Third Edition

CHEMISTRY

The Central Science

THEODORE L. BROWN

University of Illinois

H. EUGENE LeMAY, JR.

University of Nevada

Prentice-Hall, Inc.

Englewood Cliffs, N.J. 07632

Library of Congress Cataloging in Publication Data

Brown, Theodore L.
Chemistry: the central science.

Includes index.
1. Chemistry. I. LeMay, H. Eugene (Harold Eugene), (date). II. Title.
QD31.2.B78 1985 540 84-8413
ISBN 0-13-128950-0

Development editor: Raymond Mullaney
Editorial/production supervision: Karen J. Clemments
Interior design: Levavi & Levavi
Art direction and cover design: Janet Schmid
Manufacturing buyer: Raymond Keating
Page layout: Gail Collis
Cover photograph: "Rainbow" (© *Geoff Gove, The Image Bank*)

Printed in the United States of America

10 9 8 7 6 5 4 3 2 1

ISBN 0-13-128950-0 01

Prentice-Hall International, Inc., *London*
Prentice-Hall of Australia Pty. Limited, *Sydney*
Editora Prentice-Hall do Brasil, Ltda., *Rio de Janeiro*
Prentice-Hall Canada Inc., *Toronto*
Prentice-Hall of India Private Limited, *New Delhi*
Prentice-Hall of Japan, Inc., *Tokyo*
Prentice-Hall of Southeast Asia Pte. Ltd., *Singapore*
Whitehall Books Limited, *Wellington, New Zealand*

**To those from whom we have learned,
and to our wives and children**

Contents

3 STOICHIOMETRY 58

4 ENERGY RELATIONSHIPS: THE FIRST LAW OF THERMODYNAMICS 95

5 ELECTRONIC STRUCTURES OF ATOMS: BASIC CONCEPTS 128

9 GASES 259

10 LIQUIDS, SOLIDS, AND INTERMOLECULAR FORCES 293

11 SOLUTIONS 336

12 CHEMISTRY OF THE ENVIRONMENT 376

16 AQUEOUS EQUILIBRIA: FURTHER CONSIDERATIONS 506

17 CHEMICAL THERMODYNAMICS 540

18 ELECTROCHEMISTRY 565

19 NUCLEAR CHEMISTRY 603

Preface

The preface is almost always the last part of a textbook to be written. Only when the book has assumed its final form can the author declare to the reader what he thinks he has accomplished. Thus, what is for students the beginning is, in one sense, for authors the culmination of a long, often difficult journey. In this preface we want to give you some feeling for the ideas and concepts that have informed our writing of the text. We also wish to suggest how you can best make use of this book in your study of chemistry.

You will notice that this is the third edition of *Chemistry: The Central Science*. The first two editions have been very successful; apparently many teachers of chemistry (and many students as well, to judge from letters we have received) felt that we had developed a good book. But, as with any human effort, there is bound to be room for improvement. During the past few years, with the help of our publisher we have asked a great many teachers and students for suggestions as to how the book might be improved. In addition, we have watched carefully for the need to make changes to keep the book up to date. This third edition represents the end result of more than 2 years of work. The changes we have made are numerous, and some of them are important, but they need not concern you. What you should know is that we have adhered to the general approach and style that distinguished the first two editions. Our aim has been to present chemistry to you in a clear, readable fashion. We have tried continually to keep in mind the audience for whom the book is intended—you, the student.

Most of you are studying chemistry because it has been declared an essential part of the curriculum in which you are enrolled. That curriculum may be agriculture, dental hygiene, electrical engineering, geology, microbiology, metallurgy, paleontology, or one of many other related areas of study. It is fair to ask why it is that so many diverse areas of study should all relate in an essential way to chemistry. The answer is that chemistry is, by its nature, the *central* science. In any area of human activity that deals with some aspect of the material world, there must inevitably be a concern for the fundamental character of the materials involved—their endurance, their interactions with other materials, and their changes under a given set of conditions. This is true whether the materials involved are a polymer used to coat electronics components, the color used by a Renaissance painter, or the blood cells of a child born with sickle-cell anemia. It is very likely that chemistry plays an important role in the profession to which you now aspire, or may decide later to pursue. You will be a better professional, a more creative and knowledgeable person, if you understand the chemical concepts applicable to your work and are able to apply these concepts as needed.

The relationship of chemistry to professional goals is important, and this factor provides reason enough for you to study chemistry. There is, however, an even more important reason. Because chemistry is so central and so intimately involved in almost every aspect of our contact with the material world, this science is an integral part of our culture. The involvement of chemistry in our lives goes much deeper than the well-known advertising slogan, "Better things for better living through chemistry." In addition to all the obvious ways in which we use the products of chemical research and production—plastic bags, children's toys, counter tops, weed and insect killers, photographic films—we indirectly use thousands of chemical products via the foods we eat, the cars we drive, the medical care we receive, and so forth. During the past several years, we have become increasingly aware that our use of chemicals has had a profound and frightening effect on our environment. Indeed, many scientists are convinced that we have so intensely polluted this planet and so unthinkingly sowed the seeds of future pollution that the fate of civilization is all but sealed. Whether this is so remains to be seen; however, if you are to be a responsible citizen, you will surely need to be informed on many complex issues involving chemistry and the use of chemicals. Because vested interests have a powerful stake in public policy, the public often is presented with conflicting information and claims. You can more fully appreciate and analyze the complex issues put before you if you understand the fundamental principles involved and keep them in mind during your reading and study.

With all of these considerations in mind, you should now be impatient and eager to begin your study of chemistry. Now that you are ready to go, we should say something about how this book can best help you. You might first take a few minutes to glance through the table of contents. The particular sequence of chapters that we have chosen is one that we feel promotes a natural unfolding of the science of chemistry. However, the order in which the chapters of the book are covered in the classroom will be determined by your instructor. You should not be disturbed if the order is not the same as the order in the book. The book has been written so as to make allowance for alternative chapter orders and, in some instances, for the complete omission of certain chapters. Notice that some materials interspersed throughout the book deal with the chemical aspects of the world in which we live: the air, the earth, and the waters on the earth's surface. In these materials we have attempted to connect the chemical facts and principles introduced in other, usually earlier, chapters to the familiar (and sometimes not so familiar) aspects of our surroundings on earth. Your instructor, the person who will guide you through this book, may not feel that there is sufficient time to cover some or all of these environmental topics. We suggest that you read them anyway; they will help you appreciate the many ways in which chemical concepts and observations are related to contemporary life.

If you should at some point encounter a term or concept you are expected to know but can't remember, use the index at the back of the book. A good index is a rarity; we have worked hard to make your index in this book as complete and accurate as possible. Use it often. (Remember the index also when you later use the book as a reference, after having finished the course. It can help you find what you want more quickly than any other means.)

The difficulties that many chemistry students experience often can be traced to faulty exposition and confusing explanations in their text. This book has been worked on very thoroughly by many people to ensure that it is as clear, concise, and free of confusion as possible. However, you may find that a single reading of a chapter will not suffice if you are to use the book effectively as a learning tool. We suggest that you read every assigned chapter as early as possible, preferably before the material is covered in lecture. This will make you aware of important concepts and terms even before they are treated by the lecturer. Later, you will need to go through the assigned sections of the book much more carefully, making sure that you understand the new terms and problems put before you. We have inserted a great many *sample exercises* into the text, so that you might have clearly worked-out examples of problem solving of various types. You should study these exercises carefully, noting every aspect of them, especially if numerical problem solving is involved.

The review section at the end of each chapter is an integrated package designed to help you determine whether you have in fact learned all the material assigned you in each chapter. The *summary* points out the highlights of the chapter; sometimes we say things a little differently in the summary in order to add an extra element of understanding to what you have gotten from the chapter itself. The *key terms* that you should know are also collected for your convenience. The *learning goals* are placed at the end of the chapter to enable you to test yourself. You should make sure that you can meet each learning goal. This can best be done if you state a definition and then check it, write a formula and then check it, or solve a problem and then check it. It may happen, of course, that your instructor will not have covered part of the material in a chapter. You can then skip over the learning goals for this material, but you should still read the complete summary and learn all the key terms. By learning even nonrequired terms and concepts you can expand your chemical vocabulary with little effort.

The *exercises* at the back of each chapter are designed to test your understanding of the materials covered in the chapter. They are grouped according to topic, except for a number of additional exercises. The purpose of the additional exercises is to test your ability to solve a problem when it is not clearly identified as to topic. Also, some of the questions in this category require the application of material from more than one topic area. Problems marked with brackets are, in general, a little more difficult to solve than the others. We have prepared a solutions manual that contains detailed answers to all the end-of-chapter exercises; you should consult this manual only after working out problems on your own.

Finally, you should note that there are several appendices following Chapter 26. These are designed to aid you in various ways. You should get acquainted with what is there by glancing through them before the course gets under way. In particular, note that answers are provided to many of the end-of-chapter exercises. Color question numbers in the text indicate that the answer to the question is in the answer section following the appendices.

Your instructor may have elected to have you purchase the *Student's Guide* designed for use with the text. This guide, written by Professor James C. Hill, of California State University, Sacramento, is a nicely organized and well-written supplement to the text. You will find it filled with helpful ideas, problem-solving techniques, and fresh insights into the materials presented in the text. We are very happy that Jim has agreed to write the study guide; we feel that it is a valuable learning aid for use with the text.

Most general chemistry courses involve laboratory as well as classroom work. There is a very good reason for this. Chemistry is an experimental science; the entire theoretical structure of chemistry is based on the results of laboratory experiments. As you study chemistry, you should try to relate what you learn in the classroom and from the text to operations and observations made in the course of your laboratory work. A very fine laboratory manual for use with this text has been written by Professors John H. Nelson and Kenneth C. Kemp of the Department of Chemistry, University of Nevada, Reno. We believe that it is also an important learning tool in your study of chemistry.

During the many years that we have been practicing chemists, we have found chemistry to be an exciting intellectual challenge and an extraordinarily rich and varied part of our human cultural heritage. We hope that all the hassles you must face regarding course grades will not keep you from sharing with us some of that enthusiasm and appreciation. We have, in effect, been engaged by your instructor to help you learn chemistry. We are confident that we've done that job well. In any case, we would appreciate your writing us, either to tell us of the book's shortcomings, so that we might do better, or of its virtues, so that we'll know where we have helped you most.

THEODORE L. BROWN
School of Chemical Sciences
University of Illinois
Urbana 61801

H. EUGENE LEMAY, JR.
Department of Chemistry
University of Nevada
Reno 89557

Acknowledgments

This book owes its final shape and form to the assistance and hard work of many people. Several colleagues reviewed the manuscript and helped us immensely by sharing their insights and criticizing our initial writing efforts. We would like especially to thank the following:

David L. Adams, North Shore Community College
George Brubaker, Illinois Institute of Technology
Ronald J. Clark, Florida State University
A. Wallace Cordes, University of Arkansas
Lawrence M. Epstein, University of Pittsburgh
Daniel T. Haworth, Marquette University
James C. Hill, California State University
Edwin M. Larsen, University of Wisconsin, Madison
John D. Petersen, Clemson University
George H. Schenk, Wayne State University
Robert P. Stewart, Jr., Miami University, Oxford
Charles R. Ward, University of North Carolina, Wilmington
Charles A. Wilkie, Marquette University

We deeply appreciate the assistance of the following members of Prentice-Hall's College Division: Betsy Perry and Nancy Forsyth, our Chemistry Editors, and Tim Moore, Marketing Manager, each of whom provided guidance, support, and counsel throughout the project; Raymond Mullaney, Director, Product Development Department, who has been with this project from its beginnings; Karen J. Clemments, Senior Production Editor, who has guided this edition through production; and Janet Schmid, Senior Designer.

Finally, special thanks are due to Becky Lazaro and Mary Kaylor, who so ably typed manuscript copy from our rough drafts.

1 Introduction: Some Basic Concepts

Perhaps the only thing permanent about our world is change. All around us are numerous examples of change in ourselves and our environment. Trees change color in autumn, iron rusts, snow melts, paint peels, seeds become flowers, and logs burn. We grow up, we grow old. Living plants and animals undergo continual change, and even dead plants and animals continue to change as they decay. Such changes have long fascinated people and have prompted them to look more closely at nature's working in hopes of better understanding themselves and their environment.

Understanding change is closely tied to understanding the nature and composition of **matter.** Matter is the physical material of the universe; it is anything that occupies space and has mass. Chemistry is the science that is concerned primarily with matter and the changes that it undergoes. Therefore, as we begin our study of chemistry, our primary focus will be on matter. First, however, let's sketch a somewhat broader picture of chemistry.

Chemistry is a changing science. Therefore, the questions that chemists seek to answer are constantly changing also. Because of this, we might define chemistry as what chemists do. In many regards this definition is unsatisfactory. Nevertheless, it does suggest that chemistry itself changes as chemists absorb new information from other fields, tackle new problems, or reexamine old ones in new ways. One of the important activities of chemists is the synthesis of new materials or the improvement in the ways of making old ones. This aspect of chemistry has had great impact on our lives; chemists have synthesized new fibers, medicines, fertilizers, pesticides, and structural materials. Many new chemicals never find any commercial use but are nevertheless important to chemists in answering subtle questions about matter and its changes. In designing ways to synthesize new materials, it is useful to know the factors that determine how fast and to what extent the required changes proceed. Such knowledge allows chemists to improve, avoid, or control many changes of matter. This knowledge is necessary, for example, in devising ways to clean up automobile exhaust or to make fertilizers at lower cost. Chemists are also interested in determining the identity and concentration of substances. Such analysis may involve determining the quality of a soap in a manufacturing operation, the concentration of a pollutant in the air, the amount of gold in a potential ore, the amount of mercury in a lake, the identity of the substances in some physiologically active mixture, or the chemicals resulting from the utilization of a drug in the body. Chemists are interested not only in determining what things are made of, but also in discovering the ways their composition and structure are related to properties. For example, what makes a particular substance poisonous or sweet or hard or explosive?

The intent of this text is to introduce you to basic chemical facts and theories, not as ends in themselves but as means to help you understand the material world and to recognize the constraints and opportunities it provides. We hope that this text will provide not only a firm foundation for further scientific studies, whether they be in chemistry or some other field, but also a background to enable you to evaluate scientific information found in news media and periodicals. In the remainder of this chapter we consider some background material useful to your studies—the metric system, uncertainty in measurement, and problem solving in chemistry. We also briefly explore the historical and philosophical background of chemistry and the scientific approach to problems.

1.1 THE EMERGENCE OF CHEMISTRY: A HISTORICAL PERSPECTIVE

Chemistry has two roots. First, it is rooted in the craft traditions such as metallurgy, brewing, tanning, and dyeing, which provided a practical understanding of how matter behaves. Second, it can also be traced to the philosophers of ancient Greece who concerned themselves with questions of the basic nature of matter. The growth of chemistry is in some respects a reflection of the practical problems overcome in the course of human society's cultural and technical development. In addition, however, chemistry is an expression of humankind's innate curiosity and desire to understand its surroundings without regard to the practical application of that understanding.

Metallurgy, the science and art of obtaining and working with metals, exemplifies the development of chemical knowledge through the craft traditions. This craft developed for many years and achieved considerable sophistication without any theoretical framework that would explain metallurgical operations and guide their development. Developments were made largely through trial and error and through accidental discoveries. For the most part, the early pattern of discovery of metals followed their ease of recovery from ores, the earthy mixtures that are mined as sources of metals. Gold was one of the first metals used because it is found in nature in an uncombined, metallic state, for example, as gold nuggets. Copper, which is more abundant, was not used until about 3500 B.C. At that time processes for obtaining copper metal from its ores were discovered. This discovery was undoubtedly accidental. It could have occurred when some copper ore was dropped or thrown into the coals of a fire. Copper metal can be obtained by heating copper ore with charcoal. Methods for obtaining iron, which is much more abundant than copper, did not develop until about 1500 B.C.

People probably attempted to understand changes in matter even before they began to use these changes to their advantage. We know that many early explanations presumed the existence of supernatural powers. In contrast to this approach, the early Greeks sought to understand matter and its changes purely on the basis of logic. However, they were not especially concerned with using these ideas as guides to improve their crafts. Modern science differs from the approach of the Greeks; it depends not only on logic but also on the systematic gathering of facts and on careful observations. Furthermore, the ideas of science are widely used to guide the development of our technologies.

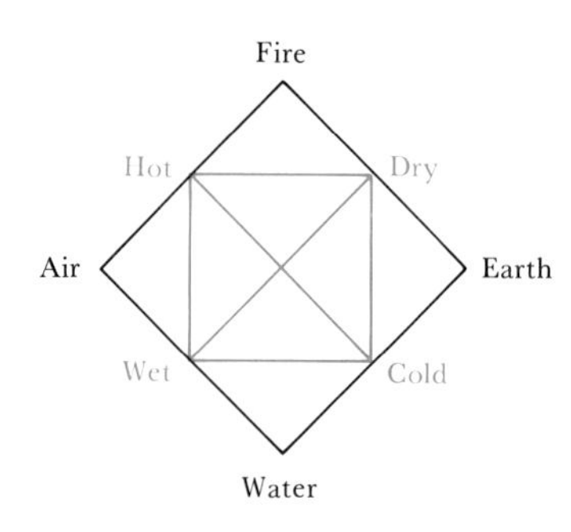

FIGURE 1.1 Schematic representation of the four elements of the Greeks and the four properties associated with these elements.

The ideas of the early Greek philosophers such as Anaximander (sixth century B.C.) and Empedocles (fifth century B.C.) found their most lasting expression in the teachings and writings of Plato (427–347 B.C.) and Aristotle (384–322 B.C.). The early Greek philosophers had proposed that all of nature was composed of four elements: fire, earth, air, and water. This idea provides a logical explanation for many observations. For example, a green log can burn, producing smoke (air) and flame (fire), leaving behind ashes (earth), and perhaps even giving a fleeting glimpse of sap (water). The idea of four elements was extended by emphasizing the basic properties of the elements. These were coldness, hotness, dryness, and wetness. Each element, in its ideal form, had two associated properties as shown schematically in Figure 1.1. For example, water was the wet, cold element. The Greeks believed that an element could be changed into any other element by altering its properties. For example, water could be changed to earth by replacing the property of wetness with the property of dryness.

This concept and its associated logic persisted for over 1000 years, influencing thought through the Middle Ages. However, by the time of the Middle Ages the list of elements and their associated properties had grown. Operating in this framework of logic the alchemists sought to convert common or ordinary metals into gold (Figure 1.2). Their attempts were based on the idea that addition of the essential quality or property of "nobility" to these metals would cause them to "grow" into gold. In the course of their futile attempts to bring about such changes the alchemists discovered new chemicals and developed new ways for working with them.

One of the important steps in the transition from alchemy to chemistry occurred in 1661 when Robert Boyle (1627–1691) published a book entitled *The Skeptical Chemist*. Boyle rejected the Greek concept of elements; in its place he proposed that an element be defined as a substance that cannot be decomposed into simpler substances. As we shall see in Section 2.2, this is essentially the modern definition of an element.

An additional hindrance to the development of chemistry was a theory developed by the German physician G. E. Stahl (1660–1734) to explain combustion, the act of burning. Stahl proposed that all flammable substances contain a component called *phlogiston* after the Greek word for fire. According to this widely held theory, materials lose phlogiston to air when they burn.

FIGURE 1.2 *The Alchemist,* a seventeenth-century painting by the Flemish painter David Teniers (1582–1649). (*Bettmann Archive*)

The Birth of Modern Chemistry

In 1772, a wealthy French nobleman named Antoine Lavoisier (Figure 1.3) began experimenting with combustion. By weighing objects before and then after combustion, Lavoisier observed that burning objects gain weight. Furthermore, he observed that when combustion is carried out in a closed container, there is no change in weight; the weight of all of the substances in the container is the same before and after combustion even though they change form. There was no evidence for the loss of phlogiston. Instead, the experiments indicated that when a substance burns it gains something from the air. The weight gained by the burning sample was the same as the weight lost by the air. Thus there is no net change in weight when combustion is carried out in a closed container.

On the basis of these and other observations, Lavoisier proposed that the role of air in combustion is not to carry off phlogiston but to provide oxygen. When an object burns, oxygen is removed from air and becomes incorporated into the burning object. Indeed, there is no such thing as phlogiston. Because of his reliance on carefully controlled experiments and his use of quantitive measurements, Lavoisier is generally considered the father of modern chemistry.

Air is only 20 percent oxygen; consequently, combustion is much more vigorous in pure oxygen than in air. The fire that broke out on the *Apollo 1* rocket in January 1967 and resulted in the deaths of astronauts Grissom, White, and Chaffee burned so very fast because the space capsule was filled with pure oxygen. Because of that incident, a mixture of 60 percent oxygen and 40 percent nitrogen has been used in space capsules ever since. A fire may be extinguished by placing a blanket over it because the blanket excludes the oxygen. People in a closed room can be suffocated by a fire because it removes oxygen from air.

1.2 THE SCIENTIFIC APPROACH

The fundamental activity of science is making careful observations. These may be of both a qualitative and quantitative nature and often involve controlled experiments. Scientists seek general relations that will unify their observations. A concise verbal statement or a mathematical equation that summarizes a broad variety of observation and experience is known as a scientific **law.** A familiar example is the law of gravity; it summarizes our experience that what goes up must come down. We also seek to understand our laws. A tentative explanation is called a **hypothesis.** A hypothesis is useful only if it can be used to make predictions that can be tested by further experiments and thereby verified or refuted. A hypothesis that continually withstands such tests is called a **theory.** A theory may serve to unify a broad area and may provide a basis for explaining many laws. Such is the case with the atomic theory of matter, which we will begin to examine in Chapter 2.

There is no fail-proof, step-by-step scientific method that scientists use. The approaches of various scientists depend on their temperament, circumstances, and training. Rarely will two scientists approach the same problem in the same way. The scientific approach involves doing one's utmost with one's mind to understand the workings of nature. Just because we can spell out the results of science so concisely or neatly in textbooks does not mean that scientific progress is smooth, certain, and

FIGURE 1.3 Antoine Lavoisier, after a painting by Louis David. Lavoisier (1743–1794) conducted studies that led to the downfall of the phlogiston theory and the birth of modern chemistry. Unfortunately, Lavoisier's career was cut short by the French Revolution. He was not only a member of the French nobility but also a tax collector. He was guillotined in 1794 during the final months of the Reign of Terror. He is now generally considered to be the father of modern chemistry because of his reliance on carefully controlled experiments and his use of quantitative measurements. The tribunal that sentenced him to death, however, declared that France had no need for scientists. The great mathematician Lagrange, then living in Paris, remarked: "It took but a moment to cut off that head, though a hundred years perhaps will be required to produce another like it." (*Courtesy Burndy Library*)

predictable. The path of any scientific study is likely to be irregular and uncertain; progress is often slow and many promising leads turn out to be dead ends. Through the course of our studies we will see that serendipity (fortunate accidental discovery) has played an important role in the development of science. What we will often miss discussing are the doubts, conflicts, clashes of personalities, and revolutions of perception that have led to our present ideas.

Selection of Theories

No hypothesis or theory can ever be exposed to all the possible tests necessary for absolute verification. However, a hypothesis or theory can be disproved by obtaining experimental results inconsistent with it. Scientific advance depends on such disproofs to eliminate faulty hypotheses and theories. In the absence of such disproofs, scientists often choose between hypotheses or theories by comparing the ability of each to explain the evidence at hand; the one that explains the facts better is chosen.

It has been suggested that scientific progress is more rapid when scientists are open and imaginative enough to formulate several alternative hypotheses to explain their observations. Experiments can then be designed to test these alternatives, thereby excluding some of them. This approach involves a continual search for alternative hypotheses. A person who works with only a single hypothesis can become strongly attached to it. Research can then become a strenuous and devoted attempt to force nature into the conceptual boxes supplied by that hypothesis.

In the end it is the collective judgment of the scientific community that effectively decides among theories. Therefore, one of the most important activities of a scientist is public disclosure of scientific results through publication. The authority of science does not rest ultimately on the individual who has done the work, but rather on whether others can repeat the work and obtain the same results or extend the work in a self-consistent fashion.

1.3 MEASUREMENT AND THE METRIC SYSTEM

Lavoisier's studies of combustion (Section 1.1) should impress upon us the importance of quantitative measurements. This idea is simply common sense to us now, although it has not always been so. Consider a person who is ill. Using only sense perception we may conclude that this person is running a fever, but we are not certain, and another person may disagree with our conclusion. To tell accurately, we use a thermometer, and much can depend upon the result of our measurement—for example, whether the person's temperature is 98.6°F or 102°F. This example suggests three points about measurement. First, our five senses are the most important tools we have, but they are limited. We therefore resort to various instruments to extend and quantify our sense perceptions. Second, an advantage of quantitative data is that they allow different people to obtain the same results, thus avoiding many arguments based on opinions. Third, measurements depend on a standard of reference. For instance, there is a considerable difference between 102°F and 102°C. (The meaning of °F and °C will be discussed shortly.)

The standards used in science are those of the **metric system.** This is the system of weights and measures used throughout most of the world; the United States is also moving toward adopting it in many facets of society. The weights of most canned products in the grocery store are now given in grams as well as in ounces; there are even highway signs that show distance in both miles and kilometers (Figure 1.4).

According to international agreement reached in 1960, certain basic metric units and units derived from them are to be preferred in scientific use. The preferred units are known as **SI units** after the French *Système International d'Unités*. The seven basic units of the SI system are given in Table 1.1. All other SI units of measure are derived from these base units. For example, speed, which is the ratio of distance to elapsed time, has units of meters per second, or m/s.

Adoption of SI units is an attempt to further systematize the metric system. Non-SI units are to be progressively discouraged and with time phased out. However, until SI units are fully adopted by practicing

FIGURE 1.4 Road sign along an interstate highway in Michigan, showing distance in metric and English-system units. (*Michigan Department of State Highways and Transportation*)

TABLE 1.1 Basic SI units

Physical quantity	Name of unit	Abbreviation
Mass	Kilogram	kg
Length	Meter	m
Time	Second	s or sec
Electric current	Ampere	A
Temperature	Kelvin	K
Luminous intensity	Candela	cd
Amount of substance	Mole	mol

scientists, it is necessary to be aware of both SI units and the non-SI units that are still in use. Whenever we first encounter a non-SI unit in the text, the proper SI unit will also be given.

The SI system employs a series of prefixes to indicate decimal fractions or multiples of various units. For example, the prefix milli- represents a 10^{-3} fraction of a unit: a milligram is 10^{-3} gram, a millimeter is 10^{-3} meter, and so forth. The prefixes most commonly encountered in chemistry are given in Table 1.2.* In using the SI system and in working problems throughout this text, it is important to have a comfortable familiarity with exponential notation. If you are unfamiliar with exponential notation or want to review it, refer to Appendix A.1.

TABLE 1.2 Selected prefixes used in the SI system

Prefix	Abbreviation	Meaning	Example
Mega-	M	10^6	1 megameter (Mm) = 1×10^6 m
Kilo-	k	10^3	1 kilometer (km) = 1×10^3 m
Deci-	d	10^{-1}	1 decimeter (dm) = 0.1 m
Centi-	c	10^{-2}	1 centimeter (cm) = 0.01 m
Milli-	m	10^{-3}	1 millimeter (mm) = 0.001 m
Micro-	μ[a]	10^{-6}	1 micrometer (μm) = 1×10^{-6} m
Nano-	n	10^{-9}	1 nanometer (nm) = 1×10^{-9} m
Pico-	p	10^{-12}	1 picometer (pm) = 1×10^{-12} m

[a]This is the Greek letter mu (pronounced "mew").

Length

The basic SI unit of length is the meter (m). From the comparisons between metric and English-system measurements given in Table 1.3 we can see that the meter is only slightly longer than a yard. A diagrammatic comparison between metric and English-system measures of length is made in Figure 1.5. We will consider interconversion of English and metric system measures more closely in Section 1.5. For the moment it is more important that we clearly understand the use of the prefixes that are given in Table 1.2.

*It is interesting that the monetary system of the United States is decimal: 0.01 of a dollar is a cent (centidollar) while 0.001 of a dollar (a tenth of a cent) is a mill (millidollar). Extending this usage, we could theoretically refer to $1000 as a kilodollar or kilobuck.

TABLE 1.3 Metric-English system equivalents

Length	Mass	Volume
1 meter = 1.094 yards	1 kilogram = 2.205 pounds	1 liter = 1.06 quarts
2.54 centimeters = 1 inch	453.6 grams = 1 pound	1 cubic foot = 28.32 liters

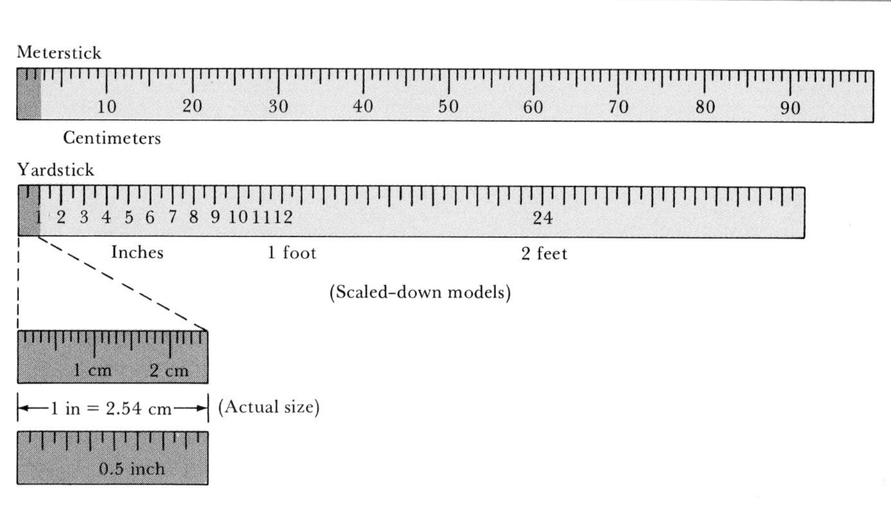

FIGURE 1.5 Comparison of common metric measures of length.

Volume

The measure for volume is a derived unit based on the fundamental SI unit of length cubed, m^3. The cubic meter, m^3, is the volume of a cube that is 1 m on each edge. Related units such as the cubic centimeter, cm^3 (sometimes written cc), or cubic decimeter, dm^3, are also used. Another common measure of volume is the liter (L), a volume roughly the size of a quart (Table 1.3). A liter is the volume occupied by 1 cubic decimeter, dm^3. There are 1000 mL in a liter, and each milliliter is the same volume

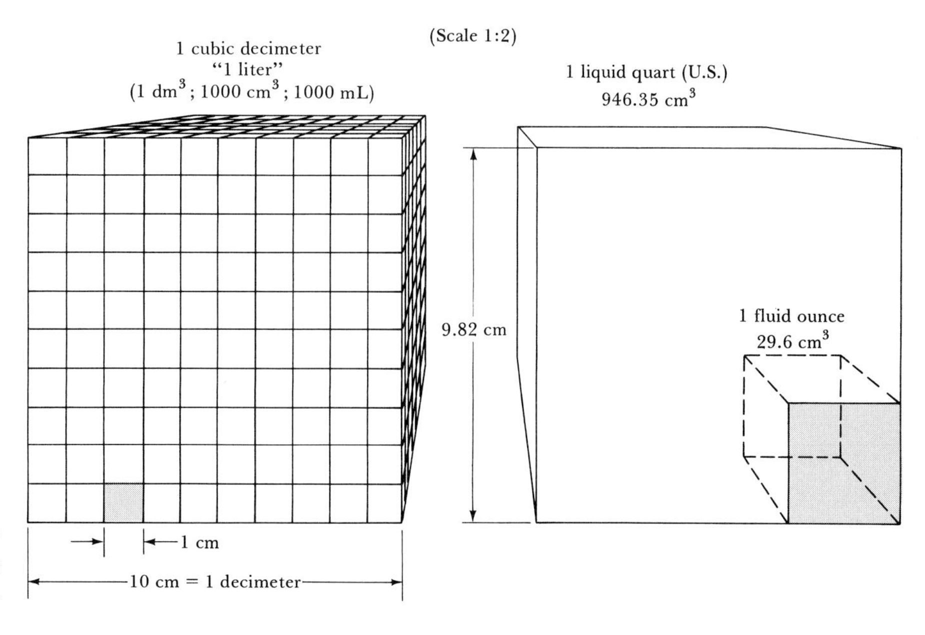

FIGURE 1.6 Comparison of common measures of volume.

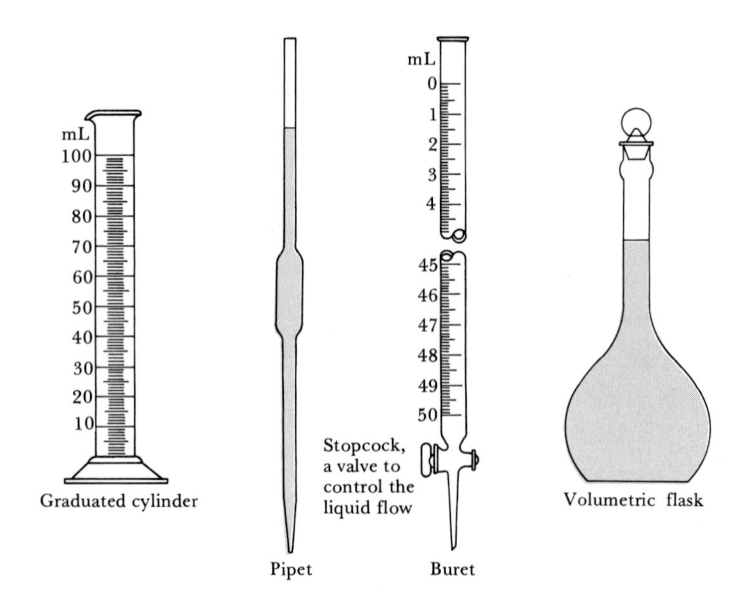

FIGURE 1.7 Common devices used in chemistry laboratories to measure volume.

as a cubic centimeter (Figure 1.6). Thus milliliter and cubic centimeter are commonly used interchangeably in expressing volume. The liter is the first metric unit that we have encountered that is not an SI unit.

The devices most frequently used in chemistry to measure volume are illustrated in Figure 1.7. Pipets and burets allow delivery of liquids with more accurately known volumes than do graduated cylinders. Volumetric flasks are used to prepare accurately a designated volume of solution.

Mass

The basic SI unit of mass* is the kilogram (kg). As shown in Table 1.3 and Figure 1.8, a kilogram is equal to 2.2 lb. The unit of mass used most frequently in chemistry is the gram (g), which is $\frac{1}{1000}$ of a kilogram. The mass of an object is determined by balancing it against a set of known masses using a device known as a balance. Several types of common laboratory balances are shown in Figure 1.9.

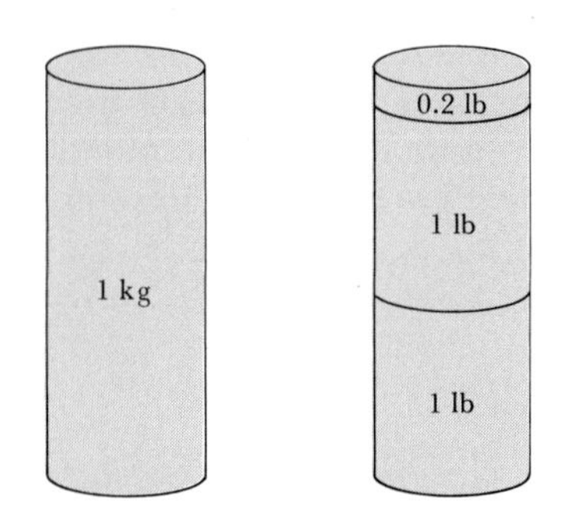

FIGURE 1.8 Comparison of common measures of mass.

Density

Density is a quantity widely employed by chemists to identify substances. It is defined as the amount of mass in a unit volume of the substance:

$$\text{Density} = \text{mass/volume} \quad [1.1]$$

Density is commonly expressed in units of grams per cubic centimeter (g/cm^3 or $g\ cm^{-3}$). The densitites of some common substances are listed in Table 1.4.

TABLE 1.4 Densities of some selected substances

Substance	Density (g/cm^3)
Air	0.001
Balsa wood	0.16
Water	1.00
Table salt	2.16
Iron	7.9
Gold	19.32

*Mass and weight are often incorrectly thought to be the same. Mass is a measure of the amount of material in an object; the weight of that object, however, depends not only on its mass but also on the attractive force of gravity. In outer space, where gravitational forces are very weak, an astronaut may be weightless, but he is not massless. In fact, he has the *same* mass as he has on earth. Nevertheless, it is common practice to use the terms mass and weight interchangeably.

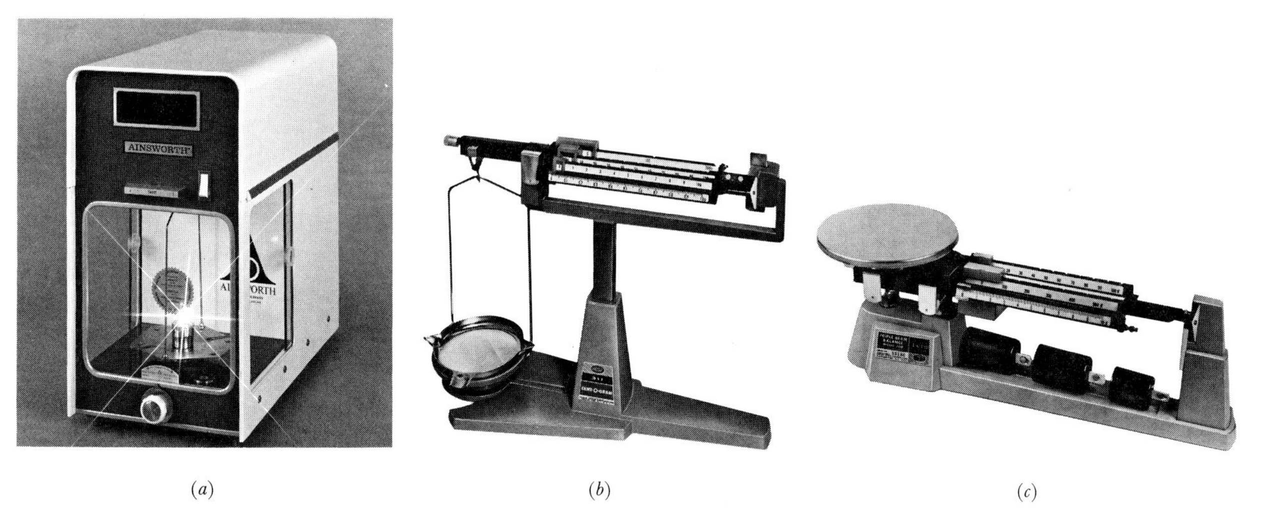

(*a*) (*b*) (*c*)

FIGURE 1.9 Three common types of laboratory balances: (*a*) analytical balance; (*b*) triple-beam balance of the stirrup type; (*c*) triple-beam platform balance. (*a, Denver Instrument Company, Ainsworth Division; b and c, Ohaus Scale Corp.*)

SAMPLE EXERCISE 1.1

(a) Calculate the density of mercury if 1.00×10^2 g occupies a volume of 7.36 cm^3. (b) Calculate the mass of 65.0 cm^3 of mercury.

Solution:

(a) $$\text{Density} = \frac{\text{mass}}{\text{volume}} = \frac{1.00 \times 10^2 \text{ g}}{7.36 \text{ cm}^3} = 13.6 \text{ g/cm}^3$$

(b) Solving Equation 1.1 for mass gives

$$\text{mass} = \text{volume} \times \text{density}:$$
$$\text{mass} = (65.0 \text{ cm}^3)(13.6 \text{ g/cm}^3) = 884 \text{ g}.$$

The terms *density* and *weight* are sometimes confused. A person who says that iron weighs more than air generally means that iron has a higher density than air; 1 kg of air has the same mass as 1 kg of iron, but the iron is confined to a smaller volume, thereby giving it a higher density.

Temperature

The temperature scales commonly employed in scientific studies are the Celsius (or centigrade) and Kelvin scales; the kelvin is the SI unit of temperature. The Celsius scale is based on assignment of 0°C to the freezing point of water and 100°C to its boiling point at sea level. The corresponding temperatures in the Fahrenheit scale are 32°F and 212°F. There are 100° between the freezing point and boiling point of water in the Celsius scale, whereas there are 180° between these points on the Fahrenheit scale. Consequently, the Celsius and Fahrenheit scales are related as follows:*

*Using Equation 1.2, it can be shown that −40°C = −40°F. This fact permits another simple method of converting between Celsius and Fahrenheit scales. To convert a temperature from Fahrenheit to Celsius add 40°, multiply the result by $\frac{5}{9}$, and then subtract 40°. To convert from Celsius to Fahrenheit add 40°, multiply the result by $\frac{9}{5}$, and then subtract 40°.

$$°C = \frac{100}{180}(°F - 32°) = \frac{5}{9}(°F - 32°) \quad [1.2]$$

The Kelvin scale is based on the properties of gases, and its origins will be considered more fully in Chapter 9. Zero on this scale corresponds to −273.15°C, and the size of a kelvin is the same as a degree Celsius. The Kelvin and Celsius scales are therefore related by Equation 1.3.

$$K = °C + 273.15 \quad [1.3]$$

[According to the SI convention, a degree sign (°) is not used with the Kelvin scale. Thus we write 273 K and not 273°K.] Further comparisons between the Celsius, Kelvin, and Fahrenheit scales are made in Figure 1.10 and Table 1.5.

SAMPLE EXERCISE 1.2

If a weather forecaster predicts that the temperature for the day will reach 30°C, what is the predicted temperature (a) in K; (b) in °F?

Solution:

(a) $K = 30 + 273 = 303$

(b)
$$°C = \frac{5}{9}(°F - 32°)$$
$$30° = \frac{5}{9}(°F - 32°)$$
$$\left(\frac{9}{5}\right)(30°) = °F - 32°$$
$$54° + 32° = °F$$
$$86° = °F$$

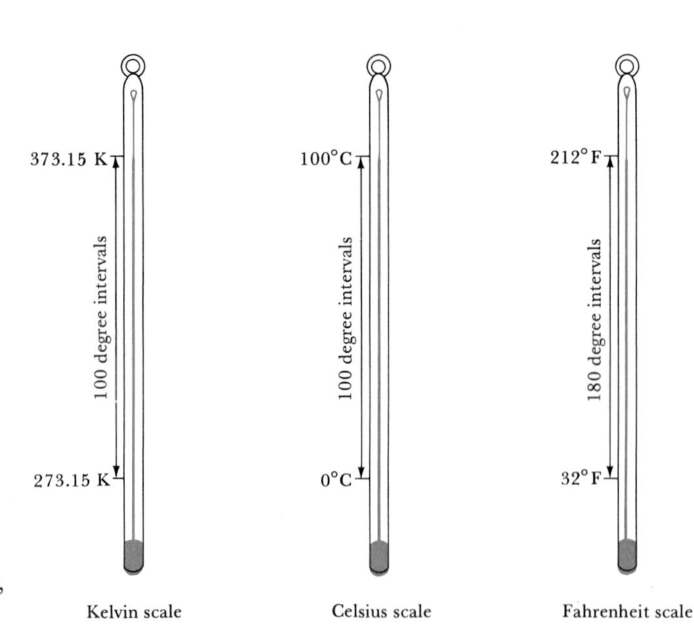

FIGURE 1.10 Comparison of the Fahrenheit, Celsius, and Kelvin temperature scales.

TABLE 1.5 Some comparisons of Fahrenheit, Celsius, and Kelvin temperatures

Absolute zero	−460°F	−273°C	0 K
Freezing point of water	32°F	0°C	273 K
Average room temperature	68°F	20°C	293 K
Normal body temperature	98.6°F	37°C	310 K
Boiling point of water	212°F	100°C	373 K

We sense temperature as a measure of the hotness or coldness of an object. Indeed, temperature determines the direction of heat flow; heat always flows spontaneously from a substance at high temperature to one at low temperature. Thus we feel the influx of energy when we touch a hot stove, and we know that the stove is at a higher temperature than our hand.

Temperature is an intensive property, meaning that its value does not depend on the amount of material chosen. Thus, two samples of a liquid may have the same temperature though one has a volume of one cup and the other sample fills a bathtub. Density is another example of an intensive property; the density of mercury, as calculated in Sample Exercise 1.1, is 13.6 g/cm^3 whether one has 100 g or 1000 g. By contrast, both volume and mass are extensive properties, because they depend on the amount of material. Similarly, the *heat content* of a sample is an extensive property. In some types of solar heating systems for homes, air is heated by passage through solar panels exposed to the sun. The heated air is then passed through a large bed of stones. Because the heat content of the stones is large, the heat stored in them during the day is sufficient to heat the house during the night. Note the important distinction between heat content, an extensive property, and temperature, an intensive property.

SAMPLE EXERCISE 1.3

Consider the following description, labeling each property or characteristic as intensive or extensive: "The yellow sample is solid at 25°C. It weighs 6.0 g and has a density of 2.3 g/cm^3."

Solution: Mass is an extensive property; color, physical state (that is, solid), temperature, and density are intensive properties.

1.4 UNCERTAINTY IN MEASUREMENT

All measurements have some degree of uncertainty; how great the uncertainty is depends on both the measuring device and the skill of its operator. For example, the magnitude of the uncertainty associated with weighing an object will depend on the type of balance employed. On a triple-beam platform balance (Figure 1.9), the mass of a sample substance can be measured to the nearest 0.1 g; mass differences less than this cannot be detected on this balance. We might therefore indicate the mass of a dime measured on this balance as 2.2 ± 0.1 g; the ± 0.1 (read plus or minus 0.1) is a measure of the uncertainty of the measurement. It is important to have some indication of how precisely any measurement is made; the $\pm$ notation is one way to accomplish this. We could measure the mass of the dime more precisely on an analytical balance, to the nearest 0.0001 g if we are careful. We might therefore report the mass as 2.2405 ± 0.0001 g. It is common to drop the $\pm$ notation with the understanding that *there is uncertainty of at least one unit in the last digit of the measured quantity;* that is, measured quantities are often reported in such a way that only the last digit is uncertain. All of the digits, including the uncertain one, are called significant digits or, more commonly, **significant figures.** The number 2.2 has two significant figures, while the number 2.2405 has five significant figures.

SAMPLE EXERCISE 1.4

What is the difference between 4.0 g and 4.00 g?

Solution: Many people would say there is no difference, but a scientist would note the difference in the number of significant figures between the two measurements. The number 4.0 has two significant figures while 4.00 has three. This implies that the second measurement has been made more precisely. A mass of 4.0 g indicates that the mass of the sample must be between 3.95 g and 4.05 g, closer to 4.0 g than to 3.9 g or 4.1 g. A mass of 4.00 g means that the sample must have a mass between 3.995 and 4.005 g, closer to 4.00 g than to 3.99 g or 4.01 g.

The following rules apply to determining the number of significant figures in a measured quantity:

1. All nonzero digits are significant—457 cm (three significant figures); 0.25 g (two significant figures).
2. Zeros between nonzero digits are significant—1005 kg (four significant figures); 1.03 cm (three significant figures).
3. Zeros to the left of the first nonzero digit in a number are not significant; they merely indicate the position of the decimal point—0.02 g (one significant figure); 0.0026 cm (two significant figures).
4. When a number ends in zeros that are to the right of the decimal point, they are significant—0.0200 g (three significant figures); 3.0 cm (two significant figures).
5. When a number ends in zeros that are not to the right of a decimal point, the zeros are not necessarily significant—130 cm (two or three significant figures); 10,300 g (three, four, or five significant figures). The way to remove this ambiguity is described below.

Use of standard exponential notation avoids the potential ambiguity of whether the zeros at the end of a number are significant (rule 5). For example, a mass of 10,300 g can be written in exponential notation showing three, four, or five significant figures:

1.03×10^4 g (three significant figures)

1.030×10^4 g (four significant figures)

1.0300×10^4 g (five significant figures)

In these numbers all the zeros to the right of the decimal point are significant (rules 2 and 4).

The rules we have stated apply to nonintegral measured quantities; for these cases the number of significant figures is indicative of the associated uncertainty of the measurement. It is important to distinguish these nonintegral numbers from exact, integral ones, those that are defined or that result from counting. For example, there are exactly 3 ft in a yard, exactly four people in my immediate family, exactly 1000 g in a kilogram, and exactly 12 eggs in a dozen eggs. These are examples of exact numbers, numbers with no associated uncertainty. They can be considered to have an infinite number of significant figures. The number 1 in the metric-English system equivalents given in Table 1.3 is an exact number.

Combining Measured Quantities in Calculations

In carrying measured quantities through calculations the rule used is that the precision of the result is limited by the least precise measurement. *In multiplication and division the result must be reported as having no more significant figures than the measurement with the fewest significant figures.* When the result contains more than the correct number of significant figures it must be rounded off.

For example, the area of a rectangle whose edge lengths are 6.221 cm and 5.2 cm should be reported as 32 cm^2:

$$\text{Area} = (6.221\ \text{cm})(5.2\ \text{cm}) = 32.3492\ \text{cm}^2 \longrightarrow \text{round off to } 32\ \text{cm}^2$$

We round off to two significant figures because 5.2 cm has only two significant figures.

In rounding off numbers, the following rules are followed (each example is rounded to two digits):

1. If the leftmost digit to be removed is more than 5, the preceding number is increased by 1; 2.376 rounds to 2.4.
2. If the leftmost digit to be removed is less than 5, the preceding number is left unchanged; 7.248 rounds to 7.2.
3. If the leftmost digit to be removed is 5, the preceding number is not changed if it is even and is increased by 1 if it is odd; 2.25 rounds to 2.2; 4.35 rounds to 4.4.

The rule used to determine the number of significant figures in multiplication and division cannot be used for *addition and subtraction.* For these operations, *the result should be reported to the same number of decimal places as that of the term with the least number of decimal places.* The following is an example:

This number limits	20.4	⟵ one decimal place
the number of significant	1.322	⟵ three decimal places
figures in the result ⟶	83	⟵ zero decimal places
	104.722	⟶ round off to 105

SAMPLE EXERCISE 1.5

How many significant figures are there in each of the following numbers: (a) 4.003; (b) 6.023×10^{23}; (c) 5000; (d) the sum $15.3 + 0.2334$; (e) the product (16)(5.7793)?

Solution: (a) Four; the zeros are significant figures. (b) Four; the exponential term does not add to the number of significant figures. (c) One, two, three, or four. In this case, the ambiguity could have been avoided by using standard exponential notation. Thus, 5×10^3 has only one significant figure; 5.00×10^3 has three. (d) Three; the sum, expressed to the proper three significant figures, is 15.5. (e) Two; the product, expressed to the proper two significant figures, is 92.

SAMPLE EXERCISE 1.6

A gas at 25°C exactly fills a container previously determined to have a volume of $1.05 \times 10^3\ cm^3$. The container plus gas are weighed and found to have a mass of 837.6 g. The container, when emptied of all gas, has a mass of 836.2 g. What is the density of the gas at 25°C?

Solution: The mass of the gas is just the difference in the two masses: $(837.6 - 836.2)\ g = 1.4\ g$. Notice that this figure has only two significant figures, even though the individual masses each have four. Note also that there is one significant figure to the right of the decimal place. This means that there is an uncertainty of 0.1 g in the measured 1.4 g mass of the gas. Since there are 14 tenths of a gram in 1.4 g, we can say that the mass of gas is known to within approximately 1 part in 14. On the other hand, the volume is known to about one part in a hundred, since the uncertainty in this quantity is in the second decimal place in the number 1.05. That is, the volume is in the range from $1.04 \times 10^3\ cm^3$ to $1.06 \times 10^3\ cm^3$. We have that

$$\text{Density} = \frac{\text{mass}}{\text{volume}}$$

The density is thus given by $1.4\ g/1.05 \times 10^3\ cm^3$. It can be expressed as $0.0013\ g/cm^3$ or $1.3\ mg/cm^3$. There are two significant figures in this quantity, corresponding to the smaller number of significant figures in the two numbers that form the ratio.

The example above illustrates an important point about significant figures; they are always related to the uncertainties in our ability to measure or estimate something. Whenever you do an experiment or evaluate someone else's experimental results, the uncertainty in the measured values is as important as the number itself.

In conducting laboratory experiments, it is important to choose the apparatus you employ so as to achieve the precision your experiment requires. For example, suppose you need to measure out 10 mL of a liquid to a precision of about 1 part in 50, for use as a reagent in a reaction. One-fiftieth of 10 is $10/50 = 0.2$. This means you must measure out 10.0 ± 0.2 mL of the liquid. It would be a mistake to employ a 100-mL graduated cylinder as illustrated in Figure 1.7. The markings on this cylinder correspond to 1-mL intervals; it would be impossible to estimate between them to a precision of 0.2 mL. On the other hand, a 10-mL graduated cylinder with markings every 0.1 or 0.2 mL would be sufficiently precise.

It is also important to have a feeling for significant figures when using a calculator, because they ordinarily display more digits than are significant. As an example, suppose that a student precisely weighs a nickel, and finds it to have a mass of 4.9566 g. He looks up the density of the alloy from which the coin is made and finds it listed as $8.8\ g/cm^3$. The volume of the nickel he weighed is then:

$$\text{Volume} = \frac{\text{mass}}{\text{density}} = \frac{4.9566\ \text{g}}{8.8\ \text{g/cm}^3}$$

Carrying out this division on a calculator, the student obtains 0.56325. This must be rounded off to two significant figures: 0.56 cm^3. In future calculations, answers will be given to the proper number of significant figures; the round-off process will not be shown.

1.5 DIMENSIONAL ANALYSIS—AN APPROACH TO PROBLEM SOLVING

Before we go on, perhaps a word of caution is in order. Sometimes students have little difficulty reading their chemistry text or following the lecture and yet have difficulty on exams. In some instances the problem is lack of familiarity with terms. Often the problem is that the students have a passive but not an active understanding of the material. They can see how someone else has worked a problem, but they are unable to work any on their own. An active understanding involves being able to use the material in new situations, including especially working problems that are not identical to those used as sample exercises in the text. It is important to use the problems at the end of each chapter to test yourself to determine how well you are able to use the material in the chapter. Colored numbers indicate problems whose answers can be found at the back of the book. Bracketed numbers indicate problems of above-average difficulty. Wherever possible we have used an approach that can be referred to as **dimensional analysis** in solving problems. If you develop a facility with this approach, which is illustrated in the following discussion, your work will be much easier. If you need a review of basic mathematics, refer to Appendix A.

A number reported for a measured quantity is meaningless unless its units are specified. If units are treated as algebraic quantities, they can be carried through all calculations and will indicate whether the calculation has been performed correctly. This approach is illustrated in the following examples and in the sample exercises that follow.

Consider the conversion of mass from pounds to kilograms. If a man weighs 175 lb, what is his mass in kilograms? From Table 1.3 we have the following relationship: 1 kg = 2.205 lb. We can therefore write the following equalities:

$$1 = \frac{1\ \text{kg}}{2.205\ \text{lb}};\quad 1 = \frac{2.205\ \text{lb}}{1\ \text{kg}}$$

These equalities, which can be read as 1 kg per 2.205 lb and 2.205 lb per kilogram, are referred to as **unit conversion factors.** Multiplication of a quantity by these factors changes the units in which the quantity is expressed but not its value. To convert pounds to kilograms we choose the unit conversion factor that cancels pounds:

$$?\ \text{kg} = (175\ \cancel{\text{lb}})\left(\frac{1\ \text{kg}}{2.205\ \cancel{\text{lb}}}\right) = 79.4\ \text{kg}$$

Now consider a more complex conversion of units, the calculation of the number of inches in 3.00 km. We can begin by writing the equality we are working toward:

$$? \text{ in.} = 3.00 \text{ km}$$

From the relations shown in Tables 1.2 and 1.3 and from our basic knowledge of the English system we can write the following equalities:

$$1 \text{ km} = 1000 \text{ m}; \ 1 \text{ m} = 1.094 \text{ yd}; \ 1 \text{ yd} = 36 \text{ in.}$$

Therefore,

$$\frac{1000 \text{ m}}{1 \text{ km}} = 1; \ \frac{1.094 \text{ yd}}{1 \text{ m}} = 1; \ \frac{36 \text{ in.}}{1 \text{ yd}} = 1$$

If we multiply 3.00 km by these factors we have

$$\begin{aligned} ? \text{ in.} &= (3.00 \text{ km})\left(\frac{1000 \text{ m}}{1 \text{ km}}\right)\left(\frac{1.094 \text{ yd}}{1 \text{ m}}\right)\left(\frac{36 \text{ in.}}{1 \text{ yd}}\right) \\ &= 1.18 \times 10^5 \text{ in.} \end{aligned}$$

Each conversion factor is applied so as to cancel the units of the preceding factor. This converts kilometers successively to meters to yards to inches. Because we are left with the proper units, we know that the problem has been correctly set up. (There are exactly 1000 m in a kilometer and exactly 36 in. in a yard. The number of significant figures in the result, in this case three, is thus determined by the number of significant figures in the quantity being converted.)

Because 1 m = 1.904 yd, it is also true that 1 m/1.094 yd = 1. If this factor is applied instead of its reciprocal as above, the result is

$$\begin{aligned} ? \text{ in.} &= (3.00 \text{ km})\left(\frac{1000 \text{ m}}{1 \text{ km}}\right)\left(\frac{1 \text{ m}}{1.094 \text{ yd}}\right)\left(\frac{36 \text{ in.}}{1 \text{ yd}}\right) \\ &= 9.88 \times 10^4 \frac{\text{m}^2 \text{ in.}}{\text{yd}^2} \end{aligned}$$

Clearly the units do not cancel to give the desired units, so we know that we have not obtained the conversion we want.

SAMPLE EXERCISE 1.7

You have to pour 2.0 cubic yards (yd^3) of concrete for a patio. What is this volume in cubic meters (m^3)?

Solution:

$$\begin{aligned} ? \text{ m}^3 &= (2.0 \text{ yd}^3)\left(\frac{1 \text{ m}}{1.094 \text{ yd}}\right)\left(\frac{1 \text{ m}}{1.094 \text{ yd}}\right)\left(\frac{1 \text{ m}}{1.094 \text{ yd}}\right) \\ &= (2.0 \text{ yd}^3)\left(\frac{1 \text{ m}}{1.094 \text{ yd}}\right)^3 \\ &= (2.0 \text{ yd}^3)\left(\frac{1 \text{ m}^3}{1.309 \text{ yd}^3}\right) \\ &= 1.5 \text{ m}^3 \end{aligned}$$

SAMPLE EXERCISE 1.8

You are approaching a city and see a sign indicating a speed limit of 40 km/h. What is the corresponding speed in miles per hour?

Solution:

$$? \frac{\text{mi}}{\text{h}} = \left(40 \frac{\text{km}}{\text{h}}\right)\left(\frac{1000 \text{ m}}{1 \text{ km}}\right)\left(\frac{1.094 \text{ yd}}{1 \text{ m}}\right)\left(\frac{1 \text{ mi}}{1760 \text{ yd}}\right)$$
$$= 25 \frac{\text{mi}}{\text{h}}$$

SAMPLE EXERCISE 1.9

The acid in an automobile battery (a solution of sulfuric acid) has a density of 1.2 g/cm^3. What is the mass (in grams) of 200 mL (2.00×10^2 mL) of this acid?

Solution:

$$? \text{ g} = (2.00 \times 10^2 \text{ mL})\left(\frac{1 \text{ cm}^3}{1 \text{ mL}}\right)\left(1.2 \frac{\text{g}}{\text{cm}^3}\right)$$
$$= 2.4 \times 10^2 \text{ g}$$

Notice that density can be thought of as a unit conversion factor for converting volume to mass or vice versa.

SAMPLE EXERCISE 1.10

In Sample Exercise 1.1, the density of mercury was found to be 13.6 g/cm^3. Convert this to the basic SI units of kg/m^3.

Solution: Because the units in both numerator and denominator are to be changed, we need to employ two different unit conversion factors. These are:

$$1 = \frac{1 \text{ kg}}{1000 \text{ g}}; \; 1 = \frac{100 \text{ cm}}{1 \text{ m}}$$

$$\text{Density} = \left(\frac{13.6 \text{ g}}{1 \text{ cm}^3}\right)\left(\frac{100 \text{ cm}}{1 \text{ m}}\right)^3\left(\frac{1 \text{ kg}}{1000 \text{ g}}\right)$$
$$= 1.36 \times 10^4 \text{ kg/m}^3$$

Note that the first unit conversion factor is cubed to provide the correct dimension of volume.

Dimensional analysis cannot be used on all problems that we will work, and you should feel free to abandon it and work problems stepwise if you wish. However, you should *always* carry units throughout all of your calculations, making sure that they cancel properly. Whenever you finish a calculation, look at both the units and magnitude of your answer and ask yourself whether your answer makes any sense. This will help you to avoid making some embarrassingly simple errors.

FOR REVIEW

Summary

We have defined chemistry as the study of the properties, composition, and changes of **matter**. Chemistry has two origins: (1) the craft traditions such as metallurgy and (2) the more philosophical search for basic understanding of matter. Modern chemistry rests upon certain scientific **laws** arrived at through both qualitative observations and quantitative measurements. **Hypotheses** are devised to provide a tentative explanation for the laws. If the hypotheses are successful they become **theories.**

Measurements are made using the **metric system,**

which is based on the decimal system. Uncertainties associated with measurements can be expressed by use of significant figures. We have seen that when measured quantities are carried through calculations, it is important to keep track of units.

Learning goals

Having read and studied this chapter, you should be able to:

1 Use the metric system and list the basic SI units and the common prefixes.
2 Interconvert metric and English-system measurements using dimensional analysis.
3 Convert temperatures between the Fahrenheit, Celsius, and Kelvin scales.
4 Perform calculations involving density.
5 Determine the number of significant figures in a derived quantity.

Key terms

Among the more important terms and expressions used for the first time in this chapter are the following:

Density (Section 1.3) is mass per unit volume.

A hypothesis (Section 1.2) is a trial idea or explanation; a tentative theory.

An intensive property (Section 1.3) is one that is independent of the amount of material under consideration; an extensive property (Section 1.3) depends on the amount of material being considered.

A scientific law (Section 1.2) is a concise verbal or mathematical statement that summarizes one or more observed relationships between physical quantities.

Mass (Section 1.3) is a measure of the amount of "stuff" in an object. It measures the resistance of a stationary object to being moved. In SI units, mass is measured in kilograms.

Matter (introduction) is the physical material (the "stuff") of the universe; it is anything that occupies space and has mass.

Significant figures (Section 1.4) are the digits that indicate the precision with which a measurement has been made—those digits of a measured number that have uncertainty only in the last digit.

A theory (Section 1.2) is an explanation of a set of related observations.

EXERCISES*

Introductory concepts

1.1 Which of the following are matter: (a) air; (b) heat; (c) dust; (d) light; (e) water; (f) steak?

1.2 Examine your surroundings and try to list five synthetic and five nonsynthetic substances that you can see.

1.3 What is the difference between a scientific theory and a scientific law?

1.4 What is the principal reason for adopting one theory over a conflicting one?

1.5 How did Boyle's definition of an element place that concept on an experimental basis?

Metric system; SI units

1.6 What are the basic SI units appropriate to express the following: (a) the length of a racetrack; (b) the volume of a swimming pool; (c) the mass of a bar of gold; (d) the area of the state of Rhode Island; (e) the length in time of a lunar cycle?

1.7 What word prefixes indicate the following multipliers: (a) 1×10^3; (b) 0.1; (c) 1×10^6; (d) 1×10^{-6}; (e) 1×10^{-12}?

*Colored exercise number indicates answers are given in the "Answers to Selected Exercises" section of the text. Brackets indicate more difficult questions.

1.8 Fill in the blank with the missing unit: (a) 3.2×10^{-6} g = 3.2 ______; (b) 8.8×10^{-9} s = 8.8 ______; (c) 5.7×10^{-3} L = 5.7 ______; (d) 1.23×10^{-12} m = 1.23 ______.

1.9 Indicate whether the following units measure length, area, volume, or mass: (a) cm^3; (b) km^2; (c) mg; (d) dm.

1.10 (a) 4.53 kg is ______ mg; (b) 6.50 m is ______ cm; (c) 2.57 g is ______ kg; (d) 35 s is ______ μs; (e) 25.3 pm is ______ nm.

1.11 (a) 3×10^2 km = ______ m; (b) 82 ms = ______ μs; (c) 476 nm = ______ mm; (d) 326 g = ______ kg; (e) 1327 cm = ______ m.

1.12 It is sometimes said that aluminum is light whereas lead is heavy. How might this comparison be stated in a scientifically more precise fashion?

1.13 (a) A gem-quality ruby has a mass of 5.2 g and a volume of 1.3 cm^3. What is its density? (b) A piece of ebony wood has a mass of 15.3 g and a volume of 12.8 cm^3. What is its density?

1.14 The density of magnesium is 1.74 g/cm^3. (a) What is the volume of a piece of magnesium whose mass is 5.75 g? (b) What is the mass of a piece of magnesium having a volume of 10.3 cm^3?

1.15 Chloroform, a liquid once used as an anesthetic, has a density of 1.49 g/mL. (a) What is the volume of 5.00 g of chloroform? (b) What is the mass of 5.00 mL of chloroform?

1.16 A cube composed of one of the substances listed in Table 1.4 measures 2.00 cm on each edge and has a mass of 63 g. (a) What is its density? (b) What is the identity of the substance?

1.17 Make the following temperature conversions: (a) 25 °C to °F; (b) 98 °F to °C; (c) −25 °C to K; (d) 525 K to °C; (e) 75.0 °F to K.

1.18 Gallium metal has one of the largest liquid ranges of any substance. It melts at 30 °C and boils at 1983 °C. What are its melting and boiling points in °F?

Significant figures

1.19 Indicate which of the following are exact numbers: (a) the number of eggs in a dozen; (b) the area of Texas; (c) the number of days in September; (d) the number of centimeters in an inch; (e) the number of ounces in a pound; (f) the number of players on a baseball team.

1.20 Indicate the number of significant figures in each of the following: (a) 1302.1; (b) 43.55; (c) 0.00388; (d) 7.12×10^{-2}; (e) 5.0×10^{2}; (f) 1200.

1.21 How many significant figures are there in each of the following numbers: (a) 0.34×10^{6}; (b) 0.058; (c) 3000; (d) 218; (e) 3.02×10^{-3}; (f) 6.300?

1.22 Round off each of the following numbers to four significant figures: (a) 4,567,985; (b) 6.3375×10^{3}; (c) 0.00238866; (d) 0.98758; (e) 0.322589×10^{-3}.

1.23 Express the following numbers in appropriate exponential notation: (a) 1245; (b) 65,000 to show three significant figures; (c) 59,750 to show four significant figures; (d) 0.00456.

1.24 Carry out the following operations and round off the answers to the appropriate number of significant figures: (a) 3.22×0.17; (b) $4568/1.3$; (c) $1.987/(3.46 \times 10^{8})$; (d) $0.0003/162$; (e) $(12.3 + 0.092)/8.3$; (f) $328 \times (0.125 + 5.43)$.

1.25 Carry out the following mathematical operations and round off to the appropriate number of significant figures: (a) $(16.788 - 15.990)/118.9$; (b) $3.4 \times 2.668/1012$; (c) $15.67 + 0.8896 + 2.0 + 1.2 \times 10^{-4}$; (d) $(4.43 \times 1.254) + 0.18$.

Conversion of units; dimensional analysis

1.26 (a) What is the engine piston displacement in liters of an engine whose displacement is listed as 320 in.^{3}? (b) If an auto is traveling at 62 mph, what is its speed in km/h? (c) A car weighing 4.85×10^{3} lb has a mass of how many kilograms? (d) What is a person's height in meters if that height is 5 ft 10 in.? (e) How many liters of gas can be held in a gas tank whose capacity is 12.5 U.S. gallons?

1.27 (a) If a person weighs 155 lb, what is the person's mass in kilograms? (b) If you are 145 cm tall, what is your height in inches? (c) If an automobile is able to travel 22.0 mi on a gallon of gasoline, what is the gas mileage in km/L? (There are 1.61 km in a mile.) (d) In 1978 the world's record for the 100-m dash was 9.9 s. To cover 100 m in 9.9 s, what must the average speed be in miles per hour? (e) A wine barrel whose capacity is 31 gal can contain how many liters of wine?

1.28 Convert the measures in the following statements to an appropriate metric unit. (a) The longest road tunnel is St. Gotthard, Switzerland-Italy: 10 mi 120 yd. (b) The biggest dam is at New Cornelia Tailings, United States: 274,026,000 yd^3 of material used. (c) The deepest mine is the Western Deep Levels Gold Mine, South Africa: 11,647 ft. (d) The great pyramid of Cheops is 481 ft high and 755 ft on each side at the base and is built of 2,300,000 blocks weighing 2.5 tons each. (e) The 1928 Indianapolis 500-mi race was won by Lou Meyer, with an average speed of 99.482 mph.

1.29 A metal block has dimensions of 4.5 cm × 12.54 cm × 1.25 cm. Calculate its volume in cubic centimeters; in cubic meters.

1.30 A cylindrical container of radius r and height h has a volume of $\pi r^2 h$. (a) Calculate the volume in cubic centimeters of a cylinder with radius 6.5 cm and a height 28.6 cm. (b) Calculate the volume in cubic meters of a cylinder that is 8.0 ft high and 20.0 in. in diameter. (c) Calculate the mass in kilograms of water required to fill the cylinder of part (b) if the dimensions listed are the inside dimensions. The density of water is 1.00 g/cm^3.

1.31 An organic liquid that occupies a volume of 3.47 L has a mass of 4.268 kg. What is its density in g/cm^3?

1.32 The maximum allowable concentration of carbon monoxide in urban air is 10 mg/m^3 over an 8-h period. At this level, what mass of carbon monoxide is present in a room that is 2.5 × 15 × 40 m in dimensions?

Additional exercises

1.33 What is the difference between an intensive and an extensive property?

1.34 Which of the following are intensive properties: color; mass; density; melting point; volume?

1.35 Determine the number of (a) centimeters in 1 km; (b) picoseconds in 1 ms; (c) micrograms in 34.2 mg; (d) kilograms in 3.05×10^{5} g.

1.36 Convert the quantities in the following statements to metric units. (a) Women of medium frame, those 5 ft 3 in. in height, should weigh in the range 110–122 lb. (b) A man 5 ft 9 in. in height, body weight about 154 lb, should consume about 2.5 oz of protein each day (1 oz = 28.35 g).

1.37 If Jules Verne expressed the title of his famous book *Twenty Thousand Leagues Under the Sea* in basic SI units, what would the title be? (1 league = 3.45 mi; 1 mi = 1609 m).

1.38 (a) A fathom, used as a measure of water depths, is defined as 1.8288 m, exactly. What is the depth in fathoms of Lake Superior, which is 1302 ft at maximum depth? (b) A furlong, a measure of distance used in horse racing, is defined as 201.168 m. What is the distance in miles of the Kentucky Derby, which is a 10.000-furlong race?

1.39 What is the mass of 1 dm^3 of mercury, density 13.6 g/cm^3?

1.40 A graduated cylinder weighs 204.58 g. Some liquid is added, and the volume is read as 58.3 mL. The mass of cylinder and liquid is 251.65 g. Calculate the density of the liquid.

1.41 The density of air at ordinary atmospheric pressure and 25°C is 1.19 g/L. What is the weight in kilograms of the air in a room that measures $8.2 \times 13.5 \times 2.75$ m?

1.42 Water has a density of 1.00 g/cm^3 whereas gasoline has a density of 0.64 g/cm^3. What mass of liquid, in grams, fills a 16.5-gal tank if the liquid is (a) water; (b) gasoline?

1.43 What is the radius, in centimeters, of a metal sphere whose mass is 2.00×10^2 g if the metal is: (a) iron, density = 7.86 g/cm^3; (b) aluminum, density = 2.70 g/cm^3; (c) lead, density = 11.3 g/cm^3. The volume of a sphere is given by the equation $V = 4\pi r^3/3$.

1.44 A graduated cylinder on a balance has a mass of 57.832 g. An organic liquid, toluene, with a density of 0.866 g/cm^3 is added until the combined mass reads 87.127 g. What is the volume of the liquid in the graduated cylinder?

1.45 A graduated cylinder contains 20.0 mL of water. An irregularly shaped object is placed in the cylinder and the water level rises to the 31.2-mL mark. If the object has a mass of 49.7 g, what is its density?

1.46 An organic liquid that occupies a volume of 3.47 L has a weight of 4.268 kg. What is the density of this liquid in kilograms/liter?

1.47 A 600-mL bottle of methanol is sold in a discount store as a gasoline additive to prevent freezing, for $0.39. The density of methanol is 0.79 g/cm^3 at room temperature. What is the price of the methanol per kilogram?

1.48 A few years ago, a cartoon pictured a thief making his getaway, gun in one hand and a bucket of gold dust in the other. If the bucket had a volume of 8 qt and was full of gold whose density is 19.3 g/cm^3, what was its mass? Comment on the thief's strength.

1.49 (a) What is the temperature, in °C, of an animal whose body has a temperature of 100.6°F? (b) On August 24, 1960, the temperature at the Vostok Station in Antarctica was recorded as −127°F. What is this temperature in °C? (c) Helium has the lowest boiling point of any liquid, 4 K. What is the boiling point in °C? in °F?

1.50 Is the use of significant figures in each of the following statements appropriate? Why or why not? (a) The 1976 circulation of *Reader's Digest* magazine was 17,887,299. (b) In the United States, 1.4 million persons have the surname Brown. (c) The average annual rainfall in San Diego, California, is 20.54 in. (d) The population of East Lansing, Michigan, in 1979 was 51,237.

1.51 The recommended adult dose of Elixophyllin, a drug used to treat asthma, is 6 mg/kg of body weight. Calculate the dose, in milligrams, for a 170-lb person.

2 Atoms, Molecules, and Ions

Our present understanding of the changes we see around us—such as the melting of ice and the burning of wood—is intimately tied to our understanding of the nature and composition of matter. For example, before we can hope to understand what is happening when ice melts we must know what ice is—what it is composed of. It is possible to resolve or separate matter into a great variety of different **pure substances.** These are materials or portions of matter whose composition and intrinsic properties are uniform throughout. For example, seawater can be separated into several different pure substances, the most abundant being water and ordinary table salt (sodium chloride).

In this chapter we examine the composition of matter. We attempt to answer many fundamental questions: What types of pure substances are there? How can matter be separated into pure substances? Can pure substances be broken into simpler components? How do substances differ at the microscopic or atomic level? How do we represent the compositions of substances and how do we name them?

There is much more in the answers to these questions than can be contained in a single chapter. You should regard this chapter as, in part, an introduction. At this point you need to master some basic ideas and concepts and acquire a chemical vocabulary to facilitate laboratory work and further study of the text.

2.1 STATES OF MATTER

As a starting point in answering the questions we have posed, it is useful to note that matter exists in three states: gas (also known as vapor), liquid, or solid. A **gas** has neither a shape of its own nor a fixed volume. It takes the shape and volume of any container into which it is introduced. It can be compressed to fit a small container; it will expand to occupy a large one. Air is a gas.

A **liquid** has no specific shape; it assumes the shape of the portion of any container that it occupies. It does not expand to fill the entire container; it has a specific volume. Furthermore, a liquid is only slightly compressible. Water and gasoline are common liquids.

A solid has a firmness that is not associated with either gases or liquids. It has a fixed volume and shape. Like liquids, solids are not compressible to any appreciable degree. Numerous objects around us are solids—nails, coins, salt, and sugar to name a few.

The state of a substance depends on temperature and pressure. Above 100°C, water exists as a gas, known as steam. Between 0°C and 100°C, it exists as a liquid. Below 0°C, it exists as a solid—ice.

Changes of state, such as the change of ice to liquid water, are examples of physical changes. Physical changes are ones that do not involve creation of new substances; they involve no change in the composition of the specimen of matter under consideration. Chemical changes, also called chemical reactions, involve conversion of one substance into another. For example, when hydrogen burns in air, it undergoes a chemical change in which it is converted to water. Every pure substance has a unique set of properties or characteristics that allows us to recognize it and distinguish it from other substances. Chemical properties are those properties that refer to the way a substance is able to change into other substances (its reactivity, how it "reacts"). The physical properties of a substance are those that do not involve a change in the chemical identity of the substance.

SAMPLE EXERCISE 2.1

Chlorine is a greenish-yellow gas with a density of 3.21 g/L. It can be changed to a liquid by cooling to −34.6°C; it reacts explosively with sodium to form sodium chloride (table salt). Which of these properties are physical properties and which are chemical?

Solution: The color and density of chlorine and the temperature at which it changes state, from a gas to a liquid, are all physical properties. They do not involve a change of chlorine into any other substance. The ability of chlorine to react explosively with sodium is a chemical property. In reacting with sodium, chlorine is changed into a different substance, sodium chloride.

2.2 ELEMENTS, COMPOUNDS, AND MIXTURES

All specimens of matter can be classified either as pure substances or as mixtures of two or more substances. Most matter around us consists of mixtures. Mixtures are characterized by variable composition and by the fact that they can be separated by physical means. That is, mixtures can be separated by taking advantage of differences in physical properties such as boiling points. For example, we can recognize that blood is a mixture because its composition may vary in many ways, such as in its iron content. Furthermore, blood can be separated into two components, packed cells and plasma, by centrifugation, a physical method of separation.

In some mixtures, such as that of water and clay, the components are readily distinguished. Such mixtures are said to be heterogeneous. In other cases the mixture may be uniform throughout, or homogeneous. Homogeneous mixtures are known as solutions. For example, when salt and water are mixed, a homogeneous mixture, or solution, forms; the salt is said to dissolve in the water.

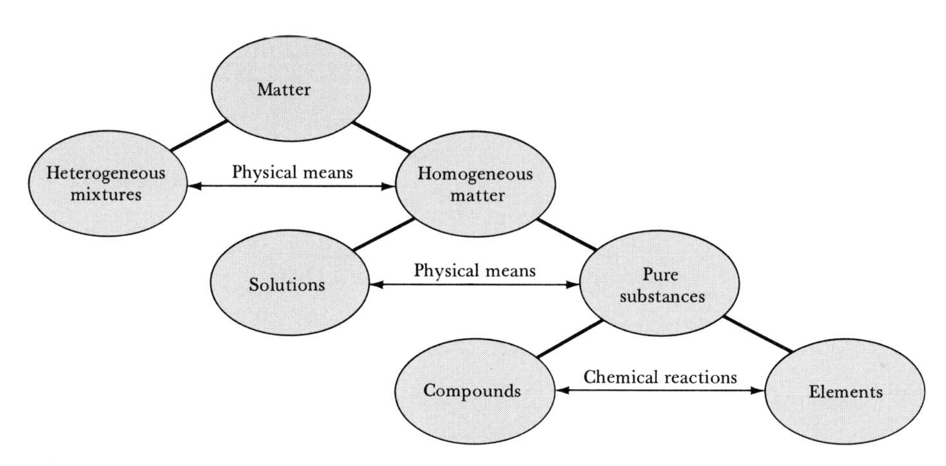

FIGURE 2.1 Classification of matter.

Pure Substances

Pure substances are homogeneous, but unlike solutions they have a constant, invariable composition. Each substance also has a distinct set of intrinsic properties by which we can identify it. (The composition and properties of solutions vary with the relative amounts of the components from which the solution is formed.)

There are two classes of substances: elements and compounds. Elements are those substances that cannot be decomposed into simpler substances. Compounds are substances that may be decomposed (by chemical means) into two or more elements. The classification of matter into mixtures, substances, compounds, and elements is summarized in Figure 2.1.

Elements

Elements are the basic substances out of which all matter is composed. In light of the seemingly endless variety in our world, it is perhaps surprising that there are only 106 known elements. Not all of these are of equal importance or abundance. Ninety percent, by weight, of the portion of the earth to which we have access for raw materials is composed of only five elements: oxygen, silicon, aluminum, iron, and calcium. Over 90 percent of the human body is composed of just three elements: oxygen, carbon, and hydrogen. At the other extreme, about 20 elements are either found in nature in only minute traces or are prepared in the laboratory and are therefore available in only very small quantities.

Some of the more familiar elements are listed in Table 2.1, together with the chemical symbols used to denote them. All of the known elements are listed on the inside front cover of this text. It can be seen that the abbreviation or symbol for an element consists of one or two letters with the first letter capitalized. These symbols are usually derived from the English name (first and second columns of Table 2.1), but sometimes they are derived instead from a foreign name (third column). You will need to know these symbols and to learn others as we encounter them in the text.

TABLE 2.1 Some common elements and their symbols

Carbon (C)	Aluminum (Al)	Copper (Cu, from *cuprum*)
Fluorine (F)	Barium (Ba)	Iron (Fe, from *ferrum*)
Hydrogen (H)	Calcium (Ca)	Lead (Pb, from *plumbum*)
Iodine (I)	Chlorine (Cl)	Mercury (Hg, from *hydrargyrum*)
Nitrogen (N)	Helium (He)	Potassium (K, from *kalium*)
Oxygen (O)	Magnesium (Mg)	Silver (Ag, from *argentum*)
Phosphorus (P)	Platinum (Pt)	Sodium (Na, from *natrium*)
Sulfur (S)	Silicon (Si)	Tin (Sn, from *stannum*)

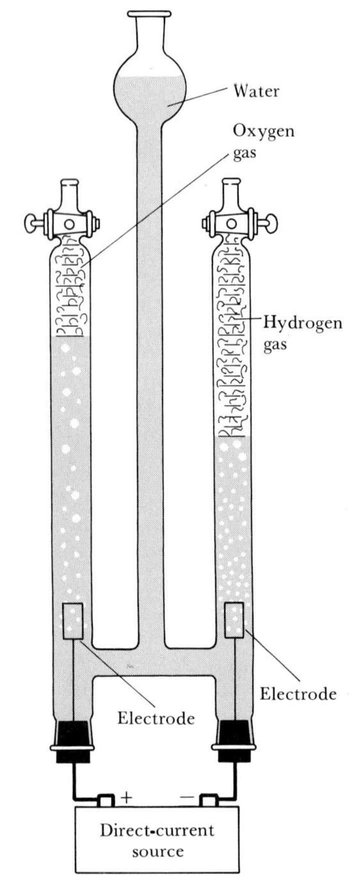

FIGURE 2.2 Decomposition of the compound water into the elements hydrogen and oxygen by passing a direct electrical current through it.

Compounds

Compounds are substances composed of two or more elements united chemically in definite proportions by mass. We can gain a clearer understanding of the distinctions between elements, compounds, and mixtures by examining a common substance, water. With the discovery of methods of generating electricity, it was found that water could be decomposed into the elements hydrogen and oxygen, as shown in Figure 2.2. This decomposition clearly indicates that water is not an element. On the other hand, it is not merely a mixture of hydrogen and oxygen either. The properties of water are clearly unique and much different from those of its constituent elements as seen in Table 2.2. Furthermore, the composition of water is not variable; pure water consists of 11 percent hydrogen and 89 percent oxygen by mass, regardless of its source.

The observation that the elemental composition of a pure compound is always the same is known both as the **law of constant composition** and the **law of definite proportions.** Although this law has been known for over 150 years, the general belief persists among some people that there is a fundamental difference between compounds prepared in the laboratory and the corresponding compounds found in nature. However, a pure compound has the same composition and properties regardless of source. Both chemists and nature must use the same elements and operate under the same natural laws. Differences in composition and properties between substances indicate that the compounds are not the same or that at least one is impure. Harmful chemicals such as strychnine are made by nature as well as by people. Nature is capable of making many compounds, especially those found in living systems, that chemists are not yet able to prepare. Chemists, on the other hand, have succeeded in forming many substances, for example, synthetic fibers and certain pesticides, not found in nature.

TABLE 2.2 Comparison of water, hydrogen, and oxygen

	Water	Hydrogen	Oxygen
Physical state[a]	Liquid	Gas	Gas
Normal boiling point	100°C	−253°C	−183°C
Density[a]	1.00 g/mL	0.090 g/L	1.43 g/L
Combustible?	No	Yes	No

[a] At room temperature and atmospheric pressure.

SAMPLE EXERCISE 2.2

Identify the following as element, compound, or mixture: milk, gold, table salt, ink.

Solution: Gold (Au) is an element (refer to the table of elements on the inside front cover of the text). Table salt is a compound that we have now mentioned several times. It is composed of the elements sodium (Na) and chlorine (Cl). Both milk and ink are mixtures and are recognized as such by their variable compositions.

Many procedures have been developed to separate (or partly separate) mixtures into their component substances. One of the most common separation procedures is *filtration*. Filtration is used to separate solids from liquids as illustrated in Figure 2.3. Other examples of separations based on filtration abound outside the lab: air and oil filters in automobiles, air filters on furnaces in our homes, filters in coffee makers, and so forth.

Another familiar separation procedure is *distillation*. This procedure is based on differences in the volatilities of substances (that is, differences in the ease with which substances form gases). It is the procedure by which the "moonshiner" obtains whiskey using a "still," and by which large petrochemical plants achieve the separation of petroleum into gasoline, diesel fuel, lubricating oil, and so forth. Figure 2.4 shows a simple laboratory distillation apparatus. Imagine that the solution in the distilling flask is a salt–water mixture. Water, of course, vaporizes at a much lower temperature than salt, so the water boils off, leaving the salt in the distilling flask. The water is condensed by cooling elsewhere in the system and collected. The liquid obtained by condensation in a distillation is known as the distillate.

A number of separation procedures are based on differences in the degree to which various substances are adsorbed onto the surface of an inert material. (An inert material is one that does not undergo a chemical change.) The difference between *ad*sorption and *ab*sorption should be noted: *Adsorption* means adherence to a surface, whereas *absorption* denotes passage into the interior; water is absorbed by a sponge. Separations that utilize adsorption differences are known as *chromatographic procedures*. The name *chromatography* arose from its use in separating pigments; it means literally "graphing of colors," For example, a solution containing the colored pigments of a leaf may be washed through a column packed with alumina (aluminum oxide). The various components will move through the column at different speeds due to differ-

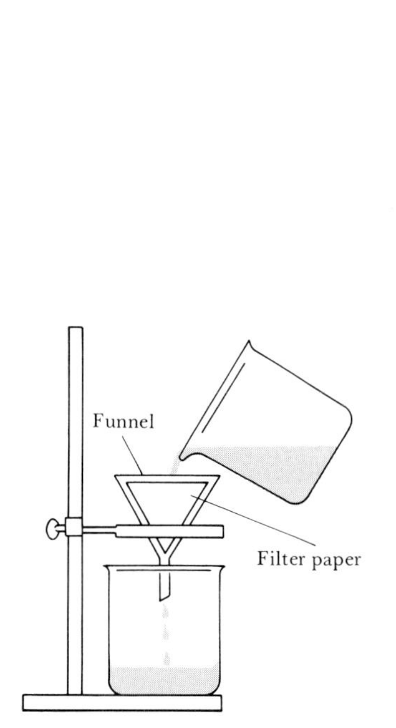

FIGURE 2.3 Separation of a liquid and solid by filtration.

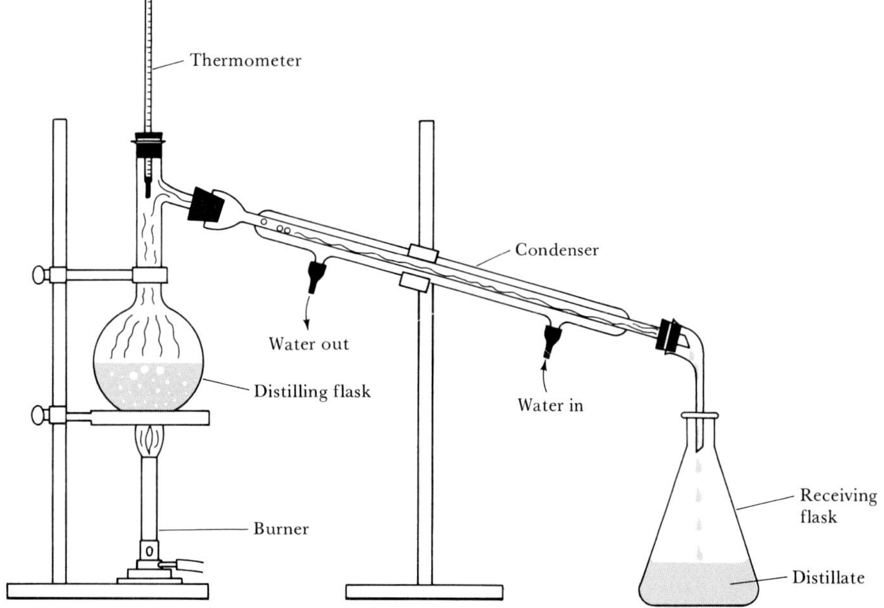

FIGURE 2.4 Simple laboratory distillation setup. Cool water circulating through the jacket of the condenser causes the liquid to condense.

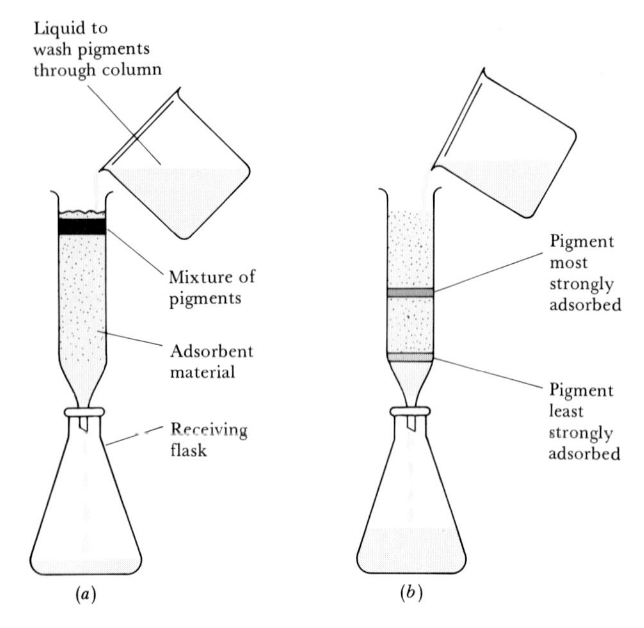

FIGURE 2.5 Separation by column chromatography: (*a*) initial stage; (b) after some time has elapsed.

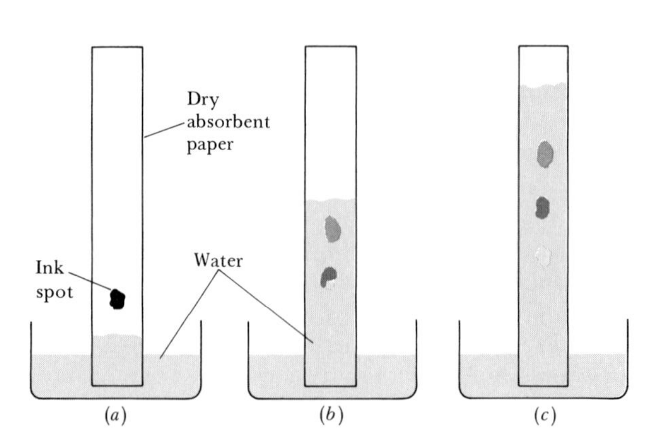

FIGURE 2.6 Separation of ink into components by paper chromatography. (*a*) Water begins to move up the paper. (*b*) Water moves past the ink spot, lifting different components of the ink at different rates. (*c*) Water has separated the ink into three different components.

ences in the degree to which they are adsorbed. This type of separation, illustrated in Figure 2.5, is known as **column chromatography.** By using various instruments instead of visual inspection to determine the location of the components, it is not necessary to restrict the technique to colored materials.

When the adsorbent material is paper and the solution containing the mixture moves upward through the paper, the technique is called **paper chromatography.** As the liquid moves upward on the paper, it carries the mixture along. The components that are adsorbed most strongly to the paper move most slowly. The technique is illustrated in Figure 2.6.

2.3 THE ATOMIC THEORY

The classification of matter as mixtures and substances, compounds and elements, is central to modern chemistry. It helps us to systematize many chemical facts. However, it also raises a number of questions. Why is one element different from another? Why is a compound different from a mixture? Why do elements combine to form compounds? These questions are connected with the facts that we have already discussed. We need a theory to help us explain these facts. We therefore shift our attention now to a discussion of our present theory of matter, the atomic theory. We will find that this theory provides answers to the questions we have raised.

The seeds of the atomic theory go back at least to the time of the ancient Greeks. The Greeks pondered a seemingly abstract question: Can matter be divided endlessly into smaller and smaller pieces, or is it composed of some ultimate particle that cannot be further divided? The main line of Greek thought, following the views of Plato and Aristotle, was that matter is continuous. However, some Greek philosophers, notably Democritus, disagreed with this view and argued that matter was composed of small indivisible particles that Democritus called *atomos,* meaning indivisible. This atomic concept was also central to the natural philosophy of the Roman poet and philosopher Lucretius, who lived in the first century B.C. He wrote a famous poem, *De Rerum Natura* (On the Nature of Things), in which he elaborated at length on the atomic view of matter.

Even if it were granted that matter is atomic in nature, the question arises how the atoms of different substances differ from one another. Lucretius suggested that the atoms of substances that have a bitter taste have barbs on their surfaces that scrape the tongue, whereas the atoms of substances with a bland taste must have a smooth surface. Not much improvement in the atomic view of matter occurred in the 18 centuries following Lucretius. The philosophical ideas of Plato and Aristotle, neither of whom accepted the atomistic view of matter, held sway in European thought for many centuries. Even though the atomic idea was occasionally revived, early proponents of the particulate theory of matter relied largely on intuition to support their views. During this long period, however, there was a thin, intermittent stream of experimental work. Much of it was prompted by erroneous notions, such as the alchemical belief that common metals such as lead might be transformed into precious metals. Nevertheless, experience of how chemical substances react with one another accumulated, and more quantitative methods of studying chemical reactions were developed. The way was prepared for a new and more meaningful statement of an atomic theory. It came, in the early years of the nineteenth century, from John Dalton, an English schoolteacher (Figure 2.7). Dalton's atomic theory, published in the period 1803–1807, was strongly tied to experimental observation. His efforts were so successful that his theory has dominated our thinking since his time and has had to undergo little revision.

The basic postulates of Dalton's theory were as follows:

1 Each element is composed of extremely small particles called atoms.
2 All atoms of a given element are identical.

FIGURE 2.7 John Dalton (1766–1844) was the son of a poor English weaver. Because he was a Quaker, Dalton's quiet life-style stands in contrast to the life-styles of Priestley and Lavoisier. Dalton began teaching at the age of 12; he spent most of his years in Manchester, where he taught both grammar school and college. His lifelong interest in meteorology led him to study gases and hence to chemistry and eventually to the atomic theory. It was perhaps because Dalton's training was not in chemistry that he was able to approach problems with a viewpoint different from that of chemists of the time. (*Library of Congress*)

3 Atoms of different elements have different properties (including different masses).
4 Atoms of an element are not changed into different types of atoms by chemical reactions; atoms are neither created nor destroyed in chemical reactions.
5 Compounds are formed when atoms of more than one element combine.
6 In a given compound, the relative number and kind of atoms are constant.

This theory provides us with a mental picture of matter. As represented schematically in Figure 2.8, we visualize an element as being composed of tiny particles called atoms. **Atoms** are the basic building blocks of matter; they are the smallest units of an element that can combine with other elements. Compounds involve atoms of two or more elements combined in definite arrangements. Mixtures do not involve the intimate interactions between atoms that are found in compounds.

Dalton's theory embodies several simple laws of chemical combination that were known at the time. Because atoms are neither created nor destroyed in the course of chemical reactions (postulate 4), it is readily evident that matter is neither created nor destroyed in such reactions. Thus we have the **law of conservation of matter,** which was discovered by Lavoisier. This law is one of the principal topics of Chapter 3. The law of constant composition, which was cited in Section 2.2, is explained by postulate 6: In a given compound the relative number and kind of atoms is constant. A third law discovered by Dalton and consistent with his theory is the **law of multiple proportions:** When two elements form more than one compound, for a fixed mass of one element the masses of the second element are related to each other by small whole numbers. For example, the substances water and hydrogen peroxide both consist of the elements hydrogen and oxygen. In water there are 8.0 g of oxygen for each gram of hydrogen, whereas in hydrogen peroxide there are 16.0 g of oxygen for each gram of hydrogen. The ratio of the masses of oxygen that combine with a gram of hydrogen in these compounds is in the ratio of the small whole number two: Hydrogen peroxide has twice as much oxygen per unit mass of hydrogen as water does. Using the atomic theory, we understand this to mean that hydrogen peroxide contains twice as many oxygen atoms per hydrogen atom as does water. We now know that water contains one oxygen atom for each two hydrogen atoms,

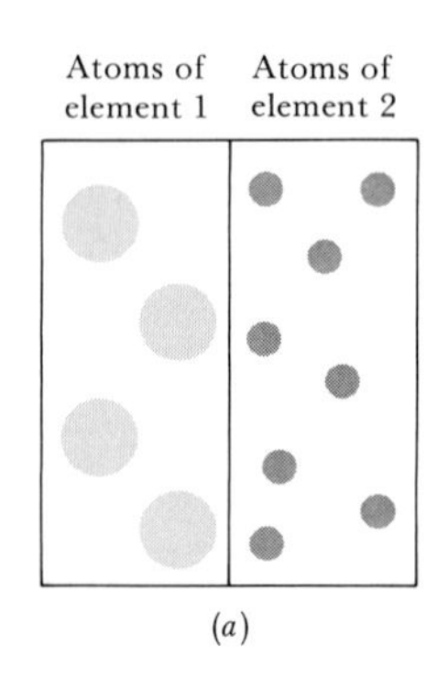

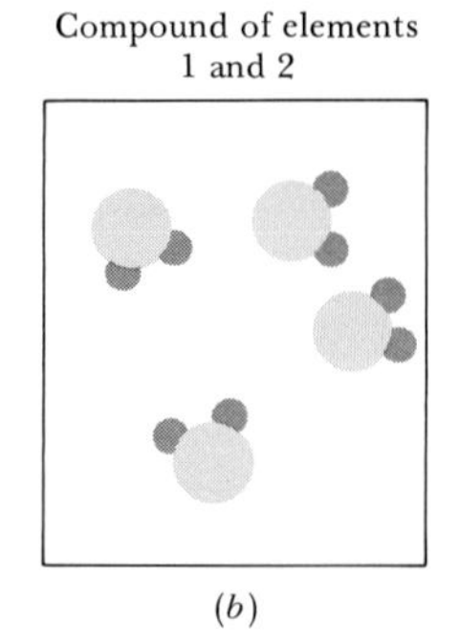

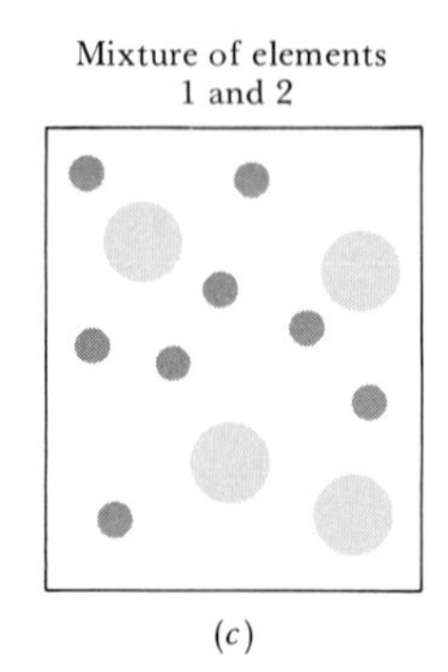

FIGURE 2.8 Difference between elements, compounds, and mixtures as visualized through Dalton's atomic theory. (*a*) Elements are composed of small particles called atoms. All atoms of a given element are identical; atoms of different elements are different. (*b*) Compounds involve atoms of two or more elements combined in definite arrangements. (*c*) Mixtures have variable compositions. There is no restriction on the relative numbers of atoms of elements 1 and 2.

whereas hydrogen peroxide contains two oxygen atoms for each two hydrogen atoms.

Thus we see that the atomic theory ties together many observations and helps us explain them. To our earlier question of what makes one element different from another we can now answer that they have different types of atoms. However, this explanation only begs a further question: How are the atoms of different elements different from each other? We need to consider the structure of the atom to answer this question. This topic is taken up in the next section. We will see that as we begin to understand the structure of the atom, we will begin to understand many more aspects of matter.

2.4 THE STRUCTURE OF THE ATOM

Our present understanding of atomic structure is very precise. Chemists and physicists are able to use a wide range of sophisticated instruments to measure in great detail all the properties of individual atoms. It is therefore difficult to imagine that only 100 years ago scientists knew very little about atoms beyond what was contained in Dalton's atomic theory. Dalton and his contemporaries viewed the atom as an indivisible object. However, data slowly accumulated to indicate that the atom has a substructure of smaller particles. We will consider here just a few of the most important experiments that led to our present model of atomic structure.

Cathode Rays and Electrons

In the mid-1800s a number of investigators began to study electrical discharge through partially evacuated tubes. Radiation is produced within such tubes when voltages become high enough to permit current flow (about 1000 volts). This radiation became known as **cathode rays** because it emanated from the negative electrode, or cathode. This and a number of additional facts suggested that the radiation consisted of a stream of negatively charged particles that were named "electrons." For example, the rays travel in straight lines in the absence of magnetic or electric fields. However, they are deflected by magnetic and electric fields in a manner expected for negatively charged particles. The behavior of a negatively charged particle in a magnetic field and in an electric field is shown in Figure 2.9. The movement of the rays can be deter-

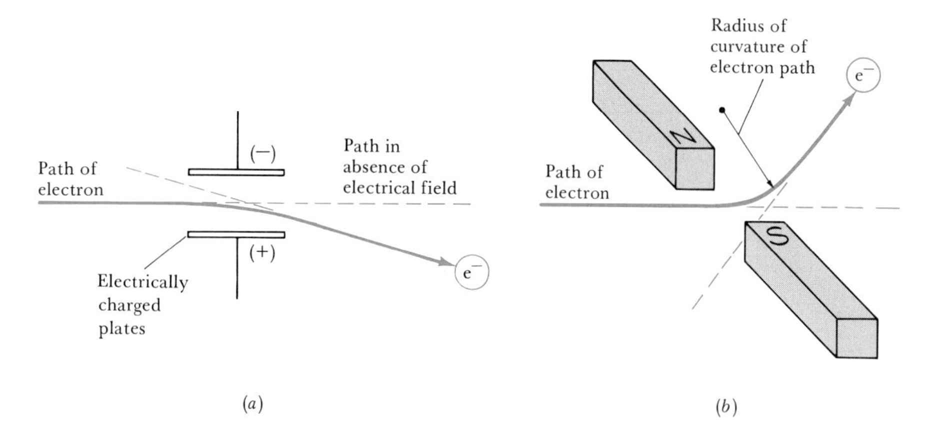

FIGURE 2.9 The path of an electron (*a*) in an electric field, and (*b*) in a magnetic field.

FIGURE 2.10 J. J. Thomson (1856–1940). (*Copyright Cavendish Laboratory, University of Cambridge*)

mined because of their ability to cause certain materials, including glass, to give off light, or fluoresce. In fact, a television picture tube is a cathode-ray tube; the television picture results from fluorescence from the television screen. Because the rays (electrons) were found to be independent of the nature of the cathode material, it was deduced that they are a basic component of all matter.

In 1897 the British physicist J. J. Thomson (Figure 2.10) was able to measure the ratio of the electrical charge to the mass of the electron using a cathode-ray tube such as that shown schematically in Figure 2.11. When only the magnetic field is turned on, the electron strikes point A of the tube. When the magnetic field is off and the electric field is on, the electron strikes point C. When both the magnetic and electric fields are off or when they are balanced so as to cancel each other's effects, the electron strikes point B. By carefully and quantitatively determining the

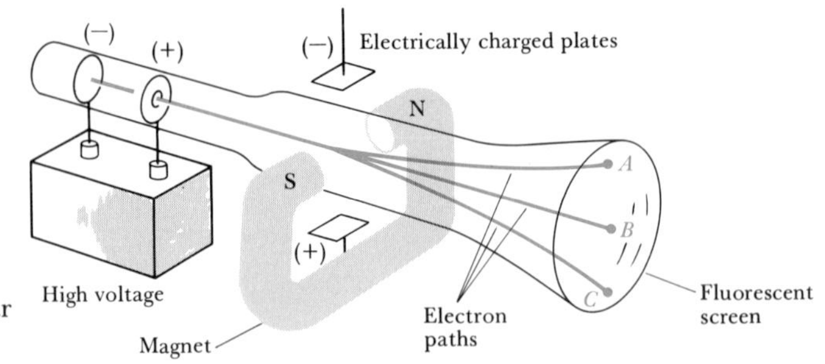

FIGURE 2.11 Cathode-ray tube with perpendicular magnetic and electric fields.

effects of magnetic and electric fields on the motion of the cathode rays, Thomson was able to determine the charge-to-mass ratio of 1.76×10^8 coulombs per gram.*

The external force exerted on an electron moving in a magnetic field is given by Hev, where H is the strength of the magnetic field, e is the charge on the electron, and v is its velocity. The motion of the electron is determined by the balance of this force and the tendency of the electron to continue its straight-line motion (the centrifugal force on the electron). This latter force is given by the relation mv^2/r, where m is the mass of the electron, v is its velocity, and r is the radius of the curved path taken by the electron as it moves through the magnetic field (refer to Figure 2.9). Thus the path taken by the electron must be consistent with the equation

$$Hev = \frac{mv^2}{r}$$

By rearranging this equation, we see that the radius of the path of the electron moving through the magnetic field is given by $r = mv^2/Hev = mv/He$. The larger the radius, the smaller the deflection of the particle from the straight-line path that it would take in the absence of the magnetic field. Thus, the more massive the particle, the greater its velocity, the smaller the magnetic field, and/or the smaller the charge on the particle, the greater the tendency of the particle to continue its straight-line motion.

The relation given above can also be arranged into the form $e/m = v/rH$. The velocity of the electron can be determined by balancing the effects of the magnetic field against a perpendicular electric field so that the electron moves in a straight line. The magnitude of the force exerted by the electric field on the electron is given by Ee where E is the strength of the electric field. For the electron moving in a straight line through the balanced magnetic and electric fields

$$Hev = Ee$$

so that

$$v = \frac{E}{H}$$

Thus, from the first relation, $e/m = v/rH = E/H^2r$, where all of the quantities on the right side of the equation can be determined experimentally. Thus, Thomson's experiment permitted determination of the ratio e/m for the electron.

In 1909 Robert Millikan of the University of Chicago determined the charge on the electron by measuring the effect of an electric field on the rate at which charged oil droplets fall under the influence of gravity. The apparatus that he used is shown schematically in Figure 2.12. The rate at which the droplets fall in air is determined by their size and mass. By watching a particular droplet, Millikan could measure its rate of fall and calculate from this its mass. The experiment was arranged so that a

*The coulomb (C) is the SI unit for electrical charge.

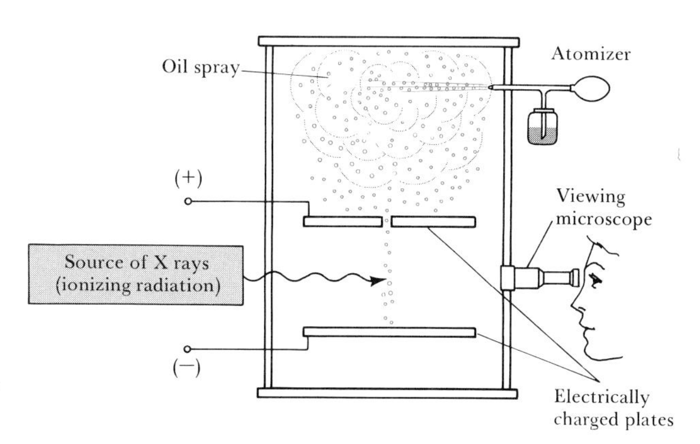

FIGURE 2.12 Schematic represention of Millikan's apparatus for studying the rate of fall of oil droplets.

source of radioactivity was near the droplets. The radioactive source caused charges to form. The charged particles floating around in the air would often become attached to an oil droplet. When an electrical charge was applied to the plates, charged oil droplets in the region between the plates would be acted upon by the electric field. Their fall could be accelerated, retarded, or even reversed, depending on the charge on the droplet and the polarity of the voltage applied to the plates. By carefully measuring the effects of the electrical field on the movements of many droplets, Millikan found that the charge on the oil drops was always an integral multiple of 1.60×10^{-19} C, which he deduced was the charge of the electron. The mass of the electron, 9.11×10^{-28} g, was then calculated by combining Millikan's value of the charge with Thomson's charge-to-mass ratio:

$$\text{Mass} = \frac{1.60 \times 10^{-19}\ \text{C}}{1.76 \times 10^{8}\ \text{C/g}} = 9.11 \times 10^{-28}\ \text{g}$$

Radioactivity

The discovery of radioactivity by the French scientist Henri Becquerel in 1896 provided additional evidence for the complexity of the atom. Becquerel's imagination had been captured by W. C. Roentgen's discovery of X rays, which had been reported in January 1896. Roentgen had been quick to grasp the practical importance of his discovery, and within a short time X rays had been used in medicine. Members of the international scientific community also sensed that this was something big. Becquerel was well aware that certain substances upon exposure to sunlight become luminous, a phenomenon referred to as fluorescence. He sought to determine whether such fluorescent substances gave off X rays. In his initial experiments, Becquerel chose to work with a fluorescent uranium mineral. He placed this in the sunlight over a photographic plate that had been carefully wrapped to protect it from the direct radiation of the sun. When the plate was developed, he found the image of the mineral on the plate. Toward the end of February 1896, Becquerel incorrectly reported that penetrating rays, presumably X rays, could be induced by sunlight and emitted as part of fluorescence. However, the weather turned bad, and Becquerel had to postpone further studies. While the sun stayed behind the clouds, Becquerel kept the mineral and the wrapped photographic plate in a desk drawer. On March 1, 1896, he decided to develop the plate, not expecting to find any images. He was surprised to find very intense silhouettes. Becquerel concluded correctly this time that the mineral was producing a spontaneous radiation and referred to this phenomenon as **radioactivity**. At Becquerel's suggestion, Marie Sklodowska Curie (Figure 2.13) and her husband, Pierre, began their famous experiments to isolate the radioactive components of the mineral, called pitchblende.

Further study of the nature of radioactivity, principally by the British scientist Ernest Rutherford, revealed three types of radiation—alpha (α), beta (β), and gamma (γ) radiation. Each type differed in electrical behavior and penetrating ability. The behavior of these three types of radiation in an electric field is shown in Figure 2.14.

FIGURE 2.13 Marie Sklodowska Curie (1867–1934). When M. Curie presented her doctoral thesis, it was described as the greatest single contribution of any doctoral thesis in the history of science. Among other things, two new elements, polonium and radium, had been discovered. In 1903, Becquerel, M. Curie, and her husband, Pierre, were jointly awarded the Nobel Prize in physics. In 1911 M. Curie won a second Nobel Prize, this time in chemistry. Irene Curie, daughter of Marie and Pierre Curie, was also a scientist. She and her husband, Frederic Joliot, shared the 1935 Nobel Prize in chemistry for their work in artificial production of radioactive substances by bombarding certain elements with particles. (*Culver Pictures*)

It was possible to show that the β radiation was identical to a stream of high-speed electrons. The particles of this radiation were called β particles. In units of the charge of the electron, each β particle has a charge of $1-$. The α rays were found to consist of positively charged particles with a charge of $2+$. The α particles are comparatively much more massive than the β particles. Rutherford was able to show that the α particles combined with electrons in the surroundings to form atoms of helium. He thus concluded that the α rays consist of the positively charged core of the helium atom. The γ rays are high-energy radiation like X rays; they do not consist of particles. The α rays have low penetrating power; they are stopped by paper. The β particles have about 100 times greater penetrating ability, and the γ rays about 1000 times greater penetrating ability than α rays. The comparative properties of the three types of radiation are summarized in Table 2.3.

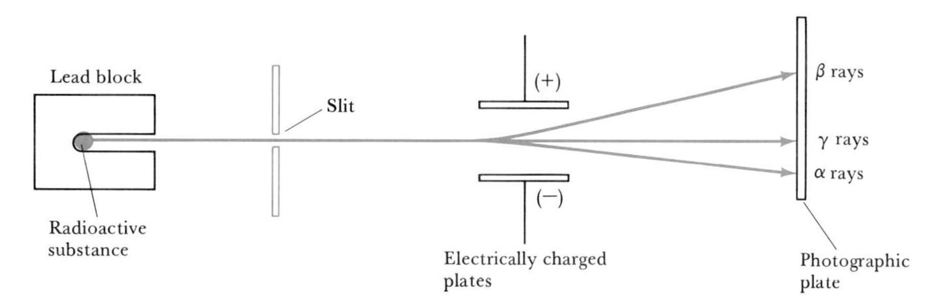

FIGURE 2.14 Behavior of alpha (α), beta (β), and gamma (γ) rays in an electric field.

TABLE 2.3 Summary of the properties of alpha, beta, and gamma rays

	Type of radiation		
	α	β	γ
Charge	2+	1−	0
Mass	6.64×10^{-24} g	9.11×10^{-28} g	0
Relative penetrating power	1	100	1000
Identity	$^4_2$He nuclei	Electrons	High-energy radiation

Rutherford and the Nuclear Atom

By 1909 Rutherford had firmly established that α rays consisted of particles with a 2+ charge. He then began to study the ways these α particles interact with matter. By this time it was well accepted that the atom was electrical in nature and contained electrons. The prevalent model of the atom, as developed by J. J. Thomson, pictured the atom as a cloud of positive charge in which negatively charged electrons were embedded like seeds in a watermelon.

In 1910 Rutherford and his co-workers performed an experiment that led to the downfall of Thomson's model. Rutherford was studying the manner of scattering of a narrow beam of α particles as they passed through a thin gold foil. He had found slight scattering, on the order of 1 degree, which was consistent with Thomson's model. One day Hans Geiger, an associate of Rutherford's, suggested that Ernest Marsden, a 20-year-old undergraduate working in their laboratory, get some experience in conducting such experiments. Rutherford suggested that Marsden see if α particles were scattered through large angles. In his own words:

> I may tell you in confidence that I did not believe they would be since we knew that the α particle was a very massive particle with a great deal of energy. . . . Then I remember two or three days later Geiger coming to me in great excitement and saying, "We have been able to get some α particles coming backwards." . . . It was quite the most incredible event that has ever happened to me in my life. It was almost as if you fired a 15-inch shell into a piece of tissue paper and it came back and hit you.

What Rutherford and his co-workers had observed was that the vast majority of α particles passed directly through the foil without deflection. Only a few underwent deflection, some even bouncing back in the direction from which they had come, as shown in Figure 2.15.

By 1911 Rutherford was able to explain these observations by postulating that all of the positive charge and most of the mass of the atom resides in a very small, extremely dense region, which he called the **nucleus.** Most of the total volume of the atom is merely space in which electrons move around the nucleus. In the α-scattering experiment, most α particles pass directly through the foil because they do not encounter the nucleus; they merely pass through the empty space of the atom. An occasional α particle, however, comes into the close vicinity of a nucleus. The repulsion between the highly charged gold nucleus and an α particle

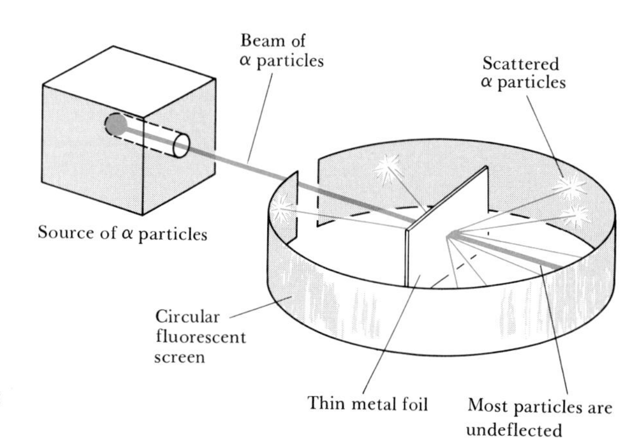

FIGURE 2.15 Rutherford's experiment on the scattering of α particles.

is strong enough to deflect the less massive α particle, as depicted in Figure 2.16.

The Modern View of Atomic Structure

Since the time of Rutherford, physicists have learned a great deal about the detailed structure of atomic nuclei. In the course of these discoveries the list of subnuclear particles has grown long and continues to increase. As chemists we can take a very simple view of the atom, because only three subatomic particles, the **proton, neutron,** and **electron,** have a bearing on chemical behavior.

Protons have a positive charge; neutrons are uncharged; electrons have a negative charge. The charges on the proton and electron are equal in magnitude. Because atoms have no net electrical charge, there are equal numbers of electrons and protons in an atom. The protons and neutrons reside together in a very small volume within the atom known

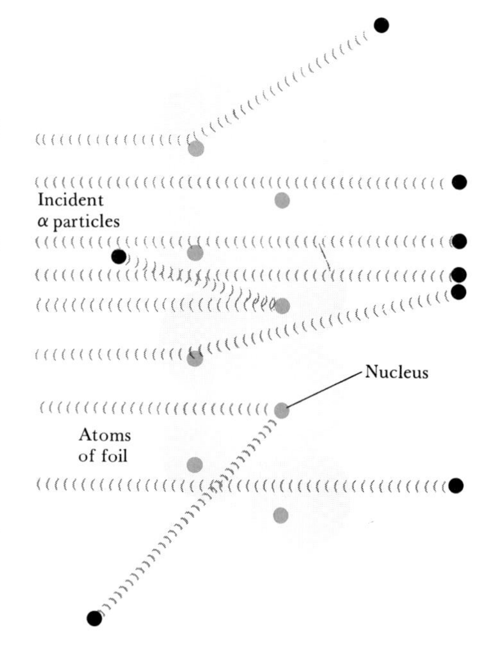

FIGURE 2.16 Rutherford's model explaining his experiment with scattering of α particles. The gold foil is actually several thousand atoms thick. The α particles are deflected backward only when they collide directly with, or pass very close to, the much heavier, positively charged gold nuclei. According to Coulomb's law, like charges repel one another. The α particles, with 2+ charge, and the gold nuclei, with 79+ charge, undergo a strongly repulsive interaction when the α particle closely approaches a gold nucleus. The less massive α particle is deflected from its path by the repulsive interaction. Only a small fraction of the particles are strongly deflected, because the gold nucleus has a small volume.

TABLE 2.4 Comparison of the proton, neutron, and electron

Particle	Charge	Mass
Proton	Positive (1+)	1.67×10^{-24} g
Neutron	None (neutral)	1.67×10^{-24} g
Electron	Negative (1−)	9.11×10^{-28} g

as the **nucleus.** Most of the rest of the atom is space in which the electrons move. The electrons are attracted to the nucleus and kept from flying off completely free in space by the attraction that exists between particles of unlike electrical charge (coulombic or electrostatic attraction). The charges and masses of the subatomic particles are summarized in Table 2.4.

Two bodies of the same charge repel each other. A positively charged body and a negatively charged one will be attracted to each other. The force of the interaction (F) between two charged bodies is given by Coulomb's law: $F = Q_1Q_2/d^2$ where Q_1 and Q_2 are the magnitudes of the charge on bodies 1 and 2, and d is the distance between them. The formula indicates that doubling the magnitude of one of the charges will double the force of attraction or repulsion. Doubling the distance of separation between the charges will reduce the force to one-fourth its previous value.

Atoms have diameters on the order of 1–5 Å. The angstrom unit (Å), which is 10^{-10} m, is widely used to indicate dimensions on the atomic scale. For example, the diameter of a chlorine atom is 2.0 Å. In time the angstrom may be replaced by a more acceptable SI unit such as the picometer or nanometer. In these units the diameter of a chlorine atom is 200 pm or 0.20 nm. For the most part we will employ angstroms throughout the text to indicate atomic and molecular dimensions.

SAMPLE EXERCISE 2.3

The diameter of a U.S. penny is 19 mm. How many chlorine atoms would fit side by side along this diameter?

Solution: The diameter of a single chlorine atom is 2.0 Å. We convert units as follows (the symbol for chlorine is Cl):

$$\frac{\text{No. of Cl atoms}}{\text{penny}} = \left(\frac{1 \text{ Cl atom}}{2.0 \text{ Å}}\right)\left(\frac{10^{10} \text{ Å}}{1 \text{ m}}\right) \times \left(\frac{1 \text{ m}}{10^3 \text{ mm}}\right)\left(\frac{19 \text{ mm}}{1 \text{ penny}}\right) = 9.5 \times 10^7 \frac{\text{Cl atoms}}{\text{penny}}$$

This exercise helps to illustrate how very small atoms are in relationship to ordinary dimensions.

The diameters of atomic nuclei are on the order of 10^{-4} Å, only a small fraction of the diameter of the atom as a whole. If the atom were scaled upward in size so that the nucleus were 2 cm in diameter (about the diameter of a penny), the atom would have a diameter of 200 m (about twice the length of a football field.) The proton and neutron have approximately equal mass, but the electron weighs only about $\frac{1}{1835}$ as much as either of these. Therefore the tiny nucleus carries most of the

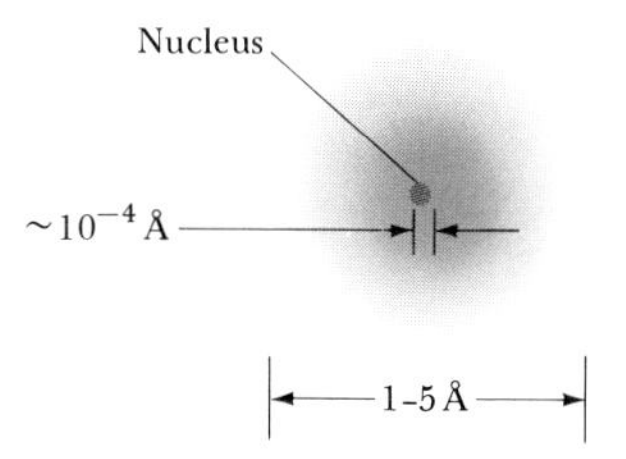

FIGURE 2.17 Schematic cross-sectional view through the center of an atom. The nucleus, which contains positive protons and neutral neutrons, is the location of virtually all the mass of the atom. The rest of the atom is mainly the space in which the light, negatively charged electrons move.

mass of the atom. Indeed, the density of the nucleus is on the order of 10^{13}–10^{14} g/cm^3. A matchbox full of material of such density would weigh over $2\frac{1}{2}$ billion tons. Astrophysicists have suggested that the matter in the interior of a collapsed star may reach approximately this density.

An illustration of the atom that incorporates the features we have just discussed is shown in Figure 2.17. The electrons, which take up most of the volume of the atom, play the major role in chemical reactions. The significance of representing the region containing the electrons as an indistinct cloud will become clear in later chapters when we consider the energies and spatial arrangements of the electrons.

The identity of an element depends on the number of protons in the nucleus of an atom of that element. In fact we may define an element as a substance whose atoms all have the same number of protons. Thus all atoms of the element carbon have six protons and six electrons. Most also have six neutrons although some have more. Atoms of a given element that differ in number of neutrons, and consequently in mass, are called **isotopes.** The symbol $^{12}_{6}C$ or simply ^{12}C (read "carbon twelve," carbon-12) is used to represent the carbon atom with six protons and six neutrons. The number of protons, which is called the **atomic number,** is shown by the subscript. Since all atoms of a given element have the same atomic number, this subscript is redundant and hence often omitted. The superscript is called the **mass number** and is the total number of protons plus neutrons in the atom. Some carbon atoms contain six protons and eight neutrons and are consequently represented as ^{14}C (read "carbon fourteen"). Normally, subscripts and superscripts are used with the symbol for an element only when reference is made to a particular isotope of that element. Three isotopes of oxygen and their chemical symbols are shown schematically in Figure 2.18. The term **nuclide** is applied in a general way to a nucleus with a specified number of protons and neutrons. For example, the nucleus of $^{16}_{8}O$ is referred to as the $^{16}_{8}O$ nuclide. (We will have more to say about the isotopic compositions of the elements in Section 3.7.)

SAMPLE EXERCISE 2.4

How many protons, neutrons, and electrons are there in ^{197}Au?

Solution: According to the list of elements given in the front inside cover of this text, gold has an atomic number of 79. Consequently ^{197}Au has 79 protons, 79 electrons, and 197 − 79 = 118 neutrons.

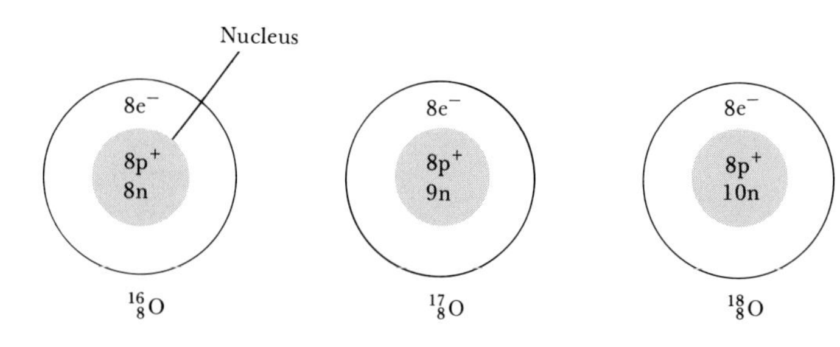

FIGURE 2.18 Distribution of subatomic particles in three isotopes of oxygen (p^+ = proton, n = neutron, e^- = electron). The shaded circle represents the nucleus; the eight electrons are in space surrounding the nucleus, as illustrated in Figure 2.17.

SAMPLE EXERCISE 2.5

Write the nuclear isotope symbols for the three isotopes of hydrogen, with mass numbers of 1, 2, and 3.

Solution: Because all three of the hydrogen isotopes must have the same number of protons, 1, the three symbols are: ${}^{1}_{1}H$; ${}^{2}_{1}H$; ${}^{3}_{1}H$.

On the atomic level gold, oxygen, and carbon differ in terms of the number of protons, neutrons, and electrons their respective atoms contain. These subatomic particles, however, are common to all substances. We can therefore state that an atom is the smallest representative sample of an element, because breaking the atom into subatomic particles destroys its identity.

In order to change a base or common metal like lead, atomic number 82, to gold, atomic number 79, requires removal of three protons from the nucleus of the lead atom. Because the nucleus is extremely small and buried in the heart of the atom, and because of the very strong binding forces between particles in the nucleus, this removal is exceedingly difficult. The energies required to cause changes in the nucleus are enormously greater than the energies associated with even the most vigorous chemical reactions. Thus, we still agree with Dalton that atoms of an element are not changed into different types of atoms by chemical reactions. Therein lies the futility of the alchemists' attempts to change base metals to gold.

2.5 THE PERIODIC TABLE: A PREVIEW

Dalton's atomic theory, and the various empirical laws that it helped to explain (Section 2.3), set the stage for a vigorous growth in chemical experimentation during the early part of the nineteenth century. As the body of chemical observations grew, and the list of known elements expanded, attempts were made to find regularities in chemical behavior. These efforts culminated in the development of the periodic table in 1869. We will have much to say about the periodic table in later chapters, but it is so important and useful that you should become acquainted with it now.

Many elements show very strong similarities to each other. For example, lithium (Li), sodium (Na), and potassium (K) are all soft, very reactive metals. The elements helium (He), neon (Ne), and argon (Ar) are very nonreactive gases. If the elements are arranged in order of increasing atomic number, their chemical and physical properties are found to show a repeating or periodic pattern. For example, each of the soft, reactive metals, lithium, sodium, and potassium, comes immediately after one of the nonreactive gases, helium, neon, and argon, as shown in Figure 2.19. The arrangement of elements in order of increasing atomic number, with elements having similar properties placed in vertical columns, is known as the **periodic table.** The periodic table is shown in Figure 2.20 and is also given on the inside front cover of your text for easy reference. In most classrooms where chemistry is taught, large periodic tables are hung on the walls—a testimony to their usefulness. You may notice slight variations in periodic tables from one book to another, or between the lecture hall and the text. These are matters of style or the particular information included; there are no fundamental differences.

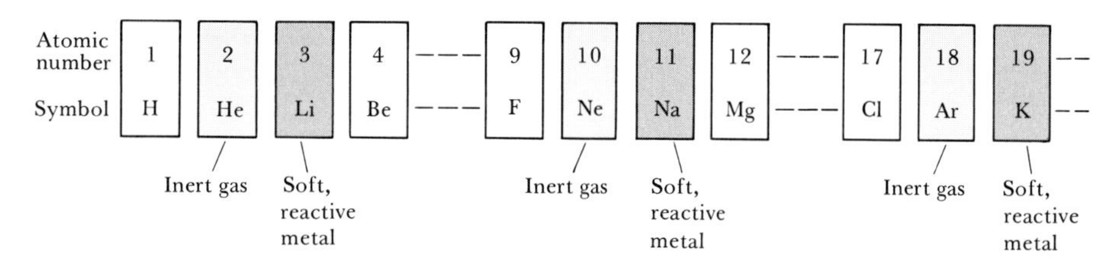

FIGURE 2.19 Arrangement of elements by atomic number to illustrate the periodic or repeating pattern in properties that is the basis of the periodic table.

The elements in a column of the periodic table are known as a **family** or **group.** They are identified as group 1A, 2A, and so forth, as shown at the top of the periodic table. For example, three familiar elements that have similar properties are copper (Cu), silver (Ag), and gold (Au), which occur together in group 1B. Some groups are also described by a family name. The members of group 1A—lithium, sodium, potassium, rubidium (Rb), cesium (Cs), and francium (Fr)—are known as the **alkali metals.** The members of group 2A—berylium (Be), magnesium (Mg), calcium (Ca), strontium (Sr), barium (Ba), and radium (Ra)—are known as the **alkaline earth metals.** The members of group 7A—fluorine (F), chlorine (Cl), bromine (Br), iodine (I), and astatine (At)—are known as the **halogens.** The members of group 8A—helium (He), neon (Ne), argon (Ar), krypton (Kr), xenon (Xe), and radon (Rn)—are known as the **noble gases, inert gases,** or **rare gases.**

1A	2A	3B	4B	5B	6B	7B	8B	8B	8B	1B	2B	3A	4A	5A	6A	7A	8A
1 H																	2 He
3 Li	4 Be											5 B	6 C	7 N	8 O	9 F	10 Ne
11 Na	12 Mg											13 Al	14 Si	15 P	16 S	17 Cl	18 Ar
19 K	20 Ca	21 Sc	22 Ti	23 V	24 Cr	25 Mn	26 Fe	27 Co	28 Ni	29 Cu	30 Zn	31 Ga	32 Ge	33 As	34 Se	35 Br	36 Kr
37 Rb	38 Sr	39 Y	40 Zr	41 Nb	42 Mo	43 Tc	44 Nu	45 Rh	46 Pd	47 Ag	48 Cd	49 In	50 Sn	51 Sb	52 Tc	53 I	54 Xe
55 Cs	56 Ba	57 La	72 Hf	73 Ta	74 W	75 Re	76 Os	77 Ir	78 Pt	79 Au	80 Hg	81 Tl	82 Pb	83 Bi	84 Po	85 At	86 Rn
87 Fr	88 Ra	89 Ac	104 Unq	105 Unp	106 Unh												

58 Ce	59 Pr	60 Nd	61 Pm	62 Sm	63 Eu	64 Gd	65 Tb	66 Dy	67 Ho	68 Er	69 Tm	70 Yb	71 Lu
90 Th	91 Pa	92 U	93 Np	94 Pu	95 Am	96 Cm	97 Bk	98 Cf	99 Es	100 Fm	101 Md	102 No	103 Lw

Metals
Metalloids
Nonmetals

FIGURE 2.20 Periodic table of the elements, showing the division of elements into metals, metalloids, and nonmetals.

We will learn in Chapters 5 and 6 that the elements in a family of the periodic table have similar properties because they have the same type of arrangement of electrons at the periphery of their atoms. However, we need not wait until then to make good use of the periodic table; after all, the table in pretty much its modern form was invented by chemists who knew nothing of the electronic structures of atoms! We can use the table, as they intended, to correlate the behaviors of elements and to aid in remembering many facts. You will find it helpful to refer to the periodic table frequently in studying the remainder of this chapter.

SAMPLE EXERCISE 2.6

Which of the following elements would you expect to show the greatest similarity in chemical and physical properties: Li, Be, F, S, Cl?

Solution: The elements F and Cl should be most alike because they are in the same family (group 7A, the halogen family).

One pattern that is evident when elements are arranged in the periodic table is the grouping together of the **metallic elements.** These elements, which are grouped together on the left side of the periodic table, share many characteristic properties, such as luster and high electrical and heat conductivity. The metallic elements are separated from the **nonmetallic elements** by the diagonal line that runs across the right side of the periodic table from boron (B) to astatine (At). Note that the majority of the elements are metallic. The nonmetals, which occupy the upper right side of the periodic table, are gases, liquids, or crystalline solids. They lack those physical characteristics that distinguish the metallic elements. Many of the elements that lie along the line that separates metals from nonmetals, such as antimony (Sb), possess properties intermediate between those of metals and nonmetals. These elements are often referred to as **semimetals** or **metalloids.**

2.6 MOLECULES AND IONS

We have seen that the atom is the smallest representative sample of an element. However, only the noble gas elements are normally found in nature as isolated atoms. Most matter is composed of molecules or ions, which are formed from atoms.

Molecules

Molecules are composed of combinations of tightly bound atoms. The resultant assembly or package of atoms behaves in many ways as a single object, just as a television set composed of many parts can be recognized as a single object. The nature of the forces (bonds) that bind the atoms together will be examined in Chapter 7.

When elements exist in molecular form, they contain only one type of atom. For instance, the element oxygen, as it is normally found in air, consists of molecules composed of pairs of oxygen atoms. This molecular form of oxygen is represented by the **chemical formula** O_2 ("oh two"). The subscript in this formula indicates that two oxygen atoms are present in each molecule. The molecule is said to be **diatomic.** Oxygen also exists in a form known as ozone, which consists of three bound oxygen atoms. Correspondingly, the chemical formula for ozone is O_3. Ozone and "normal" oxygen exhibit quite different chemical properties.

The elements that normally occur as diatomic molecules are hydrogen, oxygen, nitrogen, and the halogens. Their location in the periodic table is shown in Figure 2.21. When we speak of the substance hydrogen, we mean H_2 unless we indicate explicitly otherwise. Likewise, when we speak of oxygen, nitrogen, or any of the halogens, we are referring to O_2, N_2, F_2, Cl_2, Br_2, or I_2. Thus the properties of oxygen and hydrogen listed earlier in Table 2.2 are those of O_2 and H_2. In other forms, these elements behave much differently.

Molecules of compounds contain more than one type of atom. For example, a molecule of water consists of two hydrogen atoms and one oxygen atom. It is therefore represented by the chemical formula H_2O (read "aitch two oh"). Another compound composed of these same elements is hydrogen peroxide, H_2O_2. These two compounds have quite different properties. A number of common molecules are shown in Figure 2.22. Pay close attention to how the chemical formula of each molecule reflects its composition. Note also that these substances are composed only of nonmetallic elements. Most molecular substances that we will encounter contain only nonmetals.

SAMPLE EXERCISE 2.7

Which of the molecules shown in Figure 2.22 are compounds and which are elements?

Solution: The compounds are H_2O, CO_2, CO, H_2O_2, CH_4, and C_2H_4. The elements are O_2, O_3, and H_2 (one type of atom).

Chemical formulas that indicate the *actual* numbers and types of atoms in a molecule are called **molecular formulas.** (The formulas in Figure 2.22 are molecular formulas.) Chemical formulas that give only the *relative* number of atoms of each type in a molecule are called **empirical** or **simplest formulas.** The subscripts in an empirical formula are always smallest whole-number ratios; conversely, the subscripts in a molecular formula are always integral multiples (that is 1×, 2×, 3×, and so forth) of the subscripts in the empirical formula of the substance. For example, the empirical formula for hydrogen peroxide is HO; its molec-

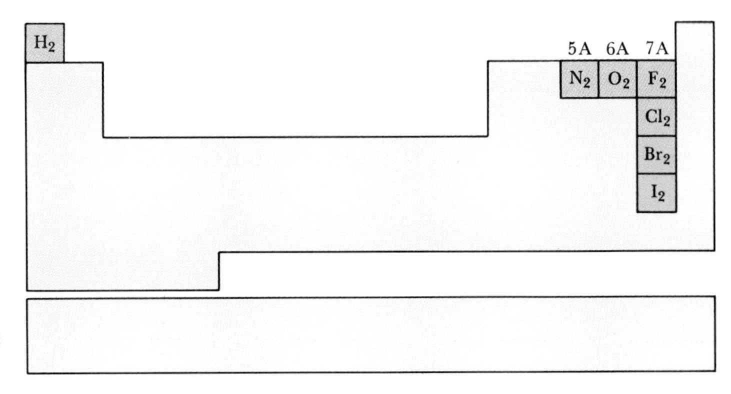

FIGURE 2.21 Elements that exist as diatomic molecules at room temperature.

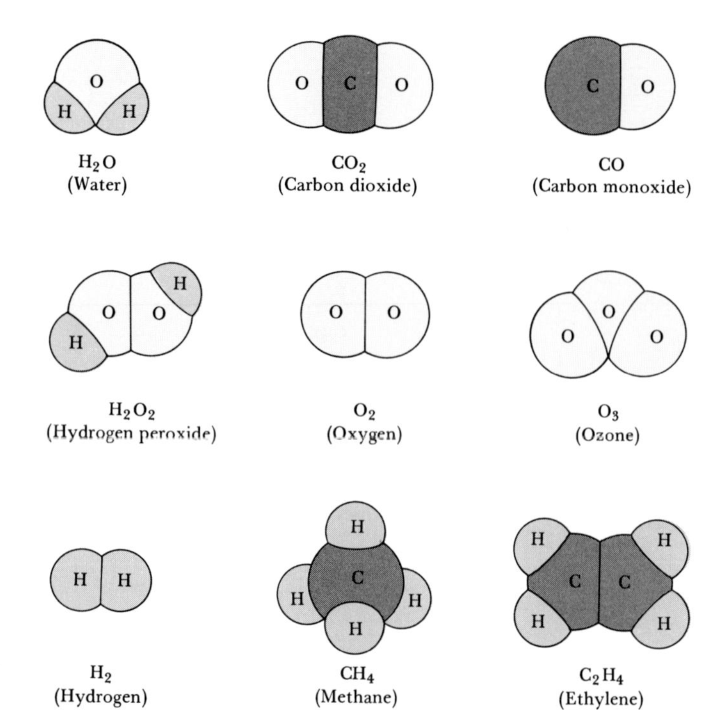

FIGURE 2.22 Representations of some simple, common molecules.

ular formula is H_2O_2. The empirical formula for ethylene is CH_2; its molecular formula is C_2H_4. For some substances the empirical formula and molecular formula are identical, as in the case of water, H_2O.

SAMPLE EXERCISE 2.8

Write the empirical formulas for the following molecules: (a) glucose, a substance also known as blood sugar and as dextrose, whose molecular formula is $C_6H_{12}O_6$; (b) nitrous oxide, a substance used as an anesthetic and commonly called laughing gas, whose molecular formula is N_2O.

Solution: (a) The empirical formula has subscripts that are smallest whole-number ratios. The smallest ratios are obtained by dividing each subscript by the largest common factor, in this case 6. The resultant, empirical formula is CH_2O. (b) Because the subscripts in N_2O are already the lowest integral numbers, the empirical formula for nitrous oxide is the same as its molecular formula, N_2O.

Molecular formulas are preferred over empirical formulas because they provide more information. Only empirical formulas, however, can be written for substances that exist as three-dimensional structures in which there are no discrete atoms or molecules. (As we shall see shortly, this situation includes ionic substances.) For example, the element carbon normally exists in extended structures rather than as isolated atoms or molecules. Because the empirical formula for any element is simply the symbol for that element, carbon is represented by its symbol, C. As a further example, boron nitride consists of a three-dimensional array of boron and nitrogen atoms; it is represented by its empirical formula BN.

Often the formulas of molecules are written so as to show how atoms are joined together in a molecule. For example, the formulas for water and hydrogen peroxide can be written as follows:

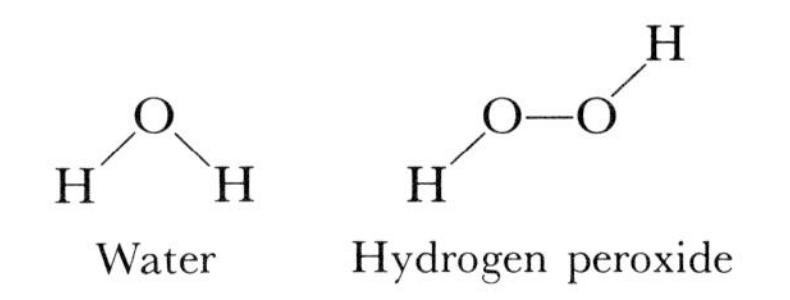

Such formulas are known as structural formulas. The lines between the symbols for the elements represent the bonds that hold the respective atoms together. These formulas indicate which atoms are attached to which; however, they do not necessarily tell anything about the shapes of molecules.

Ions

Whereas the nucleus of an atom is unchanged by ordinary chemical processes, atoms readily gain or lose electrons. If electrons are removed or added to a neutral atom, a charged particle called an ion is formed. For example, the sodium atom, which has 11 protons and 11 electrons, can lose an electron. The resulting ion has 11 protons and 10 electrons; its net charge is consequently 1+. This ion is symbolically represented as Na^+. The net charge on an ion is represented by a superscript; +, 2+, and 3+ mean a net charge resulting from loss of one, two, or three electrons, respectively. The superscripts −, 2−, and 3− represent net charges resulting from gain of one, two, or three electrons, respectively. The formation of the Na^+ ion from a Na atom is shown schematically below:

$$11p^+ \; 11e^- \xrightarrow[\text{electron}]{\text{Lose one}} 11p^+ \; 10e^-$$

Na atom — Na^+ ion

Chlorine, with 17 protons and 17 electrons, can gain an electron in chemical reactions, producing the Cl^- ion:

$$17p^+ \; 17e^- \xrightarrow[\text{electron}]{\text{Gain one}} 17p^+ \; 18e^-$$

Cl atom — Cl^- ion

In general, metal atoms lose electrons most readily, whereas nonmetal atoms tend to gain electrons.

In addition to simple ions such as Na^+ and Cl^-, there exist polyatomic ions such as NO_3^- (nitrate ion) and SO_4^{2-} (sulfate ion). These ions consist of atoms joined together as in a molecule, but having a net positive or negative charge. We will consider further examples of polyatomic ions in Section 2.7.

The chemical properties of ions are greatly different from those of the atoms from which they are derived. The change of an atom or molecule to an ion is like that from Dr. Jekyll to Mr. Hyde: Although the body may be essentially the same (plus or minus a few electrons), the personality is much different.

Many atoms gain or lose electrons so as to end up with the same number of electrons as the closest noble gas. The members of the noble gas family are chemically very nonreactive and form very few compounds. We might deduce that this is because they have very stable electron arrangements. Nearby elements can obtain these same stable arrangements by losing or gaining electrons. For example, loss of one electron from an atom of sodium leaves it with the same number of electrons as the neutral neon atom (atomic number 10). Similarly, when chlorine gains an electron it ends up with 18, the same as argon (atomic number 18). We will content ourselves with this simple observation in explaining the formation of ions until later chapters in which we consider chemical bonding.

SAMPLE EXERCISE 2.9

Predict the charges expected for the most stable ions of barium and oxygen.

Solution: Refer to the periodic table. Barium has atomic number 56. The nearest noble gas is xenon, atomic number 54. Barium can obtain the stable arrangement of 54 electrons by losing two of its electrons, thereby forming the Ba^{2+} ion.

Oxygen has atomic number 8. The nearest noble gas is neon, atomic number 10. Oxygen can obtain this stable electron arrangement by gaining two electrons, thereby forming an ion of 2– charge, O^{2-}.

Gains and losses of electrons that result in the formation of ions occur in transactions between different kinds of atoms. The compounds that result from these electron transfers are composed of positively charged and negatively charged ions. The positively charged ions are usually metal ions; the negatively charged ones are usually nonmetal ions. Consequently, ionic compounds tend to be composed of metals combined with nonmetals. For example, NaCl, ordinary table salt, is such a compound; it consists of equal numbers of Na^+ and Cl^- ions.

The ions in ionic compounds are arranged in three-dimensional structures. The arrangement of Na^+ and Cl^- ions in NaCl is shown in Figure 2.23. Because there is no discrete molecule of NaCl, we are able to write

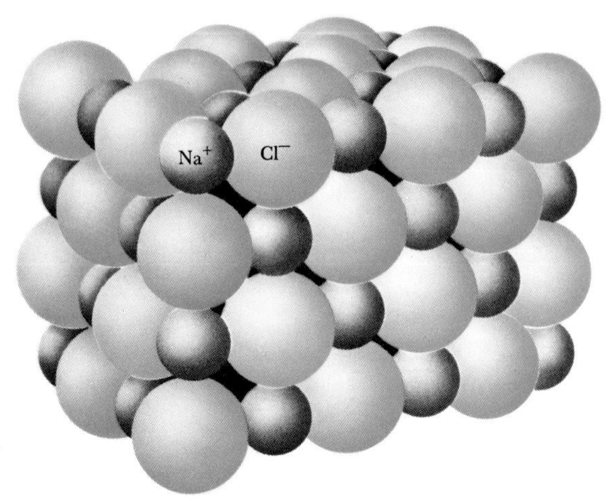

FIGURE 2.23 Arrangement of ions in sodium chloride (NaCl).

only an empirical formula for this substance. Indeed, only empirical formulas can be written for any ionic compound.

It is a simple matter to write the empirical formula for an ionic compound if we know the charges of the ions of which it is composed. Chemical compounds are always electrically neutral. Consequently, the ions in an ionic compound always occur in such a ratio that the total positive charge is equal to the total negative charge. Thus there is one Na^+ to one Cl^- giving NaCl, one Ba^{2+} to two Cl^- giving $BaCl_2$, and so forth.

SAMPLE EXERCISE 2.10

What are the empirical formulas of the compounds formed by (a) Al^{3+} and Cl^- ions; (b) Al^{3+} and O^{2-} ions; (c) Mg^{2+} and NO_3^- ions?

Solution: (a) It requires three Cl^- ions to balance the charge of one Al^{3+} ion. Thus the formula is $AlCl_3$. (b) It requires two Al^{3+} ions to balance the charge or three O^{2-} ions (that is, the total positive charge is +6 and the total negative charge is −6). Thus the formula is Al_2O_3. (c) Two NO_3^- ions are needed to balance the charge of one Mg^{2+}. Thus the formula is $Mg(NO_3)_2$. In this case the formula for the entire negative ion must be enclosed in parentheses so that it is clear that the subscript 2 applies to all the atoms of that ion.

2.7 NAMING OF INORGANIC COMPOUNDS

As you proceed in your study of this text, and in chemistry laboratory work, you will need to refer to specific chemical substances by name. We present here some of the basic rules for naming simple compounds. You may not have immediate use for some of the rules, but they are gathered here in one place for your convenience whenever a question of **nomenclature**, or naming of substances, arises.

There are now about 4 million known chemical substances. Naming them all would be a hopelessly complicated task if each had a special name independent of all the others. Many important substances that have been known for a long time, such as water (H_2O) and ammonia (NH_3), do have individual, traditional names. For most substances, however, we rely upon a set of rules that lead to an informative, systematic name for each substance.

One of the earliest classification schemes in chemistry was the distinction between inorganic and organic compounds. Organic compounds contain carbon, usually in combination with hydrogen, oxygen, nitrogen, and sulfur. Organic compounds were first associated only with plants and animals. However, a great number of organic compounds have now been prepared that do not occur in nature. We will discuss the chemistry and naming of organic compounds in Chapter 25; however, we will have many occasions throughout the text to illustrate chemical principles with organic compounds as examples. In this section we will consider the basic rules for naming inorganic compounds, those that the early chemists associated with the nonliving portion of our world. Let us first consider the naming of ionic compounds.

Ionic Compounds

The names of ionic compounds are based on the names of the ions of which they are composed. For example, NaCl is called sodium chloride after the Na^+ or sodium ion, and the Cl^- or chloride ion. The positive ion is always named first and listed first in writing the formula for the compound. The negative ion is named and written last. To see how the names of these ions arise, consider first the naming of positive ions, also called **cations** (pronounced CAT -ion). Ions may be monatomic (composed of a single atom) or polyatomic (formed from two or more atoms). Monatomic cations are most commonly formed from metallic elements. These ions take the name of the element itself:

Na^+ sodium ion $\quad$ Zn^{2+} zinc ion $\quad$ Al^{3+} aluminum ion

If an element can form more than one positive ion, the positive charge of the ion is indicated by a Roman numeral in parentheses following the name of the metal:

Fe^{2+}	iron(II) ion	Cu^+	copper(I) ion
Fe^{3+}	iron(III) ion	Cu^{2+}	copper(II) ion

At this stage you have no way of knowing which elements commonly exist in more than one charge state. This need not be a source of difficulty; if there is any doubt in your mind, use the Roman numeral designation of charge as part of the name. It is never wrong to use this form of charge designation, even though it may sometimes be unnecessary.

An older method still widely used for distinguishing between two differently charged ions of a metal is to use the endings *-ous* or *-ic*; these endings represent the lower and higher charged ions, respectively. They are used together with the root of the Latin name of the element:

Fe^{2+}	ferrous ion	Cu^+	cuprous ion
Fe^{3+}	ferric ion	Cu^{2+}	cupric ion

The only common polyatomic cations are those given below:

NH_4^+ ammonium ion $\quad$ Hg_2^{2+} mercury(I) or mercurous ion

The name mercury(I) ion is given to Hg_2^{2+} because it can be considered to consist of two Hg^+ ions. Mercury also occurs as the monatomic Hg^{2+} ion, which is known as the mercury(II) or mercuric ion.

Negative ions are called **anions** (pronounced AN -ion). Monatomic anions (those derived from a single atom) are most commonly formed from atoms of the nonmetallic elements. They are named by dropping the ending of the name of the element and adding the ending *-ide*:

H^-	hydride ion	O^{2-}	oxide ion	N^{3-}	nitride ion
F^-	fluoride ion	S^{2-}	sulfide ion	P^{3-}	phosphide ion

Only a few common polyatomic ions end in *-ide*:

OH^- hydroxide ion $\quad$ CN^- cyanide ion $\quad$ O_2^{2-} peroxide ion

Table 2.5 lists the most common cations and anions. Notice that there are many polyatomic anions containing oxygen. Anions of this kind are

TABLE 2.5 Common ions

Positive ions (cations)	Negative ions (anions)
Ammonium (NH_4^+)	Acetate ($C_2H_3O_2^-$)
Copper(I) or cuprous (Cu^+)	Bromide (Br^-)
Hydrogen (H^+)	Chloride (Cl^-)
Silver (Ag^+)	Chlorate (ClO_3^-)
Sodium (Na^+)	Cyanide (CN^-)
Potassium (K^+)	Fluoride (F^-)
	Hydride (H^-)
Barium (Ba^{2+})	Hydrogen carbonate (HCO_3^-) or bicarbonate
Calcium (Ca^{2+})	Hydrogen sulfate (HSO_4^-) or bisulfate
Chromium(II) or chromous (Cr^{2+})	Hydroxide (OH^-)
Cobalt(II) or cobaltous (Co^{2+})	Iodide (I^-)
Copper(II) or cupric (Cu^{2+})	Nitrate (NO_3^-)
Iron(II) or ferrous (Fe^{2+})	Perchlorate (ClO_4^-)
Lead(II) or plumbous (Pb^{2+})	Permanganate (MnO_4^-)
Magnesium (Mg^{2+})	
Manganese(II) or manganous (Mn^{2+})	Carbonate (CO_3^{2-})
Mercury(II) or mercuric (Hg^{2+})	Chromate (CrO_4^{2-})
Tin(II) or stannous (Sn^{2+})	Oxide (O^{2-})
Zinc (Zn^{2+})	Peroxide (O_2^{2-})
	Sulfate (SO_4^{2-})
Aluminum (Al^{3+})	Sulfide (S^{2-})
Chromium(III) or chromic (Cr^{3+})	Sulfite (SO_3^{2-})
Iron(III) or ferric (Fe^{3+})	
	Phosphate (PO_4^{3-})

referred to as oxyanions. A particular element such as sulfur may form more than one oxyanion. When this occurs, there are rules for indicating the relative numbers of oxygen atoms in the anion. When an element has two oxyanions, the name of the one that contains more oxygen ends in *-ate*; the name of the one with less oxygen ends in *-ite*:

NO_2^-	nitrite ion	SO_3^{2-}	sulfite ion
NO_3^-	nitrate ion	SO_4^{2-}	sulfate ion

When the series of anions of a given element extends to three or four members, as with the oxyanions of the halogens, prefixes are also employed. The prefix *hypo-* indicates less oxygen, whereas the prefix *per-* indicates more oxygen:

ClO^-	hypochlorite ion (less oxygen than chlorite)
ClO_2^-	chlorite ion
ClO_3^-	chlorate ion
ClO_4^-	perchlorate ion (more oxygen than chlorate)

Notice that if you memorize the rules just indicated, you need only know the name for one oxyanion in a series to deduce the names for the other members.

SAMPLE EXERCISE 2.11

The formula for the selenate ion is SeO_4^{2-}. Write the formula for the selenite ion.

Solution: The selenite ion should have one less oxygen than the selenate ion; hence, SeO_3^{2-}.

Because many names of ions predate the establishment of systematic rules, there are many exceptions to the rules. For example, the permanganate ion is MnO_4^-; we thus expect that the manganate ion should be MnO_3^-, but this ion is unknown. The name *manganate* is given to the species MnO_4^{2-}.

Many polyatomic anions that have high charges readily add one or more hydrogen ions to form anions of lower charge. These ions are named by prefixing the word *hydrogen* or *dihydrogen,* as appropriate, to the name of the hydrogen-free anion. An older method, which is still used, is to use the prefix *bi-*:

HCO_3^-	hydrogen carbonate (or bicarbonate) ion
HSO_4^-	hydrogen sulfate (or bisulfate) ion
$H_2PO_4^-$	dihydrogen phosphate ion

We are now in a position to combine the names of cations and anions to name and write the formulas for ionic compounds. The following examples illustrate the relationship between formula and name:

barium bromide	$BaBr_2$	mercurous chloride	Hg_2Cl_2
copper(II) nitrate	$Cu(NO_3)_2$	aluminum oxide	Al_2O_3

SAMPLE EXERCISE 2.12

Name the following compounds: (a) K_2SO_4; (b) $Ba(OH)_2$; (c) $FeCl_3$.

Solution: (a) This compound is composed of K^+ and SO_4^{2-} ions. Because K^+ is called the potassium ion and SO_4^{2-} is called the sulfate ion, the name of the compound is potassium sulfate. (b) This compound is composed of Ba^{2+} and OH^- ions. Ba^{2+} is the barium ion; OH^- is the hydroxide ion. Thus the compound is called barium hydroxide. (c) This compound is composed of Fe^{3+}, which is called iron(III) or ferric ion, and chloride ions, Cl^-. The compound is iron(III) chloride or ferric chloride.

SAMPLE EXERCISE 2.13

Write the chemical formulas for the following compounds: (a) calcium carbonate; (b) stannous fluoride; (c) iron(II) perchlorate.

Solution: (a) The calcium ion is Ca^{2+}; the carbonate ion is CO_3^{2-}. Because of the charges of each ion, there will be one Ca^{2+} ion for each CO_3^{2-} in the compound, giving the empirical formula $CaCO_3$. (b) The stannous ion, also known as the tin(II) ion, is Sn^{2+}. The fluoride ion is F^-. Two F^- ions are needed to balance the positive charge of Sn^{2+}, giving the formula SnF_2. This compound, incidentally, is a common ingredient in many toothpastes. (c) The iron(II) ion is Fe^{2+}; the perchlorate ion is ClO_4^-. Two ClO_4^- ions are required to balance the charge on one Fe^{2+}, giving $Fe(ClO_4)_2$.

These examples should indicate the importance of remembering the charges of the common ions. Indeed, you may find it necessary to memorize many of the entries in Table 2.5.

Acids

There is an important class of compounds known as acids that are named in a special way. These compounds will be discussed further in Section 3.3 and then extensively considered in Chapter 15. An acid may be defined as a substance that yields one or more hydrogen ions (H^+) when dissolved in water. Anions whose names end in *-ide* have associated acids that have a *hydro-* prefix and an *-ic* ending, as in these examples:

Anion	Corresponding acid
Cl^- (chloride)	HCl (hydrochloric acid)
S^{2-} (sulfide)	H_2S (hydrosulfuric acid)

Many of the most important acids are derived from oxyanions. If the anion has an *-ate* ending, the corresponding acid is given an *-ic* ending. Anions whose names end in *-ite* have associated acids whose names end in *-ous*. Prefixes in the name of the anion are retained in the name of the acid. These rules are illustrated by the oxyacids of chlorine:

Anion	Corresponding acid
ClO^- (hypochlorite)	HClO (hypochlorous acid)
ClO_2^- (chlorite)	$HClO_2$ (chlorous acid)
ClO_3^- (chlorate)	$HClO_3$ (chloric acid)
ClO_4^- (perchlorate)	$HClO_4$ (perchloric acid)

SAMPLE EXERCISE 2.14

Name the following acids: (a) HCN; (b) HNO_3; (c) H_2SO_4; (d) H_2SO_3.

Solution: (a) The anion from which this acid is derived is CN^-, the cyanide ion. Because this ion has an *-ide* ending, the acid is given a *hydro-* prefix and an *-ic* ending: hydrocyanic acid. Incidentally, only water solutions of HCN are referred to as hydrocyanic acid; the pure compound is called hydrogen cyanide. (b) Because NO_3^- is the nitrate ion, HNO_3 is called nitric acid (the *-ate* ending of the anion is replaced with an *-ic* ending in naming the acid). (c) Because SO_4^{2-} is the sulfate ion, H_2SO_4 is called sulfuric acid. (d) Because SO_3^{2-} is the sulfite ion, H_2SO_3 is sulfurous acid (the *-ite* ending of the anion is replaced with an *-ous* ending).

Molecular Compounds

The procedures for naming binary (two element) molecular compounds are similar to those for ionic compounds. In these molecular compounds it is possible to associate a more positive nature with one element in the molecule and a more negative nature with the other element. (In Chapter 7 we will consider how the more positive atom is selected.) The element with the more positive nature is named first and also appears first in the chemical formula. The second element is named with an *-ide* ending. For example, the name for HCl is hydrogen chloride. (This is the name used when referring to the pure compound; water solutions of HCl are referred to as hydrochloric acid.)

TABLE 2.6
Prefixes used in naming binary compounds formed between nonmetals

Prefix	Meaning
Mono-	1
Di-	2
Tri-	3
Tetra-	4
Penta-	5
Hexa-	6

Often a pair of elements can form several different molecular compounds. For example, carbon and oxygen form CO and CO_2. To distinguish these compounds from one another, the prefixes given in Table 2.6 are used to denote the numbers of atoms of each element present. Thus CO is called carbon monoxide, and CO_2 is called carbon dioxide. A few additional examples follow:

NF_3	nitrogen trifluoride
N_2O_4	dinitrogen tetroxide
SO_3	sulfur trioxide

FOR REVIEW

Summary

Matter exists in three states: gas, liquid, and solid. Most matter consists of a mixture of substances. Mixtures can be either homogeneous or heterogeneous; homogeneous mixtures are called solutions. Mixtures can be resolved into two types of pure substances: compounds and elements. Each substance has a unique set of chemical and physical properties that can be used to identify it.

Atoms are the basic building blocks of matter; they are the smallest units of an element that can combine with other elements. Atoms are composed of a nucleus (containing protons and neutrons) and electrons that move around the nucleus. We considered some of the historically significant experiments that led to this model of the atom: Thomson's experiments on the behavior of cathode rays (a stream of electrons) in magnetic and electric fields; Millikan's oil-drop experiment; Becquerel's and Rutherford's studies of radioactivity; and Rutherford's studies of the scattering of α particles by thin metal foils.

Elements can be classified by atomic number, the number of protons in the nucleus of an atom. All atoms of a given element have the same atomic number. The mass number of an atom is the sum of the number of protons and neutrons. Atoms of the same element that differ in mass number are known as isotopes.

The periodic table is an arrangement of elements in order of increasing atomic number with elements with similar properties placed in vertical columns. The elements in a vertical column are known as a periodic family or group. The metallic elements, which comprise the majority of the elements, are on the left side of the table; the nonmetallic elements are located on the right side.

Atoms can combine to form molecules. Atoms can also either gain or lose electrons, thereby forming charged particles called ions.

Each element has a one- or two-letter symbol. These symbols are combined to represent compounds, as in the formula H_2O for water. Three types of formulas were considered: simplest (or empirical) formulas, molecular formulas, and structural formulas. We also discussed the basic rules for naming inorganic compounds.

Learning goals

Having read and studied this chapter, you should be able to:

1 Differentiate between the three states of matter.
2 Distinguish between elements, compounds, and mixtures.
3 Give the chemical symbols for the elements discussed in this chapter.
4 Distinguish between the physical and chemical properties of a substance.
5 Describe the composition of an atom in terms of protons, neutrons, and electrons.
6 Give the approximate size, relative mass, and charge of an atom, proton, neutron, and electron.
7 Write the chemical symbol for an element (for example $^{12}_{6}C$), having been given its mass number and atomic number, and perform the reverse operation.
8 Write the symbol and charge for an atom or ion, having been given the number of protons, neutrons, and electrons, and perform the reverse operation.
9 Describe the properties of the electron as seen in cathode rays. Describe the means by which J. J. Thomson determined the ratio e/m for the electron.
10 Describe Millikan's oil-drop experiment and in-

dicate what property of the electron he was able to measure.

11 Cite the evidence from studies of radioactivity for the existence of subatomic particles.

12 Describe the experimental evidence for the nuclear nature of the atom.

13 Use the periodic table to predict the charges of monatomic ions.

14 Write the simplest formula for a compound, having been given the charges of the ions of which it is composed.

15 Write the name of a simple inorganic compound, having been given its chemical formula, and perform the reverse operation.

Key terms

Among the more important terms and expressions used for the first time in this chapter are the following:

Alpha (α) particles (Section 2.4) are helium nuclei (consisting of two protons and two neutrons) emitted from the nuclei of certain radioactive atoms.

The atomic number (Section 2.4) of an element is the number of protons in the nucleus of one of its atoms.

Beta (β) particles (Section 2.4) are electrons emitted from the nuclei of certain radioactive atoms.

A chemical change (Section 2.1) is a process in which one or more substances are converted into other substances.

Chemical properties (Section 2.1) are those properties of a substance that describe its composition and reactivity—how a substance "reacts" or changes into other substances.

A compound (Section 2.2) is a substance composed of two or more elements united chemically in definite proportions by mass. A compound can consist of either molecules or of ions.

An electron (Section 2.4) is a negatively charged subatomic particle found outside the atomic nucleus. It forms a part of all atoms. It has a mass about $\frac{1}{1835}$ times that of a proton.

An element (Section 2.2) is a substance whose atoms all have the same atomic number; there are 106 known elements. Elements cannot be separated into simpler substances by ordinary chemical means.

An empirical formula or simplest formula (Section 2.6) is a chemical formula that shows the kinds of atoms and their relative numbers in a substance.

Gamma (γ) radiation (Section 2.4), or gamma rays, is a form of radiation similar to X rays that is emitted by radioactive atoms.

Ions (Section 2.6) are electrically charged atoms or groups of atoms (polyatomic ions). Ions can be positively or negatively charged, depending on whether electrons are lost (positive) or gained (negative) from the atoms.

Isotopes (Section 2.4) are atoms of the same element containing different numbers of neutrons and therefore having different masses.

The mass number (Section 2.4) is the sum of the number of protons and neutrons in the nucleus of a particular atom.

A molecular formula (Section 2.6) is a chemical formula that indicates the actual number of atoms of each element in one molecule of a substance.

A molecule (Section 2.6) is a chemical combination of two or more atoms.

A neutron (Section 2.4) is a neutral particle found in the nucleus of an atom; it has approximately the same mass as a proton.

A physical change (Section 2.1) is a change (such as a phase change) that occurs with no change in chemical properties.

Physical properties (Section 2.1) are those properties (such as color and freezing point) that can be measured without changing the composition of a substance.

A proton (Section 2.4) is a positively charged subatomic particle found in the nucleus of any atom.

Radioactivity (Section 2.4) is the spontaneous disintegration of an unstable atomic nucleus with accompanying emission of alpha, beta, or gamma radiation.

A solution (Section 2.2) is a homogeneous mixture.

A structural formula (Section 2.6) is a formula that shows not only the number and kind of atoms in a molecule, but also the arrangement of the atoms.

EXERCISES

Substances; states of matter

2.1 Which of the following are heterogeneous mixtures, which are pure substances, and which are solutions: (a) wood; (b) wine; (c) salt; (d) a Bufferin or Anacin tablet; (e) gold bars in a Swiss bank; (f) milk?

2.2 Which of the following are chemical and which are physical processes: (a) forming copper wire from a bar of copper; (b) tarnishing of silver; (c) burning of hydrogen; (d) souring of milk; (e) melting of iron; (f) boiling of water; (g) heating the filament of an incandescent lamp to provide illumination?

2.3 Characterize the following as chemical or physical changes: (a) melting of ice; (b) hard-boiling of an egg; (c) dissolving salt in water; (d) evaporation of water from a lake; (e) rusting of iron; (f) decomposition of water into hydrogen and oxygen; (g) pulverizing rocks.

2.4 Consider the following description of the element sodium: "Sodium is silver-white and soft. It is a good conductor of electricity. It can be prepared by passing electricity through molten sodium chloride. The metal boils at 883°C; the vapor is violet-colored. Sodium metal tarnishes rapidly in air. It burns on heating in air or in an atmosphere of bromine vapor." Indicate which of the properties in this description are physical and which are chemical.

2.5 Identify each of the following substances as gases, liquids, or solids under ordinary conditions: (a) mercury; (b) iron; (c) oxygen; (d) aluminum; (e) alcohol; (f) hydrogen; (g) water; (h) chlorine; (i) helium.

2.6 In 1807 Humphry Davy passed an electric current through molten potassium hydroxide and isolated a bright, shiny, very reactive substance. He claimed the discovery of a new element, which he named potassium. In those days, before the advent of modern instruments, what was the basis on which one could claim that a substance was an element?

2.7 A solid white substance A is heated strongly. It decomposes to form a new white substance B and a gas. The gas has exactly the same properties as the product obtained when a carbon rod is burned in an excess of oxygen. What can we say about whether solids A and B and the gas are elements or compounds?

2.8 Write the chemical symbol for each of the following elements: (a) oxygen; (b) hydrogen; (c) nitrogen; (d) sodium; (e) potassium; (f) magnesium, (g) gold.

2.9 What chemical elements are represented by the following chemical symbols: (a) Pb; (b) Cs; (c) S; (d) Al; (e) He; (f) Fe; (g) Ca?

[2.10] Suggest a method of separating each of the following mixtures: (a) sand from water; (b) salt from water; (c) alcohol from water; (d) the color pigments in a leaf; (e) the components of air; (f) the components of gasoline.

Dalton's atomic theory; atomic structure

2.11 When the elements hydrogen and bromine are caused to react, a gas is formed. The composition of this product gas is the same, no matter what the relative amounts of hydrogen and bromine from which it is formed. How is this observation explained in terms of Dalton's atomic theory? What "law" does it illustrate?

2.12 Ascorbic acid, vitamin C, can be isolated from the juices of citrus fruits. It can also be synthesized in the laboratory in a series of reactions beginning with sorbose, a readily obtained sugar. In what respects should the two types of vitamin C differ in their nutritional value? Bearing in mind the material of Section 2.2, if the two materials did differ, in what respects would they be different?

2.13 A chemistry student prepared a series of compounds containing only nitrogen and oxygen:

Compound	Mass of nitrogen	Mass of oxygen
A	16.8	19.2
B	17.1	39.0
C	33.6	57.3

Show how these data obey the law of multiple proportions.

2.14 The tracks of energetic particles such as α and β rays through gases can be followed in a device called the Wilson cloud chamber. The tracks appear as streaks. The tracks of a few α particles through air are shown in Figure 2.24. Explain why the paths of the α particles are mostly long and account for the sudden change in direction of one particle, as shown.

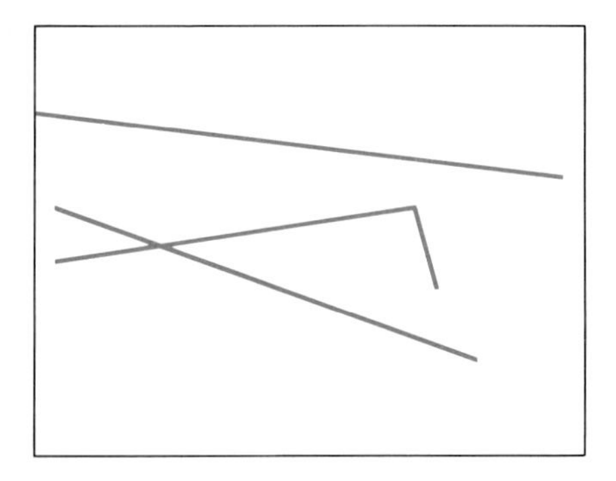

FIGURE 2.24

2.15 In a physics laboratory a student carried out the Millikan oil-drop experiment, using several oil droplets for her measurements, and calculated the charges on the drops. She obtained the following data:

Droplet	Calculated charge (C)
A	1.60×10^{-19}
B	3.15×10^{-19}
C	4.81×10^{-19}
D	6.31×10^{-19}

What is the significance of the fact that the droplets carried different charges? What conclusion can the student draw from these data regarding the charge of the electron? What value (and to how many significant figures) should she report for the electronic charge?

2.16 The diameter of the iridium (Ir) atom is estimated to be 2.7 Å. Express this distance in picometers (pm); in nanometers (nm). What number of Ir atoms end to end would form a line 1 mm long?

2.17 What restrictions on the numbers of protons, neutrons, and electrons apply to each of the following:

(a) atoms of a particular isotope of an element; (b) different isotopes of the same element; (c) nuclides with the same mass number?

2.18 Indicate the number of protons, neutrons, and electrons present in each of the following atoms: (a) ^{13}C; (b) ^{55}Mn; (c) ^{97}Mo.

2.19 Determine the number of protons, neutrons, and electrons present in each of the following nuclei: (a) ^{6}Li; (b) ^{57}Fe; (c) ^{27}Al; (d) ^{19}F.

2.20 Write the correct symbol, with both superscript and subscript, for each of the following (use the list of elements on the inside front cover): (a) the nuclide of potassium with mass 39; (b) the nuclide of chlorine with mass 35; (c) the nuclide of silicon with mass 29; (d) two isotopes of sulfur, one with 16 neutrons, the other with 18 neutrons.

2.21 Fill in the gaps in the following table:

Symbol	^{79}Se		
Protons			56
Electrons		35	
Neutrons			
Mass number		80	137

The periodic table; molecules and ions

2.22 Using the table on the inside front cover as a guide, write the symbol for each of the following elements; locate the element in the periodic table and indicate whether it is a metal, metalloid, or nonmetal: (a) manganese; (b) bromine; (c) chromium; (d) germanium; (e) hafnium; (f) selenium; (g) argon.

2.23 Locate each of the following elements on the periodic table, indicate whether it is a metal, metalloid, or nonmetal, and give the name of the element: (a) K; (b) Kr; (c) S; (d) Sb; (e) C; (f) Ca; (g) H.

2.24 Which pair in each of the following groups of elements would you expect to be the most similar in chemical and physical properties: (a) Ca, Si, I, P, Sr, Sc; (b) Mg, Al, S, I, Sb, Ga? Explain.

2.25 Write the empirical formulas corresponding to each of the following molecular formula. (a) C_4H_8; (b) C_3H_8; (c) P_4O_{10}; (d) N_2O_4; (e) C_6H_6; (f) $C_4H_2O_4$.

2.26 Each of the following elements is capable of forming an ion in chemical reactions. Refer to the periodic table and predict the charge found on the most stable ion formed by each: (a) K; (b) F; (c) Ba; (d) S; (e) Al; (f) Sc.

2.27 Predict the formula for the ionic compound formed between the following ions: (a) Ca^{2+} and Br^-; (b) NH_4^+ and SO_4^{2-}; (c) Mg^{2+} and NO_3^-; (d) Na^+ and S^{2-}; (e) K^+ and OH^-.

2.28 Predict the formula for the ionic compound formed between each of the following pairs of elements: (a) Li, O; (b) Sr, S; (c) Cs, Br; (d) Cl, Mg; (e) Sc, F.

2.29 From the group of elements F, Kr, Ca, Cs, W, Sb, and S, select the one that (a) should form an ion with a 2− charge; (b) is least reactive chemically; (c) should form an ion with a 1+ charge; (d) is a metalloid.

2.30 Based on their compositions, predict whether each of the following compounds is molecular or ionic: (a) N_2O; (b) Na_2O; (c) CaO; (d) CO; (e) ScF_3; (f) NF_3; (g) KBr.

Naming inorganic compounds

2.31 Name the following ions: (a) Cr^{2+}; (b) Sr^{2+}; (c) I^-; (d) S^{2-}; (e) Ti^{3+}; (f) OH^-; (g) Cs^+; (h) NO_3^-; (i) NH_4^+.

2.32 Name the following compounds: (a) $KClO_3$; (b) Na_3PO_4; (c) $Ba(HSO_3)_2$; (d) $Zn(OH)_2$; (e) $Ca(CN)_2$; (f) $Fe(NO_3)_3$; (g) NH_4Cl; (h) BaO_2; (i) $LiHCO_3$; (j) $CuSO_4$; (k) NaH.

2.33 Write the formulas for each of the following compounds: (a) calcium chloride; (b) barium hydroxide; (c) potassium cyanide; (d) copper(I) bromide; (e) chromium(III) oxide; (f) ammonium sulfate; (g) potassium permanganate; (h) sodium phosphate; (i) zinc nitrate; (j) mercury(II) sulfide; (k) sodium hydrogen carbonate; (l) silver chlorate; (m) lead(II) sulfate; (n) aluminum acetate.

2.34 Name each of the following acids: (a) HBr; (b) H_2SO_3; (c) $HBrO_3$; (d) H_3PO_4; (e) H_2Se.

2.35 Write the correct formula for each of the following acids: (a) nitrous acid; (b) sulfuric acid; (c) hydroiodic acid; (d) hydrocyanic acid; (e) carbonic acid.

2.36 Name each of the following binary molecular compounds: (a) SCl_2; (b) XeO_3; (c) N_2O_5; (d) SiF_4; (e) Cl_2O_7.

2.37 Write the correct formula for each of the following binary molecular compounds: (a) selenium dioxide; (b) carbon tetrachloride; (c) tetraphosphorus hexaoxide; (d) hydrogen selenide; (e) iodine pentafluoride.

2.38 Write the chemical formula for each substance mentioned in the following word descriptions (use the inside front cover to find the symbols for the elements you don't know). (a) Zinc carbonate can be heated to form zinc oxide and water. (b) On treatment with hydrofluoric acid, silicon dioxide forms silicon tetrafluoride and water. (c) Sulfur dioxide reacts with water to form sulfurous acid. (d) The substance hydrogen phosphide is commonly called phosphine. (e) Perchloric acid reacts with cadmium to form cadmium(II) perchlorate. (f) Vanadium(III) bromide is a colored solid.

2.39 Assume that you encounter the following phrases in your reading. What is the chemical formula for each compound mentioned? (a) Potassium chlorate is used as a laboratory source of oxygen. (b) Sodium hypochlorite is used as a household bleach. (c) Ammonia is important in the synthesis of fertilizers such as ammonium nitrate. (d) Hydrofluoric acid is used to etch glass. (e) The smell of rotten eggs is due to hydrogen sulfide. (f) When hydrochloric acid is added to sodium bicarbonate (baking powder), carbon dioxide gas forms.

Additional Exercises

2.40 Distinguish between the members of the following pairs: (a) homogeneous and heterogeneous; (b) gas and liquid; (c) atomic number and mass number; (d) chemical and physical properties; (e) Ca and Ca^{2+}; (f) hydrochloric acid and chloric acid; (g) iron(II) and iron(III); (h) sodium carbonate and sodium bicarbonate; (i) H_2O and H_2O_2; (j) metal and nonmetal; (k) H and He; (l) chloride and chlorate.

2.41 Distinguish between the members of the following pairs: (a) alpha and beta particles; (b) proton and neutron; (c) ion and atom; (d) molecular formula and empirical formula; (e) chemical change and physical change; (f) solid and liquid; (g) O and O^{2-}; (h) C and Ca; (i) sodium sulfate and sodium sulfite; (j) sodium hydrogen phosphate and sodium dihydrogen phosphate; (k) cuprous chloride and cupric chloride; (l) O_2 and O_3.

2.42 Many ions and compounds have very similar names and there is great potential of confusing them. Write the correct chemical formulas to distinguish between (a) calcium sulfide and calcium hydrogen sulfide; (b) hydrobromic acid and bromic acid; (c) aluminum nitride and aluminum nitrite; (d) iron (II) oxide and iron (III) oxide; (e) ammonia and ammonium ion; (f) potassium sulfite and potassium bisulfite; (g) mercurous chloride and mercuric chloride; (h) chloric acid and perchloric acid.

2.43 List as many properties as possible that would allow you to distinguish among a solid, liquid, and gas.

2.44 Suppose you were given a sample of a homogeneous liquid. What would you do to determine whether it is a solution or a pure substance?

2.45 Classify each of the following as element, compound, or mixture: (a) air; (b) cement; (c) magnesium; (d) ink; (e) soda pop; (f) sodium oxide; (g) sulfur; (h) seawater.

2.46 List as many properties as possible that would allow you to distinguish between (a) silver and aluminum; (b) water and alcohol.

2.47 Identify each of the following elements if a neutral atom of each (a) contains 4 protons; (b) has an atomic number of 12; (c) contains 11 electrons.

2.48 Identify each of the following: (a) an element of group 2A that has 20 electrons in the neutral atom; (b) an unreactive element with 36 protons in the nucleus; (c) an element with atomic mass between 70 and 90 that tends to form ions of 1− charge; (d) a metallic element that forms 2+ ions having 28 electrons.

2.49 Fill in the gaps in the following table:

Symbol	${}^{12}_{6}C$	${}^{17}_{8}O^{2-}$			
Protons	6		12		8
Neutrons			13	12	10
Electrons		10		10	10
Net charge	0	2−	0	1+	

2.50 Fill in the gaps in the following table:

Symbol	${}^{37}Cl^-$	${}^{40}Ar$	
Protons			21
Neutrons			23
Electrons			
Net charge			3+

2.51 Fill in the gaps in the following table:

Symbol	${}^{9}_{4}Be^{2+}$	
Atomic number		35
Mass number		80
Net charge		0

2.52 Each of the following radioactive nuclides is used in medical science. Write the chemical symbol for each, showing the mass number and atomic number: (a) cobalt-60; (b) iodine-131; (c) sulfur-35.

2.53 (a) Explain why ${}^{18}O$ and ${}^{16}O$ have essentially identical chemical properties. (b) Explain why the symbols for an alpha particle and a beta particle could be given as ${}^{4}_{2}\alpha$ and ${}^{0}_{-1}\beta$.

2.54 Describe the contributions to atomic theory made by the following persons: (a) Dalton; (b) Rutherford; (c) Thomson; (d) Millikan; (e) Becquerel.

2.55 Describe what is meant by the term "family" in speaking of the elements and the periodic table.

2.56 The formula of the compound formed between cadmium and oxygen is CdO. Predict the formula of the compound formed between (a) cadmium and chlorine; (b) cadmium and sulfur; (c) zinc and oxygen.

2.57 The formula of the compound formed by aluminum and oxygen is Al_2O_3. Predict the formula of the compound between (a) gallium and oxygen; (b) aluminum and sulfur; (c) gallium and sulfur; (d) aluminum and fluorine.

2.58 Which of the following are elements and which are compounds: (a) I_2; (b) SO_2; (c) S_8; (d) NH_3; (e) P_4; (f) H_2O_2?

2.59 Write the empirical formula for each of the following substances: (a) acetic acid, $C_2H_4O_2$; (b) dichloroethylene, $C_2H_2Cl_2$; (c) nicotine, $C_{10}H_{14}N_2$; (d) tetraphosphorus decasulfide, P_4S_{10}; (e) mercurous chloride, Hg_2Cl_2.

2.60 Give the chemical name for each of the following compounds: (a) HNO_3; (b) SF_6; (c) Na_2O; (d) Na_2CO_3; (e) $CoCl_2$; (f) $BaSO_4$; (g) H_2O_2; (h) NO_2.

2.61 Write the formula for each of the following: (a) ammonium nitrate; (b) aluminum hydroxide; (c) hydrobromic acid; (d) perchloric acid; (e) potassium chlorate; (f) cobalt(II) carbonate; (g) calcium hydride; (h) silver sulfide; (i) iron(III) chromate.

2.62 Many familiar substances have common, unsystematic names. In each of the following cases, give the correct systematic name: (a) saltpeter (KNO_3); (b) soda ash (Na_2CO_3); (c) lime (CaO); (d) baking soda ($NaHCO_3$); (e) lye ($NaOH$); (f) muriatic acid (HCl); (g) milk of magnesia ($Mg(OH)_2$); (h) dry ice (CO_2); (i) ammonia (NH_3).

3 Stoichiometry

Antoine Lavoisier (Section 1.1) was among the first to draw conclusions about chemical processes from careful, quantitative observations. His work laid the basis for the law of conservation of mass, one of the most fundamental laws of chemistry. In this chapter, we will consider many practical problems based on the law of conservation of mass. These problems involve the quantitative relationships between substances undergoing chemical changes. The study of these quantitative relationships is known as **stoichiometry** (pronounced stoy-key-AHM-uh-tree), a word derived from the Greek words *stoicheion* ("element") and *metron* ("measure").

3.1 LAW OF CONSERVATION OF MASS

Studies of countless chemical reactions have shown that the total mass of all substances present after a chemical reaction is the same as the total mass before the reaction. This observation is embodied in the **law of conservation of mass:** There are no *detectable* changes in mass in any chemical reaction.* More precisely, *atoms are neither created nor destroyed during a chemical reaction;* instead, they merely exchange partners or become otherwise rearranged. The simplicity with which this law can be stated should not mask its significance. As with many other scientific laws, this law has implications far beyond the walls of the scientific laboratory.

The law of conservation of mass reminds us that we really can't throw anything away. If we discharge wastes into a lake to get rid of them, they are diluted and seem to disappear. However, they are part of the envi-

*In Chapter 19, we will discuss the relationship between mass and energy summarized by the equation $E = mc^2$ (E is energy, m is mass, and c is the speed of light). We will find that whenever an object loses energy it loses mass, and whenever it gains energy it gains mass. These changes in mass are too small to detect in chemical reactions. However, for nuclear reactions, such as those involved in a nuclear reactor or in a hydrogen bomb, the energy changes are enormously larger; in these reactions there are detectable changes in mass.

ronment. They may undergo chemical changes or remain inactive; they may reappear as toxic contaminants in fish or in water supplies or lie on the bottom unnoticed. Whatever their fates, the atoms are not destroyed.

The law of conservation of mass suggests that we are converters, not consumers. In drawing upon nature's storehouse of iron ore to build the myriad iron-containing objects used in modern society, we are not reducing the number of iron atoms on the planet. We may, however, be converting the iron to less useful, less available forms from which it will not be practical to recover it later. For example, consider the millions of old washing machines that lie buried in dumps. Of course, if we expend enough energy, we can bring off almost any chemical conversions we choose. We have learned in recent years, however, that energy itself is a limited resource. Whether we like it or not, we must learn to conserve all our energy and material resources.

3.2 CHEMICAL EQUATIONS

We have seen (in Sections 2.2 and 2.6) that chemical substances can be represented by symbols and formulas. These chemical symbols and formulas can be combined to form a kind of statement, called a **chemical equation**, that represents or describes a chemical reaction. For example, the combustion of carbon involves a reaction with oxygen (O_2) in the air to form gaseous carbon dioxide (CO_2). This reaction is represented as

$$C + O_2 \longrightarrow CO_2 \qquad [3.1]$$

We read the + sign to mean "reacts with" and the arrow as "produces." Carbon and oxygen are referred to as **reactants** and carbon dioxide as the **product** of the reaction.

It is important to keep in mind that a chemical equation is a description of a chemical process. Before you can write a complete equation you must know what happens in the reaction or be prepared to predict the products. In this sense, a chemical equation has qualitative significance; it identifies the reactants and products in a chemical process. In addition, a chemical equation is a quantitative statement; it must be consistent with the law of conservation of mass. This means that the equation must contain equal numbers of each type of atom on each side of the equation. When this condition is met the equation is said to be **balanced.** For example, Equation 3.1 is balanced because there are equal numbers of carbon and oxygen atoms on each side.

A slightly more complicated situation is encountered when methane (CH_4), the principal component of natural gas, burns and produces carbon dioxide (CO_2) and water (H_2O). The combustion is "supported by" oxygen (O_2), meaning that oxygen is involved as a reactant. The unbalanced equation is

$$CH_4 + O_2 \longrightarrow CO_2 + H_2O \qquad [3.2]$$

The reactants are shown to the left of the arrow, the products to the right. Notice that the reactants and products both contain one carbon atom. However, the reactants contain more hydrogen atoms (four) than the products (two). If we place a coefficient 2 in front of H_2O, indicating formation of two molecules of water, there will be four hydrogens on each side of the equation:

$$CH_4 + O_2 \longrightarrow CO_2 + 2H_2O \quad [3.3]$$

Before we continue to balance this equation, let's make sure that we clearly understand the distinction between a coefficient in front of a formula and a subscript in a formula. Refer to Figure 3.1. Notice that changing a subscript in a formula, such as from H_2O to H_2O_2, changes the identity of the chemical involved. The substance H_2O_2, hydrogen peroxide, is quite different from water. *The subscripts in the chemical formulas should never be changed in balancing an equation.* On the other hand, placing a coefficient in front of a formula merely changes the amount and not the identity of the substance; $2H_2O$ means two molecules of water, $3H_2O$ means three molecules of water, and so forth. Now let's continue balancing Equation 3.3. There are equal numbers of carbon and hydrogen atoms on both sides of this equation; however, there are more oxygen atoms among the products (four) than among the reactants (two). If we place a coefficient 2 in front of O_2 there will be equal numbers of oxygen atoms on both sides of the equation:

$$CH_4 + 2O_2 \longrightarrow CO_2 + 2H_2O \quad [3.4]$$

The equation is now balanced. There are four oxygen atoms, four hydrogen atoms, and one carbon atom on each side of the equation. The balanced equation is shown schematically in Figure 3.2.

Now, let's look at a slightly more complicated example, analyzing stepwise what we are doing as we balance the equation. Combustion of octane (C_8H_{18}), a component of gasoline, produces CO_2 and H_2O. The balanced chemical equation for this reaction can be determined by using the following four steps.

First, the reactants and products are written in the unbalanced equation

$$C_8H_{18} + O_2 \longrightarrow CO_2 + H_2O \quad [3.5]$$

Before a chemical equation can be written the identities of the reactants and products must be determined. In the present example this information was given to us in the verbal description of the reaction.

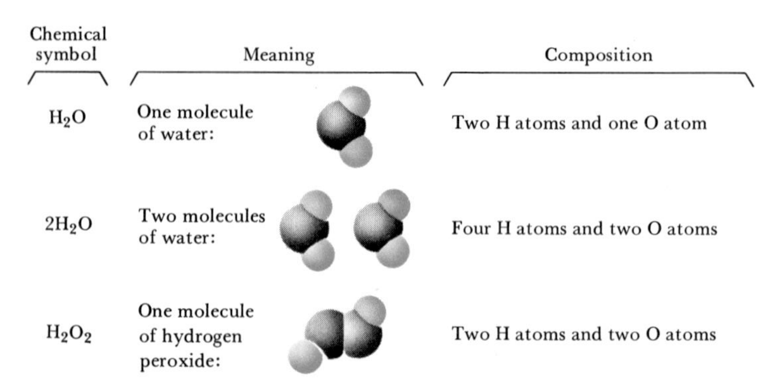

FIGURE 3.1 Illustration of the difference in meaning between a subscript in a chemical formula and a coefficient in front of the formula. Notice that the number of atoms of each type (listed under composition) is obtained by multiplying the coefficient and the subscript associated with each element in the formula.

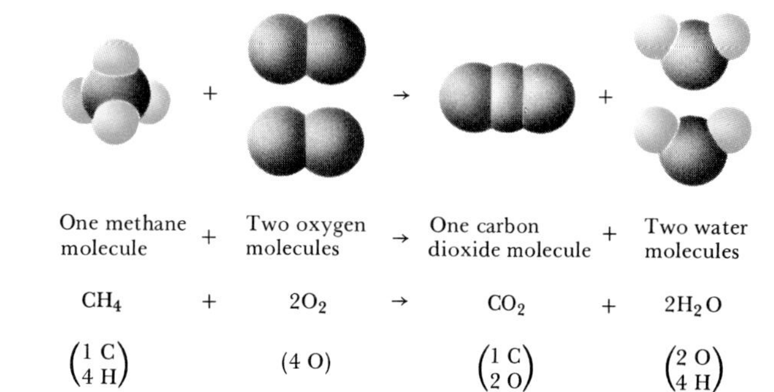

FIGURE 3.2 Balanced chemical equation for the combustion of CH_4. The drawings of the molecules involved call attention to the conservation of atoms through the reaction.

Second, the number of atoms of each type on each side of the equation is determined. In the reaction above there are 8C, 18H, and 2O among the reactants, and 1C, 2H, and 3O among the products; clearly, the equation is not balanced, because the number of atoms of each type differs from one side of the equation to the other.

Third, to balance the equation, coefficients are placed in front of the chemical formulas to indicate different quantities of reactants and products, so that the same number of atoms of each type appears on both sides of the equation. To decide what coefficients to try first, it is often convenient to focus attention on the molecule with the most atoms, in this case C_8H_{18}. This molecule contains 8C, all of which must end up in CO_2 molecules. Therefore, we place a coefficient 8 in front of CO_2. Similarly, the 18H end up as $9H_2O$. At this stage the equation reads

$$C_8H_{18} + O_2 \longrightarrow 8CO_2 + 9H_2O \qquad [3.6]$$

Although the C and H atoms are now balanced, the O atoms are not; there are 25O atoms among the products but only 2 among the reactants. It takes $12.5O_2$ to produce 25O atoms among the reactants:

$$C_8H_{18} + 12.5O_2 \longrightarrow 8CO_2 + 9H_2O \qquad [3.7]$$

However, this equation is not in its most conventional form, because it contains a fractional coefficient. Therefore, we must go on to the next step.

Fourth, for most purposes a balanced equation should contain the smallest possible whole-number coefficients. Therefore, we multiply each side of the equation above by 2, removing the fraction and achieving the following balanced equation:

$$2C_8H_{18} + 25O_2 \longrightarrow 16CO_2 + 18H_2O \qquad [3.8]$$

16C, 36H, 50O	16C, 36H, 50O
Reactants	Products

The atoms are inventoried below the equation to show graphically that the equation is indeed balanced. You might note that although atoms are conserved, molecules are not—the reactants contain 27 molecules while the products contain 34. All in all, this approach to balancing equations is largely trial and error. It is much easier to verify that an equation is balanced than actually to balance one, so practice in balancing equations is essential.

It should also be noted that the physical state of each chemical in a chemical equation is often indicated parenthetically using the symbols (g), (l), (s), and (aq) to indicate gas, liquid, solid, and aqueous (water) solution, respectively. Thus the balanced equation above can be written

$$2C_8H_{18}(l) + 25O_2(g) \longrightarrow 16CO_2(g) + 18H_2O(l) \quad [3.9]$$

Sometimes an upward arrow (↑) is employed to indicate the escape of a gaseous product, whereas a downward arrow (↓) indicates a precipitating solid (that is, a solid that separates from solution during the reaction). Often the conditions under which the reaction proceeds are indicated above the arrow between the two sides of the equation. For example, the temperature or pressure at which the reaction occurs could be so indicated. The symbol Δ is often placed above the arrow to indicate the addition of heat.

SAMPLE EXERCISE 3.1

Balance the following equation:

$$Na(s) + H_2O(l) \longrightarrow NaOH(aq) + H_2(g)$$

Solution: A quick inventory of atoms reveals that there are equal numbers of Na and O atoms on both sides of the equation, but that there are two H atoms among reactants and three H atoms among products. To increase the number of H atoms among reactants, we might place a coefficient 2 in front of H_2O:

$$Na(s) + 2H_2O(l) \longrightarrow NaOH(aq) + H_2(g)$$

Now we have four H atoms among reactants but only three H atoms among the products. The H atoms can be balanced with a coefficient 2 in front of NaOH:

$$Na(s) + 2H_2O(l) \longrightarrow 2NaOH(aq) + H_2(g)$$

If we again inventory the atoms on each side of the equation, we find that the H atoms and O atoms are balanced but not the Na atoms. However, a coefficient 2 in front of Na gives two Na atoms on each side of the equation:

$$2Na(s) + 2H_2O(l) \longrightarrow 2NaOH(aq) + H_2(g)$$

If the atoms are inventoried once more we find two Na atoms, four H atoms, and two O atoms on each side of the equation. The equation is therefore balanced.

3.3 CHEMICAL REACTIONS

Our discussion in Section 3.2 focused on how to balance chemical equations given the reactants and products for the reactions. You were not asked to predict the products for a reaction. Students sometimes ask how the products are determined. For example, how do we know that sodium metal (Na) reacts with water (H_2O) to form H_2 and NaOH as shown in Sample Exercise 3.1? These products are identified by experiment. As the reaction proceeds, there is a fizzing or bubbling where the sodium is in contact with the water (if too much sodium is used the reaction is quite violent, so small quantities would be used in our experiment). If the gas is captured, it can be identified as H_2 from its chemical and physical properties. After the reaction is complete, a clear solution remains. If this is evaporated to dryness, a white solid will remain. From its properties this solid can be identified as NaOH. However, it is not necessary to perform an experiment every time we wish to write a reaction. We can predict what will happen if we have seen the reaction or a similar one before. So far we have seen too little chemistry to predict the products for many reactions. Nevertheless, even now you should be able to make some predictions. For example, what would you expect to happen when potassium metal is added to water? We have just discussed the reaction of sodium metal with water, for which the balanced chemical equation is

$$2Na(s) + 2H_2O(l) \longrightarrow 2NaOH(aq) + H_2(g) \qquad [3.10]$$

Because sodium and potassium are in the same family of the periodic table (the alkali metal family, family 1A), we would expect them to behave similarly, producing the same types of products. Indeed, this prediction is correct, and the reaction of potassium metal with water is

$$2K(s) + 2H_2O(l) \longrightarrow 2KOH(aq) + H_2(g) \qquad [3.11]$$

You can readily see that it will be helpful in your study of chemistry if you are able to classify chemical reactions into certain types. We have just considered two examples of a type we might call reaction of an active metal with water. Let's briefly consider here a few of the more important and common types you will be encountering in your laboratory work and in the chapters ahead.

Combustion in Oxygen

We have already encountered three examples of combustion reactions: the combustion of carbon, Equation 3.1; of methane, Equation 3.4; and of octane (C_8H_{18}), Equation 3.8. Combustion is a rapid reaction that usually produces a flame. Most of the combustions we observe involve O_2 as a reactant. From the examples we have already seen it should be easy to predict the products of the combustion of propane C_3H_8. We expect that combustion of this compound would lead to carbon dioxide and water as products, by analogy with our previous examples. That expectation is correct; propane is the major ingredient in LP (liquid propane) gas, used for cooking and home heating. It burns in air as described by the balanced equation

$$C_3H_8(g) + 5O_2(g) \longrightarrow 3CO_2(g) + 4H_2O(l) \qquad [3.12]$$

If we looked at further examples, we would find that combustion of compounds containing oxygen atoms as well as carbon and hydrogen (for example, CH_3OH) also produces CO_2 and H_2O.

Acids, Bases, and Neutralization

Acids are substances that increase the H^+ ion concentration in aqueous solution. For example, hydrochloric acid, which we often represent as $HCl(aq)$, exists in water as $H^+(aq)$ and $Cl^-(aq)$ ions. Thus the process of dissolving hydrogen chloride in water to form hydrochloric acid can be represented as follows:

$$HCl(g) \xrightarrow{H_2O} HCl(aq)$$

or [3.13]

$$HCl(g) \xrightarrow{H_2O} H^+(aq) + Cl^-(aq)$$

The H_2O given above the arrows in these equations is to remind us that the reaction medium is water. Pure sulfuric acid is a liquid; when it dissolves in water it releases H^+ ions in two successive steps:

$$H_2SO_4(l) \xrightarrow{H_2O} H^+(aq) + HSO_4^-(aq) \qquad [3.14]$$

$$HSO_4^-(aq) \xrightarrow{H_2O} H^+(aq) + SO_4^{2-}(aq) \qquad [3.15]$$

Thus, although we frequently represent aqueous solutions of sulfuric acid as $H_2SO_4(aq)$, these solutions actually contain a mixture of $H^+(aq)$, $HSO_4^{2-}(aq)$, and $SO_4^{2-}(aq)$.

Bases are compounds that increase the hydroxide ion, OH^-, concentration in aqueous solution. A base such as sodium hydroxide does this because it is an ionic substance composed of Na^+ and OH^- ions. When NaOH dissolves in water, the cations and anions simply separate in the solution:

$$NaOH(s) \xrightarrow{H_2O} Na^+(aq) + OH^-(aq) \qquad [3.16]$$

Thus, although aqueous solutions of sodium hydroxide might be written as $NaOH(aq)$, sodium hydroxide exists as $Na^+(aq)$ and $OH^-(aq)$ ions. Many other bases such as $Ca(OH)_2$ are also ionic hydroxide compounds. However, NH_3 (ammonia) is a base although it is not a compound of this sort.

It may seem odd at first glance that ammonia is a base, because it contains no hydroxide ions. However, we must remember that the definition of a base is that it *increases* the concentration of OH^- ions in water. Ammonia does this by a reaction with water. We can represent the dissolving of ammonia gas in water as follows:

$$NH_3(g) + H_2O(l) \longrightarrow NH_4^+(aq) + OH^-(aq) \qquad [3.17]$$

Solutions of ammonia in water are often labeled ammonium hydroxide, NH_4OH, to remind us that ammonia solutions are basic. (Ammonia is referred to as a weak base, which means that not all the NH_3 that dissolves in water goes on to form NH_4^+ and OH^- ions; but that is a matter for Chapter 15 and need not concern us here.)

Acids and bases are among the most important compounds in industry and in the chemical laboratory. Table 3.1 lists several acids and bases and the amount of each compound produced in the United States each year. You can see that these substances are produced in enormous quantities.

Solutions of acids and bases have very different properties. Acids have

TABLE 3.1 U.S. production of some acids and bases, 1982

Compound	Formula	Annual production (kg)
Acids:		
Sulfuric	H_2SO_4	3.0×10^{10}
Phosphoric	H_3PO_4	7.7×10^9
Nitric	HNO_3	6.9×10^9
Hydrochloric	HCl	2.3×10^9
Bases:		
Sodium hydroxide	$NaOH$	8.3×10^9
Calcium hydroxide	$Ca(OH)_2$	1.3×10^{10}
Ammonia	NH_3	1.4×10^{10}

a sour taste, whereas bases have a bitter taste.* Acids can change the colors of certain dyes in a specific way that differs from the effect of a base. For example, the dye known as litmus is changed from blue to red by an acid, and from red to blue by a base. In addition, acidic and basic solutions differ in chemical properties in several important ways. When a solution of an acid is mixed with a solution of a base, a neutralization reaction occurs. The products of the reaction have none of the characteristic properties of either the acid or base. For example, when a solution of hydrochloric acid is mixed with precisely the correct quantity of a sodium hydroxide solution, the result is a solution of sodium chloride, a simple ionic compound possessing neither acidic nor basic properties. (In general, such ionic products are referred to as salts.) The neutralization reaction can be written as follows:

$$HCl(aq) + NaOH(aq) \longrightarrow H_2O(l) + NaCl(aq) \qquad [3.18]$$

When we write the reaction as we have here, it is important to keep in mind that the substances shown as (aq) are present in the form of the separated ions, as discussed above. Notice that the acid and base in Equation 3.18 have combined to form water as a product. The general description of an acid-base neutralization reaction in aqueous solution, then, is that *an acid and base react to form a salt and water.* Using this general description we can predict the products formed in any acid-base neutralization reaction.

*Tasting chemical solutions is, of course, not a good practice. However, we have all had acids such as ascorbic acid (vitamin C), acetylsalicylic acid (aspirin), and citric acid (in citrus fruits) in our mouths, and we are familiar with the characteristic sour taste. It differs from the taste of soaps, which are mostly basic.

SAMPLE EXERCISE 3.2

Write a balanced equation for the reaction of hydrobromic acid, HBr, with barium hydroxide, $Ba(OH)_2$.

Solution: The products of any acid-base reaction are a salt and water. The salt is that formed from the cation of the base, $Ba(OH)_2$, and the anion of the acid, HBr. The charge on the barium ion is 2+ (see Table 2.5), and that on the bromide ion is 1−. Therefore, to maintain electrical neutrality, the formula for the salt must be $BaBr_2$. The unbalanced equation for the neutralization reaction is therefore

$$HBr(aq) + Ba(OH)_2(aq) \longrightarrow H_2O(l) + BaBr_2(aq)$$

To balance the equation we must provide two molecules of HBr to furnish the two Br^- ions and to supply the two H^+ ions needed to combine with the two OH^- ions of the base. The balanced equation is thus

$$2HBr(aq) + Ba(OH)_2(aq) \longrightarrow 2H_2O(l) + BaBr_2(aq)$$

Precipitation Reactions

One very important class of reactions occurring in solution is the precipitation reaction, in which one of the reaction products is insoluble. We will concern ourselves in this brief introduction with reactions between acids, bases, or salts in aqueous solution. As a simple example, consider the reaction between hydrochloric acid solution and a solution of the salt silver nitrate, $AgNO_3$. When the two solutions are mixed, a finely divided white solid forms. Upon analysis this solid proves to be silver chloride, AgCl, a salt that has a very low solubility* in water. The reaction as just described can be represented by the equation

$$HCl(aq) + AgNO_3(aq) \longrightarrow AgCl(s) + HNO_3(aq) \quad [3.19]$$

The formation of a precipitate in a chemical equation may be represented by a following (*s*), by a downward arrow following the formula for the solid, or by underlining the formula for the solid. You are reminded once again that substances indicated by (*aq*) may be present in the solution as separated ions.

The following equations provide further examples of precipitation reactions:

$$Pb(NO_3)_2(aq) + Na_2CrO_4(aq) \longrightarrow PbCrO_4(s) + 2NaNO_3(aq) \quad [3.20]$$

$$CuCl_2(aq) + 2NaOH(aq) \longrightarrow Cu(OH)_2(s) + 2NaCl(aq) \quad [3.21]$$

Notice that in each equation the positive ions (cations) and negative ions (anions) exchange partners. Reactions that fit this pattern of reactivity, whether they be precipitation reactions, neutralization reactions, or reactions of some other sort, are called **metathesis reactions** (muh-TATH-uh-sis; Greek, "to transpose").

*Solubility will be considered in some detail in Chapter 11. It is a measure of the amount of substance that can be dissolved in a given quantity of solvent (see Section 3.9).

SAMPLE EXERCISE 3.3

When solutions of sodium phosphate and barium nitrate are mixed, a precipitate of barium phosphate forms. Write a balanced equation to describe the reaction.

Solution: Our first task is to determine the formulas of the reactants. The sodium ion is Na^+ and the phosphate ion is PO_4^{3-}; thus sodium phosphate is Na_3PO_4. The barium ion is Ba^{2+} and the nitrate ion is NO_3^-; thus barium nitrate is $Ba(NO_3)_2$. The Ba^{2+} and PO_4^{3-} ions combine to form the barium phosphate precipitate, $Ba_3(PO_4)_2$. The other ions, Na^+ and NO_3^-, remain in solution and are represented as $NaNO_3(aq)$. The *unbalanced* equation for the reaction is thus

$$Na_3PO_4(aq) + Ba(NO_3)_2(aq) \longrightarrow Ba_3(PO_4)_2(s) + NaNO_3(aq)$$

Because the NO_3^- and PO_4^{3-} ions maintain their identity through the reaction, we can treat them as units in balancing the equation. There are two (PO_4) units on the right, so we place a coefficient 2 in front of Na_3PO_4. This then gives six Na atoms on the left, necessitating a coefficient of 6 in front of $NaNO_3$. Finally, the presence of six (NO_3) units on the right requires a coefficient of 3 in front of $Ba(NO_3)_2$:

$$2Na_3PO_4(aq) + 3Ba(NO_3)_2(aq) \longrightarrow Ba_3(PO_4)_2(s) + 6NaNO_3(aq)$$

A balanced equation implies a quantitative relation between the reactants and the products involved in a chemical reaction. Thus complete combustion of a molecule of C_3H_8 requires exactly five molecules of O_2, no more and no less, as shown in Equation 3.12. Although it is not possible to count directly the number of molecules of each type in any reaction, this count can be made indirectly if the mass of each molecule is known. Indeed, this indirect approach is the one taken to obtain quantitative information about the amounts of substances involved in any chemical transformation. Therefore, before we can pursue the quantitative aspects of chemical reactions further, we must explore the concept of atomic and molecular weights.

3.4 ATOMIC AND MOLECULAR WEIGHTS

Dalton's atomic theory led him and other scientists of the time to a new problem. If it is true that atoms combine with one another in the ratios of small whole numbers to form compounds, what *are* the ratios with which they combine? Atoms are too small to be measured individually by any means available in the early nineteenth century. However, if one knew the *relative* masses of the atoms, then by measuring out convenient quantities in the laboratory, one could determine the relative numbers of atoms in a sample. Consider a simple analogy: Suppose that oranges are on the average four times heavier than plums; the number of oranges in 48 kg of oranges will then be the same as the number of plums in 12 kg of plums. Similarly, if you knew that oxygen atoms were on the average 16 times more massive than hydrogen atoms, then you would know that the number of oxygen atoms in 16 g of oxygen is the same as the number of hydrogen atoms in 1 g of hydrogen. Thus the problem of determining the combining ratios becomes one of determining the relative masses of the atoms of the elements.

This is all very well, but there was great difficulty in getting started. Since atoms and molecules can't be seen, there was no simple way to be sure about the relative numbers of atoms in *any* compound. Dalton thought that the formula for water was HO. However, the French scientist Gay-Lussac showed in a brilliant set of measurements that it required two volumes of hydrogen gas to react with one volume of oxygen to form two volumes of water vapor. This observation was inconsistent with Dalton's formula for water. Furthermore, if oxygen were assumed to be a monatomic gas, as Dalton did, one could obtain two volumes of water vapor only by splitting the oxygen atoms in half, which of course violates the concept of the atom as indivisible in chemical reactions.

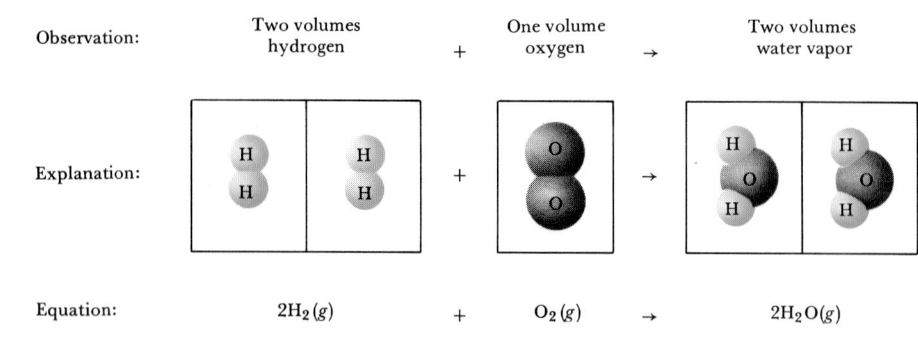

FIGURE 3.3 Gay-Lussac's experimental observation of combining volumes shown together with Avogadro's explanation of this phenomenon.

The Italian physicist Amedeo Avogadro suggested in 1811 that Gay-Lussac's results could be explained if it were assumed that both hydrogen and oxygen exist in the gas phase as diatomic molecules H_2 and O_2. Avogadro also proposed that equal volumes of gases, if measured at the same temperature and pressure, contain equal numbers of molecules. (When we talk about gases in Chapter 9 we will see the basis for Avogadro's hypothesis, which is entirely correct.) The way in which Avogadro's hypothesis explains Gay-Lussac's observations is illustrated in Figure 3.3. The key here is in assuming that the reacting gases are diatomic. Thus it is possible to obtain two volumes of water vapor from one volume of oxygen gas. Because the combining volume of hydrogen is twice that of oxygen, the formula for water must be H_2O, not HO as Dalton had suggested.

SAMPLE EXERCISE 3.4

How many liters of NH_3 will be produced when 2 L of N_2 combine with 6 L of H_2 (all volumes measured at the same temperature and pressure)?

Solution: The balanced equation is

$$N_2(g) + 3H_2(g) \longrightarrow 2NH_3(g)$$

From the coefficients it can be deduced that each liter of N_2 combines with 3 L of H_2 to produce 2 L of NH_3. Thus 4L of NH_3 will form from 2 L of N_2 and 6 L of H_2.

It required several years for Avogadro's ideas to gain acceptance, but by about 1860 matters had pretty well straightened themselves out. In the meantime chemists such as Berzelius (Figure 3.4) had been making painstaking measurements of the masses of the elements that combined with one another and had established the atomic weights* of many elements with quite good precision.

Because different types of atoms contain different numbers of subatomic particles, they differ in mass. Atomic weights were originally assigned on a relative scale with hydrogen given a value of 1. Modern atomic weights are based on the assignment of an atomic mass of exactly

*The term *atomic weight* is a bit of a misnomer; what is really meant is atomic mass. However, the term atomic weight has become so commonly used that there is little point in attempting to change. We will use the expressions "atomic weight" or "molecular weight" but will refer to individual atoms or molecules in terms of mass.

FIGURE 3.4 Jons Jakob Berzelius (1779–1848), Swedish chemist. Berzelius is one of the truly great figures in nineteenth-century science. He developed the modern system of symbols and formulas in chemistry and discovered several elements. Berzelius was an exceptionally gifted experimentalist. His determinations of the atomic weights of several elements were the most accurate known for many years. (*Bettmann Archive*)

12 for the ^{12}C isotope of carbon (Section 2.5). We refer to the units of these atomic weights as atomic mass units (amu). Thus ^{12}C has an atomic weight of exactly 12 amu. The atomic weights of elements are reported as average values, reflecting the relative abundances of each isotope of each element. Naturally occurring chlorine is 75.53 percent ^{35}Cl, which has a mass of 34.969 amu, and 24.47 percent ^{37}Cl, which has a mass of 36.966 amu. The average atomic weight for chlorine can be calculated to four significant figures from this information:

$$\begin{aligned} \text{Av AW}^* &= (75.53\%)(34.969 \text{ amu}) + (24.47\%)(36.966 \text{ amu}) \\ &= 26.41 \text{ amu} + 9.05 \text{ amu} \\ &= 35.46 \text{ amu} \end{aligned}$$

The last digit in this calculation is uncertain; the accepted value for the atomic weight of chlorine, to five significant figures, is 35.453 amu. The atomic weights of the elements are listed below the symbol for the element on the periodic table found on the inside front cover of this text. They are also listed in the table of elements on the inside front cover.

The formula weight of a substance is merely the sum of the atomic weights of each atom it contains. For example, H_2SO_4, sulfuric acid, has a formula weight of 98.0 amu:

$$\begin{aligned} \text{FW} &= 2(\text{AW of H}) + \text{AW of S} + 4(\text{AW of O}) \\ &= 2(1.0 \text{ amu}) + 32.0 \text{ amu} + 4(16.0 \text{ amu}) \\ &= 98.0 \text{ amu} \end{aligned}$$

*The abbreviation AW is used for atomic weight, MW for molecular weight, and FW for formula weight. The SI abbreviation for amu is simply u, as in 35.46 u. We will not be using this abbreviation.

Here we have rounded off the atomic weights so that our result has three significant figures. We will round off the atomic weights in this way for most problems.

If the chemical formula of a substance is its molecular formula, then the formula weight is called the **molecular weight.** For example, the molecular formula for glucose (the sugar transported by the blood to body tissues to satisfy energy requirements) is $C_6H_{12}O_6$. The molecular weight of glucose is therefore

$$6(12.0 \text{ amu}) + 12(1.0 \text{ amu}) + 6(16.0 \text{ amu}) = 180.0 \text{ amu}$$

With ionic substances such as NaCl, which exist as three-dimensional arrays of ions (Figure 2.23), it is really inappropriate to speak of molecules. Thus we cannot write molecular formulas and molecular weights for such substances. The formula weight of NaCl is

$$23.0 \text{ amu} + 35.5 \text{ amu} = 58.5 \text{ amu}$$

SAMPLE EXERCISE 3.5

Calculate the formula weight of (a) sucrose, $C_{12}H_{22}O_{11}$ (table sugar); (b) calcium nitrate, $Ca(NO_3)_2$.

Solution: (a) By adding the weights of the atoms in sucrose we find it to have a formula weight of 342.0 amu:

$$\begin{aligned} 12\text{C atoms} &= 12(12.0 \text{ amu}) = 144.0 \text{ amu} \\ 22\text{H atoms} &= 22(1.0 \text{ amu}) = 22.0 \text{ amu} \\ 11\text{O atoms} &= 11(16.0 \text{ amu}) = \underline{176.0 \text{ amu}} \\ & \qquad\qquad\qquad\quad 342.0 \text{ amu} \end{aligned}$$

(b) If a chemical formula has parentheses, the subscript outside the parentheses is a multiplier for all atoms inside. Thus for $Ca(NO_3)_2$ we have

$$\begin{aligned} 1\text{Ca atom} &= 1(40.1 \text{ amu}) = 40.1 \text{ amu} \\ 2\text{N atoms} &= 2(14.0 \text{ amu}) = 28.0 \text{ amu} \\ 6\text{O atoms} &= 6(16.0 \text{ amu}) = \underline{96.0 \text{ amu}} \\ & \qquad\qquad\qquad\quad 164.1 \text{ amu} \end{aligned}$$

Percentage Composition from Formulas

It is occasionally necessary to calculate the percentage composition of a compound (that is, the percentage by mass or weight contributed by each element in the substance). For example, we may wish to compare the calculated composition of a substance with that found experimentally in order to verify the purity of the compound. If you know the chemical formula, it is a straight-forward matter to calculate percentage composition. The calculation of such percentages is illustrated in Sample Exercise 3.6.

SAMPLE EXERCISE 3.6

Calculate the percentage composition of $C_{12}H_{22}O_{11}$.

Solution: In general, the percentage of a given element in a compound is given by

$$\frac{\text{Total mass of that element}}{\text{Formula weight of compound}} \times 100$$

The formula weight of $C_{12}H_{22}O_{11}$ is 342 amu (Sample Exercise 3.5). Therefore, the percentage composition is

$$\%C = \frac{12(12.0 \text{ amu})}{342 \text{ amu}} \times 100 = 42.1\%$$

$$\%H = \frac{22(1.0 \text{ amu})}{342 \text{ amu}} \times 100 = 6.4\%$$

$$\%O = \frac{11(16.0 \text{ amu})}{342 \text{ amu}} \times 100 = 51.5\%$$

The same elemental composition is obtained from the empirical formula of a substance as from its molecular formula. In Section 3.7 we shall see how the experimentally determined percentage composition of a compound can be used to calculate the empirical formula of the compound.

3.5 THE MASS SPECTROMETER

The most direct and accurate means for determining atomic and molecular weights is provided by the **mass spectrometer.** This instrument is illustrated schematically in Figure 3.5. A gaseous sample is introduced into the instrument at *A* and then bombarded by a stream of high-energy electrons at *B*. Collisions between the electrons and the atoms or molecules of the gas produce positive ions. These ions (of mass m and charge e) are accelerated toward a negatively charged wire grid (*C*). After they pass through the grid, they encounter two slits, which produce a narrow beam of the ions. This beam of ions is then passed between the poles of a magnet, which forces the ions into a circular path whose radius depends on the charge-to-mass ratio of the ions, e/m. Ions of smaller e/m ratio follow a curved path of larger radius than those having a larger e/m ratio. Consequently, ions with equal charges but different masses are separated from each other. By continuously changing either the strength of the magnetic field or the accelerating voltage on the negatively charged grid, ions of varying e/m ratios can be caused to enter the detector at the end of the instrument. Because the ions are principally singly charged, the separation is primarily on the basis of mass. A graph of the intensity of signal from the detector versus the mass of the ion is called a **mass spectrum.**

The mass spectrometer provided the first unambiguous evidence for the existence of isotopes. Suppose we allow a stream of mercury vapor to enter the mass spectrometer. If all the atoms of mercury were in fact identical, all of them upon ionization would possess the same e/m ratio and would therefore appear at the same place in the mass spectrum. But the mass spectrum actually observed for mercury is as shown in Figure 3.6. It is clear that there are several kinds of atoms in mercury, differing slightly in their masses.

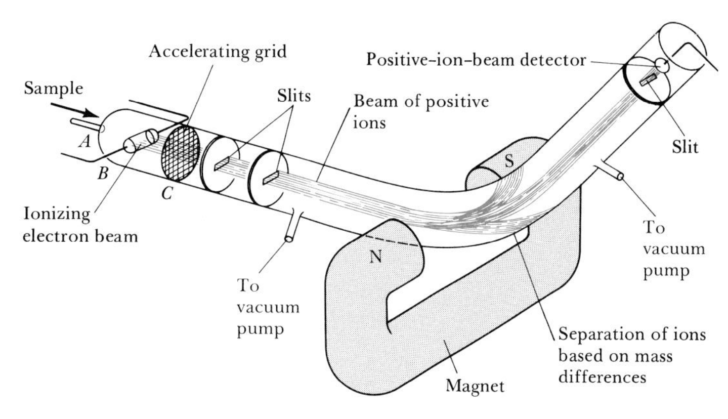

FIGURE 3.5 Schematic diagram of a modern mass spectrometer.

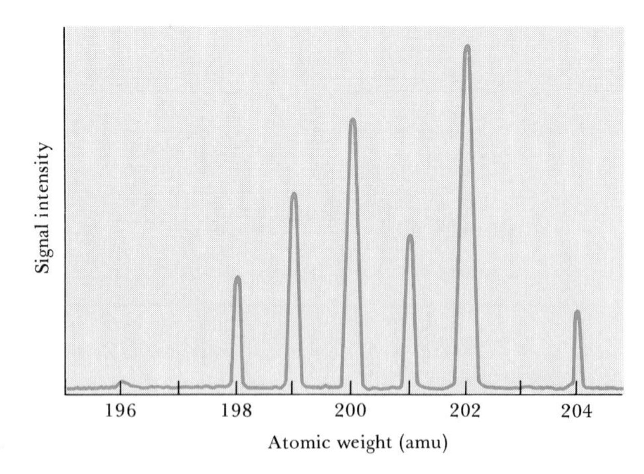

FIGURE 3.6 Mass spectrum of Hg^+ ions in mercury vapor.

TABLE 3.2 Mass spectrum of mercury

Mass number	Atomic weight (amu)	Fractional abundance
196	195.965	0.0014
198	197.967	0.10039
199	198.967	0.1683
200	199.968	0.2312
201	200.970	0.1323
202	201.970	0.2979
204	203.973	0.0685

Using the mass spectrometer, it is possible to measure relative values of the ratio e/m for ions with great accuracy. It is also possible to measure with high accuracy the relative numbers of the different isotopes of an element. As we have noted previously, the atomic weight is the weighted average of the masses of all the isotopes of that element. For example, from the mass spectrum of mercury, Figure 3.6, we obtain the data shown in Table 3.2. The atomic weight of mercury, 200.59 amu, is the sum of the masses of the various isotopes, each multiplied by its fractional abundance:

$$\begin{aligned} \text{AW} &= (0.0014)(195.965 \text{ amu}) + (0.10039)(197.967 \text{ amu}) + \cdots \\ &\qquad + (0.0685)(203.973 \text{ amu}) \\ &= 200.59 \text{ amu} \end{aligned}$$

3.6 THE CHEMICAL MOLE

We indicated earlier that the concept of atomic weights is important because it permits us to count atoms indirectly, by weighing samples. A convenient number of atoms, molecules, or formula units is that number whose mass in grams is equal numerically to the atomic weight, molecular weight, or formula weight. This quantity is called the **mole,** abbreviated mol.* Thus 1 mol of ^{12}C atoms is the number of ^{12}C atoms in exactly 12 g of ^{12}C. A mole of any element is defined as the quantity of that element, as it occurs in nature, that contains the same number of atoms as there are in exactly 12 g of ^{12}C. It has been determined experimentally that the number of atoms in this quantity of ^{12}C is 6.022×10^{23}. This number is given a special name, **Avogadro's number,** in honor of Amedeo Avogadro, who proposed that equal volumes of gases at the same temperature and pressure contain equal numbers of molecules (Section 3.4). For most purposes, we will use 6.02×10^{23} for Avogadro's number throughout the text; this number should be committed to memory.

A mole of ions, molecules, or anything else contains Avogadro's number of these objects:

*The term *mole* comes from the Latin word *moles* meaning "a mass." The term *molecule* is the diminutive form of this word and means "a small mass."

$$1 \text{ mol } ^{12}\text{C atoms} = 6.02 \times 10^{23} \ ^{12}\text{C atoms}$$

$$1 \text{ mol } H_2O \text{ molecules} = 6.02 \times 10^{23} \ H_2O \text{ molecules}$$

$$1 \text{ mol } NO_3^- \text{ ions} = 6.02 \times 10^{23} \ NO_3^- \text{ ions}$$

The concept of a mole as being 6.02×10^{23} of something is analogous to the concept of a dozen as 12 of something or a gross as 144 of something. Because a ^{24}Mg atom has twice the mass of a ^{12}C atom, 1 mol of ^{24}Mg has twice the mass of 1 mol of ^{12}C. A ^{12}C atom has a mass of exactly 12 amu, whereas a ^{24}Mg atom has a mass of 24.0 amu. Because 1 mol of ^{12}C atoms weighs 12 g (by definition), 1 mol of ^{24}Mg atoms must weigh 24.0 g. In fact, a mole of atoms of any element has a mass in grams numerically equal to the atomic weight of a single atom:

One ^{12}C atom has a mass of 12 amu; 1 mol ^{12}C weighs 12 g.

One ^{24}Mg atom has a mass of 24.0 amu; 1 mol ^{24}Mg weighs 24.0 g.

One Au atom has a mass of 197 amu; 1 mol Au weighs 197 g.

We can generalize this idea to include molecules and ions: The mass of a mole of formula units of any substance (that is, 6.02×10^{23} of them) is always equal to the formula weight expressed in grams:

One H_2O molecule has a mass of 18.0 amu; 1 mol H_2O weighs 18.0 g.

One NO_3^- ion has a mass of 62.0 amu; 1 mol NO_3^- weighs 62.0 g.

One NaCl unit has a mass of 58.5 amu; 1 mol NaCl weighs 58.5.

Further examples of mole relationships are shown in Table 3.3.

The first entries in Table 3.3, those for N and N_2, point out the importance of stating the chemical form of a substance exactly when we use the mole concept. Suppose you read that 1 mol of nitrogen is produced in a particular reaction. You might interpret this statement to mean 1 mol of nitrogen atoms (14.0 g). Unless otherwise stated, what was probably meant is 1 mol of nitrogen molecules, N_2 (28.0 g), because N_2 is the usual chemical form of the element. However, to avoid ambiguity it is always best to state explicitly the chemical form being discussed.

TABLE 3.3 Mole relationships

Name	Formula	Formula weight (amu)	Mass of 1 mol of formula units (g)	Number and kind of particles in 1 mol
Atomic nitrogen	N	14.0	14.0	6.02×10^{23} N atoms
Molecular nitrogen	N_2	28.0	28.0	6.02×10^{23} N_2 molecules $2(6.02 \times 10^{23})$ N atoms
Silver	Ag	107.9	107.9	6.02×10^{23} Ag atoms
Silver ions	Ag^+	107.9[a]	107.9	6.02×10^{23} Ag^+ ions
Barium chloride	$BaCl_2$	208.2	208.2	6.02×10^{23} $BaCl_2$ units 6.02×10^{23} Ba^{2+} ions $2(6.02 \times 10^{23})$ Cl^- ions

[a] Recall that the electron has negligible mass; thus ions and atoms have essentially the same mass.

SAMPLE EXERCISE 3.7

What is the mass of 1 mol of glucose, $C_6H_{12}O_6$?

Solution: By adding the weights of the atoms in glucose, we find it to have a formula weight of 180 amu:

$$\begin{aligned} 6\text{C atoms} &= 6(12.0\ \text{amu}) = 72.0\ \text{amu} \\ 12\text{H atoms} &= 12(1.0\ \text{amu}) = 12.0\ \text{amu} \\ 6\text{O atoms} &= 6(16.0\ \text{amu}) = \underline{96.0\ \text{amu}} \\ & \qquad\qquad\qquad\quad 180.0\ \text{amu} \end{aligned}$$

Hence 1 mol of $C_6H_{12}O_6$ weighs 180 g. Glucose is sometimes called dextrose; it is also known as blood sugar. It is found widely in nature, occurring, for example, in honey and fruits. Other types of sugars used as food must be converted into glucose in the stomach or liver before they are used by the body as energy sources. Because glucose requires no conversion, it is often given intravenously to patients who need immediate nourishment.

SAMPLE EXERCISE 3.8

How many C atoms are there in 1 mol of $C_6H_{12}O_6$?

Solution: There are 6.02×10^{23} $C_6H_{12}O_6$ molecules in 1 mol. Each molecule contains 6C atoms; hence there are $6(6.02 \times 10^{23})$C atoms:

$$\begin{aligned} \text{C atoms} &= (1\ \text{mol}\ C_6H_{12}O_6) \\ &\times \left(\frac{6.02 \times 10^{23}\ \text{molecules}}{1\ \text{mol}}\right)\left(\frac{6\text{C atoms}}{1\ \text{molecule}}\right) \\ &= 3.61 \times 10^{24}\ \text{C atoms} \end{aligned}$$

To illustrate how the mole concept and Avogadro's number allow us to interconvert masses and number of particles, let's calculate the number of copper atoms in a penny. A penny weighs 3 g, and we'll assume that it is 100 percent copper:

$$\begin{aligned} \text{Cu atoms} &= (3\ \text{g Cu})\left(\frac{1\ \text{mol Cu}}{63.5\ \text{g Cu}}\right)\left(\frac{6.02 \times 10^{23}\ \text{Cu atoms}}{1\ \text{mol Cu}}\right) \\ &= 3 \times 10^{22}\ \text{Cu atoms} \end{aligned}$$

Notice how we were able to use dimensional analysis (Section 1.5) in a straightforward manner to go from grams to numbers of atoms; the conversion sequence is grams $\longrightarrow$ moles $\longrightarrow$ atoms.

We might reflect momentarily on the number of copper atoms in a penny, 3×10^{22}. This is a tremendously large number. It becomes more impressive when we realize that the entire United States could be covered to a depth of 30 mi with this number of ice cubes, each 1 in. on an edge. We should remember that Avogadro's number is even larger.

SAMPLE EXERCISE 3.9

How many moles of glucose, $C_6H_{12}O_6$, are there in
(a) 538 g
(b) 1.00 g of this substance?

Solution:
(a) One mol of $C_6H_{12}O_6$ weighs 180 g (Sample Exercise 3.7). Therefore, there must be more than 1 mol in 538 g.

Moles $C_6H_{12}O_6$ =

$$(538\ \text{g}\ C_6H_{12}O_6)\left(\frac{1\ \text{mol}\ C_6H_{12}O_6}{180\ \text{g}\ C_6H_{12}O_6}\right)$$

$$= 2.99\ \text{mol}$$

(b) In this case there must be less than 1 mol.

$$\text{Moles } C_6H_{12}O_6 = (1.00 \text{ g } C_6H_{12}O_6)\left(\frac{1 \text{ mol } C_6H_{12}O_6}{180 \text{ g } C_6H_{12}O_6}\right) = 5.56 \times 10^{-3} \text{ mol}$$

The conversion of mass to moles and of moles to mass is frequently encountered in calculations using the mole concept. Notice that the number of moles is always the mass divided by the mass of 1 mol (the formula weight expressed in grams).

SAMPLE EXERCISE 3.10

What is the mass, in grams, of 0.433 mol of $C_6H_{12}O_6$?

Solution: Because this is less than 1 mol, the mass will be less than 180 g, the mass of 1 mol.

$$\text{Grams } C_6H_{12}O_6 = (0.433 \text{ mol } C_6H_{12}O_6)\left(\frac{180 \text{ g } C_6H_{12}O_6}{1 \text{ mol } C_6H_{12}O_6}\right) = 77.9 \text{ g}$$

Notice that the mass of a certain number of moles of a substance is always the number of moles times the mass of 1 mol.

SAMPLE EXERCISE 3.11

How many glucose molecules are there in 5.23 g of $C_6H_{12}O_6$?

Solution: In this case we need to carry out unit conversions in the order grams $\longrightarrow$ moles $\longrightarrow$ molecules. Because the mass we begin with corresponds to less than a mole, there should be fewer than 6.02×10^{23} molecules.

$$\text{Molecules } C_6H_{12}O_6 = (5.23 \text{ g } C_6H_{12}O_6)\left(\frac{1 \text{ mol } C_6H_{12}O_6}{180 \text{ g } C_6H_{12}O_6}\right) \times \left(\frac{6.02 \times 10^{23} \; C_6H_{12}O_6 \text{ molecules}}{1 \text{ mol } C_6H_{12}O_6}\right) = 1.75 \times 10^{22} \text{ molecules}$$

SAMPLE EXERCISE 3.12

What is the mass, in grams, of 1.00×10^{23} molecules of $C_6H_{12}O_6$?

Solution: In this problem, we need to carry out unit conversions in the order molecules $\longrightarrow$ moles $\longrightarrow$ grams.

$$\text{Grams } C_6H_{12}O_6 = (1.00 \times 10^{23} \text{ molecules}) \times \left(\frac{1 \text{ mol}}{6.02 \times 10^{23} \text{ molecules}}\right)\left(\frac{180 \text{ g}}{1 \text{ mol}}\right) = 29.9 \text{ g}$$

3.7 EMPIRICAL FORMULAS FROM ANALYSES

Before we use the mole concept to determine the masses of substances involved in chemical reactions, let's see how it is used in deducing the formulas of chemical substances. As an example, let us determine the formula of a compound formed between mercury and chlorine. This colorless solid is found on analysis to consist of 73.9 percent mercury and 26.1 percent chlorine by mass. This means that if we had a 100-g sample of the solid, the sample would contain 73.9 g of mercury (Hg) and 26.1 g of chlorine (Cl). We divide each of these weights by the appropriate atomic weight to obtain the number of moles of each element in 100 g:

$$73.9 \text{ g Hg}\left(\frac{1 \text{ mol Hg}}{200.6 \text{ g Hg}}\right) = 0.368 \text{ mol Hg}$$

$$26.1 \text{ g Cl}\left(\frac{1 \text{ mol Cl}}{35.5 \text{ g Cl}}\right) = 0.735 \text{ mol Cl}$$

We then divide the larger number of moles by the smaller to obtain the ratio 1.99 mole of Cl/mole of Hg. Because of experimental errors, the results of an analysis may not lead to exact integers for the ratios of moles. The ratio obtained in this case is very close to 2; the formula for the compound is thus $HgCl_2$. This is the simplest, or empirical, formula because it uses as subscripts the smallest set of integers that express the correct ratios of atoms present (Section 2.6).

SAMPLE EXERCISE 3.13

Phosgene, a poison gas used during World War I, contains 12.1 percent C, 16.2 percent O, and 71.7 percent Cl. What is the empirical formula of phosgene?

Solution: For simplicity, we may assume that we have 100 g of material. The number of moles of each element is then

$$\text{Moles C} = (12.1 \text{ g})\left(\frac{1 \text{ mol C}}{12.0 \text{ g}}\right) = 1.01 \text{ mol C}$$

$$\text{Moles O} = (16.2 \text{ g})\left(\frac{1 \text{ mol O}}{16.0 \text{ g}}\right) = 1.01 \text{ mol O}$$

$$\text{Moles Cl} = (71.7 \text{ g})\left(\frac{1 \text{ mol Cl}}{35.5 \text{ g}}\right) = 2.02 \text{ mol Cl}$$

The simplest ratio, found by dividing each number by the smallest, 1.01, is C : O : Cl = 1 : 1 : 2 and the empirical formula is $COCl_2$. Because other experiments show that the molecular weight of the phosgene molecule is 99 amu, $COCl_2$ is also the molecular formula.

When a compound containing carbon and hydrogen is combusted in an apparatus such as that shown in Figure 3.7, the carbon of the compound is converted to CO_2, and all the hydrogen to H_2O. The amount of CO_2 produced can be measured by determining the mass increase in the CO_2 absorber. Similarly, the amount of H_2O formed is determined from the increase in mass of the water absorption tube. As an example, let's consider an analysis of a sample of ascorbic acid (vitamin C). Combustion of 1.000 g of ascorbic acid produces 1.500 g of CO_2 and 0.405 g of

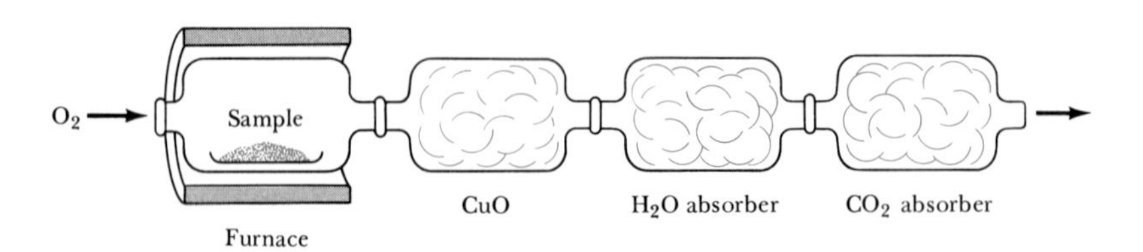

FIGURE 3.7 Apparatus for determining the percentages of carbon and hydrogen in a compound. Copper oxide serves to oxidize traces of carbon and carbon monoxide to carbon dioxide, and hydrogen to water. Magnesium perchlorate, $Mg(ClO_4)_2$, is used to absorb water, whereas sodium hydroxide, NaOH, absorbs carbon dioxide.

H_2O. From these two bits of experimental information we must calculate the quantities of C and H in the 1.000-g sample of ascorbic acid:

$$(1.500 \text{ g } CO_2)\left(\frac{1 \text{ mol } CO_2}{44.00 \text{ g } CO_2}\right)\left(\frac{1 \text{ mol C}}{1 \text{ mol } CO_2}\right)\left(\frac{12.01 \text{ g C}}{1 \text{ mol C}}\right) = 0.409 \text{ g C}$$

$$(0.405 \text{ g } H_2O)\left(\frac{1 \text{ mol } H_2O}{18.0 \text{ g } H_2O}\right)\left(\frac{2 \text{ mol H}}{1 \text{ mol } H_2O}\right)\left(\frac{1.01 \text{ g H}}{1 \text{ mol H}}\right) = 0.045 \text{ g H}$$

From other experiments it can be established that ascorbic acid contains only C, H, and O. Thus the amount of oxygen in the compound must be

$$1.000 \text{ g} - (0.409 \text{ g} + 0.045 \text{ g}) = 0.546 \text{ g}$$

From these data, we can now proceed to calculate the number of moles of each element present in 1 g of ascorbic acid:

$$\text{Moles C} = (0.409 \text{ g C})\left(\frac{1 \text{ mol C}}{12.0 \text{ g C}}\right) = 0.0341 \text{ mol C}$$

$$\text{Moles H} = (0.045 \text{ g H})\left(\frac{1 \text{ mol H}}{1.01 \text{ g H}}\right) = 0.045 \text{ mol H}$$

$$\text{Moles O} = (0.546 \text{ g O})\left(\frac{1 \text{ mol O}}{16.0 \text{ g O}}\right) = 0.0341 \text{ mol O}$$

The relative number of moles of each element can be found by dividing by the smallest number, 0.0341. The ratio of C:H:O is thus 1:1.32:1, which is the same as 3:4:3. The empirical formula is thus $C_3H_4O_3$. In order to determine the molecular formula an experiment must be performed to determine molecular weight. The molecular weight of ascorbic acid is 176 amu. The formula weight for $C_3H_4O_3$ is 3(12.0 amu) + 4(1.0 amu) + 3(16.0 amu) = 88.0 amu. Thus there are two of these formula units in the molecule, and the molecular formula is consequently $C_6H_8O_6$.

3.8 QUANTITATIVE INFORMATION FROM BALANCED EQUATIONS

The mole concept provides a key to placing the quantitative information available in a balanced chemical equation on a practical, macroscopic level. Consider the following balanced equation:

$$2H_2(g) + O_2(g) \longrightarrow 2H_2O(l) \qquad [3.22]$$

The coefficients tell us that two molecules of H_2 react with each molecule of O_2 to form two molecules of H_2O. Therefore, $2(6.02 \times 10^{23})$ molecules of H_2 will react with 6.02×10^{23} molecules of O_2 to form $2(6.02 \times 10^{23})$ molecules of H_2O. This is the same as saying that 2 mol of H_2 react with 1 mol of O_2 to form 2 mol of H_2O. The point is that the coefficients in a balanced equation can be interpreted *both* as the *relative numbers of molecules* (or formula units) involved in a reaction *and* as the *relative number of moles*. These interpretations are summarized in Table 3.4. Notice that 4.04 g of H_2 will react with each 32.00 g of O_2 to form 36.04 g

TABLE 3.4 Interpretations of equations

	$2H_2$	+	O_2	$\longrightarrow$	$2H_2O$
Molecular ratio:	2 molecules	React with	1 molecule	To form	2 molecules
	$2(6.02 \times 10^{23})$ molecules	React with	6.02×10^{23} molecules	To form	$2(6.02 \times 10^{23})$ molecules
Mole ratio:	2 mol	React with	1 mol	To form	2 mol
Weight ratio:	2(2.02) g = 4.04 g	React with	32.00 g	To form	2(18.02) g = 36.04 g

of H_2O, because these are the masses of 2 mol of H_2, 1 mol of O_2, and 2 mol of H_2O, respectively. Notice also that the sum of the masses of the reactants equals the mass of the product as it must in any chemical reaction according to the law of conservation of mass. The quantities 2 mol of H_2, 1 mol of O_2, and 2 mol of H_2O, which are related by Equation 3.22, are called stoichiometrically equivalent quantities. We can represent this as

$$2 \text{ mol } H_2 \simeq 1 \text{ mol } O_2 \simeq 2 \text{ mol } H_2O$$

where the symbol $\simeq$ is taken to mean "stoichiometrically equivalent to." These stoichiometric relations can be used to give conversion factors to relate quantities of reactants and products in a chemical reaction. For example, the number of moles of H_2O produced from 1.57 mol of O_2 can be calculated as follows:

$$\text{Moles } H_2O = (1.57 \text{ mol } O_2)\left(\frac{2 \text{ mol } H_2O}{1 \text{ mol } O_2}\right)$$
$$= 3.14 \text{ mol } H_2O$$

As a different example, consider the following reaction:

$$2CuFeS_2(s) + 5O_2(g) \longrightarrow 2Cu(s) + 2FeO(s) + 4SO_2(g) \qquad [3.23]$$

This equation describes a process in the smelting of copper using chalcopyrite ($CuFeS_2$) as the mineral source of the copper. Using the mole concept we can calculate the mass of Cu that can be produced from 1.00 g of chalcopyrite. From the coefficients in Equation 3.23 we can write the following stoichiometric relationship: 2 mol $CuFeS_2 \simeq$ 2 mol Cu. In order to use this relationship, however, the quantity of $CuFeS_2$ must be converted to moles. Because 1 mol $CuFeS_2$ = 183 g $CuFeS_2$, we have

$$\text{Moles } CuFeS_2 = (1.00 \text{ g } CuFeS_2)\left(\frac{1 \text{ mol } CuFeS_2}{183 \text{ g } CuFeS_2}\right)$$
$$= 5.46 \times 10^{-3} \text{ mol } CuFeS_2$$

The stoichiometric factor from the balanced equation, 2 mol $CuFeS_2 \simeq$ 2 mol Cu, can then be used to calculate moles of Cu:

$$\text{Moles Cu} = (5.46 \times 10^{-3}\ \text{mol CuFeS}_2)\left(\frac{2\ \text{mol Cu}}{2\ \text{mol CuFeS}_2}\right)$$
$$= 5.46 \times 10^{-3}\ \text{mol Cu}$$

Finally, the mass of the Cu, in grams, can be calculated using the relationship 1 mol Cu = 63.5 g Cu:

$$\text{Grams Cu} = (5.46 \times 10^{-3}\ \text{mol Cu})\left(\frac{63.5\ \text{g Cu}}{1\ \text{mol Cu}}\right)$$
$$= 0.347\ \text{g Cu}$$

These steps, of course, can be combined in a single sequence of factors:

$$\text{Grams Cu} = (1.00\ \text{g CuFeS}_2)\left(\frac{1\ \text{mol CuFeS}_2}{183\ \text{g CuFeS}_2}\right)\left(\frac{2\ \text{mol Cu}}{2\ \text{mol CuFeS}_2}\right)\left(\frac{63.5\ \text{g Cu}}{1\ \text{mol Cu}}\right)$$
$$= 0.347\ \text{g Cu}$$

We can similarly calculate the amount of SO_2 produced in the production of this quantity of copper using the relationship 2 mol $CuFeS_2 \simeq$ 4 mol SO_2, which also comes from the coefficients in Equation 3.23:

$$\text{Grams SO}_2 = (1.00\ \text{g CuFeS}_2)\left(\frac{1\ \text{mol CuFeS}_2}{183\ \text{g CuFeS}_2}\right)\left(\frac{4\ \text{mol SO}_2}{2\ \text{mol CuFeS}_2}\right)\left(\frac{64.1\ \text{g SO}_2}{1\ \text{mol SO}_2}\right)$$
$$= 0.701\ \text{g SO}_2$$

It is interesting to note that the mass of SO_2 produced in this reaction is approximately twice the mass of the copper. Consequently, considerable air pollution from sulfur dioxide is often generated in the vicinity of copper smelters (Figure 3.8).

SAMPLE EXERCISE 3.14

How much water is produced in the combustion of 1.00 g of glucose, $C_6H_{12}O_6$:

$$C_6H_{12}O_6(s) + 6O_2(g) \longrightarrow 6CO_2(g) + 6H_2O(l)$$

Solution: This problem can be solved by a stepwise conversion of grams of $C_6H_{12}O_6$ to moles of $C_6H_{12}O_6$ to moles of H_2O to grams of H_2O. From the balanced equation we have 1 mol $C_6H_{12}O_6 \simeq$ 6 mol H_2O. Therefore, we have

$$\text{Grams H}_2\text{O} = (1.00\ \text{g C}_6\text{H}_{12}\text{O}_6)\left(\frac{1\ \text{mol C}_6\text{H}_{12}\text{O}_6}{180\ \text{g C}_6\text{H}_{12}\text{O}_6}\right)$$
$$\times \left(\frac{6\ \text{mol H}_2\text{O}}{1\ \text{mol C}_6\text{H}_{12}\text{O}_6}\right)\left(\frac{18.0\ \text{g H}_2\text{O}}{1\ \text{mol H}_2\text{O}}\right)$$
$$= 0.600\ \text{g H}_2\text{O}$$

We may note that an average person ingests 2 L of water daily and eliminates 2.4 L. The difference is produced in metabolism of foodstuffs as above. (Metabolism is a general term used to describe all the processes of a living animal or plant.) The desert rat (kangaroo rat) is able to take great advantage of its metabolic water to help it survive in the dry desert. In fact, it apparently never drinks water.

FIGURE 3.8 Extended exposure to high concentrations of SO_2 and other pollutants can cause extensive damage to plants, animals, and structural materials. This photograph, taken in the Chest Creek Watershed, Clearfield County, Pennsylvania, shows the effects of air and water pollution. Timber in the center background has been killed by fumes from the burning mine-refuse pile at upper center. Stream pollution by mine acid and coal sedimentation is shown at center, whereas soil erosion is evident in the foreground. (*USDA-SCS*)

Limiting Reagent

In many situations an excess of one or more substances is available for chemical reaction. Some will therefore be left over when the reaction is complete. For example, consider again the combustion of hydrogen:

$$2H_2(g) + O_2(g) \longrightarrow 2H_2O(l)$$

Suppose that 2 mol of H_2 and 2 mol of O_2 are available for reaction. The balanced eqution tells us that only 1 mol of O_2 is required to completely consume 2 mol of H_2, thereby forming 2 mol of H_2O; 1 mol of O_2 will therefore be left over at the end of the reaction. This situation is represented schematically in Figure 3.9. The substance that is completely consumed in a reaction is called the **limiting reagent**, because it determines or limits the amount of product formed. In our present example, H_2 is the limiting reagent.

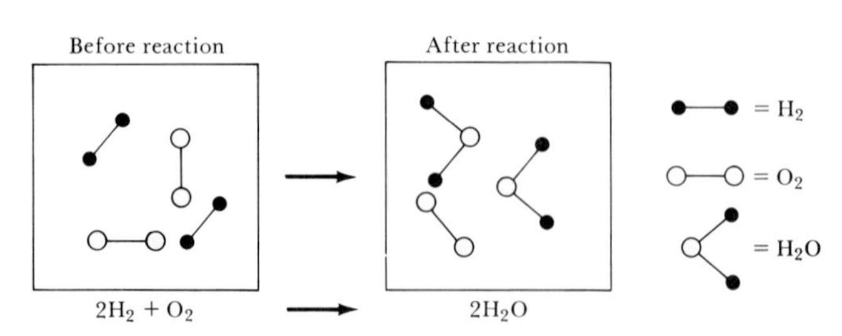

FIGURE 3.9 Diagram showing the complete utilization of a limiting reagent in a reaction. Because the H_2 is completely consumed, it is the limiting reagent in this case. Because there is a stoichiometric excess of O_2, some is left over at the end of the reaction.

SAMPLE EXERCISE 3.15

Part of the SO_2 that is introduced into the atmosphere ends up being converted to sulfuric acid, H_2SO_4. The net reaction is

$$2SO_2(g) + O_2(g) + 2H_2O(l) \longrightarrow 2H_2SO_4(l)$$

How much H_2SO_4 can be prepared from 5.0 mol of SO_2, 1.0 mol of O_2, and an unlimited quantity of H_2O?

Solution: The number of moles of O_2 needed for complete consumption of 5.0 mol of SO_2 is

$$\text{Moles } O_2 = (5.0 \text{ mol } SO_2)\left(\frac{1 \text{ mol } O_2}{2 \text{ mol } SO_2}\right) = 2.5 \text{ mol } O_2$$

This quantity of O_2 is not available; therefore, all of the SO_2 cannot be consumed; O_2 must be the limiting reagent. We use the quantity of the limiting reagent, O_2, to calculate the quantity of H_2SO_4 prepared:

$$\text{Moles } H_2SO_4 = (1.0 \text{ mol } O_2)\left(\frac{2 \text{ mol } H_2SO_4}{1 \text{ mol } O_2}\right) = 2.0 \text{ mol } H_2SO_4$$

We might note that in forming 2.0 mol of H_2SO_4, 2.0 mol of SO_2 are required. Therefore, 3.0 mol of SO_2 is left over.

Another approach to the problem of the limiting reagent is to calculate the number of moles of product that could be formed from each of the given amounts of reagents, assuming they were all completely consumed. The reagent that leads to the smallest amount of product is the limiting reagent.

SAMPLE EXERCISE 3.16

Consider the precipitation reaction described in Sample Exercise 3.3. Suppose that a solution containing 3.50 g of Na_3PO_4 were mixed with a solution containing 6.40 g of $Ba(NO_3)_2$. How many grams of $Ba_3(PO_4)_2$ can be formed?

Solution: The balanced equation for the reaction is

$$2Na_3PO_4(aq) + 3Ba(NO_3)_2(aq) \longrightarrow Ba_3(PO_4)_2(aq) + 6NaNO_3(aq)$$

From this equation we have the following stoichiometric relations:

$$2 \text{ mol } Na_3PO_4 \simeq 3 \text{ mol } Ba(NO_3)_2 \simeq 1 \text{ mol } Ba_3(PO_4)_2$$

The mass of 1 mol of each substance can be found by determining the formula weight for each substance. The results are as follows:

$$1 \text{ mol } Na_3PO_4 = 164 \text{ g } Na_3PO_4$$
$$1 \text{ mol } Ba(NO_3)_2 = 261 \text{ g } Ba(NO_3)_2$$
$$1 \text{ mol } Ba_3(PO_4)_2 = 602 \text{ g } Ba_3(PO_4)_2$$

Let us now calculate the amount of product that could be formed from each of our given amounts of reactants, assuming that each is the limiting reagent:

$$\text{Grams } Ba_3(PO_4)_2 = (3.50 \text{ g } Na_3PO_4) \times \left(\frac{1 \text{ mol } Na_3PO_4}{164 \text{ g } Na_3PO_4}\right)\left(\frac{1 \text{ mol } Ba_3(PO_4)_2}{2 \text{ mol } Na_3PO_4}\right) \times \left(\frac{602 \text{ g } Ba_3(PO_4)_2}{1 \text{ mol } Ba_3(PO_4)_2}\right) = 6.42 \text{ g } Ba_3(PO_4)_2$$

and

$$\text{Grams } Ba_3(PO_4)_2 = (6.40 \text{ g } Ba(NO_3)_2) \times \left(\frac{1 \text{ mol } Ba(NO_3)_2}{261 \text{ g } Ba(NO_3)_2}\right)\left(\frac{1 \text{ mol } Ba_3(PO_4)_2}{3 \text{ mol } Ba(NO_3)_2}\right) \times \left(\frac{602 \text{ g } Ba_3(PO_4)_2}{1 \text{ mol } Ba_3(PO_4)_2}\right) = 4.92 \text{ g } Ba_3(PO_4)_2$$

These calculations show that $Ba(NO_3)_2$ is the limiting reagent, and that mixing 3.50 g of Na_3PO_4 with 6.40 g of $Ba(NO_3)_2$ will yield at most 4.92 g of $Ba_3(PO_4)_2$.

The quantity of product that is calculated to form when all of the limiting reagent reacts is called the *theoretical yield.* The amount of product actually obtained in a reaction is called the *actual yield.* The actual yield is almost always less than the theoretical yield. There are many reasons for this difference. For example, part of the reactants may not react, or they may react in a way different from that desired (side reactions). In addition, it is not always possible to recover all of the reaction product from the reaction mixture. The *percent yield* of a reaction relates the actual yield to the theoretical (calculated) yield:

$$\text{Percent yield} = \frac{\text{actual yield}}{\text{theoretical yield}} \times 100$$

For example, in the experiment described in Sample Exercise 3.16, we calculated that 4.92 g of $Ba_3(PO_4)_2$ should form when 3.50 g of Na_3PO_4 is mixed with 6.40 g of $Ba(NO_3)_2$. This is the theoretical yield of $Ba_3(PO_4)_2$ in the reaction. If the actual yield turned out to be 4.70 g, the percent yield would be

$$\frac{4.70\ \text{g}}{4.92\ \text{g}} \times 100 = 95.5\%$$

3.9 MOLARITY AND SOLUTION STOICHIOMETRY

We have seen how it is possible to employ the concept of the mole to determine the quantities of substances that react with one another or to determine the quantity of product that can be obtained from a given mass of reactant. However, it often happens that chemicals are employed not as solid materials but in the form of solutions, particularly aqueous solutions. For example, the reaction described in Sample Exercise 3.16 involves aqueous solutions. It is a great convenience to be able to measure out quantities of dissolved substances by measuring out volumes of the solutions, rather than by weighing out solids or liquids each time and then dissolving these in water.

In discussing solutions it is often convenient to call one component the **solvent** and the others **solutes.** The component of a solution whose physical state is preserved when the solution is formed is known as the solvent. For example, when sodium chloride (a solid) is mixed with water, the resultant solution is a liquid. Consequently, water is referred to as the solvent and sodium chloride as the solute. If all components of a solution are in the same state, the one present in greatest amount is called the solvent.

The term **concentration** is used to designate the amount of solute dissolved in a given quantity of solvent or solution. The method for expressing concentration that is most useful for discussing solution stoichiometry is **molarity.** The molarity (symbol M) of a solution is defined as the number of moles of solute in a liter of solution (soln):

$$\text{Molarity} = \frac{\text{moles solute}}{\text{volume of soln in liters}} \qquad [3.24]$$

A 1.50 molar solution (written 1.50 M) contains 1.50 mol of solute in every liter of solution. To make a liter of 0.150 M sucrose, $C_{12}H_{22}O_{11}$, in water requires 0.150 mol of $C_{12}H_{22}O_{11}$. This quantity of solid sucrose (51.3 g) is first dissolved in less than a liter of water. The resulting solution is then diluted to a total volume of 1 L. A volumetric flask, which is a flask calibrated to contain a precise volume of liquid, is used for this purpose. The operation is shown in Figure 3.10.

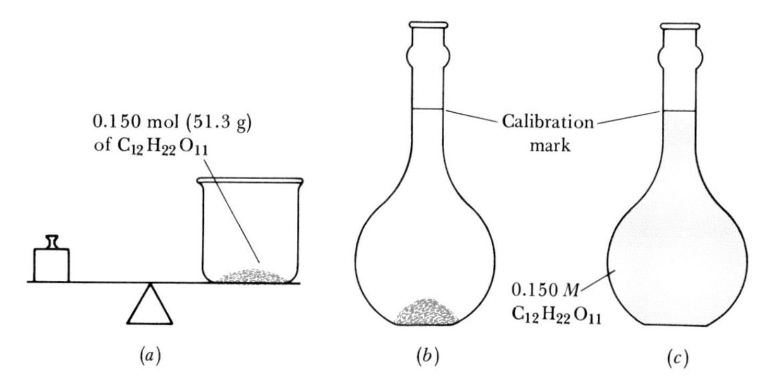

FIGURE 3.10 Procedure for preparation of 1 L of a 0.150 *M* solution of $C_{12}H_{22}O_{11}$. (*a*) Weigh out 0.150 mol (51.3 g) of $C_{12}H_{22}O_{11}$ (MW = 342 amu). (*b*) Add the $C_{12}H_{22}O_{11}$ to a 1-L volumetric flask. (*c*) Add water to dissolve, then more water until the solution reaches the calibration mark.

SAMPLE EXERCISE 3.17

Calculate the molarity of a solution made by dissolving 23.4 g of sodium sulfate (Na_2SO_4) in enough water to form 125 mL of solution.

Solution:

$$\text{Molarity} = \frac{\text{moles } Na_2SO_4}{\text{liters soln}}$$

$$\text{Moles } Na_2SO_4 = (23.4 \text{ g } Na_2SO_4)\left(\frac{1 \text{ mol } Na_2SO_4}{142 \text{ g } Na_2SO_4}\right)$$

$$= 0.165 \text{ mol } Na_2SO_4$$

$$\text{Molarity} = \left(\frac{0.165 \text{ mol } Na_2SO_4}{125 \text{ mL soln}}\right)\left(\frac{1000 \text{ mL soln}}{1 \text{ L soln}}\right)$$

$$= \frac{1.32 \text{ mol } Na_2SO_4}{\text{L soln}}$$

$$= 1.32\ M$$

One advantage of the expression for concentration in molarity is that it allows us to measure out a solution volume of known concentration and readily calculate the number of moles of solute dispensed. Molarity can be used to interconvert volume and moles just as density can be used to interconvert volume and mass (Sample Exercise 1.9). Calculation of the number of moles of HNO_3 in 2.0 L of 0.200 *M* HNO_3 solution illustrates conversion of volume to moles:

$$\text{Moles } HNO_3 = (2.0 \text{ L soln})\left(\frac{0.200 \text{ mol } HNO_3}{1 \text{ L soln}}\right)$$

$$= 0.40 \text{ mol } HNO_3$$

Notice how dimensional analysis can be used in this conversion if we express molarity as moles HNO_3/liters soln. Notice also that to obtain moles we multiplied liters and molarity: moles = liters × *M*. This same expression for moles can be obtained directly by algebraic rearrangement of Equation 3.24. The use of molarity to convert moles to volume can be illustrated by calculating the volume of 0.30 M HNO_3 solution required to supply 2.0 mol of HNO_3:

$$\text{Liters soln} = (2.0 \text{ mol HNO}_3)\left(\frac{1 \text{ L soln}}{0.30 \text{ mol HNO}_3}\right)$$

$$= 6.7 \text{ L soln}$$

In this case we needed to apply the reciprocal of molarity to convert moles to volume: liters = moles × $1/M$.

SAMPLE EXERCISE 3.18

How many grams of Na_2SO_4 are required to make 350 mL of 0.50 M Na_2SO_4?

Solution: Because

$$M\,\text{Na}_2\text{SO}_4 = \frac{\text{moles Na}_2\text{SO}_4}{\text{liters soln}}$$

$$\text{Moles Na}_2\text{SO}_4 = (0.350 \text{ L soln})\left(\frac{0.50 \text{ mol Na}_2\text{SO}_4}{\text{L soln}}\right)$$

$$= 0.175 \text{ mol Na}_2\text{SO}_4$$

Because each mole of Na_2SO_4 weighs 142 g, the required number of grams of Na_2SO_4 is

$$(0.175 \text{ mol Na}_2\text{SO}_4)\left(\frac{142 \text{ g Na}_2\text{SO}_4}{1 \text{ mol Na}_2\text{SO}_4}\right)$$

$$= 24.8 \text{ g Na}_2\text{SO}_4$$

It is also possible to work this problem by direct conversion of milliliters to liters to moles to grams. In doing so, we use molarity as a conversion factor between volume and moles:

$$\text{Grams Na}_2\text{SO}_4 = (350 \text{ mL soln})\left(\frac{1 \text{ L soln}}{1000 \text{ mL soln}}\right)$$

$$\times \left(\frac{0.50 \text{ mol Na}_2\text{SO}_4}{\text{L soln}}\right)\left(\frac{142 \text{ g Na}_2\text{SO}_4}{1 \text{ mol Na}_2\text{SO}_4}\right)$$

$$= 24.8 \text{ g Na}_2\text{SO}_4$$

Dilution

It is often convenient to make a solution of a certain concentration from a more concentrated solution. For example, suppose you need to prepare a liter of 0.10 M HNO_3 solution from a solution of 1.0 M HNO_3. The desired solution will contain 0.10 mol of HNO_3. Therefore, you need to remove 0.10 mol of HNO_3 from the 1.0 M solution. There is 0.10 mol of HNO_3 in 100 mL of the 1.0 M solution. Thus 100 mL is withdrawn from the 1.0 M HNO_3 solution using a pipet and is added to a 1-L volumetric flask. It is then diluted to 1 L as shown in Figure 3.11.*

When more solvent is added to a solution, thereby diluting it, the number of moles of solute remains unchanged:

$$\text{Moles solute before dilution} = \text{moles solute after dilution} \qquad [3.25]$$

Because number of moles = M × liters, we can write

$$\text{(Initial molarity)(initial volume)} = \text{(final molarity)(final volume)} \qquad [3.26]$$
$$M_{\text{initial}}V_{\text{initial}} = M_{\text{final}}V_{\text{final}}$$

*In diluting an acid or base, the acid or base should be added to water, then further diluted by addition of more water. Adding water directly to concentrated acid or base can cause spattering because of the intense heat generated.

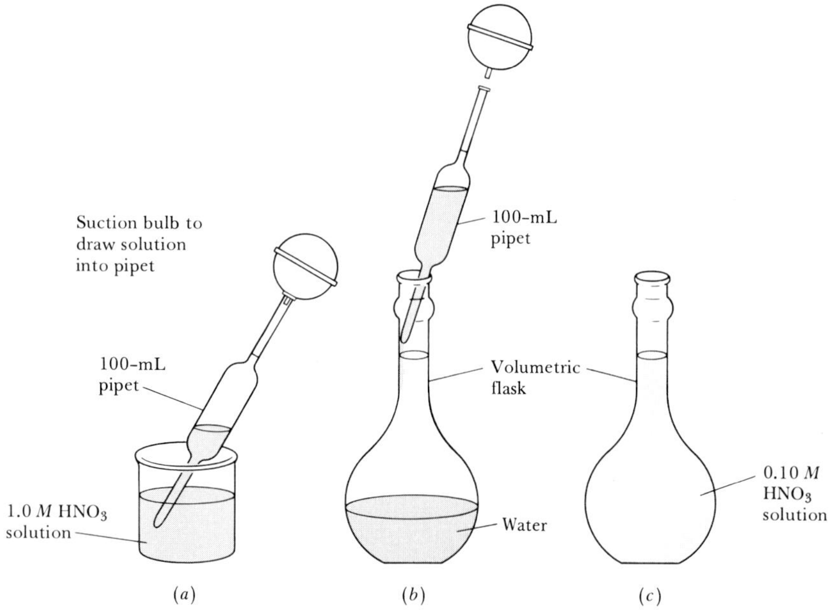

FIGURE 3.11 Procedure for preparation of 1 L of 0.10 M HNO_3 by dilution of 1.0 M HNO_3. (*a*) Draw 100 mL of the 1.0 M solution into a pipet. (*b*) Add this amount of 1.0 M HNO_3 to a small amount of water in a 1-L volumetric flask. (*c*) Add additional water to dilute the solution to a total volume of 1 L. The result is a 0.10 M HNO_3 solution.

SAMPLE EXERCISE 3.19

How much 3.0 M H_2SO_4 would be required to make 500 mL of 0.10 M H_2SO_4?

Solution: Using Equation 3.26, $M_{\text{initial}}V_{\text{initial}} = M_{\text{final}}V_{\text{final}}$, we can write

$$V_{\text{initial}} = \frac{M_{\text{final}}V_{\text{final}}}{M_{\text{initial}}} = \frac{(0.10\ M)(500\ \text{mL})}{3.0\ M} = 17\ \text{mL}$$

We see that if we start with 17 mL of 3.0 M H_2SO_4 and dilute it to a total volume of 500 mL, the desired 0.10 M solution will be obtained.

Titration

The procedure by which a solution of known concentration is added to another solution until the chemical reaction between the two solutes is complete is known as a titration. Titrations are widely used in chemistry to analyze the compositions of mixtures. The solution whose concentration is known is called the standard solution. In titrations, the standard solution is slowly added from a buret to a solution that contains a known volume or known mass of solute. The latter solution is commonly referred to as the unknown. The point at which stoichiometrically equivalent quantities of substances have been brought together is known as the equivalence point of the titration. Some titrations are carried out by measuring out a known volume of a standard solution and then adding the solution of unknown concentration until the equivalence point is reached.

In order to titrate an unknown with a standard solution, there must be some way to determine when the equivalence point of the titration has been reached. In acid-base titrations, organic dyes known as acid-base

FIGURE 3.12 Procedure for titration of an unknown base against a standardized solution of H_2SO_4. (*a*) A known quantity of base is added to a flask. (*b*) An acid-base indicator is added, and standardized H_2SO_4 is added from a buret. (*c*) Equivalence point is signaled by a color change of the indicator. The concentrations and volumes shown correspond to the example discussed in Sample Exercise 3.21.

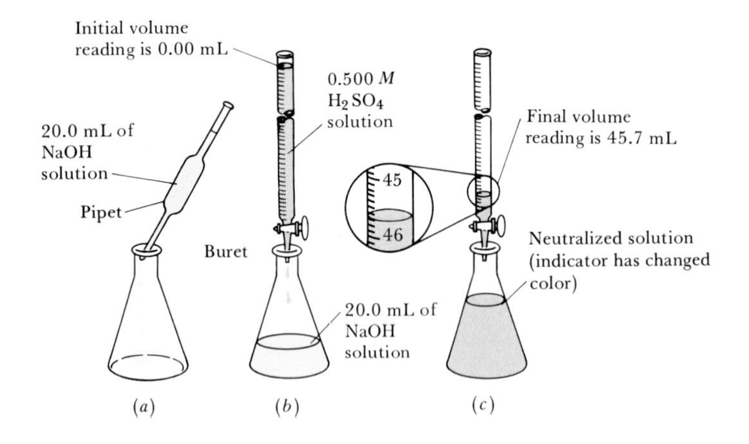

indicators are used for this purpose. For example, the dye known as phenolphthalein is colorless in acidic solution but is red in basic solution. If phenolphthalein is added to an unknown solution of acid, the solution will be colorless. Standard base can then be added from the buret until the solution barely turns from colorless to red. This indicates that the acid has been neutralized, and the drop of base that caused the solution to become colored has no acid to react with. The solution therefore becomes basic, and the dye turns red. The experimental procedure for the titration of NaOH and H_2SO_4 is summarized in Figure 3.12.

SAMPLE EXERCISE 3.20

What volume of 0.500 *M* NaCl is required to react completely with 0.200 mol of $Pb(NO_3)_2$? The chemical equation for this reaction is

$$2NaCl(aq) + Pb(NO_3)_2(aq) \longrightarrow 2NaNO_3(aq) + PbCl_2(s)$$

Solution: According to the reaction equation

$$2 \text{ mol NaCl} \simeq 1 \text{ mol Pb(NO}_3)_2$$

Therefore,

$$\text{Moles NaCl} = (0.200 \text{ mol Pb(NO}_3)_2)\left(\frac{2 \text{ mol NaCl}}{1 \text{ mol Pb(NO}_3)_2}\right) = 0.400 \text{ mol NaCl}$$

Because

$$\text{Liters NaCl soln} = \frac{\text{mol NaCl}}{M \text{ NaCl soln}}$$

$$\text{Liters NaCl soln} = \frac{0.400 \text{ mol NaCl}}{0.500 \text{ mol NaCl/L soln}} = 0.800 \text{ L}$$

Therefore, 800 mL of the 0.500 *M* NaCl solution could be measured out and added to the 0.200 mol of $Pb(NO_3)_2$. This problem can also be solved by direct conversion of moles $Pb(NO_3)_2$ to moles NaCl to volume of NaCl solution:

$$\text{Liters NaCl soln} = (0.200 \text{ mol Pb(NO}_3)_2) \times \left(\frac{2 \text{ mol NaCl}}{1 \text{ mol Pb(NO}_3)_2}\right)\left(\frac{1 \text{ L NaCl soln}}{0.500 \text{ mol NaCl}}\right) = 0.800 \text{ L}$$

SAMPLE EXERCISE 3.21

One method used commercially to peel potatoes is to soak them in a solution of NaOH for a short time and then spray off the peel after the potatoes are removed from the NaOH. The concentration of NaOH is normally in the range 3 to 6 *M*. The NaOH is analyzed periodically to determine its ability to peel potatoes rapidly. In one such analysis, 45.7 mL of 0.500 *M* H_2SO_4 is required to react completely with a 20.0-mL sample of NaOH solution:

$$H_2SO_4(aq) + 2NaOH(aq) \longrightarrow 2H_2O(l) + Na_2SO_4(aq)$$

What is the concentration of the NaOH solution?

Solution:

Moles H_2SO_4 =

$$(45.7 \text{ mL soln})\left(\frac{1 \text{ L soln}}{1000 \text{ mL soln}}\right)\left(0.500\frac{\text{mol } H_2SO_4}{\text{L soln}}\right)$$
$$= 2.28 \times 10^{-2} \text{ mol } H_2SO_4$$

According to the balanced equation, 1 mol $H_2SO_4 \simeq 2$ mol NaOH. Therefore,

Moles NaOH =

$$(2.28 \times 10^{-2} \text{ mol } H_2SO_4)\left(\frac{2 \text{ mol NaOH}}{1 \text{ mol } H_2SO_4}\right)$$
$$= 4.56 \times 10^{-2} \text{ mol NaOH}$$

Knowing the number of moles of NaOH present in 20.0 mL of solution allows us to calculate the concentration of this solution:

$$M \text{ NaOH} = \frac{\text{mol NaOH}}{\text{L soln}} =$$
$$\left(\frac{4.56 \times 10^{-2} \text{ mol NaOH}}{20.0 \text{ mL soln}}\right)\left(\frac{1000 \text{ mL soln}}{1 \text{ L soln}}\right)$$
$$= 2.28\frac{\text{mol NaOH}}{\text{L soln}} = 2.28\ M$$

SAMPLE EXERCISE 3.22

The quantity of Cl^- in a water supply is determined by titrating the sample against $AgNO_3$:

$$AgNO_3(aq) + Cl^-(aq) \longrightarrow AgCl(s) + NO_3^-(aq)$$

What mass of chloride ion is present in a 10.0-g sample of the water if 20.2 mL of 0.100 M $AgNO_3$ is required to react with all the chloride in the sample?

Solution: We must first determine the number of moles of $AgNO_3$ required in the titration:

$$(20.2 \text{ mL soln})\left(\frac{1 \text{ L soln}}{1000 \text{ mL soln}}\right)\left(0.100\frac{\text{mol } AgNO_3}{\text{L soln}}\right)$$
$$= 2.02 \times 10^{-3} \text{ mol } AgNO_3$$

From the balanced equation we see that 1 mol $AgNO_3 \simeq 1$ mol Cl^-. Therefore, the sample must contain 2.02×10^{-3} mol of Cl^-:

Moles Cl^- =

$$(2.02 \times 10^{-3} \text{ mol } AgNO_3)\left(\frac{1 \text{ mol } Cl^-}{1 \text{ mol } AgNO_3}\right)$$
$$= 2.02 \times 10^{-3} \text{ mol } Cl^-$$

The number of moles of Cl^- can then be converted to grams:

Grams Cl^- =

$$(2.02 \times 10^{-3} \text{ mol } Cl^-)\left(\frac{35.5 \text{ g } Cl^-}{1 \text{ mol } Cl^-}\right)$$
$$= 7.17 \times 10^{-2} \text{ g } Cl^-$$

We might note that the weight percentage of Cl^- in the water is

$$\% \ Cl^- = \frac{7.17 \times 10^{-2} \text{ g } Cl^-}{10.0 \text{ g soln}} \times 100 = 0.717\%$$

Chloride ion is one of the major ions in water and sewage. Ocean water contains 1.92 percent Cl^-. Whether or not water containing Cl^- exhibits a salty taste depends on the other ions present. If the accompanying ions are Na^+ ions, a salty taste may be detected with as little as 0.03 percent Cl^-.

FOR REVIEW

Summary

The **law of conservation of mass** states that there are no detectable changes in mass in any chemical reaction. This indicates that there are the same number of atoms of each type present after a chemical reaction as there were before the reaction. A **balanced equation** shows equal numbers of each type of atom on each side of the equation and is thereby consistent with the law of conservation of mass. We discussed how equations are balanced by placing coefficients in front of the chemical formulas for the substances involved in the reaction. We have also seen how it is possible to predict the products of simple reactions by analogy to known reactions and by use of the periodic table. Among the reaction types seen in this chapter are the following: (1) combustion in oxygen, in which an organic compound reacts with oxygen, forming carbon dioxide, water, and possibly other products, depending on the composition of the compound; (2) neutralization reaction, in which an acid and base react to form water and a salt; and (3) precipitation reaction, in which one of the products of a reaction between two substances in solution is insoluble in the solution.

The coefficients in a balanced equation can be interpreted as either the relative number of formula units involved in the reaction or the relative number of moles. A mole of any substance is Avogadro's number (6.022×10^{23}) of formula units of that substance. The mass of a mole of atoms, molecules, or ions is the formula weight expressed in grams. For example, a single molecule of H_2O weighs 18 amu; a mole of H_2O weighs 18 g. The empirical formula, or simplest formula, of a substance expresses the composition in terms of the smallest possible set of whole-number subscripts denoting the relative numbers of atoms. We have seen how the mole concept can be used to determine the empirical formula of a compound and to calculate the quantities involved in chemical reactions. In dealing with reactions between substances in solution, it is convenient to employ the concept of solution concentration. Molarity is defined as the number of moles of solute per liter of solution. Molarity serves as a conversion factor for interconverting solution volume and number of moles of solute.

Learning goals

Having read and studied this chapter, you should be able to:

1. Balance chemical equations.
2. Predict the products of a chemical reaction, having seen a suitable analogy.
3. Calculate the formula weight of any substance.
4. Interconvert number of moles, mass in grams, and number of atoms, ions, or molecules.
5. Calculate the empirical formula of a compound, having been given appropriate analytical data such as elemental percentages or the quantity of CO_2 and H_2O produced by combustion.
6. Calculate the molecular formula, having been given the empirical formula and molecular weight.
7. Calculate the mass of a particular substance produced or used in a chemical reaction (mass-mass problems).
8. Determine the limiting reagent in a reaction.
9. Define molarity.
10. Solve problems involving interconversions among molarity, solution volume, and number of moles of solute.
11. Explain what is meant by the term *titration.*
12. Calculate concentration or mass of solute in a sample from titration data.

Key terms

Among the more important terms and expressions used for the first time in this chapter are the following:

An acid (Section 3.3) is a substance that produces an excess of H^+ ions when it dissolves in water.

The atomic mass unit (amu) (Section 3.4) is based on the value of exactly 12 for the isotope of carbon with six protons and six neutrons in the nucleus.

Atomic weight (Section 3.4) or atomic mass is the average mass of the atoms of an element in atomic mass units. It is numerically equal to weight in grams of one mole of the element.

Avogadro's number (Section 3.6) is the number of ^{12}C atoms in exactly 12 g of ^{12}C, 6.022×10^{23}.

A balanced chemical equation (Section 3.2) is one that satisfies the law of conservation of mass; it contains equal numbers of atoms of each element on both sides of the equation.

A base (Section 3.3) is a substance that produces an excess of OH^- ions when it dissolves in water.

A combustion reaction (Section 3.3) is one that proceeds with evolution of heat and usually also a flame. Most combustion involves reaction with oxygen, as in the burning of a match.

Solution concentration (Section 3.9) denotes the number of moles of a solute present in a given quantity of solvent.

The law of conservation of mass (Section 3.1) states that atoms are neither created nor destroyed during a chemical reaction.

The empirical formula (Section 3.7) of any substance expresses the simplest whole-number ratio of the elements present in that substance. For example, the empirical formula of glucose, $C_6H_{12}O_6$, is CH_2O.

The equivalence point (Section 3.9) is the point in a titration at which the added solute just completely reacts with the solute present in solution.

Formula weight (Section 3.4) is the weight of the

collection of atoms represented by a chemical formula. For example, the formula weight of NO_2 (46.0 amu) is the sum of the weights of one nitrogen atom and two oxygen atoms. If the formula is the molecular formula of the substance, the formula weight is the molecular weight of the substance.

An indicator (Section 3.9) is a substance added to a solution to indicate by a color change the point at which the added solute has just reacted with all the solute present in solution.

A limiting reagent (Section 3.8) is the reactant present in smallest stoichiometric quantity in a mixture of reactants. The amount of product that can form is limited by the complete consumption of the limiting reagent.

The mass spectrometer (Section 3.5) is an instrument used to measure the precise masses and relative amounts of atomic and molecular ions.

A metathesis reaction (Section 3.3) is one in which two substances react to form two new substances through an exchange of the component parts of each reacting substance. For example AB + CD ⟶ AC + BD. Precipitation and neutralization reactions are examples of metathesis reactions.

Molarity (Section 3.9) is the concentration of a solution expressed as moles of solute per liter of solution; abbreviated *M*.

A mole (Section 3.6) is a collection of Avogadro's number (6.022×10^{23}) of objects; for example, a mole of H_2O is 6.022×10^{23} H_2O molecules.

A neutralization reaction (Section 3.3) is one in which an acid and a base react in stoichiometrically equivalent amounts. The product of a neutralization reaction is water and a salt.

A precipitation reaction (Section 3.3) is one occurring between substances in solution in which one of the products is insoluble.

A salt (Section 3.3) is an ionic compound that can be formed by an acid-base neutralization reaction.

A titration (Section 3.9) is the process of reacting a solution of unknown concentration with one of known concentration (a standard solution). The procedure is commonly used to determine the concentrations of solutions of unknown concentrations.

EXERCISES

Conservation of mass; chemical equations

3.1 Which of the following equations, as written, are consistent with the law of conservation of mass?

(a) $CHF_3(g) + O(g) \longrightarrow OH(g) + CF_3(g)$
(b) $CCl_4(g) + O_2(g) \longrightarrow COCl_2(g) + Cl_2(g)$
(c) $H_2SO_4(aq) + 2KOH(aq) \longrightarrow K_2SO_4(aq) + H_2O(l)$
(d) $4C_3H_5NO(g) + 19O_2(g) \longrightarrow 12CO_2(g) + 10H_2O(g) + 4NO_2(g)$

3.2 Balance the following equations:

(a) $Al(s) + Cl_2(g) \longrightarrow AlCl_3(s)$
(b) $P_2O_3(s) + H_2O(l) \longrightarrow H_3PO_3(aq)$
(c) $Ca(OH)_2(aq) + HBr(aq) \longrightarrow CaBr_2(aq) + H_2O(l)$
(d) $Al(s) + S_8(s) \longrightarrow Al_2S_3(s)$
(e) $Mg_2C_3(s) + H_2O(l) \longrightarrow Mg(OH)_2(s) + C_3H_4(g)$

3.3 Balance the following equations:

(a) $C(s) + F_2(g) \longrightarrow CF_4(g)$
(b) $N_2O_5(s) + H_2O(l) \longrightarrow HNO_3(l)$
(c) $C_2H_4(g) + O_2(g) \longrightarrow CO_2(g) + H_2O(g)$
(d) $La_2O_3(s) + H_2O(l) \longrightarrow La(OH)_3(s)$
(e) $HCl(g) + CaO(s) \longrightarrow CaCl_2(s) + H_2O(g)$
(f) $Pb(NO_3)_2(aq) + NaCl(aq) \longrightarrow PbCl_2(s) + NaNO_3(aq)$

3.4 Write a balanced chemical equation to correspond to each of the following word descriptions. (a) Phosphine, $PH_3(g)$, is combusted in air to form gaseous water and solid diphosphorus pentoxide. (b) Barium metal reacts with methyl alcohol, $CH_3OH(l)$, to form hydrogen gas and dissolved barium methoxide, $Ba(OCH_3)_2$. (c) Boron sulfide, $B_2S_3(s)$, reacts violently with water to form dissolved boric acid, H_3BO_3, and hydrogen sulfide gas, H_2S. (d) Copper metal reacts with hot concentrated sulfuric acid to form aqueous copper(II) sulfate, sulfur dioxide gas, and liquid water. (e) When ammonia gas, NH_3, is passed over hot liquid sodium metal, hydrogen is released and sodium amide, $NaNH_2$, is formed as a solid product.

3.5 Write balanced chemical equations to correspond to each of the following verbal descriptions. (a) Cyanic acid, HCNO, is quite unstable. The gas reacts with water to form ammonia and gaseous carbon dioxide. (b) When solid mercury(II) nitrate is heated, it decomposes to form solid mercury(II) oxide and gaseous nitrogen dioxide and oxygen. (c) When liquid phosphorus trichloride is added to water, it reacts violently to form aqueous phosphorous acid and aqueous hydrogen chloride. (d) When solid potassium nitrate is heated, it decomposes to solid potassium nitrite, and oxygen gas is evolved. (e) When hydrogen sulfide (H_2S) gas is passed over solid hot iron(III) hydroxide, it reacts to form solid iron(III) sulfide and gaseous water.

Chemical reactions

3.6 Balance the following equations and classify each reaction as one of the following types: combustion, neutralization, or precipitation:

(a) $Ba(OH)_2(aq) + HCl(aq) \longrightarrow BaCl_2(aq) + H_2O(l)$

(b) $CO(g) + O_2(g) \longrightarrow CO_2(g)$

(c) $CaCl_2(aq) + Na_2CO_3(aq) \longrightarrow CaCO_3(s) + NaCl(aq)$

(d) $CH_3SH(g) + O_2(g) \longrightarrow CO_2(g) + H_2O(l) + SO_2(g)$

(e) $Ni(NO_3)_2(aq) + KOH(aq) \longrightarrow Ni(OH)_2(S) + KNO_3(aq)$

3.7 Complete and balance each of the following equations:

(a) $C_4H_{10}(g) + O_2(g) \longrightarrow$ ______(g) + ______(l)

(b) $C_2H_5OH(l) + O_2(g) \longrightarrow$ ______(g) + ______(l)

(c) $Al(OH)_3(s) + HNO_3(aq) \longrightarrow$ ______(aq) + ______(l)

(d) $Cu(OH)_2(s) + HCl(aq) \longrightarrow$ ______(aq) + ______(l)

(e) $AgNO_3(aq) + H_2SO_4(aq) \longrightarrow Ag_2SO_4(s) +$ ______(aq)

(f) $CaCl_2(aq) + Na_3PO_4(aq) \longrightarrow Ca_3(PO_4)_2(s) +$ ______(aq)

3.8 Drawing upon the following list of substances as reactants—$BaCl_2$, C_2H_5OH, Na_2CrO_4, O_2, H_2O, KOH, H_2SO_4, $Ca(OH)_2$, $Pb(NO_3)_2$, HNO_3, and K—write balanced equations for each of the following: (a) a combustion reaction; (b) formation of insoluble $PbCrO_4$ from solution; (c) formation of hydrogen gas; (d) neutralization of sulfuric acid; (e) solid $Ca(OH)_2$, which has a low solubility in water, is dissolved by reaction with another reagent; (f) formation of insoluble $BaSO_4$ from solution.

[3.9] We can sometimes extend our ability to write balanced equations by recognizing the analogies between compounds. From the hints provided, write complete balanced equations for the following. (a) Combustion of nitromethane, $CH_3NO_2(g)$, leads to $NO_2(g)$ as one of the products. (b) Reaction of potassium with liquid ammonia is very much like reaction of this metal with water. (c) Fluorine, like oxygen, can support combustion; for example, methane, $CH_4(g)$, can be made to "burn" in an atmosphere of F_2. (d) Reaction of a metal with an acid solution is like reaction of active metals with water, except that a salt of the metal rather than the hydroxide is the product. For example, Zn reacts with dilute HCl solution.

Atomic and molecular weights; the mass spectrometer

3.10 Calculate the formula weight of each of the following: (a) C_3H_8; (b) SO_2; (c) SiF_4; (d) $(NH_4)_2SO_4$; (e) CH_3OH.

3.11 Calculate the formula weight of each of the following: (a) $KBrO_3$; (b) Cr_2O_3; (c) $KMnO_4$; (d) $Ca(OH)_2$; (e) CH_3OCH_3.

3.12 Determine the formula weights of each of the following substances: (a) thionyl chloride, $SOCl_2$; (b) thallium(I) oxalate, $Tl_2(C_2O_4)$; (c) coumarin, $C_9H_6O_2$; (d) coniferin, $C_{16}H_{22}O_8 \cdot 2H_2O$; (e) xenon tetrafluoride, XeF_4.

3.13 Calculate the weight percentage of each element in each of the following compounds: (a) H_2SO_4; (b) $K_2Cr_2O_7$; (c) $Mg(NO_3)_2$; (d) C_2H_5OH; (e) $Ca(HCO_3)_2$.

3.14 Each of the following compounds is used as a nitrogen source in agricultural fertilizers: NH_3; $NaNO_3$; $(NH_4)_2SO_4$. Calculate the percentage nitrogen in each.

3.15 Gay-Lussac found that one volume of hydrogen gas reacts with one volume of chlorine gas to form *two* volumes of hydrogen chloride as product. Is this observation consistent with an assumption that hydrogen and chlorine gas are monatomic? Explain. What does this experiment tell us directly about the formula for hydrogen chloride? Explain.

3.16 We know that chlorine and fluorine are both diatomic gases. One volume of chlorine gas reacts with three volumes of fluorine gas to yield two volumes of product, with all gases measured at the same temperature and pressure. What is the formula for the product?

3.17 The element magnesium consists of three isotopes with masses 23.9924, 24.9938, and 25.9898 amu. These three isotopes are present in nature to the extent of 78.6, 10.1, and 11.3 percent by mass, respectively. From these data calculate the average atomic mass of magnesium.

3.18 The element neon consists of three isotopes with masses 19.99, 20.99, and 21.99 amu. These three isotopes are present in nature to the extent of 90.92, 0.25, and 8.83 percent, respectively. From these data, calculate the atomic weight of neon.

[3.19] The element silver consists in nature of two isotopes, ^{107}Ag with atomic mass 106.905 amu, and ^{109}Ag with atomic mass 108.905 amu. The accepted atomic weight of Ag is 107.870. From this calculate the relative amounts of ^{107}Ag and ^{109}Ag in nature.

3.20 The mass spectrum of a sample of lead oxide contains ions of the formula PbO^+. The lead oxide has been prepared from isotopically pure ^{16}O. The ion masses seen and their relative intensities are listed as follows:

PbO^+ ion mass	Fractional intensity
220.002	0.0137
222.056	0.2630
223.050	0.2080
224.055	0.5153

The mass of ^{16}O is 15.9948. Calculate the average atomic weight of lead in this sample.

3.21 In his determination of the atomic weight of the element zinc, Berzelius determined in 1818 that the weight ratio of Zn to O in the oxide of zinc was 4.032. He thought that the formula of the compound was ZnO_2. Assuming an atomic weight of 16.00 for oxygen, what value does this give for the atomic weight of zinc? By comparing this with the presently accepted value, what can you say about Berzelius's assumption? If it was in error, what should it have been?

[3.22] Stas reported in 1865 that he had reacted a weighed amount of pure silver with nitric acid and had recovered all the silver as pure silver nitrate, $AgNO_3$. The weight ratio of Ag to $AgNO_3$ was found to be 0.634985. Using only this ratio and the presently accepted values of the atomic weights of silver and oxygen, calculate the atomic weight of nitrogen. Compare this calculated atomic weight with the currently accepted value.

The chemical mole

3.23 (a) What is the mass of 1 mol of calcium nitrate, $Ca(NO_3)_2$? (b) How many moles of $Ca(NO_3)_2$ are there in 2.50 g of this substance? (c) What is the mass, in grams, of 0.325 mol of $Ca(NO_3)_2$? (d) How many Ca^{2+} ions are there in 8.73 g of $Ca(NO_3)_2$?

3.24 Determine the number of (a) moles of SO_2 in 10.5 g of this substance; (b) moles of C_2H_2 in 25.7 g of this substance; (c) moles of caffeine, $C_8H_{10}N_4O_2$, in 0.788 g of caffeine; (d) grams of NaCN in 0.358 mol of this substance; (e) grams of ozone, O_3, in 2.33 mol of O_3; (f) grams of vitamin C, ascorbic acid, $C_6H_8O_6$, in 0.0594 mol of ascorbic acid.

3.25 What is the mass of each of the following: (a) 3.00 mol of CO_2; (b) 3.58×10^{22} atoms of Ar; (c) 1.83×10^{24} molecules of HCl; (d) 0.0080 mol of ethylene C_2H_4?

3.26 Complete each of the following. (a) 8.2 g of solid iodine contains ______ I_2 molecules. (b) 6.5×10^{17} Fe atoms has a total mass of ______ g. (c) The number of silver atoms in 8.0 g of AgCl is ______. (d) 8.0×10^{-5} moles of glucose, $C_6H_{12}O_6$, has a mass of ______ g. (e) The number of moles of H_2O in 55.7 g is ______.

3.27 An average snowflake contains 1.7×10^{18} H_2O molecules. (a) How many moles of H_2O does it contain? (b) How many grams of water?

3.28 The molecule pyridine, C_5H_5N, was found to adsorb on the surfaces of certain metal oxides. A 5.0-g sample of finely divided zinc oxide, ZnO, was found to adsorb 0.068 g of pyridine. How many pyridine molecules are adsorbed? What is the ratio of pyridine molecules to formula units of zinc oxide? If the surface area of the oxide is 48 m^2/g, what is the average area of surface per adsorbed pyridine molecule?

3.29 The allowable concentration level of vinyl chloride, C_2H_3Cl, in the atmosphere in a chemical plant is 2.05×10^{-6} g per liter. How many moles of vinyl chloride in each liter does this represent? How many molecules per liter?

3.30 It requires about 25 μg minimum of tetrahydrocannabinol, THC, the active ingredient in marijuana, to produce intoxication from smoking substances containing THC. The molecular formula of THC is $C_{21}H_{30}O_2$. How many molecules of THC does this 25 μg represent? How many moles?

Empirical formulas

3.31 (a) A sample of a compound contains 0.36 mol of hydrogen and 0.090 mol of carbon. What is the empirical formula of this compound? (b) A sample of a compound contains 11.66 g of iron and 5.01 g of oxygen. What is the empirical formula of this compound? (c) What is the empirical formula of hydrazine, which contains 87.5% N and 12.5% H?

3.32 What are the empirical formulas of compounds with the following compositions: (a) 40.0 percent C, 6.7 percent H, 53.3 percent O; (b) 10.4 percent C, 27.8 percent S, 61.7 percent Cl; (c) 32.79 percent Na, 13.02 percent Al, 54.19 percent F; (d) 83.0 percent I, 7.85 percent C, 9.15 percent N?

3.33 In a laboratory, 1.55 g of an organic compound containing carbon, hydrogen, and oxygen is combusted for analysis, as illustrated in Figure 3.7. Combustion resulted in 1.45 g of CO_2 and 0.890 g of H_2O. What is the empirical formula?

3.34 Cyclopropane, a substance used with oxygen as a general anesthetic, contains only two elements, carbon and hydrogen. When 1.00 g of this substance is completely combusted, 3.14 g of CO_2 and 1.29 g of H_2O are produced. What is the empirical formula of cyclopropane?

3.35 (a) The empirical formula of dichloroethane, a substance used for dry cleaning, is CH_2Cl. If its molecular weight is 99 amu, what is its molecular formula? (b) Butyric acid, a compound produced when butter becomes rancid, has the empirical formula C_2H_4O. If its molecular weight is 88 amu, what is its molecular formula? (c) The elemental analysis of acetylsalicylic acid (aspirin) is 60.0 percent C, 4.48 percent H, and 35.5 percent O. If the molecular mass of this substance is 180.2 amu, what is its molecular formula?

3.36 The characteristic odor of pineapple is due to ethyl butyrate, a compound containing carbon, hydrogen, and oxygen. Combustion of 2.78 mg of ethyl butyrate leads to formation of 6.32 mg of CO_2 and 2.58 mg of H_2O. What is the empirical formula? The properties of the compound suggest that the molecular weight should be between 100 and 150. What is the likely molecular formula?

3.37 Strontium hydroxide is isolated as a hydrate, which means that a certain number of water molecules are included in the solid structure. The formula of the hydrate can be written as $Sr(OH)_2 \cdot xH_2O$, where x indicates the number of moles of water in the solid per mole of $Sr(OH)_2$. When 6.85 g of the hydrate is dried in an oven, 3.13 g of anhydrous $Sr(OH)_2$ is formed. What is the value for x?

[3.38] An oxybromate compound, $KBrO_x$, where x is unknown, is analyzed and found to contain 52.92 percent Br. What is the value for x?

[3.39] Fungal laccase, a blue protein found in wood-rotting fungi, is approximately 0.39 percent copper. If a laccase molecule contains four copper atoms, what is its approximate molecular weight?

Chemical calculations involving equations

3.40 (a) What weight of NH_3 is formed when 5.38 g of Li_3N reacts with water according to the equation

$$Li_3N(s) + 3H_2O(l) \longrightarrow 3LiOH(s) + NH_3(g)$$

(b) What mass of $CaCO_3(s)$ is required to produce 2.87 g $CO_2(g)$ according to the reaction

$$CaCO_3(s) \longrightarrow CaO(s) + CO_2(g)$$

(c) What mass of CO_2 is formed when 9.53 g of C_2H_5OH is produced in the fermentation of sugar according to the reaction

$$C_6H_{12}O_6(aq) \longrightarrow 2C_2H_5OH(aq) + 2CO_2(g)$$

(d) What mass of Mg is required to react with excess $CuSO_4$ to form 1.89 g Cu_2O according to the reaction

$$2Mg(s) + 2CuSO_4(aq) + H_2O(l) \longrightarrow 2MgSO_4(aq) + Cu_2O(s) + H_2(g)$$

3.41 White phosphorus, P_4, burns in excess oxygen to form diphosphorus pentoxide, P_2O_5. Write a balanced chemical equation for this combustion. (a) How many moles of P_2O_5 are formed by combustion of 6.00 g of P_4? (b) When P_2O_5 is added to excess water, it eventually hydrolyzes to H_3PO_4:$P_2O_5(s) + 3H_2O(l) \longrightarrow 2H_3PO_4(aq)$. How many grams of H_3PO_4 are formed from 6.00 g of P_4?

3.42 Magnesium oxide reacts with gaseous phosphorus pentachloride to form solid magnesium chloride and solid diphosphorus pentoxide as products. Write a balanced equation for the reaction. Assuming an excess of $MgO(s)$ and complete reaction, what mass of PCl_5 is required to produce 4.50 kg of diphosphorus pentoxide?

3.43 A sample of aluminum carbide, $Al_4C_3(s)$, is added to a dilute acid solution. The following reaction goes to completion:

$$Al_4C_3(s) + HCl(aq) \longrightarrow AlCl_3(aq) + CH_4(g)$$

Balance this equation. In one particular experiment, the methane produced was collected and found to have a mass of 1.754 g. What mass of $Al_4C_3(s)$ was added to the acid solution? What mass of $AlCl_3$ is formed in solution as the other product?

Limiting reactant; theoretical yields

3.44 (a) In the reaction

$$Mg_2Si(s) + 4H_2O(l) \longrightarrow 2Mg(OH)_2(s) + SiH_4(g)$$

how many moles of SiH_4 are formed by complete reaction of 3.95 g of Mg_2Si? (b) Calculate the weight of water necessary to react with all the Mg_2Si. (c) In a second experiment involving the same reaction, 42.5 mg of Mg_2Si is reacted with 27.0 mg of water. What mass of SiH_4 is formed, assuming that the limiting reagent is completely reacted?

3.45 A mass of 1.84 g of NaOH is dissolved in water, and 3.05 g of $H_2S(g)$ is bubbled in. The following reaction occurs:

$$H_2S(g) + 2NaOH(aq) \longrightarrow Na_2S(aq) + 2H_2O(l)$$

What mass of Na_2S is produced, assuming that the limiting reagent is completely consumed?

3.46 What mass of $AgBr(s)$ is formed when a solution containing 3.45 g of KBr is mixed with a solution containing 7.28 g of $AgNO_3$?

3.47 Ethylene, C_2H_4, burns in air according to the following equation:

$$C_2H_4(g) + 3O_2(g) \longrightarrow 2CO_2(g) + 2H_2O(l)$$

How many grams of CO_2 will be formed when a mixture of 2.93 g of C_2H_4 and 4.29 g of O_2 is ignited assuming that the above reaction is the only one to occur?

3.48 A student reacts benzene, C_6H_6, with bromine, Br_2, in an attempt to prepare bromobenzene, C_6H_5Br:

$$C_6H_6 + Br_2 \longrightarrow C_6H_5Br + HBr$$

(a) What is the theoretical yield of bromobenzene in this reaction when 30.0 g of benzene reacts with an excess of Br_2? (b) Dibromobenzene, $C_6H_4Br_2$, is produced as a by-product in the synthesis of bromobenzene. If the actual yield of bromobenzene was 56.7 g, what was the percentage yield?

[3.49] Azobenzene ($C_{12}H_{10}N_2$) is an important intermediate in the manufacture of dyes. It can be prepared by the reaction between nitrobenzene ($C_6H_5NO_2$) and triethylene glycol ($C_6H_{14}O_6$) in the presence of zinc and potassium hydroxide:

$$2C_6H_5NO_2 + 4C_6H_{14}O_4 \xrightarrow[KOH]{Zn} C_{12}H_{10}N_2 + 4C_6H_{12}O_4 + 4H_2O$$

(a) What is the theoretical yield of azobenzene when 115 g of nitrobenzene and 327 g of triethylene glycol are allowed to react? (b) If the reaction yields 55 g of azobenzene, what is the percent yield of azobenzene?

Molarity; solution stoichiometry

3.50 Calculate the molarity of each of the following solutions: (a) 5.44 g of Na_2CrO_4 in 250 mL of solution; (b) 0.088 mol of H_2SO_4 in 1.5 L of solution; (c) 85.6 kg of NaF in 3.45×10^8 L of solution; (d) 7.6×10^{-3} g of MgF_2 in 100 mL of solution.

3.51 Calculate the mass, in grams, of solute present in each of the following solutions: (a) 240 mL of 0.112 M CdF_2; (b) 125 mL of 0.180 M HNO_3; (c) 5.63 L of 0.0224 M NH_4CN; (d) 5.6×10^{12} m^3 of 2.2×10^{-8} M tetrachlorobiphenyl, $Cl_4C_{12}H_6$.

3.52 What volume of solution is required to obtain (a) 5.50 g of NH_4Cl from a 0.600 M solution of NH_4Cl; (b) 3.25 mol of H_2SO_4 from a 6.55 M solution of H_2SO_4; (c) 0.088 g of As from a 0.020 M solution of H_3AsO_4; (d) 2.50 g of Cr from a 0.0600 M solution of $K_2Cr_2O_7$?

3.53 (a) What mass of Na_2SO_4 is required to prepare 250.0 mL of 0.100 M Na_2SO_4? (b) What volume of 1.000 M KNO_3 must be diluted with water to prepare 500.0 mL of 0.250 M KNO_3?

3.54 Describe how you would prepare 100.0 mL of 0.1000 M glucose solution starting with (a) solid glucose, $C_6H_{12}O_6$; (b) 1.000 L of 2.000 M glucose solution.

3.55 Calculate the molarity of the solution produced by mixing (a) 50.0 mL of 0.200 M NaCl and 100.0 mL of 0.100 M NaCl; (b) 24.5 mL of 1.50 M NaOH and 20.5 mL of 0.850 M NaOH. (Assume that the volumes are additive.)

3.56 A bottle of perchloric acid, $HClO_4$ is analyzed to have a concentration of 2.06 M. (a) Describe how you would use this solution to prepare 1.00 L of 0.100 M perchloric acid solution. (b) What volume of the original solution is required to react completely with 1.65 L of 0.780 M KOH? (c) What volume of the more dilute solution is needed to react completely with 39.5 mL of 0.125 M Na_2CO_3 according to the reaction

$$2HClO_4(aq) + Na_2CO_3(aq) \longrightarrow 2NaClO_4(aq) + CO_2(g) + H_2O(l)$$

(d) A volume of 54.6 mL of the 0.100 M $HClO_4$ solution is required to neutralize 25.0 mL of an NaOH solution of unknown molarity. What is the concentration of the NaOH solution?

3.57 In the laboratory, 6.67 g of $Sr(NO_3)_2$ is dissolved in enough water to form 0.750 L. A 0.100-L sample is withdrawn from this stock solution and titrated with a 0.0460 M solution of Na_2CrO_4. What volume of Na_2CrO_4 solution is required to precipitate all the $SrCrO_4$?

3.58 Exactly 26.3 mL of a 0.110 M HCl solution was required to neutralize all the base in 50.0 mL of a RbOH solution. What is the molarity of the RbOH solution? What mass of RbOH is required to make up 1.00 L of a solution of this concentration?

3.59 A sample of zinc nitrate is known to contain some zinc chloride as an impurity. When 2.87 g of the material is dissolved in water and titrated with 0.0130 M $AgNO_3$, it requires 2.94 mL to precipitate all the chloride as insoluble AgCl. What is the weight percentage of zinc chloride in the zinc nitrate?

3.60 A sample of solid $Ca(OH)_2$ is allowed to stand in contact with water at 30°C for a long time, until the solution contains as much dissolved $Ca(OH)_2$ as it can hold. A 100-mL sample of this solution is withdrawn and titrated with 5.00×10^{-2} M HBr. It requires 48.8 mL of the acid solution for neutralization. What is the concentration of the $Ca(OH)_2$ solution? What is the solubility of $Ca(OH)_2$ in water, at 30°C, in grams of $Ca(OH)_2$ per 100 mL of solution?

[3.61] Calculate the molarity of OH^-, Cl^-, and Ca^{2+} present in a solution made by mixing equal volumes of 0.200 M $Ca(OH)_2$ and 0.300 M HCl.

Additional exercises

3.62 Balance the following equations:

(a) $Li_3N(s) + H_2O(l) \longrightarrow NH_3(g) + LiOH(aq)$

(b) $C_3H_7OH(l) + O_2(g) \longrightarrow CO_2(g) + H_2O(g)$

(c) $PBr_3(l) + H_2O(l) \longrightarrow H_3PO_3(aq) + HBr(aq)$

(d) $Mg_3B_2(s) + H_2O(l) \longrightarrow Mg(OH)_2(aq) + B_2H_6(g)$

(e) $CCl_4(g) + O_2(g) \longrightarrow CCl_2O(g) + Cl_2(g)$

(f) $La(NO_3)_3(aq) + Ba(OH)_2(aq) \longrightarrow La(OH)_3(s) + Ba(NO_3)_2(aq)$

3.63 Given the following reactions:

$$2HCl(aq) + CaCO_3(s) \longrightarrow CaCl_2(aq) + CO_2(g) + H_2O(l)$$

$$Zn(s) + 2HCl(aq) \longrightarrow H_2(g) + ZnCl_2(aq)$$

$$Na_2O(s) + H_2O(l) \longrightarrow 2NaOH(aq)$$

predict what will happen in the following cases and write a balanced chemical equation for each reaction. (a) Hydrochloric acid is added to solid $BaCO_3$. (b) Nitric acid is added to solid $CaCO_3$. (c) Solid potassium oxide, K_2O, is added to water. (d) Zinc metal is added to hydrobromic acid, HBr. (e) Barium oxide, $BaO(s)$ is added to water.

3.64 The mass spectrum of carbon indicates that it contains 98.89 percent ^{12}C (atomic mass = exactly 12 amu) and 1.11 percent ^{13}C (atomic mass = 13.003 amu). Calculate the average atomic weight of carbon.

3.65 Which of the following compounds, all used as fertilizers, contains the highest percentage by weight of phosphorus: (a) $Ca_5(PO_4)_3F$; (b) $Ca(H_2PO_4)_2 \cdot H_2O$; (c) $(NH_4)_2HPO_4$?

3.66 The alloy Co_5Sm is used to form the permanent magnets in very lightweight headsets such as in the Sony Walkman and similar units. What is the percent by weight of samarium in this alloy?

[3.67] One of the earliest accurate formula weight measurements involved measurement of the weight ratio of $KClO_3$ to KCl, based on decomposition of $KClO_3$: $2KClO_3(s) \longrightarrow 2KCl(s) + 3O_2(g)$. In 1911 Stähler and Meyer measured this ratio and found it to be 1.64382. Using only this ratio and the presently accepted atomic weight of oxygen, calculate the formula weight for KCl. Compare this calculated formula weight with the presently accepted value.

3.68 Calculate the number of moles in the following: (a) 15.0 g of Na_2CO_3; (b) 44.2 g of Kr; (c) 65 kg of H_2; (d) 2.3×10^{-7} g of $HgCl_2$.

3.69 Halothane, a widely used anesthetic, contains carbon, hydrogen, chlorine, bromine, and fluorine. Its elemental analysis yields the following results: carbon 12.2 percent, hydrogen 0.51 percent, fluorine 28.9 percent, chlorine 18.0 percent, and bromine 40.4 percent. What is the empirical formula of this substance?

3.70 Calculate the empirical formula and molecular formula of each of the following substances: (a) epinephrine (adrenaline), a hormone secreted in the bloodstream in times of danger or stress: 59.0 percent C, 7.1 percent H, 26.2 percent O, and 7.7 percent N; MW about 183 amu; (b) nicotine, a component of tobacco; 74.1 percent C, 8.6 percent H, and 17.3 percent N; MW = 160 ± 5 amu; (c) ethylene glycol, the substance used as the primary component of most antifreeze solutions: 38.7 percent C, 9.7 percent H, and 51.6 percent O; MW = 62.1 amu; (d) caffeine, a stimulant found in coffee; 49.5 percent C, 5.15 percent H, 28.9 percent N, and 16.5 percent O; MW about 195 amu.

3.71 The koala bear dines exclusively on eucalyptus leaves. Its digestive system detoxifies the eucalyptus oil, a poison to other animals. The chief constituent in eucalyptus oil is a substance called eucalyptol, which contains 77.87 percent C, 11.76 percent H, and the remainder O. (a) What is the empirical formula of this substance? (b) What is its molecular formula if the substance has a molecular mass of 154.2 amu?

3.72 Butadiene, a substance used to manufacture certain types of synthetic rubber, contains only two elements, carbon and hydrogen. When 1.00 g of this substance is completely combusted, 3.26 g of CO_2 and 0.998 g of H_2O are produced. What is the simplest formula of butadiene?

[3.73] Under a certain set of laboratory conditions, 2.1 g of Na reacts with water to form 1.14 L of H_2 gas. When 3.4 g of another alkali metal is reacted with water under the same conditions, 497 mL of hydrogen gas is evolved. Which of the other alkali metals was reacted?

3.74 Many "antacids" contain $Al(OH)_3$. (a) Write the equation for the neutralization of HCl in stomach acid by $Al(OH)_3$. (b) How many moles of HCl are neutralized by 5.0 g of $Al(OH)_3$?

3.75 What mass of ZnO is obtained by heating 75 kg of $ZnSO_3(s)$ according to the following equation:

$$ZnSO_3(s) \longrightarrow ZnO(s) + SO_2(g)$$

How many grams of $ZnSO_3$ are required to form 2.0 kg of SO_2 in this reaction?

3.76 A mixture of 3.50 g of H_2 and 26.0 g of O_2 is caused to react to form H_2O. How much H_2, O_2, and H_2O remain after reaction is complete?

3.77 A particular coal contains 2.8 percent sulfur by weight. When this coal is burned, the sulfur appears as $SO_2(g)$. This SO_2 is reacted with CaO to form $CaSO_3(s)$. If the coal is burned in a power plant that uses 2000 tons of coal per day, what is the daily production of $CaSO_3$?

3.78 Several samples of a particular brand of cigar were analyzed and found to contain on the average 8.00×10^{-6} g of iron, Fe, per cigar. The quantity of iron remaining in the ashes and butts of smoked cigars was found to average 5.92×10^{-6} g of iron. Assuming that the missing iron was lost from the cigar during smoking as gaseous iron pentacarbonyl, $Fe(CO)_5$, what weight of iron pentacarbonyl is formed on the average during smoking of each cigar?

3.79 In the Solvay process for forming soda ash (sodium carbonate), sodium bicarbonate is formed at one stage in the reaction:

$$NH_4HCO_3(aq) + NaCl(aq) \longrightarrow NaHCO_3(s) + NH_4Cl(aq)$$

Assuming complete precipitation, what mass of $NaHCO_3(s)$ could be formed from a solution containing excess NaCl and 4.60 kg of NH_4HCO_3?

[3.80] Chloromycetin is an antibiotic with the formula $C_{11}H_{12}O_5N_2Cl_2$. A 1.03-g sample of an ophthalmic ointment containing chloromycetin was chemically treated so as to convert its chlorine into Cl^- ions. The Cl^- was then precipitated as AgCl. If the AgCl weighed 0.0129 g, calculate the weight percentage of chloromycetin in the sample.

3.81 Adipic acid, $C_6H_{10}O_4$, is a raw material used for the production of nylon. Adipic acid is made industrially by oxidation of cyclohexane, C_6H_{12}:

$$5O_2 + 2C_6H_{12} \longrightarrow 2C_6H_{10}O_4 + 2H_2O$$

(a) If 2.00×10^2 kg of C_6H_{12} reacts with an unlimited supply of O_2, what is the theoretical yield of adipic acid? (b) If the reaction produces 2.95×10^2 kg of adipic acid, what is the percent yield for the reaction?

3.82 Aspirin, $C_9H_8O_4$, is produced from salicylic acid, $C_7H_6O_3$, and acetic anhydride, $C_4H_6O_3$:

$$C_7H_6O_3 + C_4H_6O_3 \longrightarrow C_9H_8O_4 + HC_2H_3O_2$$

(a) How much salicylic acid is required to produce 1.5×10^2 kg of aspirin, assuming that all of the salicylic acid is converted to aspirin? (b) How much salicylic acid would be required if only 80 percent of the salicylic acid is converted to aspirin? (c) What is the theoretical yield of aspirin if 185 kg of salicylic acid is allowed to react with 125 kg of acetic anhydride? (d) If the situation described in part (c) produces 182 kg of aspirin, what is the percentage yield?

3.83 Calculate the molarity of each of the following solutions: (a) 35.0 g of H_2SO_4 in 600 mL of solution; (b) 42.0 mL of 0.550 *M* HNO_3 solution diluted to a volume of 0.500 L; (c) a solution formed by dissolving 2.56 g of NaBr in water to form 65.0 mL of solution; (d) a solution formed by mixing 35 mL of 0.50 *M* KBr solution with 65 mL of 0.36 *M* KBr solution (assume a total volume of 100 mL).

3.84 Describe how you would prepare each of the following solutions: (a) 0.300 L of 0.35 *M* KOH solution, beginning with solid KOH; (b) 1.00 L of a 0.500 *M* $AgNO_3$ solution, starting with solid $AgNO_3$; (c) 1 liter of a 0.200 *M* solution of H_2SO_4, starting with a concentrated H_2SO_4 solution of unknown concentration and a quantity of 0.150 *M* NaOH solution.

3.85 Using modern analytical techniques it is possible to detect sodium in concentrations as low as 50 picograms per milliliter. What is this detection limit expressed in (a) molarity of Na; (b) atoms of Na per cubic centimeter?

3.86 If it requires 64.0 mL of 1.06 *M* NaOH solution to neutralize 10.0 mL of the H_2SO_4 solution from an auto battery, what is the molarity of the H_2SO_4 solution?

[3.87] The arsenic in a 1.22-g sample of a pesticide was converted to AsO_4^{3-} by suitable chemical treatment. It was then titrated using Ag^+ to form Ag_3AsO_4 as a precipitate. It if took 25.0 mL of 0.102 *M* Ag^+ to reach the equivalence point in this titration, what is the percentage of arsenic in the pesticide?

3.88 What volume of 0.20 *M* HBr is required to react with 18.0 g of zinc according to the equation

$$2HBr(aq) + Zn(s) \longrightarrow ZnBr_2(aq) + H_2(g)$$

[3.89] A solid sample of $Zn(OH)_2$ is added to 0.400 L of a 0.550 *M* solution of HBr. The solution that remains is still acidic. It is then titrated with 0.500 *M* NaOH solution, and 165 mL of the NaOH solution is required to reach the equivalence point. What was the mass of $Zn(OH)_2$ added to the HBr solution?

4 Energy Relationships: The First Law of Thermodynamics

In Chapters 2 and 3 we have considered the classification of matter into elements and compounds, the structure of matter in terms of atoms, ions, and molecules, and the changes of matter in chemical reactions. However, an important feature of chemical reactions is that they also involve changes in energy. Chemistry is therefore concerned with both matter and energy. Our focus in this chapter is on energy and its involvement in chemical processes. Most energy produced in modern society comes from chemical reactions, especially combustion of coal, petroleum products, and natural gas.

An ample supply of energy at reasonable cost is essential to the working of modern society. This fact lies at the heart of a most critical problem confronting the human race: Where will the energy come from to sustain civilization in the decades to come? In the short century since humankind learned how to utilize fossil fuels as a major energy source—first coal, then oil and gas—we have used up a substantial part of all the reserves that exist. Furthermore, it is becoming increasingly evident that burning a large fraction of all the fossil fuel stores that remain could result in disastrous changes in world climate (see Section 12.4). In the meantime, the demands for energy from an ever-increasing world population continue to grow. In the United States, we have been faced with political and economic crises resulting from an excessive dependence on foreign sources of oil and from the lack of a sound long-term energy plan.

The realization that energy may be in short supply, or may become excessively expensive, has in recent years brought about a keener interest in conserving energy. In addition, more attention has been paid to how we in the United States can more effectively utilize coal, which is relatively plentiful. To do this we must convert coal to more useful forms such as synthetic gasoline, synthetic gas, or liquid fuel oils. In addition, more attention will be given to converting plant matter into useful forms of fuel such as alcohols. Measures such as these will be the keystones of energy utilization in the decades to come, until we can arrive at long-range solutions to the problems of nuclear energy use or can develop nuclear fusion as an energy source (see Sections 19.7 and 19.8). Thus chemistry is the key to effective use of the resources on which we will be most dependent for the foreseeable future.

4.1 THERMODYNAMICS

Our aim in the study of chemistry is to be able to explain a great variety of chemical observations in terms of a relatively small number of theories and general principles. In doing this we need to employ concepts relating to the nature of matter, as introduced in Chapters 2 and 3. In addition, however, we need to understand the role of energy. This chapter is really an introduction to **thermodynamics,** the science of heat and energy, and the laws governing the conversion of heat into mechanical, electrical, and other forms of energy. Thermodynamics is concerned with **macroscopic properties**—that is, properties such as temperature, pressure, and volume—that apply to matter in the bulk form. It does not depend on any particular theories about the forces that operate between individual atoms or molecules, nor does it depend on our knowledge of the nature of subatomic particles such as the electron or proton. In fact, most of thermodynamics was developed at a time when scientists knew little or nothing of the existence of such particles.

To study the changes with which thermodynamics is concerned, it is important to focus our attention on a limited and well-defined part of the universe. That portion of the universe we single out for study is called the **system;** everything else is called the **surroundings.** As an example, suppose we wish to study the properties of a gas contained in a cylinder, as illustrated in Figure 4.1. We expect that the properties of the gas will not depend on the specific materials of which the cylinder and piston are constructed. The system in this case is just the gas; the cylinder, piston, and everything beyond them are the surroundings.

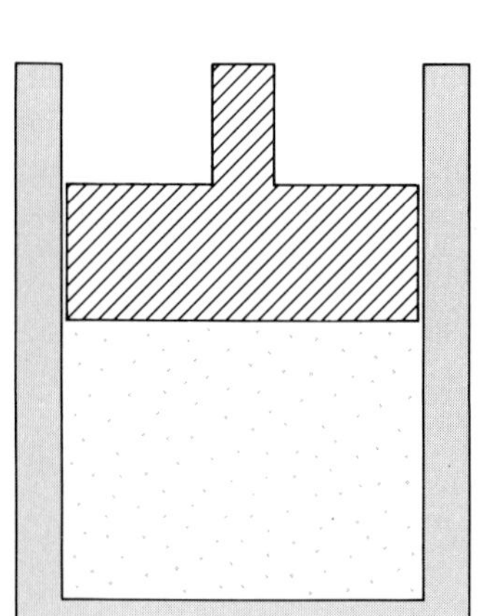

FIGURE 4.1 Gas in a cylinder, as an example of a system. If we are interested in studying the properties of the gas, the cylinder and piston would normally be considered part of the surroundings.

Thermodynamics is based on three laws that summarize our observations of how nature behaves. These laws embody the results of much systematic study of the properties of matter. In this chapter we will be concerned with the **first law of thermodynamics,** which is a statement of our experience that we cannot derive energy from nothing. The first law tells us that in any process that occurs in nature the energy of the universe is conserved. The energy gained or lost by the system we study is exactly matched by a counterbalancing loss or gain by the rest of the universe. In the material that follows we will be studying the changes that occur in properties such as the temperature and pressure of a system during a process. We will relate those changes to the exchanges of energy, in the form of heat and work, that the system undergoes with its surroundings. The first law of thermodynamics will serve as our guiding principle. Let's begin by considering the nature of energy and the various forms that it can assume.

4.2 THE NATURE OF ENERGY

Because energy is not tangible, as are material objects, it is a little difficult to define it completely. One definition of energy is that it is the capacity to do work or transfer heat. But what is work? The most familiar kind of work, **mechanical work** (w), is defined and measured by the product of the net force (f) and the distance (d) through which that force moves:

$$w = f \times d \qquad [4.1]$$

Energy is required to accomplish work: for example, to lift a book against the force of gravity, or to separate a positively charged ion from a negatively charged one, thus overcoming the attractive force that exists between two particles. The energy of the system changes when work is done on it or when the system does work on its surroundings. For example, let the ram of the pile driver shown in Figure 4.2 be our system. When the ram is lifted against the force of gravity, work is done on the ram. The ram has a different energy because of its new position. We call this kind of energy **potential energy,** the energy stored in a system by virtue of its position or the positions of its parts.

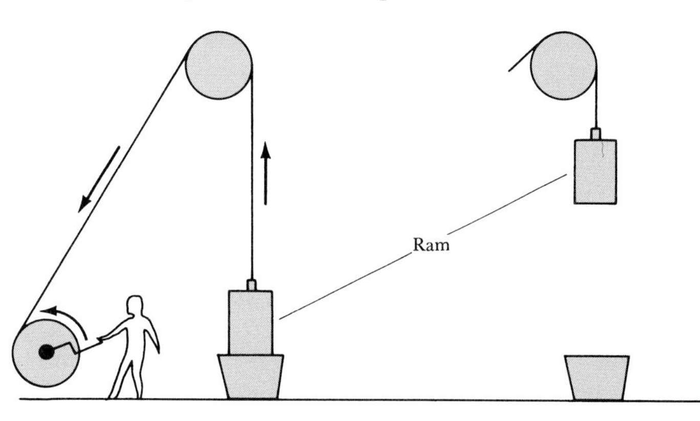

FIGURE 4.2 The ram of the pile driver has potential energy that depends on its height and mass. The energy to increase the potential energy of the ram is provided by the person operating the crank.

Work can also be done to impart **kinetic energy** to a system. Kinetic energy is the energy possessed by an object by virtue of its motion with respect to other matter. For example, a baseball pitcher does work on a baseball by imparting kinetic energy to it in a pitch. The magnitude of the kinetic energy of an object is related to its mass (m) and velocity or speed, (v):

$$E_k = \tfrac{1}{2}mv^2 \qquad [4.2]$$

As you can see from this expression, the greater the mass of an object and the higher its speed, the larger its kinetic energy.

In addition to the mechanical forms of energy we have been considering, systems may possess other forms of energy, such as thermal energy, radiant energy (light), chemical energy, and electrical energy. In later chapters we will have occasion to discuss all these forms of energy in more detail. Whatever the form, the units of energy are the same. The SI unit of energy is the **joule** (J); $1\ \text{J} = 1\ \text{kg-m}^2/\text{s}^2$. Traditionally, energy changes accompanying chemical reactions have been expressed in units of calories (cal). A **calorie** is the amount of energy required to raise the temperature of 1 g of water by 1°C, from 14.5°C to 15.5°C.* A kilocalo-

*The temperature interval 14.5 to 15.5°C is specified because the energy required to raise the temperature of water by 1°C is slightly different for different temperatures. However, it is constant to three significant figures, 1.00 cal, over the entire range of liquid water, from 0 to 100°C.

rie, 1000 cal, is the same as the Calorie (capitalized), used in expressing the energy values of foods. One calorie is the same amount of energy as 4.184 J. In much engineering work it is common to use the British thermal unit (Btu); 1 Btu = 1.05 kJ. In most places in this text the joule or kilojoule is used as the unit of energy.

SAMPLE EXERCISE 4.1

What is the kinetic energy in joules and calories of a 6.0-kg object moving at a speed of 5.0 m/s?

Solution: To evaluate the kinetic energy we employ Equation 4.2:

$$E_k = \tfrac{1}{2}mv^2$$
$$= (\tfrac{1}{2})(6.0\ \text{kg})(5.0\ \text{m/s})^2 = 75\ \text{kg-m}^2/\text{s}^2 = 75\ \text{J}$$

To obtain the kinetic energy in calories, we employ the conversion factor 1 cal = 4.184 J:

$$(75\ \text{J})\left(\frac{1\ \text{cal}}{4.184\ \text{J}}\right) = 18\ \text{cal}$$

Thermal energy is the energy that an object possesses by virtue of the kinetic energies of the atoms or molecules of which it is composed. The atoms and molecules that make up an object are in constant motion, moving through space as in a gas or liquid, or vibrating about their positions in a solid. These various kinds of kinetic energies are often quite complex, and we will not be concerned about them in detail. It is important only to realize that they give rise to what we call the thermal energy of a system, and that the thermal energy increases as the temperature of the system increases. The term *heat* has a different meaning; **heat** refers to the flow of energy between two bodies that are at different temperatures. We speak of heat flow from a hot body to a cold body. Such a heat flow corresponds to a transfer of energy, usually but not necessarily as thermal energy.

Heat and work are both forms of energy, and both are means by which a system can exchange energy with its surroundings. In general, energy is transferred whenever two systems, or a system and its surroundings, at different temperatures are placed in mechanical or thermal contact with one another. An exchange of work involves a mechanical or electrical link between the system and its surroundings. Heat flow, by contrast, occurs solely because of a temperature difference between the system and its surroundings. The precise character of the energy transfer will vary with the detailed nature of the system. To illustrate, consider the cylinder of an internal combustion engine just after combustion of the gasoline, as illustrated in Figure 4.3. Our system consists of the hot gas under pressure in the cylinder. The gas is hotter than its surroundings, and it is under high pressure. There is a very rapid expansion of the gas as the piston is driven downward. In this expansion the gas does work on its surroundings, because it moves the piston and parts of the engine connected to it. Because the gas is hotter than its surroundings, there will also be heat flow directly from the hot gas to the walls of the cylinder. In the expansion, therefore, there is a net transfer of energy from the system (the hot gas) to the surroundings, in the form of both heat flow and work done. The energy source for both the work and heat flow is the **internal energy** of the gas, that is, the total of all the kinetic energies of the gas

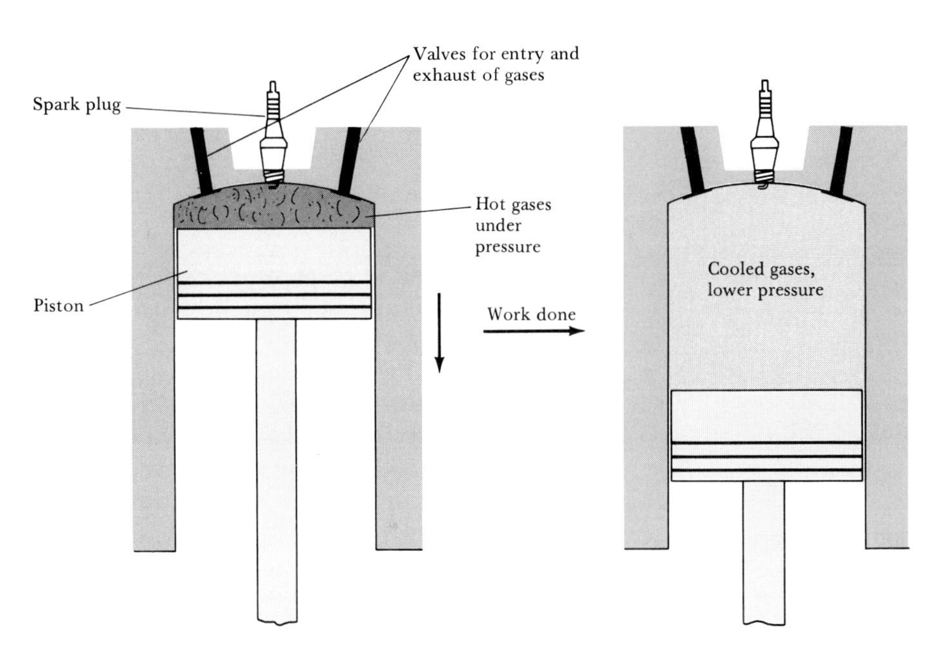

FIGURE 4.3 Conversion of internal energy into work. Expansion of the hot gases formed in the cylinder by combustion forces the piston to move through a distance. The gas does work on its surroundings; in the process, the temperature of the gas is reduced.

particles and the potential energies of interaction between them. Since this energy is the source of the heat transferred and work done, the first law tells us that the internal energy of the gas must decrease. In the expansion, therefore, the temperature of the gas decreases.

Engineers concerned with engine design want to minimize energy that appears as heat, because it increases the engine temperature without producing useful work. In practice, only about 10 to 15 percent of the chemical energy released as heat when the fuel is burned is converted into the work of moving the automobile. The remainder is transferred to the environment as waste heat, through the exhaust and the car's cooling system. The question of how efficiently a temperature difference between two systems (in this case, between the hot gas and the surroundings) can be employed to produce useful work is important. However, it falls within the province of the second law of thermodynamics, which we will consider in Chapter 17.

Our discussion has led us to see that any system interacts with its surroundings in two general ways: (1) it may do work on its surroundings or have work done on it by its surroundings, and (2) it may exchange heat with its surroundings. We can use these ideas to write a very useful algebraic expression of the first law of thermodynamics. Let's first define the **change in the internal energy**, which we represent as ΔE,* as the difference between the internal energy of the system at the completion of a process and that at the beginning:

$$\Delta E = E_{\text{final}} - E_{\text{initial}}$$

*The symbol Δ is commonly used to denote *change*. For example, a change in volume can be represented by ΔV.

According to the first law, when a system undergoes any chemical or physical change, the accompanying change in its internal energy, ΔE, is given by the heat (q) added to the system, less any work (w) that the system does on its surroundings:

$$\Delta E = q - w \qquad [4.3]$$

You will need to be careful about the signs attached to the quantities in this equation. Notice that heat *added to the system* is assigned a positive sign. Work w *done by the system* is also assigned a positive sign. As you use Equation 4.3 you can keep these sign conventions straight by using your physical intuition. Heat added *to* the system, or work done *on* the system, both increase the internal energy of the system, and thus make ΔE positive. For example, if the system absorbs 50 J of heat and has 20 J of work done on it in a particular process, $q = 50\text{ J}$ and $w = -20\text{ J}$. Then $\Delta E = 50\text{ J} - (-20\text{ J}) = 70\text{ J}$. Notice that work done on the system has a negative sign; the negative sign before w in Equation 4.3 converts this into a positive contribution to ΔE. Heat flow from the system to the surroundings would have associated with it a negative sign, and would result in a negative contribution to ΔE.

SAMPLE EXERCISE 4.2

The gas in the cylinder illustrated in Figure 4.1 is allowed to expand against a constant force acting on the piston. It is determined that the gas does 240 J of work on the surroundings in this expansion, and at the same time absorbs 160 J of heat from the surroundings, through the walls of the cylinder. What is the change in internal energy of the gas?

Solution: We can calculate the change in internal energy, ΔE, by use of Equation 4.3. The quantity q is positive when heat is added to the system. The quantity w, which represents work done *by* the system, is positive when work is done by the system on its surroundings, as in the present case. Thus we have

$$\Delta E = q - w = 160 - 240 = -80\text{ J}$$

Notice that, although w is positive, the negative sign before it ensures that the internal energy decreases when the system does work on its surroundings.

We usually have no way of knowing the precise value of the internal energy when the system is in a given state. However, we do know that it has a fixed value for a given set of conditions of temperature, pressure, and any other variables that might affect it. We also know that the total internal energy is proportional to the quantity of matter in the system; energy is an extensive property (Section 1.3). As a simple example, suppose we define our system as 50 g of water in a beaker on the bench at 25°C, as in Figure 4.4. Our system could have arrived at that state by cooling of 50 g of water from 100°C, or by the melting of 50 g of ice, and subsequent warming of the water to 25°C. The internal energy of the 50 g of water is the same in either case. The internal energy is an example of a **state function,** a property of a system that is determined by specifying its state (temperature, pressure, location, and so forth). *The value of a state function does not depend on the particular previous history of the sample, but only on its present condition.*

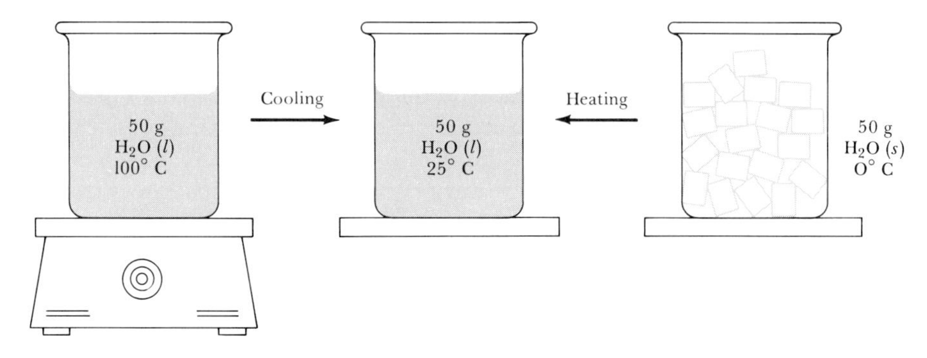

FIGURE 4.4 Illustration of the fact that internal energy, a state function, depends only on the present state of the system and not on the path by which it arrived at that state. The internal energy of 50 g of water at 25 °C is the same whether the water is cooled from a higher temperature to 25 °C, or is obtained by melting of 50 g of ice, and subsequent warming to 25 °C.

You may well ask: What is an example of something that is *not* a state function? The work done by a system in a given process is not a state function. Rather, it depends on the manner in which the process is carried out. As an example, let's consider a flashlight battery as our system, and let the change in the system be the complete discharge of the battery, at constant temperature. If the battery is discharged in a flashlight [Figure 4.5(*a*)], no mechanical work is accomplished. All of the energy lost from the battery appears as radiant energy and heat. If the battery is used in a mechanical toy [Figure 4.5(*b*)], the same change in state of the battery produces mechanical work and heat. The change in state of the system, and thus the change in ΔE, is the same in both cases. However, the amount of work done in the two cases is different. We see from this example that although ΔE is always the same for a given change in the system, the way in which the change is performed will determine the relative contributions of q and w, the means by which energy is transferred.

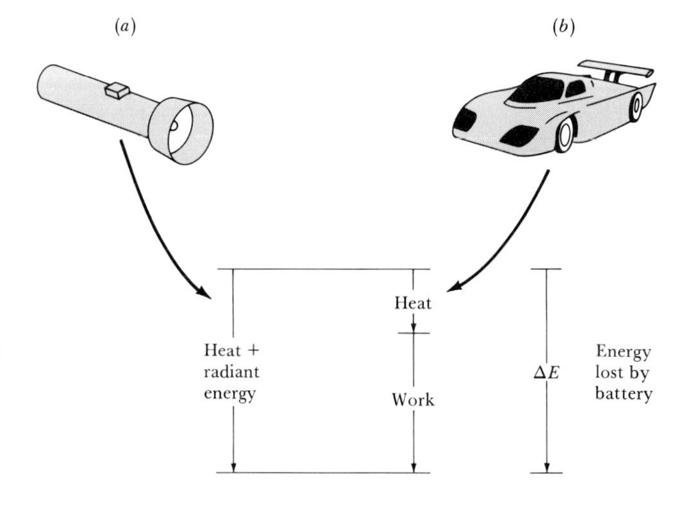

FIGURE 4.5 When a flashlight battery is discharged in lighting a flashlight, all the energy of the battery appears as radiant energy and heat; no work is done. When the battery is used in the toy car, work is done in moving the car from place to place. Thus the work done by the system (the flashlight battery) is not a state function, because its magnitude depends on the particular path by which the system gets from its initial to its final state.

4.3 CHEMICAL CHANGES AT CONSTANT PRESSURE; ENTHALPY

Although there are many varieties of work, in chemistry we are concerned primarily with electrical work and the work done by expanding gases. In this chapter we restrict ourselves to the latter type of work, illustrated by the earlier example of an expanding gas moving a piston. This kind of work is especially important in the case of chemical reactions that occur at atmospheric pressure, and where a gas is produced or consumed in the course of the reaction. As a further example, let's consider the simple physical process of converting a mole of water from liquid to gaseous form at 100°C. Figure 4.6 shows an experimental arrangement in which this process might be carried out. We assume that the piston is weightless, but that it is acted upon by the earth's atmosphere. When heat is applied to the system, the water passes from the liquid to the gaseous state. The heat supplied to the system goes to increase its internal energy by a certain amount, and to do work against the force of the atmosphere. This work is called pressure-volume work. We will discuss the properties of gases in more detail in Chapter 9. At this point all we need note is that the work done when a gas expands or contracts against a constant external pressure is given by the product of the constant pressure, P, times the change in volume, ΔV:

$$w = P\,\Delta V \tag{4.4}$$

For this special case, where the work done by the system is pressure-volume work only, we have $\Delta E = q_P - w = q_P - P\,\Delta V$. (Here the subscript P on q denotes that the process occurs under constant pressure conditions.) Rearranging gives

$$q_P = \Delta E + P\,\Delta V \tag{4.5}$$

This equation tells us that the heat added to bring about any change at constant pressure, q_P, is the sum of the internal energy change plus the pressure-volume work.

When we heat liquids or solids at constant pressure they do not expand by very much. However, we know that gases do expand upon heating under constant pressure conditions. Therefore, the quantity $P\,\Delta V$ has a significant magnitude only when there is a change in the number of moles of gas in the process. In our example above, there is

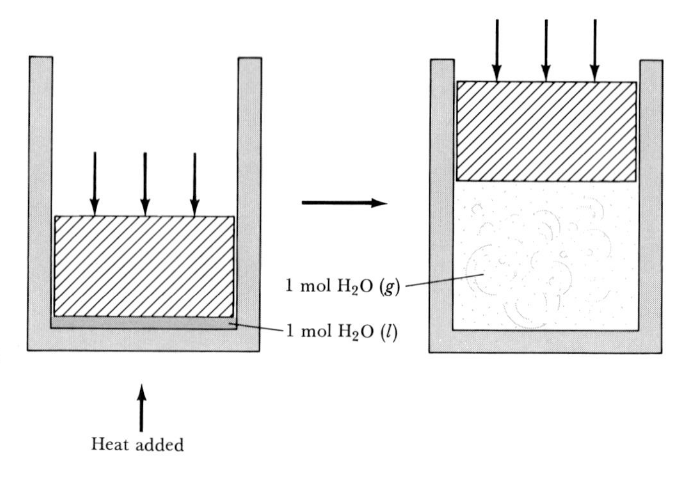

FIGURE 4.6 Conversion of a mole of water from the liquid to vapor state. The system does work on its surroundings as gas is formed and exerts a force on the piston. The pressure P (force per unit area) causes the piston to move through a volume ΔV. The work done by the system on the surroundings is given by $P\Delta V$.

such a change, because we are converting a mole of *liquid* water to a mole of *gaseous* water. (That is, we go from 0 to 1 mol of gas.) By contrast, in the process of converting a mole of solid water to a mole of liquid water, that is, in the melting of a mole of ice, there is very little pressure-volume work, because the volumes of a mole of ice and a mole of liquid water are very nearly the same. Thus ΔV, and consequently $P\,\Delta V$, is very small for this process, and q_P and ΔE are essentially equal.

Enthalpy

Most chemical reactions, including those that occur in living systems, take place under the essentially constant pressure of the earth's atmosphere. For this reason it is useful to define a new state function, called enthalpy, represented by the symbol $\boldsymbol{H}$, that is particularly suited for discussing such processes. Enthalpy is defined by the equation

$$H = E + PV \qquad [4.6]$$

For processes occurring at constant pressure, the *change* in enthalpy, ΔH, is related to the changes in other properties of the system as follows:

$$\Delta H = \Delta E + P\,\Delta V \qquad [4.7]$$

By comparing this with Equation 4.5 you can see that the enthalpy change for a process occurring at constant pressure is identical with q_P, the heat added to the system under constant pressure, when the only type of work involved is pressure-volume work: $\Delta H = q_P$.

When a chemical reaction occurs, the enthalpies of the products will generally differ from those of the reactants. Thus there is an overall change in the enthalpy of the system in going from reactants to products. The enthalpy change for a reaction (ΔH_{rxn}) is given by the enthalpy of the products minus that of the reactants:

$$\Delta H_{rxn} = H(\text{products}) - H(\text{reactants}) \qquad [4.8]$$

For a given set of reactants and products, H(products) and H(reactants) will be proportional to the total mass of material present. That is, *enthalpy is an extensive property,* like internal energy and volume.

To get a feeling for the sort of quantity enthalpy is, consider the combustion of a piece of magnesium ribbon:

$$2\text{Mg}(s) + \text{O}_2(g) \longrightarrow 2\text{MgO}(s) + \text{heat} + \text{light} \qquad [4.9]$$

You have almost surely observed this reaction. Camera flash bulbs are filled with magnesium ribbon and oxygen gas; passage of an electrical current through the magnesium ribbon causes it to ignite, producing heat and light. In this process the system consists initially of just the magnesium and oxygen. In the course of the reaction that takes place, no new material is added. However, chemical change produces a new product, MgO, and results in the release of considerable energy. Figure 4.7 shows diagrammatically the loss of enthalpy by the system as reaction proceeds.

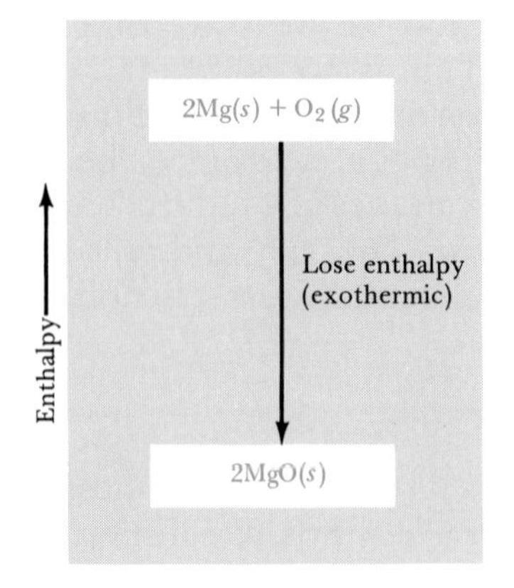

FIGURE 4.7 Enthalpy change in the exothermic reaction of magnesium with oxygen. Heat is given off by the system to its surroundings, thereby decreasing the enthalpy of the system.

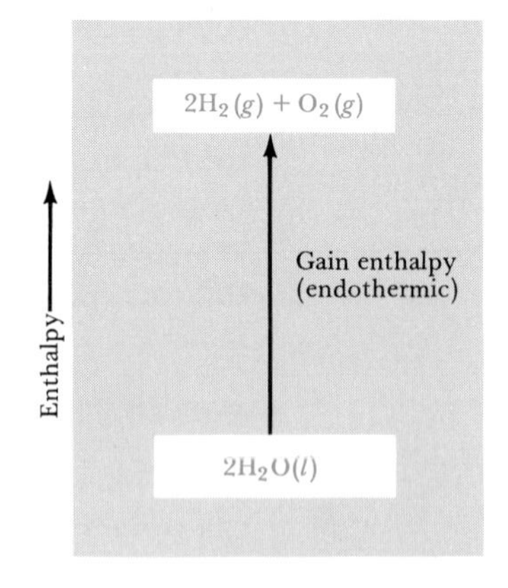

FIGURE 4.8 Enthalpy change in the endothermic decomposition of water to its elements. Heat is absorbed by the system, thereby increasing the enthalpy of the system.

An exothermic process is one in which the system transfers heat to its surroundings. When such processes occur at constant pressure the enthalpy of the products is less than that of the reactants, and ΔH is consequently negative in sign. The combustion of magnesium ribbon in an oxygen atmosphere is an example of an exothermic process. A process in which heat is absorbed from the surroundings is termed endothermic; ΔH for such a process is positive in sign. An example of an endothermic process is the decomposition of water into its elements:

$$\text{Energy} + 2H_2O(l) \longrightarrow 2H_2(g) + O_2(g) \qquad [4.10]$$

As we noted in Chapter 2, this reaction can be carried out by applying electrical energy to the system. The enthalpy change of the system for this process is depicted in Figure 4.8. Note that the total enthalpy of the system is increased.

To explore further how we might use the concept of enthalpy, let's consider the combustion of methane to form carbon dioxide and water. It is found experimentally that 802 kJ of heat is produced when 1 mol of CH_4 is burned in a constant-pressure system. We can express this fact as follows:

$$CH_4(g) + 2O_2(g) \longrightarrow CO_2(g) + 2H_2O(g) \qquad \Delta H = -802 \text{ kJ} \qquad [4.11]$$

The negative sign for ΔH tells us that this is an exothermic process. The coefficients in the balanced equation are taken to represent the number of moles of reactants giving the associated enthalpy change. As we have noted earlier, the amount of heat associated with a reaction is directly proportional to the amounts of substances involved. Thus combustion of 1 mol of CH_4 with 2 mol of O_2 produces 802 kJ of heat, whereas combustion of 2 mol of CH_4 with 4 mol of O_2 produces 1604 kJ.

SAMPLE EXERCISE 4.3

When barium chlorate, $Ba(ClO_3)_2$, is added to water at 25 °C, the salt dissolves and the solution temperature decreases below 25 °C. Is the process of dissolving exothermic or endothermic? What is the sign of ΔH?

Solution: The fact that the solution temperature decreases as the salt dissolves indicates that the process is one that requires absorption of heat from the surroundings. Thus it is an endothermic process. To maintain the temperature of the system, consisting of salt plus water, at 25 °C, it is necessary to add heat; thus ΔH is positive, as is always the case for an endothermic process.

SAMPLE EXERCISE 4.4

How much heat is produced when 4.50 g of methane gas is burned in a constant-pressure system?

Solution: According to Equation 4.11, 802 kJ is produced when a mole of CH_4 is burned. A mole of CH_4 has a mass of 16.0 g.

$$\text{Heat produced} = (4.50 \text{ g } CH_4) \times \left(\frac{1 \text{ mol } CH_4}{16.0 \text{ g } CH_4}\right)\left(802 \frac{\text{kJ}}{\text{mol } CH_4}\right) = 226 \text{ kJ}$$

Another important feature of ΔH is that the enthalpy change for a reaction is equal in magnitude but opposite in sign to ΔH for the reverse reaction. For example:

$$CO_2(g) + 2H_2O(g) \longrightarrow CH_4(g) + 2O_2(g) \qquad \Delta H = 802 \text{ kJ} \qquad [4.12]$$

If more energy were produced by combustion of CH_4 than was required for the reverse reaction, it would be possible to use these processes to create an unlimited supply of energy. Some CH_4 could be combusted, and that portion of the energy necessary to reform CH_4 could be saved. The rest could be used to do work of some type. After CH_4 is reformed, it could again be combusted, and so forth, continually supplying energy. This, of course, is clearly contrary to our experience—such behavior does not obey the law of conservation of energy. The observed situation is shown in Figure 4.9.

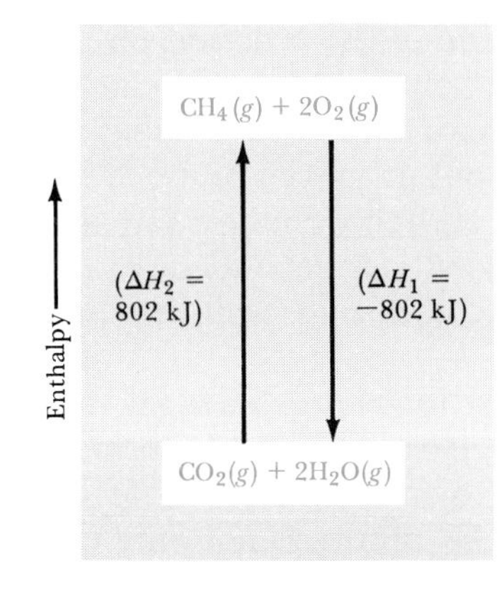

FIGURE 4.9 If a substance is combusted and then reformed from its combustion products, the total enthalpy change in the system must be zero ($\Delta H_1 + \Delta H_2 = 0$). In general, the net energy change in the system for a series of reactions that eventually regenerate the original substances in their original states is zero.

The enthalpy change of a reaction also depends on the state of the reactants and products. If the product in the combustion of methane, Equation 4.11, were liquid H_2O instead of gaseous H_2O, ΔH would be −890 kJ instead of −802 kJ. More heat is available for transfer to the surroundings because 88 kJ is released when 2 mol of gaseous water is condensed to the liquid state:

$$2H_2O(g) \longrightarrow 2H_2O(l) \qquad \Delta H = -88 \text{ kJ} \qquad [4.13]$$

Therefore, the states of the reactants and products must be specified. In addition, we will generally assume that the reactants and products are both at the same temperature, usually 25 °C, unless otherwise indicated.

The enthalpy change associated with a given chemical process is often of great significance. In the sections that follow we will consider some ways in which we can evaluate this important quantity. As we shall see, ΔH for a reaction can be either directly determined by experiment or calculated from a knowledge of the enthalpy changes associated with other reactions by invoking the first law. Let's begin by discussing how we make use of the concept of a state function to carry out calculations of enthalpy changes.

4.4 HESS'S LAW

Because enthalpy is a state function, the enthalpy change, ΔH, associated with any chemical process depends only on the amount of matter that undergoes change, and on the nature of the initial state of the reactants and final state of the products. This means that if a particular reaction can be carried out in one step or in a series of steps, the sum of the enthalpy changes associated with the individual steps must be the same as the enthalpy change associated with the one-step process. For example, the enthalpy change for the combustion of methane to form carbon dioxide and liquid water can be calculated from ΔH for the condensation of water vapor and ΔH for combustion to methane to form gaseous water:

$$CH_4(g) + 2O_2(g) \longrightarrow CO_2(g) + 2H_2O(g) \qquad \Delta H = -802\text{ kJ}$$

$$\text{(Add)} \qquad 2H_2O(g) \longrightarrow 2H_2O(l) \qquad \Delta H = -88\text{ kJ}$$

$$CH_4(g) + 2O_2(g) + 2H_2O(g) \longrightarrow CO_2(g) + 2H_2O(l) + 2H_2O(g) \qquad \Delta H = -890\text{ kJ}$$

Net equation:

$$CH_4(g) + 2O_2(g) \longrightarrow CO_2(g) + 2H_2O(l) \qquad \Delta H = -890\text{ kJ}$$

To obtain the net equation, the sum of the reactants of the two equations are placed on one side of the arrow, and the sum of the products on the other. Because $2H_2O(g)$ occurs on both sides of the arrow, it can be canceled like an algebraic quantity that is on both sides of an equal sign.

Hess's law states that if a reaction is carried out in a series of steps, ΔH for the reaction will be equal to the sum of the enthalpy changes for each step. The overall enthalpy change for the process is independent of the number of steps or the particular nature of the path by which the reaction is carried out.

Hess's law provides a useful means of calculating energy changes that are difficult to measure directly. For instance, it is not possible to measure directly the heat of combustion of carbon to form carbon monoxide. Combustion of 1 mol of carbon with $\frac{1}{2}$ mol of O_2 produces not only CO but also CO_2, leaving some carbon unreacted. However, the heat of the reaction forming CO can be calculated as shown in Sample Exercise 4.5.

SAMPLE EXERCISE 4.5

The heat of combustion of C to CO_2 is -393.5 kJ/mol of CO_2, whereas that for combustion of CO to CO_2 is -283.0 kJ/mol of CO_2. Calculate the heat of combustion of C to CO.

Solution: You need first to write out the two combustion reactions and then turn the combustion reaction for CO around, so that CO is the product. The two equations can then be added, as follows:

$$2C(s) + 2O_2(g) \longrightarrow 2CO_2(g) \qquad \Delta H = -2(393.5) = -787.0 \text{ kJ}$$
$$2CO_2(g) \longrightarrow 2CO(g) + O_2(g) \qquad \Delta H = 566.0 \text{ kJ}$$
$$2C(s) + O_2(g) \longrightarrow 2CO(g) \qquad \Delta H = -221.0 \text{ kJ}$$

Note that it is necessary to multiply the first equation through by 2 to obtain a proper canceling of terms. This requires that ΔH for the process also be multiplied by 2, as shown. Remember that when reactions are turned around, the sign of ΔH for the reversed process is the opposite of the original value.

The heat of combustion of C(s) to form CO(g) is $\frac{1}{2}(-221.0 \text{ kJ}) = -110.5 \text{ kJ}$ *per mole of CO formed.*

SAMPLE EXERCISE 4.6

Given the following reactions and their respective enthalpy changes:

$$C_2H_2(g) + \tfrac{5}{2}O_2(g) \longrightarrow 2CO_2(g) + H_2O(l) \qquad \Delta H = -1299.6 \text{ kJ/mol } C_2H_2$$
$$C(s) + O_2(g) \longrightarrow CO_2(g) \qquad \Delta H = -393.5 \text{ kJ/mol C}$$
$$H_2(g) + \tfrac{1}{2}O_2(g) \longrightarrow H_2O(l) \qquad \Delta H = -285.9 \text{ kJ/mol } H_2$$

calculate ΔH for the reaction

$$2C(s) + H_2(g) \longrightarrow C_2H_2(g)$$

Solution: Because we want to end up with C_2H_2 we turn the first equation around; the sign of ΔH is therefore changed. Because we need to start with 2C(s), we multiply the second equation and its ΔH by 2. We then add the resultant equations and their enthalpy changes in accordance with Hess's law:

$$2CO_2(g) + H_2O(l) \longrightarrow C_2H_2(g) + \tfrac{5}{2}O_2(g) \qquad \Delta H = 1299.6 \text{ kJ}$$
$$2C(s) + 2O_2(g) \longrightarrow 2CO_2(g) \qquad \Delta H = -787.0 \text{ kJ}$$
$$H_2(g) + \tfrac{1}{2}O_2(g) \longrightarrow H_2O(l) \qquad \Delta H = -285.9 \text{ kJ}$$
$$2C(s) + H_2(g) \longrightarrow C_2H_2(g) \qquad \Delta H = 226.7 \text{ kJ}$$

When the three equations are added, there are $2CO_2$, $\frac{5}{2}O_2$, and H_2O on both sides of the arrow. These are canceled in writing the net equation.

The first law of thermodynamics, in the form of Hess's law, teaches us that we can never expect to obtain more (or less) energy from a chemical reaction by changing the method of carrying out the reaction. For example, for the reaction of methane, CH_4, and oxygen, O_2, to form CO_2 and H_2O, we may envision the reaction to occur either directly or with the initial formation of CO, which is then subsequently combusted. This set of choices is illustrated in Figure 4.10. Because ΔH is a state function,

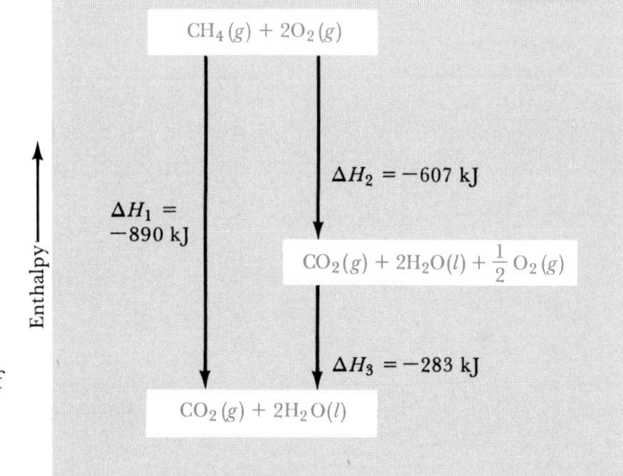

FIGURE 4.10 The quantity of heat generated by combustion of CH_4 is independent of whether the reaction takes place in one or more steps ($\Delta H_1 = \Delta H_2 + \Delta H_3$).

either path produces the same change in the enthalpy content of the system. That is, $\Delta H_1 = \Delta H_2 + \Delta H_3$. Again, we may note that if this were not so, it would be possible to create energy continuously, in conflict with the first law of thermodynamics.

4.5 HEATS OF FORMATION

By using the methods we have just discussed, we can calculate the enthalpy changes for many reactions from a few tabulated values. It is convenient to summarize many of these data in terms of **standard heats of formation.*** The heat of formation of a compound, ΔH_f, is the enthalpy change involved in forming a mole of that compound from its constituent elements. A standard enthalpy change, $\Delta H°$, is one that takes place with all reactants and products in their **standard states.** That is, all substances are in the forms most stable at the particular temperature of interest and at standard atmospheric pressure (see Chapter 9). The temperature usually chosen for purposes of tabulating data is 25°C. For example, the standard heat of formation, ΔH_f°, for ethanol (C_2H_5OH) is the enthalpy change for the following reaction:

$$2C(\text{graphite}) + 3H_2(g) + \tfrac{1}{2}O_2(g) \longrightarrow C_2H_5OH(l) \qquad [4.14]$$

The elemental source of oxygen is O_2, not O or O_3, because O_2 is the stable form of oxygen at 25°C and standard atmospheric pressure. Similarly, the elemental source of carbon is graphite and not diamond, because the former is the stable (lowest energy) form at 25°C and standard atmospheric pressure. The conversion of graphite to diamond requires the addition of energy as shown in Equation 4.15:

$$C(\text{graphite}) \longrightarrow C(\text{diamond}) \qquad \Delta H° = 1.88\ \text{kJ} \qquad [4.15]$$

A few standard heats of formation are given in Table 4.1; a more complete table is provided in Appendix D. *By convention the standard heat of formation of the stable form of any element is zero.*

The standard enthalpy change for any reaction can be found by summing the heats of formation of all reaction products, taking care to multiply each molar heat of formation by the coefficient of that substance in the balanced equation, then subtracting a similar sum for all the heats of formation of all the reactants:

$$\Delta H^\circ_{\text{rxn}} = \sum n\,\Delta H_f^\circ(\text{products}) - \sum m\,\Delta H_f^\circ(\text{reactants}) \qquad [4.16]$$

The symbol Σ (sigma) means "the sum of" and n and m are the stoichiometric coefficients of the chemical equation. For example, $\Delta H°$ for the combustion of glucose, Equation 4.17, is given by Equation 4.18:

$$C_6H_{12}O_6(s) + 6O_2(g) \longrightarrow 6CO_2(g) + 6H_2O(l) \qquad [4.17]$$

$$\Delta H^\circ_{\text{rxn}} = [6\,\Delta H_f^\circ(CO_2) + 6\,\Delta H_f^\circ(H_2O)] - [\Delta H_f^\circ(C_6H_{12}O_6) + 6\,\Delta H_f^\circ(O_2)] \qquad [4.18]$$

*It would be more precise to refer to heats of formation as "enthalpies of formation." However, the usual practice is to use the more general term "heat of formation."

TABLE 4.1 Standard heats of formation, ΔH_f°, at 25 °C

Substance	Formula	ΔH_f° (kJ/mol)
Acetylene	$C_2H_2(g)$	226.7
Ammonia	$NH_3(g)$	−46.19
Benzene	$C_6H_6(l)$	49.04
Calcium carbonate	$CaCO_3(s)$	−1207.1
Calcium oxide	$CaO(s)$	−635.5
Carbon dioxide	$CO_2(g)$	−393.5
Carbon monoxide	$CO(g)$	−110.5
Diamond	$C(s)$	1.88
Ethane	$C_2H_6(g)$	−84.68
Ethanol	$C_2H_5OH(l)$	−277.7
Ethylene	$C_2H_4(g)$	52.30
Hydrogen bromide	$HBr(g)$	−36.23
Hydrogen chloride	$HCl(g)$	−92.30
Hydrogen fluoride	$HF(g)$	−268.6
Hydrogen iodide	$HI(g)$	25.9
Glucose	$C_6H_{12}O_6(s)$	−1260
Methane	$CH_4(g)$	−74.85
Methanol	$CH_3OH(l)$	−238.6
Silver chloride	$AgCl(s)$	−127.0
Sodium bicarbonate	$NaHCO_3(s)$	−947.7
Sodium carbonate	$Na_2CO_3(s)$	−1130.9
Sodium chloride	$NaCl(s)$	−411.0
Sucrose	$C_{12}H_{22}O_{11}(s)$	−2221
Water	$H_2O(l)$	−285.8
Water vapor	$H_2O(g)$	−241.8

Using the heats of formation recorded in Table 4.1, this process gives

$$\begin{aligned}\Delta H^\circ_{\text{rxn}} &= \left[(6\text{ mol CO}_2)\left(-393.5\,\frac{\text{kJ}}{\text{mol CO}_2}\right)\right.\\ &\qquad \left.+ (6\text{ mol H}_2\text{O})\left(-285.9\,\frac{\text{kJ}}{\text{mol H}_2\text{O}}\right)\right]\\ &\qquad - \left[(1\text{ mol C}_6\text{H}_{12}\text{O}_6)\left(-1260\,\frac{\text{kJ}}{\text{mol C}_6\text{H}_{12}\text{O}_6}\right)\right.\\ &\qquad \left.+ (6\text{ mol O}_2)\left(0\,\frac{\text{kJ}}{\text{mol O}_2}\right)\right]\\ &= -2816\text{ kJ}\end{aligned}$$

The general relationship expressed in Equation 4.16 follows directly from the fact that ΔH is a state function. This fact allows calculation of the enthalpy change for any reaction from the energies required to convert the initial reactants into elements and then combine the elements into the desired products. This reaction pathway is shown in Figure 4.11 for the combustion of glucose.

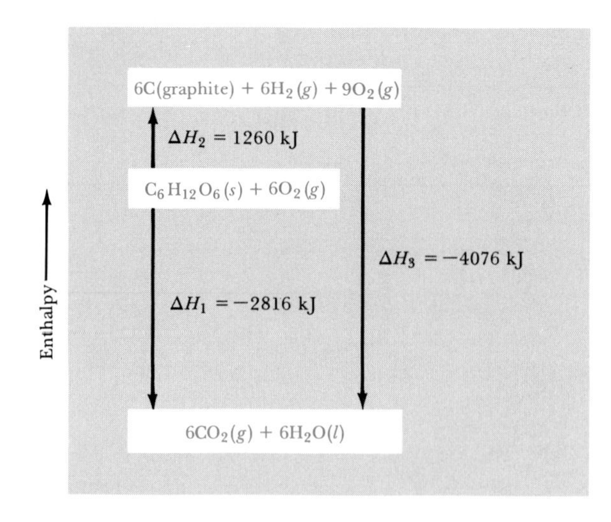

FIGURE 4.11 Because ΔH is a state function, $\Delta H_1 = \Delta H_2 + \Delta H_3$. Note that ΔH_2 is $-\Delta H_f^\circ(C_8H_{12}O_6)$, while ΔH_3 is $\Delta H_f^\circ(H_2O)$ + $6\,\Delta H_f^\circ(CO_2)$. This is the same result as given in Equation 4.18, because $\Delta H_f^\circ(O_2)$ is zero.

SAMPLE EXERCISE 4.7

Compare the quantity of heat produced by combustion of 1.00 g of glucose ($C_6H_{12}O_6$) with that produced by 1.00 g of sucrose ($C_{12}H_{22}O_{11}$).

Solution: The example worked in the text gave $\Delta H^\circ = -2816$ kJ for the combustion of a mole of glucose. The molecular weight of glucose is

$$\left(-2816\,\frac{\text{kJ}}{\text{mol}}\right)\left(\frac{1\ \text{mol}}{180\ \text{g}}\right) = -15.6\,\frac{\text{kJ}}{\text{g}}$$

180 amu. Therefore, the heat produced per gram is

For sucrose:

$$C_{12}H_{22}O_{11}(s) + 12O_2(g) \longrightarrow 12CO_2(g) + 11H_2O(l)$$

$$\begin{aligned}\Delta H^\circ_{rxn} &= [12\,\Delta H_f^\circ(CO_2) + 11\,\Delta H_f^\circ(H_2O)] \\ &\quad - [\Delta H_f^\circ(C_{12}H_{22}O_{11}) + 12\,\Delta H_f^\circ(O_2)] \\ &= 12(-393.5\ \text{kJ}) + 11(-285.9\ \text{kJ}) \\ &\quad - (-2221\ \text{kJ}) \\ &= (-4722 - 3145 + 2221)\ \text{kJ} \\ &= -5646\ \text{kJ/mol}\ C_{12}H_{22}O_{11}\end{aligned}$$

The molecular weight of sucrose is 342 amu. Therefore, the heat produced per gram of sucrose is

$$\left(-5646\,\frac{\text{kJ}}{\text{mol}}\right)\left(\frac{1\ \text{mol}}{342\ \text{g}}\right) = -16.5\,\frac{\text{kJ}}{\text{g}}$$

Both sucrose and glucose are carbohydrates. As a rule of thumb the energy obtained from the combustion of a gram of carbohydrate is 17 kJ (4 kcal or 4 Cal).

SAMPLE EXERCISE 4.8

The standard enthalpy change for the reaction

$$CaCO_3(s) \longrightarrow CaO(s) + CO_2(g)$$

is 178.1 kJ. From the values for the standard heats of formation given in Table 4.1, calculate the standard heat of formation of $CaCO_3(s)$.

Solution: We have that the standard enthalpy change in the reaction is

$$\Delta H^\circ_{rxn} = [\Delta H_f^\circ(CaO) + \Delta H_f^\circ(CO_2)] - \Delta H_f^\circ(CaCO_3)$$

Inserting the known values, we have

$$178.1\ \text{kJ} = -635.5\ \text{kJ} - 393.5\ \text{kJ} - \Delta H_f^\circ(CaCO_3)$$

$$\Delta H_f^\circ(CaCO_3) = -1207.1\ \text{kJ/mol}$$

4.6 MEASUREMENTS OF ENERGY CHANGES; CALORIMETRY

Nearly all chemical reactions result in either the evolution or absorption of heat. Measurements of the amount of heat evolved or absorbed provide the data needed to calculate important enthalpy values such as the standard heats of formation listed in Table 4.1. Let's consider how the enthalpy changes that accompany chemical reactions can be measured.

In general, we measure quantities of heat by measuring the amounts of heat that flow from one object to another. The actual temperature change experienced by a body when it absorbs a certain amount of heat is determined by its **heat capacity**—the energy required to raise the temperature of the body by 1 °C. An object of large heat capacity absorbs more heat when it undergoes a certain temperature change than does an object of smaller heat capacity. For example, it requires considerably more heat to raise the temperature of the water in an outdoor swimming pool from 15 °C to 25 °C than it does to produce the same temperature change in a fish tank. The heat capacity of an object can be expressed as the ratio $q/\Delta T$, where q is the total heat flow into or out of an object and ΔT is the temperature change produced by the heat flow. The **molar heat capacity**, C, is defined as the amount of heat required to produce a change of 1 °C in exactly 1 mol of a substance.*

In general, the molar heat capacity, C, moles of substance, n, and temperature change, ΔT are related to q, the heat added or evolved, through the expression

$$q = n \times C \times \Delta T \qquad [4.19]$$

The term **specific heat** refers to the heat required to produce a temperature change of 1 °C in 1 *gram* of a substance. The specific heats of several substances are listed in Table 4.2. Notice that water has the highest specific heat of all the substances listed. We will learn about the reasons for this when we discuss water in Section 10.4.

*In some cases it is important to specify whether the process occurs under constant-pressure or constant-volume conditions. Then the heat capacity is labeled C_P or C_V, to denote constant-pressure or constant-volume conditions, respectively. For liquids and solids, the two values are essentially the same.

TABLE 4.2 Specific heats of some selected substances

Compound	Temperature (°C)	Specific heat (J/°C-g)
$H_2O(l)$	15	4.184
$H_2O(s)$	−11	2.03
Al(s)	20	0.89
$C(s)$	20	0.71
$Fe(s)$	20	0.45
$Hg(l)$	20	0.14
$CaCO_3(s)$	0	0.85
$MgO(s)$	0	0.87
$HgS(s)$	0	0.21

SAMPLE EXERCISE 4.9

The molar heat capacity of iron(III) oxide, Fe_2O_3, is 120 J/K-mol. (a) What is the specific heat of this substance? (b) What quantity of heat is required to increase the temperature of a 2.0-kg brick of Fe_2O_3 from 120°C to 380°C?

Solution: (a) The specific heat of Fe_2O_3 is the heat required to raise 1 g of the substance by 1°C. Thus

$$\left(\frac{120\ \text{J}}{\text{K-mol Fe}_2\text{O}_3}\right)\left(\frac{1\ \text{mol Fe}_2\text{O}_3}{160\ \text{g Fe}_2\text{O}_3}\right)$$

$$= 0.75\ \text{J/K-g Fe}_2\text{O}_3$$

(b) To obtain the total heat capacity of the brick (let's call it C_T), we can simply multiply the mass of the brick by the specific heat of Fe_2O_3:

$$C_T = 2.0 \times 10^3\ \text{g Fe}_2\text{O}_3\left(\frac{0.75\ \text{J}}{\text{K-g Fe}_2\text{O}_3}\right) = 1.5\ \text{kJ/K}$$

The total heat flow into the brick, q, is the heat capacity of the brick, C_T, times the temperature change: $q = C_T \times \Delta T$. In this case, $\Delta T =$ 380°C − 120°C = 260 K.

$$q = (1.5\ \text{kJ/K})260\ \text{K} = 390\ \text{kJ}$$

Alternatively, we could use Equation 4.19 directly, converting kilograms of Fe_2O_3 to moles in the calculation:

$$q = (2.0 \times 10^3\ \text{g Fe}_2\text{O}_3)\left(\frac{1\ \text{mol Fe}_2\text{O}_3}{160\ \text{g Fe}_2\text{O}_3}\right)$$

$$\times \left(\frac{120\ \text{J}}{\text{K-mol}}\right)(260\ \text{K}) = 390\ \text{kJ}$$

The fact that liquid water has an exceptionally high specific heat has many important consequences. As an example, the heat collected in solar panels in solar heating systems may be stored in the form of hot water or heated rocks. Note that the specific heats of typical minerals such as MgO or $CaCO_3$ are much lower than that for water. The specific heats for these substances are typical of inorganic solids. Thus a much larger mass of rocks is required to provide the same thermal storage capacity as a given amount of water.

The measurement of heat effects is known as **calorimetry.** The techniques and equipment employed in calorimetry depend on the nature of the process being studied. One of the most important reaction types studied by means of calorimetry is combustion (Section 3.3). A compound, usually an organic compound, is allowed to react completely with excess oxygen. Equation 4.11 is an example of such a reaction. Combustion reactions are most conveniently studied by means of a **bomb calorimeter,** a device shown schematically in Figure 4.12. The substance to be studied is placed in a small cup within a sealed vessel called a bomb. The bomb, which is designed to withstand high pressures, has an inlet valve for adding oxygen and also has electrical contacts to initiate the combustion reaction. After the sample has been placed in the bomb, the bomb is sealed and pressurized with oxygen. It is then placed in the calorimeter, which is essentially an insulated container, and covered with an accurately measured quantity of water. When all of the components within the calorimeter have come to the same temperature, the combustion reaction is initiated by passing an electrical current through a fine wire that is in contact with the sample. The sample ignites when the wire gets hot.

Heat is evolved when combustion occurs. This heat is absorbed by the calorimeter contents, causing a rise in the temperature of the water. The temperature of the water is very carefully measured before reaction, and

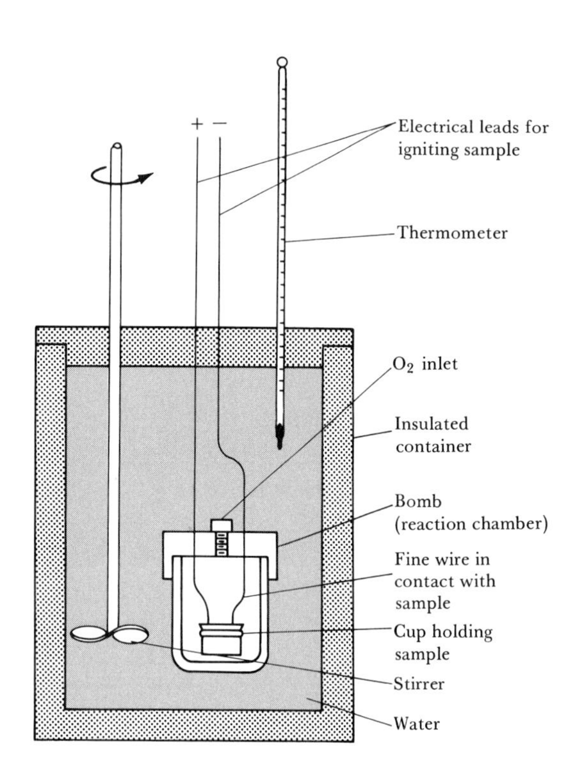

FIGURE 4.12 Bomb calorimeter, in which reactions occur at constant volume.

then again shortly after reaction, when the contents of the calorimeter have again arrived at a common temperature. In keeping with the first law of thermodynamics, the heat evolved in the combustion of the sample is absorbed by its surroundings, that is, the calorimeter contents.

To calculate the heat of combustion from the measured temperature increase in the calorimeter, it is necessary to know the heat capacity of the calorimeter. This is normally ascertained by combusting a sample that gives off a known quantity of heat. For example, it is known that combustion of exactly 1 g of benzoic acid, $C_7H_6O_2$, in a bomb calorimeter, produces 26.38 kJ of heat. Suppose that 1 g of benzoic acid is combusted in our calorimeter, and it causes a temperature increase of 5.022°C. The heat capacity of the calorimeter is then given by 26.38 kJ/5.022°C = 5.252 kJ/°C. Once we know the value of the heat capacity of the calorimeter, we can measure temperature changes produced by other reactions, and from these we can calculate the heat, q, evolved in the reaction:

$$q_{\text{evolved}} = -C_{\text{calorimeter}} \times \Delta T \qquad [4.20]$$

SAMPLE EXERCISE 4.10

The temperature of a bomb calorimeter is raised 1.17°C when 1.00 g of hydrazine, N_2H_4, is burned in it. If the calorimeter has a heat capacity of 16.53 kJ/°C, what is the quantity of heat evolved? What is the heat evolved upon combustion of a mole of N_2H_4?

Solution: The quantity of heat evolved is just the product of the temperature change times the heat capacity (Equation 4.20):

$$-\left(\frac{16.53\text{ kJ}}{1°\text{C}}\right)1.17°\text{C} = -19.3\text{ kJ heat evolved}$$

Since this is the amount of heat that results from combustion of 1.00 g of hydrazine, the amount released by combustion of 1 mol of N_2H_4 is

$$\left(\frac{-19.3\text{ kJ}}{1\text{ g } N_2H_4}\right)\left(\frac{32.0\text{ g } N_2H_4}{1\text{ mol } N_2H_4}\right) = -619\text{ kJ/mol } N_2H_4$$

In general, the change in internal energy, ΔE, in any process is given by Equation 4.3, $\Delta E = q - w$. However, under the constant-volume conditions of the bomb calorimeter, there is no pressure-volume work or other work done, so $w = 0$. Thus $\Delta E = q_V$, where q_V is the heat absorbed by the system (the bomb contents) in the process occurring at constant volume. The reactions studied in a bomb calorimeter are normally exothermic processes, so q_V is negative and so is ΔE.

The values of ΔE obtained in bomb calorimeter experiments can readily be converted to ΔH values, using Equation 4.7, $\Delta H = \Delta E + P\Delta V$, by calculating the value of $P\Delta V$ for reaction at 1 atm pressure. We need not concern ourselves here with how that is done.* The important point is that the heat evolved upon combustion of a sample at a constant pressure of 1 atm can be calculated from the bomb calorimeter data. These values are termed **heats of combustion.** The heat of combustion is normally given for the reaction that leads to $CO_2(g)$ and $H_2O(l)$ as products.

For many reactions, such as those occurring in solution, it is a simple matter to control pressure so that ΔH can be measured directly. One constant-pressure calorimeter is shown in Figure 4.13. Such simple, "coffee cup" calorimeters are often used in freshman chemistry labs to illustrate the principles of calorimetry. Because the calorimeter is not sealed, the reaction occurs under the essentially constant pressure of the atmosphere. The heat change of the reaction is determined from the temperature increase of a known quantity of solution in the calorimeter, as shown in Sample Exercise 4.11.

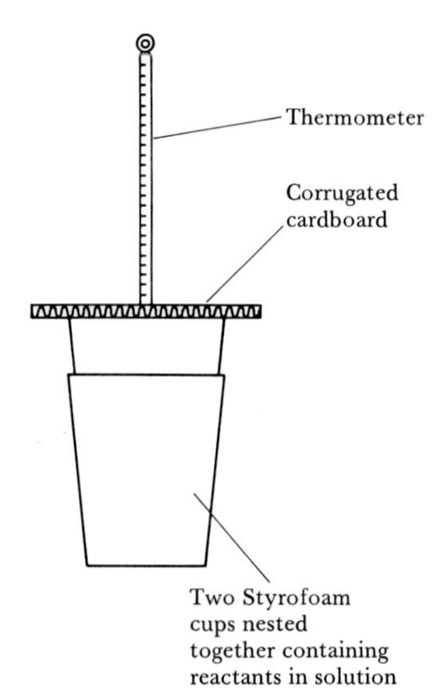

FIGURE 4.13 "Coffee-cup" calorimeter, in which reactions occur at constant pressure.

SAMPLE EXERCISE 4.11

When 50 g of 1.0 *M* HCl (containing 0.050 mol of HCl) and 50 g of 1.0 *M* NaOH (containing 0.050 mol of NaOH) are mixed in a "coffee cup" calorimeter, the temperature of the resultant solution increases from 21.0°C to 27.5°C. Assuming that it takes 4.18 J to increase the temperature of 1.00 g of the solution by 1.00°C, calculate the heat of the reaction

$$HCl(aq) + NaOH(aq) \longrightarrow H_2O(l) + NaCl(aq)$$

Solution: We can ignore the heat capacity of the plastic material from which the cup is made. Thus the total heat capacity of the calorimeter is just the heat capacity of the fluid contents:

$$100\text{ g}\left(\frac{4.18\text{ J}}{°\text{C-g}}\right) = 418\text{ J/°C}$$

The temperature rise of 6.5°C then represents 6.5°C × 418 J/°C = 2.7 kJ. To put this on a molar basis, we need to know the heat produced per mole of NaOH and HCl:

$$\text{Heat evolved per mole} = -2.7\text{ kJ/0.050 mol} = -54\text{ kJ/mol}$$

*To employ Equation 4.7 you need to know about the pressure-volume properties of gases, a subject discussed in Chapter 9. It works out that the $P\Delta V$ term is equal to $RT\Delta n$, where Δn is the total number of moles of gas in the products minus the total number of moles of gas in the reactants; R is the molar gas constant (8.314 J/K-mol), and T is the temperature in degrees kelvin.

4.7 FUEL VALUES OF FUELS AND FOODS

Most common chemical reactions used to produce heat are combustion reactions. The energy released when a fuel or food is combusted is known as its **fuel value.** Because all heats of combustion are exothermic, it is common to report fuel values without their associated negative sign. Furthermore, because fuels and foods are usually mixtures, fuel values are reported on a gram rather than a mole basis. For example, the fuel value of octane, C_8H_{18}, a component of gasoline, is the heat produced by combustion of 1 g of this substance as shown in Equation 4.21:

$$2C_8H_{18}(l) + 25O_2(g) \longrightarrow 16CO_2(g) + 18H_2O(g) \qquad [4.21]$$

Note that in this reaction the product water is considered to be in the gaseous form. This is appropriate because under the conditions that octane would be used as a fuel, water would be vaporized. The enthalpy change for this reaction is $\Delta H = -10{,}920$ kJ. Because each mole of C_8H_{18} weighs 114 g, the fuel value of octane is 47.9 kJ/g:

$$\left(\frac{10{,}920 \text{ kJ}}{2 \text{ mol } C_8H_{18}}\right)\left(\frac{1 \text{ mol } C_8H_{18}}{114 \text{ g } C_8H_{18}}\right) = 47.9 \frac{\text{kJ}}{\text{g } C_8H_{18}}$$

In accordance with the first law of thermodynamics, the fuel value of a substance is the same no matter how or where it reacts as long as the products of the reaction are the same. Thus a bomb calorimeter can often be used to measure the fuel values of foods. Unquestionably, this procedure is much easier than having to measure the heat given off in an engine or in our bodies.

Foods

Most of the energy our bodies need comes from carbohydrates and fats. Carbohydrates are decomposed in the stomach into glucose, $C_6H_{12}O_6$. Glucose is soluble in blood and is known as blood sugar. It is transported by the blood to cells, where it reacts with O_2 in a series of steps, eventually producing $CO_2(g)$, $H_2O(l)$, and energy:

$$C_6H_{12}O_6(s) + 6O_2(g) \longrightarrow 6CO_2(g) + 6H_2O(l) \qquad \Delta H^\circ = -2816 \text{ kJ}$$

The breakdown of carbohydrates is rapid, so their energy is quickly supplied to the body. However, the body stores a very small amount of carbohydrates. The average fuel value of carbohydrates, assuming $H_2O(l)$ as product, is 17 kJ/g (4 kcal/g).

Like carbohydrates, fats produce CO_2 and H_2O in both their metabolism and their combustion in a bomb calorimeter. The reaction of stearin, $C_{57}H_{110}O_6$, a typical fat, is as follows:

$$2C_{57}H_{110}O_6(s) + 163O_2(g) \longrightarrow 114CO_2(g) + 110H_2O(l) \qquad \Delta H^\circ = -75{,}520 \text{ kJ}$$

The chemical energy from foods that is not used either to maintain body temperature or for muscular activity or for the organization of the atoms in food into body parts is stored in the form of fats. Fats are well suited to

TABLE 4.3 Fuel values and compositions of some common foods

	Approximate composition (%)			Fuel value	
	Protein	Fat	Carbohydrate	kJ/g	kcal/g
Apples (raw)	0.4	0.5	13	2.5	0.59
Beer[a]	0.3	0	1.2	1.8	0.42
Bread (white, enriched)	9	3	52	12	2.8
Cheese (cheddar)	28	37	4	20	4.7
Eggs	13	10	0.7	6	1.4
Fudge	2	11	81	18	4.4
Green beans (frozen)	1.9	—	7.0	1.5	0.38
Hamburger	22	30	—	15	3.6
Milk	3.3	4.0	5.0	3.0	0.74
Peanuts	26	39	22	23	5.5

[a]Beers typically contain 3.5 percent ethanol, which has fuel value.

serve as the body's energy reserve for at least two reasons: (1) they are insoluble in water, which permits their storage in the body; and (2) they produce more energy per gram than either proteins or carbohydrates, which makes them efficient energy sources on a weight basis. The average fuel value of fats, assuming $H_2O(l)$ as product, is 38 kJ/g (9 kcal/g).

In the case of proteins, metabolism in the body produces less energy than combustion in a calorimeter because the products are different. Proteins contain nitrogen, which is released in the bomb calorimeter as N_2. In the body this nitrogen ends up mainly as urea, $(NH_2)_2CO$. Proteins are used by the body mainly as building materials for organ walls, skin, hair, muscle, and so forth. On the average, the metabolism of proteins produces 17 kJ/g (4 kcal/g).

The food values for a variety of common foods are shown in Table 4.3. The amount of energy the body requires varies considerably depending on such factors as body weight, age, and muscular activity. When a person is doing average work, his or her daily energy requirement is about 10,000 to 13,000 kJ (2500 to 3000 kcal). This is about the same amount of energy as that consumed by a 100-W light bulb operating for a 24-h period.

SAMPLE EXERCISE 4.12

It is estimated that for a person of average weight, running or jogging requires consumption of about 100 Cal/mile. What weight of hamburger provides the fuel value requirements for running 3 miles?

Solution: Recall that the dietary Calorie is equivalent to 1 kcal. The running requires 300 Calories, or 300 kcal. Using data from Table 4.3:

$$\text{Hamburger required} = 300\text{ kcal} \times \left(\frac{1\text{ g hamburger}}{3.6\text{ kcal}}\right)$$

$$= 83\text{ g hamburger}$$

Thus the calories consumed in running 3 miles are replaced by less hamburger than is present in a "quarter-pounder."

TABLE 4.4 Fuel values and compositions of some common fuels

	Approximate elemental composition (%)			
	C	H	O	Fuel value (kJ/g)
Wood (pine)	50	6	44	18
Anthracite coal (Pennsylvania)	82	1	2	31
Bituminous coal (Pennsylvania)	77	5	7	32
Charcoal	100	0	0	34
Crude oil (Texas)	85	12	0	45
Gasoline	85	15	0	48
Natural gas	70	23	0	49
Hydrogen	0	100	0	142

Fuels

Several common fuels are compared in Table 4.4. Notice that an increase in the percentage of carbon or hydrogen in the fuel increases the fuel value. For example, the fuel value of bituminous coal is greater than that of wood because of its greater carbon content.

Coal, oil, and natural gas, which are presently our major sources of energy, are known as **fossil fuels.** All are thought to have formed over millions of years from the decomposition of plants and animals. All are presently being depleted far more rapidly than they are being formed. **Natural gas** consists of gaseous hydrocarbons, compounds of hydrogen and carbon. It varies in composition but contains primarily methane (CH_4), with small amounts of ethane (C_2H_6), propane (C_3H_8), and butane (C_4H_{10}). **Oil,** which is also known as petroleum, is a liquid composed of hundreds of compounds. Most of these compounds are hydrocarbons, with the remainder being mainly organic compounds containing sulfur, nitrogen, or oxygen. **Coal,** which is solid, contains hydrocarbons of high molecular weight as well as compounds containing sulfur, oxygen, and nitrogen. The sulfur in oil and coal is important from the standpoint of air pollution, as we shall discuss in Chapter 12.

Hydrogen (H_2) is a very attractive fuel because of its high fuel value and because its combustion produces water, a "clean" chemical that produces no negative environmental effects. However, hydrogen cannot be used as a primary energy source because there is so little H_2 in nature. Most hydrogen is produced by the decomposition of water or hydrocarbons. This decomposition requires energy; in fact, because of heat losses, more energy must be used to generate hydrogen than can be reclaimed when the hydrogen is subsequently used as a fuel. However, should large, cheap sources of energy become available because of technical advances, perhaps in areas such as nuclear or solar power generation, a portion of this energy could be used to generate hydrogen. The hydrogen could then serve as a convenient energy carrier. It would be cheaper to transport hydrogen using existing gas pipelines than to transport electrical energy; hydrogen is both portable and storable. Because present industrial technology is based on combustible fuels, hydrogen could replace oil and natural gas as these fuels become scarcer and more expensive.

4.8 ENERGY USAGE: TRENDS AND PROSPECTS

The average energy consumption per person in the United States each day amounts to about 8.8×10^5 kJ. This amount of energy is about 100 times greater than the individual food-energy requirement. Until re-

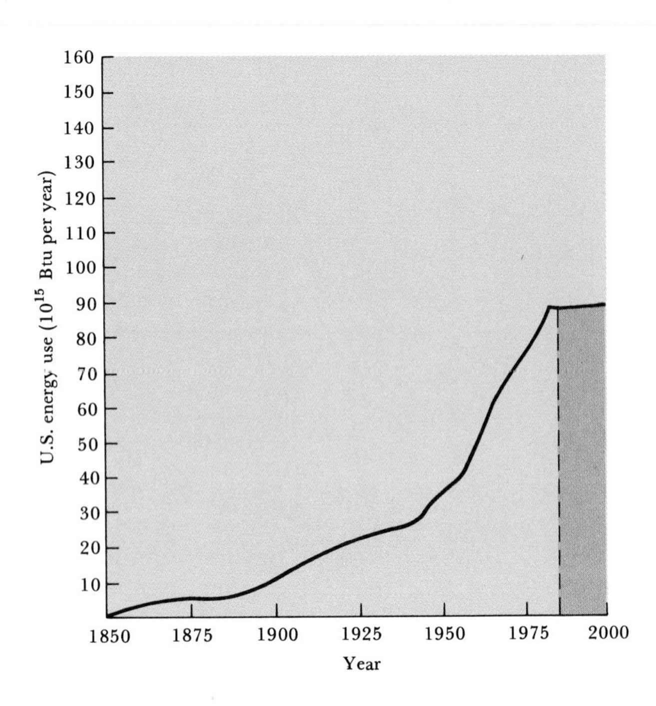

FIGURE 4.14 Energy consumption in the United States since 1850, and projected to the year 2000 (1 Btu = 1.05 kJ).

cently, total energy consumption in the United States has been increasing each year, as illustrated in Figure 4.14. We are a very energy-intensive society; currently, the United States accounts for nearly 30 percent of the world's annual production of energy. However, the increasing cost of energy, particularly of oil, has led to significant conservation measures in recent years. Per capita consumption of energy in 1982 was 16 percent lower than in 1973, the year in which it reached an all-time high.

During the past century the fossil fuels—first coal, later oil and gas—have become the major energy sources in our society. Together they account for more than 80 percent of national energy consumption. Eventually, such fossil fuel energy sources must decline in importance as we exhaust the readily available supplies. Current projections are that 80 percent of all the known and estimated oil reserves will have been consumed before the children born during the 1980s will have lived out their lives. It is clear that alternative energy sources must be developed to replace those fossil fuels if the 8 billion people expected as the world's population by 2025 are to have a decent standard of living.

Of the various sources of energy that could serve as alternatives to fossil fuels, only nuclear and solar energies are potentially capable of furnishing sufficient quantities of energy to satisfy the world's needs. Nuclear power supplied about 4 percent of the gross energy consumed in the United States in 1982; it will furnish about 8 percent in 1990. Aspects of the use of nuclear reactors for energy production are discussed in Chapter 19. In the paragraphs that follow we'll consider some aspects of coal and solar radiation as energy sources.

Coal is the most abundant fossil fuel; it constitutes 80 percent of the fossil fuel reserves of the United States and 90 percent of those of the world. However, the use of coal presents a number of problems. Coal is a complex mixture of substances, and it contains components that produce air pollution. Because it is a solid, recovery from its underground deposits is expensive and often dangerous. Coal deposits are not always close to the locations where energy use is high, so there are often substantial shipping costs to add to the total cost of coal use.

Some experts feel that coal could be used most effectively if it were converted to a gaseous form, often called **syngas**, for synthetic gas. In such a conversion the sulfur is removed, thereby decreasing air pollution when the syngas is burned. Syngas could be easily transported in pipelines and could supplement our diminishing supplies of natural gas. Gasification of coal requires addition of hydrogen to coal. Typically, the coal is pulverized and treated with superheated steam. The product contains a mixture of CO, H_2, and CH_4, all of which can be used as fuels. However, conditions are maintained to maximize production of CH_4. A simplified schematic showing some of the reactions that occur is given in Figure 4.15.

The syngas produced in coal gasification can also be used to make chemicals. Methyl alcohol, CH_3OH, and acetic anhydride, $(CH_3CO)_2O$, are examples of chemicals produced in large quantities from synthesis gas (Figure 4.16).

Solar energy is the world's largest energy source. The solar energy that falls on only 0.1 percent of the land area of the United States is equivalent to all the energy that this nation currently uses. The problem with the use of solar energy is that it is dilute (that is, distributed over an area) and fluctuates with time and weather conditions. The effective use of solar energy will depend on the development of some means of storing the energy collected for use at a later time. Any practical means for doing this will almost certainly involve use of a chemical process that can be made to proceed in the direction in which it is endothermic, by applica-

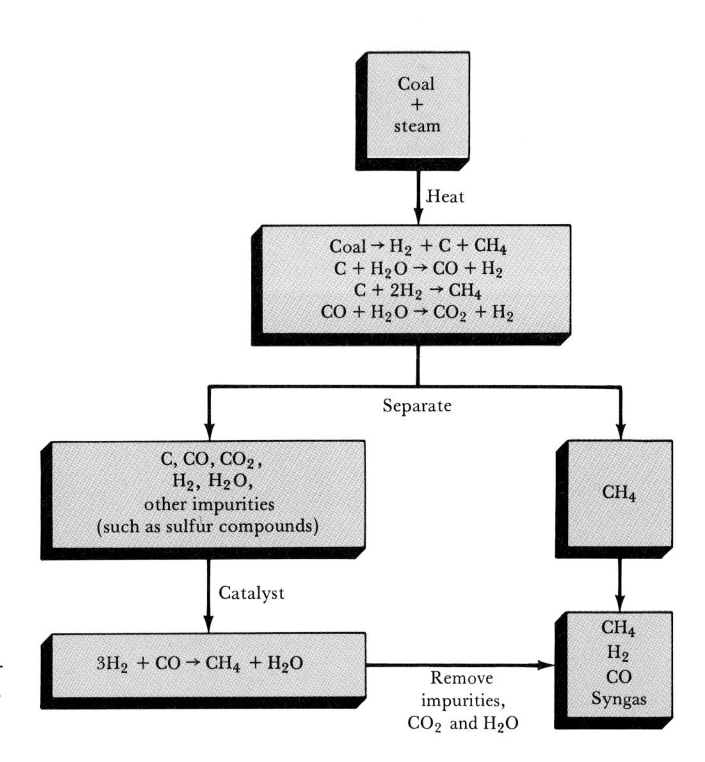

FIGURE 4.15 Basic processes involved in the gasification of coal to form synthetic gas (syngas). A catalyst is a substance that increases the speed of a reaction without itself being consumed in the reaction.

FIGURE 4.16 View of a large, new plant for conversion of coal to acetic anhydride, $(CH_3CO)_2O$. This plant is designed to process 900 tons of high-sulfur-content Appalachian coal daily, for both feedstock and to furnish plant power. Acetic anhydride is used in the manufacture of photographic film, synthetic fibers, cigarette filters, and coatings. (*Eastman Kodak*)

tion of heat. The reaction can then be made to proceed in the reverse direction, with the evolution of heat, at a later time. One such reaction is the following:

$$CH_4(g) + H_2O(g) + \text{heat} \rightleftharpoons CO(g) + 3H_2(g) \qquad [4.22]$$

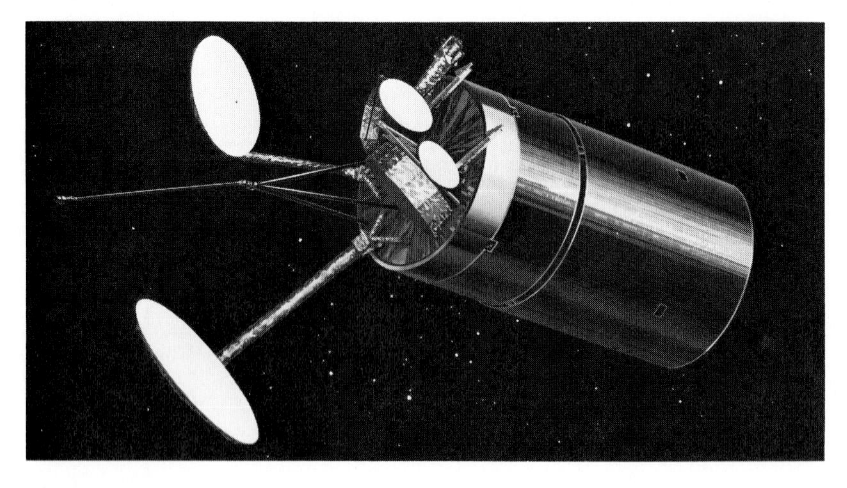

FIGURE 4.17 The Intelsat VI communications satellite, launched in July 1984, is 6.6 m (about 22 ft) in height and 2.2 m (about 7 ft) in diameter. The solar array around the body of the satellite furnishes 990 watts of direct current power at the beginning of its life. The white objects are the antenna reflectors used in the transmission and reception of signals. The satellite is used to distribute cable television programming to the contiguous United States, Alaska, and Hawaii. (*Courtesy Hughes Aircraft Corporation*)

This reaction can be made to proceed in the forward direction at high temperatures, which can be obtained directly in a solar furnace. The CO and H_2 formed in the reaction could then be reacted later, with evolution of heat that can be used to produce useful work.

Solar energy can be converted directly into electricity by use of photovoltaic devices, sometimes called solar cells. The efficiencies of solar energy conversion by use of such devices have increased dramatically during the past few years as a result of intensive research efforts. Photovoltaics are vital to the generation of power for satellites, as illustrated in Figure 4.17. However, for large-scale generation of useful energy at the earth's surface, they are not yet practical because of high unit cost. Even if the costs are reduced, some means must be found to store the energy produced by the solar cells, because the sun shines only intermittently and during only part of the day at any place. Once again, the solution to this problem will almost certainly be to use the energy to run a chemical reaction in the direction in which it is endothermic.

FOR REVIEW

Summary

This chapter has been about energy and the first law of thermodynamics. Thermodynamics is the study of heat and energy, and the rules that govern the conversions between heat and various forms of energy. To study the thermodynamic properties of matter, we define some specific amount of matter as our system and study the interactions between this system and its surroundings. The system may exchange heat with its surroundings, or do work on the surroundings, or have work done on it by the surroundings. Work comes in many forms; mechanical work (w) is measured by the product of a force (f) acting through a distance (d): $w = f \times d$. The thermal energy of a system is the energy associated with the motions of the atoms or molecules of which it is constituted. Thermal energy increases with increasing temperature. The term heat has a special significance in thermodynamics; it refers to the flow of energy between two bodies at different temperatures when they are placed in thermal contact.

The joule (J) is the SI unit of energy. $1\ J = 1\ kg\text{-}m^2/s^2$. The calorie (cal) is also often used; $1\ cal = 4.184\ J$.

The internal energy, E, of a system is all the energy possessed by the atomic, molecular, or ionic units of which the system is composed. We do not in general know the magnitude of this internal energy; we are concerned primarily with *changes* in the internal energy, ΔE, that accompany changes in the system. The first law of thermodynamics tells us that the change in internal energy is given by the heat added to the system during the change, q, less the work done by the system on the surroundings: $\Delta E = q - w$. The internal energy is a state function; that is, the value of the internal energy for the system depends only on the condition of the system, as described by variables such as temperature, pressure, and so forth. It does not depend on the details of how the system came to be in that state.

The enthalpy, H, of a system, is a thermodynamic state function defined as $E + PV$, where P is the pressure acting on the system, and V is its volume. Enthalpy is a most useful function for dealing with changes that occur at constant pressure; then $\Delta H = \Delta E + P\Delta V$. For endothermic processes, those that involve absorption of heat, ΔH is positive. For exothermic processes, those that involve transfer of heat from the system to the surroundings, ΔH is negative.

The enthalpy change for a reaction that proceeds in a given direction is equal in magnitude but opposite in sign to that for the reaction when it proceeds in the reverse direction. Because enthalpy is a state function, the overall enthalpy change for a given chemical process is the same whether the process is carried out in one step or in a series of steps, as stated in Hess's law. This means that we can calculate the enthalpy changes for many reactions that cannot be carried out directly, by making use of known enthalpy values for reactants and products. By convention, the enthalpies of formation of elements in their standard states (that is, in their stable form under 1 atm pressure) are defined as zero. The standard enthalpy of formation of a substance is defined as the enthalpy of formation of that substance from the elements, with all reactants and products in their standard states. The standard state refers to conditions of 1 atm pressure and each reactant and product in its stable form (gas, liquid, or solid) at the temperature in question, usually 298 K. Enthalpies of formation can be used to calculate changes in ΔH for other reactions:

$$\Delta H^\circ_{rxn} = \Sigma\, n\, \Delta H^\circ_f(\text{products}) - \Sigma\, m\, \Delta H^\circ_f(\text{reactants})$$

The term calorimetry refers to experiments in which the heats evolved or absorbed in chemical and physical changes are measured experimentally. A calorimeter is a device for making such measurements, by provision for insulating the system and a fixed portion of the surroundings, so that accurate temperature changes can be measured. A bomb calorimeter is used to measure the heat evolved when a substance is combusted in oxygen at constant volume. The data so obtained can be employed to calculate the heat of combustion, the enthalpy change that occurs on oxidation of a substance at 1 atm pressure. In such measurements it is necessary to employ values of heat capacity; the molar heat capacity refers to the heat required to raise the temperature of 1 mol of a substance by 1°C; the specific heat is the heat required to raise the temperature of 1 g of substance by 1°C.

The fuel values of foods and fuels are defined as the heat energy released when a gram of food or fuel is combusted. In this chapter, we considered briefly the fuel values of some common foods and fuels, including various forms of fossil fuels. Finally, we have considered current and projected rates of fuel consumption, and the prospects for various sources of energy in the future.

Learning goals

Having read and studied this chapter, you should be able to:

1. Provide examples of different forms of energy, and describe the relationship between heat and other forms of energy.
2. Know the important units in which energy is expressed, and be able to convert from one to another.
3. Provide a definition of the first law of thermodynamics, and describe how the first law relates to our everyday experience.
4. Describe how the change in internal energy of a system is related to the exchanges of heat and work between the system and its surroundings.
5. Describe what is meant by the term *state function.*
6. Provide a definition of enthalpy, H. You should also be able to relate the enthalpy change in a process occurring at constant pressure to the change in internal energy and volume during the process.
7. Describe what is meant by the term *standard state.*
8. Be able to sketch an energy diagram such as that shown in Figure 4.10, given the enthalpy changes in the processes involved.
9. Be able to state Hess's law, and show how it can be applied to calculate the enthalpy change in a process, given the enthalpy changes in other processes that could be combined to yield the reaction of interest.
10. Describe the basis on which standard heats of formation are defined.
11. Be able to calculate the enthalpy change in a reaction occurring at constant pressure, given the standard enthalpies of formation of each reactant and product.
12. Describe the process by which heats of combustion are measured in a bomb calorimeter.
13. Provide definitions of the terms *molar heat capacity* and *specific heat.*
14. Calculate the heat capacity of a calorimeter, given the data regarding temperature changes and quantities of heat involved. You should also be able to calculate the heat evolved or absorbed in a process from a knowledge of the heat capacity of the system and the measured temperature change in a process.
15. Define the term *heat of combustion.*
16. Indicate what is meant by the term *fuel value.*
17. Describe the major sources of energy on which humankind must depend, and the likely availability of these for the foreseeable future.
18. Describe the process by which coal is converted to syngas; indicate some of the uses that might be made of syngas.

Key terms

Among the more important terms and expressions used for the first time in this chapter are the following:

A bomb calorimeter (Section 4.6) is a device for measurements of heat evolved in combustion of a substance under constant-volume conditions.

Calorimetry (Section 4.6) refers to the experimental measurements of heat effects in chemical and physical processes.

An endothermic process (Section 4.3) is one in which a system absorbs heat from its surroundings.

Energy (Section 4.2) is the ability to do work or to transfer heat.

The enthalpy, H (Section 4.3) is defined by the relationship $H = E + PV$. The enthalpy change, ΔH, for a reaction that occurs at constant pressure is the heat evolved or absorbed in the reaction, q_P.

An exothermic process (Section 4.3) is one in which a system loses heat to its surroundings.

The first law of thermodynamics (Section 4.1) is a statement of our experience that energy is conserved in any process. We can express the first law in many ways. One of the more useful expressions is that the change in internal energy, ΔE, of a system in any process is equal to the heat, q, added *to* the system, less the work, w, done *by* the system on its surroundings: $\Delta E = q - w$.

Heat (Section 4.2) refers to the flow of energy from a body at higher temperature to one at lower temperature when they are placed in thermal contact.

The heat capacity (Section 4.6) is defined as the quantity of heat required to cause a 1°C change in temperature. The molar heat capacity is the heat required to raise the temperature of 1 mol of a substance by 1°C. The specific heat is the heat required to raise the temperature of 1 g of a substance by 1°C.

The heat of combustion (Section 4.6) is defined as the heat evolved in the reaction of a given quantity of a pure substance with oxygen at 1 atm pressure, to yield fully oxidized products. For example, the heat of combustion of methyl alcohol, CH_3OH, is the heat evolved in the reaction:

$$CH_3OH(l) + \tfrac{3}{2}O_2(g) \longrightarrow CO_2(g) + 2H_2O(l)$$

under 1 atm pressure and at some defined temperature, usually 25°C.

The heat of formation (Section 4.5) of a substance is defined as the heat evolved when a substance is formed from the elements in their standard states.

Hess's law (Section 4.4) states that the heat evolved in a given process can be expressed as the sum of the heats of several processes that, when added, yield the process of interest.

Internal energy, E (Section 4.2), is the total energy possessed by a system. When a system undergoes a change, the change in internal energy, ΔE, is defined as the heat, q, added to the system, less the work, w, done by the system on its surroundings: $\Delta E = q - w$.

The joule, J (Section 4.2), is the SI unit of energy, 1 kg-m^2/s^2. A related unit is the calorie; 4.184 J = 1 cal.

Kinetic energy (Section 4.2) is the energy that an object possesses by virtue of its motion with respect to another object.

Macroscopic properties (Section 4.1) are those that apply to matter in the bulk form. Microscopic properties, on the other hand, are those that are possessed by the individual atoms, molecules, or ions that make up the system. Thermodynamics is concerned with macroscopic properties of systems.

Mechanical work, w (Section 4.2), is defined and measured by the product of the net force, f, and distance, d, through which that force acts: $w = f \times d$.

Potential energy (Section 4.2) is the energy that an object possesses as a result of its position with respect to another object, or by virtue of its composition.

A standard state (Section 4.5) for a substance is defined as the pure material in the physical state (gas, liquid, or solid) most stable at the temperature in question, under 1 atm pressure. Standard states are most usually referred to 25°C.

The surroundings (Section 4.1) are everything that lies outside the system that we study.

Syngas (Section 4.8) is a mixture of gases, mainly H_2 and CO, that results from heating coal in the presence of steam. The mixture can be used as a fuel or for synthesis of other substances.

A system (Section 4.1) is a portion of the universe that we single out for study. We must be careful to state exactly what the system contains and what transfers of energy it may have with its surroundings.

Thermal energy (Section 4.2) refers to the motional energies of all the atoms and molecules that make up a system. Thermal energy increases with increasing temperature.

EXERCISES

Energy; thermodynamics; the first law

4.1 Suppose that one wishes to measure the heat required to cause melting of a certain quantity of ice. In setting up such an experiment, what constitutes the system, in a thermodynamic sense?

4.2 Water at the top of a waterfall possesses potential energy. Describe ways in which this potential energy can be converted into other forms of energy or work.

4.3 Indicate which of the following is independent of the path by which a change occurs: (a) the change in potential energy when a book is transferred from table to shelf; (b) the heat evolved when a cube of sugar is oxidized to $CO_2(g)$ and $H_2O(g)$; (c) the work accomplished in burning a gallon of gasoline.

4.4 Calculate the kinetic energy, in joules, of a 28-g bullet moving with a speed of 3.9×10^4 cm/s. What happens to this kinetic energy when the bullet slams into an impenetrable steel plate?

4.5 Calculate the kinetic energy, in kilojoules, of an automobile weighing 1450 kg, moving at a speed of 90 km/h. Also express this energy in calories.

4.6 Occasionally, people try to patent machines that are claimed to perform work without a corresponding consumption of energy or heat flow from another system. How do you evaluate these claims in light of the first law of thermodynamics? Explain.

4.7 Calculate the change in internal energy of a system in each of the following processes. (a) A gas expands very rapidly, so that there is no heat exchange with the surroundings; in the expansion it does 450 J of work on the surroundings. (b) 200 g of water is heated from 30°C to 40°C, a process that requires approximately 8360 J of heat. (c) A gas contracts as it is cooled; it has 300 J of work done on it and loses 146 J of heat to its surroundings.

4.8 Using a bicycle, a butane lighter, and a rubber band to create your examples, illustrate as many forms of energy as you can think of.

Enthalpy; heats of formation

4.9 When a mole of dry ice, $CO_2(s)$, is converted to $CO_2(g)$ at 1 atmosphere pressure, −78°C, the heat absorbed by the system is greater than the increase in internal energy of the CO_2. Why is this so? What has happened to the remaining energy?

4.10 A gas is confined to a cylinder as illustrated in Figure 4.1 under a pressure of 1 atmosphere. When the temperature of the gas is increased, 660 J of heat is added, and the gas expands, doing 140 J of work on the surroundings. What are the values for ΔH and ΔE in this process?

[4.11] A dry cell battery is connected to a heater that is immersed in an insulated water bath. As the dry cell discharges, the heater delivers 1300 J of energy to the water plus heater, raising the temperature. The dry cell undergoes a slight temperature increase during the time of discharge. Consider the dry cell and the water plus heater as two separate systems. What kinds of changes have taken place in the internal energy and enthalpy of each of these systems, neglecting any exchanges of heat that each may have with its surroundings?

4.12 Classify the following processes as exothermic or endothermic: (a) sublimation (that is, the conversion of solid to gas) of ice; (b) reaction of potassium with water; (c) dissolution of potassium nitrate in water, causing a temperature drop; (d) decomposition of water into H_2 and O_2.

4.13 Combustion of a mole of liquid acetone, C_3H_6O, results in liberation of 1790 kJ. What is the heat evolved upon combustion of 24 g of acetone?

4.14 For the reaction

$$2K(s) + 2H_2O(l) \longrightarrow 2KOH(aq) + H_2(g)$$
$$\Delta H = -389 \text{ kJ}.$$

Calculate the heat change that occurs when 6.00 g of potassium reacts with excess water.

4.15 Nitrogen trioxide is an unstable species that may be involved in many reactions of nitrogen oxides. The standard enthalpy of formation of $NO_3(g)$ is 70.9 kJ/mol. The standard enthalpy of formation of $N_2O_5(g)$ is 11.3 kJ/mol. Using these values and those in Appendix D, calculate the enthalpy change in each of the following reactions:

$$N_2O_5(g) \longrightarrow NO_2(g) + NO_3(g)$$
$$NO_3(g) \longrightarrow \tfrac{1}{2}N_2(g) + \tfrac{3}{2}O_2(g)$$
$$NO_2(g) \longrightarrow \tfrac{1}{2}N_2(g) + O_2(g)$$

4.16 The standard enthalpies of formation of gaseous propyne, C_3H_4, propylene, C_3H_6, and propane, C_3H_8, are +185.4, +20.4, and −103.8 kJ/mol, respectively. Calculate the heat evolved per mole on combustion of each substance to yield $CO_2(g)$ and $H_2O(g)$. Also calculate the heat evolved on combustion of 1 kg of each substance. Which is the most efficient fuel in terms of heat evolved per unit mass?

4.17 From the following heats of reaction:

$$N_2(g) + 2O_2(g) \longrightarrow 2NO_2(g) \qquad \Delta H = 67.6 \text{ kJ}$$
$$NO(g) + \tfrac{1}{2}O_2(g) \longrightarrow NO_2(g) \qquad \Delta H = -56.6 \text{ kJ}$$

calculate the heat of the reaction

$$N_2(g) + O_2(g) \longrightarrow 2NO(g)$$

4.18 Using Appendix D, calculate the enthalpy change in each of the following reactions:

(a) $3H_2(g) + N_2(g) \longrightarrow 2NH_3(g)$
(b) $3H_2(g) + P_2(g) \longrightarrow 2PH_3(g)$
(c) $Na(s) + 2H_2O(g) \longrightarrow 2NaOH(s) + H_2(g)$
(d) $2Fe(s) + \tfrac{3}{2}O_2(g) \longrightarrow Fe_2O_3(s)$
(e) $PH_3(g) + 3HCl(g) + \tfrac{1}{2}O_2(g) \longrightarrow POCl_3(l) + 3H_2(g)$
(f) $CaCO_3(s) \longrightarrow CaO(s) + CO_2(g)$
(g) $2NOCl(g) \longrightarrow 2NO(g) + Cl_2(g)$

4.19 From the following data for three prospective fuels, calculate which could provide the most energy per unit volume.

Fuel	Density (g/cm³) at 20°C	Molar heat of combustion (kJ/mol)
Nitroethane, $C_2H_5NO_2(l)$	1.052	−1348
Ethanol, $C_2H_5OH(l)$	0.789	−1371
Diethyl ether, $(C_2H_5)_2O(l)$	0.714	−2727

4.20 Draw a diagram analogous to Figure 4.11 for the following reactions:

$$2SO_2(g) + O_2(g) \longrightarrow 2SO_3(g) \qquad \Delta H = -196 \text{ kJ}$$
$$2S(s) + 3O_2(g) \longrightarrow 2SO_3(g) \qquad \Delta H = -790 \text{ kJ}$$

and the reaction

$$S(s) + O_2(g) \longrightarrow SO_2(g)$$

Label the enthalpy changes for the processes shown in your diagram.

4.21 By looking in a table of thermodynamic data such as Appendix D, how can you tell when an element is not in its standard state? Find two examples in Appendix D of thermodynamic data for elements that are not in their standard states at 25°C.

4.22 From the data in Appendix D calculate the following quantities at 25°C: (a) the enthalpy of vaporization of methyl alcohol, CH_3OH; (b) the enthalpy of vaporization of bromine; (c) the enthalpy of dissociation of $I_2(g)$ to form $I(g)$; (d) the enthalpy of conversion of solid nickel, $Ni(s)$, to $Ni(g)$; (e) the enthalpy of oxidation of $Zn(g)$ to form $ZnO(s)$.

4.23 The standard heat of formation of dihydrothiophene, $C_4H_6S(l)$, is 53.1 kJ/mol. The standard heat of formation of thiophene, $C_4H_4S(l)$, is 80.3 kJ/mol. Calculate the enthalpy change in the reaction

$$C_4H_4S(l) + H_2(g) \longrightarrow C_4H_6S(l)$$

4.24 The thermite reaction, which involves reaction of aluminum metal with a metal oxide, is highly exothermic. Calculate the heat evolved for reaction of a mole of $MnO_2(s)$ with excess $Al(s)$ according to the reaction

$$3MnO_2(s) + 4Al(s) \longrightarrow 3Mn(s) + 2Al_2O_3(s)$$

Calculate the heat evolved per gram of a stoichiometric mixture of reactants.

4.25 Calcium carbide, CaC_2, reacts with water to form acetylene, C_2H_2, and $Ca(OH)_2$. From the following heat of reaction data, and the data in Appendix D, calculate ΔH_f° for $CaC_2(s)$:

$$CaC_2(s) + 2H_2O(l) \longrightarrow Ca(OH)_2(s) + C_2H_2(g) \qquad \Delta H^\circ = -127.2 \text{ kJ}$$

Calorimetry; fuel values

4.26 (a) A beaker contains 150 g of ethanol at 18°C. What quantity of heat, in joules, is required to raise the temperature of the ethanol to 35°C, assuming that none of the heat is lost to the surroundings? The specific heat of ethanol is 2.4 J/°C-g. (b) A swimming pool contains 340 m^3 of water. What quantity of heat is required to raise the temperature of the water from 17°C to 24°C, assuming that none of the heat supplied is lost to the surroundings? The density of water is 1.00 g/mL.

4.27 A house is being designed to have passive solar energy features. Brickwork is to be incorporated into the interior of the house to act as a heat absorber. Each brick weighs approximately 1.8 kg. The specific heat of the brick is 0.85 J/°C-g. How many bricks will need to be incorporated into the interior of the house to provide the same total heat capacity as 1000 gal of water?

4.28 A small "coffee cup" calorimeter of the type shown in Figure 4.13 contains 105 g of water at 21.8°C. A 88-g block of nickel metal is heated to 100°C and then placed in the water in the calorimeter. The contents of the calorimeter come to an average temperature of 27.9°C. What is the specific heat of nickel metal? What is the heat capacity per mole of this element? (The heat capacity of the water in the calorimeter is 4.18 J/°C-g.)

4.29 A 50.0-mL sample of a 1.00 *M* solution of $CuSO_4$ is mixed in a "coffee cup" calorimeter with 50.0 mL of 2.00 *M* KOH solution. Both solutions are initially at the same temperature. The temperature of the mixture in the calorimeter rises 6.2°C when the solutions are mixed. From these data calculate ΔH for the process

$$CuSO_4(aq;\ 1\ M) + 2KOH(aq;\ 2\ M) \longrightarrow Cu(OH)_2(s) + K_2SO_4(aq;\ 0.5\ M)$$

Assume that the specific heat and density of the solution following mixing of the reagents is the same as that of pure water.

[4.30] (a) When a 0.235-g sample of benzoic acid is combusted in a bomb calorimeter, a 1.642°C rise in temperature is observed. When a 0.265-g sample of caffeine, $C_8H_{10}O_2N_4$, is burned, a 1.525°C rise in temperature is measured. Using the value 26.38 kJ/g for the heat of combustion of benzoic acid, calculate the heat of combustion per mole of caffeine at constant volume. (b) Assuming that there is an uncertainty of 0.002°C in each temperature reading, and that the masses of samples are measured to 0.001 g, what is the estimated uncertainty in the value calculated for the molar heat of combustion of caffeine?

[4.31] The "law" of Dulong and Petit—really an empirical rule announced by them in 1819—states that the molar heat capacities of all solid elements are the same, the value being 27 J/°C-mol. This means that the heat capacity per gram times the atomic weight should equal 27. The rule was very useful in the nineteenth century in obtaining some idea of the atomic weight of a newly discovered element. Use a handbook of chemical data to test the rule on the elements with atomic numbers 22, 24, 26, 27, 32, 42, 47, 50, 53, 56, 74, 78, 79, and 82. Compute the molar heat capacity in each case, compare the mean value for all these elements with 27 J/°C-mol, and calculate the average deviation from the mean of the results.

[4.32] When 4.00 g of KCl is added to a calorimeter of the type illustrated in Figure 4.18, the temperature is observed to change from 25.818°C to 24.886°C. The heater

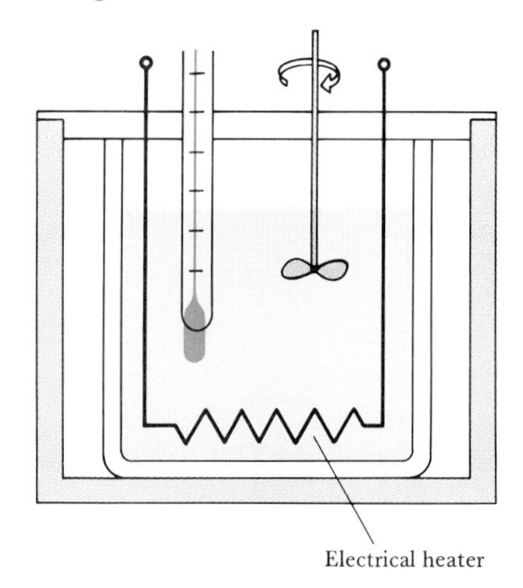

FIGURE 4.18

in the calorimeter has a resistance of 6.00 ohms. A current of 4.00 amperes is passed through the heater for a period of 20.0 s. As a result, the temperature in the calorimeter is observed to increase by 2.290°C. The heat dissipated in the electrical heater is given by the formula $q = i^2Rt$, where i is the current in amperes, R the resistance in ohms, t the time the current flows in seconds, and q is given in joules. From these data calculate the molar heat of solution of KCl in water at approximately 25°C.

4.33 The total caloric value of 100 g of fresh pineapple is 52 Cal, due mainly to carbohydrate. However, the pineapple also contains 0.4 g of protein and 0.4 g of fat. Given these data, calculate the percentage of the pineapple that is water.

4.34 Dry red beans contain 62 percent carbohydrate, 22 percent protein, 1.5 percent fat, and the remainder water. Calculate the calorie content of 100 g of beans.

4.35 The heat of combustion of ethanol, $C_2H_5OH(l)$, is -1371 kJ/mol. A 12-oz (355-mL) bottle of beer contains 3.7 percent ethanol by weight. Assuming the density of the beer to be 1.0 g/cm^3, what caloric content does the alcohol in a bottle of beer represent?

[4.36] Aspirin is produced commercially from salicylic acid, $C_7O_3H_6$. A large shipment of salicylic acid is contaminated with boric oxide, which, like salicylic acid, is a white powder. The heat of combustion of salicyclic acid at constant volume is known to be -3.00×10^3 kJ/mol. Boric oxide, because it is fully oxidized, does not burn. When a 3.556-g sample of the contaminated salicyclic acid is burned in a bomb calorimeter, the temperature increases 2.556°C. From previous measurements, the heat capacity of the calorimeter is known to be 13.62 kJ/°C. What is the amount of boric oxide in the sample, in terms of percent by weight?

4.37 In 1977, the "proved" U.S. reserves of natural gas were estimated to be 2.09×10^{11} ft^3. Assuming the fuel value of the gas to be 980 kJ/ft^3, calculate the total fuel value of this quantity of natural gas. What weight of anthracite coal has an equivalent total fuel value?

4.38 Calculate the standard enthalpy change in Equation 4.22. A large mirror 6 m in diameter can capture 25 kJ/s of solar energy at optimal operation. A 65 percent efficiency in converting this solar energy to run a reaction in the forward direction was obtained in some experiments. What mass of methane undergoes reaction in an hour under these optimal conditions?

Additional exercises

4.39 Distinguish between the following terms: (a) internal energy and enthalpy; (b) heat of combustion and heat of formation; (c) work and heat; (d) calorie and joule; (e) heat of combustion at constant pressure and heat of combustion at constant volume.

4.40 The following reactions are of importance in the underground burning of coal. Calculate the enthalpy change in each case.

(a) $C(s) + O_2(g) \longrightarrow CO_2(g)$
(b) $CO(g) + \frac{1}{2}O_2(g) \longrightarrow CO_2(g)$
(c) $CO_2(g) + C(s) \longrightarrow 2CO(g)$
(d) $H_2O(g) + C(s) \longrightarrow CO(g) + H_2(g)$

4.41 It is estimated that U.S. consumption of petroleum in 1986 will be 18.2 million barrels per day. A barrel of crude oil has an estimated energy value of 5.8 million Btu. Calculate the total energy value of the U.S. annual consumption of petroleum in 1986, in quads (1 quad = 1×10^{15} Btu).

4.42 Given the following reactions and their associated enthalpy changes:

$PH_3(g) \longrightarrow P(g) + 3H(g)$ $\quad \Delta H = 965$ kJ
$O_2(g) \longrightarrow 2O(g)$ $\quad \Delta H = 490$ kJ
$2H(g) + O(g) \longrightarrow H_2O(g)$ $\quad \Delta H = -930$ kJ
$2P(g) + 5O(g) \longrightarrow P_2O_5(s)$ $\quad \Delta H = -3382$ kJ

calculate ΔH for the combustion of 1 mol of $PH_3(g)$ to give $H_2O(g)$ and $P_2O_5(s)$ as products.

4.43 Give at least one example of each of the following kinds of energy conversions: (a) chemical energy into thermal energy; (b) mechanical energy into electrical energy; (c) solar energy into chemical energy; (d) kinetic energy into potential energy; (e) chemical energy into work.

4.44 Given the following reactions and their associated enthalpy changes:

$\frac{1}{2}H_2(g) + \frac{1}{2}Br_2(g) \longrightarrow HBr(g)$ $\quad \Delta H = -36$ kJ
$\frac{1}{2}H_2(g) \longrightarrow H(g)$ $\quad \Delta H = 218$ kJ
$\frac{1}{2}Br_2(g) \longrightarrow Br(g)$ $\quad \Delta H = 112$ kJ

calculate ΔH for the reaction

$H(g) + Br(g) \longrightarrow HBr(g)$

4.45 Calculate the enthalpy changes for the following reactions. Indicate which are endothermic and which are exothermic.

(a) $6H_2(g) + P_4(s) \longrightarrow 4PH_3(g)$
(b) $3Fe(s) + 2O_2(g) \longrightarrow Fe_3O_4(s)$
(c) $2HF(g) + Br_2(l) \longrightarrow 2HBr(g) + F_2(g)$
(d) $CaO(s) + CO_2(g) \longrightarrow CaCO_3(s)$

4.46 The heat of formation of $CH_3OH(g)$ is -201.3 kJ/mol. The heat of vaporization of $CH_3OH(l)$ at 25°C is 37.4 kJ/mol. Write the equations for these processes and calculate ΔH for the reaction

$2C(s) + O_2(g) + 4H_2(g) \longrightarrow 2CH_3OH(l)$

[4.47] Ammonia, NH_3, boils at -33°C; at this temperature, it has a density of 0.81 g/cm^3. The heat of formation of $NH_3(g)$ is -46.2 kJ/mol, and the heat of vaporization of $NH_3(l)$ is 4.6 kJ/mol. Calculate the heat evolved when 1 L of liquid NH_3 is burned in air to give $N_2(g)$ and $H_2O(g)$. How does this compare with the heat evolved upon complete combustion of a liter of liquid methyl alcohol, CH_3OH, density at 25°C = 0.792 g/cm^3, and heat of formation $\Delta H_f^\circ(CH_3OH(l)) = -239$ kJ/mol?

4.48 The heat of combustion at constant volume of azobenzene, $C_{12}H_{10}N_2$, is 6476 kJ/mol. What is the temperature rise in a bomb calorimeter with a heat capacity of 8.12 kJ/°C when 1.80 g of azobenzene is combusted?

4.49 When a highly elastic rubber ball having a mass of 72.0 g is dropped from a height of 3.600 m onto a thick, hard surface, it bounces to a height of 3.200 m. Describe the net change in potential energy that has occurred at that moment. What form has this energy difference taken? (The potential energy of a mass in the earth's gravitational field is given by $E = mgh$, where m is mass, g is the gravitational-acceleration constant, 9.81 m/s^2, and h is the height above some reference level.)

4.50 A large coal-burning electric power plant requires 1000 tons per hour of coal for peak operation. The coal has a fuel value of 13,600 Btu/lb. Calculate the total heat generated per hour by coal combustion, in units of kilojoules (1 Btu = 1.05 kJ). The coal contains 1.22 percent sulfur, which we can assume to be present as elemental sulfur. What heat is evolved per hour as a result of combustion of this sulfur to $SO_2(g)$?

[4.51] A recent U.S. government survey of coal reserves places the quantity of coal recoverable by means of current technology at about 2.5×10^{11} tons. Assuming that this coal has an average fuel value of 31 kJ/g, calculate the total fuel value of the coal. Assuming that the U.S. use of energy stabilizes at about 94 quads per year (1 quad = 1.05×10^{15} kJ), how many years of energy supply does this coal represent? If the average composition of the coal is $C_{135}H_{97}O_9NS$, what mass of CO_2 in kilograms will be produced by this combustion?

5 Electronic Structures of Atoms: Basic Concepts

In earlier chapters, we did not address the question of why atoms form molecules or ions. Why, for example, do hydrogen and oxygen atoms "stick" together to form the familiar water molecule, H_2O? Before we can answer this question and many others like it, we must know more about atoms themselves. It is especially important to know how electrons are arranged in atoms. We refer to the electron arrangement of an atom as its electronic structure. The goal of this chapter is to give you an understanding of the electronic structure of the simplest atom, the hydrogen atom. In subsequent chapters we will apply these ideas to other atoms and then to the bonding between atoms. A great deal of our understanding of electronic structure has come from studies of the properties of light or radiant energy. We begin our study, then, by considering the characteristics of radiant energy.

5.1 RADIANT ENERGY

Different kinds of radiant energy—such as the warmth from a glowing fireplace, the light reflected off snow in the mountains, and the X rays used by a dentist—*seem* very different from one another. Yet they share certain fundamental characteristics. All types of radiant energy, also called **electromagnetic radiation,** move through a vacuum at a speed of 2.9979250×10^8 m/s, the "speed of light." This is one of the most accurately known physical constants. For our purposes, however, it will be sufficient to use 3.00×10^8 m/s.

All radiant energy has wavelike characteristics analogous to those of waves that move through water. A cork bobbing on water as waves pass by is not swept along with the wave. Rather, it moves up and down with the wave motion. Water waves are the result of energy imparted to the water, perhaps by the dropping of a stone, the movement of a boat, or the force of wind on the water surface. This energy is expressed as the up-and-down movement of the water.

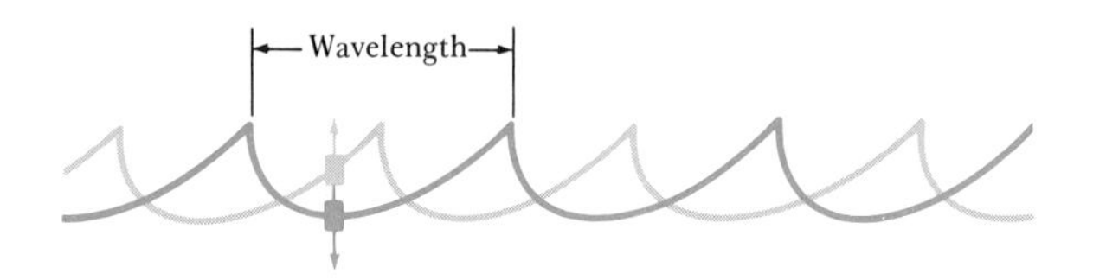

FIGURE 5.1 Characteristics of water waves. The distance between corresponding points on each wave is called the wavelength. The number of times per second that the cork bobs up and down is called the frequency.

If we look at a cross section of a water wave (Figure 5.1), we see that it is periodic in character. That is, the wave form repeats itself at regular intervals. The distance between identical points on successive waves is called the **wavelength.** As the wave passes, the cork floating on the surface moves up and down. The number of times per second that the cork moves through a complete cycle of upward and then downward motion is called the **frequency** of the wave.

In a similar way, radiant energy has a characteristic frequency and wavelength associated with it. These are illustrated schematically in Figure 5.2. The frequency of the radiation is the number of cycles that occur in 1 s as the radiation flows past a given point. Short wavelengths correspond to high frequency, and long wavelengths to low frequency, as shown in Figure 5.2. The wavelength and frequency are thus related.

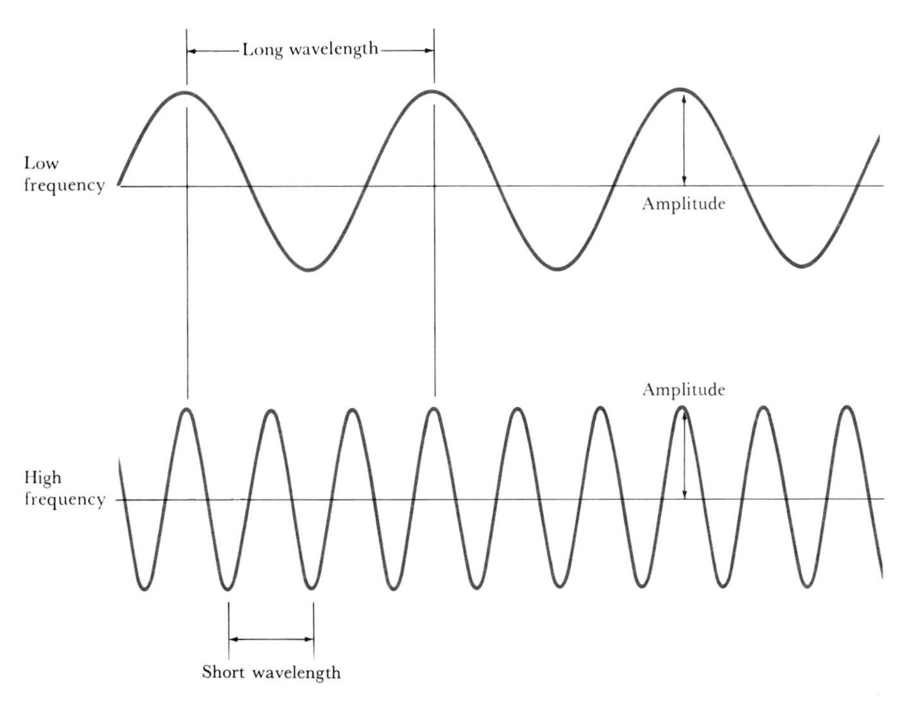

FIGURE 5.2 Wave characteristics of radiant energy. Unlike water waves, radiant energy is not associated with the movement of matter. Instead, it is associated with periodic changes in the electric and magnetic fields that accompany the wave. That is why it is referred to as electromagnetic radiation. Radiant-energy waves move through space with the speed of light. The frequency of the high-frequency radiation in this illustration is three times that of the low-frequency radiation. In the time that the low-frequency, long-wavelength radiation goes through one cycle, the high-frequency, short-wavelength radiation goes through three cycles. The *amplitude* of the wave refers to its intensity. Here both waves have the same amplitude.

The product of frequency, ν (nu), and wavelength, λ (lambda), equals the speed of light, c:

$$\nu\lambda = c \quad [5.1]$$

Wavelength is expressed in units of length per cycle; the unit of length chosen depends on the type of radiation as shown in Table 5.1. Figure 5.3 shows the ranges of wavelength that characterize the various types of radiant energy. Frequency is given in units of hertz (Hz), which is cycles per second. For example, the frequency of an AM radio station might be written as 810 kilohertz (kHz), or 810,000 cycles/s. Because it is understood that cycles are involved, the units are normally given simply as /s or s^{-1}.

SAMPLE EXERCISE 5.1

The yellow light given off by a sodium lamp has a wavelength of 589 nm. What is the frequency of this radiation?

Solution: We can rearrange Equation 5.1 to give $\nu = c/\lambda$. We insert the value for c and λ, and then convert nm to m. This gives us

$$\nu = \left(\frac{3.00 \times 10^8\ \text{m/s}}{589\ \text{nm}}\right)\left(\frac{10^9\ \text{nm}}{1\ \text{m}}\right)$$
$$= 5.09 \times 10^{14}/\text{s}$$

Note that frequency has units of /s or s^{-1}.

Radiations of different wavelengths affect matter differently. For example, overexposure of a part of your body to infrared radiation may cause a "heat burn," overexposure to visible and near-ultraviolet light causes sunburn and suntan, and overexposure to X radiation causes tissue damage, possibly even cancer. These diverse effects are due to differences in the energy of the radiation. Radiations of high frequency and short wavelength are more energetic than radiation of lower frequency and longer wavelength. The quantitative relation between frequency and energy was developed at the turn of the century in the far-reaching and revolutionary *quantum theory* of Max Planck (Figure 5.4).

TABLE 5.1 Wavelength units for electromagnetic radiation

Unit	Symbol	Length (m)	Type of radiation
Ångstrom	Å	10^{-10}	X ray
Nanometer (or millimicron)[a]	nm (mμ)	10^{-9}	Ultraviolet, visible
Micrometer (or micron)	μ	10^{-6}	Infrared
Millimeter	mm	10^{-3}	Infrared
Centimeter	cm	10^{-2}	Microwaves
Meter	m	1	TV, radio

[a]The millimicron is the same size unit as the nanometer; use of the latter is preferred.

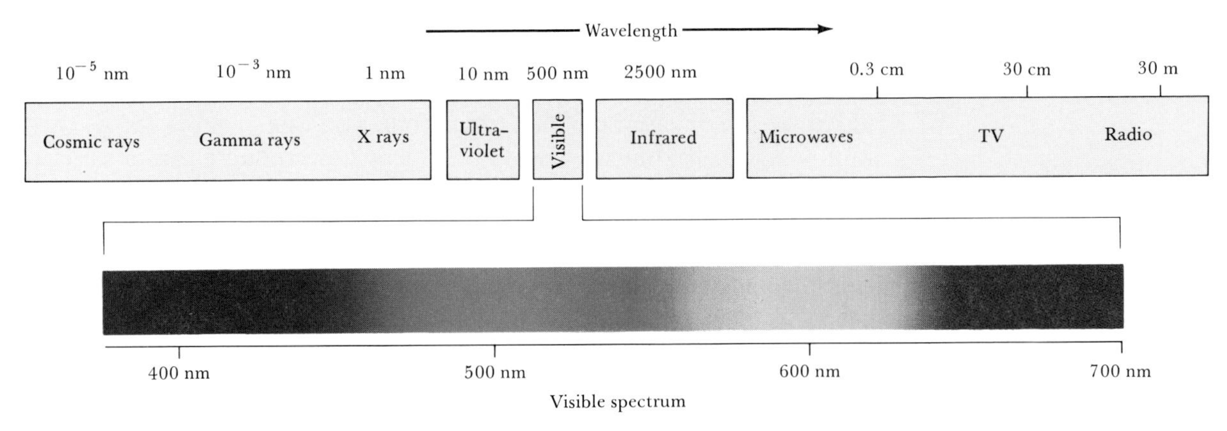

FIGURE 5.3 Wavelengths of electromagnetic radiation characteristic of various regions of the electromagnetic spectrum. Notice that color can be expressed quantitatively by wavelength or frequency.

5.2 THE QUANTUM THEORY

The quantum theory is concerned with the rules that govern the gain or loss of energy from an object. Planck's contribution was to see that when we deal with gain or loss of energy from objects in the atomic or subatomic size range, the rules *seem* to be different from those that apply when we are dealing with energy gain or loss from objects of ordinary dimensions. A very crude and fanciful analogy might best illustrate what is involved. Imagine a large dump truck loaded with perhaps 20 tons of fine-grained sand. Let's say that the amount of sand on the truck is measured by driving it onto a supersensitive scale that measures the weight of a 30-ton object to the nearest pound. With this as our measuring device, the gain or loss of a few grains of sand from the truck would be too small to be measured; a spoonful of sand might be added or deleted with no change in scale reading.

Now imagine, if you will, a tiny little truck, operated by people the

FIGURE 5.4 Max Planck (1858–1947), physicist. Born in Kiel, Germany, Planck was the son of a law professor at the University of Kiel. When he announced his intention to study physics, Planck was warned that all the major discoveries had already been made in this field. Nevertheless, Planck became a physicist, and in 1892 was named professor of physics at the University of Berlin. In 1900 he presented a paper before the Berlin Physical Society which launched the greatest intellectual revolution in the history of science. (*Library of Congress*)

size of tiny mites. For this little truck a full load of sand would consist of perhaps a dozen grains of sand of the same size as those carried by the large truck. In this microscopic world, the load on the truck can be added to or decreased only by rolling on or off one or more full grains of sand. On a scale that weighs this tiny truck, even one grain of sand, the smallest piece attainable, represents a substantial fraction of the full load and is easily measurable.

In our analogy the sand represents energy. An object of ordinary, or macroscopic, dimensions, like the dump trucks we have seen on highways, contains energy in so many tiny pieces that the gain or loss of individual pieces is completely unnoticed. On the other hand, an object of atomic dimensions, such as our imaginary little truck, contains such a small amount of energy that the gain or loss of even the smallest possible piece makes a substantial difference. The essence of Planck's quantum theory is that there *is* such a thing as a smallest allowable gain or loss of energy. Even though the amount of energy gained or lost at one time may be very tiny, there is a limit to how small it may be. Planck termed the smallest allowed increment of energy gained or lost a **quantum.** In our analogy, a single grain of sand represents a quantum of sand.

You should keep in mind that the rules regarding the gain or loss of energy are always the same, whether we are concerned with objects on the size scale of our ordinary experience or with microscopic objects. However, it is only when dealing with matter at the atomic level of size that the impact of the quantum restriction is evident. Humans, being creatures of macroscopic dimensions, had no reason to suppose that the quantum restriction existed until they devised means of observing the behavior of matter at the atomic level. The major tool for doing this at the time of Planck's work was by observation of the radiant energy absorbed or emitted by matter. An object can gain or lose energy by absorbing or emitting radiant energy. Planck assumed that the amount of energy gained or lost at the atomic level by absorption or emission of radiation had to be a whole number multiple of a constant times the frequency of the radiant energy. Let us call this amount of energy gained or lost ΔE. Then, according to Planck's theory,

$$\Delta E = h\nu,\ 2h\nu,\ 3h\nu,\ \text{etc.} \qquad [5.2]$$

The constant h is known as Planck's constant. It has the value 6.63×10^{-34} joule second, or J-s. The smallest increment of energy at a given frequency, $h\nu$, is called a *quantum* of energy.

SAMPLE EXERCISE 5.2

Calculate the smallest increment of energy (that is, the quantum of energy) that an object can absorb from yellow light whose wavelength is 589 nm.

Solution: We obtain the magnitude of a quantum of energy from Equation 5.2, $\Delta E = h\nu$. The value of Planck's constant is given both in the text above and in the table of physical constants on the inside back cover of the text, $h = 6.63 \times 10^{-34}$ J-s. The frequency, ν, is calculated from the given wavelength as shown in Sample Exercise 5.1: $\nu = c/\lambda = 5.09 \times 10^{14}/\text{s}$. Thus we have

$$\Delta E = h\nu = (6.63 \times 10^{-34}\ \text{J-s})(5.09 \times 10^{14}/\text{s}) = 3.37 \times 10^{-19}\ \text{J}$$

Planck's theory tells us that an atom or molecule emitting or absorbing radiation whose wavelength is 589 nm cannot lose or gain energy by radiation except in multiples of 3.37×10^{-19} J. It cannot, for example, gain 5.00×10^{-19} J from this radiation because this amount is not a multiple of 3.37×10^{-19} J.

At this point you may be wondering about the practical applications of Planck's quantum theory. A few years after Planck presented his theory, scientists began to see its applicability to a great many experimental observations. It soon became apparent that Planck's theory had within it the seeds of a revolution in the way the physical world is viewed. Let's consider a few applications that are of special importance for chemistry.

The Photoelectric Effect

Light shining on a clean metallic surface can cause the surface to emit electrons; this phenomenon, known as the photoelectric effect, can be demonstrated as shown in Figure 5.5. For each metal there is a minimum frequency of light below which no electrons are emitted, regardless of how intense the beam of light. In 1905, Albert Einstein (1879–1955) used the quantum theory to explain the photoelectric effect. He assumed that the radiant energy striking the metal surface is a stream of tiny energy packets. Each energy packet, called a photon, is a quantum of energy $h\nu$. Thus radiant energy itself is considered to be quantized. *Photons of high-frequency radiation have high energies, whereas photons of lower-frequency radiation have lower energy:*

$$E_{\text{photon}} = h\nu \qquad [5.3]$$

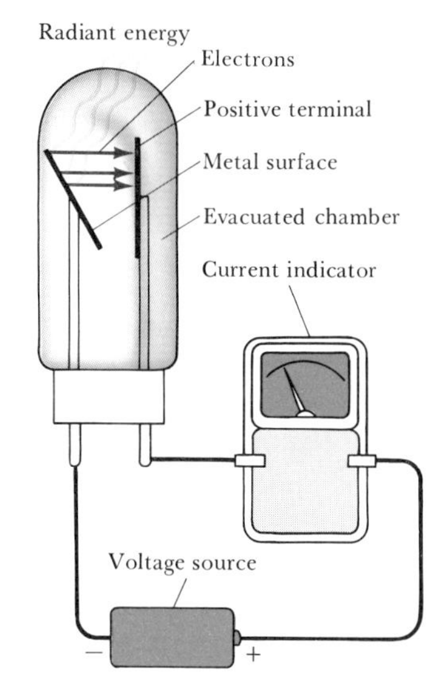

FIGURE 5.5 The photoelectric effect. When photons of sufficiently high energy strike the metal surface in the tube, electrons are emitted from the metal. The electrons are drawn toward the other electrode, which is a positive terminal. As a result, current flows in the circuit. If the energy of a photon, $h\nu$, is too small, no electrons are freed and no current flows in the circuit. Photocells in automatic doors use the photoelectric effect to generate the electrons that operate the door-opening circuits.

When the photons are absorbed by the metal, their energy is transferred to an electron in the metal. A certain amount of energy is required for the electron to overcome the attractive forces that hold it within the metal. Otherwise it cannot escape from the metal surface, even if the light beam is quite intense. If a photon does have sufficient energy, the electron is emitted. If a photon has more than the minimum energy required to free an electron, the excess appears as the kinetic energy of the emitted electron. Thus the kinetic energy of the emitted electron, E_k, equals the energy supplied by the photon, $h\nu$, minus the binding energy holding the electron in the metal, E_b:

$$E_k = h\nu - E_b \qquad [5.4]$$

Line Spectra

A particular source of radiant energy may emit a single wavelength, as in the light from a laser (Figure 5.6), or may contain many different wavelengths, as in the radiations from a light bulb or a star. Radiation composed of a single wavelength is termed monochromatic. When the radiation from a source such as a star is separated into its monochromatic components, a spectrum is produced. This separation of radiations of differing wavelengths can be achieved by dispersing them in a prism, as

FIGURE 5.6 Monochromatic light being emitted from the discharge tube of an argon-gas laser. (*Laser Analytics Division, Spectra-Physics*)

shown in Figure 5.7. When the white light from a light bulb is passed through a prism it is dispersed into a continuous range of colors; violet merges into blue, blue into green, and so forth with no blank spots. This rainbow of colors, containing light of all wavelengths, is called a **continuous spectrum.** The most familiar example of a continuous spectrum is the rainbow, produced by the dispersal of sunlight by raindrops or mist.

Not all emitters of light radiate all colors or wavelengths. For example, when hydrogen is placed under reduced pressure in a tube such as that depicted in Figure 5.8 and a high voltage is applied, light is emitted. (If, instead of hydrogen, neon gas were placed in the tube, the familiar red-orange glow of neon lights would be produced; sodium vapor would produce the yellow radiation of many modern streetlights.) When the light coming from such tubes is separated into its monochromatic components, only certain colors or wavelengths of light are found to be present. A spectrum containing radiation of only specific wavelengths is called a **line spectrum** (see Figure 5.8).

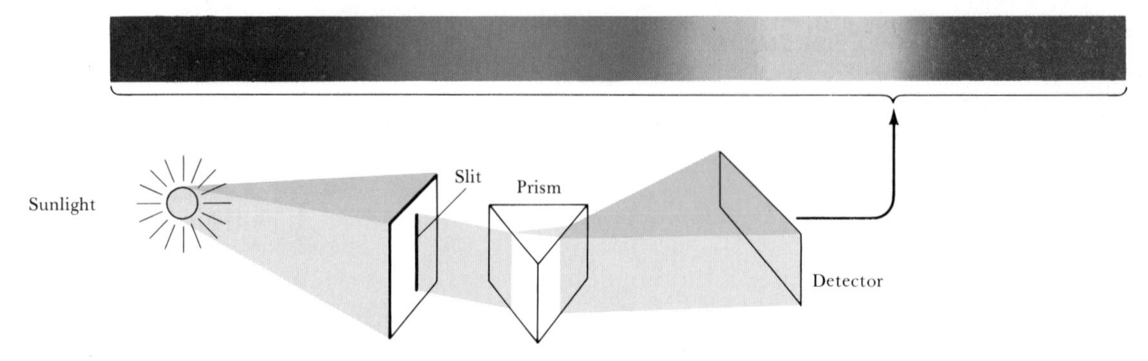

FIGURE 5.7 Production of a continuous visible spectrum by passing a narrow beam of white light through a prism. The white light could be sunlight or light from an incandescent lamp.

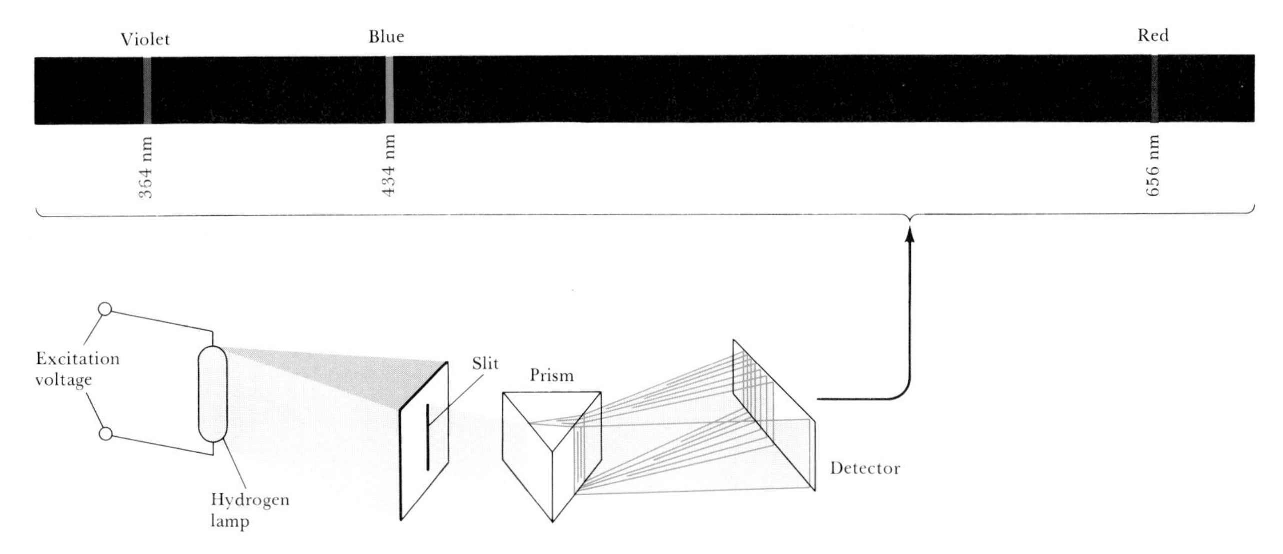

FIGURE 5.8 Production of the emission line spectrum of hydrogen by passing a narrow beam of light emitted by a hydrogen lamp through a prism. The lamp emits light when hydrogen atoms are excited in an electrical discharge.

The spectrum of the radiation emitted by a substance is called an emission spectrum. The line spectrum of hydrogen shown in Figure 5.8 is an emission spectrum. Substances also exhibit absorption spectra. When continuous electromagnetic radiation, such as that from a lightbulb, is passed through a substance, certain wavelengths may be absorbed. The spectrum of the radiation that passes through is called an absorption spectrum. The absorption spectrum of hydrogen consists of what looks like a continuous spectrum interrupted by black lines at 656.3 nm, 486.3 nm, and the other wavelengths found in the emission spectrum. The absorption spectrum of hydrogen is a line spectrum that is complementary to the emission spectrum of this substance.

Each element has a characteristic line spectrum that can be used to identify it. Figure 5.9 shows the emission spectrum of the sun. Notice that there are several dark bands in the spectrum. Because of its high temperature the sun emits a continuous spectrum of radiation. However, elements that are present in the outer regions of the sun, where the temperatures are not so high, absorb radiation at characteristic wavelengths. Absorptions due to hydrogen, iron, and sodium atoms give rise to the dark bands evident in Figure 5.9. It was from a similar spectrum of our sun that helium was first discovered in 1868. Some of the absorption lines of the sun's spectrum could not be matched with those of any then known element; it was concluded that the sun contained an element previously unknown on earth. This element was named helium after *helios,* the Greek word for the sun. Helium was subsequently isolated and characterized in the laboratory in 1895.

FIGURE 5.9 Spectrum of the sun's radiation. Note the presence of dark lines, called Fraunhofer lines, due to absorption of radiation by hydrogen and other atoms at the outer limits of the sun. (*Courtesy Bausch & Lomb*)

Although it was not realized at the time by the scientists who gathered the data, line spectra provide beautiful applications of the quantum theory. In 1914, the Danish physicist Niels Bohr (1885–1962) incorporated Planck's theory to explain the line spectrum of hydrogen. The basic idea introduced by Bohr was that the absorptions and emissions of light by hydrogen atoms correspond to energy changes of electrons within the atoms. The fact that only certain frequencies are absorbed or emitted by an atom tells us that only certain energy changes are possible. (We say that the energy changes within an atom are quantized.) Further details of Bohr's model are described in the next section.

5.3 BOHR'S MODEL OF HYDROGEN

Bohr's model of the hydrogen atom took into account two important developments that were relatively new at that time. The first was Rutherford's experiments , which established the nuclear nature of the atom (Section 2.6). The other was Einstein's work, which showed that radiant energy could be thought of as a stream of discrete bundles of energy called photons. We are not concerned here with the details of the Bohr treatment, because it is not strictly correct. However, it does introduce an important concept: the quantization of the energy of electrons in atoms.

Bohr's model consists of a series of postulates that may be summarized as follows:

1. The electron of a hydrogen atom moves about the central proton in a circular orbit. However, an electron in an atom cannot have just any energy; only orbits of certain radii, and having certain energies, are permitted. An electron in one of these orbits is said to be in an "allowed" energy state.
2. In the absence of radiant energy, an electron in an atom remains indefinitely in one of the allowed energy states. When radiant energy is present, however, the atom may absorb energy. When this happens, the electron undergoes a change from one allowed energy state to another. The frequency of the radiant energy absorbed (ν) corresponds exactly to the energy difference (ΔE) between two of the allowed energies: $\Delta E = h\nu$.

We need not concern ourselves with the details of how Bohr used quantum theory to calculate the energies of the electron. The main point is that he was able to calculate a set of allowed energies. Each of these allowed energies corresponds to a circular path of different radius. In Bohr's model, each allowed orbit was assigned an integer n, known as the **principal quantum number**, that may have values from 1 to infinity. The radius of the electron orbit in these energy states varies as n^2:

$$\text{Radius} = n^2(5.3 \times 10^{-11}\ \text{m}) \qquad [5.5]$$

Thus the larger the value of n, the farther the electron from the nucleus.

The energy of the electron depends on the orbit it occupies:

$$E_n = -R_H\left(\frac{1}{n^2}\right) \qquad [5.6]$$

The constant R_H in Equation 5.6 is called the Rydberg constant; it has the value of 2.18×10^{-18} J. From Equation 5.6 we have that the energy of the electron is -2.18×10^{-18} J when the electron is in the orbit closest to the nucleus, $n = 1$. When it is in the second orbit, $n = 2$, its energy is

$$E_2 = (-2.18 \times 10^{-18}\,\text{J})\left(\frac{1}{2^2}\right) = -5.45 \times 10^{-19}\,\text{J}$$

The orbital radii and energies are illustrated for $n = 1, n = 2$, and $n = 3$ in Figure 5.10. Another more common way of representing the allowed energies is shown in Figure 5.11.

The negative sign in Equation 5.6 denotes stability relative to some reference state. In other words, the more negative the value for energy, the more stable the system is. (One way to remember this convention is to think of a ball rolling around on a surface with many hills and valleys. The ball will naturally come to rest in a valley; it is more stable when its potential energy is lowest.) The reference, or zero-energy, state for the electron in hydrogen is chosen to be that in which the electron is completely separated from the nucleus. This, of course, corresponds to an infinitely large value for the principal quantum number n:

$$E_\infty = (-2.18 \times 10^{-18}\,\text{J})\left(\frac{1}{\infty^2}\right) = 0$$

The energy of the electron in any other orbit is then negative relative to this reference state. The lowest energy, or most stable, state, with $n = 1$, is known as the ground state. When the electron is in a higher energy

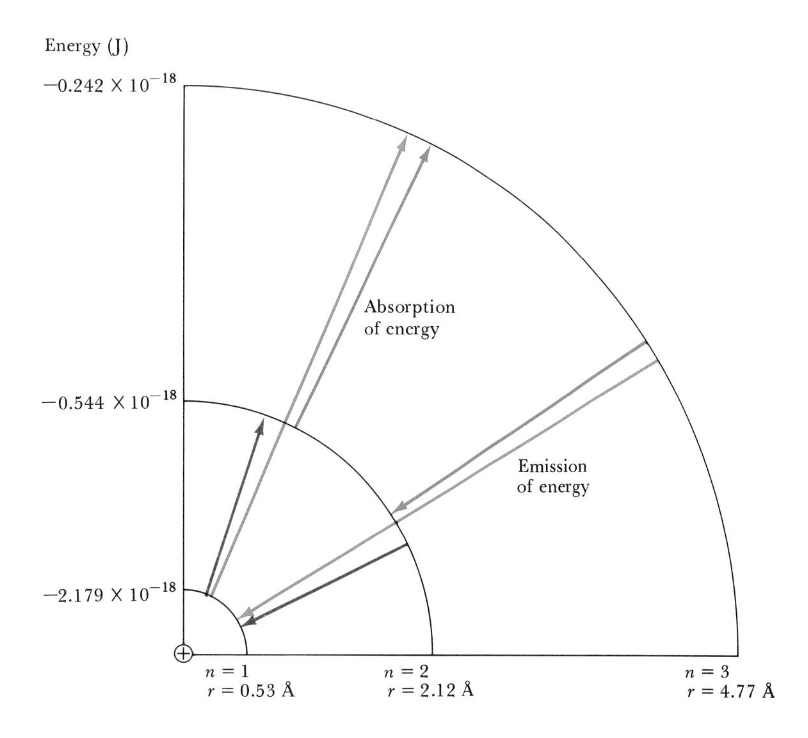

FIGURE 5.10 Radii and energies of the three lowest-energy orbits in the Bohr model of hydrogen. The arrows refer to transitions of the electron from one allowed energy state to another. When the transition takes the electron from a lower- to a higher-energy state, absorption occurs. When the transition is from a higher- to a lower-energy state, emission occurs.

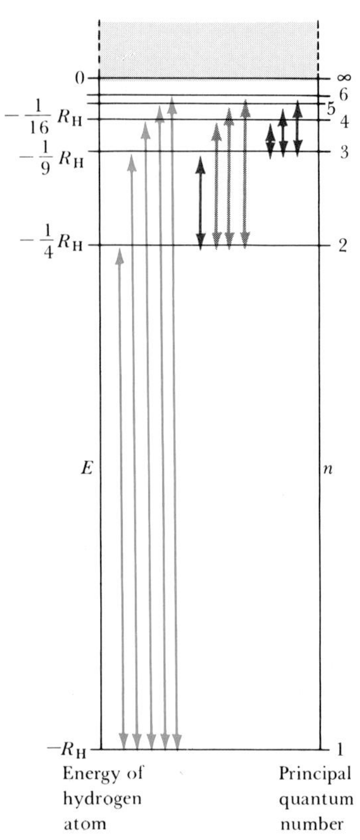

FIGURE 5.11 Energy levels in the hydrogen atom from the Bohr model. The arrows refer to transitions of the electron from one allowed energy state to another, as described in Figure 5.10. Only the lowest six energy levels are shown.

orbit, that is, $n = 2$ or higher, the atom is said to be in an electronically **excited state.**

Bohr's model of the hydrogen atom quantitatively explained the observed line spectra of that substance. The absorptions or emissions in the line spectra correspond to transitions of the electron from one orbit to another. Radiant energy is absorbed when the electron moves from one orbit to another having a larger radius; it requires energy to pull the electron away from the nucleus. Conversely, energy is emitted when the electron moves from a larger orbit to another having a smaller radius. The changes in energy, ΔE, are given by the difference between the energy of the final state of the electron, E_f, and the initial state, E_i:

$$\Delta E = E_f - E_i$$

Substituting the expression for the energy of the electron, Equation 5.6, gives

$$\Delta E = \left(\frac{-R_H}{n_f^2}\right) - \left(\frac{-R_H}{n_i^2}\right) = R_H\left(\frac{1}{n_i^2} - \frac{1}{n_f^2}\right)$$

Because $\Delta E = h\nu$, we have

$$\Delta E = h\nu = R_H\left(\frac{1}{n_i^2} - \frac{1}{n_f^2}\right) \qquad [5.7]$$

In this expression, n_i and n_f represent the quantum numbers for the initial and final states, respectively. Notice that when the final state quantum number (n_f) is larger than the initial state quantum number(n_i), the term in parentheses is positive, and ΔE is positive. This means that the system has absorbed a photon and thus increased in energy (an endothermic process). When n_i is larger than n_f, as happens in emission, energy is given off; ΔE is negative.

Complete removal of the electron from a hydrogen atom, corresponding to a transition from the $n = 1$ (ground) state to the $n = \infty$ state, is known as ionization. This is represented as

$$H(g) \longrightarrow H^+(g) + e^-$$

The energy required for ionization from the ground state is called the **ionization energy.**

SAMPLE EXERCISE 5.3

Calculate the frequency of the hydrogen line that corresponds to the transition of the electron from the $n = 4$ to the $n = 2$ states.

Solution: We employ Equation 5.7, substituting $n_i = 4$ and $n_f = 2$ because these are the quantum numbers for the initial and final orbits, respectively:

$$\Delta E = h\nu = R_H\left(\frac{1}{n_i^2} - \frac{1}{n_f^2}\right)$$

$$\nu = \frac{R_H}{h}\left(\frac{1}{n_i^2} - \frac{1}{n_f^2}\right)$$

$$= \frac{2.18 \times 10^{-18}\,\text{J}}{6.63 \times 10^{-34}\,\text{J-s}}\left(\frac{1}{4^2} - \frac{1}{2^2}\right)$$

$$= \frac{2.18 \times 10^{-18}\,\text{J}}{6.63 \times 10^{-34}\,\text{J-s}}\left(\frac{1}{16} - \frac{1}{4}\right)$$

$$= \frac{2.18 \times 10^{-18}\,\text{J}}{6.63 \times 10^{-34}\,\text{J-s}}\left(-\frac{3}{16}\right)$$

$$= -6.17 \times 10^{14}/\text{s}$$

The negative sign simply indicates that the light is emitted. This value for ν can be used to calculate the wavelength of the radiation, so we can compare it with the hydrogen spectrum shown in Figure 5.8:

$$\lambda = \frac{c}{\nu} = \frac{3.00 \times 10^{8}\,\text{m/s}}{6.17 \times 10^{14}/\text{s}}$$

$$= 4.86 \times 10^{-7}\,\text{m} = 486\,\text{nm}$$

All of the lines shown in Figure 5.8 correspond to transitions of an electron from higher orbits to the $n = 2$ orbit.

SAMPLE EXERCISE 5.4

Calculate the energy required for ionization of an electron from the ground state of the hydrogen atom.

Solution: The ionization energy may be written as the difference between the final and initial state energies. We have $n_f = \infty$, $n_i = 1$.

$$\Delta E = E_f - E_i = R_H\left(\frac{1}{1^2} - \frac{1}{\infty^2}\right)$$

$$= R_H(1 - 0)$$

This is just equal to R_H, $2.18 \times 10^{-18}\,\text{J}$.

It is often useful to express this energy on a molar basis. To do this we simply multiply by Avogadro's number:

$$\left(2.18 \times 10^{-18}\,\frac{\text{J}}{\text{atom}}\right)\left(6.02 \times 10^{23}\,\frac{\text{atoms}}{\text{mol}}\right) \times \left(\frac{1\,\text{kJ}}{1000\,\text{J}}\right) = 1.31 \times 10^{3}\,\text{kJ/mol}$$

Bohr's model was very important because it introduced the idea of quantized energy states for electrons in atoms. This feature is incorporated into our current model of the atom. However, Bohr's model was adequate for explaining only atoms and ions with one electron, such as H, He^{+}, or Li^{2+}. It was not adequate to account for the atomic spectra of other atoms or ions, except in a rather crude way. Consequently, Bohr's model was eventually replaced by a new way of viewing atoms that is called **quantum mechanics** or **wave mechanics.** This newer model maintains the concept of quantized energy states, but in addition still further applications of Planck's quantum theory enter the picture.

5.4 MATTER WAVES

In the years following Bohr's development of a model for the hydrogen atom, the dual nature of radiant energy had become a familiar concept. Depending on the experimental circumstances, radiation might appear to have either a wavelike or particlelike (photon) character. Louis de Broglie, a young man working on his Ph.D. thesis in physics at the Sorbonne, in Paris, made a rather daring, intuitive extension of this idea. If radiant energy could under appropriate conditions behave as though it were a stream of particles, could not matter under appropriate conditions possibly show the properties of a wave? Suppose that the electron in orbit around the nucleus of a hydrogen atom could be thought of as a wave, with a characteristic wavelength, as illustrated in Figure 5.2. De Broglie suggested that the electron in its circular path about the nucleus has associated with it a particular wavelength. He went on to propose that the characteristic wavelength of the electron or any other particle depends on its mass, m, and velocity, v:

$$\lambda = \frac{h}{mv} \quad [5.8]$$

(h is Planck's constant). The quantity mv for any object is called its momentum. De Broglie used the term matter waves to describe the wave characteristics of material particles.

Because de Broglie's hypothesis is perfectly general, any object of mass m and velocity v would give rise to a characteristic matter wave. However, it is easy to see from Equation 5.8 that the wavelength associated with an object of ordinary size, such as a golf ball, is so tiny as to be completely out of the range of any possible observation. This is not so for electrons, because their mass is so small.

SAMPLE EXERCISE 5.5

What is the characteristic wavelength of an electron with a velocity of 5.97×10^6 m/s? (The mass of the electron is 9.11×10^{-28} g.)

Solution: The value of Planck's constant, h, is 6.63×10^{-34} J-s (1 J = 1 kg-m²/s²).

$$\lambda = \frac{h}{mv}$$

$$= \frac{6.63 \times 10^{-34}\ \text{J-s}}{(9.11 \times 10^{-28}\ \text{g})(5.97 \times 10^6\ \text{m/s})} \times \left(\frac{1\ \text{kg-m}^2/\text{s}^2}{1\ \text{J}}\right)\left(\frac{10^3\ \text{g}}{1\ \text{kg}}\right)$$

$$= 1.22 \times 10^{-10}\ \text{m} = 0.122\ \text{nm}$$

By comparing this value with the wavelengths of electromagnetic radiations shown in Figure 5.3, we see that the characteristic wavelength is about the same as that of X rays.

Within a few years after de Broglie published his theory, the wave properties of the electron were demonstrated experimentally. Electrons were diffracted by crystals, just as X rays, which are definitely radiant energy, are diffracted. (We shall have more to say about the X-ray diffraction experiment in Chapter 10.)

The technique of electron diffraction has been highly developed. In the electron microscope, the wave characteristics of electrons are used to obtain electron diffraction pictures of tiny objects. The electron microscope is an important technique for studying surface phenomena at the very highest magnifications. An example of an electron microscope picture is shown in Figure 5.12. Pictures such as this are powerful demonstrations that tiny particles of matter can indeed behave as waves.

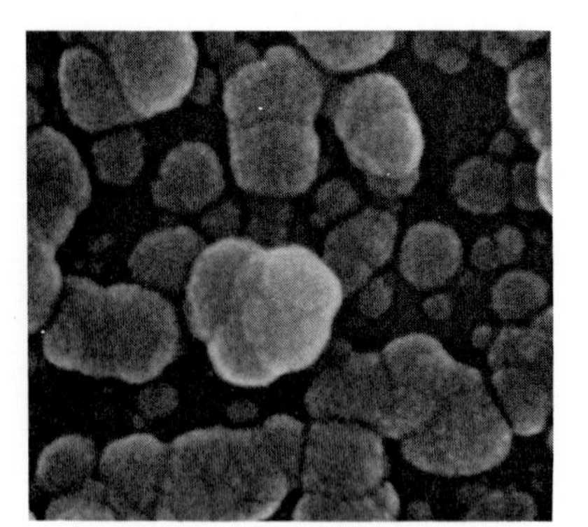

FIGURE 5.12 Gold particles on silicon. The particles are shown magnified over 100,000 times. (*Photograph courtesy of International Scientific Instruments, Inc.*)

The Uncertainty Principle

Discovery of the wave properties of matter raised new and interesting questions. If a subatomic particle can exhibit the properties of a wave phenomenon, is it possible to say precisely just where that particle is located? One can hardly speak of the precise location of a wave. The amplitude, or intensity, of a wave can be defined at a certain point, as illustrated in Figure 5.2, but the wave as a whole extends in space. Its location is therefore not defined precisely, at least not in the same sense that one can define the location of a particle. We might logically expect to be able to measure not only a particle's location, but also its direction and speed of motion.

However, the German physicist Werner Heisenberg* concluded that there is a fundamental limitation on just how precisely we can hope to know both the location and the momentum of a particle. Just as in the case of quantum effects, the limitation becomes important only when we deal with matter at the subatomic level, that is, with masses as small as that of an electron. Heisenberg's principle is called the **uncertainty principle.** When applied to the electrons in an atom, this principle states that it is inherently impossible for us to know both the exact momentum of the electron and its location in space. Thus it is not appropriate to imagine the electrons as moving in well-defined circular orbits about the nucleus, always at the same radius.

De Broglie's hypothesis and Heisenberg's uncertainty principle set the stage for a new and more broadly applicable theory of atomic structure. In this new approach, any attempt to define precisely the instantaneous location and momentum of the electrons is abandoned. The wave nature of the electron is recognized, and its behavior is described in terms appropriate to waves.

5.5 THE QUANTUM-MECHANICAL DESCRIPTION OF THE ATOM

The mathematics employed to determine electron energy levels in even so simple a system as the hydrogen atom by quantum-mechanical methods are quite advanced. There is therefore no point in presenting any mathematical details here. We can, however, understand the idea content of quantum mechanics with the aid of a qualitative description.

The approach of quantum mechanics involves a mathematical description of the wave properties of the electron. The equation involved also takes into account the attraction of the electron for the nucleus and the kinetic energy of the electron. Then certain conditions are imposed on the possible solutions to the resultant mathematical equation to make them physically reasonable. The result is a series of solutions that describe the allowed energy states of the electron. These solutions, usually represented by the symbol ψ (the Greek lowercase letter *psi*), are called **wave functions.**†

A wave function provides information about an electron's location in space when it is in a certain allowed energy state. The allowed energy states are the same as those predicted by the Bohr model. However, the Bohr model suggests that the electron is in a circular orbit of some particular radius about the nucleus. In the quantum-mechanical model for hydrogen, it is not so simple to describe the electron's location. The uncertainty principle suggests that if we know the momentum of the electron with high accuracy, our knowledge of its location is very uncer-

*Heisenberg (1901–1976) was one of the leading physicists of the twentieth century. He received the Nobel Prize in physics in 1932.

†In mathematics, the term *function* is used to describe a relationship between two variables—for example, time and distance. Thus the distance (d) covered by an auto moving at a constant speed (s) is a *function* of the amount of time (t) it has been in motion: $d = st$.

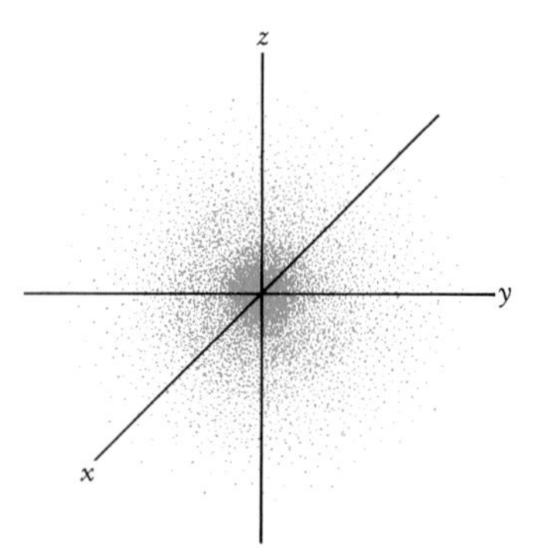

FIGURE 5.13 Electron density distribution in the ground state of the hydrogen atom.

tain. Thus, for an individual electron, we cannot hope to specify its location around the nucleus. Rather we must be content with a kind of statistical knowledge. In the quantum-mechanical model we therefore speak of the *probability* that the electron will be in a certain region of space at a given instant. As it turns out, the square of the wave function (ψ^2) at a given point in space represents the probability that the electron will be found at that location. If the value of ψ^2 is summed over all the points in space around the nucleus, its value must equal 1. This is true because a probability of 1 is certainty, and it is certain that the electron must be somewhere in the space around the nucleus.

One way of representing the probability of finding the electron in various regions of an atom is shown in Figure 5.13. In this figure the density of the dots represents the probability of finding the electron; the region of high probability is represented by a high density of dots. These regions correspond to relatively large values for ψ^2. It is common to refer to the electron density in certain regions of space, this being another way of expressing probability; regions where there is a high probability of finding the electron are said to be regions of high electron density. In Section 5.6, we will say more about the ways in which we can represent electron density.

Quantum Numbers and Orbitals

The complete quantum-mechanical solution to the hydrogen atom problem yields a set of wave functions and a corresponding set of energies. Each wave function represents an allowed solution. The term orbital denotes an allowed energy state for the electron. It also refers to the probability function that defines the distribution of electron density in space. Thus an orbital has both a characteristic energy and a characteristic shape. For example, the lowest energy orbital in the hydrogen atom has an energy of -2.18×10^{-18} J, and the shape illustrated in Figure 5.13. Note that an *orbital* (quantum-mechanical model) is not the same as an *orbit* (Bohr model).

The Bohr model introduced a single quantum number (n) to describe an orbit. The quantum-mechanical model introduces three quantum numbers (n, l, and m_l) to describe an orbital. We will consider what information we obtain from each of these and how they are interrelated:

1 The principal quantum number (n) can have integral values of 1, 2, 3, and so forth, just as in the Bohr model. This quantum number relates to the average distance of the electron from the nucleus and is the quantum-mechanical equivalent of Bohr's principal quantum number. In the hydrogen atom, orbitals possessing the same principal quantum number are of the same energy; as in the Bohr model $E_n = -R_H(1/n^2)$. We shall see in Chapter 6 that this is not the case for atoms having many electrons.

2 The second quantum number is known as the azimuthal quantum number, and is given the symbol l. It defines the shape of the orbital. (We will consider these shapes in Section 5.6.) The possible values for l are limited by the value for n; for each value of n, l can have integral values from 0 to $n - 1$. For example, if $n = 3$, l can have values of 0,

TABLE 5.2 Relationship among values of n, l, and m_l through $n = 4$

n	l	Subshell designation	m_l	Number of orbitals in subshell
1	0	1*s*	0	1
2	0	2*s*	0	1
	1	2*p*	1, 0, −1	3
3	0	3*s*	0	1
	1	3*p*	1, 0, −1	3
	2	3*d*	2, 1, 0, −1, −2	5
4	0	4*s*	0	1
	1	4*p*	1, 0, −1	3
	2	4*d*	2, 1, 0, −1, −2	5
	3	4*f*	3, 1, 2, 0, −1, −2, −3	7

1, or 2, but cannot equal 3 or higher. The value of l is generally designated by the letters *s*, *p*, *d*, *f*, and *g*, corresponding to l values of 0, 1, 2, 3, and 4, respectively.*

l	0	1	2	3	4
Designation of orbital	*s*	*p*	*d*	*f*	*g*

Table 5.2 shows the possible values of l for each value of n up through $n = 4$.

A collection of orbitals with the same value of n is referred to as an **electron shell.** One or more orbitals with the same set of n and l values is referred to as a **subshell.** For example, the shell with $n = 3$ is composed of three subshells, $l = 0$, 1, and 2 (the allowed values of l for $n = 3$). These subshells are called the 3*s*, 3*p*, and 3*d* subshells, respectively; in each of these designations, the number indicates the value for the principal quantum number (n) and the letter corresponds to the value of l.

3 The third quantum number, labeled m_l, is called the **magnetic** or **orientational quantum number.** It describes the orientation of the orbital in space. This quantum number may have integral values ranging from l to $-l$. Thus, when $l = 0$, m_l must be 0. When $l = 1$, m_l can have values of 1, 0, or −1.

The possible values of the three quantum numbers through $n = 4$ are summarized in Table 5.2. Notice that the allowed subshells are 1*s*, 2*s*, 2*p*, 3*s*, 3*p*, 3*d*, and so forth. Also note that each *s* subshell contains one orbital ($m_l = 0$), and each *p* subshell contains three orbitals ($m_l = 1, 0, -1$); similarly, each *d* subshell contains five orbitals, and each *f* subshell contains seven orbitals. Figure 5.14 shows the number and relative energies

*The letters *s*, *p*, *d*, and *f* came from the words *sharp, principal, diffuse,* and *fundamental.* These words were used to describe certain features of spectra before quantum mechanics was developed.

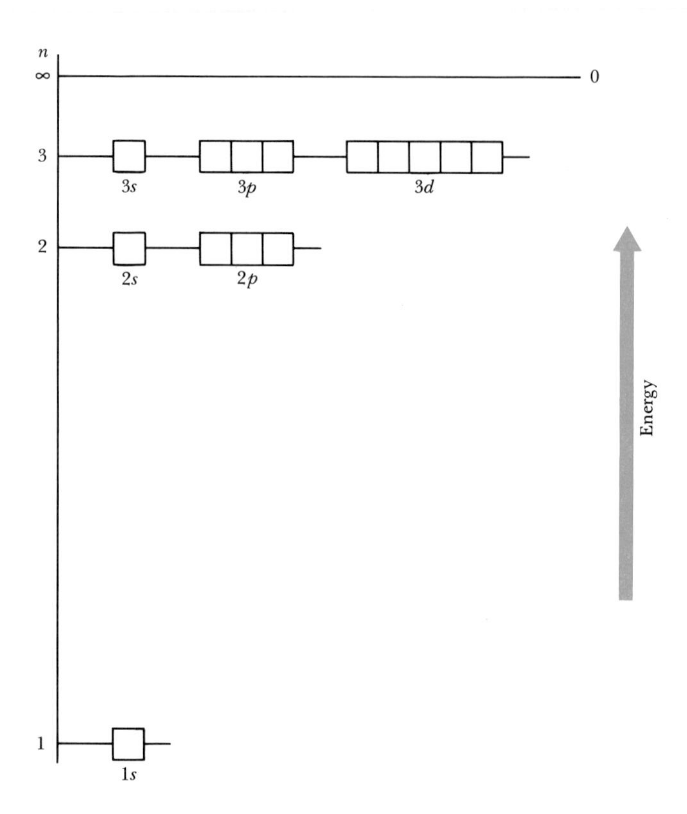

FIGURE 5.14 Orbital energy levels in the hydrogen atom and in hydrogenlike ions (those containing just one electron). Note that all orbitals with the same value for the principal quantum number, n, have the same energy. This is true only in one-electron systems.

of all hydrogen atom orbitals through $n = 3$. Each box represents an orbital; orbitals of the same subshell, such as the 2*p*, are grouped together. When the electron is in the lowest energy orbital (the 1*s* orbital), the hydrogen atom is said to be in its ground state. When it is in any other orbital, the atom is in an excited state. At ordinary temperatures essentially all hydrogen atoms are in their ground states. The electron may be promoted to an excited state orbital by absorption of a photon of appropriate energy.

SAMPLE EXERCISE 5.6

What are the values of n, l, and m_l for orbitals in the 3*d* subshell?

Solution: The number given in the designation of the subshell is the principal quantum number. Thus $n = 3$. The letter represents the value for l; *s* orbitals have $l = 0$, *p* have $l = 1$, and *d* have $l = 2$; thus $l = 2$ for the 3*d* subshell. The values of m_l can vary from l to $-l$; thus m_l can equal 2, 1, 0, -1, or -2. Consequently, there are five 3*d* orbitals.

SAMPLE EXERCISE 5.7

Without referring to Table 5.2, determine the number of orbitals for which $n = 4$; indicate the values of n, l, and m_l for each of these orbitals.

Solution: For $n = 4$, the possible values of l are 0, 1, 2, and 3. These correspond to the 4*s*, 4*p*, 4*d*, and 4*f* subshells. There is one 4*s* orbital ($n = 4$, $l = 0$, $m_l = 0$); there are three 4*p* orbitals ($n = 4$, $l = 1$, $m_l = 1, 0, -1$; there are five 4*d* orbitals ($n = 4$, $l = 2$, $m_l = 2, 1, 0, -1, -2$); there are seven 4*f* orbitals ($n = 4$, $l = 3$, $m_l = 3, 2, 1, 0, -1, -2, -3$).

5.6 REPRESENTATIONS OF ORBITALS

In our discussion of orbitals we have so far emphasized their energies. But the wave function also provides information about the electron's location in space when it is in a particular allowed energy state. We need to examine the ways that we can visualize or picture the orbitals.

The *s* Orbitals

The lowest energy (most stable) orbital, the 1*s* orbital, is spherically symmetric, as shown in Figure 5.13. Figures of this type, showing electron density, are one of the several ways we have to help us visualize orbitals. This figure indicates that the probability of finding the electron around the nucleus decreases as we move away from the nucleus in any direction. When the probability function (ψ^2) for the 1*s* orbital is graphed as a function of the distance from the nucleus (r), it rapidly approaches zero at large distance, as shown in Figure 5.15. This effect indicates that the electron, which is drawn toward the nucleus by electrostatic attraction, is unlikely ever to get very far from the nucleus.

If we similarly consider the 2*s* and 3*s* orbitals of hydrogen, we find that they are also spherically symmetric. Indeed, *all s* orbitals are spherically symmetric. The manner in which the probability function (ψ^2) varies with r for the 2*s* and 3*s* orbitals is shown in Figure 5.15. Notice that for the 2*s* orbital, ψ^2 goes to zero and then increases again in value before finally approaching zero at a larger value of r. The intermediate regions where ψ^2 goes to zero are called **nodal surfaces** or simply **nodes.** The number of nodes increases with increasing value for the principal quantum number (n). The 3*s* orbital possesses two nodes, as illustrated in Figure 5.15. Notice also that as n increases the electron is more and more likely to be located farther from the nucleus. That is, the size of the orbital increases as n increases.

The most widely used method of representing orbitals is to display a boundary surface that encloses some substantial fraction, say 90 percent, of the total electron density for the orbital. For the *s* orbitals these con-

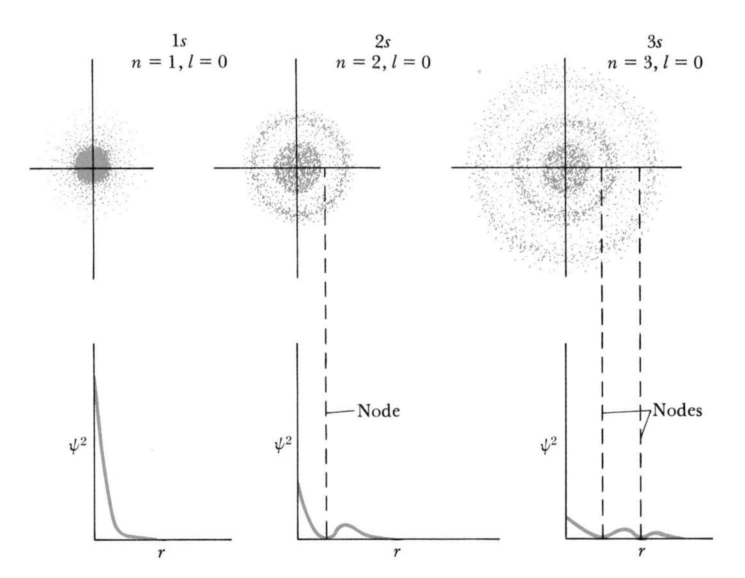

FIGURE 5.15 Electron-density distributions in 1*s*, 2*s*, and 3*s* orbitals. The lower part of the figure shows how the electron density, represented by ψ^2, varies as a function of distance from the nucleus. In the 2*s* and 3*s* orbitals, the electron-density function drops to zero at certain distances from the nucleus. The spherical surfaces around the nucleus at which ψ^2 is zero are called nodes.

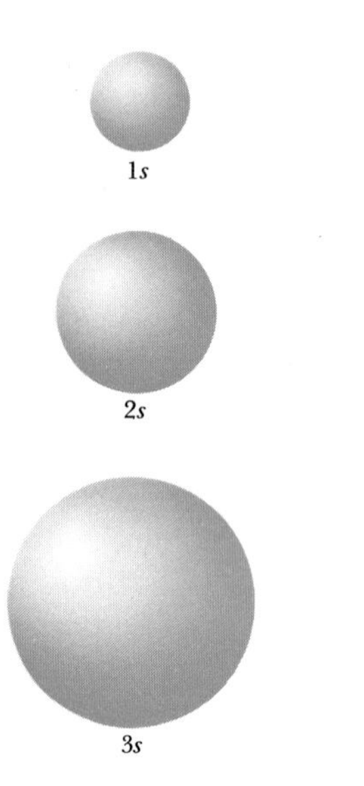

FIGURE 5.16 Contour representations of the 1s, 2s, and 3s orbitals. The spherical surfaces connect points of equal value of ψ^2. The surface encloses 90 percent of total ψ^2 for each orbital.

tour representations are merely spheres. The contour or boundary surface representations of the 1s, 2s, and 3s orbitals are shown in Figure 5.16. They have the same shape, but they differ in size. The fact that there are nodes within the 2s and 3s surfaces is lost in these representations. This is not a serious disadvantage; it turns out that for most qualitative discussions the most important features of orbitals are their size and shape. These features are adequately represented by the contour diagrams.

The *p* Orbitals

The distribution of electron density for a 2*p* orbital is shown in Figure 5.17. As we can see from this figure, the electron density is not distributed in a spherically symmetric fashion as in an *s* orbital. Instead, the electron density is concentrated on two sides of the nucleus separated by a node at the nucleus; we often say that this orbital has two lobes. It is useful to recall that we are making no statement of how the electron is moving within the orbital; Figure 5.17 portrays the *averaged* distribution of the 2*p* electron in space.

There are three *p* orbitals in each shell beginning with $n = 2$. For example, there are three 2*p* orbitals, three 3*p* orbitals, and so forth. The orbitals of a given principal quantum number have the same size and shape but differ from each other in orientation. The contour surfaces of the three 2*p* orbitals are shown in Figure 5.18. It is convenient to label these as the $2p_x$, $2p_y$, and $2p_z$ orbitals. The letter subscript indicates the axis along which the orbital is oriented. As it turns out, there is no necessary connection between one of these subscripts and a particular value of m_l. To explain why this is so would require discussion of material beyond the scope of an introductory text.

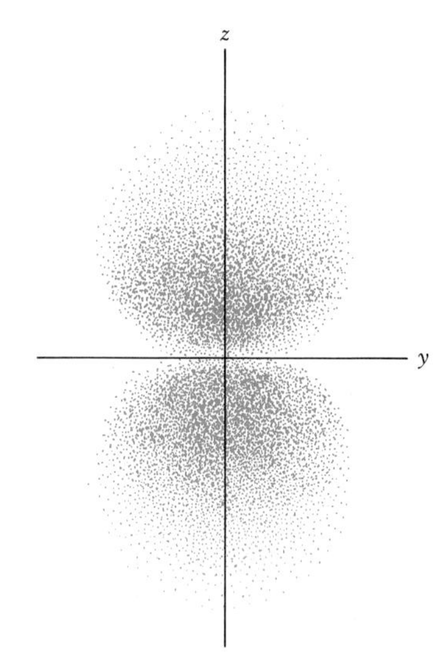

FIGURE 5.17 Electron-density distribution in a 2*p* orbital.

Just as with the *s* orbitals, the distance from the nucleus to the center of the electron density moves outward as we go from the 2*p* to 3*p* to 4*p* orbitals. In other words, orbital size increases with increase in the principal quantum number (n). In accurate contour representations of the 3*p* and higher *p* orbitals, there are small regions of electron density separating the major lobes. These details of the shapes of the *p* orbitals are not of major chemical importance. The general overall shape of the *p* orbitals is more important. We shall therefore always represent the *p* orbitals as shown in Figure 5.18, regardless of the value for the principal quantum number (n).

The *d* and *f* Orbitals

When $l = 2$, the orbital is a *d* orbital. There are no *d* orbitals with n lower than 3. This follows from the fact that l can never be larger than $n - 1$. There are five equivalent 3*d* orbitals corresponding to the five possible values for m_l: 2, 1, 0, −1, and −2. Similarly, there are five 4*d* orbitals, and so forth. Just as in the case of the *p* orbitals, the differing values of m_l correspond to different orientations of orbitals in space. The most useful representations of the 3*d* orbitals are shown in Figure 5.19. Notice that

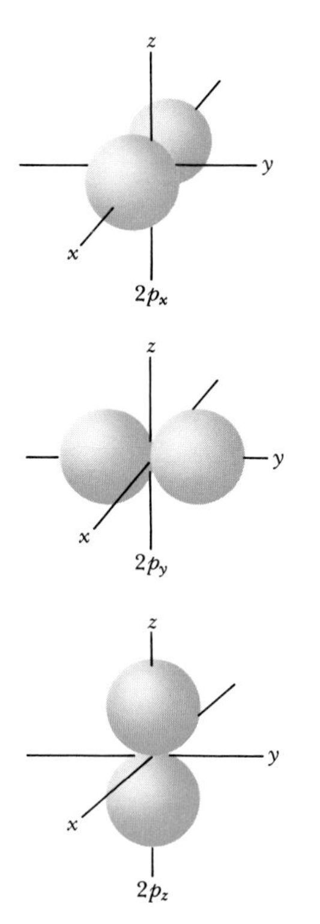

FIGURE 5.18 Contour representations of the three $2p$ orbitals. The three orbitals with differing orientations correspond to different values of the orientational quantum number, m_l. There is no necessary connection between a particular orientation and a particular value for m_l. The only important point to remember is that because there are three possible values for m_l, there are three p orbitals with differing orientations.

four of the orbitals are of the same shape but have differing orientations. The fifth orbital, labeled the d_{z^2}, has a different shape. It is not possible to represent the five d orbitals as having the same shape in an ordinary x, y, z axis system. Although the fifth d orbital looks different, it has the same energy in an atom as any of the other four orbitals.

The representations of higher d orbitals are very much like those for the $3d$. The contour representations shown in Figure 5.19 are commonly employed for all d orbitals, regardless of major quantum number.

There are seven equivalent f orbitals (for which $l = 3$) for each value of n of 4 or greater. The f orbitals are difficult to represent in three-dimensional contour diagrams. We shall have no need to concern ourselves with orbitals having values for l greater than 3.

As we shall see in a later chapter, an understanding of the number and shapes of atomic orbitals is important to a proper understanding of the molecules formed by combining atoms. *You should commit to memory the orbital representations shown in Figures 5.16, 5.18, and 5.19.*

The orbitals we have been describing are those of a hydrogen atom. In the atoms of other elements, there are many electrons moving about a single nucleus. These electrons are attracted by the nucleus and at the same time are repelled by one another. The resulting distribution of electron density and the energies of the allowed energy states of the electrons are thus the product of extremely complex forces. Nevertheless, as we shall see in the following chapter, the electronic structures of atoms having many electrons can be built up by the progressive addition of electrons to orbitals that are very much like those of the hydrogen atom.

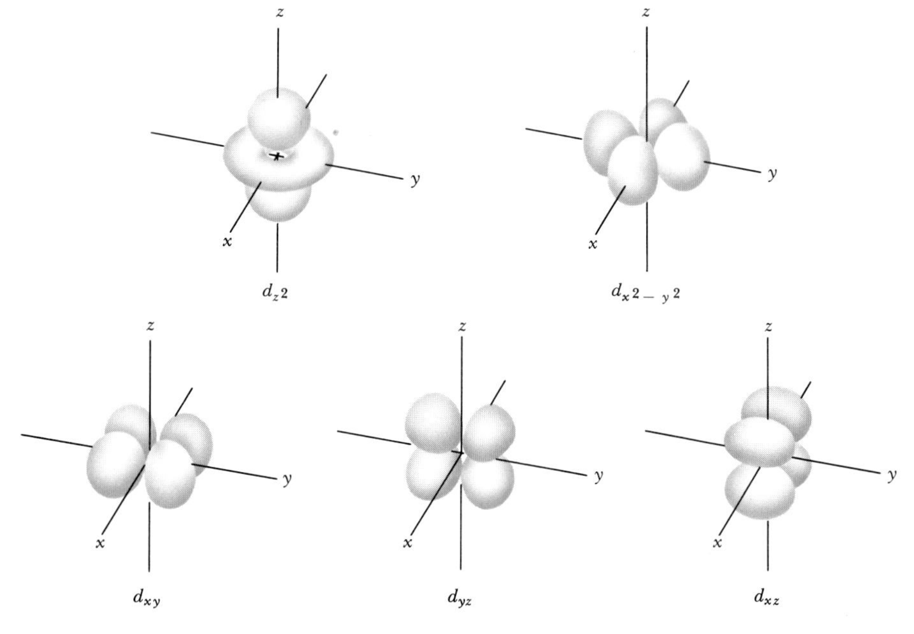

FIGURE 5.19 Contour representations of the five $3d$ orbitals.

FOR REVIEW

Summary

Radiant energy moves through a vacuum at the "speed of light," $c = 3.00 \times 10^8$ m/s; it has wavelike characteristics that allow it to be described in terms of wavelength (λ) and frequency (ν); these are interrelated: $c = \lambda\nu$. The dispersion of radiation into its component wavelengths produces a spectrum. If all wavelengths are present, the spectrum is said to be continuous; if only certain wavelengths are present, it is called a line spectrum.

The quantum theory describes the minimum amount of radiant energy that an object can gain or lose, $E = h\nu$; this smallest quantity is called a quantum. A quantum of radiant energy is called a photon. The quantum theory was used to explain the photoelectric effect and the line spectrum of the hydrogen atom. The absorptions or emissions of light by an atom, which produce its line spectrum, correspond to energy changes of electrons within the atom; the energy of the electron in an atom is quantized.

Electrons exhibit wave properties and can be described by a wavelength, $\lambda = h/mv$. Discovery of the wave properties of the electron led to the uncertainty principle, which indicates that the position and momentum of a particle as light as an electron can be determined simultaneously with only limited accuracy.

The ideas described above culminated in our current model of the electronic structures of atoms, in which we speak of the probability of the electron being found at a particular point in space. Although the positions of the electrons are defined in this averaged sense, their energies are precisely known. Each allowed state of an electron in an atom corresponds to a particular set of values for three quantum numbers. Each such allowed energy state is termed an orbital. An orbital is described by a combination of an integer and letters, corresponding to the three values for the quantum numbers. The principal quantum number (n) is indicated by the integers 1, 2, 3, ... This quantum number relates most directly to the size and energy of an orbital. The azimuthal quantum number (l) is indicated by the letters s, p, d, f, and so on, corresponding to values of l of 0, 1, 2, 3, ... The l quantum number defines the shape of the orbital. The magnetic or orientational quantum number (m_l) describes the orientation of the orbital in space. For example, the three $3p$ orbitals are designated $3p_x$, $3p_y$, and $3p_z$, the subscript letters indicating the axis along which the orbital is oriented.

Restrictions in the values of the three quantum numbers gives rise to the following allowed subshells:

$1s$
$2s$, $2p$
$3s$, $3p$, $3d$
$4s$, $4p$, $4d$, $4f$
⋮

There is one orbital in an s subshell, three in a p subshell, five in a d subshell, and seven in an f subshell. The contour representations are the most generally useful way to visualize the spatial characteristics of the orbitals.

Learning goals

Having read and studied this chapter, you should be able to:

1 Describe the wave properties and characteristic speed of propagation of radiant energy (electromagnetic radiation).

2 Use the relationship $\lambda\nu = c$, which relates the wavelength (λ) and frequency (ν) of radiant energy to its speed (c).

3 Describe how the wavelength and frequency differ in the various parts of the electromagnetic spectrum, such as the infrared, visible, and ultraviolet.

4 Explain the essential feature of Planck's quantum theory, namely, that the smallest increment, or quantum, of radiant energy of frequency ν that can be emitted or absorbed is $h\nu$, where h is Planck's constant.

5 Explain how Einstein accounted for the photoelectric effect by considering the radiant energy to be a stream of particlelike photons striking a metal surface. In other words, you should be able to explain all the observations about the photoelectric effect using Einstein's model.

6 Explain what is meant by the term *spectrum* and by the expression *line spectrum* in referring to the light emitted or absorbed by an atom.

7 List the assumptions made by Bohr in his model

of the hydrogen atom. Most important, you should be able to explain how Bohr's model relates to Planck's quantum theory.

8 Explain the concept of an allowed energy state and how this concept is related to the quantum theory.

9 Calculate the energy differences between any two allowed energy states of the electron in hydrogen.

10 Explain the concept of ionization energy.

11 Calculate the characteristic wavelength of a particle from a knowledge of its mass and velocity.

12 Describe the uncertainty principle and explain the limitations it places on our ability to define simultaneously the location and momentum of a subatomic particle, particularly an electron.

13 Explain the concepts of *orbital, electron density*, and *probability* as used in the quantum-mechanical model of the atom; explain the physical significance of ψ^2.

14 Describe the three quantum numbers used to define an orbital in an atom and list the limitations placed on the values each may have.

15 Describe the correspondence between letter designations and values for the azimuthal quantum number (l).

16 Describe the shapes of s, p, and d orbitals.

Key terms

Among the more important terms and expressions used for the first time in this chapter are the following:

The **azimuthal quantum number**, l (Section 5.5), is one of the quantum numbers that specifies an atomic orbital. The value for l defines the shape of the orbital. The values allowed for l are restricted by the value for the principal quantum number (n); l may take on integral values from 0 to $n - 1$.

The **electron density** (Section 5.5) at a particular point in space in an atom is the probability that the electron will be found in the region immediately around that point. This probability is expressed by ψ^2, the square of the wave function.

The **ionization energy** (Section 5.3) for the hydrogen atom is the energy required to move an electron from its lowest energy, or most stable, orbit to a point infinitely far from the nucleus. In effect, this involves removing the electron from the atom.

A **line spectrum** (Section 5.2) contains radiation only of certain specific wavelengths.

Matter wave (Section 5.4) is the term applied to the wave characteristics of a subatomic particle.

A **node** (Section 5.6), as applied to electron density in atoms, is the locus of points (for example, a plane or a spherical surface) at which the electron density is zero. For example, the node in a 2s orbital (Figure 5.15) is a spherical surface.

An **orbit** (Section 5.3) in the Bohr theory of the hydrogen atom is any one of the allowed energy states of the electron. Each orbit corresponds to a different value for the principal quantum number (n)

An **orbital** (Section 5.5) represents an allowed energy state of an electron in the modern, quantum-mechanical model of the atom. The term *orbital* is also used to describe the spatial distribution of the electron. An orbital is defined by specifying the values of three quantum numbers: n, l, and m_l.

The **orientational quantum number**, m_l (Section 5.5), describes the orientation of an orbital in space. This quantum number may have integral values ranging from l to $-l$. For each value of l there are thus $2l + 1$ values for m_l.

A **photon** (Section 5.2) is a quantum, or smallest increment, of radiant energy, $h\nu$.

The **principal quantum number**, n (Section 5.5), is the quantum number that relates most directly to the size and energy of an atomic orbital. It may take on integer values of 1, 2, 3, ...

A **quantum** (Section 5.2) is the smallest increment of radiant energy that may be absorbed or emitted. The magnitude of the quantum of radiant energy is $h\nu$.

Radiant energy, or **electromagnetic radiation** (Section 5.1), is a form of energy possessing wave character and being propagated through space with a characteristic speed 3.00×10^8 m/s.

The term **spectrum** (Section 5.2) refers to the distribution among various wavelengths of the light emitted or absorbed by an object. A continuous spectrum contains radiation distributed over many wavelengths.

A **subshell** (Section 5.5) is designated by a particular set of values for the quantum numbers n and l. For example, we speak of the 2p subshell ($n = 2$, $l = 1$), which is composed of three orbitals ($2p_x$, $2p_y$, and $2p_z$).

The **uncertainty principle** (Section 5.4) states that there is an inherent uncertainty in the precision with which we can simultaneously specify the location and momentum of a particle. It is of importance only for the lightest particles such as the electron.

A **wave function** (Section 5.5) is a mathematical description of an allowed energy state, or orbital, for an electron in the quantum-mechanical model for an atom. It is usually symbolized by the Greek letter ψ.

EXERCISES

Radiant energy

5.1 List the following types of electromagnetic radiation in order of increasing wavelength: (a) radiation from a room heater; (b) radiation from an FM station; (c) the green light from a traffic signal; (d) cosmic radiation from outer space; (e) X rays used in medical diagnosis.

5.2 (a) What is the wavelength of radiation whose frequency is 6.24×10^{13}/s? (b) What is the frequency of radiation whose wavelength is 2.20×10^{-6} nm? (c) What distance does light travel in 1.00 h?

5.3 The planetary space probe *Voyager II* should pass close to Neptune in 1989. At that time it will be 2.82×10^9 mi from Earth. How long will it take for the pictures transmitted from *Voyager II* to reach Earth?

5.4 Excited barium atoms emit radiation of 455-nm wavelength. What is the frequency of this radiation? From Figure 5.3 indicate the color of the radiation.

5.5 (a) A television relay transmitter operates at a frequency of 927.9 MHz. Calculate the wavelength of the signal from the transmitter. (b) What is the frequency of radiation with a wavelength of 1.27 m?

5.6 What is the wavelength of radiation with a frequency of (a) 4.3×10^{15} s^{-1}; (b) 6.68×10^9/s?

Quantum theory

5.7 Compare FM radio and visible-light radiations with respect to each of the following: (a) frequency; (b) speed; (c) wavelength; (d) ability to penetrate glass; (e) ability to penetrate a wooden wall; (f) energy of the radiation.

5.8 (a) Calculate the smallest increment of energy (a quantum) that can be emitted or absorbed at a wavelength of 667 nm. (b) Calculate the energy of a photon of frequency 4.5×10^{12}/s. (c) What frequency of radiation has photons of energy 2.15×10^{-18} J?

5.9 Calculate and compare the energy of an ultraviolet-light photon of wavelength 106 nm with that of an infrared photon of wavelength 44 μm.

5.10 In astronomy it is often necessary to be able to detect just a few photons, because the light signals from distant stars are so weak. A photon detector receives a signal of total energy 4.05×10^{-18} J from radiation of 540-nm wavelength. How many photons have been detected?

5.11 Excited chromium atoms strongly emit radiation of 427 nm. What color is this radiation? What is the energy, in kilojoules per photon, of radiation of this wavelength? What is the energy in kilojoules per mole?

5.12 The minimum energy required to remove an electron from nickel metal is 8.05×10^{-19} J. What is the maximum wavelength of radiation that will provide photons of at least this energy? What happens to the photoemission if the intensity of radiation of this wavelength is doubled?

5.13 Radiation that impinges on chemical substances may cause rupture of a chemical bond. If a minimum energy of 332 kJ/mol is required to break a carbon-chlorine bond in a plastic material, what is the longest wavelength of radiation that possesses the necessary energy?

5.14 Potassium metal requires radiation with wavelength shorter than 540 nm before it can emit an electron from its surface via the photoelectric effect. (a) What is the minimum energy required to produce this photoelectron? (b) What is the frequency of radiation of the minimum required energy? (c) If potassium is irradiated with light of 440-nm wavelength, what is the maximum possible kinetic energy of the emitted electron?

5.15 Indicate whether each of the following statements is true or false. Where it is false, correct it. (a) The speed of light increases in proportion to its frequency. (b) 10 Å = 1 nm. (c) Photons of light of 400-nm wavelength are of higher energy than photons of light of 500-nm wavelength. (d) A photon of energy 2.80×10^{-19} J corresponds to visible light. (e) The energy of a photon is inversely proportional to its wavelength. (f) The wavelength of a photon is proportional to its frequency.

Bohr's model; matter waves

5.16 Calculate the radius of the orbit of an electron in hydrogen in the $n = 4$ orbit.

5.17 Indicate whether energy is emitted or absorbed when the following electronic transitions occur in hydrogen: (a) $n = 3$ to $n = 5$; (b) from an orbit with radius 8.48 Å to one with radius 2.12 Å; (c) ionization of an electron from the $n = 3$ state.

5.18 What wavelength of light is absorbed when an electron moves from the $n = 2$ to $n = 5$ states in hydrogen? Compare this with the wavelengths observed in the emission spectrum of hydrogen (Figure 5.8).

5.19 A hydrogen emission line in the infrared region, at 1875.6 nm, corresponds to a transition from a higher n level to the $n = 3$ level. What is the value of n for the higher-energy level?

5.20 Calculate the shortest and longest possible wavelengths for hydrogen emission lines that involve the $n = 5$ level as the final state.

5.21 The Li^{2+} ion has only one electron. Would you expect the ionization energy for Li^{2+} to be larger or smaller than that for H? Explain.

5.22 For one-electron ions, the energy of the electron is given by the equation $E_n = -Z^2R_H(1/n^2)$, where Z is the atomic number of the nucleus. What is the energy required to remove the remaining electron from the He^+ ion? Refer to Sample Exercise 5.4 and compare this value with that for the hydrogen atom. Explain the reason for the difference.

5.23 (a) Calculate the radius of the electron orbit in hydrogen when $n = 3$. (b) The radius of the electron orbit for one-electron ions of nuclear charge Z greater than 1 is given by the equation $r = n^2(0.53 \times 10^{-8}\text{ cm})/Z$. Calculate the radius of the electron orbit for $n = 1$ in Li^{2+}.

5.24 What properties of an electron are involved in the electron microscope, used to obtain highly magnified images of biological molecules and particles on surfaces?

5.25 Compare the characteristic wavelengths of an electron and a proton, assuming that they are moving at the same speed.

5.26 Neutron diffraction has become an important technique for determining the structures of molecules. Calculate the velocity of a neutron that has a characteristic wavelength of 0.880 Å. (Refer to the inside back cover for information regarding the mass of the neutron.)

5.27 Indicate whether each of the following statements is true or false. Correct those that are false. (a) The characteristic wavelength of an elementary particle depends on its charge. (b) The n_1-to-n_3 transition in hydrogen requires more energy than does the n_2-to-n_5 transition. (c) Two photons of wavelength 400 nm have the same total energy as one photon of wavelength 200 nm. (d) Photons of visible light possess higher energy than photons of infrared radiation. (e) Emission of a photon of wavelength 486 nm results from an electron transition from the $n = 2$ to $n = 4$ state in hydrogen.

Wave functions; orbitals; quantum numbers

5.28 (a) Give the values for n, l, and m_l for each orbital in the $4d$ subshell; (b) for each orbital in the $n = 3$ shell.

5.29 How do the three $3p$ orbitals differ from one another? How do the $2p$ and $3p$ orbitals differ?

5.30 Sketch the contour representations for the following orbitals: (a) s; (b) p_z; (c) d_{xy}; (d) $d_{x^2-y^2}$.

5.31 Which of the following are incorrect designations for an atomic orbital: $3f$, $3d$, $2p$, $4s$, $4f$, $2d$?

5.32 What is the physical significance of the square of the wave function?

5.33 What characteristics of an orbital are determined by (a) the principal quantum number, n; (b) the azimuthal quantum number, l; (c) the orientational quantum number, m_l?

5.34 Indicate whether each of the following statements is true or false. Correct those that are false. (a) The energy of an electron in hydrogen depends only on the principal quantum number, n. (b) The energies of electrons in H and He^+ are the same when the principal quantum number n is the same. (c) The number of orbitals in a subshell of azimuthal quantum number l is the same regardless of the value of the major quantum number n. (d) The set $n = 4$, $l = 3$, and $m_l = 3$ is a permissible set of quantum numbers for an electron in hydrogen. (e) The contour representation of the $3p_z$ orbital looks much like that for the $3d_{z^2}$ orbital.

5.35 Indicate the number of orbitals that can have each of the following designations: (a) $n = 5$; (b) $3p$; (c) $3f$; (d) $4d_{xz}$.

Additional exercises

5.36 A high-powered laser is pulsed on for a period of 100 ns. During that time it emits 5.00×10^{21} photons of wavelength 1.05 μm. What is the total energy emitted during the pulse?

5.37 (a) What is the characteristic frequency of a photon of energy 2.15×10^{-19} J? (b) What is the energy of a photon of wavelength 482 nm?

5.38 Chlorophyll absorbs blue light, $\lambda = 460$ nm, and emits red light, $\lambda = 660$ nm. Calculate the net energy change in the chlorophyll system when a single photon of 460-nm wavelength is absorbed, and a photon of 660-nm wavelength is emitted.

5.39 (a) The distance that light can travel in 365.25 days is called a light-year. What is this distance in kilometers? (b) A parsec is an astonomical unit of distance equal to 3.26 light-years. What is this distance in kilometers?

[5.40] Using Appendix D, calculate the energy change for the process $O_2(g) \longrightarrow 2O(g)$. This process, called photodissociation, occurs in the earth's upper atmosphere upon absorption of solar radiation. Calculate the maximum wavelength that a photon may have if it is to have sufficient energy to cause this dissociation.

5.41 (a) How many photons of high-energy ultraviolet light, wavelength 460 Å, are required to supply 1 J? (b) How many photons of infrared radiation of 46-μm wavelength are required to supply 1 J?

5.42 In the photoelectric effect, the minimum energy required to eject an electron from a metal is called the work function. Rubidium metal will emit electrons when the wavelength of radiation is 574 nm or less. (a) Calculate the work function for rubidium. (b) If rubidium is irradiated with 420-nm light, what is the maximum kinetic energy of the emitted electrons?

5.43 When the light from a neon light is dispersed in a prism to form a spectrum, it is found that the spectrum is not continuous; rather, it consists of several sharp lines, each of a specific frequency. Explain in general terms why a line spectrum is produced.

5.44 Draw the contour representations for all of the orbitals with quantum numbers $n = 3$, $l = 2$.

5.45 How many orbitals are there on an atom that have the following designations associated with them: (a) $4d$; (b) $3p_x$; (c) s; (d) $3s$?

5.46 Which of the following transitions in a He^+ ion would require the lowest energy photon: (a) $2s \longrightarrow 3p$; (b) $1s \longrightarrow 4s$; (c) $4p \longrightarrow 5p$?

5.47 Predict how the curves for the plots of ψ^2 versus r would appear for He^+ as compared with those for H shown in Figure 5.16. Explain your answer.

5.48 Write the set of quantum numbers allowed for each of the following orbitals: (a) 3*s*; (b) 2*p*; (c) 5*d*.

5.49 State the values that may be assumed by each of the following quantum numbers: (a) n; (b) l; (c) m_l.

5.50 Deuterium is the isotope of hydrogen with a nuclear mass of 2 (Section 2.6). What differences, if any, would you expect to see in the emission spectrum of deuterium as compared with that for hydrogen?

5.51 State the essential idea involved in the Heisenberg uncertainly principle. How does it relate to the idea of a probability distribution for the electron in the space around the nucleus in an atom?

5.52 How does the spectrum of the sun, Figure 5.9, relate to the quantum theory?

6 Electronic Structure: Periodic Relationships

We might say that the modern era in chemistry had its beginnings with Dalton's atomic theory. It happened that at this same time several other notable discoveries occurred to provide new impetus to experimental studies in chemistry. Many of the elements were isolated for the first time during the 50-year period following Dalton's work (1803–1807). As the quantity of chemical information grew, more attention was given to the possibilities of classifying it in useful ways. The most important product of these attempts at classification is the periodic table, which we introduced in Chapter 2.

In Chapter 5, we saw how the work of physicists led to our present understanding of atomic structure, at least as far as the hydrogen atom is concerned. The application of the concepts outlined there to atoms containing more than one electron proved to be possible in principle, but difficult in practice. While physicists were wrestling with this problem, chemists were making significant progress in understanding atomic structure from another point of view. In brief, they asked what sorts of inferences about the arrangements of electrons in atoms they might be able to make by using the periodic table and the known chemical behavior of the elements. One of the chief contributors to this development was the American chemist Gilbert N. Lewis (1875–1946). Even before Bohr had proposed his theory of the hydrogen atom, Lewis had suggested that electrons in atoms are arranged in shells.

In this chapter we will see why there is a close connection between the periodic table and the ways in which electrons are arranged in atoms. Our first goal will be to use the concepts developed in Chapter 5 to understand the electronic structures of many-electron atoms. We will then relate these electronic structures to the periodic table and to the chemical behavior of the elements.

6.1 ORBITALS IN MANY-ELECTRON ATOMS

The electronic structures of atoms with two or more electrons can be described in terms of orbitals like those we have described for hydrogen. Thus we can continue to use orbital designations such as $1s$, $2p_x$, and so forth. Furthermore, each of these orbitals has the same shape as those for hydrogen.

To describe the electronic structure of an atom we need to know how the orbitals are occupied by electrons. In an atom in its ground state, the electrons are always found in the lowest-energy orbitals available. We therefore need to consider the relative energies of the orbitals.

The energy of any atom or ion with only one electron depends only on the principal quantum number, n, and on the nuclear charge, Z:

$$E_n = -R_H\left(\frac{Z^2}{n^2}\right) \quad [6.1]$$

Notice that Equation 5.6, which applies to hydrogen, is just a special case of this equation, for which $Z = 1$. For many-electron atoms, the energy also depends on the azimuthal quantum number, l; that is, the energy differs for the different subshells. To understand why the s, p, d, and f subshells have different energies in a many-electron atom, we need to consider the forces that operate between the electrons, and the way in which these forces are influenced by the shapes of the orbitals. In a many-electron atom each electron is acted upon not only by the attractive force of the nucleus, but also by the repulsive forces of all the other electrons in the atom. In principle, it is possible to calculate the net force acting on each electron. In practice, this problem is almost impossibly complicated for any but the simplest atoms. However, we can get a good idea of what is involved by simplifying the problem somewhat. Let's focus on one electron at a time, and consider how it interacts with the *average* environment created by the nucleus and all the other electrons in the atom. Because of their opposite charge, electrons that are relatively close to the nucleus screen or shield the nucleus from the electrons farther out. That is, there is a decrease in the effective positive charge acting on an electron because of the other electrons between it and the nucleus. We refer to this effect as the **screening effect**; the resultant positive charge that the electron experiences is called the **effective nuclear charge**. The effective nuclear charge that a particular electron experiences, Z_{eff}, depends on the number of protons in the nucleus, Z, minus the average number of electrons that are between it and the nucleus, S:

$$Z_{eff} = Z - S \quad [6.2]$$

This idea is illustrated in Figure 6.1. At some particular instant, the electron shown is at a distance r from the nucleus. Assume that the averaged motion of all the other electrons results in a spherical electron distribution. All the electron density within the sphere of radius r shields the electron from the nucleus; that electron density has the same effect as it would if concentrated at the nucleus. Therefore, the electron density between the nucleus and the electron on which we have focused cancels a certain amount of the nuclear charge. Of course each electron is moving about; as it moves closer to or farther from the nucleus, the effective

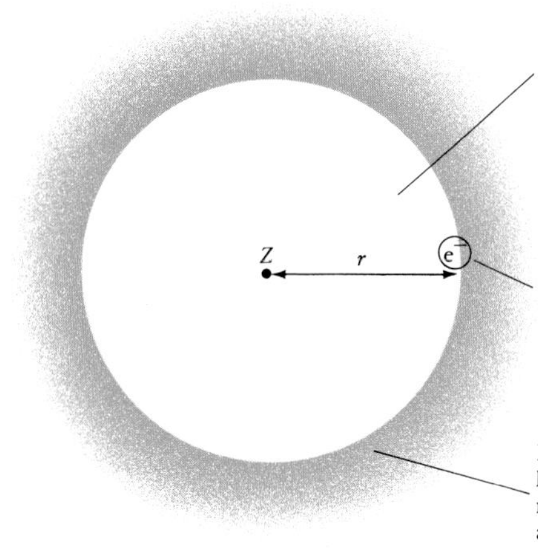

Electronic charge, S, within sphere of radius r counters the nuclear charge, so $Z_{eff} = Z - S$

Z r e^-

Electron of interest at radius r from nucleus

Electrons outside sphere or radius r have no effect on value of effective nuclear charge experienced by electron at radius r

FIGURE 6.1 Shielding of the nuclear charge from an electron by other electrons in an atom. As an example, if the nuclear charge were 5, and the sphere of radius r contained three electrons, the effective nuclear charge would be $5 - 3 = 2$.

nuclear charge it experiences increases or decreases. However, we are interested in its *average* behavior.

The manner in which the probability function (ψ^2) varies as we move outward from the nucleus differs for orbitals in different subshells. Consider the orbitals for which $n = 3$. The $3s$ electron distribution extends closer to the nucleus than does the $3p$; the $3p$, in turn, extends closer than does the $3d$. As a result, the other electrons of the atom (that is, the $1s$, $2s$, and $2p$) shield the $3s$ electrons from the nucleus less effectively than they shield the $3p$ or $3d$. Thus the $3s$ electrons experience a larger Z_{eff} than do the $3p$ electrons, and these in turn experience a larger Z_{eff} than do the $3d$ electrons.

The energy of an electron depends on the effective nuclear charge, Z_{eff} in a way analogous to Equation 6.1:

$$E_n \propto -Z_{eff}^2 \qquad [6.3]$$

where the symbol $\propto$ is read as "proportional to." Since Z_{eff} is larger for the $3s$ electrons, they have a lower energy (that is, they are more stable) than the $3p$, which in turn are lower in energy than the $3d$. As a result, the energy-level diagram for the orbitals of many-electron atoms is like that shown in Figure 6.2. Keep in mind that this is a *qualitative* energy-level diagram; the exact energies and their spacings differ from one atomic species to another. In all cases, however, the relative energies of the orbitals through $n = 3$ are as shown. Notice that all orbitals of a given subshell, such as the $3d$ orbitals, have the same energy. In scientific jargon, orbitals that have the same energy are said to be **degenerate.**

SAMPLE EXERCISE 6.1

Based on the energy-level diagram of Figure 6.2, would you expect the average distance from the nucleus of a $3d$ electron to be greater or less than that of a $2p$? Explain.

Solution: The energy of the $2p$ orbitals is considerably lower than for a $3d$. This indicates that the average attractive interaction of the $2p$ electron with the nucleus is much greater than for an electron in a $3d$ orbital. The increased attractive interaction is due to a smaller average distance of the $2p$ electron from the nucleus.

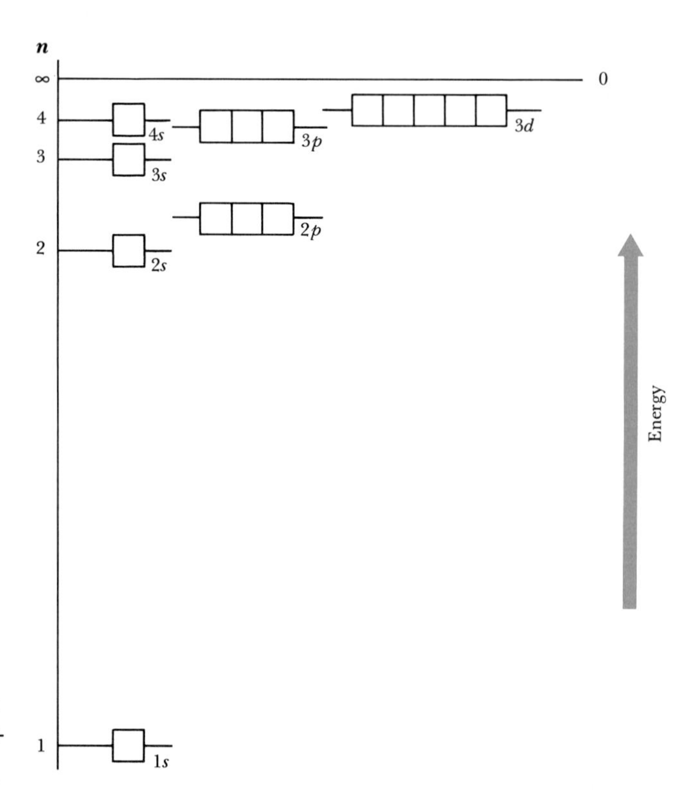

FIGURE 6.2 Ordering of orbital energy levels in many-electron atoms. Note that orbitals with the same value for the principal quantum number, *n*, but differing values of azimuthal quantum number, *l*, differ in energy.

With this much background we now have an idea of the ordering of the energies of atomic orbitals in many-electron atoms. We must now consider what rules govern the placement of electrons into these orbitals. However, before the rules can be stated in their most useful form, we must learn about a fourth, and final, quantum number.

6.2 ELECTRON SPIN AND THE PAULI EXCLUSION PRINCIPLE

When the atomic spectra of atoms were first studied, certain complicating features of the spectra were noted. Lines that were at first thought to be single lines were found under high resolution to be closely spaced pairs. The only way in which these extra splittings could be accounted for was to introduce a new quantum number in addition to the three quantum numbers we have already encountered. This new quantum number is associated with the electron itself. The electron behaves as though it were spinning on its own axis and thereby acting like a tiny magnet. It was necessary to define an **electron-spin quantum number** (m_s), which has values of $+\frac{1}{2}$ or $-\frac{1}{2}$, corresponding to the two possible orientations of the electron spin in a magnetic field.

In 1921, Otto Stern and Walter Gerlach succeeded in actually separating a beam of atoms into two groups according to the orientation of the electron spin. Their experiment is diagrammed in Figure 6.3. Let us assume that the beam of atoms is hydrogen. We saw in Section 2.6 that charged particles are deflected upon moving through a magnetic field. In the Stern-Gerlach experiment, since the atoms moving through the magnetic field are neutral, any deflection of the beam cannot be due to charge on the atoms. However, the magnet due to the electron's spin

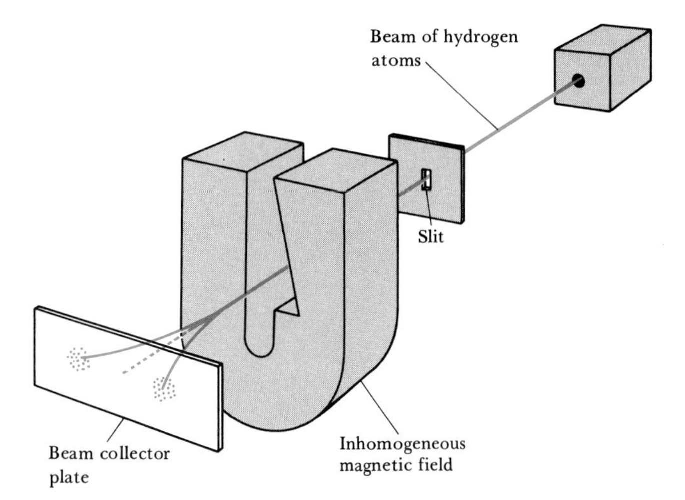

FIGURE 6.3 Diagrammatic illustration of the Stern-Gerlach experiment. A beam of hydrogen atoms is allowed to pass through an inhomogeneous magnetic field. Atoms in which the electron-spin quantum number, m_s is $+\frac{1}{2}$ are deflected in one direction, whereas those in which m_s is $-\frac{1}{2}$ are deflected in the other.

does interact with the magnetic field, causing the atom to be deflected from its straight-line path. The direction in which the atom is deflected depends on the orientation of the spin of the electron. We expect that there will be equal numbers of electrons with each of the two possible orientations, as is found to be the case. The presence of electron spin turns out to be important in determining the electronic structures of atoms. The first to recognize this was a German physicist, Wolfgang Pauli. In 1924 he spelled out what has become known as the **Pauli exclusion principle.** This principle declares that no two electrons in an atom can have the same set of four quantum numbers, n, l, m_l, and m_s. As we will see in the next section, the Pauli exclusion principle dictates that each orbital can hold a maximum of two electrons. This restriction provides the key to one of the great problems of chemistry—an explanation of the structure of the periodic table of the elements.

6.3 THE PERIODIC TABLE AND ELECTRON CONFIGURATIONS

The essential feature of the periodic table is that when elements are arranged in order of increasing atomic number, similarities in physical and chemical properties are repeated at regular intervals (Section 2.7). Elements with similar properties such as the alkali metals, group 1A, are placed together in columns called groups or families.* The horizontal rows of the periodic table are called rows or periods. The first row consists of only two members, H and He. The next two rows, starting with Li and Na, contain eight members each. The following rows, beginning with K, Rb, and Cs, contain 18, 18, and 32 members, respectively. Why do certain elements possess chemical similarities, and why does the periodic table have its particular shape with rows containing 2, 8, 18, 18, and 32 members? The answers to both questions are found in the nature of the arrangements or configurations of electrons in atoms—that is, in how electrons populate the available orbitals.

The atomic number corresponds not only to the number of protons in the nucleus of an atom, but also to the number of electrons in that atom.

*Hydrogen is generally placed in family 1A or 7A. It shows a few remote resemblances to both families but is really not a member of either.

Thus the periodic table represents an ordering of the elements not only according to nuclear charge, but also according to the number of electrons about the nucleus. Gilbert N. Lewis, who was mentioned in the introduction, reasoned that if the chemical properties of the elements are repeated at intervals, then their electron configurations must be repeating in some way. He suggested that electrons in atoms are arranged in shells, and that once a shell of electrons is filled, additional electrons must go into a new shell. He further reasoned that the noble gases, which are chemically very nonreactive, have closed or filled shells of electrons. Chemical behavior might then be understood in terms of the tendency of an atom to achieve a closed shell by gaining, losing, or sharing electrons. For example, we saw in Section 2.9 that sodium, atomic number 11, loses one electron in chemical reactions to form the sodium ion, Na^+. This ion has the stable, closed-shell electron arrangement of the noble gas neon, atomic number 10. These ideas of Lewis generate additional questions: Why do certain electron configurations occur periodically? Why are electrons arranged in shells?

To answer these questions and to account in more detail for the way in which various atomic properties vary with atomic number, we must apply the ideas about orbitals that we have learned about in this and the preceding chapter. Our goal is to understand how electrons are arranged in each element of the periodic table. When you have mastered this material, you will be able to describe the arrangement of electrons in any element. This is an important skill; as we have implied, electron arrangements are the key to the chemical behavior of the elements. So let's consider how the electrons in each element populate orbitals, starting with hydrogen and then moving through the periodic table with increasing atomic number.

The Electron Configurations of the Elements

The most stable, or ground, state of an atom will be that in which all the electrons are in the lowest possible energy states. If there were no restrictions on the possible values for the quantum numbers of the electrons, all the electrons would crowd into the 1*s* orbital, because this is lowest in energy. The Pauli principle, however, tells us that there are limits on the quantum numbers that the electrons may have. The 1*s* orbital corresponds to values $n = 1$, $l = 0$, and $m_l = 0$. The electron that occupies this orbital can have a spin quantum number (m_s) of $+\frac{1}{2}$ or $-\frac{1}{2}$. If one electron has quantum numbers $n = 1, l = 0, m_l = 0$, and $m_s = +\frac{1}{2}$, then a second electron can have $n = 1, l = 0, m_l = 0$, and $m_s = -\frac{1}{2}$. Thus the exclusion principle requires that there can be at most *two electrons* in the 1*s* orbital, or for that matter *in any single atomic orbital*. In hydrogen there is one electron in the 1*s* orbital. In helium, atomic number 2, there are two electrons in this orbital. These electrons must have opposite values for the electron-spin quantum number (m_s). The electrons are said to be paired. Because of this pairing, the magnetic properties of the electrons effectively cancel. If the Stern-Gerlach experiment (Figure 6.3) were performed on a beam of helium atoms, no separation of the beam by the magnetic field would occur as it does with hydrogen.

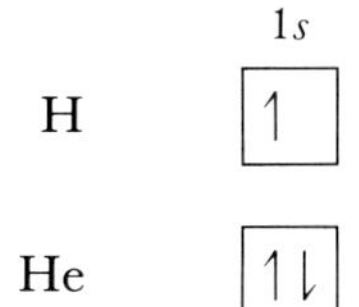

A particular arrangement of electrons in the orbitals of an atom is referred to as an **electron configuration.** It is convenient to have a shorthand notation for representing the electron configuration. This is done by writing the symbol for each subshell occupied by an electron, with a superscript to indicate the number of electrons occupying that subshell. For hydrogen the electron configuration is $1s^1$; for helium it is $1s^2$. Another way of representing the electron configurations of these atoms is shown at left. In this kind of diagram each orbital is represented by a box, and each electron by a half-arrow. A half-arrow pointing upward (↿) represents an electron spinning in one direction ($m_s = +\frac{1}{2}$), whereas a downward half-arrow (⇂) represents an electron spinning in the opposite direction ($m_s = -\frac{1}{2}$). We shall refer to representations of this type as **orbital diagrams.**

The two electrons present in helium complete the filling of orbitals with principal quantum number $n = 1$. Helium therefore possesses a very stable electron configuration, as reflected in its chemical inertness. The electron configurations of lithium and of several elements that follow it in the periodic table are shown in Table 6.1. Recall that a maximum of two electrons can be placed in each orbital. Thus, for lithium, with three electrons, the third electron cannot enter the 1*s* orbital, but must be placed in the next most stable orbital, the 2*s* (refer to Figure 6.2). The change in principal quantum number for the third electron represents a large jump in energy, and a corresponding jump in the average distance of the electron from the nucleus. We may say that it represents the start of a new shell of electrons. As you can see by examining the periodic table, lithium represents the start of a new row of the periodic table. It is the first member of the alkali metals family (group 1A).

The element that follows lithium is beryllium; its electron configuration is $1s^22s^2$ (Table 6.1). Boron, atomic number 5, has an electron configuration $1s^22s^22p^1$. The fifth electron must be placed in a 2*p* orbital, because the 2*s* orbital is filled. Because each of the three 2*p* orbitals are of equal energy, it doesn't matter which 2*p* orbital is occupied. With the

TABLE 6.1 Electron configurations of several lighter elements

Element	Total electrons	Orbital diagram 1s	2s	2p			3s	Electron configuration
Li	3	↿⇂	↿					$1s^22s^1$
Be	4	↿⇂	↿⇂					$1s^22s^2$
B	5	↿⇂	↿⇂	↿				$1s^22s^22p^1$
C	6	↿⇂	↿⇂	↿	↿			$1s^22s^22p^2$
Ne	10	↿⇂	↿⇂	↿⇂	↿⇂	↿⇂		$1s^22s^22p^6$
Na	11	↿⇂	↿⇂	↿⇂	↿⇂	↿⇂	↿	$1s^22s^22p^63s^1$

next element, carbon, we come to a new situation. We know that the sixth electron must go into a $2p$ orbital, where there is already one electron. However, does this new electron go into the $2p$ orbital that already has one electron, or into one of the others? This question is answered by **Hund's rule,** which states that electrons occupy degenerate orbitals singly to the maximum extent possible, and with their spins parallel. In the case of carbon, then, the sixth electron goes into one of the other $2p$ orbitals, and with its spin in the same orientation as the other $2p$ electron. Hund's rule is based on the fact that electrons repel one another because they have the same electrical charge. By occupying different orbitals, the electrons remain as far as possible from one another in space, thus minimizing electron-electron repulsions. When electrons must occupy the same orbital, the repulsive interaction between the paired electrons is greater than between electrons in different, equivalent orbitals.

Neon, the last member of the second row, has ten electrons. Two electrons fill the $1s$ orbital, two electrons fill the $2s$ orbital, and the remaining six electrons fill the $2p$ orbitals. The electron configuration is thus $1s^22s^22p^6$ (Table 6.1). In neon, all of the orbitals with $n = 2$ are filled. The filling of the $2s$ and $2p$ orbitals by the eight electrons that they can hold represents a very stable configuration. As a result, neon is chemically quite inert. Lewis, in his model for the electron configurations of elements, noted that the octet of electrons (eight) in the outermost shell of an atom or ion represents an especially stable arrangement.

Sodium, atomic number 11, marks the beginning of a new row of the periodic table. Sodium has a single $3s$ electron beyond the stable configuration of neon. We can abbreviate the electron configuration of sodium as follows:

$$\text{Na} \qquad [\text{Ne}]3s^1$$

The symbol [Ne] represents the electron configuration of the ten electrons of neon, $1s^22s^22p^6$. Writing the electron configuration in this manner helps us focus attention on the outermost electrons of the atom. The outer electrons are the ones largely responsible for the chemical behavior of an element. For example, we can write the electron configuration of lithium as follows:

$$\text{Li} \qquad [\text{He}]2s^1$$

By comparing this with the electron configuration for sodium, it is easy to appreciate why lithium and sodium are so similar chemically: They have the same type of outer-shell electron configuration. All the members of the alkali metal family (group 1A) have a single s electron beyond an inner-core noble-gas configuration. The outer-shell electrons are often referred to as **valence-shell electrons.**

SAMPLE EXERCISE 6.2

Draw the orbital diagram representation for the electron configuration of oxygen, atomic number 8.

Solution: The ordering of orbitals is as shown in Figure 6.2. Two electrons each go into the $1s$ and $2s$

orbitals. This leaves four electrons for the three $2p$ orbitals. Following Hund's rule, we put one electron into each $2p$ orbital until all three have one each. The fourth electron must then be paired up with one of the three electrons already in a $2p$ orbital, so that the correct representation is

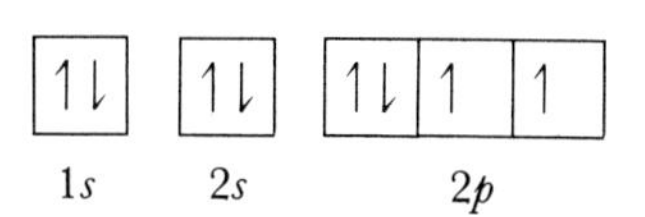

The corresponding electron configuration is written $1s^22s^22p^4$ or $[He]2s^22p^4$. The $1s^2$ or [He] electrons are the inner-shell or core electrons of the oxygen atom. The $2s^22p^4$ electrons are the outer-shell or valence electrons.

SAMPLE EXERCISE 6.3

What is the characteristic outer-shell electron configuration of the group 7A elements, the halogens?

Solution: The first member of the halogen family is fluorine, atomic number 9. The abbreviated form of the electronic configuration for fluorine is

$$\text{F} \qquad [He]2s^22p^5$$

Similarly, the abbreviated form of the electron configuration for chlorine, the second halogen, is

$$\text{Cl} \qquad [Ne]3s^23p^5$$

From these two examples we see that the characteristic outer-shell electron configuration of a halogen is ns^2np^5, where n ranges from 2 in the case of fluorine to 5 in the case of iodine.

The rare-gas element argon marks the end of the row started by sodium. The configuration for argon is $1s^22s^22p^63s^23p^6$. The element following argon in the periodic table is potassium (K), atomic number 19. In all its chemical properties, potassium is very obviously a member of the alkali metal family. The experimental facts about the properties of potassium leave no doubt that the outermost electron of this element occupies an s orbital. But this means that the highest-energy electron has *not* gone into a $3d$ orbital, which we might naïvely have expected it to do. In this case the ordering of energy levels is such that the $4s$ orbital is lower in energy than the $3d$ (see Figure 6.2).

Following complete filling of the $4s$ orbital (this occurs in the calcium atom), the next set of equivalent orbitals to become filled is the $3d$. (You'll find it helpful as we go along to refer often to the periodic table on the inside front cover.) Beginning with scandium, and extending through zinc, electrons are added to the five $3d$ orbitals until they are completely filled. Thus the fourth row of the periodic table is ten elements wider than the previous rows because of the insertion of the elements known as the **transition metals.** Note the position of these ten elements in the periodic table (inside front cover).

In accordance with Hund's rule, electrons are added to the $3d$ orbitals singly until all five orbitals have one electron each. Additional electrons are then placed in the $3d$ orbitals with spin pairing until the shell is completely filled. The orbital diagram representations and electron configurations of two transition elements are as follows:

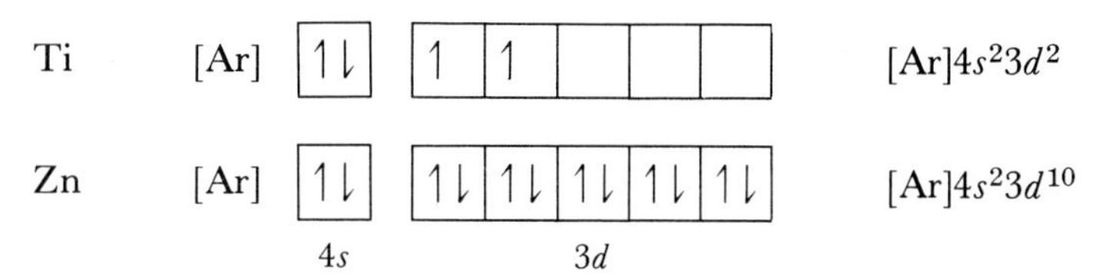

The $3d$ and $4s$ orbital energies are very close together. Occasionally, an electron may be moved from one of these types of orbitals to another. For example, we might expect chromium to have the outer electron configuration $4s^23d^4$, but it is actually $4s^13d^5$. This anomalous behavior is partly due to the special stability associated with precisely half-filled sets of degenerate orbitals. Apparently, there is just enough gain in stability in arriving at this arrangement to cause the electron to move from a $4s$ to a $3d$ orbital.

Upon completion of the $3d$ transition series, the $4p$ orbitals begin to be occupied, until the completed octet of outer electrons is again arrived at in krypton (Kr), atomic number 36. Krypton is another of the rare gases. Rubidium (Rb) marks the beginning of the fifth row of the periodic table. This row is in every respect like the preceding one, except that the value for n is one greater. The sixth row of the table begins similarly to the preceding one: one electron in the $6s$ orbital of cesium (Cs) and two electrons in the $6s$ orbital of barium (Ba). The next element, lanthanum (La), represents the start of the third series of transition elements. But with cerium (Ce), element 58, a new set of orbitals, the $4f$, enter the picture. The energies of the $5d$ and $4f$ orbitals are very close. For lanthanum itself, the $5d$ orbital energy is just a little lower than the $4f$. However, for the elements immediately following lanthanum, the $4f$ orbital energies are a little lower, so that the highest energy electrons go into the $4f$ orbitals.

There are seven equivalent $4f$ orbitals, corresponding to the seven allowed values of m_l, ranging from 3 to -3. Thus it requires 14 electrons to fill the $4f$ orbitals completely. The 14 elements corresponding to the filling of the $4f$ orbitals are elements 58 to 71, known as the **rare-earth,** or **lanthanide,** elements. In order not to make the periodic table unduly wide, the rare-earth elements are set together below the other elements. The properties of the rare-earth elements are all quite similar, and they occur together in nature. For many years it was virtually impossible to separate them from one another.

Following completion of the rare-earth series, the third transition element series is completed, followed by filling of the $6p$ orbitals. This brings us to radon (Rn), heaviest of the rare-gas elements. The final row of the periodic table begins as the one before it. The **actinide** elements involve completion of the $5f$ electron orbitals. This series consists mainly of elements not found in nature, those that have been synthesized in nuclear reactions.

Using the Periodic Table to Write Electron Configurations

Our rather brief survey of electron configurations of the elements has taken us through the entire periodic table. You will find that a familiarity with the general structure of the table will enable you to write down the electronic configuration of any element. There are a few instances in which minor shifts of an electron or two from one orbital to another occur, when orbitals have closely similar energies. We have given as one example the case of chromium, which possesses an outer electron configuration $4s^13d^5$, rather than the $4s^23d^4$ we might have expected. Another

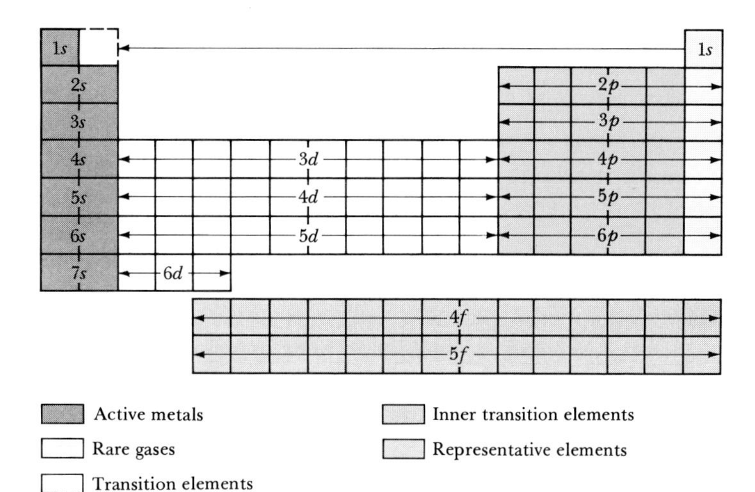

FIGURE 6.4 Block diagram of the periodic table showing the groupings of the elements according to the type of orbital being filled with electrons. The arrangement of elements in this figure is the same as in the periodic table.

interesting case occurs with copper and its congeners* silver and gold. In copper the configuration is found to be $4s^13d^{10}$. Evidently, the stability associated with completing the *d* subshell causes the electron to move from *ns* to $(n - 1)d$. There are a few other similar instances among the transition elements, lanthanides, and actinides. Although these minor departures from the expected are interesting, they are not of great chemical significance.

We've seen that the periodic table is structured so that elements with the same outer-shell electron configuration are arranged in columns. The elements can be grouped also in terms of the *type* of orbital into which the electrons are placed. These different groupings are indicated by the shadings on the outline of the periodic table shown in Figure 6.4. The first two groups of elements on the left contain the **active metals**, with outermost *s* electrons. The **transition elements** are those for which the *d* orbitals are incomplete and being filled. The **representative elements** are those for which the outermost *p* orbitals are being filled. The **rare-gas**, or noble-gas, elements are those for which the octet of outermost electrons has been attained. The two series of elements for which the *f* orbitals are being filled are sometimes called the **inner transition elements.**

*A congener is an element in the same family or group of the periodic table as another.

SAMPLE EXERCISE 6.4

Write the electron configuration for the element bismuth, atomic number 83.

Solution: We can do this by simply moving across the periodic table one row at a time and writing the occupancies of the orbitals corresponding to each row (refer to Figure 6.4):

First row	$1s^2$
Second row	$2s^22p^6$
Third row	$3s^23p^6$
Fourth row	$4s^23d^{10}4p^6$
Fifth row	$5s^24d^{10}5p^6$
Sixth row	$6s^24f^{14}5d^{10}6p^3$

Total: $1s^22s^22p^63s^23p^63d^{10}4s^24p^64d^{10}4f^{14}5s^25p^65d^{10}6s^26p^3$

Note that 3 is the lowest possible value that n may have for a d orbital, and that 4 is the lowest possible value of n for an f orbital.

The total of the superscripted numbers should equal the atomic number of bismuth, 83. It does not matter a great deal precisely in which order the orbitals are listed. They may be listed, as shown above, in the order of increasing major quantum number. However, it is also possible to list them in the sequence read from the periodic table: $1s^22s^22p^63s^23p^64s^23d^{10}4p^65s^24d^{10}5p^66s^24f^{14}5d^{10}6p^3$.

It is a simple matter to write the abbreviated electron configuration of an element using the periodic table. First locate the element of interest (in this case element 83) and then move backward until the first noble gas is encountered (in this case Xe, element 54). Thus the inner core is [Xe]. The outer electrons are then read from the periodic table as before. Moving from Xe to Cs, element 55, we find ourselves in the sixth row. Moving across this row to Bi gives us the outer electrons. The complete electron configuration is thus: [Xe]$6s^24f^{14}5d^{10}6p^3$.

TABLE 6.2 The electron configurations of the elements

Atomic number	Symbol	Electron configuration	Atomic number	Symbol	Electron configuration	Atomic number	Symbol	Electron configuration
1	H	$1s^1$	36	Kr	[Ar]$4s^23d^{10}4p^6$	71	Lu	[Xe]$6s^24f^{14}5d^1$
2	He	$1s^2$	37	Rb	[Kr]$5s^1$	72	Hf	[Xe]$6s^24f^{14}5d^2$
3	Li	[He]$2s^1$	38	Sr	[Kr]$5s^2$	73	Ta	[Xe]$6s^24f^{14}5d^3$
4	Be	[He]$2s^2$	39	Y	[Kr]$5s^24d^1$	74	W	[Xe]$6s^24f^{14}5d^4$
5	B	[He]$2s^22p^1$	40	Zr	[Kr]$5s^24d^2$	75	Re	[Xe]$6s^24f^{14}5d^5$
6	C	[He]$2s^22p^2$	41	Nb	[Kr]$5s^14d^4$	76	Os	[Xe]$6s^24f^{14}5d^6$
7	N	[He]$2s^22p^3$	42	Mo	[Kr]$5s^14d^5$	77	Ir	[Xe]$6s^24f^{14}5d^7$
8	O	[He]$2s^22p^4$	43	Tc	[Kr]$5s^24d^5$	78	Pt	[Xe]$6s^14f^{14}5d^9$
9	F	[He]$2s^22p^5$	44	Ru	[Kr]$5s^14d^7$	79	Au	[Xe]$6s^14f^{14}5d^{10}$
10	Ne	[He]$2s^22p^6$	45	Rh	[Kr]$5s^14d^8$	80	Hg	[Xe]$6s^24f^{14}5d^{10}$
11	Na	[Ne]$3s^1$	46	Pd	[Kr]$4d^{10}$	81	Tl	[Xe]$6s^24f^{14}5d^{10}6p^1$
12	Mg	[Ne]$3s^2$	47	Ag	[Kr]$5s^14d^{10}$	82	Pb	[Xe]$6s^24f^{14}5d^{10}6p^2$
13	Al	[Ne]$3s^23p^1$	48	Cd	[Kr]$5s^24d^{10}$	83	Bi	[Xe]$6s^24f^{14}5d^{10}6p^3$
14	Si	[Ne]$3s^23p^2$	49	In	[Kr]$5s^24d^{10}5p^1$	84	Po	[Xe]$6s^24f^{14}5d^{10}6p^4$
15	P	[Ne]$3s^23p^3$	50	Sn	[Kr]$5s^24d^{10}5p^2$	85	At	[Xe]$6s^24f^{14}5d^{10}6p^5$
16	S	[Ne]$3s^23p^4$	51	Sb	[Kr]$5s^24d^{10}5p^3$	86	Rn	[Xe]$6s^24f^{14}5d^{10}6p^6$
17	Cl	[Ne]$3s^23p^5$	52	Te	[Kr]$5s^24d^{10}5p^4$	87	Fr	[Rn]$7s^1$
18	Ar	[Ne]$3s^23p^6$	53	I	[Kr]$5s^24d^{10}5p^5$	88	Ra	[Rn]$7s^2$
19	K	[Ar]$4s^1$	54	Xe	[Kr]$5s^24d^{10}5p^6$	89	Ac	[Rn]$7s^26d^1$
20	Ca	[Ar]$4s^2$	55	Cs	[Xe]$6s^1$	90	Th	[Rn]$7s^26d^2$
21	Sc	[Ar]$4s^23d^1$	56	Ba	[Xe]$6s^2$	91	Pa	[Rn]$7s^25f^26d^1$
22	Ti	[Ar]$4s^23d^2$	57	La	[Xe]$6s^25d^1$	92	U	[Rn]$7s^25f^36d^1$
23	V	[Ar]$4s^23d^3$	58	Ce	[Xe]$6s^24f^15d^1$	93	Np	[Rn]$7s^25f^46d^1$
24	Cr	[Ar]$4s^13d^5$	59	Pr	[Xe]$6s^24f^3$	94	Pu	[Rn]$7s^25f^6$
25	Mn	[Ar]$4s^23d^5$	60	Nd	[Xe]$6s^24f^4$	95	Am	[Rn]$7s^25f^7$
26	Fe	[Ar]$4s^23d^6$	61	Pm	[Xe]$6s^24f^5$	96	Cm	[Rn]$7s^25f^76d^1$
27	Co	[Ar]$4s^23d^7$	62	Sm	[Xe]$6s^24f^6$	97	Bk	[Rn]$7s^25f^9$
28	Ni	[Ar]$4s^23d^8$	63	Eu	[Xe]$6s^24f^7$	98	Cf	[Rn]$7s^25f^{10}$
29	Cu	[Ar]$4s^13d^{10}$	64	Gd	[Xe]$6s^24f^75d^1$	99	Es	[Rn]$7s^25f^{11}$
30	Zn	[Ar]$4s^23d^{10}$	65	Tb	[Xe]$6s^24f^9$	100	Fm	[Rn]$7s^25f^{12}$
31	Ga	[Ar]$4s^23d^{10}4p^1$	66	Dy	[Xe]$6s^24f^{10}$	101	Md	[Rn]$7s^25f^{13}$
32	Ge	[Ar]$4s^23d^{10}4p^2$	67	Ho	[Xe]$6s^24f^{11}$	102	No	[Rn]$7s^25f^{14}$
33	As	[Ar]$4s^23d^{10}4p^3$	68	Er	[Xe]$6s^24f^{12}$	103	Lr	[Rn]$7s^25f^{14}6d^1$
34	Se	[Ar]$4s^23d^{10}4p^4$	69	Tm	[Xe]$6s^24f^{13}$	104	Rf	[Rn]$7s^25f^{14}6d^2$
35	Br	[Ar]$4s^23d^{10}4p^5$	70	Yb	[Xe]$6s^24f^{14}$	105	Ha	[Rn]$7s^25f^{14}6d^3$

SAMPLE EXERCISE 6.5

Draw the orbital diagram representation for zirconium, atomic number 40; show only those electrons beyond the krypton inner core.

Solution: Zirconium has four electrons beyond the nearest noble gas, krypton, atomic number 36. Examining the periodic table, we see that zirconium is a transition element from the fifth row of the table. This means that its outermost electrons are in 5*s* and 4*d* orbitals. Two electrons occupy the 5*s* orbital; two must be placed in the five 4*d* orbitals. As indicated by Hund's rule, the 4*d* electrons occupy separate orbitals. Thus we have

Zr	[Kr]	↿⇂	↿	↿			
		5*s*	4*d*				

A complete list of the electron configurations of the elements is contained in Table 6.2. You can use this table to check your answers as you practice writing electron configurations. We have written these configurations as they would be read off the periodic table. However, they are sometimes written with orbitals of a given principal quantum number gathered together. Thus we might write the electron configuration of arsenic (atomic number 33) as $[Ar]3d^{10}4s^24p^3$ instead of $[Ar]4s^23d^{10}4p^3$ as shown in Table 6.2.

The configurations of many of the heavier elements are not known for certain. The configurations must be deduced by analysis of atomic spectra. These methods are extremely complicated, especially for the heavier elements, in which the energy levels are closely spaced.

As we have seen, the periodic table arises from the periodic nature of electron configurations. However, the periodic table developed in an entirely empirical manner and was used by generations of chemists before there was any knowledge about electron configurations. It is instructive to consider briefly the historical development of the periodic table.

During the earliest years of the nineteenth century, many new elements were discovered in a short period of time. By 1830 there were about 56 known elements. The identification of many new elements and the development of their descriptive chemistry naturally led to various attempts at classification. The classification process that occurred in chemistry in those years is common to the development of all science. As the quantity and variety of data increase, some means of orderly classification is sought, simply as a means of managing the large number of facts at hand. When a workable classification scheme is found, it may form the basis for development of a theory that accounts for the regularities observed. In 1869, Dmitri Mendeleev in Russia (Figure 6.5) and Lothar Meyer in Germany, working quite independently of each other, published very similar schemes for classification of the elements. Their tables of the elements were the forerunners of the modern periodic table. Mendeleev arranged the elements in order of increasing atomic weight and observed that when this is done, similar chemical and physical properties recur periodically. By arranging the elements so that those with similar characteristics are in vertical groupings, Mendeleev constructed the periodic table.

Although Meyer and Mendeleev came to essentially the same conclusion about the periodicity of properties, Mendeleev must be given credit for more vigorously advancing his ideas and stimulating much new work in chemistry. By sticking to his notion that elements of similar characteristics must be listed in groups, he was forced to leave several spaces in his table blank. For example, arsenic (As) was the element of next highest atomic weight after zinc (Zn). But its placement immediately after Zn in the table would have required that it fall under aluminum (Al). This, however, did not make sense in terms of its properties. Rather, it clearly belonged under phosphorus, as shown in Figure 6.6. This meant that in the table there were two blank spaces that Mendeleev boldly predicted would be filled by as yet undiscovered elements. He gave these elements the names eka-aluminum and eka-silicon and suggested that they might be found in nature with other members of their respective families. For example, eka-aluminum might be found in certain ores containing aluminum, because according to the periodic law it was likely to have properties similar to those of aluminum.

By noting that the properties of elements within a vertical family varied in a regular way with increasing atomic weight, Mendeleev was able to predict the properties of the unknown elements. Thus the properties of eka-silicon should be intermediate between those of silicon (Si) and tin (Sn). In 1871 Mendeleev predicted the properties for eka-silicon, and not many years later, in 1886, the element germanium (Ge) was discovered. That the element germanium was the eka-silicon predicted by Mendeleev is shown by the data listed in Table 6.3.

FIGURE 6.5 Dmitri Mendeleev. Mendeleev rose from very poor beginnings to a position of great eminence in nineteenth-century science. He was born in Siberia, the youngest child in a family of at least 14. His mother endured great personal sacrifice to make it possible for him to enroll in a university in Saint Petersburg. Mendeleev proved to be a brilliant student in sciences and mathematics and eventually was able to study in France and Germany. He spent most of his career as a professor of chemistry in the University of Saint Petersburg. Despite his eminence as a scientist, he was often in trouble because of his liberal, unorthodox opinions. (*Library of Congress*)

	B 10.81	C 12.01	N 14.01
	Al 26.98	Si 28.09	P 30.97
Zn 65.37	?	?	As 74.92
Cd 112.41	In 114.82	Sn 118.69	Sb 121.75

FIGURE 6.6 Portion of Mendeleev's periodic table showing the symbol for the element and the modern value for atomic mass.

TABLE 6.3 Comparison of the properties for eka-silicon predicted by Mendeleev with the known properties of the element germanium

	Mendeleev's prediction for eka-silicon	Observed properties of germanium
Appearance	Gray	Grayish white
Atomic weight	72	72.59
Density (g/cm^3)	5.5	5.35
Specific heat (J/g-K)	0.31	0.31
Formula of oxide	XO_2	GeO_2
Density of oxide (g/cm^3)	4.7	4.70
Formula of chloride	XCl_4	$GeCl_4$
Density of chloride (g/cm^3)	1.9	1.84

6.4 ELECTRON SHELLS IN ATOMS

We have seen how the electron configuration of an atom may be built up by adding electrons to orbitals of successively higher energy, in accordance with the Pauli exclusion principle and Hund's rule. How do our results compare with Lewis's idea of electron shells? Consider the noble gases helium, neon, and argon, whose electron configurations are as follows:

He	$1s^2$
Ne	$1s^22s^22p^6$
Ar	$1s^22s^22p^63s^23p^6$

Very accurate calculations of the total electronic charge distribution in these atoms can be made using large computers. These distributions are shown in Figure 6.7. The quantity plotted on the vertical axis is called the radial electron density. It corresponds to the probability of the electron being located at a particular distance from the nucleus. As Figure 6.7 shows, the radial electron density does not fall off continuously as we move away from the nucleus. Rather it shows maxima corresponding to distances at which there are higher probabilities of finding electrons. These maxima correspond to the traditional idea of shells of electrons; however, these shells are diffuse and overlap considerably.

Helium shows a single shell, neon two, and argon three. Each of these maxima is due mainly to electrons in the atom that have the same value for the principal quantum number n. Thus for helium the $1s$ electrons possess a maximum in radial electron density at about 0.3 Å. In argon, the maximum in the $1s$ radial electron density occurs at only 0.05 Å. The second maximum, which occurs at larger radial distance, is due to both the $2s$ and $2p$ electrons. The third maximum is due to $3s$ and $3p$ electrons.

The reason for the smaller radial distance of the orbital of the $1s$ electrons in the heavier atom is clear when we recall that the nuclear charge of helium is only 2, whereas that for argon is 18. The $1s$ electrons are the innermost electrons of the atom. The electrons of quantum number $n = 2$ and greater, present in elements beyond helium, therefore do

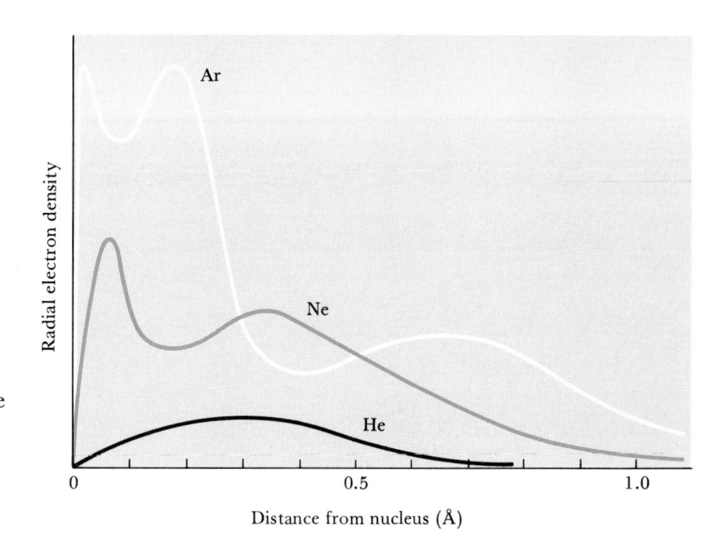

FIGURE 6.7 Radial electron-density graphs for the first three rare-gas elements, He, Ne, and Ar. The maxima that occur in the radial electron density correspond to electrons with the same value of principal quantum number n.

not do much to shield the 1*s* electrons from the increasing nuclear charge. As a result, the size of the 1*s* orbital shrinks steadily as nuclear charge increases. Thus the calculations show that in many-electron atoms the inner electrons are pulled with ever-increasing force into the region around the nucleus as the nuclear charge increases.

6.5 IONIZATION ENERGY

Now that we have some understanding of the electronic structures of atoms, we can examine some properties of atoms that are strongly dependent on electron configuration. We will consider three that provide important insights into the chemical behavior of atoms: ionization energy, electron affinity, and atomic size.

Recall from our discussion in Chapter 5 that the ionization energy (I) is the energy required to remove an electron from a gaseous atom or ion. The first ionization energy for an element, I_1, is therefore that required for the process shown in Equation 6.4:

$$M(g) \longrightarrow M(g)^{+} + e^{-} \quad [6.4]$$

where M is a gaseous, neutral atom. The second ionization energy, I_2, is then the energy for the removal of the second electron, Equation 6.5:

$$M(g)^{+} \longrightarrow M(g)^{2+} + e^{-} \quad [6.5]$$

Successive ionization energies are defined in a similar manner. The values of successive ionization energies are known for many elements. Values for the elements sodium through argon are listed in Table 6.4. Remember that a larger value of I corresponds to tighter binding of the electron to the atom or ion.

As we might expect, each successive removal of an electron requires more energy. The reason for this is that the positive nuclear charge that provides the attractive force remains the same, whereas the number of electrons, which produce repulsive interaction, steadily decreases. For example, the electronic configuration for silicon (Si) is $1s^2 2s^2 2p^6 3s^2 3p^2$. If we look at the successive ionization energies for silicon given in Table 6.4, we see a steady increase from 780 kJ/mol to 4350 kJ/mol for the four

TABLE 6.4 Successive values of ionization energies (I) for the elements sodium through argon (kJ/mol)[a]

Element	I_1	I_2	I_3	I_4	I_5	I_6	I_7
Na	490	4560		(Inner-shell electrons)			
Mg	735	1445	7730				
Al	580	1815	2740	11,600			
Si	780	1575	3220	4350	16,100		
P	1060	1890	2905	4950	6270	21,200	
S	1005	2260	3375	4565	6950	8490	27,000
Cl	1255	2295	3850	5160	6560	9360	11,000
Ar	1525	2665	3945	5770	7230	8780	12,000

[a] Although the ionization energies are given here in units of kJ/mol, they are also often given in units of electron volts; 1 electron volt is equal to 96.49 kJ/mol.

values of I that correspond to loss of the four valence-shell electrons with principal quantum number $n = 3$. The fifth electron, however, requires considerably more energy for removal, 16,100 kJ/mol. This sharp increase in ionization energy reflects the fact that the fifth electron is an inner-shell electron. This electron is in a $2p$ orbital and consequently penetrates closer to the nucleus than do the $3s$ and $3p$ electrons. The $2p$ electron in silicon not only has a smaller average distance r from the nucleus, but it also experiences a larger effective nuclear charge because it penetrates the charge distribution of the other electrons.

The effect of change in the principal quantum number can be seen in other comparisons as well. For example, if we compare Mg and Al, we see from Table 6.4 that the energies required to remove first one and then two electrons from the two metals are not so very different. Yet the energy required to remove the third electron from Mg, 7730 kJ/mol, is much greater than the energy required to remove a third electron from Al, 2740 kJ/mol. The difference in energies must therefore arise predominately from the fact that a third electron removed from Mg is an inner-shell $2p$ electron; in contrast, the third electron removed from Al is an outer-shell $3s$ electron.

These and similar ionization energy data thus support the idea that only the outermost electrons, those beyond the noble-gas core, are involved in the sharing and transfer of electrons that give rise to chemical change. The reason for this is that the inner electrons are too tightly bound to the nucleus to be lost from the atom or even shared with another atom.

SAMPLE EXERCISE 6.6

As can be seen in Table 6.4, the energy required to remove an electron from P^{4+} is 6270 kJ/mol, as compared with 16,000 kJ/mol for removal of an electron from Si^{4+}. Account for the large difference.

Solution: The outer electron configuration of phosphorus is $3s^23p^3$. After removal of four of these electrons, the highest-energy electron remaining is a $3s$. The outer electron configuration of silicon is $3s^23p^2$. After removal of these four electrons the highest-energy electron remaining is a $2p$. It requires considerably less energy to remove the $3s$ electron, which lies largely outside the $1s^22s^22p^6$ core of electrons, than to remove an electron from the $2p$ level, as would be necessary for silicon.

Periodic Trends in Ionization Energies

It is of interest to observe how the first ionization energies, I_1, vary with atomic number. Figure 6.8 shows a graph of I_1 versus atomic number. It is evident that there is an overall periodicity in this property. Overlooking for the moment the lesser displacements, there is a gradual increase in I_1 with atomic number in any one horizontal row. Thus the alkali metals show the lowest ionization energy in each row and the rare-gas elements the highest. A few simple considerations help to explain this trend. Proceeding along any horizontal row of the table, the electrons that are added to counterbalance the increasing nuclear charge do not completely shield the outermost electrons from the nucleus. Thus the effective nuclear charge increases steadily. For example, the inner $1s^2$

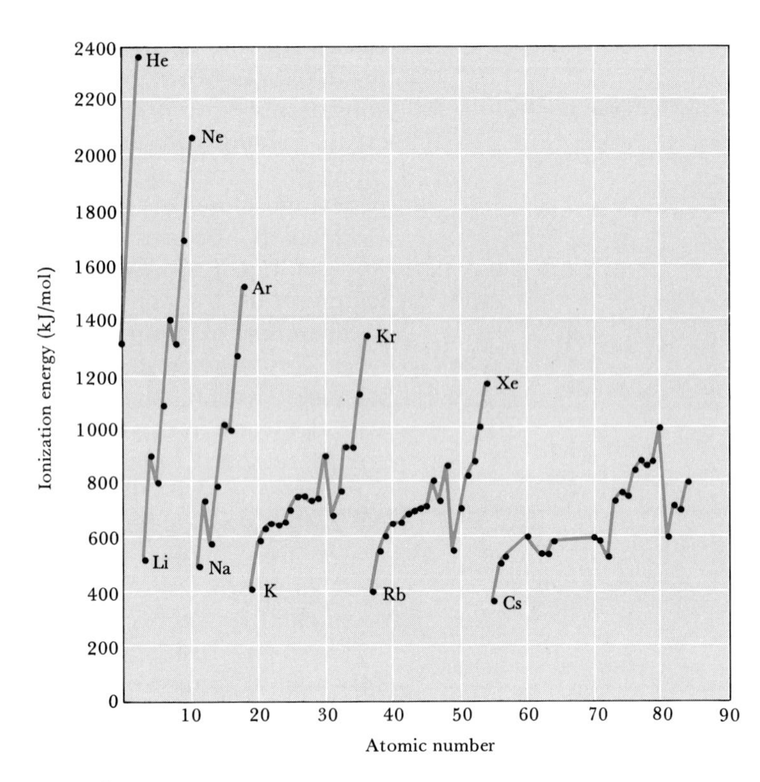

FIGURE 6.8 Ionization energy versus atomic number.

electrons of lithium ($1s^22s^1$) shield the outer $2s$ electron from the 3+ charged nucleus. Consequently, the outer electron experiences an effective nuclear charge of about 1+. For beryllium ($1s^22s^2$), the effective nuclear charge experienced by each outer $2s$ electron is larger; in this case the inner $1s^2$ electrons are shielding a 4+ nucleus, and each $2s$ electron only partially shields the other from the nucleus. As the effective nuclear charge increases, the electron becomes harder to remove. The irregularities within a given row are primarily associated with the enhanced stability of filled or half-filled subshells.

For elements in any column or family of elements, there is a gradual decrease in ionization energy with increasing atomic number. For example, it requires more energy to remove an electron from a lithium atom than from a potassium atom. When we compare elements of a vertical column, we are comparing elements with the same outer electron configurations but differing values for n, the principal quantum number. As n increases so also does the average distance of the electron from the nucleus. As its average distance from the nucleus increases, the electron

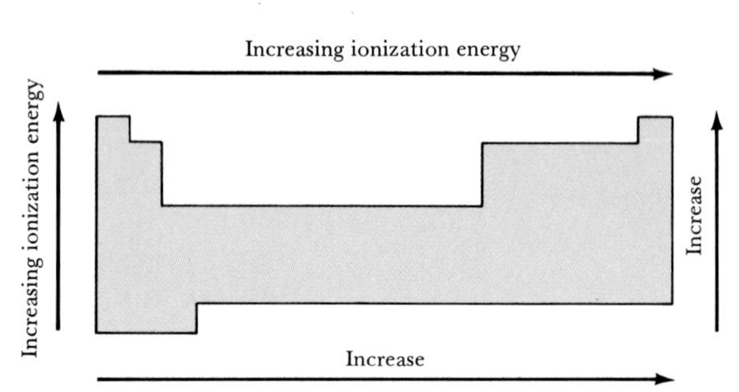

FIGURE 6.9 Variation of the first ionization energy for the elements in relation to the periodic table.

becomes easier to remove. Therefore, if all other factors are the same, ionization energy decreases with increasing atomic radius. In terms of the periodic table, the ionization energies vary in the manner shown in Figure 6.9.

6.6 ELECTRON AFFINITIES

Atoms not only lose electrons to form positively charged ions but also gain them to form negatively charged ones. The ionization energy measures the energy changes associated with removing electrons from gaseous atoms. The energy change that occurs when an electron is added to a gaseous atom or ion is called the electron affinity, E.* The process may be represented for a neutral atom as

$$M(g) + e^- \longrightarrow M^-(g) \qquad [6.6]$$

For an ion of 1+ charge, the equation is

$$M^+(g) + e^- \longrightarrow M(g) \qquad [6.7]$$

For most neutral and for all positively charged species, energy is evolved when the electron is added; E is thus negative in sign. The electron affinities of neutral atoms are quite difficult to measure, and accurate values are now known for only about 50 elements. Figure 6.10 shows the

*We have defined electron affinity in such a way that a negative E is associated with an exothermic process. Thus the more negative the value for E, the greater the attraction for electrons. However, E is sometimes defined as the energy given off when an electron is added to a gaseous atom or ion. In references that define E in this second way, the more positive the E value, the greater the attraction for electrons.

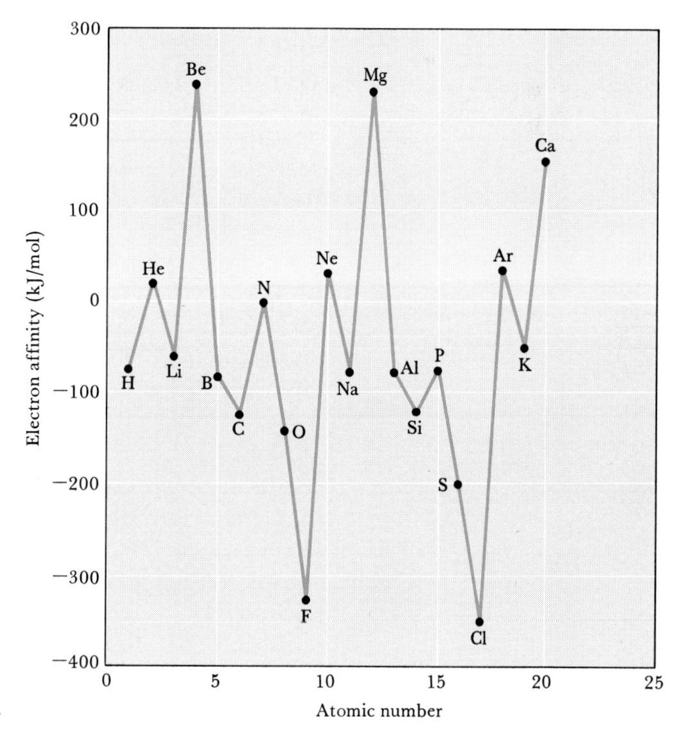

FIGURE 6.10 Electron affinity versus atomic number.

variation of electron affinities for the first 20 elements in the periodic table.

In general, electron affinities become more negative (stronger attraction for an electron) as we move from left to right across any row of the periodic table. The most negative electron affinities are found with the halogens, group 7A. The halogens have an outer electron configuration of the type ns^2np^5. The addition of a single electron gives the stable configuration characteristic of the noble gases. The noble gases, which have filled outer *s* and *p* subshells, have no attraction for an additional electron; energy is required to add an electron. Similarly, energy is required to add an electron to the members of the alkaline earth family, group 2A; each member of this family has a filled outer *s* subshell.

SAMPLE EXERCISE 6.7

Although electron affinities generally become more negative as we move from left to right across any row of the periodic family, the electron affinity of N is more positive than that of C (see Figure 6.10). Rationalize this observation in terms of the electron configurations of N and C.

Solution: The orbital diagram for the outer-shell electrons of C is as follows:

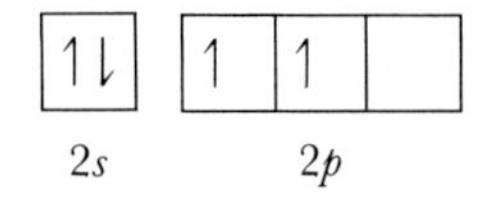

There is an empty $2p$ orbital available to receive the added electron. Nitrogen has a half-filled $2p$ subshell; there is already one electron in each $2p$ orbital. Consequently, when an extra electron is added it must be placed into an orbital that is already occupied by an electron. The resultant electron-electron repulsion causes the nitrogen atom to have less attraction for an extra electron than does carbon. In general, atoms with filled or half-filled subshells have more positive electron affinities (less electron attraction) than elements on either side of them in the periodic table.

Table 6.5 gives the electron affinities of the halogens. Notice that the values differ very little within the family. As we proceed from fluorine to iodine the added electron is going into a *p* orbital of increasing major quantum number. The average distance of the electron from the nucleus steadily increases, and electron-nuclear attraction should thus steadily decrease. If this were all that is involved, fluorine would have the highest electron affinity. But the orbitals that hold the outermost electrons of the halogen are increasingly spread out as we proceed from fluorine to iodine. The electron-electron repulsions between these electrons and the added one therefore decrease with increasing atomic weight of the halogen. A lower electron-nuclear attraction is thus counterbalanced by lower electron-electron repulsion. The overall result is that the electron affinities differ very little among the halogens. Other families likewise show little variation in electron affinities among their members.

Note (Table 6.5) that addition of one electron to an oxygen atom results in evolution of energy; that is, it leads to a more stable species than the originally separated atom and electron. However, addition of a second electron to form O^{2-} requires energy, even though it results in a completed octet of electrons about the nucleus; the sign of E is therefore positive. In this case, and for S^{2-} also, the electron-electron repulsions outweigh the electron-nuclear attractions. However, such ions are common in crystalline solids because they are stabilized by electrostatic interactions with cations.

TABLE 6.5 Electron affinities of some elements

Element	Ion formed	E (kJ/mol)
H	H^-	−73
F	F^-	−332
Cl	Cl^-	−349
Br	Br^-	−325
I	I^-	−295
O	O^-	−141
O	O^{2-}	+710
S	S^{2-}	+375

6.7 ATOMIC SIZES

One conclusion we can draw from the quantum-mechanical model is that an atom does not have a sharply defined boundary that determines its size. This is evident in Figure 6.7; the electron-density distributions illustrated do not end sharply; rather, they simply drop off with increasing distance from the nucleus, approaching zero at large distance. Given this state of affairs we might well ask whether it makes sense to speak of an atomic radius for an atom. Certainly such a concept would be difficult to define for an isolated atom. Suppose, however, that two atoms form a chemical bond between them, as in Br_2. The distance between the centers of the two bromine atoms in Br_2 can be thought of as the sum of two bromine atom radii. The Br—Br distance in Br_2 is 2.286 Å; we might then say that, to the nearest 0.01 Å, the radius of the bromine atom is 1.14 Å. Similarly, in compounds containing carbon-carbon bonds, the C—C distance is found to be very close to 1.54 Å. We can thus assign an atomic radius of 0.77 Å to carbon. If our concept of an atomic radius is going to be very useful, these atomic radii should remain pretty much the same when the atom is bound to another element. Thus the distance between carbon and bromine in a C—Br bond should be 1.14 + 0.77 Å, or 1.91 Å. It turns out that carbon-bromine bonds in various compounds do in fact exhibit about this C—Br distance. Atomic

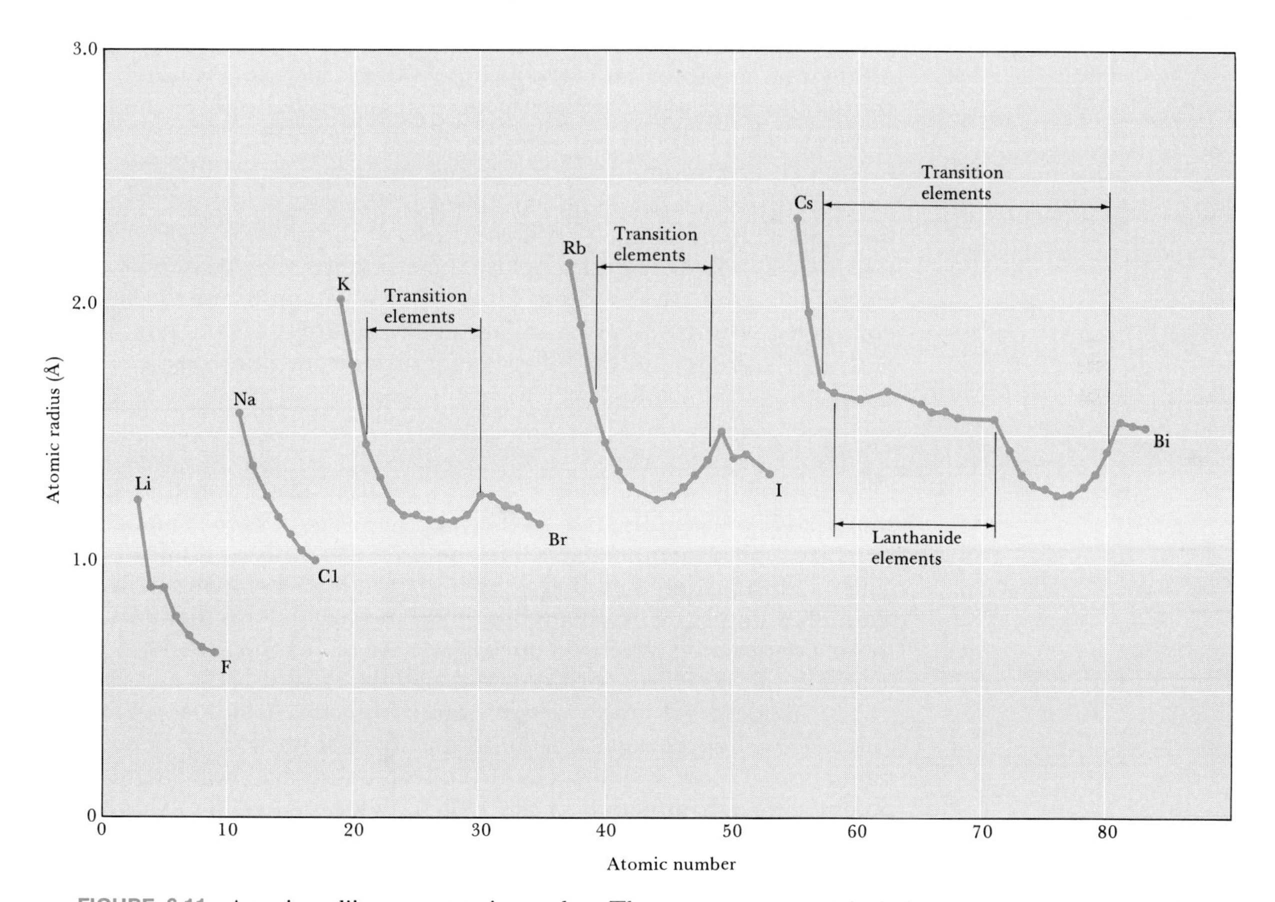

FIGURE 6.11 Atomic radii versus atomic number. The rare gases are not included in this graph because there is no simple way of relating their radii to those of the other elements on the basis of solid-state structure determinations. Gaps in the graph are due to lack of experimental data.

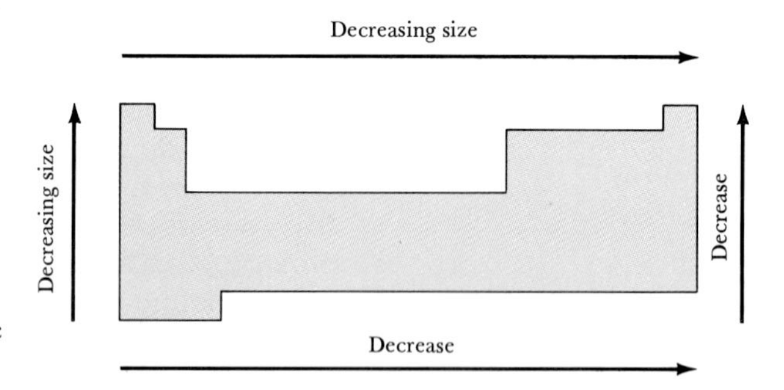

FIGURE 6.12 General periodic trends in atomic size.

radii are not so easy to evaluate for metallic elements, but a list of atomic radii has been assembled for many elements, by considering a large body of experimental data. These atomic radii are graphed in Figure 6.11 as a function of atomic number.

Provided we keep in mind that there are uncertainties in these values because of the means by which they are obtained, we can discern some interesting trends in the data.

It is evident that atomic radii show periodic variation. The atomic size generally decreases as we move from the alkali metals (group 1A) toward the halogens (group 7A). Furthermore, we find that among the elements of any one group or family of the periodic table—for example, among the alkali metals—the radius increases regularly with increasing atomic number. These trends are summarized in Figure 6.12.

We can understand these variations using the same line of reasoning employed in discussing the variations in ionization energy. Proceeding from left to right across a horizontal row of the periodic table, the effective nuclear charge experienced by the outermost electrons increases as a result of incomplete shielding. Thus the orbital containing the electron is contracted. On the other hand, in any vertical row, orbital size increases with increasing value for the principal quantum number.

6.8 THE GROUP 1 AND GROUP 2 METALS

Our discussion of ionization energies, electron affinity, and atomic size should give some indication of the way the periodic table can be used to organize and remember facts. Not only do elements in a family possess general similarities, but there are also trends in behavior as we move through a family or from one family to another. In this section we want to use the periodic table and our knowledge of electron configurations to examine the chemistry of the alkali metals, family 1A.

The alkali metal family gets its name from the Arabic word meaning *ashes*. Many compounds of sodium and potassium, the two most abundant alkali metals, were isolated from wood ashes by early chemists. The names soda ash and potash are still sometimes used for Na_2CO_3 and K_2CO_3, respectively.

The alkali metals are all shiny, soft, metallic solids of relatively low density. Some of their physical properties are given in Table 6.6. Notice how their melting points and densities vary in a fairly regular manner with increasing atomic number.

TABLE 6.6 Some physical and chemical properties of the alkali metals

Element	Symbol	Atomic number	Atomic weight	Melting point (°C)	Density (g/cm^3)	Formula of hydroxide	Formula of chloride
Lithium	Li	3	6.939	181	0.53	LiOH	LiCl
Sodium	Na	11	22.9898	98	0.97	NaOH	NaCl
Potassium	K	19	39.102	63	0.86	KOH	KCl
Rubidium	Rb	37	85.47	39	1.53	RbOH	RbCl
Cesium	Cs	55	132.905	29	1.87	CsOH	CsCl

The alkali metals are all very reactive, readily losing one electron to form ions with a 1+ charge:

$$M \longrightarrow M^+ + e^- \qquad [6.8]$$

(The symbol M in this equation and others through this section represents any one of the alkali metals.) This behavior corresponds to their ns^1 outer electron configurations and their low ionization energies. As we have noted in Section 6.5, the ease with which the elements lose electrons increases as we move down a family. Consequently, Cs is especially reactive. Of course, Fr is even more reactive, but it is an extremely rare, radioactive element; Cs is therefore the heaviest alkali metal generally encountered in the laboratory.

Owing to their reactivities, the alkali metals exist in nature only as compounds. The metals can be obtained by passing an electric current through a molten salt. Humphry Davy first prepared sodium and potassium in 1807 by passing a current through the molten sodium and potassium hydroxides. Today, sodium is prepared commercially by electrolysis of molten NaCl. The electrical energy is used to force an electron onto Na^+ and remove one from Cl^-:

$$Na^+ + e^- \longrightarrow Na \qquad [6.9]$$

$$2Cl^- \longrightarrow Cl_2 + 2e^- \qquad [6.10]$$

Although the metals have a silvery luster, their surfaces quickly lose this luster and become white when exposed to air. This behavior is due to reaction with oxygen in the air to form oxides:

$$4M(s) + O_2(g) \longrightarrow 2M_2O(s) \qquad [6.11]$$

All except lithium can also form peroxides, M_2O_2; potassium, rubidium, and cesium are able to form superoxides, MO_2. (The superoxide ion has a charge of 1−; O_2^-.)

The oxides of the alkali metals react readily with water to form hydroxides:

$$M_2O(s) + H_2O(l) \longrightarrow 2M^+(aq) + 2OH^-(aq) \qquad [6.12]$$

These hydroxides are water soluble and completely ionized into M^+ and OH^- ions in aqueous solution. They are the most common laboratory bases.

The alkali metals also combine directly with water, generating hydrogen gas and forming hydroxides:

$$2M(s) + 2H_2O(l) \longrightarrow 2M^+(aq) + 2OH^-(aq) + H_2(g) \qquad [6.13]$$

This reaction is most violent in the case of the heavier members of the family, in keeping with their weaker hold on the single outer-shell electron.

The reactivity of the alkali metals permits them to react directly with most elements. Some further examples follow:

$$2M(s) + H_2(g) \longrightarrow 2MH(s) \qquad [6.14]$$

$$2M(s) + S(s) \longrightarrow M_2S(s) \qquad [6.15]$$

$$2M(s) + Cl_2(g) \longrightarrow 2MCl(s) \qquad [6.16]$$

In all cases the alkali metal exists as 1+ ions in these compounds.

Almost all compounds of the alkali metals are soluble in water. Consequently, sodium, the most abundant of the alkali metals, is found as Na^+ in seawater. It reaches the sea through the dissolving action of water moving through the ground following rains.

The alkali metal salts are colorless, as are their aqueous solutions. Color is produced when an electron in an atom is excited from one energy level to another by visible radiation. Alkali metal ions, having lost their outermost electrons, have no electrons that can be excited by visible radiation.

The Alkaline Earth Elements

The elements of group 2A are called the alkaline earths. A few properties of these elements are listed in Table 6.7. Like the group 1A elements, the group 2A elements are also reactive; however, they lose their electrons less readily than their alkali metal neighbors. As we have noted in Section 6.5, the ease with which the elements lose electrons decreases as we move across the periodic table from left to right. Beryllium and magnesium, the lightest members of the family, are least reactive. Indeed, the chemical reactivity of magnesium is low enough to permit its use as a structural metal. It is nonetheless reasonably reactive; the metal is protected from loss of electrons to substances such as O_2 and H_2O in the environment by a thin and compact surface coat of water-insoluble

TABLE 6.7 Some physical and chemical properties of the alkaline earth metals

Element	Symbol	Atomic number	Atomic mass	Melting point (°C)	Density	Formula of hydroxide	Formula of chloride
Beryllium	Be	4	9.01	1283	1.85	$Be(OH)_2$	$BeCl_2$
Magnesium	Mg	12	24.30	650	1.74	$Mg(OH)_2$	$MgCl_2$
Calcium	Ca	20	40.08	851	1.54	$Ca(OH)_2$	$CaCl_2$
Strontium	Sr	38	87.62	757	2.58	$Sr(OH)_2$	$SrCl_2$
Barium	Ba	56	137.33	704	3.65	$Ba(OH)_2$	$BaCl_2$

MgO. Like the alkali metals, the alkaline earth elements are found in nature only as compounds.

Each group 2A element has two valence-shell electrons beyond the electron configuration of the nearest rare-gas element. The metals have higher melting points and are both harder and denser than their alkali metal neighbors. This difference in properties results from the fact that two valence-shell electrons are employed in the bonding between alkaline earth metal atoms in the pure element, as compared with just one electron in the alkali metals. Finally, note that, in keeping with their ns^2 outer electron configuration, the alkaline earth metals form 2+ ions, as reflected in the formulas shown in Table 6.7 for the hydroxides and chlorides.

Comparison of A and B Groups

The label attached to each family in the periodic table has either an A or B designation.* Elements of the same group number but different letter labels may share certain characteristics in common, but in general there is not a close connection. Let's consider the distinction between A and B elements in terms of electron configuration. As an example, locate the group 1A and 1B elements on the periodic table. In both families of group 1 elements the outer electron configuration involves a single *s* electron. In the case of the alkali metals (group 1A), this *s* electron is outside a stable noble-gas core of electrons. For example, the single $4s$ electron in potassium is outside a set of eight electrons associated with filled $3s$ and $3p$ subshells, $3s^23p^6$. In the corresponding group 1B element, copper, the single electron is outside filled $3s$, $3p$, and $3d$ subshells, $3s^23p^63d^{10}$. The presence of the additional ten *d* electrons in copper, and of course the increase of ten in nuclear charge that goes along with them, has a profound effect on the chemical behavior of the single *s* electron. Because these *d* electrons only partially shield the nucleus, the *s* electron of the 1B group element experiences a larger effective nuclear charge than does the *s* electron of the 1A group. As a result, the 1B elements have higher ionization energies than do the corresponding 1A elements. Table 6.8 lists the ionization energies for the group 1 and 2 elements. Compare the values for the A and B groups.

Because the 1B members have a much stronger hold on their outer-shell electron, these metals are much less reactive than are the 1A metals. In fact, the 1B metals, often referred to as the *coinage metals,* can be found in elemental form in nature and are widely used in jewelry.

*The designation of groups as "A" or "B" is purely arbitrary. In fact, some periodic tables use a different system from that used in this text; however, the system used in this text is presently the most common one for designating groups.

TABLE 6.8 Comparative values of ionization energies for group 1A, 1B, 2A, and 2B elements (kJ/mol)

1A	I_1	1B	I_1	2A	I_1	I_2	2B	I_1	I_2
K	418	Cu	859	Ca	589	1145	Zn	907	1733
Rb	403	Ag	730	Sr	549	1064	Cd	867	1630
Cs	375	Au	890	Ba	503	965	Hg	994	1805

FOR REVIEW

Summary

Our major concern in this chapter has been the relationship between electron configurations and the properties of atoms, especially as organized by the periodic table. We saw that the electron configurations of many-electron atoms can be written by placing electrons into orbitals in the following order:

1*s*, 2*s*, 2*p*, 3*s*, 3*p*, 4*s*, 3*d*, 4*p*, . . .

Subshells with a given principal quantum number, such as the 3*s*, 3*p*, and 3*d* subshells, do not have the same energies. This fact can be understood in terms of the screening effect and the average distance of an electron from the nucleus in each of these subshells.

The Pauli exclusion principle places a limit of two on the number of electrons that may occupy any one atomic orbital. These two electrons differ in their electron-spin quantum number, m_s. As the electrons populate orbitals of equal energy, they do not pair up until each orbital contains one electron; this observation is called Hund's rule. Using the relative energies of the orbitals, the Pauli exclusion principle, and Hund's rule, it is possible to write the electron configuration of any atom. When we do so, we see that the elements in any given family in the periodic table have the same type of electron arrangements in their outermost, incomplete shells. For example, the electron configurations of fluorine and chlorine, which are both members of the halogen family, are $[He]2s^22p^5$ and $[Ne]3s^23p^5$, respectively. This periodicity in electron configurations, summarized in Figure 6.4, permits us to write the electron configuration of an element from its position in the periodic table.

Many properties of atoms that have chemical significance exhibit periodic character. Among the most important of these are atomic radii, ionization energy, and electron affinity. Electron configurations and the periodic table help us to understand trends in these properties. We also illustrated how electron configurations help us to organize and understand some of the chemistry of the alkali and alkaline earth metals.

Learning goals

Having read and studied this chapter, you should be able to:

1. List the factors that determine the energy of an electron in a many-electron atom. You should be able to explain the fact that electrons with the same value of principal quantum number (n) but differing values of the azimuthal quantum number (l) possess different energies.
2. Explain the concepts of effective nuclear charge and the screening effect as they relate to the energies of electrons in atoms.
3. State the Pauli exclusion principle and Hund's rule and illustrate how they are used in writing the electronic structures for the elements.
4. Explain the basis on which the modern periodic table is constructed.
5. Define the term "group" or "family" in terms of electron configuration.
6. Describe the various blocks of elements in the periodic table in terms of the type of orbital being occupied by electrons in that block (*s*, *p*, *d*, and *f* blocks).
7. List the names and give the locations in the periodic table for the active metals (*s*-block), representative elements (*p*-block), transition metals (*d*-block), and inner transition metals (*f*-block).
8. Write the electron configuration for any element once you know its place in the periodic table.
9. Write the orbital diagram representation for electron configurations of atoms.
10. Relate the historical development of the periodic table, with emphasis on the logical process employed by Mendeleev and Meyer in placing the elements in the table.
11. Explain the effect of increasing nuclear charge on the radial density function in many-electron atoms.
12. Explain the general variations in first ionization energies among the elements, as depicted in Figure 6.8. You should also be able to explain the observed changes in values of the successive ionization energies for a given atom.
13. Explain the variation in atomic radii with atomic number. You should be able to relate this variation, as shown in Figure 6.11, to the variation in corresponding values of the first ionization energies.
14. Explain the concept of electron affinity and its relationship to ionization energy.

Key terms

Among the more important terms and definitions used for the first time in this chapter are the following:

The active metals (Section 6.3), groups 1A and 2A, are those in which electrons occupy only the s orbitals of the valence shell.

The effective nuclear charge (Section 6.1) is the charge at the nucleus experienced by an electron in a many-electron atom. This charge is not the full nuclear charge, because there is some shielding of the nuclear charge by other electrons in the atom. How much shielding occurs depends on the average distance of the electron from the nucleus, compared with the other electrons in the atom.

The electron affinity (Section 6.6) is the energy change that occurs when an electron is added to a gaseous atom or ion.

An electron configuration (Section 6.3) is a particular arrangement of electrons in the orbitals of an atom.

Electron spin (Section 6.2) is a property of the electron that makes it behave as though it were a tiny magnet. Associated with the electron spin is a spin quantum number (m_s), which may have values of $+\frac{1}{2}$ or $-\frac{1}{2}$.

Hund's rule (Section 6.3) states that electrons must occupy degenerate orbitals one at a time until all orbitals have at least one electron, before pairing of electrons in the orbitals occurs. Note carefully that the rule applies only to orbitals that are degenerate, which means that they have the same energy.

The inner transition elements (Section 6.3), or lanthanides and actinides, are those in which the $4f$ or $5f$ orbitals are partially occupied.

The Pauli exclusion principle (Section 6.2) states that no two electrons in an atom may have all four quantum numbers, n, l, m_l, and m_s, the same. As a consequence of this principle, there can be no more than two electrons in any one atomic orbital.

The representative elements (Section 6.3) are those in which the p orbitals are partially occupied.

The screening effect (Section 6.1) is the effect of inner electrons in decreasing the nuclear charge experienced by outer electrons. (Also called the shielding effect.)

Transition elements (Section 6.3) are those in which the d orbitals are partially occupied.

Valence-shell electrons (Section 6.3) are the electrons in the outermost shell of an atom.

EXERCISES

Energies of orbitals

6.1 How does the average distance from the nucleus of an electron in a $2s$ electron in neon compare with that for a $2p$ electron? Explain.

6.2 Which of the following electrons in each of the following sets experiences the largest effective nuclear charge in a many-electron atom: (a) $2s$, $2p$; (b) $3d$, $4d$; (c) $3d$, $2p$.

6.3 Explain why the effective nuclear charge experienced by a $3s$ electron in magnesium is larger than that experienced by a $3s$ electron in sodium.

6.4 What quantum numbers must be the same in order that orbitals in a many-electron atom be degenerate (have the same energy)?

6.5 Suppose that we could follow the forces acting on an individual electron in its motion in a many-electron atom. What change occurs in the effective nuclear charge as the electron moves closer to the nucleus? Explain.

6.6 If the ionization energy of hydrogen is 1312 kJ/mol, what is the ionization energy of He^{+}?

6.7 Which of the quantum numbers assignable to an electron are related to the energy of the electron in a hydrogen atom? Which are related to the energy of an electron in a many-electron atom?

Electron spin; the Pauli principle

6.8 What is the total number of electrons in an atom that may have the following quantum numbers: (a) $n = 4$; (b) $n = 3$, $l = 2$, $m_l = -1$; (c) $n = 3$, $l = 2$, $m_s = +\frac{1}{2}$.

6.9 What is the maximum number of electrons that can be put into each of the following subshells: (a) $4d$; (b) $4f$; (c) $5f$; (d) $2d$?

6.10 List the possible values of the four quantum numbers for each electron in the beryllium atom.

6.11 Indicate whether each of the following statements is true or false. If the statement is false, correct it. (a) The m_l value for an electron in a $5d$ orbital can have any value less than 5. (b) As many as 9 electrons with principal quantum number $n = 3$ can have m_s value of $+\frac{1}{2}$. (c) The set of quantum numbers $n = 3$, $l = 0$, $m_l = -2$, $m_s = -\frac{1}{2}$ is allowable. (d) As a general rule, the maximum number of electrons that can occupy all the orbitals of major quantum number n is $2n^2$.

Electron configurations of the elements

6.12 Write the electron configurations for the following atoms using the appropriate noble-gas inner core for abbreviation: (a) Ca; (b) Ge; (c) Br; (d) Co; (e) Eu.

6.13 Write the complete electron configuration for the following atoms without referring to any data other than the periodic table: (a) In; (b) Se; (c) La; (d) Ag; (e) Cl. Do not use the noble-gas inner core abbreviation, and list the orbitals in the order in which they are filled.

6.14 Draw the orbital diagrams for the electrons beyond the appropriate noble-gas inner core for each of the following elements: (a) As; (b) Mn; (c) Sn; (d) Lu; (e) Te.

6.15 If it were possible to do the Stern-Gerlach experiment with a beam of sodium atoms, would the beam diverge into two? What would you expect to occur with a beam of magnesium atoms? Explain.

6.16 Referring only to Figure 6.4, identify the family or type of element with each of the following electron configurations (we use the symbol X to represent one of the noble-gas elements): (a) $[X]ns^2np^3$; (b) $[X]ns^1$; (c) $[X]ns^2(n-1)d^m$; (d) $[X]ns^2(n-2)f^m$.

6.17 Classify each of the following in terms of the groupings illustrated in Figure 6.4: (a) Sc; (b) Cr; (c) Se; (d) Sm; (e) K; (f) Ar.

6.18 Arrange the following group of orbitals in the order in which they fill with electrons: $4d$; $5p$; $4f$; $6p$; $5s$; $6s$; $3p$.

6.19 How many unpaired electrons are there in an element of each of the following groups: (a) 6A; (b) 5B; (c) 8A; (d) 2B?

6.20 When sodium atoms are excited they emit line spectra that are similar in many respects to those produced by excited hydrogen atoms. Which electron or electrons are responsible for the observed line spectra?

Periodicity and atomic properties

6.21 Name the element (or elements) that fits each of the following descriptions: (a) alkaline earth metal with two $5s$ valence-shell electrons; (b) element with the most negative electron affinity; (c) element with a half-filled $4p$ subshell; (d) element with the lowest ionization energy; (e) all the elements that possess two unpaired $3p$ electrons.

6.22 How do each of the following properties vary as we move from left to right across the periodic table: (a) electron affinity; (b) atomic size; (c) ionization energy; (d) nuclear charge?

6.23 Based on their positions in the periodic table, select the atom with the larger ionization energy from each of the following pairs: (a) B, Cl; (b) N, P; (c) Mg, Cs; (d) O, N; (e) Ga, Ge.

6.24 Order the following elements in the order of increasing atomic radius: Mg, C, Kr, S, K, Cl, Co.

6.25 In each of the following pairs, indicate which element has the larger ionization energy, larger size, and more negative electron affinity: (a) P, Cl; (b) Al, Ga; (c) Cs, La; (d) La, Hf. In each case, provide an explanation in terms of electron configuration and effective nuclear charge.

6.26 Explain, in terms of electron configurations, why hydrogen exhibits properties similar to both Li and F.

6.27 Compare the effective nuclear charges operating on the $3s$ electrons in Mg^+ and Al^{2+}. What evidence for your response is to be found in Table 6.4?

6.28 Although SiO_2, which can be thought of as containing Si^{4+} ions, is common, AlO_2 is not a known compound. Explain this result in terms of the data in Table 6.4.

6.29 Addition of an electron to Na(g) is a slightly exothermic process, whereas addition of an electron to Mg(g) is strongly endothermic. Explain this difference in terms of the ground-state electron configurations of the two elements.

6.30 The electron affinity of chlorine is strongly negative; that is, addition of an electron to Cl is an exothermic process. On the other hand, addition of an electron to Ar is an endothermic process. Account for the difference in terms of the electron configurations of the two elements.

Group 1A and 2A elements

6.31 Write balanced equations for each of the following processes. (a) Potassium is added to water. (b) Barium is added to water. (c) Calcium oxide dissolves in water. (d) Na_2O is added to water. (e) Potassium burns in air. (f) Sodium vapor reacts with bromine vapor.

6.32 Compare the elements potassium and calcium with respect to the following properties: (a) electron configuration; (b) typical ionic charge; (c) first ionization energy; (d) formula of hydroxide; (e) melting point of the metal; (f) atomic radius; (g) electron affinity. Account for the differences in the two elements.

6.33 The electron affinity of copper is -123 kJ/mol. Compare this number with that for potassium, and account for the difference. The electron affinity for zinc is not known. Predict roughly what value it should have.

6.34 Explain why it requires less energy to remove an electron from Sr than from Cd.

[6.35] The first ionization energies of the group 1A elements K, Rb, and Cs decrease steadily with increasing atomic mass (Table 6.8). However, in group 1B, Au has a higher value for I_1 than does Ag. Provide an explanation for this observation.

Additional exercises

6.36 By examining the modern periodic table, find as many examples as you can of violations of Mendeleev's periodic law that the chemical and physical properties of the elements are periodic functions of their atomic *weights*. Why do these violations occur? State the modern version of the periodic law.

6.37 How many unpaired electrons do each of the following atoms possess: (a) Zn; (b) P; (c) Co; (d) Se; (e) Gd; (f) Cu? (Use Table 6.2.)

6.38 Without looking at Table 6.2, write the electron configuration of each of the following elements: (a) Ne; (b) As; (c) Mo; (d) Be; (e) Pd. When you have completed writing your description of the electron configurations, compare your work with the configurations shown in Table 6.2. Any differences should be readily explained in terms of the near-degeneracy of different sets of orbitals.

6.39 Which of the electrons in each of the following sets experiences the larger effective nuclear charge in a many-electron atom: (a) $3p$, $3s$; (b) $4p$, $5s$; (c) $4p$, $3d$; (d) $5f$, $5d$?

6.40 In a chlorine atom, which electrons experience the largest effective nuclear charge? Which experience the lowest?

6.41 Why do the horizontal rows (periods) of the periodic table not all contain the same number of elements?

6.42 Explain how the existence of elements with very low chemical reactivity at atomic numbers 2, 10, 18, 35, and 54 are consistent with the Pauli exclusion principle.

[6.43] What is the lowest value of the principal quantum number n for which there can be a g subshell? How many electrons are required to fill a g subshell? Are there any elements known whose ground-state electron configurations contain electrons in the g subshell?

6.44 Draw the orbital diagram for each of the following elements: (a) Ge; (b) Ni; (c) B; (d) Kr; (e) In.

6.45 Identify the specific element or group of elements that can have the following electron configurations: (a) $1s^22s^22p^4$; (b) $[Ar]4s^13d^{10}$; (c) [noble gas]$ns^2(n-1)d^{10}np^3$; (d) $[Kr]5s^24d^2$; (e) $1s^22s^22p^6$.

6.46 Magnesium not only has a higher first ionization energy than both its neighbors, sodium or aluminum, it has a strongly endothermic electron affinity. Explain these characteristics.

[6.47] Make a graph of the second ionization energies, I_2, listed in Table 6.4, as a function of atomic number. Account for the general trend in the series from Mg through Ar. Account for the exceptionally large value for Na. Suggest why the value for Al is larger than one might expect from the general trend.

6.48 Name the element or elements that fit each of the following descriptions: (a) noble-gas element with a completed $4p$ subshell; (b) alkaline earth metal with $6s$ valence-shell electrons; (c) nonmetallic elements with ns^2np^4 outer-shell electron configurations; (d) element in the row beginning with Rb that possesses the highest ionization energy; (e) elements with six unpaired valence-shell electrons.

6.49 From the data in Table 6.5 calculate the energy change in the process

$$O^-(g) + e^- \longrightarrow O^{2-}(g)$$

Compare this value with that for the process

$$F(g) + e^- \longrightarrow F^-(g)$$

What is the essential difference in these two processes, and how does this difference account for the difference in the two energy changes?

6.50 Draw an orbital diagram for an excited state of the boron atom, in which there are three unpaired valence shell electrons.

[6.51] The element technetium, atomic number 43, is not observed in nature because it happens to be radioactive. Assuming that it can be synthesized in a nuclear reactor in quantity, predict some of its characteristic properties. These should include electron configuration, density, melting point, and formulas of oxides. Consult a handbook of chemistry for information on related elements.

6.52 Draw the ground-state orbital diagram for aluminum. Draw an orbital diagram for an excited state of this atom that possesses three unpaired electrons.

[6.53] There are certain similarities in properties that exist between the first member of any periodic family and the element located one position below it and to the right in the periodic table. For example, in some ways Li resembles Mg, Be resembles Al, and so forth. This observation is known as the diagonal relationship. In terms of what we have learned in this chapter, offer a possible explanation for this relationship.

[6.54] Where would you go to look up the electron affinities of the singly charged positive ions of the elements? Explain.

7 Basic Concepts of Chemical Bonding

Deep within the earth below the city of Detroit and beneath the rolling plains of Kansas lie enormous deposits of a white mineral, halite. This substance, also known as sodium chloride, NaCl, was deposited in these and other places millions of years ago, when extensive primordial seas dried upon the changing surface of the earth. Sodium chloride is the most abundant dissolved substance present in seawater and is found in human body tissues in large quantities. However, it is most familiar to us as ordinary table salt. This substance consists of sodium ions and chloride ions, Na^+ and Cl^-.

Water, H_2O, is another very abundant substance. We drink it, swim in it, and use it as a cooling agent. It is essential to life as we know it. This substance is composed of molecules.

Why are some substances composed of ions while others are composed of molecules? The key to this question is found in the electronic structures of the atoms involved and in the nature of the chemical forces within the compounds. In this chapter and the next we shall examine the relationships between electronic structure, chemical bonding forces, and the properties of substances. As we do this, we shall find it useful to classify chemical forces into three broad groups: (1) ionic bonds, (2) covalent bonds, and (3) metallic bonds.

The term **ionic bond** refers to the electrostatic forces that exist between particles of opposite charge. As we shall see, ions may be formed from atoms by transfer of one or more electrons from one atom to another. Ionic substances generally result from the interaction of metals from the far left side of the periodic table with the nonmetallic elements from the far right side (excluding the rare gases, group 8A).

The **covalent bond** results from a sharing of electrons between two atoms. The most familiar examples of covalent bonding are seen in the interactions of nonmetallic elements with one another.

Metallic bonds are found in solid metals such as copper, iron, and aluminum. In the metals, each metal atom is bonded to several neighboring atoms. The bonding electrons are relatively free to move throughout the three-dimensional structure. Metallic bonds give rise to such typical metallic properties as high electrical conductivity and luster. We will postpone further discussion of metallic bonding until Chapter 22.

7.1 LEWIS SYMBOLS AND THE OCTET RULE

The term **valence** is commonly used in discussions of both ionic and covalent bonding. The valence of an element is a measure of its capacity to form chemical bonds. Originally, it was determined by the number of hydrogen atoms with which an element combined. Thus the valence of oxygen is 2 in H_2O; in CH_4 the valence of carbon is 4.

We speak of **valence electrons** when referring to the electrons that take part in chemical bonding. These electrons are the ones residing in the outermost electron shell of the atom, the valence shell. **Electron-dot symbols** (also known as **Lewis symbols,** after G. N. Lewis) are a simple and convenient way of showing the valence electrons of atoms and keeping track of them in the course of bond formation. The electron-dot symbol for an element consists of the chemical symbol for the element plus a dot for each valence electron. For example, sulfur has an electron configuration $[Ne]3s^23p^4$; its electron-dot symbol therefore shows six valence electrons:

$$\cdot\ddot{\underset{\cdot\cdot}{S}}\cdot$$

TABLE 7.1 Electron-dot symbols

Element	Electron configuration	Electron-dot symbols
Li	$[He]2s^1$	$Li\cdot$
Be	$[He]2s^2$	$\cdot Be\cdot$
B	$[He]2s^22p^1$	$\cdot\dot{B}\cdot$
C	$[He]2s^22p^2$	$\cdot\dot{\underset{\cdot}{C}}\cdot$
N	$[He]2s^22p^3$	$\cdot\dot{\underset{\cdot}{N}}:$
O	$[He]2s^22p^4$	$:\dot{\underset{\cdot}{O}}:$
F	$[He]2s^22p^5$	$\cdot\ddot{\underset{\cdot}{F}}:$
Ne	$[He]2s^22p^6$	$:\ddot{\underset{\cdot\cdot}{Ne}}:$

Other examples are shown in Table 7.1.

The number of valence electrons of any active metal or representative element is the same as the column number of the element in the periodic table. Thus the electron-dot symbols for both oxygen and sulfur, members of family 6A, show six dots.

The way atoms gain, lose, or share electrons can often be viewed as an attempt on the part of the atoms to achieve the same number of electrons as the noble gas closest to them in the periodic table. The noble gases, you will recall, have very stable electron arrangements, as evidenced by their high ionization energies, low affinity for electrons, and general lack of chemical reactivity. Because all noble gases (except He) have eight valence electrons, many atoms undergoing reactions also end up with eight valence electrons. This observation has led to what is known as the **octet rule.** Of course, because He has only two electrons, atoms near it in the periodic table, such as H, will generally tend to obtain two electrons. As we shall see, there are many exceptions to the octet rule. Nevertheless, it provides a useful framework for introducing many important concepts of bonding.

7.2 IONIC BONDING

When sodium metal is brought into contact with chlorine gas, Cl_2, a violent reaction ensues. The product of that reaction is sodium chloride, NaCl, a substance composed of Na^+ and Cl^- ions:

$$2Na(s) + Cl_2(g) \longrightarrow 2NaCl(s)$$

These ions are arranged throughout the solid NaCl in a regular three-dimensional array, as shown in Figure 7.1.

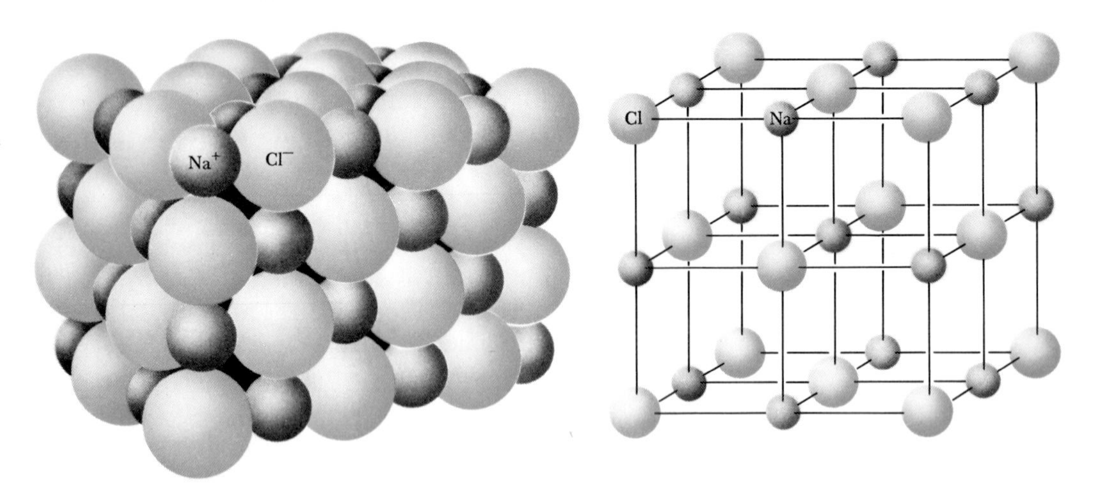

FIGURE 7.1 Two ways of representing the crystal structure of sodium chloride. The structure on the left shows the ions in their correct sizes relative to the distances between them. The larger spheres represent the chloride ions. The structure on the right emphasizes the symmetry of the structure, whereas the one on the left better illustrates the way the ions are packed together in the solid.

The formation of Na^+ from Na and of Cl^- from Cl_2 indicates that an electron has been lost by a sodium atom and gained by a chlorine atom. Such electron transfer to form oppositely charged ions is favored when the atoms involved differ greatly in their attraction for electrons. Our example of NaCl is rather typical for ionic compounds; it involves a metal of low ionization energy and a nonmetal with a high affinity for electrons. Using electron-dot symbols (and showing a chlorine atom rather than the Cl_2 molecule) we can represent this reaction as follows:

$$\text{Na}\cdot + \cdot\ddot{\underset{..}{\text{Cl}}}: \longrightarrow \text{Na}^+ + :\ddot{\underset{..}{\text{Cl}}}:^-$$

Each ion has an octet of electrons, the octet on Na^+ being the $2s^22p^6$ electrons that lie below the single $3s$ valence electron of the Na atom.

Energetics of Ionic Bond Formation

The formation of ionic compounds is not merely the result of low ionization energies and high affinities for electrons, although these factors are very important. The formation of an ionic compound from the elements is always an exothermic process; the compound forms because it is more stable (lower in energy) than its elements. Much of the stability of NaCl results from the packing of the oppositely charged Na^+ and Cl^- ions together as shown in Figure 7.1. A measure of just how much stabilization results from this packing is given by the lattice energy. This quantity is the energy required for 1 mol of the solid ionic substance to be separated completely into ions far removed from one another. We can write the process as

$$\text{NaCl}(s) \longrightarrow \text{Na}^+(g) + \text{Cl}^-(g) \quad [7.1]$$

To get a picture of this process, imagine that the lattice shown in Figure 7.1 expands from within, so that the spaces between the ions grow larger and larger, until the ions are very far apart. The energy that would be required for that to occur for a lattice containing 1 mol of Na^+ and 1 mol of Cl^- ions is the lattice energy.

The lattice energy for NaCl(*s*) amounts to 785 kJ/mol. This is a very large amount of energy, and it accounts for the fact that sodium chloride is a stable, solid substance with a high melting point. We see from the structure of NaCl (Figure 7.1) that each sodium ion is surrounded by six nearest-neighbor chloride ions of opposite charge. Similarly, each chloride ion is surrounded by six sodium ions. The attractive force between each ion and its nearest neighbors of opposite charge provides much of the stabilizing lattice energy. Furthermore, each ion also experiences repulsive interactions with ions of like charge in the lattice and is attracted to other ions of opposite charge in addition to its nearest neighbors. The lattice energy is the result of all such electrostatic interactions, taken over the entire lattice. Ionic substances may have arrangements of ions that differ from that shown in Figure 7.1. The lattice arrangements for a few other common ionic substances are shown in Figure 7.2. In each, the ions are arranged to maximize the attractive forces between ions of opposite charge while minimizing the repulsive forces between ions

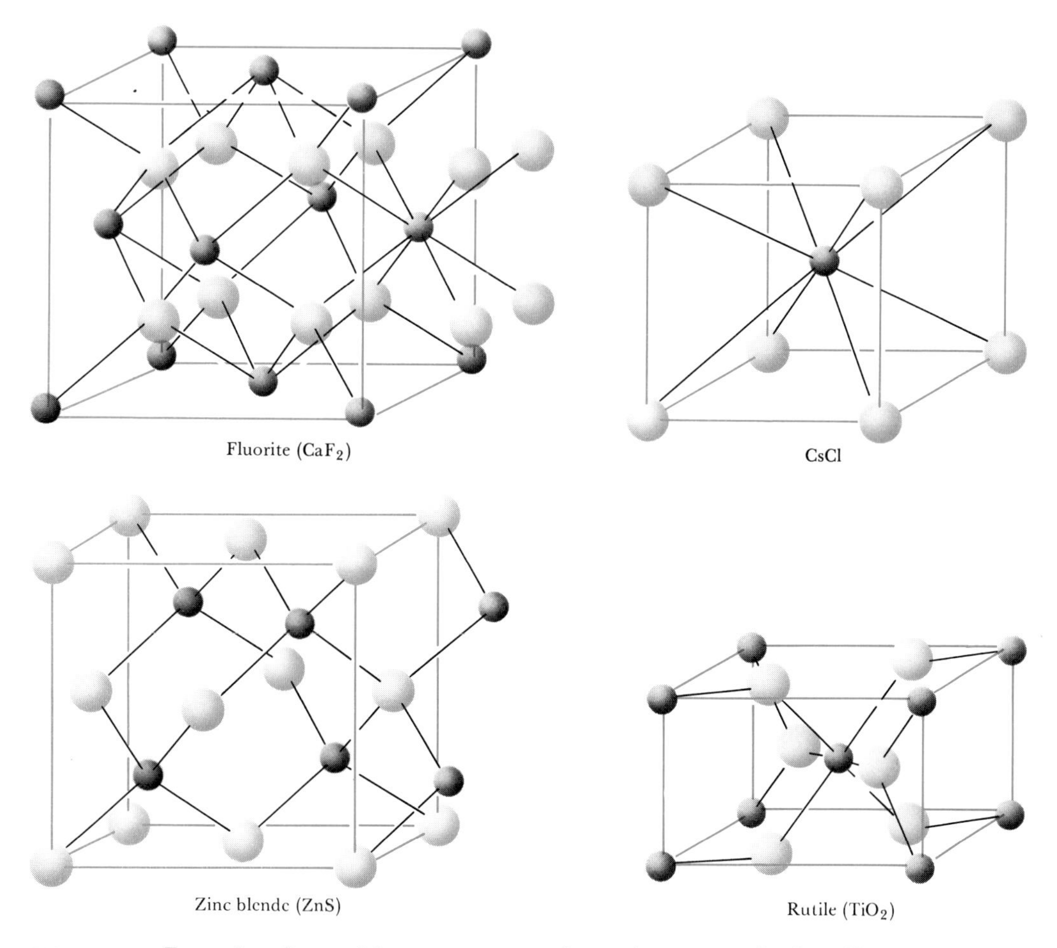

FIGURE 7.2 Examples of several important types of crystal structure. In these illustrations, the color spheres represent the anions, the gray spheres the cations.

of like charge. The overall structure consequently depends on the charges of the ions and their relative sizes. In all these substances, the forces between ions lead to brittle, crystalline solids with high melting points, properties characteristic of ionic compounds.

The potential energy of two interacting charges is given by Equation 7.2:

$$E = k\frac{Q_1Q_2}{d} \qquad [7.2]$$

Q_1 and Q_2 are the magnitudes of the charges on the particles in coulombs and d is the distance between their centers in meters. The constant k has the value 8.99×10^9 J-m/C^2. As Equation 7.2 indicates, the attractive interaction between two oppositely charged ions increases as the magnitudes of their charges increase and as the distance between their centers decreases. Thus for a given arrangement of ions, the lattice energy increases as the charges on the ions increase and as their radii decrease. (The decrease in radii allows ions to approach more closely to one another.)

Now if lattice energy increases as the charges of the ions increase, why doesn't sodium lose two electrons to form Na^{2+}? The second electron would have to come from the inner shell of the sodium atom. We can see from the large value for the second ionization energy of Na (Table 6.4), that it would require too much energy to form the Na^{2+} ion in chemical compounds. Thus sodium and the other group 1A metals are found in ionic substances only as 1+ ions. Similarly, addition of a second electron to a chloride ion to form a hypothetical Cl^{2-} is never observed. The other group 7A elements (the halogens) are also found only as the 1− ions F^-, Br^-, or I^- in ionic substances.

Magnesium, an element of group 2A, also forms an ionic compound with chlorine, of composition $MgCl_2$. In this instance, the metal achieves a rare-gas configuration by loss of two electrons. In any ionic compound there must be charge balance. Thus two Cl^- anions are formed for each Mg^{2+} cation. It requires much less energy to remove two electrons from Mg (Table 6.4) than from Na because both electrons are in the valence shell of Mg. Of course, it requires more energy to remove two electrons from magnesium than is required for removing just one. This energy is more than recovered, however, in the increased lattice energy of $MgCl_2$, which comes from the higher charge on the metal ion. Thus magnesium and the other group 2A metals do not form ionic compounds in which the metal has a 1+ charge; there is no known MgCl. Similarly, the group 6A elements are found in ionic compounds as O^{2-}, S^{2-}, and so forth, in which the ion possesses a rare-gas configuration. In the formation of the ionic solid MgO, both magnesium and oxygen attain the rare-gas configuration, by transfer of two electrons:

$$\mathrm{Mg} + :\ddot{\mathrm{O}}: \longrightarrow \mathrm{Mg}^{2+} + \left[:\ddot{\underset{..}{\mathrm{O}}}:\right]^{2-}$$

Thus there is a balance between ionization energies and lattice energies that determines the magnitude of the positive charge on a metal ion.

Similarly, the negative charge on a nonmetal is determined by the balance between electron affinities and lattice energies. You may recall from Section 6.6 that the electron affinity of O^{2-} is positive. That is, energy is required to add two electrons to an oxygen atom to form O^{2-}. Once again, however, this energy requirement is more than made up for by the large lattice energy resulting from thc 2+ and 2− charges of the Mg^{2+} and O^{2-} ions.

SAMPLE EXERCISE 7.1

Which substance would have the higher lattice energy, NaF or MgO? Explain.

Solution: Magnesium oxide, MgO, would have the higher lattice energy. The electrostatic attraction between oppositely charged ions increases with the charge on the ion. For this reason, it would require a greater amount of energy to separate a mole of Mg^{2+} ions and a mole of O^{2-} ions to infinite separation, as compared with separating a mole of Na^+ and a mole of F^-. In this case, the effect of larger charge is much more important than any slight differences in ionic radii in the two compounds.

SAMPLE EXERCISE 7.2

Predict the formula of the compound formed between aluminum (Al) and fluorine; between aluminum and oxygen.

Solution: Aluminum, with atomic number 13, has three electrons beyond the inert-gas configuration. It might then be expected to lose three electrons to form the Al^{3+} ions. Each fluorine atom attains the noble-gas configuration by accepting one electron to form F^-. As we saw in Chapter 2, the formula of an ionic compound depends on the charges on the ions. The formula may be derived by the principle of electroneutrality: The total charge of the cations must balance that of the anions. Three F^- ions are required to balance the charge of one Al^{3+} ion; the expected formula is thus AlF_3.

In forming a compound with oxygen, each aluminum atom again loses three electrons to form Al^{3+}. Each oxygen atom accepts two electrons to form O^{2-}, thereby achieving the noble-gas configuration (an octet of valence-shell electrons). Two Al^{3+} balance the charge of three O^{2-}; the formula for aluminum oxide is therefore Al_2O_3.

Transition Metal Ions

Ionic bond theory correctly predicts the charges found on many simple ions, based on the notion that attainment of a rare-gas configuration leads to maximum stability. For some elements, however, the rule must be modified, and for others it is not applicable at all. For example, metals of group 1B (Cu, Ag, Au) are observed to occur often as the 1+ ions (as in CuBr and AgCl). Silver possesses a $4d^{10}5s^1$ outer electron configuration. In forming Ag^+ the $5s$ electron is lost. This leaves a completely filled shell of 18 electrons in the $n = 4$ level. Because it is a completed shell, it is somewhat like a rare-gas arrangement. Similarly, the group 2B elements most commonly are seen as the 2+ ions (Zn^{2+}, Cd^{2+}, Hg^{2+}) in ionic compounds. The valence-shell s electrons are lost in forming the ions, leaving an electronic arrangment consisting of 18 electrons in the highest occupied level.

For most of the transition metals, the attainment of a rare-gas configuration by loss of electrons is not feasible; that would require the loss of too many electrons. The outer electron configurations of these elements are either $(n - 1)d^x ns^2$ or $(n - 1)d^x ns^1$, where n is 4, 5, or 6, and x may vary from 1 to 10. In forming ions the transition metals lose the valence-shell s electrons first, then as many d electrons as are required to form an ion of particular charge. Most of the transition metals are found in more than one charge state. For example, the element chromium is found in compounds as Cr^{2+} or Cr^{3+}. There are no simple rules to tell which charge state of a transition metal ion will exist in a particular case.

SAMPLE EXERCISE 7.3

Write the electron configuration for the Co^{2+} ion; for the Co^{3+} ion.

Solution: Cobalt (atomic number 27) has an electron configuration $[Ar]4s^2 3d^7$. To form a 2+ ion, two electrons must be removed. As discussed in the text above, the $4s$ electrons are removed before the $3d$. Consequently, the Co^{2+} ion has an electron configuration of $[Ar]3d^7$. To form Co^{3+} requires the removal of an additional electron; the electron configuration for this ion is $[Ar]3d^6$.

This is a good point at which to review Table 2.5, which lists common ions. Recall that positively charged ions are called **cations**, whereas negatively charged ones are called **anions**. Notice that some ions are polyatomic. Examples of polyatomic cations include the vanadyl ion, VO^{2+}, and the familiar ammonium ion, NH_4^+. However, most polyatomic ions are anions. Examples include the carbonate ion, CO_3^{2-}, found in many mineral deposits, and the brightly colored yellow chromate ion, CrO_4^{2-}.

In polyatomic ions, two or more atoms are bound together by predominantly covalent bonds. They form a stable grouping that carries a charge, either positive or negative. We will examine the covalent bonding forces in these ions in Chapter 8. For now, it is simply necessary for you to realize that the group of atoms as a whole acts as a charged species in forming an ionic compound with an ion of opposite charge.

SAMPLE EXERCISE 7.4

The dichromate ion, $Cr_2O_7^{2-}$, is readily obtained in the form of its ammonium salt. Write the formula for ammonium dichromate.

Solution: The charge on the dichromate ion is 2−, that on the ammonium ion is 1+. We therefore require two ammonium ions to balance the charge of the dichromate ion. The formula for the salt is thus $(NH_4)_2Cr_2O_7$.

This salt is very soluble in water and is thus a convenient source of the dichromate ion. However, it decomposes violently on heating, to yield chromium(III) oxide and nitrogen gas:

$$(NH_4)_2Cr_2O_7(s) \longrightarrow Cr_2O_3(s) + N_2(g) + 4H_2O(g)$$

7.3 SIZES OF IONS

The radii of ions are interesting because they help us understand better how the electrons in ions are attracted to the central nuclear charge. They are also important in many practical ways. For example, the sizes of ions are important in determining the lattice energy in an ionic solid. They also determine how easily the ions can be removed from water in water-softening devices. As another example, many metal ions are im-

TABLE 7.2 Radii (Å) of ions with rare-gas electron configurations

Group 1A		Group 2A		Group 3A		Group 6A		Group 7A	
Li^+	0.68	Be^{2+}	0.30			O^{2-}	1.45	F^-	1.33
Na^+	0.98	Mg^{2+}	0.65	Al^{3+}	0.45	S^{2-}	1.90	Cl^-	1.81
K^+	1.33	Ca^{2+}	0.94	Sc^{3+}	0.68	Se^{2-}	2.02	Br^-	1.96
Rb^+	1.48	Sr^{2+}	1.10	Y^{3+}	0.90	Te^{2-}	2.22	I^-	2.19
Cs^+	1.67	Ba^{2+}	1.31						

portant in biological reactions. Biological systems are often very specific; they work well with one metal ion but not with another, even though it has the same charge and seems very similar. It often happens that only a small difference in ionic size is sufficient to cause one metal ion to be biologically active and another not to be so.

When ions are drawn close together, electron-electron repulsions eventually become as large as the attractive forces. The repulsive forces place a limit on the distance of closest approach. We may thus think of ions as having characteristic radii that determine their distances from other ions. From X-ray diffraction studies of ionic solids, which will be described in Section 10.5, it is possible to determine the distances between ions. Using data for a large number of structures, chemists have analyzed these distances to obtain a set of ionic radii.

The radii for several ions with rare-gas configurations are listed in Table 7.2. As we might expect, the radii of cations are smaller than the radii of the corresponding neutral atoms from which they are derived, as illustrated in Figure 7.3. For example, the radius of the potassium atom

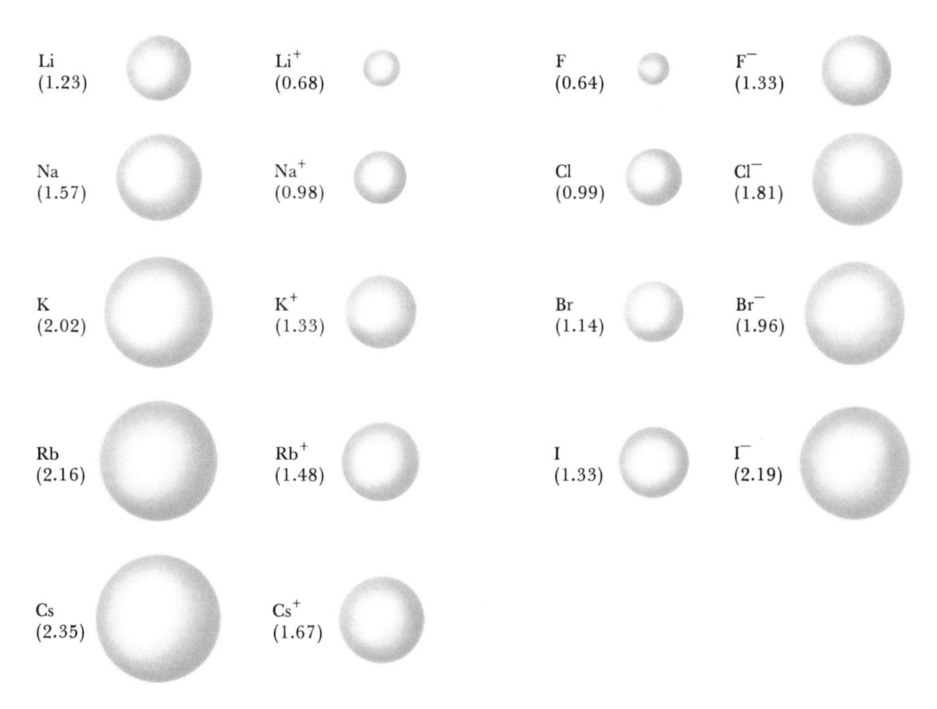

FIGURE 7.3 Relative sizes of atoms and ions. The values in parentheses are radii (Å).

is 2.2 Å, whereas the radius of K^+ is 1.33 Å. Positive ions are formed by removing one or more electrons from the outermost region of the atom. Thus the most spatially extended orbitals are vacated. In addition, removal of an electron decreases the total electron-electron repulsions. On the other hand, the radii of anions are larger than those of the corresponding neutral atoms. For example, the atomic radius of Br^- is 1.96 Å, whereas for bromine atom it is 1.14 Å. Electrons added to atoms to form negative ions with the rare gas configuration go into *p* orbitals that are already partially filled. The increased electron-electron repulsions caused by the additions result in a larger average distance from the nucleus. The electrons spread out in space to minimize their interactions with one another.

SAMPLE EXERCISE 7.5

Based on the data given in Figure 7.3, how would you compare the effective nuclear charge experienced by a 4*p* electron in Br^- with the effective nuclear charge experienced by a 4*p* electron in Br? Explain.

Solution: The effective nuclear charge experienced by a 4*p* electron in Br^- is smaller than in Br. The *actual* nuclear charge is the same in both cases. The only difference is that an extra electron has been added to Br to form Br^-. This extra electron, which goes into the singly occupied 4*p* orbital of Br, acts to some extent to shield the nucleus from the other 4*p* electrons. Thus they experience a smaller effective nuclear charge. This effect, along with the increased electron-electron repulsions, causes Br^- to have a larger radius than Br.

The effect of varying nuclear charge on ionic radii is seen in the variation in radius in an isoelectronic series of ions. The term "isoelectronic" means that the ions possess the same number and arrangement of electrons. For example, in the series O^{2-}, F^-, Na^+, Mg^{2+}, and Al^{3+}, there are ten electrons arranged in the neon electron configuration about each nucleus. The nuclear charge in this series increases steadily in the order listed. With the number of electrons remaining constant, the radius of the ion decreases as the nuclear charge increases, attracting the electrons more strongly toward the nucleus:

Increasing nuclear charge ⟶

O^{2-}	F^-	Na^+	Mg^{2-}	Al^{3+}
1.45 Å	1.33 Å	0.98 Å	0.65 Å	0.45 Å

We saw in Section 6.7 that as the principal quantum number of an orbital increases, the average distance of the electron from the nucleus increases also. The relative radial extensions of the 1*s*, 2*s*, and 3*s* orbitals, shown in Figure 5.15, provide a good example. Thus, for ions of the same charge in any one family, the ionic radius increases with increasing period, that is, as the principal quantum number of the outermost occupied orbital increases.

One further comparison worth keeping in mind is the relative sizes of ions from the A and B subgroups. You may recall from the discussion in Section 6.8 that the outermost *s* electron of the group 1B elements experiences a higher effective nuclear charge. This happens because the ten *d* electrons added in going from a group 1A element to a group 1B element in the same row (for example, in going from K to Cu) do not completely shield the valence *s* electron from the nucleus. They also do not completely shield one another from the nucleus. As a result, not only is the atom of the group 1B element smaller, the ion formed by removal of the valence *s* electron is smaller also. For example, the radius of Cu^+, 0.96 Å, is less than that for the corresponding group 1A ion, K^+, radius 1.33 Å.

SAMPLE EXERCISE 7.6

The radius of Zn^{2+} is 0.74 Å, as compared with 0.99 Å for Ca^{2+}. Account for the smaller radius of Zn^{2+}.

Solution: Zinc occurs in the same horizontal row of the periodic table as Ca, but it has ten additional electrons, and a nuclear charge ten larger than for Ca. As a result of incomplete shielding of the added nuclear charge by the added 3*d* electrons, the effective nuclear charge experienced by all the electrons in the $n = 3$ shell in Zn^{2+} is considerably greater than in Ca^{2+}. As a result, the radius of Zn^{2+} is smaller.

7.4 COVALENT BONDING

We have seen that ionic substances possess several characteristic properties. They are usually brittle substances with high melting points. They are usually also crystalline, meaning that the solids have flat surfaces that make characteristic angles with one another. Ionic crystals can often be cleaved; that is, they break apart along smooth, flat surfaces. The characteristics of ionic substances result from the ionic forces that maintain the ions in a rigid, well-defined, three-dimensional arrangement such as one of those illustrated in Figures 7.1 and 7.2.

The vast majority of chemical substances do not have the characteristics of ionic materials; we need only think of water, gasoline, banana peelings, hair, antifreeze, and plastic bags as examples. Most of the substances with which we come in daily contact tend to be gases, liquids, or solids with low melting points; many vaporize readily—for example, mothball crystals. Many in their solid forms are plastic rather than rigidly crystalline—for example, paraffin or plastic bags.

For the very large class of substances that do not behave like ionic substances, a different model for the bonding between atoms is required. G. N. Lewis reasoned that an atom might acquire a rare-gas electron configuration by sharing electrons with other atoms. A chemical bond formed by sharing a pair of electrons is called a **covalent bond.**

The hydrogen molecule, H_2, furnishes the simplest possible example of a covalent bond. Using electron-dot symbols, formation of the H_2 molecule by combination of two hydrogen atoms can be represented as

$$H\cdot + \cdot H \longrightarrow H:H$$

The shared pair of electrons provides each hydrogen atom with two electrons in its valence shell (the 1*s*) orbital, so that in a sense it has the electron configuration of the rare gas helium (the shared electrons are counted with both atoms). Similarly, when two chlorine atoms combine to form the Cl_2 molecule,

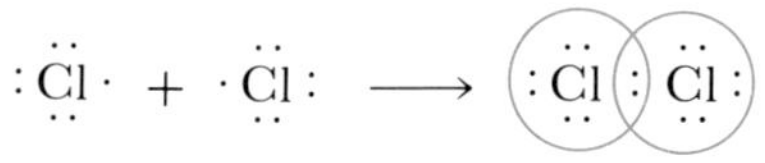

$$:\ddot{\underset{..}{Cl}}\cdot + \cdot\ddot{\underset{..}{Cl}}: \longrightarrow (:\ddot{\underset{..}{Cl}}(:)\ddot{\underset{..}{Cl}}:)$$

each chlorine atom, by sharing the bonding electron pair, acquires eight electrons (an octet) in its valence shell. It thus achieves the rare-gas electron configuration of argon. The structures shown above for H_2 and Cl_2 are called Lewis structures. In writing Lewis structures, it is the usual practice to show each electron pair shared between atoms as a line, and the unshared electron pairs as dots. Thus the Lewis structures for H_2 and Cl_2 are shown as follows:

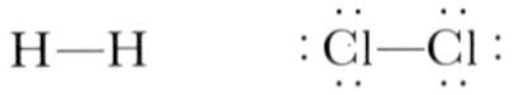

$$\text{H—H} \qquad :\ddot{\underset{..}{Cl}}\text{—}\ddot{\underset{..}{Cl}}:$$

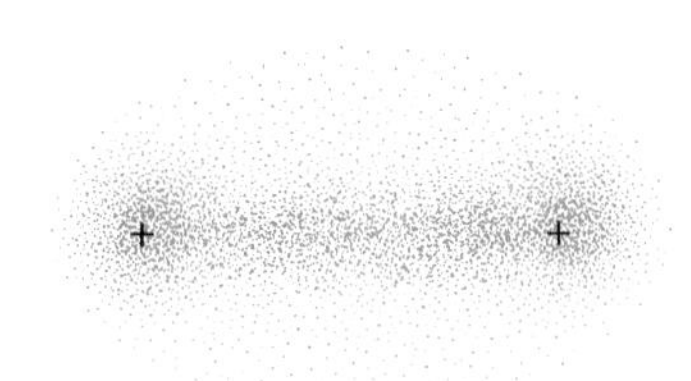

FIGURE 7.4 Electron distribution in the H_2 molecule.

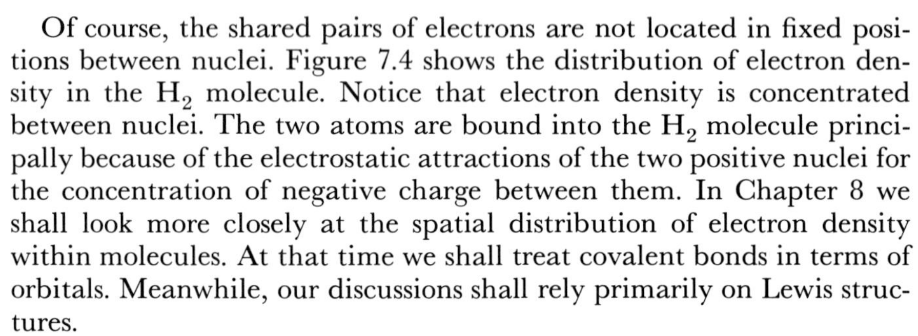

Of course, the shared pairs of electrons are not located in fixed positions between nuclei. Figure 7.4 shows the distribution of electron density in the H_2 molecule. Notice that electron density is concentrated between nuclei. The two atoms are bound into the H_2 molecule principally because of the electrostatic attractions of the two positive nuclei for the concentration of negative charge between them. In Chapter 8 we shall look more closely at the spatial distribution of electron density within molecules. At that time we shall treat covalent bonds in terms of orbitals. Meanwhile, our discussions shall rely primarily on Lewis structures.

In the Lewis model, the valence of an element is associated with the number of electron pairs shared to complete the octet of electrons. Because the number of valence electrons is the same as the group number for the nonmetals, one might predict that the 7A elements, such as F, would form one covalent bond to achieve an octet; 6A elements such as O would form two covalent bonds; 5A elements such as N would form three covalent bonds; and 4A elements such as C would form four covalent bonds. These predictions are borne out in many compounds. For example, consider the simple hydrides of the nonmetals of the second row (period) of the periodic table:

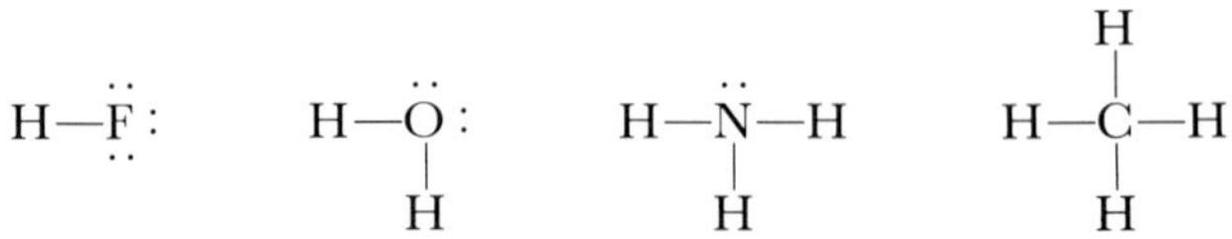

$$\text{H—}\ddot{\underset{..}{F}}: \qquad \text{H—}\ddot{O}:\ (\text{with H below O}) \qquad \text{H—}\ddot{N}\text{—H}\ (\text{with H below N}) \qquad \text{H—C—H}\ (\text{with H above and below C})$$

Thus the Lewis model is very successful in accounting for the compositions of compounds of the nonmetals, in which covalent bonding predominates.

SAMPLE EXERCISE 7.7

Using the Lewis theory and the theory of ionic bonding described earlier, explain the formulas of the following hydrides: NaH; MgH_2; AlH_3; SiH_4; PH_3; H_2S; HCl.

Solution: The hydrides of the metallic elements are ionic compounds consisting of metallic cations and the hydride ion, H^-. These ionic substances are formed by transfer of one or more electrons from the metal to hydrogen atoms. Each hydrogen atom accepts one electron to form H^-. One hydride ion is required for each electron removed from the metal to form the rare-gas configuration. The first three compounds thus correspond to the compositions Na^+H^-; $Mg^{2+}2H^-$; $Al^{3+}3H^-$. The remaining compounds are best formulated as covalent, in which an electron pair is shared between the central atom and each hydrogen. Si, an element of group 4A, requires four electrons to attain the rare-gas configuration of eight valence-shell electrons. Phosphorus requires three, sulfur two, and chlorine one. The formulas of the hydrides are in accord with the number of electrons needed.

Multiple Bonds

The sharing of a pair of electrons constitutes a single covalent bond, generally referred to simply as a single bond. In many molecules, atoms attain complete octets by sharing more than one pair of electrons between them. When two electron pairs are shared, two lines are drawn, representing a double bond. A triple bond corresponds to the sharing of three pairs of electrons. Such multiple bonding is found, for example, in the N_2 molecule:

$$:\dot{\underset{\cdot}{N}}\cdot + \cdot\dot{\underset{\cdot}{N}}: \longrightarrow :N:::N: \quad (\text{or } :N{\equiv}N:)$$

Because each nitrogen atom possesses five electrons in its valence shell, the sharing of three electron pairs is required to achieve the octet configuration. The properties of N_2 are in complete accord with this Lewis structure. Nitrogen gas is a diatomic gas with exceptionally low reactivity. The low chemical reactivity results from the very stable nitrogen-nitrogen bond. Study of the structure of N_2 reveals that the nitrogen atoms are separated by only 1.10 Å. The short N—N bond distance is a result of the triple bond between the atoms. From structure studies of many different substances in which nitrogen atoms share one or two electron pairs, it has been learned that the average distance between bonded nitrogen atoms varies with the number of shared electron pairs:

N—N	N=N	N≡N
1.47 Å	1.24 Å	1.10 Å

As a general rule, the distance between bonded atoms decreases as the number of shared electron pairs increases.

Carbon dioxide, CO_2, provides a further example of a molecule containing multiple bonds:

$$:\dot{\underset{\cdot}{O}}: + \cdot\dot{\underset{\cdot}{C}}\cdot + :\dot{\underset{\cdot}{O}}: \longrightarrow \ddot{\underset{\cdot\cdot}{O}}::C::\ddot{\underset{\cdot\cdot}{O}} \quad (\text{or } \ddot{\underset{\cdot\cdot}{O}}{=}C{=}\ddot{\underset{\cdot\cdot}{O}})$$

7.5 DRAWING LEWIS STRUCTURES

We have seen in the foregoing section some simple examples of Lewis structures. It is actually fairly easy to draw the Lewis structures for most compounds and ions formed from nonmetallic elements. Doing so is a key skill that is important in mastering the material in this chapter and the next.

In drawing Lewis structures it is a good idea to follow a regular procedure. We'll first outline the procedure, then go through several examples to show their application.

1 Add up the number of valence-shell electrons. (Use the periodic table as necessary to help you determine the number of valence-shell electrons on each atom.) If the species involved is neutral, the total number of valence-shell electrons is just the sum of the valence-shell electrons from each atom. If the ion is negatively charged, add the charge of the ion to the total electron count. If it is positively charged, subtract the charge of the ion. For example, the number of valence-shell electrons in N_2O is $(2 \times 5) + 6 = 16$. In CO_3^{2-} it is $4 + (3 \times 6) + 2 = 24$.

2 Write the symbol for the atoms involved so as to show which atoms are connected to which. This is often very easy; for example, in CH_4, we know that the hydrogens must be connected to the carbon. But when the formula is H_3PO_4, or N_2O, or $H_4C_2O_4$, is isn't so immediately obvious which atoms are bonded to which. In these cases you must have more information before you can draw the Lewis structure. Often the atoms are written in the order in which they are connected in the molecule or ion, as in HCN. Also, when a central atom has a group of other atoms bonded to it, it is usually written first, as in CO_3^{2-} or CCl_4.

3 Once you have an idea of the arrangement of atoms, draw in a single bond between each pair of atoms that are presumed to be bonded to one another. Then put the remaining electrons on the atoms as unshared pairs, in an attempt to achieve an octet about each atom. (Hydrogen is, of course, the exception, because it always has just two electrons in its valence-shell orbital.)
   (a) If placing all the valence electrons in this manner leads to an octet around each atom, and uses up all the valence-shell electrons, your Lewis structure is complete.
   (b) If the octet rule cannot be satisfied in this manner because there are not enough electron pairs, then one or more of the electron pairs that were placed as unshared pairs must be used to form double or triple bonds.
   (c) If it turns out that there are too many electron pairs to complete an octet with single bonds between atoms, then the octet rule is not obeyed for one or more atoms in the compound. We will learn more about exceptions to the octet rule in Section 7.7.

These rules are illustrated in the following sample exercise.

SAMPLE EXERCISE 7.8

Draw Lewis structures for the following molecules: (a) PCl_3; (b) HCN; (c) ClO_3^-.

Solution: (a) Phosphorus (group 5A) has five valence electrons and each chlorine (group 7A) has seven. The total number of valence-shell electrons is therefore $5 + (3 \times 7) = 26$. There are various ways the atoms might be arranged. However, in binary (two element) compounds such as PCl_3, the first element listed in the chemical formula is generally surrounded by the remaining atoms. Thus we begin with a skeleton structure that shows single bonds between phosphorus and each chlorine:

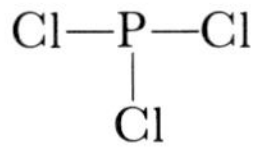

Cl—P—Cl
|
Cl

(It is not important that we place the atoms in exactly this arrangement; Lewis structures are not drawn to show geometry. However, it is important to show correctly which atoms are bonded to which.) The three single bonds account for six electrons. When the remaining 20 electrons are then placed in pairs, they lead to an octet of electrons about each atom:

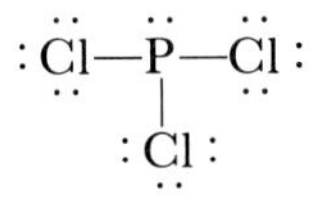

:Cl—P—Cl:
|
:Cl:

(Remember that the bonding electrons are counted for both atoms.)

(b) Hydrogen has one valence-shell electron, carbon (group 4A) has four, and nitrogen (group 5A) has five. The total number of valence-shell electrons is therefore 1 + 4 + 5 = 10. Again there are various ways we might choose to arrange the atoms. Because hydrogen can accommodate only one electron pair, it always has only one single bond associated with it in any compound. This fact causes us to reject C—H—N as a possible arrangement. The remaining two possibilities are H—C—N and H—N—C. The first is the arrangement found experimentally. You might have guessed this to be the atomic arrangement because the formula is written with the atoms in this order. Thus we begin with a skeleton structure that shows single bonds between hydrogen, carbon, and nitrogen:

H—C—N

These two bonds account for four electrons. If we then place the remaining six electrons around carbon and nitrogen, we cannot achieve an octet:

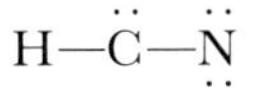

H—C—N

We next try a double bond between C and N. Again we are unable to achieve an octet around these atoms. Finally, we try a triple bond. The single bond between H and C and the triple bond between C and N together account for eight electrons. The remaining two electrons can be placed on nitrogen to give an octet around both C and N atoms:

H—C≡N:

(c) Chlorine has seven valence electrons, and oxygen (group 6A) has six. An extra electron is added to account for the fact that the ion has a 1− charge. The total number of valence-shell electrons is therefore 7 + (3 × 6) + 1 = 26. After putting in the single bonds and distributing the unshared electron pairs, we have

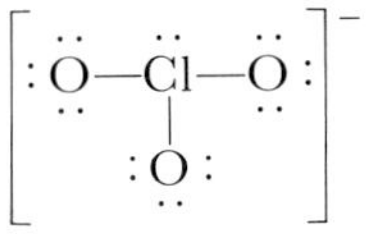

[:O—Cl—O: | :O:]$^-$

(For oxyanions such as ClO_3^-, SO_4^{2-}, NO_3^-, CO_3^{2-}, and so forth, the oxygen atoms surround the central nonmetal atom.)

The concept of **formal charge** is sometimes used as an aid in deciding between alternative Lewis structures. The formal charge is largely a means of "bookkeeping" the valence-shell electrons. To establish the formal charge on any atom in a molecule or ion, we assign electrons to the atom as follows:

1 All bonding electrons are divided equally between the atom and the atoms to which it forms bonds.

2 All nonbonding electrons are assigned entirely to the atom on which they are found.

The formal charge is defined as *the number of valence-shell electrons in the isolated atom, minus the number of electrons assigned to the atom in the Lewis structure*. Let's illustrate these rules by calculating the formal charge on the central atom in the second-row nonmetal hydrides:

	Hydride			
	H—F: (with lone pairs)	H—O—H (O with lone pairs)	H—N(—H)—H (N with lone pair)	CH_4 (H—C—H with H above and below)
Bonding electrons assigned	1	2	3	4
Nonbonding electrons assigned	6	4	2	0
Total electrons assigned	7	6	5	4
Electrons in isolated atom	7	6	5	4
Formal charge	0	0	0	0

Note that the formal charge on the central atom is zero in all cases.

To see how the idea of formal charge can help in making a distinction between alternative Lewis structures, consider the cyanate ion, NCO^-. There are three possible orders for the atoms in this ion. For each we can write a Lewis structure that yields an octet about each atom. For each structure we can then calculate the formal charge on each atom. The results are as follows:

	$[\ddot{N}=C=\ddot{O}]^-$	$[\ddot{C}=\ddot{O}=\ddot{N}]^-$	$[\ddot{O}=N=\ddot{C}]^-$
Formal charge	−1 0 0	−2 +2 −1	0 +1 −2

Because the ion as a whole has a charge of 1−, the formal charges of all atoms must sum to 1−. As a general rule the most stable Lewis structure will be that in which the atoms bear the smallest formal charges. The Lewis structure on the left is clearly superior to the other two in producing the smallest variations in formal charge among the atoms. This suggests that the arrangement shown at the left is the preferred structure for the ion; indeed, it is the observed structure.

Although the concept of formal charge is useful in helping to decide between alternative Lewis structures, you must keep in mind that *the formal charges do not represent real charges on the atoms.* Other factors that we will be discussing in the material ahead contribute to determine the actual net charges on atoms in molecules and ions.

7.6 RESONANCE FORMS

We sometimes encounter substances in which the known arrangement of atoms is not adequately described by a single Lewis structure. The structural chemistry of nonmetallic elements affords several examples. Consider ozone, O_3, about which we shall have much to say in Chapter 12. This fascinating substance consists of bent molecules with both O—O distances the same:

1.278 Å 1.278 Å
O O O
117°

Because each oxygen atom contributes 6 valence-shell electrons, the ozone molecule has 18 valence-shell electrons. In writing the Lewis structure, we find that we must have one double bond to attain an octet of electrons about each atom:

$$:\ddot{\underset{\cdot\cdot}{O}}—\ddot{O}=\ddot{O}:$$

But this structure cannot by itself be correct, because it requires that one O—O bond be different from the other, contrary to the observed structure. However, in drawing the Lewis structure we could just as easily have put the O=O bond on the left:

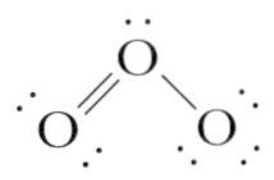

The two alternative Lewis structures for ozone are equivalent except for the placement of electrons. Equivalent Lewis structures of this sort are called resonance forms. To properly describe the structure of ozone, we write both Lewis structures and indicate that the real molecule is described by an average of the structures suggested by the two resonance forms:

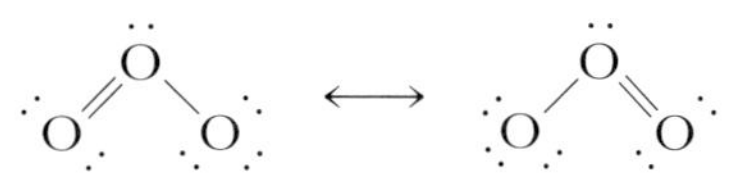

The double-headed arrow is used to indicate that the structures shown are resonance forms.

The fact that we must write more than one Lewis structure to describe a molecule or ion does not imply anything especially different about these species. You must not suppose that the molecule really exists in two or more different forms and oscillates rapidly between them. There is only one form of the molecule, that which is observed experimentally. The fact that we need to write two or more different resonance forms is simply a limitation of the use of Lewis structures in describing the electron distributions in molecules.

One rule that must be followed in writing resonance forms is that the arrangement of the nuclei must be the same in each structure. That is, the same atoms must be bonded to one another in all the structures, so that the only differences are in the arrangements of electrons.

As an additional example of resonance forms, let us draw the Lewis structure for the nitrate ion, NO_3^-, one of the most commonly encountered anions. We find that three equivalent Lewis structures are required in this instance:

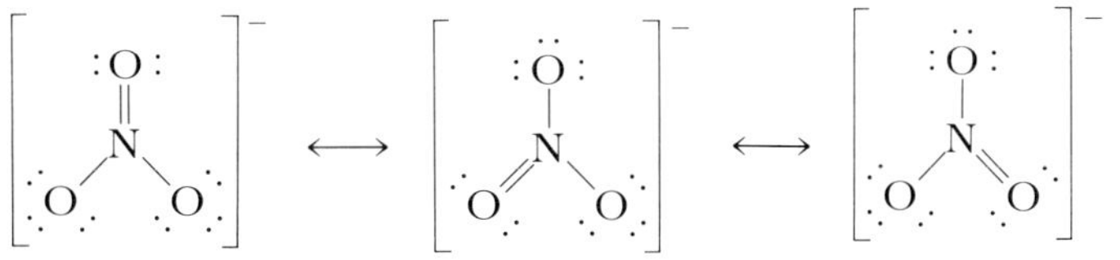

Note that the arrangement of nuclei is the same in each structure; only the placement of electrons differs. All three Lewis structures taken together adequately describe the nitrate ion, which is observed to be planar (all atoms in the same plane), with all three N—O distances equal.

In the examples of resonance structures we have seen so far, all the resonance forms have the same importance in contributing to the overall description of the molecule or ion. In the example in Sample Exercise 7.9, the contributing resonance structures are not equally important.

SAMPLE EXERCISE 7.9

The molecule chlorine dioxide, ClO_2, is bent, with a central chlorine atom bound to two oxygen atoms. The two Cl—O distances are observed to be equal. Describe the ClO_2 molecule in terms of three possible resonance forms.

Solution: The chlorine atom has 7 valence-shell electrons, and oxygen has 6. We therefore have a total of 19 electrons ($2 \times 6 + 7$) to place in this molecule. Note that ClO_2 has an odd number of electrons. This is an unusual situation; it means that one electron must remain unpaired.

We draw the arrangement of atoms in accord with the experimental facts and put one single bond between chlorine and each oxygen.

```
   Cl
  /  \
 O    O
```

This leaves us with 15 electrons to place. We put electron pairs on the atoms to achieve an octet on each atom, or as close to it as we can get. Because the molecule possesses an odd number of electrons, we can be sure that at least one atom will not possess an octet of electrons. The odd electron must go on one or the other of the atoms, so that three possible structures result:

```
  Cl             Cl             Cl
 /  \    <-->   /  \    <-->   /  \
O    O         O    O         O    O
```

Two of these are equivalent, but the other is different, in that the odd electron is on chlorine rather than oxygen. To find out how much weight to attach to this third resonance structure in comparison with the other two, we would have to have experimental information on how the odd electron is distributed in the real ClO_2 molecule.

7.7 EXCEPTIONS TO THE OCTET RULE

In the Lewis theory, attention is focused on attainment of an octet of electrons about each atom. There are many molecules and ions, however, in which the octet rule is not obeyed. These exceptions are of three types.

One fairly small group consists of molecules in which there is an odd number of electrons. In the vast majority of molecules the number of electrons is even, and complete pairing of spins occurs. But in molecules such as ClO_2, NO, and NO_2, the number of electrons is odd. For example, NO contains $5 + 6 = 11$ valence electrons. Obviously, under these conditions complete pairing of all electrons is impossible, and an octet around each atom cannot be achieved.

A second possible failure to obey the octet rule comes when there are fewer than eight electrons about an atom. This is also a relatively rare situation. One example is boron trifluoride, BF_3, a planar molecule with three fluorine atoms around each boron, as shown in Figure 7.5. The properties of BF_3 appear to be best described in terms of single bonds between B and F. The Lewis structure, also shown in the figure, indicates that there are only six electrons about the boron. One of the valence-shell orbitals on the boron must therefore be vacant. The chemical behavior of this molecule reflects this vacancy. It is attracted to other molecules in which there is an atom with an unshared pair of electrons that can be donated to the boron. For example, it reacts with ammonia, NH_3, to form the compound BF_3NH_3:

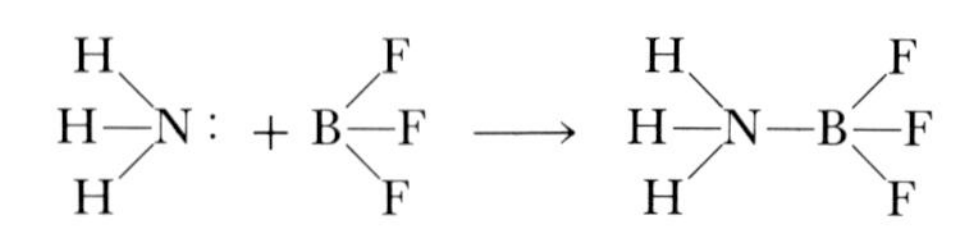

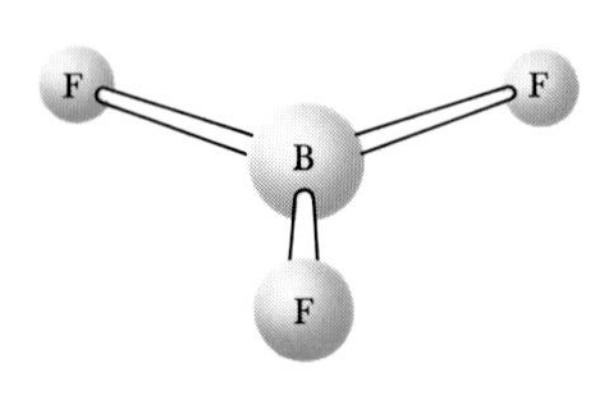

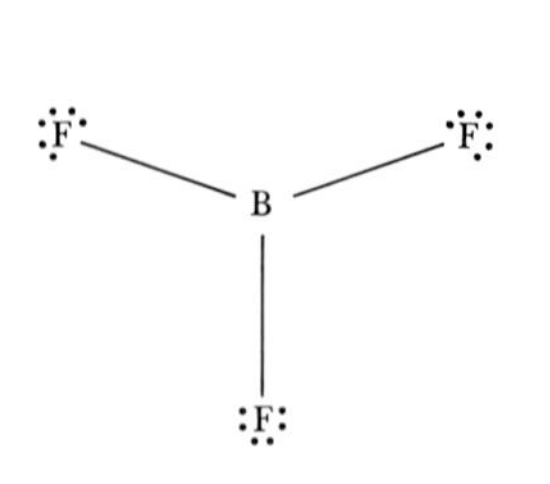

FIGURE 7.5 Geometrical structure and Lewis structure for boron trifluoride, BF_3.

A bond such as the one between N and B, in which one of the two atoms involved has been the source of the electron pair, is often called a **coordinate covalent**, or **dative, bond.** We should remember that electrons do not "belong" to certain atoms. However, in accounting for the overall placement of electrons we can see that both electrons in the B—N bond can be associated originally with the nitrogen. We shall again consider this type of reaction in discussing acid-base chemistry in Chapter 15.

A third and most numerous class of substances in which the octet rule is not obeyed is that in which there are *more* than eight electrons in the valence shell about an atom. As an example, consider PCl_5. This compound may be made by reaction of PCl_3 with chlorine gas:

$$PCl_3(g) + Cl_2(g) \longrightarrow PCl_5(g)$$

When we draw the Lewis structure for this molecule we are forced to place ten electrons about the central phosphorus atom:

Cl—P(—Cl)(—Cl)(—Cl)—Cl

Among the many other examples of molecules and ions with "expanded" valence shells are PF_5, AsF_6^-, SF_4, and ClF_3. We shall see in Chapter 8 how such structures can be accounted for by the use of additional atomic orbitals. Even without referring to these considerations, however, it is possible to see some logical pattern in their occurrences. We note that:

1. The occurrences of expanded valence shells increase with increasing size of the central atom. There are no instances of such molecules among nonmetals of the second row. For example, the NF_3 molecule is a stable species, but NF_5 is unknown. On the other hand, both PF_3 and PF_5 are well-characterized molecules.
2. Expanded valence shells occur most often when the central atom is bonded to the smallest and most strongly electron-attracting atoms, such as F, and Cl, and O.

These observations suggest that expanded valence shells may occur when size considerations allow more atoms to crowd about a central atom than are required by the octet rule. For example, consider the pentahalides of phosphorus. Phosphorus pentafluoride, PF_5, is a very stable species. Phosphorus pentachloride, PCl_5, is reasonably stable, but dissociates in the vapor phase at 300°C into PCl_3 and Cl_2. Phosphorus pentabromide, PBr_5, dissociates in the vapor state even more readily; the compound PI_5 is not known. We see from these observations that the stability of the pentahalide decreases as the size of the halogen atom increases and as its electron attraction decreases.

7.8 STRENGTH OF COVALENT BONDS

The stability of a molecule can be related to the strengths of the covalent bonds it contains. The strength of a covalent bond between two atoms is determined by the energy required to break that bond. The bond-dissociation energy is the enthalpy change, ΔH, required to break a particular bond in a mole of gaseous substance. For example, the dissociation energy for the bond between chlorine atoms in the Cl_2 molecule is the energy required to dissociate a mole of Cl_2 into chlorine atoms:

$$:\ddot{\underset{..}{Cl}}-\ddot{\underset{..}{Cl}}:(g) \longrightarrow 2:\ddot{\underset{..}{Cl}}\cdot(g) \qquad \Delta H = D(\text{Cl—Cl}) = 242\ \text{kJ}$$

We have used the designation D(bond type) in this equation and elsewhere to represent bond-dissociation energies.

It is a relatively easy matter to assign bond energies to bonds in diatomic molecules. As we have seen in the example above, the bond energy is just the energy required to break the diatomic molecule into its component atoms. However, for bonds that occur only in polyatomic molecules (such as the C—H bond) we must often utilize average bond energies. For example, the enthalpy change for the process shown below (called "atomization") can be used to define an average bond strength for the C—H bond:

$$\text{H}-\underset{\text{H}}{\overset{\text{H}}{\text{C}}}-\text{H}(g) \longrightarrow \cdot\dot{\underset{.}{\text{C}}}\cdot(g) + 4\text{H}\cdot(g) \qquad \Delta H = 1660\ \text{kJ}$$

Since there are four equivalent C—H bonds in methane, the heat of atomization is equal to the total bond energies of the four C—H bonds. Thus the average C—H bond energy, as determined from the reaction above, is $D(\text{C—H}) = (1660/4)\ \text{kJ/mol} = 415\ \text{kJ/mol}$.

The average C—H bond energy is not the same as the energy required for the process $CH_4(g) \longrightarrow H(g) + CH_3(g)$; that process requires 435 kJ. Use of that bond-dissociation energy in thermochemical calculations will naturally give different results than the use of the average bond energy. The successive bond-dissociation energies for CH_4 are as follows:

$$CH_4 \longrightarrow CH_3 + H \qquad D(CH_3\text{—}H) = 435\ \text{kJ}$$
$$CH_3 \longrightarrow CH_2 + H \qquad D(CH_2\text{—}H) = 464\ \text{kJ}$$
$$CH_2 \longrightarrow CH + H \qquad D(CH\text{—}H) = 422\ \text{kJ}$$
$$CH \longrightarrow C + H \qquad D(C\text{—}H) = 339\ \text{kJ}$$

The sum of these energies is 1660 kJ, which yields the average of 415 kJ cited above.

The bond energy for a given set of atoms, say C—H, depends on the remainder of the molecule of which it is a part. However, the variation from one molecule to another is generally small. This supports the idea that the bonding electron pairs are localized between atoms. If we consider C—H bond strengths in many other molecules, we find that the average strength is 413 kJ/mol, which compares closely with the 415 kJ/mol value calculated from CH_4.

Table 7.3 lists several average bond energies. Notice that the bond energy is always a positive quantity; energy is always required to break chemical bonds. Conversely, energy is given off when a bond forms be-

TABLE 7.3 Average bond energies (kJ/mol)

Single bonds							
C—H	413	N—H	391	O—H	463	F—F	155
C—C	348	N—N	163	O—O	146		
C—N	293	N—O	201	O—F	190	Cl—F	253
C—O	358	N—F	272	O—Cl	203	Cl—Cl	242
C—F	485	N—Cl	200	O—I	234		
C—Cl	328	N—Br	243			Br—F	237
C—Br	276			S—H	339	Br—Cl	218
C—I	240	H—H	436	S—F	327	Br—Br	193
C—S	259	H—F	567	S—Cl	253		
		H—Cl	431	S—Br	218	I—Cl	208
Si—H	323	H—Br	366	S—S	266	I—Br	175
Si—Si	226	H—I	299			I—I	151
Si—C	301						
Si—O	368						
Multiple bonds							
C=C	614	N=N	418	O_2	495		
C≡C	839	N≡N	941				
C=N	615			S=O	323		
C≡N	891			S=S	418		
C=O	799						
C≡O	1072						

tween two gaseous atoms or molecular fragments. Of course, the greater the bond energy, the stronger the bond.

Referring to Table 7.3, compare the bond energies for the C—C, C=C, and C≡C bonds. Notice that the bond strength increases as the number of pairs of electrons shared between the atoms increases. We noted in Section 7.4 that the distance between atoms decreases as we move from a single to a double to a triple bond. For carbon-carbon bonds we have the following average bond lengths and bond strengths:

C—C	C=C	C≡C
1.54 Å	1.34 Å	1.20 Å
348 kJ/mol	614 kJ/mol	839 kJ/mol

In general, longer bonds have lower bond energies. Consequently, stronger bonds are generally associated with smaller atoms.

A molecule with strong chemical bonds generally has less tendency to undergo chemical change than does one with weak bonds. This relationship between strong bonding and chemical stability helps explain the chemical form in which many elements are found in nature. For example, Si—O bonds are among the strongest ones that silicon forms. It is not surprising therefore that SiO_2 and other substances containing Si—O bonds (silicates) are so common; it is estimated that over 90 percent of the earth's crust is composed of SiO_2 and silicates. We will have more to say about these compounds in Chapter 22.

Bond Energies and Chemical Reactions

A knowledge of bond energies is helpful in understanding why some reactions are exothermic (negative ΔH) while others are endothermic (positive ΔH). An exothermic reaction results when the bonds in the product molecules are stronger than those in the reactants. Consider the following reaction:

$$\text{H}-\underset{\text{H}}{\overset{\text{H}}{\underset{|}{\overset{|}{\text{C}}}}}-\text{H}(g) + \text{Cl}-\text{Cl}(g) \longrightarrow \text{H}-\underset{\text{H}}{\overset{\text{H}}{\underset{|}{\overset{|}{\text{C}}}}}-\text{Cl}(g) + \text{H}-\text{Cl}(g)$$

In the course of this reaction, 1 mol of C—H bonds and 1 mol of Cl—Cl bonds must be broken (reactants). Using average bond energies, we can estimate that this requires 655 kJ:

$$\Delta H(\text{bond breakage}) = D(\text{C—H}) + D(\text{Cl—Cl}) = (413 + 242)\text{ kJ} = 655\text{ kJ}$$

However, 1 mol of C—Cl bonds and 1 mol of H—Cl bonds are formed in the products; this produces 759 kJ:

$$\Delta H(\text{bond formation}) = -D(\text{C—Cl}) - D(\text{H—Cl}) = (-328 - 413)\text{ kJ} = -759\text{ kJ}$$

The overall energy change for the reaction is the sum of the energy required to break the pertinent bonds in the reactants and the energy given up in forming the new bonds in the products:

$$\begin{aligned} \Delta H_{rxn} &= \Delta H(\text{bond breakage}) + \Delta H(\text{bond formation}) \\ &= 655\text{ kJ} - 759\text{ kJ} = -104\text{ kJ} \end{aligned} \qquad [7.3]$$

The reaction is exothermic because the bonds in the products (especially the H—Cl bond) are stronger than those in the reactants (especially the Cl—Cl bond).

The example above illustrates how bond energies can be used to estimate heats of reactions using Equation 7.3. In practice, we seldom calculate ΔH in this way, because it can be determined more accurately from heats of formation (Section 4.5). However, we might use bond energies to estimate ΔH for a reaction if the heat of formation of some reactant or product is not known. Whenever one uses bond energies in this fashion, it is important to remember that they refer to gaseous molecules and that they are often just averaged values.

SAMPLE EXERCISE 7.10

Using Table 7.3, estimate ΔH for the following reaction (where we show explicitly the bonds involved in the reactants and products):

$$\text{H}-\underset{\text{H}}{\overset{\text{H}}{\underset{|}{\overset{|}{\text{C}}}}}-\underset{\text{H}}{\overset{\text{H}}{\underset{|}{\overset{|}{\text{C}}}}}-\text{H}(g) + \tfrac{7}{2}\text{O}_2(g) \longrightarrow 2\text{O}=\text{C}=\text{O}(g) + 3\text{H}-\text{O}-\text{H}(g)$$

$$C_2H_6(g) + \tfrac{7}{2}O_2(g) \longrightarrow 2CO_2(g) + 3H_2O(g)$$

Solution: Among the reactants, we must break six C—H bonds and a C—C bond in C_2H_6; we also break $\frac{7}{2}O_2$ bonds. Among the products, we form four C═O bonds (two in each CO_2) and six O—H bonds (two in each H_2O). Using Equation 7.3 and data from Table 7.3 we have

$$\begin{aligned}\Delta H &= 6D(\text{C—H}) + D(\text{C—C}) + \tfrac{7}{2}D(\text{O}_2)\\ &\qquad - 4D(\text{C=O}) - 6D(\text{O—H})\\ &= 6(413\text{ kJ}) + 348\text{ kJ} + \tfrac{7}{2}(495\text{ kJ}) - 4(799\text{ kJ})\\ &\qquad - 6(463\text{ kJ})\\ &= 4558\text{ kJ} - 5974\text{ kJ} = -1416\text{ kJ}\end{aligned}$$

This estimate can be compared with the value of −1428 kJ calculated from more precise thermochemical data; the agreement is excellent.

7.9 BOND POLARITY; ELECTRONEGATIVITIES

The electron pairs shared between two different atoms are not necessarily shared equally. We can visualize two extreme cases in the degree to which electron pairs are shared. On the one hand, we have bonding between two identical atoms, as in Cl_2 or N_2, where the electron pairs must be equally shared. At the other extreme, illustrated by NaCl, there will be essentially no sharing of electrons. We know that in this case the compound is best described as composed of Na^+ and Cl^- ions. The 3*s* electron of the Na atom is, in effect, transferred completely to chlorine. The bonds occurring in most covalent substances fall somewhere between these extremes. In practice, then, we describe bonds as either ionic or covalent depending on which extreme the bond most closely resembles.

The concept of **bond polarity** is useful in describing the sharing of electrons between atoms. A **nonpolar bond** is one in which the electrons are shared equally between two atoms. In a **polar covalent bond,** one of the atoms exerts a greater attraction for the electrons than the other. If the difference in relative abilities to attract electrons is large enough an ionic bond is formed.

The ability of an atom to attract electrons to itself in a chemical bond is referred to as **electronegativity.** Electronegativity can be related to electron affinity and ionization energy, properties that reflect the tendency of an isolated, gaseous atom to gain or lose an electron. In practice, numerical estimates of electronegativity are based on a variety of other properties. For example, Linus Pauling (1901–), who first developed the concept of electronegativity, based his scale on bond-energy relationships. We will not be concerned in detail with how the electronegativity values are obtained, but rather with using the concept in discussing chemical bonding.

Figure 7.6 shows electronegativity values for many of the elements. Notice that the most electronegative element is fluorine, with an electronegativity of 4.0. The least electronegative element, cesium, has an electronegativity of 0.79. The values for all other elements lie between these extremes. The values listed for the transition elements are those for the 2+ state. When the element is in a higher charge state, its electronegativity is higher.

Note that in a horizontal row of the table there is a more or less steady increase in electronegativity in moving from left to right, that is, from the most metallic to the most nonmetallic elements. Notice also that, with a few exceptions, there is an overall decrease in electronegativity with increasing atomic number in any one group of the periodic table.

1A	2A	3B	4B	5B	6B	7B	8B			1B	2B	3A	4A	5A	6A	7A
H 2.2																
Li 1.0	Be 1.6											B 1.8	C 2.5	N 3.0	O 3.4	F 4.0
Na 0.93	Mg 1.3											Al 1.6	Si 1.9	P 2.2	S 2.6	Cl 3.2
K 0.82	Ca 1.0	Sc 1.4	Ti 1.5	V 1.6	Cr 1.7	Mn 1.6	Fe 1.8	Co 1.9	Ni 1.9	Cu 2.0	Zn 1.6	Ga 1.8	Ge 2.0	As 2.2	Se 2.6	Br 3.0
Rb 0.82	Sr 0.9	Y 1.2	Zr 1.3	Nb 1.6	Mo 2.2	Tc —	Ru 2.2	Rh 2.3	Pd 2.2	Ag 1.9	Cd 1.7	In 1.8	Sn 1.8	Sb 2.0	Te 2.1	I 2.7
Cs 0.79	Ba 0.9								Pt 2.3	Au 2.5	Hg 2.0	Tl 2.0	Pb 2.3	Bi 2.0	Po —	

FIGURE 7.6 Electronegativities of the elements. The values for the transition elements are those for the 2+ valence state.

This is what we might expect, because we know that ionization energies tend to decrease with increasing atomic number in a group, and electron affinities don't change very much. You do not need to memorize numerical values for electronegativity. However, you should know the periodic trends so you can predict which of two elements is more electronegative.

Keep in mind that electronegativities are *approximate* measures of the *relative* tendencies of these elements to attract electrons to themselves in a chemical bond. The electronegativity varies with the type of chemical environment in which an element is situated. We have noted, for example, that the electronegativities of transition metals vary with their valence. Similarly, the electronegativity of chlorine in its bonding to phosphorus in PCl_3 is likely to be different from its value in the chlorate ion, ClO_3^-, in which it is in a much different bonding situation. Variations of this sort are not so large as to render the concept of electronegativity useless, but we must avoid placing too much reliance on precise values for electronegativities. So long as we remain aware of their limitations, electronegativity values provide a useful guide to polarities in chemical bonds.

The electronegativity difference between two atoms is a measure of the polarity of the bond between them; the greater the difference in electronegativity, the more polar the bond. The shared electron pair has a greater probability of being located on the more electronegative of the two atoms. For example, the electronegativity difference between H and F is 4.0 − 2.2 = 1.8. Consequently, the sharing of electrons is unequal (the bond is polar), with the more electronegative fluorine attracting electron density off the hydrogen. We represent this situation in the following two ways:

$$\overset{\delta+}{H}—\overset{\delta-}{F} \qquad \text{or} \qquad \overset{\longmapsto}{H—F}$$

The $\delta+$ and $\delta-$ are meant to represent partial positive and negative charges, respectively. (The symbol δ is the Greek lowercase letter delta.) The arrow represents the pull of electron density off the hydrogen by the fluorine, leaving the hydrogen with a partial positive charge; the head of the arrow points in the direction in which the electrons are attracted, toward the more electronegative atom. We will consider the molecular consequences of bond polarities in Chapter 8 after we have discussed molecular shapes.

SAMPLE EXERCISE 7.11

Which of the following bonds is more polar: (a) B—Cl or C—Cl; (b) P—F or P—Cl? Indicate in each case which atom has the partial negative charge.

Solution: (a) The difference in the electronegativities of boron and chlorine is $3.2 - 1.8 = 1.4$; the difference between carbon and chlorine is $3.2 - 2.5 = 0.7$. Consequently, the B—Cl bond is the more polar; the chlorine atom carries the partial negative charge because it has a higher electronegativity. We should be able to reach this same conclusion without using a table of electronegativities; instead, we can rely on periodic trends. Because boron is to the left of carbon in the periodic table, we would predict that it has a lower attraction for electrons. Chlorine, being on the right side of the table, has a strong attraction for electrons. The most polar bond will be the one between the atoms having the lowest attraction for electrons (boron) and the highest attraction (chlorine).

(b) Because fluorine is above chlorine in the periodic table, we would predict it to be more electronegative. Consequently, the P—F bond will be more polar than the P—Cl bond. You should compare the electronegativity differences for the two bonds to verify this prediction.

7.10 OXIDATION NUMBERS

In light of our discussion of polar covalent bonds, it is useful to examine the similarities between reactions in which electrons are completely transferred and those in which only partial shifts of electron density occur. The reaction between sodium and chlorine atoms to form NaCl involves transfer of an electron from sodium to chlorine. We can imagine the overall reaction as the result of two separate processes:

$$\text{Na}\cdot \longrightarrow \text{Na}^+ + e^- \qquad [7.4]$$

$$:\ddot{\underset{..}{\text{Cl}}}\cdot + e^- \longrightarrow [:\ddot{\underset{..}{\text{Cl}}}:]^- \qquad [7.5]$$

Each of these processes is called a **half-reaction.** Adding these two half-reactions together gives the overall formation of the ionic species from the neutral atoms:

$$\text{Na} + \text{Cl} \longrightarrow \text{Na}^+ + \text{Cl}^-$$

Processes such as the one shown in Equation 7.4, which involve loss of electrons, are called **oxidations.** Processes such as the one shown in Equation 7.5, which involve gain of electrons, are called **reductions.** A substance that has lost electrons is said to be oxidized; one that gains electrons is said to be reduced.

The extent of oxidation-reduction, if any, is not so clear in the reaction between hydrogen atoms and chlorine atoms. The product in this case,

HCl, is a gaseous substance whose normal boiling point is $-84°C$. Because ionic compounds are solids at room temperature, the bonding in HCl is best described as polar covalent. Because chlorine is more electronegative than hydrogen, the electrons in the H—Cl bond are displaced toward chlorine:

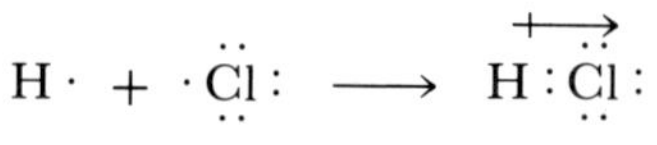

$$H\cdot + \cdot\ddot{\underset{..}{Cl}}: \longrightarrow \overset{+\!\!\longrightarrow}{H:\ddot{\underset{..}{Cl}}:}$$

The H atom in HCl therefore carries a somewhat positive charge and the Cl atom a somewhat negative one. So far as keeping track of electrons is concerned, it would seem a reasonable simplification to assign the shared pair of electrons completely to the more electronegative chlorine atom:

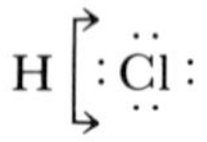

$$H\left[:\ddot{\underset{..}{Cl}}:\right.$$

Chlorine would then have eight valence-shell electrons, one more than the neutral chlorine atom. By assigning electrons in this way, we have in effect given a 1− charge to the chlorine. Hydrogen, stripped of its electron, is assigned a charge of 1+.

Charges assigned in this fashion are called **oxidation numbers** or **oxidation states.** The oxidation number of an atom may vary from compound to compound. However, in each case it is the charge that results when the electrons in a covalent bond are assigned to the more electronegative atom; it is the charge that an atom would possess if the bonding were ionic. When electrons are shared between like atoms as in H_2, they are divided equally between those atoms. Thus in H_2 each hydrogen atom is assigned one electron; the oxidation number of hydrogen in H_2 is therefore zero. For monoatomic ions, such as Fe^{2+}, the oxidation number is just the charge of the ion (2+ in this example).

Although it is possible to determine oxidation numbers using Lewis structures as shown above for HCl and H_2, we seldom do this in practice. It is generally easier to determine oxidation numbers using the following set of rules:

1 The oxidation number of a substance in the elemental state is zero. Thus, for Cl_2, N_2, Na, or P_4, the oxidation number of each atom is zero, because electrons involved in bonding must be shared equally between atoms.

2 In a compound, the more electronegative elements are assigned negative oxidation numbers; the less electronegative elements are assigned positive oxidation numbers. The magnitude of the oxidation number corresponds more or less to the valence, or number of shared electron-pair bonds. For example, hydrogen always has an oxidation number of -1 when it is bonded to a less electronegative element, for example, in NaH; it is $+1$ when it is bonded to a more electronegative element. Thus, in HCl, hydrogen is assigned an oxidation number of $+1$, and chlorine an oxidation number of -1.

3 In any molecule or ion, the sum of the positive and negative oxidation numbers must equal the overall charge on the species. For example, in

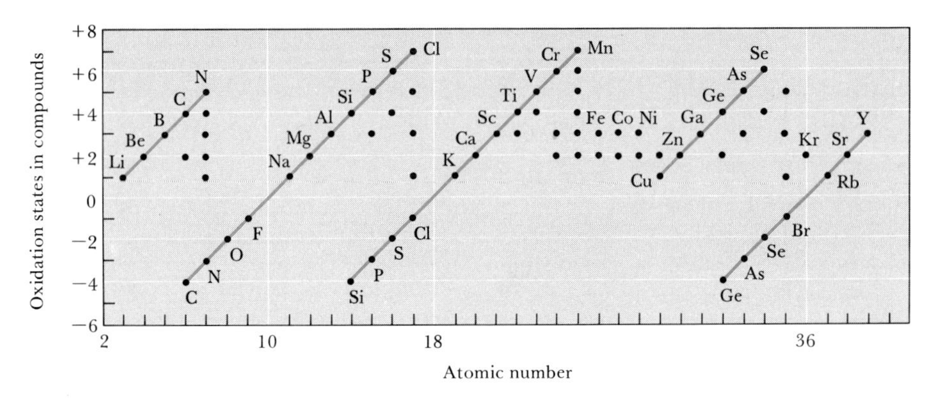

FIGURE 7.7 Common oxidation numbers for the first 39 elements in the periodic table. Notice that the maximum and minimum oxidation states (through which lines have been drawn for emphasis) are a periodic function of atomic number.

OF_2, fluorine is assigned an oxidation number of -1 because it is the more electronegative element. Oxygen must then have an oxidation number of $+2$. In AlO_2^-, if oxygen is assigned an oxidation number of -2, then Al must have an oxidation number of $+3$ in order that the overall charge come out $1-$.

The periodic table provides us with many guidelines to the assignment of oxidation numbers. As shown in Figure 7.7 there is a periodicity in oxidation numbers. All the alkali metals (group 1A) have oxidation numbers of $+1$ in their compounds with other elements. There is simply no other commonly observed means of chemical bonding to other elements for these metals other than loss of an electron to form the $+1$ ion. The elements of group 2A are always found in the $+2$ oxidation state. In group 3A, the most commonly encountered element, Al, is always found in the $+3$ oxidation state.

The most electronegative element, F, is always found in the -1 oxidation state. Other nonmetals are found in negative oxidation states unless combined with a more electronegative atom. Oxygen is nearly always in the -2 oxidation state. The only common exception to this general rule occurs in peroxides. In the peroxide ion, O_2^{2-}, and in molecular peroxides such as H_2O_2, oxygen has an oxidation state of -1. Oxygen is more electronegative than any element with which it forms oxyanions (NO_3^-, ClO_4^-, and so forth). Consequently, the central atom in these oxyanions always has a positive oxidation state.

SAMPLE EXERCISE 7.12

Predict the maximum and minimum oxidation states for sulfur.

Solution: Sulfur has six valence electrons. Assignment of all of these to a more electronegative atom leads to the maximum oxidation state for sulfur, $+6$. When sulfur has a share of an octet of electrons and all are assigned to the sulfur, the resultant oxidation state is -2, the lowest that sulfur exhibits. As shown in Figure 7.7, sulfur has other common oxidation states between these extremes.

SAMPLE EXERCISE 7.13

Assign oxidation numbers to each element in the following compounds: (a) $PbCl_2$; (b) NO_3^-; (c) $KMnO_4$; (d) BaO_2.

Solution: (a) The most electronegative element in $PbCl_2$ is Cl; because only one electron can be added to complete the octet around Cl, its oxidation state is -1. Halogens always have an oxidation state of -1 in binary compounds (compounds containing only two different elements) with metals. The sum of the oxidation states of all of the atoms must equal zero (rule 3). Letting x equal the oxidation state of Pb, we have $x + 2(-1) = 0$. Consequently, the oxidation state of Pb, x, must be $+2$.

(b) Oxygen has a common oxidation state of -2. In this case the sum of the oxidation numbers is -1, the net charge on the NO_3^- ion. Letting x equal the oxidation state of N, we have $x + 3(-2) = -1$. From this we conclude that the oxidation state of nitrogen is $+5$.

(c) The sum of the oxidation numbers of all the atoms is zero. Potassium, an alkali metal, has an oxidation state of $+1$. The oxidation state of O is -2. If we let x equal the oxidation state of Mn, we have $1 + x + 4(-2) = 0$. Thus the oxidation state of Mn is $+7$.

(d) The sum of the oxidation states of all the atoms is zero. Ba, a group 2A metal, must have an oxidation state of $+2$. As a result, oxygen cannot have an oxidation state of -2. We conclude that this compound is a peroxide, oxygen having an oxidation state of -1: $+2 + 2(-1) = 0$.

It is important to keep in mind that the oxidation numbers, or oxidation states, do not correspond to real charges on the atoms, except in the special case of simple ionic substances. Nevertheless, they furnish a useful means of organizing chemical facts, especially in the case of metallic elements. Their most frequent use is in naming compounds and in balancing chemical equations for reactions in which changes in oxidation numbers occur.

Oxidation Numbers and Nomenclature

The rules for naming simple inorganic compounds were discussed in Chapter 2. This is a good point at which to review those rules. Now that the concept of oxidation number has been explained, we can add another widely used method for naming binary compounds. You learned in Section 2.10 that the name of a binary compound consists of the name of the less electronegative element first, followed by the name of the more electronegative element, modified to have an *-ide* ending. The following examples are illustrative:

BaSe	barium selenide
ZnS	zinc sulfide
MgH_2	magnesium hydride

When one or the other of the elements involved has more than one possible oxidation state, the number of atoms may be included in the name as a Greek prefix (1 = mono-, 2 = di-, 3 = tri-, 4 = tetra-, 5 = penta-, 6 = hexa-, 7 = hepta-), *or* the oxidation state is indicated using a Roman numeral, as in these examples:

MnO_2	manganese dioxide	*or* manganese(IV) oxide
Mn_2O_3	dimanganese trioxide	*or* manganese(III) oxide
P_2O_5	diphosphorus pentoxide	*or* phosphorus(V) oxide

P_2O_3	diphosphorus trioxide	*or* phosphorus(III) oxide
$SnCl_4$	tin tetrachloride	*or* tin(IV) chloride
$SnCl_2$	tin dichloride	*or* tin(II) chloride

The use of roman numerals to indicate the oxidation state of the element has the disadvantage that it gives only the simplest formula for the compound, and not the molecular formula. For example, the compounds NO_2 and N_2O_4 are both nitrogen(IV) oxide, and are not distinguished by this name. However, the names nitrogen dioxide and dinitrogen tetroxide do distinguish them. This shortcoming is often unimportant because many compounds are referred to merely by their simplest formula. As an example, phosphorus(V) oxide normally exists as P_4O_{10} molecules, but it is nearly always represented by its simplest formula, P_2O_5.

Oxidation-Reduction Reactions

Many chemical reactions involve atoms that undergo changes in oxidation numbers. Atoms that lose electrons (increase in oxidation number) are said to be oxidized. Atoms that gain electrons (decrease in oxidation number) are said to be reduced. Oxidation and reduction are always discussed together because whenever one substance loses electrons another must gain them. We will take up the topic of balancing oxidation-reduction (or redox) equations in Chapter 19.

SAMPLE EXERCISE 7.14

Is the following reaction an oxidation-reduction reaction? If so, indicate which atom is oxidized and which is reduced:

$$Zn(s) + 2HCl(aq) \longrightarrow ZnCl_2(aq) + H_2(g)$$

Solution: To decide whether this is an oxidation-reduction reaction, we must determine whether any atom undergoes a change in oxidation state. Zinc starts with an oxidation state of 0 and ends up as +2 in $ZnCl_2$. Consequently, the reaction is an oxidation-reduction reaction in which zinc is oxidized. But what is reduced? Hydrogen starts with an oxidation state of +1 in HCl and ends up as zero in H_2. It has gained electrons and has therefore been reduced. The chlorine maintains an oxidation state of −1 throughout the reaction; it does not undergo either oxidation or reduction.

7.11 BINARY OXIDES

With even such simple ideas about bonds as those discussed in this chapter, we can organize and understand many chemical facts. Before taking a closer look at covalent bonding, let's apply some of what we have learned to a discussion of the binary compounds of oxygen.

Oxygen is a very important element that is abundant in air and is found in numerous compounds in the environment. The chemistry of the element is dominated by its high ionization energy and strong atttraction for electrons. The term "oxidation," which we have defined as an increase in oxidation number, was originally applied only to reactions of oxygen, O_2. Because of its high electronegativity, oxygen almost invariably gains electrons in its reactions. It thereby oxidizes (removes electrons) from the substance with which it reacts. For example, O_2

reacts with magnesium to produce magnesium oxide, MgO (Mg in a +2 oxidation state):

$$2Mg(s) + O_2(g) \longrightarrow 2MgO(s)$$

With the exception of the noble gases He, Ne, Ar, and Kr, oxygen forms compounds with every element. Many elements form more than one oxide. For example, iron forms FeO, Fe_2O_3, and Fe_3O_4, while chlorine forms Cl_2O, Cl_2O_3, ClO_2, Cl_2O_6, and Cl_2O_7. The oxides of the metals are generally ionic substances, and those of the nonmetals are generally covalent.

Ionic Oxides

Because of their low ionization energies, metals generally combine with oxygen to form ionic solids in which oxygen is present as the oxide ion. Because the oxide ion, O^{2-}, has a high charge and is relatively small, metal oxides tend to have high lattice energies. Consequently, the melting points of these ionic oxides are typically high (around 2000°C), and most are rather insoluble in water.

When a metal oxide does dissolve to any extent in water, the aqueous O^{2-} ion reacts to form hydroxide ions:

$$:\ddot{\underset{\cdot\cdot}{O}}:^{2-} + H:\ddot{\underset{\cdot\cdot}{O}}:H \longrightarrow :\ddot{\underset{\cdot\cdot}{O}}:H^- + :\ddot{\underset{\cdot\cdot}{O}}:H^-$$

This reaction involves the transfer of H^+ (hydrogen without electrons) from H_2O to the O^{2-} ion.

The solubility of ionic oxides in water is strongly dependent on the size and charge of the associated cation and is greatest for large cations carrying a low charge. The oxides of the alkali metals, such as Na_2O, are quite soluble, as are the oxides of the heavier alkaline earths (especially Ba). They therefore react readily with water to form hydroxides as illustrated by the following examples:

$$Na_2O(s) + H_2O(l) \longrightarrow 2NaOH(aq)$$

$$BaO(s) + H_2O(l) \longrightarrow Ba(OH)_2(aq)$$

As we move from left to right across a period in the periodic table, the solubilities of the metal oxides generally decrease. This trend corresponds to the increasing charge and decreasing size of the metal ions. As we move down a family, the solubilities generally increase; this corresponds to the increasing size of the cation.

Many metal oxides that are water insoluble dissolve in acid solutions owing to the chemical reaction that occurs between O^{2-} ions and H^+ ions to form H_2O. For example, iron(III) oxide, Fe_2O_3, is insoluble in water but dissolves in acids to form soluble salts:

$$Fe_2O_3(s) + 6HCl(aq) \longrightarrow 2FeCl_3(aq) + 3H_2O(l)$$

Metal oxides that dissolve in water to give OH^--containing solutions as well as those that dissolve in acid solutions are called **basic oxides.**

Covalent Oxides

Nonmetals have sufficiently high electronegativities to keep oxygen from removing electrons to form O^{2-} ions. Instead, these elements form molecules containing polar covalent bonds with oxygen. These molecular substances are typically gases, liquids, or solids with low melting points. A notable exception is silicon dioxide, SiO_2, which we shall consider in more detail in Section 10.5. This substance, also known as silica, is relatively hard and has a high melting point (approximately 1600°C).

Whereas metals tend to form basic oxides, nonmetals tend to form acidic ones. **Acidic oxides** dissolve in water to form acidic solutions or dissolve in solutions of bases like NaOH. For example, chlorine(VII) oxide, Cl_2O_7, dissolves in water, reacting to form perchloric acid, $HClO_4$:

$$Cl_2O_7(l) + H_2O(l) \longrightarrow 2HClO_4(aq)$$

Carbon dioxide reacts in a similar way to form an acidic (H^+-containing) solution:

$$CO_2(g) + H_2O(l) \longrightarrow H_2CO_3(aq)$$

This accounts for the acidity of carbonated water. Some nonmetal oxides, CO_2 included, dissolve more readily in aqueous solutions of NaOH than they do in water. Carbon dioxide dissolves in such basic solutions to form hydrogen carbonate ions, HCO_3^-, and carbonate ions, CO_3^{2-}:

$$CO_2(g) + NaOH(aq) \longrightarrow NaHCO_3(aq)$$

$$CO_2(g) + 2NaOH(aq) \longrightarrow Na_2CO_3(aq) + H_2O(l)$$

As the concentration of OH^- increases, less HCO_3^- and more CO_3^{2-} forms. Notice that there is no oxidation-reduction occurring in these reactions. Carbon has the oxidation state of +4 in CO_2, HCO_3^-, and CO_3^{2-}.

We have seen that the ionic oxides, formed with metallic elements in low oxidation states, are basic in character; that is, they dissolve in water to form basic solutions. The oxides of the nonmetallic elements, especially in high oxidation states, form acidic solutions. However, metallic oxides involving metals in high oxidation states are also acidic. For example, Mn_2O_7 forms a strongly acidic solution of the acid $HMnO_4$ on dissolving in water. This behavior is related to the fact that the metal-oxygen bond has more covalent character when the metal is in a high oxidation state. We will have more to say about the acid-base character of oxides in later chapters, but it is useful to keep in mind two simple general rules: (1) For a given oxidation state, the acidity of an oxide increases with increasing electronegativity of the element involved. (2) For any given element, the acidity of the oxide increases with increasing oxidation state of the element.

Amphoteric Oxides

Certain oxides that are virtually insoluble in water are soluble in both acids and bases. Such oxides are said to be **amphoteric**. As shown in Figure 7.8, most of the elements that form amphoteric oxides lie near the

Increasing acidic character →

Increasing base character ↓

1A	2A	3A	4A	5A	6A	7A
Li_2O	BeO	B_2O_3	CO_2	N_2O_5		F_2O
Na_2O	MgO	Al_2O_3	SiO_2	P_2O_5	SO_3	Cl_2O_7
K_2O	CaO	Ga_2O_3	GeO_2	As_2O_5	SeO_3	Br_2O_7
Rb_2O	SrO	In_2O_3	SnO_2	Sb_2O_5	TeO_3	I_2O_7
Cs_2O	BaO	Tl_2O_3	PbO_2	Bi_2O_5	PoO_3	At_2O_7

FIGURE 7.8 Simplest formulas of the oxides of the representative elements in their maximum oxidation states. Those that are basic are shown without shading, those that are amphoteric with light shading, and those that are acidic with dark shading.

diagonal line in the periodic table that divides metals from nonmetals. For example, aluminum oxide, Al_2O_3, is amphoteric and reacts with both acidic and basic solutions:

$$Al_2O_3(s) + 6HCl(aq) \longrightarrow 2AlCl_3(aq) + 3H_2O(l)$$

$$Al_2O_3(s) + 6NaOH(aq) \longrightarrow 2Na_3AlO_3(aq) + 3H_2O(l)$$

Chromium(III), which has the same charge as the aluminum ion and has a virtually identical ionic radius, also forms an amphoteric oxide, Cr_2O_3. We will consider amphoterism again, in Chapter 16, when we consider acids and bases in more detail.

SAMPLE EXERCISE 7.15

Write a balanced chemical equation for the reaction that occurs when (a) $CaO(s)$ is added to water; (b) $SO_2(g)$ is bubbled through water; (c) $SO_2(g)$ is bubbled through an aqueous solution of KOH.

Solution: (a) Metal oxides can react with water to form metal hydroxide solutions. In this case we have

$$CaO(s) + H_2O(l) \longrightarrow Ca(OH)_2(aq)$$

(b) Nonmetal oxides are acidic oxides. To the extent to which they dissolve they react with water to form acidic solutions. In this case we have

$$SO_2(g) + H_2O(l) \longrightarrow H_2SO_3(aq)$$

(c) Many acidic oxides, including SO_2, have low solubilities in water. However, they dissolve in hydroxide solutions to form salts of their oxyanions. In this case,

$$SO_2(g) + KOH(aq) \longrightarrow KHSO_3(aq)$$

or

$$SO_2(g) + 2KOH(aq) \longrightarrow K_2SO_3(aq) + H_2O(l)$$

FOR REVIEW

Summary

In this chapter, we have dealt with the interactions between atoms that lead to formation of chemical bonds. **Ionic bonding** results from the complete transfer of electrons from one atom to another, with formation of a three-dimensional lattice of charged particles. The stabilities of ionic substances result from the powerful electrostatic attractive forces between an ion and all the surrounding ions of opposite charge. These interactions are measured by the **lattice energy.**

Covalent bonding results from the sharing of electrons between atoms. The rules that govern this sharing are based on the stability of the rare-gas electron configuration (the **octet rule**). We can represent shared electron-pair structures of molecules by means of **Lewis structures,** which show the sharing of electron pairs between atoms. The sharing of one pair of electrons produces a **single bond;** the sharing of two or three pairs of electrons between atoms produces **double** and **triple bonds,** respectively.

It sometimes happens that a single Lewis structure is inadequate to represent a particular molecule but that an average of two or more Lewis structures does form a satisfactory representation. In these cases, the Lewis structures are referred to as **resonance forms.** It also sometimes happens that the octet rule is not obeyed; this situation occurs mainly when a large atom is surrounded by small, electronegative atoms like F, O, or Cl. In such instances the large atom often has more than an octet of electrons.

The strengths of covalent bonds increase with the number of electron pairs shared between two atoms. In single bonds, the bond strengths are generally higher between atoms of smaller size.

It is important to recognize that even in covalent bonding, electrons may not be shared equally between two atoms. **Electronegativity** is a measure of the ability of an atom to compete with other atoms for the electrons shared between them. Highly electronegative elements strongly attract electrons. The electronegativities of the elements, which show a regular periodic relationship, are an important guide to chemical behavior. We shall be using the concept of electronegativity often throughout the text. The difference in electronegativities of bonded atoms is used to determine the polarity of a bond.

Another application of electronegativity is in the assignment of **oxidation numbers,** formal whole-number charges assigned to atoms in molecules and ions. Although the oxidation numbers do not represent the real charges on atoms except in simple ionic substances, they are of great value in helping us to organize chemical facts and are an aid in the balancing of equations and in the naming of compounds.

Oxidation may be defined as a process in which an atom undergoes an increase in oxidation number. **Reduction** is a process in which an element undergoes a decrease in oxidation number. In an **oxidation-reduction reaction,** both oxidation and reduction occur in such a manner as to balance the total increases and decreases in oxidation numbers.

Several ideas presented in this chapter were used in discussing the chemistry of binary oxides. Metal oxides tend to be ionic substances that are either soluble in water to give basic solutions or that can be dissolved by acids; they are called **basic oxides.** Nonmetal oxides, by contrast, are covalent substances. They tend to be **acidic oxides,** which dissolve in water to produce acidic solutions or which dissolve in bases. Some oxides are borderline; they are called **amphoteric oxides.**

Learning goals

Having read and studied this chapter, you should be able to:

1 Determine the number of valence electrons for any atom and write its Lewis symbol.

2 Describe the origin of the energy terms that lead to stabilization of ionic lattices.

3 Describe the more commonly observed arrangements of ions in lattices, as illustrated in Figures 7.1 and 7.2.

4 Predict on the basis of the periodic table the probable formulas of ionic substances formed between common metals and nonmetals.

5 Write the electron configurations of ions.

6 Describe the effects of gain or loss of electrons on atomic radii in producing ionic radii.

7 Explain the concept of an isoelectronic series and the origin of changes in ionic radius within such a series.

8 Describe the basis of the Lewis theory and predict the valence of common nonmetallic elements from their position in the periodic table.

9 Write the Lewis structures for molecules and ions containing covalent bonds, using the periodic table.

10 Write resonance forms for molecules or polyatomic ions that are not adequately described by a single Lewis structure.

11 Explain the significance of electronegativity and

in a general way relate the electronegativity of an element to its position in the periodic table.

12 Predict the relative polarities of bonds using either the periodic table or electronegativity values.

13 Relate bond energies to bond strengths and use bond energies to estimate ΔH for reactions.

14 Assign oxidation numbers to atoms in molecules and ions.

15 Give the meaning of the terms *oxidation, reduction,* and *oxidation-reduction reactions.*

16 Determine whether oxidation-reduction has occurred in a reaction; if it has, be able to identify the substance that is oxidized and the one that is reduced.

17 Assign acceptable names to simple inorganic compounds and ions.

18 Describe the general differences in physical properties between substances with ionic bonds and those with covalent bonds.

19 Describe how the water solubility of a metal oxide is related to cation size and charge.

20 Write balanced chemical equations for the reactions of oxides with water, metallic oxides with acids, and nonmetal oxides with bases.

Key terms

Among the more important terms and expressions used for the first time in this chapter are the following:

An **amphoteric oxide** (Section 7.11) is a water-insoluble oxide that dissolves in either an acidic or basic solution.

Bond energy (Section 7.8) is the enthalpy change, ΔH, required to break a chemical bond when a substance is in the gas phase.

Bond polarity (Section 7.9) is a measure of the difference in ability of the two atoms in a chemical bond to attract electrons.

A **cation** (Section 7.2) is a positively charged ion; an **anion** is negatively charged.

A **coordinate covalent bond** (Section 7.7) is a covalent bond in which the electrons are furnished by only one of the bonded atoms.

A **covalent bond** (Section 7.4) is a bond formed between two or more atoms by a sharing of electrons.

A **double bond** (Section 7.4) is a covalent bond involving two electron pairs.

Electronegativity (Section 7.9) is a measure of the ability of an atom that is bonded to another atom to attract electrons to itself.

The **formal charge** (Section 7.5) on an atom is defined as the number of valence-shell electrons in an isolated atom, minus the number of electrons assigned to that atom in a molecule or compound. In assigning electrons, nonbonded electron pairs are assigned entirely to the atom; shared electron pairs are divided equally between the sharing atoms.

A **half-reaction** (Section 7.10) is half of an overall oxidation-reduction reaction, corresponding to either the oxidation half or the reduction half.

An **ionic bond** (Section 7.2) is a bond formed on the basis of the electrostatic forces that exist between oppositely charged species in solid lattices made up of ions. The ions are formed from atoms by transfer of one or more electrons.

An **isoelectronic series** (Section 7.3) is a series of atoms, ions, or molecules having the same number of electrons.

Lattice energy (Section 7.2) is the energy required to separate completely the ions in an ionic solid.

A **Lewis structure** (Section 7.4) is a representation of covalent bonding in a molecule that is drawn using Lewis symbols. Covalently shared electron pairs are shown as lines, and unshared electron pairs are shown as a pair of dots. Only the valence-shell electrons are shown.

The **octet rule** (Section 7.1) states that bonded atoms tend to possess or share a total of eight valence-shell electrons.

Oxidation (Section 7.10) is the half of an oxidation-reduction process that corresponds to an increase in oxidation number.

Oxidation number (Section 7.10) or oxidation state is a positive or negative whole number assigned to an atom in a molecule or ion on the basis of a set of formal rules; to some degree it reflects the positive or negative character of that atom.

A **polar covalent bond** (Section 7.9) is a covalent bond in which the electrons are not shared equally.

Reduction (Section 7.10) is the half of an oxidation-reduction process that corresponds to a decrease in oxidation number.

Resonance forms (Section 7.6) are individual Lewis structures in cases where two or more Lewis structures are equally good descriptions of a single molecule. The resonance structures in such an instance are "averaged" to give a correct description of the real molecule.

A **triple bond** (Section 7.4) is a covalent bond involving three electron pairs.

Valence (Section 7.1) may be defined as the capacity of an atom for entering into chemical combination with other atoms. Ionic valence is equal to the number of electrons gained or lost in forming the ionic species. Covalence is equal to the number of electrons from an atom that are involved in shared electron-pair bonds with other atoms.

Valence electrons (Section 7.1) are the outer-shell electrons of an atom; these are the ones the atom uses in bonding.

EXERCISES

Valence electrons, Lewis symbols, ionic bonding

7.1 In each of the following examples of an electron-dot symbol, indicate the group of the periodic table in which the element X belongs: (a) $\dot{\underset{\cdot}{X}}\cdot$; (b) $\cdot X \cdot$; (c) $\cdot\ddot{\underset{\cdot}{X}}\cdot$.

7.2 Write the Lewis symbol for each of the following elements: (a) sulfur, S; (b) scandium, Sc; (c) silicon, Si; (d) argon, Ar.

7.3 Write the Lewis symbol for each of the following atoms or ions: (a) S^{2-}; (b) Ca^{2+}; (c) Br; (d) Ca.

7.4 What change must occur in the electron configurations of each of the following elements if it is to achieve a configuration that obeys the octet rule: (a) Cl; (b) Mg; (c) N; (d) Rb?

7.5 Write the electron configuration of each of the following ions. Which possess noble-gas configurations: (a) Ca^{2+} (b) Br^-; (c) Mn^{2+}; (d) Ti^{4+}; (e) Zn^{2+}?

7.6 Write the outer electron configuration for each of the following ions: (a) Mn^{3+}; (b) Sr^{2+}; (c) Mo^{3+}; (d) Te^{2-}; (e) Tl^+; (f) Cu^{2+}.

7.7 Predict the chemical formula of the ionic compound formed between each of the following pairs of elements: (a) Sc, O; (b) Mg, Br; (c) Ba, S; (d) Ti, Cl.

7.8 Write the chemical formula of the ionic compound formed between the following cations and anions: (a) Zn^{2+}, $SO_4{}^{2-}$; (b) Ca^{2+}, $PO_4{}^{3-}$; (c) VO^{2+}, Cl^-.

7.9 Which of the following compounds contains ions that do not have a noble-gas electron configuration: (a) CuCl; (b) CdO; (c) $TiCl_4$; (d) MoO_3?

7.10 It requires energy to remove two electrons from Ca to form the Ca^{2+} ion. It also requires energy to form the O^{2-} ion by adding two electrons to an oxygen atom. Why, then, is CaO stable relative to the free elements?

7.11 The energy required to evaporate solid argon is very low; on the other hand, the energy required to evaporate RbCl, formed from the two elements on each side of Ar, is quite high. What accounts for the difference?

7.12 Explain the following trends in lattice energies: (a) CaS > KCl; (b) LiF > CsBr; (c) MgO > MgS; (d) MgO > BaO.

7.13 The ionic compound MY has a higher lattice energy than MX. What does this suggest about the relative radii of the X and Y anions?

7.14 Indicate whether each of the following formulas is a likely formula for a stable ionic compound, and give an explanation for your answer: (a) Rb_2O; (b) BaCl; (c) MgF_3; (d) $ScBr_3$; (e) Na_3N.

7.15 Describe electron configurations other than a completed octet in the valence shell that are relatively stable arrangements often found in ions.

Sizes of ions

7.16 Refer to Figure 7.3. Why are the anions of the halogens larger than the neutral atoms from which they are derived?

7.17 Explain the following variation in atomic or ionic radii: (a) $Br^- > Kr > Rb^+$; (b) $N^{3-} > O^{2-} > F^-$.

7.18 Based on the data of Figure 7.3, how would you compare the effective nuclear charge experienced by a $3p$ electron in K with that of a $3p$ electron in K^+? Explain.

7.19 Compare the effective nuclear charge experienced by a $3p$ electron in Cl^- with that experienced by a $3p$ electron in K^+. Explain.

7.20 Arrange the members of each of the following sets in the order of increasing size: (a) Li^+, Rb^+, K^+; (b) Br^-, Na^+, Mg^{2+}; (c) Ar, Cl^-, S^{2-}, K^+; (d) Al^{3+}, Cl^-, Ar.

7.21 Identify a neutral atom that is isoelectronic with each of the following: (a) Br^-; (b) I^+; (c) Sr^{2+}; (d) Ga^{3+}.

Lewis Structures; resonance structures

7.22 Label each of the following compounds as ionic or covalent: (a) CO_2; (b) BaO; (c) H_2O; (d) SO_3; (e) Fe_2O_3. Mention at least one property of each substance that supports your answer.

7.23 Draw Lewis structures for each of the following compounds: (a) SiH_4; (b) CO_2; (c) NF_3; (d) $TeCl_2$; (e) ONCl; (f) ClOH.

7.24 Draw Lewis structures for each of the following ions: (a) $NO_3{}^-$; (b) $PO_4{}^{3-}$; (c) $ClO_2{}^-$; (d) $BH_4{}^-$; (e) $CrO_4{}^{2-}$; (f) $AlCl_4{}^-$.

7.25 Draw Lewis structures for each of the following: indicate which have atoms that do not obey the octet rule: (a) BF_3; (b) NO_2; (c) $NO_2{}^-$; (d) $BeCl_2$; (e) SO_2.

7.26 Draw resonance structures for each of the following molecules or ions: (a) $NO_3{}^-$; (b) ClO_3; (c) SeO_2; (d) HNO_3 (H bound to an oxygen).

7.27 The oxalate ion, $C_2O_4{}^{2-}$, has the following structure:

```
O.       .O
  C···C
O'       'O
```

Draw the Lewis structures for this ion.

7.28 Which of the following compounds do not obey the octet rule: (a) GeH_4; (b) BCl_3; (c) Cl_2O; (d) ClO_2; (e) $TeCl_4$; (f) XeF_4; (g) BeF_2? In each case, indicate the nature of the departure from the octet rule.

7.29 Predict the relative N—O bond lengths in the following compounds and ions: NO, $NO_3{}^-$, $NO_2{}^-$. Provide an explanation for your answer based on Lewis structures.

7.30 SF_6 is a very stable substance, used as an electrically insulating gas in high-voltage tranformers. Yet SCl_6 is quite unstable, and has never been isolated. Explain the difference in stabilities.

Bond energies

7.31 Using Hess's law (Section 4.4), show how the Si—H bond dissociation energies in SiH_4, SiH_3, SiH_2, and SiH are related to the average Si—H bond dissociation energy in SiH_4.

7.32 Using the bond energies tabulated in Table 7.3, estimate ΔH for each of the following reactions:

(a) $H_3C{-}O{-}H + H{-}I \longrightarrow H_3C{-}I + H{-}O{-}H$ (structural formulas: H—C(H)(H)—O—H + H—I ⟶ H—C(H)(H)—I + H—O—H)

(b) $H{-}CCl_3 + O \longrightarrow Cl{-}C({=}O){-}Cl + H{-}Cl$ (structural formulas: H—C(Cl)(Cl)—Cl + O ⟶ Cl—C(=O)—Cl + H—Cl)

(c) $C{\equiv}O + H{-}O{-}H \longrightarrow H{-}H + O{=}C{=}O$

7.33 Using the data of Table 7.3, calculate the enthalpy change in each of the following reactions: (a) $HCN + 3H_2 \longrightarrow CH_4 + NH_3$; (b) $HC{\equiv}CH + 2Cl_2 \longrightarrow HCl_2C{-}CCl_2H$; (c) $CH_3OH + NH_3 \longrightarrow CH_3NH_2 + H_2O$.

7.34 Based on data given in this chapter, what is the general relationship between bond order (the number of bonding electron pairs in the bond) and the bond energy?

[7.35] The enthalpy of formation of NH(*g*) at 25°C is 360 kJ/mol. From this value, and using the data in Table 7.3, calculate the N—H bond dissociation energy in NH.

Bond polarities; electronegativities

7.36 Explain what is meant by the statement that the H—F bond is polar covalent.

7.37 Without looking at the table of electronegativities (refer to a periodic table instead), arrange the members of each of the following sets in order of increasing electronegativity: (a) O, P, S; (b) Mg, Al, Si; (c) S, Cl, Br; (d) C, Si, N.

7.38 Which of the following bonds is polar: (a) B—Cl; (b) Cl—Cl; (c) Hg—Sb; (d) As—F; (e) Co—C? Indicate the more electronegative atom in each polar bond.

7.39 Arrange the bonds in each of the following sets in the order of increasing polarity: (a) C—S, B—F, N—O; (b) Pb—Cl, Pb—Pb, Pb—C; (c) H—F, O—F, Be—F.

7.40 Based on the data in Sections 6.5 and 6.6, explain why the electronegativity of fluorine is higher than that for chlorine.

7.41 What is the general rule regarding the variation in electronegativity in moving from left to right in a horizontal row of the periodic table? Explain the origin of the trend.

7.42 In a given family of the periodic table, what is the general relationship between electronegativity and size?

Oxidation numbers; oxidation-reduction

7.43 Calculate the oxidation numbers of all elements in each of the following compounds or ions: (a) N_2O; (b) PBr_3; (c) HPO_3^{2-}; (d) P_4O_6; (e) ClF_3; (f) K_2O_2; (g) NH_2^-.

7.44 What are the maximum and minimum oxidation states exhibited by each of the following elements: (a) Cr; (b) Al; (c) Br; (d) Cs; (e) As?

7.45 Provide an appropriate name for each of the following ions or compounds (refer to Table 2.5 as necessary): (a) NaH_2PO_4; (b) Cr_2O_3; (c) V_2O_5; (d) ClO_3^-; (e) BrO^-; (f) CoF_3; (g) Cr^{2+}; (h) HSO_3^-.

7.46 Name each of the following compounds (refer to Table 2.5 as necessary): (a) Cl_2O; (b) $NaMnO_4$; (c) MnO; (d) N_2O_3; (e) $Ba(ClO_3)_2$.

7.47 Give the chemical formula for each of the following substances: (a) chlorine trifluoride; (b) diphosphorus pentoxide; (c) sodium hypochlorite; (d) iron(III) fluoride; (e) molybdenum(VI) oxide.

7.48 Indicate which of the following are oxidation-reduction reactions. In each oxidation-reduction reaction identify the element that is oxidized and the element that is reduced:

(a) $Na_2CO_3(s) + 2HCl(aq) \longrightarrow 2NaCl(aq) + H_2O(l) + CO_2(g)$

(b) $2KI(aq) + F_2(g) \longrightarrow 2KF(aq) + I_2(s)$

(c) $MnO_2(s) + 4HCl(aq) \longrightarrow MnCl_2(aq) + Cl_2 + 2H_2O(l)$

(d) $CaS(s) + 2HCl(aq) \longrightarrow CaCl_2(aq) + H_2S(g)$

(e) $2PbO_2(s) \longrightarrow 2PbO(s) + O_2$

(f) $O_2(g) + 4HCl(g) \longrightarrow 2Cl_2(g) + 2H_2O(g)$

7.49 Indicate the oxidation states of the elements that undergo a change in oxidation state in each of the following reactions:

(a) $S(s) + 2F_2(g) \longrightarrow SF_4(g)$

(b) $2CuSO_4(aq) + 4KI(aq) \longrightarrow 2CuI(s) + 2K_2SO_4(aq) + I_2(s)$

(c) $NH_3(g) + 3Cl_2(g) \longrightarrow NCl_3(g) + 3HCl(g)$

(d) $I_2(aq) + SO_2(aq) + 2H_2O(l) \longrightarrow 2HI(aq) + H_2SO_4$

(e) $2PbS(s) + 3O_2(g) \longrightarrow 2PbO(s) + 2SO_2(g)$

[7.50] Calculate the formal charge on the indicated atom in each of the following molecules or ions: (a) sulfur in SO_2; (b) fluorine in BF_3; (c) phosphorus in $OPCl_3$; (d) phosphorus in PO_4^{3-}; (e) iodine in ICl_3; (f) boron in BH_4^-.

[7.51] Use the concept of formal charge to choose the more likely skeleton structure in each of the following cases: (a) NNO or NON; (b) HCN or HNC; (c) NOBr or ONBr.

Binary oxides

7.52 Write the formula of an oxide formed when each of the following elements forms a compound with oxygen:

(a) Mg; (b) N; (c) Ge; (d) Cr; (e) F; (f) S. In each case, indicate the oxidation state of the element combined with oxygen.

7.53 Indicate the type of product formed in each of the following cases. (a) An active metal reacts with oxygen. (b) An active metal oxide reacts with water. (c) A nonmetal oxide reacts with water. (d) An insoluble metal oxide dissolves on reaction with hydrochloric acid.

7.54 Metallic oxides are generally high-melting solids, whereas nonmetal oxides are more commonly gases, liquids, or low-melting solids. What accounts for this difference?

7.55 Write balanced chemical reactions for each of the following reactions. (a) Barium metal reacts with oxygen to form barium peroxide. (b) Gaseous dichlorine heptoxide dissolves in water to form an acidic solution. (c) Gaseous sulfur dioxide dissolves in aqueous sodium hydroxide solution. (d) Solid chromium(III) oxide dissolves in hydrochloric acid solution. (e) Solid strontium oxide is added to water. (f) Solid diphosphorus pentoxide is added to water.

7.56 Which of the following pairs would you expect to be less soluble in water: (a) Al_2O_3 or Na_2O; (b) SiO_2 or CO_2? Explain your answer in each case.

Additional exercises

7.57 Select the ions or atoms from each of the following sets that are isoelectronic with each other: (a) K^+, Ca^{2+}, Rb^+; (b) Cu^{2+}, Ca^{2+}, Sc^{3+}; (c) S^{2-}, Se^{2-}, Ar; (d) F^-, Ne, Na^+.

7.58 Suggest an explanation for the observation that the radii of anions are generally larger than those of cations.

7.59 The lattice energy of LiH is 858 kJ/mol, whereas that for MgH_2 is 2790 kJ/mol. Account for the large difference in these two quantities.

7.60 Explain the following trend in lattice energies: LiH, 858 kJ/mol; NaH, 782 kJ/mol; KH, 699 kJ/mol; RbH, 674 kJ/mol.

7.61 How is the tendency of a pair of elements to form an ionic compound related to their electronegativity difference?

7.62 From the ionic radii given in Table 7.2, calculate the potential energy of each of the following ion pairs, assuming that they are separated by the sum of their ionic radii (the magnitude of the electronic charge is given on the inside back cover): (a) Mg^{2+}, O^{2-}; (b) Na^+, Br^-.

[7.63] From the ionic radii given in Figure 7.3, calculate the potential energy of a Na^+ and Cl^- ion pair that are just touching (the magnitude of the electronic charge is given on the inside back cover). Calculate the energy of a mole of such pairs. How does this value compare with the lattice energy of NaCl? Explain the difference.

7.64 A white solid, melting at 1115°C, insoluble in water but slightly soluble in aqueous NaOH, is likely to be which of the following: SrO; GeO_2; SeO_2; N_2O_3?

7.65 What type of bond would you expect to be found in each of the following substances: (a) K_2S; (b) SCl_2; (c) MnF_3; (d) B_2H_6; (e) Cr; (f) P_4S_{10}?

7.66 How would you expect the energy of an ionic interaction between positive ion A and negative ion B to be affected by the following changes? (a) The charge on A is doubled. (b) The charge on B is doubled. (c) The charges on both A and B are doubled. (d) The radii of both A and B are simultaneously doubled.

7.67 Write the electronic configurations for the following molybdenum ions: Mo^{2+}; Mo^{4+}; Mo^{6+}. Write the formulas for oxides in which molybdenum has each of these oxidation states.

7.68 In which of the following compounds is there the possibility of one or more unpaired electrons: (a) NO; (b) CdO; (c) CoF_3; (d) $KMnO_4$; (e) Cl_2O; (f) $SrCl_2$; (g) $BeCl_2(g)$; (h) $Na_2(g)$?

7.69 Write the Lewis structures for each of the following compounds or ions (the diagrams show the atom-atom connections):

(a) Thiosulfate ion, $S_2O_3^{2-}$

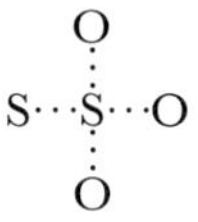

```
    O
    :
S···S···O
    :
    O
```

(b) Cyanic acid, HCNO

```
H···N···C···O
```

(c) Acetic acid, CH_3COOH

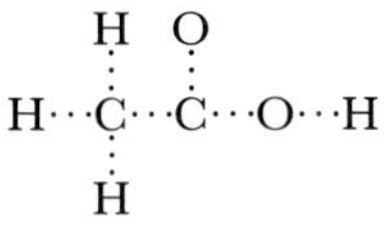

```
    H   O
    :   :
H···C···C···O···H
    :
    H
```

(d) Selenous acid, H_2SeO_3

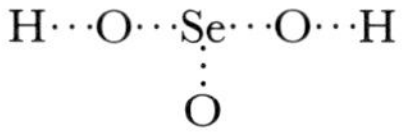

```
H···O···Se···O···H
         :
         O
```

(e) Dinitrogen tetroxide, N_2O_4

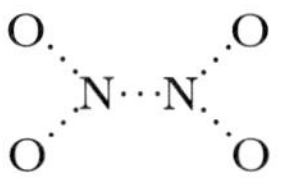

```
O.       .O
  ·N···N·
O'       'O
```

(f) Dinitrogen trioxide, N_2O_3

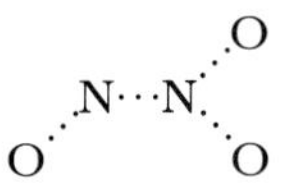

```
          .O
  .N···N·
O'       'O
```

7.70 Write Lewis structures for each of the following ions: (a) NO_2^-; (b) HSO_3^-; (c) $S_2O_6^{2-}$ (S—S single bond); (d) MnO_4^-; (e) ClO^-; (f) OCN^-. Indicate those for which resonance structures must be drawn to represent the structure properly.

7.71 Estimate ΔH for each of the following reactions:

(a) $HBr(g) + 2F_2(g) \longrightarrow BrF_3(g) + HF(g)$

(b) $2NCl_3(g) \longrightarrow N_2(g) + 3Cl_2(g)$

(c) $CO(g) + 2H_2(g) \longrightarrow CH_3OH(g)$
(d) $C_2H_4(g) + F_2(g) \longrightarrow C_2H_4F_2(g)$

7.72 Indicate the oxidation state of the metal in each of the following compounds: (a) $NiO(OH)$; (b) CoF_3; (c) K_2MnO_4; (d) $Na_2Cr_2O_7$; (e) MgH_2; (f) H_2GeO_2.

7.73 Give the chemical formula for each of the following compounds: (a) sodium hypchlorite, used in household bleaches; (b) calcium dihydrogen phosphate, used in soft drinks to add tartness; (c) calcium nitrite, used as a protective additive in cement formulations; (d) molybdenum(IV) sulfide, used as a lubricant.

7.74 What do you predict for the relative C—C bond lengths in the series, C_2H_6, C_2H_4, C_2H_2? Explain your answer.

7.75 Sodium hydroxide solutions are best stored in plastic bottles rather than in glass containers. Glass is made primarily from SiO_2. Why is storage of a strong base such as NaOH in glass containers not a good procedure?

7.76 Given the following bond dissociation energies, calculate the average bond energy for the Ti—Cl bond.

	ΔH (kJ/mol)
$TiCl_4(g) \longrightarrow TiCl_3(g) + Cl(g)$	335
$TiCl_3(g) \longrightarrow TiCl_2(g) + Cl(g)$	423
$TiCl_2(g) \longrightarrow TiCl(g) + Cl(g)$	444
$TiCl(g) \longrightarrow Ti(g) + Cl(g)$	519

7.77 Write the Lewis structure for each of the following substances: (a) nitrous oxide, N_2O, known as laughing gas and used as an inhalation anesthetic in dental surgery; (b) ethyl alcohol, CH_3CH_2OH, the alcohol of alcoholic beverages (contains a C—C bond); (c) formaldehyde, H_2CO, used to preserve biological samples; (d) urea, H_2NCONH_2, which is excreted in the urine of mammals (contains two N—C bonds).

7.78 Write a set of reactions for the formation of each of the following compounds, starting with the materials indicated: (a) $CaSO_3(s)$ beginning with $CaO(s)$ and elemental sulfur; (b) H_3AsO_4, beginning with $As_2O_5(s)$; (c) $BaSO_4(s)$, beginning with $BaO(s)$ and $SO_3(g)$.

[7.79] Using the bond energies in Table 7.3 and the following bond distances, construct a graph of bond energies versus bond distances. What conclusions can you reach from your graph? Bond distances are as follows: C—C (1.54 Å), Si—Si (2.35 Å), N—N (1.45 Å), O—O (1.48 Å), S—S (2.05 Å), F—F (1.42 Å), Cl—Cl (1.99 Å), Br—Br (2.28 Å), I—I (2.67 Å), C—F (1.35 Å), C—O (1.43 Å), and S—Br (2.27 Å).

7.80 From Figure 7.7 it is evident that some metallic elements, such as Cr, Mn, and others, can exhibit more than one oxidation number, whereas others, such as Ca, exhibit only one. What distinguishes these two groups of metallic elements?

8 Geometries of Molecules; Molecular Orbitals

Molecules of different substances have diverse shapes; some molecules are flat, others long and thin. Some large molecules such as proteins are found to be globular clusters, whereas others are coiled. The shape of a molecule, together with the strength and polarity of its bonds, largely determines the physical and chemical properties of that substance. Molecular shape is especially important in biochemical reactions, in which only molecules that are of a certain shape and size may be able to enter into a reaction. For example, small changes in the shapes and sizes of drug molecules may alter activity or may reduce the toxic side effects of a drug that is otherwise beneficial.

Diversity in molecular shape arises because atoms attach to one another in various geometric arrangements. The overall molecular shape of a molecule is determined by its *bond angles,* the angles made by the lines joining the nuclei of the atoms in the molecule (the internuclear axes). The distance between the nuclei of two bonded atoms is called the *bond distance*. Figure 8.1 shows the bond angles and bond distances that define the shape and size of the SF_6 molecule. It is possible to determine bond-distance and bond-angle information about a molecule from various experimental techniques. We will not discuss these techniques here, but you should simply accept the fact that by using them the chemist has acquired a great deal of information about molecular geometries. Our concern is with how these geometries can be explained.

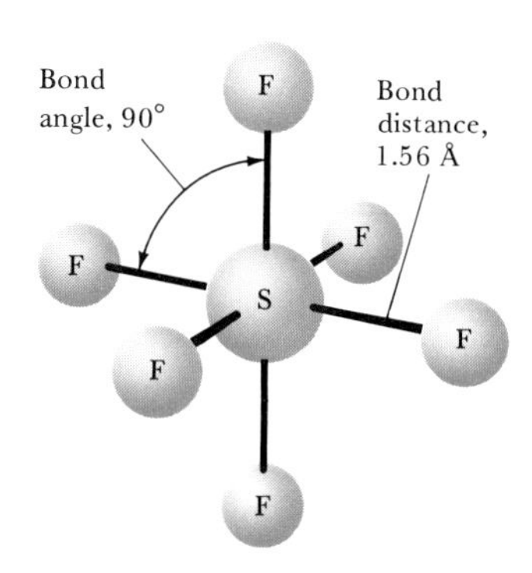

FIGURE 8.1 Geometrical structure of the SF_6 molecule, showing bond-distance and bond-angle information.

In Chapter 7, we employed the simple and empirical Lewis model to account for the formulas of covalent compounds. The Lewis structures that we draw are not geometrical figures. They simply describe the bonding connections between atoms in a two-dimensional representation. For example, the Lewis structure for methane, CH_4, does not tell us whether the molecule is planar or tetrahedral. To develop a model that explains three-dimensional structure, we must take into account the electrostatic interactions between electron pairs.

8.1 THE VALENCE-SHELL ELECTRON-PAIR REPULSION (VSEPR) MODEL

We have seen that an atom is bonded to other atoms in a molecule by electron pairs that occupy its valence-shell atomic orbitals. These electron pairs are attracted to the central nucleus by the electron-nucleus attractive force. At the same time, though, they are repelled by the other electron pairs about the central atom. As a result, the electron pairs remain as far apart from one another as possible, while maintaining their distance from the central nucleus.

You will remember from Chapter 7 that the most commonly observed electronic arrangement about an atom in covalent compounds is an octet, or four pairs, of electrons. We might then ask how four electron pairs could arrange themselves about a central atom so as to minimize electron-pair repulsions, while maintaining their distance from the central nucleus. Two possibilities are shown in Figure 8.2. As it turns out, the tetrahedral arrangement is the one that has the electron pairs as far apart as possible for a given distance from the central nucleus. (This can be demonstrated using some rather lengthy trigonometric calculations, but we won't go through these.) We therefore consider that four electron pairs about an atom will be arranged so that each pair occupies a region of space toward the vertex of a tetrahedron. If each of these four electron pairs bonds the central atom to another atom—for example, as in methane—then the observed arrangement of bonded atoms about the central atom is tetrahedral.

We can extend the idea we have just discussed to include cases with differing numbers of electron pairs about the central atom. Table 8.1 lists differing numbers of electron pairs in the valence shell and the arrangement that corresponds to maximum separation of the pairs. Our model correctly predicts the structures of molecules with expanded valence shells; PF_5 and PCl_5 are observed, for example, to be trigonal bipyramids, as shown in Table 8.1. Similarly, SF_6 is found to be octahedral, as predicted. Study Table 8.1 closely, paying attention to the electron-pair orientations associated with each number of electron pairs. Also notice the names (such as tetrahedral and trigonal bipyramidal) associated with each geometry.

The model for molecular geometries we have been discussing is often called the **valence-shell electron-pair repulsion (VSEPR) model.** It is based on the idea that the arrangement of electron pairs around a central atom is determined by the repulsions between the electron pairs around that atom. The locations of bonded atoms, in turn, are determined by the locations of the electron pairs, because the bonding electrons lie along the internuclear axes.

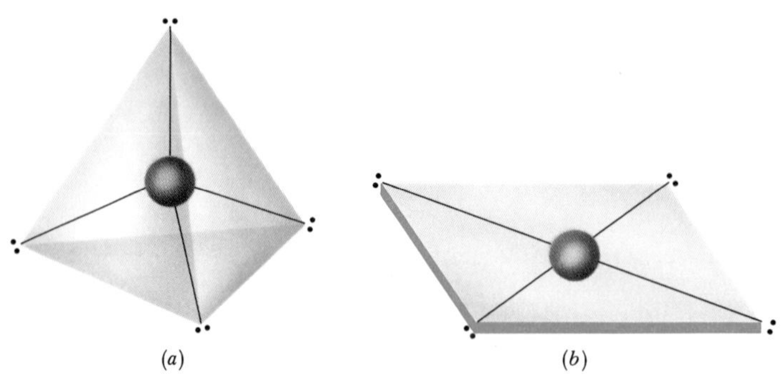

FIGURE 8.2 Two possible arrangements of four electron pairs around a central atom: (*a*) tetrahedral; (*b*) square planar. For a given distance of the electron pairs from the central atom, the repulsions are lower for the tetrahedral arrangement than for any other, such as the square planar.

TABLE 8.1 Geometries of electron-pair distributions about a central atom as a function of the number of electron pairs

Number of electron pairs	Arrangement of electron pairs	Predicted bond angles	Example
2	 Linear	180°	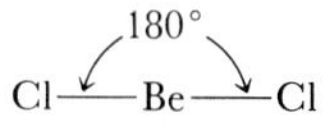 Cl—Be—Cl (180°)
3	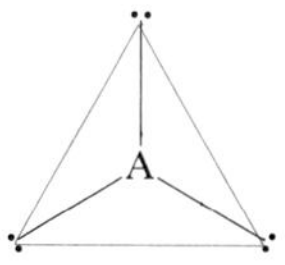 Trigonal planar	120°	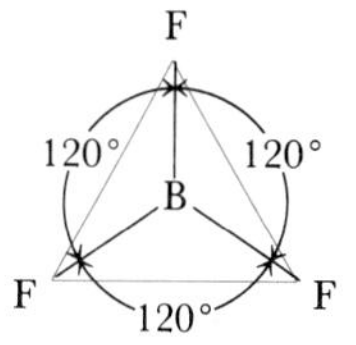 BF_3 (120°, 120°, 120°)
4	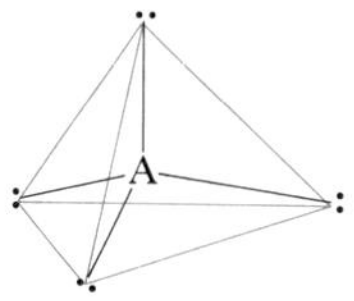 Tetrahedral	109.5°	CH_4 (109.5°) 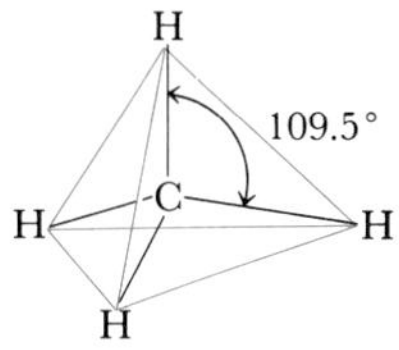
5	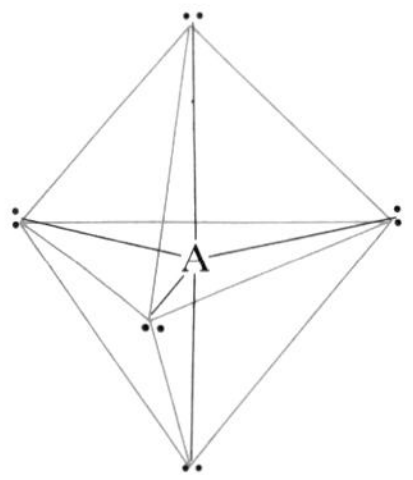 Trigonal bipyramidal	120° 90°	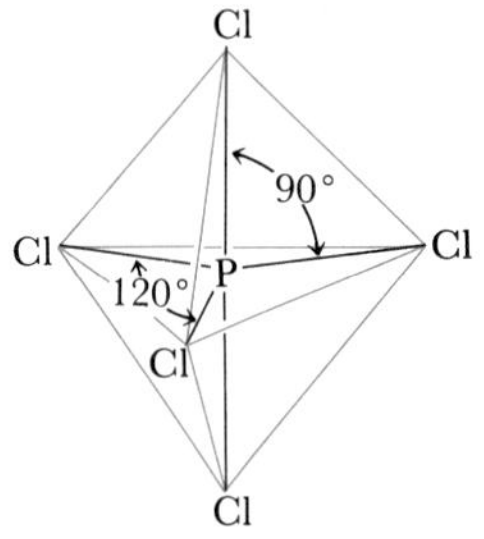 PCl_5 (90°, 120°)
6	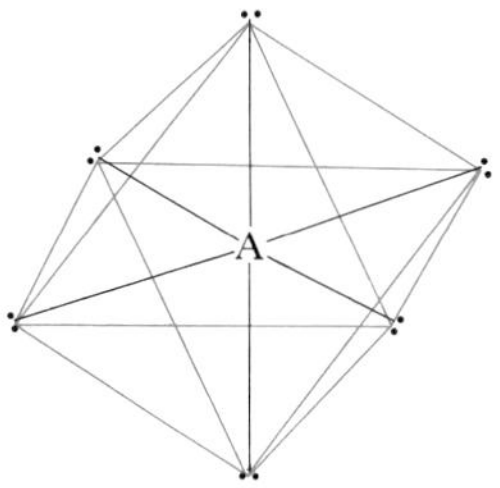 Octahedral	90°	SF_6 (90°, 90°) 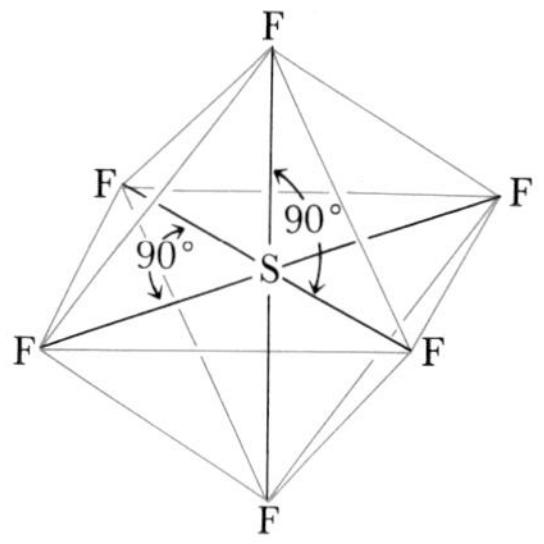

It often happens that the electron pairs about a central atom are not all shared with other atoms. For example, remember that in the Lewis structure for water there are two unshared electron pairs about the oxygen:

$$\mathrm{H{-}\ddot{O}{:}} \quad (\text{with H bonded below O})$$

These unshared electron pairs are as important as are shared pairs in determining the structure. However, when the structure of a molecule is studied experimentally, only the positions of the atoms are observed, not the positions of electrons. The effects of unshared electron pairs on the geometry must therefore be inferred from what we see of the structures of molecules.

Usually, unshared electron pairs aren't shown explicitly in writing chemical formulas. Before we can use the VSEPR model to predict the geometry of a molecule we must determine whether it has any unshared valence-shell electrons. The presence of such unshared electrons is revealed when we draw the Lewis structure for the molecule. The Lewis structures and molecular geometries of CH_4, NH_3, H_2O, and HF are shown in Table 8.2. In each of these molecules there is an octet of electrons about the central atom; however, the number of unshared electrons varies. The VSEPR model predicts that the four electron pairs will be arranged in a tetrahedral fashion (see Table 8.1). Consequently, CH_4 can be described as tetrahedral, NH_3 as pyramidal, and H_2O as nonlinear (also referred to as bent or angular). When we describe molecular shape in this way, we are focusing on the positions of the atoms, not the electron pairs.

TABLE 8.2 Distribution between shared and unshared electron pairs in hydrides of the second-row nonmetals

Lewis structure	Number of shared pairs	Number of unshared pairs	Structure
H H—C—H H	4	0	H, C, H, H, H; 109.5° Tetrahedral
H—N̈—H H	3	1	N, H, H, H; 107° Trigonal pyramidal
H—Ö: H	2	2	O, H, H; 104.5° Nonlinear
H—F̤̈:	1	3	F, H Linear

It is possible to extend the VSEPR model a little further. We can imagine that each electron pair around a central atom has a certain volume requirement, but that the volume requirement of an unshared pair of electrons is larger than for a pair that is shared between two atoms. This is so because the shared pair is held by the attractive forces of two nuclear centers. This should have the effect of contracting the spatial distribution of the electrons. If we use this idea we conclude that the HNH angle in NH_3 should be a little less than the tetrahedral angle of 109.5° found in CH_4, and indeed it is, as shown in Table 8.2. Similarly, the HOH angle in H_2O is reduced from the tetrahedral angle. In the third-row hydrides, the corresponding angles in PH_3 and H_2S are reduced even more.

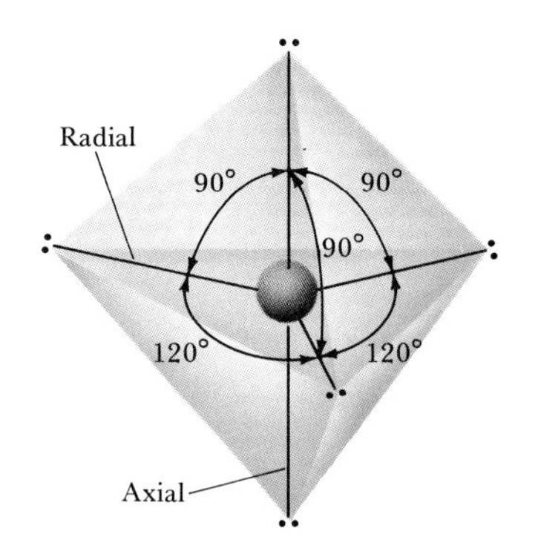

FIGURE 8.3 Trigonal bipyramidal arrangement of five electron pairs about a central atom. There are two geometrically distinct types of electron pairs, the two axial and three radial pairs. The repulsions experienced by the axial pairs are greater. For this reason, unshared electron pairs tend to occupy the radial positions.

Let's apply the idea that unshared electron pairs have a larger volume requirement to molecules with five electron pairs in the valence shell. In the trigonal bipyramid, shown in Figure 8.3, electron pairs in the axial and radial positions are not equivalent. In the axial position, an electron pair is repelled by three other electron pairs at 90° angles from it. For an electron pair in the radial position, there are only two such 90° interactions and two others much farther removed, at 120°. As a result, the repulsions from other electron pairs are larger for a pair located in an axial position that for a pair located in a radial position. Thus we predict that electron pairs with larger volume requirements will go into the radial locations.

The molecule SF_4 represents an example of a molecule with four shared electron pairs and one unshared pair. In the alternative VSEPR structures, shown in Figure 8.4, the unshared electron pair could be in either an axial or radial (equatorial) location. From the observed geometrical structure, also shown in Figure 8.4, we can infer that the unshared electron pair resides in a radial position, as predicted. Note that the "axial" S—F bonds are slightly bent back, suggesting that the unshared electron pair, with its greater volume requirement, is pushing them back.

Using the idea that unshared electron pairs are important in determining the observed geometry, we can predict molecular geometries for various numbers of shared and unshared electron pairs about a central atom. Table 8.2 outlines these possibilities for the case of four electron pairs. In Table 8.3 are listed the geometries to be expected for five or six

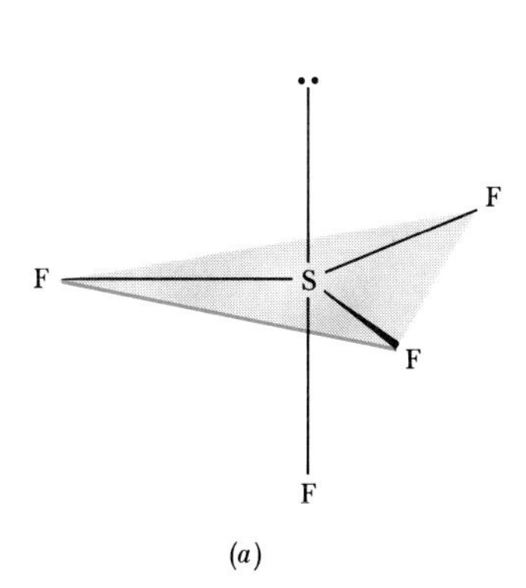

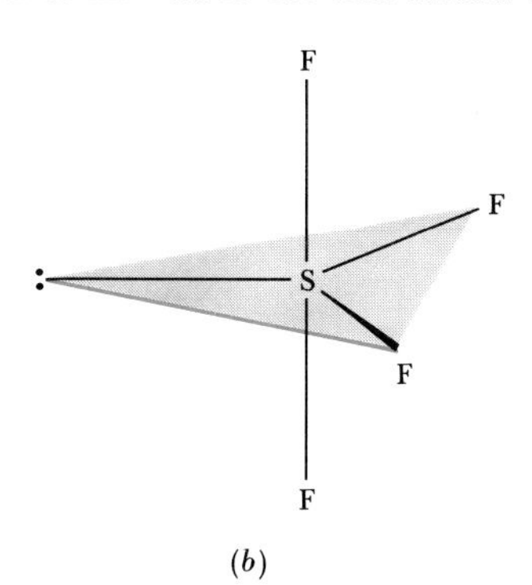

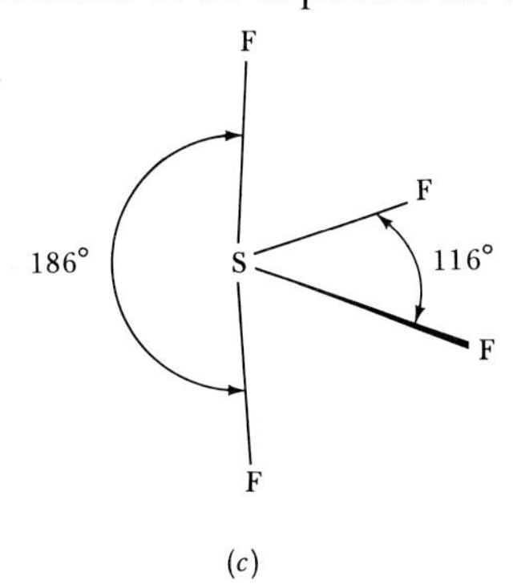

FIGURE 8.4 Alternative VSEPR structures for SF_4 and the observed structure: (*a*) unshared pair in axial position; (*b*) unshared pair in radial position; (*c*) observed structure.

TABLE 8.3 Relationship between numbers of shared and unshared electron pairs and observed molecular shapes for molecules with more than an octet of electrons

Number of electron pairs about central atom	Geometrical arrangement of electron pairs	Number of bonding pairs	Number of unshared pairs	Formula	Observed molecular shape	Exampl
5	Trigonal bipyramidal	5	0	AB_5	Trigonal bipyramidal	PCl_5
		4	1	AB_4	Seesaw	SF_4
		3	2	AB_3	T-shaped	ClF_3
		2	3	AB_2	Linear	XeF_2
6	Octahedral	6	0	AB_6	Octahedral	SF_6
		5	1	AB_5	Square pyramidal	BrF_5
		4	2	AB_4	Square planar	XeF_4

electron pairs. The guiding rule in arriving at these predictions is that the unshared electron pairs have a larger volume requirement than do shared electron pairs and that this determines in some cases where they are placed. Some of the possible arrangements have been omitted from the table because there are no known molecules with those arrangements.

The VSEPR model is very useful in predicting the structures of molecules. The predictions of the model are basically contained in Tables 8.1, 8.2, and 8.3. When the VSEPR model is applied to predicting the structures of molecules of ions containing multiple bonds, one additional rule is required: *Multiple bonds are to be considered the same as single bonds for the purpose of determining the overall geometry*. However, we can imagine that the double bond will occupy more space than a single bond, and will therefore have some effect on bond angles. These ideas are illustrated in Sample Exercise 8.1.

In summary, to use the VSEPR model to predict geometry, you should proceed in the following way:

1 Count the number of valence electrons and write out the Lewis structure for the molecule or ion.
2 From the Lewis structure determine the number of unshared electron pairs and number of bonds around each central atom. (Remember to count multiple bonds as a single bond.)
3 Determine the arrangement of bonds and unshared electron pairs that minimizes electron-pair repulsions (Table 8.1, 8.2, and 8.3).
4 Describe the shape of the molecule in terms of the positions of the atoms, which are determined by the locations of the bonding electron pairs. That is, we determine the arrangement of the electrons about the central atom for the purpose of predicting the atomic positions.

SAMPLE EXERCISE 8.1

Using the VSEPR model, predict the geometrical structures of the following: (a) $SnCl_3^-$; (b) CO_2; (c) IF_5; (d) ethylene, C_2H_4.

Solution: (a) The Lewis structure for the $SnCl_3^-$ ion is as follows:

$$\left[:\ddot{\text{Cl}}-\ddot{\text{Sn}}-\ddot{\text{Cl}}:\right]^- \text{ (with } :\ddot{\text{Cl}}: \text{ bonded below Sn)}$$

The four electron pairs around Sn should thus be disposed at the corners of a tetrahedron. One of the corners is occupied by the unshared electron pair. The geometrical arrangement of the atoms is thus pyramidal:

Sn (with unshared pair above), bonded to Cl, Cl, Cl

(b) The Lewis structure for carbon dioxide is

$$\ddot{\text{O}}=\text{C}=\ddot{\text{O}}$$

The rule that each multiple bond should be treated the same as a single electron pair means that, in effect, we have two electron pairs in the valence shell. The molecule should therefore be linear (Table 8.1). The CO_2 molecule is in fact observed to be linear.

(c) The Lewis structure for IF_5 is

I (with one unshared pair) bonded to F, F, F, F, F

The iodine has six pairs of valence-shell electrons around it, five of them bonding pairs. The electron pairs should point toward the corners of an octahedron. The geometrical arrangement of atoms is therefore square pyramidal (Table 8.3).

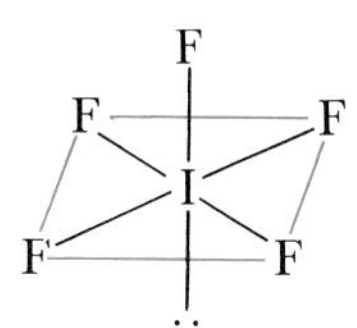

(d) The Lewis structure for ethylene is

H₂C=CH₂: H and H bonded to each C, C=C

The multiple bond between carbon atoms is treated as a single electron pair, so we have effectively three electron pairs in the valence shell. The three bonds about each carbon are distributed in a plane, with approximately 120° bond angles (Table 8.1). The molecule is planar because of the presence of the double bond; we'll learn more about that in a following section:

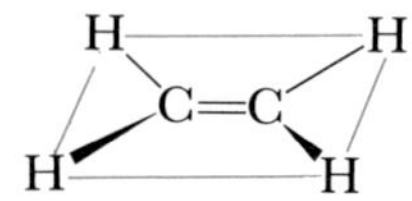

The experimentally determined H—C—H bond angle is 117°. Thus the spatial demands of the C=C double bond appear to force the C—H bonds together to a slight extent, thereby reducing the H—C—H angle from the "ideal" 120°.

8.2 DIPOLE MOMENTS

The shape of a molecule and the polarity of its bonds together determine the charge distribution in the molecule. A diatomic molecule such as HF, which contains a polar covalent bond (Section 7.9), has a concentration of negative charge on one end and a concentration of positive charge on the other end. A molecule whose centers of positive and negative charges do not coincide is described as being polar. The degree of polarity is measured by the **dipole moment** of the molecule. The dipole moment, μ, is defined as the product of the charge at either end of the dipole, Q, times the distance, r, between the charges, : $\mu = Qr$. Thus the dipole moment increases as the quantity of charge which is separated increases, or as the distance between the positive and negative centers increases.

Dipole moments are generally reported in units of debye, D; a debye is 3.33×10^{-30} C-m. Let's consider what the dipole moment value tells us about the separation of charge in a polar molecule. As an example, the dipole moment of HCl is 1.03 D. The H—Cl bond distance in this molecule is 1.36 Å. From these data, we can calculate the value of Q in the formula for the dipole moment, assuming that the charges are centered on the H and Cl atoms. We have

$$\mu = Qr$$

$$(1.03\ \text{D})\left(3.33 \times 10^{-30}\ \frac{\text{C-m}}{\text{D}}\right) = Q \times (1.36\ \text{Å})\left(10^{-10}\ \frac{\text{m}}{\text{Å}}\right)$$

$$Q = \frac{3.43 \times 10^{-30}\ \text{C-m}}{1.36 \times 10^{-10}\ \text{m}}$$

$$= 2.54 \times 10^{-20}\ \text{C}$$

Note from the inside back cover that the charge of the electron is 1.602×10^{-19} C. Thus, as a percentage of the electronic charge, Q is $100 \times (2.54 \times 10^{-20})/(1.602 \times 10^{-19}) = 16\%$. If the H—Cl were ionic, there would be a full + charge on H and a full − charge on Cl. Thus, Q would be 100 percent of the electronic charge. In reality, Q is less than this because the H—Cl bond is only partially ionic. The dipole moments of the hydrogen halides are listed in Table 8.4. Notice that the dipole moment decreases as the electronegativity difference decreases.

The dipole moment of a molecule containing more than two atoms depends both on bond polarity and molecular geometry. To have a dipole moment, a polyatomic molecule must have polar bonds. However, even if polar bonds are present, the molecule itself might not have a

TABLE 8.4 Some properties of hydrogen halides

Compound	Electronegativity difference	Dipole moment (D)
HF	1.8	1.91
HCl	1.0	1.03
HBr	0.8	0.79
HI	0.5	0.38

dipole moment if the bonds are arranged so their polarities cancel. This situation is found in the linear CO_2 molecule:

$$\overset{\longleftarrow\!\!+\;\;+\!\!\longrightarrow}{O{=}C{=}O}$$

Remember that the arrow points toward the more electronegative atom (Section 7.9). Because oxygen has a greater electronegativity than carbon, the bonds are polar, with electron density concentrating on the oxygen atoms. However, the centers of negative and positive charges are found at the same point, on carbon; consequently, the molecule has no net dipole moment. Figure 8.5 shows several further examples of both polar and nonpolar molecules, all of which have polar bonds. Notice that the symmetric arrangement of bonds in the nonpolar molecules leaves them without positive and negative sides; molecules with dipole moments have positive and negative ends, and this allows them to align themselves between the charged plates of a capacitor. Notice also that unshared electrons as well as polar bonds can contribute to molecular polarity.

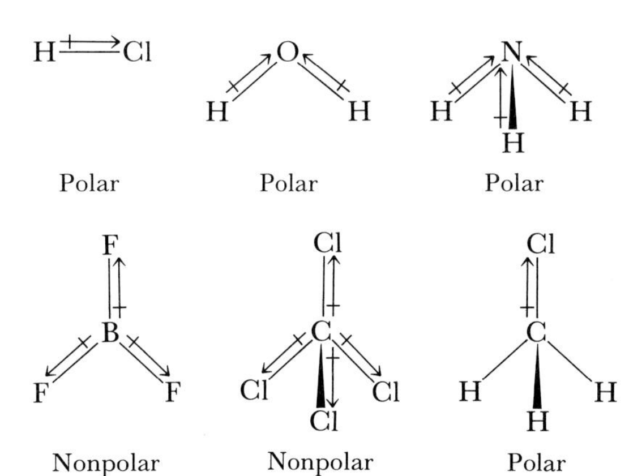

FIGURE 8.5 Examples of molecules with polar bonds. Some of these molecules are nonpolar because their bond polarities cancel each other.

SAMPLE EXERCISE 8.2

Predict whether the following molecules are polar or nonpolar: (a) ICl; (b) SO_2; (c) SF_6.

Solution: (a) Chlorine is a more electronegative element than iodine. Consequently, ICl will be polar with chlorine as the negative end:

$$\overset{+\longrightarrow}{\text{I—Cl}}$$

All diatomic molecules with polar bonds are polar molecules.

(b) Oxygen is more electronegative than sulfur; the molecule therefore has the polar bonds necessary for the molecule to be polar. The molecule has the following resonance forms:

$$:\ddot{\text{O}}\text{—}\ddot{\text{S}}\text{=}\ddot{\text{O}} \longleftrightarrow \ddot{\text{O}}\text{=}\ddot{\text{S}}\text{—}\ddot{\text{O}}:$$

The nonlinear shape of the molecule results from the trigonal planar arrangement of electron pairs around sulfur (Table 8.1), there being one unshared pair of electrons on the sulfur atom. Because of the molecular shape, the bond polarities do not cancel and the molecule is polar:

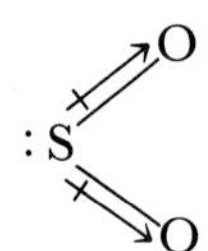

(It is unlikely that the unshared electron pair on sulfur completely cancels the effect of the polar S—O bonds, though it certainly does so to some extent.)

(c) Fluorine is more electronegative than sulfur. The bond dipoles therefore point toward fluorine. The six S—F bonds are arranged in a symmetric octahedral fashion around the central sulfur (there are no unshared valence-shell electron pairs on sulfur):

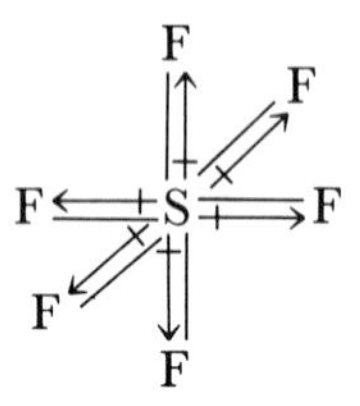

The bond dipoles cancel each other; the molecule does not have a negative and positive side. It is nonpolar with an overall dipole moment of zero.

8.3 HYBRID ORBITALS AND MOLECULAR SHAPE

The VSEPR theory provides a simple model for predicting the shapes of molecules. However, it does not explain why bonds exist between atoms. The shared and unshared pairs are taken as given; the model is used simply to deduce the shape of the molecule or ion. In developing a theory of covalent bonding, chemists have also approached the problem from another direction. Suppose we take the formula and geometrical structure of the molecule as given. How can we account for the observed geometries in terms of the atomic orbitals used by the atoms in forming bonds to one another?

In the Lewis theory, covalent bonding occurs when atoms share electrons. Such sharing involves a concentration of electron density between nuclei. This buildup of electron density occurs when a valence atomic orbital of one atom merges or overlaps with that of another atom. The orbitals are then said to share a region of space or to **overlap.** We can think of the shared electron pair as occupying an orbital that consists in part of the atomic orbitals contributed by each atom. This idea is illustrated in Figure 8.6, which shows the coming together of two hydrogen atoms to form the H_2 molecule. The overlap of the 1*s* orbital on one H atom with that on the other increases as the atoms draw near. The potential energy decreases because the electrons are simultaneously attracted to both nuclei. At very short internuclear distances the potential energy begins to increase due to increasing repulsions between nuclei and between electrons. The internuclear distance at the minimum in this potential energy curve corresponds to the observed bond distance. Thus the observed bond distance is a compromise between increased overlap of the atomic orbitals, which draws the atoms together, and the nuclear-nuclear and electron-electron repulsions, which force them apart.

FIGURE 8.6 Formation of the stable H_2 molecule by the overlap of two hydrogen 1*s* orbitals. The minimum in the energy surface, at r_e, represents the equilibrium H—H bond distance in H_2.

In extending this model of bond formation to polyatomic molecules, we want a model that localizes, as much as possible, the electronic charge in the regions between the atoms. In constructing such a model we must work with the valence-shell atomic orbitals of the atoms. At first this would appear to be a difficult task. Let's consider methane, CH_4, as an example. This molecule has a tetrahedral arrangement of four hydrogen atoms around a central carbon, as illustrated in Table 8.2. We know that the carbon has a 2*s* and three 2*p* atomic orbitals available for bonding to hydrogens. Each hydrogen has, of course, just its 1*s* atomic orbital. The carbon valence-shell orbitals do not have the directional characteristics we want; the 2*s* orbital is spherically symmetric, and the three 2*p* orbitals are oriented along the *x*, *y*, and *z* axes (Figure 5.18). We cannot therefore employ these orbitals directly in forming the four equivalent bonds to hydrogen. The solution to our problem lies in the theory of **hybridization.**

The process of hybridization corresponds to a mathematical mixing of the valence-shell atomic orbitals. It is not important to understand the mathematical details; however, we can understand the physical significance of the process. In effect, we want to describe the four orbitals directed from carbon toward the hydrogen atoms as an appropriate mixture of the 2*s* and three 2*p* orbitals.

We can imagine a stepwise procedure of getting a carbon atom initially in its ground state ready for bonding to four hydrogen atoms. The electronic configuration of the ground state carbon atom, in orbital diagram representation, is as follows:

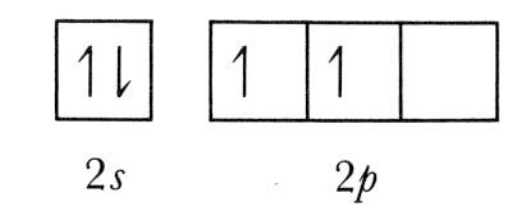

Notice that a carbon atom in its ground state could at most form two bonds to two other atoms, because there are just two unpaired electrons available. To obtain the capability for forming four bonds, one of the 2*s* electrons must be "promoted" to the vacant 2*p* orbital:

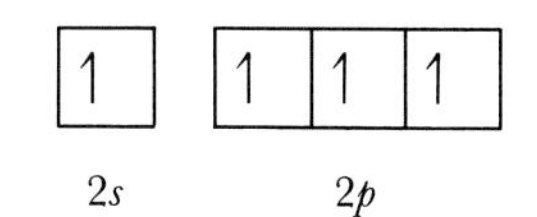

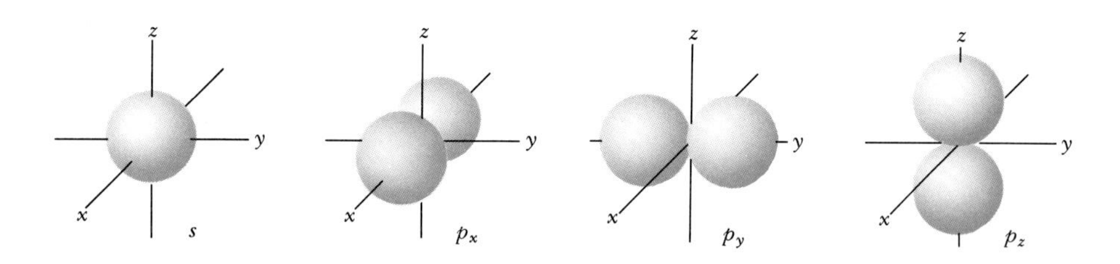

Hybridize to form four sp^3 hybrid orbitals

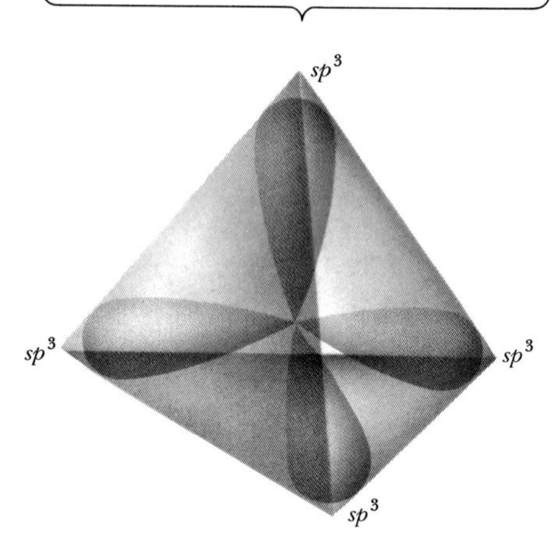

FIGURE 8.7 Formation of four sp^3 hybrid orbitals from a set of one *s* and three *p* orbitals.

Because the 2*p* orbital is of higher energy than the 2*s*, this promotion costs energy. Next, the mathematical functions that describe the 2*s* and 2*p* orbitals can be combined with one another. The combination yields four equivalent orbitals called **hybrid orbitals,** which point toward the corners of a tetrahedron (Figure 8.7). Each one of these hybrid orbitals consists of a certain amount of the 2*s* orbital and a certain amount of the 2*p* orbitals. Because the four hybrid orbitals are made up of one 2*s* and three 2*p* orbitals, they are labeled sp^3 hybrids. (Notice that the superscript 3 relates to the relative proportions of *s* and *p* orbitals and not to the number of electrons they contain.) When the 2*s* and 2*p* orbitals are mixed so as to give the four hybrid orbitals we obtain

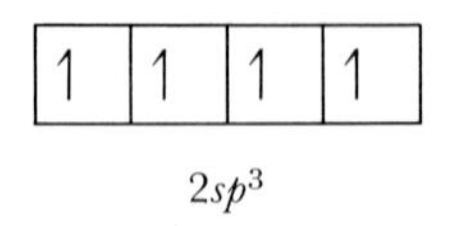

The carbon is now "prepared" to form four equivalent bonds with tetrahedral geometry. The interaction of the four sp^3 hybrid orbitals from carbon with four hydrogen atoms to form methane, CH_4, is shown in Figure 8.8(*a*).

The first of the steps outlined above, promotion of the 2*s* electron, requires energy. The second does not, because it merely means an averaging to give four equivalent orbitals, each with energy equal to the average of the original set. The overall process does require energy. The reason it occurs is that the system more than recovers this energy in bond

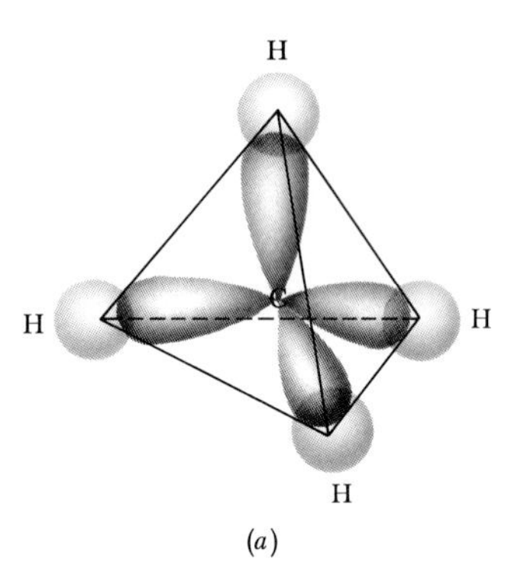

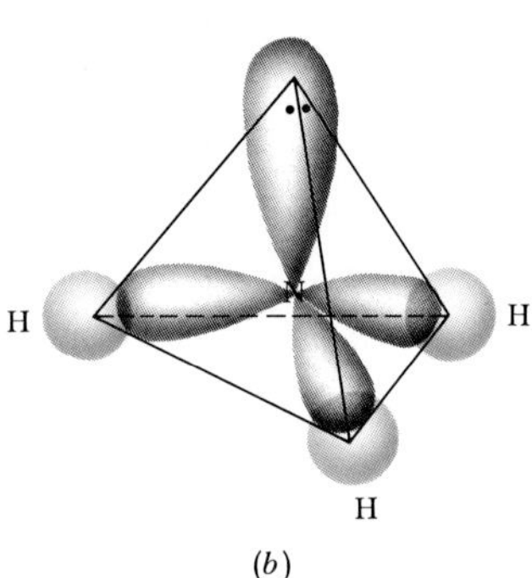

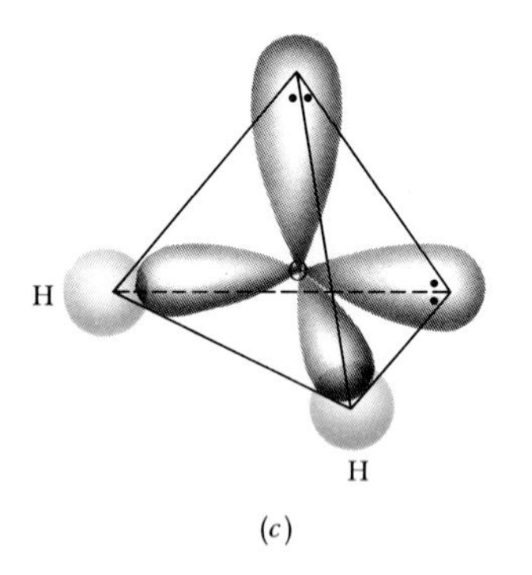

FIGURE 8.8 Formation of methane (*a*), ammonia (*b*), and water (*c*) by overlap of hydrogen 1*s* orbitals with the sp^3 hybrid orbitals of the central atom.

formation. As we've seen, a ground-state carbon atom could at most form two bonds to hydrogen, using the two electrons in its $2p$ orbitals. With promotion and hybridization, four bonds are possible. Furthermore, hybridization permits greater orbital overlap, because the hybrid orbitals point directly toward the bonded atom.

In molecules such as NH_3 and H_2O, the central atom has about it four electron pairs in an approximately tetrahedral arrangement. The orbitals used by the central atom can be thought of as sp^3 hybrid orbitals. In NH_3 one of the sp^3 hybrid orbitals contains the unshared electron pair, the other three are employed in the bonds to hydrogen, Figure 8.8(*b*). In H_2O two of the orbitals contain unshared pairs, two are employed in bonds to hydrogen, Figure 8.8(*c*). In these examples, the hybrid orbitals employed by the central atom are not pure sp^3 hybrids, because the bond angles about the central atom are not exactly the 109.5° tetrahedral angle, as shown in Table 8.2. In the mathematical formulation of hybrid orbitals, it is possible for the contribution of the *s* orbital to be a little greater in one hybrid orbital, a little smaller in another, in order to make the angles between the hybrid orbitals come as close as possible to the observed bond angles. We shall make no attempt to follow through with these refinements but shall instead employ just the idealized equivalent hybrid set.

Various combinations of atomic orbital sets can be mixed, or hybridized, to obtain different geometries of orbitals about a central atom. Table 8.5 shows several of the more important hybrid orbital combinations and the geometries to which they correspond. The hybrid orbital sets are illustrated in Figure 8.9. Note that expanded valence shells about atoms, that is, those in which there are more than eight electrons in the valence shell, can be accommodated by mixing in *d* orbitals along with *s* and *p*.

The purpose in formulating hybrid orbitals is to provide a convenient model in which we can imagine the electrons to be localized in the region between two atoms. The picture of hybrid orbitals has limited predictive value; that is, we cannot say in advance that in NH_3 the nitrogen uses essentially sp^3 hybrid orbitals. Once given the molecular geometry, however, we can employ the concept of hybridization to describe the atomic orbitals employed by the central atom in bonding.

TABLE 8.5 Geometrical arrangements characteristic of hybrid orbitals

Atomic orbital set	Hybrid orbital set	Geometrical arrangement	Examples
s,p	sp	Linear (180° angle)	$Be(CH_3)_2$, $HgCl_2$
s,p,p	sp^2	Trigonal planar (120° angles)	BF_3
s,p,p,p	sp^3	Tetrahedral (109.5° angles)	CH_4, $AsCl_4^+$, $TiCl_4$
d,s,p,p[a]	dsp^2	Square planar (90° angles)	$PdBr_4^{2-}$
d,s,p,p,p[a]	dsp^3	Trigonal bipyramidal (120° and 90° angles)	PF_5
d,d,s,p,p,p[a]	d^2sp^3	Octahedral (90° angles)	SF_6, $SbCl_6^-$

[a]Depending on the particular element involved, the *d* orbital that mixes with the *s* and *p* may be of major quantum number one lower or of the same major quantum number. This has no effect on the geometrical characteristics of the resulting hybrid set.

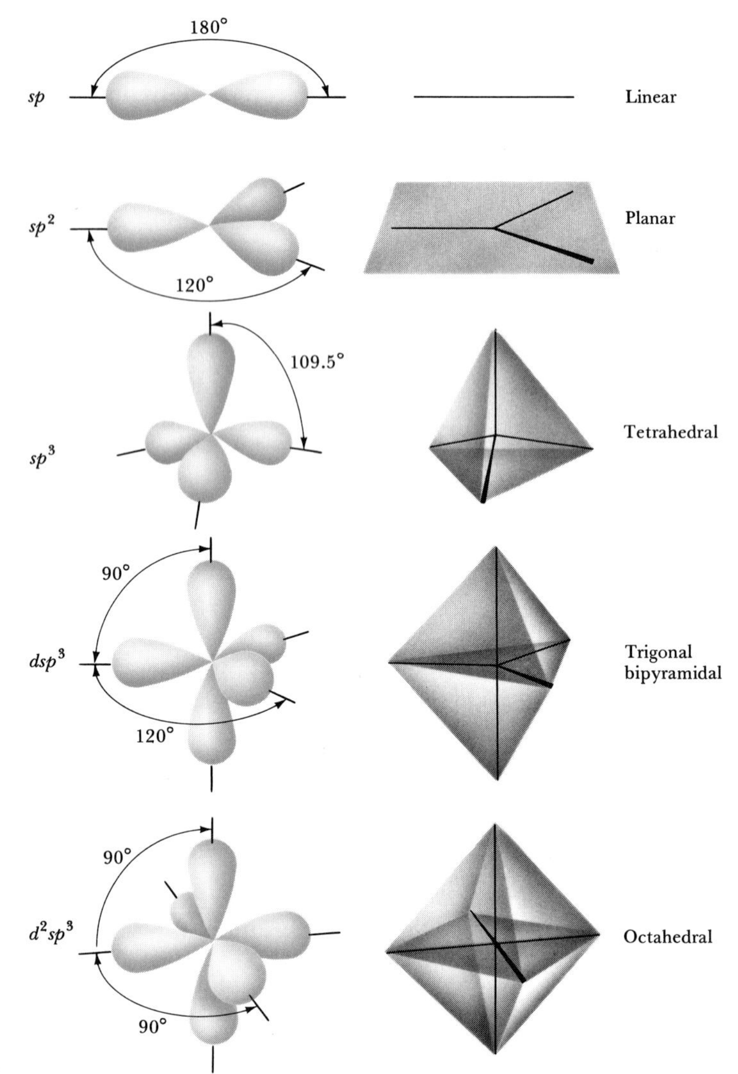

FIGURE 8.9 Geometrical arrangements characteristic of hybrid orbital sets.

SAMPLE EXERCISE 8.3

The molecule BeH_2 is known to be linear. Account for the bonding in BeH_2 in terms of the hybrid orbitals employed by Be in bonding to the two hydrogen atoms.

Solution: The orbital diagram for the Be atom is as follows:

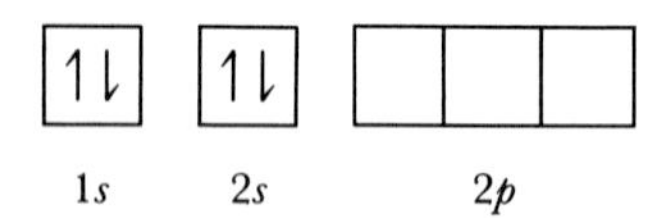

The Be atom in its ground state is incapable of forming bonds with other atoms, because all the electrons are paired. However, suppose one of the electrons is promoted to the $2p$ orbital:

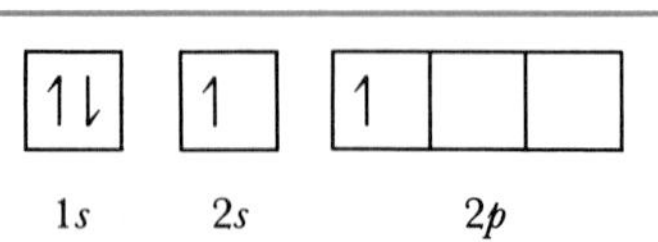

and the $2s$ and one of the $2p$ orbitals are mixed to form two *sp* hybrid orbitals:

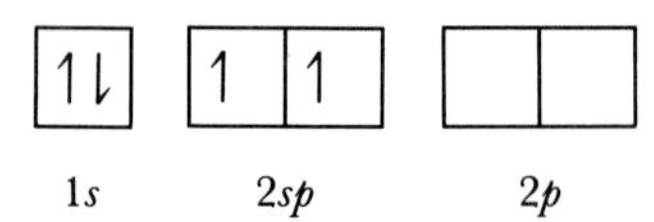

The electrons in the two *sp* hybrid orbitals can form shared electron pair bonds with two hydrogen atoms. The *sp* hybrid orbitals are directed at 180° angles from one another, Figure 8.9, so the BeH_2 molecule is linear.

SAMPLE EXERCISE 8.4

Indicate the hybridization of orbitals employed by the central atom in each of the following: (a) NH_2^-; (b) SF_4 (see Figure 8.4).

Solution: (a) The Lewis structure for NH_2^- is

$$[\mathrm{H}:\ddot{\underset{..}{\mathrm{N}}}:\mathrm{H}]^-$$

From the VSEPR model we conclude that the four electron pairs around N should be arranged in a tetrahedral fashion. Such a tetrahedral arrangement is characteristic of sp^3 hybridization (Figure 8.9); two of the hybrid orbitals contain unshared electron pairs, the other two contain pairs shared with hydrogen.

(b) As shown in Figure 8.4, there are 10 valence-shell electrons around sulfur in SF_4. With an expanded octet of ten electrons, the use of a *d* orbital on the sulfur is indicated. The trigonal-bipyramidal arrangement of valence-shell electron pairs shown in Figure 8.4 corresponds to dsp^3 hybridization (Figure 8.9). One of the hybrid orbitals contains an unshared electron pair; the other four are involved in bonding to fluorine.

8.4 HYBRIDIZATION IN MOLECULES CONTAINING MULTIPLE BONDS

The concept of hybridization may be applied also to molecules containing multiple bonds. For example, we have seen (Sample Exercise 8.1) that ethylene possesses a carbon-carbon double bond and is a planar molecule. The planar arrangement of three bonds and 120° bond angles about each carbon suggests that the hybrid orbital set it uses to bond to the other carbon and the two hydrogens is sp^2 (Table 8.5). Because the valence orbitals on carbon consist of a 2*s* and *three* 2*p* orbitals, one 2*p* orbital remains unused after forming the sp^2 hybrid set. This is illustrated in Figure 8.10. We notice that the carbon atoms of ethylene are bonded together through overlap of sp^2 hybrid orbitals. The resultant electron density is concentrated symmetrically between the nuclei along the line joining them. Bonds of this kind are called **σ (sigma) bonds.** The bonds between carbon and hydrogen are also σ bonds. Because the hydrogen uses a 1*s* orbital that is spherical, each bond is symmetrical with respect to that particular C—H bond axis.

The formation of the σ bonds gives three electron pairs about carbon in the sp^2 hybrid orbitals, and an electron on each carbon in the 2*p* orbital that is perpendicular to the plane of the molecule. The *p* orbitals shown on the two carbon atoms in Figure 8.10 can overlap with one another in a sideways fashion. By means of this overlap a second C—C covalent bond is formed, as shown in Figure 8.11. Each carbon has achieved an octet of electrons. The second C—C bond differs from the first, because it results from concentrating electron density above and below the C—C bond axis. This kind of bond is called a **π (pi) bond.**

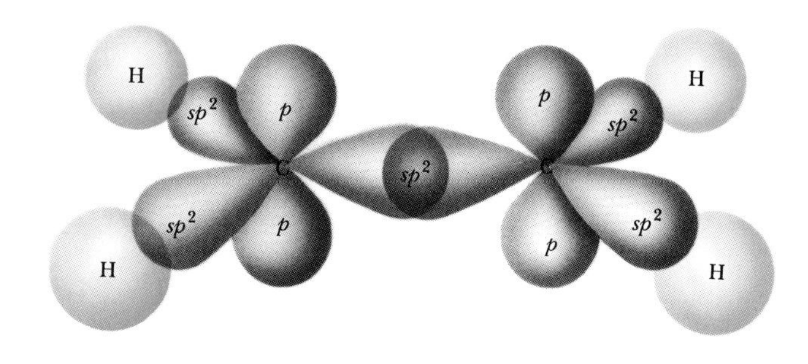

FIGURE 8.10 Hybridization of carbon orbitals in ethylene. The σ bond framework, formed from sp^2 hybrid orbitals on the carbon atoms, determines the observed geometrical structure of the molecule.

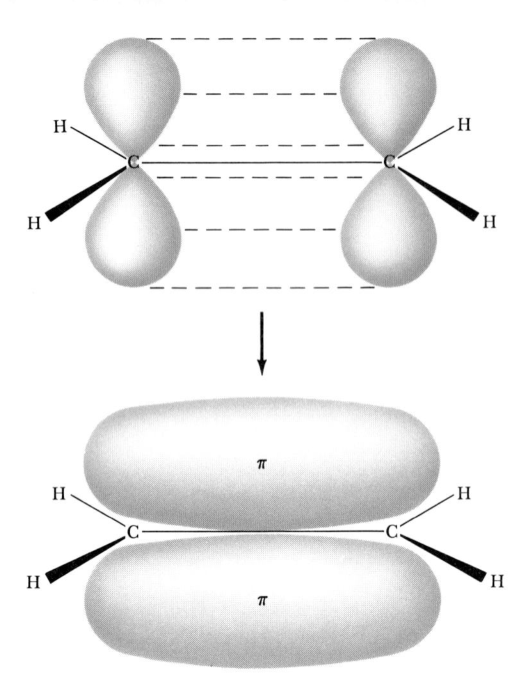

FIGURE 8.11 Formation of the π bond in ethylene by overlap of the $2p$ orbitals on each carbon atom. Note that the centers of charge density in the π bond are above and below the bond axis, whereas in σ bonds the centers of charge density lie on the bond axes.

It is useful to consider the hybridization about carbon in ethylene in terms of an orbital diagram:

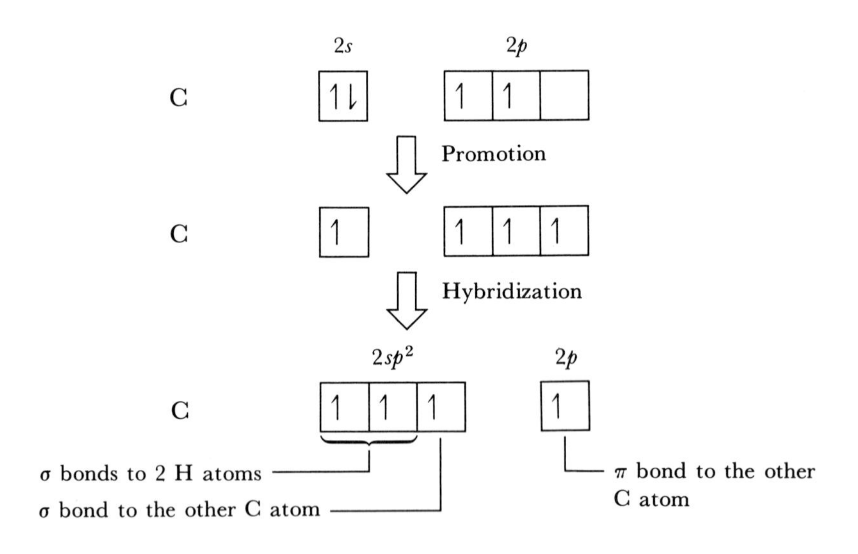

The first step in our imaginary process of preparing the carbon atom for bonding is again the promotion of one of the $2s$ electrons to the vacant $2p$ orbital. We then form an sp^2 hybrid orbital set from the $2s$ and two of the $2p$ orbitals. The orbitals of the hybrid set are used in bonding as indicated above. By electron pairing in each orbital, carbon gains an octet of electrons in its valence-shell orbitals. Because two pairs of electrons are shared between carbon atoms in ethylene, as compared with only one pair in an ordinary C—C single bond, the C—C bond distance is shorter in ethylene than, for example, in ethane, C_2H_6:

$$\underset{\text{Ethylene}}{\mathrm{H_2C{=}CH_2}} \qquad \underset{\text{Ethane}}{\mathrm{H_3C{-}CH_3}}$$

Ethylene C—C distance = 1.34 Å   Ethane C—C distance = 1.54 Å

According to the hybrid orbital picture, the double bond between carbons in ethylene consists of one σ and one π bond. This picture of the bonding in ethylene is in good accord with its chemical properties. For example, ethylene reacts with bromine, Br_2, to form dibromoethane:

$$\underset{\text{Ethylene}}{\mathrm{H_2C{=}CH_2}} + \mathrm{Br_2} \longrightarrow \underset{\text{Dibromoethane}}{\mathrm{BrH_2C{-}CH_2Br}} \qquad [8.1]$$

In this reaction the double bond is opened, and a bromine adds to each carbon. The C—C π bond is converted to two C—Br σ bonds.

Acetylene, C_2H_2, is a linear molecule containing a triple bond: H—C≡C—H. Each carbon may be visualized as using *sp* hybrid orbitals in forming σ bonds with the other carbon and with a hydrogen. Each carbon then has two remaining valence *p* orbitals at right angles to the axis of the *sp* hybrid set (Figure 8.12). These overlap to form a pair of π bonds. Thus the triple bond can be thought of as formed from one σ and two π bonds. All double bonds consist of a σ and a π bond, while all triple bonds consist of a σ and two π bonds.

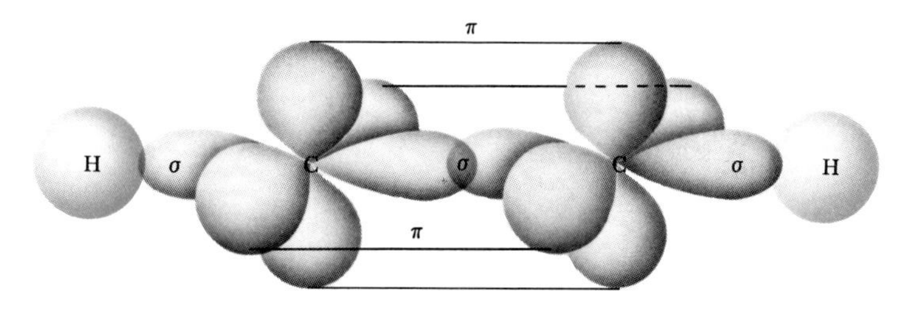

FIGURE 8.12 Formation of two π bonds in acetylene from the overlap of two sets of carbon 2*p* orbitals.

Delocalized Orbitals

In each of the molecules that we've discussed in this chapter, the bonding electrons are *localized;* by this we mean that the σ and π electrons are associated totally with the two atoms forming the bond. There are some molecules, however, in which the π electrons are free to move over several atoms; we say that such electrons are *delocalized* over those atoms. Benzene, C_6H_6, is an example of such a molecule. This molecule has the following resonance forms:

(Two Kekulé structures of benzene, C_6H_6, with alternating C=C double bonds, connected by ⟷)

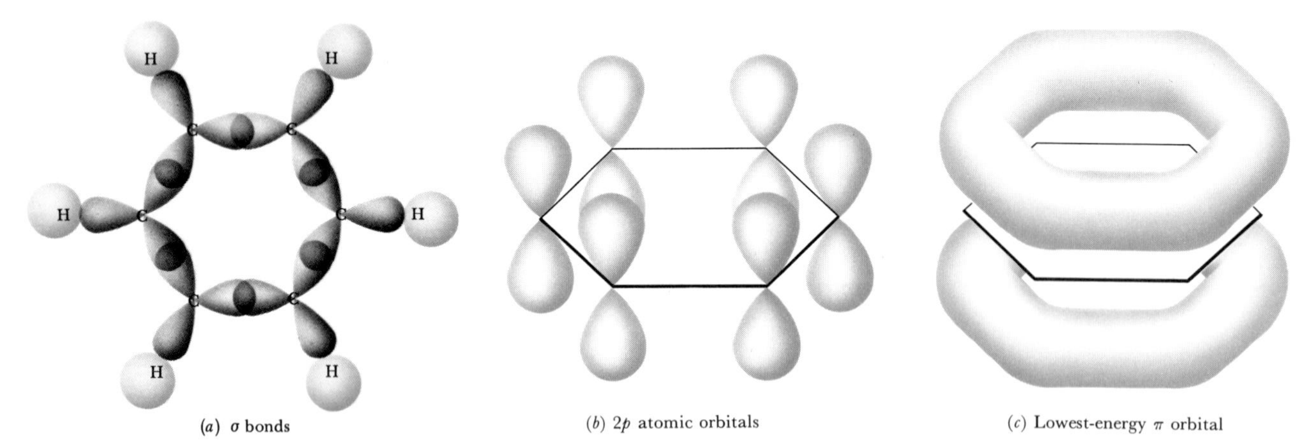

FIGURE 8.13 The σ and π bond networks in benzene, C_6H_6. (*a*) The σ bonds all lie in the molecular plane and are formed from carbon sp^2 hybrid orbitals. (*b*) The π bond network is formed from overlap of the $2p$ orbital on each carbon atom with the $2p$ orbitals of each of its neighbors. Six π molecular orbitals result. The lowest-energy bonding π molecular orbital is illustrated in (*c*).

All C—C bonds in benzene are of equal length; the C—C distance is 1.395 Å, intermediate between the values for C—C single bonds (1.54 Å) and C=C double bonds (1.34 Å). The bond angles around each carbon are 120°.

To describe benzene in terms of hybridization of carbon orbitals, we follow the procedure of setting up a hybrid orbital set consistent with the skeletal structure for the molecule. Because each carbon is surrounded by three atoms at 120° angles in a plane, the appropriate hybrid set (Table 8.5) is sp^2, as shown in Figure 8.13(*a*). This leaves a p orbital on each carbon perpendicular to the plane of the benzene ring. The situation is very much like that in ethylene, except that now we have six carbon $2p$ orbitals, in a cyclic arrangement [Figure 8.13(*b*)]. The six carbon $2p$ atomic orbitals interact with one another to form π orbitals. Each of the $2p$ orbitals overlaps with two others, one on each adjacent carbon atom, to form a kind of doughnut of electron density above and below the plane of the benzene right, as illustrated in Figure 8.13(*c*).

Because there is one electron from each $2p$ orbital, there is a total of three electron pairs in the π orbitals formed in this manner. The electrons in the π orbitals of benzene are said to be *delocalized* in the sense that any one electron is free to move around the entire circle of carbon atoms. This delocalization of the π electrons gives benzene a special stability. For example, this substance does not react readily with bromine, as does ethylene.

It is common to represent benzene and related molecules by leaving off the hydrogens attached to the carbon and showing only the carbon-carbon framework. The presence of π electrons may be shown using one of the Lewis structures or by placing a circle in the center of the carbon ring. Thus we can represent benzene as

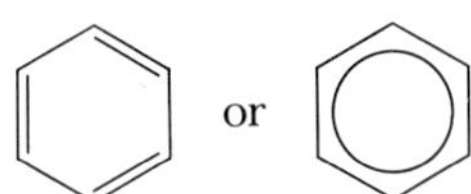

Benzene and many related molecules are referred to as aromatics. The name "aromatics" arose because some of the compounds related to benzene possess a spicy fragrance. However, benzene and many other aromatics do not have a pleasant smell. Several aromatic compounds are depicted in Figure 8.14.

Many of the aromatic hydrocarbons are carcinogenic, that is, cancer causing. Benzo[*a*]pyrene,* shown in Figure 8.14, is among the most potent carcinogens. It is found in significant quantities in urban atmospheres and, most especially, in cigarette smoke. It is regarded as a major cause of lung cancer. The major sources of benzo[*a*]pyrene in urban atmospheres are coal-burning power plants, engines that burn gasoline and diesel fuel, and incinerators used for refuse disposal. Studies show that for nonsmokers, the incidence of lung cancer among urban dwellers is higher than for rural populations with the same age profile. In both groups the incidence of lung cancer is much higher for cigarette smokers than for nonsmokers.

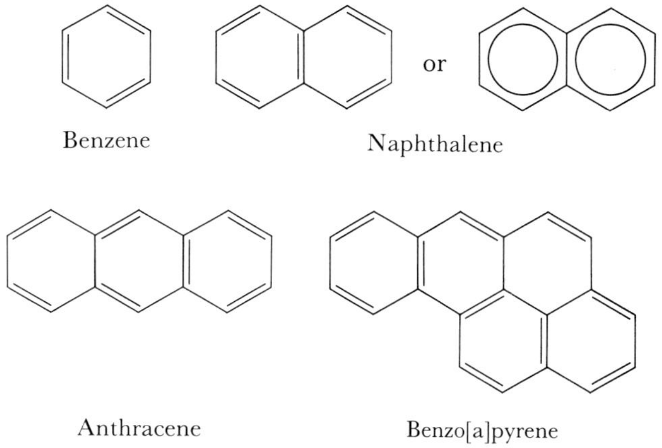

Benzene

Naphthalene

or

Anthracene

Benzo[a]pyrene

FIGURE 8.14 Several aromatic organic structures. Each vertex represents a carbon atom. A hydrogen is attached to each carbon that has only three bonds in the structures as shown.

General Conclusions

On the basis of all the examples we've seen, we can formulate a few general conclusions that are helpful in using the concept of hybrid orbitals to discuss molecular structures.

1. Every pair of bonded atoms shares one or more pairs of electrons. In every bond at least one pair of electrons is localized in the space between the atoms, in a σ bond. The appropriate set of hybrid orbitals used to form the σ bonds between an atom and its neighbors is determined by the observed geometry of the molecule. The relationship between hybrid orbital set and geometry about an atom is given in Table 8.5.
2. The electrons in σ bonds are localized in the region between two bonded atoms and do not make a significant contribution to the bonding between any other two atoms.
3. When atoms share more than one pair of electrons, the additional pairs are in π bonds. The centers of charge density in a π bond lie above and below the bond axis.
4. π bonds may extend over more than two bonded atoms. Electrons in π bonds that extend over more than two atoms are said to be delocalized.

*The symbol [a] in the name for benzo[*a*]pyrene is used to designate the manner in which the six-membered rings are connected together.

SAMPLE EXERCISE 8.5

Formaldehyde, which is a planar molecule, has the following Lewis structure:

```
H
 \
  C=O:
 /
H
```

Describe the bonding in formaldehyde in terms of an appropriate set of hybrid orbitals at the carbon atom.

Solution: Using the VSEPR model, we would predict the bond angles around C to be about 120° (trigonal-planar geometry). The presence of three bonds in a plane about a central atom suggests sp^2 hybrid orbitals for the σ bonds (Table 8.5). There remains a $2p$ orbital on carbon, perpendicular to the plane of the three σ bonds.

Oxygen is also predicted from the VSEPR model to have two unshared electron pairs and the O—C σ bond in a trigonal plane, with approximately 120° angles between them. In this case also, there remains a $2p$ orbital on oxygen, perpendicular to the plane of the three σ bonds. This orbital overlaps with the similarly oriented $2p$ orbital on carbon to form a π bond between carbon and oxygen, as illustrated in Figure 8.15.

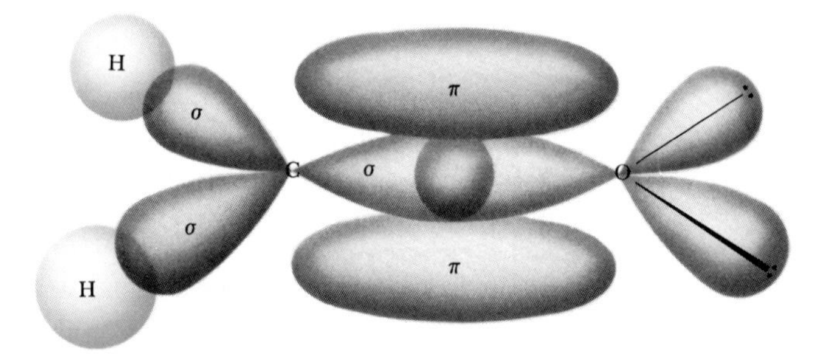

FIGURE 8.15 Formation of σ and π bonds in formaldehyde.

8.5 MOLECULAR ORBITALS

The models of covalent bonding and molecular geometries discussed in this and the preceding chapter are very useful. They provide a nice way of relating the formulas and structures of molecules to the electron configurations of the atoms involved. For example, we can understand why methane has the formula CH_4, and why the arrangement of C—H bonds about the central carbon is tetrahedral. But in all of this discussion, we have sidestepped a rather important question: Why do atoms combine to form covalent bonds in the first place? The answer to this question has to be expressed in terms of energy.

We have seen that electrons in atoms exist in allowed energy states called atomic orbitals. The quantum theory tell us that, in a similar way, electrons exist in molecules in allowed energy states that are called **molecular orbitals.** Because molecules are more complex than atoms, it is no surprise that molecular orbitals are more complex than atomic orbitals.

One of the most useful ways of viewing molecular orbitals is to imagine that they are formed by combining the atomic orbitals of the atoms making up the molecule. By examining the formation of molecular orbitals in this fashion, we can understand why some molecules are more stable and others less stable than the separated atoms. There are various rules and restrictions for how atomic orbitals can be combined to form molecular orbitals:

1. When a set of atomic orbitals is combined to form molecular orbitals, the number of molecular orbitals formed is equal to the number of atomic orbitals in the set.
2. The average energy of the molecular orbitals formed by combining a set of atomic orbitals is approximately equal to the average energy of the atomic orbitals. However, some of the molecular orbital energies are lower than the energies of the starting atomic orbitals, while others are higher.
3. The Pauli principle is obeyed for molecular orbitals just as for atomic orbitals; there can be at most two electrons in each molecular orbital, and they must have their spins paired.
4. Atomic orbitals combine most effectively with other atomic orbitals of similar energy.
5. The greater the amount of overlap between two atomic orbitals, the more stable the resultant molecular orbital. If the overlap is zero, no molecular orbital results.
6. When a molecular orbital is formed by overlap of two nonequivalent atomic orbitals, the bonding molecular orbital contains a greater contribution from the atomic orbital of lower energy. Conversely, the antibonding molecular orbital contains a greater contribution from the higher energy atomic orbital. (Bonding and antibonding orbitals will be explained shortly.)

Let us now illustrate these rules by considering the molecular orbital descriptions of simple diatomic molecules. We begin with two hydrogen atoms coming together to form H_2. The $1s$ orbital on each H atom interacts with the $1s$ orbital on the other atom to form two molecular orbitals (rule 1).

One of the possible interactions or combinations leads to a positive overlap that concentrates electron density between the nuclei. This interaction produces the **bonding** sigma orbital, labeled σ, shown at the bottom of Figure 8.16. This combination is lower in energy (more stable) than the separated atoms. The second molecular orbital that results from the interaction of two $1s$ orbitals is shown on the top of Figure 8.16. This

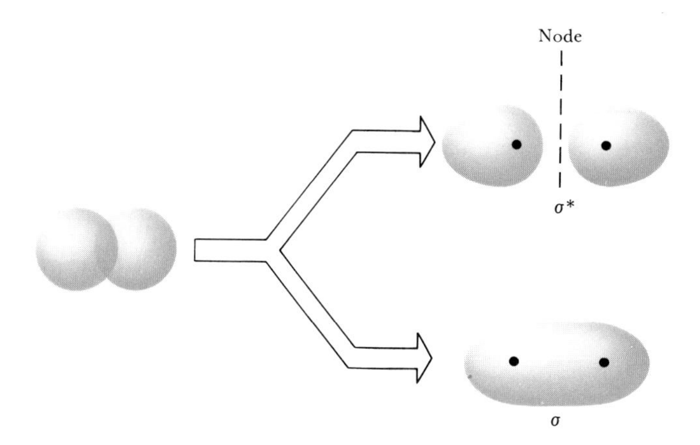

FIGURE 8.16 Contour diagram of the wave functions for the σ and σ^* molecular orbitals in H_2. Note the presence of a nodal plane midway between the atoms in the σ^* antibonding orbital.

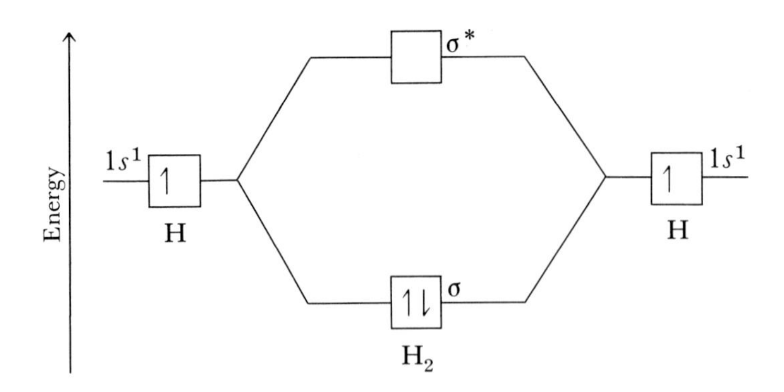

FIGURE 8.17 Energy-level diagram for the molecular orbitals in the H_2 molecule.

combination leads to negative overlap that concentrates the electron density outside the region between the nuclei. This molecular orbital is the antibonding sigma orbital, labeled σ^*. It is called an antibonding orbital because it is of higher energy (less stable) than the isolated atoms.

The interactions between two 1*s* orbitals to form σ and σ^* molecular orbitals can be represented by an orbital energy-level diagram as shown in Figure 8.17. Because each isolated hydrogen atom contains one electron, there are two electrons in the H_2 molecule. We use the Pauli exclusion principle (rule 3) in placing these two electrons into the molecular orbitals. Consequently, they occupy the lowest energy orbital available to them, the σ orbital, with their spins paired. The energy of the two electrons in the σ orbital is lower than that of the electrons in the isolated 1*s* orbitals; consequently, the H_2 molecule is more stable than the two separate hydrogen atoms.

Now let's consider the coming together of two helium atoms to form the He_2 molecule. Again, we have two 1*s* orbitals interacting to form a bonding σ and antibonding σ^* pair of molecular orbitals. In this case, however, each helium atom contributes two electrons, so that the total number to be placed is four. As Figure 8.18 shows, two are placed in the σ orbital, and the other two must be placed in the σ^* orbital. Using rule 2 we recognize that the σ^* orbital is destablized to the same extent that the σ orbital is stabilized. Thus the two electrons that must go into the σ^* orbital destabilize the He_2 molecule just as much as the two electrons in the σ orbital stabilize it. We predict from this model that there is no net stabilization in He_2. Laboratory studies have shown that there is no significant tendency for two helium atoms to bond together to form He_2.

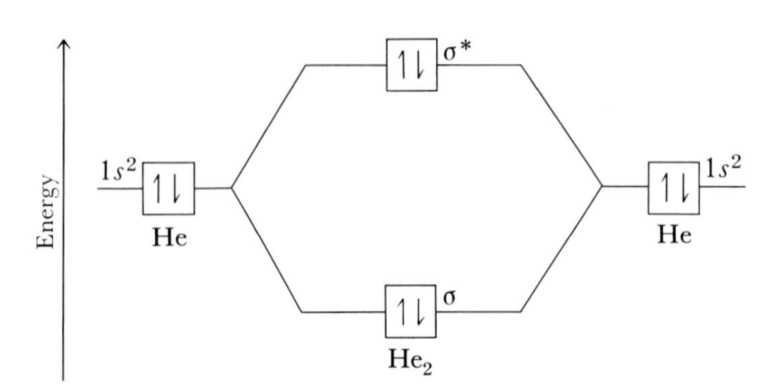

FIGURE 8.18 Energy-level diagram for the molecular orbitals in the He_2 molecule.

From these two examples of the molecular-orbital model, we see that H_2 is stable because electrons can be accommodated in bonding molecular orbitals. The He_2 molecule, on the other hand, is not stable because there are as many electrons in antibonding orbitals as in bonding orbitals.

SAMPLE EXERCISE 8.6

Would you expect the polyatomic ion He_2^+ to be stable relative to a separated He atom and He^+ ion? Explain.

Solution: The energy-level diagram for this system is shown in Figure 8.19. Two helium 1*s* orbitals interact to form bonding and antibonding molecular orbitals. In He_2^+ we have a total of three electrons. Two of these are placed in the bonding orbital, the third in the antibonding orbital, as shown in Figure 8.19.

The stability gained from two electrons in the bonding orbital is greater than the destabilization due to one electron in the antibonding orbital. The He_2^+ molecular ion is therefore predicted to be stable relative to its dissociation products. It has been shown in laboratory studies that He_2^+ is stable relative to separated He and He^+.

In considering the formation of Li_2 from two Li atoms, a new factor enters the picture. The lithium atom has the electron configuration $1s^22s$. In combining two lithium atoms to form the Li_2 molecule we must consider the possible interactions of the 1*s* orbital on one lithium atom with the 2*s* orbital on the other. Such an interaction is theoretically possible, but rule 4 tells us that it will not be of importance. The energy difference between the 1*s* and 2*s* orbitals is simply too great for a strong interaction. As a result, the energy-level diagram for the Li_2 molecule is as shown in Figure 8.20. Notice that the molecular orbitals are labeled with subscripts to indicate the set of atomic orbitals from which they are formed. The electrons in the σ_{1s} and σ^*_{1s} orbitals make no net contribution to the bonding. Their average energy is just the energy of the 1*s* electrons in the isolated Li atoms. All of the bonding in the Li_2 molecule is thus due to the 2*s* electrons. This example illustrates the general rule that *filled atomic subshells do not contribute to bonding in molecule formation.* This rule means that whenever an atom has a completed *s*, *p*, or *d* level, the electrons in those atomic orbitals do not contribute to bond formation. It also means that all filled shells do not contribute to bonding. We need only consider the valence-shell electrons. This conclusion is equivalent to

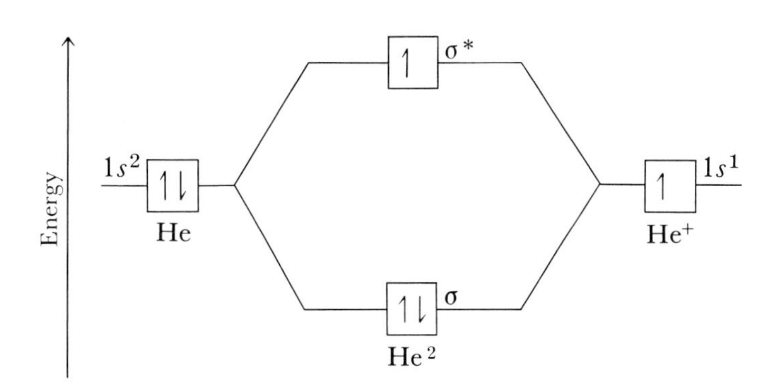

FIGURE 8.19 Energy-level diagram for the molecular orbitals in He_2^+.

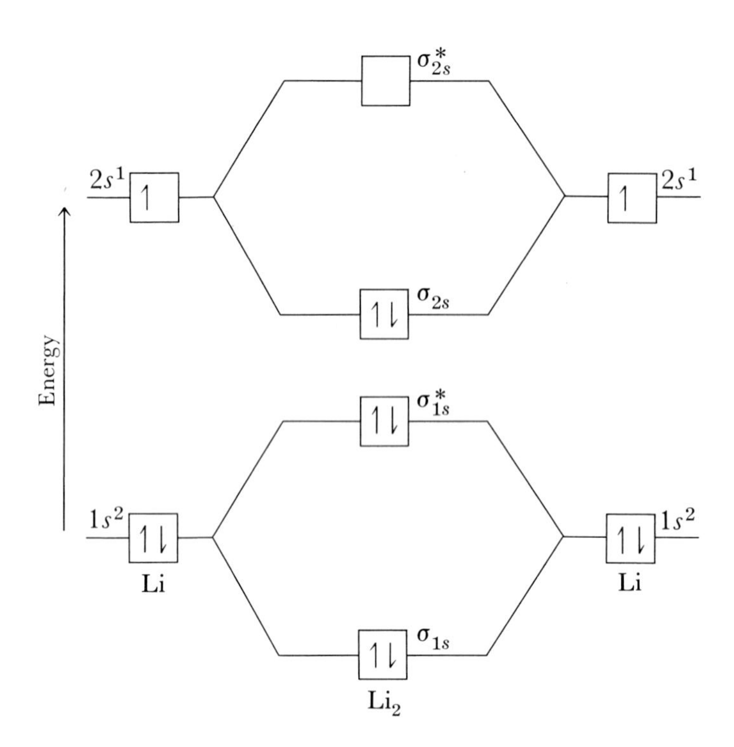

FIGURE 8.20 Energy-level diagram for the Li_2 molecule.

the assumption we make when we draw Lewis structures that show only the valence-shell electrons.

Using this rule, we can account for the bonding in the other alkali metal diatomic molecules. The valance-shell part of the molecular-orbital diagram for Na_2 or K_2 would look just like the valence-shell part for Li_2 (see Figure 8.21). In Figure 8.21, n represents the major quantum number of the valence-shell s orbital. There is no need to show any of the electrons below the valence shell, because they do not contribute to bonding.

Although Li_2 and the other alkali metals are similar to H_2 in terms of the energy-level diagram, the change in major quantum number is of great importance. Table 8.6 shows the bond-dissociation energies of H_2 and the alkali metal diatomic molecules through Cs_2. (The alkali metals are not normally in the form of diatomic molecules, but they can be studied as diatomic molecules in the gas phase at high temperatures.)

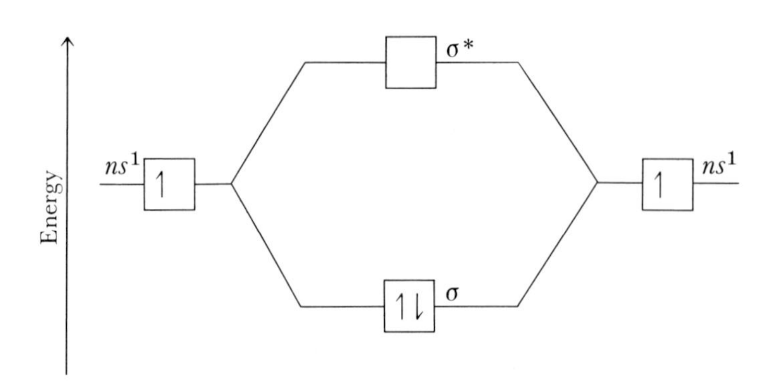

FIGURE 8.21 Energy-level diagram for alkali metal diatomic molecules.

TABLE 8.6 Bond-dissociation energies for H_2 and the alkali metals in diatomic molecules

Molecule	Major quantum number of s orbital	Bond distance (Å)	Dissociation energy (kJ/mol)
H_2	1	0.75	430
Li_2	2	2.67	100
Na_2	3	3.08	72.4
K_2	4	3.92	49.4
Rb_2	5	4.2	45.2
Cs_2	6	4.7	42.6

The bond-dissociation energy decreases steadily in this series. The major factor responsible for this effect is a decrease in the extent to which the valence-shell s orbitals overlap (rule 5). The inner-shell electrons set a limit on how close the nuclei can draw together, because of repulsions between filled shells. In addition, the s orbitals become more spatially extended with an increase in major quantum number. The overall result is a decrease in the extent of overlap of the s orbitals with increasing major quantum number for the diatomic molecules of Table 8.6.

SAMPLE EXERCISE 8.7

Beryllium, the fourth element of the periodic table, does not form a stable diatomic molecule. Provide a reason for this in terms of molecular orbital formation.

Solution: The electron configuration for Be is $1s^2 2s^2$. The energy-level diagram for Be_2 involves interactions of the $1s$ and $2s$ orbitals, just as for Li_2, as shown above. The four valence-shell electrons in Be_2, however, completely fill both σ_{2s} and σ_{2s}^* orbitals, to leave a net bonding of zero.

Overlaps of *p* and *s* Atomic Orbitals

When we consider the formation of diatomic molecules for the elements beyond beryllium, we need to take account of the ways in which the p orbitals on the atoms may combine with one another and with s orbitals. Figure 8.22 shows the contour diagrams for the molecular orbitals formed by combining s and p atomic orbitals in all the possible ways that lead to the formation of molecular orbitals. These combinations are of two kinds. When the orbitals that make up the combination concentrate electron density about the bond axis, the molecular orbital formed is of the σ type. When the orbitals that make up the combination concentrate electron density above and below the bond axis, the molecular orbital formed from the atomic orbitals is labeled π. (You will remember that in Section 8.4 we used these same designations in describing bonds formed from overlaps of hybrid atomic orbitals.)

Both a bonding and an antibonding combination is obtained in each case. Note that the electron density in the bonding orbitals tends to be concentrated in the region between the nuclei, whereas in the antibonding orbital it is concentrated in the regions in back of the nuclei.

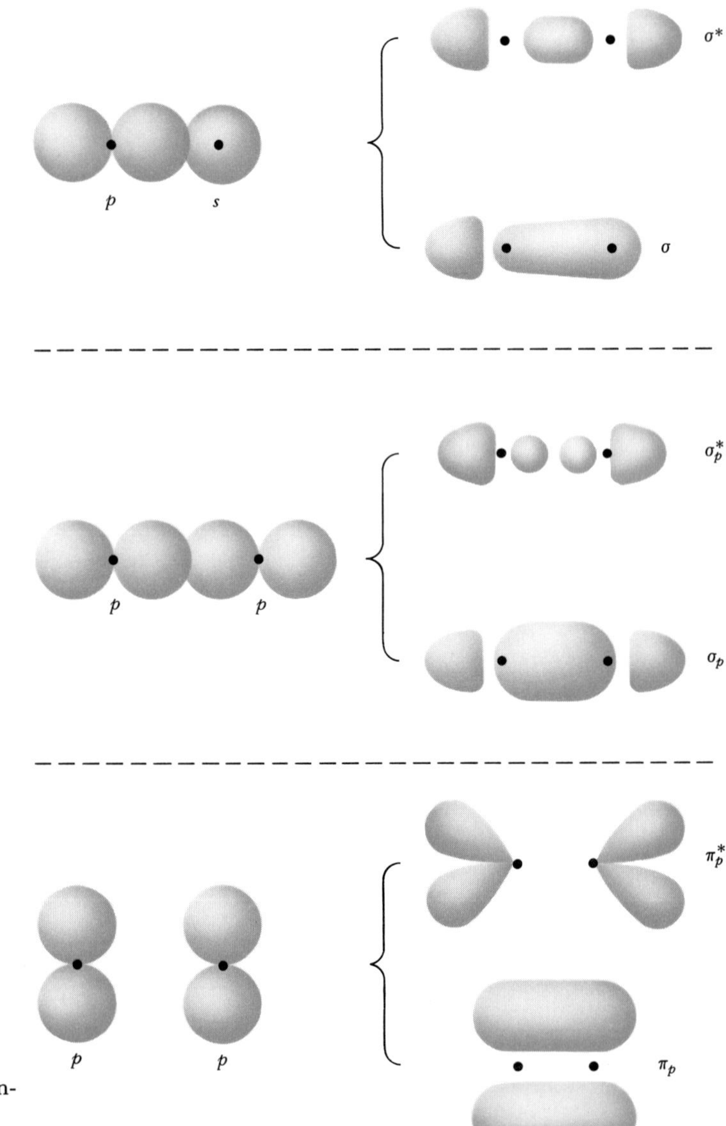

FIGURE 8.22 Contour diagrams for molecular orbitals formed by overlap of *s* and *p* orbitals. These diagrams are designed to illustrate the general shapes of bonding and antibonding orbitals and are not accurate representations.

8.6 MOLECULAR-ORBITAL DIAGRAMS FOR DIATOMIC MOLECULES

We are now ready to consider the molecular-orbital energies in a diatomic molecule formed from two atoms of an element from the second row of the periodic table. Without specifying which element is involved, let us imagine that two atoms of a second-row element come together to form a molecule. The atomic orbitals on each atom interact with those on the other atom, in accordance with the rules described above. From these interactions, a set of molecular orbitals results, as shown in Figure 8.23. We might expect that the 2*s* orbitals would interact exclusively with one another and the 2*p* orbitals with other 2*p* orbitals. Because there are two possible ways in which the 2*p* orbitals can interact, there are both σ and π molecular orbitals formed in this case. The only complication in this simple picture comes from the fact that the σ_{2s} and σ_{2p}

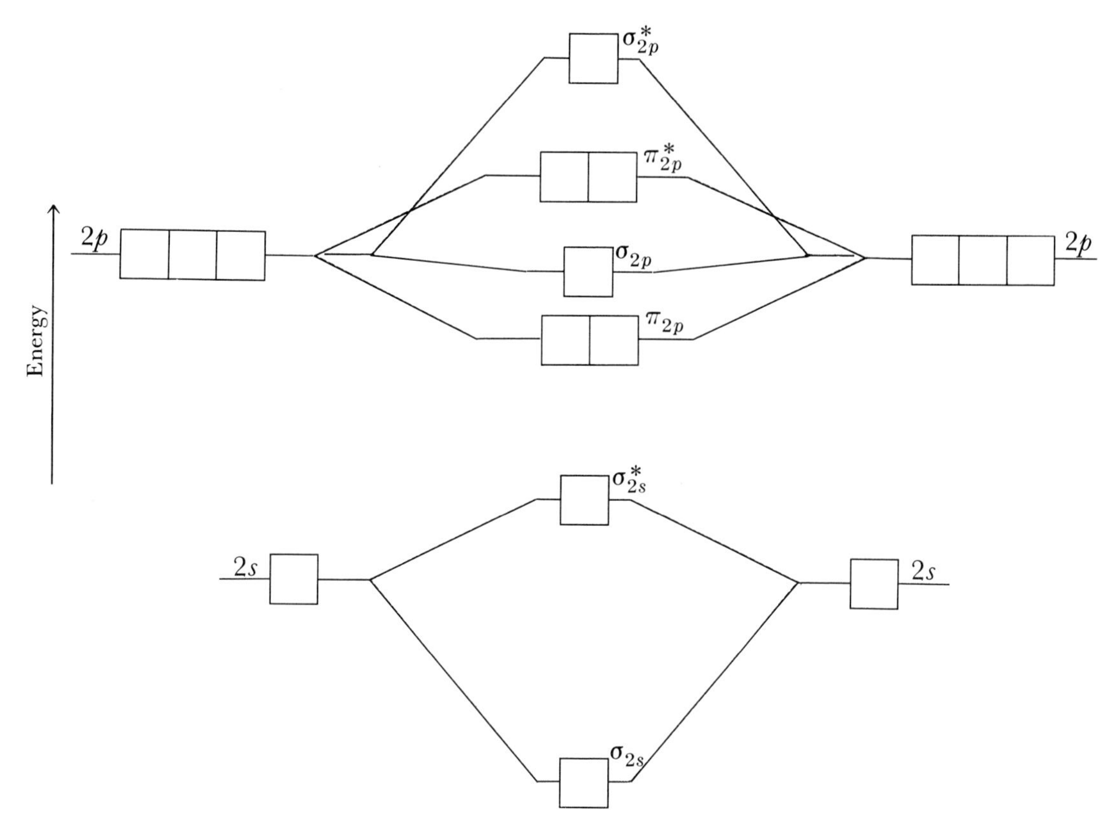

FIGURE 8.23 General energy-level diagram for molecular orbitals of second-row diatomic molecules.

orbitals can interact with one another, as can the σ^*_{2s} and σ^*_{2p}. As a result, the σ_{2p} and σ^*_{2p} orbitals, formed from the $2p$ orbitals, are pushed a little upward in energy; the σ_{2s} and σ^*_{2s} orbitals, formed from the $2s$ orbitals, are pushed downward. These changes in energy result from a certain amount of mixing of the σ_{2s} and σ_{2p} orbitals, and of the σ^*_{2s} and σ^*_{2p} orbitals. Note also that the energy splitting between bonding and antibonding orbitals is larger for the σ_{2p} orbitals than it is for the π_{2p} orbitals. This occurs because the overlap of p orbitals is greater when they are oriented along the axis with respect to one another in σ fashion than when they are oriented in π fashion (rule 5).

Using this basic energy-level diagram we can readily deduce the electronic configuration of any of the second-row diatomic molecules. Some of these molecules are familiar substances, for example, O_2 or N_2. Others are known only in the vapor state at high temperature or under other unusual conditions. In all cases, however, the electronic structures are known from experiments. Bond order in these molecules is defined as the excess of bonding electron pairs over antibonding pairs. Thus in Li_2, with two valence electrons, the bond order is one. The molecule Be_2, with four valence-shell electrons, should have a net bond order of zero, because the four electrons are placed in the σ_{2s} and σ^*_{2s} orbitals. However, B_2 should be stable, with a net bonding pair, for a bond order of one. As shown in Table 8.7, the two bonding electrons occupy the pair of

TABLE 8.7 Electronic configuration and some experimental data for several second-row diatomic molecules

	B_2	C_2	N_2	O_2	F_2
σ^*_{2p}	[ ]	[ ]	[ ]	[ ]	[ ]
π^*_{2p}	[ ][ ]	[ ][ ]	[ ][ ]	[↑][↑]	[↑↓][↑↓]
σ_{2p}	[ ]	[ ]	[↑↓]	[↑↓]	[↑↓]
π_{2p}	[↑][↑]	[↑↓][↑↓]	[↑↓][↑↓]	[↑↓][↑↓]	[↑↓][↑↓]
σ^*_{2s}	[↑↓]	[↑↓]	[↑↓]	[↑↓]	[↑↓]
σ_{2s}	[↑↓]	[↑↓]	[↑↓]	[↑↓]	[↑↓]
Bond order	One	Two	Three	Two	One
Bond-dissociation energy (kJ/mol)	290	620	941	495	155
Bond distance (Å)	1.59	1.31	1.10	1.21	1.43
Ionization energy (kJ/mol)	—	1150	1495	1205	1700

π orbitals of equal energy. Hund's rule operates here just as for electrons in atomic orbitals: Electrons occupy degenerate orbitals singly and with spins parallel until all degenerate orbitals have been occupied. Thus B_2 is predicted to have two unpaired electrons, in agreement with experimental results. Pairing of the π electrons comes in C_2, which has a net bond order of two. With nitrogen, the bond order reaches its maximum, three. In view of this high bond order, the exceptional stability of the N_2 molecule and its high bond-dissociation energy are understandable.

The next molecule in our series, O_2, is especially interesting. We have a total of 12 valence electrons to place. As shown in Table 8.7, the last two of these must go into the π^*_{2p} orbitals with spins parallel. Thus the bond order in O_2 is two. Because of the unpaired electrons, molecular oxygen is **paramagnetic.** (A paramagnetic substance is one that is drawn into a magnetic field.) Oxygen is the only commonly available, simple molecule with this property. The paramagnetism of O_2 can be demonstrated by observing the effect of a magnetic field on a tube containing liquid oxygen, as shown in Figure 8.24. When the field is applied, the tube is moved laterally as the sample is drawn into the magnetic field. The prediction of the paramagnetism of O_2 is a most elegant achievement of molecular-orbital theory. The simple Lewis theory (Section 7.4) does not account for the paramagnetism, and there is no other bonding model for O_2 in which this property is explained so naturally.

In the next molecule in our series, F_2, the electrons in the π^*_{2p} orbitals are paired; the net bond order is one. In the hypothetical Ne_2 molecule, all valence molecular orbitals would be filled, for a net bond order of zero. The Ne_2 molecule is nonexistent. It is interesting to compare the electronic structures of the diatomic molecules with their observable properties. One of the most important measures of bond strength is the

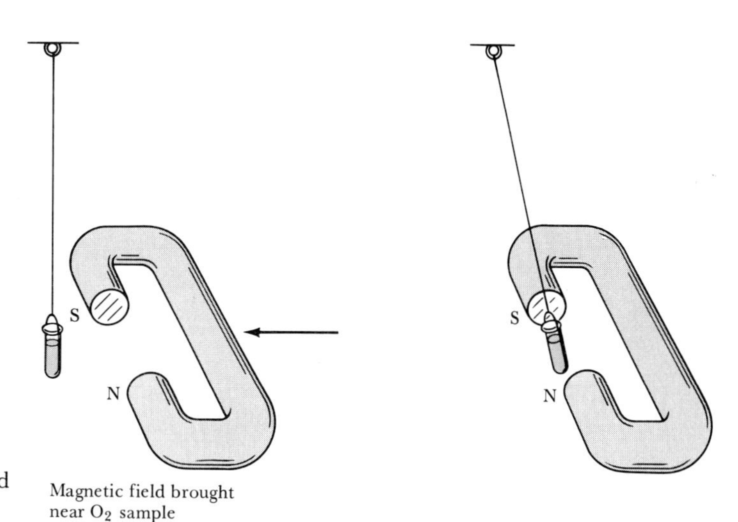

FIGURE 8.24 Illustration of an experiment that shows the paramagnetic character of O_2. When the magnetic field is applied, the sample of liquid O_2 is drawn into the field.

bond-dissociation energy—that is, the energy required to separate the two atoms of the molecule to a very large distance from one another (Section 7.8). Another important characteristic of a bond is the distance separating the bonded atoms. We have seen in previous discussions (Section 7.4) that bond distances are smaller in the case of multiple bonds. A third property of the diatomic molecules that relates to bonding is the **ionization energy**—that is, the energy required to remove the highest energy electron from the molecule.

The bond-dissociation energies, bond distances, and ionization energies for the diatomic molecules are listed in Table 8.7. Note that the bond-dissociation energies of molecules with the same bond order are not the same. This should not surprise us, because overlaps differ, and many other factors contribute to the total energy of the molecule. Still, it is roughly true that bond-dissociation energy increases with bond order. Similarly, bond length decreases with increasing bond order.

In addition to the neutral diatomic molecules listed in Table 8.7, there are several known diatomic ions, such as N_2^+ and O_2^+. These species can be produced in the gas phase and their properties studied, even though they are too reactive to permit isolation. We'll see in Chapter 12 that N_2^+ and O_2^+ are important components of the earth's upper atmosphere. It is possible, using the energy-level diagram for molecular orbitals, to predict some of the properties of these ions.

SAMPLE EXERCISE 8.8

Predict the following properties of O_2^+: (a) number of unpaired electrons, (b) bond order, (c) bond-dissociation energy, and (d) bond length.

Solution: O_2^+ has one electron less than O_2. The electron removed from O_2 to form O_2^+ is one of the two unpaired π^* electrons. O_2^+ should therefore have just one unpaired electron left. Because the electron removed has come from an antibonding orbital, the bond order in O_2^+ is larger than for O_2; it is in fact intermediate between O_2 and N_2. Counting a single electron as contributing a bond order of $\frac{1}{2}$, the bond order is $2\frac{1}{2}$. The bond-dissociation energy and bond length should be about midway between that for O_2 and N_2, say, 720 kJ/mol and 1.15 Å, respectively. The observed properties of O_2^+ are: number of unpaired electrons, one; bond length, 1.123 Å; dissociation energy, 625 kJ/mol.

In principle, the energy-level diagram shown in Figure 8.23 is applicable to the diatomic molecules of the third- and higher-row elements. Except for the halogens, however, in which only a single bond is possible, the formation of diatomic molecules among the other elements is not the most stable form of bonding. The molecules P_2 and S_2 have been observed in the high-temperature vapors of these elements, but at lower temperatures other forms of the element are more stable (Section 8.7). Because P—P and S—S single bonds are quite stable, the major reason for the relative instability of the diatomic molecules in these two cases seems to be that their π bonds are not very strong.

Polar Covalent Molecular Orbitals

The concept of molecular orbitals can also be used to understand molecule formation between two different atoms. When the atoms are not very different, the energy-level diagram of Figure 8.23 can be used with some modification. The energy levels of one of the atoms are shifted with respect to the other. Such a diagram should be applicable to such molecules as NO, CO, CN, and so forth.

When the two atoms forming a molecule are appreciably different, extensive changes need to be made in the energy-level diagram. In many instances, it is possible to make simplifying assumptions that work quite well. For example, consider HF, in which the two atoms have quite different electronegativities. This is a simple case because hydrogen has only a single orbital for interaction with the orbitals of fluorine. The energy of the hydrogen $1s$ orbital is higher that that of the highest valence-shell orbital of fluorine, the $2p$, as shown in Figure 8.25. It is *much* higher than the fluorine $2s$ orbital. Thus we need consider only the interaction of the hydrogen $1s$ orbital with the $2p$. Of the three $2p$ orbitals, only one is of correct symmetry to interact with the hydrogen $1s$. The interaction leads to a bonding and antibonding pair of orbitals, as shown in Figure 8.25. The other orbitals of fluorine are relatively unperturbed by molecular formation.

The bonding orbital occupied by a pair of electrons in HF is a polar covalent bond. This means that the contributions of the two atomic

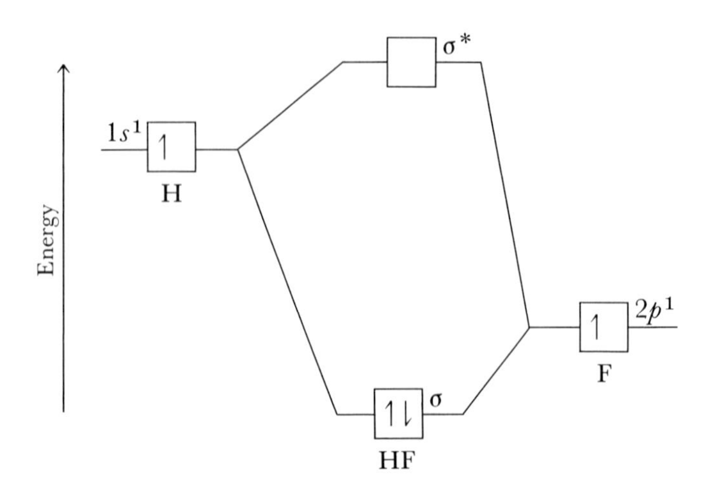

FIGURE 8.25 Energy-level diagram for HF molecular orbitals.

orbitals to the molecular orbitals are not the same. In the bonding orbital, the fluorine $2p$ orbital is more important than the hydrogen $1s$. As a result, the bonding electron pair is shifted more toward the fluorine atom than toward hydrogen. In the antibonding orbital, the contribution of the hydrogen $1s$ orbital is greater than that of the fluorine $2p$. Hydrogen fluoride provides an example of rule 6 in the list of rules for formation of molecular orbitals.

8.7 STRUCTURES OF THE NONMETALLIC ELEMENTS

Our discussion of molecular shape and bonding has provided the background for us to examine the structures of some common nonmetals. As we do so, we will also take the opportunity to introduce a few properties of these elements, especially ones that reflect molecular structure and bonding.

The halogens, family 7A, all consist of diatomic molecules, with formulas of the type X_2. Fluorine has a pale yellow color while chlorine is greenish-yellow. Bromine is a dark red liquid that boils at 58°C. Iodine is a dark violet solid with an almost metallic appearance; it readily forms a violet vapor when gently heated. The diatomic structure of the halogens is maintained not only in the vapor phase, but also in the liquid and solid phases. The diatomic nature of the halogens reflects the fact that each halogen atom has seven valence electrons. Covalent bonding with a single bond between the halogen atoms therefore provides an octet of electrons around each:

$$:\ddot{\underset{..}{X}}-\ddot{\underset{..}{X}}:$$

Oxygen, the first member of family 6A, also occurs as a diatomic molecule, O_2. This substance is familiar as a common and important component of the atmosphere. The short O—O bond distance (1.21 Å) and the relatively high bond-dissociation energy (495 kJ/mol) of O_2 suggest that it possesses a double bond. The following Lewis structure is consistent with this observation:

$$\ddot{\underset{..}{O}}=\ddot{\underset{..}{O}}$$

However, this Lewis structure leads to the incorrect prediction that all valence electrons are paired. In fact, the O_2 molecule is paramagnetic; it contains two unpaired electrons. Although we can write useful Lewis structures for most molecules, the Lewis structure for O_2 fails to account for the paramagnetism of the molecule. We have seen, in Section 8.6, that molecular-orbital theory correctly predicts both oxygen's bond order of two and its paramagnetism.

Oxygen also occurs as a gaseous triatomic molecule, O_3, known as ozone. This substance is responsible for the pungent odor often detected around electric motors. In contrast to O_2, which is essential for life, O_3 is toxic. Ozone and O_2 are allotropes of oxygen. Allotropes are different forms of the same element in the same state (in this case, both gases). Ozone is less stable and more reactive than O_2. In particular, it is a much stronger oxidizing agent. Although O_2 reacts directly with most ele-

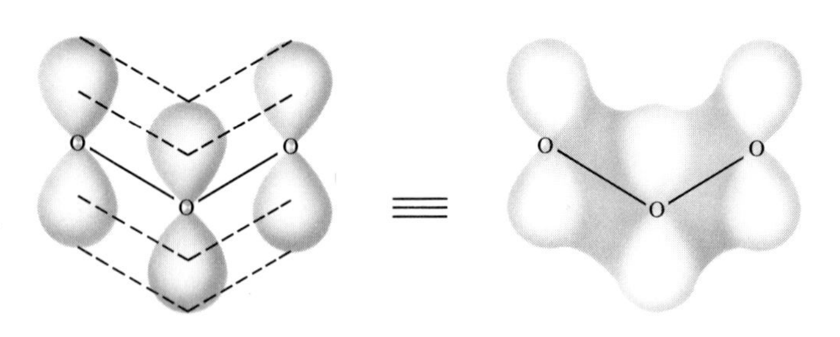

FIGURE 8.26 Delocalized π bond in ozone formed by overlap of $2p$ orbitals on each of the oxygen atoms.

ments, high temperatures are often required; O_3 is more likely to react readily at lower temperatures.

The ozone molecule may be represented by the following resonance forms:

$$:\ddot{O}=\ddot{O}-\ddot{O}: \longleftrightarrow :\ddot{O}-\ddot{O}=\ddot{O}:$$

The molecule is bent, as we would predict using the VSEPR model; the bond angle is found experimentally to be 116.8° and the O—O bond distance is 1.278 Å. The σ bonds between the oxygen atoms involve approximately sp^2 hybridization of the valence-shell electrons on the central oxygen atom. A π bond that is delocalized over the three atoms is formed by sideways overlap of p orbitals on each atom (Figure 8.26). The bond-dissociation energy of the molecule, corresponding to the reaction $O_3(g) \longrightarrow O_2(g) + O(g)$, is 107 kJ/mol. This is much lower than the bond energy of O_2. The lower bond energy in O_3 explains why O_3 is more reactive than O_2.

Sulfur, the second member of family 6A, exists in several allotropic forms. The most stable and common allotrope at room temperature is a yellow solid with molecular formula S_8. The S_8 molecule consists of an eight-membered ring of sulfur atoms as shown in Figure 8.27. The S—S—S bond angle is 107.8°, close to the tetrahedral angle associated with sp^3 hybridization. Each sulfur atom in this structure achieves an octet of electrons by bonding to two other sulfur atoms by single σ-type bonds. By contrast, we have seen that oxygen satisfies its bonding capacity by forming π bonds in the O_2 and O_3 molecules. π bonds are generally more common between smaller atoms like O, N, and C than between larger ones like S, P, and Si.

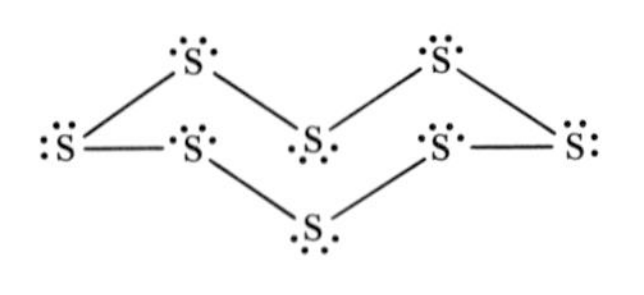

FIGURE 8.27 Structure of S_8 molecules as found in the most common allotropic form of sulfur at room temperature.

Nitrogen, the lightest member of family 5A, exists in the earth's atmosphere as the very stable N_2 molecule. The Lewis structure for N_2 indicates that each nitrogen achieves an octet by forming a triple bond:

The bond order of three, arising from a σ and two π bonds, is also predicted by molecular-orbital theory. The very strong, nonpolar N—N bond results in exceptionally low chemical reactivity. The N—N dissociation energy is 941 kJ/mol and the nitrogen atoms are separated by only 1.10 Å.

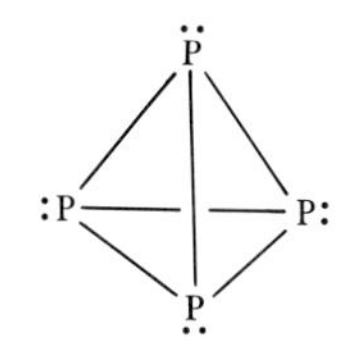

FIGURE 8.28 Structure of P_4 molecules as found in white phosphorus, a common allotropic form of the element.

In contrast to nitrogen, the next member of family 5A, phosphorus, does not exist as a diatomic molecule, except at high temperatures in the vapor state. Phosphorus forms weak P—P π bonds and consequently seeks to achieve its octet by utilizing only σ bonds; because each atom has five valence electrons, three σ bonds are required. When phosphorus condenses from the vapor state it forms so-called white phosphorus. This reactive allotrope consists of P_4 molecules formed by a tetrahedral arrangement of phosphorus atoms, as shown in Figure 8.28. Each atom in this structure is bonded to three other phosphorus atoms by single bonds. However, this geometry requires that each P—P—P bond angle be only 60°. This angle is much smaller than that in any other molecule that we have discussed thus far. The 60° bond angle requires that the bonding electron pairs in the molecule be crowded rather closely together. The high reactivity of the P_4 molecule is probably due to the strain introduced by electron-electron repulsions.

The heavier elements of group 5A, arsenic, antimony, and bismuth, show increasingly metallic properties with increasing period. The allotropy of arsenic is somewhat similar to that of phosphorus, but for antimony and bismuth the structures of the elements are more complex and possess many of the characteristics of metals. These elements are frequently referred to as metalloids, to indicate that they are on the borderline between metals and nonmetals in their properties. The group 5A elements exemplify the rule that *in any one group, the metallic characteristics of the elements increase with increasing atomic number.* In the periodic table shown on the inside front cover, the heavy line that runs diagonally from just below B to Po represents the approximate dividing line between metallic and nonmetallic elements.

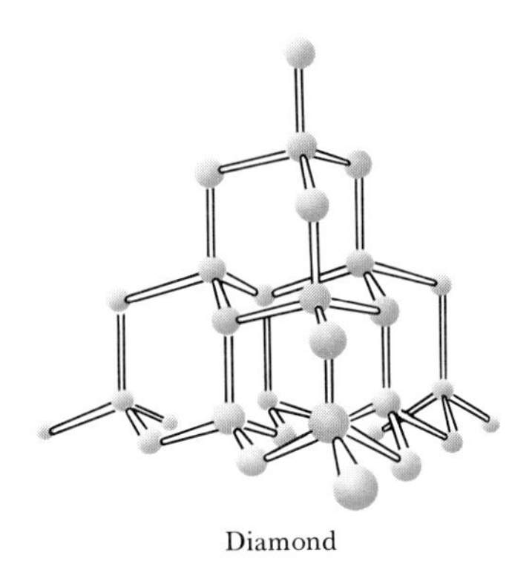

FIGURE 8.29 Structure of diamond, a major allotropic form of carbon.

Carbon, the lightest of the group 4A elements, exists in two major allotropic forms, graphite and diamond. The structure of diamond is shown in Figure 8.29. You can see that each carbon atom is surrounded by four other carbon atoms arranged at the corners of a tetrahedron. Each carbon can therefore be described as using sp^3 hybrid orbitals in forming σ bonds with each of its four neighboring carbon atoms. The stability of the diamond lattice comes from the fact that the carbon atoms are interconnected in a three-dimensional array of strong carbon-carbon single bonds. Diamond is a very hard and brittle material. In fact, industrial-grade diamonds are employed in the blades of saws for the most demanding cutting jobs. The melting point of diamond, above 3500°C, is higher than that for any other element. It is exceptionally inert chemically. However, when heated to about 1800°C in the absence of air, diamond converts to graphite, the other allotropic form of the element. Diamonds also burn at about 900°C when heated in air or in oxygen. (Naturally, there is not much interest in carrying out experiments of this kind.)

The structure of graphite is shown in Figure 8.30. The carbon atoms are arranged in layers; each carbon atom is surrounded by three other carbon atoms, all at the same distance of separation. The distance between adjacent carbon atoms in the plane is 1.42 Å, very close to the C—C distance in benzene (1.395 Å). The distance between adjacent layers is 3.41 Å, too great a distance for a covalent bond to exist. Graph-

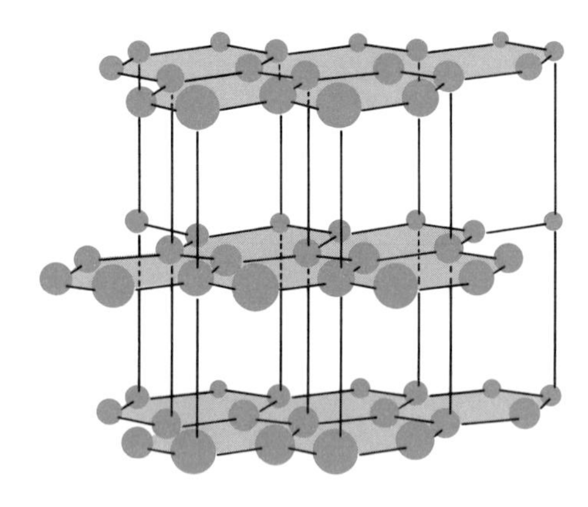

FIGURE 8.30 Structure of graphite.

ite occurs in gray, shiny plates and is usually found in masses of thin, easily separated sheets (see Figure 8.31). The layers easily slide past one another when rubbed, giving the substance a greasy feel. Graphite has been used as a lubricant and in making the "lead" in pencils.

The properties of graphite are due to the nature of the bonding between carbon atoms. Because the individual layers in graphite are not directly bonded together, they slide past one another easily and are readily separated. We can understand the bonding within each layer by supposing that each carbon is bonded to its three neighbors by sp^2 hybrid

FIGURE 8.31 Piece of graphite photographed with an electron microscope (magnification about 15 million times). The bright bands are layers of carbon atoms that are only 3.41 Å apart. (*P. A. Marsch and A. Voet, J. M. Huber Corp.*)

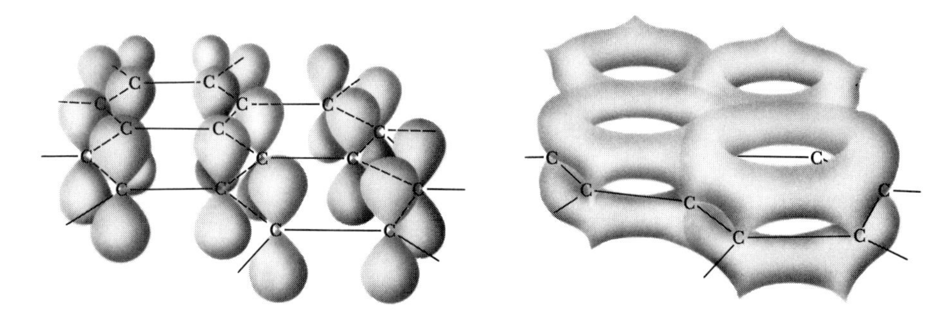

FIGURE 8.32 Formation of the π bonds in graphite.

σ bonds. The remaining $2p$ orbital is employed in π bonding to the same three atoms, as illustrated in Figure 8.32. But because those three neighboring atoms are also involved in π bonding to two other carbons, the network of π bonds extends over essentially the entire plane. The electrons that occupy the π bonds are free to move from one bond to the next. Thus graphite represents an extension to an entire plane of the kind of delocalization we saw in benzene. In benzene the π electrons are free to move about the circumference of the ring. In graphite they are free to move over the entire plane. Because of this freedom of motion, graphite is a good conductor of heat and electricity in directions along the planes of carbon atoms. (If you have ever taken apart a flashlight battery, you know that the central electrode in the battery is made of graphite.) Although graphite readily conducts heat and electricity along the planes, it is an insulator in the direction normal to (perpendicular to) the planes. This is so because there is no means by which electrons can move easily from one plane to the other.

The structures of elemental silicon and germanium are the same as for diamond. No graphitelike allotrope of these elements is known. This observation parallels our expectation that π bonds between these larger atoms will be of low strength. Tin exists in two allotropic forms. White tin is the more metallic in character; this is the more stable form at room temperature and above. However, at lower temperatures, the element converts to a gray form that has the diamond structure. The conversion of tin from the metallic form to a gray powder was a great nuisance when tin was used for construction in cold climates. This phenomenon was referred to as "tin disease." Tin is another example of an element on the borderline between metallic and nonmetallic, with perhaps more metallic character in its properties. Lead, the heaviest element of group 4A, is quite markedly metallic in character.

FOR REVIEW

Summary

In this chapter, we've applied the basic principles of chemical bonding to several important areas of chemical structure and behavior. The three-dimensional structures of molecules are determined by the distances between bonded atoms and by the directions of chemical bonds with respect to one another around a particular atom. The valence-shell electron-pair repulsion (VSEPR) model explains these relative directions in terms of the repulsions that exist between electron pairs. According to this model, electron pairs around an atom orient themselves so as to minimize electrostatic repulsions; that is, they remain as far apart as possible. By recognizing that unshared electron pairs take up more space (exert greater repulsive forces) than shared electron pairs, it is possible to account for the departures of bond angles from the idealized values and to explain many other aspects of molecular structure. The shape of a molecule and the bond polarities determine whether or not a molecule will be polar. The degree of polarity of a molecule is measured by its dipole moment.

The Lewis model for covalent bonding introduced in Chapter 7 can be extended to account very nicely for the geometrical properties of molecules. We can imagine that the atoms in a molecule are bonded to one another by electron pairs that occupy pairs of overlapping atomic orbitals. The extent to which the atomic orbitals share the same region of space, called overlap, is important in determining the amount of stability that results from bond formation. The bonds directed along the internuclear axes are called σ bonds. It is possible to formulate orbitals on an atom that are directed toward each of the other atoms surrounding it by forming hybrid orbitals. These orbitals are made up of mixtures of the familiar *s*, *p*, and *d* atomic orbitals. Depending on the particular number of other atoms bonded to an atom and their arrangement in space, a particular set of hybrid orbitals can be formulated that has the necessary directional characteristics. For example, sp^3 hybrid orbitals are directed toward the corners of a tetrahedron.

In addition to the σ bonds, which determine the geometry of the bonding around a particular atom, there may be also π bonds constructed from remaining, unhybridized atomic orbitals. Thus double bonds, consisting of a σ and a π bond, or a triple bond, consisting of a σ and two π bonds, may be formed. In some molecules the π bonds may extend, or be delocalized, over several atoms. Delocalization of the π electrons in a cyclic structure, such as in benzene, or throughout a plane, as in graphite, leads to a special stability.

The coming together of atoms to form molecules may be viewed also as the coming together of atomic orbitals to form molecular orbitals. Atomic orbitals may combine with one another in various ways. The rules for combining atomic orbitals on atoms to form molecular orbitals allow us to account very well for the observed properties of the diatomic molecules formed by the first several elements of the periodic table. The molecular-orbital model is particularly impressive in explaining the fact that the O_2 molecule contains two unpaired electrons. Many of the ideas presented in this chapter are important in understanding the structures of the nonmetallic elements.

Learning goals

Having read and studied this chapter, you should be able to:

1. Relate the number of electron pairs in the valence shell of an atom in a molecule to the geometrical arrangement around that atom.
2. Explain why unshared electron pairs exert a greater repulsive interaction on other pairs than do shared electron pairs.
3. Predict the geometrical structure of a molecule or ion from its Lewis structure.
4. Predict from the molecular shape and the electronegativities of the atoms involved whether a molecule can have a dipole moment.
5. Explain the concept of hybridization and its relationship to geometrical structure.
6. Assign a hybridization to the valence orbitals of an atom in a molecule, knowing the number and geometrical arrangement of the atoms to which it is bonded.
7. Formulate the bonding in a molecule in terms of σ bonds and π bonds, from its Lewis structure.
8. Explain the concept of delocalization in π bonds.
9. Explain the concept of orbital overlap.
10. Describe how molecular orbitals are formed by overlap of atomic orbitals.
11. Explain the relationship between bonding and antibonding molecular orbitals.
12. Construct the molecular-orbital energy-level diagram for a diatomic molecule or ion built from elements of the first or second row and predict the bond order and number of unpaired electrons.
13. Describe the structures of the common nonmetallic elements.
14. Relate the structures and bonding of nonmetallic elements to their properties.

Key terms

Among the more important terms and expressions used for the first time in this chapter are the following:

An allotrope (Section 8.7) is one of the forms of an element, when that element is capable of existing in more than one form in the same state. For example, O_2 and O_3 are allotropes of oxygen.

An antibonding molecular orbital (Section 8.5) is a molecular orbital in which electron density is concentrated outside the region between the two nuclei of bonded atoms. Such orbitals, designed as σ^* or π^*, are less stable (of higher energy) than bonding molecular orbitals.

A bonding molecular orbital (Section 8.5) is one in which the electron density is concentrated in the internuclear region. The energy of a bonding molecular orbital is lower than the energy of the separate atomic orbitals from which it forms.

Bond order (Section 8.6) is expressed as the number of bonding electron pairs shared between two atoms, less the number of antibonding electron pairs.

The dipole moment (Section 8.2) is a measure of the separation between centers of positive and negative charges in polar molecules.

Hybridization (Section 8.3) refers to the mixing of different types of atomic orbitals to produce a set of equivalent hybrid orbitals.

A molecular orbital (Section 8.5) is an allowed state for an electron in a molecule. A molecular orbital is entirely analogous to an atomic orbital, which is an allowed state for an electron in an atom. A molecular orbital may be classified as σ or π, depending on the disposition of electron density with respect to the internuclear axis.

The term overlap (Section 8.3) refers to the extent to which atomic orbitals on different atoms share the same region of space to form a molecular orbital. When overlap is large, a strong bond may be formed.

Paramagnetism (Section 8.6) is a property that a substance may possess if it contains one or more unpaired electrons. A paramagnetic substance is drawn into a magnetic field.

A pi (π) bond (Section 8.4) is a covalent bond in which electron density is concentrated above and below the line joining the bonding atoms.

A sigma (σ) bond (Section 8.4) is a covalent bond in which electron density is concentrated along the internuclear axis.

The valence-shell electron-pair repulsion (VSEPR) model (Section 8.1) accounts for the geometrical arrangements of shared and unshared electron pairs around a central atom in terms of the repulsions between electron pairs.

EXERCISES

The VSEPR model

8.1 Describe the geometrical arrangements that are characteristic for each of the following numbers of electron pairs about a central atom: (a) 3; (b) 4; (c) 5; (d) 6.

8.2 What is the geometrical shape characteristic of a molecule or ion AX_n that has the following numbers of unshared electron pairs and A—X bonds, respectively: (a) 0, 4; (b) 2, 2; (c) 1, 4; (d) 2, 4; (e) 0, 2?

8.3 Predict the geometry of each of the following molecules or ions: (a) NF_3; (b) BeH_4^{2-}; (c) KrF_2; (d) $GaCl_3$; (e) $SOCl_2$; (f) SO_2Cl_2.

8.4 Explain why the HNH bond angle in ammonia is smaller than the HCH bond angle in methane.

8.5 Use the VSEPR model to predict the geometry of each of the following molecules or ions: (a) BH_3; (b) SiH_4; (c) SiF_6^{2-}; (d) $SnCl_2$; (e) ICl_2^-; (f) SO_3^{2-}; (g) $AsCl_4^-$.

8.6 Use the VSEPR model to predict the arrangement of valence-shell electron pairs around the central atom in each of the following molecules or ions, and also predict the shape of the species: (a) ICl_3; (b) NO_2; (c) CO_3^{2-}; (d) ClO_3^-; (e) $GaCl_4^-$; (f) SCl_2; (g) PCl_4^+; (h) $SeOCl_2$.

8.7 In each of the following cases, indicate which of the two structures proposed is likely to be correct, and give the reason for your answer:

(a) 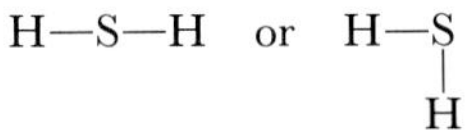 H—S—H or H—S(—H) [S bonded to H, with second H below S]

(b) 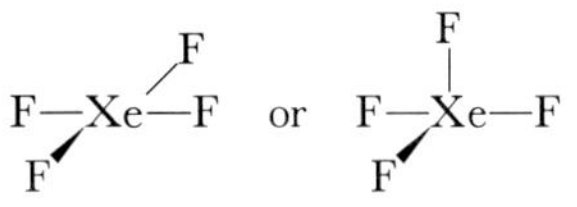 XeF_4 structures: F—Xe—F with F above and F wedged below, or F—Xe—F with F directly above and F wedged below

(c) 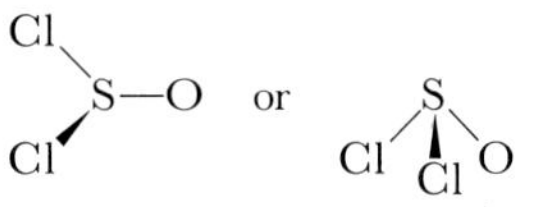 Cl, Cl—S—O (planar) or S with Cl, Cl, O arranged pyramidally

(d) 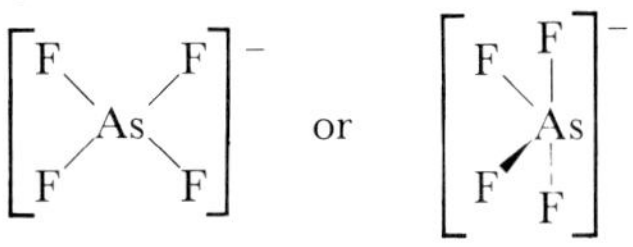 $[AsF_4]^-$: square planar As with four F, or As with F above, F wedged, F below

8.8 The three species NO_2^+, NO_2, and NO_2^- all have a central nitrogen atom. The ONO bond angles in the three species are 180°, 134°, and 115°, respectively. Explain this variation in bond angle.

Dipole moments

8.9 Indicate whether each of the following molecules is capable of possessing a dipole moment: (a) CCl_4; (b) NF_3; (c) SO_3; (d) CS_2; (e) SCl_2; (f) GeI_2.

8.10 In spite of the larger electronegativity difference between the bonded atoms, $BeCl_2(g)$ has no dipole moment, whereas SCl_2 does possess one. What is the origin of this difference in properties?

8.11 Predict whether the following molecules are polar or nonpolar: (a) PH_3; (b) $SiCl_4$; (c) BF_3; (d) IF; (e) C_2H_4 (ethylene); (f) SCO; (g) CH_2Cl_2.

8.12 The bond length in HBr is 1.41 Å. From the dipole moment (Table 8.4) calculate the percentage ionic character in the H—Br bond, assuming that the partial charges are centered on H and Br.

8.13 How is the dipole moment of a diatomic molecule A—B expected to vary as a function of the electronegativity difference between A and B?

Hybrid orbitals

8.14 What energy changes are involved in the conceptual rearrangement of electrons in an atom to prepare it for forming hybrid orbitals?

8.15 Without referring to the tables in the chapter, indicate the designation for the hybrid orbitals formed from each of the following combinations of atomic orbitals: (a) one *s* and two *p*; (b) one *d*, one *s*, and three *p*; (c) two *d*, one *s*, and three *p*. What bond angles are associated with each?

8.16 Without referring to tables in the chapter, indicate the geometrical character of each of the following hybrid orbital sets: (a) sp^2d; (b) sp; (c) d^2sp^3; (d) sp^2.

8.17 Indicate the hybrid orbital set employed by the central atom in each of the following compounds or ions: (a) CH_3^+; (b) PF_5; (c) $AlCl_4^-$; (d) $TiCl_4$; (e) $SOCl_2$; (f) BrF_3; (g) ClO_2^-; (h) XeF_4; (i) AsF_4^+.

8.18 Consider the Lewis structure for glycine, the simplest amino acid.

```
       H :O:
 ..    |  ||   ..
H2N—C—C—O—H
       |       ..
       H
```

(a) What are the approximate bond angles about each of the two carbon atoms, and what are the hybridizations of the orbitals on each of them? (b) What are the hybridizations of the orbitals on the two oxygens and the nitrogen atom, and what are the approximate bond angles at the nitrogen? (c) What is the total number of σ bonds in the entire molecule, and the total number of π bonds?

8.19 (a) What kind of hybrid orbitals are used by each carbon atom in the molecule shown?

```
          H  O  H  H
          |  ||  |  |
H—C≡C—C—C—C=C—Cl
          |
          H
```

(b) How many π bonds does this molecule possess?

8.20 The geometrical structure of the nitrate ion, NO_3^-, can be accounted for either in terms of Lewis structures, using resonance forms, or in terms of delocalized π bonding. Explain the structure of NO_3^- ion in both of these terms.

Molecular orbitals

8.21 How do bonding and antibonding molecular orbitals differ with respect to (a) energies; (b) the spatial distribution of electron density?

8.22 Using Figure 8.23 as a guide, sketch the molecular-orbital energy-level diagram for each of the following: (a) B_2^+; (b) Li_2^+; (c) C_2^+; (d) Ne_2^{2+}. In each case indicate whether the stability of the species increases or decreases upon addition of one electron.

8.23 Show with drawings the form of the bonding and antibonding orbitals formed from overlap of (a) two *p* orbitals; (b) two *s* orbitals; (c) an *s* and a *p* orbital.

8.24 The acetylide ion, C_2^{2-}, occurs in calcium carbide. Draw the molecular-orbital energy-level diagrams for both C_2^{2-} and C_2 and compare them with regard to their expected relative bond energies, bond lengths, and magnetism.

8.25 Neglecting the difference in electronegativities, sketch the energy-level diagram for the CN^- ion. What is the bond order in this ion? How would the bond order be affected by removal of an electron to form CN?

8.26 Which of the following species would you expect to be paramagnetic: (a) NO; (b) O_2^+; (c) C_2; (d) CO?

8.27 The S_2 molecule, seen in the gas phase under special conditions, has a dissociation energy of 425 kJ/mol. Compare this with the value for O_2, and sketch a possible energy-level diagram for the S_2 molecule.

8.28 Predict the relative stabilities of the species N_2^+, N_2, and N_2^-. Predict also the variation in bond length in the series.

Structures of the nonmetallic elements

8.29 Among the second-row nonmetallic elements the internuclear bond distances vary as follows: carbon (diamond), 1.54 Å; nitrogen (N_2), 1.09 Å; oxygen (O_2), 1.208 Å; fluorine (F_2), 1.417 Å. Account for these variations in terms of the concepts of bond order and effective nuclear charge.

8.30 The molecule P_2 exists in phosphorus vapor. It has a bond-dissociation energy of 489 kJ/mol. Describe the structure of this molecule in terms of molecular-orbital theory. Compare its bond-dissociation energy with that for N_2, and account for the difference. Why doesn't phosphorus exist in this form at ordinary temperatures and pressures?

8.31 The following table provides data regarding O—O bond distances and dissociation energies:

Compound	O—O distance (Å)	Bond-dissociation energy (kJ/mol)
H_2O_2	1.48	213
O_2	1.21	495
O_3	1.28	107 (for $O_3 \longrightarrow O_2 + O$)

(a) Explain in terms of the structures of the molecules involved why the O—O distance in O_3 is intermediate between that for O_2 and H_2O_2. (b) Explain why the energy requirement for rupture of the O—O bond in O_3 is lower than for either H_2O_2 or O_2.

8.32 Account for each of the following: (a) the hardness of diamond; (b) the low boiling point of elemental bromine as compared with elemental carbon; (c) the fact that white phosphorus readily burns in air, whereas elemental sulfur is unreactive.

8.33 In black phosphorus, a solid form of the element, each phosphorus atom is bonded to three others in a sheet-like arrangement. The P—P bond distances of 2.23 Å are regarded as a normal single bond P—P distance. (a) What term is used to describe the fact that phosphorus exists in two different solid forms? (b) Black phosphorus is much more stable than white phosphorus. Account for the difference. (c) Draw a Lewis structure representation of a portion of the sheet structure in black phosphorus. What approximate value would you predict for the P—P—P bond angles in the structure?

Additional exercises

8.34 Using VSEPR theory, predict the geometry of each of the following: (a) AsO_3^{3-}; (b) OCN^-; (c) N_3^-; (d) $GaCl_2^+$; (e) HCN.

8.35 What experimental evidence can you cite to support the molecular-orbital formulation of the bonding in O_2?

8.36 The H—P—H bond angle in PH_3 is 93°; in PH_4^+ it is 109.5°. Account for the difference.

[8.37] The Lewis structure for allene is

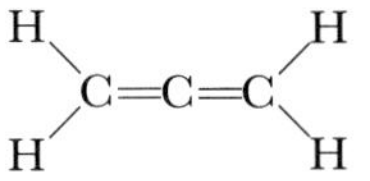

H   H
C=C=C
H   H

Make a sketch of the structure of this molecule that is analogous to Figure 8.11. In addition, answer the following two questions: (a) Is the molecule planar? (b) Does it have a dipole moment?

8.38 Aluminum chloride exists in the gas phase as $AlCl_3$ molecules. It reacts with excess chloride ion in solution to form $AlCl_4^-$ ions. Describe the structure of each of these species.

8.39 Which of the following molecules will have a dipole moment: (a) IF; (b) SCl_2; (c) BCl_3; (d) HCN; (e) HC≡CH; (f) $Cl_2C{=}CH_2$?

[8.40] Dichlorobenzene, $Cl_2C_6H_4$, can have one of the following three structures:

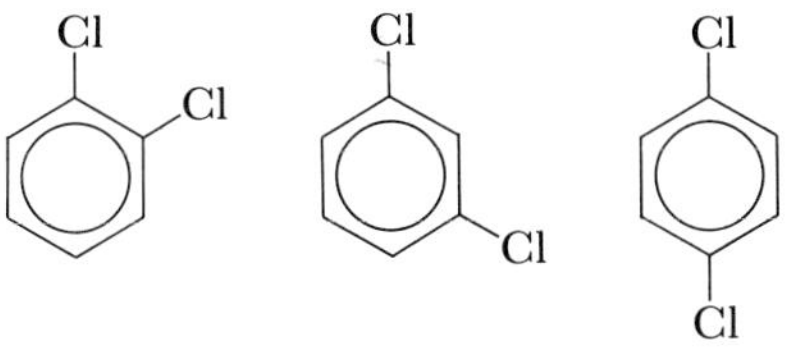

Cl Cl Cl

Cl Cl Cl

Assuming that the value for a C—Cl bond dipole moment is 1.70 D, and that the value for the C—H bond dipole moment is zero, and assuming that there are no direct influences of the bond dipoles on one another, calculate the approximate dipole moments to be expected for the three structures. (Such substances, which have the same molecular formulas but different atomic arrangements, are called *isomers*.) The three isomers are found experimentally to have dipole moments of O, 1.72, and 2.50 D. Assign the dipole moments to the three isomers.

8.41 Indicate the hybrid orbital set employed by the atom shown in boldface type in each of the following structures: (a) **S**Cl_2; (b) F_2**C**$=CH_2$; (c) **Br**F_4^+; (d) **Tl**Cl_3; (e) **Se**O_3^{2-}; (f) **Sn**Cl_6^{2-}.

8.42 Cumene hydroperoxide, for which the structure is

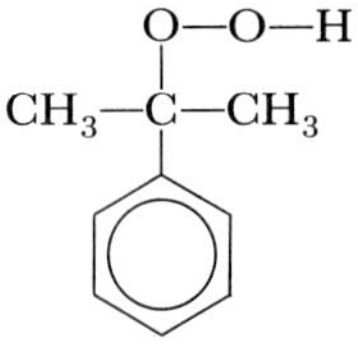

O—O—H
CH_3—C—CH_3

is an intermediate in the formation of phenol, C_6H_5OH, an important industrial chemical. Indicate the hybrid orbital set employed by all the carbon and oxygen atoms of cumene hydroperoxide. Using the data in Table 7.3, estimate the enthalpy change in the oxidation of cumene to the hydroperoxide:

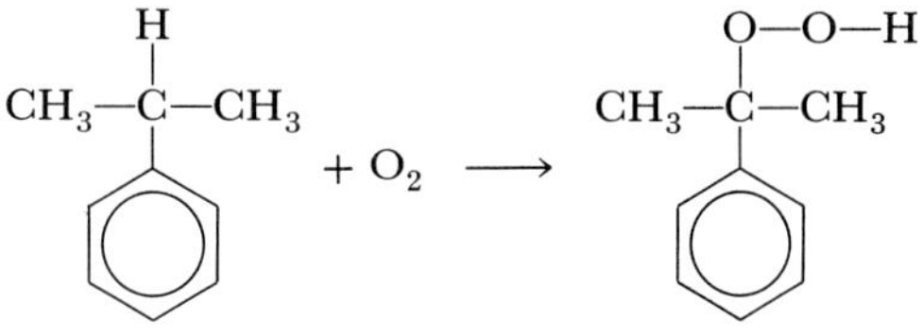

H
CH_3—C—CH_3 + O_2 ⟶ CH_3—C—CH_3 (with O—O—H)

8.43 Indicate the number of orbitals on a particular atom available for π bonding when the following hybrid orbital sets are used to define the molecular framework: (a) sp; (b) sp^2; (c) d^2sp^3; (d) dsp^2.

8.44 What change in the hybridization of orbitals at the central atom occurs in each of the following reactions?

(a) $GaCl_4^- \longrightarrow GaCl_3 + Cl^-$
(b) $PCl_5 \longrightarrow PCl_3 + Cl_2$
(c) $SF_6 \longrightarrow SF_4 + F_2$

8.45 When applying the VSEPR model we count double and triple bonds as a single pair of electrons. Present an argument as to why this is justified.

8.46 Describe the molecular shape and bonding in the unstable molecule diimine, HN=NH: (a) Draw the Lewis structure. (b) Use the VSEPR model to predict the molecular geometry. (c) Indicate the type of hybrid orbitals employed by nitrogen. (d) Sketch the structure using σ and π bond orbitals, and label them. (e) Experimentally, it is found that the dipole moment of the molecule is zero.

Is this consistent with your structure? Is there a form of the molecule, with the same Lewis structure, that could have a nonzero dipole moment?

8.47 Draw the Lewis structure and describe the state of hybridization of the nitrogen atoms in the azide ion, N_3. The N—N bond distances in this ion are both 1.15 A. By comparison with the data for representative N—N bond distances presented in Section 7.4, is this distance consistent with your Lewis structure?

8.48 The structure of indium triiodide is shown in Figure 8.33. It is evident from this structure that the molecular formula is In_2I_6. Draw the Lewis structure for this molecule. What is the hybridization about In? The In_2I_6 molecules dissociate in the gas phase into InI_3 molecules. Draw the Lewis structure for this molecule and indicate the hybridization about In. What characteristic of the electronic structure of iodine in this compound is of importance in forming In_2I_6? What characteristic of In in InI_3 is of importance?

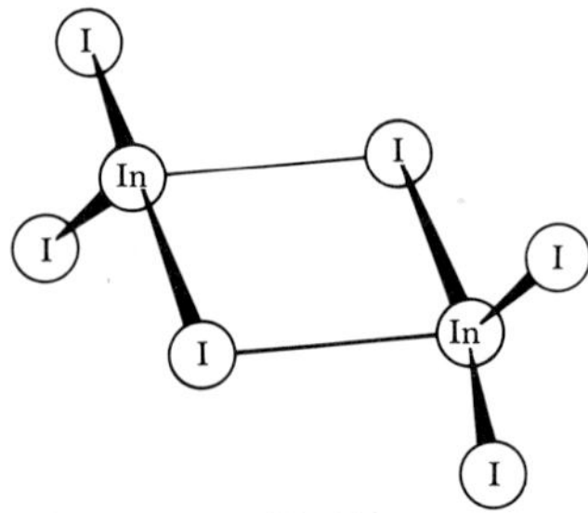

FIGURE 8.33 Structure of In_2I_6.

[8.49] In each of the following pairs of molecules, one is stable and well known, the other is unstable or unknown. Identify the unstable member in each pair and explain why it is the less stable compound: (a) CH_4, CH_5; (b) HI, NI_3; (c) OF_6, SF_6; (d) HC≡CH, HSi≡SiH; (e) PO_3, SO_3.

8.50 Using the molecular orbital diagram for homonuclear diatomic molecules and ions, determine the bond order in each of the following species, and indicate the expected order of bond-dissociation energies: CN; NO; BC; OF^+.

8.51 The ions O_2^-, O_2^{2-}, and O_2^+ occur in several compounds. Compare these three ions with O_2 by listing the four in order of increasing bond length.

8.52 Indicate the expected order of bond-dissociation energies in the series H_2, Ca_2, K_2, Li_2. Explain your reasoning.

8.53 Predict the nitrogen-oxygen bond orders in the series NO^-, NO, NO^+. Predict the order of N—O bond distances in this same series.

8.54 Which of the following molecules should have the lowest ionization energy: NO, N_2, O_2^+? Explain your answer.

[8.55] Telluric acid, $Te(OH)_6$, is formed by reaction of TeO_2 with hydrogen peroxide in aqueous solution. Write a balanced chemical equation for this process. Indicate the oxidation numbers of all species that undergo oxidation or reduction in this reaction. Draw the Lewis structure for telluric acid. What hybridization is employed by Te in this compound? Fusion of $Te(OH)_6$ with NaOH gives Na_6TeO_6. What is the structure of the TeO_6^{6-} ion?

8.56 The nitrogen-nitrogen distances in N_2H_4, N_2F_2, and N_2 are 1.45, 1.25, and 1.10 Å, respectively. Account for this variation in bond distances.

8.57 In addition to the S_8 ring structure shown in Figure 8.27, liquid sulfur also exists in less stable allotropic forms with S_6, S_5, and even S_4 rings. Suggest a reason why these allotropes are less stable than S_8.

[8.58] The element tellurium is readily oxidized to form a Te_4^{2+} species. This ion consists of a square of four Te atoms. Draw a plausible set of Lewis structures (resonance forms are involved) for this ion. Should the square be planar or nonplanar?

8.59 Liquid AsF_3 is a conducting liquid. The conductivity is thought to be due to the presence of ions formed in the following reaction:

$$2AsF_3 \longrightarrow AsF_2^+ + AsF_4^-$$

Predict the geometries of these two ions; indicate the hybridization employed by As in each.

8.60 Indicate the hybridization of the central atoms and the overall geometrical structures of each of the following molecules or ions: (a) SiH_3^-; (b) B_2Cl_4; (c) SOF_2; (d) C_2O_3 (atoms are bonded in the order OCCCO); (e) NSF_3.

8.61 The molecules in each of the following sets have the same molecular formula, but have different shapes: (a) SiF_4, SF_4, XeF_4; (b) BF_3, NF_3, ClF_3. Indicate the shape of each molecule, and explain the origin of the differing shapes.

9 Gases

In the past several chapters we have learned about the electronic structures of atoms and about how atoms come together to form molecules or ionic substances. We commonly observe matter, however, not on the atomic and molecular level, but as a solid, liquid, or gas. In the next few chapters we will be considering some of the important characteristics of these states of matter. We will be interested in learning why substances are found in one or the other state; what forces operate between atoms, ions, or molecules in these states; what transitions may occur between states; and about some of the characteristic properties of matter in each state.

9.1 CHARACTERISTICS OF GASES

Many familiar substances exist at ordinary temperature and pressure as gases. These include several elements (H_2, N_2, O_2, F_2, Cl_2, and the noble gases), and a great variety of compounds. Table 9.1 lists a few of the more common gaseous compounds. Under appropriate conditions, substances that are ordinarily liquids or solids can also exist in the gaseous state,

TABLE 9.1 Some common compounds that are gases

Formula	Name	Characteristics
HCN	Hydrogen cyanide	Very toxic, slight odor of bitter almonds
HCl	Hydrogen chloride	Toxic, corrosive, choking odor
H_2S	Hydrogen sulfide	Very toxic, odor of rotten eggs
CO	Carbon monoxide	Toxic, colorless, odorless
CO_2	Carbon dioxide	Colorless, odorless
CH_4	Methane	Colorless, odorless, flammable
N_2O	Nitrous oxide	Colorless, sweet odor, "laughing gas"
NO_2	Nitrogen dioxide	Red-brown, irritating odor
NH_3	Ammonia	Colorless, pungent odor
SO_2	Sulfur dioxide	Colorless, irritating odor

where they are often referred to as *vapors*. The substance H_2O, for example, is familiar to us as liquid water, ice, or water vapor. Frequently, a substance exists in all three separate states of matter, or phases, at the same time. A Thermos flask containing a mixture of ice and water at 0°C has a certain pressure of water vapor in the gas phase over the liquid and solid phases.

Normally, the three states of matter differ very obviously from one another. Gases differ dramatically from solids and liquids in several respects. A gas expands to fill its container. Consequently, the volume of a gas is given simply by specifying the volume of the container in which it is held. Volumes of solids and liquids, on the other hand, are not determined by the container. The corollary of this is that gases are highly compressible. When pressure is applied to a gas, its volume readily contracts. Liquids and solids, on the other hand, are not very compressible at all. Great pressures must be applied to cause the volume of a liquid or solid to diminish by even as little as 5 percent.

Two or more gases form homogeneous mixtures in all proportions, regardless of how different the gases may be. Liquids, on the other hand, often do not form homogeneous mixtures. For example, when water and gasoline are poured into a bottle, the water vapor and gasoline vapors above the liquids form a homogeneous gas mixture. The two liquids, by contrast, remain largely separate; each dissolves in the other to only a slight extent.

The characteristic properties of gases arise because the individual molecules of a gas are relatively far apart. In a liquid, the individual molecules are close together and take up perhaps 70 percent of the total space. By comparison, in the air we breathe, the molecules take up only about 0.1 percent of the volume. In liquids, the molecules are constantly in contact with neighbors. They experience attractive forces for one another; this is what keeps the liquid together. However, when a pair of molecules come close together, repulsive forces prevent any closer approach. These attractive and repulsive forces differ from one substance to another. The result is that different liquids behave differently. By contrast, the molecules of a gas are well separated and are not much influenced by one another. As we shall see in more detail later, gas molecules are in constant motion, and they frequently collide. On the average, though, they remain fairly far apart. For example, in air, the average distance between molecules is about 10 times as great as the sizes of the molecules themselves. Each molecule thus tends to behave as though the others weren't there. The relative degree of isolation of the molecules causes different gases to behave similarly, even though they are made up of different molecules.

9.2 PRESSURE

Among the most readily measured properties of a gas are its temperature, volume, and pressure. It is not surprising, therefore, that many early studies of gases focused on relationships between these properties. We have already discussed volume and temperature (Section 1.3). Let us now consider the concept of pressure.

In general terms, **pressure** carries with it the idea of a force, something that tends to move something else in a given direction. Pressure is, in

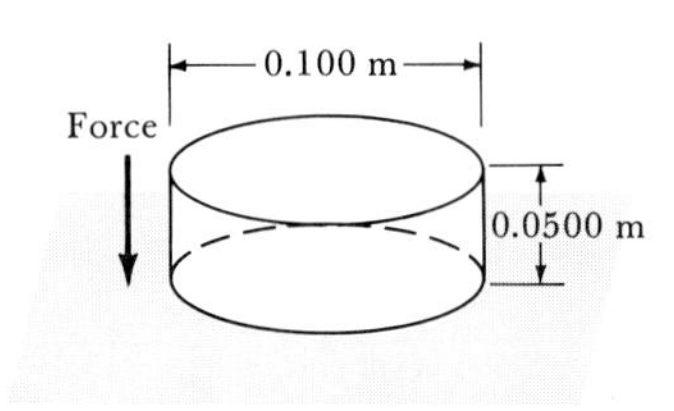

FIGURE 9.1 Aluminum cylinder resting on a flat surface. The diameter of the cylinder is 0.100 m and its height is 0.0500 m. The area of its base is thus $A = \pi r^2 = \pi(0.0500 \text{ m})^2 = 7.85 \times 10^{-3} \text{ m}^2$. Its volume is $V = \pi r^2 h = \pi(0.0500 \text{ m})^2(0.0500 \text{ m}) = 3.93 \times 10^{-4} \text{ m}^3$. The density of aluminum is $2.70 \text{ g/cm}^3 = 2.70 \times 10^3 \text{ kg/m}^3$. The mass of the cylinder is then $M = d \times V = (2.70 \times 10^3 \text{ kg/m}^3) \times (3.93 \times 10^{-4} \text{ m}^3) = 1.06 \text{ kg}$. Calculation of the pressure exerted by this cylinder on the surface upon which it rests is described in the text.

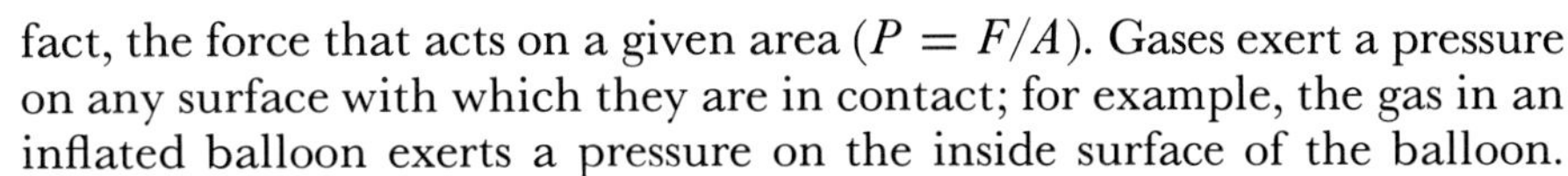

fact, the force that acts on a given area ($P = F/A$). Gases exert a pressure on any surface with which they are in contact; for example, the gas in an inflated balloon exerts a pressure on the inside surface of the balloon.

To understand better the concept of pressure and the units in which it is expressed, consider the aluminum cylinder illustrated in Figure 9.1. Because of the gravitational force, this cylinder exerts a downward force upon the surface on which it rests. According to Newton's second law of motion, the force exerted by an object is the product of its mass, m, times its acceleration, a: $F = ma$. The acceleration due to the gravitational force of earth is 9.81 m/s^2. The mass of the cylinder is 1.06 kg; thus the force with which the earth attracts it is

$$(1.06 \text{ kg})(9.81 \text{ m/s}^2) = 10.4 \text{ kg-m/s}^2 = 10.4 \text{ N}$$

A kg-m/s^2 is the SI unit for force; it is called the newton, abbreviated N: $1 \text{ N} = 1 \text{ kg-m/s}^2$. The cylinder has a cross-sectional area of $7.85 \times 10^{-3} \text{ m}^2$; thus the pressure exerted by the cylinder is

$$P = \frac{F}{A} = \frac{10.4 \text{ N}}{7.85 \times 10^{-3} \text{ m}^2} = 1.32 \times 10^3 \text{ N/m}^2$$

A N/m^2 is the standard unit of pressure in SI units. It is given the name pascal (abbreviated Pa) after Blaise Pascal (1623–1662), a French mathematician and scientist: $1 \text{ Pa} = 1 \text{ N/m}^2$.

Like the aluminum cylinder in our example above, the earth's atmosphere is also attracted toward earth by gravitational attraction. A column of air 1 m^2 in cross section extending through the atmosphere has a mass of roughly 10,000 kg and produces a resultant pressure of about 100 kPa:

$$P = \frac{F}{A} = \frac{(10{,}000 \text{ kg})(9.81 \text{ m/s}^2)}{1 \text{ m}^2} = 1 \times 10^5 \text{ Pa} = 1 \times 10^2 \text{ kPa}$$

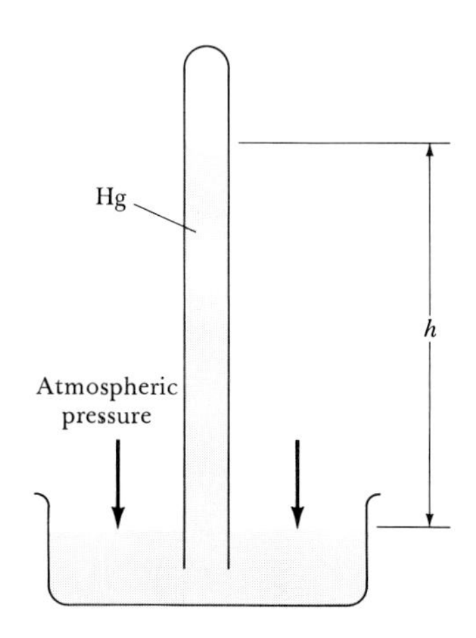

FIGURE 9.2 Mercury barometer. The space in the tube above mercury is nearly a vacuum; a negligible amount of mercury vapor occupies this space.

Of course, the actual atmospheric pressure at any location depends on altitude and weather conditions.

Atmospheric pressure can be measured by use of a mercury barometer such as that illustrated in Figure 9.2. Such a barometer is formed by filling a glass tube more than 76 cm long, which is closed on one end, with mercury and inverting it in a dish of mercury. Care must be taken that no air gets into the tube. When the tube is inverted in this manner, some of the mercury runs out, but a column remains.

The mercury surface outside the tube experiences the full force of the earth's atmosphere over each unit area. However, the atmosphere is not in contact with the mercury surface within the tube. The atmosphere pushes the mercury up the tube until the pressure due to the weight of the mercury column balances the atmospheric pressure. Thus the height of the mercury column fluctuates as the atmospheric pressure fluctuates. The standard atmospheric pressure, which corresponds to the typical pressure at sea level, is defined as the pressure sufficient to support a column of mercury 760 mm in height. This pressure, which corresponds to 1.01325×10^5 Pa, is used to define another unit in common use, the atmosphere (abbreviated atm):

$$1 \text{ atm} = 760 \text{ mm Hg} = 1.01325 \times 10^5 \text{ Pa} = 101.325 \text{ kPa}$$

One mm Hg pressure is also referred to as a torr, after the Italian scientist Evangelista Torricelli (1608–1647), who invented the barometer: 1 mm Hg = 1 torr.

In this text we will ordinarily express gas pressure in units of atm or mm Hg. However, you should be able to convert gas pressures from one set of units to another.

SAMPLE EXERCISE 9.1

(a) Convert 0.605 atm to millimeters of mercury (mm Hg). (b) Convert 3.5×10^{-4} mm Hg to atmospheres.

Solution: (a) Because 1 atm = 760 mm Hg, conversion of atm to mm Hg is made by multiplying the number of atm by the factor 760 mm Hg/1 atm:

$$(0.605 \text{ atm})\left(\frac{760 \text{ mm Hg}}{1 \text{ atm}}\right) = 460 \text{ mm Hg}$$

Notice that the units cancel in the required manner.

(b) To convert from mm Hg to atm, we must multiply by the conversion factor 1 atm/760 mm Hg:

$$(3.5 \times 10^{-4} \text{ mm Hg})\left(\frac{1 \text{ atm}}{760 \text{ mm Hg}}\right) = 4.6 \times 10^{-7} \text{ atm}$$

Convert a pressure of 735 mm Hg to kPa.

Solution: From the material discussed above we know that 1 atm = 101.3 kPa = 760 mm Hg. Thus the conversion factor we want is of the form 101.3 kPa/760 mm Hg. We use this conversion factor to convert the pressure given:

$$(735 \text{ mm Hg})\left(\frac{101.3 \text{ kPa}}{760 \text{ mm Hg}}\right) = 98.0 \text{ kPa}$$

In countries that use the metric system, for example, Canada, atmospheric pressure is expressed in weather reports in units of kPa.

A device called a *manometer,* whose principle of operation is similar to that of a barometer, can be used to measure the pressures of enclosed gases. Figure 9.3(*a*) shows a closed-tube manometer, a device normally used to measure pressures below atmospheric pressure. The pressure is just the difference in the heights of the mercury levels in the two arms.

An open-tube manometer such as that pictured in Figure 9.3(*b*) and (*c*) is often employed to measure gas pressures that are near atmospheric pressure. The difference in the heights of the mercury levels in the two arms of the manometer relate the gas pressure to atmospheric pressure. If the pressure of the enclosed gas is the same as atmospheric pressure, the levels in the two arms are equal. If the pressure of the enclosed gas is greater than atmospheric pressure, mercury is forced higher in the arm exposed to the atmosphere, as in Figure 9.3(*b*). If atmospheric pressure exceeds the gas pressure, the mercury is higher in the arm exposed to the gas as in Figure 9.3(*c*).

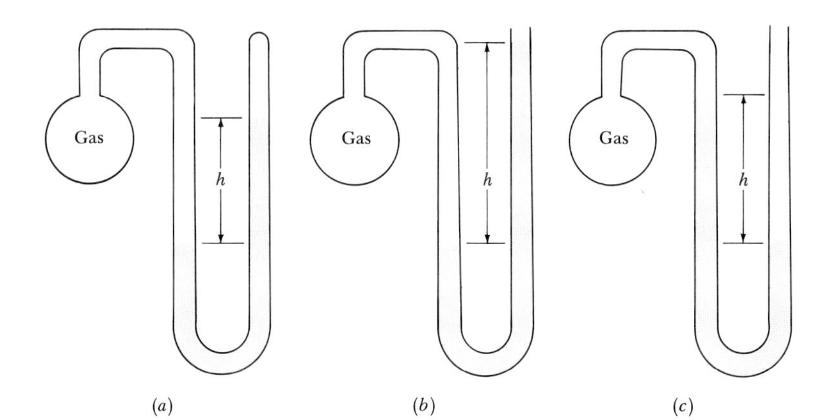

FIGURE 9.3 Closed-end manometer (*a*) and open-end manometers (*b*) and (*c*). In (*b*) gas pressure exceeds atmospheric pressure; in (*c*) gas pressure is less than atmospheric pressure.

SAMPLE EXERCISE 9.3

Consider a container of gas with an attached open-tube manometer. The manometer is not filled with mercury but rather with another nonvolatile liquid, dibutylphthalate. The density of mercury is 13.6 g/mL; that of dibutylphthalate is 1.05 g/mL. If the conditions are as shown in Figure 9.3(*b*) with $h = 12.2$ cm when atmospheric pressure is 0.964 atm, what is the pressure of the enclosed gas in mm Hg?

Solution: Converting atmospheric pressure to mm Hg, we have

$$(0.964 \text{ atm})\left(\frac{760 \text{ mm Hg}}{1 \text{ atm}}\right) = 733 \text{ mm Hg}$$

The pressure associated with a 12.2-cm column of dibutylphthalate is equivalent to a mercury column of

$$(12.2 \text{ cm})\left(\frac{1.05 \text{ g/mL}}{13.6 \text{ g/mL}}\right) = 0.94 \text{ cm} = 9.4 \text{ mm}$$

If the situation is like that in Figure 9.3(*b*), the pressure of the enclosed gas exceeds atmospheric pressure by this amount:

$$P = 733 \text{ mm Hg} + 9 \text{ mm Hg} = 742 \text{ mm Hg}$$

9.3 THE GAS LAWS

Experiments with a large number of gases reveal that the four variables temperature (T), pressure (P), volume (V), and quantity of gas in moles (n) are sufficient to define the state or condition of many gaseous substances. The first relationship between these variables was found in 1662 by Robert Boyle (1627–1691). Boyle's law states that *the volume of a fixed quantity of gas maintained at constant temperature is inversely proportional to the gas pressure.* That is, as pressure increases, volume decreases, as shown in Figure 9.4; doubling the pressure results in the gas volume decreasing to

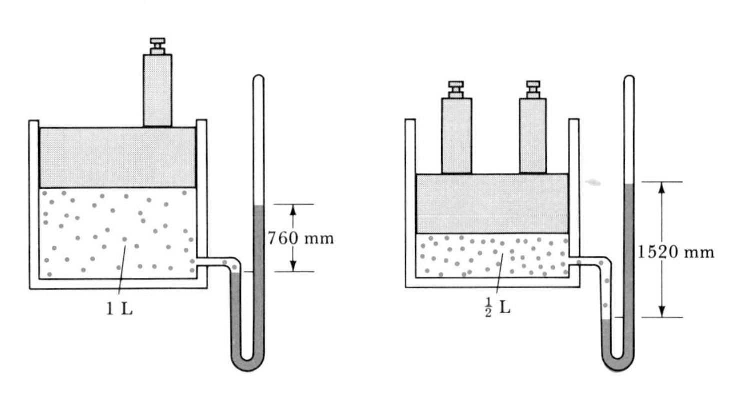

FIGURE 9.4 Gas pressure versus volume. An illustration of Boyle's law.

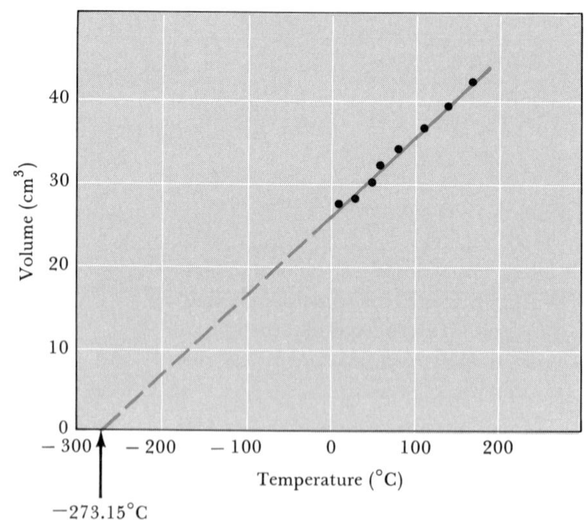

FIGURE 9.5 Volume of an enclosed gas as a function of temperature at constant pressure.

one-half its original value. This relationship can be expressed as $V \propto 1/P$ (where $\propto$ is read "proportional to"), or as $V = c/P$ (where c, the proportionality constant, depends on temperature and the amount of gas). Boyle's law expresses the important fact that a gas is compressible.

The relationship between gas volume and temperature was discovered in 1787 by Jacques Charles (1746–1823), a French scientist. Charles found that the volume of a fixed quantity of gas at constant pressure increases in a linear fashion with temperature. Some typical data are shown in Figure 9.5. Notice that the extrapolated (extended) line through the data points passes through −273°C. Note also that the gas is predicted to have zero volume at this temperature. Of course, this condition is never fulfilled, because all gases liquefy or solidify before reaching −273°C. In 1848, William Thomson (1824–1907), a British physicist whose title was Lord Kelvin, proposed the idea of an absolute temperature scale, now known as the Kelvin scale, with −273°C = 0 K. In terms of this scale, **Charles's law** can be stated as follows: *The volume of a fixed amount of gas maintained at constant pressure is directly proportional to absolute temperature.* This relationship can be expressed as $V \propto T$ or as $V = cT$, where the proportionality constant, c, depends on the pressure and amount of gas. Thus doubling absolute temperature, say from 200 K to 400 K, causes the gas volume to double.

The relationship between gas volume and the quantity of gas follows from the work of the French scientist Joseph Louis Gay-Lussac (1778–1850) and the Italian scientist Amedeo Avogadro (1776–1856). Gay-Lussac is one of those extraordinary figures in the early history of modern science who can truly be called an adventurer. He was interested in lighter-than-air balloons and in 1804 made an ascent to 23,000 ft. This exploit set the altitude record for several decades, but Gay-Lussac had other reasons for making the flight: He tested the variation of the earth's magnetic field and sampled the composition of the atmosphere as a function of elevation.

To control lighter-than-air balloons properly, Gay-Lussac needed to know more about the properties of gases. He therefore was led to carry

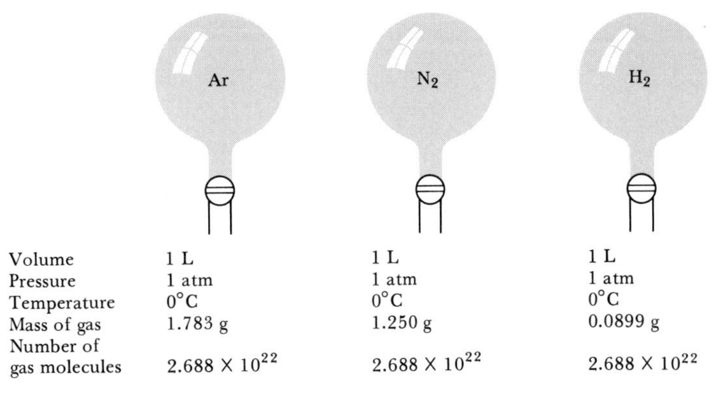

FIGURE 9.6 Comparison illustrating Avogadro's hypothesis. Note that argon gas consists of argon atoms; we can regard these as one-atom molecules. Each gas has the same volume, temperature, and pressure and thus contains the same number of molecules. Because a molecule of one substance differs in mass from a molecule of another, the mass of gas in the three containers differ.

out several experiments. The most important of these led to his discovery in 1808 of the law of combining volumes (Section 3.4). Recall that this law states that the volumes of gases that react with one another at the same pressure and temperature are in the ratios of small whole numbers.

It was most especially Gay-Lussac's work that led Avogadro in 1811 to propose what is known as Avogadro's hypothesis: *Equal volumes of gases at the same temperature and pressure contain equal numbers of molecules.* We saw in Chapter 3 the importance of both Gay-Lussac's and Avogadro's work in setting the stage for a correct appreciation of atomic weights. Let us now consider how their results can help us understand the nature of the gaseous state. Suppose that we have three 1-L bulbs containing H_2, N_2, and Ar, respectively (Figure 9.6), and that each gas is at the same pressure and temperature. According to Avogadro's hypothesis, those bulbs contain *equal numbers of gaseous particles,* although the masses of the substances in the bulbs differ greatly.

Avogadro's law follows from Avogadro's hypothesis: *The volume of a gas maintained at constant temperature and pressure is directly proportional to the quantity of gas:* $V \propto n$ or $V = cn$. Thus doubling the number of moles of gas will cause the volume to double if T and P remain constant.

9.4 THE IDEAL-GAS EQUATION

In the preceding section, we examined three historically important gas laws:

Boyle's law:	$V \propto \frac{1}{P}$	(constant n, T)
Charles's law:	$V \propto T$	(constant n, P)
Avogadro's law:	$V \propto n$	(constant P, T)

These three relationships are special cases of a more general gas law:

$$V \propto \frac{nT}{P}$$

If we call the proportionality constant R, we have

$$V = R\left(\frac{nT}{P}\right)$$

TABLE 9.2 Numerical values of the gas constant, R, in various units

Units	Numerical value
Liter-atm/K-mol	0.08206
Calories/K-mol	1.987
Joules/K-mol[a]	8.314

[a] SI units.

Rearranging, we have this relationship in its more familiar form:

$$PV = nRT \qquad [9.1]$$

This equation is known as the **ideal-gas equation.** The term R is called the **gas constant.** The numerical value of R depends on the units chosen for the four variables in the equation. Temperature, T, must be expressed in an absolute temperature scale, normally the Kelvin scale. The quantity of gas, n, is normally expressed in moles. The units chosen for pressure, P, and volume, V, are most often atm and liters, respectively. However, other units could be used. Table 9.2 shows the numerical values for R in a few of the more important units. The last two values given, which include calories and joules among the units, arise because the product $P \times V$ has energy units (Section 4.2).

SAMPLE EXERCISE 9.4

Using the ideal-gas equation, calculate the volume of exactly 1 mol of gas at 0°C (273.15 K) and exactly 1 atm pressure.

Solution: Rearranging Equation 9.1 to solve for V gives

$$V = \frac{nRT}{P}$$

Inserting the numerical values for each term gives

$$V = \frac{(1 \text{ mol})(0.08206 \text{ L-atm/mol-K})(273.15 \text{ K})}{1 \text{ atm}}$$

$$= 22.41 \text{ L}$$

The conditions 0°C and 1 atm are referred to as the **standard temperature and pressure** (abbreviated **STP**). Many properties of gases are tabulated for these conditions. The volume calculated in Sample Exercise 9.4, 22.41 L, is known as the *molar volume* of an ideal gas at STP.

SAMPLE EXERCISE 9.5

A flashbulb of volume 2.6 cm^3 contains O_2 gas at a pressure of 2.3 atm and a temperature of 26°C. How many moles of O_2 does the flashbulb contain?

Solution: Because we know volume, temperature, and pressure, the only unknown quantity in the ideal-gas equation (Equation 9.1) is the number of moles, n. Solving Equation 9.1 for n gives

$$n = \frac{PV}{RT}$$

The values of all quantities involved are tabulated and changed to units consistent with those for R:

$$P = 2.3 \text{ atm}$$
$$V = 2.6 \text{ cm}^3 = 2.6 \times 10^{-3} \text{ L}$$
$$n = ?$$
$$R = 0.0821 \text{ L-atm/mol-K}$$
$$T = 26°\text{C} = 299 \text{ K}$$

Thus we have

$$n = \frac{(2.3 \text{ atm})(2.6 \times 10^{-3} \text{ L})}{(0.0821 \text{ L-atm/mol-K})(299 \text{ K})}$$
$$= 2.4 \times 10^{-4} \text{ mol}$$

The fact that Equation 9.1 is called the *ideal*-gas equation correctly suggests that there may be conditions where gases don't exactly obey this equation. For example, we might calculate the quantity of a gas, n, for given conditions of P, V, and T and find it to differ somewhat from the measured quantity under these conditions. Ordinarily, however, the difference between ideal and real behavior is so small that it can be ignored. We will examine deviations from ideal behavior later, in Section 9.9.

Relationship Between the Ideal-Gas Equation and the Gas Laws

The simple gas laws, such as Boyle's law, which were discussed in Section 9.3, are special cases of the ideal-gas equation. For example, when the temperature and quantity of gas are held constant, n and T have fixed values. Thus the product nRT is the product of three constants and must itself be a constant:

$$PV = nRT = \text{constant} = c \qquad \text{or} \qquad PV = c \tag{9.2}$$

Thus we have Boyle's law. Equation 9.2 expresses the fact that even though the individual values of P and V can change, their product PV remains constant.

Similarly, when n and P are constant, the ideal-gas equation gives the following relationship:

$$V = \left(\frac{nR}{P}\right)T = \text{constant} \times T \qquad \text{or} \qquad \frac{V}{T} = c \tag{9.3}$$

As we have noted earlier, this relationship was first observed by Charles and is known as Charles's law.

SAMPLE EXERCISE 9.6

The pressure of nitrogen gas in a 12.0-L tank at 27°C is 2300 lb/in.^{2}. What volume would the gas in this tank have at 1 atm pressure (14.7 lb/in.^{2}) if the temperature remains unchanged?

Solution: Let us begin by making up a table that lists the initial and final values for the pressure, temperature, and volume of the gas. (Always convert temperature to the Kelvin scale.)

	Temperature (K)	Pressure (lb/in.^{2})	Volume (L)
Initial	300	2300	12.0
Final	300	14.7	?

If the quantity of gas (n) and the temperature (T) do not change, the product PV must remain constant (see Equation 9.2). Thus, if we have two different sets of conditions for the same quantity of gas at constant temperature, we can write

$$P_1V_1 = P_2V_2$$

In our example, P_1 is 2300 lb/in.^{2}, P_2 is 14.7 lb/in.^{2}, V_1 is 12 L, and V_2 is unknown. Inserting all the known quantities and solving for V_2, we obtain

$$V_2 = \frac{P_1V_1}{P_2} = \frac{(2300 \text{ lb/in.}^2)(12.0 \text{ L})}{14.7 \text{ lb/in.}^2} = 1880 \text{ L}$$

We can also approach this problem from a slightly different viewpoint. We recognize that if the pressure decreases, volume must increase. Assuming that there is an inverse proportionality between pressure and volume, we look for a pressure factor that will increase the volume when the pressure decrease occurs. This pressure factor should be the ratio of the higher to the lower pressure:

$$\begin{aligned} V &= (12.0 \text{ L})(\text{pressure factor}) \\ &= (12.0 \text{ L})\left(\frac{2300 \text{ lb/in.}^2}{14.7 \text{ lb/in.}^2}\right) \\ &= 1880 \text{ L} \end{aligned}$$

SAMPLE EXERCISE 9.7

A large natural-gas storage tank is arranged so that the pressure is maintained at 2.2 atm. On a cold day in December, when the temperature is $-15°C$ (4°F), the volume of gas in the tank is 28,500 ft^3. What is the volume of the same quantity of gas on a warm July day when the temperature is 31°C (88°F)?

Solution: After converting temperatures to degrees Kelvin, we have the following values for initial and final conditions:

	Temperature (K)	Pressure (atm)	Volume (ft^3)
Initial	258	2.2	28,500
Final	304	2.2	?

Because pressure and quantity of gas are constant, the quotient V/T must be constant (Equation 9.3). It follows that

$$\frac{V_1}{T_1} = \frac{V_2}{T_2}$$

Inserting known quantities and solving for V_2 gives

$$V_2 = V_1 \times \frac{T_2}{T_1} = (28{,}500\ ft^3)\left(\frac{304\ K}{258\ K}\right) = 33{,}600\ ft^3$$

We could also take an intuitive approach to this problem. When a gas is heated at constant pressure, it expands. Thus we need to multiply the gas volume at the lower temperature by a factor greater than 1. That factor is the ratio of temperatures, as shown above.

We know that when a confined gas is heated at constant volume, the pressure increases. For example, a popcorn kernel bursts open under the pressure of steam that forms within the kernel when it is heated in oil. We could make quantitative measurements of the change in pressure of a confined gas by placing the gas in a steel container fitted with a pressure gauge, and then varying the temperature. We would find that the pressure increases linearly with absolute temperature, perhaps as shown by the sample data labeled A in Figure 9.7. If the experiment were repeated with a different-sized sample of the same gas, we might obtain the results labeled B in the figure. Note that in both cases the extrapolated pressure at 0 K is zero.

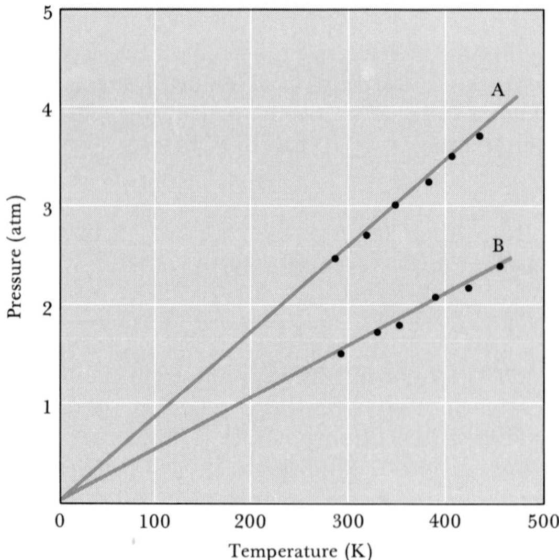

FIGURE 9.7 Variation of gas pressure with temperature under constant-volume conditions.

If both n and V in Equation 9.1 are fixed, the pressure varies with temperature as expressed in Equation 9.4:

$$P = \left(\frac{nR}{V}\right)T = \text{constant} \times T \qquad [9.4]$$

Thus the ideal-gas equation predicts a linear relationship between pressure and absolute temperature, extrapolating to zero pressure at 0 K. Again, we must remind ourselves that real gases lose their gaseous properties before absolute zero is reached.

SAMPLE EXERCISE 9.8

Why do the two samples of gas for which data are shown in Figure 9.7 show two different linear relationships?

Solution: Inspection of Equation 9.4 shows us that the linear relationship between P and T passes through the origin, and has slope nR/V. (You may wish to review linear equations, Appendix A.4.) In our example, both R and V are the same for the two samples, but the number of moles of gas, n, is different. The slope of the pressure versus temperature relationship for a given volume is proportional to the amount of gas that is confined. There is more gas in sample A.

SAMPLE EXERCISE 9.9

The gas pressure in an aerosol can is 1.5 atm at 25°C. Assuming that the gas inside obeys the ideal-gas equation, what would the pressure be if the can were heated to 450°C?

Solution: Let us proceed, as in Sample Exercise 9.7, by writing down the initial and final conditions of temperature, pressure, and volume that the problem gives us. (Remember that we must convert temperatures to degrees kelvin).

	Volume	Pressure (atm)	Temperature (K)
Initial	V_1	1.5	298
Final	V_1	P_2	723

Because the pressure increases in proportion to the absolute temperature, the pressure at the higher temperature, P_2, will equal P_1 times a ratio of absolute temperatures that is greater than 1:

$$P_2 = (1.5 \text{ atm})\left(\frac{723 \text{ K}}{298 \text{ K}}\right) = 3.6 \text{ atm}$$

It is evident from this example why aerosol cans carry the warning not to incinerate.

9.5 DALTON'S LAW OF PARTIAL PRESSURES

The pressure of a gas under conditions of constant volume and temperature is directly proportional to the number of moles of gas:

$$P = \left(\frac{RT}{V}\right)n = \text{constant} \times n \qquad [9.5]$$

Suppose that the gas with which we are concerned is not a single kind of gas particle but is rather a mixture of two or more different substances. We might expect that the total pressure exerted by the gas mixture is the sum of pressures due to the individual components. Each of the individual components, if present alone under the same temperature and volume conditions as the mixture, would exert a pressure that we term the **partial pressure.** John Dalton was the first to observe that the *total pressure of a mixture of gases is just the sum of the pressures that each gas would exert if it were present alone:*

$$P_t = P_1 + P_2 + P_3 + \cdots \qquad [9.6]$$

Each of the gases obeys the ideal-gas equation. Thus we can write

$$P_1 = n_1\left(\frac{RT}{V}\right), \quad P_2 = n_2\left(\frac{RT}{V}\right), \quad P_3 = n_3\left(\frac{RT}{V}\right), \quad \text{etc.}$$

All of the gases experience the same temperature and volume. Therefore, by substituting into Equation 9.6, we obtain

$$P_t = \frac{RT}{V}(n_1 + n_2 + n_3 + \cdots) \qquad [9.7]$$

That is, the total pressure at constant temperature and volume is determined by the total number of moles of gas present, whether that total represents just one substance or a mixture.

SAMPLE EXERCISE 9.10

If a 0.20-L sample of O_2 at 0°C and 1.0 atm pressure and a 0.10-L sample of N_2 at 0°C and 2.0 atm pressure are both placed in a 0.40-L container at 0°C, what is the total pressure in the container?

Solution: We can solve this problem by calculating the pressure that each gas would exert if it were alone in the 0.40-L container. Because the quantity of each gas and the temperature are constant, the only quantities that vary are pressure and volume. Using the same type of approach employed in Sample Exercise 9.6, we have

$$P_1V_1 = P_2V_2$$

$$P_2 = P_1\left(\frac{V_1}{V_2}\right)$$

$$P_{O_2} = (1.0\text{ atm})\left(\frac{0.20\text{ L}}{0.40\text{ L}}\right) = 0.5\text{ atm}$$

$$P_{N_2} = (2.0\text{ atm})\left(\frac{0.10\text{ L}}{0.40\text{ L}}\right) = 0.5\text{ atm}$$

According to Dalton's law, the total pressure is the sum of each of the partial pressures exerted by the gases individually:

$$P_t = P_{O_2} + P_{N_2} = 0.5\text{ atm} + 0.5\text{ atm} = 1.0\text{ atm}$$

SAMPLE EXERCISE 9.11

What pressure, in atm, is exerted by a mixture of 2.00 g of H_2 and 8.00 g of N_2 at 273 K in a 10.0-L vessel?

Solution: The pressure depends on the total moles of gas (Equation 9.7). Calculating the moles of H_2 and N_2, we have

$$n_{H_2} = (2.00\text{ g }H_2)\left(\frac{1\text{ mol }H_2}{2.02\text{ g }H_2}\right) = 0.990\text{ mol }H_2$$

$$n_{N_2} = (8.00\text{ g }N_2)\left(\frac{1\text{ mol }N_2}{28.0\text{ g }N_2}\right) = 0.256\text{ mol }N_2$$

Using Equation 9.7 gives us

$$P_t = \frac{RT}{V}(n_{H_2} + n_{N_2})$$

$$= \frac{(0.0821\text{ L-atm/mol-K})(273\text{ K})}{10.0\text{ L}} \times (0.990\text{ mol} + 0.256\text{ mol})$$

$$= 2.79\text{ atm}$$

This total pressure is the sum of the partial pressures of H_2 and N_2.

9.6 TYPICAL PROBLEMS INVOLVING GASES

We have two reasons for being interested in gases. In the first place, the properties of the gaseous state are important for our understanding of the nature of matter on the atomic and molecular level. We will deal with this aspect of gases in later sections. A second, very practical reason for studying gases is that they may be reactants and products in chemical reactions. Just as we must learn to deal with pure solid or liquid reagents, or with solutions of reagents, so we must also learn to express quantities of gases under various conditions in terms of volume and pressure. In this section we will review several typical problems involving gases that might be encountered in the chemical laboratory or related situations.

It is possible to vary P, V, and T for a given quantity of gas. From the ideal-gas equation we have

$$\frac{PV}{T} = nR = \text{constant}$$

Thus as long as the total quantity of gas, n, is constant, PV/T is a constant. If we represent the initial and final conditions of pressure, temperature, and volume by subscripts 1 and 2, respectively, we can write the following expression:

$$\frac{P_1V_1}{T_1} = \frac{P_2V_2}{T_2} \quad [9.8]$$

SAMPLE EXERCISE 9.12

A quantity of helium gas occupies a volume of 16.5 L at 78°C and 45.6 atm. What is its volume at STP?

Solution: It is best to begin problems of this sort by writing down all we know of the initial and final values of temperature, pressure, and volume. (Remember that you must always convert temperatures to the absolute temperature scale, K.)

	Pressure (atm)	Volume (L)	Temperature (K)
Initial	45.6	16.5	351
Final	1 (exactly)	V_2	273

Putting the quantities into Equation 9.8, we have

$$\frac{(45.6\ \text{atm})(16.5\ \text{L})}{351\ \text{K}} = \frac{(1\ \text{atm})(V_2)}{273\ \text{K}}$$

$$V_2 = \left(\frac{45.6\ \text{atm}}{1\ \text{atm}}\right)\left(\frac{273\ \text{K}}{351\ \text{K}}\right)(16.5\ \text{L}) = 585\ \text{L}$$

We can also approach this problem in a more intuitive fashion. The new volume will be a product of the old volume times a ratio of pressures times a ratio of temperatures. The decrease in pressure from 45.6 atm to 1 atm will cause an increase in volume. Thus the pressure ratio is greater than 1: 45.6 atm/1 atm. The temperature of the gas has decreased; this change leads to a smaller volume. Thus the temperature ratio is less than 1: 273 K/351 K. The resultant equation is the same one that is used above:

$$V_2 = (16.5\ \text{L})\left(\frac{45.6\ \text{atm}}{1\ \text{atm}}\right)\left(\frac{273\ \text{K}}{351\ \text{K}}\right) = 585\ \text{L}$$

Note that the previous problem (as well as Sample Exercises 9.6 to 9.9) involved the effects of changing conditions upon a variable such as volume. In such problems R is not needed; it cancels from the calculations. In the remaining examples we will need to employ R. For example, we can use the ideal-gas equation to calculate the magnitude of one of the variables, given values for the other three. This sort of calculation we performed earlier, in Sample Exercises 9.4 and 9.5.

Molecular Weights and Gas Densities

Some of the most useful calculations using the ideal-gas equation involve measurements and calculation of the gas density. Density has the units of mass per unit volume. We can arrange the gas equation to obtain

$$\frac{n}{V} = \frac{P}{RT}$$

Now n/V has the units of moles per liter. Suppose that we multiply both sides of this equation by molecular weight ($\mathscr{M}$), which is the number of grams in 1 mol of a substance:

$$\frac{n(\mathscr{M})}{V} = \frac{P\mathscr{M}}{RT} \qquad [9.9]$$

But the product of the quantities n/V and $\mathscr{M}$ equals density, because the units multiply as follows:

$$\frac{\text{Moles}}{\text{Liter}} \times \frac{\text{grams}}{\text{mole}} = \frac{\text{grams}}{\text{liter}}$$

Thus the density of the gas is given by the expression on the right in Equation 9.9:

$$d = \frac{P\mathscr{M}}{RT} \qquad [9.10]$$

or, rearranging,

$$\mathscr{M} = \frac{dRT}{P} \qquad [9.11]$$

Sample Exercises 9.13 and 9.14 illustrate the uses of these relations.

SAMPLE EXERCISE 9.13

What is the density of carbon dioxide gas at 745 mm Hg and 65°C?

Solution: The molecular weight of carbon dioxide, CO_2, is $12.0 + (2)(16.0) = 44.0$ g/mol. If we are to use 0.0821 L-atm/K-mol for R, we must convert pressure to atmospheres. We have then

$$d = \frac{\left(\frac{745}{760}\text{ atm}\right)(44.0\text{ g/mol})}{(0.0821\text{ L-atm/K-mol})(338\text{ K})}$$
$$= 1.56\text{ g/L}$$

The problem can be turned around a bit to determine the molecular weight of a gas from its density as shown in Sample Exercise 9.14.

SAMPLE EXERCISE 9.14

A large flask fitted with a stopcock is evacuated and weighed; its mass is found to be 134.567 g. It is then filled to a pressure of 735 mm Hg at 31 °C with a gas of unknown molecular weight, and then reweighed; its mass is 137.456 g. The flask is then filled with water and again weighed; its mass is now 1067.9 g. Assuming that the ideal-gas equation applies, what is the molecular weight of the unknown gas? (The density of water at 31 °C is 0.997 g/cm^3.)

Solution: First we must determine the volume of the flask. This is given by the difference in weights of the empty flask and the flask filled with water, divided by the density of water at 31 °C, 0.997 g/cm^3:

$$V = \frac{1067.9\text{ g} - 134.6\text{ g}}{0.997\text{ g/cm}^3} = 936\text{ cm}^3$$

The next task is to calculate the number of moles of gas that will occupy this volume at the temperature and pressure indicated.

$$n = \frac{PV}{RT}$$

We must be careful to insert all quantities in the appropriate units. When this is done we obtain

$$n = \frac{\left(\frac{735}{760}\text{ atm}\right)(0.936\text{ L})}{(0.0821\text{ L-atm/K-mol})(304\text{ K})} = 0.0363\text{ mol}$$

We now have the number of moles of gas. We know also that this number of moles weighs 137.456 g − 134.567 g = 2.889 g.

This molecular weight is then simply the mass divided by the number of moles which that mass represents:

$$\frac{2.889\text{ g}}{0.0363\text{ mol}} = 79.6\text{ g/mol}$$

This problem can also be solved by more direct use of Equation 9.11. Because the mass of the gas is 137.456 g − 134.567 g = 2.889 g, its density is 2.889 g/0.936 L = 3.087 g/L. Using Equation 9.11, we have

$$\mathcal{M} = \frac{dRT}{P}$$

$$= \frac{(3.087\text{ g/L})(0.0821\text{ L-atm/mol-K})(304\text{ K})}{(735/760)\text{ atm}}$$

$$= 79.7\text{ g/mol}$$

(The difference between the 79.6 reported in the first method of solution and the 79.7 g/mol obtained this time is due to the round-off procedures used in the intermediate calculations in the first method.)

Quantities of Gases Involved in Chemical Reactions

An experiment that often comes up in the course of laboratory work is the determination of the number of moles of gas collected from a chemical reaction. Sometimes this gas is collected over water. For example, solid potassium chlorate ($KClO_3$) may be decomposed by heating it in a test tube with an arrangement as shown in Figure 9.8. The balanced equation for the reaction is

$$2KClO_3(s) \longrightarrow 2KCl(s) + 3O_2(g)$$

The oxygen gas is collected in a bottle that is initially filled with water and inverted in a water pan.

The volume of gas collected is measured by raising or lowering the bottle as necessary until the water levels inside and outside the bottle are the same. When this condition is met, the pressure inside the bottle is

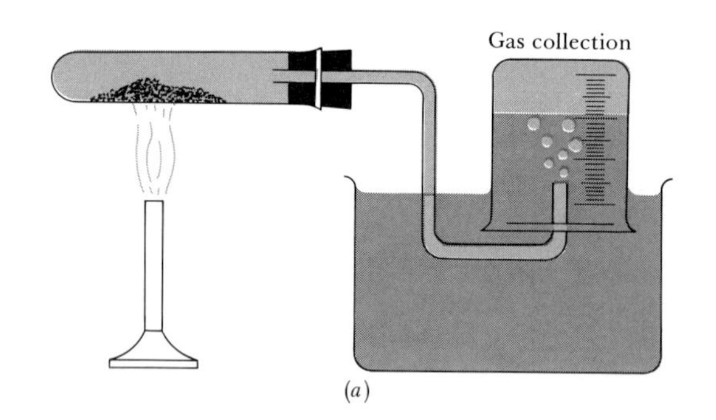

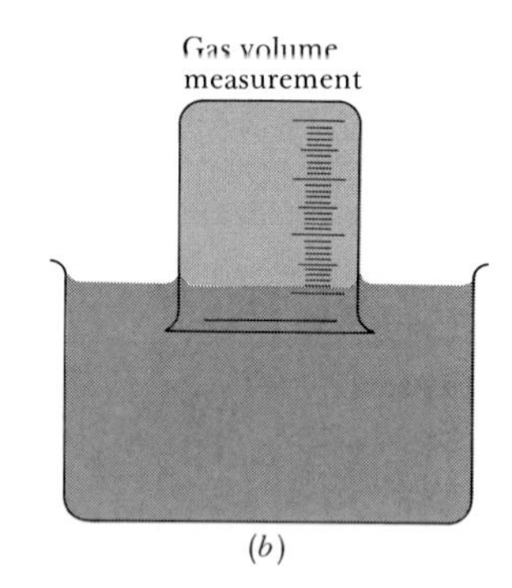

FIGURE 9.8 (*a*) Collection of a gas over water. (*b*) When the gas has been collected, the bottle is raised or lowered to equalize pressures inside and outside before measuring the volume of the gas collected.

equal to the atmospheric pressure outside. But the total pressure inside is the sum of the pressure of gas collected and the water vapor in equilibrium with liquid water:

$$P_{total} = P_{gas} + P_{H_2O} \qquad [9.12]$$

The pressure exerted by water vapor, P_{H_2O}, at various temperatures is shown in Appendix C.

SAMPLE EXERCISE 9.15

Suppose that 0.200 L of oxygen gas is collected over water as in Figure 9.8 at a temperature of 26°C and a pressure of 750 mm Hg. What volume would the O_2 gas collected occupy when dry, at the same temperature and pressure?

Solution: The pressure of O_2 gas in the vessel is the difference between the total pressure, 750 mm Hg, and the vapor pressure of water at 26°C, 25 mm (Appendix C):

$$P_{O_2} = 750 - 25 = 725 \text{ mm Hg}$$

If the water were dried from the gas sample, while at the same time maintaining the same total pressure, the O_2 pressure after drying would be 750 mm Hg. The corrected O_2 volume would thus be less for the dry gas, because its partial pressure is greater than for the wet gas:

$$V_{O_2} = (0.200 \text{ L})\left(\frac{725 \text{ mm Hg}}{750 \text{ mm Hg}}\right) = 0.193 \text{ L}$$

The collection of a gas over water is often done in an experiment to determine the number of moles of gaseous product. To compute the number of moles of gas collected, a correction must be applied for the partial pressure of water vapor in the collection bottle. We shall go through a complete analysis of such an experiment in Sample Exercise 9.16 to show how the stoichiometric relationships come together.

SAMPLE EXERCISE 9.16

A 2.55-g sample of ammonium nitrite (NH_4NO_2) is heated in a test tube connected as shown in Figure 9.8. The ammonium nitrite is expected to decompose according to the equation

$$NH_4NO_2(s) \longrightarrow N_2(g) + 2H_2O(g)$$

If it does decompose in this way, what volume of N_2 will be collected in the flask? The water and gas temperature are 26°C, and the barometric pressure is 745 mm Hg.

Solution: We begin by calculating the number of moles of N_2 gas formed:

$$2.55 \text{ g } NH_4NO_2\left(\frac{1 \text{ mol } NH_4NO_2}{64.0 \text{ g } NH_4NO_2}\right) \times \left(\frac{1 \text{ mol } N_2}{1 \text{ mol } NH_4NO_2}\right) = 0.0398 \text{ mol } N_2$$

To predict the volume of N_2 gas that is to be collected, we might be tempted simply to calculate the volume that would be occupied by 0.0398 mol of N_2 at 745 mm Hg. But we must also take into account the partial pressure of water vapor at 26°C, because the gas within the bottle is saturated with water vapor. From a table of water-vapor pressure versus temperature in the laboratory manual or in a chemistry handbook (see also Appendix C), we can determine that the vapor pressure of water at 26°C is 25 mm Hg. The pressure of nitrogen gas in the flask when the water levels inside and out have been equalized is thus 745 − 25 = 720 mm Hg. We must calculate the predicted volume of gas using this pressure (720/760 = 0.947 atm). (Pressure must be expressed in atm when we employ the value for the gas constant, R, expressed in L-atm/mol-K.) Rearranging Equation 9.1, we obtain

$$V = \frac{nRT}{P}$$

Inserting all known quantities in correct units, we obtain

$$V = \frac{(0.0398 \text{ mol})(0.0821 \text{ L-atm/K-mol})(299 \text{ K})}{0.947 \text{ atm}}$$
$$= 1.03\text{L}$$

9.7 KINETIC-MOLECULAR THEORY

The ideal-gas equation describes *how* gases behave, but it doesn't explain *why* they behave as they do. For example, why does a gas expand when heated at constant pressure, or why does its pressure increase when the gas is compressed at constant temperature? To understand the physical properties of gases, we need a model that helps us picture what happens to gas particles as experimental conditions such as pressure or temperature change. Such a model, known as the **kinetic-molecular theory**, was developed over a period of about 100 years, culminating in 1857 when Rudolf Clausius (1822–1888) published a complete and satisfactory form of the theory. This theory permits us not only to explain qualitatively the behavior of gases, but also to derive quantitatively the ideal-gas equation.

The kinetic-molecular theory is summarized by the following statements:

1. Gases consist of large numbers of molecules that are in continuous, random motion. (The word "molecule" is used here to designate the smallest particle of any gas; some gases, such as the noble gases, consist of uncombined atoms.)
2. The volume of all the molecules of the gas is negligible compared to the total volume in which the gas is contained.
3. Attractive and repulsive forces between gas molecules are negligible.
4. Energy can be transferred between molecules during collisions, but the *average* kinetic energy of the molecules does not change with time, as long as the temperature of the gas remains constant. In other words, the collisions are perfectly elastic.
5. The average kinetic energy of the molecules is proportional to absolute temperature. At any given temperature the molecules of all gases have the same average kinetic energy.

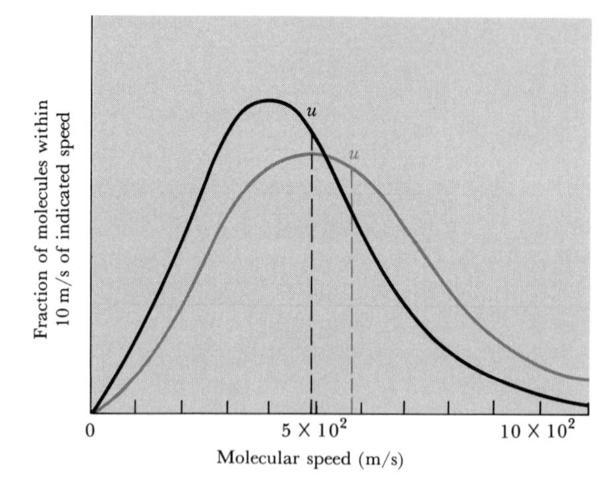

FIGURE 9.9 Distribution of molecular speeds for nitrogen at 0°C (black line) and 100°C (colored line).

The kinetic-molecular theory gives us an understanding of both pressure and temperature at the molecular level. The pressure of a gas is caused by collisions of the molecules with the walls of the container; it is determined both by the frequency of collisions per unit area and by the impulse imparted per collision (that is, by how frequently and by how "hard" the molecules strike the walls). The absolute temperature of a substance is a measure of the average kinetic energy of its molecules; absolute zero is the temperature at which the average kinetic energy of the molecules would be zero.

The idea that average kinetic energy and temperature are proportional provides a particularly important insight into matter. Let's consider this idea further. The molecules of a gas move at varying speeds. At one instant some of them are moving rapidly, others slowly. Figure 9.9 illustrates the distribution of molecular speeds within nitrogen gas at 0°C (black line) and at 100°C (colored line). Notice that at higher temperatures the distribution curve has shifted toward higher speeds.

Figure 9.9 also shows the value of the **root-mean-square** (rms) **speed,** u, of the molecules at each temperature. This quantity represents the square root of the average squared speeds of the molecules. The rms speed is not the same as the average speed. The difference between the two, however, is so small that for most purposes they can be considered equal.* The rms speed is important because the average kinetic energy of the gas molecules, ϵ, is related directly to u^2:

$$\epsilon = \frac{1}{2} mu^2 \qquad [9.13]$$

where m is the mass of the molecule. When the temperature of a gas increases, the average kinetic energy of the gas molecules increases pro-

*To illustrate the difference between rms speed and average speed, suppose that we have four objects with speeds of 4.0, 6.0, 10.0, and 12.0 m/s. Their average speed is $\frac{1}{4}(4.0 + 6.0 + 10.0 + 12.0) = 8.0$ m/s. However, the rms speed, u, is

$$\sqrt{\tfrac{1}{4}(4.0^2 + 6.0^2 + 10.0^2 + 12.0^2)} = \sqrt{74.0} = 8.6 \text{ m/s}$$

In general, the average speed equals $0.921 \times u$. Thus the average speed is directly proportional to the rms speed, and the two are in fact nearly equal.

portionally. The relationship between average kinetic energy and temperature that is obtained from kinetic-molecular theory is given in Equation 9.14, where R is the ideal gas constant, N is Avogadro's number, and T is the absolute temperature:

$$\epsilon = \frac{1}{2}mu^2 = \frac{3RT}{2N} \qquad [9.14]$$

SAMPLE EXERCISE 9.17

What is the average kinetic energy of a nitrogen molecule at 27°C?

Solution: Using Equation 9.14, we have

$$\epsilon = \frac{3RT}{2N} = \frac{3(8.314\ \text{J/K-mol})(300\ \text{K})}{2(6.022 \times 10^{23}/\text{mol})} = 6.21 \times 10^{-21}\ \text{J}$$

The average kinetic energy of a gas molecule at 27°C will have this same value for any other gaseous substance.

The empirical observations of gas properties as expressed in the various gas laws are readily understood in terms of the kinetic-molecular theory. The following examples are illustrative.

1. *Effect of a volume increase at constant temperature:* The fact that temperature remains constant means that the average kinetic energy of the gas molecules remains unchanged. This in turn means that the rms speed of the molecules, u, is unchanged. However, if the volume is increased, the molecules must move a longer distance between collisions; consequently, there are fewer collisions per unit time with the container walls, and pressure decreases. Thus the model accounts in a simple way for Boyle's law.
2. *Effect of a temperature increase at constant volume:* An increase in temperature means an increase in the average kinetic energy of the molecules, and thus an increase in u. If there is no change in volume, there will be more collisions with the walls per unit time. Furthermore, there will be a larger change in momentum on each collision (the molecules strike the walls harder). Hence the model explains the observed pressure increase.

SAMPLE EXERCISE 9.18

A sample of O_2 gas initially at STP is transferred from a 2-L container to a 1-L container at constant temperature. What effect does this change have on (a) the average kinetic energy of O_2 molecules; (b) the average speed of the O_2 molecules; (c) the total number of collisions of O_2 molecules with the container walls in a unit time; (d) the number of collisions of O_2 molecules with a unit area of container wall in a unit time?

Solution: (a) The average kinetic energy of O_2 molecules is determined only by temperature. The average kinetic energy is not changed by the compression of O_2 from 2 L to 1 L at constant temperature. (b) If the average kinetic energy of the O_2 molecules doesn't change, neither does u (see Equation 9.12). Both the average and rms speeds remain constant. (c) The total number of collisions with the container walls in a unit time must increase, because the molecules are moving within a smaller volume but with the same average speed as before. Under these conditions they must encounter a wall more frequently. (d) The number of collisions with a unit area of wall increases, because the total number of collisions with the walls is higher and the area of wall is smaller than before.

The Ideal-Gas Equation

Beginning with the postulates of the kinetic-molecular theory it is possible to derive the ideal-gas equation. Rather than proceed through a derivation, let's consider in somewhat qualitative terms how the ideal-gas equation might follow. As we have seen (Section 9.2), pressure is force per unit area. The total force of the molecular collisions on the walls, and hence the pressure produced by these collisions, depends both on how strongly the molecules strike the walls (impulse imparted per collision) and on the rate at which these collisions occur:

$$P \propto \text{impulse imparted per collision} \times \text{rate of collisions} \qquad [9.15]$$

The impulse imparted by a collision of a molecule with a wall depends on the momentum of the molecule; that is, it depends on the product of its mass and speed, mu. The rate of collisions is proportional to both the number of molecules per unit volume, n/V, and their speed, u. If there are more molecules in a container, there will be more frequent collisions with the container walls. As the molecular speed increases or the volume of the container decreases, the time required for molecules to transverse the distance from one wall to another is reduced, and the molecules collide more frequently with the walls. Thus we have

$$P \propto mu \times \frac{n}{V} \times u \propto \frac{nmu^2}{V} \qquad [9.16]$$

From Equation 9.14 we have that $mu^2 \propto T$. (This proportionality also follows from the basic idea that the average kinetic energy is proportional to temperature.) Making this substitution into Equation 9.16, we have

$$P \propto \frac{n(mu^2)}{V} \propto \frac{nT}{V} \qquad [9.17]$$

Let us now convert the proportionality sign to an equal sign by expressing n as the number of moles of gas, and then insert a proportionality constant, R, the molar gas constant:

$$P = \frac{nRT}{V} \qquad [9.18]$$

This expression, of course, is the familiar ideal-gas equation.

9.8 MOLECULAR EFFUSION AND DIFFUSION; GRAHAM'S LAW

We have already made reference to the fact that the molecules of a gas do not all move at the same speed. Instead, the molecules are distributed over a range of speeds, as shown for nitrogen at two different temperatures in Figure 9.9. The distribution of molecular speeds depends on the mass of the gas molecules and on temperature. The average kinetic energy of the molecules in any gas is determined only by temperature; thus the quantity $\frac{1}{2}mu^2$, which is kinetic energy, must have the same value for

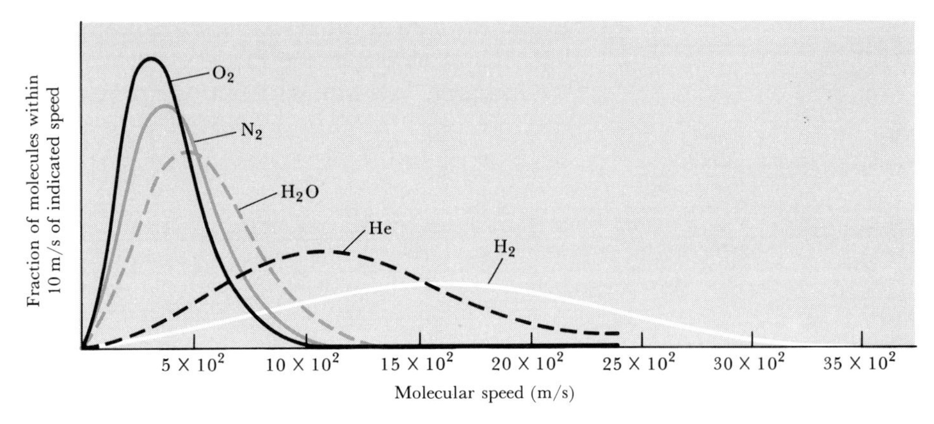

FIGURE 9.10 Distribution of molecular speeds for different gases at 25°C.

two gases at the same temperature, though their masses may differ. This in turn means that molecules of larger mass must have smaller average speeds. From the kinetic-molecular theory it can be shown that the rms speed, u, is given by Equation 9.19. Note that u is proportional to the square root of the absolute temperature, and *inversely* proportional to the square root of the molecular weight, $\mathcal{M}$:

$$u = \sqrt{\frac{3RT}{\mathcal{M}}} \quad [9.19]$$

This relationship tells us that at any given temperature lighter molecules have higher rms speeds. In fact, the entire distribution of molecular speeds is skewed to higher values for gases of lower molecular weights, as shown for several gases in Figure 9.10.

The dependence of molecular speeds on mass has several interesting consequences. For example, the rate at which a gas is able to escape through a tiny hole, as when a gas escapes through a hole in a balloon, depends on the molecular mass of the gas. This process is known as **effusion.** Effusion is related to, but it not quite the same as, **diffusion.** The latter term refers to the spread of one substance throughout a space, or throughout a second substance. For example, the molecules of a perfume diffuse through a room.

Graham's Law of Effusion

In about 1830, Thomas Graham discovered that the effusion rates of gases are inversely related to the square roots of their molecular weights. Assume that we have two gases at the same initial pressure contained in identical containers, each with an identical pinhole in one wall. Let the rate of effusion be called r. **Graham's law** states that

$$\frac{r_1}{r_2} = \sqrt{\frac{\mathcal{M}_2}{\mathcal{M}_1}} \quad [9.20]$$

Equation 9.20 compares the *rates* of effusion of two different gases; the lighter gas effuses more rapidly.

Graham's law follows from our previous discussion if we assume that the rate of effusion is proportional to the rms speed of the molecules. Because R and T are constant, we have from Equation 9.19:

$$\frac{r_1}{r_2} = \frac{u_1}{u_2} = \sqrt{\frac{3RT/\mathcal{M}_1}{3RT/\mathcal{M}_2}} = \sqrt{\frac{\mathcal{M}_2}{\mathcal{M}_1}}$$

SAMPLE EXERCISE 9.19

(a) Calculate the rms speed, in m/s, of an O_2 molecule at 27°C. (b) If an unknown gas effuses at a rate that is only 0.468 times that of O_2 at the same temperature, what is the molecular weight of the unknown gas?

Solution: (a) Using Equation 9.19, we have

$$u = \sqrt{\frac{3RT}{\mathcal{M}}}$$

$$= \sqrt{\frac{(3)(8.314\ \text{J/K-mol})(300\ \text{K})}{32.0\ \text{g/mol}} \times \frac{10^3\ \text{g}}{1\ \text{kg}}}$$

$$= 484\ \text{m/s}$$

The factor 10^3 g/kg is needed to convert mass to SI units consistent with the rest of the units in the problem. Recall (Section 4.2) that 1 J = 1 kg-m^2/s^2.

(b) Using Equation 9.21, we have

$$\frac{r_x}{r_{O_2}} = \sqrt{\frac{\mathcal{M}_{O_2}}{\mathcal{M}_x}}$$

Thus

$$\frac{r_x}{r_{O_2}} = 0.468 = \sqrt{\frac{32.0\ \text{g/mol}}{x}}$$

Solving for x yields

$$\frac{32.0\ \text{g/mol}}{x} = (0.468)^2 = 0.219$$

$$x = \frac{32.0\ \text{g/mol}}{0.219} = 146\ \text{g/mol}$$

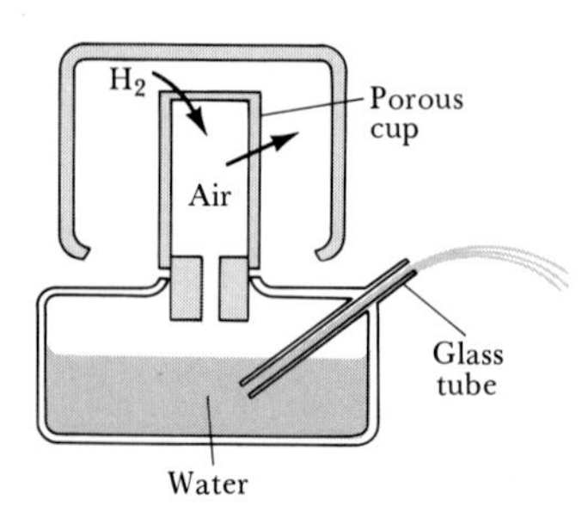

FIGURE 9.11 Hydrogen-fountain demonstration of the greater rate of diffusion of hydrogen compared with air. A large container filled with H_2 gas is placed over the porous cup containing air. Hydrogen diffuses into the cup more rapidly than the molecules of air diffuse outward. As a result the pressure inside the vessel increases and water is pushed out the glass tube, which is open to the outside.

The explanation for diffusive flow is more complicated than that for effusion, because it involves the effects of molecular collisions, whereas effusion does not. Nevertheless, the dependence of diffusion rate on molecular mass is also given by Equation 9.20: The ratio of the rates of diffusion of two gases under identical experimental conditions is inversely proportional to the square root of the ratio of molecular masses.

The fact that both effusion and diffusion rates are greater for lighter gases has many interesting applications. One popular lecture demonstration, the hydrogen fountain, is illustrated in Figure 9.11. It makes use of the fact that the diffusion of hydrogen through the walls of the porous cup is faster than diffusion of atmospheric gases out of the cup through the wall. Thus excess pressure builds up in the enclosure. Another example is seen in the behavior of toy balloons that have been filled with helium. Helium diffuses outward through the balloon surface more rapidly than the atmospheric gases diffuse in, so the balloon collapses much more rapidly than it would if filled with air.

In the course of the effort during World War II to develop the atomic bomb, it was necessary to separate the relatively low-abundance uranium isotope ^{235}U (0.7 percent) from the much more abundant ^{238}U (99.3 percent). This was done by converting the uranium into a volatile compound, UF_6, which boils at 56°C. The gaseous UF_6 was allowed to diffuse from one chamber into a second through a porous barrier. Because of the slight difference in molecular weights, the relative rates of passage through the barrier for $^{235}UF_6$ and $^{238}UF_6$ are not exactly the same. The ratio of diffusion rates is given by the square root of the ratio of molecular weights, Equation 9.20:

$$\frac{r_{235}}{r_{238}} = \sqrt{\frac{352.04}{349.03}} = 1.0043$$

Thus the gas initially appearing on the opposite side of the barrier would be very slightly enriched in the lighter molecule. The diffusion process was repeated thousands of times, leading to a nearly complete separation of the two nuclides of uranium.

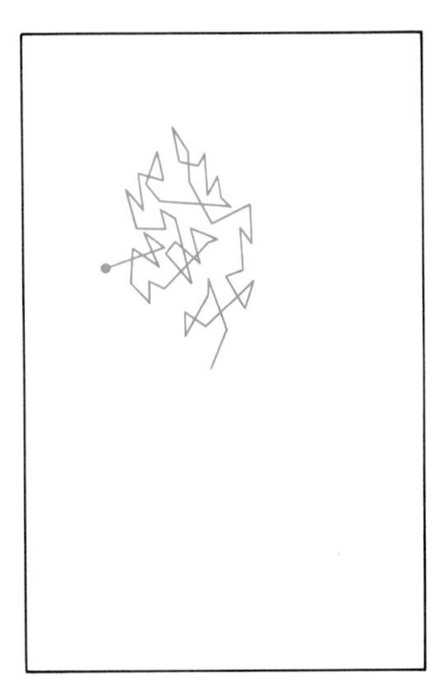

FIGURE 9.12 Schematic illustration of the diffusion of a gas molecule. For the sake of clarity all the other gas molecules in the container are not shown. The path of the molecule of interest begins at the dot. Each short segment of line represents travel between collisions. The path traveled by the molecule is often described as a "random walk."

Mean Free Path

We can see from the horizontal scale of Figure 9.10 that the speeds of molecules are really quite high. Translated into more familiar units, the average speed of N_2 at room temperature, 515 m/s, corresponds to 1850 km/h, or 1150 mi/h. Yet we know that molecules don't travel through the atmosphere from one place to another at these speeds. For example, if a vial of a substance possessing a strong odor is opened at one end of a room, it is some time, perhaps a few minutes, before the odor is noted by a person at the other end. Therefore, the movement of gas molecules in the atmosphere is much slower than the molecular speeds would suggest. We must ascribe these slower diffusion rates to collisions between molecules. The collisions are in fact very frequent for a gas at atmospheric pressure—about 10^{10} times per second for each molecule. The paths of the gas molecules are therefore interrupted very often. Diffusion of gas molecules, depicted in Figure 9.12, thus consists of a random motion, first in this direction, then that, at one instant at high speed, the next at low speed, with no overall direction to the motion of the gas as a whole except when there is a pressure difference.

The average distance traveled by a molecule between collisions is referred to as the mean free path. This distance depends on the effective radius of the molecules, because larger molecules are more likely to undergo a collision. It depends also on the number of molecules in a unit volume—the larger the number of molecules per unit volume, the more likely is collision. Table 9.3 lists some values of experimentally determined mean free paths for several gases at STP. As you can see from these values, the molecules of a gas at STP do not travel very far before undergoing collision. By contrast, at an elevation of about 100 km in the earth's atmosphere, the mean free paths of nitrogen and oxygen molecules are on the order of 10 cm, more than a million times longer than at the surface.

TABLE 9.3 Mean free paths for several gases at 0°C, 1 atm

Gas	Mean free path (nm)
Carbon dioxide	39.7
Carbon monoxide	58.4
Argon	63.5
Nitrogen	60.0
Oxygen	64.7
Hydrogen	112.3
Helium	179.8

Thermal conductivity is an important property of gases. It is a measure of the rate at which heat energy can be transferred through a gas. The thermal conductivity of a gas depends on the average speed of its molecules and on their mean free path. Those gases whose molecules move fastest and with the largest mean free path have the highest thermal conductivities. Among the gases in Table 9.3, carbon dioxide has the highest molecular weight and thus has the lowest average speed per molecule for a given temperature. It also has the shortest mean free path of the gases listed. We can conclude that carbon dioxide has the lowest thermal conductivity of any of the gases in the table. On the other hand, helium should have the highest thermal conductivity.

9.9 NONIDEAL GASES; DEPARTURES FROM THE IDEAL-GAS EQUATION

Although the ideal-gas equation is a very useful description of gases, all real gases fail to obey the relationship to a greater or lesser degree. The extent to which a real gas departs from ideal behavior may be seen by slightly rearranging the ideal-gas equation:

$$\frac{PV}{RT} = n \qquad [9.21]$$

For a mole of gas, $n = 1$; the quantity PV/RT should therefore equal 1 if the gas is ideal. Figure 9.13 shows the quantity PV/RT plotted as a function of pressure for a few gaseous substances, compared with the expected behavior of an ideal gas. It is clear from the figure that real gases are simply not ideal. However, the pressures shown are very high; at more ordinary pressures, below 10 atm, the deviations from ideal behavior are not so large, and the ideal-gas equation can be used without serious error.

Thus we see that deviations from ideal behavior tend to be larger at higher pressures than at lower ones. Temperature also has an effect. Gases tend to show significant deviations from ideal behavior at temperatures near their liquefaction points; that is, the deviations increase as temperature decreases, becoming significant near the temperature at which the gas is converted into a liquid.

We can understand these pressure and temperature effects on nonideality by considering two factors that are considered negligible in the kinetic-molecular theory: (1) the molecules of a gas possess finite volumes, and (2) at short distances of approach they exert attractive forces upon one another.

When gases are contained at relatively low pressure, say 1 atm, the volume of the space they occupy is very large in comparison with the

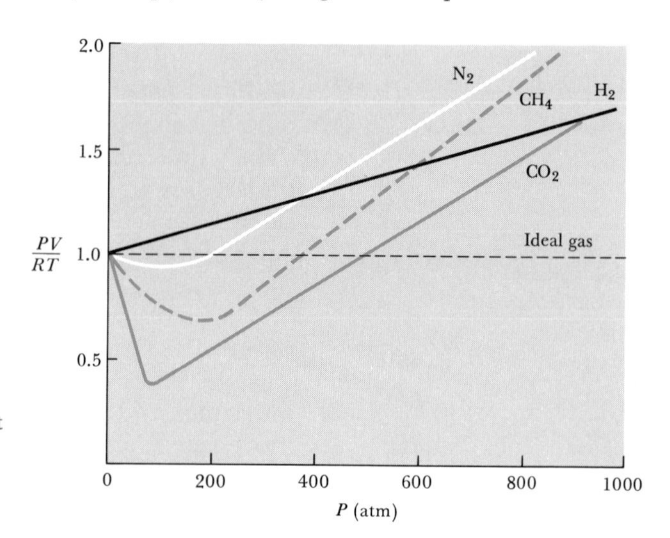

FIGURE 9.13 PV/RT versus pressure for several gases at 300 K. The data for CO_2 pertain to a temperature of 313 K, because CO_2 liquefies under high pressure at 300 K.

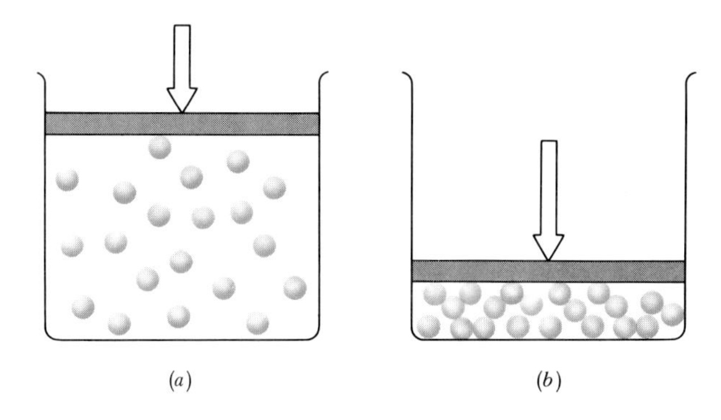

FIGURE 9.14 Illustration of the effect of the finite volume of gas molecules on the properties of a real gas at high pressure. In (*a*), at low pressure, the volume of the gas molecules is small compared with the container volume. In (*b*), at high pressure, the volume of the gas molecules themselves is a large fraction of the total space available.

volumes of the gas molecules themselves. At increasingly high pressures, however, the volume taken up by the molecules becomes a larger fraction of the total. This effect is illustrated in Figure 9.14. Thus, at pressures of several hundred atmospheres, the free volume in which the gas molecules can move is considerably smaller than the volume of the container. As a result, the value of V that *should* be used in the product PV is smaller than the volume of the container. Hence, when we use the container volume in the ideal-gas equation we obtain a product PV that is larger than it should be. The data for H_2 in Figure 9.13 illustrate this situation. Notice that PV/RT increases steadily with increasing pressure. This is due entirely to the finite volume of the H_2 molecules.

The attractive forces between molecules come into play at short distances, when the molecules undergo collisions with one another. Because of these attractive forces, the molecules tend to stick together upon collision. If we could stop the action at any one instant in a real gas we might see something like that illustrated in Figure 9.15. Some of the molecules, having come together in collisions, are "paired up" for short times, because of the attractive forces. This means that the total number of particles present in the gas is decreased. Thus PV/RT is smaller than it should be for an ideal gas.

From the data illustrated in Figure 9.13 we can guess that the attractive forces between molecules are greatest for CO_2. The product PV

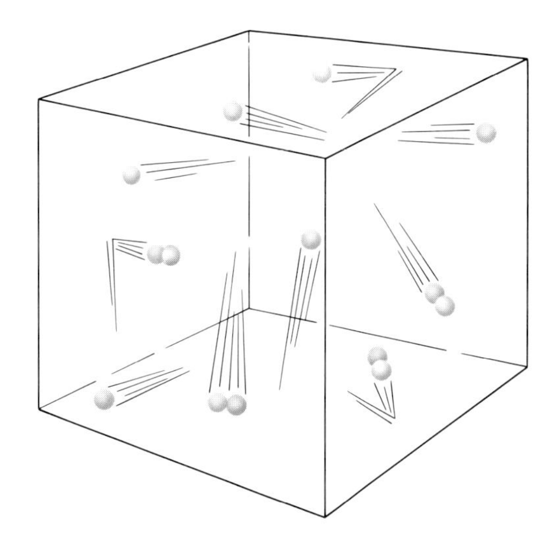

FIGURE 9.15 Illustration of the effects of intermolecular attractive forces. Upon collision, molecules may form pairs that are held together for short periods by the attractive forces between them. Thus the total number of independent particles in the gas and the quantity PV/RT are reduced.

shows a substantital negative departure from the ideal-gas relationship over a wide range of pressure. The attractive forces are less important for CH_4, and less important still for N_2. Because there is very little attractive interaction between H_2 molecules the quantity PV/RT for this gas is continuously larger than that expected for an ideal gas.

Temperature has no effect on the finite volumes of the gas molecules, but it does determine how *effective* attractive forces are. As the gas is cooled, the motional energies decrease while intermolecular attractions remain constant. In a sense, cooling of a gas deprives molecules of the energy they need to overcome their mutual attractive influence. Thus the molecules stick together more upon collision. As we noted above in discussing pressure effects, the resultant pairing of molecules causes PV/RT to be smaller for a real gas than for the ideal gas.

The Van Der Waals Equation

Engineers and scientists who work with gases at high pressures often cannot use the ideal-gas equation to predict the pressure-volume properties of gas, because departures from the ideal-gas behavior are too large. Various equations of state have been developed to predict more realistically the pressure-volume behavior of real gases. These equations, while more realistic, are also considerably more complicated than the simple ideal-gas equation (Equation 9.1). Equation 9.22 shows the van der Waals equation, named after Johannes van der Waals, who presented it in 1873.

$$\left(P + \frac{an^2}{V^2}\right)(V - nb) = nRT \qquad [9.22]$$

This equation differs from the ideal-gas equation by the presence of two correction terms; one corrects the volume, another modifies the pressure. The term nb in the expression $(V - nb)$ is a correction for the finite volume of the gas molecules; the van der Waals constant b, different for each gas, has units of liters/mole. It is a measure of the actual volume occupied by the gas molecules. Values of b for several gases are listed in

TABLE 9.4 Van der Waals constants for gas molecules

Substance	a (L^2-atm/mol^2)	b (L/mol)
He	0.0341	0.02370
Ne	0.211	0.0171
Ar	1.34	0.0322
Kr	2.32	0.0398
Xe	4.19	0.0510
H_2	0.244	0.0266
N_2	1.39	0.0391
O_2	1.36	0.0318
Cl_2	6.49	0.0562
CO_2	3.59	0.0427
CH_4	2.25	0.0428
CCl_4	20.4	0.1383

Table 9.4. Note that b increases with an increase in mass of the molecule or in the complexity of its structure.

The correction to the pressure takes account of the intermolecular attractions between molecules. Notice that it consists of the constant a, different for each gas, times the quantity $(n/V)^2$. The units of n/V are moles/liter. This quantity is squared because the number of molecular pairings, as illustrated in Figure 9.15, is proportional to the square of the number of molecules per unit volume. Values of the van der Waals constant a are listed in Table 9.4 for several gases. Notice that a increases with an increase in molecular weight and with an increase in complexity of molecular structure.

SAMPLE EXERCISE 9.20

Compare the van der Waals constants a and b for N_2 and CO_2 and account for the differences.

Solution: We note that the constant b, which is a measure of molecular volume, is larger for CO_2: 0.0427 L/mol compared with 0.0391 L/mol for N_2. Because CO_2 is a larger molecule, consisting of three atoms rather than two, the larger value of b for CO_2 is expected. Furthermore, because CO_2 is both larger and more complex than N_2, we expect that the attractive forces between CO_2 molecules upon collision will be greater than the analogous attractive forces for N_2. Thus the value for a should be larger for CO_2 than for N_2, as observed. In effect, under the same conditions of temperature and pressure, there are more molecular pairs per unit volume of CO_2 than for N_2.

Note also that the relative values of a and b are consistent with the relative magnitudes of the departures from the ideal-gas equation, as illustrated in Figure 9.13.

To get some feeling for the magnitudes of the departures from ideal behavior, let's calculate these departures for CO_2 at STP.

SAMPLE EXERCISE 9.21

Calculate the correction terms to pressure and volume for CO_2 at STP, using the data in Table 9.4, and compare with the ideal-gas values for P and V.

Solution: From Equation 9.22 we see that the volume correction term is given by nb. Since $n = 1$, nb equals 0.0427 L, which is to be compared with 22.4 L, the molar volume of an ideal gas at STP. The correction to volume is thus

$$\frac{0.0427}{22.4} \times 100 = 0.191\% \simeq 0.2\%$$

The correction to pressure is given by an^2/V^2. Inserting the value of a from Table 9.4, $n = 1$, and $V = 22.4$ L, we obtain

$$\frac{an^2}{V^2} = \frac{\left(\dfrac{3.59\ \text{L}^2\text{-atm}}{\text{mol}^2}\right) 1\ \text{mol}^2}{(22.4\ \text{L})^2} = 0.007\ \text{atm}$$

The required correction to pressure is thus $0.007 \times 100 = 0.7\%$. We conclude that the ideal-gas law is obeyed by CO_2 at STP conditions to within 1 percent.

FOR REVIEW

Summary

Many substances are capable of existing in any one of the three states of matter—solid, liquid, or gas. This chapter has been concerned with the gaseous state. To describe the state or condition of a gas, it is necessary to specify four variables: pressure, temperature, volume, and quantity of gas. Volume is usually measured in liters (L), and temperature in the

Kelvin scale. Pressure is defined as the force per unit area. It is expressed in SI units as pascals, Pa ($1 \text{ Pa} = 1 \text{ N/m}^2 = 1 \text{ kg/m-s}^2$), or more commonly in millimeters of mercury (mm Hg). One standard atmosphere pressure equals 101.325 kPa, or 760 mm Hg. A barometer is often used to measure the atmospheric pressure. A monometer can be used to measure the pressure of enclosed gases.

The ideal-gas equation, $PV = nRT$, is the equation of state for an ideal gas. Most gases at pressures of about 1 atm and temperatures of 300 K and above obey the ideal-gas equation reasonably well. We can use the ideal-gas equation to calculate variations in one variable when one or more of the others are changed. For example, for a constant quantity of gas at constant temperature, the pressure of the gas is inversely proportional to the volume (Boyle's law). Similarly, for a constant quantity of gas at constant pressure, the volume of a gas is directly proportional to temperature (Charles's law). In gas mixtures, the total pressure is the sum of the partial pressures that each gas would exert if it were present alone under the same conditions (Daltons' law of partial pressures). In all applications of the ideal-gas equation we must remember to convert temperatures to the absolute temperature scale, Kelvin.

It is important to be able to use the ideal-gas equation to solve problems involving gases as reactants or products in chemical reactions. From the gas density, d, under given conditions of pressure and temperature, it is possible to calculate the molecular weight of the gas: $\mathcal{M} = dRT/P$. In calculating the quantity of gas collected over water, correction must be made for the partial pressure of water vapor in the container.

The kinetic-molecular theory accounts for the properties of an ideal gas in terms of a set of assumptions about the nature of gases. Briefly these assumptions are that molecules are in ceaseless, chaotic motion; that the volume of gas molecules is negligible in relation to the volume of their container; that the gas molecules have no attractive forces for one another; and finally, that the average kinetic energy of the gas molecules is proportional to absolute temperature.

The molecules of a gas do not all have the same kinetic energy at a given instant. Their speeds are distributed over a wide range; the distribution varies with the molecular weight of the gas and with temperature. The root-mean-square (rms) speed, u, varies in proportion to the square root of absolute temperature, and inversely with the square root of molecular weight: $u = (3RT/\mathcal{M})^{1/2}$. It follows that the rate at which a gas escapes (effuses) through a tiny hole is inversely proportional to the square root of its molecular weight (Graham's law). Molecules in a real gas possess finite volume and thus undergo frequent collisions with one another. These frequent collisions limit the rate at which a gas molecule can diffuse through a space occupied by other gas molecules and determine the thermal conductivity of a gas.

Departures from ideal behavior increase in magnitude as pressure increases and as temperature decreases. The extent of nonideality of a real gas can be seen by examining the quantity PV/RT for 1 mol of the gas, as a function of pressure; this quantity is exactly 1 for an ideal gas at all pressures. Real gases depart from the ideal behavior because the molecules possess finite volume (leads to $PV/RT > 1$), or because the molecules experience attractive forces for one another upon collision (leads to $PV/RT < 1$). The van der Waals equation is an equation of state for gases that attempts to correct the ideal-gas equation to take account of two properties of real gases.

Learning goals

Having read and studied this chapter, you should be able to:

1 Describe the general characteristics of gases as compared with other states of matter and list the ways in which gases are distinctly different.

2 List the variables that are required to define the state of a gas.

3 Define atmosphere, millimeters of mercury, and kilopascals, the most important units in which pressure is expressed. You should also understand the principle of operation of a barometer.

4 Explain the way in which pressure, volume, and temperature are related in the ideal-gas equation. That is, you should remember Equation 9.1.

5 Solve problems involving changes in the condition or state of a gas. You should be able to explain how one variable is affected by a change in another, when the other variables are maintained constant.

6 Calculate the quantity of a gas under a given set of conditions that is required as a reactant or formed as product in a chemical reaction.

7 Correct for the effects of water vapor pressure in calculating the quantity of a gas collected over water.

8 Explain the concept of gas density and describe how it is related to temperature, pressure, and molecular weight.

9 Calculate molecular weight, given gas density under defined conditions of temperature and pressure. You should also be able to calculate gas

density under stated conditons, knowing molecular weight.

10 List and explain the assumptions on which the kinetic theory of gases is based.

11 Describe graphically how gas molecules are distributed over a range of speeds and how that distribution changes with temperature.

12 Describe how the relative rates of diffusion or effusion of two gases depend on their relative molecular weights (Graham's law).

13 Explain the concept of mean free path and how it relates to the rates of diffusion of molecules in the gas state and to thermal conductivity.

14 Explain the origin of deviations shown by real gases from the relationship $PV/RT = 1$ for a mole of ideal gas.

15 List the two major factors responsible for deviations of gases from ideal behavior.

16 Explain the origins of the correction terms to P and V that appear in the van der Waals equation of state for a gas.

Key terms

Among the more important terms and expressions used for the first time in this chapter are the following:

According to **Avogadro's hypothesis** (Section 9.3), equal volumes of gases at the same temperature and pressure contain equal numbers of molecules.

A **barometer** (Section 9.2) is a device for measuring atmospheric pressure in terms of the height of a liquid column sustained by that pressure.

According to **Boyle's law** (Section 9.3), at constant temperature, the product of the volume and pressure of a given amount of gas is a constant.

According to **Charles's law** (Section 9.3), (1) at constant pressure, the volume of a given quantity of gas is proportional to absolute temperature, and (2) at constant volume, the pressure of a given quantity of gas is proportional to absolute temperature.

Dalton's law of partial pressures (Section 9.5) states that the total pressure of a mixture of gases is just the sum of the pressures that each gas would exert if it were present alone.

Diffusion (Section 9.8) refers to the rate at which a substance spreads into and throughout a space. Thus a gas might diffuse throughout a room, or atmospheric oxygen might diffuse through the waters of a lake. **Effusion** refers to the rate at which a gas escapes through an orifice or hole.

The **gas constant,** R (Section 9.4), is the constant of proportionality in the ideal-gas equation.

Graham's law (Section 9.8) states that the relative rate of effusion of two gases is inversely proportional to the square root of the ratio of their molecular weights.

The **ideal-gas equation** (Section 9.4) is an equation of state for gases that embodies Boyle's law, Charles's law, and Avogadro's hypothesis in the form $PV = nRT$.

The **kinetic-molecular theory of gases** (Section 9.7) consists of a set of assumptions about the nature of gases. These assumptions, when translated into mathematical form, yield the ideal-gas equation.

The **mean free path** (Section 9.8) in a gas sample is the average distance traveled by a gas molecule between collisions.

Pressure (Section 9.2) is a measure of the force exerted on a unit area. In work with gases, pressure is most commonly expressed in units of atmospheres (atm) or of millimeters of mercury (mm Hg)—760 mm Hg = 1 atm; in SI units, pressure is expressed in pascals (Pa).

The **root-mean-square** (rms) **speed,** u (Section 9.7), of a gas is given by the square root of the average of the squared speeds of the gas molecules.

The **standard atmospheric pressure** (Section 9.2) is defined as 760 mm Hg, or in SI units, 101.325 kPa.

Standard temperature and pressure (STP) (Section 9.4), 0°C and 1 atm pressure, are frequently used reference conditions for a gas.

The **thermal conductivity** of a gas (Section 9.8) is a measure of the rate at which heat energy can be transferred through it.

The **torr** (Section 9.2) is a unit of pressure (1 torr = 1 mm Hg).

The **van der Waals equation** (Section 9.9) is an equation of state for real gases containing terms that correct for the existence of attractive forces between molecules and for their finite volumes.

EXERCISES

Introduction; pressure

9.1 Compressibility, which is the change in volume of a substance in response to a change in pressure, is much greater for gases than for liquids or solids. Explain.

9.2 Perform the following conversions: (a) 706 mm Hg to atm; (b) 366 kPa to atm; (c) 165 mm Hg to torr; (d) 0.933 atm to mm Hg; (e) 0.897 atm to kPa; (f) 598 mm Hg to kPa.

9.3 Perform the following conversions: (a) 0.322 atm to mm Hg; (b) 897 mm Hg to atm; (c) 0.573 atm to kPa; (d) 50.4 kPa to atm; (e) 388 mm Hg to kPa; (f) 5.0×10^{-2} mm Hg to torr

9.4 (a) If the weather report for Montreal states that the temperature is 15°C and the barometric pressure is 99.8 kPa, what is the temperature in degrees Fahrenheit and pressure in inches of mercury? (b) If the weather in Chicago is reported to be 48°F with a barometric pressure of 29.50 in. Hg, what is the temperature in degrees Celsius and pressure in kilopascals?

9.5 (a) A plastic bucket has a flat bottom of area 3.2×10^3 cm^2. When filled with water it weighs 9.60 kg. What pressure, in pascals, does the bucket exert on a flat surface on which it rests? (b) A phonograph stylus resting on a record has an effective mass of 2.5 g. The area of the stylus making contact with the record averages 2.7×10^{-4} mm^2. What pressure, in pascals, does the stylus exert on the record? (c) Calculate the pressure, in pascals, exerted on a tabletop by a cube of aluminum (density = 2.70 g/cm^3) measuring 5.00 cm on each edge.

9.6 An open-tube manometer containing mercury is connected to a container of gas. What is the pressure of the enclosed gas, in mm Hg, in each of the following situations? (a) The mercury in the arm attached to the gas is 2.5 cm higher than the one open to the atmosphere; atmospheric pressure is 0.933 atm. (b) The mercury in the arm attached to the gas is 5.3 cm lower than the one open to the atmosphere; atmospheric pressure is 0.897 atm.

9.7 Suppose that the mercury in the barometer of Figure 9.2 were replaced by a liquid metal alloy with a density of 5.73 g/cm^3. The density of mercury is 13.6 g/cm^3. What height of column of liquid alloy would be supported by 1.00 atm pressure?

9.8 A laboratory experiment calls for collecting O_2 gas over water as shown in Figure 9.8. The water is at 25°C, at which temperature its density is 1.00 g/mL and the vapor pressure is 23.8 mm Hg. Atmospheric pressure is 735.5 mm Hg. If the water level in the collection vessel is 5.0 cm higher than the water level outside, what is the pressure of the O_2 gas?

[9.9] Water-well pumps, such as the familiar pitcher pump, which depend on creating a vacuum to lift water to the surface, cannot lift water from a depth greater than about 32 ft. Why is this so?

Ideal-gas equation

9.10 The ideal-gas equation is called an "equation of state." What does this expression mean?

9.11 Which of the following statements, if any, are false? Correct any false statements. (a) At constant temperature and volume, pressure is inversely proportional to the number of moles of gas. (b) At constant volume the pressure of a given amount of gas increases in proportion to the absolute temperature. (c) An increase in volume of a given quantity of gas at constant pressure arises from an increase in absolute temperature. (d) At constant volume the pressure of a gas is inversely proportional to temperature.

9.12 For an ideal gas, calculate (a) the pressure of the gas if 3.32 mol occupies 5.28 L at 38°C; (b) the volume occupied by 0.105 mol at −23°C and 5.00 atm; (c) the number of moles in 0.500 L at 27°C and 725 mm Hg; (d) the temperature at which 0.670 mol occupies 25.0 L at 2.45 atm.

9.13 Assuming a gas to behave ideally, calculate (a) its pressure if 5.27×10^{-2} mol occupies 1.53 mL at −23°C; (b) its volume if 5.23 mol has a pressure of 0.886 atm and a temperature of 37°C; (c) the quantity of gas, in moles, if 1.00 L at −15°C has a pressure of 1.25 atm; (d) the temperature, in kelvin, at which 3.92×10^{-3} mol occupies 52.5 mL at 745 mm Hg.

9.14 Many gases are shipped in high-pressure containers. If a steel tank whose volume is 50.0 L contains O_2 gas at a total pressure of 1500 kPa at 23°C, what mass of oxygen does it contain? What volume would the gas occupy at STP?

9.15 (a) A neon sign is made of glass tubing whose inside diameter is 2.0 cm and whose length is 4.0 m. If the sign contains neon at a pressure of 1.5 mm Hg at 35°C, how many grams of neon are contained in the sign? (b) A good laboratory vacuum system can maintain a vacuum of 10^{-6} mm Hg. How many molecules will be present in 1.00 L at 22°C at this pressure? (c) A typical ranch house has a volume of 13,000 ft^3. In winter the house has a water-vapor pressure of only 4 mm Hg. A humidifier is installed to increase the water-vapor pressure to 9 mm Hg, which represents about 40 percent humidity. What mass of water vapor must the humidifier produce to increase the humidity to the desired level? Assume that the house temperature averages 22°C.

The gas laws

9.16 Starting with the ideal-gas equation, derive Avogadro's law.

9.17 (a) A fixed quantity of gas is compressed at constant temperature from a volume of 566 mL to 308 mL. If the initial pressure was 622 mm Hg, what is the final pressure? (b) A fixed quantity of gas is allowed to expand at constant temperature from 1.45 L to 4.37 L. If the initial pressure was 1.25 atm, what is the final pressure? (c) What is the final volume of a gas if a 3.00-L sample is heated from 23°C to 350°C at constant pressure? (d) If a gas originally at 15°C and having a volume of 233 mL is compressed to 150 mL while its pressure is held constant, what is its final temperature? (e) A gas exerts a pressure of 137 kPa at 27°C. The temperature of the gas is increased to 148°C with no significant volume change. What is the gas pressure at the higher temperature?

9.18 The density of a particular gas in a cylinder with a volume of 6.50 L is found to be 1.45 g/L. The gas is compressed at constant temperature until the volume is 3.20 L. What is the density of the gas under the new conditions?

9.19 At 25°C and 1.00 atm, a gas occupies a volume of 0.500 L. How many liters will it occupy (a) at 0°C and 0.305 atm; (b) at STP?

9.20 The volume of a sample of chlorine, Cl_2, is 3.24 L at 577 torr and 23°C. (a) What volume will the Cl_2 occupy at 127°C and 780 torr? (b) What volume will the Cl_2 occupy at STP? (c) At what temperature will the volume be 5.00 L if the pressure is 5.00×10^2 mm Hg? (d) At what pressure will the volume be 5.00 L if the temperature is 87°C?

9.21 Suppose that you are responsible for filling a 12.0-L weather balloon with 1.00 atm pressure of helium each day and releasing it. Assuming no waste, how many 25-L tanks of helium at a pressure of 2250 lb/in.^{2} would you require in a year (all temperatures about 300 K)? (1 atm = 14.7 lb/in.^{2}.)

9.22 The Rankine temperature scale is an absolute temperature scale with 0 K = 0°R. The size of each degree on the Rankine scale is the same as that on the Fahrenheit scale. What temperature, in °R, corresponds to (a) 273 K; (b) 55°F; (c) −15°C?

[9.23] A tire was inflated to a gauge pressure of 30 psi (pounds per square inch) while at 15°C. After driving for some time, the tire was heated by friction to 60°C. If atmospheric pressure is 14.0 psi, what is the gauge pressure of the tire at the higher temperature? (Gauge pressure is the pressure by which atmospheric pressure is exceeded.)

Dalton's law of partial pressures

9.24 Consider a mixture composed of 1.5 g of H_2, 85.2 g of O_2, and 17.0 g of Ar confined to a volume of 3.20 m^3 at 88°C. Calculate (a) the partial pressure of H_2; (b) the total pressure in the vessel.

9.25 A mixture of cyclopropane gas, C_3H_6, and oxygen, O_2, is widely used as an anesthetic. (a) How many moles of each gas are present in a 1.00-L container at 23°C if the partial pressure of cyclopropane is 150 mm Hg and that of oxygen is 585 mm Hg? (b) What is the partial pressure of C_3H_6 in a mixture of 2.00 g of C_3H_6 and 20.0 g of O_2 if the total pressure in the vessel is 750 mm Hg?

9.26 Consider the arrangement of bulbs shown in Figure 9.16. Each of the bulbs contains a gas at the pressure shown. What is the pressure of the system when all the stopcocks are opened, assuming that the temperature remains constant? (We can neglect the volume of the capillary tubing connecting the bulbs.)

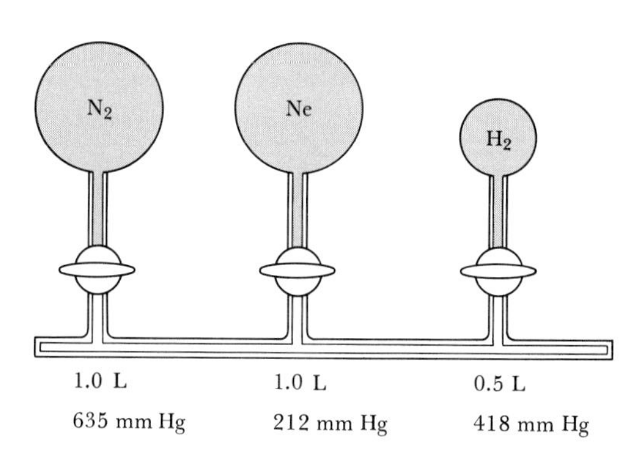

FIGURE 9.16

9.27 A quantity of N_2 gas originally held at 2.00 atm pressure in a 1.00-L container at 20°C is transferred to a 10.0-L container at 25°C. A quantity of O_2 gas originally at 1.50 atm and 23°C in a 5.00-L container is transferred to this same container. What is the total pressure in the new container?

Density and molecular weight

9.28 (a) Calculate the density of argon gas at STP. (b) Calculate the density of a gas at 27°C and 0.870 atm if the gas has a molecular weight of 34.1 g/mol. (c) Calculate the density of CO_2 at 0.980 atm and −5°C.

9.29 (a) Calculate the molecular weight of a gas if 0.608 g occupies 750 mL at 385 mm Hg and 35°C. (b) Calculate the molecular weight of a gas that has a density of 1.84 g/L at 52°C and 600 mm Hg.

9.30 (a) Cyanogen is 46.2 percent carbon and 53.8 percent nitrogen by mass. At 25°C and 750 mm Hg, 1.05 g of cyanogen occupies 0.500 L. What is the molecular formula of cyanogen? (b) Benzene is 92.3 percent carbon and 7.7 percent hydrogen by mass. At 120°C and 698 mm Hg, 0.555 g of benzene occupies 0.250 L. What is the molecular formula of benzene?

9.31 In the Dumas bulb technique for determining the molecular weight of an unknown liquid, one vaporizes the sample of a liquid that boils below 100°C in a boiling-water bath and determines the mass of vapor required to just fill the bulb (see Figure 9.17). From the following data, calculate the molecular weight of the unknown liquid: mass of unknown vapor, 0.912 g; volume of bulb, 342 cm^3; pressure, 742 mm Hg; temperature, 99°C.

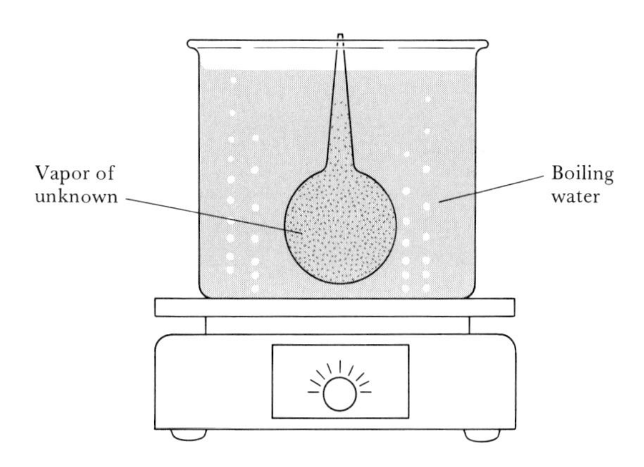

FIGURE 9.17

[9.32] A gaseous mixture of He and N_2 is found to have a density of 0.550 g/L at 25°C and 715 mm Hg. What is the percent He (by mass) in the mixture?

Quantities of gases in chemical reactions

9.33 Calculate the volume of O_2 at 710 mm Hg and 36°C required to react completely with 6.50 g of CuS according to the following equation:

$$CuS(s) + 2O_2(g) \longrightarrow CuSO_4(s)$$

9.34 A piece of solid magnesium is reacted with dilute hydrochloric acid to form hydrogen gas:

$$Mg(s) + 2HCl(aq) \longrightarrow MgCl_2(aq) + H_2(g)$$

What volume of H_2 is collected over water at 28°C (see Appendix C) by reaction of 1.25 g of Mg with 50.0 mL of 0.10 *M* HCl? The barometer records an atmospheric pressure of 748 mm Hg.

9.35 A 1.86-g sample of lead nitrate, $Pb(NO_3)_2$, is heated in an evacuated cylinder with a volume of 1.62 L. The salt decomposes when heated:

$$2Pb(NO_3)_2(s) \longrightarrow 2PbO(s) + 4NO_2(g) + O_2(g)$$

Assuming complete decomposition, what is the pressure in the cylinder after decomposition and cooling to a final temperature of 30°C?

9.36 Assume that a single cylinder of an automobile engine has a volume of 500 cm^3. (a) If the cylinder is full of air at 70°C and 0.980 atm, how many moles of O_2 are present? (The mole fraction of O_2 in dry air is 0.2095.) (b) How many grams of C_8H_{18} could be combusted by this quantity of O_2 assuming complete combustion with formation of CO_2?

9.37 When nitroglycerin, $C_3H_5N_3O_9$, explodes, all of the products are gases: N_2, CO_2, NO, and H_2O. If 5.00 g of nitroglycerin explodes, what total pressure is exerted, in atm, in a 1.00-L container at 1500°C, assuming that the container can withstand such pressures?

[9.38] A mixture is known to contain only KCl and $KClO_3$. When 9.08 g of the mixture is heated, 1.60 L of O_2 gas is collected over H_2O at 23°C and 0.966 atm total pressure. Assuming that all oxygen has been expelled from the $KClO_3$ in the mixture, forming KCl, calculate the percent (by mass) $KClO_3$ in the original mixture.

Kinetic-molecular theory; Graham's law

9.39 What change or changes in the state of a gas bring about each of the following effects? (a) The number of impacts per unit time on a given container wall increases. (b) The average energy of impact of molecules with the wall of the container decreases. (c) The average distance between gas molecules increases. (d) The average speed of molecules in the gas mixture is increased.

9.40 Suppose that we have two 1-L flasks, one containing N_2 at STP, the other containing SF_6 at STP. How do these systems differ with respect to (a) the average kinetic energies of the molecules; (b) the total number of collisions occurring per unit time with the container walls; (c) the shapes of the distribution curves of molecular speeds; (d) the relative rates of effusion through a pinhole leak?

9.41 Vessel A contains H_2 gas at 0°C and 1 atm. Vessel B contains O_2 gas at 20°C and 0.5 atm. The two vessels have the same volume. (a) Which vessel contains more molecules? (b) Which contains the most mass? (c) In which vessel is the average kinetic energy of molecules higher? (d) In which vessel is the average speed of molecules higher?

9.42 What experimental observations justify the assumption that the volumes of gas molecules are negligible compared with the volume of the gas container?

9.43 A sample of N_2 gas is initially at 1 atm pressure and 300 K. It is heated to 600 K and its volume is simultaneously doubled. What effect do these changes have on (a) the rms speed of the N_2 molecules; (b) the average kinetic energy of the N_2 molecules; (c) the number of collisions per unit time with a unit area of the container wall?

9.44 Calculate the rms speed, in m/s, for each of the following cases: (a) Cl_2 at 273 K; (b) H_2 at 273 K; (c) H_2 at 127°C.

9.45 What is the ratio of the average kinetic energies of H_2 and N_2 molecules at 25°C? What is the ratio of the rms speeds of the molecules of the two gases?

9.46 Which gas in each of the following pairs should have the higher thermal conductivity: (a) CO or He; (b) N_2 or CO_2; (c) Ar or O_2?

9.47 Which gas will effuse faster, NH_3 or CO_2? What are their relative rates of effusion?

9.48 Calculate the ratio of rates of effusion of (a) He and Ne; (b) SO_2 and SO_3; (c) SF_4 and SF_6.

9.49 A gas of unknown molecular mass is allowed to effuse through a small opening under constant pressure conditions. It required 88 s for 1 L of the gas to effuse. Under identical experimental conditions it required 28 s for 1 L of O_2 gas to effuse. Calculate the molecular weight of the unknown gas. (Remember that the faster the rate of effusion, the shorter the time required for effusion of 1 L; that is, rate and time are inversely proportional.)

9.50 Suppose that a nozzle system is being designed for a rocket in which the thrust will come from reaction of H_2 with F_2, a highly exothermic reaction. The two substances will be effused into the reaction zone through a series of tiny holes, all the same size. What should be the ratio of the number of holes for F_2 to those for H_2 to maintain a 1:1 stoichiometry in the reaction zone?

Departures from ideal-gas behavior

9.51 List two conditions under which deviations from ideal behavior are observed. Give two reasons for such deviations.

9.52 Which of the following gases would you expect to show the largest negative departure from the PV/RT relationship expected for an ideal gas: C_2H_6, N_2, He, or CH_4? For which of these gases should the correction for finite volume of the gas molecules be largest?

9.53 For each of the following pairs of gases, indicate which you would expect to deviate more from the PV/RT relationship expected for an ideal gas: (a) O_2 or UF_6; (b) BF_3 or $SiCl_4$; (c) CO_2 or SO_2.

9.54 It turns out that the van der Waals constant b is equal to four times the total volume actually occupied by the molecules of a mole of gas. Using this figure, calculate the fraction of the volume in a container actually occupied by Ar atoms (a) at STP; (b) at 100 atm pressure and 0°C.

(Assume for simplicity that the ideal-gas equation still holds.)

9.55 Calculate the pressure that CCl_4 will exert at 80°C if 1.00 mol occupies 30.0 L assuming that (a) CCl_4 behaves as an ideal gas; (b) it obeys the van der Waals equation. (Values for the van der Waals constants are given in Table 9.4.)

[9.56] Calculate the density of He at 100 atm pressure and 150 K using (a) the ideal-gas equation; (b) van der Waals equation.

[9.57] Figure 9.18 shows the PV/RT behavior for CH_4 as a function of temperature. Explain why the negative departures from the ideal-gas relationship decrease with increasing temperature. (A rather more subtle question: Why is the positive departure at high pressures larger when the gas is at lower temperatures?)

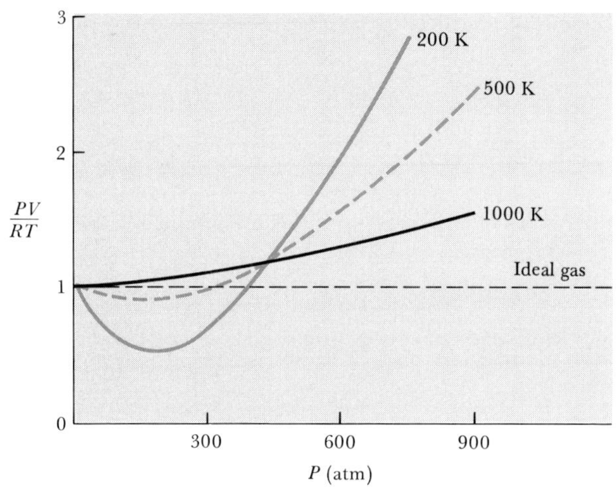

FIGURE 9.18

Additional exercises

9.58 Calculate the pressure, in pascals, exerted on the fluid in a hypodermic syringe if the diameter of the plunger is 1.0 cm and a downward force of 3.0 N is applied.

9.59 Suppose that a cylinder such as the one shown in Figure 9.1 had a radius of 5.00 cm and a height of 10.0 cm. What pressure, in pascals, does it exert on the table on which it stands if the cylinder is composed of copper (density $= 8.94\ g/cm^3$)?

9.60 At 25°C, the density of mercury is 13.6 g/mL while that of water is 1.00 g/mL. On a day when the atmospheric pressure is 735 mm Hg, how high would a column in a water-filled barometer be? (a) Ignore the vapor pressure of water. (b) Include the vapor pressure of water.

9.61 Suppose that the manometer illustrated in Figure 9.3(c) contains mineral oil (density $0.752\ g/cm^3$) as fluid in place of mercury, which has a density of $13.6\ g/cm^3$. The oil has no significant partial pressure of its own. When h is 17.5 cm and the atmospheric pressure is 0.980 atm, what is the pressure of the enclosed gas?

9.62 The airship *Hindenburg* had a volume of 2×10^8 L. (a) How many grams of hydrogen would have been needed to fill the airship at a pressure of 0.9 atm and temperature of 18°C? (b) How many grams of H_2O are produced by complete combustion of this quantity of H_2?

[9.63] An ideal gas at 650 mm Hg occupies a bulb of unknown volume. A certain portion of this gas is withdrawn and found to occupy $2.05\ cm^3$ at STP. The pressure of the gas remaining in the bulb is 590 mm Hg. Assuming that the pressure measurements of the gas in the bulb were made at the same temperature, 23°C, what is the volume of the bulb?

9.64 If the partial pressure of ozone, O_3, in the stratosphere is 3.0×10^{-3} atm and the temperature is 250 K, how many ozone molecules are present per liter volume?

9.65 A person exhales about 800 g of CO_2 a day. Find the pressure exerted by this quantity of CO_2 in a room measuring $3\ m \times 3\ m \times 2.5\ m$ if the temperature is 21°C.

9.66 A bubble of helium gas is trapped in a cavity of $0.172\ cm^3$ volume inside a mineral sample at a partial pressure of 125 atm at 18°C. What pressure does the helium exert after release from the mineral and storage in a bulb of 125-cm^3 volume at 23°C?

9.67 A student obtains the following data for the volume of air as a function of temperature at constant pressure:

Temperature (°C)	16	55	85	103	126	163
Volume (cm^3)	31	35	38	40	43	47

Make a graph of the data. Is Charles's law obeyed? Extrapolate the data carefully to zero volume. At which temperature does this intercept occur? What special significance does this temperature have?

9.68 Magnesium can be used as a "getter" in evacuated enclosures, to react with the last traces of oxygen. (The magnesium is usually heated by passing an electric current through a wire or ribbon of the metal.) If an enclosure of 0.283 L has a partial pressure of O_2 of 6.5×10^{-6} mm Hg at 27°C, what mass of magnesium will react according to the following equation?

$$2Mg(s) + O_2(g) \longrightarrow 2MgO(s)$$

[9.69] A mixture of methane, CH_4, and acetylene, C_2H_2, occupies a certain volume at a total pressure of 70.5 mm Hg. The sample is burned, forming CO_2 and H_2O. The H_2O is removed and the remaining CO_2 found to have a pressure of 96.4 mm Hg at the same volume and temperature as the original mixture. What fraction of the gas was acetylene?

9.70 Calcium hydride, CaH_2, reacts with water to produce H_2 gas:

$$CaH_2(s) + 2H_2O(l) \longrightarrow 2H_2(g) + Ca^{2+}(aq) + 2OH^-(aq)$$

This reaction is used to generate H_2 to inflate life rafts and for similar uses where a simple compact means of H_2 generation is desired. Assuming complete reaction with water, how many grams of CaH_2 are required to fill a balloon to a total pressure of 1.12 atm at 15°C if its volume is 5.50 L?

9.71 A sample of Ar is collected over water in a 1.00-L flask at 20°C. The total pressure of the wet gas in the flask is 735 mm Hg. When the volume of the wet gas is reduced to 0.500 L while still in contact with H_2O, the total pressure is 1452 mm Hg. Why doesn't a decrease to one-half volume cause the pressure to precisely double to 1470 mm Hg?

9.72 The density of dry air at 30.0°C, 720 mm Hg, is 1.104 g/L. Calculate the average molecular weight of the air.

[9.73] A glass vessel fitted with a stopcock has a mass of 337.428 g when evacuated. When filled with Ar it has a mass of 339.712 g. When evacuated and refilled with a mixture of Ne and Ar, under the same conditions of temperature and pressure, it weighs 339.146 g. What is the mole percentage of Ne in the gas mixture?

9.74 Calculate the density of an equimolar mixture of He and Ar at 82°C and 600 mm Hg.

9.75 Derive Equation 9.18, $u = \sqrt{3RT/\mathcal{M}}$, starting with Equation 9.14.

9.76 Suppose it has been suggested that a relatively rare isotope of carbon, ^{13}C, could be separated from the more abundant ^{12}C by using a diffusion process similar to that described in the text for UF_6, but using either CO or CO_2. Calculate the relative rates of diffusion for ^{12}CO and ^{13}CO and similarly for $^{12}CO_2$ and $^{13}CO_2$. Which substance would give the greater degree of separation?

9.77 A balloon made of rubber permeable to small molecules is filled with hydrogen, H_2. This balloon is then placed in a box that contains pure helium, He. Will the balloon expand or contract? Explain.

[9.78] Scandium metal reacts with excess hydrochloric acid to produce H_2 gas. When 2.25 g of scandium is treated in this way, it is found that 2.41 L of H_2 measured at 100°C and 722 mm Hg pressure is liberated. Write the balanced chemical equation for the reaction that occurred.

[9.79] Consider the experiment illustrated in Figure 9.19. A gas is confined in the leftmost cylinder under 1 atm pressure, and the other cylinder is evacuated. When the stopcock is opened, the gas expands to fill both cylinders. Only a very small temperature change is noted when this expansion occurs. Explain how this observation relates to assumption 4 of the kinetic-molecular theory (Section 9.7).

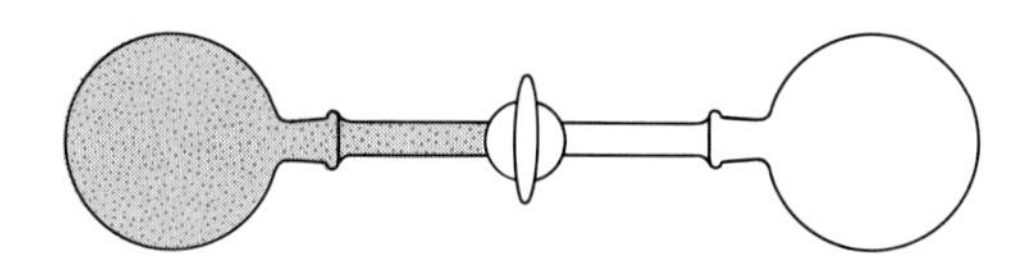

FIGURE 9.19

10 Liquids, Solids, and Intermolecular Forces

The water vapor in air (which we recognize as humidity), the water in a lake, and the ice in a glacier are all forms of the same chemical substance, H_2O. All have the same chemical properties; however, their physical properties differ greatly. The physical properties of any substance depend on the physical state of the substance. Some properties characteristic of each state of matter are listed in Table 10.1. In Chapter 9 we discussed the gaseous state in some detail. We now turn our attention to the physical properties of liquid and solids. You will remember that we could explain the most important physical properties of gases in terms of the kinetic-molecular theory. This same theory, with some modifications, can also help us to understand the characteristics of liquids and solids. In this chapter we will also examine those properties of the molecules themselves that are important in determining whether a substance is a gas, liquid, or solid under a given set of conditions. A key concept in comparing the properties of substances is that of intermolecular forces, the attractive forces responsible for keeping molecules together in liquids and solids. In this chapter we will be exploring the relationships among structure, intermolecular forces, and physical properties.

10.1 THE KINETIC-MOLECULAR DESCRIPTION OF LIQUIDS AND SOLIDS

In the kinetic-molecular theory a gas is viewed as a collection of widely separated molecules in constant, chaotic motion. The average kinetic energy of the molecules is much larger than the energy associated with the attractive forces between them. Indeed, in the derivation of the ideal-gas equation attractive forces are ignored altogether.

In liquids, the attractive forces between molecules have energies comparable to the kinetic energies of the molecules. The attractive forces are thus able to hold the molecules in close proximity; however, the intermolecular attractions are not sufficiently strong (relative to the kinetic energies of molecules) to keep the molecules from moving in a more or less chaotic fashion. Because the molecules in a liquid are close together, liquids have much larger densities than gases. They also have definite

TABLE 10.1 Some characteristic properties of the states of matter

Gas	1. Assumes both the volume and shape of container. 2. Is compressible. 3. Diffusion occurs rapidly. 4. Flows readily.
Liquid	1. Assumes the shape of the portion of the container it occupies. 2. Does not expand to fill container. 3. Is virtually incompressible. 4. Diffusion occurs slowly. 5. Flows readily.
Solid	1. Retains its own shape and volume. 2. Is virtually incompressible. 3. Diffusion occurs extremely slowly. 4. Does not flow.

volumes, independent of the shape or size of their container. Because there is so little free space between molecules, liquids are much less compressible than gases. Because the molecules are free to move relative to each other, liquids do not have definite shape; they can be poured, and they flow to assume the shape of their container.

In solids, the intermolecular attractions are sufficiently strong relative to kinetic energies so that the molecules are virtually locked in place. Each molecule takes up a certain position relative to its neighbors, often in a highly regular pattern that extends through the solid. Solids that possess highly ordered structures are said to be crystalline. The transition from a liquid to a crystalline solid is rather like the change that occurs on a military parade ground when the troops are called to formation. Solids, like liquids, are not very compressible because the molecules are close together, with little free space available between them. Because the particles of a solid are not free to undergo long-range movement (translational motion), solids are rigid. Although translational motion is restricted, the molecules within a solid may undergo vibrational motion; that is, they may move back and forth periodically about the positions they occupy.

Figure 10.1 illustrates schematically the comparisons among the three states of matter based on the kinetic-molecular model. The particles that

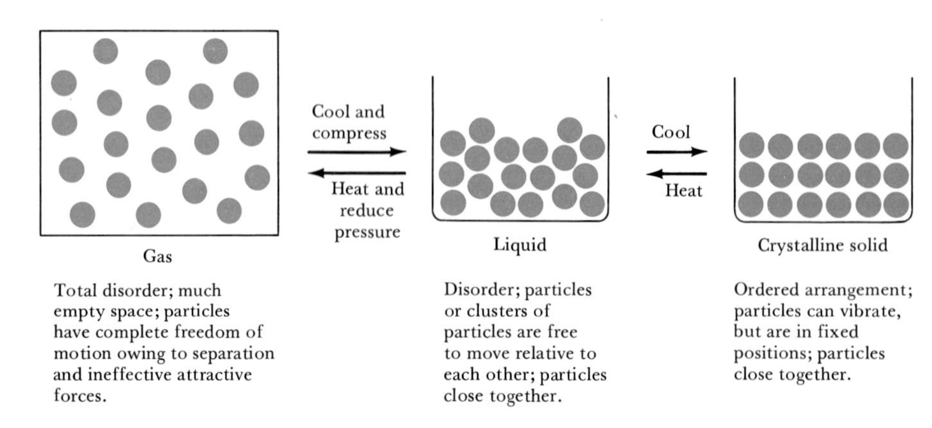

FIGURE 10.1 Molecular-level comparison of gases, liquids, and solids. The density of particles in the gas phase is exaggerated compared with the usual situation.

compose the substance can be individual atoms, as in Ar; molecules, as in H_2O; or ions, as in NaCl. For convenience, we will use the term "molecule" to refer to all these possibilities. The key factor that determines the physical state of a substance is the average kinetic energy of the molecules relative to the average energy of the attractive forces between them. Substances that exist as gases at room temperature possess weaker intermolecular attractions than do those that are liquids; those that are solids possess stronger intermolecular attractions. Conversions from one state to another can be brought about by heating or cooling, thereby changing the average kinetic energy of the molecules. For example, NaCl, which is a solid at room temperature, melts at 804°C and boils at 1465°C when heated under 1 atm pressure. Ar, which is a gas at room temperature, can be liquefied at −186°C and solidified at −189°C when the pressure is 1 atm. Before we examine liquids, solids, or intermolecular forces in more detail, let's consider the interconversions that can occur between the states of matter.

10.2 EQUILIBRIA BETWEEN PHASES

Imagine an experiment in which we begin with a quantity of ethyl alcohol, also called ethanol, in a closed container, with a large space above the liquid. Suppose that we could somehow begin the experiment with all the ethanol in liquid form. The pressure in the container under these conditions would be zero, as illustrated in Figure 10.2(*a*). We would find that the pressure in the container quickly rises as a function of time. After a short time the pressure would attain a constant value, termed the **vapor pressure** of liquid ethanol [Figure 10.2(*b*)].

We can account for these observations by using the kinetic-molecular theory. We know that the molecules of liquid ethanol are in constant motion. At any given temperature some of the molecules possess more than the average thermal energy, some less. That is, there is a distribution of molecular energies in the liquid similar to the distribution in a gas, illustrated in Figure 9.9.

At any instant some of the molecules at the surface of the liquid possess sufficient energy to escape from the attractive forces of their neighbors. Transfer of molecules into the gas phase is called **vaporization** or **evaporation.** The movement of molecules from the liquid to the gas phase goes on continuously. However, as the number of gas-phase molecules increases, the probability increases that a molecule from the gas phase will strike the liquid surface and stick there. We call this process

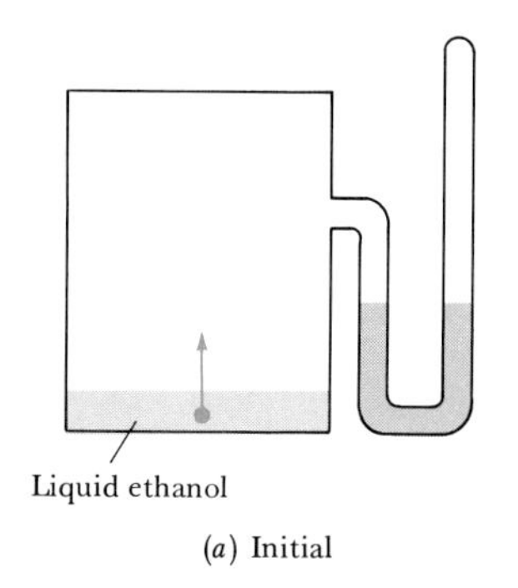

FIGURE 10.2 Schematic representation of the vaporization of ethanol in a closed container. At equilibrium the rate of condensation and the rate of vaporization are equal. This produces a stable vapor pressure, one that does not change with time as long as the temperature remains constant.

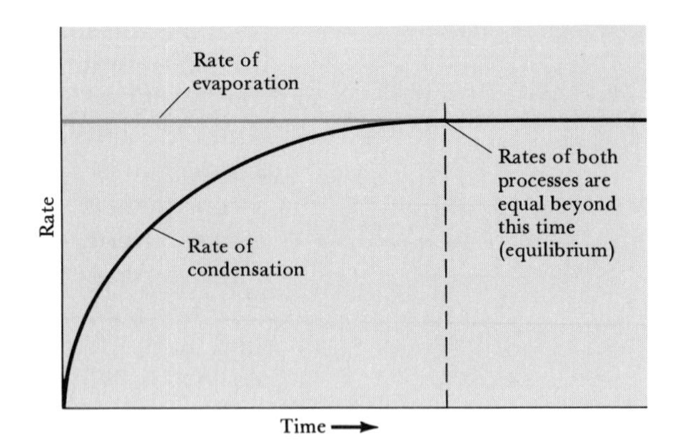

FIGURE 10.3 Comparison of the rates of evaporation and condensation of a liquid as a function of time as the system approaches equilibrium.

condensation. Eventually, the two opposing processes occur at an equal rate, as illustrated in Figure 10.3. The pressure in the gas phase at this point becomes constant.

The system we have just discussed provides a simple example of dynamic equilibrium, in which two opposing processes take place at equal rates. To the observer it may appear that nothing is taking place; in our example of ethyl alcohol, the pressure remains constant at some value determined by the temperature of the liquid. In fact, a great deal is happening; molecules continuously pass from the liquid state to the gas state, and from the gas state to liquid state. All equilibria between matter in different phases possess this dynamic character.

Although equilibria can exist simultaneously between all three states of matter, we usually encounter equilibria occurring between just two states, as in our example of the equilibrium between liquid and gaseous ethanol. Figure 10.4 summarizes the possible two-phase equilibria and the terms associated with them. Some of these are more familiar than others, but a little thought will bring to mind several examples of each possible kind of phase equilibrium.

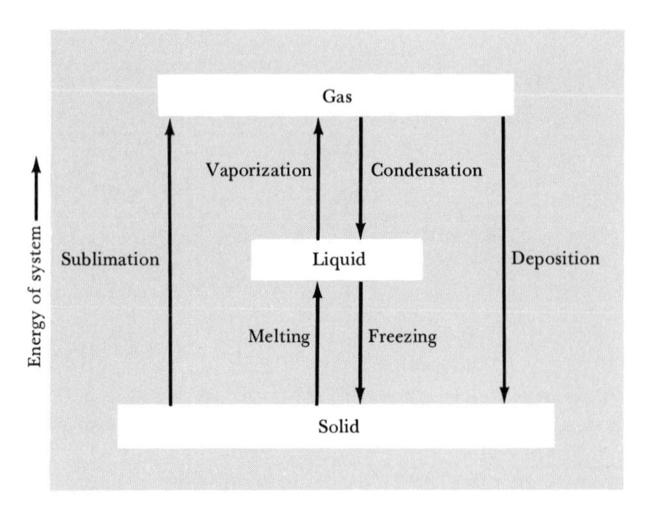

FIGURE 10.4 Possible phase changes and the terms associated with them.

SAMPLE EXERCISE 10.1

Using water as the substance, give examples of all the possible equilibria between two different phases.

Solution: *Vaporization* of liquid water occurs from lakes or other water bodies in the summer months. *Condensation* of water vapor occurs on a cold window surface or on the lawn in the early morning hours. Ice *melts* in the spring, and lakes *freeze* in winter. When air moves over snow or ice at temperatures below freezing, the solid *sublimes*. The formation of snow crystals in clouds in the winter provides an example of *deposition*.

The change of matter from one state to another is called a phase change. Each of the phase changes diagramed in Figure 10.4, when it occurs at constant temperature, has associated with it an enthalpy change. As we have already seen, energy is required to overcome the attractive forces between molecules in vaporization, the phase change from the liquid to the gaseous state. Similarly, we expect that sublimation will require energy; the attractive forces that bind the molecules to one another in the solid must be overcome to produce a gas of widely separated particles. Melting also requires energy; the regularity of the solid state, in which molecules are packed to attain the most effective intermolecular interactions, is disrupted when the solid melts to form the liquid state. Thus vaporization, sublimation, and melting are all endothermic processes; an input of heat is required for them to proceed. For example, an input of 2.26 kJ is required to cause the vaporization of 1 g of water at 100°C. The enthalpies of vaporization, sublimation, and melting are thus positive quantities.

Condensation is just the reverse of the process of vaporization. Similarly, deposition is the reverse of sublimation, and freezing is the reverse of melting. Thus the enthalpies of condensation, deposition, and freezing are the same magnitude but opposite in sign from the enthalpies of vaporization, sublimation, and melting, respectively. The melting process is also referred to as fusion; thus the enthalpy of melting is frequently called the enthalpy of fusion, or heat of fusion. Incidentally, the freezing point of a substance is identical to its melting point; the two differ only in the temperature direction from which the phase change is approached. In other words, the freezing point is a property of the liquid, whereas the melting point is a property of the solid.

ΔH for Phase Changes

Figure 10.5 shows how the enthalpy of 1 mol of H_2O changes as it is heated from -25°C to 125°C. The vertical jumps in enthalpy at the melting point, 0°C (*BC*), and the boiling point, 100°C (*DE*), correspond to the molar enthalpy of fusion and the molar enthalpy of vaporization, respectively. Temperature remains constant during phase changes, because the added energy is used to overcome attractive forces between molecules. The enthalpy changes of the pure ice, liquid water, and water vapor phases as they are heated depend on the molar heat capacities of these phases (Section 4.6).

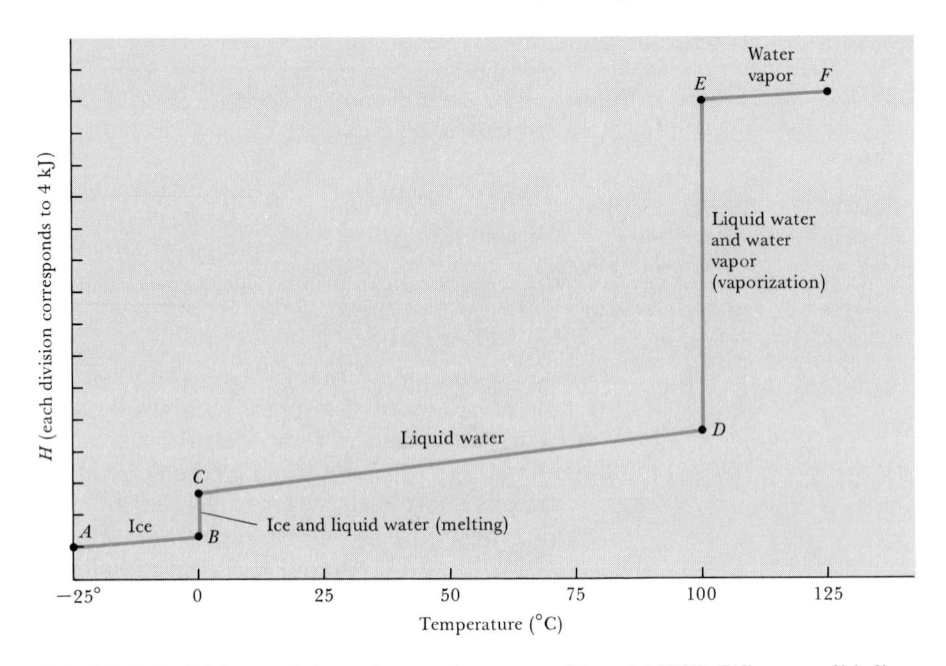

FIGURE 10.5 Molar enthalpy of water between −25 and 125 °C. When a solid, liquid, or gas is heated, its enthalpy always increases. However, its temperature does not change during phase changes such as that associated with the melting ($B \longrightarrow C$) or vaporization ($D \longrightarrow E$) processes. The molar heat capacity of ice is 37.6 J/mol-°C; the molar heat capacity of liquid water is 75.3 J/mol-°C; the molar heat capacity of water vapor is 33.1 J/mol-°C. The molar enthalpy of fusion is 6.02 kJ/mol, and the molar enthalpy of vaporization is 40.67 kJ/mol.

SAMPLE EXERCISE 10.2

Calculate the enthalpy change associated with converting 1.00 mol of ice at −25 °C to water vapor at 125 °C. Use the data given in the caption for Figure 10.5.

Solution: The heat required to bring the ice from −25 °C to 0 °C is given by

$$\Delta H = nC_p\,\Delta T$$
$$= (1.00\text{ mol})\left(37.6\frac{\text{J}}{\text{mol-°C}}\right)(25\text{ °C})$$
$$= 940\text{ J} = 0.94\text{ kJ}$$

The heat required to melt 1.00 mol of ice at 0 °C is given by the molar enthalpy of fusion, 6.02 kJ. Heating liquid water from 0 °C to 100 °C then requires

$$\Delta H = nC_p\,\Delta T$$
$$= (1.00\text{ mol})\left(75.3\frac{\text{J}}{\text{mol-°C}}\right)(100\text{ °C})$$
$$= 7530\text{ J} = 7.53\text{ kJ}$$

The heat required to vaporize the water at 100 °C is given by the molar enthalpy of vaporization, 40.67 kJ. Finally, to heat the water vapor from 100 °C to 125 °C requires

$$\Delta H = nC_p\,\Delta T$$
$$= (1.00\text{ mol})\left(33.1\frac{\text{J}}{\text{mol-°C}}\right)(25\text{ °C})$$
$$= 830\text{ J} = 0.83\text{ kJ}$$

Thus the total enthalpy change is 0.94 kJ + 6.02 kJ + 7.53 kJ + 40.67 kJ + 0.83 kJ = 55.99 kJ.

10.3 PROPERTIES OF LIQUIDS

We have seen that molecules may escape from the surface of a liquid into the gas phase by vaporization or evaporation. The weaker the attractive forces between molecules in the liquid phase, the more readily vaporiza-

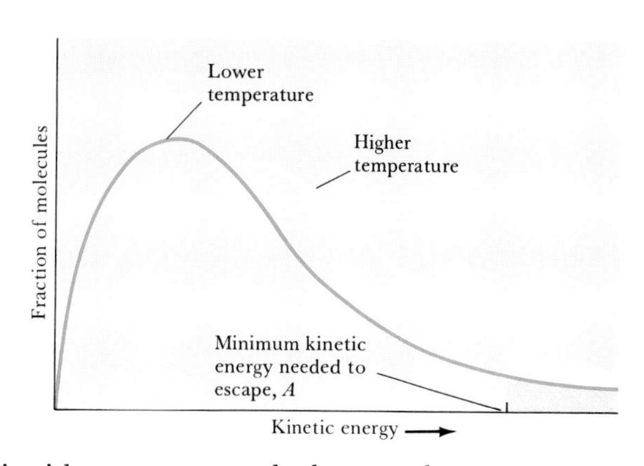

FIGURE 10.6 Distribution of kinetic energies of surface molecules of a hypothetical liquid at two different temperatures. The same minimum kinetic energy, A, is needed to escape at both temperatures. This energy reflects the magnitude of the attractive forces between molecules. The fraction of molecules having sufficient kinetic energy to escape the liquid (the shaded area) is larger at the higher temperature.

tion occurs. When two liquids are compared, the one that evaporates more readily is described as being the more volatile. For example, alcohol (ethanol) is more volatile than water. The rate of vaporization increases with increasing temperature. Figure 10.6 provides a graphic explanation for this behavior. In that figure the distribution of kinetic energies of the particles at the surface of the liquid is compared at two temperatures with the attractive forces at the surface. Notice that the distribution curves are like those shown earlier for gases (Figure 9.9); average kinetic energy increases with increasing temperature. Therefore, as the temperature of a liquid is increased, there is an increase in the number of particles having sufficient kinetic energy to escape from the surface. Loss of particles with high kinetic energy causes the average kinetic energy of the particles remaining in the liquid to decrease. Anyone getting out of a swimming pool, particularly on a windy day, has experienced the cooling that accompanies evaporation. The molar enthalpies of vaporization (ΔH_v) of some common liquids at their boiling points are summarized in Table 10.2.

TABLE 10.2 Enthalpies of vaporization of some common liquids at their boiling points

Substance	Formula	ΔH_v (kJ/mol)	Boiling point (°C)
Benzene	C_6H_6	30.8	80.2
Ethanol	C_2H_5OH	39.2	78.3
Ether	$C_2H_5OC_2H_5$	26.0	34.6
Mercury	Hg	59.3	356.9
Methane	CH_4	10.4	−164
Water	H_2O	40.7	100

The rate of metabolic heat generation is such that an adult human's body temperature would rise 1 to 30°F per hour (depending on the level of activity) if the heat were not dissipated. This rise would result in a heat stroke at 106°F and death at 110 to 112°F. Because heat does not flow spontaneously from colder to hotter objects, cooling by simple conductive heat loss from the body is not possible when the outside temperature is greater than body temperature. Our bodies rely for cooling on the heat absorbed by evaporation of water; we are cooled by sweating. Thus one of the significant roles played by water in our bodies is that of a cooling agent. We may note that water is particularly well suited for this role; it takes more energy to vaporize a gram of water than to vaporize an equal mass of any known liquid substance. Comparing water and mercury (Table 10.2) we have, for water,

$$\left(40.7\ \frac{kJ}{mol}\right)\left(\frac{1\ mol}{18.0\ g}\right) = 2.26\ kJ/g$$

For mercury we have

$$\left(59.3\ \frac{kJ}{mol}\right)\left(\frac{1\ mol}{200.5\ g}\right) = 0.296\ kJ/g$$

Note also that when the surrounding air is humid and thus contains many water molecules, the net rate of evaporation of water is slower, and we feel more uncomfortable.

Vapor Pressure

Liquids vary greatly in their vapor pressures at any given temperature. Gasoline or water evaporates fairly quickly when exposed to moving air, whereas engine oil exposed to the air at room temperature seems not to be volatile at all. The volatility of a liquid is determined by the magnitude of the intermolecular forces that constrain the molecules to remain together in the liquid. (We will consider the origins of intermolecular forces in Section 10.4.) Volatility increases with increasing temperature, as the average kinetic energy of the molecules increases in relation to the intermolecular forces. Figure 10.7 depicts the variation in vapor pressure with temperature for four common substances that differ greatly in volatility. Note that in all cases the vapor pressure increases quite nonlinearly with increasing temperature.

As we shall see in a few of the end-of-chapter exercises, it is usually possible to represent the dependence of the vapor pressure, P, on temperature as a log function of the form

$$\log P = \frac{-\Delta H_v}{2.303RT} + C \qquad [10.1]$$

In this equation T is absolute temperature, R is the gas constant (Table 9.2), ΔH_v is the enthalpy of vaporization, and C is a constant. This equation, called the Clausius-Clapeyron equation, tells us that a graph of $\log P$ versus $1/T$ should be linear. The slope of the line can be used to calculate the enthalpy of vaporization.

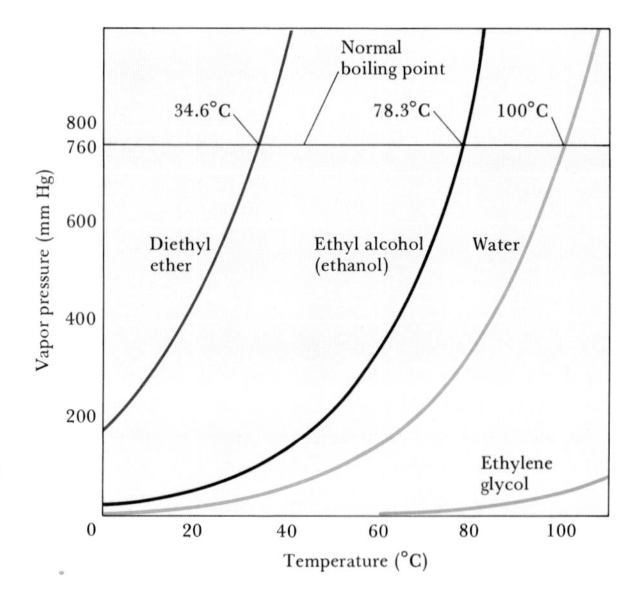

FIGURE 10.7 Equilibrium vapor pressures of four common liquids shown as a function of temperature. The temperature at which the vapor pressure is 760 mm Hg is the normal boiling point of the liquid.

Boiling Points

A liquid is said to boil when vapor bubbles form in the interior of the liquid; this condition occurs when the vapor pressure equals the external pressure acting on the liquid's surface. Consequently, the boiling point of a liquid depends on pressure. From Figure 10.7 it can be seen that the boiling point of water at 1 atm pressure (760 mm Hg) is 100°C. The boiling point of a liquid at 1 atm pressure is called its normal boiling point. At 650 mm Hg water boils at 96°C.

SAMPLE EXERCISE 10.3

What is the boiling point of ethanol at 400 mm Hg?

Solution: From Figure 10.7 it can be seen that the boiling point must be about 64°C, as compared to the normal boiling point of 78.3°C.

The fact that boiling points vary with pressure is utilized in a pressure cooker. The time required to cook food depends on the temperature. As long as water is present, the maximum temperature of the food is the boiling point of water. Pressure cookers are sealed and allow steam to escape only when it exceeds a predetermined pressure; the pressure above the water can therefore increase above atmospheric pressure. The higher pressure causes water to boil at a higher temperature, thereby allowing the food to get hotter. It therefore cooks more rapidly. The effect of pressure on boiling point also explains why it takes longer to cook food at higher elevations than at sea level; at higher altitudes the atmospheric pressure is lower, and water boils at a lower temperature.

Critical Temperature and Pressure

We have seen that the transition from the gaseous to the liquid state is promoted by the intermolecular attractive forces that cause condensation. On the other hand, condensation is opposed by the kinetic energies of the molecules, which keep them in independent motion.

As the temperature is raised, the kinetic energies of molecules increase in relation to intermolecular attractions, and a gas becomes more difficult to liquefy. Consequently, the pressure required to condense a gas to a liquid increases. Finally, a temperature is reached at which no amount of pressure, however great, causes the gas to pass from the gaseous state to the liquid state. The highest temperature at which a substance can exist as a liquid is called its critical temperature. The critical pressure is the pressure required to bring about liquefaction at this critical temperature. The critical temperatures and pressures of substances are of considerable practical importance to engineers and others working with gases. The relative values of these quantities among different substances also provide some indication of the relative magnitudes of intermolecular forces. Table 10.3 shows the critical temperatures and pressures for the noble gas elements and a few substances composed of nonpolar diatomic molecules. The critical temperature and pressure increase in the same way as the boiling point.

TABLE 10.3 Molecular masses, normal boiling points, enthalpies of vaporization, and critical temperatures and pressures of the noble gas elements and a few substances composed of nonpolar diatomic molecules

Substance	Molecular mass (amu)	Normal boiling point (K)	ΔH_v (kJ/mol)[a]	Critical temperature (K)	Critical pressure (atm)
He	4	4.2	0.081	5.2	2.26
Ne	20	27	1.76	44.4	25.9
Ar	40	87	6.52	151	48
Kr	84	121	9.03	210	54
Xe	131	164	12.64	290	58
Rn	222	211	16.78	377	62
H_2	2	20	0.903	33.2	12.8
N_2	28	77	5.58	126	33.5
O_2	32	90	6.82	154	49.7
Cl_2	71	239	20.40	417	76.1

[a] Measured at the temperature of the normal boiling point.

At ordinary pressures, a substance above its critical temperature behaves as an ordinary gas. However, as it is subjected to increasing pressure, up to several hundred atmospheres, its character changes. Like a gas it still expands to fill the confines of its container. Its density, however, comes to approximate that of a liquid. For example, the critical temperature of water is 374.4°C and its critical pressure is 217.7 atm. At this temperature and pressure, the density of water is 0.4 g/mL. Thus it is perhaps more appropriate to speak of a substance under such conditions as a *supercritical fluid* rather than as a gas.

Like liquids, supercritical fluids can behave as solvents, dissolving a wide range of substances. This ability forms the basis of a system for separating the components of mixtures, known as supercritical fluid extraction. The solvent power of a supercritical fluid increases as its density increases. Conversely, lowering its density (either by decreasing pressure or increasing temperature) causes the supercritical fluid and the dissolved material to separate. With skillfull manipulation of temperature and pressure, it is possible to separate the components of very complicated mixtures.

The process of supercritical fluid extraction is now under extensive study in the chemical, food, drug, and energy industries. A process to remove caffeine from green coffee beans by extraction with supercritical carbon dioxide has been in commercial operation for several years. (The critical temperature of CO_2 is 31.1°C and its critical pressure is 73.0 atm.) At the proper temperature and pressure, the CO_2 removes caffeine from the beans, but leaves the flavor and aroma components, producing decaffeinated coffee. Other applications of supercritical CO_2 extraction include removal of nicotine from tobaccos and removal of oil from potato chips, producing a lower calorie product that is less greasy but has the same flavor and texture.

Viscosity

Some liquids literally flow like molasses, whereas others flow quite easily. The resistance of liquids to flow is referred to as their **viscosity.** The larger the viscosity, the more slowly the liquid flows. Liquids such as molasses or motor oils are relatively viscous; water and organic liquids such as carbon tetrachloride are not. Viscosity can be measured by measuring how long it takes a certain amount of a liquid to flow through a thin tube, under gravitational force. In another method, the rate of fall of steel spheres through the liquid is measured. The spheres fall more slowly through the more viscous liquids.

Viscosity is related to the ease with which individual molecules of the liquid can move with respect to one another. It thus depends on the attractive forces between molecules, and on whether there are structural

TABLE 10.4 Viscosities and surface tensions of some common liquids at 20°C, and of water at several temperatures.

Substance	Formula	Viscosity ($N\text{-}s/m^2$)[a]	Surface tension, γ (J/m^2)
Benzene	C_6H_6	0.65×10^{-3}	2.89×10^{-2}
Ethanol	C_2H_5OH	1.20×10^{-3}	2.23×10^{-2}
Ether	$C_2H_5OC_2H_5$	0.23×10^{-3}	1.70×10^{-2}
Glycerin	$C_3H_8O_3$	1490×10^{-3}	6.34×10^{-2}
Mercury	Hg	1.55×10^{-3}	46×10^{-2}
Water at:	H_2O		
20°C		1.00×10^{-3}	7.29×10^{-2}
40°C		0.652×10^{-3}	6.99×10^{-2}
60°C		0.466×10^{-3}	6.70×10^{-2}
80°C		0.356×10^{-3}	6.40×10^{-2}

[a]This is the SI unit: $1\ N\text{-}s/m^2 = 1\ kg/m\text{-}s$.

features that cause the molecules to become entangled. Viscosity decreases with increasing temperature, because at higher temperature the greater average kinetic energy of the molecules more easily overrides the attractive forces between molecules. The viscosities of some common liquids are listed in Table 10.4.

Surface Tension

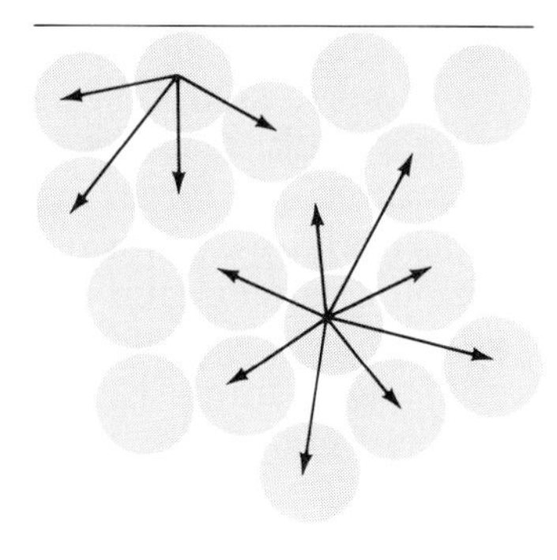

FIGURE 10.8 Molecular-level view of the unbalanced intermolecular forces on the surface of a liquid that result in surface tension.

Liquids have a tendency to assume a minimum surface area. This minimum is achieved when the liquid has a spherical shape. For example, when water is placed on a waxy surface, it "beads up," forming distorted spheres. The state of minimum energy is that state of minimum surface area; energy must therefore be supplied to increase the surface area. The energy required to increase the surface area of a liquid by a unit amount is known as its **surface tension.** The origin of surface tension is an imbalance of forces at the surface of the liquid, as shown in Figure 10.8. There is a net inward pull on the surface that contracts the surface and makes the liquid behave almost as if it had a skin. This effect permits a carefully placed needle to float on the surface of water or some insects to "walk" on water even though their densities are greater than that of water (Figure 10.9).

FIGURE 10.9 Surface tension permits an insect such as the water strider to "walk" on water. (*© 1977. Michael P. Godomski, National Audubon Society, Photo Researchers, Inc.*)

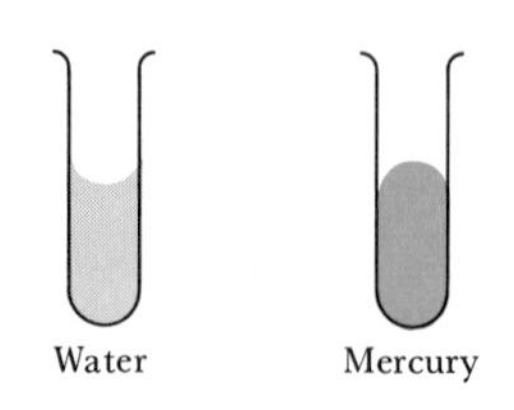

FIGURE 10.10 Comparison between the shape of the meniscus of water in a glass tube and the shape of the mercury meniscus.

The forces between like molecules that are reflected in a substance's vapor pressure, boiling point, heat of vaporization, viscosity, and surface tension are called cohesive forces. The forces between unlike substances, such as water and glass, are called adhesive forces. In a glass tube, the adhesive forces between water and glass are sufficiently strong relative to the cohesive forces for water to form a concave-upward surface (Figure 10.10). Such a curved surface on a liquid is known as a meniscus. For mercury, cohesive forces are greater than the adhesive forces with glass; mercury does not adhere to glass and a concave-downward meniscus is observed (Figure 10.10).

When a liquid adheres to or wets the walls of a tube as water adheres to glass, the liquid is drawn up the tube. This phenomenon is known as capillary rise, or capillary action. Wetting of the walls of a tube tends to increase the surface area of the liquid. Surface tension tends to reduce this area, and consequently the liquid is drawn up the tube. One method of determining surface tension involves measuring the height to which a liquid rises in a tube of known radius. The surface tensions of several common liquids are given in Table 10.4. The surface tension of $7.26 \times 10^{-2}\ J/m^2$ for water at 20°C indicates that an energy of $7.26 \times 10^{-2}\ J$ must be supplied to increase the surface area of a given amount of water by 1 m^2. Surface tension generally decreases with increasing temperature, as illustrated by the data for water in Table 10.4.

10.4 INTERMOLECULAR ATTRACTIVE FORCES

We have made repeated reference to the influence of the strengths of intermolecular forces of attraction on the physical properties of liquids. Indeed, the physical properties of matter ultimately depend on intermolecular attractive forces. For example, enthalpies of vaporization, boiling points, surface tensions, and viscosities of liquids generally increase in magnitude as intermolecular attractive forces increase. Such forces are also directly related to properties of solids, such as enthalpy of fusion, melting point, and hardness. It is therefore desirable to understand the origins and relative strengths of these forces. Rather than continuing to refer to them in a vague way, then, let's examine these forces before moving on to consider the properties of solids.

We have already seen that covalent bonds hold atoms together *within* molecules (Chapter 7). The fact that a molecular substance such as H_2O exists in the solid and liquid phases indicates that there are also attractive forces *between* molecules. These are the attractive forces that we have referred to as intermolecular forces. The magnitudes of intermolecular forces vary over a wide range. However, they are generally much smaller than those of the covalent bonds that bind atoms together in the same molecule. For example, only 16 kJ/mol is required to overcome the intermolecular attractions between HCl molecules in liquid HCl in order to vaporize it. By contrast, the energy required to dissociate HCl into H and Cl atoms is 431 kJ/mol.

In the introduction to Chapter 7, we noted that there are three types of strong chemical forces: ionic bonds, covalent bonds, and metallic bonds. We now consider five relatively weaker types of forces: ion-dipole attractions, ion-induced dipole attractions, dipole-dipole attractions, dispersion forces, and hydrogen bonds. We will refer to these weak forces

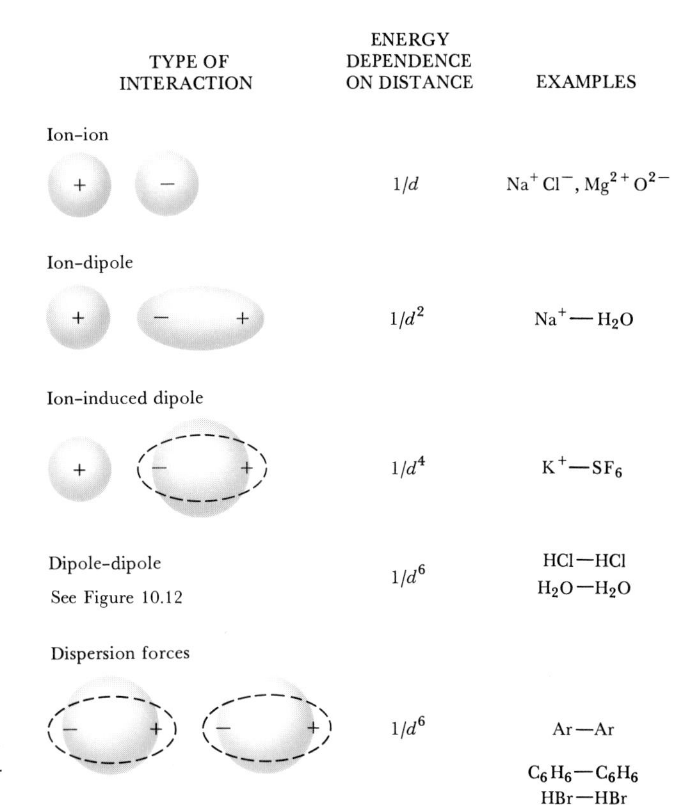

FIGURE 10.11 Illustration of the various types of intermolecular forces. Note that the ion-induced dipole and dispersion forces depend on a distortion (dashed lines) of the electron distribution in a nonpolar atom or molecule.

collectively as intermolecular forces even though they may involve atoms and ions as well as molecules.

You will recall that the energy of interaction between two ions depends on the magnitude of the charge on each ion, Q_1 and Q_2, and on the distance between the centers on the ions, d: $E = Q_1Q_2/d$ (Section 7.2).* Each of the intermolecular forces that we have mentioned can also be considered to be essentially electrostatic in nature. The energies of these interactions can be expressed as functions that depend on how far apart the particles are. The distance dependence is important, because it determines how close two particles must approach before the interaction between them becomes significant. An intermolecular force that varies as $1/d^2$, say, where d is the distance between centers, will operate over a much larger range of distance than an intermolecular attraction that varies as $1/d^6$. As an example, let us suppose that $d = 1$. Then $1/d^2$ and $1/d^6$ both equal 1. Now let the distance d double, to 2. Then $1/d^2$ equals 0.25, whereas $1/d^6$ equals only 0.016. It is clear that as the distance d increases the intermolecular force that varies as $1/d^6$ decreases much more rapidly than the intermolecular force that varies as $1/d^2$. The classes of intermolecular force we are about to discuss are illustrated in Figure 10.11. Hydrogen bonds, which are not included in this figure, will be defined and described later in this section.

*The equation $E = Q_1Q_2/d$ holds only in vacuum. If the ions are immersed in some medium, the attraction, or repulsion, is decreased by a factor ε, known as the dielectric constant: $E = Q_1Q_2/\varepsilon d$. The magnitude of ε is characteristic of the medium ($\varepsilon = 1$ in vacuum; otherwise, $\varepsilon > 1$); the dielectric constant for H_2O is 81.7.

Ion-Dipole Forces

An **ion-dipole** force exists between an ion and a neutral polar molecule that possesses a permanent dipole moment. Recall that polar molecules are those to which a positive and a negative end can be assigned owing to molecular shape and to charge separation caused by unequal sharing of electrons (Section 8.2). A simple example is HCl (μ = 1.03 D). The end of the dipole possessing an opposite charge from that of the ion is attracted to the ion, as illustrated in Figure 10.11. The energy of the interaction between an ion and a dipole depends on the charge on the ion, Q, and the magnitude of the dipole moment of the dipole, μ: $Q\mu/d^2$, where d is the distance from the center of the ion to the midpoint of the dipole. Ion-dipole forces are especially important in solutions of ionic substances in polar liquids, as, for example, in a solution of NaCl in water. We will have more to say about such solutions later (Section 11.2).

SAMPLE EXERCISE 10.4

Predict the relative strengths of the ion-dipole attractions between H_2O and the following ions: Na^+, K^+, Mg^{2+}.

Solution: The factors to consider are the sizes and charges of the ions. The ions vary in radius in the order $K^+ > Na^+ > Mg^{2+}$ (Section 7.3). Because Mg^{2+} is smallest and has the highest charge, the Mg^{2+}-H_2O interaction is strongest. Because K^+ is largest, and has a charge of +1, it has the weakest interaction. Thus the relative strengths of ion-dipole attraction vary in the order $Mg^{2+} > Na^+ > K^+$.

Ion-Induced Dipole Forces

An ion may distort the electron density in a nearby atom or molecule, thereby inducing a dipole moment. For example, if the ion is positively charged, the electrons of the nearby molecule are drawn toward it. In the process, there is a separation of positive and negative charges as shown in Figure 10.11. That is, the electron cloud of the molecule is distorted so that it acquires an **induced dipole moment.** The attractive energy of interaction between an ion and an induced dipole varies as $1/d^4$. The strength of the interaction also depends on the charge of the ion and the **polarizability** of the molecule, that is, the extent to which its electron density can be distorted. The polarizability of a molecule depends on how tightly it holds electrons. In general, the more electrons in a molecule, and the more spread out or diffuse the electron cloud, the greater the polarizability of the molecule.

Dipole-Dipole Forces

Dipole-dipole forces exist between polar molecules. The sign and magnitude of the interaction varies with the relative orientations of the two dipoles, as illustrated in Figure 10.12. In crystalline solids, the dipolar molecules tend to orient themselves to maximize the attractive interactions between ends of unlike charge [Figure 10.12(*a*)]. However, other factors, such as the shapes of the molecules and location of the dipole

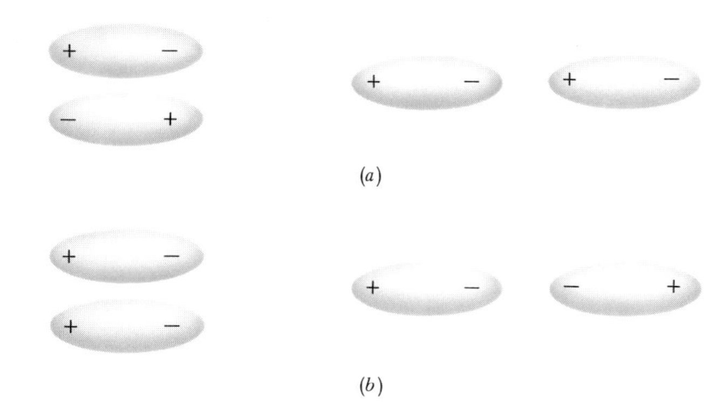

FIGURE 10.12 Variation in the dipole-dipole interaction with orientation. In (*a*) the dipoles are aligned so as to produce an attractive interaction. In (*b*) the interactions are repulsive.

within the molecule, also play a role in determining the orientations of the molecules.

In liquids, the dipolar molecules are free to move with respect to one another; they will sometimes be in an orientation that is attractive, sometimes in an orientation that is repulsive. The net effect, averaged over time, is an attractive energy that varies as μ^4/d^6, where d is the distance between the centers of the dipoles and μ is the dipole moment. Thus dipole-dipole forces in a liquid are significant only between polar molecules that are very close together. Nevertheless, we find clear evidence that *for molecules of approximately equal molecular mass, and size, intermolecular attractions increase with increasing polarity*. Thus the boiling points of the organic liquids listed in Table 10.5 increase with increasing magnitude of the dipole moment.

London Dispersion Forces

Table 10.3 contains the masses, normal boiling points, and enthalpies of vaporization of several substances composed either of atoms or simple, nonpolar molecules. The fact that these substances can be liquefied, some of them at quite high absolute temperatures, tells us that there must be attractive interactions between the molecules. Yet none of the various types of attractive forces we have discussed can apply here. The solution to this longstanding puzzle was provided in 1930 by Fritz London, who applied the new theory of quantum mechanics to the problem. London's contribution was essentially to distinguish what we see on the average from what we would see if we could freeze the charge distribution in a collection of molecules at any particular instant. In a collection

TABLE 10.5 Molecular masses, dipole moments, and boiling points of several simple organic substances

Substance	Molecular mass (amu)	Dipole moment, μ (D)	Boiling point (K)
Propane, $CH_3CH_2CH_3$	44	0.0	231
Dimethyl ether, CH_3OCH_3	46	1.3	249
Methyl chloride, CH_3Cl	50	2.0	249
Acetaldehyde, CH_3CHO	44	2.7	293
Acetonitrile, CH_3CN	41	3.9	355

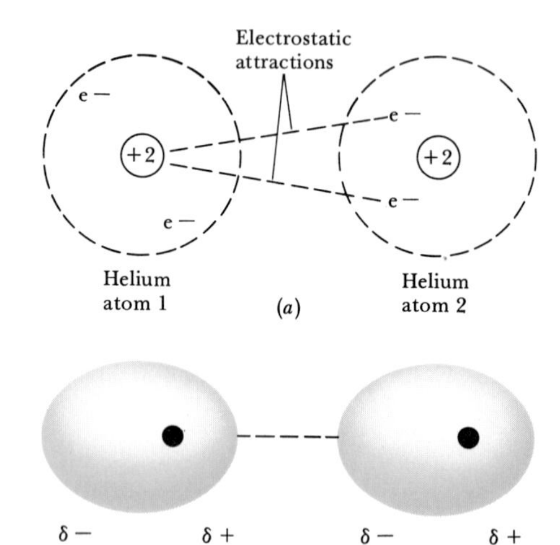

FIGURE 10.13 Two schematic representations of an instantaneous dipole on two adjacent helium atoms, showing the electrostatic attraction between them.

of helium atoms, the average distribution of electronic charge about each nucleus is spherically symmetrical. But the electrons are in constant motion. Because each electron constantly experiences repulsive interactions with other electrons on the same atom, *and on other adjacent atoms,* the motion of any one electron is at least partly determined by the motions of all its near neighbors. Suppose it happens that at some instant the electrons in a given helium atom are slightly displaced, so that the atom possesses an *instantaneous dipole moment.* This instantaneous dipole moment would induce a similar dipole moment on an adjacent atom, because of the somewhat synchronized motions of the electrons, as illustrated in Figure 10.13. The result is an attractive interaction between the two atoms, called the London dispersion force.

London's analysis showed that dispersion forces between two molecules should vary as $1/d^6$, where, as usual, d is the distance between the molecular centers (Figure 10.11). Thus this force is significant only when the molecules are close together.

The strength of the attractive dispersion forces depends on how easily the electron cloud is distorted or polarized. In general, the larger the molecule, the farther its electrons are from the nuclei, and consequently the greater its polarizability. Therefore, the magnitude of the London dispersion forces increases with increase in molecular size. Because molecular size and mass generally parallel each other, it is often suggested that *dispersion forces increase in magnitude with increasing molecular mass.* The truth of this generalization is evident in Table 10.3; the boiling points and enthalpies of vaporization of the noble gas elements increase steadily with increase in atomic mass. A similar trend is evident among the diatomic molecules. However, we must expect that the shapes of the molecules involved will also play a role. Note, for example, that although O_2 has a slightly lower mass than Ar, it has a slightly higher boiling point and enthalpy of vaporization. As another example, *n*-pentane and neopentane, illustrated in Figure 10.14, have the same molecular formula, C_5H_{12}, yet the boiling point of *n*-pentane* is 27°C higher than that of

*The *n* in *n*-pentane is an abbreviation for the word "normal." A normal hydrocarbon is one whose carbon atoms are arranged in a straight chain.

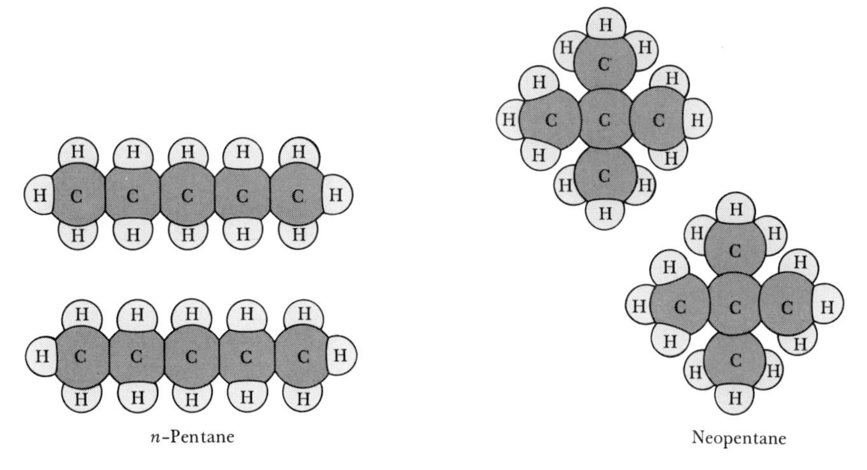

FIGURE 10.14 Illustration of the effect of molecular shape on intermolecular attraction. The boiling point of *n*-pentane is 36.2°C, whereas that of neopentane is 9.5°C.

neopentane. The difference can be traced to the different shapes of the two molecules. The overall attraction between molecules is greater in the case of *n*-pentane because the molecules are able to come in contact over the entire length of the molecule. In the case of neopentane, less contact is possible between molecules.

SAMPLE EXERCISE 10.5

Which of the following substances is most likely to exist as a gas at room temperature and normal atmospheric pressures: P_4O_{10}, Cl_2, AgCl, or I_2?

Solution: In essence, the question asks which substance has the weakest intermolecular attractive forces, because the weaker these forces, the more likely the substance is to exist as a gas at any given temperature and pressure. We should therefore select Cl_2, because this is both a nonpolar molecule and also has the lowest molecular weight. In fact, Cl_2 does exist as a gas at room temperature and normal atmospheric pressure, whereas the others are solids. Of the other substances, AgCl is least likely to be a gas because it exists as Ag^+ and Cl^- ions with very strong ionic bonds holding the ions within the solid.

Dispersion forces operate between all molecules, whether they are polar or nonpolar. For example, in comparing HCl and HBr we find that HCl has the larger dipole moment ($\mu = 1.03$ D, compared to $\mu = 0.79$ D for HBr). HCl is also smaller, which permits the centers of the dipoles to approach each other more closely. Thus the dipole-dipole forces between HCl molecules are stronger than those between HBr molecules. However, dispersion forces are stronger between the more massive HBr molecules, whose electrons are more polarizable. The fact that the boiling point of HBr (206.2 K) is higher than that of HCl (189.5 K) suggests that the *overall* attractive forces are stronger in the case of HBr. In other cases, it may be the more polar molecule that has the stronger overall attractive forces. It is difficult to make generalizations about the relative strengths of intermolecular attractions unless we restrict ourselves to comparing molecules of either similar size and shape or similar polarity and shape. If molecules are of similar size and shape, dispersion forces are approximately equal, and therefore attractive forces increase with increasing polarity. If molecules are of similar polarity and shape, attractive forces tend to increase with increasing molecular mass, because of increasing dispersion forces.

Hydrogen Bonding

Figure 10.15 shows the boiling points of the simple hydrides of the group 4A and 6A elements as a function of molecular mass. In general, the boiling points increase with increasing molecular mass, owing to increased dispersion forces. The notable exception of this trend is H_2O. Clearly, the boiling point of this substance is much higher than would be expected on the basis of its molecular mass. A similar abnormally high boiling point is observed when the boiling point of NH_3 is compared with those of the other group 5A element hydrides, or when HF is compared with the hydrides of the other group 7A elements. These compounds, H_2O, NH_3, and HF, also have many other unusual characteristics that distinguish them from other substances of similar molecular mass and polarity. For example, water has a high melting point, high heat capacity, and a high heat of vaporization. Each of these properties indicates that the intermolecular forces between H_2O molecules are abnormally strong.

The origin of these strong intermolecular attractions is the **hydrogen bond,** a special type of intermolecular attraction that exists between the hydrogen atom in a polar bond and an electronegative atom on an adjacent molecule. This intermolecular attractive force is most important in substances in which a hydrogen atom is bonded to nitrogen, oxygen, or fluorine. The electronegativity of hydrogen is 2.2, much less than the electronegativity of nitrogen (3.0), oxygen (3.4), or fluorine (4.0). As a result, the bond between hydrogen and any of these three elements is

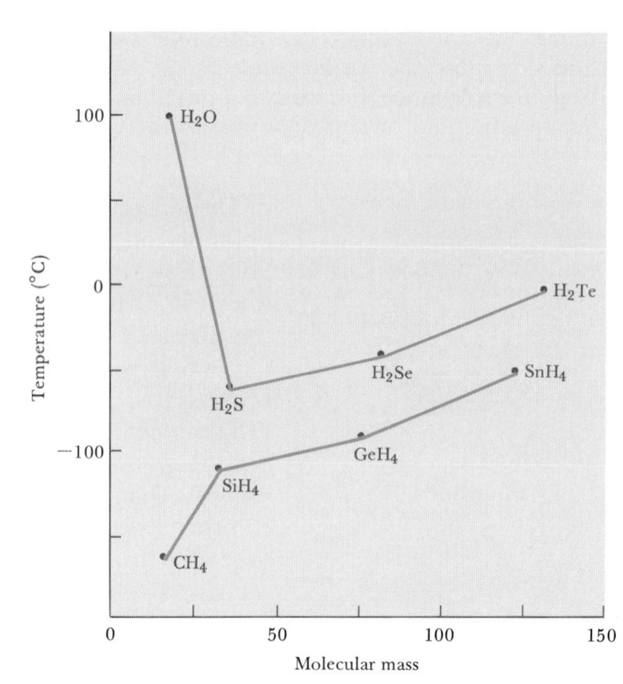

FIGURE 10.15 Boiling points of the group 4A and 6A hydrides as a function of molecular mass.

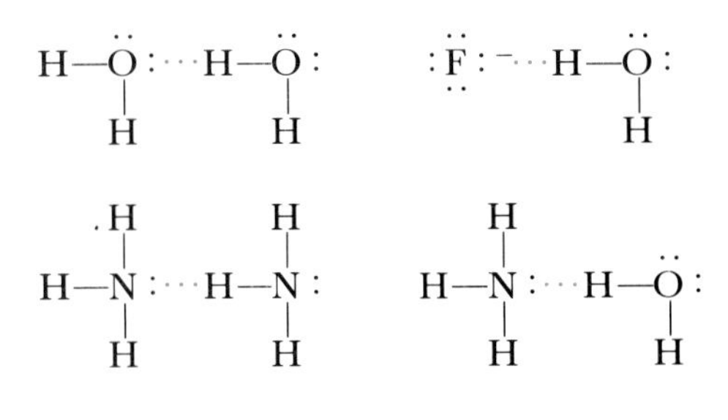

FIGURE 10.16 Examples of hydrogen bonding. Solid lines represent covalent bonds; dotted lines represent hydrogen bonds.

quite polar, with hydrogen at the positive end. We can think of each bond as having a bond dipole moment. The bond dipole moments have been estimated for these three bonds as follows:

Bond:	N—H	O—H	F—H
	⟵+	⟵+	⟵+
Dipole moment (D):	1.0	1.6	2.3

Each of these bond dipoles is capable of interacting with an unshared electron pair on the nitrogen, oxygen, or fluorine atom of an adjacent molecule. It is this electrostatic interaction between the X—H bond dipole of one molecule (where X is an electronegative atom) and the unshared electron pair of an electronegative atom in another molecule that we call hydrogen bonding. Several examples are shown in Figure 10.16. In each case the hydrogen bond is represented by a dotted line.

The energies of hydrogen bonds vary from about 4 or 5 kJ/mol to 25 kJ/mol or so. Thus they are not more than a few percent of the energies of ordinary chemical bonds (see Table 7.3). Nevertheless, hydrogen bonding is generally much stronger than dipole-dipole or dispersion forces, and it has important consequences for the properties of many substances, including those in biological systems.

Hydrogen bonding is responsible, for example, for the rather open structure for ice (Figure 10.17), which causes it to have a *lower* density than liquid water. By contrast, for most substances the solid phase is *more* dense than the liquid. This unusual property of water has some important consequences. First of all, it permits ice to float on water (Figure

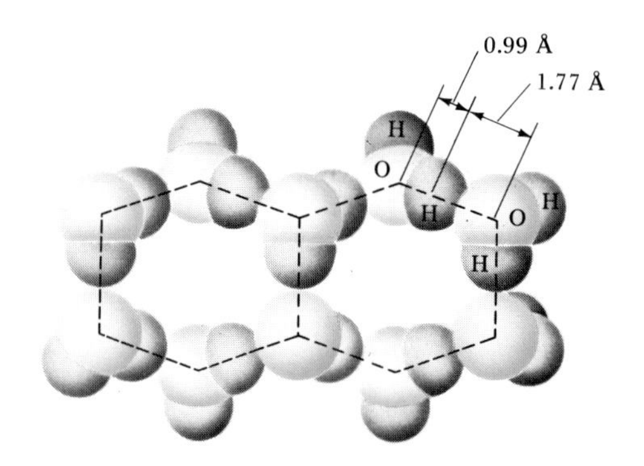

FIGURE 10.17 Arrangement of water molecules in ice. Each hydrogen atom on one water molecule is oriented toward a nonbonding pair of electrons on an adjacent water molecule. Distances between the centers of the bonded atoms are shown. In the full structure each oxygen atom is hydrogen bonded to two O—H groups.

FIGURE 10.18 Because ice has a lower density than liquid water, it floats. (*Naval Photographic Center, Naval Station*)

10.18). When ice forms in cold weather, it covers the top of the water, thereby insulating the water below. If ice were more dense than water, ice forming at the top of a lake would fall to the bottom, and the lake could freeze solid. Most aquatic life could not survive under these circumstances. The expansion of water upon freezing is also what causes water pipes to break in freezing weather. The low density of ice compared to water can be understood in terms of hydrogen-bonding interactions between water molecules. The interactions in the liquid are random. However, when water freezes, the molecules assume the ordered arrangement shown in Figure 10.17. This structure, which extends in all directions in space, permits the maximum number of hydrogen-bonding interactions between the H_2O molecules. Because the structure has large hexagonal holes, ice is more open and less dense than the liquid. Application of pressure depresses the melting point of ice because the pressure causes the solid to revert more readily to the more dense liquid.

SAMPLE EXERCISE 10.6

Which of the following can form hydrogen bonds with H_2O: (a) $CH_3\ddot{\underset{..}{O}}CH_3$; (b) CH_4; (c) NH_4^+?

Solution: The water molecule can interact with the electron pair on an electronegative atom in another molecule or with the H—X bond dipole of another molecule:

$$:\ddot{\underset{\displaystyle\overset{|}{H}}{O}}\text{—H}\cdots:\text{X} \quad \text{or} \quad \text{H—}\ddot{\underset{\displaystyle\overset{|}{H}}{O}}:\cdots\text{H—X}$$

(a) In CH_3OCH_3 there is an electronegative atom, O, with unshared electron pairs to form hydrogen bonds with H_2O. (b) In CH_4 there is no strongly electronegative atom, so no hydrogen bonding is possible. (c) In NH_4^+ there are polar N—H bonds that can interact with the unshared electron pairs of H_2O, forming hydrogen bonds.

SAMPLE EXERCISE 10.7

List the substances $BaCl_2$, H_2, CO, HF, and Ne in order of increasing boiling points.

Solution: The boiling point reflects the attractive forces in the liquid. These are stronger for ionic substances than for molecular ones, so $BaCl_2$ has the highest boiling point. The intermolecular forces of the remaining substances depend on molecular weight, polarity, and hydrogen bonding. The other molecular weights are H_2 (2), CO (28), HF (20), and Ne (20). The boiling point of H_2 should be the lowest, because it is nonpolar and has the lowest molecular weight. The molecular weights of CO, HF, and Ne are roughly the same. HF has hydrogen bonding, so it has the highest boiling point of the three. CO, which is slightly polar and has the highest molecular weight, is next. Ne, which is nonpolar, comes last of these three. The predicted boiling points are therefore

$$H_2 < Ne < CO < HF < BaCl_2$$

The actual normal boiling points are H_2 (20 K), Ne (27 K), CO (83 K), HF (293 K), and $BaCl_2$ (1813 K).

If the hydrogen bond is indeed the result of an electrostatic interaction between the X—H bond dipole and an unshared electron pair on another atom, Y, then the strength of hydrogen bonding should increase as the X—H bond dipole increases. Thus, if Y is the same, we would expect hydrogen-bonding strengths to increase in the series

$$N—H \cdots Y < O—H \cdots Y < F—H \cdots Y$$

This is indeed found to be the case. But what property of Y is important? The atom Y must possess an unshared electron pair that attracts the positive end of the X—H dipole. The electron pair must not be too diffuse in space; if the electrons occupy too large a volume the X—H dipole does not experience a strong, directed attraction. For this reason, we find that hydrogen bonding is not very strong unless Y is one of the smaller atoms N, O, or F. Among these three elements, we find that hydrogen bonding is stronger when the electron pair is not attracted too strongly to its own nuclear center. The ionization energy of the electrons on Y is a good measure of this aspect. For example, the ionization energy of an unshared-pair electron on nitrogen in a covalent molecule is less than the corresponding value for oxygen. Nitrogen is thus a better donor of the electron pair to the X—H bond. For a given X—H, hydrogen-bond strength increases in the order

$$X—H \cdots F < X—H \cdots O < X—H \cdots N$$

When X and Y are the same, the energy of hydrogen bonding increases in the order

$$N—H \cdots N < O—H \cdots O < F—H \cdots F$$

When the Y atom carries a negative charge, the electron pair is especially able to form strong hydrogen bonds. The hydrogen bond in the $F—H \cdots F^-$ ion is among the strongest known; the reaction

$$F^-(g) + HF(g) \longrightarrow FHF^-(g)$$

has a ΔH value of about -155 kJ/mol.

10.5 SOLIDS

Solids are rigid; they cannot be poured like liquids or compressed like gases. Solids such as quartz or diamond possess highly regular crystalline shapes or cleavage planes as illustrated in Figure 10.19. These facts suggest a regular atomic arrangement within the solid. Indeed, a statement of this regular arrangement is sometimes included as part of the definition of solids. For our purposes, we shall divide materials that we would ordinarily call solids because of their rigidity into two groups: crystalline solids and amorphous solids. **Crystalline solids** are characterized by a regular three-dimensional arrangement of atoms. **Amorphous solids** lack this regular atomic-level organization. Familiar materials of the second type include substances such as rubber and glass, which are composed of large or complicated molecules. In some texts, amorphous solids are referred to as supercooled liquids because they have the molecular dis-

FIGURE 10.19 Crystalline solids come in a wide variety of forms: (*a*) Pyrite, (*b*) calcite, (*c*) selenite, (*d*) quartz. Cleavage planes are most evident in large crystals such as calcite and quartz. (*Parts a-c from Pfizer, Inc.; part d © Louise K. Berman; Photo Researchers, Inc.*)

order of liquids. In fact, glass is capable of flowing, as revealed by a careful examination of the window panes of very old houses. The panes are thicker at the bottom than at the top because the glass has flowed under the continued influence of the force of gravity. In this section our focus is primarily on crystalline solids.

Bonding in Solids

The particles of a crystalline solid, whether they be atoms, molecules, or ions, assume positions that maximize attractive forces between them. We can classify solids according to the types of particles and the types of forces that hold the particles in their positions. Table 10.6 summarizes solids in this fashion. It is important to study this table carefully because it contains a great deal of information about solids, relating atomic-level arrangements and forces to the bulk properties of the crystal.

Solid argon and methane are examples of atomic solids and molecular solids, respectively. Intermolecular forces consist of dispersion forces, dipole-dipole forces, and hydrogen bonds. Because these forces are weak, such solids normally have relatively low melting points and exhibit trends in melting points that are similar to those discussed for the boiling points of molecular substances. Most substances that are gases or liquids at room temperature form molecular solids at low temperature.

The properties of molecular solids depend not only on the magnitudes of the forces that operate between molecules but also on the abilities of

TABLE 10.6 Crystal classifications

Crystal classification	Form of unit particles	Forces between particles	Properties	Examples
Atomic	Atoms	London dispersion forces	Soft, very low melting point, poor thermal and electrical conductors	Rare gases—Ar, Kr
Molecular	Polar or nonpolar molecules	London dispersion, dipole-dipole forces, hydrogen bonds	Fairly soft, low to moderately high melting point, poor thermal and electrical conductors	Methane, CH_4; sugar, $C_{12}H_{22}O_{11}$; dry ice, CO_2
Ionic	Positive and negative ions	Electrostatic attraction	Hard and brittle, high melting point, poor thermal and electrical conductors	Typical salts—for example, NaCl, $Ca(NO_3)_2$
Covalent (network)	Atoms that are connected in covalent-bond network	Covalent bond	Very hard, very high melting point, poor thermal and electrical conductors	Diamond, C; quartz, SiO_2.
Metallic	Atoms	Metallic bond	Soft to very hard, low to very high melting point, excellent thermal and electrical conductors, malleable and ductile	All metallic elements—for example, Cu, Fe, Al, W

the molecular units to pack efficiently in three dimensions. For example, consider the melting and boiling points of several chlorine-substituted benzene molecules, as given in Figure 10.20. These are all planar molecules; the chlorine atoms have a larger radius than the carbon or hydrogen atoms that make up the benzene molecule. The attractive forces between molecules increase with an increase in molecular mass, as evidenced by the increase in boiling points with increasing number of chlorine atoms. The melting points, however, do not vary in the same smooth manner. For example, chlorobenzene melts at a much lower temperature than benzene itself, because it is a less symmetrical molecule and cannot pack as efficiently in the solid state. When all other factors are about equal, more symmetrical molecules melt at a higher temperature than do unsymmetrical ones.

Ionic solids have higher melting points than do atomic and molecular solids, because ionic bonds are stronger than intermolecular forces. Ionic solids are also harder and fracture when struck rather than simply deforming.

In covalent or network solids the lattice units are joined by covalent bonds. Such materials are consequently much stronger than molecular solids. Diamond, whose structure and bonding was discussed in Section 8.7, is an example of this type of lattice (see Figure 8.30).

Bonding in metals differs from the bonding in other solids. Each atom in a metallic lattice typically has 8 to 12 atoms adjacent to it, reflecting a close-packing arrangement. The bonding is too strong to be due to London-dispersion forces, and yet there are not enough valence electrons for ordinary covalent bonds between the atoms. We can visualize the valence electrons as delocalized, somewhat as outlined for the π electrons in graphite (Section 8.7). This model allows us to picture metals as composed of metal atoms held together by electrons that are dis-

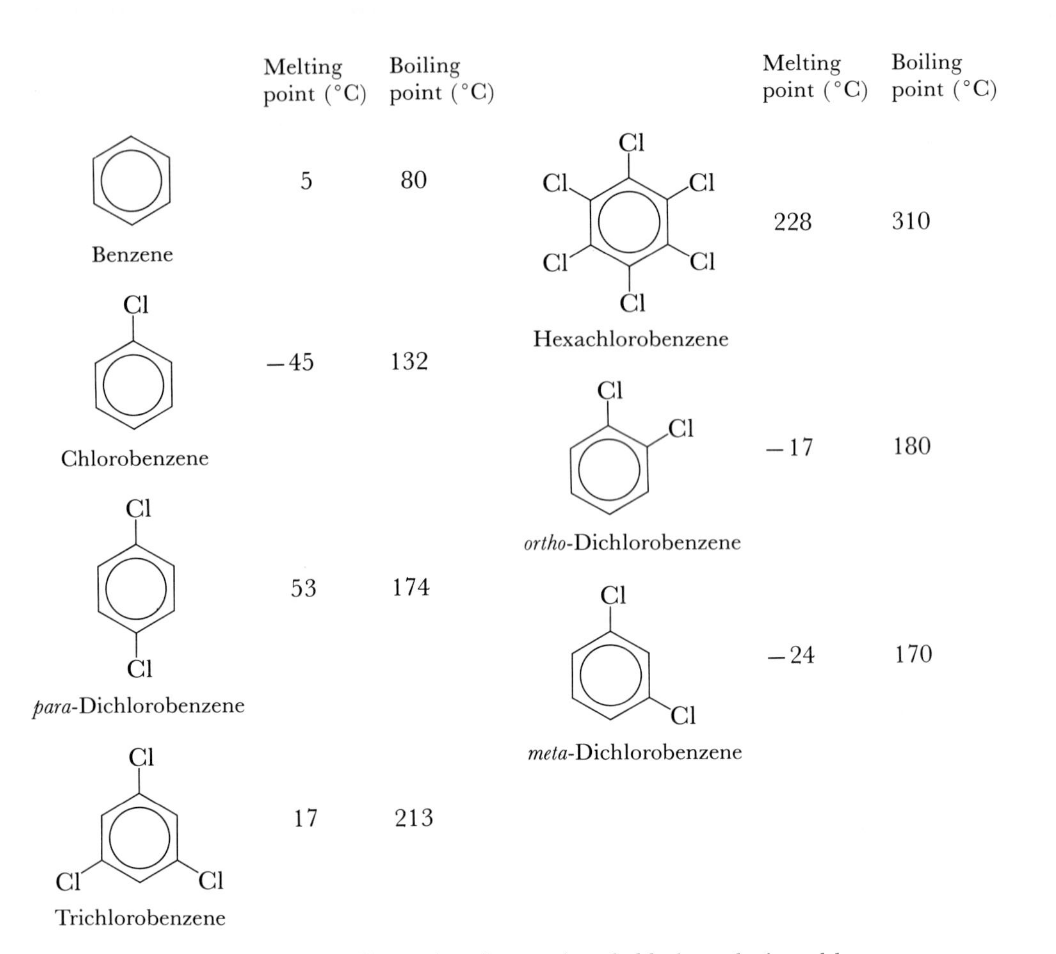

FIGURE 10.20 Melting and boiling points for a series of chlorine-substituted benzenes.

tributed throughout all the spaces between the atoms. The electrons are free to move through the orbitals that extend over the entire metal. However, they maintain a uniform average distribution. Metals vary greatly in the strength of metallic bonding, as evidenced by such physical properties as melting and boiling points. For example, in platinum, with a melting point of 1770°C and a boiling point of 3824°C, the metallic bonding is very strong; in cesium, melting point 29°C and boiling point 678°C, the metallic bonding is comparatively weak. The properties and structures of metals will be examined more closely in Chapter 23.

Crystal Lattices and Unit Cells

The order characteristic of crystalline solids allows us to convey a picture of an entire crystal by looking at only a small part of it. That is, a small, repeating unit of the solid, known as the unit cell, can be selected and the solid regarded as being built by repeating these unit cells in three dimensions. The unit cell can therefore be regarded as the basic "building block" of the crystal. A simple two-dimensional example is afforded by a sheet of wallpaper, such as that shown in Figure 10.21. There are several ways of choosing the repeat pattern or unit cell of the design, but the

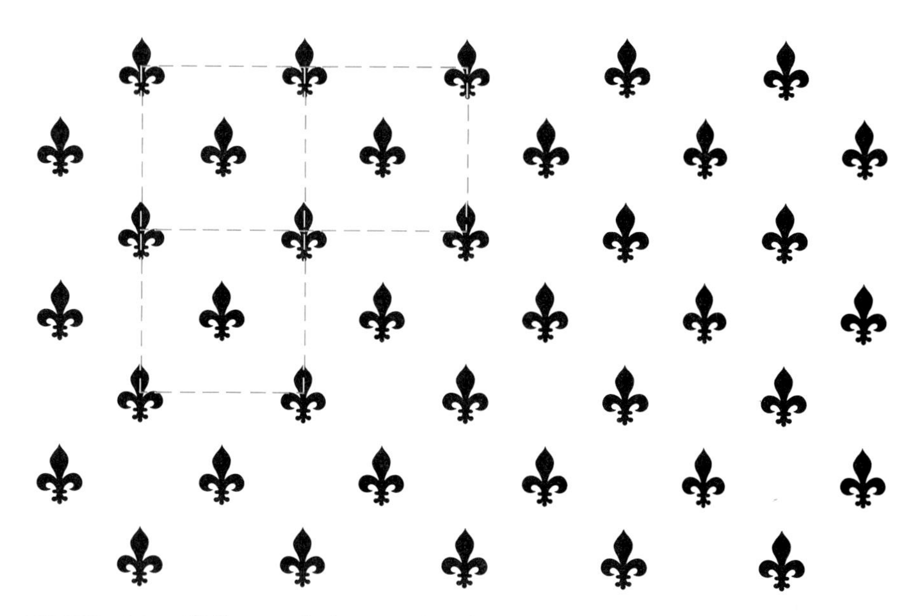

FIGURE 10.21 Wallpaper design showing the repeat pattern or unit cell of the design.

choice is usually taken to be the smallest one that shows clearly the symmetry characteristic of the entire pattern.

In representing a crystalline solid it is convenient to simplify the structure so as to concentrate on its three-dimensional, or spatial characteristics. This objective can be accomplished by representing the crystal as an array of points, each of which represents an identical environment within the crystal. Such an array of points is called a crystal lattice. The crystal lattice gives an abstract representation that provides a clear picture of the shape and size of the unit cell. The crystal structure may then be imagined to be formed by arranging the contents of the unit cell on this imaginary network of points.

Figure 10.22 shows a simple crystal lattice and its associated unit cell. In general, unit cells are parallelepipeds (that is, six-sided figures whose

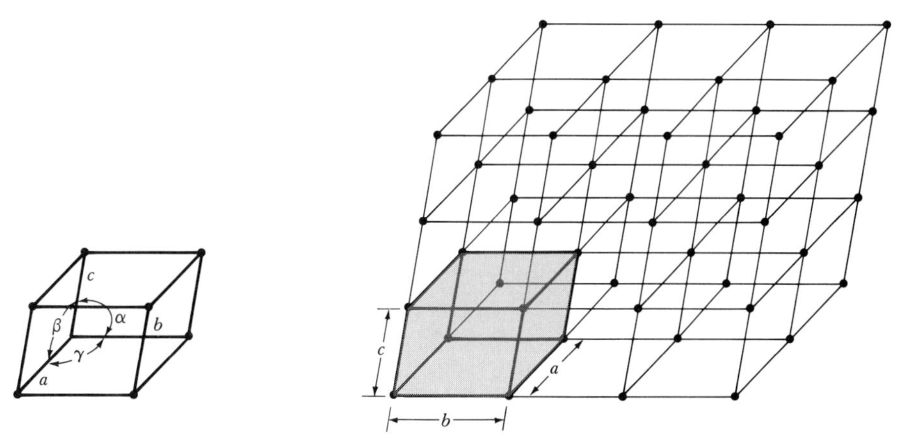

FIGURE 10.22 Simple crystal lattice and its associated unit cell. Angle α is the angle between the b and c axes; β is between the a and c axes; and γ is between the a and b axes.

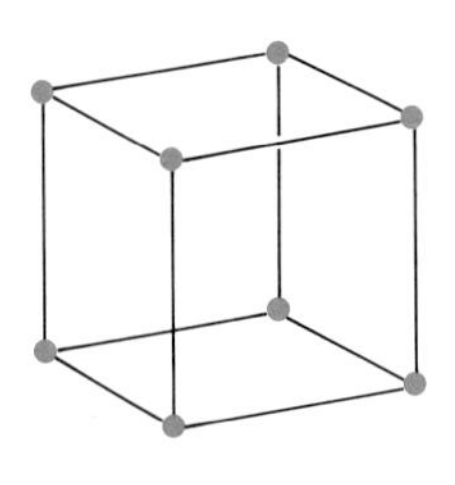

Simple cubic

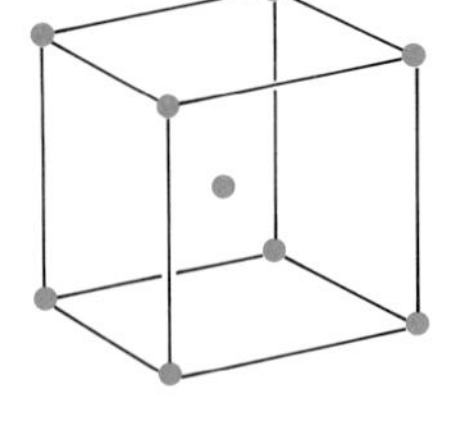

Body-centered cubic

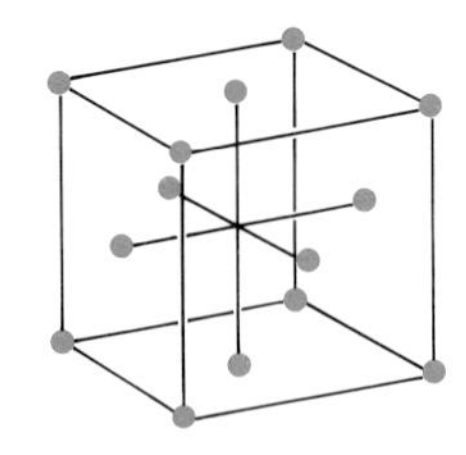

Face-centered cubic

FIGURE 10.23 The three types of unit cells of cubic lattices.

faces are parallelograms). Each unit cell can be described in terms of the lengths of the edges of the cell (a, b, and c) and by the angles between these edges (α, β, and γ), as shown in Figure 10.20. The lattices of millions of compounds can be described in terms of the seven basic types of unit cells described in Table 10.7. The simplest of these is the cubic unit cell. There are three kinds of cubic unit cells, as illustrated in Figure 10.23. When lattice points are at the corners only, the unit cell is described as **simple cubic.** When lattice points also occur at the center of the unit cell, the cell is known as **body-centered cubic.** A third type of cubic cell has lattice points at each corner as well as at the center of each face; this arrangement is known as **face-centered cubic.**

Figure 10.24 shows a portion of the crystal structure of sodium chloride. Two ways of locating the lattice points so they are in identical environments are shown. In Figure 10.22(a) the points are centered on Cl^-, whereas in Figure 10.22(b) they are centered on Na^+. In either case the structure is seen to possess a lattice with a face-centered-cubic unit cell.

Figure 10.22 shows "exploded" views of portions of the NaCl structure in which the ions have been moved apart so that the symmetry of the structure can be seen more clearly. In this representation, no attention is

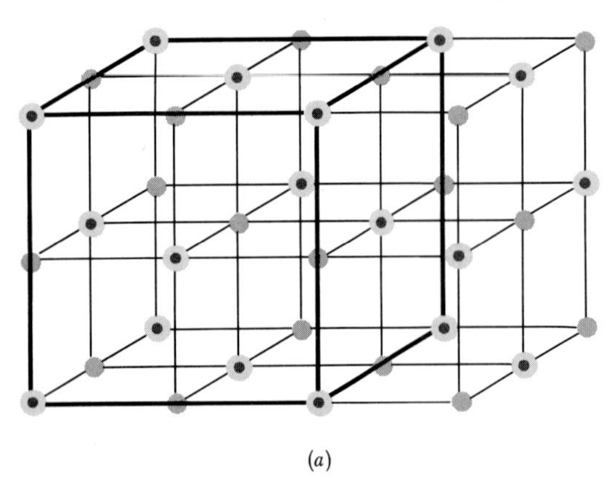

(a)

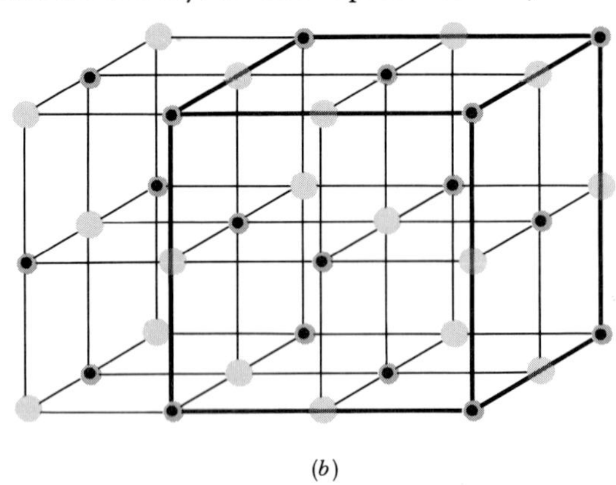

(b)

FIGURE 10.24 Portion of the crystal structure of NaCl illustrating two ways of positioning its crystal lattice. Gray spheres represent Na^+ ions, blue spheres represent Cl^- ions, and dots represent lattice points. Heavy lines are used to define the face-centered unit cell of each lattice. In (a) the lattice points are centered on Cl^- ions. In (b) the lattice points are centered on Na^+ ions. However, the lattice points need not be centered on ions at all. The only restriction on where lattice points are placed is that the points must be in positions that have identical environments. Regardless of where the lattice points are placed in the NaCl structure, the result is a face-centered-cubic lattice.

TABLE 10.7 Basic unit cells of crystal lattices

System	Edge lengths	Angles[a]
Cubic	$a = b = c$	$\alpha = \beta = \gamma = 90°$
Tetragonal	$a = b \neq c$	$\alpha = \beta = \gamma = 90°$
Orthorhombic	$a \neq b \neq c$	$\alpha = \beta = \gamma = 90°$
Monoclinic	$a \neq b \neq c$	$\alpha = \beta = 90° \neq \gamma$
Triclinic	$a \neq b \neq c$	$\alpha \neq \beta \neq \gamma$
Rhombohedral	$a = b = c$	$\alpha = \beta = \gamma \neq 90°$
Hexagonal	$a = b \neq c$	$\alpha = \beta = 90°; \gamma = 120°$

[a] $\neq$ should be read as "not restricted by symmetry to be equal to."

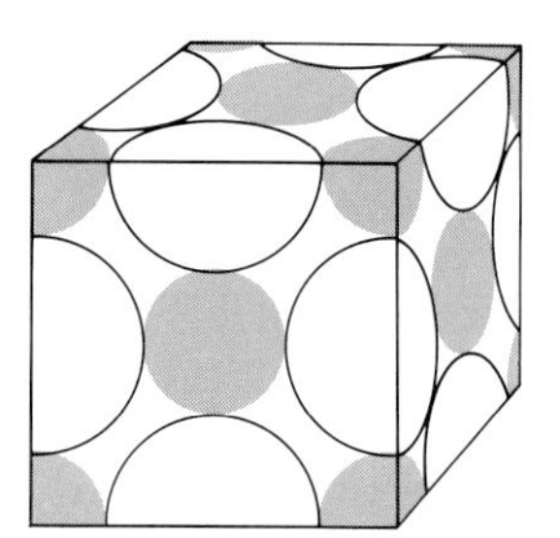

FIGURE 10.25 Space-filling representation of the unit cell of sodium chloride. The Cl^- ions are the larger spheres, the Na^+ ions are the smaller ones. Notice that only a portion of most ions actually lies within the boundaries of the single unit cell.

paid to the relative sizes of the Na^+ and Cl^- ions. In contrast, Figure 10.25 provides a representation that shows the relative sizes of the ions and how they fill the unit cell. Notice that the particles at the corners, edges, and faces do not lie wholly within the unit cell. Instead, these particles are shared by other unit cells. A particle at a corner is shared by eight unit cells, one at the edge is shared by four, and one at the center of a face is shared by two. The total cation-to-anion ratio of the unit cell must be the same as that for the entire crystal. Thus, if we sum all the Na^+ and Cl^- ions within the unit cell of NaCl, there must be one Na^+ for each Cl^-. Similarly, the unit cell for $CaCl_2$ would have one Ca^{2+} for each two Cl^-, and so forth for other crystals. In Sample Exercises 10.8 and 10.9, the contents of the NaCl unit cell are determined and used to calculate the density of the solid.

SAMPLE EXERCISE 10.8

Determine the net number of Na^+ and Cl^- ions in the NaCl unit cell (Figure 10.25).

Solution: There is $\frac{1}{8}$ of a Na^+ on each corner (each Na^+ on a corner is shared by eight cubes which intersect at that point). There is $\frac{1}{2}$ of a Na^+ on each face, $\frac{1}{4}$ of a Cl^- on each edge, and a whole Cl^- in the center of the cube. We therefore have the following:

$$Na^+: \quad (\tfrac{1}{8}\,Na^+ \text{ per corner})(8 \text{ corners}) = 1\,Na^+$$
$$(\tfrac{1}{2}\,Na^+ \text{ per face})(6 \text{ faces}) = 3\,Na^+$$
$$Cl^-: \quad (\tfrac{1}{4}\,Cl^- \text{ per edge})(12 \text{ edges}) = 3\,Cl^-$$
$$(1\,Cl^- \text{ per center})(1 \text{ center}) = 1\,Cl^-$$

Thus the unit cell contains $4Na^+$ and $4Cl^-$. This result agrees with the compound's stoichiometry: one Na^+ for each Cl^-.

SAMPLE EXERCISE 10.9

If the unit cell of NaCl is 5.64 Å on an edge, calculate the density of NaCl.

Solution: The volume of the unit cell is $(5.64\ \text{Å})^3$. Because each unit cell contains $4Na^+$ and $4Cl^-$ (Sample Exercise 10.8), its mass is

$$4(23.0 \text{ amu}) + 4(35.5 \text{ amu}) = 234.0 \text{ amu}$$

The density is mass/volume:

$$\text{Density} = \frac{234.0 \text{ amu}}{(5.64\ \text{Å})^3}\left(\frac{1 \text{ g}}{6.02 \times 10^{23} \text{ amu}}\right) \times \left(\frac{1\ \text{Å}}{10^{-8} \text{ cm}}\right)^3$$
$$= 2.17 \text{ g/cm}^3$$

This value agrees with that found by simple density measurements, 2.165 g/cm^3. Thus the size and contents of the unit cell are consistent with the macroscopic density of the substance.

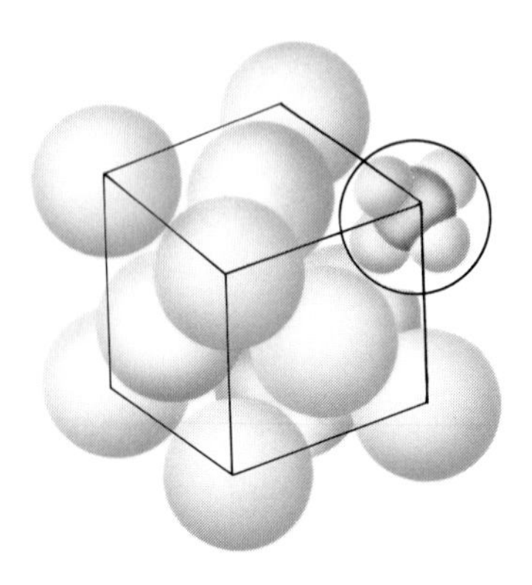

FIGURE 10.26 Unit cell of solid methane. Each large sphere represents a CH_4 molecule, as shown in the upper right.

Close Packing

The structures adopted by crystalline solids are those that bring particles into closest contact, to maximize the attractive forces between them. In many cases the particles that make up the solids are spherical or approximately so. Such is the case for atoms in metallic and atomic solids. Many molecules can also be approximated as spheres, as seen for CH_4 in Figure 10.26. It is therefore instructive to consider how equal-sized spheres can pack most efficiently (that is, with the minimum amount of empty space).

> "There are several ways," Dr. Breed said to me, "in which certain liquids can crystallize—can freeze—several ways in which their atoms can stack and lock in an orderly, rigid way."
>
> That old man with spotted hands invited me to think of the several ways in which cannonballs might be stacked on a courthouse lawn, of the several ways in which oranges might be packed into a crate.
>
> "So it is with atoms in crystals, too; and two different crystals of the same substance can have quite different physical properties."*

The most efficient packing arrangements for spheres are referred to as closest-packed structures. These structures are quite common, accounting, for example, for about two-thirds of the structures of the metallic elements. The most efficient arrangement of a layer of equal-sized spheres can be seen in Figure 10.27(*a*). Each sphere is surrounded by six

*Excerpt from *Cat's Cradle,* by Kurt Vonnegut, Jr. Copyright © 1963 by Kurt Vonnegut, Jr. Reprinted by permission of Delacorte Press/Seymour Lawrence and Donald C. Farber for Kurt Vonnegut, Jr.

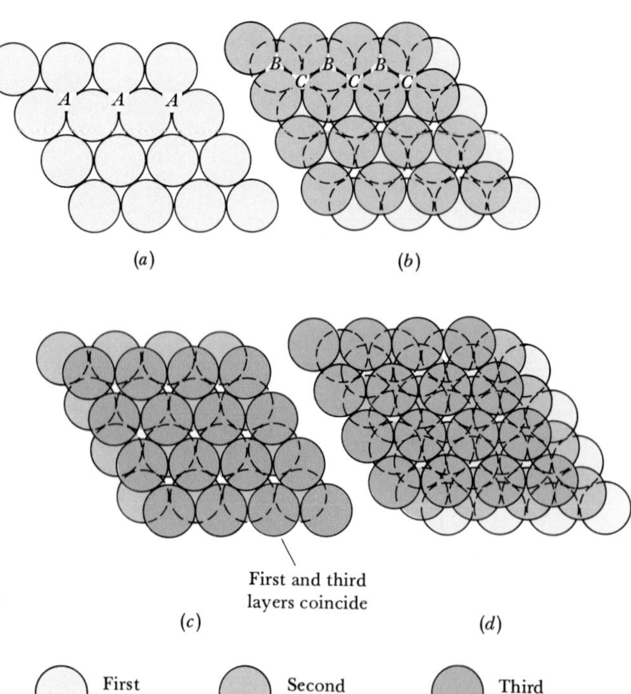

FIGURE 10.27 Closest packing of equal-sized spheres. (*a*) One layer. (*b*) Two superimposed layers. (*c*) Three superimposed layers in hexagonal close-packed arrangement. (*d*) Three superimposed layers in cubic close-packed arrangement.

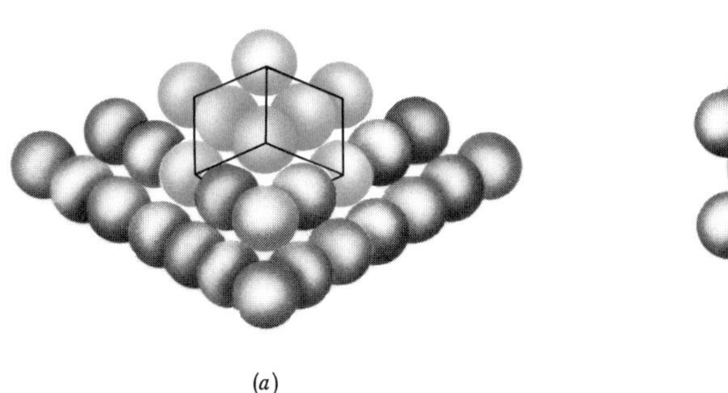

(a)

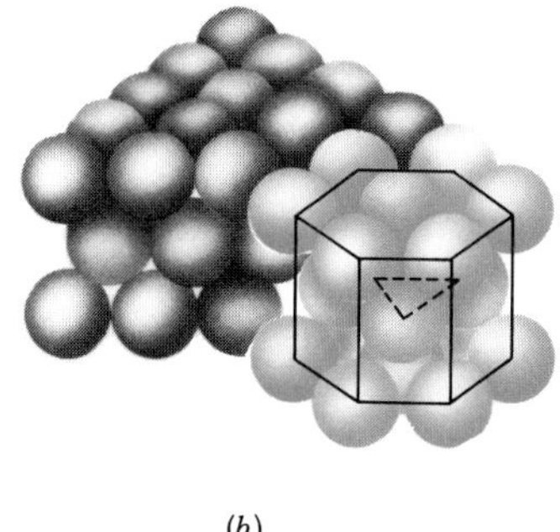

(b)

FIGURE 10.28 Cubic close-packing structure (*a*), and hexagonal close-packing structure (*b*), drawn so as to show the unit cell for each arrangement.

others in the layer. The most efficient arrangement of the spheres in a second layer is in the depressions of the first layer, labeled *A* in Figure 10.27(*a*). This is shown in Figure 10.27(*b*). The spheres of the third layer sit in depressions in the second layer. However, there are two types of depressions, and they lead to two different structures. If the spheres of the third layer are placed immediately above those of the first layer, in the positions labeled *B* in Figure 10.27(*b*), the structure known as the **hexagonal close-packed structure** results—Figure 10.27(*c*). If we consider more than three layers of spheres, the arrangement of the layers in this structure can be represented as *ABABAB*. . . . The second type of close-packed structure results if the third-layer spheres are placed in the positions labeled *C* in Figure 10.27(*b*). The resultant structure is known as the **cubic close-packed structure**—Figure 10.27(*d*). In this case the stacking sequence can be represented as *ABCABC*. . . . Although it is not obvious from Figure 10.27, the cubic-close-packed arrangement of spheres has a face-centered cubic unit cell. In Figure 10.28, the close-packed structures are viewed from a perspective that permits the unit cells to be seen more clearly. In both of the close-packed structures, each sphere has 12 equidistant nearest neighbors: six in one plane, three above that plane, and three below. The spheres are therefore said to have a **coordination number** of 12.

When unequal-sized spheres are packed in a lattice, the large particles sometimes assume one of the close-packed arrangements with small particles occupying the holes between the large spheres. For example, in Li_2O the oxide ions assume a cubic close-packed structure, whereas the Li^+ ions occupy small cavities that exist between oxide ions. We shall consider this particular view of crystals further in Chapter 23 when we discuss the structures of metal oxides.

X-ray Study of Crystal Structure

Much of what we know about the internal molecular-level regularity of crystalline solids is revealed by their interaction with X rays. X rays, you will recall, are electromagnetic waves of short wavelength and high energy (Section 5.1). They were first discovered in 1895 by the German physicist Wilhelm Roentgen. However, it was not until 1913 that they

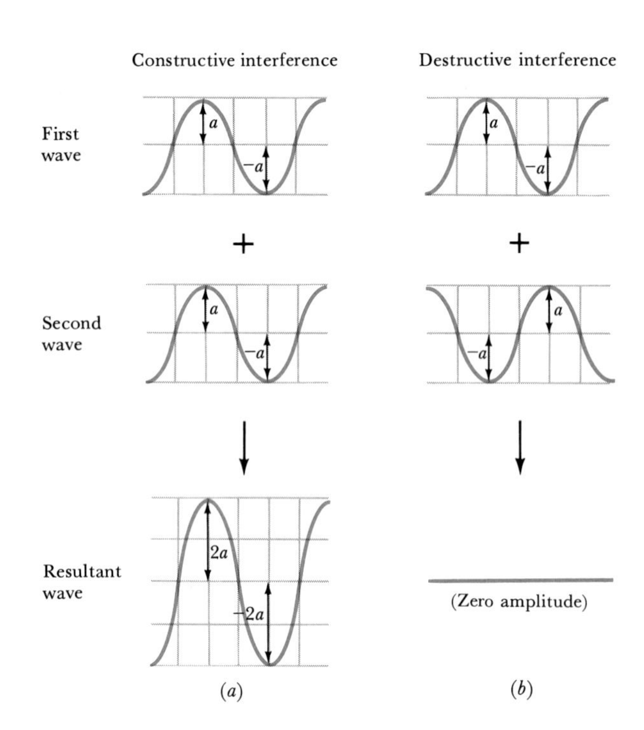

FIGURE 10.29 (*a*) Constructive interference (waves in phase), and (*b*) destructive interference (waves out of phase).

were used to determine the location of atoms within a crystalline solid. In that year, the Englishman W. L. Bragg and his father, W. H. Bragg, determined the location of the zinc and sulfur atoms in a ZnS crystal from a mathematical analysis of the X-ray diffraction pattern of this substance.

Diffraction is a phenomenon that results from scattering of light waves by a regular arrangement of points or lines. It is observed when the wavelength of the light rays is comparable to the distances that separate the points or lines. The scattered waves then interfere with each other when they come together. If the waves are in phase (that is, their peaks and troughs coincide), their amplitudes add, producing a more intense wave. This situation, shown in Figure 10.29(*a*), is termed constructive interference. If the two waves are out of phase, as shown in Figure 10.29(*b*), destructive interference occurs; the resultant wave then has no intensity.

A crystalline solid is a regular array of particles. These particles are typically separated by distances of 1 to 4 Å (that is, 0.1 to 0.4 nm). A glance at Figure 5.3 reminds us that the wavelengths of X rays are in the range of these short distances. Thus when X rays impinge on a crystalline solid, the rays are scattered by the particles of the crystal.* The diffraction pattern of the scattered rays can be examined by allowing them to fall on photographic film, as illustrated in Figure 10.30. Rays that are in phase produce spots on the film. The various angles at which the X rays are scattered can be determined from measurements made on the photographic film. To see how the angle of scattering from a crystal affects different rays, we need to look in detail at the paths taken by two rays of

*Each atom absorbs some of the energy of the X rays and then reemits it in all directions, thereby scattering the X rays.

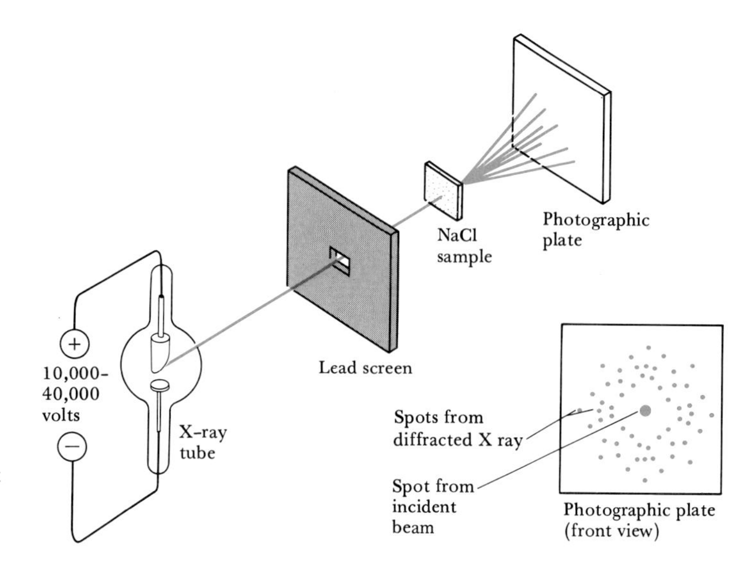

FIGURE 10.30 X-ray diffraction pattern for NaCl and the experimental method by which it is obtained. The pattern shown is based on use of nonmonochromatic X rays.

equal wavelength that are scattered by adjacent atoms. The geometrical situation is illustrated in Figure 10.31.

The incoming rays are in phase at AB. Wave ACA' is scattered or reflected by an atom in the first layer of the solid, whereas wave BEB' is reflected by an atom in the second layer. If these two waves are to be in phase at $A'B'$, the extra distance covered by BEB' must be a whole-number multiple of the wavelength, λ. In Figure 10.29 (*b*), the extra distance, DEF, is 2λ. Now notice that the triangle CDE is a right triangle. Using trigonometry, it can be shown that the distance DE is $d \sin \theta$, where d is the distance between the planes and θ is the angle between the incoming wave and the plane. Because the distance DEF is twice DE, we have

$$DEF = 2\lambda = 2d \sin \theta$$

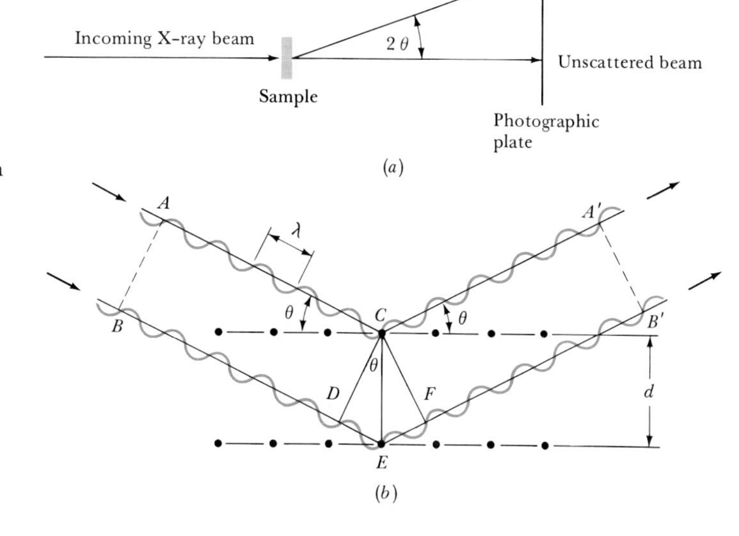

FIGURE 10.31 Scattering of X rays by atoms in parallel planes. (*a*) The experimental arrangement with the X-ray beam scattered at an angle of 2θ from the unscattered beam. (*b*) The atomic-level view of the atoms and waves. The incoming X rays of wavelength λ are diffracted by atoms, which are represented by dots. The atoms are arranged in planes that are separated by a distance d. The incoming X rays make an angle of θ with the planes and are scattered at an angle of 2θ from the unscattered beam, which is not shown in (*b*).

It can be shown that the general equation for constructive interference is

$$n\lambda = 2d \sin \theta \qquad \text{where } n = 1, 2, 3, 4, \ldots \qquad [10.2]$$

This relationship, known as the **Bragg law,** allows determination of the spacing between planes from the known wavelength of the light and experimentally determined values of θ at which constructive interference occurs.

SAMPLE EXERCISE 10.10

When a crystal of calcite ($CaCO_3$) is irradiated with monochromatic X rays whose wavelength is 0.1540 nm, a first-order ($n = 1$) reflection occurs at 14.69°. What is the distance between the planes responsible for this reflection?

Solution: We wish to solve for the quantity d in Equation 10.2. Upon rearranging the equation, we have

$$d = \frac{n\lambda}{2 \sin \theta}$$

$$= \frac{(1)(0.1540\ \text{nm})}{2 \sin (14.69°)} = \frac{(1)(0.1540\ \text{nm})}{2(0.2536)} = 3.304\ \text{nm}$$

Crystal Defects

Although we have generally pictured crystalline solids as being composed of perfectly ordered arrays of particles, this is merely a useful abstraction like that of an ideal gas. Real crystals contain imperfections whose number and type can play an important role in determining the properties of the solid. For example, a crystal lattice that has many sites where particles are missing (vacancies) can be more readily deformed than can a perfect crystal lattice of the same substance. It is not hard to appreciate that structural imperfections can occur readily. In forming a solid, the structural units can be thought of as a mob that has been required to assume a military drill formation. The faster crystal formation occurs, the greater the chance for defects.

Amorphous Solids

Not all solids have a regular arrangement of particles. For example, the material obtained when quartz, SiO_2, is melted and then rapidly cooled is amorphous. Quartz has a three-dimensional structure like that of diamond (Section 8.7). When quartz is melted (at approximately 1600°C) it becomes a viscous, tacky liquid. Although the silicon-oxygen network remains largely intact, many Si—O bonds are broken, and the orderliness of the quartz is lost. If the melt is rapidly cooled, the atoms are unable to return to their orderly arrangement, and an **amorphous solid** known as quartz glass or silica glass results. The lack of molecular-level regularity in this glass is shown in Figure 10.32, where its structure is compared with that of quartz.

Even when a solid lacks any long-range order, there may be small regions of the solid where particles are arranged in an orderly fashion.

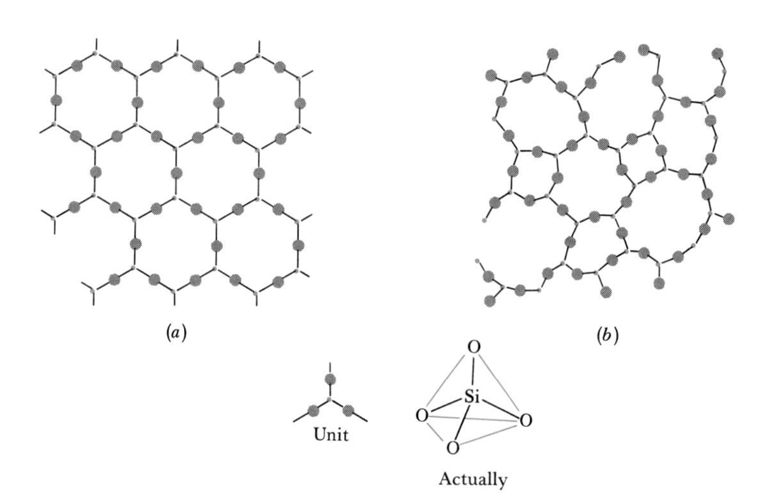

FIGURE 10.32 Schematic comparison of (*a*) crystalline SiO_2 (quartz) and (*b*) amorphous SiO_2 (quartz glass). The small dots represent silicon atoms; the large ones represent oxygen atoms. The structure is actually three-dimensional and not planar as drawn. The unit shown as the basic building block (silicon and three oxygens) actually has four oxygens, the fourth coming out of the plane of the paper and capable of bonding to other silicon atoms.

The extent of such ordering is referred to as the *degree of crystallinity* of the solid. Synthetic *polymers,* large molecules composed of many molecular parts fused together, often have such ordered regions, as shown in Figure 10.33.

The simplest synthetic polymer is polyethylene, a waxy-feeling substance used in packaging films, wire insulation, and molded articles. Polyethylene is formed by causing ethylene, C_2H_4, molecules to bond together to form chains. Typically 700 to 2000 C_2H_4 molecules combine to form a chain containing 1400 to 4000 carbon atoms:

$$H_2C{=}CH_2 \longrightarrow \cdots{-}CH_2{-}CH_2{-}CH_2{-}CH_2{-}CH_2{-}\cdots$$

Ethylene Polyethylene

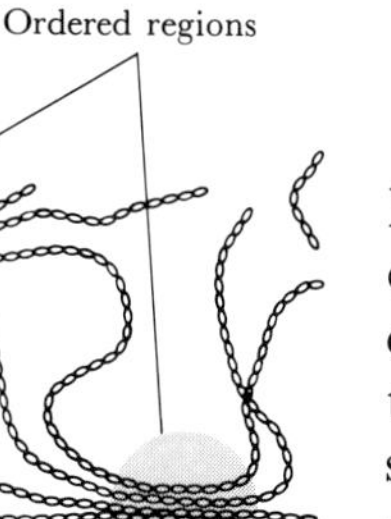

FIGURE 10.33 Solid composed of large, flexible molecules. This solid lacks the long-range order characteristic of crystalline solids, but nonetheless contains small regions where the molecules are arranged in an orderly fashion.

The properties of a plastic* such as polyethylene are determined by at least three factors: (1) the length of the polymer chain, (2) the degree of crystallinity, and (3) the extent of bonding between chains. As the length of a polymer chain increases, intermolecular attractive forces increase, thus making the polymer mechanically stronger and harder. Mechanical strength and hardness also increase as the degree of crystallinity increases. The regular arrangement of chains in crystallites permits closer approach of the molecules, thereby increasing intermolecular attractions.

The effect of the degree of crystallinity on the properties of polyethylene can be seen in Table 10.8. The degree of crystallinity depends on the conditions of the polymerization.

*Although the term "plastic" has come to mean a certain type of synthetic material, the term is most precisely used to describe any material that changes shape when a force is exerted and maintains its distortion upon removal of the force. Materials such as rubber that distort but return to their original shape when the force is removed are called elastomers.

TABLE 10.8 Properties of polyethylene as a function of crystallinity

	DEGREE OF CRYSTALLINITY				
	55%	62%	70%	77%	85%
Melting point (°C)	109	116	125	130	133
Density (g/cm^3)	0.92	0.93	0.94	0.95	0.96
Stiffness[a]	25	47	75	120	165
Yield stress[a]	1700	2500	3300	4200	5100

[a]These tests reflect the increased mechanical strength of the polymer with increased crystallinity. The physical units for the stiffness test are psi $\times$ 10^{-3} (psi = pounds per square inch), while those for the yield stress test are psi. Discussion of the exact meaning and significance of these tests is beyond the scope of this text.

To soften a plastic or make it more pliable a substance with low molecular weight known as a *plasticizer* may be added during the course of fabrication. The plasticizer molecules occupy positions between polymer strands, thereby interfering with intermolecular forces between chains and lowering the degree of crystallinity. The oily film that develops on the insides of the windows of new cars left standing in the sun is due in part to loss of plasticizers from the plastics in the cars' interiors. Continual loss of the plasticizer leaves the plastics brittle, causing them eventually to crack.

In contrast to the weakening of bonds between chains by use of a plasticizer, the bonds can be strengthened by replacing the intermolecular bonds with covalent bonds between the chains. These bonds are known as cross-links. The process of vulcanizing rubber, which increases its rigidity, involves cross-linking of polymer chains.

10.6 PHASE DIAGRAMS

The physical state of a substance depends not only on inherent intermolecular attractive forces, but also on temperature and pressure. Now that we have examined each state, let us conclude our discussions by considering the temperatures and pressures at which the various phases of a substance can exist. Such information can be summarized in a **phase**

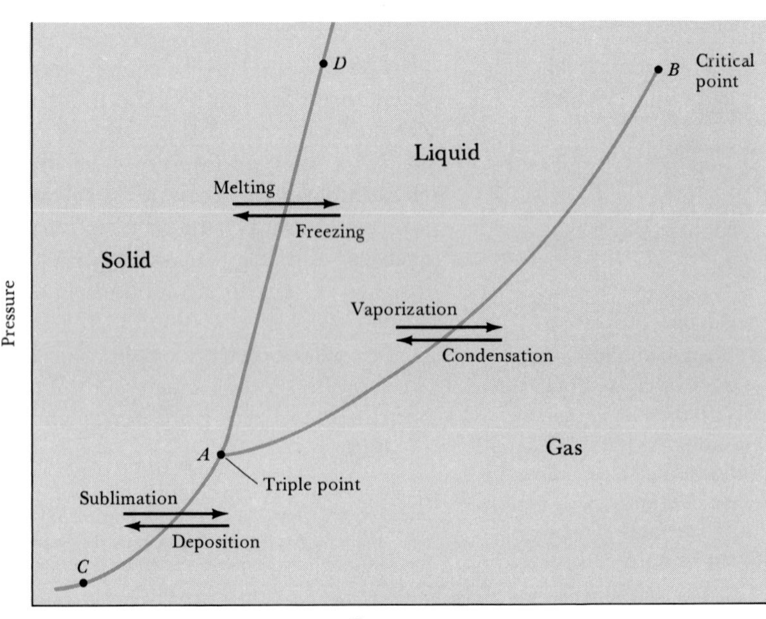

FIGURE 10.34 General shape for a phase diagram of a system exhibiting three phases: gas, liquid, and solid.

diagram. The general form for such a diagram for a single substance that exhibits three phases is shown in Figure 10.34. This diagram contains three important curves, each of which represents the conditions of temperature and pressure at which the various phases can coexist at equilibrium.

1 The line from *A* to *B* is the vapor-pressure curve of the liquid. It represents the equilibrium between the liquid and gas phases at various temperatures. This curve ends at *B*, the critical point. The temperature at this point is the critical temperature (Section 10.3), the temperature above which the substance cannot exist as a liquid regardless of how much pressure is applied. At this temperature the liquid and gas phases become indistinguishable. The pressure at the critical temperature is the critical pressure. Beyond the critical point the substance is described as a supercritical fluid. The point on curve *AB* where the equilibrium vapor pressure is 1 atm is, of course, the normal boiling point of the substance.
2 The line *AC* represents the variation in the vapor pressure of the solid as a function of temperature.
3 The line from *A* through *D* represents the change in melting point of the solid with increasing pressure. This line normally slopes slightly to the right as pressure increases; most solids expand upon melting, and increasing pressure therefore favors formation of the more dense solid phase. Thus higher temperatures are required to melt the solid at higher pressures. The melting point at 1 atm pressure is the normal melting point. (Remember that this is also the normal freezing point.)

Point *A*, where the three curves intersect, is known as the triple point. All three phases are at equilibrium at this temperature and pressure. Any other point on the three curves represents an equilibrium between two phases. Any point on the diagram that does not fall on a line corresponds to conditions under which only one phase is present. Notice that the gas phase is the stable phase at low pressures and high temperatures. The conditions under which the solid phase is stable extend to low temperatures and high pressures. The stability range for liquids lies between the other two regions.

Figure 10.35 shows the phase diagrams of H_2O and CO_2. Notice that the solid-liquid equilibrium (melting point) line of CO_2 is normal; its melting point increases with increasing pressure. On the other hand, the melting point of H_2O *decreases* with increasing pressure. Water is among the very few substances whose liquid form is more compact than the solid (Section 10.4).

The triple point of H_2O (0.0098°C and 4.58 mm Hg) is much lower than that of CO_2 (−56.4°C and 5.11 atm). For CO_2 to exist as a liquid, the pressure must exceed 5.11 atm. Consequently, solid CO_2 does not melt but rather sublimes when heated at 1 atm. Thus CO_2 does not have a normal melting point; instead, it has a normal sublimation point, −78.5°C. The fact that CO_2 sublimes rather than melting as it absorbs energy at ordinary pressures makes solid CO_2 (commonly called "dry ice") a very convenient coolant.

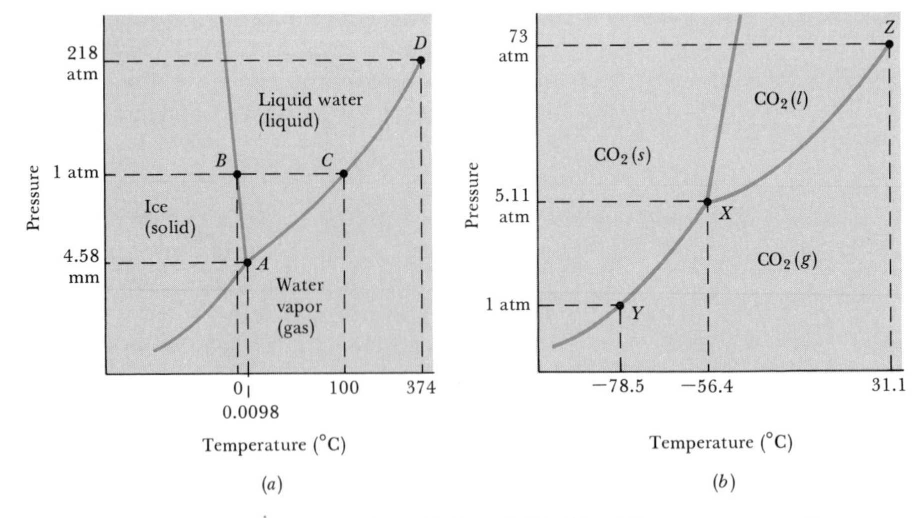

FIGURE 10.35 Phase diagrams of (*a*) H_2O and (*b*) CO_2. The axes are not drawn to scale in either case. In (*a*), for water, the triple point *A* (0.0098°C, 4.58 mm Hg), the normal melting (or freezing) point *B* (0°C, 1 atm), the normal boiling point *C* (100°C 1 atm), and the critical point *D* (374.4°C, 217.7 atm) are shown. In (*b*), for carbon dixoide, the triple point *X* (−56.4°C, 5.11 atm), the normal sublimation point *Y* (−78.5°C, 1 atm), and the critical point *Z* (31.1°C, 73.0 atm) are shown.

SAMPLE EXERCISE 10.11

Referring to Figure 10.36, describe any changes in the phases present when H_2O is (a) kept at 0°C while the pressure is increased from that at point 1 to that at point 5 (vertical colored line); (b) kept at 1.00 atm while the temperature is increased from that at point 6 to that at point 9 (horizontal colored line).

Solution: (a) At 1 the H_2O exists totally as a vapor. At 2 a solid-vapor equilibrium exists. Above that pressure, at point 3, all the H_2O is converted to a solid. At 4 some of the solid melts and an equilibrium between solid and liquid is achieved. At still higher pressures, all the H_2O melts, so that only the liquid phase is present at point 5.

(b) At point 6 the H_2O exists entirely as a solid. When the temperature reaches 4, the solid begins to melt and an equilibrium condition occurs between the solid and the liquid phases. At a yet higher temperature, 7, the solid has been converted entirely to a liquid. When point 8 is encountered, vapor forms and a liquid-vapor equilibrium is achieved. Upon further heating, to point 9, the H_2O is converted entirely to the vapor phase.

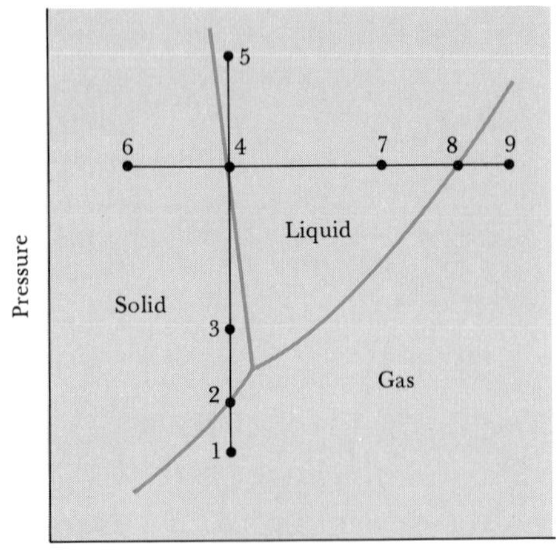

FIGURE 10.36 Phase diagram of H_2O (refer to Sample Exercise 10.11).

The effect of pressure on the melting point of water is partially responsible for making ice skating possible. Because skates have very thin edges, a 130-lb person can exert a pressure equivalent to 500 atm on the ice. Consequently, the ice below the blade melts (the friction between the blade and ice also facilitates melting). The resultant film of water under the blade allows movement of the skater over the ice.

Sublimation of H_2O, which occurs at pressures below 4.58 mm Hg, is used in the freeze-drying procedure for dehydrating foods. The food is frozen and then introduced into a vacuum chamber. The water is thereby removed from the food through sublimation.

FOR REVIEW

Summary

In this chapter we have concerned ourselves with the properties of liquids and solids. We can understand the physical properties of matter in these states in terms of the kinetic-molecular theory, which we used earlier to explain the properties of gases. In a liquid the intermolecular forces keep molecules in close proximity, though they retain freedom to move with respect to one another. The free volume in a liquid is thus small, and liquids are not very compressible. The particles that make up a solid are even more restrained than in a liquid; they occupy specific locations in a three-dimensional arrangement. Thus solids retain both their shape and their volume. Any substance may exist in more than one state of matter, or phase. The equilibria between phases are dynamic; that is, there is continuous transfer of particles from one phase to the other. Equilibrium in such a dynamic system occurs when the rates of transfer between the phases are equal. The change of matter from one state of matter to another is termed a phase change. Conversions of a solid to a liquid (melting), solid to a gas (sublimation), or liquid to a gas (vaporization) are all endothermic processes; that is, the enthalpies of melting, sublimation, or vaporization are all positive. The reverse processes, conversion of a liquid to a solid (freezing), gas to a solid (deposition), or a gas to a liquid (condensation) are all exothermic; thus the enthalpy changes for these processes are all negative.

The vapor pressures of liquids increase nonlinearly with temperature. Boiling occurs when the vapor pressure equals the externally applied pressure. The normal boiling point is the temperature at which the vapor pressure of the liquid equals 1 atm.

Physical properties of liquids, such as their vapor pressure, normal boiling point, viscosity, and surface tension, are related to the intermolecular forces between molecules.

The intermolecular forces that keep the particles of a liquid or solid together are essentially electrostatic in nature. They include ion-ion, ion-dipole, ion-induced dipole, dipole-dipole, and London dispersion forces. The relative importance of each of these contributions to the intermolecular attractions depends on the character of each substance. In general, the intermolecular attractive forces depend on the polarizability, polarity, and shape of the molecule. The magnitudes of London dispersion forces between nonpolar molecules of similar shape and complexity tend to increase with increasing molecular mass. Hydrogen bonding is an important source of intermolecular attractions in compounds containing O—H, N—H, and F—H bonds, and in mixtures containing these. The unusual properties of water, such as its high boiling point and high heat of vaporization, are due to the extensive O—H···O hydrogen bonding in both the liquid and solid forms.

Solids are characterized on a macroscopic level by their rigidity, and on the atomic level by the relatively fixed nature of the particles. Solids whose particles are arranged in a regularly repeating pattern are said to be crystalline, while those whose particles show no such order are said to be amorphous.

Solids may be classified according to the type of bonding between the units of the solid as atomic or molecular, ionic, covalent network, or metallic. The magnitude of the forces operating between the units of the lattice ranges from very small, as with the rare gases, to very large, as in diamond (a covalent network structure), MgO (ionic), or W (metallic).

The essential structural features of any crystalline solid (crystal) can be represented by its unit cell, the smallest part of the crystal that can, by simple displacement, reproduce the three-dimensional structure. The three-dimensional patterns of crystals can also be represented by their crystal lattices. The points in a crystal lattice represent positions in the structure where there are identical environments (although these positions might be centered on atoms, ions, or molecules, this is not necessary).

In many solid structures the particles have a close-packing arrangement, in which spherical particles are arranged so as to leave the minimal amount of free volume. Two closely related forms of close packing, cubic and hexagonal, are possible. In both, each close-packed unit is in contact with 12 equivalent neighbors.

The structures of crystalline solids can be determined by observations of X-ray diffraction patterns. The Bragg law, $n\lambda = 2d \sin \theta$, expresses the angles at which constructive interference of the diffracted rays occurs, in terms of the distance d between planes in the crystal.

Crystalline solids have imperfections or defects that cause them to have less than perfect order. On the other hand, amorphous solids often have some small regions where the particles are arranged in an orderly fashion. The extent to which these ordered regions occur is expressed by the percent crystallinity of the solid.

The equilibria between the solid, liquid, and gas phases of a substance as a function of temperature and pressure are displayed on a phase diagram. Equilibrium between any two phases on such a diagram is indicated by a line. The point on such a diagram at which all three phases coexist in equilibrium is called the triple point.

Learning goals

Having read and studied this chapter, you should be able to:

1. Distinguish between gases, liquids, and solids on a molecular level.
2. Employ the kinetic-molecular theory and the concept of intermolecular attractions to explain the properties of each phase, such as surface tension, viscosity, vapor pressure, and boiling and melting points.
3. Explain the nature of the equilibria that may exist between phases. Account for the enthalpy changes that accompany phase changes.
4. Calculate the heat absorbed or evolved when a given quantity of a substance changes from one condition to another, given the needed heat capacities and enthalpy changes associated with phase changes.
5. Describe the manner in which the vapor pressure of a substance changes with temperature and the relationship between the pressure on the surface of a liquid and the boiling point of that liquid.
6. Explain the meaning of the terms *critical temperature* and *critical pressure* and account for the variation in critical temperatures of different substances in terms of intermolecular forces.
7. Describe the various types of intermolecular attractive forces, indicate how each arises, and indicate the manner in which each varies with distance.
8. Predict, for any particular substance of known structure, which types of intermolecular forces may be operative and which particular type is of major importance.
9. Describe the nature of the hydrogen bond and distinguish those molecular systems in which hydrogen bonding is likely to be important.
10. Distinguish between crystalline and amorphous solids.
11. Predict the type of solid (atomic, molecular, ionic, covalent network, or metallic) formed by a substance and predict its general properties.
12. Determine the net contents in a cubic unit cell given a drawing or verbal description of the cell, and use this information together with the atomic weights of the atoms in the cell and the cell dimensions to calculate the density of the substance.
13. Explain the origin of the diffraction patterns obtained when X rays impinge on a crystal.
14. Use Bragg's law to calculate spacings in a crystal given the appropriate experimental data.
15. Draw a phase diagram of a substance, given appropriate data.
16. Use a phase diagram to predict what phases are present at any given temperature and pressure.

Key terms

Among the more important terms and expressions used for the first time in this chapter are the following:

An amorphous solid (Section 10.5) is a solid whose molecular arrangement lacks a regular and long-range pattern.

The boiling point (Section 10.3) of a liquid is the temperature at which its vapor pressure equals the external pressure. The normal boiling point is the temperature at which the liquid boils when the external pressure is 1 atm (that is, the temperature at which the vapor pressure of the liquid is 1 atm).

The Bragg law Section 10.5) relates the angles at which X rays are scattered from a crystal to the spacing between the layers of particles.

Capillary action (Section 10.3) is the term used to describe the process by which a liquid rises in a tube because of a combination of adhesion with the walls of the tube and cohesion between liquid particles.

Close packing (Section 10.5) refers to the most efficient packing of spheres in a three-dimensional array. There are two closely similar forms, cubic close packing and hexagonal close packing.

Condensation (Section 10.2) is a phase change in which a gas is converted to a liquid.

Critical pressure (Section 10.3) is the pressure at which a gas at its critical temperature is converted to the liquid state.

Critical temperature (Section 10.3) is the highest temperature at which it is possible to convert the gaseous form of a substance to a liquid. The critical temperature increases with an increase in the magnitude of intermolecular forces.

The crystal lattice (Section 10.5) of a solid is an imaginary network of points on which the repeating unit of the structure (the contents of the unit cell) may be imagined to be laid down so that the structure of the crystal is obtained. Each point represents an identical environment in the crystal.

A crystalline solid (Section 10.5) (or simply a crystal) is a solid whose internal arrangement of atoms, molecules, or ions shows a regular repetition in any direction through the solid.

A dynamic equilibrium (Section 10.2) is a state of balance in which opposing processes occur at the same rate.

Hydrogen bonds (Section 10.4) are intermolecular attractions between molecules containing hydrogen bonded to oxygen, nitrogen, or fluorine.

Intermolecular forces (Section 10.4) are the short-range attractive forces operating between the particles that make up the units of a liquid or solid substance. These same forces also cause gases to liquefy or solidify at low temperatures.

London dispersion forces (Section 10.4) are intermolecular forces resulting from attractions between induced dipoles.

The melting point (Section 10.2) of a solid (or the freezing point of a liquid) is the temperature at which solid and liquid phases coexist in equilibrium. The normal melting point is the melting point at 1 atm pressure.

A meniscus (Section 10.3) is the curved upper surface of a liquid column.

A phase change (Section 10.2) represents the conversion of a substance from one state of matter to another. The phase changes we consider are melting and freezing (solid $\leftrightarrow$ liquid), sublimation and deposition (solid $\leftrightarrow$ gas), and vaporization and condensation (liquid $\leftrightarrow$ gas).

A phase diagram (Section 10.6) is a graphic representation of the equilibria between the solid, liquid, and gaseous phases of a substance as a function of temperature and pressure.

The polarizability (Section 10.4) of a molecule describes the ease with which its electron cloud is distorted by an outside influence, thereby inducing a dipole moment.

Surface tension (Section 10.3) is the intermolecular, cohesive attraction that causes a liquid surface to become as small as possible.

The triple point (Section 10.6) of a substance is the temperature at which solid, liquid, and gas phases coexist in equilibrium.

A unit cell (Section 10.5) is the smallest portion of a crystal that reproduces the structure of the entire crystal when repeated in different directions in space. It is the repeating unit or "building block" of the crystal lattice.

Vaporization (Section 10.2), or evaporation, is a phase change in which a liquid is converted to a gas.

Vapor pressure (Section 10.2) is the pressure exerted by a vapor in equilibrium with its liquid or solid phase.

Viscosity (Section 10.3) is a measure of the resistance of fluids to flow.

X-ray diffraction (Section 10.5) refers to the scattering of X rays by the units of a regular crystalline solid. The scattering patterns obtained can be used to deduce the arrangements of particles in the solid lattice.

EXERCISES

Kinetic-molecular theory of solids and liquids; phase changes

10.1 State how the following physical properties differ for the three states of matter—solid, liquid, and gas: (a) density; (b) rate of diffusion; (c) compressibility; (d) ability to flow.

10.2 Suppose that a drop of liquid bromine is added to a 1-L vessel containing air and another drop is added to a 1-L vessel containing water (in which the bromine dissolves). Compare the relative rates at which bromine would diffuse through the 1-L volume in each case. Explain in terms of the kinetic-molecular theory.

10.3 Explain why heat is always liberated when a gas condenses to a liquid and when a liquid freezes to a solid.

10.4 Ethyl chloride, C_2H_5Cl, has a normal boiling point of 12°C. When liquid C_2H_5Cl under pressure is sprayed on a surface at atmospheric pressure, the surface is cooled considerably. Explain.

10.5 The molar enthalpy of sublimation of a solid is always larger than the molar heat of vaporization of the corresponding liquid. Explain.

10.6 The enthalpy of melting of ice at 0°C is 6.01 kJ/mol, and the enthalpy of vaporization of water at 0°C is 44.86 kJ/mol. Calculate the enthalpy of sublimation of ice at 0°C.

10.7 (a) How much heat is produced when 75.0 g of steam at 135°C is converted to ice at −20.0°C? (b) How much heat is required to convert 30.0 g of ice at −10.0°C to steam at 105.0°C?

10.8 For many years drinking water has been cooled in hot climates by the evaporation of water from the surface of canvas bags or porous clay pots. How many grams of water can be cooled from 35°C to 20°C by the evaporation of 1.0 g of water?

10.9 How much water would have to evaporate from the skin per minute to take away all the 250 kJ/h of heat generated by a 65-kg person while asleep (this is called the basal metabolism).

10.10 Ethyl alcohol melts at −114°C and boils at 78°C. The enthalpy of fusion at −114°C is 105 J/g, and the enthalpy of vaporization at 78°C is 870 J/g. If the heat capacity of solid ethyl alcohol is taken to be 0.97 J/g-°C, and that for the liquid 2.3 J/g-°C, how much heat is required to convert 10.0 g of ethyl alcohol at −120°C to the vapor phase at 78°C?

Properties of liquids

10.11 Explain how each of the following affects the vapor pressure of a liquid: (a) surface area; (b) temperature; (c) intermolecular attractive forces; (d) volume of liquid.

10.12 Explain the following observations. (a) Raindrops tend to have a nearly spherical form. (b) A tin can filled with steam at 100°C and sealed collapses when it is cooled. (c) It takes longer to boil eggs at high altitudes than at lower ones. (d) Water may be kept cool in hot climates by storing it in a slightly porous canvas bag. (e) When heated above 279°C, carbon disulfide, CS_2, cannot be liquefied regardless of how great the pressure exerted on the gas.

10.13 Suppose that the atmospheric pressure at a high mountain camp is 500 mm Hg. Using Figure 10.7, determine the temperature at which diethyl ether, ethanol, and water will boil at this pressure.

10.14 Describe some of the molecular characteristics of a substance that might lead to high viscosity.

10.15 Indicate what change, if any, should occur in each of the following properties as a result of an increase in the strength of intermolecular forces: (a) vapor pressure; (b) normal boiling points; (c) normal melting point; (d) surface tension; (e) viscosity; (f) heat of fusion; (g) heat of vaporization; (h) molecular weight.

[10.16] Graph the surface tension of water (Table 10.4), as a function of temperature. What is the equation for the relationship between temperature and surface tension for this substance?

[10.17] Test the applicability of the Clausius-Clapeyron equation (Equation 10.1), using the following vapor pressure versus temperature data for mercury:

Temperature (°C)	Vapor pressure of Hg (mm Hg)
50.0	0.01267
60.0	0.02524
70.0	0.04825
80.0	0.0880
90.0	0.1582

If the Clausius-Clapeyron equation is obeyed, use the slope of the line to calculate ΔH_v for mercury in this temperature range.

[10.18] We might guess that the Clausius-Clapeyron equation (Equation 10.1), would be applicable also to the vapor pressure data for a solid. Use this equation to estimate the heat of sublimation of ice from the following data:

Temperature (°C)	Vapor pressure (mm Hg)
−20.0	0.640
−16.0	1.132
−12.0	1.632
−8.0	2.326
−4.0	3.280
0.0	4.579

Intermolecular forces

10.19 Indicate some of the ways in which intermolecular forces differ from the *intra*molecular forces that hold atoms together *within* molecules.

10.20 Using the thermodynamic data listed in Appendix D, calculate ΔH for the following processes at 25°C:

$$Br_2(l) \longrightarrow Br_2(g)$$
$$Br_2(g) \longrightarrow 2Br(g)$$

Discuss the relative magnitudes of these enthalpy changes in terms of the forces involved in each case.

10.21 What is the nature of the major attractive intermolecular force in each of the following: (a) $MgO(s)$; (b) $I_2(s)$; (c) $HF(l)$; (d) $CH_3CN(l)$ (dipole moment 3.9 D)?

10.22 Indicate all the various types of intermolecular attractive forces that may operate in each of the following: (a) $CH_3OH(l)$; (b) $Xe(l)$; (c) $H_2S(l)$; (d) $ClF(l)$; (e) $Ca(NO_3)_2(s)$. In each case indicate, if you can, the type of attractive force that makes the major contribution.

10.23 List the following compounds in the expected order of increasing energy of the hydrogen-bonding interaction between molecules: H_2S; CH_3NH_2; C_6H_5OH (phenol).

10.24 What are the required structural features for a substantial hydrogen-bonding contribution to the intermolecular attractive forces?

10.25 How would you expect the heats of vaporization to vary among the members of the following sets of compounds: (a) PH_3, AsH_3, SbH_3; (b) CH_4, SiH_4, GeH_4; (c) CH_4, CF_4, CCl_4, CBr_4; (d) CH_4, NH_3, H_2O?

10.26 Indicate which of the following substances (a) is most likely to exist as a crystalline solid at room temperature, and (b) would be the least readily liquefied gas: HF, PCl_3, F_2, $FeCl_2$, SO_2.

10.27 From the following list of substances, indicate which is expected to have the highest boiling point and which the lowest: CO_2, Ar, CF_4; RbCl; SiF_4. Explain your choices.

10.28 Predict the order of increasing viscosity among the following liquids, all at a common temperature: (a) propanol, $CH_3CH_2CH_2OH$; (b) propane, $CH_3CH_2CH_3$; (c) propane-1,3-diol, $HOCH_2CH_2CH_2OH$. Explain your reasoning.

10.29 The boiling points of the fluorides of second-row elements of the periodic table are: LiF, 1717°C; BeF_2, 1175°C; BF_3, −101.0°C; CF_4, −128°C; NF_3, −120°C; OF_2, −145°C; F_2, −188°C. Account for this variation in terms of the nature and strengths of the intermolecular forces that operate.

10.30 In dichloromethane, CH_2Cl_2 ($\mu = 1.60$ D), the dispersion-force contribution to the intermolecular attractive forces is about five times larger than the dipole-dipole contribution. How would you expect the relative importance of the two kinds of intermolecular attractive forces to vary (a) in dibromomethane ($\mu = 1.43$ D); (b) in difluoromethane ($\mu = 1.93$ D)?

10.31 The critical temperatures for the hydrogen halides vary as follows: HF, 188°C; HCl, 51°C; HBr, 90°C; HI, 151°C. Explain this observed variation in terms of the intermolecular attractive forces operative in each case.

Solids

10.32 Indicate the type of crystal (atomic, molecular, metallic, covalent, or ionic) each of the following would form upon solidification: (a) O_2; (b) H_2S; (c) Ag; (d) KCl; (e) Si; (f) $Al_2(SO_4)_3$; (g) Ne; (h) SiO_2; (i) NH_3; (j) MgO; (k) NaOH; (l) CH_4.

10.33 Compare the following group of substances with respect to hardness, electrical conductivity, and melting point: $BaCl_2$; Ni; SCl_2; C (diamond).

10.34 For each of the following pairs of substances, predict which will have the higher melting point and indicate why: (a) $CuBr_2$, Br_2; (b) CO_2, SiO_2; (c) S, Cr; (d) CsBr, CaF_2.

10.35 Why does hexachlorobenzene (Figure 10.18) have higher melting and boiling points than benzene?

10.36 What is the net number of lattice points associated with one of each of the following types of unit cells: (a) simple cubic; (b) body-centered cubic; (c) face-centered cubic; (d) simple monoclinic?

10.37 The unit cell of cesium chloride, CsCl, may be chosen as a cube with Cl^- ions on the corners and Cs^+ in the center of the cube. Why is it improper to refer to the crystal lattice of CsCl or the unit cell of that lattice as being body-centered cubic?

10.38 The unit cell of zinc blende (ZnS) is given in Figure 7.2. If the unit cell shown is cubic, how might the lattice of this structure best be described?

10.39 In the face-centered cubic unit cells possessed by many metals, the atom in the center of each face is in contact with the corner atoms as shown in Figure 10.37. (a) Copper crystallizes in a face-centered cubic lattice. If the edge of the unit cell is 3.61 Å, calculate the atomic radius of Cu. (b) Nickel crystallizes in a face-centered cubic lattice. If the radius of a nickel atom is 1.24 Å, what is the length of each edge of the unit cell?

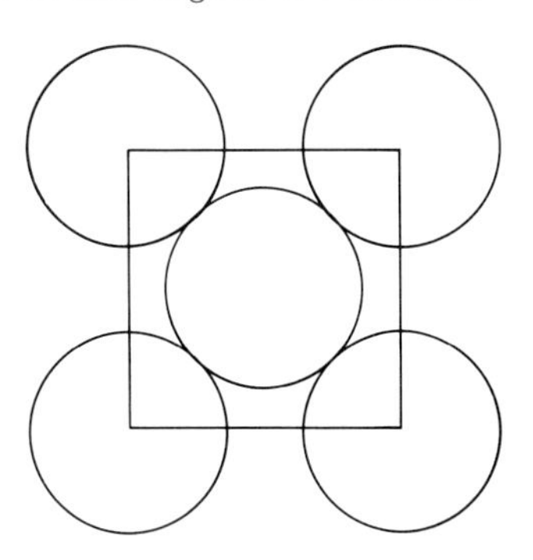

FIGURE 10.37

10.40 (a) Potassium fluoride has the NaCl type of crystal structure. The density of KF at 25°C is 2.468 g/cm^3. Calculate the dimensions of the KF unit cell. (b) Silver chloride, which has the NaCl type of crystal structure, has a density of 5.57 g/cm^3 at 25°C. What is the length of the edge of the AgCl unit cell?

10.41 An element crystallizes in a body-centered cubic lattice. The edge of the unit cell is 2.86 Å and the density of the crystal is 7.92 g/cm^3. Calculate the atomic weight of the element.

10.42 (a) Neon crystallizes in a face-centered cubic lattice. The edge of the unit cell is 452 pm; the atomic weight of neon is 20.2. Calculate the density of crystalline neon. (b) Potassium crystallizes in a body-centered cubic lattice whose unit cell measures 533 pm on each edge. Calculate the density of potassium.

10.43 (a) The unit cell of aluminum metal is cubic with an edge length of 4.05 Å. Determine the type of unit cell (simple, face-centered, or body-centered) if the metal has a density of 2.70 g/cm^3. (b) Platinum has a density of 21.5 g/cm^3. How many atoms are in the unit cell if it is cubic with an edge length of 3.914 Å? Is the cell simple cubic, body-centered cubic, or face-centered cubic?

10.44 KCl has the same structure as NaCl. The length of the unit cell is 628 pm. The density of KCl is 1.984 g/cm^3, and its formula mass is 74.55. Using this information, calculate Avogadro's number.

10.45 (a) In a three-dimensional, close-packed array of equal-sized spheres, what is the coordination number of each sphere? (b) What is the coordination number of each sphere in a simple-cubic structure? (c) What is the coordination number of each sphere in a body-centered cubic lattice?

[10.46] The free space in a structure composed of

equal-sized spheres may be found by subtracting the volumes of the spheres in a unit cell from the volume of the cell. (a) Calculate the percentage of free space in each of the three types of cubic unit cells if each contains equal-sized spheres. In the simple cubic cell the spheres touch along the unit-cell edge. In the face-centered cubic cell they touch along the diagonal of each face. In the body-centered cubic cell they touch along the body diagonal. (b) Arrange the cubic unit cells in order of increasing packing efficiency.

10.47 In the diffraction of a crystal using X rays with a wavelength of 1.54 Å, a first-order ($n = 1$) reflection was obtained at an angle of 12.5°. What is the distance between the planes of particles that causes this diffraction?

10.48 The first-order ($n = 1$) diffraction of X rays from crystal planes separated by 2.81 Å occurs at 11.8°. What is the wavelength of the X rays?

10.49 At what angle will first-order ($n = 1$) diffraction occur when X rays whose wavelength is 154 pm are diffracted by planes separated by 205 pm?

Phase diagrams

10.50 The normal melting and boiling points of xenon are −112°C and −108°C, respectively, and its triple point is at −121°C at a pressure of 282 mm Hg. Sketch the phase diagram for xenon, showing the three points given above and indicating the areas in which each phase is stable.

10.51 The normal melting and boiling points of sulfur dioxide, SO_2, are −72.7°C and −10.0°C, respectively. Its triple point is at −75.5°C and 1.65×10^{-3} atm, and its critical point is at 157°C and 78 atm. Sketch the phase diagram for SO_2 showing the four points given above and indicating the areas in which each phase is stable.

10.52 Refer to Figure 10.35 and describe all the phase changes that would occur in each of the following cases. (a) Water vapor originally at 1.0×10^{-3} atm and −0.10°C is slowly compressed at constant temperature until the final pressure is 10 atm. (b) Water originally at −10°C and 0.30 atm is heated at constant pressure until the temperature is 80.0°C.

10.53 Refer to Figure 10.35 and describe the phase changes (and the temperatures at which they occur) when CO_2 is heated from −80°C to −20°C at (a) a constant pressure of 3 atm; (b) a constant pressure of 6 atm.

10.54 Briefly explain why the melting-point curve in the phase diagram for water slopes to the left with increasing pressure, whereas that for carbon dioxide slopes to the right. Which is the more typical behavior for substances?

Additional exercises

10.55 Explain the phenomenon of critical temperature in terms of the kinetic-molecular theory.

10.56 Calculate the heat required to convert 10.0 g of propanol, C_3H_7OH, from a solid at −135°C into a vapor at 120°C. The normal melting point and boiling point of C_3H_7OH are −127°C and 97°C, respectively. The heat of fusion is 5.18 kJ/mol while the heat of vaporization is 41.7 kJ/mol. The heat capacities of the solid, liquid, and gas states are 142, 170, and 108 J/mol-°C, respectively.

10.57 Account for the observed variation in viscosities of the series of straight-chain hydrocarbons, as listed below:

Compound	Formula	Viscosity (N-s/m²)
Pentane	C_5H_{12}	0.225×10^{-3}
Hexane	C_6H_{14}	0.313×10^{-3}
Heptane	C_7H_{16}	0.397×10^{-3}
Octane	C_8H_{18}	0.546×10^{-3}

10.58 Explain the folowing observed temperature variation in viscosity of glycerol, $HOCH_2CH_2(OH)CH_2OH$: 20°C, 1.49 N-s/m²; 25°C, 0.942 N-s/m²; 30°C, 0.622 N-s/m².

10.59 The molar heats of sublimation of benzene, C_6H_6, and the monohalobenzenes, C_6H_5X, are as follows:

Compound	Formula	ΔH_s (kJ/mol)
Benzene	C_6H_6	33.85
Fluorobenzene	C_6H_5F	34.61
Chlorobenzene	C_6H_5Cl	41.04
Bromobenzene	C_6H_5Br	44.43
Iodobenzene	C_6H_5I	49.58

Account for this variation in terms of the nature of the intermolecular forces that are likely to be most important.

10.60 The enthalpy of vaporization of water at 0°C is 44.86 kJ/mol, whereas at 100°C the enthalpy of vaporization of water is 40.65 kJ/mol. Account for the fact that water has a lower enthalpy of vaporization at the higher temperature.

[10.61] From the following data for the vapor pressure of $NH_3(l)$ versus temperature, use the Clausius-Clapeyron equation to estimate the molar heat of vaporization of $NH_3(l)$.

Temperature (°C)	Vapor pressure of $NH_3(l)$ (atm)
−60.0	0.2161
−54.0	0.3167
−50.0	0.4034
−44.0	0.5693
−40.0	0.7083
−34.0	0.9676

Note that in working this problem you do *not* need to change the units of pressure. Why is this so?

[10.62] The critical temperature and pressure of $CClF_3$ (Freon 13) are 29°C and 39 atm, respectively. For $CClH_3$ (methyl chloride) the corresponding values are 143°C and 66 atm. What do these values tell us about the relative intermolecular attractive forces between molecules in the two substances? Are the results surprising? If so, why?

10.63 The critical temperatures of the boron trihalides vary as follows: BF_3, $-12°C$; BCl_3, $179°C$; BBr_3, $300°C$. Explain this variation in terms of the nature of the intermolecular attractive forces in these systems.

10.64 A clean, very dry silica or glass surface has many Si—OH groups on it. Describe the nature of the adhesive forces that might operate between such a surface and CH_3OH.

10.65 Hydrogen peroxide, H_2O_2, melts at $-0.4°C$ and boils at 151°C. In comparison with water, do these data suggest that hydrogen bonding might be important? Draw a diagram that shows the nature of the hydrogen-bonding interactions that could occur between molecules in this substance.

[10.66] The normal boiling points and molar heats of vaporization at the boiling points vary among the hydrogen halides as follows:

Compound	Boiling point (K)	ΔH_v(kJ/mol)
HF	292	7.5
HCl	188	16.1
HBr	207	17.6
HI	238	19.7

Account for the variation in boiling point among the compounds. In light of your explanation, how can you account for the abnormally low heat of vaporization of HF? (*Hint:* The density of HF vapor just above the boiling point indicates a molecular mass much greater than 20.)

10.67 Account for the following observations. (a) SO_2 is a gas, whereas MnO_2 is a solid at room temperature. (b) I_2 crystals are soft and easily crushed, whereas NaI crystals are hard and brittle.

10.68 Amorphous silica is found to have a density of about 2.2 g/cm^3, whereas the density of crystalline quartz is 2.65 g/cm^3. Account for this difference in densities.

[10.69] Assume that 3.000 g of H_2O is introduced into an evacuated flask whose volume is 1.000 L. If the temperature of the water and flask is 30.0°C, what mass of water will evaporate? The vapor pressure of water at 30.0°C is 31.82 mm Hg.

[10.70] Chromium crystallizes in a body-centered cubic unit cell whose edge length is 288.4 pm. If the atoms touch along the body diagonal of the unit cell, calculate the atomic radius of a Cr atom.

10.71 The unit cell of diamond is cubic. The density is 3.52 g/cm^3 and the volume of the unit cell is 0.0454 nm^3. Calculate the net number of carbon atoms in the cell.

10.72 Irradiation of polyethylene with X rays introduces cross-linking into the material. What effect would this have on the properties of the polyethylene?

10.73 From which of the following materials would you expect to obtain well-defined X-ray diffraction patterns: (a) a sugar crystal; (b) KBr; (c) liquid water; (d) pure iron; (e) ice; (f) a section of a rubber stopper?

10.74 A crystal has a simple cubic lattice with a unit cell length of 4.06 Å. Using X rays of 1.78 Å, calculate the first three diffraction angles at which the X rays are diffracted by the planes that parallel the unit cell faces.

10.75 The triple point of benzene is 5°C, 21 mm Hg. The density of solid benzene is 1.005 g/cm^3, whereas that of the liquid is 0.894 g/cm^3. The normal boiling point of benzene is 80°C; its critical point is 289°C, 48 atm. Sketch the phase diagram.

11 Solutions

Very few of the materials that we encounter in everyday life are pure substances; most are mixtures. Many of these mixtures are homogeneous; that is, their components are uniformly intermingled on a molecular level. We have seen that homogeneous mixtures are called solutions (Sections 2.2 and 3.9). Examples of solutions abound in the world around us. The air we breathe is a homogeneous mixture of several gaseous substances. The familiar metal brass is a solution of zinc in copper. The oceans are a solution of many dissolved substances in water. The fluids that run through our bodies are solutions, carrying a great variety of essential nutrients, salts, and so forth.

Solutions may be gaseous, liquid, or solid; examples of each kind are given in Table 11.1. Recall that the solvent is the component whose phase is retained when the solution forms (Section 3.11); if all components are in the same phase, the one in greatest amount is called the solvent. Other components are called solutes. Liquid solutions are the most common, and it is on this type of solution that we focus our attention in this chapter.

In Chapter 10 we considered the various types of intermolecular forces that exist between molecular and ionic particles. In this chapter we will

TABLE 11.1 Examples of solutions

State of solution	State of solvent	State of solute	Example
Gas	Gas	Gas	Air
Liquid	Liquid	Gas	Oxygen in water
Liquid	Liquid	Liquid	Alcohol in water
Liquid	Liquid	Solid	Salt in water
Solid	Solid	Gas	Hydrogen in platinum
Solid	Solid	Liquid	Mercury in silver
Solid	Solid	Solid	Silver in gold (certain alloys)

see that these forces are also involved in the interactions between solutes and solvents. We will examine the solution process, the factors that determine the amount of solute that can dissolve in a given quantity of solvent, and some properties of the solutions that result. We will be particularly concerned with aqueous solutions of ionic substances, because of their central importance in chemistry and in our daily lives. Near the end of the chapter we will consider a type of mixture, known as a colloid, that is on the borderline between heterogeneous mixtures and solutions. Before we examine these topics, however, it is useful to discuss ways of describing the concentrations of solutions, that is, the amount of solute dissolved in a given quantity of solvent or solution.

11.1 WAYS OF EXPRESSING CONCENTRATION

The concentration of a solution can be expressed either qualitatively or quantitatively. The terms dilute and concentrated are used to describe a solution qualitatively. A solution with a relatively small concentration of solute is said to be **dilute;** one with a large concentration is said to be **concentrated.**

Several quantitative expressions of concentration are employed in chemistry. One of the simplest is the **weight percentage.** The weight percentage of a component of a solution is given by Equation 11.1. If a solution of hydrochloric acid contains 36 percent HCl by weight, it has 36 g of HCl for each 100 g of solution:

$$\text{Wt \% of component} = \frac{\text{mass of component in soln}}{\text{total mass of soln}} \times 100 \qquad [11.1]$$

SAMPLE EXERCISE 11.1

A solution is made containing 6.9 g of $NaHCO_3$ per 100 g of water. What is the weight percentage of solute in this solution?

Solution:

$$\text{Wt \% of solute} = \frac{\text{mass solute}}{\text{mass soln}} \times 100$$

$$= \frac{6.9\ \text{g}}{6.9\ \text{g} + 100\ \text{g}} \times 100 = 6.5\%$$

Notice that the mass of solution is the sum of the mass of solvent and the mass of solute. The weight percentage of solvent in this solution is $(100 - 6.5)\% = 93.5\%$.

Several concentration expressions are based on the number of moles of one or more components of the solution. Three are commonly used in chemistry: mole fraction, molarity, and molality. The **mole fraction** of a component of a solution is given by Equation 11.2.

$$\text{Mole fraction of component} = \frac{\text{moles component}}{\text{total moles all components}} \qquad [11.2]$$

This concept was used in Chapter 9 when we discussed Dalton's law of partial pressures. The symbol X is commonly used for mole fraction, with a subscript to indicate the component on which attention is being focused. For example, the mole fraction of HCl in a hydrochloric acid solution can be represented as X_{HCl}. The mole fractions of all components of a solution will total 1.

SAMPLE EXERCISE 11.2

Calculate the mole fraction of HCl in a solution of hydrochloric acid containing 36 percent HCl by weight.

Solution: Assume that there is 100 g of solution (you can verify for yourself that assuming any other quantity will not change the result, though it can make the arithmetic more difficult). The solution therefore contains 36 g of HCl and 64 g of H_2O.

$$\text{Moles HCl} = (36 \text{ g HCl})\left(\frac{1 \text{ mol HCl}}{36.5 \text{ g HCl}}\right) = 0.99 \text{ mol HCl}$$

$$\text{Moles } H_2O = (64 \text{ g } H_2O)\left(\frac{1 \text{ mol } H_2O}{18 \text{ g } H_2O}\right) = 3.6 \text{ mol } H_2O$$

$$X_{HCl} = \frac{\text{moles HCl}}{\text{moles } H_2O + \text{moles HCl}} = \frac{0.99}{3.6 + 0.99} = \frac{0.99}{4.6} = 0.22$$

Molarity (M), which we discussed in Section 3.9, is defined as the number of moles of solute in a liter of solution:

$$\text{Molarity} = \frac{\text{moles solute}}{\text{liters soln}} \qquad [11.3]$$

SAMPLE EXERCISE 11.3

(a) Calculate the molarity of an ascorbic acid ($C_6H_8O_6$, vitamin C) solution prepared by dissolving 1.80 g in enough water to make 125 mL of solution. (b) How many milliliters of this solution contain 0.0100 mol of ascorbic acid?

Solution:

(a) $$M = \frac{\text{mol } C_6H_8O_6}{\text{L soln}} = \left(\frac{1.80 \text{ g } C_6H_8O_6}{125 \text{ mL soln}}\right)\left(\frac{1 \text{ mol } C_6H_8O_6}{176 \text{ g } C_6H_8O_6}\right)\left(\frac{10^3 \text{ mL soln}}{1 \text{ L soln}}\right) = 0.0818 \frac{\text{mol } C_6H_8O_6}{\text{L soln}} = 0.0818\, M$$

(b) Rearranging Equation 11.3, solving for liters, we have

$$\text{Liters soln} = \text{moles solute} \times \frac{1}{\text{molarity}} = (0.0100 \text{ mol } C_6H_8O_6) \times \left(\frac{1 \text{ L soln}}{0.0818 \text{ mol } C_6H_8O_6}\right) = 0.122 \text{ L}$$

$$\text{mL soln} = (0.122 \text{ L})\left(\frac{10^3 \text{ mL}}{1 \text{ L}}\right) = 122 \text{ mL}$$

The **molality** (m) of a solution is defined as the number of moles of solute in a kilogram of solvent:

$$\text{Molality} = \frac{\text{moles solute}}{\text{mass of solvent in kilograms}} \qquad [11.4]$$

Notice the difference between molarity and molality. These two ways of expressing concentration are similar enough to be easily confused. Molality is defined in terms of the mass of solvent, whereas molarity is defined in terms of the volume of solution. A 1.50 molal (written 1.50 *m*) solution contains 1.50 mol of solute for every kilogram of solvent. (When water is the solvent, the molality and molarity of a dilute solution are numerically about the same, because 1 kg of solvent is nearly the same as 1 kg of solution, and 1 kg of the solution has a volume of about 1 L.)

The molality of a given solution does not vary with temperature, because masses do not vary with temperature. Molarity, however, changes with temperature because of the expansion or contraction of the solution.

SAMPLE EXERCISE 11.4

What is the molality of a solution made by dissolving 5.0 g of toluene (C_7H_8) in 225 g of benzene (C_6H_6)?

Solution:

$$\text{Molality} = \frac{\text{moles } C_7H_8}{\text{kilograms } C_6H_6} =$$

$$\left(\frac{5.0 \text{ g } C_7H_8}{225 \text{ g } C_6H_6}\right)\left(\frac{1 \text{ mol } C_7H_8}{92.0 \text{ g } C_7H_8}\right)\left(\frac{1000 \text{ g } C_6H_6}{1 \text{ kg } C_6H_6}\right)$$

↑ Convert grams C_7H_8 to moles

↑ Convert grams C_6H_6 to kilograms

$$= 0.24\ m$$

For most purposes in ordinary chemical laboratory work molarity is the most useful expression of concentration. However, the other ways we have just considered find use in special situations, and you should be familiar with them. Sample Exercise 11.5 shows by way of example how the various expressions of concentration are related.

SAMPLE EXERCISE 11.5

Given that the density of a solution of 5.0 g of toluene and 225 g of benzene (Sample Exercise 11.4) is 0.876 g/mL, calculate the concentration of the solution in (a) molarity; (b) mole fraction of solute; (c) weight percentage of solute.

Solution: (a) The total mass of the solution is equal to the mass of the solvent plus the mass of the solute:

$$\text{Mass soln} = 5.0 \text{ g} + 225 \text{ g} = 230 \text{ g}$$

The density of the solution is used to convert the mass of the solution to its volume:

$$\text{Milliliters soln} = (230 \text{ g})\left(\frac{1 \text{ mL}}{0.876 \text{ g}}\right) = 263 \text{ mL}$$

Density must be known in order to interconvert molarity and molality, because one is on a mass basis whereas the other is on a volume basis. The number of moles of solute must be known to calculate either molarity or molality:

$$\text{Moles } C_7H_8 = (5.0 \text{ g } C_7H_8)\left(\frac{1 \text{ mol } C_7H_8}{92.0 \text{ g } C_7H_8}\right)$$

$$= 0.054 \text{ mol}$$

Molarity is moles of solute per liter of solution:

$$\text{Molarity} = \frac{\text{moles } C_7H_8}{\text{liter soln}}$$

$$= \left(\frac{0.054 \text{ mol } C_7H_8}{263 \text{ mL soln}}\right)\left(\frac{1000 \text{ mL soln}}{1 \text{ L soln}}\right)$$

$$= 0.21\ M$$

Compare this value with the molality of the same solution calculated in Sample Exercise 11.4.

(b) The mole fraction of solute is expressed as

$$X_{C_7H_8} = \frac{\text{moles } C_7H_8}{\text{moles } C_7H_8 + \text{moles } C_6H_6}$$

We have already calculated the number of moles of C_7H_8.

$$\text{Moles } C_6H_6 = (225 \text{ g } C_6H_6)\left(\frac{1 \text{ mol } C_6H_6}{78.0 \text{ g } C_6H_6}\right)$$

$$= 2.88 \text{ mol}$$

$$X_{C_7H_8} = \frac{0.054 \text{ mol}}{0.054 \text{ mol} + 2.88 \text{ mol}} = \frac{0.054}{2.93}$$

$$= 0.018$$

(c) The weight percentage of solute is calculated as follows:

$$\text{Wt \% } C_7H_8 = \frac{5.0 \text{ g } C_7H_8}{5.0 \text{ g } C_7H_8 + 225 \text{ g } C_6H_6} \times 100$$

$$= \frac{5.0 \text{ g}}{230 \text{ g}} \times 100 = 2.2\%$$

The concentration units that we have just discussed are the ones that we will use through the rest of this chapter and elsewhere in this text. However, one further concentration expression, called **normality**, may be encountered in other places. Normality (abbreviated N) is defined as the number of **equivalents** of solute per liter of solution.

$$\text{Normality} = \frac{\text{equivalents solute}}{\text{liters soln}} \qquad [11.5]$$

An equivalent is defined according to the type of reaction being examined. For acid-base reactions, an equivalent of an acid is the quantity that supplies 1 mol of H^+; an equivalent of a base is the quantity reacting with 1 mol of H^+. In an oxidation-reduction reaction, an equivalent is the quantity of substance that gains or loses 1 mol of electrons. The masses of 1 equivalent of several substances are given in Table 11.2. An equivalent is always defined in such a way that 1 equivalent of reagent A will react with 1 equivalent of reagent B. For example, in an oxidation-reduction reaction, 31.6 g (1 equivalent) of $KMnO_4$ will react with 67.0 g (1 equivalent) of $Na_2C_2O_4$ (refer to Table 11.2). Similarly, in an acid-base reaction, 49.0 g of H_2SO_4 is stoichiometrically equivalent to 26.0 g of $Al(OH)_3$.

Where $KMnO_4$ is reduced to Mn^{2+}, thereby gaining five electrons, we have

$$1 \text{ mol } KMnO_4 = 5 \text{ equivalents of } KMnO_4$$

TABLE 11.2 Equivalent-mass relationships

Reactant	Product	Reaction type	Mass of 1 mol of reactant (g)	Mass of 1 equivalent of reactant (g)
$KMnO_4$	Mn^{2+}	Reduction (5 e^-)	158.0	158.0/5 = 31.6
$KMnO_4$	MnO_2	Reduction (3 e^-)	158.0	158.0/3 = 52.7
$Na_2C_2O_4$	CO_2	Oxidation (2 e^-)	134.0	134.0/2 = 67.0
H_2SO_4	SO_4^{2-}	Acid (2 H^+)	98.0	98.0/2 = 49.0
$Al(OH)_3$	Al^{3+}	Base (3 OH^-)	78.0	78.0/3 = 26.0

Therefore, if 1 mol of $KMnO_4$ is dissolved in sufficient water to form 1 L of solution, the concentration of the solution can be expressed as either 1 *M* or 5 *N*. Normality is always a whole-number multiple of molarity. In oxidation-reduction reactions, the whole number is the number of electrons gained or lost by one formula unit of the substance. In acid-base reactions the whole number is the number of H^+ or OH^- available in a formula unit of the substance. If H_2SO_4 is used as an acid, forming SO_4^{2-} as a product, it will lose *two* H^+ ions. Thus its normality will be *twice* its molarity. For example, a 0.255 *M* solution of H_2SO_4 will be $2 \times 0.255 = 0.510\ N$.

SAMPLE EXERCISE 11.6

How many milliliters of 0.200 *N* $KMnO_4$ solution are required to oxidize 25.0 mL of 0.120 *N* $FeSO_4$ solution?

Solution: At the equivalence point, equal numbers of equivalents of $KMnO_4$ and $FeSO_4$ must react. Rearranging Equation 11.5, solving for equivalents, we have

$$\text{Equivalents solute} = \text{normality} \times \text{liters soln}$$

Thus the number of equivalents of $FeSO_4$ is

$$\text{Equivalents } FeSO_4 = \left(0.120 \frac{\text{equivalents}}{\text{L}}\right)(0.0250\ \text{L}) = 3.00 \times 10^{-3}\ \text{equivalent}$$

This must also be the number of equivalents of $KMnO_4$ reacted. Thus rearranging Equation 11.5, solving for liters, we have

$$\text{Liters soln} = \text{equivalents solute} \times \frac{1}{N}$$

The volume of $KMnO_4$ used is therefore

$$\text{Liters soln} = (3.00 \times 10^{-3}\ \text{equivalent})\left(\frac{1\ \text{L}}{0.200\ \text{equivalent}}\right) = 0.0150\ \text{L} = 15.0\ \text{mL}$$

This exercise illustrates one advantage of using normality over molarity in stoichiometric calculations: We do not need to refer to a completely balanced chemical equation.

11.2 THE SOLUTION PROCESS

A solution is formed when one substance disperses uniformly throughout another. With the exception of gas mixtures, all solutions involve substances in a condensed phase. We learned in Chapter 10 that substances in the liquid and solid state experience intermolecular attractive forces that hold the individual particles together. Intermolecular forces also operate between a solute particle and the solvent that surrounds it.

Any of the various kinds of intermolecular forces that we discussed in Chapter 10 can operate between solute and solvent particles in a solution. As a general rule, we expect solutions to form when the attractive forces between solute and solvent are comparable in magnitude with those that exist between the solute particles themselves or between solvent particles themselves. For example, the ionic substance NaCl dissolves readily in water. When NaCl is added to water, the water molecules orient themselves on the surface of the NaCl crystals as shown in Figure 11.1. The positive end of the water dipole is oriented toward the Cl^- ions, while the negative end of the water dipole is oriented toward the Na^+ ions. The ion-dipole attractions between Na^+ and Cl^- ions and water molecules are sufficiently strong to pull these ions from their positions in the crystal. Notice that the corner Na^+ ion is held in the crystal by only three adjacent Cl^- ions. In contrast, an Na^+ ion on the edge of

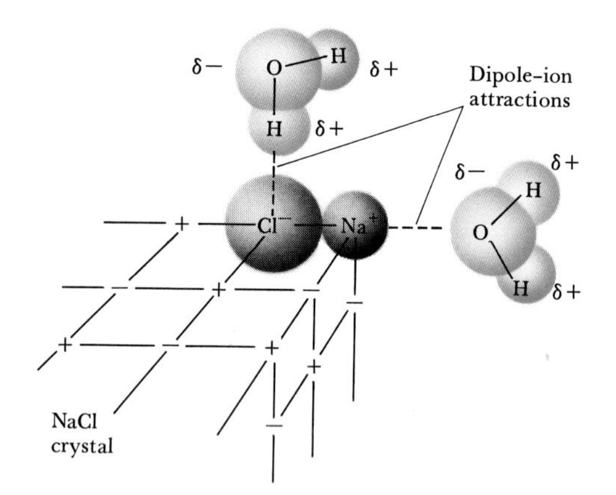

FIGURE 11.1 Interactions between H_2O molecules and the Na^+ and Cl^- ions of a NaCl crystal.

the crystal has four nearby Cl^- ions, and a Na^+ ion in the interior of the crystal has six surrounding Cl^- ions. The corner Na^+ ion is therefore particularly vulnerable to removal from the crystal. Once this Na^+ ion has been removed, adjacent Cl^- ions are similarly exposed and are therefore removed more easily than before.

Once removed from the crystal, the Na^+ and Cl^- ions are surrounded by water molecules as shown in Figure 11.2. Such interactions between solute and solvent molecules are known as **solvation.** When the solvent is water it is known as **hydration.**

Frequently, hydrated ions remain in crystalline salts that are obtained by evaporation of water from aqueous solutions. Common examples include $FeCl_3 \cdot 6H_2O$ [iron(III) chloride hexahydrate] and $CuSO_4 \cdot 5H_2O$ [copper(II) sulfate pentahydrate]. The $FeCl_3 \cdot 6H_2O$ consists of $Fe(H_2O)_6{}^{3+}$ and Cl^- ions; the $CuSO_4 \cdot 5H_2O$ consists of $Cu(H_2O)_4{}^{2+}$ and $SO_4(H_2O)^{2-}$ ions. Water molecules can also occur in positions in the crystal lattice that are not specifically associated with either a cation or anion. $BaCl_2 \cdot 2H_2O$ (barium chloride dihydrate) is an example. Compounds such as $FeCl_3 \cdot 6H_2O$, $CuSO_4 \cdot 5H_2O$, and $BaCl_2 \cdot 2H_2O$, which contain a salt and water combined in definite proportions, are known as *hydrates;* the water associated with them is called *water of hydration*.

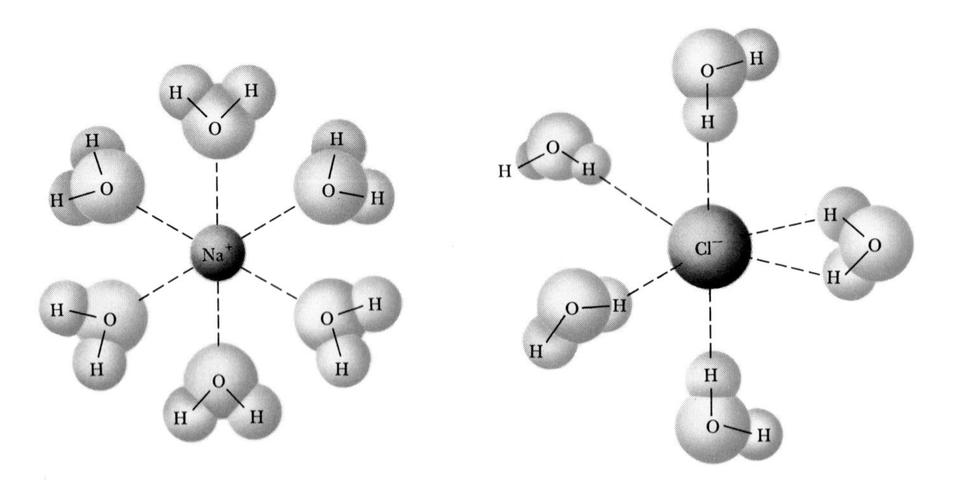

FIGURE 11.2 Hydrated Na^+ and Cl^- ions. The negative ends of the water dipole point toward the positive ion. The positive ends of the water dipole point toward the negative ion. We do not know whether one or both positive hydrogens are oriented toward the negative ion.

Sodium chloride dissolves in water because the water molecules interact with the Na^+ and Cl^- ions sufficiently strongly to overcome the attraction between Na^+ and Cl^- ions in the crystal. To form a solution, the water molecules must also separate to make room for the ions. Therefore, we can visualize three types of interaction that take place in the solution process: (1) solute-solute interactions; (2) solute-solvent interactions; and (3) solvent-solvent interactions. The enthalpy changes accompanying each of these interactions when a salt dissolves in water are shown in Figure 11.3. Energy is required in processes (*a*) and (*b*) to overcome solute-solute and solvent-solvent attractions. Energy is released in process (*c*), when solute and solvent interact with one another. As shown in Figure 11.3, the net solution process can be either exothermic or endothermic. For example, ammonium nitrate, NH_4NO_3, dissolves by an endothermic reaction ($\Delta H = 26.4$ kJ/mol). It has consequently been used to make instant ice packs, which are used to treat athletic injuries. The solid NH_4NO_3 is placed inside a thin-walled plastic bag. This, in turn, is sealed inside a thicker-walled bag together with some water. The small bag can be broken by kneading the larger bag. The resultant solution gets quite cold. By contrast, some solution processes are exothermic; for example, when sodium hydroxide, NaOH, is added to water the resultant solution can get quite hot.

Figure 11.3 shows us that the heat of solution is the net result of enthalpy terms that are rather large when ionic solutions are involved. The overall heat may be exothermic or slightly endothermic. In both examples shown in Figure 11.3, however, there is a large exothermic heat term due to the attractive interaction of the ions with the solvent molecules—the solvation term (*c*). We can understand from a diagram of this kind why NaCl does not dissolve in a nonpolar liquid such as gasoline. The hydrocarbon molecules of the gasoline could experience only weak ion-induced dipole attractions (see Figure 10.11), and these would not go very far toward compensating for the energies required to separate the ionic particles from one another.

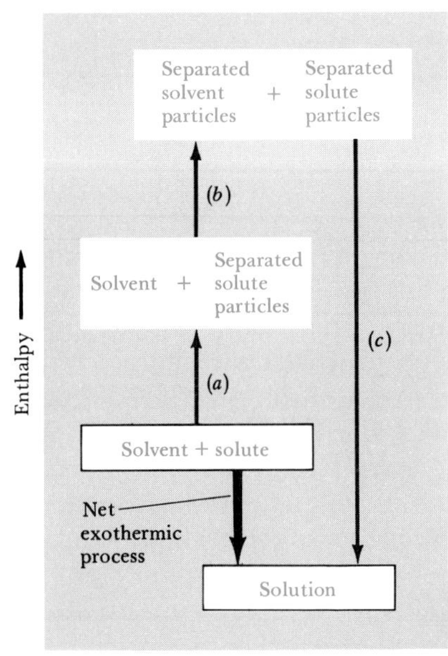

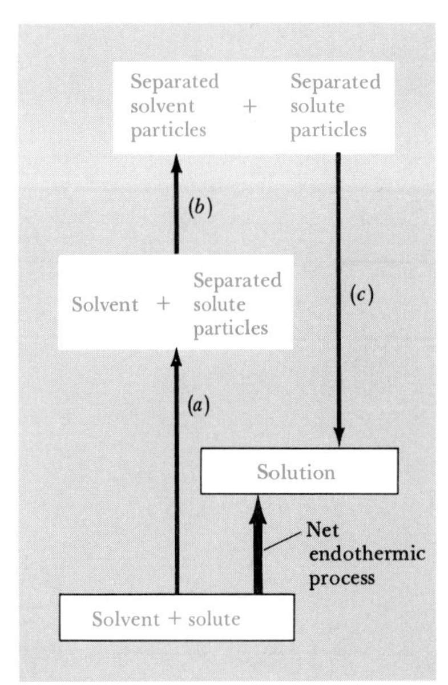

FIGURE 11.3 Analysis of the enthalpy changes accompanying the solution process: process (*a*) represents the enthalpy required to separate solute particles; process (*b*) represents the enthalpy required to separate solvent particles; process (*c*) represents the enthalpy released when solute and solvent particles interact with each other. The figure on the left shows a net exothermic heat of solution, whereas the one on the right shows a net endothermic heat of solution.

By reasoning in a similar way we can understand why a polar liquid such as water does not dissolve to a great extent in a nonpolar liquid such as carbon tetrachloride, CCl_4. The water molecules experience strong attractive hydrogen-bonding interactions with one another (Section 10.4); these attractive forces would need to be overcome to disperse the water molecules throughout the nonpolar liquid. But there are no compensating attractive forces of comparable magnitude between water and carbon tetrachloride molecules. Consider the other possible case, the dissolving of carbon tetrachloride as a solute in water as solvent. The attractive forces between carbon tetrachloride molecules are mainly of the London-dispersion-force type, since the CCl_4 molecules are nonpolar. There would thus not be much loss in attractive energy by dispersing the CCl_4 molecules in water as solvent. In this case, however, water molecules must separate from one another to some degree, to make a space for the solute CCl_4 molecules. This separation costs energy that is not recovered in the form of the attractive interactions between water and the CCl_4 molecules. Thus solution formation does not occur readily in this case, either.

Finally, let us consider yet another example that raises a new idea. When two nonpolar substances such as CCl_4 and hexane, C_6H_{14}, are mixed, they readily dissolve in one another in all proportions. The attractive forces between molecules in both these cases are of the London-dispersion-force type. We note that CCl_4 boils at 77°C, whereas C_6H_{14} boils at 69°C. Thus it is reasonable to suppose that the magnitudes of the attractive forces between molecules are comparable in the two substances. This means that when they are mixed there is little or no energy change upon solution formation. Yet this process occurs spontaneously; that is, it occurs to an appreciable extent without any extra input of energy from outside the system.

Experience tells us that two distinct factors are involved in processes that occur spontaneously. The most obvious has to do with energy. If you let go of a book you have in hand, it falls to the floor. The book is acted upon by the force of gravity. At its initial height it has a potential energy higher than when it is on the floor. Unless it is restrained, the book falls and thereby loses energy. This leads us to the first basic principle identifying spontaneous processes and influencing their direction: *Processes in which the energy content of the system decreases tend to occur spontaneously*. Spontaneous processes tend to be exothermic. Change occurs in the direction that leads to a lower energy content.

On the other hand, we can think of processes that do not result in a lower energy, or that may even be endothermic, and yet still occur spontaneously. The mixing of CCl_4 and C_6H_{14} provides a simple example. All such processes are characterized by an increase in the disorder, or randomness, of the system. To illustrate, suppose that we could suddenly remove a barrier that separated 500 mL of CCl_4 from 500 mL of C_6H_{14}, as in Figure 11.4(*a*). Before the barrier is removed each liquid occupies a volume of 500 mL. We know that we can find all the CCl_4 molecules in the 500 mL to the left, all the C_6H_{14} molecules in the 500 mL to the right of the barrier. When equilibrium has been established after removal of the barrier, the two liquids together occupy a volume of about

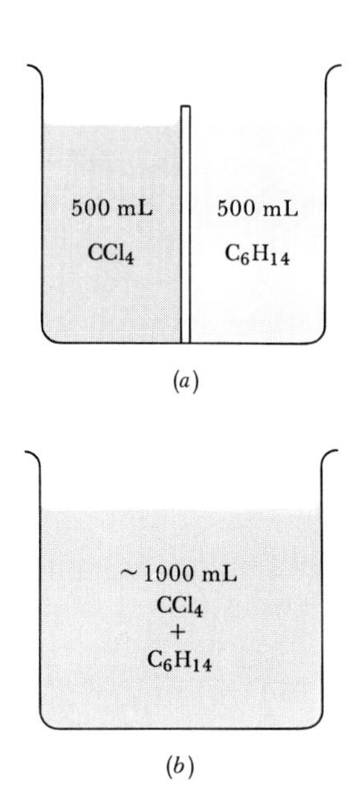

FIGURE 11.4 Formation of a homogeneous solution between CCl_4 and C_6H_{14} upon removal of a barrier separating the two liquids. The solution in (*b*) is more disordered, or random in character, than the separate liquids before solution formation (*a*).

1000 mL.* Formation of a homogeneous solution has resulted in an increased disorder, or randomness, in that the molecules of each substance are now distributed in a volume twice as large as that which they occupied before mixing. This example illustrates our second basic principle: *Processes in which the disorder of the system increases tend to occur spontaneously.*

When molecules of different types are mixed, an increase in disorder occurs spontaneously unless the molecules are restrained by sufficiently strong intermolecular forces or by physical barriers. Thus, because of the strong bonds holding the sodium and chloride ions together, sodium chloride does not spontaneously dissolve in gasoline. On the other hand, gases spontaneously expand unless restrained by their containers; in this case intermolecular forces are too weak to restrain the molecules. We shall discuss spontaneous processes again in Chapter 17; at that time we shall consider the balance between the tendency toward lower energy and toward increased disorder in greater detail. For the moment we need to be aware that formation of a solution is always favored by the increase in disorder that accompanies mixing. Consequently, a solution will form unless solute-solute or solvent-solvent interactions are too strong relative to the solute-solvent interactions.

In all of our discussions of solutions we must be careful to distinguish the *physical* process of solution formation from chemical processes that lead to a solution. For example, zinc metal is dissolved on contact with hydrochloric acid solution. The following chemical reaction occurs:

$$Zn(s) + 2HCl(aq) \longrightarrow H_2(g) + ZnCl_2(aq) \qquad [11.6]$$

In this instance the chemical form of the substance being dissolved is changed. If the solution is evaporated to dryness, $Zn(s)$ is not recovered as such; instead $ZnCl_2(s)$ is recovered. In contrast, when $NaCl(s)$ is dissolved in water, it can be recovered by evaporation of its solution to dryness. Our focus throughout this chapter is on solutions from which the solute can be recovered unchanged from the solution.

Solubility

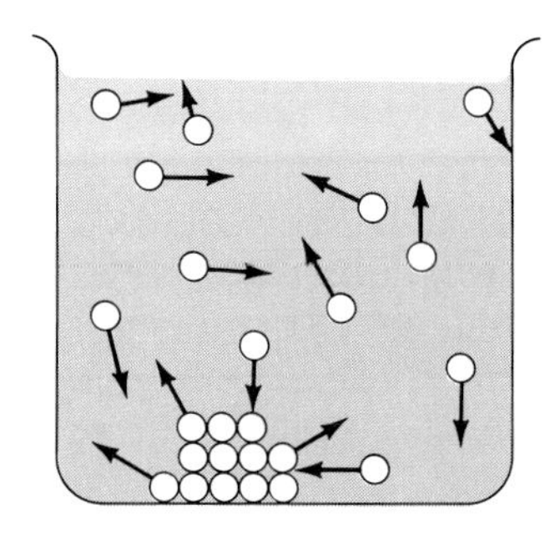

FIGURE 11.5 Movement of solute particles in solvent containing excess solute. Both dissolution and crystallization occur.

As the solution process involving a solid solute proceeds, and the concentration of solute particles in solution increases, the chances of their colliding with the surface of the solid increases (Figure 11.5). Such a collision may result in the solute particle becoming attached to the solid. This process, which is the opposite of the solution process, is called **crystallization**. Thus two opposing processes occur in a solution in contact with undissolved solute. This situation is represented in Equation 11.7 by use of a double arrow:

$$\text{Solute} + \text{solvent} \underset{\text{crystallize}}{\overset{\text{dissolve}}{\rightleftharpoons}} \text{solution} \qquad [11.7]$$

*There may be a slight change in the total volume upon mixing, but that is unimportant for our example.

When the rates of these opposing processes become equal, there will be no further net increase in the amount of solute in solution. This is an example of **dynamic equilibrium**, similar to that discussed in Section 10.2, where the processes of evaporation and condensation were considered. A solution that is in equilibrium with undissolved solute is said to be **saturated.** Additional solute will not dissolve if added to a saturated solution. The amount of solute needed to form a saturated solution in a given quantity of solvent is known as the **solubility** of that solute. For example, the solubility of NaCl in water at 0°C is 35.7 g per 100 mL of water. This is the maximum amount of NaCl that can be dissolved in water to give a stable equilibrium solution at that temperature. If less solute is added than the equilibrium amount, the solution is said to be **unsaturated.** In some cases it is possible to prepare solutions that contain more solute than the equilibrium amount. Such solutions, which are said to be **supersaturated,** are unstable, and under the proper conditions solute will crystallize from them to give saturated solutions. Sodium acetate, $NaC_2H_3O_2$, forms supersaturated solutions very easily. The solubility of this substance at 0°C is 119 g per 100 mL of H_2O. It is more soluble at higher temperatures. If a hot solution containing more than 119 g of $NaC_2H_3O_2$ per 100 mL of H_2O is slowly cooled to 0°C, the excess solute remains dissolved; the resultant solution is supersaturated. Addition of a small crystal of $NaC_2H_3O_2$ gives a surface on which crystallization can start, and crystallization occurs until the concentration of the solution drops to the saturation level.

11.3 FACTORS AFFECTING SOLUBILITY

As the discussion in the preceding section indicates, the extent to which one substance dissolves in another depends on the nature of both the solute and the solvent. It also depends on temperature and, at least for gases, on pressure.

Solubilities and Molecular Structure

TABLE 11.3
Solubilities of several gases in water at 20°C, with 1 atm gas pressure

Gas	Solubility (M)
N_2	6.9×10^{-4}
CO	1.04×10^{-3}
O_2	1.38×10^{-3}
Ar	1.50×10^{-3}
Kr	2.79×10^{-3}

Our discussion of the factors that lead to formation of solutions enables us to understand many observations regarding solubilities. As a simple example, consider the data in Table 11.3 for the solubilities of various simple gases in water. Note that solubility increases with increasing molecular mass. The attractive forces between the gas and solvent molecules are mainly of the London-dispersion-force type, which increase with increasing size and molecular mass of the gas molecules. When chemical reaction occurs between the gas and solvent, much higher gas solubilities result. We will encounter instances of this in later chapters, but as a simple example, the solubility of Cl_2 in water under the same conditions given in Table 11.3 is 0.102 M. This is much higher than would be predicted from the trends in the table, based just on molecular mass. We can infer from this that the dissolving of Cl_2 in water is accompanied by some kind of chemical process. The use of chlorine as a bactericide in municipal water supplies and swimming pools is based on its chemical reaction with water (Section 12.6, Equation 12.18).

Polar liquids tend to dissolve readily in polar solvents. For example,

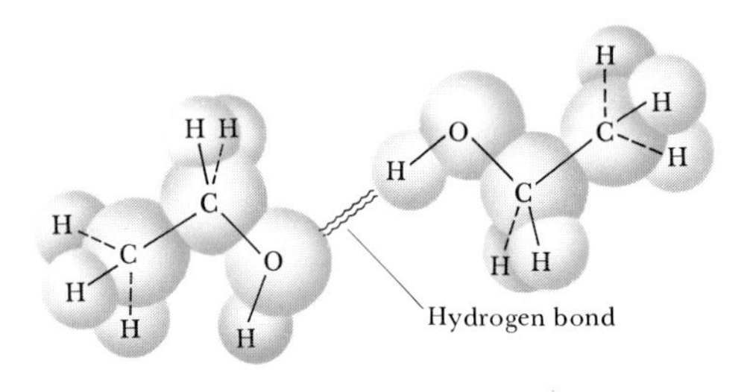

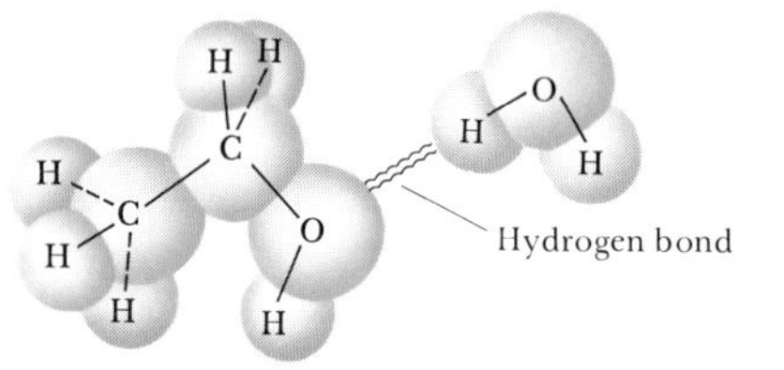

FIGURE 11.6 Hydrogen-bonding interactions between ethanol molecules and between water and ethanol molecules.

$$\overset{\overset{\large O}{\|}}{CH_3CCH_3}$$
Acetone

$$CH_3CH_2\overset{\overset{\large O}{\|}}{C}CH_2CH_3$$
Diethyl ketone

acetone, a polar molecule whose structure is shown at left, mixes in all proportions with water. Pairs of liquids that mix in all proportions are said to be **miscible,** while liquids that do not mix are termed **immiscible.** Water and hexane, C_6H_{14}, for example, are immiscible. Diethyl ketone (see structure at left), which is similar to acetone but has a higher molecular mass, dissolves in water to the extent of about 47 g per liter of water at 20°C, but is not completely miscible.

Hydrogen-bonding interactions between solute and solvent may lead to high solubility. For example, water is completely miscible with ethanol, CH_3CH_2OH. The CH_3CH_2OH molecules are able to form hydrogen bonds with water molecules as well as with each other. This is shown in Figure 11.6. Because of this hydrogen-bonding ability, the solute-solute, solvent-solvent, and solute-solvent forces are not appreciably different within a mixture of CH_3CH_2OH and H_2O. There is no significant change in the environment of the molecules as they are mixed. The increase in disorder accompanying mixing therefore plays a significant role in formation of the solution.

The number of carbon atoms in an alcohol affects its solubility in water, as shown in Table 11.4. As the length of the carbon chain increases, the OH group becomes an ever smaller part of the molecule and the molecule becomes more like a hydrocarbon. The solubility of the

TABLE 11.4 Solubilities of some alcohols in water

Alcohol	Solubility in H_2O (mol/100 g H_2O at 20°C)
CH_3OH (methanol)	∞[a]
CH_3CH_2OH (ethanol)	∞
$CH_3CH_2CH_2OH$ (propanol)	∞
$CH_3CH_2CH_2CH_2OH$ (butanol)	0.11
$CH_3CH_2CH_2CH_2CH_2OH$ (pentanol)	0.030
$CH_3CH_2CH_2CH_2CH_2CH_2OH$ (hexanol)	0.0058
$CH_3CH_2CH_2CH_2CH_2CH_2CH_2OH$ (heptanol)	0.0008

[a]The infinity symbol indicates that there is no real limit to the solubility of this alcohol in water.

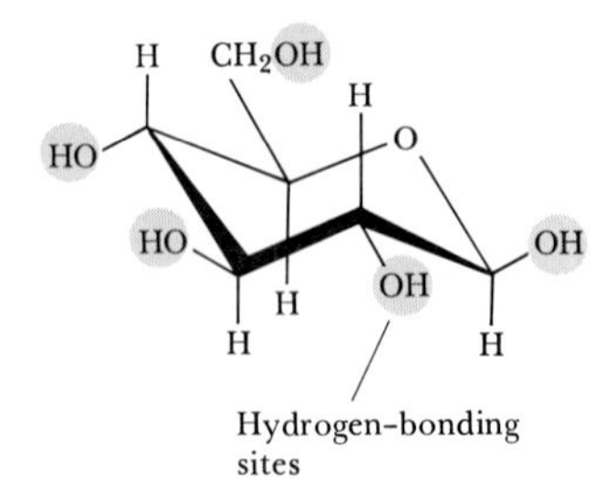

FIGURE 11.7 Structure of glucose. Colored spheres indicate sites capable of hydrogen bonding with water.

alcohol decreases correspondingly. If the number of OH groups along the carbon chain increases, more solute-water hydrogen bonding is possible and solubility generally increases. In the case of glucose, $C_6H_{12}O_6$, there are five OH groups on a six-carbon framework that make the molecule very soluble in water (83 g dissolve in 100 mL of water at 17.5 °C). The glucose molecule is shown in Figure 11.7.

Examination of pairs of substances such as those listed in the preceding paragraphs has led to an important generalization: *Substances with similar intermolecular attractive forces tend to be soluble in one another.* This generalization is often simply stated as "*likes dissolve likes.*" Nonpolar substances are soluble in nonpolar solvents, whereas ionic and polar solutes are soluble in polar solvents. Network solids like diamond and quartz are not soluble in either polar or nonpolar solvents because of the strong bonding forces within the solid.

SAMPLE EXERCISE 11.7

Predict whether each of the following substances is more likely to dissolve in carbon tetrachloride, CCl_4, which is a nonpolar solvent, or in water: C_7H_{16}, $NaHCO_3$, HCl, I_2.

Solution: Both C_7H_{16} and I_2 are nonpolar. We would therefore predict that they would be more soluble in CCl_4 than in H_2O. On the other hand, $NaHCO_3$ is ionic and HCl is polar covalent. Water would be a better solvent than CCl_4 for these two substances.

Vitamins B and C are water soluble; the A, D, E, and K vitamins are soluble in nonpolar solvents and in the fatty tissue of the body (which is nonpolar). Because of their water solubility, vitamins B and C are not stored to any appreciable extent in the body, and so foods containing these vitamins should be included in the daily diet. In contrast, the fat-soluble vitamins are stored in sufficient quantities to keep vitamin-deficiency diseases from appearing even after a person has subsisted for a long period on a vitamin-deficient diet. With the ready availability of vitamin supplements, cases of hypervitaminosis, illness caused by an excessive amount of vitamins, are now being seen by physicians in this country. Because only the fat-soluble vitamins are stored, they are the ones for which true hypervitaminosis has been observed. The different solubility patterns of the water-soluble vitamins and the fat-soluble ones can be rationalized in terms of the structures of the molecules. The chemical structures of vitamin A (retinol) and of vitamin C (ascorbic acid) are shown below and at the top of p. 349.

Note that the vitamin A molecule is an alcohol with a very long carbon chain. It is nearly nonpolar, and because the OH group is such a small part of the molecule, the molecule mainly resembles a hydrocarbon. The situation is like that of the long-chain alcohols listed in Table 11.4. In contrast, the vitamin C molecule is smaller and has more OH groups that can form hydrogen bonds with water. It is somewhat like the glucose example discussed above.

Vitamin A

Vitamin C

```
                 O—H  H
                 |    |
        O        C————C—O—H
      /   \    / \    |
O=C         C     H   H
   \       / \
    C====C    H
   /      \
H—O        O—H
```

Effect of Pressure on Solubility

The solubility of a gas in any solvent is increased as the pressure of the gas over the solvent increases. By contrast, the solubilities of solids and liquids are not appreciably affected by pressure. We can understand the effect of pressure on the solubility of a gas by considering the equilibrium that operates, illustrated in Figure 11.8. Suppose that we have a gaseous substance distributed between the gas and solution phases. When equilibrium is established, the rate at which gas molecules enter the solution equals the rate at which solute molecules escape from the solution to enter the gas phase. The small arrows in Figure 11.8(*a*) represent the rates of these opposing processes. This is another example of dynamic equilibrium, which we have encountered before (Section 10.2). Now suppose that we exert added pressure on the piston and compress the gas above the solution, as shown in Figure 11.8(*b*). If we reduce the volume to half its original value, the pressure of the gas would increase to about twice its original value. But this would mean that the rate at which gas molecules strike the surface to enter the solution phase would increase. Thus the solubility of the gas in the solution would increase until there is again an equilibrium; that is, until the rates at which gas molecules enter the solution equals the rate at which solute molecules escape from the solvent, as indicated by the arrows in Figure 11.8(*b*). This means that the solubility of the gas should increase in direct proportion to the pressure. The relationship between pressure and solubility is expressed in terms of a simple equation known as **Henry's law:**

$$C_g = kP_g \qquad [11.8]$$

where C_g is the solubility of the gas in the solution phase, P_g is the pressure of the gas over the solution, and k is a proportionality constant,

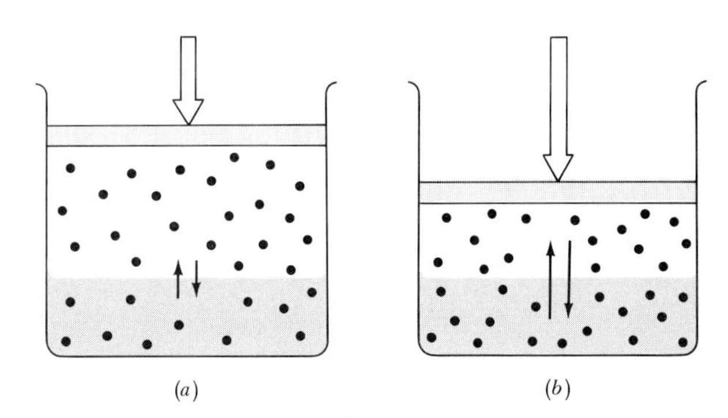

FIGURE 11.8 Effect of pressure on the solubility of a gas. When the pressure is increased, as in (*b*), the rate at which gas molecules enter the solution increases; the result is that the concentration of solute molecules at equilibrium increases in proportion to the pressure.

known as the Henry's law constant. As an example, the solubility of pure nitrogen gas in water at 25 °C and 0.78 atm pressure is 5.3×10^{-4} *M*. If the partial pressure of N_2 is doubled, Henry's law predicts that the solubility in water is also doubled to 1.06×10^{-3} *M*.

The effect of pressure on solubility is utilized in the production of carbonated beverages such as champagne, beer, and soda pop. These are bottled under a carbon dioxide pressure slightly greater than 1 atm. When the bottles are opened to the air, the partial pressure of CO_2 above the solution is decreased, and CO_2 bubbles out of the solution.

Deep-sea divers rely on compressed air for their oxygen supply. According to Henry's law, the solubilities of gases increase with pressure. If a diver is suddenly exposed to atmospheric pressure, where the solubility of gases is less, bubbles form in the bloodstream and in other fluids of the body. These bubbles affect nerve impulses and give rise to the disease known as "the bends," or decompression sickness. Nitrogen is the main problem because it has the highest partial pressure in air and because it can be removed only through the respiratory system. Oxygen is consumed in metabolism. Substitution of helium for nitrogen minimizes this effect, because helium has a much lower solubility in biological fluids than does N_2. Cousteau's divers on *Conshelf III* used a mixture of 98 percent helium and 2 percent oxygen. At the high pressures (10 atm) experienced by the divers, this percentage of oxygen gives an oxygen partial pressure of about 0.2 atm, which is the partial pressure in normal air at 1 atm. If the oxygen partial pressures become too great, the urge to breathe is reduced, CO_2 is not removed from the body, and this leads to CO_2 poisoning.

Effect of Temperature on Solubility

The solubilities of several common gases in water as a function of temperature are graphed in Figure 11.9. These solubilities correspond to a constant pressure of 1 atm of gas over the solution. Note that in general, solubility decreases with increasing temperature. If a glass of cold tap

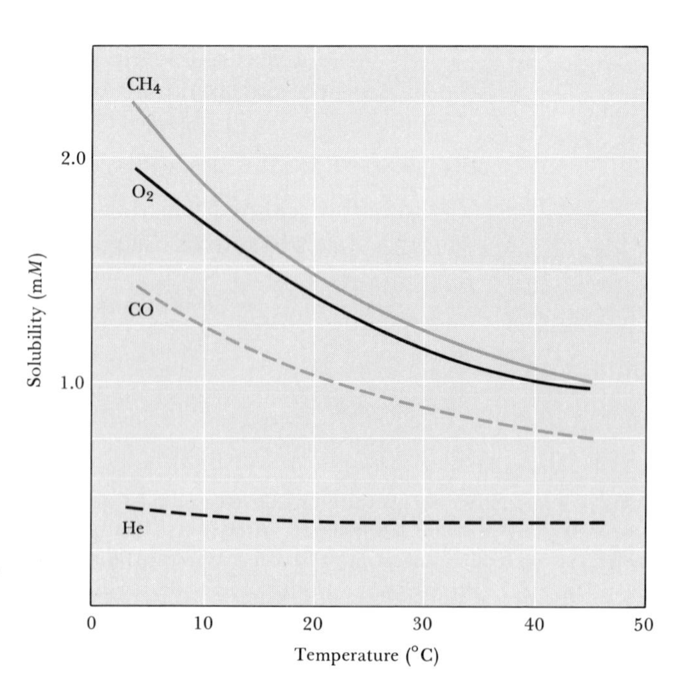

FIGURE 11.9 Solubilities of several gases in water as a function of temperature. Note that solubilities are in units of millimoles per liter, for a constant pressure of 1 atm in the gas phase.

water is warmed, bubbles of air are seen on the side of the glass. Similarly, a carbonated beverage like soda pop goes "flat" as it is allowed to warm; as the temperature of the solution increases, CO_2 escapes from the solution. The decreased solubility of O_2 in water as temperature increases is one of the effects of the thermal pollution of lakes and streams. The effect is particularly serious in deep lakes, because warm water is less dense than cold water. It therefore tends to remain on top of the cold water, at the surface. This situation impedes the dissolving of oxygen into the deeper layers, thus stifling the respiration of all aquatic life needing oxygen. Fish may suffocate and die in these circumstances.

Figure 11.10 shows the effect of temperature on the solubilities of several ionic substances in water. Note that for most solids the solubility increases with increasing temperature. The effect of temperature on solubility depends on the enthalpy change for the solution process. For solutes that dissolve with endothermic heats of solution, solubility increases as temperature increases. We can understand these effects in terms of a principle first put forward by Henri Louis Le Châtelier (1850–1936), a French industrial chemist. **Le Châtelier's principle** can be stated as follows: *If a system at equilibrium is disturbed by a change in temperature, pressure, or the concentration of one of the components, the system will tend to shift its equilibrium position so as to counteract the effect of the disturbance.* Let us consider a solution in equilibrium with undissolved solute. Assume that the process of solution is endothermic, that is, that heat is absorbed from the surroundings as the solution is formed. Then the equilibrium can be represented as follows:

$$\text{Solute} + \text{solvent} + \text{heat} \rightleftharpoons \text{solution} \qquad [11.9]$$

Now according to Le Châtelier's principle, if we add heat to this system, the equilibrium will shift in such a direction as to undo the effect of the

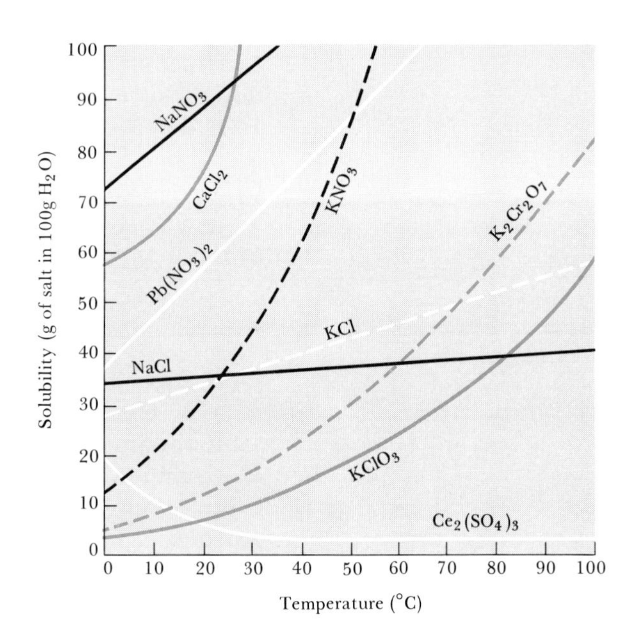

FIGURE 11.10 Solubilities of several common ionic solids shown as a function of temperature.

added heat. That is, it will shift in the direction in which heat is absorbed—to the right. Thus, an increase in temperature, which represents an addition of heat to the system, results in increased solubility. On the other hand, if the heat of solution were negative—that is, if the solution process were exothermic—solubility would decrease with increasing temperature. This case is illustrated by the data for $Ce_2(SO_4)_3$ in Figure 11.10.

Le Châtelier's principle is very useful in making predictions about the effects of changes on equilibrium systems. We will examine it more closely in Section 14.3.

SAMPLE EXERCISE 11.8

Use Le Châtelier's principle to predict the effect of increased gas pressure on the solubility of a gas.

Solution: We can answer this by reference to Figure 11.8. The equilibrium system is disturbed by an increase in the gas pressure. According to Le Châtelier's principle, the system will respond by shifting the equilibrium so as to return the pressure toward its original value. This happens when more of the gas dissolves in the solution phase. Thus we predict that solubility will increase with increasing pressure, in agreement with our previous conclusion.

11.4 ELECTROLYTE SOLUTIONS

Solutes can be classified according to whether or not they exist in aqueous solutions as ions. Those that do are called electrolytes. A familiar example is sodium chloride whose solid consists of Na^+ and Cl^- ions. This substance dissolves in water to give hydrated Na^+ and Cl^- ions (Section 11.2). A solution labeled 1.0 *M* NaCl actually contains 1.0 mol of Na^+ and 1.0 mol of Cl^- per liter of solution. Because all NaCl present in solution is in the form of ions, NaCl is said to be completely ionized and is referred to as a strong electrolyte. Not surprisingly, ionic compounds form electrolyte solutions.

Molecular compounds that contain highly polar covalent bonds may also dissolve in water to form ions. For example, hydrogen chloride gas, HCl, dissolves readily in water to form an aqueous solution that we know as hydrochloric acid. The HCl in that solution exists entirely as H^+ and Cl^- ions. (We will take a closer look at this ionization process in Section 15.1.) Other compounds ionize incompletely on dissolving in water. An example is $HgCl_2$, which participates in the following equilibrium:

$$HgCl_2(aq) \rightleftharpoons HgCl^+(aq) + Cl^-(aq) \qquad [11.10]$$

Only a small fraction of the $HgCl_2$ that is dissolved is present in solution as ions. Compounds such as this one, which ionize only partially in solution, are called weak electrolytes. Finally, those substances that undergo no ionization in solution are called nonelectrolytes. Examples include nonpolar gases such as O_2 and most organic compounds, such as glucose, $C_6H_{12}O_6$, and ethanol, C_2H_5OH. These substances maintain their molecular structures rather than ionizing in solution.

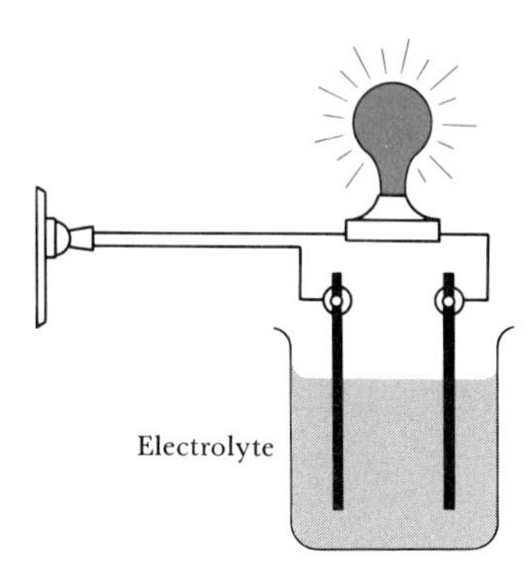

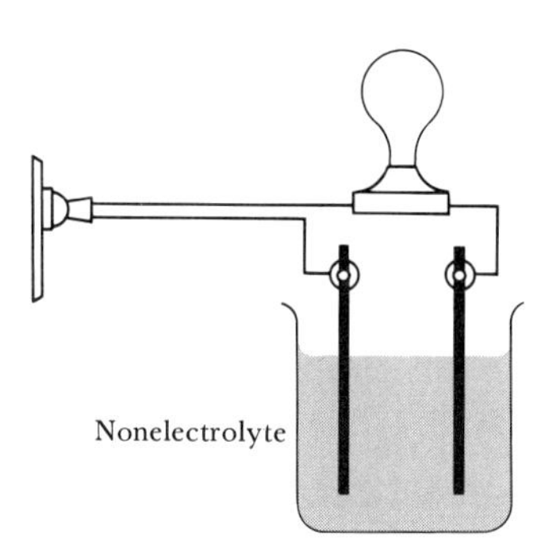

FIGURE 11.11 Simple device that can be used to distinguish between electrolyte and nonelectrolyte solutions.

Whether or not a substance forms an electrolyte solution can be determined experimentally by testing the ability of the solution to conduct an electric current. Pure water itself is a poor conductor. However, solute ions carry electrical charge through aqueous solutions. Thus a device such as that shown in Figure 11.11 could be used to distinguish whether or not ions are present if the solutions are not too dilute. If ions are present, they are able to complete the electric circuit, thereby permitting the light bulb to glow.

It is very important to distinguish solubility from whether a substance is a strong or weak electrolyte. For example, silver chloride, AgCl, is only very slightly soluble in water. Therefore, if we shake some solid AgCl with water for a long time until equilibrium is established, most of the AgCl will remain as an insoluble solid. All of the AgCl that does dissolve, however, is present in solution as Ag^+ and Cl^- ions. We say, therefore, that AgCl is a strong electrolyte.

In Section 3.3 we introduced acids, bases, and salts. You should review that section at this time. We can distinguish acids and bases according to the degree to which they ionize in solution. Acids and bases that are completely ionized in solution are referred to as **strong acids** and **strong bases.** Those that are partly ionized are referred to as **weak acids** and **weak bases.** The terms weak acid and weak base have no relation to how reactive the acid or base is. Hydrofluoric acid, HF, is a weak electrolyte and is therefore called a weak acid. A 0.1 *M* solution of HF is 8 percent ionized. However, HF is a very reactive acid that vigorously attacks many substances, including glass.

The following generalizations are useful in recognizing which substances are strong electrolytes and which weak:

1 Most **salts** (that is, ionic compounds) are strong electrolytes. A few of the heavy metals form weak electrolytes, as in the example of $HgCl_2$ cited in Equation 11.10.

2 Most **acids** are weak electrolytes. The common strong acids are HCl, HBr, HI, HNO_3, H_2SO_4, and $HClO_4$.

3 The common strong **bases** are the hydroxides of Li, Na, K, Rb, and Cs (the alkali metals, group 1A), and those of Ca, Sr, and Ba (the heavy alkaline earths, group 2A). NH_3 is a weak electrolyte.

SAMPLE EXERCISE 11.9

Classify each of the following substances as nonelectrolyte, weak electrolyte, or strong electrolyte: $CaCl_2$, HNO_3, CH_3OH (methanol), $HC_2H_3O_2$ (acetic acid), KOH, H_2O_2.

Solution: Only one of the substances, $CaCl_2$, is a salt. It is a strong electrolyte. Two of the substances, HNO_3 and $HC_2H_3O_2$, are acids; HNO_3 is a common strong acid (strong electrolyte); because $HC_2H_3O_2$ is not a common strong acid, our best guess would be that it is a weak acid (weak electrolyte). This is correct. There is one base, KOH. It is one of the common strong bases (a strong electrolyte) because it is a hydroxide of an alkali metal. The remaining compounds, CH_3OH and H_2O_2, are neither acids, bases, nor salts. They are nonelectrolytes.

SAMPLE EXERCISE 11.10

What is the concentration of all species in a 0.1 M solution of $Ba(NO_3)_2$?

Solution: $Ba(NO_3)_2$ is a salt and as such we expect it to be a strong electrolyte. It ionizes into Ba^{2+} and the polyatomic nitrate ion, NO_3^-. A 0.1 M solution of $Ba(NO_3)_2$ is 0.1 M in Ba^{2+} and 0.2 M in NO_3^- [since 1 mol of $Ba(NO_3)_2$ supplies 1 mol of Ba^{2+} and 2 mol of NO_3^-].

11.5 COLLIGATIVE PROPERTIES

Many properties of a solution depend not only on the concentration of solute but also on its particular nature. For example, the density of a solution depends on the identity of both solute and solvent as well as on the concentration of solute. However, for certain kinds of solutes, a number of physical properties of the solution depend on the number of solute particles present in solution but not on the identity of the solute. Such properties are called colligative properties. They include vapor-pressure lowering, boiling-point elevation, freezing-point depression, and osmotic pressure.

Vapor-Pressure Lowering

If a nonvolatile solute is added to a solvent, the vapor pressure of the solvent is decreased. For example, solutions of sugar or salt both have lower vapor pressures than does pure water. Experiments have shown that the vapor-pressure lowering depends on the concentration of solute particles. For example, 1.0 mol of a nonelectrolyte such as glucose will produce essentially the same vapor-pressure lowering in a given quantity of water as will 0.5 mol of NaCl. There is 1.0 mol of particles in both solutions, because 0.5 mol of NaCl dissolves, giving 0.5 mol of Na^+ and 0.5 mol of Cl^-. Quantitatively, the vapor pressure of solutions containing nonvolatile solutes is given by Raoult's law, Equation 11.11, where P_A is the vapor pressure of the solution, X_A is the mole fraction of the solvent, and P_A° is the vapor pressure of the pure solvent:

$$P_A = X_A P_A^\circ \qquad [11.11]$$

For example, the vapor pressure of water is 17.5 mm Hg at 20°C. Imagine holding the temperature constant while adding glucose, $C_6H_{12}O_6$, to the water so that the resulting solution has $X_{H_2O} = 0.80$ and $X_{C_6H_{12}O_6} = 0.20$. According to Equation 11.11, the vapor pressure of water over the solution will be 14 mm Hg:

$$P_{H_2O} = (0.80)(17.5 \text{ mm Hg}) = 14 \text{ mm Hg}$$

The vapor-pressure lowering, ΔP_A, is $P_A^\circ - P_A$. Using Equation 11.11, we can write this as

$$\begin{aligned}\Delta P_A &= P_A^\circ - X_A P_A^\circ \\ &= P_A^\circ(1 - X_A) \qquad [11.12]\end{aligned}$$

But since $X_A + X_B = 1$, where X_B is the mole fraction of solute particles, we can write

$$X_B = 1 - X_A$$
$$\Delta P_A = X_B P_A^\circ \qquad [11.13]$$

In our example, the vapor pressure will be lowered by 3.5 mm Hg:

$$\Delta P_A = (0.20)(17.5 \text{ mm Hg}) = 3.5 \text{ mm Hg}$$

It is easy to explain this effect qualitatively. At a given temperature, the average kinetic energy of all particles in the solution is the same. Essentially the same number of particles are at the surface in both the solution and pure solvent. However, in the case of a nonvolatile solute, only a fraction of these particles are volatile and have sufficient kinetic energy to escape from the solution.

SAMPLE EXERCISE 11.11

Calculate the vapor-pressure lowering caused by the addition of 100 g of sucrose, $C_{12}H_{22}O_{11}$, to 1000 g of water if the vapor pressure of the pure water at 25°C is 23.8 mm Hg.

Solution:

$$X_{C_{12}H_{22}O_{11}} = \frac{\text{moles } C_{12}H_{22}O_{11}}{\text{moles } H_2O + \text{moles } C_{12}H_{22}O_{11}}$$

$$\text{Moles } C_{12}H_{22}O_{11} = (100 \text{ g})\left(\frac{1 \text{ mol } C_{12}H_{22}O_{11}}{342 \text{ g } C_{12}H_{22}O_{11}}\right) = 0.292 \text{ mol}$$

$$\text{Moles } H_2O = (1000 \text{ g})\left(\frac{1 \text{ mol } H_2O}{18.0 \text{ g } H_2O}\right) = 55.5 \text{ mol}$$

$$X_{C_{12}H_{22}O_{11}} = \frac{0.292}{55.5 + 0.292} = \frac{0.292}{55.8}$$

$$\Delta P_A = \left(\frac{0.292}{55.8}\right)(23.8 \text{ mm Hg}) = 0.125 \text{ mm Hg}$$

Boiling-Point Elevation

We saw in Section 10.3 that the vapor pressure of a liquid increases with increasing temperature; it boils when its vapor pressure is the same as the external pressure over the liquid. Because nonvolatile solutes lower the vapor pressure of the solution, a higher temperature is required to cause the solution to boil. Figure 11.12 shows the variation in vapor pressure of a pure liquid with temperature (the black line), as compared with a solution (the colored line).* Because the vapor pressure of the solution is lower than that of the solvent at all temperatures, in accordance with

*Both lines refer to the usual laboratory situation, in which we have a constant total gas pressure of 1 atm over the system.

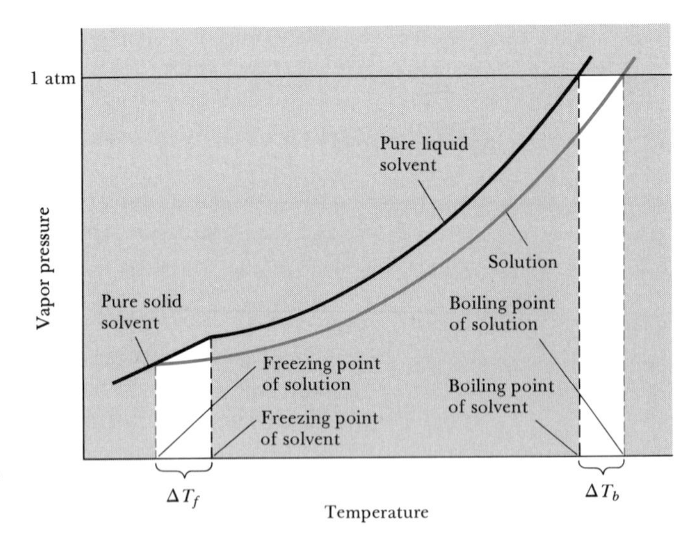

FIGURE 11.12 Vapor-pressure versus temperature curves of a pure solvent and a solution of a nonvolatile solute, under a constant total pressure of 1 atm. The vapor pressure of the solid solvent is unaffected by the presence of solute if the solid solvent freezes out without containing a significant concentration of solute, as is usually the case.

Raoult's law, a higher temperature is required to attain a vapor pressure of 1 atm. The increase in boiling point, ΔT_b (relative to the boiling point of the pure solvent), is directly proportional to the number of solute particles per mole of solvent molecules. We know that molality expresses the number of moles of solute per 1000 g of solvent, which represents a fixed number of moles of solvent. Thus ΔT_b is proportional to molality, as shown in Equation 11.14:

$$\Delta T_b = K_b m \qquad [11.14]$$

The magnitude of K_b, which is called the **molal boiling-point-elevation constant,** depends on the solvent. Some typical values for several common solvents are given in Table 11.5.

For water, K_b is 0.52°C/m; therefore, a 1 m aqueous solution of sucrose or any other aqueous solution that is 1 m in nonvolatile solute particles will boil at a temperature 0.52°C higher than pure water. It is important to notice that the boiling-point elevation is proportional to the number of solute particles present in a given quantity of solution. When NaCl dissolves in water 2 mol of solute particles, 1 mol of Na^+ and 1 mol of Cl^-, are formed for each mole of NaCl that dissolves. Thus a 1 m solution of NaCl in water causes a boiling-point elevation twice as large as a 1 m solution of a nonelectrolyte such as sucrose.

TABLE 11.5 Molal boiling-point-elevation and freezing-point-depression constants

Solvent	Normal boiling point (°C)	K_b (°C/m)	Normal freezing point (°C)	K_f (°C/m)
Water, H_2O	100.0	0.52	0.0	1.86
Benzene, C_6H_6	80.1	2.53	5.5	5.12
Carbon tetrachloride, CCl_4	76.8	5.02	−22.3	29.8
Ethanol, C_2H_5OH	78.4	1.22	−114.6	1.99
Chloroform, $HCCl_3$	61.2	3.63	−63.5	4.68

Freezing-Point Depression

The freezing point corresponds to the temperature at which the vapor pressures of the solid and liquid phases are the same. The freezing point of a solution is lowered because the solute is not normally soluble in the solid phase of the solvent. For example, when aqueous solutions freeze, the solid that separates out is almost always pure ice. Thus the vapor pressure of the solid is unaffected by the presence of solute. On the other hand, if the solute is nonvolatile, the vapor pressure of the solution is reduced in proportion to the mole fraction of solute (Equation 11.13). This means that the temperature at which the solution and solid phases will have the same vapor pressure is reduced, as illustrated in Figure 11.12. Note that the point that represents the equilibrium between solid and liquid lies to the left of the corresponding point for pure solvent, at a lower temperature. Like the boiling-point elevation, the decrease in freezing point, ΔT_f, is directly proportional to the molality of the solute:

$$\Delta T_f = K_f m \qquad [11.15]$$

The values of K_f, the molal freezing-point-depression constant, for several common solvents are given in Table 11.5. For water K_f is 1.86°C/m; therefore, a 0.5 m aqueous solution of NaCl or any aqueous solution that is 1 m in nonvolatile solute particles will freeze 1.86°C lower than pure water. The freezing-point lowering caused by solutes explains the use of antifreeze (Sample Exercise 11.12) and the use of calcium chloride, $CaCl_2$, to melt ice on roads during winter.

SAMPLE EXERCISE 11.12

Calculate the freezing point and the boiling point of a solution of 100 g of ethylene glycol ($C_2H_6O_2$) in 900 g of H_2O.

Solution:

$$\text{Molality} = \frac{\text{moles } C_2H_6O_2}{\text{kilograms } H_2O} =$$

$$\left(\frac{100 \text{ g } C_2H_6O_2}{900 \text{ g } H_2O}\right)\left(\frac{1 \text{ mol } C_2H_6O_2}{62.0 \text{ g } C_2H_6O_2}\right)\left(\frac{1000 \text{ g } H_2O}{1 \text{ kg } H_2O}\right) = 1.79\, m$$

$$\Delta T_f = K_f m = \left(1.86 \frac{°C}{m}\right)(1.79\, m) = 3.33°C$$

$$\text{Freezing point} = (\text{normal f.p. of solvent}) - \Delta T_f = 0.00°C - 3.33°C = -3.33°C$$

$$\Delta T_b = K_b m = \left(0.52 \frac{°C}{m}\right)(1.79\, m) = 0.93°C$$

$$\text{Boiling point} = (\text{normal b.p. of solvent}) + \Delta T_b = 100.00°C + 0.93°C = 100.93°C$$

Ethylene glycol is the main component of antifreeze. It keeps the fluid in the cooling system of a car from freezing by lowering the freezing point of water. Because it also raises the boiling point of water, it is also useful in the summer to prevent what antifreeze advertisements call "boil-over."

SAMPLE EXERCISE 11.13

List the following solutions in order of their expected freezing points: 0.050 m $CaCl_2$; 0.15 m NaCl; 0.10 m HCl; 0.050 m $HC_2H_3O_2$; 0.10 m $C_{12}H_{22}O_{11}$.

Solution: First notice that $CaCl_2$, NaCl, and HCl are strong electrolytes, $HC_2H_3O_2$ is a weak electrolyte, and $C_{12}H_{22}O_{11}$ is a nonelectrolyte. The molality of each solution in total particles is as follows:

$0.050\ m\ CaCl_2$ $(0.15\ m$ in particles) $0.15\ m\ NaCl$ $(0.30\ m$ in particles) $0.10\ m\ HCl$ $(0.20\ m$ in particles) $0.050\ m\ HC_2H_3O_2$ (between 0.050 and $0.10\ m$ in particles) $0.10\ m\ C_{12}H_{22}O_{11}$ $(0.10\ m$ in particles)	The freezing points are expected to run from $0.15\ m$ NaCl (lowest freezing point) to $0.10\ m$ HCl to $0.050\ m\ CaCl_2$ to $0.10\ m\ C_{12}H_{22}O_{11}$ to 0.050 $HC_2H_3O_2$ (highest freezing point).

Because NaCl is a strong electrolyte, the expected freezing-point depression of a $0.100\ m$ aqueous solution of NaCl is $(0.200\ m) \times (0.186°C/m) = 0.372°C$. However, the measured value is slightly less than this, 0.348°C. Although strong electrolytes like NaCl give nearly the effect expected from the number of ions present in solution, the effect is always slightly less than that of a corresponding number of nonelectrolyte particles. This is due to the residual electrostatic attraction between ions, even when separated by solvent molecules. This attraction makes solutions of electrolytes behave as though their ion concentrations were less than they actually are.

Osmosis

Certain materials, including many membranes in biological systems, are semipermeable, meaning that they allow some particles to pass through, but not others. **Osmosis** is the net movement of solvent molecules but not solute particles through a semipermeable membrane from a more dilute solution into a more concentrated one. Consider two solutions separated by a semipermeable membrane as shown in Figure 11.13. The net migration of solvent occurs from the right arm to the left arm of the apparatus as if the solutions were trying to equalize their concentrations. Pressure can be applied to the left arm of the apparatus, as shown in Figure 11.14, to prevent the osmosis. The pressure required to stop osmosis from a pure solvent into a solution is known as the **osmotic pressure**, π, of the solution. The osmotic pressure is related to concentration as shown in Equation 11.16:

$$\pi = MRT \qquad [11.16]$$

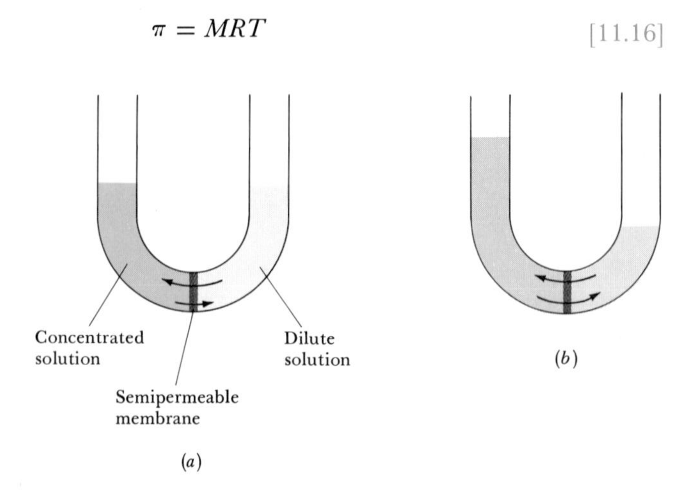

FIGURE 11.13 Osmosis: (*a*) net movement of solvent from the solution with low solute concentration into the solution with high solute concentration; (*b*) osmosis stops when the column of solution on the left becomes high enough to exert sufficient pressure to stop the osmosis.

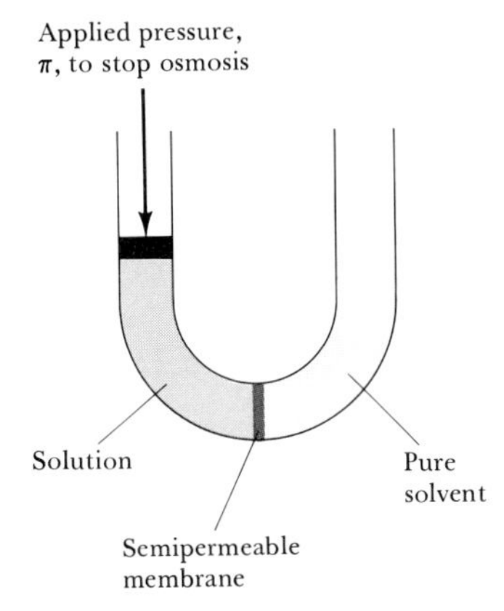

FIGURE 11.14 Applied pressure on the left of apparatus stops net movement of solvent from the right side of the semipermeable membrane. This applied pressure is known as the osmotic pressure of the solution.

where M is molarity, R is the ideal gas constant, and T is the temperature on the Kelvin scale.

If two solutions of identical osmotic pressure are separated by a semipermeable membrane, no osmosis will occur. The two solutions are said to be **isotonic.** If one solution is of lower osmotic pressure, it is described as being **hypotonic** with respect to the more concentrated solution. The more concentrated solution is said to be **hypertonic** with respect to the dilute solution.

Osmosis plays a very important role in living systems. For example, the membranes of red blood cells are semipermeable. Placement of red blood cells in a solution that is hypertonic relative to the intracellular solution (the solution within the cells) causes water to move out of the cell as shown in Figure 11.15. This causes the cell to shrivel, a process known as **crenation.** Placement of the cell in a solution that is hypotonic relative to the intracellular fluid causes water to move into the cell. This causes rupturing of the cell, a process known as **hemolysis.** Persons needing replacement of body fluids or nutrients who cannot be fed orally are administered solutions by intravenous (or IV) infusion, meaning slow addition to the veins. To prevent crenation or hemolysis of red blood cells, the IV solutions must be isotonic with the intracellular fluids of the cells.

SAMPLE EXERCISE 11.14

The average osmotic pressure of blood is 7.7 atm at 25°C. What concentration of glucose, $C_6H_{12}O_6$, will be isotonic with blood?

Solution:

$$\pi = MRT$$

$$M = \frac{\pi}{RT} = \frac{7.7 \text{ atm}}{\left(0.082 \dfrac{\text{L-atm}}{\text{K-mol}}\right)(298 \text{ K})}$$

$$= 0.31\ M$$

In clinical situations, the concentrations of solutions are generally expressed in terms of weight percentages. The weight percentage of a 0.31 M solution of glucose is 5.3 percent. The concentration of NaCl that is isotonic with blood is 0.16 M since NaCl ionizes to form two particles, Na^+ and Cl^- (a 0.155 M solution of NaCl is 0.310 M in particles). A 0.16 M solution of NaCl is 0.9 percent in NaCl. Such a solution is known as a physiological saline solution.

There are many interesting examples of osmosis. A cucumber placed in concentrated brine loses water via osmosis and shrivels into a pickle. A carrot that has become limp because of water loss to the atmosphere can

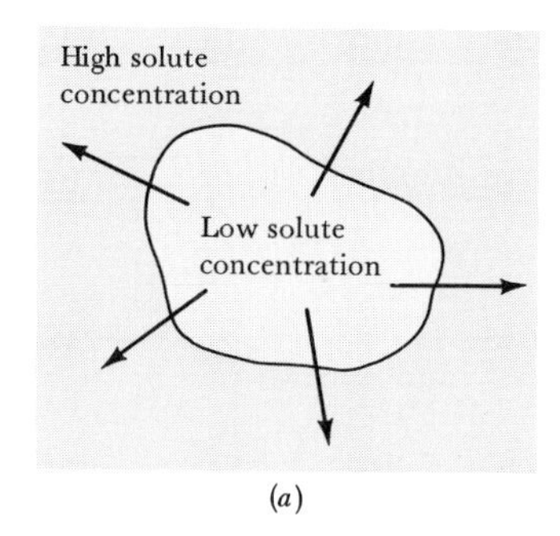

(a)

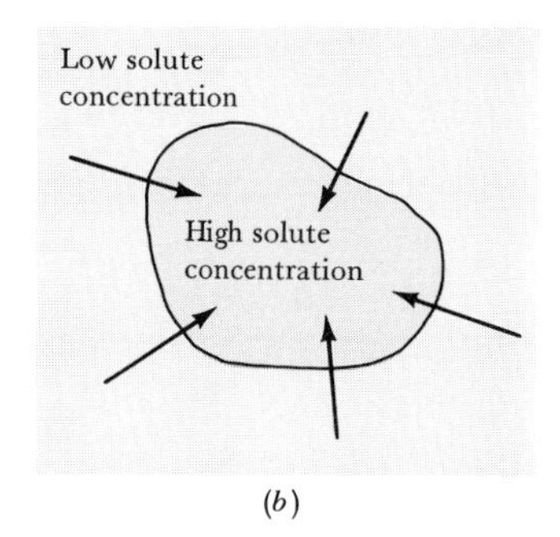

(b)

FIGURE 11.15 Osmosis through the semipermeable membrane of a red blood cell: (*a*) crenation caused by movement of water from the cell; (*b*) hemolysis caused by movement of water into the cell.

be placed in water. Water moves into the carrot through osmosis, making it firm once again. People eating a lot of salty food experience water retention in tissue cells and intercellular spaces because of osmosis. The resultant swelling or puffiness is called edema. Movement of water from soil into plant roots and subsequently into the upper portions of the plant is due at least in part to osmosis. The preservation of meat by salting and of fruit by adding sugar protects against bacterial action. Through the process of osmosis, a bacterium on salted meat or candied fruit loses water, shrivels, and dies.

In osmosis, water moves from an area of high water concentration (low solute concentration) into an area of low water concentration (high solute concentration). Such movement of a substance from an area where its concentration is high to an area where it is low is spontaneous. Biological cells transport not only water, but other select materials through their membrane walls. This permits entry of nutrients and allows for disposal of waste materials. In some cases, substances must be moved from an area of low concentration to one of high concentration. This movement is called *active transport*. It is not spontaneous but requires expenditures of energy by the cell.

Determination of Molecular Weight

The colligative properties of solutions provide a useful means of experimentally determining molecular weights. Any of the four colligative properties could be used to determine molecular weight. The procedures are illustrated in Sample Exercises 11.15 and 11.16.

SAMPLE EXERCISE 11.15

A solution of an unknown nonvolatile nonelectrolyte was prepared by dissolving 0.250 g in 40.0 g of CCl_4. The normal boiling point of the resultant solution was increased by 0.357°C. Calculate the molecular weight of the solute.

Solution: Using Equation 12.21, we have

$$\text{Molality} = \frac{\Delta T_b}{K_b} = \frac{0.357°\text{C}}{5.02°\text{C}/m} = 0.0711\ m$$

Thus the solution contains 0.0711 mol of solute per kilogram of solvent. The solution was prepared from 0.250 g of solute and 40.0 g of solvent. The number of grams of solute in a kilogram of solvent is therefore

$$\frac{\text{Grams solute}}{\text{Kilograms CCl}_4} = \left(\frac{0.250\text{ g solute}}{40.0\text{ g CCl}_4}\right)\left(\frac{1000\text{ g CCl}_4}{1\text{ kg CCl}_4}\right) = \frac{6.25\text{ g solute}}{1\text{ kg CCl}_4}$$

Notice that a kilogram of solvent contains 6.25 g, which from the ΔT_b measurement must be 0.0711 m. Therefore,

$$0.0711\text{ mol} = 6.25\text{ g}$$

$$1\text{ mol} = \frac{6.25\text{ g}}{0.0711} = 87.9\text{ g}$$

Therefore, $\mathscr{M} = 87.9$ amu

SAMPLE EXERCISE 11.16

The osmotic pressure of 0.200 g of hemoglobin in 20.0 mL of solution is 2.88 mm Hg at 25°C. Calculate the molecular weight of hemoglobin.

Solution:

$$\pi = MRT$$

$$M = \frac{\pi}{RT} = \frac{2.88 \text{ mm Hg}}{\left(0.0821 \frac{\text{L-atm}}{\text{K-mol}}\right)(298 \text{ K})}\left(\frac{1 \text{ atm}}{760 \text{ mm Hg}}\right)$$

$$= 1.55 \times 10^{-4} \frac{\text{mol}}{\text{L}}$$

A liter of solution contains

$$\left(\frac{0.200 \text{ g solute}}{20.0 \text{ mL soln}}\right)\left(\frac{100 \text{ mL soln}}{1 \text{ L soln}}\right) = 10.0 \text{ g/L}$$

Therefore,

$$1.55 \times 10^{-4} \frac{\text{mol}}{\text{L}} = 10.0 \frac{\text{g}}{\text{L}}$$

$$1 \text{ mol} = \frac{10.0 \text{ g}}{1.55 \times 10^{-4}} = 6.45 \times 10^4 \text{ g}$$

Therefore, $\mathscr{M} = 64{,}500$ amu

Osmotic pressure is particularly useful for measuring the molecular weights of high-molecular-weight substances.

11.6 REACTIONS IN AQUEOUS SOLUTION

The main thrust of this chapter has been on the physical properties of solutions. Of course, chemists are also interested in solutions as media for chemical reactions. Particles in solution are free to move throughout the solution volume, and their resultant collisions often lead to chemical reaction. Because of the importance of water as a solvent in our natural environment, we devote this section to a brief examination of chemical reactions in aqueous solution.

Net Ionic Equations

In Section 3.3 we discussed two important reaction types, neutralization and precipitation. We should now examine these and related reactions in more detail, keeping in mind that some solutes are electrolytes, whereas others are not. Consider first the neutralization reaction between hydrochloric acid, HCl, and sodium hydroxide, NaOH. In the past we have written the equation for this reaction in the following form:

$$HCl(aq) + NaOH(aq) \longrightarrow H_2O(l) + NaCl(aq) \quad [11.17]$$

Equations written in this fashion, showing the complete chemical formulas of reactants and products, are often called **molecular equations.** Of course, the term is a bit of a misnomer in this case because all of the substances involved in the reaction except H_2O are ionic. Because HCl, NaOH, and NaCl are strong electrolytes, we could write the chemical equation so as to stress the fact that they are completely ionized in solution:

$$H^+(aq) + Cl^-(aq) + Na^+(aq) + OH^-(aq) \longrightarrow H_2O(l) + Na^+(aq) + Cl^-(aq) \quad [11.18]$$

An equation written in this form, with all soluble strong electrolytes shown as they exist in solution, as ions, is known as an **ionic equation.** Notice that Na^+ and Cl^- appear on both sides of the equation. Ions that appear in identical chemical forms on both sides of an ionic equation are known as **spectator ions.** When the spectator ions are omitted from an ionic equation, the resultant equation is called the **net ionic equation.** In our present example, omitting Na^+ and Cl^- gives

$$H^+(aq) + OH^-(aq) \longrightarrow H_2O(l) \quad [11.19]$$

This net ionic equation expresses the essential feature of the neutralization reaction between any strong acid and strong base: H^+ and OH^- ions combine to form H_2O.

Because net ionic equations can illustrate the similarities between a large number of reactions involving electrolytes, they are widely used. In writing net ionic equations the following conventions are used:

1. Only soluble strong electrolytes are written in ionic form.
2. The chemical formulas of soluble weak electrolytes and nonelectrolytes, as well as all insoluble substances, are written in "molecular" form.
3. Spectator ions are omitted.

SAMPLE EXERCISE 11.17

Write the net ionic equation for the complete neutralization of phosphoric acid, H_3PO_4, by sodium hydroxide, NaOH.

Solution: The molecular equation for this reaction is

$$H_3PO_4(aq) + 3NaOH(aq) \longrightarrow 3H_2O(l) + Na_3PO_4(aq)$$

H_3PO_4 is not one of the common strong acids listed in Section 11.4, where rules were given for predicting which substances are strong electrolytes. We would therefore predict that it is a weak electrolyte, a prediction that is indeed correct. Both NaOH and Na_3PO_4 are strong electrolytes. Thus the ionic equation for the reaction is

$$H_3PO_4(aq) + 3Na^+(aq) + 3OH^-(aq) \longrightarrow 3H_2O(l) + 3Na^+(aq) + PO_4^{3-}(aq)$$

Elimination of the spectator ion, Na^+, gives the net ionic equation,

$$H_3PO_4(aq) + 3OH^-(aq) \longrightarrow 3H_2O(l) + PO_4^{3-}(aq)$$

SAMPLE EXERCISE 11.18

When aqueous solutions of barium nitrate, $Ba(NO_3)_2$, and sodium chromate, Na_2CrO_4, are mixed, a precipitate of barium chromate, $BaCrO_4$, forms while sodium nitrate, $NaNO_3$, remains dissolved in the solution. Write the net ionic equation for the reaction.

Solution: The balanced molecular equation is

$$Ba(NO_3)_2(aq) + Na_2CrO_4(aq) \longrightarrow BaCrO_4(s) + 2NaNO_3(aq)$$

All reactants and products involved in the reaction are ionic compounds ("salts") and are thus strong electrolytes. However, in writing the ionic equation, the formula for $BaCrO_4$ is not written in ionic form because it is a solid:

$$Ba^{2+}(aq) + 2NO_3^-(aq) + 2Na^+(aq) + CrO_4^{2-}(aq) \longrightarrow BaCrO_4(s) + 2Na^+(aq) + 2NO_3^-(aq)$$

Elimination of the spectator ions, Na^+ and NO_3^-, gives the net ionic equation,

$$Ba^{2+}(aq) + CrO_4^{2-}(aq) \longrightarrow BaCrO_4(s)$$

This equation summarizes the general chemical fact that soluble barium salts will react with soluble chromate salts to form a precipitate of $BaCrO_4$. Thus the same net ionic equation would describe the reaction between $BaCl_2(aq)$ and $K_2CrO_4(aq)$.

Solubility Rules

In order to predict whether precipitation will occur when two solutions are mixed, we must have more knowledge of solubilities than presented so far in this text. The rule "likes dissolve likes" allows us to predict

TABLE 11.6 Solubility rules for common ionic compounds in water[a]

	Mainly water soluble
NO_3^-	All nitrates are soluble.
$C_2H_3O_2^-$	All acetates are soluble.
ClO_3^-	All chlorates are soluble.
Cl^-	All chlorides are soluble except $AgCl$, Hg_2Cl_2, and $PbCl_2$.
Br^-	All bromides are soluble except $AgBr$, Hg_2Br_2, $PbBr_2$, and $HgBr_2$.
I^-	All iodides are soluble except AgI, Hg_2I_2, PbI_2, and HgI_2.
SO_4^{2-}	All sulfates are soluble except $CaSO_4$, $SrSO_4$, $BaSO_4$, $PbSO_4$, Hg_2SO_4, and Ag_2SO_4.
	Mainly water insoluble
S^{2-}	All sulfides are insoluble except those of the 1A and 2A elements and $(NH_4)_2S$.
CO_3^{2-}	All carbonates are insoluble except those of the 1A elements and $(NH_4)_2CO_3$.
SO_3^{2-}	All sulfites are insoluble except those of the 1A elements and $(NH_4)_2SO_3$.
PO_4^{3-}	All phosphates are insoluble except those of the 1A elements and $(NH_4)_3PO_4$.
OH^-	All hydroxides are insoluble except those of the 1A elements, $Ba(OH)_2$, $Sr(OH)_2$, and $Ca(OH)_2$.

[a]The following cations are considered: those of the 1A and 2A families, NH_4^+, Ag^+, Al^{3+}, Cd^{2+}, Co^{2+}, Cr^{3+}, Cu^{2+}, Fe^{2+}, Fe^{3+}, Hg_2^{2+}, Hg^{2+}, Mn^{2+}, Ni^{2+}, Pb^{2+}, Sn^{2+}, and Zn^{2+}

correctly that a substance like AgCl is more likely to be soluble in water than in a nonpolar solvent. However, it doesn't permit us to predict the extent to which AgCl will dissolve in water. In fact, only 1.3×10^{-5} mol of AgCl dissolves in a liter of water at 25°C. For all practical purposes, the compound is insoluble. In our discussions, any substance whose solubility is less than 0.01 mol/L at 25°C will be referred to as being insoluble. We shall consider solubility from a quantitative point of view in Chapter 16. In the meantime, we will rely on the general solubility rules for common ions that are given in Table 11.6. These rules are organized according to anions. Examination of the entries reveals an additional important generalization concerning cations: *All common salts of the alkali metals (group 1A) and of the ammonium ion, NH_4^+, are water soluble.*

SAMPLE EXERCISE 11.19

Write balanced molecular, ionic, and net ionic equations for the precipitation reactions (if any) that occur when solutions of the following compounds are mixed: (a) $BaCl_2$ and Na_2SO_4; (b) KCl and Na_2SO_4.

Solution: (a) Both $BaCl_2$ and Na_2SO_4 are soluble and ionize to give Ba^{2+}, Cl^-, Na^+, and SO_4^{2-} ions. The possible precipitation products are $BaSO_4$ and NaCl. The $BaSO_4$ is insoluble according to the rule given in Table 11.6 for SO_4^{2-} compounds. Therefore, the molecular equation is

$$BaCl_2(aq) + Na_2SO_4(aq) \longrightarrow BaSO_4(s) + 2NaCl(aq)$$

The ionic equation is

$$Ba^{2+}(aq) + 2Cl^-(aq) + 2Na^+(aq) + SO_4^{2-}(aq) \longrightarrow BaSO_4(s) + 2Na^+(aq) + 2Cl^-(aq)$$

Both $Na^+(aq)$ and $Cl^-(aq)$ are spectator ions. The net ionic equation is

$$Ba^{2+}(aq) + SO_4^{2-}(aq) \longrightarrow BaSO_4(s)$$

(b) Both reactants are soluble and ionize in solution. There are no possible insoluble salts resulting from reaction. Both NaCl and K_2SO_4 are soluble. Therefore, there is no reaction; the solutes merely mix in the solution.

Formation of Gases

The neutralization and precipitation reactions we have just considered proceed essentially to completion. The "driving force," that is, the factor that causes the reaction to occur, is formation of water in the first instance and formation of an insoluble salt in the second.

An additional source of "driving force" for completion of reactions in aqueous solution is formation of a volatile nonelectrolyte. The net ionic equations for three common examples of this reaction type are as follows:

1 A carbonate reacting with an acid to form CO_2 and H_2O:

$$CO_3^{2-}(aq) + 2H^+(aq) \longrightarrow CO_2(g) + H_2O(l) \qquad [11.20]$$

2 A sulfite reacting with an acid to form SO_2 and H_2O:

$$SO_3^{2-}(aq) + 2H^+(aq) \longrightarrow SO_2(g) + H_2O(l) \qquad [11.21]$$

3 A sulfide reacting with an acid to form H_2S:

$$MnS(s) + 2H^+(aq) \longrightarrow Mn^{2+}(aq) + H_2S(g) \qquad [11.22]$$

SAMPLE EXERCISE 11.20

Write balanced complete ionic and net ionic equations for any reactions that occur when the following compounds are mixed: (a) $FeCO_3(s)$ and $HCl(aq)$; (b) $NiS(s)$ and $HCl(aq)$.

Solution: (a) This reaction is similar to Equation 11.20, but the reactant carbonate is a solid, insoluble in water.

Complete ionic:

$$FeCO_3(s) + 2H^+(aq) + 2Cl^-(aq) \longrightarrow Fe^{2+}(aq) + 2Cl^-(aq) + H_2O(l) + CO_2(g)$$

Net ionic:

$$FeCO_3(s) + 2H^+(aq) \longrightarrow Fe^{2+}(aq) + H_2O(l) + CO_2(g)$$

(b) This reaction is of the type of Equation 11.22. Most sulfides will react with acids even though they are insoluble. Similarly, solid carbonates and hydroxides will generally react with acids, as exemplified in part (a).

Complete ionic:

$$NiS(s) + 2H^+(aq) + 2Cl^-(aq) \longrightarrow Ni^{2+}(aq) + 2Cl^-(aq) + H_2S(g)$$

Net ionic:

$$NiS(s) + 2H^+(aq) \longrightarrow Ni^{2+}(aq) + H_2S(g)$$

Notice that in writing the net ionic equation, in both (a) and (b), both solids and gases are included. The only species that goes through the reaction unchanged is the Cl^- ion.

Chemical Reactions: A Summary

This is a good point at which to summarize the important things we have learned so far about various types of chemical change that might occur upon mixing substances in solution. It is of course possible that no reaction at all will occur when solutions of ionic substances are mixed. For example, mixing aqueous solutions of calcium chloride and potassium nitrate leads to no chemical change. Both salts are strong electrolytes, neither is notably acidic or basic, and no insoluble salt is formed by any combination of the ions present. We do observe a net chemical change when one or more of the following occur:

1 *A precipitate is formed.* Several examples of this type are given above.
2 *A gas is formed.* Equations 11.20, 11.21, and 11.22 provide examples of this reaction type.
3 *A nonelectrolyte is formed.* The neutralization reaction is an example of this type of reaction, since $H_2O(l)$ is a nonelectrolyte.
4 *A weak electrolyte is formed.* As an example of this reaction type, consider the reaction of sodium acetate, $NaC_2H_3O_2$, with hydrochloric acid. The net ionic equation describes formation of acetic acid, a weak electrolyte:

$$H^+(aq) + C_2H_3O_2^-(aq) \rightleftharpoons HC_2H_3O_2(aq) \quad [11.23]$$

We will encounter many more examples of this reaction type in Chapters 15 and 16, when we discuss aqueous equilibria involving acids, bases, and other substances in aqueous solution.
5 *One or more electrons are transferred.* You will recall that we saw examples of oxidation-reduction equations in Section 7.10. Clearly, when substances undergo changes in their oxidation numbers, a chemical reaction has occurred. As we proceed through the text from this point, oxidation-reduction reactions occurring in an aqueous medium will be written in the form of net ionic equations.*
6 *One or more covalent bonds are made or broken.* Many of the reactions in this category—for example, the combustion reactions discussed in Section 3.3—do not occur in an aqueous medium. However, many other such reactions do occur in water. For example, an alcohol such as ethanol is oxidized by potassium dichromate to form acetaldehyde:

$$2H^+(aq) + 3C_2H_5OH(aq) + Cr_2O_7^{2-}(aq) \longrightarrow 3C_2H_4O(aq) + Cr_2O_3(s) + 4H_2O(l) \quad [11.24]$$

Notice that in this reaction we not only have new covalent bonds formed, we also have changes in oxidation numbers and formation of a nonelectrolyte. This example points up the fact that any particular reaction might contain elements of more than one of the categories we have just defined. The main point you should keep in mind is that when one or more of the "driving forces" for reaction that we have just outlined is present, a reaction occurs, as opposed to simply a mixing with no net chemical change.

11.7 COLLOIDS

When finely divided clay particles are dispersed through water, they do not remain suspended but eventually settle out of the water. The dispersed clay particles are much larger than molecules and consist of many thousands or even millions of atoms. In contrast, the dispersed particles of a solution are of molecular size. Between these extremes is the situation in which dispersed particles are larger than molecules, but not so large that the components of the mixture separate under the influence of gravity. These intermediate types of dispersions or suspensions are called **colloidal dispersions** or simply **colloids.** Thus colloids are on the dividing

*The techniques used to balance oxidation-reduction equations are described in Section 18.1.

TABLE 11.7 Types of colloids

Phase of colloid	Dispersing (solventlike) substance	Dispersed (solutelike) substance	Colloid type	Example
Gas	Gas	Gas	—	None (all are solutions)
Gas	Gas	Liquid	Aerosol	Fog
Gas	Gas	Solid	Aerosol	Smoke
Liquid	Liquid	Gas	Foam	Whipped cream
Liquid	Liquid	Liquid	Emulsion	Milk
Liquid	Liquid	Solid	Sol	Paint
Solid	Solid	Gas	Solid foam	Marshmallow
Solid	Solid	Liquid	Solid emulsion	Butter
Solid	Solid	Solid	Solid sol	Ruby glass

line between solutions and heterogeneous mixtures. Like solutions, colloids can be either gases, liquids, or solids. Examples of each are listed in Table 11.7.

The size of the dispersed particle is the property used to classify a mixture as a colloid. Colloid particles range in diameter from approximately 10 to 2000 Å, whereas solute particles are smaller. The colloid particle may consist of many atoms, ions, or molecules or may even be a single giant molecule. For example, the hemoglobin molecule, which carries oxygen in blood, has molecular dimensions of 65 Å × 55 Å × 50 Å and a molecular weight of 64,500 amu.

Even though colloid particles are so small that the dispersion appears uniform even under a microscope, they are large enough to scatter light very effectively. Consequently, most colloids appear cloudy or opaque unless they are very dilute. Furthermore, because they scatter light, a light beam can be seen as it passes through a colloidal suspension as shown in Figure 11.16. This scattering of light by colloidal particles is known as the **Tyndall effect.** Such light scattering makes it possible to see the light beam coming from the projection housing in a smoke-filled theater or the light beam from an automobile on a dusty dirt road.

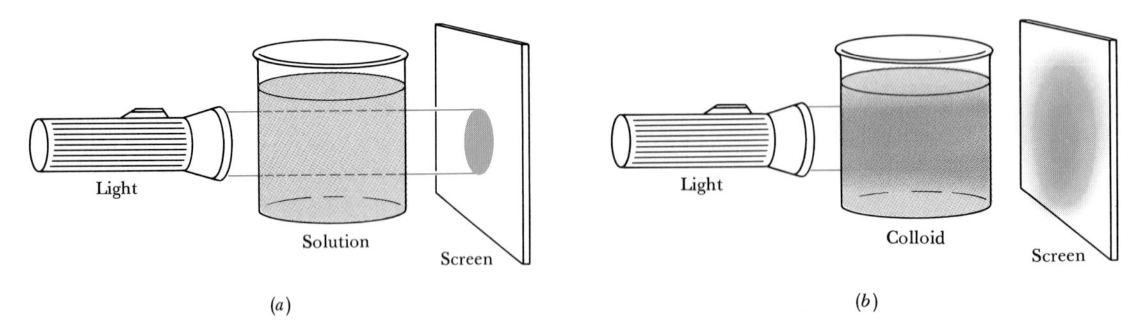

FIGURE 11.16 Comparison of the action of light on a true solution and a colloid. The light beam is shown in (*a*) as passing through the solution without scattering. However, the particles of a colloid scatter the light, so that the outline of the beam is visible, as illustrated by the colored section in (*b*). The beam is diffused, and thus may not appear distinctly on the screen.

FIGURE 11.17 Examples of hydrophilic groups at the surface of a giant molecule (macromolecule) that help keep the macromolecule suspended in water.

Hydrophilic and Hydrophobic Colloids

The most important colloids are those in which the dispersing medium is water. Such colloids are frequently referred to as hydrophilic (water loving) or hydrophobic (water hating). Hydrophilic colloids are most like the solutions that we have previously examined. In the human body, the extremely large molecules that make up such important substances as enzymes and antibodies are kept in suspension by interaction with surrounding water molecules. The molecules fold so that polar or charged groups can interact with water molecules at the periphery of the molecules. These hydrophilic groups generally contain oxygen or nitrogen. Some examples are shown in Figure 11.17.

In the genetic disease known as sickle-cell anemia, hemoglobin molecules are abnormal, in that they have lower solubility, especially in the unoxygenated form. Consequently, as much as 85 percent of the hemoglobin in the red blood cells crystallizes from solution. This distorts the cells into a sickle shape as shown in Figure 11.18. These clog the capillaries, thus causing gradual deterioration of the vital organs. The disease is hereditary, and if both parents carry the defective genes, it is likely that the child will possess only abnormal hemoglobin. Such children seldom survive more than a few years after birth. The reason for the insolubility of hemoglobin in sickle-cell anemia can be traced to a change in one part of the molecule. The normal atom group shown at right is replaced at one place in the hemoglobin molecule by

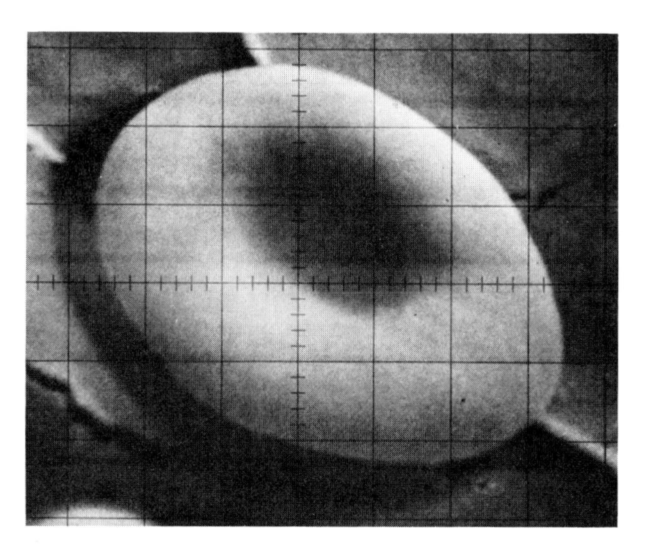

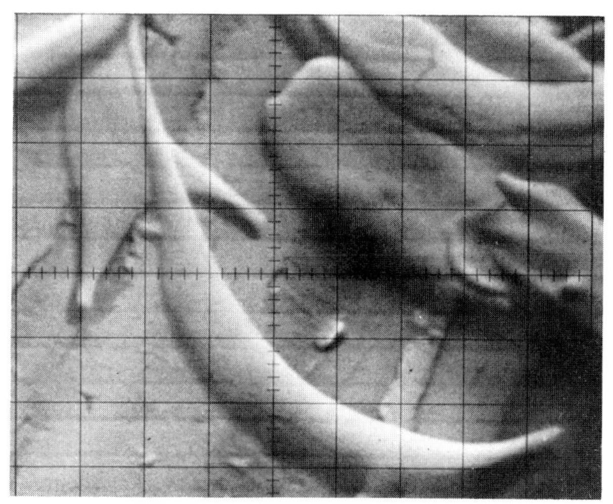

FIGURE 11.18 Electron micrograph showing a normal red blood cell (left; magnification 10,000 times) and sickled red blood cells (right; magnification 5000 times). (*Courtesy Philips Electronic Instruments, Inc.*)

the abnormal combination. The normal group of atoms has a polar group that can interact with water, thereby enhancing the solubility of the hemoglobin. The abnormal group of atoms is nonpolar (hydrophobic), and its presence leads to the aggregation of this defective form of hemoglobin into particles too large to remain suspended in biological fluids.

$$-CH_2-CH_2-\overset{\overset{\displaystyle O}{\|}}{C}-OH$$
Normal

$$-\underset{\underset{\displaystyle CH_3}{|}}{CH}-CH_3$$
Abnormal

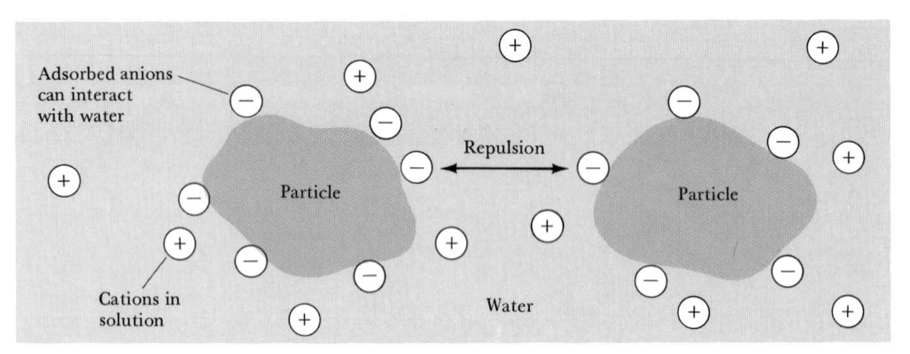

FIGURE 11.19 Schematic representation of the stabilization of a hydrophobic colloid by adsorbed ions.

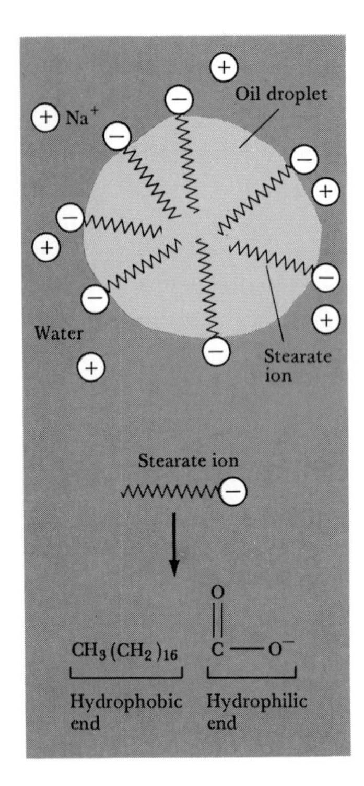

FIGURE 11.20 Stabilization of an emulsion of oil in water by stearate ions.

$$\underbrace{CH_3(CH_2)_{16}}_{\text{Hydrophobic end}}\underbrace{\overset{\overset{\displaystyle O}{\|}}{C}O^-Na^+}_{\text{Hydrophilic end}}$$

Sodium stearate

Hydrophobic colloids can be prepared in water only if they are stabilized in some way. Otherwise, their natural lack of affinity for water causes them to separate from the water. Hydrophobic colloids can be stabilized by adsorption of ions on their surface as shown in Figure 11.19. These adsorbed ions can interact with water, thereby stabilizing the colloid. At the same time the mutual repulsion between colloid particles with adsorbed ions of the same charge keeps the particles from colliding and thereby getting larger. Hydrophobic colloids can also be stabilized by the presence of other hydrophilic groups on their surfaces. For example, small droplets of oil are hydrophobic. They do not remain suspended in water; instead, they separate, forming an oil slick on the surface of the water. Addition of sodium stearate, whose structure is shown at left, or any similar substance having one end that is hydrophilic (polar or charged) and one that is hydrophobic (nonpolar), will stabilize a suspension of oil in water as shown in Figure 11.20. The hydrophobic ends of the stearate ions interact with the oil droplet, whereas the hydrophilic ends point out toward the water with which they interact.

The cleansing action of a soap or detergent is due to its ability to stabilize hydrophobic colloids. These cleansing agents contain substances such as sodium stearate. The action of such molecules in removing dirt from a piece of cloth is illustrated in Figure 11.21.

Colloidal dispersions of liquids in liquids are called *emulsions* (Table 11.7). An agent that stabilizes an emulsion is known as an *emulsifying agent*. Soaps and detergents are emulsifying agents for dispersal of oil in water. The fats that we eat must also be dispersed before they can be digested. The emulsifying agents that accomplish this task are known as bile salts. Because these substances are secreted by the gallbladder, people who have their gallbladders removed must restrict their intake of fats and oils.

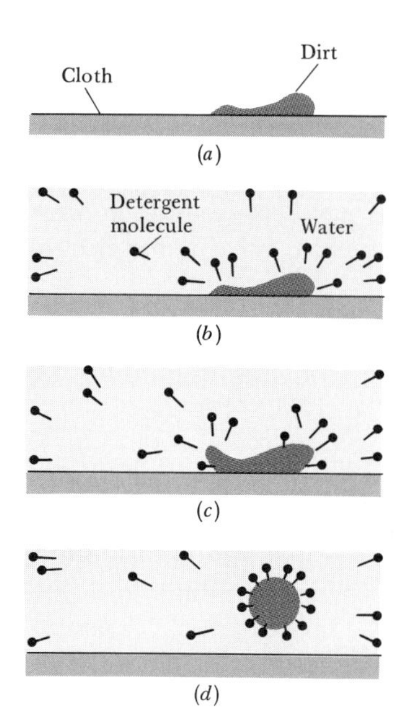

FIGURE 11.21 Schematic representation of the action of detergent ions on dirt. (*a*) Dirt on a piece of cloth. (*b*) Detergent molecules in water attach their hydrophobic ends to the dirt and (*c*) begin to lift it. (*d*) Detergent surrounds the dirt and holds it in suspension so that it can be washed away.

Removal of Colloidal Particles

It is often desirable to remove colloidal particles from a dispersing medium, as in the removal of smoke from stacks or butterfat from milk. Because colloidal particles are so small, they cannot be removed by simple filtration. The separation can often be accomplished by causing the colloidal particles to increase in size, a process referred to as coagulation. The resultant, larger particles can then be removed by filtration or by merely allowing them to settle out of the dispersing medium. Coagulation is normally accomplished either by heating or by adding an electrolyte to the mixture. When a colloidal dispersion is heated, the increased motion of particles increases their collisions. The particles increase in size as they stick together following collision. The addition of electrolytes causes neutralization of the surface charges of the particles, thereby removing the electrostatic repulsions that inhibit their coming together. The effect of electrolytes is seen in the depositing of suspended clay in a river as it mixes with salt water. This results in the formation of river deltas wherever rivers empty into oceans or other salty bodies of water.

Semipermeable membranes can also be used to separate ions from colloidal particles; the ions can pass through the membrane, whereas the colloid particles cannot. This type of separation is known as dialysis. This process is used in the purification of blood in artificial kidney machines. This process is used in the purification of blood in artificial kidney machines. Our kidneys are responsible for removing the waste products of metabolism from blood. In the kidney machine, blood is circulated through a dializing tube immersed in a washing solution. That solution is isotonic in ions that must be retained by the blood but is lacking the waste products. Wastes therefore dialize out of the blood.

FOR REVIEW

Summary

Solutions are homogeneous mixtures of atoms, ions, or molecules. The relative amounts of solute and solvent in a solution can be described qualitatively (dilute or concentrated solutions) or quantitatively (weight percentage, molarity, molality, normality, and mole fraction were discussed).

The extent to which a solute will dissolve in a particular solvent depends on the relative magnitudes of solute-solute, solute-solvent, and solvent-solvent attractive forces as well as on the changes in disorder accompanying mixing. The rule "likes dissolve likes" was found to be useful in rationalizing solubilities. It is possible to change the solubility of a solute by changing temperature. If the solution process is endothermic, an increase in temperature promotes solubility. With a gas, solubility increases in proportion to gas pressure. These effects can be understood in terms of Le Châtelier's principle.

Substances that exist in solution as ions are called electrolytes. Those substances that are completely ionized in solution are called strong electrolytes. Reactions occur between electrolytes if an insoluble substance, a gas, a weak electrolyte, or a nonelectrolyte can form. Net ionic equations focus attention on the particular species that actually undergo some change during the reaction.

The presence of a solute in a solvent lowers the vapor pressure and the freezing point and increases the boiling point of the solvent. These changes are termed colligative properties. The magnitude of the change depends on the total concentration of solute particles in solution and not on their characteristics. Osmotic pressure is the pressure that must be applied to a solution to prevent transfer of solvent molecules from a pure solvent through a semipermeable membrane. It is a colligative property because it is proportional to the total concentration of solute particles. Colligative properties can be used to determine the molecular weights of nonvolatile nonelectrolytes.

True solutions can be differentiated from colloids on the basis of particle size. Colloids play an important role in many chemical and biological systems.

Learning goals

Having read and studied this chapter, you should be able to:

1. Define molarity, molality, mole fraction, normality, and weight percentage and calculate concentrations in any of these concentration units.
2. Convert concentration in one concentration unit into any other (given the density of the solution where necessary).
3. Give definitions of the qualitative terms used to describe solutions: dilute, concentrated, saturated, unsaturated, and supersaturated.
4. Describe the solution process, including the bonds made and broken when a substance dissolves.
5. Describe the energy changes that occur in the solution process in terms of the attractive forces that operate in the solvent and solute and the attractive forces between solvent and solute.
6. Describe the role of disorder in the solution process.
7. Rationalize the solubilities of substances in various solvents in terms of their molecular structures and intermolecular forces.
8. Discuss the effects of pressure and temperature on solubilities of gases.
9. Relate the effect of temperature on the solubilities of salts in water to the enthalpy change in the solution process.
10. State Le Châtelier's principle and apply it to rationalize the effects of temperature and pressure on solubilities.
11. Predict which substances are electrolytes and which are nonelectrolytes; predict which electrolytes are strong electrolytes and which are weak.
12. Predict the course of reactions involving formation of a precipitate, gas, or nonelectrolyte, and write net ionic equations for these reactions.
13. Describe the effect of solute concentration on solvent vapor pressure; state Raoult's law.
14. Explain the origin of the colligative properties—boiling-point elevation, freezing-point lowering, and osmotic pressure—of solutions.
15. Determine the molecular weight of a solute from the magnitude of the effect of a known concentration of solute on one of the colligative properties of a solvent.
16. Explain the difference between the magnitude of changes in colligative properties caused by electrolytes compared to those caused by nonelectrolytes.
17. Describe how a colloid differs from a true solution.

Key terms

Among the more important terms and expressions used for the first time in this chapter are the following:

Colligative properties (Section 11.5) are those properties of a solvent (vapor-pressure lowering, freezing-point lowering, boiling-point elevation, osmotic pressure) that depend on the total concentration of solute particles present.

Colloids (Section 11.7) are particles that are larger than normal molecules, but that are nevertheless small enough to remain suspended in a dispersing medium indefinitely.

Dialysis (Section 11.7) is the separation of small solute particles from colloid particles by means of a semipermeable membrane.

An electrolyte (Section 11.4) is a solute that gives a solution containing ions; such a solution will conduct an electrical current. Strong electrolytes (strong acids, strong bases, and most common salts) are completely ionized in solution. Weak electrolytes are partially ionized in solution.

Henry's law (Section 11.3) states that the solubility of a gas in a liquid, C_g, is proportional to the pressure of gas over the solution: $C_g = kP_g$.

Le Châtelier's principle (Section 11.3) states that any attempt to change the conditions of a system at equilibrium results in a shift in the position of the equilibrium in the direction that tends to offset the change.

Molality (Section 11.1) is the concentration of a solution expressed as moles of solute per kilogram of solvent; abbreviated *m*.

Mole fraction (Section 11.1) is the ratio of the number of moles of one component of a solution to the total number of moles of all substances present in the solution; abbreviated *X*, with a subscript identifying the component.

A net ionic equation (Section 11.6) is an equation for a reaction involving ions that is obtained by eliminating spectator ions (those ions that go through the reaction unchanged and appear on both sides of the overall ionic equation).

Normality (Section 11.1) is the concentration of a solution expressed as equivalents of solute per liter of solution; abbreviated *N*. An equivalent is defined according to the type of reaction being considered; 1 equivalent of a given reactant will react with 1 equivalent of a second reactant.

Osmosis (Section 11.5) is the net movement of solvent through a semipermeable membrane toward the solution with greatest solute concentration. The osmotic pressure of a solution is the pressure that must be applied to a solution to stop osmosis from pure solvent into the solution.

Raoult's law (Section 11.5) states that the partial pressure of a solvent over a solution, P_A, is given by the vapor pressure of the pure solvent, P_A°, times the mole fraction of solvent in the solution, X_A: $P_A = X_A P_A^\circ$.

A saturated solution (Section 11.2) is a solution in which undissolved solute and dissolved solute are in equilibrium. The solubility of a substance is the amount of solute that dissolves to form a saturated solution. Solutions containing less solute than this are said to be unsaturated, whereas those containing more are said to be supersaturated.

Solvation (Section 11.2) is the clustering of solvent molecules around a solute particle. When the solvent is water, this clustering is known as hydration.

The Tyndall effect (Section 11.7) describes the visible path of a light beam passing through a colloidal dispersion caused by scattering of light by the colloid particles.

The weight percentage (Section 11.1) of a solution is the number of grams of solute it contains in each 100 g of solution.

EXERCISES

Concentrations of solutions

11.1 Calculate the weight percentage of solute in each of the following solutions: (a) 1.00g of NaCl in 50.0 g of H_2O; (b) 14.0 g of benzene, C_6H_6, in 25.0 g of carbon tetrachloride, CCl_4; (c) 3.55 g of $Ba(NO_3)_2$ in 100.0 g of H_2O.

11.2 Calculate the mole fraction of methyl alcohol, CH_3OH, in each of the following solutions: (a) 6.00 g of CH_3OH dissolved in 855 g of H_2O; (b) 5.00 g of CH_3OH dissolved in 865 g of CCl_4; (c) 6.00 g of CH_3OH and 6.00 g of C_2H_5OH dissolved in 865 g of H_2O.

11.3 Calculate the molarity of each of the following solutions: (a) 1.00 g of KBr in 1.00 L of solution; (b) 4.00 g of NaOH in 0.450 L of solution; (c) 3.58 g of $Ca(NO_3)_2 \cdot 4H_2O$ in 0.400 L; (d) 50.0 mL of 0.300 *M* HCl diluted to 1.00 L; (e) 20.0 mL of 0.150 *M* $Al(NO_3)_3$ diluted to 50.0 mL.

11.4 Calculate the molality of each of the following solutions: (a) 14.0 g of benzene, C_6H_6, dissolved in 30.0 g of carbon tetrachloride, CCl_4; (b) 3.50 g of NaI dissolved in 0.200 L of water; (c) 5.85 g of KNO_3 dissolved in 0.250 L of water.

11.5 For each of the following solutions, calculate the weight percentage, mole fraction, and molality of the solute: (a) 3.6 g of KI in 340 g of water; (b) 35.2 g of ethylene glycol, $C_2H_4(OH)_2$, in 65.0 g of water; (c) 5.00 g of methanol, CH_3OH, in 45.0 g of ethanol, C_2H_5OH.

11.6 A solution containing 66.0 g of acetone, C_3H_6O, and 46.0 g of H_2O has a density of 0.926 g/mL. Calculate (a) the weight percentage; (b) the mole fraction; (c) the molality; (d) the molarity of H_2O in this solution.

11.7 A sulfuric acid solution containing 571.6 g of H_2SO_4 per liter of solution has a density of 1.329 g/mL. Calculate (a) the weight percentage; (b) the mole fraction; (c) the molality; (d) the molarity of H_2SO_4 in this solution.

11.8 Given that the density of a 0.790 *M* aqueous solution of KI is 1.093 g/mL, calculate its molality.

11.9 Calculate the number of moles of solute present in each of the following solutions: (a) 256 mL of 0.358 *M* $Ca(NO_3)_2$; (b) 4.00×10^2 L of 0.0655 *M* HBr; (c) 450 g of an aqueous solution that is 0.565 percent NaCl by weight; (d) 35.2 g of an aqueous solution that is 0.0505 percent K_2SO_4 by weight; (e) 0.100 kg of 0.25 *m* HCl; (f) 158 g of 0.0327 *m* $Al(NO_3)_3$.

11.10 Calculate the number of grams of $La(NO_3)_3$ present in 1.00×10^2 g of water in each of the following solutions: (a) a 0.050 *m* solution; (b) an aqueous solution in which $X_{La(NO_3)_3} = 0.0105$; (c) a solution that is 1.30 percent $La(NO_3)_3$ by weight.

11.11 Describe how you would prepare each of the following aqueous solutions starting with solid KBr: (a) 2.40 L of 2.00×10^{-2} *M* KBr; (b) 150 g of 0.420 *m* KBr; (c) 2.5 L of a solution that is 14 percent KBr by weight; (d) a 0.200 *M* solution of KBr that would contain just enough KBr to precipitate 26.0 g of AgBr from a solution containing 0.44 mol of $AgNO_3$.

11.12 Concentrated nitric acid has a density of 1.42 g/mL and is 69 percent HNO_3 by weight. What is the molarity of this solution?

11.13 (a) In a certain reaction Sn^{4+} is reduced to Sn^{2+}. What is the normality of a 0.36 *M* solution of Sn^{4+}? What is the molarity of a 0.42 *N* solution of Sn^{4+}? (b) What is the molarity of a 0.105 *N* solution of H_3PO_4 reacted with NaOH solution, in which all three hydrogens of the H_3PO_4 react?

11.14 In each of the following reactions indicate the normality of a 0.1 *M* solution of the reagent underlined:

(a) $H_3PO_4(aq) + 3NaOH(aq) \longrightarrow Na_3PO_4(aq) + 3H_2O(l)$

(b) $BaCl_2(aq) + H_2SO_4(aq) \longrightarrow BaSO_4(s) + 2HCl(aq)$

(c) $MnS(s) + 2HCl(aq) \longrightarrow MnCl_2(aq) + H_2S(g)$

The solution process; solubility and structure

11.15 Indicate the type of solute-solvent intermolecular attractive force (Section 10.4) that should be most important in each of the following solutions: (a) KCl in water; (b) benzene, C_6H_6, in carbon tetrachloride, CCl_4; (c) HF in water; (d) acetonitrile, CH_3CN, in acetone, C_6H_6O (see Table 10.5).

11.16 Offer an explanation, in terms of the intermolecular forces involved, for each of the following observations. (a) $CoCl_2$ is soluble to the extent of 540 g per kilogram of ethyl alcohol. (b) Water is completely miscible with dioxane, for which the structural formula is

H_2C——CH_2
O　　　O
H_2C——CH_2

However, water is not soluble in cyclohexane, for which the structural formula is

H_2C——CH_2
CH_2　　　CH_2
H_2C——CH_2

(c) Chloroform, $CHCl_3$, is soluble in water to the extent of only 1 g per 100 g of water, but is miscible with ethyl alcohol.

11.17 Judging from the nature of solute-solvent interactions, rank the following substances in order of increasing solubility in water: CH_3OH, CH_4, NaF, CH_3F.

11.18 Br_2 is more soluble than I_2 in CCl_4. Suggest an explanation for this observation.

11.19 Explain in terms of the enthalpy changes in the various steps involved (Figure 11.3) why KBr is not soluble in CCl_4.

11.20 Which of the following pairs of liquids would you expect to be miscible and which nonmiscible: (a) H_2O and $HOCH_2CH(OH)CH_3OH$; (b) $C(CH_2OH)_4$ and C_7H_{16}; (c) Hg(l) and NaI(l); (d) CH_2Cl_2 and $CH_3CH_2OCH_2CH_3$. Explain in each case.

11.21 Which ion in each of the following pairs would you expect to be more strongly hydrated (a) Na^+ or K^+; (b) Cl^- or Br^-; (c) Fe^{2+} or Fe^{3+}; (d) Ca^{2+} or Al^{3+}?

Effect of temperature and pressure on solubility

11.22 State Henry's law. From the data listed in Table 11.3, which has the larger Henry's law constant in water at 20°C, N_2 or O_2?

11.23 The Henry's law constant for helium gas in water at 30°C is 3.7×10^{-4} M/atm; that for N_2 at 30°C is 6.0×10^{-4} M/atm. If the two gases are each present at 0.50 atm pressure, calculate the solubility of each gas.

11.24 The partial pressure of O_2 in air at sea level is 0.21 atm. Using the data in Table 11.3 together with Henry's law, calculate the molar solubility of O_2 in the surface water of a lake saturated with O_2 at 20°C.

[11.25] In terms of the kinetic-molecular theory, explain why the solubility of gases in liquids generally decreases with an increase in temperature.

11.26 The dissolving of $CuCl_2$ is accompanied by the evolution of 46.4 kJ/mol $CuCl_2$. Using Le Châtelier's principle, predict the effects of temperature on the solubility of $CuCl_2$.

11.27 Using Figures 11.9 and 11.10, predict whether the formation of a saturated solution of each of the following compounds is an exothermic or an endothermic process: (a) $NaNO_3$; (b) CH_4; (c) $Ce_2(SO_4)_3$; (d) $KClO_3$.

Electrolyte solutions

11.28 Classify each of the following substances as a nonelectrolyte, weak electrolyte, or strong electrolyte in water; (a) LiOH; (b) HClO; (c) $HClO_4$; (d) $Ba(NO_3)_2$; (e) O_2; (f) glucose, $C_6H_{12}O_6$; (g) HF; (h) ethyl alcohol, C_2H_5OH; (i) $NaC_2H_3O_2$; (j) NH_3.

11.29 Identify those substances in Exercise 11.28 that are acids, those that are bases, those that are salts, and those that are none of these three types of compounds. How do these classifications relate to whether a substance is an electrolyte or nonelectrolyte?

11.30 For each of the following solutions, indicate whether the substance exists entirely in molecular form, entirely in ionic form, or as a mixture of molecular and ionic forms in the solutions indicated: (a) $CaCl_2$ in water; (b) $HC_2H_3O_2$ in water; (c) acetone, $(CH_3)_2CO$, in water; (d) NaOH in water; (e) CH_3OH in C_2H_5OH; (f) $Ba(NO_3)_2$ in water; (g) HNO_3 in water.

11.31 Indicate the total concentration of all solute species present in each of the following aqueous solutions: (a) 0.10 M NaOH; (b) 0.35 M $CaBr_2$; (c) 0.14 M C_2H_5OH; (d) a mixture of 50.0 mL of 0.10 M $KClO_3$ and 50.0 mL of 0.20 M Na_2SO_4.

11.32 Indicate the total concentration of each ion present in the solution formed by mixing (a) 10.0 mL of 0.100 M HCl and 20.0 mL of 0.150 M HCl; (b) 25.0 mL of 0.200 M Na_2SO_4 and 15.0 mL of 0.100 M NaCl; (c) 5.00 g of KCl in 45.0 mL of 0.600 M $CaCl_2$ (assume no volume change).

Colligative properties

11.33 Which of the following are colligative properties: (a) heat of vaporization of a liquid; (b) freezing-point lowering of a solution; (c) variation of solubility of a solute with temperature; (d) osmotic pressure of a solution; (e) vapor pressure of a solution; (f) density of a solution; (g) boiling point of a solution?

11.34 Calculate the vapor pressure of water above a solution prepared by adding (a) 20.00 g of lactose,

$C_{12}H_{22}O_{11}$, to 91.50 g of water at 338 K; (b) 5.00 g of sodium sulfate, Na_2SO_4, to 95.0 g of water at 338 K; (c) 46.5 g of glycerin, $C_3H_8O_3$, to 148 g of water at 338 K. (The vapor pressure of pure water is given in Appendix C.)

11.35 Using Table 11.5, calculate the freezing and boiling points of each of the following solutions: (a) 1.2*m* glucose in ethanol; (b) 3.5 g of CCl_4 in 123 g of benzene; (c) 1.5 g of KNO_3 in 33.6 g of water; (d) 1.86 g of Li_2CrO_4 in 60.0 g of water; (e) 4.30 g of naphthalene, $C_{10}H_8$, in 95.2 g of CCl_4.

11.36 List the following aqueous solutions in order of increasing boiling point: 0.030*m* glycerin; 0.020*m* KBr; 0.030*m* benzoic acid ($HC_7H_5O_2$).

11.37 Calculate the osmotic pressure exerted by each of the following aqueous solutions: (a) 45.0 g of glycerin, $C_3H_5(OH)_3$, in 0.500 L of solution at 23°C; (b) 0.64 g of aspirin ($C_9H_8O_4$) in 200 mL of solution at 25°C; (c) 1.25 g of NaCl in 200 mL of solution at 25°C.

11.38 A solution is prepared from 26.7 g of an unknown compound and 116.2 g of acetone, C_3H_6O, at 313 K. At this temperature the vapor pressure of pure acetone is 0.526 atm, while that of the solution is 0.501 atm. Calculate the molecular weight of the unknown compound.

11.39 A dilute sugar solution prepared by mixing 8.0 g of an unknown sugar with sufficient water to form 0.200 L of solution is found to have an osmotic pressure of 2.86 atm at 25°C. What is the molecular weight of the sugar?

11.40 A sample of seawater taken from the Arctic Ocean freezes at −1.98°C; a sample taken from the middle of the Atlantic Ocean freezes at −2.08°C. What is the total molality of ionic solutes present in each of these solutions?

11.41 Camphor, an organic substance melting at 176.5°C, was at one time widely used to obtain estimates of molecular weights of organic compounds, because it has a large freezing-point-lowering constant, 37.7°C/*m*. When 0.125 g of sulfur is finely ground with 3.62 g of camphor, the resultant mixture melts at 171.4°C. What is the molecular weight of the sulfur in camphor? To what molecular formula does this correspond?

Reactions in aqueous solution

11.42 Write balanced net ionic equations for each of the following reactions. (a) Carbon dioxide gas is bubbled through an aqueous solution of sodium hydroxide, forming a solution of sodium bicarbonate. (b) Addition of solid sodium hydride, NaH, to water produces hydrogen gas and an aqueous solution of sodium hydroxide. (c) The addition of metallic zinc to an aqueous solution of copper(II) sulfate produces copper metal and a solution of zinc sulfate.

11.43 Predict whether the following compounds are soluble in water: (a) silver chlorate; (b) cesium sulfate; (c) lead iodide; (d) magnesium hydroxide; (e) barium carbonate; (f) zinc sulfide; (g) ammonium sulfate; (h) nickel acetate.

11.44 Write balanced net ionic equations for the reactions, if any, that occur between (a) ZnS(*s*) and HCl(*aq*); (b) Na_2CO_3(*aq*) and $BaCl_2$(*aq*); (c) Na_3PO_4(*aq*) and HBr(*aq*); (d) $Ba(OH)_2$(*aq*) and HCl(*aq*); (e) $Sr(C_2H_3O_2)_2$(*aq*) and $NiSO_4$(*aq*); (f) $ZnSO_3$(*aq*) and HCl(*aq*); (g) $Pb(NO_3)_2$(*aq*) and H_2S(*aq*); (h) $Fe(OH)_3$(*s*) and $HClO_4$(*aq*); (i) $Pb(ClO_3)_2$(*aq*) and $MgSO_4$(*aq*).

11.45 Write balanced net ionic equations for the reactions, if any, that occur when each of the following pairs is mixed: (a) H_2SO_4(*aq*) and $BaCl_2$(*aq*); (b) NaCl(*aq*) and $(NH_4)_2SO_4$(*aq*); (b) $AgNO_3$(*aq*) and Na_2CrO_4(*aq*); (d) KOH(*aq*) and HNO_3(*aq*); (e) $Ca(OH)_2$(*aq*) and $HC_2H_3O_2$(*aq*); (f) K_2SO_3(*s*) and H_2SO_4(*aq*).

11.46 Which of the following salts, insoluble in pure water, would you expect to dissolve (accompanying chemical reaction) upon addition of moderately concentrated HBr solution: (a) $SrCO_3$; (b) $MnSO_3$; (c) $BaSO_4$; (d) AgCl? Explain.

[11.47] Suppose you have a solution that might contain any or all of the following cations: Ni^{2+}, Ag^+, Sr^{2+}, and Mn^{2+}. Addition of HCl solution causes a precipitate to form. After filtering off the precipitate, H_2SO_4 solution is added to the resultant solution and another precipitate forms. This is filtered off, and a solution of Na_2CrO_4 is added to the resulting solution. No precipitate is observed. Which ions are present in each of the precipitates? Which ions must be absent (at least in moderate concentrations) from the original solution?

Colloids

11.48 Given a homogeneous liquid, how could you determine whether it is a solution or a colloidal dispersion?

11.49 Indicate whether each of the following is a hydrophilic or a hydrophobic colloid: (a) butterfat in homogenized milk; (b) hemoglobin in blood; (c) vegetable oil in a salad dressing.

11.50 Glucose, which has the structure shown in Figure 11.7, is not soluble in benzene. Explain how it might be possible to stabilize a colloidal dispersion of glucose in benzene using sodium stearate. Draw a sketch to show how the stabilization would occur.

11.51 It is possible to precipitate gold from an aqueous solution in such a manner that the gold particles are extremely small. Although the density of gold is 19.3 g/cm³, such a sol is stable indefinitely. In terms of the kinetic-molecular theory, how is it possible for the gold particles to remain dispersed?

11.52 Suggest an explanation for the fact that milk curdles upon the addition of acid.

Additional exercises

11.53 Calculate the molarity of each of the following solutions: (a) one that contains 10.0 g of H_2SO_4 in 500 mL of solution; (b) one made by diluting 5.0 mL of 3.0 *M* H_3PO_4 to a total volume of 25.0 mL.

11.54 The density of a sulfuric acid solution from a car battery is 1.225 g/mL. If this solution is 3.75 *M*, calculate its concentration in (a) molality; (b) mole fraction H_2SO_4; (c) weight percentage of water.

11.55 How many milliliters of concentrated hydrochloric acid, 36.5 percent by weight, density = 1.185 g/mL, are required to prepare 1.00 L of a 0.100 *M* solution?

11.56 A solution is made by dissolving 1.68 g of benzoic acid, $HC_7H_5O_2$, in 206 mL of CCl_4, density 1.59 g/mL; another is prepared by dissolving the same quantity of benzoic acid in 206 mL of C_2H_5OH, density 0.782 g/mL. (a) Calculate the mole fractions and molalities in each case. (b) Assuming that the density of each solution is the same as that of pure solvent, calculate the molarity in each case. (c) Compare the molarities and molalities of these solutions and comment on their relative magnitudes.

11.57 Caffeine has the following molecular structure:

Caffeine

It is soluble in water to the extent of 1.35 g per 100 g of water at 16°C, 46 g per 100 g of water at 65°C. (a) What structural features of the caffeine molecule account for its solubility in water? (b) Is the solution of caffeine an endothermic or exothermic process?

11.58 In terms of the concepts developed in this chapter, indicate why each of the following commercial products is formulated as it is. (a) A gas-line antifreeze additive to the gas tank consists almost entirely of methyl alcohol. (b) The solvent in a water-repellent spray for treating shoes is methylene chloride, CH_2Cl_2. (c) A wax remover for cross-country skis consists of benzene and similar aromatic hydrocarbons (Section 8.4).

11.59 Which of the following observed processes proceeds spontaneously because of an increase in the randomness, or disorder, of the process: (a) evaporation of solid I_2 crystals; (b) mixing of $Br_2(l)$ and $CCl_4(l)$; (c) precipitation of $BaSO_4(s)$ from aqueous solution; (d) alignment of iron filings in a magnetic field.

[11.60] Butylated hydroxytoluene (BHT) has the following molecular structure:

BHT

It is widely used as a preservative in a variety of foods, including dried cereals, cooking oils, and canned goods. The average person in the United States consumes about 2 mg of BHT daily. In terms of its structure, would you expect it to be readily excreted from the body or found stored in body fats? (Incidentally, BHT is not known to have any harmful properties; it is, in fact, a known antiviral agent.)

11.61 The solubility of CO_2 in water at 18°C with 1 atm CO_2 pressure over the solution is 0.0414 *M*. What is the molar solubility of CO_2 at this temperature when water is in contact with normal air at sea level, where the partial pressure of CO_2 is 0.251 mm Hg?

11.62 The following data give the solubility of radon (Rn) in water with 1 atm pressure of the gas over the solution at various temperatures: 20°C, 9.91×10^{-3} *M*; 30°C, 7.27×10^{-3} *M*; 40°C, 5.62×10^{-3} *M*. Is the solution of radon in water an exothermic or endothermic process? Explain.

11.63 A gas sample consisting of a mixture of the noble gas elements contains 3.5×10^{-6} mole fraction of radon. This gas at a total pressure of 80 atm is shaken in contact with water at 30°C. Using the data of Exercise 11.62, calculate the molar concentration of radon in the water.

11.64 Lysozyme, an enzyme that cleaves bacterial cell walls, has a molecular weight of 13,930 g/mol. What is the osmotic pressure exerted by a solution of 0.100 g of this enzyme in 150 mL of solution at 25°C?

[11.65] A constant humidity in a closed container can be achieved by placing in the chamber an aqueous solution that has the desired vapor pressure of water. Provide a recipe for preparing 1 kg of a solution of glycerin ($C_3H_9O_3$) and water that has a vapor pressure of 14.5 mm Hg at 24°C (see Appendix C).

11.66 A "canned heat" product used to warm chafing dishes consists of a homogeneous mixture of methyl alcohol, CH_3OH, and paraffin that has an average formula $C_{24}H_{50}$. What mass of CH_3OH should be added to 48.0 kg of the paraffin in formulating the mixture if the vapor pressure of the methanol over the mixture at 24°C is to be 15 mm Hg? The vapor pressure of methanol at 24°C is 177 mm Hg.

[11.67] Calculate the vapor pressure over a 1.50*m* solution of glucose at 60°C. (see Appendix C; note that you will need to convert from units of molality to units of mole fraction.)

[11.68] A lithium salt used in a lubricating grease has the formula $LiC_nH_{2n-1}O_2$. The salt is soluble in water to the extent of 0.036 g per 100 g of water at 25°C. The osmotic pressure of this solution is found to be 57.1 mm Hg. Assuming that molality and molarity in such a dilute solution are the same and that the lithium salt is completely ionized, determine an appropriate value of *n* in the formula for the salt.

[11.69] A 1.00-g sample of heroin ($C_{21}H_{23}O_5N$) "cut" with "milk sugar" (lactose, $C_{12}H_{22}O_{11}$) was dissolved in a sufficient quantity of water to form 100.0 mL of solution. If the osmotic pressure of this solution was 535 mm Hg at 25°C, what was the percent sugar present in the heroin sample?

[11.70] Pheromones are compounds secreted by the

females of many insect species to attract males. One of these compounds contains 80.78% C, 13.56% H, and 5.66% O. A solution of 1.00 g of this substance in 8.50 g of benzene freezes at 3.37°C. What is the molecular weight and molecular formula of the compound?

[11.71] Adrenaline, the hormone that triggers release of extra glucose molecules at times of stress or emergency, contains 59.0 percent C, 26.2 percent O, 7.1 percent H, and 7.6 percent N. A solution of 0.64 g of adrenaline in 36.0 g of CCl_4 causes an elevation of 0.49°C in the boiling point. What are the molecular weight and molecular formula of adrenaline?

11.72 The label on a container of antifreeze says that a mixture consisting of equal volumes of the antifreeze and of water will protect your engine to a temperature of −34°F. Assuming that the antifreeze consists entirely of ethylene glycol, $C_2H_4(OH)_2$, density 1.113 g/mL, calculate the freezing and boiling points of an equal-volume mixture of water and ethylene glycol. Compare the freezing point that you calculate with the temperature stated on the container label.

11.73 A mixture is made by adding together 50.0 mL of 0.100 M NaCl and 25.0 mL of 0.100 M $BaCl_2$ solution. Assuming that the volume of the resultant solution is the sum of the two initial solutions, calculate the concentration of each ion in the solution.

11.74 The alcoholic content of beer, wine, and hard liquor is normally given in terms of the "percent proof," which is defined as twice the percentage by volume of alcohol, C_2H_5OH, in the mixture. Calculate the number of grams of C_2H_5OH present in 0.850 L of 87-proof gin if pure alcohol has a density of 0.798 g/mL.

11.75 Antacids are often used to relieve pain and promote healing in the treatment of mild ulcers. Write balanced net ionic equations for the reactions that occur between stomach acid, HCl(aq), and each of the following substances used in various antacids: (a) $Al(OH)_3(s)$; (b) $Mg(OH)_2(s)$; (c) $MgCO_3(s)$; (d) $NaAl(CO_3)(OH)_2(s)$; (e) $CaCO_3(s)$.

11.76 Account for each of the following observations, and write a balanced net ionic chemical equation to represent what occurs. (a) A solid forms upon addition of sodium hydroxide solution to an aqueous solution of manganese chloride. (b) Limestone formations, composed mainly of $CaCO_3$, dissolve when in contact with acidic groundwater. (c) Chloric acid, $HClO_3$, was first prepared by Gay-Lussac, who reacted a barium chlorate solution with dilute H_2SO_4.

[11.77] In the apparatus shown in Figure 11.13, 20 mL of a 0.050 M solution of a nonvolatile nonelectrolyte is placed in the left arm and 20 mL of a 0.030 M solution of a nonvolatile nonelectrolyte is placed in the right arm. What are the volumes in the two arms when equilibrium is obtained?

12 Chemistry of the Environment

In previous chapters we have dealt for the most part with the principles that govern the chemical and physical behavior of matter. We are now in a position to apply these principles to an understanding of the world in which we live. In this chapter we will consider some aspects of the chemistry of our environment, focusing on the earth's atmosphere and on the aqueous environment, which we call the **hydrosphere.**

Both the atmosphere and hydrosphere of our planet make life as we know it possible. Pollution of the environment, which occurs in many different ways, is one of the most important concerns of our time. It has become evident that major reforms and much stricter standards are required if we are to preserve the quality of life in the world. As a voting citizen you will be called upon in the future to help decide bond issues and referenda that may have an impact on your health as well as on your economic security. The better you understand the chemical principles that underlie environmental issues, the better your chances of forming a sound judgment. Our intent in this chapter is therefore to provide an introduction to the nature of the earth's atmosphere and hydrosphere, and to indicate some of the ways in which pollution occurs.

12.1 EARTH'S ATMOSPHERE

Because most of us have never been very far from the earth's surface, we tend to take for granted the many ways in which the atmosphere determines the environment in which we live. In this section we will examine some of the important physical characteristics of our planet's atmosphere in light of what we know of the properties of gases.

The temperature of the atmosphere varies in a rather complex manner as a function of altitude, as shown in Figure 12.1. Just above the surface the temperature normally decreases with increasing altitude and reaches a minimum value at an elevation of about 12 km. In this region, called the **tropopause,** the temperature is about 215 K (−60°C). Above this elevation the temperature increases to about 275 K in the region of

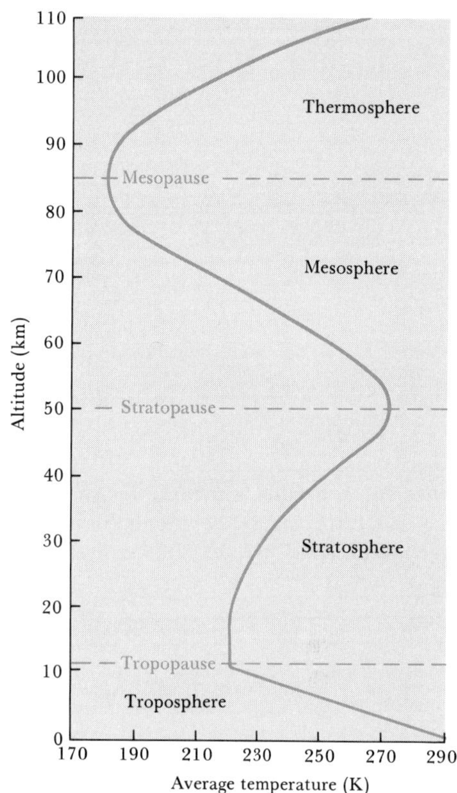

FIGURE 12.1 Temperature variations in the atmosphere at altitudes below 110 km. (*Adapted from "U.S. Standard Atmosphere, 1962." Washington, D.C.: Government Printing Office*)

50 km and then begins again to decrease. The altitude at which the temperature reaches a maximum is called the **stratopause.** Above the stratopause the temperature drops to an even lower value than at the tropopause. The region of this second temperature minimum is called the **mesopause.** Above the mesopause, the temperature rises rapidly in the region called the **thermosphere.** Note that the regions of temperature minima and maxima are denoted by the suffix *-pause*. The regions between these are denoted by the suffix *-sphere*. The boundaries between different regions are important because mixing of the atmosphere across the boundaries is relatively slow. For example, pollutant gases generated in the troposphere find their way into the stratosphere only very slowly.

The **troposphere** is the region of the atmosphere in which nearly all of us live out our entire lives. Howling winds and soft breezes, rain, sunny skies—all that we normally think of as weather occurs in this region. Even when we fly in a modern jet aircraft between distant cities, we are still in the troposphere, though we may be near the tropopause.

In contrast to the temperature changes that occur in the atmosphere, the pressure of the atmosphere decreases in a quite regular way with increasing elevation, as shown in Figure 12.2. We see from this illustration that atmospheric pressure drops off much more rapidly at lower elevations than at higher. The explanation for this characteristic of the atmosphere lies in its compressibility. Gases are very different from liquids in this regard. As a result of the atmosphere's compressibility, the pressure decreases from an average value of 760 mm Hg at sea level to 2.3×10^{-3} mm Hg at 100 km, to only 1.0×10^{-6} mm Hg at 200 km.

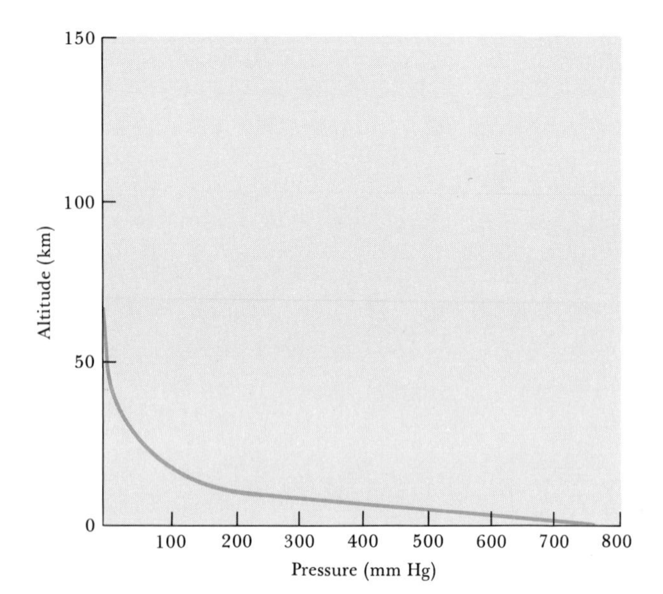

FIGURE 12.2 Variation in atmospheric pressure with altitude. At 50 km altitude the pressure has declined to about 1 mm Hg. At still higher altitudes the pressure continues to decline, although this cannot be shown on the scale of the figure. For example, at 100 km the pressure has declined to 2.3×10^{-3} mm Hg.

Composition of the Atmosphere

The atmosphere is an extremely complex system. Its temperature and pressure change over a wide range with altitude, as we have already seen. The atmosphere is subjected to bombardment by radiation and energetic particles from the sun and by cosmic radiation from outer space. This barrage of energy has profound chemical effects, especially on the outer reaches of the atmosphere. In addition, because of the earth's gravitational field, lighter atoms and molecules tend to rise to the top. As a result of all these factors, the composition of the atmosphere is not constant. However, it is useful to know the composition of the atmosphere in the region near the earth's surface. Table 12.1 shows the composition of dry air near sea level. We note that although traces of a great many

TABLE 12.1 Composition of dry air near sea level

Component[a]	Content (mole fraction)	Molecular weight
Nitrogen	0.78084	28.013
Oxygen	0.20948	31.998
Argon	0.00934	39.948
Carbon dioxide	0.000330	44.0099
Neon	0.00001818	20.183
Helium	0.00000524	4.003
Methane	0.000002	16.043
Krypton	0.00000114	83.80
Hydrogen	0.0000005	2.0159
Nitrous oxide	0.0000005	44.0128
Xenon	0.000000087	131.30

[a]Ozone, sulfur dioxide, nitrogen dioxide, ammonia, and carbon monoxide are present as trace gases in variable amounts.

substances are present, only a few dominate. The two diatomic molecules N_2 and O_2 make up about 99 percent of the entire atmosphere. Essentially all the remainder, with the exception of carbon dioxide, is made up of the monatomic rare gases.

Note that the contribution of each component of the atmosphere listed in Table 12.1 is given in terms of its mole fraction. This is simply the total number of moles of a particular component in a given sample of air, divided by the total number of moles of all the components in that sample (Section 11.1). The partial pressure of a given component in the atmosphere is given by the total atmospheric pressure times the mole fraction of that component. Using P_t to represent the total pressure, and X_1, X_2, and so on, to represent the mole fractions of the components of the gas mixture, we have:

$$\begin{aligned} P_t &= X_1P_t + X_2P_t + X_3P_t + \cdots \\ &= P_t(X_1 + X_2 + X_3 + \cdots) \end{aligned} \quad [12.1]$$

This statement follows from Dalton's law of partial pressures (Section 9.5) and Avogadro's hypothesis (Section 9.3). Of course, the mole fractions must add up to 1.

SAMPLE EXERCISE 12.1

What is the partial pressure of CO_2 in dry air when the total dry air pressure (P_t) is 735 mm Hg?

Solution: Referring to Table 12.1, we see that the mole fraction of CO_2 is 3.30×10^{-4}. This means that the fractional contribution of CO_2 to a unit total pressure of 1 would be 3.30×10^{-4}.

$$\begin{aligned} \text{Pressure } CO_2 &= (735 \text{ mm Hg } P_t) \\ &\quad \times \left(\frac{3.30 \times 10^{-4} \text{ mm Hg } CO_2}{1 \text{ mm Hg } P_t}\right) \\ &= 0.243 \text{ mm Hg } CO_2 \end{aligned}$$

In speaking of trace constituents of the atmosphere it is common to use parts per million, ppm, as the unit of concentration. In dealing with gases, a part per million refers to 1 part by volume in 1 million volume units of the whole. By virtue of the properties of gases, volume fraction and mole fraction are the same. Thus 1 part per million of a trace constituent amounts to 1 mol of that constituent in 1 million moles total of the gas; that is, it is mole fraction times 1×10^6. As an example, we can say from the data in Table 12.1 that CO_2 is present in the earth's atmosphere at a concentration of 330 ppm.

When applied to substances in solution, parts per million refers to mass. That is, it can be expressed in any of the following ways:

$$\text{ppm} = \left(\frac{\text{g solute}}{\text{g soln}}\right) \times 10^6 = \left(\frac{\text{mg solute}}{\text{kg soln}}\right) \simeq \left(\frac{\text{mg solute}}{\text{L soln}}\right) \quad [12.2]$$

The last expression is approximately true for water as solvent, because the density of water is about 1 kg/L.

Before we begin a consideration of the chemical processes that occur in various regions of the atmosphere, let's remind ourselves about some of the most important chemical properties of the two major components of the atmosphere, N_2 and O_2. We saw in Section 8.7 that the N_2 molecule possesses a triple bond between the nitrogen atoms. This very strong bond is responsible for the exceptional lack of reactivity of N_2, which undergoes reaction only under extreme conditions. The O—O bond energy in O_2 is much lower than that for N_2 (Table 8.7), and O_2 is therefore much more reactive than N_2. Oxygen reacts with many substances to form oxides, as described in Section 7.11. The oxides of nonmetals, for example, SO_2, for the most part form acidic solutions on dissolving in water. The oxides of active metals, and other metals in low oxidation states, for example, CaO, form basic solutions on dissolving in water. Many metal oxides, for example, Fe_2O_3 and Al_2O_3, are insoluble in water, but may dissolve in strongly acidic or strongly basic solutions.

12.2 THE OUTER REGIONS

Although the upper reaches of the atmosphere contain only a small fraction of the atmospheric mass, the upper atmosphere plays an important role in determining the conditions of life at the earth's surface. These upper layers form the outer bastion of defense against the hail of radiation and high-energy particles with which the planet is continually bombarded. In absorbing these assaults, the molecules and atoms of the atmosphere undergo chemical change.

Photodissociation

The sun emits radiant energy over a wide range of wavelengths. The shorter-wavelength, higher-energy radiations in the ultraviolet range of the spectrum are sufficiently energetic to cause chemical changes. We have already seen, in Section 5.2, that electromagnetic radiation can be pictured as a stream of photons. The energy of each photon is given by the relationship $E = h\nu$, where h is Planck's constant and ν is the frequency of the radiation. For a chemical change to occur when radiation falls on the earth's atmosphere, two conditions must be met. First, there must be photons with energy at least as large as that required to break a chemical bond, remove an electron, or otherwise accomplish whatever chemical process is being considered. Second, molecules must absorb these photons. This requirement means that the energy of the photons is converted into some other form of energy within the molecule.

One of the most important processes occurring in the upper atmosphere is dissociation of the oxygen molecule as a result of the absorption of a photon (**photodissociation**), as shown in Equation 12.3.

$$O_2(g) + h\nu \longrightarrow 2O(g) \qquad [12.3]$$

The minimum energy required to cause this change is determined by the dissociation energy of O_2, 495 kJ/mol. We calculate the longest-wavelength photon having sufficient energy to dissociate the O_2 molecule in Sample Exercise 12.2.

SAMPLE EXERCISE 12.2

What is the wavelength of a photon corresponding to a molar bond-dissociation energy of 495 kJ/mol?

Solution: We must first calculate the energy required on a per molecule basis and then determine the wavelength of a photon with that energy:

$$\left(495 \times 10^3 \frac{\text{J}}{\text{mol}}\right)\left(\frac{1 \text{ mol}}{6.022 \times 10^{23} \text{ molecules}}\right)$$

$$= 8.22 \times 10^{-19} \frac{\text{J}}{\text{molecule}}$$

The energy of the photon is given by $E = h\nu$. Rearranging, we have

$$\nu = \frac{E}{h} = \left(\frac{8.22 \times 10^{-19} \text{ J}}{6.625 \times 10^{-34} \text{ J-s}}\right)$$

$$= 1.24 \times 10^{15}/\text{s}$$

Recall from Section 5.1 that the product of frequency and wavelength of radiation equals the velocity of light:

$$\nu\lambda = c = 3.00 \times 10^8 \text{ m/s}$$

Therefore, rearranging this equation, we have

$$\lambda = \frac{c}{\nu} = \left(\frac{3.00 \times 10^8 \text{ m/s}}{1.24 \times 10^{15}/\text{s}}\right)\left(\frac{10^9 \text{ nm}}{1 \text{ m}}\right)$$

$$= 242 \text{ nm}$$

The calculations in Sample Exercise 12.2 tell us that any photon of wavelength *shorter* than 242 nm will have sufficient energy to dissociate the O_2 molecule. (Remember that shorter wavelength means higher energy!)

The second condition that must be met before dissociation actually occurs is that the photon must be absorbed by O_2. Fortunately for us, O_2 absorbs much of the high-energy, short-wavelength radiation from the solar spectrum before it reaches the lower atmosphere. As it does so, atomic oxygen (O) is formed. At higher elevations the dissociation of O_2 is very extensive; at 400 km only 1 percent of the oxygen is in the form of O_2; the other 99 percent is in the form of atomic oxygen. At 130 km, O_2 and O are just about equally abundant. Below this elevation O_2 is more abundant than O.

Recall from our discussion of the electronic structures of diatomic molecules that the bond-dissociation energy of N_2 is very high (Table 8.7). Thus only photons of very short wavelength possess sufficient energy to cause dissociation of this molecule. Furthermore, N_2 does not readily absorb photons, even when they do possess sufficient energy. The overall result is that very little atomic nitrogen is formed in the upper atmosphere by dissociation of N_2.

Photoionization

In 1901, Guglielmo Marconi carried out a sensational experiment, by receiving in St. John's, Newfoundland, a radio signal transmitted from Land's End, England, some 2900 km away. Because radio waves were thought to travel in straight lines, it had been assumed that radio communication over large distances on earth would be impossible. The fact that Marconi's experiment was successful suggested that the earth's atmosphere in some way substantially affected radio-wave propagation. His discovery led to intensive study of the upper atmosphere. In about 1924 the existence of electrons in the upper atmosphere was established by experimental studies.

TABLE 12.2 Ionization processes, ionization energies, and maximum wavelength of a photon capable of causing ionization

Process	Ionization energy (kJ/mol)	λ_{max} (nm)
$N_2 + h\nu \longrightarrow N_2^+ + e^-$	1495	80.1
$O_2 + h\nu \longrightarrow O_2^+ + e^-$	1205	99.3
$O + h\nu \longrightarrow O^+ + e^-$	1313	91.2
$NO + h\nu \longrightarrow NO^+ + e^-$	890	134.5

For each electron present in the upper atmosphere, there is a corresponding positively charged ion. The electrons in the upper atmosphere are a result mainly of photoionization of molecules, caused by solar radiation. For photoionization to occur, a photon must be absorbed by the molecule, and this photon must have enough energy to cause removal of the highest-energy electron. Some of the more important ionization processes occurring in the upper atmosphere, that is, above about 90 km, are shown in Table 12.2, together with the ionization energies and λ_{max}, the maximum wavelength of a photon capable of causing ionization. Photons with energies sufficient to cause ionization have wavelengths in the short, or high-energy, region of the ultraviolet. These wavelengths are completely filtered out of the radiation reaching earth as a result of absorption by the upper atmosphere.

12.3 OZONE IN THE UPPER ATMOSPHERE

At an elevation of about 90 km, most of the short-wavelength solar radiation capable of causing ionization has been absorbed. As a result, the concentration of ions and electrons drops off very rapidly at about this elevation. Radiation capable of causing dissociation of the O_2 molecule remains sufficiently intense, however, so that photodissociation of O_2 (Equation 12.3) remains important down to 30 km. The chemical processes that occur in the region below about 90 km following photodissociation of O_2 are very different than processes that occur at higher elevations. In the mesophere and stratosphere the concentration of O_2 is much greater than that of atomic oxygen. Thus, when O atoms are formed in the mesosphere and stratosphere, they undergo frequent collisions with O_2 molecules. These collisions lead to formation of ozone, O_3:

$$O(g) + O_2(g) \longrightarrow O_3^*(g) \qquad [12.4]$$

The asterisk over the O_3 denotes that the ozone molecule contains an excess of energy. Reaction of O with O_2 to form O_3 results in release of 105 kJ/mol. This energy must be gotten rid of by the O_3 molecule in a very short time, or else it will simply fly apart again into O_2 and O. This decomposition is shown in Equation 12.5 as the reverse of the process by which O_3 is formed. The double arrows, $\rightleftharpoons$, indicate the reaction is reversible; that is, that it may occur in either direction. The energy-rich O_3 molecule can get rid of the excess energy by colliding with another atom or molecule and transferring some of the excess energy to it. Let us represent the atom or molecule undergoing the collision as M. (Nearly always M is O_2 or N_2, because these are the most abundant molecules.) The transfer of energy can then be represented as in the second reaction, Equation 12.6:

$$O(g) + O_2(g) \rightleftharpoons O_3^* \qquad [12.5]$$
$$O_3^*(g) + M(g) \longrightarrow O_3(g) + M^*(g) \qquad [12.6]$$
$$O(g) + O_2(g) + M(g) \longrightarrow O_3(g) + M^*(g) \qquad [12.7]$$

The reaction in Equation 12.6 competes with the reverse process in Equation 12.5. When collisions are not very frequent, the reverse reaction in Equation 12.5 wins out; most of the O_3^* molecules formed fall apart into O and O_2 before they undergo a stabilizing collision. However, when the number of collisions per unit time is high, formation of O_3 via Equation 12.7 is favored. Because the number of molecules per unit volume is much greater at lower elevation, the frequency of stabilizing collisions is greater. However, at much lower altitudes, the solar radiation energetic enough to produce dissociation of O_2 becomes largely absorbed. The overall result of the opposing factors is a maximum in the rate of production of ozone at about 50 km altitude. The ozone molecule, once formed, does not stay around very long; ozone itself is capable of absorbing solar radiation, with the result that it is decomposed into O_2 and O. Because the energy required for this process is only 105 kJ/mol, photons of wavelength shorter than 1140 nm are sufficiently energetic. The strongest and most important absorptions are of photons with wavelengths from about 200 to 310 nm. Radiation in this wavelength range is not strongly absorbed by any species other than ozone. If it were not for the layer of ozone in the stratosphere, therefore, these short-wavelength, high-energy photons would penetrate to the earth's surface. Plant and animal life as we know it could not survive in the presence of this high-energy radiation. The "ozone shield" is thus essential for our continued well-being.

The photodecomposition of ozone reverses the reaction leading to its formation. We thus have a cyclic process of ozone formation and decomposition, summarized as follows:

$$O_2(g) + h\nu \longrightarrow O(g) + O(g)$$
$$O(g) + O_2(g) + M(g) \longrightarrow O_3(g) + M^*(g) \qquad \text{(heat released)}$$
$$O_3(g) + h\nu \longrightarrow O_2(g) + O(g)$$
$$O(g) + O(g) + M(g) \longrightarrow O_2(g) + M^*(g) \qquad \text{(heat released)}$$

The overall result of this cycle is that ultraviolet radiation from the sun is converted into heat energy. The ozone cycle in the stratosphere is responsible for the temperature rise that reaches its maximum at the stratopause, as illustrated in Figure 12.1.

The scheme described above for the life and death of ozone molecules accounts for some but not all of the known facts about the ozone layer. Many chemical reactions involving substances other than just oxygen are involved. In addition, the effects of turbulence and winds in mixing up the stratosphere must be considered. A very complicated picture results. It is quite certain, however, that the oxides of nitrogen are important in the ozone cycle.

Nitric oxide, NO, and its close relative nitrogen dioxide, NO_2, are present in the stratosphere in low concentrations. Ozone reacts with NO to form NO_2 and O_2; then NO_2 reacts with atomic oxygen to regenerate NO and form O_2. The NO is then ready again to react with O_3. The overall reaction involving NO is simply:

$$\begin{array}{lll} O_3(g) + NO(g) & \longrightarrow & NO_2(g) + O_2(g) \\ NO_2(g) + O(g) & \longrightarrow & NO(g) + O_2(g) \\ \hline O_3(g) + O(g) & \longrightarrow & 2O_2(g) \end{array} \quad [12.8]$$

We see from this sequence of reactions that NO serves the function of increasing the rate of decomposition of O_3. There is no net change in the chemical state of the NO. We have here a very simple example of a **catalyst,** a substance that has the effect of increasing the rate of a chemical reaction, without itself undergoing a net chemical change.

The overall result of ozone formation and removal reactions, coupled with atmospheric turbulence and other factors, is to produce an ozone profile in the upper atmosphere, as shown in Figure 12.3.

A few years ago it was recognized that the **chlorofluoromethanes,** principally CF_2Cl_2 and $CFCl_3$, may have an adverse effect on the ozone layer. These substances have been widely used as propellent gases in spray cans, and as refrigerant gases. They are quite inert chemically; there seems to be no relatively rapid chemical process that removes chlorofluoromethanes from the lower atmosphere. The lifetimes of these molecules in the atmosphere are thus controlled by the rate at which they diffuse into the stratosphere and become subject to the action of ultraviolet light. The action of high-energy light with wavelengths in the range 190 to 225 nm results in **photolysis,** or light-induced rupture, of a carbon-chlorine bond:

$$CF_xCl_{4-x}(g) + h\nu \longrightarrow CF_xCl_{3-x}(g) + Cl(g) \quad [12.9]$$

There may be further photochemical breakdown of the CF_xCl_{3-x} fragment. Calculations suggest that the rate of chlorine-atom formation will be maximized at an altitude of about 30 km. The atomic chlorine produced by photolysis is capable of rapid reaction with ozone to form chlorine oxide and molecular oxygen. Chlorine oxide is capable of reaction with atomic oxygen to reform atomic chlorine:

$$\begin{array}{llll} Cl(g) + O_3(g) & \longrightarrow & ClO(g) + O_2(g) & [12.10] \\ ClO(g) + O(g) & \longrightarrow & Cl(g) + O_2(g) & [12.11] \\ \hline O_3(g) + O(g) & \longrightarrow & 2O_2(g) & [12.8] \end{array}$$

This pair of reactions is analogous to those involving nitric oxide to produce the net reaction shown in Equation 12.8. In both cases the original species is regenerated. The overall result is reaction of ozone with atomic oxygen to form molecular oxygen. Many uncertainties are involved in any quantitative estimate of how much ozone destruction might be due to chlorofluoromethanes. Because rates of diffusion of molecules into the stratosphere from the earth's surface are likely to be very slow, it may be several decades before the full impact of chlorofluoromethanes is felt. Additional research on this problem is in progress in many laboratories. In the meantime, the use of the chlorofluoromethanes as aerosol propellent gases has been substantially reduced.

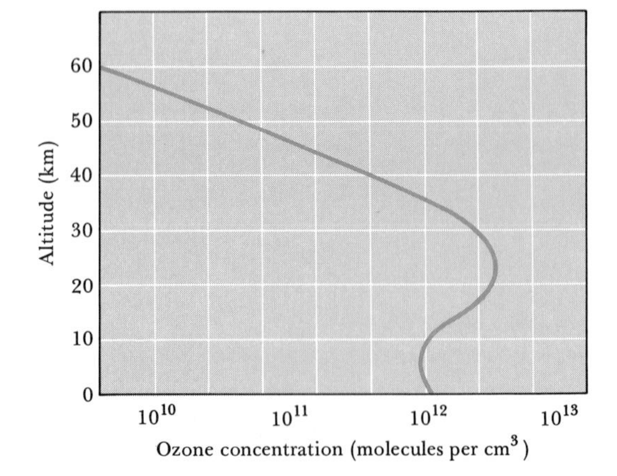

FIGURE 12.3 Variation in ozone concentration in the atmosphere as a function of altitude.

TABLE 12.3 Sources and typical concentrations of some minor atmospheric constituents

Minor constituent	Sources	Typical concentrations
Carbon dioxide (CO_2)	Decomposition of organic matter; release from the oceans; fossil-fuel combustion	330 ppm throughout troposphere
Carbon monoxide (CO)	Decomposition of organic matter; industrial processes; fuel combustion	0.05 ppm in nonpolluted air; 1 to 50 ppm in urban traffic areas
Methane (CH_4)	Decomposition of organic matter; natural-gas seepage	1 to 2 ppm throughout troposphere
Nitric oxide (NO)	Electrical discharges; internal-combustion engines; combustion of organic matter	0.01 ppm in nonpolluted air; 0.2 ppm in smog atmospheres
Ozone (O_3)	Electrical discharges; diffusion from stratosphere; photochemical smog	0 to 0.01 ppm in nonpolluted air; 0.5 ppm in photochemical smog
Sulfur dioxide (SO_2)	Volcanic gases; forest fires; bacterial action; fossil-fuel combustion; industrial processes (roasting of ores, and so on)	0 to 0.01 ppm in nonpolluted air; 0.1 to 2 ppm in polluted urban environment

12.4 CHEMISTRY OF THE TROPOSPHERE

In the preceding sections, we have described the photodissociation of oxygen and ozone. These two processes, and, to a lesser extent, other photodissociations and photoionizations, result in essentially complete absorption of all solar radiation of less than about 300 nm wavelength at the altitude of the tropopause. Because the major constituents of the atmosphere do not interact with radiation of wavelength longer than 300nm, the photochemical reactions that occur in the troposophere are entirely those of minor atmospheric constituents. Many of the minor constituents occur to only a slight extent in the natural environment but exhibit much higher local concentrations in certain areas as a result of human activities. Table 12.3 is a summary of information on several minor atmospheric constituents. Some of these are of interest for their role as air pollutants. We shall discuss here the most important characteristics of a few of these minor constituents and their chemical role as air pollutants.

Sulfur Compounds

Sulfur-containing compounds are present to some extent in the natural, unpolluted atmosphere. They originate in the bacterial decay of organic matter, in volcanic gases, and from other sources listed in Table 12.3. Some scientists think that a certain amount of sulfur dioxide may also originate in the oceans. The concentration of sulfur-containing compounds in the atmosphere as a result of distribution of material from natural sources is very small in comparison with the concentrations that may build up in urban and industrial environments as a result of human activities. Sulfur compounds, chiefly sulfur dioxide, SO_2, are among the most unpleasant and harmful of the common pollutant gases. Table 12.4 shows the concentrations of several pollutant gases in a *typical* urban environment (not one that is particularly affected by smog). According

TABLE 12.4
Concentrations of atmospheric pollutants likely to be exceeded about 50 percent of the time in a typical urban atmosphere

Pollutant	Concentration (ppm)
Carbon monoxide	10
Hydrocarbons	3
Sulfur dioxide	0.08
Nitrogen oxides	0.05
Total oxidants (ozone and others)	0.02

to these data, the level of sulfur dioxide would be 0.08 ppm or higher about half the time. This concentration is considerably lower than that of other pollutants, notably carbon monoxide. Nevertheless, sulfur dioxide is regarded as the most serious health hazard among the pollutants shown, especially for persons with respiratory difficulties. Studies of the medical case histories of large population segments in urban environments have shown clearly that those living in the most heavily polluted parts of cities have higher levels of respiratory disease and shorter life expectancies. One industrial process that may produce very high local levels of SO_2 is the roasting, or smelting, of ores. By this process a metal sulfide is oxidized, driving off SO_2, as in the following example:

$$2ZnS(s) + 3O_2(g) \longrightarrow 2ZnO(s) + 2SO_2(g) \qquad [12.12]$$

Smelting operations account for about 8 percent of the total SO_2 released in the United States. About 80 percent of the SO_2 generated comes from the combustion of coal and oil. The extent to which SO_2 emissions are a problem in the burning of fossil fuels depends on the level of sulfur concentration in the coal or oil. Oil that is burned in the power plants of electrical generating stations is the nonvolatile residue that remains after the low boiling fractions have been distilled off. Some oil, such as that from the Middle East, is relatively low in sulfur, whereas Venezuelan oil is relatively high. Because of concern about SO_2 pollution, low-sulfur oil is in greater demand and is consequently more expensive.

Coals vary considerably in their sulfur content. Much of the coal lying in beds east of the Mississippi is relatively high in sulfur content, up to 6 percent by weight. Much of the coal lying in the western states has a lower sulfur content. (This coal, however, also has a lower heat content per unit weight of coal, so that the difference in sulfur content on the basis of a unit amount of heat produced is not as large as is often assumed.)

Altogether, more than 30 million tons of SO_2 are released into the atmosphere in the United States each year. This material does a great deal of damage to both property and human health. Not all the damage, however, is caused by SO_2 itself; it is likely, in fact, that SO_3, formed by oxidation of SO_2, is the major culprit. Sulfur dioxide may be oxidized to SO_3 by any of several pathways, depending on the particular nature of the atmosphere. Once SO_3 is formed it dissolves in water droplets, forming sulfuric acid, H_2SO_4:

$$SO_3(g) + H_2O(l) \longrightarrow H_2SO_4(aq) \qquad [12.13]$$

The phenomenon of **acid rain** has been known for a long time in the Scandinavian countries and other parts of northern Europe. The dominant contributor to the acidity in the rain is sulfuric acid. The acidity has caused fish populations in many freshwater lakes to decline, and it has measurably affected other parts of the network of interdependent living things within the lakes. There is evidence that acid rain is now affecting many northern lakes in the United States. Acid rain is strongly

FIGURE 12.4 Statue of the River Nile, Rome. The extensive decay of the stonework is due to the presence of SO_2 in the atmosphere. The SO_2 reacts with limestone, $CaCO_3$, and calcium sulfate, $CaSO_4$, is eventually formed, leading to a powdering or blistering of the surface. Wind and weather certainly contribute to the decay of buildings and statuary, but the major culprit is the chemical process that destroys the limestone.

corrosive when it comes in contact with metals, paints, and similar substances. An example of this corrosive effect is shown in Figure 12.4.

Aside from the damage to human health, billions of dollars each year are lost as a result of corrosion resulting from SO_2 pollution. Obviously we all want this noxious gas removed from the environment. But removal of sulfur from coal or oil is difficult and, therefore, expensive. Rather than attempt to remove sulfur from fuel before it is burned, the sulfur dioxide formed when the fuel is combusted may be removed. There are many possible ways of doing this. One way involves blowing powdered limestone, $CaCO_3$, into the combustion chamber. The carbonate (limestone) is decomposed into lime (CaO) and carbon dioxide:

$$CaCO_3(s) \longrightarrow CaO(s) + CO_2(g) \qquad [12.14]$$

The lime then reacts with SO_2 to form calcium sulfite:

$$CaO(s) + SO_2(g) \longrightarrow CaSO_3(s) \qquad [12.15]$$

Only about half the SO_2 is removed by contact with the dry solid. It is necessary to "scrub" the furnace gas with an aqueous suspension of lime to remove the $CaSO_3$ formed, and to remove any unreacted SO_2. This process, which is illustrated in Figure 12.5, is difficult to engineer, reduces the heat effectiveness of the fuel, and leaves an enormous solid waste disposal problem. An electric power plant that would serve the needs of a population of about 150,000 people would produce about 160,000 tons per year of solid waste if it were equipped with the purification system just described. This is three times the normal fly-ash waste from a plant of this size. Various schemes may be employed to recover elemental sulfur or some other industrially useful chemical from the SO_2, but as yet no process has been found sufficiently attractive from an

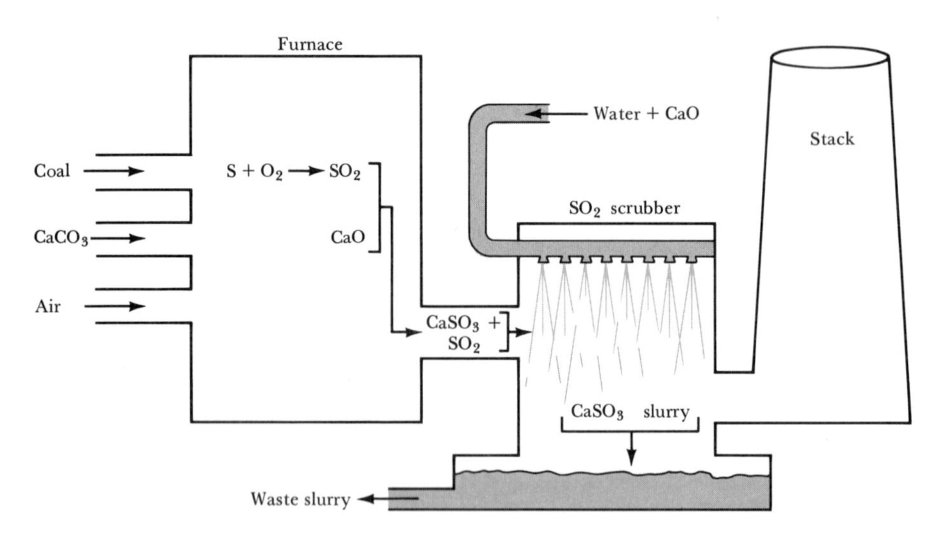

FIGURE 12.5 Common method for removing SO_2 from combusted fuel. Powdered limestone decomposes into CaO, which reacts with SO_2 to form $CaSO_3$. The $CaSO_3$ and any unreacted SO_2 enter a purification chamber where a shower of CaO and water precipitates the $CaSO_3$ into a watery residue called slurry and converts the remaining SO_2 to $CaSO_3$.

economic point of view to warrant large-scale development. Pollution by sulfur dioxide remains a major problem and will probably continue to remain so for some time.

Nitrogen Oxides; Photochemical Smog

The atmospheric chemistry of the nitrogen oxides is interesting because these substances are components of smog, a phenomenon with which city dwellers are all too familiar. The term smog refers to a particularly unpleasant condition of pollution in certain urban environments, which occurs when weather conditions produce a relatively stagnant air mass. The smog made famous by Los Angeles, but now common in many other urban areas as well, is more accurately described as a photochemical smog, because photochemical processes play an essential role in its formation.

Nitric oxide, NO, is formed in small quantities in the cylinders of internal-combustion engines via the direct combination of nitrogen with oxygen. Prior to installation of control measures, typical emission levels of NO_x were 4 g/mi. (The x is either 1 or 2; both NO and NO_2 are formed, though NO predominates.) The most recent auto-emission standards call for NO_x emission levels of less than 1 g/mi.

Nitric oxide is oxidized very slowly in air. A certain amount of NO_2 is present, however; this is formed either directly in the automobile engine or by slow oxidation of NO. The dissociation of NO_2 into NO and O requires 304 kJ/mol. This requirement corresponds to a photon wavelength of 393 nm. In sunlight, NO_2 undergoes dissociation to NO and O:

$$NO_2(g) + h\nu \longrightarrow NO(g) + O(g) \qquad [12.16]$$

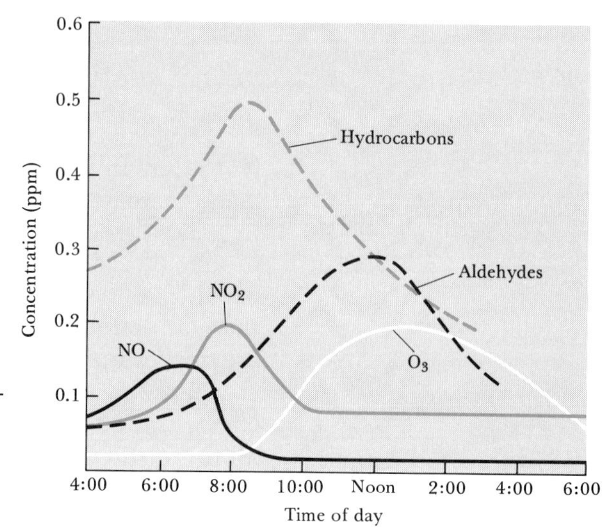

FIGURE 12.6 Concentration of smog components as a function of time of day. (*After P. A. Leighton, "Photochemistry of Air Pollution" in* Physical Chemistry: A Series of Monographs, *ed. Eric Hutchinson and P. Van Rysselberghe, vol. IX. New York: Academic Press, 1961*)

The atomic oxygen formed undergoes several possible reactions. One of these is with the abundant molecular oxygen to form ozone, as described earlier:

$$O(g) + O_2(g) + M(g) \longrightarrow O_3(g) + M^*(g) \qquad [12.7]$$

Ozone is capable of rapidly oxidizing NO to NO_2:

$$O_3(g) + NO(g) \longrightarrow NO_2(g) + O_2(g) \qquad [12.17]$$

To see the significance of these reactions for smog formation, we must look at Figure 12.6, which shows the time dependence of the concentrations of various smog components. For the moment look at just the curves for the nitrogen oxides. In the early morning hours, the NO_2 concentration is low. As auto traffic builds up, and NO is formed, oxidation by ozone (Equation 12.17) takes over. Note that the ozone level does not noticeably increase during this period.

TABLE 12.5 Typical concentrations of trace pollutants during a photochemical smog (levels are subject to wide variation from one situation to another)

Constituent	Concentration (ppm)
NO_x	0.2
NH_3	0.02
CO	40
O_3	0.5
CH_4	2
C_2H_4	0.5
Higher olefins[a]	0.25
C_2H_2 (acetylene)	0.25
Aldehydes	0.6
SO_2	0.2

[a] Higher olefins are compounds that contain a hydrocarbon chain attached to one of the carbons of the C=C bond.

In addition to nitrogen oxides and carbon monoxide, an automobile engine also emits as pollutants unburned and partially burned hydrocarbons, compounds made up entirely of carbon and hydrogen. A typical engine without emission controls emits about 10 to 15 g of such organic compounds per mile. The newest standards require that hydrocarbon emissions be less than 0.4 g/mi. Table 12.5 shows typical concentrations of trace constituents in photochemical smog. The most important organic compounds in this list for smog formation are olefins and aldehydes. An **olefin** is an organic compound, a type of hydrocarbon, containing a double bond between carbon atoms. Ethylene, C_2H_4 (Sample Exercise 8.1), is the simplest member of the series. **Aldehydes** are compounds containing a carbon-oxygen double bond on a carbon atom at the end of a hydrocarbon chain. Formaldehyde, acetaldehyde, and propionaldehyde are examples:

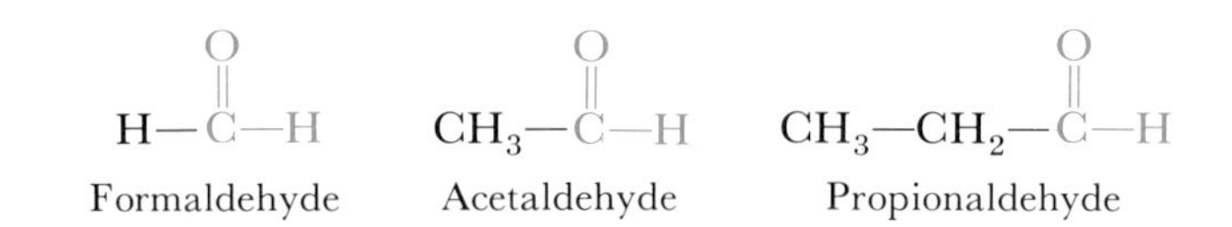

Note that in Figure 12.6 the hydrocarbon levels decrease as aldehyde levels increase. This occurs because some of the atomic oxygen produced in photodissociation of NO_2 reacts with organic compounds, eventually producing aldehydes through a complex series of reactions. Ozone is also capable of reaction with olefins to eventually yield aldehydes. Many of the compounds formed by reaction of atomic oxygen and ozone with organic compounds are **free radicals,** molecular fragments that contain an unpaired electron. They are very reactive and lead to a complex chemistry in the polluted atmosphere. One group of molecules formed in all this is the peroxyacylnitrates (PAN), especially unpleasant substances that cause eye irritation and breathing difficulties:

$$\mathrm{R{-}\overset{\overset{\displaystyle O}{\|}}{C}{-}O{-}O{-}NO_2}$$

In this diagram, R represents an organic group such as CH_3, C_6H_5, and so on.

Reduction or elimination of smog requires that the essential ingredients for its formation be removed from automobile exhaust. The stricter emission standards in effect since 1975 are designed to reduce drastically the levels of two of the major ingredients of smog: NO_x and hydrocarbons. Whether the means for meeting these standards now in use will be successful remains to be seen. Emission-control systems are not notably successful in poorly maintained autos.

Carbon Monoxide

In terms of total mass, carbon monoxide is the most abundant of all the pollutant gases. The level of CO present in fresh, nonpolluted air is small, probably on the order of 0.05 to 0.1 ppm. The estimated total amount of CO in the earth's atmosphere is about 5.2×10^{14} g. In the United States alone, however, about 1×10^{14} g of CO is produced each year. The CO is formed mostly in the incomplete combustion of fossil fuels. The major sources of CO in the United States are automobile and power-plant emissions.

Carbon monoxide is a relatively unreactive molecule, and it might be supposed that it would not be a health hazard. It does have the unusual ability, however, of binding very strongly to **hemoglobin,** the iron-containing protein that is responsible for oxygen transport in the blood. Hemoglobin consists of four protein chains loosely held together in a cluster. Each chain has within its folds a heme molecule. The structure of heme is shown in Figure 12.7. The important characteristics of heme are that the iron is situated in the center of a plane of coordinating nitrogen atoms. An oxygen molecule reacts with the iron atom to form a species called **oxyhemoglobin.** Under appropriate conditions the oxygen is released from the iron. The equilibrium between hemoglobin and oxyhemoglobin may be illustrated graphically as shown in Figure 12.8. Oxy-

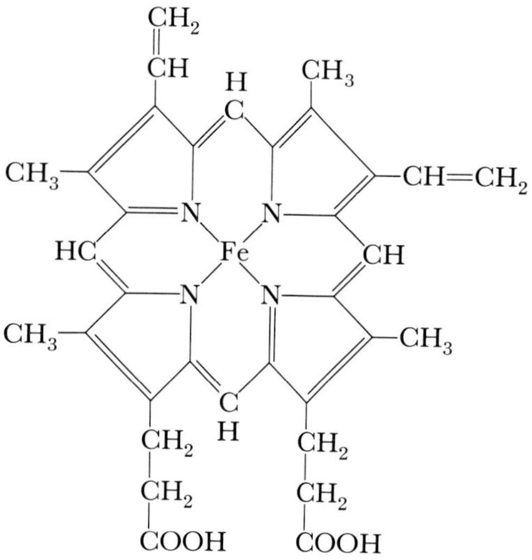

FIGURE 12.7 Structure of the heme molecule.

gen is picked up by hemoglobin in the lungs, and released in the tissues where it is needed for cell metabolism, that is, for the chemical processes occurring in the cell.

Carbon monoxide also happens to bind very strongly to the iron in hemoglobin. The complex is called carboxyhemoglobin and is represented as COHb. The affinity of human hemoglobin for CO is about 210 times greater than for O_2. As a result, a relatively small quantity of CO can inactivate a substantial fraction of the hemoglobin in the blood for oxygen transport. For example, a person breathing air that contains only 0.1 percent of CO takes in enough CO after a few hours of breathing to convert up to 60 percent of the hemoglobin into COHb, thus reducing the blood's normal oxygen-carrying capacity by 60 percent.

Under normal conditions, a nonsmoker breathing unpolluted air has about 0.3 to 0.5 percent carboxyhemoglobin, COHb, in the bloodstream. This small amount arises mainly from the production of small amounts of CO in the course of normal body chemistry, and from the small amount of CO present in clean air. Exposure to higher concentrations of CO causes the COHb level to increase. This doesn't happen instantly, but requires several hours. Similarly, when the CO level is suddenly

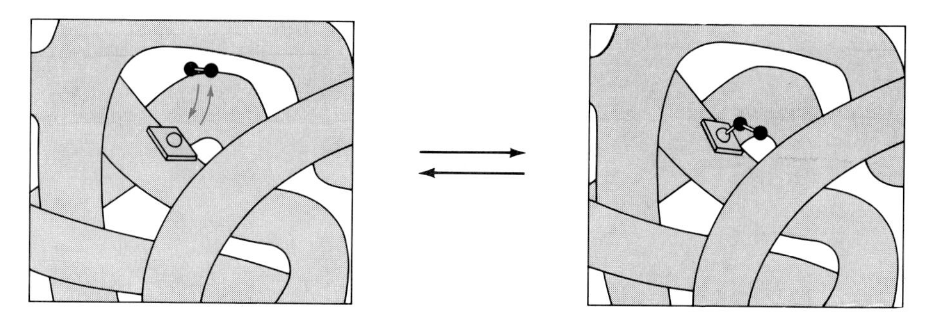

FIGURE 12.8 Equilibrium between hemoglobin and oxyhemoglobin. The gray regions represent the protein chain. The heme molecule is attached to the protein at a particular point.

TABLE 12.6 Carboxyhemoglobin (COHb) percentages in the blood of persons under various conditions

	COHb (%)
Continuous exposure, 10 ppm CO	2.0
Continuous exposure, 30 ppm CO	5.0
Nonsmokers, Chicago (1970)	2.0
Smokers, Chicago (1970)	5.8
Nonsmokers, Milwaukee (1969–1971)	1.1
Smokers, Milwaukee (1969–1971)	5.0

decreased, it requires several hours for the COHb concentration to level off at a lower value. Table 12.6 shows the percentages of COHb in blood that are typical of various groups of people.

The CO concentration in city traffic often reaches 50 ppm and may go as high as 140 ppm in traffic jams. The most serious source of carbon monoxide poisoning, however, comes from cigarette smoking. The inhaled smoke from cigarettes contains about 400 ppm of CO. The effect of smoking on COHb percentage is evident from the data in Table 12.6. A study of a group of San Francisco dockworkers presented further proof of the dramatic relationship between smoking and COHb percentage. Nonsmokers in the group averaged 1.3 percent COHb; light smokers (less than half a pack per day) averaged 3.0 percent; moderate smokers averaged 4.7 percent; and heavy smokers (two packs or more per day) averaged 6.2 percent COHb.

There is widespread evidence that chronic exposure to CO impairs performance on standardized tests. Thus it is most definitely not a good idea to smoke heavily before and during a test. In addition, motor performance is also impaired by high COHb percentages. For example, there is evidence that drivers responsible for traffic accidents have, on the average, higher than normal percentages of COHb in their blood. As has been mentioned, a chronically high level of COHb means that a certain fraction of the hemoglobin in the blood is not available for oxygen transport. This in turn means that the heart must work that much harder to ensure an adequate supply of oxygen. It is not surprising, therefore, that many medical researchers believe that chronic exposure to CO is a contributing factor in heart disease and in heart attacks.

Water Vapor, Carbon Dioxide, and Climate

We have seen how the atmosphere makes life as we know it possible on earth by screening out harmful short-wavelength radiation. In addition, the atmosphere is essential in maintaining a reasonably uniform and moderate temperature on the surface of the planet. The two atmospheric components of major importance in maintenance of the earth's surface temperature are carbon dioxide and water.

The earth is in overall thermal balance with its surroundings. This means that the planet radiates energy into space at a rate equal to the rate at which it absorbs energy from the sun. The sun is a very hot body with a temperature of about 6000 K. As seen from outer space the earth is relatively cold, with a temperature of about 254 K. The distribution of wavelengths in the radiation emitted from an object is determined by its

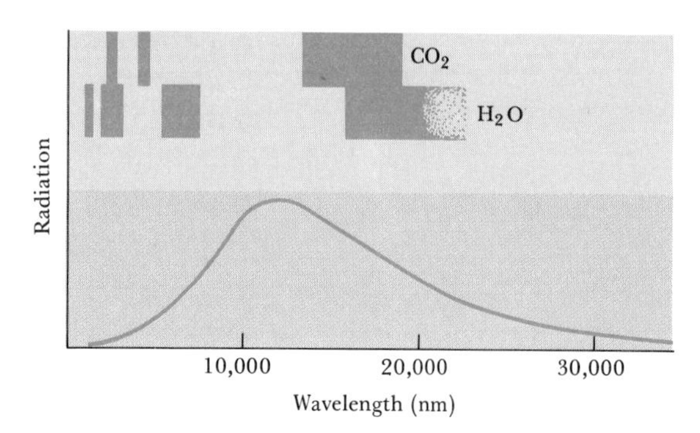

FIGURE 12.9 Long-wavelength radiation from earth compared with the absorption of infrared radiation by carbon dioxide and water.

temperature. The radiation emitted by relatively cold objects is in the low-energy, or long-wavelength, region of the spectrum. This means that the maximum in the wavelength of radiation from the earth is in the far-infrared region, around 12,000 nm (Figure 5.3). But although the troposphere is transparent to visible light, it is not at all transparent to infrared radiation. Figure 12.9 shows the distribution of radiation from the earth's surface and, on the same scale, the wavelengths absorbed by water vapor and carbon dioxide. Clearly, these atmospheric gases absorb much of the outgoing radiation from the earth's surface. It is indeed fortunate for us that they do so; they serve to maintain a livably uniform temperature at the surface by holding in, as it were, the infrared radiation from the surface.

The partial pressure of water vapor in the atmosphere varies greatly from place to place and time to time, but, in general, it is highest near the surface and drops off very sharply with increased elevation. Carbon dioxide, by contrast, is uniformly distributed throughout the atmosphere, at a concentration of about 330 ppm. Because water vapor absorbs infrared radiation so strongly, it plays the major role in maintaining the atmospheric temperature at night, when the surface is emitting radiation into space and not receiving energy from the sun. In very dry desert climates, where the water-vapor concentration is unusually low, it may be extremely hot during the day but very cold at night. In the absence of an extensive layer of water vapor to absorb and then radiate back part of the infrared radiation, the surface loses this radiation into space and cools off very rapidly.

Carbon dioxide plays a secondary, but very important, role in maintaining the surface temperature. The worldwide combustion of fossil fuels, principally coal and oil, on a prodigious scale in the modern era has materially increased the carbon dioxide level of the atmosphere. From measurements carried out over a period of time it is clear that the CO_2 concentration in the atmosphere is steadily increasing. From a knowledge of the infrared-absorbing characteristics of CO_2 and water, and using a theoretical model for the atmosphere, it has been estimated that if the CO_2 level were to double from its present level, the average surface temperature of the planet would increase 3°C. On the basis of present and expected future rates of fossil-fuel use, the atmospheric CO_2 level is expected to just about double by the year 2050. If the calculated effect of a doubling of CO_2 level on surface temperature is correct, this means that the earth's temperature will be 3°C higher within 70 years. Such a small change may seem insignificant, but it is not. Major changes in global climate could result from a temperature change of this or even smaller magnitude. Because so many factors go into determining climate, it is not possible to predict with certainty precisely what changes will occur. It is clear, however, that humanity has acquired the potential, by changing the CO_2 concentration in the atmosphere, for substantially altering the climate of the planet.

12.5 THE WORLD OCEAN

Before we begin our examination of water as a major component of the environment, we should review some of its most important characteristics as a solvent and reactant molecule. We have learned that water possesses many exceptional properties because of extensive hydrogen bonding (Section 10.4). Hydrogen bonding is responsible for the high melting and boiling points of water, and for its high heat capacity. The highly polar character of water is responsible for its exceptional ability to dissolve a wide range of ionic substances (Section 11.6). Water reacts with many substances such as oxides (Section 7.11) to form acidic or basic solutions. It is also the medium in which important types of reactions such as neutralization and precipitation occur (Section 3.3). We will see that all of these key characteristics of water come into play as we consider natural waters, beginning with the world ocean.

All of the vast layer of salty water that covers so much of the earth is connected and is of more or less constant composition. For this reason, oceanographers (scientists whose major interest is the sea) speak in terms of a world ocean, rather than of the separate oceans we learn about in geography books. The world ocean is indeed huge. Its volume is 1.35 billion cubic kilometers. It covers about 72 percent of the earth's surface. Almost all the water on earth, 97.2 percent, is in the world ocean. About 2.1 percent is in the form of ice caps and glaciers. All of the fresh water, in lakes, rivers, and ground water, amounts to only 0.6 percent. The remaining 0.1 percent is in the form of brine wells and brackish (salty) waters.

Chemical Composition of Seawater

Seawater is often referred to as saline water. The **salinity** of seawater is defined as the mass in grams of dry salts present in 1 kg of seawater. In the world ocean, the salinity varies from 33 to 37, with an average of about 35. To put it another way, this means that seawater contains about 3.5 percent dissolved salts. The list of elements present in seawater is very long. However, most are present only in very low concentrations. Table 12.7 lists the 11 ionic species present in seawater at concentrations greater than 0.001 g/kg, or 1 part per million (ppm) by weight. (To convert from g/kg to ppm, multiply by 1000; see Equation 12.2.) In a lower range of concentration, the elements nitrogen, lithium, rubidium, phosphorus, iodine, iron, zinc, and molybdenum are present in amounts ranging from 1 to 0.01 ppm. At least 50 other elements have been identified at still lower concentrations.

The sea is a vast storehouse of chemicals. Each cubic mile of seawater contains 1.5×10^{11} kg of dissolved solids. The sea is so vast that if a

TABLE 12.7 Ionic constituents of seawater present in concentrations greater than 0.001 g/kg (1 ppm) by weight

Ionic constituent	g/kg seawater	Concentration (*M*)
Chloride, Cl^-	19.35	0.55
Sodium, Na^+	10.76	0.47
Sulfate, SO_4^{2-}	2.71	0.028
Magnesium, Mg^{2+}	1.29	0.054
Calcium, Ca^{2+}	0.412	0.010
Potassium, K^+	0.40	0.010
Carbon dioxide[a]	0.106	2.3×10^{-3}
Bromide, Br^-	0.067	8.3×10^{-4}
Boric acid, H_3BO_3	0.027	4.3×10^{-4}
Strontium, Sr^{2+}	0.0079	9.1×10^{-5}
Fluoride, F^-	0.001	7×10^{-5}

[a] CO_2 is present in seawater as HCO_3^- and CO_3^{2-}

substance is present in seawater to only the extent of 1 part per billion by weight, there are still 5×10^9 kg of it in the world ocean. Nevertheless, the ocean is not used very much as a source of raw materials, because the costs of extracting the desired substances from the water are too high. Only three substances are recovered from seawater in commercially important amounts: sodium chloride, bromine, and magnesium.

Biochemical Processes in the Sea

Plants and animals present in the ocean exert an important influence on the composition of seawater. The simplest elements of the food chain are the phytoplankton, minute plants in which CO_2, water, and other nutrients are converted by photosynthesis into plant organic matter. Analysis of the composition of phytoplankton shows that carbon, nitrogen, and phosphorus are present in atomic ratios of 108:16:1, as illustrated in Figure 12.10. Thus about 108 molecules of CO_2 are required for each 16 atoms of nitrogen (usually in the form of nitrate ion) and for each single atom of phosphorus (usually present as hydrogen phosphate ion, HPO_4^{2-}). Because of its high solubility in seawater, CO_2 is always present in excess. The concentration of nitrogen or phosphorus is therefore the limiting factor in the rate of formation of organic matter via photosynthesis.

Photosynthesis occurs in the photosynthetic zone near the surface, where the sun's rays are strongest. Thus in the water extending to a depth of about 150 m, phosphate and nitrate concentrations are depleted because of the photosynthesis that has already occurred. At lower depths, dead plant and animal matter decomposes, restoring the phosphate and nitrate levels.

In the photosynthetic zone, the oxygen concentration is high because oxygen is released during photosynthesis. At lower depths, the oxygen level drops sharply because the oxygen is used up in oxidizing dead plant and animal matter. The oxygen concentration is at a minimum at about

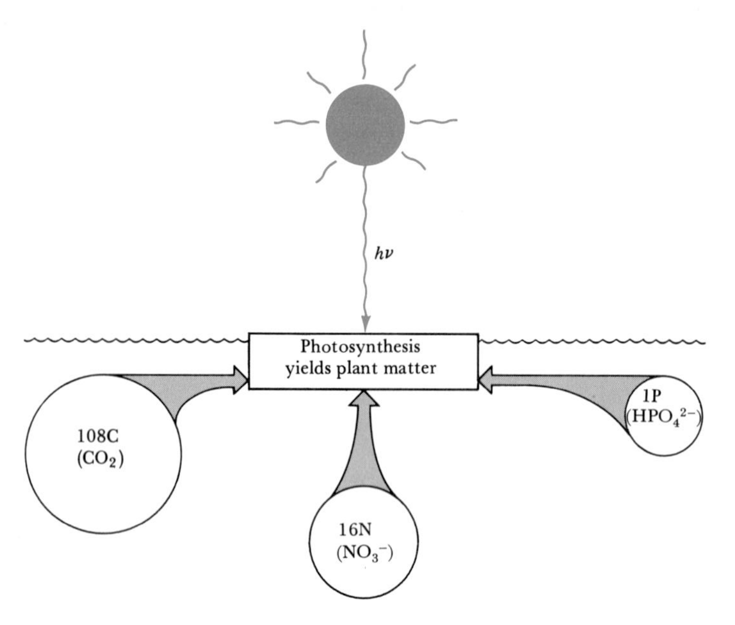

FIGURE 12.10 Schematic diagram of photosynthesis in phytoplankton at or near the surface of the sea. The relative amounts of required nutrients are indicated in the circles. The chemical forms shown are not the only ones usable, but are simply the more abundant ones.

1 km depth, the same region in which the phosphate level is restored to its highest values.

All the higher forms of life in the ocean are ultimately related to the phytoplankton at the base of the food chain. Thus the abundance of fish is coupled very closely to the conditions that determine the rate of photosynthesis. In certain regions of the ocean, where seasonal upwellings of nutrient-rich lower water occur, the rate of photosynthesis may become very high. For example, the Grand Banks in the northern Atlantic, off the coast of Newfoundland, are one of the richest fishing regions in the world ocean. The abundance of fish there is due to an upwelling of nutrient-rich deep water.

Desalination

Because of its high salt content, seawater is unfit for human consumption and indeed for most of the uses to which we put water. In the United States, the salt content of municipal water supplies is restricted by health codes to no more than about 0.05 percent. This is much lower than the 3.5 percent dissolved salts present in seawater, or the 0.5 percent or so present in brackish water found underground in some regions. The removal of salts from seawater or brackish water to the extent that the water becomes usable is called **desalination.**

There are many different ways to desalinate water, and any one of them can be made the basis of a large production plant. The challenge is to carry out the desalination with the absolute minimum energy requirement and with the least possible investment in equipment and facilities. This is important because a nation or region that must rely on desalinated water to a large extent must compete economically with other nations that may have more abundant and cheaper sources of fresh water. A small nation such as Kuwait, situated on the Persian Gulf, which has almost no freshwater resources, can afford to depend to a large extent on desalinated water only because it derives a very high per capita income from its oil production.

Water can be separated from dissolved salts by *distillation,* as described in Section 2.3. This process takes advantage of the fact that water is a volatile substance, whereas the salts are nonvolatile. The principle of distillation is simple enough, but there are many problems associated with carrying out the process on a large scale. For example, as water is distilled from a vessel containing seawater, the salts become more and more concentrated and eventually precipitate out. This causes formation of scale, which in turn causes poor heat transfer through the vessel, plugging of pipes, and so forth. The obvious solution is to discard the seawater after a certain amount of water has been distilled from it and begin again with a new batch. But unless this is done very carefully, all the heat values stored in the hot seawater will be lost, and additional heat must be then supplied to heat up the cold, incoming seawater. The heat lost represents wasted energy and higher costs. In addition, if the distillation is carried out at atmospheric pressure, the water must be heated to 100°C; at lower pressures, the boiling point of water would be lowered, and less heat would be required.

One rather successful attempt to get around some of these difficulties is called the **multistage flash-distillation** process, which is illustrated in Figure 12.11. To see how this method works, let's begin at *A* with warm seawater, which we'll call brine. The brine is pumped under pressure through the coils of a condenser in *B*, then into *C*, and then into *D*, growing hotter in each chamber. The heat comes from condensing steam that forms on the coils in each chamber. The condensed steam, which is fresh water, is collected and pumped off. In region *E* the heated brine is heated still further by steam that passes around the coils. The steam used at this point represents a large fraction of the total energy input to the system. From *E* the heated brine passes into chamber *D*, which is held at a lower pressure. Because the pressure is lower in this chamber, some of the brine is "flashed off," or distilled to form water vapor. We know that energy is required to evaporate water. As water evaporates from our

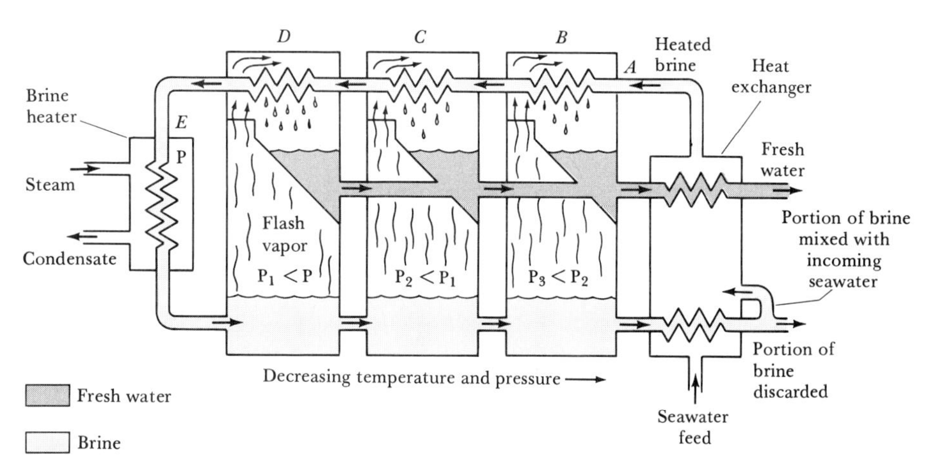

FIGURE 12.11 Schematic diagram of a multistage flash-distillation unit for desalination.

FIGURE 12.12 Multistage flash-distillation unit. This unit is capable of producing about 9 million liters of fresh water each day. (*Courtesy Aqua-Chem, Inc., Milwaukee, Wisconsin*)

bodies the surface that remains is cooled. In the same way, the brine that remains after some water has evaporated, or flashed off, is cooler. It then passes into chamber *C*, in which the pressure is a little lower still. A bit more of the water evaporates, and the brine is cooled still further. In each succeeding stage, the brine becomes more concentrated in salts and lower in temperature. At the end, a portion of the brine, which is now about 7 percent salt by weight, is mixed with incoming seawater. The other portion is discarded to prevent the salt concentration from getting too high.

A large multistage flash-distillation unit is shown in Figure 12.12. This unit is designed to produce about 9 million liters of fresh water per day.

12.6 FRESH WATER

We've seen that the total amount of fresh water on earth is not a large fraction of the total water present. What there is of it can be traced to evaporation of water from the oceans and the land, and to transpiration through the leaves of plants. The water vapor that accumulates in the atmosphere is transported via global atmospheric circulation to other latitudes, where it falls as rain or snow. The water that falls on land runs off in rivers or collects in lakes or underground caverns. Eventually, it is evaporated or carried via streams and rivers back to the oceans.

BOD and Water Quality

Fresh water, of course, isn't all water. It contains dissolved gases (principally O_2, N_2, and CO_2), a variety of cations (mainly Na^+, K^+, Mg^{2+}, Ca^{2+}, and Fe^{2+}) and a variety of anions (mainly Cl^-, SO_4^{2-}, and HCO_3^-). Suspended solids such as tiny clay particles are also likely to be present.

The amount of dissolved oxygen present is an important indicator of the quality of the water. Oxygen is necessary for fish and much other aquatic life. Cold-water fish require about 5 ppm of dissolved oxygen for survival. Water fully saturated with air at 1 atm and 20°C contains about 9 ppm of O_2. Aerobic bacteria consume dissolved oxygen as they utilize it to oxidize organic materials as food to meet their energy requirements. The organic material that the bacteria are able to oxidize is said to be **biodegradable.** This oxidation occurs by a complex set of chemical reactions, and the organic material disappears gradually. The carbon, hydrogen, oxygen, nitrogen, sulfur, and phosphorus in the biodegradable material ends up mainly as CO_2, H_2O, NO_3^-, SO_4^{2-}, and phosphates. These oxidation reactions sometimes reduce the amount of dissolved oxygen to the point where the aerobic bacteria can no longer survive. Anaerobic bacteria then take over the decomposition process, forming products such as CH_4, NH_3, H_2S, and PH_3 that contribute to the offensive odors of some polluted waters.

The amount of oxygen required to decompose all of the biodegradable organic wastes in water is called the **biological oxygen demand (BOD).** The BOD indicates the organic pollution load of the water. The standard test for such organic material is the 5-day BOD test. In this test the contaminated water is diluted in air-saturated distilled water to ensure an excess of oxygen. The amount of dissolved oxygen in the resultant solution is measured. The solution is then stored for 5 days at 20°C, after which time the amount of dissolved oxygen is again measured. The 5-day BOD, BOD_5, is calculated from the amount of dissolved oxygen consumed. The 5-day BOD is usually about three-fourths of the total BOD of the water. Drinkable water normally has a BOD_5 of 1.5 ppm O_2 or less. Raw, untreated sewage usually has a BOD_5 range of 100 to 400 ppm O_2.

Treatment of Municipal Water Supplies

The water needed for domestic uses, for agriculture, or for industrial processes is taken from naturally occurring lakes, rivers, and underground sources or from reservoirs. Much of the water that finds its way into municipal water systems is "used" water; it has already passed through one or more sewage systems or industrial plants. Consequently, it is usually necessary to treat the water before it is distributed to our faucets. Municipal water treatment usually involves five steps: coarse filtration, sedimentation, sand filtration, aeration, and sterilization. Figure 12.13 shows a typical treatment process.

After coarse filtration through a screen, the water is allowed to stand in large settling tanks in which finely divided sand and other minute particles can settle out. To aid removal of very small particles, the water

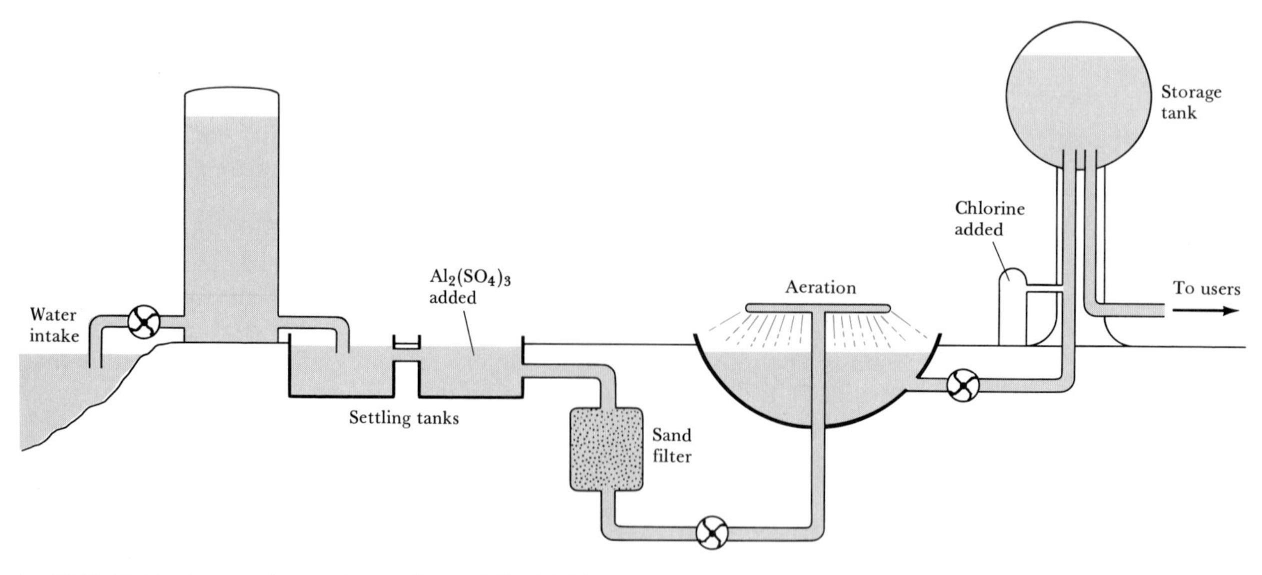

FIGURE 12.13 Common steps in treating public drinking water.

may first be made slightly basic by adding CaO, and then $Al_2(SO_4)_3$ is added. The aluminum sulfate reacts with OH^- ions to form a spongy, gelatinous precipitate of $Al(OH)_3$. This precipitate settles slowly, carrying suspended particles down with it, thereby removing nearly all finely divided matter and most bacteria. The water is then filtered through a sand bed. Following filtration the water may be sprayed into the air to hasten the oxidation of dissolved organic substances.

The final stage of the operation normally involves treating the water with a chemical agent to ensure the destruction of bacteria. Ozone, O_3, is most effective, but it must be generated at the place where it is used. Chlorine, Cl_2, is therefore more convenient to use. Chlorine can be shipped in tanks as the liquefied gas and dispensed from the tanks through a metering device directly into the water supply. The amount used depends on the presence of other substances with which the chlorine might react and on the concentrations of bacteria and viruses to be removed. The sterilizing action of chlorine is probably due not to Cl_2 itself, but to hypochlorous acid, which forms when chlorine reacts with water:

$$Cl_2(aq) + H_2O(l) \longrightarrow HClO(aq) + H^+(aq) + Cl^-(aq) \qquad [12.18]$$

Water Softening

The water treatment described thus far should remove all the substances potentially harmful to health. Sometimes additional treatment is used to reduce the concentrations of Ca^{2+} and Mg^{2+}, which are responsible for water hardness. These ions react with soaps to form an insoluble material. Although they do not form precipitates with detergents, they adversely affect the performance of such cleaning agents. In addition, mineral deposits may form when water containing these ions is heated. When water containing Ca^{2+} and bicarbonate ions is heated, some carbon dioxide is driven off. As a result, the solution becomes less acidic, and insoluble calcium carbonate forms:

$$Ca^{2+}(aq) + 2HCO_3^-(aq) \xrightarrow{\text{heat}} CaCO_3(s) + CO_2(g) + H_2O(l) \qquad [12.19]$$

The solid $CaCO_3$ coats the surfaces of hot-water systems and the insides of teakettles, thereby reducing heating efficiency. Deposits of scale can be especially serious in boilers in which water is heated under pressure in pipes running through a furnace. Formation of scale reduces the efficiency of heat transfer and may result in melting of the pipes.

Not all municipal water supplies require water softening. In those that do, the water is generally taken from underground sources in which the water has had considerable contact with limestone ($CaCO_3$) and other minerals containing Ca^{2+}, Mg^{2+}, and Fe^{2+}. The **lime-soda process** is used for large-scale municipal water-softening operations. The water is treated with "lime," CaO [or "quicklime," $Ca(OH)_2$], and "soda ash," Na_2CO_3. These chemicals cause precipitation of calcium as $CaCO_3$ and of magnesium as $Mg(OH)_2$. The role of Na_2CO_3 is to provide a source of CO_3^{2-}, if needed. If the water already contains a high concentration of bicarbonate ion, calcium can be removed as $CaCO_3$ simply by addition of $Ca(OH)_2$:

$$Ca^{2+}(aq) + 2HCO_3^-(aq) + [Ca^{2+}(aq) + 2OH^-(aq)] \longrightarrow 2CaCO_3(s) + 2H_2O(l) \qquad [12.20]$$

Lime is used only to the extent that bicarbonate is present: 1 mol of $Ca(OH)_2$ for each 2 mol of HCO_3^-. When bicarbonate is not present, addition of Na_2CO_3 causes removal of Ca^{2+} as $CaCO_3$. The carbonate ion also serves to cause precipitation of $Mg(OH)_2$:

$$Mg^{2+}(aq) + 2CO_3^{2-}(aq) + 2H_2O(l) \longrightarrow 2HCO_3^-(aq) + Mg(OH)_2(s) \qquad [12.21]$$

There are two problems with the lime-soda process. First, formation of $CaCO_3$ and $Mg(OH)_2$ may take a long time, and they may not settle out very well. Second, the solution that results after removal of the precipitates is too strongly basic. This is so because the chemicals added, $Ca(OH)_2$ and Na_2CO_3, are both bases. (We will discuss the reason that Na_2CO_3 is a base in Chapter 15.) Alum, $Al_2(SO_4)_3$, is added to remove the precipitates. Because the solution is basic, the Al^{3+} ion forms $Al(OH)_3(s)$, a gelatinous precipitate that carries finely divided solids with it out of solution. To prevent any further precipitation of $Mg(OH)_2$ or $CaCO_3$ by the ions remaining, the solution is made less basic by bubbling CO_2 through the water.

Sewage Treatment

Municipal sewage treatment is generally divided into three stages referred to as primary, secondary, and tertiary treatments. About 10 percent of sewage handled by public sewers receives no treatment, about 30 percent receives only primary treatment, while about 60 percent is subject to secondary treatment as well. Tertiary treatment is presently rare, but it is expected to become more common as communities upgrade their treatment facilities to meet federal water-pollution standards.

Primary treatment consists first of screening the incoming sewage to filter out debris and larger suspended solids. Then the sewage is passed into settling or sedimentation tanks where suspended solids, called sludge, settle out. If the water receives no secondary treatment it is then often treated with chlorine before being dumped back into the natural water system. Primary treatment removes about 60 percent of the suspended solids and 35 percent of the BOD.

Secondary treatment is based on aerobic decomposition of organic material. The most common type of secondary treatment is known as the activated-sludge method. In this method the waste from primary treatment is passed into an aeration tank, where air is blown through it as shown in Figure 12.14. This aeration results in rapid growth of aerobic bacteria that feed on the organic wastes in the water. The bacteria form a mass called activated sludge. This sludge settles out in sedimentation tanks and the liquid effluent is discharged, often after chlorination. Most of the activated sludge is returned to the aeration tank, where it aids in decomposing the organic wastes in the incoming water. After secondary treatment, about 90 percent of suspended solids and 90 percent of BOD has been removed.

Water that has received only primary and secondary treatment may contain relatively large quantities of phosphorus and nitrogen. These can cause damage to natural waters by promoting excessive growth of algae. In addition, many chemicals present in sewage are not affected by secondary treatment and simply pass through and are released to the environment. The cost of removing the many metals and organic substances that might be present in wastewater is high. As a result, very little wastewater receives a general tertiary treatment, in which such contaminants are removed.

The substances that might contaminate waste water when it is returned to the environment are derived from all the many substances flushed down toilets, ground up in garbage disposal units, and rinsed down the drains in hospitals, stores, factories, and laboratories. In addition to these effluents, natural waters receive the water from storm drains and the runoff from cattle feedlots and from farmland dosed with fertilizers, insecticides, and weed-killing chemicals.

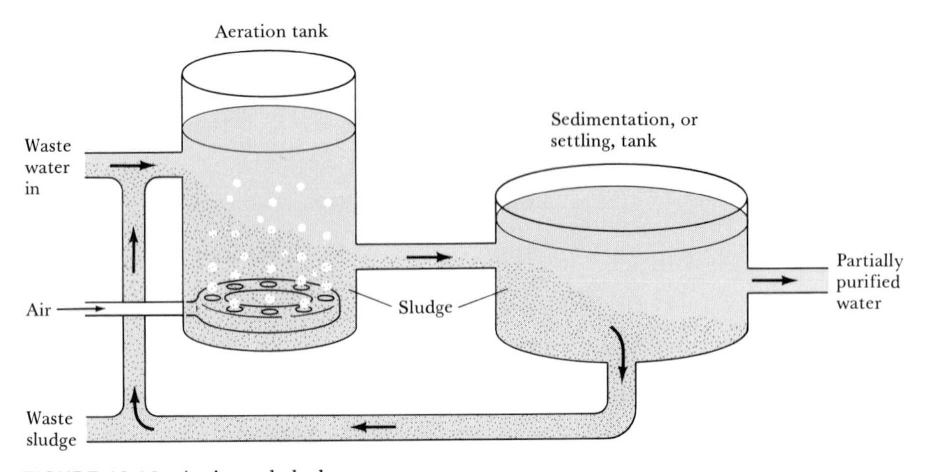

FIGURE 12.14 Activated-sludge process.

FOR REVIEW

Summary

In this chapter we've examined the physical and chemical properties of the earth's atmosphere. The complex temperature variations in the atmosphere give rise to several regions, each with characteristic properties. The lowest of these regions, the troposphere, extends from the surface to about 11 km. Above the troposphere, in order of increasing altitude, are the stratosphere, mesosphere, and thermosphere. In the upper reaches of the atmosphere, only the simplest chemical species can survive the bombardment of highly energetic particles and radiation from the sun. The average molecular weight of the atmosphere at high elevations is lower than at the earth's surface, because the lightest atoms and molecules diffuse upward and because of photodissociation. Absorption of radiation may also lead to ionization.

Ozone is produced in the mesosphere and stratosphere as a result of reaction of atomic oxygen with O_2. Ozone is itself decomposed by absorption of a photon or by reaction with an active species such as NO. Human activities could result in the addition to the stratosphere of atomic chlorine, which is capable of reacting with ozone in a catalytic cycle to convert ozone to O_2. A marked reduction in the ozone level in the upper atmosphere would have serious adverse consequences, because the ozone layer filters out certain wavelengths of ultraviolet light that are not taken out by any other atmospheric component.

In the troposphere, the lower atmosphere in which we live, the chemistry of trace atmospheric components is of major importance. Many of these minor components are pollutants; sulfur dioxide is one of the more noxious and prevalent. It is oxidized in air to form sulfur trioxide, which upon dissolving in water forms sulfuric acid. To control this source of pollution, it is necessary to prevent SO_2 from escaping from industrial operations in which it is formed. One method for doing this involves reacting the SO_2 with CaO to form calcium sulfite, $CaSO_3$.

Photochemical smog is a complex mixture of components in which both nitrogen oxides and ozone play important roles. The smog components are generated mainly in automobile engines, and smog control consists largely of controlling emissions from automobiles.

Carbon monoxide is found in high concentrations in the exhaust of automobile engines and in cigarette smoke. This compound is a health hazard because of its ability to form a strong bond with hemoglobin and thus reduce the capacity of blood for oxygen transfer from the lungs.

Carbon dioxide and water vapor are the only components of the atmosphere that strongly absorb infrared radiation. The level of carbon dioxide in the atmosphere is thus of importance in determining worldwide climate. As a result of the extensive combustion of fossil fuels (coal, oil, and natural gas) the carbon dioxide level of the atmosphere is steadily increasing.

Seawater contains about 3.5 percent by weight of dissolved salts. These salts, along with dissolved carbon dioxide, establish a buffer system that maintains the pH of seawater in the vicinity of 8.0.

The varieties of biological life that live in the sea depend on the growth of photosynthetic phytoplankton as the base of the food chain. The photosynthetic zone is near the water's surface, where the sun's rays can penetrate. The major ingredients needed for photosynthesis are CO_2 and suitable forms of nitrogen and phosphorus. Usually, the availability of one or the other of these latter elements limits the rate of photosynthesis.

Because most of the world's water is in the oceans, it is perhaps inevitable that humankind must eventually look to the seas for fresh water. Desalination refers to the removal of dissolved salts from seawater, brine, or brackish water, so as to render it fit for human consumption. Among the means by which desalination may be accomplished is distillation.

Fresh water that is available from rivers, lakes, and underground sources may require treatment to render it fit for use. The several steps which may be used in water treatment are coarse filtration, sedimentation, sand filtration, aeration, sterilization, and softening.

Wastewater treatment is applied to sewage waters or to water that has been used in an industrial operation. Municipal wastewaters are given a primary treatment to remove insoluble scum, grease, and other materials. Secondary treatment consists of aeration of sewage sludge to promote the growth of microorganisms that feed on the organic compounds present in sewage. Eventually, clear water is separated from the mass of microorganisms. The water is lower in biological oxygen demand (BOD) than before treatment. However, it may still contain many substances that are toxic to aquatic life or to humans or that cause excessive growth of algae in natural waters. The many substances that remain in waters after secondary treatment can be removed only by extensive additional processing, referred to as tertiary treatment.

Learning goals

Having read and studied this chapter, you should be able to:

1 Sketch the manner in which the atmospheric temperature varies with altitude and list the names of the various regions of the atmosphere and the boundaries between them.

2 Sketch the manner in which atmospheric pressure decreases with elevation and explain in general terms the reason for the decrease.

3 Describe the composition of the atmosphere with respect to the four most abundant components.

4 Explain what is meant by the term *photodissociation* and calculate the maximum wavelength of a photon that is energetically capable of producing photodissociation, given the dissociation energy of the bond to be broken in the process.

5 Explain what is meant by photoionization and relate the energy requirement for photoionization to the ionization energy of the species undergoing ionization.

6 Explain the presence of ozone in the mesosphere and stratosphere in terms of appropriate chemical reactions.

7 Describe how nitrogen oxides or atomic chlorine function in the stratosphere as ozone-removal agents.

8 Explain how atomic chlorine might appear in the stratosphere as a product of the chlorofluoromethanes.

9 List the names and chemical formulas of the more important pollutant substances present in the troposphere and in urban atmospheres.

10 List the major sources of sulfur dioxide as an atmospheric pollutant.

11 List the more important reactions of nitrogen oxides and ozone that occur in smog formation.

12 Explain why carbon monoxide constitutes a health hazard.

13 Explain why the concentration of carbon dioxide in the troposphere has an effect on the average temperature at the earth's surface.

14 List the more abundant ionic species present in seawater.

15 Describe how the growth of phytoplankton is related to the availability of nutrient nitrogen and phosphorus.

16 Explain the principles involved in the multistage flash-distillation process for desalination of seawater.

17 Describe the BOD test for water and explain how it is related to water purity.

18 List and explain the various stages of treatment that may be applied to a freshwater supply.

19 Describe the chemical principles involved in the lime-soda process for reducing water hardness.

20 List and explain the stages in treatment of wastewater.

Key terms

Among the more important terms and expressions used for the first time in this chapter are the following:

Acid rain (Section 12.4) refers to rainwater that has become excessively acidic because of absorption of pollutant oxides, notably SO_3, produced by human activities.

The **biological oxygen demand (BOD)** (Section 12.6) is the total capacity of a water sample for consumption of oxygen for biological oxidations. It is measured by the amount of oxygen gas taken up by a given quantity of water during a 5-day period at 20°C.

Carboxyhemoglobin (Section 12.4) is a complex formed between carbon monoxide and hemoglobin, in which CO is bound to the iron atom.

A **catalyst** (Section 12.3) is a substance that affects the rate of a chemical reaction but does not itself undergo a net, overall chemical change.

Chlorofluoromethanes (Section 12.3) are compounds of the general formula CF_xCl_{4-x}, $x = 1, 2,$ or 3, used as propellant gases in aerosol spray cans and in refrigeration units.

Desalination (Section 12.5) refers to the removal of salts from seawater, brine, or brackish water, so as to render it fit for human consumption.

Hemoglobin (Section 12.4) is an iron-containing protein responsible for oxygen transport in the blood.

The **lime-soda process** (Section 12.6) is a method for removal of Mg^{2+} and Ca^{2+} ions from water to reduce water hardness. The substances added to the water are "lime," CaO [or "quicklime," $Ca(OH)_2$], and "soda ash," Na_2CO_3, in amounts determined by the concentrations of the offending ions.

Multistage flash distillation (Section 12.5) is a method for desalination that involves distillation of saline water in several stages, for maximum possible conservation of heat stored in the water.

Photochemical smog (Section 12.4) is a complex

mixture of undesirable substances produced by the action of sunlight on an urban atmosphere polluted with automobile emissions. The major starting ingredients are nitrogen oxides and organic substances, notably olefins and aldehydes.

Photodissociation (Section 12.2) refers to the breaking of a molecule into two or more neutral fragments as a result of absorption of light.

Photoionization (Section 12.2) refers to the removal of an electron from an atom or molecule by absorption of light.

Phytoplankton (Section 12.5) are microscopic plants that abound in the water near the ocean surface. They utilize photosynthesis to consume CO_2 and suitable forms of nitrogen and phosphorus in forming plant matter. They form the base of the food chain for biological life in the oceans.

The photosynthetic zone (Section 12.5) is the region of the oceans extending from the surface to a depth of about 150 m, in which the photosynthetic growth of phytoplankton occurs.

Salinity (Section 12.5) is a measure of the salt content of seawater, brine, or brackish water. It is equal to the weight in grams of dissolved salts present in 1 kg of seawater, brine, or brackish water.

The troposphere (Section 12.1) is the region of earth's atmosphere extending from the surface to about 11 km altitude. The regions of the atmosphere extending above the troposphere are, in order of increasing altitude, the stratosphere, mesosphere, and thermosphere.

Water hardness (Section 12.6) refers to the presence in a water supply of Ca^{2+}, Mg^{2+} (and sometimes Fe^{2+}) ions. These ions form insoluble precipitates with soaps and are responsible for scale formation when the water is heated.

EXERCISES

Earth's atmosphere; the outer regions

12.1 Name the regions of the atmosphere, indicate the altitude interval for each region, and describe the variation in temperature in that region.

12.2 Name the boundaries between the regions of the atmosphere and indicate the temperature in the boundary region.

12.3 From the data in Table 12.1, calculate the partial pressures in mm Hg of argon and neon when the total pressure is 750 mm Hg.

12.4 Explain why the stratosphere, which extends from 11 km to about 50 km, contains a smaller total atmospheric mass than the troposphere, which extends from the surface to 11 km.

12.5 Describe the bonding in N_2 and O_2. Using data from earlier chapters, discuss the reasons for the relative reactivities of these two molecules in the earth's upper atmosphere.

12.6 The dissociation energy of a carbon-bromine bond is typically about 210 kJ/mol. What is the maximum wavelength of photon that can cause C—Br bond dissociation?

12.7 In CF_3Cl the C—Cl bond dissociation energy is 339 kJ/mol. In CCl_4 the C—Cl bond dissociation energy is 293 kJ/mol. What is the range of wavelengths of photons that can cause C—Cl bond rupture in one molecule, but not in the other?

12.8 What conditions need to be met for radiant energy to cause photodissociation of NO: $NO(g) \longrightarrow N(g) + O(g)$?

[12.9] Suppose that on another planet the atmosphere consisted of 20 percent Ar, 35 percent CH_4, and 45 percent O_2. What would be the average molecular weight at the surface? What would be the average molecular weight at 200 km, assuming that all the O_2 is photodissociated?

Chemistry of the stratosphere

12.10 Explain why oxygen atoms exist in the atomic state for much longer average times at 120 km elevation than at 50 km elevation.

12.11 Explain the means by which ozone, O_3, is formed in the stratosphere. What is the biological significance at the earth's surface of the ozone layer in the stratosphere?

12.12 Using the thermodynamic data in Appendix D, calculate the overall enthalpy change in each step in the catalytic cycle that converts O_3 to O_2 (see Equation 12.7).

$$NO(g) + O_3(g) \longrightarrow NO_2(g) + O_2(g)$$
$$NO_2(g) + O(g) \longrightarrow O_2(g) + NO(g)$$

12.13 The standard enthalpies of formation of ClO and ClO_2 are 101 and 102 kJ/mol, respectively. Using these data and the thermodynamic data in Appendix D, calculate the overall enthalpy change for each step in the following catalytic cycle:

$$ClO(g) + O_3(g) \longrightarrow ClO_2(g) + O_2(g)$$
$$ClO_2(g) + O(g) \longrightarrow ClO(g) + O_2(g)$$

On the basis of your results, indicate whether the ClO—ClO_2 pair is at least a possible catalyst for decomposition of ozone in the atmosphere.

12.14 Beginning with the intact chlorofluoromethane, CF_2Cl_2, write equations showing how a catalytic effect for destruction of ozone may be established in the stratosphere.

[12.15] It has recently been pointed out that there may be increased amounts of NO in the troposphere as compared with the past because of massive use of nitrogen-containing compounds in fertilizers. Assuming that NO can eventually diffuse into the stratosphere, what role might it play in affecting the conditions of life on earth? Using the index to this text, look up the chemistry of nitrogen oxides. What other chemical pathways might NO in the troposphere follow, other than diffusion into the stratosphere?

Chemistry of the troposphere

12.16 In a particular urban environment, the ozone concentration is 0.26 ppm. Assuming a temperature of 16°C and an atmospheric pressure of 750 mm Hg at the time, calculate the partial pressure of ozone and the number of O_3 molecules per cubic meter.

12.17 In a particular urban environment the NO concentration is 0.75 ppm. If the atmospheric pressure at the time is 730 mm Hg and the temperature is 20°C, calculate the partial pressure of NO, and the number of NO molecules per cubic meter.

12.18 Compare typical concentrations of CO, SO_2, and NO in nonpolluted air (Table 12.3) and urban air (Table 12.4), and indicate in each case at least one possible source of the higher values in Table 12.4.

12.19 For each of the following gases make a list of known or possible naturally occurring sources: (a) CH_4; (b) SO_2; (c) NO; (d) CO.

12.20 In a recent study carried out in Canada it was found that at a particular location far from industrial activity the sulfate in rainfall amounted to 210 mg/m^2 per year. Calculate the total mass of sulfate falling in a square mile per year.

12.21 Assuming an overall efficiency of about 30 percent, how much calcium carbonate would be required to remove the SO_2 formed in burning a ton of coal containing 2.7 percent sulfur by weight?

12.22 From Figure 12.6 we see that the concentration of NO_2 in an urban atmosphere increases to a maximum rather early in the day, then decreases to a lower, steady value. What is the explanation for this behavior?

12.23 We have noted that the affinity of carbon monoxide for hemoglobin is about 210 times that of O_2. Assume that a person is inhaling air that contains 86 ppm of CO. If all the hemoglobin leaving the lungs carries either oxygen or CO, calculate the fraction in the form of carboxyhemoglobin.

The world ocean

12.24 What is the molarity of Na^+ in a solution of NaCl whose salinity is 5 if the solution has a density of 1.0 g/mL?

12.25 Phosphorus is present in seawater to the extent of 0.07 ppm by weight (that is, 0.07 g of P per 10^6 g of H_2O). If the phosphorus is present as phosphate, PO_4^{3-}, calculate the corresponding molar concentration of phosphate.

12.26 The concentration of nitrate ion in the open sea varies as shown in Figure 12.15. Explain this variation in terms of the biochemical processes occurring in the sea.

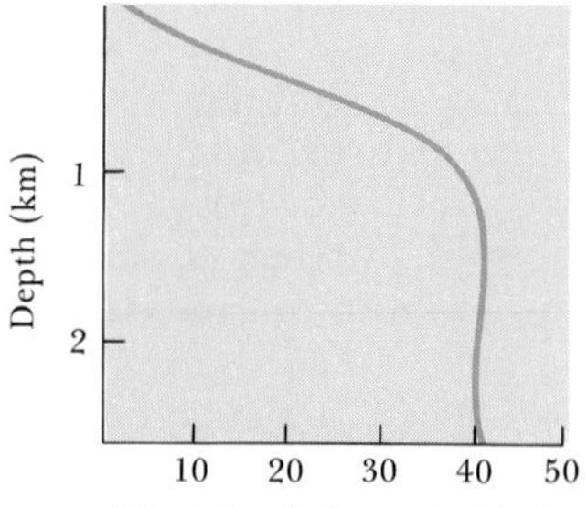

FIGURE 12.15 Concentration of nitrate ion as a function of depth.

12.27 Assuming a 10 percent efficiency of recovery, how many liters of seawater must be processed to obtain 10^8 kg of bromine in a commercial production process, assuming the bromide ion concentration listed in Table 12.7?

12.28 A first stage in recovery of magnesium from seawater is precipitation of $Mg(OH)_2$ by use of CaO:

$$Mg^{2+}(aq) + CaO(s) + H_2O(l) \longrightarrow Mg(OH)_2(s) + Ca^{2+}(aq)$$

What mass of CaO is needed to precipitate 4.0×10^7 g of $Mg(OH)_2$?

12.29 List the problems encountered in desalination by distillation, and indicate the steps that can be taken to minimize them.

[12.30] Suppose that in the large multistage flash-distillation unit shown in Figure 12.12 the efficiency of use of heat values in the steam applied to the brine heaters is 24 percent and that the steam is formed with a 32 percent efficiency from burning oil. The heat of combustion of petroleum oil is -46 kJ/g. If the unit is to produce 4 million liters of fresh water per day, and if the heat of vaporization of water in the units is 2.25 kJ/g of H_2O, how much oil must be burned each day?

Fresh water

12.31 What naturally occurring impurities are commonly present in fresh, unpolluted water?

12.32 (a) What ions are commonly responsible for the hardness of water? (b) What makes these ions objectionable?

12.33 Explain what is meant by the following terms and abbreviations: (a) BOD; (b) biodegradable; (c) aerobic decay; (d) anaerobic decomposition; (e) BOD_5.

12.34 The following organic anion is found in most detergents:

$$H_3C{-}(CH_2)_9{-}\underset{CH_3}{\overset{H}{C}}{-}C_6H_4{-}SO_3^-$$

Assume that this anion undergoes aerobic decomposition in the following manner:

$$2C_{18}H_{29}O_3S^-(aq) + 51O_2(aq) \longrightarrow 36CO_2(aq) + 28H_2O(l) + 2H^+(aq) + 2SO_4^{2-}(aq)$$

What is the total BOD of a water sample containing 1.0 g of this substance per 100 L of water?

12.35 Two 10-mL samples of wastewater were diluted to 300 mL with aerated distilled water containing sufficient inorganic nutrients to ensure complete biochemical oxidation of any biodegradable organic substances present. The samples were seeded with bacteria. One sample was then analyzed immediately for dissolved oxygen and a second was analyzed after it had been incubated for 5 days at 20°C. The initial analysis revealed the sample to contain 7.90 ppm of dissolved oxygen. The second analysis, performed 5 days later, indicated that 1.40 ppm dissolved oxygen was present. Taking sample dilution into effect, calculate the 5-day BOD of the wastewater. (Ignore any BOD associated with the dilution and preliminary treatment.)

12.36 Explain how $Al_2(SO_4)_3$ and CaO are used in the purification of water.

[12.37] In a particular water supply, the concentration of Ca^{2+} is $2.2 \times 10^{-3}\,M$, and the concentration of bicarbonate ion, HCO_3^-, is $1.3 \times 10^{-3}\,M$. What weights of $Ca(OH)_2$ and Na_2CO_3 are needed to reduce the level of Ca^{2+} to one-fourth its original level if 1.0×10^7 L of water must be treated?

Additional exercises

12.38 Describe the major factors that lead to a maximum in the ozone concentration at about 22 km elevation.

12.39 The temperature profile in the atmosphere shows a maximum at about 50 km (Figure 12.1). Explain the origin of this maximum.

12.40 It is said that atomic chlorine serves as a catalyst for the decomposition of ozone according to the reaction $O_3 + O \longrightarrow 2O_2$. What is a catalyst? How does Cl serve as a catalyst in this reaction? What overall chemical change occurs for Cl?

[12.41] In terms of the energy requirements, explain why photodissociation of oxygen is more important than photoionization of oxygen at altitudes below about 90 km.

[12.42] Experiments have been performed in which metals such as sodium or barium have been released into the atmosphere at altitudes of about 120 km. Assuming that the metals are present in the atomic form, what reactions would you expect to occur with the ionic species present? Explain.

12.43 Except for two substances, the components of the earth's atmosphere are transparent to long-wavelength, infrared radiation. What are these two substances? In what way does the absorption of infrared radiation affect the earth's climate? Explain how increased levels of infrared-absorbing substances in the atmosphere could lead to a higher average surface temperature.

12.44 The overall elemental composition of phytoplankton can be represented approximately by the formula $C_{108}H_{266}N_{16}O_{109}P$. Suppose that in a particular stretch of water the concentration of nitrogen is limiting for growth of phytoplankton; also assume that the nitrogen concentration in the zone from the surface to 100 m averages 1.0×10^{-6} mol of nitrogen atoms per liter. If half of this nitrogen is converted to phytoplankton, what is the total mass of plant matter in a square kilometer of ocean in a region from the surface to a depth of 100 m?

12.45 The average daily mass of O_2 taken up by sewage discharged in the United States is 59 g per person. How many liters of water at 9 ppm O_2 are totally depleted of oxygen in 1 day by a population of 50,000 people?

12.46 Urea, CH_4N_2O, is the end product of protein metabolism in animals. Assume that aerobic bacteria can decompose it as follows:

$$CH_4N_2O(aq) + 4O_2(aq) \longrightarrow H_2O(l) + CO_2(aq) + 2H^+(aq) + 2NO_3^-(aq)$$

What is the BOD of a body of water of total volume 3.0×10^6 L if 30 kg of urea enters?

12.47 How many moles of $Ca(OH)_2$ and of Na_2CO_3 should be added to soften 10^3 L of water in which $[Ca^{2+}] = 5.0 \times 10^{-4}\,M$ and $[HCO_3^-] = 7.0 \times 10^{-4}\,M$?

12.48 Complete and balance a chemical equation corresponding to each of the following verbal descriptions. (a) The nitric oxide molecule undergoes photodissociation in the upper atmosphere. (b) The nitric oxide molecule undergoes photoionization in the upper atmosphere. (c) Nitric oxide undergoes oxidation by ozone in the stratosphere. (d) Sulfur dioxide is formed when lead sulfide, PbS, is roasted. (e) Hypochlorous acid is formed when chlorine is added to water. (f) Hypochlorous acid reacts with aqueous ammonia to form chloramine, NH_2Cl. (g) A sample of water containing Ca^{2+} and bicarbonate ion forms a precipitate when heated.

12.49 Distinguish between primary, secondary, and tertiary sewage treatment.

13 Chemical Kinetics: Reaction Rates

Chemistry is by its very nature concerned with change. Substances with well-defined properties are converted by chemical reactions into other materials with different properties. Chemists want to know which new substances are formed from a given set of starting reactants. However, it is equally important to know how rapidly chemical reactions occur and to understand the factors that control their speeds. For example, what factors are important in determining how rapidly foods spoil? What determines the rate at which steel rusts? How does one design a rapidly setting material for dental fillings? What factors control the rate at which fuel burns in an auto engine, and how does burning rate determine the pollutant content of the engine exhaust?

The area of chemistry concerned with the speeds, or rates, at which chemical reactions occur is called **kinetics.** In this chapter we learn how to express and determine the rates at which reactions occur. We shall also learn how reaction rates are affected by variables such as concentration, temperature, and the presence of catalysts.

13.1 REACTION RATE

The speed of any event is measured by the change occurring in a given interval of time. For example, the speed of an automobile expresses its change in position in a certain time; the associated units are usually miles per hour (mi/h). Similarly, the speed or rate of a reaction is expressed as the change in the concentration of a reactant or product in a certain time; the units with which we express this speed are usually molarity per second (M/s). As an example, consider the reaction that occurs when butyl chloride, C_4H_9Cl, is placed in water. The resulting reaction produces butyl alcohol, C_4H_9OH, and hydrochloric acid:

$$C_4H_9Cl(l) + H_2O(l) \longrightarrow C_4H_9OH(aq) + HCl(aq) \qquad [13.1]$$

TABLE 13.1 Rate data for reaction of C_4H_9Cl with water

Time (s)	$[C_4H_9Cl]$ (M)	Average rate (M/s)
0	0.1000	
		1.90×10^{-4}
50	0.0905	
		1.70×10^{-4}
100	0.0820	
		1.58×10^{-4}
150	0.0741	
		1.40×10^{-4}
200	0.0671	
		1.22×10^{-4}
300	0.0549	
		1.01×10^{-4}
400	0.0448	
		0.80×10^{-4}
500	0.0368	
		0.56×10^{-4}
800	0.0200	
10,000	0	

Suppose that we begin with a 0.1000 M solution of C_4H_9Cl in water and then measure the C_4H_9Cl concentration at different times after the solution is mixed. We could in this way collect the data shown in the first two columns of Table 13.1. The average rate of the reaction over any time interval is a positive quantity given by the change in the concentration of C_4H_9Cl divided by the time in which that change occurs:

$$\text{Average rate} = -\frac{\text{change in concentration of } C_4H_9Cl}{\text{corresponding time interval}} = -\frac{\Delta[C_4H_9Cl]}{\Delta t} \quad [13.2]$$

The brackets around C_4H_9Cl in this equation indicate the concentration of that substance. The Greek capital letter delta, Δ, is read "change in"; $\Delta[C_4H_9Cl]$ is the change in the concentration of C_4H_9Cl:

$$\Delta[C_4H_9Cl] = [C_4H_9Cl]_{\text{final}} - [C_4H_9Cl]_{\text{initial}} \quad [13.3]$$

Similarly, Δt, the corresponding time interval, is the amount of time between the beginning and the end of the interval. The negative sign in Equation 13.2 indicates that the concentration of C_4H_9Cl is decreasing with time.

Notice in Table 13.1 that during the first interval of 50 s, the concentration of C_4H_9Cl decreases from 0.1000 M to 0.0905 M. Thus the average rate over this 50-s interval is

$$\text{Average rate} = -\frac{(0.0905 - 0.1000)\,M}{(50 - 0)\,\text{s}} = 1.90 \times 10^{-4}\,M/\text{s}$$

Average rates over other intervals calculated similarly are shown in the third column of Table 13.1. Notice that the average rate steadily decreases as the reaction proceeds. At some point the reaction stops; that is, there is no longer any change in concentration with time.

We can also see this decrease in rate if we display the data graphically, as in Figure 13.1. The dots represent the experimental data from the first

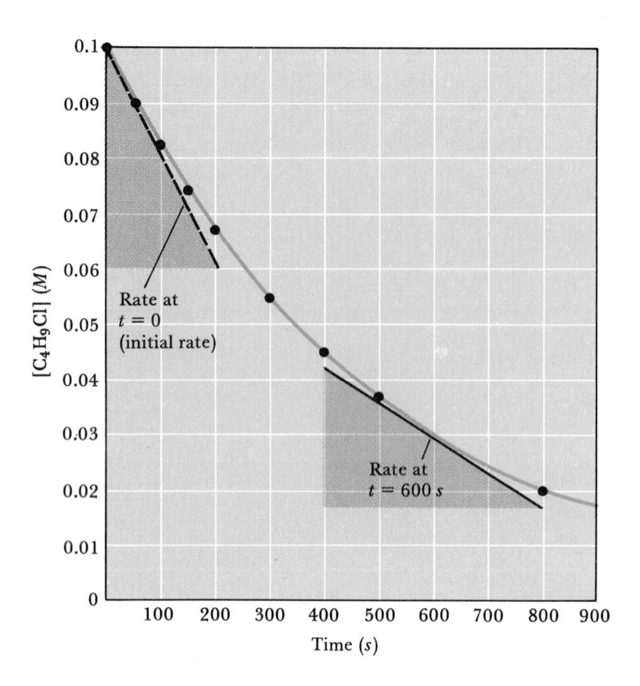

FIGURE 13.1 Concentration of butyl chloride, C_4H_9Cl, as a function of time. The dots represent the experimental data from the first two columns of Table 13.1; the colored line is the smooth curve drawn to connect the data points. The reaction rate at any time is given by the slope of the tangent to the curve at that time. The slope of the tangent is defined as the vertical change divided by the horizontal change of the tangent, that is, $\Delta[C_4H_9Cl]/\Delta t$. Tangents have been drawn that touch the curve at $t = 0$ and $t = 600$ s. The calculation of the slope at $t = 600$ s is performed in the body of the text. The slope at $t = 0$ is calculated in Sample Exercise 13.1.

two columns of Table 13.1. Using such a curve we can determine the instantaneous rate; this is the rate at a particular time as opposed to the average rate over an interval of time. The instantaneous rate is obtained from the straight-line tangent that touches the curve at the point of interest. We have drawn two such tangents on Figure 13.1, one at $t = 0$ and the other at $t = 600$ s. The slopes of these tangents give the instantaneous rates at these times.* For example, at 600 s we have

$$\text{Instantaneous rate} = -\frac{(0.017 - 0.042)\,M}{(800 - 400)\text{ s}} = 6.2 \times 10^{-5}\,M/\text{s}$$

We will usually refer to the instantaneous rate merely as the rate.

Whenever we discuss either the rate or average rate of a reaction we need to define which substance we are using as a reference. In the reaction

$$2HI(g) \longrightarrow H_2(g) + I_2(g)$$

we can measure the rate of disappearance of HI or the appearance of either H_2 or I_2. Because 2 mol of HI disappears for each mole of H_2 or I_2 that forms, the rate of disappearance of HI is twice the rate of appearance of H_2 or I_2:

$$-\frac{\Delta[HI]}{\Delta t} = 2\frac{\Delta[H_2]}{\Delta t} = 2\frac{\Delta[I_2]}{\Delta t} \qquad [13.4]$$

*You may wish to briefly review the idea of graphical determination of slopes by referring to Appendix A. If you are familiar with calculus, you may recognize that the average rate approaches the instantaneous rate as the time interval approaches zero. This limit, in the notation of calculus, is represented as $-d[C_4H_9Cl]/dt$.

SAMPLE EXERCISE 13.1

(a) Using the data in Table 13.1, calculate the average rate of disappearance of C_4H_9Cl over the time interval from 50 to 150 s. (b) Using Figure 13.1, estimate the instantaneous rate of disappearance of C_4H_9Cl at $t = 0$ (the initial rate). (c) How is the rate of disappearance of C_4H_9Cl related to the rate of appearance of C_4H_9OH?

Solution: (a) From Table 13.1 we have

$$\text{Average rate} = -\frac{\Delta[C_4H_9Cl]}{\Delta t}$$

$$= -\frac{(0.0741 - 0.0905)\,M}{(150 - 50)\text{ s}}$$

$$= 1.64 \times 10^{-4}\,M/\text{s}$$

(b) The initial rate is given by the slope of the dashed line in Figure 13.1. The slope of a straight line is given by the change in the vertical axis divided by the corresponding change in the horizontal axis. The straight line falls from $[C_4H_9Cl] = 0.100$ to $0.060\,M$ in the time change from 0 to 200 s. Thus the instantaneous rate is

$$\text{Rate} = -\frac{(0.060 - 0.100)\,M}{(200 - 0)\text{ s}}$$

$$= 2.0 \times 10^{-4}\,M/\text{s}$$

(c) Because 1 mol of C_4H_9OH forms for each mole of C_4H_9Cl that disappears,

$$-\frac{\Delta[C_4H_9Cl]}{\Delta t} = \frac{\Delta[C_4H_9OH]}{\Delta t}$$

13.2 DEPENDENCE OF REACTION RATE ON CONCENTRATIONS

The decreasing rate of reaction with passing time that is evident in Figure 13.1 is quite typical of reactions. Reaction rates diminish as the concentrations of reactants diminish. Conversely, rates generally increase when reactant concentrations are increased.

One way of studying the effect of concentration on reaction rate is to determine the way in which the rate at the beginning of a reaction depends on the starting concentrations. To illustrate this approach, consider the following reaction:

$$NH_4^+(aq) + NO_2^-(aq) \longrightarrow N_2(g) + 2H_2O(l) \qquad [13.5]$$

We might study the rate of this reaction by measuring the concentration of NH_4^+ or NO_2^- as a function of time or by measuring the volume of N_2 collected. Because of the 1:1 stoichiometry of the reaction, all of these rates will be equal. If we determine the initial reaction rate (the instantaneous rate at $t = 0$) for various starting concentrations of NH_4^+ and NO_2^- we could collect the data shown in Table 13.2. These data indicate

TABLE 13.2 Rate data for the reaction of ammonium and nitrite ions in water at 25°C

Experiment number	Initial NO_2^- concentration (M)	Initial NH_4^+ concentration (M)	Observed initial rate (M/s)
1	0.0100	0.200	5.4×10^{-7}
2	0.0200	0.200	10.8×10^{-7}
3	0.0400	0.200	21.5×10^{-7}
4	0.0600	0.200	32.3×10^{-7}
5	0.200	0.0202	10.8×10^{-7}
6	0.200	0.0404	21.6×10^{-7}
7	0.200	0.0606	32.4×10^{-7}
8	0.200	0.0808	43.3×10^{-7}

that changing either $[NH_4^+]$ or $[NO_2^-]$ changes the reaction rate. Notice that if we double $[NO_2^-]$ while holding $[NH_4^+]$ constant, the rate doubles (compare experiments 1 and 2). If $[NO_2^-]$ is increased by a factor of 4 (compare experiments 1 and 3) the rate changes by a factor of 4, and so forth. These results indicate that the rate is directly proportional to $[NO_2^-]$. When $[NH_4^+]$ is similarly varied while $[NO_2^-]$ is held constant, the rate is affected in the same manner. We conclude that the rate is also directly proportional to the concentration of NH_4^+. We can express the overall concentration dependence in the following way:

$$\text{Rate} = k[NH_4^+][NO_2^-] \qquad [13.6]$$

The proportionality constant, k, in Equation 13.6 is called the rate constant. We can evaluate the magnitude of k using the data in Table 13.2. Using the results of experiment 1, and substituting into Equation [13.6], we have

$$5.4 \times 10^{-7}\ M/\text{s} = k(0.0100\ M)(0.200\ M)$$

Solving for k gives

$$k = \frac{5.4 \times 10^{-7}\ M/\text{sec}}{(0.0100\ M)(0.200\ M)} = 2.7 \times 10^{-4}/M\text{-sec}$$

You may wish to satisfy yourself that this same value of k is obtained using any of the other experimental results given in Table 13.2. You might also note that given $k = 2.7 \times 10^{-4}/M\text{-s}$ and using Equation 13.6, we can calculate the rate for any concentration of NH_4^+ and NO_2^-. Suppose that $[NH_4^+] = 0.100\ M$ and $[NO_2^-] = 0.100\ M$; then

$$\text{Rate} = (2.7 \times 10^{-4}/M\text{-s})(0.100\ M)(0.100\ M) = 2.7 \times 10^{-6}\ M/\text{s}$$

An equation like 13.6 that relates the rate of a reaction to concentration is called a rate law. *The rate law for any chemical reaction must be determined experimentally; it cannot be predicted by merely looking at the chemical equation.* The following are some additional examples of rate laws:

$$2N_2O_5(g) \longrightarrow 4NO_2(g) + O_2(g) \qquad \text{Rate} = k[N_2O_5] \qquad [13.7]$$

$$CHCl_3(g) + Cl_2(g) \longrightarrow CCl_4(g) + HCl(g) \qquad \text{Rate} = k[CHCl_3][Cl_2]^{\frac{1}{2}} \qquad [13.8]$$

$$H_2(g) + I_2(g) \longrightarrow 2HI(g) \qquad \text{Rate} = k[H_2][I_2] \qquad [13.9]$$

The rate laws for a great many reactions have the general form

$$\text{Rate} = k[\text{reactant 1}]^m[\text{reactant 2}]^n \ldots \qquad [13.10]$$

As noted above, the proportionality constant, k, in a rate law is called the rate constant. For a given set of reactant concentrations, the reaction rate increases as k increases. The exponents m and n in Equation 13.10 are called the reaction orders and their sum is the overall reaction order. For the reaction of NH_4^+ with NO_2^-, the rate law (Equation 13.6) contains the concentration of NH_4^+ raised to the first power. Thus the reaction is said to be first order in NH_4^+. Similarly, it is also first order in NO_2^-. The overall reaction order is two.

In a great many rate laws the reaction orders are zero, one, or two. However, reaction orders can be fractional or even negative. If a reaction is zero order in a particular reactant, changing its concentration will have no influence on rate as long as some of that reactant is present. On the other hand, if the reaction is first order in a reactant, changes in the concentration of that substance will produce proportional changes in the rate; doubling the concentration will double the rate, and so forth. When the rate law is second order in a particular reactant, doubling its concentration changes the rate by a factor of $2^2 = 4$; tripling its concentration causes the rate to increase by a factor of $3^2 = 9$.

In working with rate laws, be careful not to confuse the rate constant for a reaction with the reaction rate. The rate of a reaction depends on the concentrations of reactants; the rate constant does not. As we shall see later in this chapter, the rate constant and consequently the reaction rate are affected by temperature and the presence of a catalyst.

SAMPLE EXERCISE 13.2

The initial rate of a reaction $A + B \longrightarrow C$ was measured for several different starting concentrations of A and B, with the results given below:

Experiment number	[A] (M)	[B] (M)	Initial rate (M/s)
1	0.100	0.100	4.0×10^{-5}
2	0.100	0.200	4.0×10^{-5}
3	0.200	0.100	16.0×10^{-5}

Using these data, determine (a) the rate law for the reaction; (b) the magnitude of the rate constant; (c) the rate of the reaction when [A] = 0.050 M and [B] = 0.100 M.

Solutions: (a) We may assume that the rate law has the form: rate $= k[A]^m[B]^n$, so that our task is to deduce the values of m and n. Experiments 1 and 2 indicate that the concentration of B has no influence on the reaction rate. The reaction is therefore zero order in B. Experiments 1 and 3 indicate that doubling A increases the rate fourfold. This result indicates that rate is proportional to $[A]^2$; the reaction is second order in A. The rate law is

$$\text{Rate} = k[A]^2[B]^0 = k\,[A]^2$$

This same conclusion could be reached in a more formal way by taking the ratio of the rates from two experiments:

$$\frac{\text{Rate 1}}{\text{Rate 1}} = \frac{4.0 \times 10^{-5}\,M/\text{s}}{4.0 \times 10^{-5}\,M/\text{s}} = 1$$

Using the rate law, then, we have

$$1 = \frac{\text{rate 1}}{\text{rate 2}}$$

$$= \frac{k[0.100\,M]^m[0.100\,M]^n}{k[0.100\,M]^m[0.200\,M]^n} = \frac{[0.100]^n}{[0.200]^n} = \left(\frac{1}{2}\right)^n$$

But $(\frac{1}{2})^n$ can equal 1 only if $n = 0$.

We can deduce the value of m in a similar fashion:

$$\frac{\text{Rate 1}}{\text{Rate 3}} = \frac{4.0 \times 10^{-5}\,M/\text{s}}{16.0 \times 10^{-5}\,M/\text{s}} = \frac{1}{4}$$

Using the rate law gives us

$$\frac{1}{4} = \frac{\text{rate 1}}{\text{rate 3}}$$

$$= \frac{k[0.100\,M]^m[0.100\,M]^n}{k[0.200\,M]^m[0.100\,M]^n} = \frac{[0.100]^m}{[0.200]^m} = \left(\frac{1}{2}\right)^m$$

The fact that $(\frac{1}{2})^m = \frac{1}{4}$ indicates that $m = 2$.

(b) Using the rate law and the data from experiment 1, we have

$$k = \frac{\text{rate}}{[A]^2} = \frac{4.0 \times 10^{-5}\,M/\text{s}}{(0.100\,M)^2}$$

$$= 4.0 \times 10^{-3}/M\text{-s}$$

(c) Using the rate law from part (a) and the rate constant from part (b), we have

$$\text{Rate} = k[\text{A}]^2$$
$$= (4.0 \times 10^{-3}/M\text{-s})(0.050\ M)^2$$
$$= 1.0 \times 10^{-5}\ M/\text{s}$$

Because [B] is not part of the rate law, its concentration is immaterial to the rate, provided that there is at least some B present to react with A.

13.3 RELATION BETWEEN REACTANT CONCENTRATION AND TIME

The rate law (Equation 13.10) tells us how the rate of a reaction changes as we change reactant concentrations. Such rate laws can be converted into equations that tell us what the concentrations of the reactants or products are at any time during the course of a reaction. The mathematics required involve calculus. We don't expect you to be able to perform the calculus operations; however, you should be able to use the resulting equations. We will apply this conversion to two of the simplest rate laws—those that are first order overall and those that are second order overall.

First-Order Reactions

If the rate of a reaction of the sort A $\longrightarrow$ products is first order in A, we can write the following rate law:

$$\text{Rate} = -\frac{\Delta[\text{A}]}{\Delta t} = k[\text{A}] \qquad [13.11]$$

Using calculus, this equation can be transformed into an equation that relates the concentration of A at the start of the reaction, $[\text{A}]_0$, to its concentration at any other time t, $[\text{A}]_t$:

$$\log\,[\text{A}]_t - \log\,[\text{A}]_0 = \log\frac{[\text{A}]_t}{[\text{A}]_0} = -\frac{kt}{2.30} \qquad [13.12]$$

Rearranging this equation slightly gives

$$\log\,[\text{A}]_t = \left(-\frac{k}{2.30}\right)t + \log\,[\text{A}]_0 \qquad [13.13]$$

One important fact about Equation 13.13 is that it is in the form of an equation for a straight line. Straight-line equations are of the type

$$y = ax + b \qquad [13.14]$$

(See Appendix A.4.) Equation 13.13 has this form with $y = \log\,[\text{A}]_t$, $a = -k/2.30$ (the slope), $x = t$, and $b = \log\,[\text{A}]_0$ (the intercept). Thus a graph of $\log\,[\text{A}]_t$ versus time gives a straight line with a slope of $-k/2.30$ and an intercept of $\log\,[\text{A}]_0$.

The conversion of methyl isonitrile to acetonitrile provides a simple example of a first-order reaction:

$$\underset{\text{Methyl isonitrile}}{H_3C{-}N{\equiv}C{:}} \longrightarrow \underset{\text{Acetonitrile}}{H_3C{-}C{\equiv}N{:}} \qquad [13.15]$$

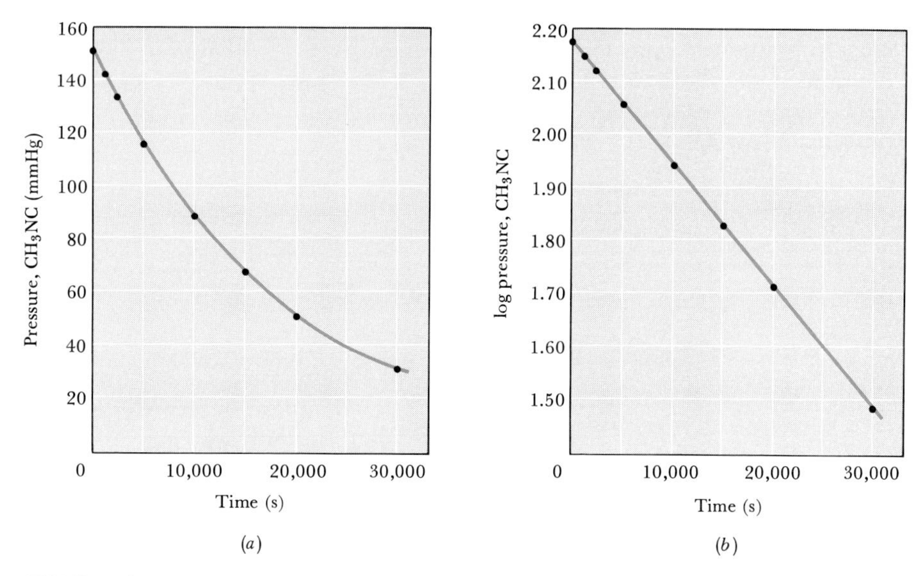

FIGURE 13.2 (*a*) Variation in the pressure of methyl isonitrile, CH_3NC, with time during the reaction $CH_3NC \longrightarrow CH_3CN$. (*b*) The data from (*a*) plotted in a linear form as log of CH_3NC pressure as a function of time.

Figure 13.2(*a*) shows how the pressure of the isonitrile varies with time as it rearranges in the gas phase at 198.9°C. Figure 13.2(*b*) shows the same data in straight-line form, as the logarithm of pressure as a function of time. Pressure is a legitimate unit of concentration for a gas because the number of moles per unit volume is directly proportional to pressure. The slope of the linear plot is -2.22×10^{-5}/s. (You should verify this for yourself, remembering that your result may vary slightly from ours because of the inaccuracies associated with reading the graph.) Because the slope equals $-k/2.30$,

$$k = -2.30(-2.22 \times 10^{-5}/\text{s}) = 5.11 \times 10^{-5}/\text{s}$$

For a first-order reaction, Equations 13.12 and 13.13 can be used to determine (1) the concentration of a reactant remaining at any time after the reaction has started, (2) the time required for a given fraction of sample to react, or (3) the time required for a reactant concentration to reach a certain level.

SAMPLE EXERCISE 13.3

The first-order rate constant for hydrolysis of a certain insecticide in water at 12°C is 1.45/yr. A quantity of this insecticide is washed into a lake in June, leading to an overall concentration of 5.0×10^{-7} g/cm^3 of water. Assuming that the effective temperature of the lake is 12°C, (a) what is the concentration of the insecticide in June of the following year; (b) how long will it take for the concentration of the insecticide to drop to 3.0×10^{-7} g/cm^3?

Solution: (a) Substituting $k = 1.45$/yr, $t = 1$ yr, and $[\text{insecticide}]_0 = 5.0 \times 10^{-7}$ g/cm^3 into Equation 13.13 gives

$$\log\,[\text{insecticide}]_t = -\frac{1.45/\text{yr}}{2.30}(1.00\text{ yr}) + \log(5.0 \times 10^{-7})$$

$$= -0.630 - 6.30 = -6.93$$

$$[\text{insecticide}]_t = 10^{-6.93} = 1.2 \times 10^{-7}\text{ g/cm}^3$$

(The concentration units for $[A]_0$ and $[A]_t$ must be the same.)

(b) Again substituting into Equation 13.13, with $[\text{insecticide}]_t = 3.0 \times 10^{-7}\ \text{g/cm}^3$, gives

$$\log(3.0 \times 10^{-7}) = -\frac{1.45/\text{yr}}{2.30}t + \log(5.0 \times 10^{-7})$$

Solving for t gives

$$t = -\frac{2.30}{1.45/\text{yr}}[\log(3.0 \times 10^{-7}) - \log(5.0 \times 10^{-7})]$$

$$= -\frac{2.30}{1.45/\text{yr}}(-6.52 + 6.30) = 0.35\ \text{yr}$$

Half-Life

The half-life of a reaction, $t_{1/2}$, is the time required for the concentration of the reactant to decrease to halfway between its initial and final values. In the cases we'll be considering, the final concentration is zero. To obtain an expression for $t_{1/2}$ for a first-order reaction, we begin with Equation 13.12. The half-life corresponds to the time when $[A]_t = \frac{1}{2}[A]_0$. Inserting these quantities into the equation, we have

$$\log\frac{\frac{1}{2}[A]_0}{[A]_0} = \left(\frac{-k}{2.30}\right)t_{1/2}$$

$$\log\tfrac{1}{2} = \left(\frac{-k}{2.30}\right)t_{1/2} \qquad [13.16]$$

$$t_{1/2} = \frac{-(2.30)\log\frac{1}{2}}{k} = \frac{0.693}{k}$$

Notice that $t_{1/2}$ is independent of the initial concentration of reactant. This result tells us that if we measure reactant concentration at *any* time in the course of a first-order reaction, the concentration of reactant will be half of that measured value at a time $0.693/k$ later. The concept of half-life is widely used in describing radioactive decay. This application is discussed in detail in Section 20.4.

The data for the first-order rearrangement of methyl isonitrile at 198.9°C are graphed in Figure 13.3. The first half-life is shown at

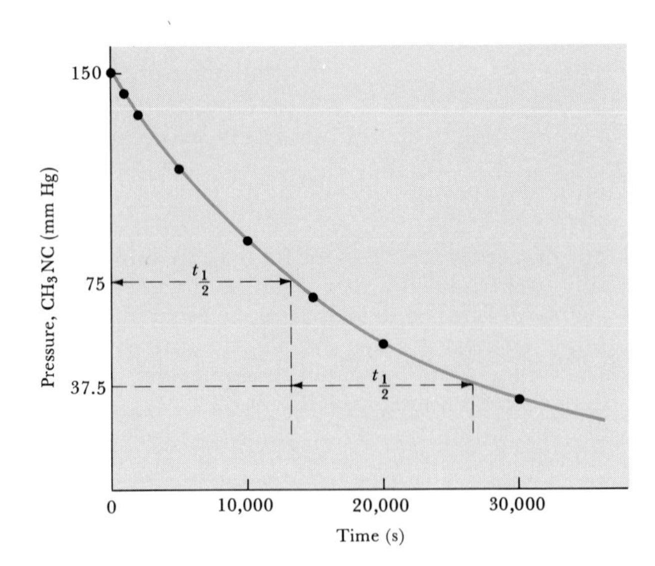

FIGURE 13.3 Pressure of methyl isonitrile as a function of time. Two successive half-lives of the rearrangement reaction, Equation 13.15, are shown.

13,320 s. At a time 13,320 s later, the isonitrile concentration has decreased to $\frac{1}{2}$ of $\frac{1}{2}$, or $\frac{1}{4}$ the original concentration. *It is a characteristic of a first-order reaction that the concentration of the reactant decreases by factors of* $\frac{1}{2}$ *in a series of regularly spaced time intervals.*

Second-Order Reactions

For a reaction that is second order in just one reactant, A, the rate is given by

$$\text{Rate} = k[\text{A}]^2$$

Relying on calculus, this rate law can be used to derive the following equation:

$$\frac{1}{[\text{A}]_t} = \frac{1}{[\text{A}]_0} + kt \qquad [13.17]$$

In this case, a linear form of the data is obtained by plotting $1/[\text{A}]_t$ versus t. The resultant line has a slope of k and an intercept of $1/[\text{A}]_0$. One way to distinguish between first- and second-order rate laws is to graph both $\log [\text{A}]_t$ and $1/[\text{A}]_t$ against t. If the $\log [\text{A}]_t$ plot is linear, the reaction is first order; if the $1/[\text{A}]_t$ plot is linear, the reaction is second order.

Using Equation 13.17, it can be shown that the half-life of a second-order reaction is given by the expression $t_{1/2} = 1/k[\text{A}]_0$. It is *not* independent of the initial concentration of reactant as is $t_{1/2}$ for a first-order reaction. Thus a constant half-life is indicative of a first-order reaction, but not a second-order one.

SAMPLE EXERCISE 13.4

The following data were obtained for the gas-phase decomposition of nitrogen dioxide at 300°C, $2NO_2(g) \longrightarrow 2NO(g) + O_2(g)$:

Time (s)	$[NO_2](M)$
0	0.0100
50	0.0079
100	0.0065
200	0.0048
300	0.0038

Is the reaction first or second order in NO_2?

Solution: To test whether the reaction is first or second order, we can construct plots of $\log [NO_2]$ and $1/[NO_2]$ against time. In doing so, we will find it useful to prepare the following table from the data given:

Time (s)	$[NO_2]$	$\log [NO_2]$	$1/[NO_2]$
0	0.0100	−2.00	100
50	0.0079	−2.10	127
100	0.0065	−2.19	154
200	0.0048	−2.32	208
300	0.0038	−2.42	263

As Figure 13.4 shows, only the plot of $1/[NO_2]$ versus time is linear. Thus the reaction obeys a second-order rate law: rate $= k[NO_2]^2$. From the slope of this straight-line graph we have that $k = 0.543/M\text{-s}$.

A reaction may also be second order by having a first-order dependence of the rate on each of two reagents, that is, rate $= k[\text{A}][\text{B}]$. It is possible to derive an expression for the variation in concentrations of A and B with time. However, we will not consider this and other more complicated rate laws in this text.

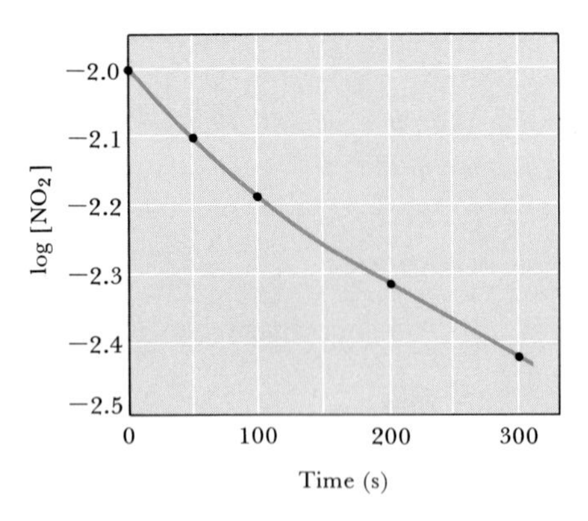

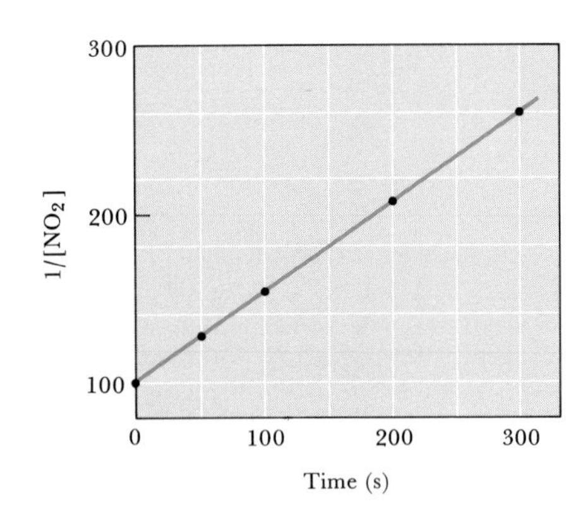

FIGURE 13.4 Kinetic data from Sample Exercise 13.4 for the reaction $2NO_2(g) \longrightarrow 2NO(g) + O_2(g)$ at 300°C. The plot of log $[NO_2]$ versus time is not linear; consequently, the reaction is not first order in NO_2. The plot of $1/[NO_2]$ versus time is linear; the reaction is second order in NO_2.

13.4 THE TEMPERATURE DEPENDENCE OF REACTION RATES

The rates of most chemical reactions increase as the temperature rises. We see examples of this generalization in many biological processes around us. The rate at which grass grows and the metabolic activity of the common housefly are both greater in warm weather than in the cold of winter. As another example, food cooks more rapidly in boiling water than in merely hot water. However, it is dangerous to place too much emphasis on the apparent rates at which biological systems operate as a function of temperature, because they are very complex and are adapted to operate optimally in a narrow temperature range. To obtain a clear understanding of how temperature affects reaction rates, we must examine simple reaction systems. As an example, let us consider the reaction about which we have already learned quite a bit, the first-order rearrangement of methyl isonitrile (Equation 13.15). Figure 13.5 shows the experimentally determined rate constant for this reaction as a function of temperature. It is evident that the rate of the reaction increases rapidly with temperature. Furthermore, the increase is nonlinear.

As we seek an explanation for this behavior, perhaps the first question to ask is, why do *any* reactions go slowly? What keeps reactions from simply occurring immediately? If the methyl isonitrile molecules are going eventually to rearrange into acetonitrile, why don't they all do it at once?

Activation Energy

We know from the kinetic-molecular theory of gases that with increasing temperature the average energy of the gas molecules increases. The fact that the rate of the methyl isonitrile reaction increases with increasing temperature suggests that perhaps the rearrangement is related to the kinetic energies of the molecules. Svante Arrhenius suggested in 1888 that before reaction can occur, a certain minimum amount of energy must be available to "propel" the molecules from one chemical state into another. The situation is rather like that shown in Figure 13.6. The boulder will be in a lower (or more stable) potential-energy state in

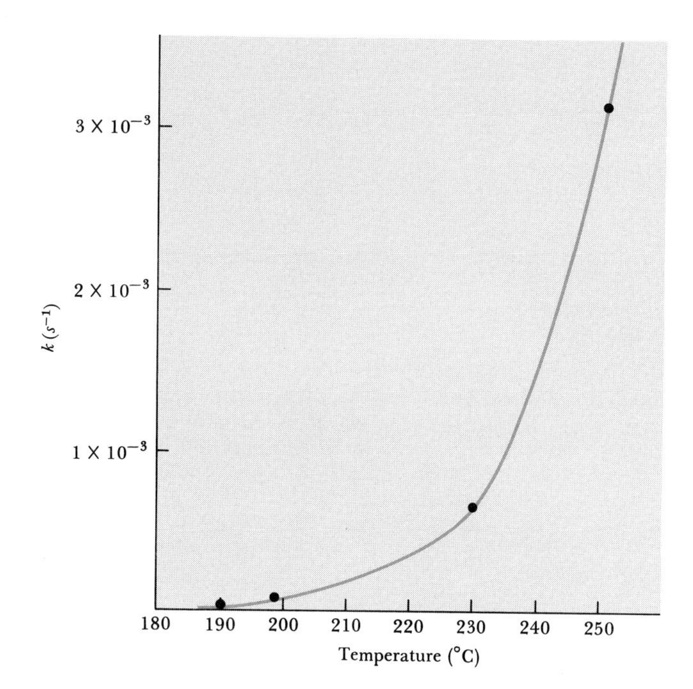

FIGURE 13.5 Variation in the first-order rate constant for rearrangement of methyl isonitrile as a function of temperature. (The four points indicated are used in connection with Sample Exercise 13.5.)

valley *B* than in valley *A*. Before it can come to rest in *B*, however, it must acquire the energy to overcome the barrier blocking its passage from the one state into the other. In the same way, molecules may require a certain minimum energy to overcome the forces that tend to keep them as they are, if they are to form the new chemical bonds that will result in a different arrangement. In our methyl isonitrile example, we might imagine that for rearrangement to occur, the N≡C portion of the molecule must turn over:

$$H_3C{-}N{\equiv}C: \longrightarrow \left[H_3C\cdots \begin{matrix} \ddot{C} \\ ||| \\ \underset{..}{N} \end{matrix} \right] \longrightarrow H_3C{-}C{\equiv}N: \qquad [13.18]$$

Even though the bonding may be more stable in the product acetonitrile than in the starting compound, energy is required to force the molecule through the relatively unstable intermediate state to the final result. The energy of the molecule as it proceeds along this reaction pathway is

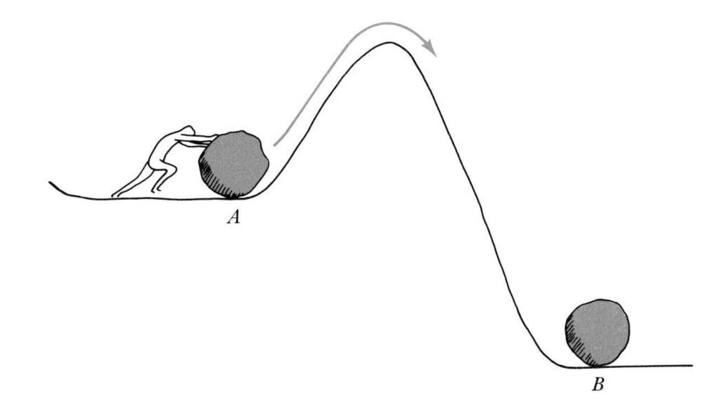

FIGURE 13.6 Illustration of the potential-energy profile for a boulder. The boulder must be moved over the energy barrier before it can come to rest in the lower-energy location, *B*.

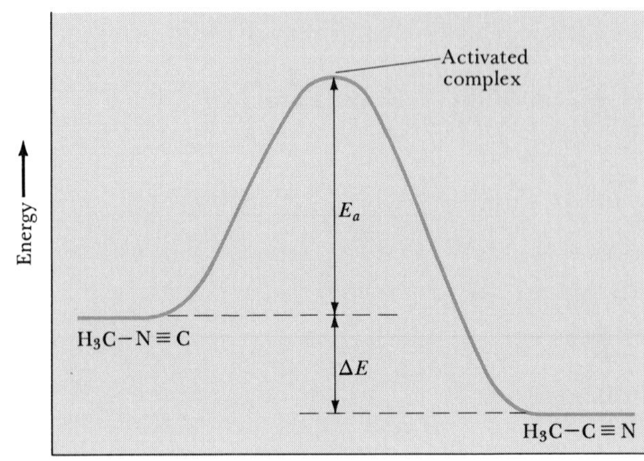

FIGURE 13.7 Energy profile for the rearrangement of methyl isonitrile. The molecule must surmount the activation-energy barrier before it can form the product, acetonitrile.

shown in Figure 13.7. Arrhenius called the energy barrier between the starting molecule and the highest energy along the reaction pathway the activation energy, E_a. The particular arrangement of atoms that has the maximum energy is often called the activated complex.

The conversion of $H_3CN{\equiv}C$ to $H_3CC{\equiv}N$ is exothermic; Figure 13.7 therefore shows the product as having a lower energy than the reactant. Notice that the reverse reaction is then endothermic; for that reaction the activation barrier is equal to the sum of ΔE and E_a for the forward reaction.

Energy is transferred between molecules through collisions. Thus, within a certain period of time, any particular isonitrile molecule might acquire enough energy to overcome the energy barrier and be converted into acetonitrile. At any given temperature only a small fraction of collisions will occur with sufficient energy to overcome the barrier to reaction. However, as shown in Figure 9.9, the distribution of molecular speeds is more spread out toward higher values when the gas is at a higher temperature. The distribution of kinetic energies of molecules changes in a similar way, as shown in Figure 13.8. This graph shows that at the higher temperature a larger fraction of molecules possesses the minimum energy needed for reaction.

In addition to the requirement that the reactant species collide with sufficient energy to begin to rearrange bonds, there is also an orientational requirement. The relative orientations of the molecules

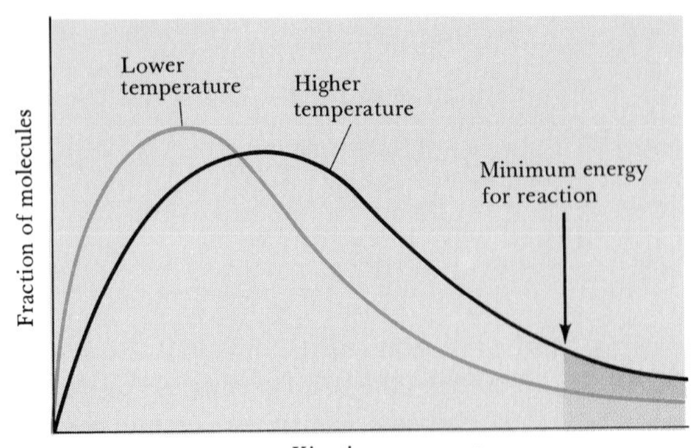

FIGURE 13.8 Distribution of kinetic energies in a sample of gas molecules at two different temperatures. At the higher temperature a larger number of molecules have higher energies. Thus a larger fraction in any one instant will have more than the minimum energy required for reaction.

during their collisions may determine whether the energy gets to the right place for reaction or whether atoms are suitably oriented to form new bonds. Thus only a fraction of the collisions possessing enough energy for reaction actually produce products.

In a mixture of H_2 and I_2 at ordinary temperatures and pressures, each molecule undergoes about 10^{10} collisions per second. If every collision between H_2 and I_2 resulted in formation of HI, the reaction would be over in much less than a second. Instead, at room temperature the reaction proceeds very slowly. Obviously, every collision does not lead to reaction. In fact, only about 1 in every 10^{13} collisions is effective. Only a small fraction of the collisions occur with suitable orientation and with sufficient energy to carry the molecule over the energy barrier to products. As the temperature increases, the number of collisions increases as does the fraction that are sufficiently energetic for reaction. With each 10°C rise in temperature, the rate of this reaction triples.

The Arrhenius Equation

Arrhenius noted that the increase in rate with increasing temperature for most reactions is nonlinear, as in the example shown in Figure 13.5. He found that most reaction-rate data obeyed the equation

$$\log k = \log A - \frac{E_a}{2.30RT} \quad [13.19]$$

where k is the rate constant. This equation is called the **Arrhenius equation.** The term E_a is the activation energy, which we have already defined; R is the gas constant (8.314 J/K-mol); and T is absolute temperature; A is constant, or nearly so, as temperature is varied. It is called the **frequency factor;** it is related to the frequency of collisions and the probability that the collisions are favorably oriented for reaction. Notice that as the magnitude of E_a increases, k becomes smaller. Thus reaction rates decrease as the energy barrier increases.

Equation 13.19 has the form of a straight line, in which one variable is $\log k$, and the other is $1/T$. The slope of the line is given by $-E_a/2.30R$; the intercept, at $1/T = 0$, is $\log k = \log A$. Thus Equation 13.19 can be used to determine E_a from a graph of $\log k$ versus $1/T$.

It is sometimes convenient to manipulate Equation 13.19 further to give the relationship between the rate constants at two different temperatures, T_1 and T_2. At T_1 we have

$$\log k_1 = \log A - \frac{E_a}{2.30RT_1}$$

At T_2,

$$\log k_2 = \log A - \frac{E_a}{2.30RT_2}$$

Subtracting $\log k_2$ from $\log k_1$ gives

$$\log k_1 - \log k_2 = \left(\log A - \frac{E_a}{2.30RT_1}\right) - \left(\log A - \frac{E_a}{2.30\,RT_2}\right)$$

Simplifying this equation and rearranging it gives

$$\log \frac{k_1}{k_2} = \frac{E_a}{2.30R}\left(\frac{1}{T_2} - \frac{1}{T_1}\right) \qquad [13.20]$$

Equation 13.20 provides a convenient means of calculating the rate constant at some temperature, T_1, when we know the activation energy and the rate constant, k_2, at some other temperature, T_2.

SAMPLE EXERCISE 13.5

The following table shows the rate constants for rearrangement of methyl isonitrile at various temperatures (these are the data that are graphed in Figure 13.5).

Temperature (°C)	k (s^{-1})
189.7	2.52×10^{-5}
198.9	5.25×10^{-5}
230.3	6.30×10^{-4}
251.2	3.16×10^{-3}

(a) From these data calculate the activation energy for the reaction. (b) What is the magnitude of the rate constant at 430.0 K?

Solution: (a) We must first convert temperatures to the absolute temperature scale, K. We then take the inverse of these temperatures, and obtain the corresponding log values for k. This gives us the following table:

T (K)	$1/T$ (K)	$\log k$
462.7	2.160×10^{-3}	−4.60
471.9	2.118×10^{-3}	−4.29
503.3	1.986×10^{-3}	−3.20
524.2	1.907×10^{-3}	−2.50

A graph of these data results in a straight line, as shown in Figure 13.9. The data points shown in the graph lie very close to the best straight line through all four points. The slope of the line is obtained by choosing two well-separated points, as shown, and reading off the coordinates of each:

$$\text{Slope} = \frac{-2.45 - (-4.45)}{0.00190 - 0.00214} = -8330$$

The numerator in this equation has no units, because logs have no units. The denominator has the units of $1/T$, that is, 1/K. Thus the overall units for the slope are K. The slope is equal to $-E_a/2.30R$. We want the value for the molar gas constant R in J/K-mol (Table 9.2), which is 8.314. Thus we obtain

$$\text{Slope} = \frac{-E_a}{2.30R}$$

$$E_a = -(\text{slope})(2.30R)$$

$$= -(-8330\ \text{K})(2.30)\left(8.31\ \frac{\text{J}}{\text{K-mol}}\right) \times \left(\frac{1\ \text{kJ}}{1000\ \text{J}}\right)$$

$$= 159\ \text{kJ/mol}$$

(b) To determine the rate constant, k_1, at 430 K, we apply Equation 13.20 with $E_a = 159$ kJ/mol, $k_2 = 2.52 \times 10^{-5}$/s, $T_2 = 462.7$ K, and $T_1 = 430.0$ K:

$$\log \frac{k_1}{2.52 \times 10^{-5}/\text{s}} = \frac{159\ \text{kJ/mol}}{(2.30)(8.31\ \text{J/K-mol})} \times \left(\frac{1}{462.7\ \text{K}} - \frac{1}{430.0\ \text{K}}\right)\left(\frac{10^3\ \text{J}}{1\ \text{kJ}}\right)$$

$$= -1.366$$

$$\frac{k_1}{2.52 \times 10^{-5}/\text{s}} = 10^{-1.366} = 4.31 \times 10^{-2}$$

$$k_1 = (2.52 \times 10^{-5}/\text{s})(4.31 \times 10^{-2})$$

$$= 1.09 \times 10^{-6}/\text{s}$$

13.5 REACTION MECHANISMS

A balanced equation for a chemical reaction indicates the substances that are present at the start of the reaction and those produced at the end. However, it provides no information about how the reaction occurs. The process by which a reaction occurs is called the reaction mecha-

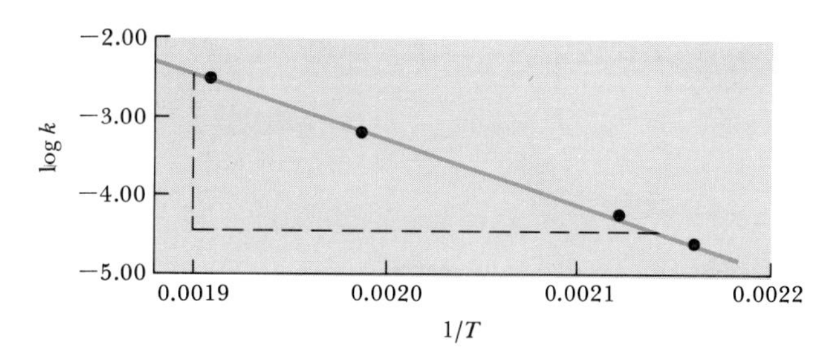

FIGURE 13.9 Log of the rate constant for rearrangement of methyl isonitrile as a function of $1/T$. The linear relationship is predicted from the Arrhenius equation.

nism. At the most sophisticated level, a reaction mechanism will describe in great detail the order in which bonds are broken and formed and the changes in relative positions of the atoms in the course of the reaction. But we will begin with more rudimentary descriptions of how reactions occur.

Elementary Reactions

We have seen (Section 13.4) that reactions take place as a result of collisions among reacting molecules. For example, the collision between methyl isonitrile, CH_3NC, and some other molecule can provide the energy to allow the CH_3NC to rearrange:

$$H_3C{-}N{\equiv}C: \longrightarrow \left[H_3C\cdots \begin{matrix} \ddot{C} \\ ||| \\ \underset{..}{N} \end{matrix} \right] \longrightarrow H_3C{-}C{\equiv}N:$$

Similarly, the reaction of O_3 and NO to form O_2 and NO_2 appears to occur as a result of a single collision involving suitably oriented and sufficiently energetic NO and O_3 molecules:

$$NO(g) + O_3(g) \longrightarrow NO_2(g) + O_2(g) \qquad [13.21]$$

Both of these processes occur in a single event or step and are called elementary reactions (or elementary steps).

The net change represented by a balanced chemical equation often occurs by a sequence of elementary reactions. For example, consider the reaction of NO_2 and CO:

$$NO_2(g) + CO(g) \longrightarrow NO(g) + CO_2(g) \qquad [13.22]$$

Below 225°C this reaction appears to proceed in two elementary steps. First, two NO_2 molecules collide and an oxygen atom is transferred from one to the other. The resultant NO_3 then transfers an oxygen atom to CO during a collision between these molecules:

$$NO_2(g) + NO_2(g) \longrightarrow NO_3(g) + NO(g) \qquad [13.23]$$

$$NO_3(g) + CO(g) \longrightarrow NO_2(g) + CO_2(g) \qquad [13.24]$$

The elementary reactions in a multistep mechanism must always add to give the chemical equation of the overall process. In the present example, the sum of the elementary reactions is

$$2NO_2(g) + NO_3(g) + CO(g) \longrightarrow NO_2(g) + NO_3(g) + NO(g) + CO_2(g)$$

Simplifying this equation by eliminating substances that appear on both sides of the arrow gives the net equation for the process, Equation 13.22. Because NO_3 is neither a reactant nor a product in the overall reaction, but is formed in one elementary reaction and consumed in the next, it is called an intermediate. Multistep mechanisms involve one or more intermediates.

The number of molecules that participate as reactants in an elementary reaction defines the molecularity of the reaction. If a single molecule is involved, the reaction is said to be unimolecular. The isomerization of methyl isonitrile (Equation 13.18) is a unimolecular process. Elementary reactions involving two reactant molecules are said to be bimolecular. The reaction between NO and O_3 (Equation 13.21) is bimolecular. Elementary reactions involving three molecules are said to be termolecular. Termolecular reactions are less probable than unimolecular or bimolecular reactions and are rarely encountered. The chance that four or more molecules will collide simultaneously with any regularity is even more remote; consequently, such collisions are never proposed as part of a reaction mechanism.

SAMPLE EXERCISE 13.6

It has been proposed that the conversion of ozone into O_2 proceeds in two steps:

$$O_3(g) \rightleftharpoons O_2(g) + O(g)$$
$$O_3(g) + O(g) \longrightarrow 2O_2(g)$$

(a) Write the equation for the overall reaction. (b) Identify the intermediate, if any. (c) Describe the molecularity of each step in the mechanism.

Solution: (a) Adding the two elementary reactions gives

$$2O_3(g) + O(g) \longrightarrow 3O_2(g) + O(g)$$

Because $O(g)$ appears in equal amounts on both sides of the equation, it can be eliminated to give the net equation for the chemical process:

$$2O_3(g) \longrightarrow 3O_2(g)$$

(b) The intermediate is $O(g)$. It is neither an original reactant nor a final product, but is formed in one step and consumed in another.

(c) The first elementary reaction involves a single reactant and is consequently unimolecular. The second step, which involves two reactant molecules, is bimolecular.

Rate Laws of Elementary Reactions

In discussing rate laws in Section 13.2, we stressed the fact that rate laws must be determined experimentally; they cannot, in general, be predicted from the coefficients of balanced chemical reactions. It is not the chemical equation for the overall process that determines the rate law; rather, it is the elementary reactions and their relative speeds.

The rate law of any elementary reaction is based directly on its molecularity. For example, consider the general unimolecular process

$$A \longrightarrow \text{products} \qquad [13.25]$$

It seems clear that as the number of A molecules increases, the number that decompose in a given interval of time will increase. Thus the rate of a unimolecular process will be first order:

$$\text{Rate} = k[\text{A}] \qquad [13.26]$$

In the case of bimolecular reactions, the rate law will be second order, as in the following examples:

$$\text{A} + \text{B} \longrightarrow \text{products} \qquad \text{Rate} = k[\text{A}][\text{B}] \qquad [13.27]$$

$$\text{A} + \text{A} \longrightarrow \text{products} \qquad \text{Rate} = k[\text{A}]^2 \qquad [13.28]$$

The second-order rate law follows from the fact that the rate of collision between A and B molecules is proportional to the concentrations of A and B.

In general, the order for each reactant in an elementary reaction is equal to its coefficient in the chemical equation for that step. Of course, we cannot tell by merely looking at a balanced chemical equation whether that reaction occurs in a single step or in a number of steps. That brings us back to the need for experimental studies.

SAMPLE EXERCISE 13.7

If the following reaction occurs in a single elementary step, predict the rate law:

$$O_3(g) + NO(g) \longrightarrow NO_2(g) + O_2(g)$$

Solution: The rate law is first order each in O_3 and NO, corresponding to the coefficients (understood to be one) in the chemical equation:

$$\text{Rate} = k[O_3][NO]$$

Rate Laws of Multistep Mechanisms

When chemical reactions occur by a mechanism involving a number of sequential elementary steps, it often happens that one step is much slower than others. The overall rate of a reaction cannot exceed the rate of the slowest elementary step of its mechanism. The slow step limits the overall reaction rate and is consequently called the **rate-determining step.** The situation is analogous to the progression of auto traffic through a tunnel. The rate of traffic flow is largely determined by the rates of the slowest cars.

When a mechanism includes a rate-determining step, the rate law for the overall reaction will be governed by the rate law for that slow step. Consider the situation where the first step of a mechanism is the rate-determining one. This situation is encountered in the reaction of NO_2 and CO mentioned earlier in this section (Equation 13.22):

$$NO_2(g) + NO_2(g) \longrightarrow NO_3(g) + NO(g) \qquad \text{(slow)}$$

$$NO_3(g) + CO(g) \longrightarrow NO_2(g) + CO_2(g) \qquad \text{(fast)}$$

The rate law found experimentally corresponds to that for the first step in the mechanism, the slow, rate-determining step:

$$\text{Rate} = k[NO_2]^2 \qquad [13.29]$$

In general, steps that occur after the rate-determining one do not affect the rate law for the overall process.

SAMPLE EXERCISE 13.8

The gas-phase reaction of nitric oxide, NO, and fluorine, F_2, to form nitrosyl fluoride, NOF, is believed to occur by the following mechanism:

$$NO(g) + F_2(g) \longrightarrow NOF(g) + F(g) \quad \text{(slow)}$$
$$NO(g) + F(g) \longrightarrow NOF(g) \quad \text{(fast)}$$

If the first step of the mechanism is rate determining, predict the rate law for the overall reaction.

Solution: The rate law for the overall reaction,

$$2NO(g) + F_2(g) \longrightarrow 2NOF(g)$$

is determined by the first step:

$$\text{Rate} = k[NO][F_2]$$

It is not so easy to derive the rate law for a mechanism in which an intermediate is a reactant in the rate-determining step. This situation arises in multistep mechanisms when the first step is *not* rate determining. Let's consider one example, the gas-phase reaction of chlorine, Cl_2, with chloroform, $CHCl_3$:

$$Cl_2(g) + CHCl_3(g) \longrightarrow HCl(g) + CCl_4(g) \qquad [13.30]$$

Experiments indicate that the rate law for this reaction is as follows:

$$\text{Rate} = k[CHCl_3][Cl_2]^{1/2} \qquad [13.31]$$

Clearly, the reaction cannot occur in a single elementary step; in that case the rate law would be first order in both $CHCl_3$ and Cl_2, corresponding to the coefficients in the balanced chemical equation for the process. To explain the rate law and other experimental features of this reaction, the following sequence of elementary reactions is proposed as the mechanism:*

$$Cl_2(g) \underset{k_{-1}}{\overset{k_1}{\rightleftharpoons}} 2Cl(g) \quad \text{(fast)} \qquad [13.32]$$

$$Cl(g) + CHCl_3(g) \xrightarrow{k_2} HCl(g) + CCl_3(g) \quad \text{(slow)} \qquad [13.33]$$

$$Cl(g) + CCl_3(g) \xrightarrow{k_3} CCl_4(g) \quad \text{(fast)} \qquad [13.34]$$

Because the second step of the mechanism is the slow, rate-determining step, the rate of the overall reaction should be governed by the rate law for that step:

$$\text{Rate} = k_2[CHCl_3][Cl] \qquad [13.35]$$

However, Cl is an intermediate generated by dissociation of Cl_2 molecules. Intermediates are normally encountered in low, unknown concentrations. Experimental rate laws are generally expressed in terms of substances present in measurable concentrations, not in terms of intermediates. Thus the experimental rate law is expressed in terms of the reactant, Cl_2, from which the intermediate Cl atoms form. To see

*The subscript 1 on k identifies this rate constant as that for the first elementary reaction of the mechanism. Similarly, k_{-1} is the rate constant for the reverse of the first reaction, and k_2 is that for the second reaction, and so forth.

how the concentration of Cl depends on the concentration of Cl_2, we assume that the first step begins to reverse itself soon after the start of the reaction, before step two has a chance to form much product. That is, both the forward and reverse reactions in Equation 13.32 are rapid relative to reaction 13.33. If we further assume that the forward and reverse reactions of the first step are approximately equal in rate (that is, they are in equilibrium), then we can write

$$\text{Rate}_1 = k_1[\text{Cl}_2] = \text{rate}_{-1} = k_{-1}[\text{Cl}]^2 \qquad [13.36]$$

Solving for Cl, we have

$$[\text{Cl}] = \left(\frac{k_1}{k_{-1}}[\text{Cl}_2]\right)^{1/2} \qquad [13.37]$$

Substituting this relationship into the rate law for the rate-determining step (Equation 13.35), we have

$$\text{Rate} = k_2[\text{CHCl}_3]\left(\frac{k_1}{k_{-1}}[\text{Cl}_2]\right)^{1/2} \qquad [13.38]$$

Combining constants and defining $k = k_2(k_1/k_{-1})^{1/2}$, we have the experimentally observed rate law:

$$\text{Rate} = k[\text{CHCl}_3][\text{Cl}_2]^{1/2}$$

The basic procedure in establishing the mechanism of a chemical reaction is first to determine the rate law experimentally. One or more elementary reactions are then postulated that account for that rate law and for other experimental observations. The fact that a mechanism gives a rate law that agrees with the observed one does not prove that that mechanism is correct. Often several mechanisms can be envisioned that give rise to the same rate law. To distinguish between such mechanisms, chemists might search for proposed intermediates or carry out other types of studies. As an example, consider the following reaction:

$$O^+(g) + NO(g) \longrightarrow NO^+(g) + O(g) \qquad [13.39]$$

This reaction occurs in the upper atmosphere. The rate law for this reaction is: rate $= k[O^+][NO]$, and the reaction is believed to occur in a single elementary step. However, we may ask whether the reaction involves the breaking of the NO bond and the formation of a new bond between N and O^+, or whether it merely involves the transfer of an electron from NO to O^+. These two possibilities are shown in Figure 13.10. The question of which of these mechanisms is operative is answered by performing the reaction using O^+ that has been highly enriched in the rare isotope ^{18}O.

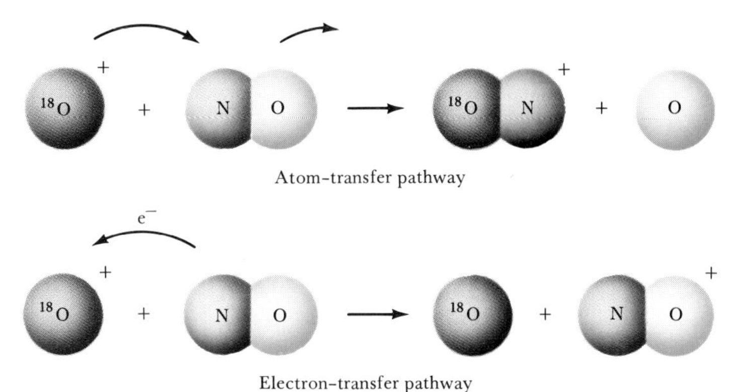

FIGURE 13.10 Alternative pathways for the reaction $O^+ + NO \longrightarrow O + NO^+$. The results of experiments using ^{18}O labeling show that the reaction proceeds according to the electron-transfer pathway.

Because this isotope is present to the extent of only 0.2 percent in nature, the NO contains only 0.2 percent $N^{18}O$. By measuring the location of ^{18}O in the reaction products, the two mechanisms can be distinguished (see Figure 13.10). The experiment indicates that no ^{18}O is incorporated into NO^+; thus the experimental results are consistent with the electron-transfer pathway but not the atom-transfer pathway.

13.6 CATALYSIS

A catalyst is a substance that acts to change the speed of a chemical reaction without itself undergoing a permanent chemical change in the process. Nearly all catalysts increase reaction rates. Catalysts are very common; most reactions occurring in the human body, the atmosphere, the oceans, or in industrial chemical processes are affected by catalysts.

If you have been exposed to chemical laboratory work, it is likely that you have carried out the reaction in which oxygen is produced by heating potassium chlorate, $KClO_3$:

$$2KClO_3(s) \xrightarrow{\Delta} 2KCl(s) + 3O_2(g) \qquad [13.40]$$

In the absence of a catalyst, $KClO_3$ does not readily decompose in this manner, even on strong heating. However, mixing black manganese dioxide, MnO_2, with the $KClO_3$ before heating causes the reaction to occur much more readily. The MnO_2 can be recovered largely unchanged from this reaction, so it is clear that the overall chemical process is still the same. Thus MnO_2 acts as a catalyst for decomposition of $KClO_3$. As another example, we know that a cube of sugar, when dissolved in water at 37°C, does not undergo oxidation at a significant rate. The sugar could be recovered essentially unchanged from the solution after several days. Yet when sugar is ingested into the human body at about 37°C it is rapidly oxidized and soon ends up mostly as carbon dioxide and water:

$$C_{12}H_{22}O_{11}(aq) + 12O_2(aq) \longrightarrow 12CO_2(aq) + 11H_2O(l) \qquad [13.41]$$

The oxidation of sugar in the biochemical system has been greatly speeded up by the presence of one or more catalysts. These biochemical catalysts are enzymes, protein molecules that act to catalyze specific biochemical reactions. (Enzymes are discussed at length in Chapter 26.)

Much industrial chemical research is devoted to the search for new and more effective catalysts for reactions of commercial importance. Extensive research efforts also are devoted to finding means of inhibiting or removing certain catalysts that promote undesirable reactions, such as those involved in corrosion of metals, aging, and tooth decay.

Homogeneous Catalysis

A catalyst that is present in the same phase as the components of a chemical reaction is known as a homogeneous catalyst. For example, a homogeneous catalyst for a reaction occurring in solution would itself be dissolved in the solution.

In Chapter 12 we considered a simple example of homogeneous catalysis, the action of NO in promoting the decomposition of ozone, O_3. The

NO acts as a catalyst by reaction with O_3 to form NO_2 and O_2. The NO_2 thus formed then reacts with atomic oxygen present in the stratosphere to reform NO and yield O_2 as the other product. The sequence of reactions and the overall result are as follows:

$$\begin{array}{rcl} NO(g) + O_3(g) & \longrightarrow & NO_2(g) + O_2(g) \\ NO_2(g) + O(g) & \longrightarrow & NO(g) + O_2(g) \\ \hline O_3(g) + O(g) & \longrightarrow & 2O_2(g) \end{array}$$

In this example, NO acts as a catalyst for O_3 decomposition because it speeds up the rate of the overall reaction without itself undergoing any net, or overall, chemical change; it is used in one step and reformed in the next.

As another example, hydrogen peroxide, H_2O_2, when dissolved in water undergoes slow decomposition, forming oxygen and water:

$$2H_2O_2(aq) \longrightarrow 2H_2O(l) + O_2(g) \quad [13.42]$$

In the absence of a catalyst, this reaction occurs at an extremely slow rate. Many different substances are capable of catalyzing the reaction; among these is bromine, Br_2. The bromine reacts with hydrogen peroxide in acidic solution, forming bromide ion and liberating oxygen:

$$Br_2(aq) + H_2O_2(aq) \longrightarrow 2Br^-(aq) + 2H^+(aq) + O_2(g) \quad [13.43]$$

If this reaction were all that were involved, bromine would not be a catalyst, because it undergoes chemical change in the reaction. It happens, however, that hydrogen peroxide reacts with bromide ion in acidic solution to form bromine:

$$2Br^-(aq) + H_2O_2(aq) + 2H^+(aq) \longrightarrow Br_2(l) + 2H_2O(l) \quad [13.44]$$

The overall sum of reaction Equations 13.43 and 13.44 is just Equation 13.42 (Add these two reactions together yourself to make sure you see that reaction Equation 13.42 results.) We see that bromine is indeed a catalyst in the reaction, because it speeds the overall reaction without itself undergoing any net, or overall, change.

On the basis of the Arrhenius expression for a chemical reaction, Equation 13.19, the rate constant k is determined by the activation energy E_a and the frequency factor A. A catalyst may affect the rate of reaction by altering the value for either E_a or A. The most dramatic catalytic effects come from lowering of E_a. As a general rule, *a catalyst lowers the overall activation energy for chemical reaction.* The lowering of E_a by a catalyst is shown schematically in Figure 13.11.

A catalyst usually lowers the overall activation energy for reaction by providing a completely different pathway for reaction. The two examples given above involve a reversible, cyclic reaction of the catalyst with the reactants. For example, in the decomposition of hydrogen peroxide, two successive reactions of H_2O_2, with bromine and then with bromide, are involved. Because these two reactions together serve as a catalytic pathway for hydrogen peroxide decomposition, *both* of these reactions

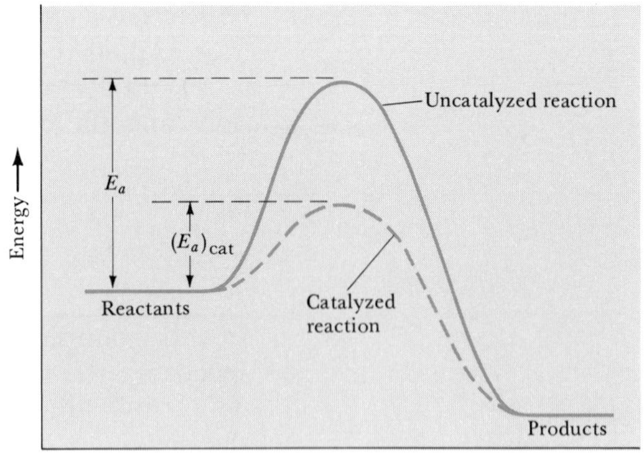

Uncatalyzed reaction
Energy
E_a
$(E_a)_{cat}$
Reactants
Catalyzed reaction
Products
Reaction pathway

FIGURE 13.11 Energy profile for a catalyzed and uncatalyzed reaction. The catalyst functions in this example to lower the activation energy for reaction. Notice that the energies of reactants and products are unchanged by the catalyst.

must have significantly lower activation energies than the uncatalyzed decomposition, as shown schematically in Figure 13.12.

The catalysts we have so far discussed (and this includes enzymes, which are discussed in more detail in Chapter 26) are homogeneous catalysts. However, a great many reactions are catalyzed by substances that exist in a different phase from the reactants.

Heterogeneous Catalysis

A **heterogeneous catalyst** exists in a different phase from the reactant molecules. For example, a reaction between molecules in the gas phase might be catalyzed by a finely divided metal oxide. In the absence of a catalyst, the reaction would occur slowly in the gas phase. However, when the catalyst is present, the reaction occurs more rapidly on the surface of the solid catalyst.

Many industrially important reactions occurring in the gas phase are catalyzed by solid surfaces. For example, hydrocarbon molecules are rearranged to form gasoline with the aid of what are called "cracking" catalysts (Section 25.2). Reactions occurring in solution may also be catalyzed by solids. Heterogeneous catalysts are often composed of finely divided metal or metal oxides. Because the catalyzed reaction occurs on

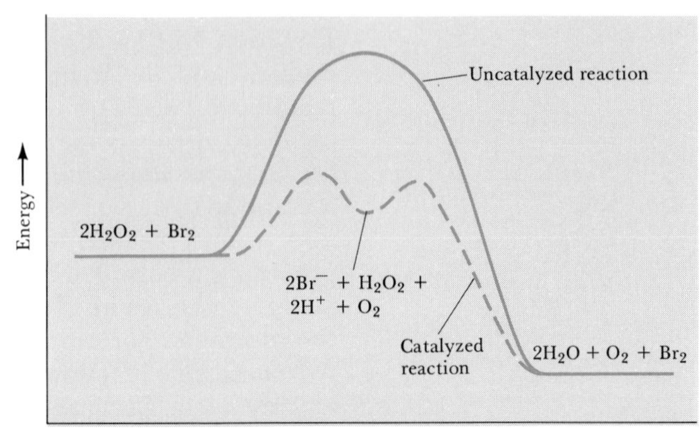

FIGURE 13.12 Energy profile for the uncatalyzed decomposition of hydrogen peroxide, and for the reaction as catalyzed by Br_2. The catalyzed reaction involves two successive steps, each of which has a lower activation energy than the uncatalyzed reaction.

the surface, special methods are often used to prepare catalysts so that they have very large surface areas.

The initial step in heterogeneous catalysis is usually **adsorption** of reactants. The term *ad*sorption should be distinguished from *ab*sorption. Adsorption refers to binding of molecules to a surface, whereas absorption refers to the uptake of molecules into the interior of another substance. Adsorption occurs because the atoms or ions at the surface of a solid are extremely reactive. Unlike their counterparts in the interior of the substance, they have unfulfilled valence requirements. The unused bonding capability of surface atoms or ions may be utilized to bond molecules from the gas or solution phase to the surface of the solid. In practice, not all the atoms or ions of the surface are reactive; various impurities may be adsorbed at the surface, and these may occupy many potential reaction sites and block further reaction. The places where reacting molecules may become adsorbed are called **active sites.** The number of active sites per unit amount of catalyst depends on the nature of the catalyst, on its method of preparation, and on its treatment before use.

As an example of heterogeneous catalysis, consider the hydrogenation of ethylene to form ethane:

$$\underset{\text{Ethylene}}{H_2C{=}CH_2} + H_2 \longrightarrow \underset{\text{Ethane}}{H_3C{-}CH_3} \qquad [13.45]$$

In the absence of a catalyst, this reaction does not occur at all readily. However, in the presence of a very finely divided metal such as nickel, palladium, or platinum, the reaction occurs rather easily at room temperature, under a few hundred atmospheres of hydrogen pressure. The mechanism by which reaction occurs is shown diagrammatically in Figure 13.13. Both ethylene and hydrogen are adsorbed at the metal surface [Figure 13.13(*a*)]. The adsorption of hydrogen results in breaking of the H—H bond and formation of two M—H bonds, where M represents the metal surface [Figure 13.13(*b*)]. The hydrogen atoms are relatively free to move about the surface. When they encounter an adsorbed ethylene, the hydrogen may become bound to the carbon [Figure 13.13(*c*)]. The carbon thus acquires four σ bonds about it, which reduces its tendency to remain adsorbed at the metal. When the other carbon also acquires a hydrogen, the ethane molecule is released from the surface [Figure 13.13(*d*)]. The active site is ready to adsorb another ethylene molecule, and thus begin the cycle again.

Heterogeneous catalysis plays a major role in the fight against urban air pollution. Two components of automobile exhausts that are involved in the formation of photochemical smog are nitrogen oxides and unburned hydrocarbons of various types (Section 12.4). In addition, automobile exhausts may contain considerable quantities of carbon monoxide. Even with the most careful attention to engine design and fuel characteristics, it is not possible under normal driving conditions to reduce the contents of these pollutants to an acceptable level in the exhaust gases coming from the engine. It is therefore necessary somehow to remove them from the exhaust gases before they are vented to the air. This removal is accomplished in the *catalytic converter*.

The catalytic converter, illustrated in Figure 13.14,

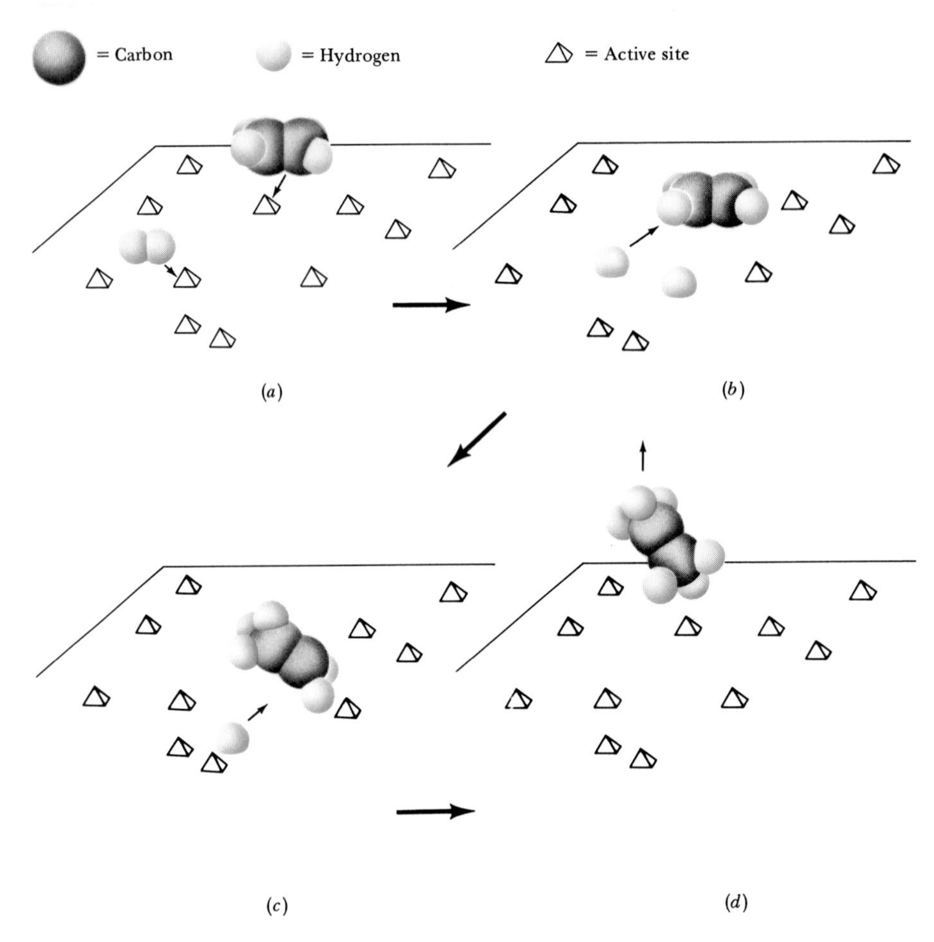

FIGURE 13.13 Mechanism for reaction of ethylene with hydrogen on a catalytic surface. (*a*) The hydrogen and ethylene are adsorbed at the metal surface. (*b*) The H—H bond is broken to give adsorbed hydrogen atoms. (*c*) These migrate to the adsorbed ethylene and bond to the carbon atoms. (*d*) As C—H bonds are formed, the adsorption of the molecule to the metal surface is decreased, and ethane is released.

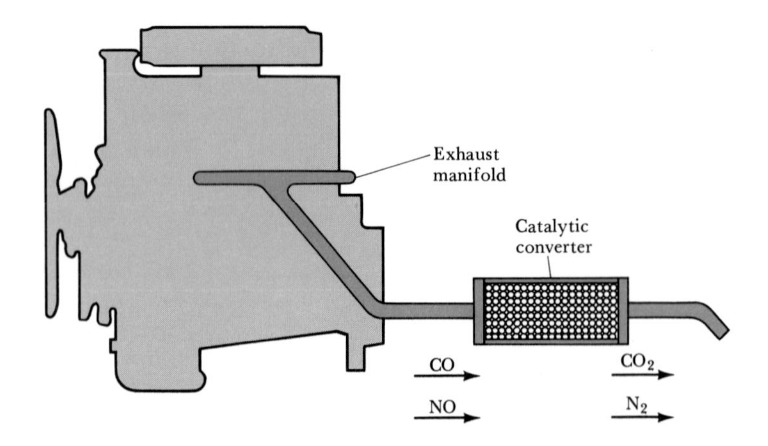

FIGURE 13.14 Illustration of the arrangement and functions of a catalytic converter.

must perform two distinct functions: (1) oxidation of CO and unburned hydrocarbons to carbon dioxide and water, and (2) reduction of nitrogen oxides to nitrogen gas:

$$\text{CO, hydrocarbons } (C_xH_y) \xrightarrow{O_2} CO_2 + H_2O$$

$$NO, NO_2 \longrightarrow N_2$$

These two functions require two distinctly different catalysts. The development of a successful catalyst system represents a very difficult challenge. The catalysts must be effective over a wide range of operating temperatures; they must continue to be active in spite of the poisoning action of various gasoline additives emitted along with the exhaust; they must be physically rugged enough to withstand gas turbulence and the mechanical shocks of driving under various conditions for thousands of miles.

Catalysts that promote the combustion of CO and hydrocarbons are, in general, the transition-metal oxides and noble metals such as platinum. As an example, a mixture of two different metal oxides such as CuO and Cr_2O_3 might be used. These materials are supported on a structure (Figure 13.15), which allows the best possible contact between the flowing exhaust gas and the catalyst surface. Either bead or honeycomb structures made from alumina, Al_2O_3, and impregnated with the catalyst may be employed. Such catalysts operate by first adsorbing oxygen gas, also present in the exhaust gas. This adsorption weakens the O—O bond in O_2, so that oxygen atoms are in effect available for reaction with adsorbed CO to form CO_2. Hydrocarbon oxidation probably proceeds somewhat similarly, with the hydrocarbons first being adsorbed by rupture of a C—H bond.

Reduction of nitrogen oxides is favored thermodynamically. That is, the decomposition of NO, for example, to yield N_2 and O_2 is favored, but the reaction is extremely slow. A catalyst is therefore necessary. The most effective catalysts are transition-metal oxides and noble metals, the same kinds of materials that catalyze the oxidation of CO and hydrocarbons. The catalysts that are most effective in one reaction, however, are usually much less effective in the other. It is therefore necessary to have two different catalytic components.

The activity of catalytic converters decreases with use, because of loss of active catalyst, cracking and fractures due to repeated heating and cooling, and poisoning of the catalysts. One of the most active poisons is the lead that comes from the tetramethyl lead, $Pb(CH_3)_4$, or tetraethyl lead, $Pb(C_2H_5)_4$, added to gasoline. Because of the severe catalyst poisoning that results from use of leaded fuels, cars built since 1975 have been engineered to discourage the use of leaded gas.

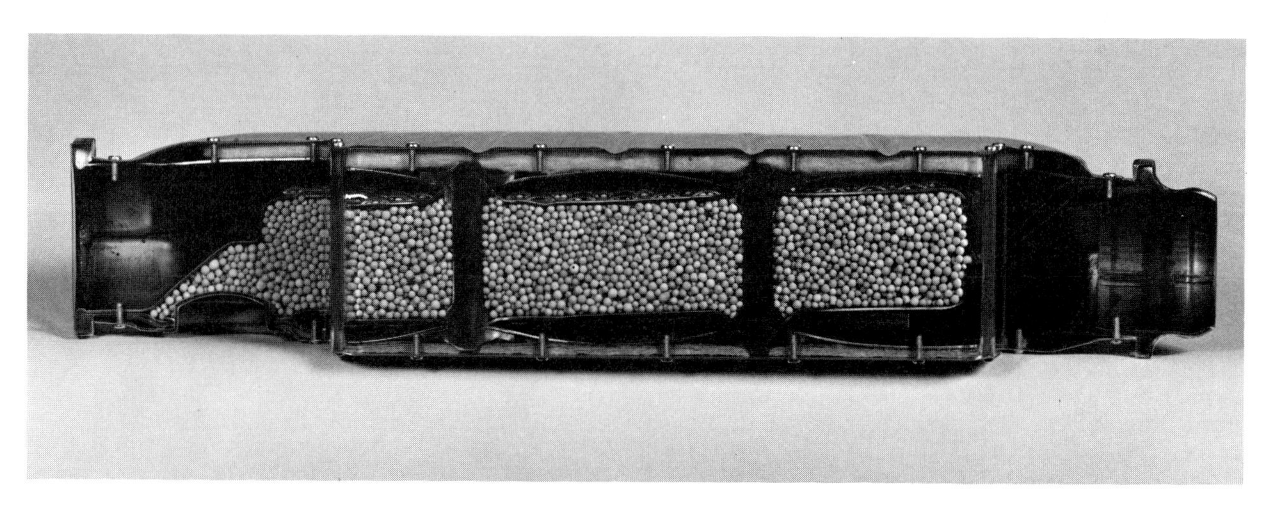

FIGURE 13.15 Cross-sectional view of a catalytic converter. The beads within the converter are impregnated with a catalyst that promotes the combustion of CO and hydrocarbons. (*General Motors Corp.*)

FOR REVIEW

Summary

In this chapter we've discussed ways of expressing reaction rates and the factors that influence these rates: namely, concentration, temperature, and catalysts. Rates are expressed as changes of concentration per unit of time; typically, for reactions in solution, the units are M/s. The quantitative relationship between rate and concentration is expressed by the rate law, which often has the following form: rate $= k[\text{reactant } 1]^m[\text{reactant } 2]^n \cdots$. The constant k in the rate law is called the rate constant; the exponents m, n, and so forth, are called reaction orders. The rate law depends not on the overall reaction, but on the mechanism by which it occurs. Thus rate laws cannot ordinarily be determined from the coefficients in the balanced chemical equation, but must be determined by experiment.

In a first-order reaction, the reaction rate is proportional to the concentration of a single reactant raised to the first power: rate $= k[\text{A}]$. In such cases $\log [\text{A}]_t = (-k/2.30)t + \log [\text{A}]_0$, where $[\text{A}]_t$ is the concentration of reactant A at time t, k is the rate constant, and $[\text{A}]_0$ is the initial concentration of reactant A. Thus a graph of $\log [\text{A}]_t$ versus time yields a straight line of slope $-k/2.30$. First-order reactions are also characterized by having a constant half-life, which is related to the rate constant: $t_{1/2} = 0.693/k$. Reactions more complex than first order yield different expressions for the rate constant and half-life. For example, for second-order reactions, rate $= k[\text{A}]^2$, the following relationship holds: $1/[\text{A}]_t = 1/[\text{A}]_0 + kt$; thus in this case a graph of $1/[\text{A}]_t$ versus time yields a straight line.

Reactions occur as a result of collisions between molecules. Collisions bring atoms together so that new chemical bonds can form. In addition, when molecules collide, part of their kinetic energies may be used to cause old bonds to break. Of course, not all collisions lead to reaction; in order for reaction to occur, molecules must collide with sufficient energy and proper orientation. The minimum energy required for a reaction to occur is called the activation energy. When the activation energy is appreciable, only a very small fraction of all collisions between reactants provides the energy required to surmount the activation-energy barrier and form the products of the reaction. By increasing the temperature of a system we can increase the fraction of molecules whose kinetic energies exceed the activation energy and thus increase the reaction rate. From the manner in which the rate constant for a reaction varies with temperature, it is possible to determine the activation energy, E_a, for a reaction using the Arrhenius equation: $\log k = \log \text{A} - E_a/2.30RT$. Accordingly, a graph of $\log k$ versus $1/T$ yields a straight line whose slope is $-E_a/2.30R$.

From a knowledge of the rate law for a reaction and with other experimental information, a picture of how the reaction proceeds may be formulated. The overall reaction may occur in a series of elementary steps or in a single elementary step. Elementary steps (or elementary reactions) are referred to as unimolecular, bimolecular, or termolecular, depending on whether one, two, or three molecules, respectively, are involved as reactants. The rate law of an elementary reaction is related directly to its molecularity. The sequence of elementary steps by which a reaction proceeds is known as its mechanism. If a reaction proceeds by a multistep mechanism, one of the elementary reactions may be much slower than the others and thereby be the rate-determining step. The rate-determining step governs the rate law for the overall process.

A catalyst speeds a reaction without itself undergoing a net chemical change. It does so by providing a different mechanism for the reaction, one having a lower activation-energy barrier. Catalysts may be either homogeneous, that is, in the same phase with the reactants, or heterogeneous, in a separate phase. Heterogeneous catalysts are particularly important in large-scale industrial chemical processes, and in applications such as the catalytic converter of an automobile.

Learning goals

Having read and studied this chapter, you should be able to:

1. Express the rate of a given reaction in terms of the variation in concentration of a reactant or product substance with time.
2. Calculate the average rate over a time interval, given the concentrations of a reactant or product at the beginning and end of that interval.
3. Calculate instantaneous rate from a graph of reactant or product concentrations as a function of time.
4. Explain the meaning of the term *rate constant* and state the units associated with rate constants for first- and second-order reactions.

5 Determine the rate law for a reaction from experimental results that show how concentration affects rate.

6 Calculate rate, rate constants, or reactant concentration, given two of these together with the rate law.

7 Use Equation 13.12 or 13.13 (for first-order reactions) or 13.17 (for second-order reactions) to determine (a) the concentration of a reactant or product at any time after a reaction has started, (b) the time required for a given fraction of sample to react, or (c) the time required for a reactant concentration to reach a certain level.

8 Use Equations 13.13 and 13.17 to determine graphically whether the rate law for a reaction is first or second order.

9 Explain the concept of reaction half-life and describe the relationship between half-life and rate constant for a first-order reaction.

10 Explain the concept of activation energy and how it relates to the variation of reaction rate with temperature.

11 Determine the activation energy for a reaction from a knowledge of how the rate constant varies with temperature (the Arrhenius equation).

12 Explain what is meant by the mechanism of a reaction using the terms *elementary steps, rate-determining step,* and *intermediate.*

13 Derive the rate law for a reaction that has a rate-determining step, given the elementary steps and their relative speeds; or, conversely, choose a plausible mechanism for a reaction given the rate law.

14 Describe the effect of a catalyst on the energy requirements for a reaction.

15 Relate the factors that are important in determining the activity of a heterogeneous catalyst.

Key terms

Among the more important terms and expressions used for the first time in this chapter are the following:

The **activated complex** (Section 13.4) is the particular arrangement of reactant and product molecules at the point of maximum energy in the rate-determining step of a reaction.

Activation energy (Section 13.4), E_a, is the minimum energy that must be supplied by the reactants in a chemical reaction to overcome the barrier to formation of products.

The **Arrhenius equation** (Section 13.4) relates the rate constant for a reaction to the frequency factor, A, the activation energy, E_a, and the temperature, T: $\log k = \log A - E_a/2.30RT$.

A **catalyst** (Section 13.6) is a substance that acts to change the speed of a chemical reaction without itself undergoing a permanent chemical change in the process.

An **elementary reaction** (Section 13.5) is a single-step **unimolecular, bimolecular,** or **termolecular** reaction. A multistep mechanism involves two or more such reactions.

An **enzyme** (Section 13.6) is a protein molecule that acts to catalyze specific biochemical reactions.

A **first-order reaction** (Section 13.3) is one in which the reaction rate is proportional to the concentration of a single reactant, raised to the first power.

The **half-life** (Section 13.3) of a reaction is the time required for the concentration of a reactant substance to decrease to halfway between its initial and final values.

Heterogeneous catalyst (Section 13.6) is a catalyst that is in a different phase from that of the reactant substances.

A **homogeneous catalyst** (Section 13.6) is a catalyst that is in the same phase as the reactant substances.

An **intermediate** (Section 13.5) is a substance formed in one elementary step of a multistep mechanism and consumed in another; it is neither a reactant nor an ultimate product of the overall reaction.

A **mechanism** (Section 13.5) for a chemical reaction is a detailed picture, or model, of how the reaction occurs; that is, the order in which bonds are broken and formed, and the changes in relative positions of the atoms as reaction proceeds.

The **rate constant** (Section 13.2) is a constant of proportionality between the reaction rate and the concentrations of reactants that appear in the rate law.

The **rate-determining step** (Section 13.5) in a chemical reaction is the slowest step in a reaction that proceeds via a series of elementary steps from reactants to products.

A **rate law** (Section 13.2) is an equation in which the reaction rate is set equal to a mathematical expression involving the concentrations of reactants (and sometimes of products also).

The **reaction order** (Section 13.2) is the sum of the powers to which all the reactants appearing in the rate expression are raised.

Reaction rate (Section 13.1) is defined in terms of the decrease in concentration of a reactant molecule or the increase in concentration of a product molecule with time. It can be expressed as either the average rate over a period of time or the instantaneous rate at a particular time.

EXERCISES

Reaction rates

13.1 The rate of disappearance of H^+ was measured for the following reaction:

$$CH_3OH(aq) + HCl(aq) \longrightarrow CH_3Cl(aq) + H_2O(l)$$

The following data were collected:

Time (min)	$[H^+]$ (M)
0	1.85
79	1.67
158	1.52
316	1.30
632	1.00

Calculate the average rate of reaction for the time interval between each measurement.

13.2 Using the data provided in Exercise 13.1, make a graph of $[H^+]$ versus time. Draw tangents to the curve at $t = 100$ and 500 min. Determine the rates at these times.

13.3 The rearrangement of methyl isonitrile, CH_3NC, was studied in the gas phase at 215°C, and the following data were obtained:

Time (s)	Pressure CH_3NC (mm Hg)
0	502
2,000	335
5,000	180
8,000	95.5
12,000	41.7
15,000	22.4

(a) Using these data, make a graph of pressure CH_3NC versus time. Draw tangents to the curve at $t = 0$, 5000, and 10,000 s. Determine the rates at these times. (b) Given that the rate law is first order in CH_3NC, determine the value of the rate constant at $t = 0$, 5000, and 10,000 s.

13.4 In each of the following reactions, how is the rate of disappearance of each reactant related to the rate of appearance of each product?

(a) $2NOCl(g) \longrightarrow 2NO(g) + Cl_2(g)$
(b) $HI(g) + CH_3I(g) \longrightarrow CH_4(g) + I_2(g)$
(c) $Ag^+(aq) + 2NH_3(aq) \longrightarrow Ag(NH_3)_2{}^+(aq)$
(d) $2H_2O_2(aq) \longrightarrow 2H_2O(l) + O_2(g)$
(e) $Cl_2(g) + 2OH^-(aq) \longrightarrow Cl^-(aq) + ClO^-(aq) + H_2O(l)$

13.5 (a) Consider the combustion of methane, CH_4, in air:

$$CH_4(g) + 2O_2(g) \longrightarrow CO_2(g) + 2H_2O(g)$$

If methane is burning at a rate of 0.20 M/s, at what rates are CO_2 and H_2O being formed? (b) If the rate of disappearance of ozone, O_3, in the reaction $2O_3(g) \longrightarrow 3O_2(g)$ is 9.0×10^{-3} atm/s, what is the rate of appearance of O_2? (c) If the rate of formation of NH_3 in the following reaction is 2.0×10^{-3} M/s, at what rates are N_2 and H_2 being consumed?

$$N_2(g) + 3H_2(g) \longrightarrow 2NH_3(g)$$

Rate laws

13.6 The decomposition of N_2O_5 in carbon tetrachloride proceeds as follows: $2N_2O_5 \longrightarrow 4NO_2 + O_2$. The rate law for this reaction is first order in N_2O_5. At 45°C, the rate constant is 6.08×10^{-4}/s. Calculate the rate of reaction when (a) $[N_2O_5] = 0.100\ M$; (b) $[N_2O_5] = 0.305\ M$.

13.7 Consider the following reaction:

$$2NO(g) + 2H_2(g) \longrightarrow N_2(g) + 2H_2O(g)$$

(a) The rate law for this reaction is first order in H_2 and second order in NO. Write the rate law. (b) If the rate constant for this reaction at 1000 K is $6.0 \times 10^4/M^2$-s, what is the reaction rate when $[NO] = 0.050\ M$ and $[H_2] = 0.010\ M$? (c) What is the reaction rate at 1000 K when the concentration of NO is doubled, to 0.10 M, while the concentration of H_2 is 0.010 M?

13.8 Consider the hypothetical reaction $A + B \longrightarrow$ products. For each of the possible rate laws listed below, indicate the reaction order with respect to A, with respect to B, and the overall reaction order: (a) rate $= k[A][B]$; (b) rate $= k[A]^2$; (c) rate $= k[A][B]^2$.

13.9 What units would each of the rate constants in Exercise 13.8 have if concentration is expressed as M and rate as M/s?

13.10 For each rate law in Exercise 13.8, indicate how the rate changes (a) if the concentration of A is doubled while the concentration of B remains constant; (b) if the concentration of B is doubled while the concentration of A is held constant.

13.11 When the concentration of a substance is doubled, what effect does it have on the rate if the order with respect to that reactant is (a) 0; (b) 1; (c) 2; (d) 3; (e) $\frac{1}{2}$?

13.12 Consider a hypothetical reaction $A + 2B \longrightarrow C$ for which the following kinetic data were obtained:

Experiment	[A] (M)	[B] (M)	$\Delta[C]/\Delta t$ (M/s)
1	0.20	0.10	7.0×10^{-5}
2	0.40	0.10	7.0×10^{-5}
3	0.20	0.20	2.8×10^{-4}

(a) Write the rate law for this reaction. (b) What is the order of the reaction with respect to A? With respect to B? What is the overall reaction order? (c) What is the numerical value for the rate constant? (d) What is the rate when [A] = 0.30 M and [B] = 0.050 M?

13.13 Consider the reaction of peroxydisulfate ion, $S_2O_8^{2-}$, with iodide ion, I^-, in aqueous solution:

$$S_2O_8^{2-}(aq) + 3I^-(aq) \longrightarrow 2SO_4^{2-}(aq) + I_3^-(aq)$$

At a particular temperature the rate of this reaction varies with reactant concentrations in the following manner:

Experiment	$[S_2O_8^{2-}]$ (M)	$[I^-]$ (M)	$-\Delta[S_2O_8^{2-}]/\Delta t$ (M/s)
1	0.038	0.060	1.4×10^{-5}
2	0.076	0.060	2.8×10^{-5}
3	0.076	0.030	1.4×10^{-5}

(a) Write the rate law for the rate of disappearance of $S_2O_8^{2-}$. (b) What is the numerical value of the rate constant for the disappearance of $S_2O_8^{2-}$? (c) What is the rate of disappearance of $S_2O_8^{2-}$ when $[S_2O_8^{2-}]$ = 0.025 M and $[I^-]$ = 0.100 M? (d) What is the rate of appearance of SO_4^{2-} when $[S_2O_8^{2-}]$ = 0.025 M and $[I^-]$ = 0.050 M?

13.14 The following data were collected for the gas-phase reaction between nitric oxide and bromine at 273°C:

$$2NO(g) + Br_2(g) \longrightarrow 2NOBr(g)$$

Experiment	[NO] (M)	$[Br_2]$ (M)	Initial rate of appearance of NOBr (M/s)
1	0.10	0.10	12
2	0.10	0.20	24
3	0.20	0.10	48
4	0.30	0.10	108

(a) Determine the rate law. (b) Calculate the value of the rate constant. (c) How is the rate of appearance of NOBr related to the rate of disappearance of Br_2? (d) What is the rate of appearance of NOBr when [NO] = 0.15 M and $[Br_2]$ = 0.25 M? (e) What is the rate of disappearance of Br_2 when [NO] = 0.075 M and $[Br_2]$ = 0.185 M?

Concentration and time; half-lives

13.15 The decomposition of a substance is found to be first order. If it takes 2.5×10^3 s for the concentration of that substance to fall to half its original value, what is the magnitude of the rate constant?

13.16 The decomposition of a substance is found to be first order. If $k = 8.6 \times 10^{-3}$/s, what is the half-life for the reaction?

13.17 The reaction

$$SO_2Cl_2(g) \longrightarrow SO_2(g) + Cl_2(g)$$

is first order in SO_2Cl_2. Using the following kinetic data, determine the magnitude of the first-order rate constant:

Time (s)	Pressure, SO_2Cl_2 (atm)
0	1.000
2,500	0.947
5,000	0.895
7,500	0.848
10,000	0.803

13.18 The rate of the gas-phase decomposition of NO_2 to form NO and O_2, $2NO_2(g) \longrightarrow 2NO(g) + O_2(g)$, is second-order in NO_2. If the concentration of NO_2 at 383°C varies with time in the following manner, what is the numerical value of the rate constant?

Time (s)	$[NO_2]$ (M)
0.0	0.100
5.0	0.017
10.0	0.0090
15.0	0.0062
20.0	0.0047

13.19 Sucrose, $C_{12}H_{22}O_{11}$, which is more commonly known as table sugar, reacts in dilute acid solutions to form two simpler sugars, glucose and fructose. Both of these sugars have the molecular formula $C_6H_{12}O_6$, though they differ in molecular structure. The reaction is

$$C_{12}H_{22}O_{11}(aq) + H_2O(l) \longrightarrow 2C_6H_{12}O_6(aq)$$

The rate of this reaction was studied at 23°C and in 0.5 M HCl; the following data were obtained:

Time (min)	$[C_{12}H_{22}O_{11}]$ (M)
0	0.316
39	0.274
80	0.238
140	0.190
210	0.146

Is the reaction first order or second order with respect to the concentration of sucrose? What is the numerical value for the rate constant?

13.20 The decomposition of N_2O_5 has been studied in carbon tetrachloride solution at 45°C. The following data are obtained under these conditions:

Time (min)	$[N_2O_5]$ (M)
0	1.40
6.67	0.10
13.33	0.87
20.00	0.68
26.67	0.53
33.33	0.42

Is the reaction first order or second order with respect to N_2O_5? What is the numerical value of the rate constant?

13.21 The first-order rate constant for the gas-phase decomposition of N_2O_5 to NO_2 and O_2 at 65°C is 4.87×10^{-3}/s. (a) If we start with 0.500 mol of N_2O_5 in a 500-mL container, how many moles of N_2O_5 will remain after 5.00 min? (b) How long will it take for the quantity of N_2O_5 to drop to 0.100 mol?

13.22 The first-order rate constant for the decomposition of a certain antibiotic in water at 20°C is 1.29/yr. (a) If a 5.0×10^{-3} M solution of this antibiotic is stored at 20°C for 1 month, what is the concentration of the antibiotic? After 1 year? (b) How long will it take for the concentration of the antibiotic to reach 1.0×10^{-3} M? (c) What is the half-life of the antibiotic at 20°C?

13.23 A reaction that is second order in A is 50 percent complete after 450 min. If $[A]_0 = 1.35$ M, what is the rate constant?

Effects of temperature; activation energy

13.24 Draw energy profiles (like those in Figures 13.7 and 13.11) for elementary reactions with the following values of E_a and ΔE: (a) $E_a = 50$ kJ/mol; $\Delta E = -30$ kJ/mol; (b) $E_a = 60$ kJ/mol; $\Delta E = 20$kJ/mol; (c) $E_a = 100$ kJ/mol; $\Delta E = 30$ kJ/mol.

13.25 (a) Which of the reactions in Exercise 13.24 would be fastest? Which would be slowest? (Assume that the reactions are of the same order, involve the same starting concentrations of reactants, and have the same magnitude for the preexponential factor, A, in the Arrhenius equation.) (b) In which case would the *reverse* reaction be the fastest? Which reverse reaction would be the slowest? Explain briefly.

13.26 Sketch the energy profile for each of the following reactions: (a) The uncatalyzed decomposition of $H_2O_2(aq)$ into $O_2(g)$ and $H_2O(l)$ for which $E_a = 75.3$ kJ/mol and $\Delta E = -98.1$ kJ/mol; (b) the reaction

$$O_3(g) + NO(g) \longrightarrow NO_2(g) + O_2(g)$$

for which $E_a = 10.3$ kJ/mol and $\Delta E = -199.6$ kJ/mol.

13.27 The rate of the reaction

$$CH_3COOC_2H_5(aq) + OH^-(aq) \longrightarrow CH_3COO^-(aq) + C_2H_5OH(aq)$$

was measured at several temperatures and the following data collected:

Temperature (°C)	k (1/M-s)
15	0.0521
25	0.101
35	0.184
45	0.332

Using these data, construct a graph of log k versus $1/T$. Using your graph, determine the value of E_a.

13.28 The temperature dependence of the rate constant for the reaction

$$CO(g) + NO_2(g) \longrightarrow CO_2(g) + NO(g)$$

is tabulated below:

Temperature (K)	k (1/M-s)
600	0.028
650	0.22
700	1.3
750	6.0
800	23

Calculate E_a and A.

13.29 The activation energy, E_a, for a reaction is 70.0 kJ/mol. What is the effect on rate if the temperature increases (a) from 25.0°C to 35.0°C; (b) from 100°C to 110°C? (Assume that E_a and A do not change with temperature.)

13.30 The gas-phase decomposition of HI into H_2 and I_2 is found to have $E_a = 182$ kJ/mol. The rate constant at 700°C is 1.57×10^{-3} M/s. What is the value of k at (a) 600°C; (b) 800°C?

13.31 The rate constant, k, for a reaction is 1.0×10^{-3}/s at 25°C. Calculate the rate constant for the same reaction at 100°C if (a) $E_a = 67.0$ kJ/mol; (b) $E_a = 134$ kJ/mol.

13.32 For the reaction

$$HI(g) + CH_3I(g) \longrightarrow CH_4(g) + I_2(g)$$

$k = 1.7 \times 10^{-2}$/s at 430 K and 9.5×10^{-2}/s at 450 K. What is the numerical value of the activation energy, E_a?

Reaction mechanisms; catalysis

13.33 The gas-phase decomposition of NO_2Cl is believed to occur in two steps:

$$NO_2Cl(g) \longrightarrow NO_2(g) + Cl(g)$$
$$NO_2Cl(g) + Cl(g) \longrightarrow NO_2(g) + Cl_2(g)$$

(a) Describe the molecularity of each step of the mechanism. (b) Write the equation for the overall reaction. (c) Identify the intermediate, if any.

13.34 Assume that each of the following reactions is an elementary reaction:

(a) $N_2O_4(g) \longrightarrow 2NO_2(g)$
(b) $2NO_2(g) \longrightarrow N_2O_4(g)$
(c) $CO(g) + Cl_2(g) \longrightarrow COCl(g) + Cl(g)$

Identify each reaction as unimolecular, bimolecular, or termolecular, and write a rate law for each.

13.35 The following mechanism has been proposed for the reaction of NO with Br_2 to form NOBr:

$$NO(g) + Br_2(g) \rightleftharpoons NOBr_2(g)$$
$$NOBr_2(g) + NO(g) \longrightarrow 2NOBr(g)$$

(a) Describe the molecularity of each step in the mechanism (including the reverse of the first reaction). (b) Write the rate law for each elementary step in the mechanism. (c) Write the chemical equation for the overall reaction. (d) Identify the intermediate, if any. (e) The observed rate law is: Rate $= k[NO]^2[Br_2]$. If the proposed mechanism is correct, what can we conclude about the relative speeds of the first and second steps?

13.36 Consider the following reaction: $H_2(g) + 2ICl(g) \longrightarrow 2HCl(g) + I_2(g)$. The rate law for this reaction is first order in both H_2 and ICl: Rate $= k[H_2][ICl]$. Which of the following mechanisms are consistent with the observed rate law?

(a) $2ICl(g) + H_2(g) \longrightarrow 2HCl(g) + I_2(g)$ (termolecular reaction)

(b) $H_2(g) + ICl(g) \longrightarrow HI(g) + HCl(g)$ (slow)
$HI(g) + ICl(g) \longrightarrow HCl(g) + I_2(g)$ (fast)

(c) $H_2(g) + ICl(g) \longrightarrow HI(g) + HCl(g)$ (fast)
$HI(g) + ICl(g) \longrightarrow HCl(g) + I_2(g)$ (slow)

(d) $H_2(g) + ICl(g) \longrightarrow HClI(g) + H(g)$ (slow)
$H(g) + ICl(g) \longrightarrow HCl(g) + I(g)$ (fast)
$HClI(g) \longrightarrow HCl(g) + I(g)$ (fast)
$I(g) + I(g) \longrightarrow I_2(g)$ (fast)

13.37 The decomposition of hydrogen peroxide is catalyzed by iodide ion. The catalyzed reaction is thought to proceed by a two-step mechanism:

$$H_2O_2(aq) + I^-(aq) \longrightarrow H_2O(l) + IO^-(aq) \quad \text{(slow)}$$
$$IO^-(aq) + H_2O_2(aq) \longrightarrow H_2O(l) + O_2(g) + I^-(aq) \quad \text{(fast)}$$

(a) Assuming that the first step of the mechanism is rate determining, predict the rate law for the overall process. (b) Write the chemical equation for the overall process. (c) Identify the intermediate, if any, in the mechanism. (d) What distinguishes an intermediate from a catalyst in a chemical reaction?

[13.38] Derive the rate law for the reaction

$$OCl^-(aq) + I^-(aq) \longrightarrow OI^-(aq) + Cl^-(aq)$$

if its mechanism proceeds as follows:

$$OCl^-(aq) + H_2O(l) \rightleftharpoons HOCl(aq) + OH^-(aq) \quad \text{(fast)}$$
$$I^-(aq) + HOCl(aq) \longrightarrow HOI(aq) + Cl^-(aq) \quad \text{(slow)}$$
$$HOI(aq) + OH^-(aq) \longrightarrow H_2O(l) + OI^-(aq) \quad \text{(fast)}$$

13.39 In older texts, catalysts were sometimes defined as substances that speed up a chemical reaction without taking part in the reaction. In what way is this definition misleading? How might the definition be modified to correct it?

13.40 Make a sketch of the energy profile for the uncatalyzed decomposition of H_2O_2 into H_2O and O_2 and the I^--catalyzed reaction that is consistent with the mechanism proposed in Exercise 13.37.

13.41 The activity of a heterogeneous catalyst is highly dependent on its method of preparation and prior treatment. Explain why this is so.

13.42 When D_2 is reacted with ethylene, C_2H_4, in the presence of a finely divided catalyst, ethane with two deuteriums, $CH_2D{-}CH_2D$, is formed. (Deuterium, D, is an isotope of hydrogen of mass 2.) There is very little ethane formed in which two deuteriums are bound to one carbon, for example, $CH_3{-}CHD_2$. Explain why this is so in terms of the sequence of steps involved in hydrogenation.

Additional exercises

13.43 (a) List three factors that can be varied to change the speed of a particular reaction. (b) Explain, on a molecular lever, how each exerts its influence.

13.44 What factors determine whether a collision will lead to chemical reaction?

13.45 Explain why the rate law for a reaction cannot in general be written using the coefficients of the balanced chemical equation as the reaction orders.

13.46 (a) How does the rate of disappearance of N_2O_5 relate to the rate of appearance of NO_2 and the rate of appearance of O_2 in the reaction

$$2N_2O_5(g) \longrightarrow 4NO_2(g) + O_2(g)$$

(b) If the rate law for this reaction is $k[N_2O_5]$, how does k differ for the disappearance of N_2O_5 and the appearance of NO_2?

13.47 A particular reaction is found to have the following rate law: rate $= k[A]^2[B]$. Which terms in this rate law are changed by each of the following changes? (a) The concentration of A is doubled. (b) A catalyst is added. (c) The concentration of A is increased by a factor of 2, whereas the concentration of B is decreased by a factor of 4. (d) The temperature is increased.

13.48 The oxidation of NO by O_3, a reaction of importance in formation of photochemical smog (Section 12.4), is first order in each of the reactants. The second-order rate constant at 28°C is $1.5 \times 10^7/M$-s. If the concentrations of NO and O_3 are each $2 \times 10^{-8}\ M$, what is the rate of oxidation of NO?

13.49 Hydrogen sulfide, H_2S, is a common and troublesome pollutant in industrial waste waters. One way to remove this substance is to treat the water with chlorine, in which case the following reaction occurs:

$$H_2S(aq) + Cl_2(aq) \longrightarrow S(s) + 2H^+(aq) + 2Cl^-(aq)$$

The rate of this reaction is first order in each reactant. The rate constant for disappearance of H_2S at 28°C is $3.5 \times 10^{-2}/M$-s. If at a given time the concentration of H_2S is 5×10^{-5} *M* and that of Cl_2 is 0.04 *M*, what is the rate of formation of Cl^-?

13.50 The anaerobic (in absence of air) breakdown of glucose to form lactic acid proceeds as follows:

$$C_6H_{12}O_6(aq) \longrightarrow 2HC_3H_5O_3(aq)$$

This reaction occurs in muscles during strenuous activity when the energy demands of the cells exceed the O_2 available for normal oxidative breakdown of glucose into CO_2 and H_2O. The lactic acid accumulates, contributing to muscle fatigue and triggering heavy breathing to supply O_2 necessary to decompose the lactic acid. Suppose that in a laboratory experiment you are assigned the task of determining the rate of the conversion of glucose to lactic acid at 25°C. Suggest some procedure that could be used in obtaining data.

13.51 The reaction of $(CH_3)_3CBr$ with hydroxide ion proceeds with the formation of $(CH_3)_3COH$:

$$(CH_3)_3CBr(aq) + OH^-(aq) \longrightarrow (CH_3)_3COH(aq) + Br^-(aq)$$

This reaction was studied at 55°C, and the following data obtained:

Experiment	$[(CH_3)_3CBr]$ (*M*)	$[OH^-]$ (*M*)	Initial rate (*M*/s)
1	0.10	0.10	1.0×10^{-3}
2	0.20	0.10	2.0×10^{-3}
3	0.10	0.20	1.0×10^{-3}

Write the rate law for this reaction and compute the rate constant.

13.52 Consider the following reaction between mercury(II) chloride and oxalate ion:

$$2HgCl_2(aq) + C_2O_4^{2-}(aq) \longrightarrow 2Cl^-(aq) + 2CO_2(g) + Hg_2Cl_2(s)$$

The initial rate of this reaction was determined for several concentrations of $HgCl_2$ and $C_2O_4^{2-}$ and the following rate data obtained:

Experiment	$[HgCl_2]$ (*M*)	$[C_2O_4^{2-}]$ (*M*)	Rate (*M*/s)
1	0.105	0.15	1.8×10^{-5}
2	0.105	0.30	7.1×10^{-5}
3	0.052	0.30	3.5×10^{-5}
4	0.052	0.15	8.9×10^{-6}

(a) What is the rate law for this reaction? (b) What is the numerical value of the rate constant? (c) What is the reaction rate when the concentration of $HgCl_2$ is 0.080 *M* and that of $C_2O_4^{2-}$ is 0.10 *M* if the temperature is the same as that used to obtain the data shown above?

13.53 Urea, NH_2CONH_2, is the end product in protein metabolism in animals. The decomposition of urea in 0.1 *M* HCl occurs according to the reaction

$$NH_2CONH_2(aq) + H^+(aq) + 2H_2O(l) \longrightarrow 2NH_4^+(aq) + HCO_3^-(aq)$$

The reaction is first order in urea. When [urea] = 0.200 *M*, the rate at 61.05°C is 8.56×10^{-5} *M*/s. (a) What is the numerical value for the rate constant, *k*? (b) What is the concentration of urea in this solution after 5.00×10^3 s if the starting concentration is 0.500 *M*? (c) What is the half-life for this reaction at 61.05°C?

13.54 The half-life of the gas-phase decomposition of N_2O_5 at 318 K is 1350 s. If the initial concentration of N_2O_5 in a reaction vessel at this temperature is 2.0 mol/L, how long after the start of the reaction will the concentration equal (a) 1.0 mol/L; (b) 0.50 mol/L; (c) 0.10 mol/L?

13.55 A sample of polluted water was oxidized with O_2 at 25°C. The percentage of organic matter in the sample that was oxidized varied with time in the following manner:

Time (days)	1	2	3	4	5	6	7	10	20
Organic matter oxidized (%)	21	37	50	60	68	75	80	90	99

Is the oxidation process first or second order? What is the rate constant for this reaction?

13.56 You wish to test the following proposed mechanism for the oxidation of HBr:

$$HBr + O_2 \longrightarrow HOOBr$$
$$HOOBr + HBr \longrightarrow 2HOBr$$
$$HOBr + HBr \longrightarrow H_2O + Br_2$$

You find that the rate is first order with respect to both HBr and O_2. You cannot detect either HOBr or HOOBr among the products. (a) If the proposed mechanism is correct, which step must be rate determining? (b) Can you either prove or disprove the mechanism from these observations?

13.57 One of the reactions that occurs in polluted air in urban areas is $2NO_2(g) + O_3(g) \longrightarrow N_2O_5(g) + O_2(g)$. The observed rate law for this process is: rate = $k[NO_2][O_3]$. The involvement of NO_3 as an intermediate has been proposed. Suggest a mechanism for this reaction.

13.58 The rate law for the reaction

$$2NO(g) + O_2(g) \longrightarrow 2NO_2(g)$$

is $k[NO]^2[O_2]$. The following mechanism has been proposed:

$$NO(g) + O_2(g) \rightleftharpoons NO_3(g) \quad \text{(fast)}$$
$$NO_3(g) + NO(g) \longrightarrow 2NO_2(g) \quad \text{(slow)}$$

Explain how the proposed mechanism leads to the observed rate law.

[13.59] The gas-phase reaction of chlorine with carbon monoxide to form phosgene, $Cl_2(g) + CO(g) \longrightarrow COCl_2(g)$, obeys the following rate law:

$$\text{Rate} = \frac{\Delta[COCl_2]}{\Delta t} = k[Cl_2]^{3/2}[CO]$$

A mechanism involving the following series of steps is consistent with the rate law:

$$Cl_2 \rightleftharpoons 2Cl$$
$$Cl + CO \rightleftharpoons COCl$$
$$COCl + Cl_2 \rightleftharpoons COCl_2 + Cl$$

Assuming that this mechanism is correct, which of the above steps is the slow, or rate-determining, step? Explain.

13.60 Indicate the molecularity for each elementary reaction of the mechanism given in Exercise 13.59. (Consider both the forward and reverse reactions.)

13.61 The reaction

$$F(g) + H_2(g) \longrightarrow HF(g) + H(g)$$

has been studied at various temperatures and has been found to have an activation energy of 22 kJ/mol. The overall change in energy in the reaction, ΔE, is -130 kJ/mol. Draw a diagram of the energy profile of the system as a function of the reaction coordinate.

13.62 The first-order rate constant for hydrolysis of a particular organic compound in water varies with temperature as follows:

Temperature (K)	Rate constant (s^{-1})
300	1.0×10^{-5}
320	5.0×10^{-5}
340	2.0×10^{-4}
355	5.0×10^{-4}

From these data calculate the activation energy in units of kJ/mol.

[13.63] The rate at which fireflies flash varies with the temperature of the insect, because a certain rate-controlling reaction proceeds at speeds that are temperature dependent. The period of the firefly flash was found to be 16.3 s at 21.0°C and 13.0 s at 27.8°C. What is the activation energy for the reaction that controls the rate of firefly flash?

13.64 The activation energy for the reaction

$$2NO_2(g) \longrightarrow 2NO(g) + O_2(g)$$

is 114 kJ/mol. If $k = 0.75/M\text{-s}$ at 600°C, what is the value of k at 500°C?

[13.65] Outline briefly the advantages you can think of for the use of catalysts as opposed to high temperatures to promote chemical reactions.

13.66 The primary focus of this chapter has been on homogeneous reactions, yet many significant processes are heterogeneous. Explain the basis for the following observations involving heterogeneous systems. (a) Bulk flour is hard to burn, yet elevators in which flour is stored have been known to explode. (b) Solids dissolve more rapidly in water when they are first crushed to give a small particle size.

14 Chemical Equilibrium

It often happens that chemical reactions do not proceed to completion. By this we mean that a mixture of reactants is not entirely converted into products. Instead, after a time, reactant concentrations no longer decrease. The reaction system at this point consists of a mixture of reactant and product substances. A chemical system in such a condition is said to be in a state of **chemical equilibrium.** We have already encountered instances of simple equilibrium conditions. For example, in a closed container, the vapor above a liquid achieves an equilibrium with the liquid phase (Section 10.2). The rate at which molecules escape from the liquid to the gas phase equals the rate at which molecules of the gas phase strike the surface and become part of the liquid. As another example, solid sodium chloride may be in equilibrium with the ions dissolved in water (Section 11.2). The rate at which ions leave the solid surface equals the rate at which other ions are removed from the liquid to become part of the solid. We know from these examples that equilibrium is not a static condition in which nothing is happening. Rather, it is dynamic; opposing processes are occurring at equal rates. In this chapter we shall consider chemical equilibrium and examine the principles on which the idea of equilibrium is based. To demonstrate the importance of equilibrium considerations and to illustrate the most basic concepts involved, we begin with a discussion of the Haber process for synthesizing ammonia.

14.1 THE HABER PROCESS

Of all the chemical reactions that humans have learned to carry out and control for their own purposes, the synthesis of ammonia from hydrogen and atmospheric nitrogen is perhaps the most important. This is especially the case in the present world situation, with food shortages becoming more critical each year. The growth of plant matter requires a substantial store of nitrogen in the soil, in a form usable by plants. The quantity of food required to feed the ever-increasing human population far exceeds that which could be produced if we relied solely on naturally available nitrogen in the soil. Huge quantities of fertilizer rich in nitrogen are required to sustain the high-yield agriculture practiced in the United States today. Great amounts of fertilizer must be used throughout the world if the gap between food supplies and human needs is to be narrowed. The only widely available source of nitrogen is the N_2 present in the atmosphere. The problem thus becomes one of "fixing" atmospheric N_2, that is, converting it to compounds that plants can use. This process is called *nitrogen fixation.*

The N_2 molecule is exceptionally unreactive. Its lack of reactivity is due in large measure to the strong triple bond between the nitrogen atoms (Section 7.4). Because this bond is so strong, there is very little tendency for the molecule to engage in chemical reactions. For this reason, the process of fixation is not easy to achieve. In nature the fixation of N_2 is carried out by a special group of nitrogen-fixing bacteria that grow on the roots of certain plants, for example, clover or alfalfa. Here we are interested in only one particular fixation reaction, the **Haber process.**

Fritz Haber, a German chemist, investigated the energy relations in the reaction between nitrogen and hydrogen and convinced himself that it should be possible to form ammonia in a reasonable yield from these two starting substances. The chemical reaction involved is

$$N_2(g) + 3H_2(g) \rightleftharpoons 2NH_3(g) \qquad [14.1]$$

The double arrow indicates the reversible character of the reaction. The NH_3 can form from N_2 and H_2, but it can also decompose into these elements.

Haber's research was of great interest to the German chemical industry. Germany was preparing for World War I, and nitrogen compounds figured heavily in the manufacture of explosives. Without a synthetic source of these nitrogen compounds Germany would be greatly handicapped. By 1913, Haber had designed a process that worked, and for the first time ammonia was produced on a large scale from atmospheric nitrogen. In the following year, World War I began.

From these unhappy beginnings as a major factor in international warfare the Haber process has become the world's principal source of fixed nitrogen. It is estimated that 16 million tons of ammonia were formed via the Haber process in the United States in 1982.* The ammonia produced in the Haber process can be applied directly to the soil. It may also be converted into ammonium salts, for example, ammonium sulfate, $(NH_4)_2SO_4$, or ammonium hydrogen phosphate, $(NH_4)_2HPO_4$, which are then used as fertilizers.

In designing the process that bears his name, Haber had to face two separate questions. First, is there a catalyst that will allow the reaction to occur at a reasonable speed under practically attainable conditions?

*The industrial fixation of nitrogen, chiefly by the Haber process, now accounts for more than 30 percent of all the nitrogen fixed on the planet.

After much long and difficult searching, Haber found a suitable catalyst; we shall return to this aspect of his work later, in Section 14.6. Second, assuming that a catalyst can be found, to what extent will nitrogen be converted into ammonia? Let us now consider this latter question, which relates to chemical equilibrium.

14.2 THE EQUILIBRIUM CONSTANT

The Haber process consists of putting together N_2 and H_2 in a high-pressure tank at a total pressure of several hundred atmospheres, in the presence of a catalyst, and at a temperature of a few hundred degrees Celsius. Under these conditions, the two gases react to form ammonia. But the reaction does not lead to complete consumption of the N_2 and H_2. Rather, at some point the reaction appears to stop, with all three components of the reaction mixture present at the same time. The manner in which the concentration of H_2, N_2, and NH_3 vary with time in this situation is shown in Figure 14.1(*a*). The condition of the system in which all concentrations have achieved steady values is referred to as chemical equilibrium. The relative amounts of N_2, H_2, and NH_3 present at equilibrium do not depend on the amounts of catalyst present. However, they do depend on the relative amounts of H_2 and N_2 with which the reaction was begun. Furthermore, if only ammonia is placed into the tank under the usual reaction conditions, at equilibrium there is again a mixture of N_2, H_2, and NH_3. The variations in concentrations as a function of time for this situation are shown in Figure 14.1(*b*). By comparing the two parts of Figure 14.1, we can see that at equilibrium the relative concentrations of H_2, N_2, and NH_3 are the same, regardless of whether the starting mixture was a 3:1 molar ratio of H_2 and N_2 or pure NH_3. Equilibrium is thus a condition of the system that can be approached from both directions. This suggests to us that equilibrium is *not* a static condition. On the contrary, both the forward reaction in Equation 14.1, in which ammonia is formed, and the reverse reaction in which H_2 and N_2 are formed from ammonia, are going on at precisely the same rates. Thus the net rate of change in the system is zero.

If we were systematically to change the relative amounts of N_2, H_2, and NH_3 in the starting mixture of gases, and then analyze the gas mixtures at equilibrium, it would be possible to determine what sort of

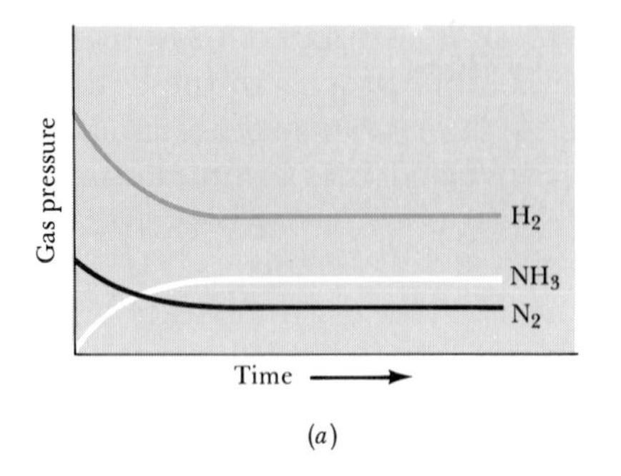

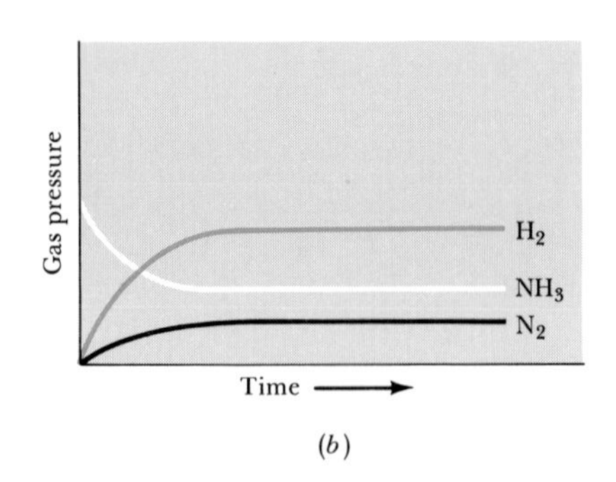

FIGURE 14.1 Variation in gas pressures in formation of the equilibrium $N_2 + 3H_2 \rightleftharpoons 2NH_3$. (*a*) The equilibrium is approached beginning with H_2 and N_2 in the ratio 3:1. (*b*) The equilibrium is approached beginning with NH_3.

"law" governs the equilibrium state. Studies of this kind were carried out on other chemical systems by chemists in the nineteenth century, long before Haber's work. In 1864, Cato Maximilian Guldberg and Peter Waage proposed their law of mass action. This law expresses the relative concentrations of reactants and products at equilibrium in terms of a quantity called the equilibrium constant. Suppose that we have the general reaction

$$jA + kB \rightleftharpoons pR + qS \quad [14.2]$$

where A, B, R, and S are the chemical species involved, and j, k, p, and q are their coefficients in the balanced chemical equation. According to the law of mass action, the equilibrium condition is expressed by the equation

$$K = \frac{[R]^p[S]^q}{[A]^j[B]^k} \quad [14.3]$$

where K is a constant, called the equilibrium constant, and the square brackets signify the *concentration* of the species within the brackets. The law of mass action applies only to a system that has attained equilibrium. In general, the equilibrium constant is given by the concentrations of all reaction products multiplied together, each raised to the power of its coefficient in the balanced equation, divided by the concentrations of all reactants multiplied together, each raised to the power of its coefficient in the balanced equation. (Remember that the convention is to write the concentration terms for the *products* in the *numerator* and those for the *reactants* in the *denominator*.)

The equilibrium constant is a true constant. Its value at any given temperature does not depend on the initial concentrations of reactants and products. It also does not matter whether there are other substances present, as long as they do not consume a reactant or product through chemical reaction. The value of the equilibrium constant does, however, vary with temperature.

As an illustration of the law of mass action, consider the gas-phase equilibrium between dinitrogen tetroxide and nitrogen dioxide:

$$N_2O_4(g) \rightleftharpoons 2NO_2(g) \quad [14.4]$$

Because NO_2 is a dark brown gas, and N_2O_4 is colorless, the amount of NO_2 in the mixture can be measured by measuring the intensity of the brown color of the gas mixture.

Following the rule given above, the equilibrium-constant expression for reaction Equation 14.4 is

$$K = \frac{[NO_2]^2}{[N_2O_4]} \quad [14.5]$$

Suppose that to determine the numerical value for K, and to verify that it is indeed constant as the concentrations of NO_2 and N_2O_4 change, three samples of NO_2 were placed in sealed glass vessels. In addition, a sample of N_2O_4 was placed in a fourth vessel. The vessels were allowed to remain at 100°C until no further change in the color of the gas was noted. The mixture of gases was then analyzed to determine the concentrations of both NO_2 and N_2O_4. The results are given in Table 14.1.

TABLE 14.1 Initial and equilibrium concentrations (molarities) of NO_2 and N_2O_4 in the gas phase at 100°C

Experiment	Initial N_2O_4 concentration (*M*)	Initial NO_2 concentration (*M*)	Equilibrium N_2O_4 concentration (*M*)	Equilibrium NO_2 concentration (*M*)	K_c
1	0.0	0.0200	0.00140	0.0172	0.211
2	0.0	0.0300	0.00280	0.0243	0.211
3	0.0	0.0400	0.00452	0.0310	0.213
4	0.0200	0.0	0.00452	0.0310	0.213

To evaluate the equilibrium constant, *K*, the equilibrium concentrations are inserted into the equilibrium-constant expression, Equation 14.5. When the concentration unit is molarity, as in the present case, we will label the equilibrium constant as K_c.

For example, using the first set of data,

$$[NO_2] = 0.0172\,M \qquad [N_2O_4] = 0.00140\,M$$

$$K_c = \frac{[NO_2]^2}{[N_2O_4]} = \frac{(0.0172)^2}{0.00140} = 0.211$$

Proceeding in the same way, the values of K_c for the other samples were calculated, as listed in Table 14.1. Note that the value for K_c is essentially constant, even though the initial concentrations vary. Furthermore, the results of experiment 4 show that equilibrium can be attained beginning with N_2O_4 as well as with NO_2. That is, equilibrium can be approached from either direction. Figure 14.2 shows how both experiments 3 and 4 result in the same equilibrium mixture even though one begins with $0.0400\,M$ NO_2 and the other with $0.0200\,M$ N_2O_4.

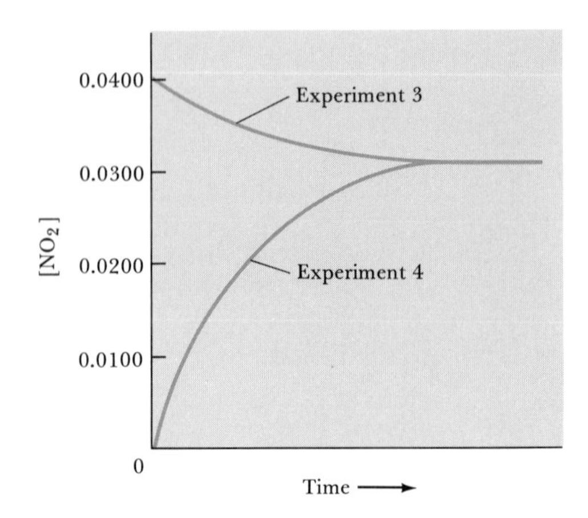

FIGURE 14.2 The same equilibrium mixture is produced starting with either $0.400\,M$ NO_2 (experiment 3) or $0.0200\,M$ N_2O_4 (experiment 4).

SAMPLE EXERCISE 14.1

Write the equilibrium-constant expression for each of the following reactions:

(a) $N_2(g) + 3H_2(g) \rightleftharpoons 2NH_3(g)$
(b) $2NH_3(g) \rightleftharpoons N_2(g) + 3H_2(g)$
(c) $2HI(g) \rightleftharpoons H_2(g) + I_2(g)$

Solution: (a) As indicated by Equation 14.3, the equilibrium-constant expression has the form of a quotient. The numerator of this quotient is obtained by multiplying the equilibrium concentrations of products, each raised to a power equal to its coefficient in the balanced equation. The denominator is obtained similarly using the equilibrium concentrations of reactants:

$$K = \frac{[NH_3]^2}{[N_2][H_2]^3}$$

(b) This reaction is just the reverse of that given in part (a). Placing products over reactants, we have

$$K = \frac{[N_2][H_2]^3}{[NH_3]^2}$$

Notice that this expression is just the reciprocal of that given in part (a). It is a general rule that the equilibrium-constant expression for a reaction written in one direction is the reciprocal of the one for the reverse reaction. This also means that the numerical value of the equilibrium constant for a reaction written in one direction is the reciprocal of the value of the equilibrium constant for the reaction written in the reverse direction.

(c) $K = \frac{[H_2][I_2]}{[HI]^2}$

One of the first tasks confronting Haber and his co-workers when they set to work on the problem of ammonia synthesis was the determination of the numerical value of the equilibrium constant for the synthesis of NH_3 at various temperatures. If the value of K for this reaction is very small, this means that the amount of NH_3 formed would be small relative to the initial amounts of N_2 and H_2 used. This situation can be described by saying that the equilibrium for

$$N_2(g) + 3H_2(g) \rightleftharpoons 2NH_3(g)$$

lies to the left, that is, toward the reactant side. Clearly, if the equilibrium were too far to the left, it would not be possible to develop a satisfactory synthesis for ammonia.

SAMPLE EXERCISE 14.2

The reaction of N_2 with O_2 to form NO might be considered as a means of "fixing" nitrogen:

$$N_2(g) + O_2(g) \rightleftharpoons 2NO(g)$$

The value for the equilibrium constant for this reaction at 25°C is $K_c = 1 \times 10^{-30}$. Describe the feasibility of this reaction for nitrogen fixation.

Solution: Because K_c is so small, very little NO will form at 25°C. The equilibrium is said to lie to the left, favoring the reactants. Consequently, this reaction is an extremely poor choice for nitrogen fixation, at least at 25°C.

The value of K can be calculated if we know the equilibrium concentrations of all reactants and products involved in the reaction. These concentrations might be obtained by a direct experimental measurement. This was the method described in constructing Table 14.1. Alternatively, the equilibrium concentrations can be calculated if we know the concentrations at the beginning of the reaction and the equilibrium concentration of at least one species. These approaches are described in Sample Exercises 14.3 and 14.4.

SAMPLE EXERCISE 14.3

In one of their experiments, Haber and co-workers introduced a mixture of hydrogen and nitrogen into a reaction vessel and allowed the system to attain chemical equilibrium at 472°C. The equilibrium mixture of gases was analyzed and found to contain 0.1207 M H_2, 0.0402 M N_2, and 0.00272 M NH_3. From these data calculate the equilibrium constant, K_c, for

$$N_2(g) + 3H_2(g) \rightleftharpoons 2NH_3(g)$$

Solution:

$$K_c = \frac{[NH_3]^2}{[N_2][H_2]^3} = \frac{(0.00272)^2}{(0.0402)(0.1207)^3} = 0.105$$

SAMPLE EXERCISE 14.4

A mixture of 5.00×10^{-3} mol of H_2 and 1.00×10^{-2} mol of I_2 are placed in a 5.00-L container at 448°C and allowed to come to equilibrium. Analysis of the equilibrium mixture shows that the concentration of HI is 1.87×10^{-3} M. Calculate K_c at 448°C for the reaction

$$H_2(g) + I_2(g) \rightleftharpoons 2HI(g)$$

Solution: The initial concentrations of H_2 and I_2 must be calculated:

$$[H_2]_i = \frac{5.00 \times 10^{-3}\text{ mol}}{5.00\text{ L}} = 1.00 \times 10^{-3}\,M$$

$$[I_2]_i = \frac{1.00 \times 10^{-2}\text{ mol}}{5.00\text{ L}} = 2.00 \times 10^{-3}\,M$$

The equilibrium concentrations of H_2 and I_2 can be calculated from these initial concentrations and the equilibrium concentration of HI. During the course of the reaction the concentration of HI changes from 0 to 1.87×10^{-3} M. The balanced equation indicates that 2 mol of HI form from each mole of H_2. Thus the amount of H_2 consumed is

$$\left(1.87 \times 10^{-3}\,\frac{\text{mol HI}}{\text{L}}\right)\left(\frac{1\text{ mol }H_2}{2\text{ mol HI}}\right) = 0.935 \times 10^{-3}\text{ mol }H_2/\text{L}$$

The equilibrium concentration of H_2 is the initial concentration minus that consumed:

$$[H_2] = 1.00 \times 10^{-3}\,M - 0.935 \times 10^{-3}\,M = 0.065 \times 10^{-3}\,M$$

The same line of argument gives the equilibrium concentration of I_2:

$$[I_2] = 2.00 \times 10^{-3}\,M - 0.935 \times 10^{-3}\,M = 1.065 \times 10^{-3}\,M$$

In calculating the equilibrium concentrations in problems of this type, it is often convenient to set up a table showing the initial concentration, the change, and the equilibrium concentration:

	$H_2(g)$	+	$I_2(g)$	$\rightleftharpoons$	$2HI(g)$
Initial:	$1.00 \times 10^{-3}\,M$		$2.00 \times 10^{-3}\,M$		$0\,M$
Change:	$-0.935 \times 10^{-3}\,M$		$-0.935 \times 10^{-3}\,M$		$+1.87 \times 10^{-3}\,M$
Equilibrium:	$0.065 \times 10^{-3}\,M$		$1.065 \times 10^{-3}\,M$		$1.87 \times 10^{-3}\,M$

The concentrations given in the problem are printed in color. The concentration changes and finally the equilibrium concentrations of H_2 and I_2 were then determined as described above.

Once we have the equilibrium concentrations of each reactant and product we can calculate the equilibrium constant:

$$K_c = \frac{[HI]^2}{[H_2][I_2]} = \frac{(1.87 \times 10^{-3})^2}{(0.065 \times 10^{-3})(1.065 \times 10^{-3})} = 50.5$$

Concentration Units and Equilibrium Constants

As we have seen, the square brackets around a chemical symbol, as in $[NH_3]$, represent the concentration of that substance. Molarity is the common concentration unit for reactions occurring in solution. For gas-phase reactions, the concentration units used are either molarity or atmospheres of pressure. When the concentration is expressed in molarity, we denote the equilibrium constant as K_c. When the units are atmospheres, we write K_p. Since the numerical values of K_c and K_p will generally be different, we must take care to indicate which we are using by means of these subscripts.

The ideal-gas equation permits us to convert between atmospheres and molarity and therefore convert between K_p and K_c:

$$PV = nRT$$

$$P = \left(\frac{n}{V}\right)RT = MRT \qquad [14.6]$$

where n/V (the number of moles per liter) is concentration in molarity, M. As a result of the relationship between pressure and molarity expressed in Equation 14.6, a general expression relating K_p and K_c can be written

$$K_p = K_c(RT)^{\Delta n} \qquad [14.7]$$

The quantity Δn in this equation is the change in the number of moles of gas upon going from reactants to products; Δn is equal to the number of moles of gaseous products minus the number of moles of gaseous reactants. For example, in the reaction

$$H_2(g) + I_2(g) \rightleftharpoons 2HI(g)$$

there are 2 mol of HI (the coefficient in the balanced equation); there are also 2 mol of gaseous reactants ($1H_2 + 1I_2$). Therefore, $\Delta n = 2 - 2 = 0$, and $K_p = K_c$ for this reaction.

SAMPLE EXERCISE 14.5

Using the value of K_c obtained in Sample Exercise 14.3, calculate K_p for

$$N_2(g) + 3H_2(g) \rightleftharpoons 2NH_3(g)$$

at 472°C.

Solution: There are 2 mol of gaseous products ($2NH_3$); and there are 4 mol of gaseous reactants ($1N_2 + 3H_2$). Therefore, $\Delta n = 2 - 4 = -2$. (Remember that Δ functions are always based on products minus reactants.) The temperature, T, is $273 + 472 = 745$ K. The value for the ideal-gas constant, R, is 0.0821 L-atm/K-mol. The value of K_c from Sample Exercise 14.3 is 0.105. We therefore have

$$K_p = \frac{P_{NH_3}^2}{P_{N_2}P_{H_2}^3} = K_c(RT)^{\Delta n}$$

$$= (0.105)(0.0821 \times 745)^{-2}$$

$$= 2.81 \times 10^{-5}$$

Concentration units can be carried through the calculation of the equilibrium constant to give units for K. For example, for the reaction $N_2O_4(g) \rightleftharpoons 2NO_2(g)$ we have $K = [NO_2]^2/[N_2O_4]$. When concentration is in molarity, the units of the equilibrium constant are $M^2/M = M$; when concentration is in atmospheres, the units are $\text{atm}^2/\text{atm} = \text{atm}$. Attaching units to the equilibrium constant has the advantage of clearly indicating the units in which concentration is expressed. Nevertheless, the more common practice is to write equilibrium constants as dimensionless quantities. We have adopted this practice in this text.

14.3 HETEROGENEOUS EQUILIBRIA

Many equilibria of importance, such as in the hydrogen-nitrogen-ammonia system, involve substances all in the same phase. Such equilibria are said to be **homogeneous.** On the other hand, it is possible for the substances in equilibrium to be in different phases, giving rise to **heterogeneous equilibria.** As an example, consider the decomposition of calcium carbonate:

$$CaCO_3(s) \rightleftharpoons CaO(s) + CO_2(g) \qquad [14.8]$$

This system involves a gas in equilibrium with two solids. If we write the equilibrium-constant expression for this process in the usual way, we obtain

$$K = \frac{[CaO][CO_2]}{[CaCO_3]} \qquad [14.9]$$

This example presents us with a problem we have not encountered previously: How do we express the concentration of a solid substance? Because the calcium carbonate or calcium oxide are present as pure solids, their "concentration" is a constant. The number of moles per liter of either of these solids is not changed, whether we have a large amount of solid present or just a little. The concentration of a pure substance, liquid or solid, can be expressed as the density divided by the molecular weight:

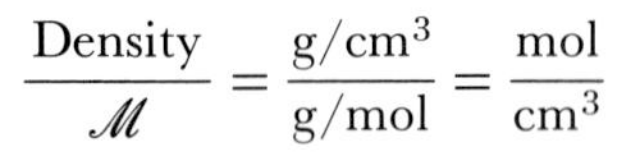

$$\frac{\text{Density}}{\mathscr{M}} = \frac{\text{g/cm}^3}{\text{g/mol}} = \frac{\text{mol}}{\text{cm}^3}$$

The density of a pure liquid or solid is a constant at any given temperature and in fact changes very little with temperature. Thus the effective concentration of a pure liquid or solid is a constant. The equilibrium-constant expression for Equation 14.8 then simplifies to

$$K = \frac{[CO_2](\text{constant 1})}{(\text{constant 2})}$$

where constant 1 is the concentration of CaO, and constant 2 is the concentration of $CaCO_3$. Moving the constants to the left-hand side of the equation, we have

$$K' = K\frac{(\text{constant 2})}{(\text{constant 1})} = [CO_2] \qquad [14.10]$$

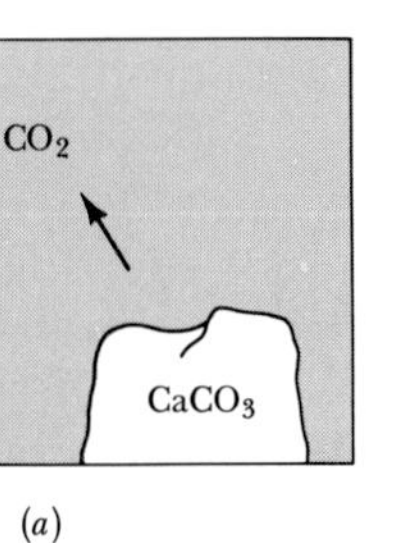

(*a*)

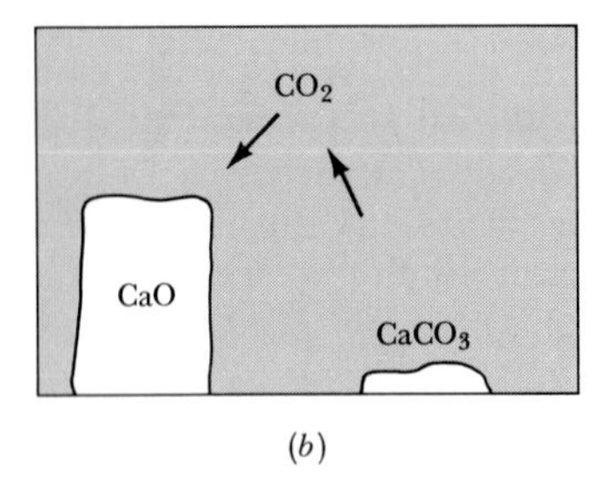

(*b*)

FIGURE 14.3 Heterogeneous equilibrium in $CaCO_3$ decomposition. Assuming the temperature in each case to be the same, the equilibrium pressure of CO_2 is the same in (*a*) as in (*b*).

As a practical matter, the overall effect is the same as if we set the concentrations of both solids equal to one in the equilibrium-constant expression.

Equation 14.10 tells us that it doesn't matter at all how much $CaCO_3$ or CaO is present, as long as there is some of each in the system. As shown in Figure 14.3, we would have the same pressure of CO_2 in the system when we have a large excess of CaO as when we have an excess of $CaCO_3$. On the other hand, if one of the three ingredients is missing, we cannot have an equilibrium.

Thus we see that any pure solids or liquids that might be involved in an equilibrium have the same effect on the equilibrium no matter how much solid or liquid is present. The "concentrations" of pure solids and liquids are incorporated into the equilibrium constant. What this means when we write equilibrium-constant expressions is that the concentrations of pure solids and liquids are absent from the expression.

SAMPLE EXERCISE 14.6

Each of the mixtures listed below was placed into a closed container and allowed to stand. Which of these mixtures are capable of attaining the equilibrium expressed by Equation 14.8: (a) pure $CaCO_3$; (b) CaO and a pressure of CO_2 greater than the value of K_c; (c) some $CaCO_3$ and a pressure of CO_2 greater than the value of K_c; (d) $CaCO_3$ and CaO?

Solution: Equilibrium can be reached in all cases except (c). In (a), $CaCO_3$ simply decomposes until the equilibrium pressure of CO_2 is attained. In (b), CO_2 combines with the CaO present until its pressure decreases to the equilibrium value. In (c), equilibrium can't be attained, because there is no way in which the CO_2 pressure can decrease so as to attain its equilibrium value. In (d), the situation is essentially the same as in (a); $CaCO_3$ decomposes until equilibrium is attained. The presence of CaO initially makes no difference.

SAMPLE EXERCISE 14.7

Write the equilibrium-constant expression for each of the following reactions:

(a) $CO_2(g) + H_2(g) \rightleftharpoons CO(g) + H_2O(l)$
(b) $SnO_2(s) + 2CO(g) \rightleftharpoons Sn(s) + 2CO_2(g)$

Solution: (a) The equilibrium-constant expression is

$$K = \frac{[CO]}{[CO_2][H_2]}$$

(Because H_2O is a pure liquid its concentration does not appear in the equilibrium-constant expression.)

(b) The equilibrium-constant expression is

$$K = \frac{[CO_2]^2}{[CO]^2}$$

(Because SnO_2 and Sn are both pure solids, they do not appear in the equilibrium-constant expression.)

14.4 USES OF EQUILIBRIUM CONSTANTS

We have seen that the magnitude of K indicates the extent to which a reaction will proceed. If K is very large, the reaction will tend to proceed far to the right; if K is very small, very little reaction will occur and the equilibrium mixture will contain mainly reactants. The equilibrium constant also allows us (1) to predict the direction in which a reaction mixture will proceed to achieve equilibrium, and (2) to calculate the concentrations of reactants and products once equilibrium has been reached.

Prediction of the Direction of Reaction

Suppose that we place a mixture of 2.00 mol of H_2, 1.00 mol of N_2, and 2.00 mol of NH_3 in a 1-L container at 472 K. Will there be a reaction between N_2 and H_2 to form more NH_3? If we insert the starting concentrations of N_2, H_2, and NH_3 into the equilibrium-constant expression, we have

$$\frac{[NH_3]^2}{[N_2][H_2]^3} = \frac{(2.00)^2}{(1.00)(2.00)^3} = 0.500$$

We have seen (Sample Exercise 14.3) that at this temperature $K_c = 0.105$. Therefore, the quotient $[NH_3]^2/[N_2][H_2]^3$ will need to change from 0.500 to 0.105 to move the system toward equilibrium. This change can happen only if $[NH_3]$ decreases and $[N_2]$ and $[H_2]$ increase. Thus the reaction proceeds toward equilibrium with the formation of N_2 and H_2 from the NH_3; the reaction proceeds from right to left.

When we substitute reactant and product concentrations into the equilibrium-constant expression as we did above, the result is known as the **reaction quotient** and is represented by the letter Q. The reaction quotient will equal the equilibrium constant, K, only if the concentrations are ones for the system at equilibrium: $Q = K$ only at equilibrium. We have seen that when the reaction quotient is larger than K, substances on the right side of the chemical equation will react to form substances on the left; the reaction moves from right to left in approaching equilibrium: if $Q > K$ the reaction moves right to left. Conversely, if $Q < K$, the reaction will move toward equilibrium with the formation of more products (from left to right). These relationships are summarized in Table 14.2.

TABLE 14.2 Effect of the relative values of Q and K on the direction of reaction

Relationship	Direction
$Q > K$	$\longleftarrow$
$Q = K$	(Equilibrium)
$Q < K$	$\longrightarrow$

SAMPLE EXERCISE 14.8

At 448°C the equilibrium constant, K_c, for the reaction

$$H_2(g) + I_2(g) \rightleftharpoons 2HI(g)$$

is 50.5. Predict the direction in which the reaction will proceed to reach equilibrium at 448°C if we start with 2.0×10^{-2} mol of HI, 1.0×10^{-2} mol of H_2, and 3.0×10^{-2} mol of I_2 in a 2.0-L container.

Solution: The starting concentrations are

$$[HI] = 2.0 \times 10^{-2}\ \text{mol}/2.0\ \text{L} = 1.0 \times 10^{-2}\ M$$
$$[H_2] = 1.0 \times 10^{-2}\ \text{mol}/2.0\ \text{L} = 5.0 \times 10^{-3}\ M$$
$$[I_2] = 3.0 \times 10^{-2}\ \text{mol}/2.0\ \text{L} = 1.5 \times 10^{-2}\ M$$

The reaction quotient is

$$Q = \frac{[HI]^2}{[H_2][I_2]} = \frac{(1.0 \times 10^{-2})^2}{(5.0 \times 10^{-3})(1.5 \times 10^{-2})} = 1.3$$

Because $Q < K_c$, [HI] will need to increase and $[H_2]$ and $[I_2]$ decrease to reach equilibrium; the reaction will proceed from left to right.

Calculation of Equilibrium Concentrations

If we know the value of the equilibrium constant for a particular reaction, we can calculate the concentrations of substances present in the reaction mixture at equilibrium. The complexity of the calculation will depend on several factors including the complexity of the balanced equation for the reaction and what concentrations we are given. The following examples illustrate the general types of problems.

SAMPLE EXERCISE 14.9

What is the partial pressure of NH_3 that is in equilibrium with N_2 and H_2 at 500°C if the equilibrium partial pressure of H_2 is 0.733 atm and that of N_2 is 0.527 atm [$K_p = 1.45 \times 10^{-5}$ at 500°C for $N_2(g) + 3H_2(g) \rightleftharpoons 2NH_3(g)$]?

Solution:

$$K_p = \frac{P^2_{NH_3}}{P_{N_2}P^3_{H_2}} = 1.45 \times 10^{-5}$$

Because K_p, P_{N_2}, and P_{H_2} are given, we can rearrange this equation and solve for P_{NH_3}:

$$P^2_{NH_3} = K_pP_{N_2}P^3_{H_2} = (1.45 \times 10^{-5})(0.527)(0.733)^3$$
$$= 3.01 \times 10^{-6}$$
$$P_{NH_3} = \sqrt{3.01 \times 10^{-6}} = 1.73 \times 10^{-3} \text{ atm}$$

SAMPLE EXERCISE 14.10

A 1-L container is filled with 0.50 mol of HI at 448°C. The value of the equilibrium constant, K_c, for the reaction

$$H_2(g) + I_2(g) \rightleftharpoons 2HI(g)$$

at this temperature is 50.5. What are the concentrations of H_2, I_2, and HI in the vessel at equilibrium?

Solution: In this case we are not given any of the equilibrium concentrations, only the starting concentrations: $[H_2] = [I_2] = 0$ and $[HI] = 0.50\,M$. The problem is related to the one we worked in Sample Exercise 14.4: We must use the balanced chemical equation to write an expression for equilibrium concentrations in terms of initial concentrations. It is useful to set up a table as in Sample Exercise 14.4. We can define x as the amount of HI that reacts forming H_2 and I_2. For each x HI that decomposes, $(x/2)$ H_2 and $(x/2)$ I_2 form

	$H_2(g)$ +	$I_2(g)$ $\rightleftharpoons$	$2HI(g)$
Initial:	$0\,M$	$0\,M$	$0.50\,M$
Change:	$(x/2)\,M$	$(x/2)\,M$	$-x\,M$
Equilibrium:	$(x/2)\,M$	$(x/2)\,M$	$(0.50 - x)\,M$

We can substitute the equilibrium concentrations into the equilibrium expression and solve for the single unknown, x:

$$K_c = \frac{[HI]^2}{[H_2][I_2]} = \frac{(0.50 - x)^2}{(x/2)^2} = 50.5$$

This equation is second order in x (it contains x^2 as the highest power of x); such equations can always be solved by use of the quadratic formula (Appendix A.3). However, a quicker solution can be obtained in this particular case by taking the square root of both sides of the equation:

$$\frac{0.50 - x}{x/2} = \sqrt{50.5} = 7.11$$

Solving for x yields

$$0.50 - x = \frac{x}{2}(7.11) = 3.56x$$
$$0.50 = x + 3.56x = 4.56x$$
$$x = \frac{0.50}{4.56} = 0.11\,M$$

Thus the equilibrium concentrations are as follows:

$$[H_2] = \frac{x}{2} = \frac{0.11\,M}{2} = 0.055\,M$$
$$[I_2] = \frac{x}{2} = \frac{0.11\,M}{2} = 0.055\,M$$
$$[HI] = 0.50\,M - 0.11\,M = 0.39\,M$$

14.5 FACTORS AFFECTING EQUILIBRIUM: LE CHÂTELIER'S PRINCIPLE

In developing his process for making ammonia from N_2 and H_2 Haber sought to know the factors that might be varied to increase the yield of NH_3. Using the values of the equilibrium constant at various temperatures he calculated the equilibrium amounts of NH_3 formed under a variety of conditions. The results of some of his calculations are shown in Table 14.3. Notice that the yield of NH_3 decreases with increasing temperature and increases with increasing pressure. These results can be qualitatively understood in terms of Le Châtelier's principle (Section 11.3). In this section we will use Le Châtelier's principle to make qualitative predictions about the response of a system at equilibrium to various changes in external conditions. We will consider three ways that a chemical equilibrium can be shifted: (1) adding or removing a reactant or product, (2) changing the pressure, and (3) changing the temperature.

TABLE 14.3 Effect of temperature and total pressure on the percentage of ammonia present at equilibrium, beginning with a 3:1 molar H_2/N_2 mixture

	Total pressure (atm)			
Temperature (°C)	200	300	400	500
400	38.7	47.8	54.9	60.6
450	27.4	35.9	42.9	48.8
500	18.9	26.0	32.2	37.8
600	8.8	12.9	16.9	20.8

Change in Reactant or Product Concentrations

A system at equilibrium is in a dynamic state; the forward and reverse processes are occurring at equal rates, and the system is in a state of balance. An alteration in the conditions of the system may cause the state of balance to be disturbed. If this occurs, the equilibrium is shifted until a new state of balance is attained. Le Châtelier's principle states that the shift will be in the direction that minimizes or reduces the effect of the change. Therefore, *if a chemical system is at equilibrium, and we add a substance (either a reactant or a product), the reaction will shift so as to reestablish equilibrium by consuming part of that added substance. Conversely, removal of a substance will result in the reaction moving in the direction that forms more of that substance.*

For example, addition of hydrogen to an equilibrium mixture of H_2, N_2, and NH_3 would cause the system to shift in such a way as to reduce the hydrogen pressure toward its original value. This can occur only if the equilibrium is shifted in the direction of forming more NH_3. At the same time, the quantity of N_2 would also be reduced slightly. This situation is illustrated in Figure 14.4. Addition of more N_2 to an equilibrium system would similarly cause a shift in the direction of forming more ammonia. On the other hand, Le Châtelier's principle tells us that if we add NH_3 to the system at equilibrium, the shift will be in such a direction as to reduce the NH_3 concentration toward its original value; that is, some of the added ammonia will decompose to form N_2 and H_2.

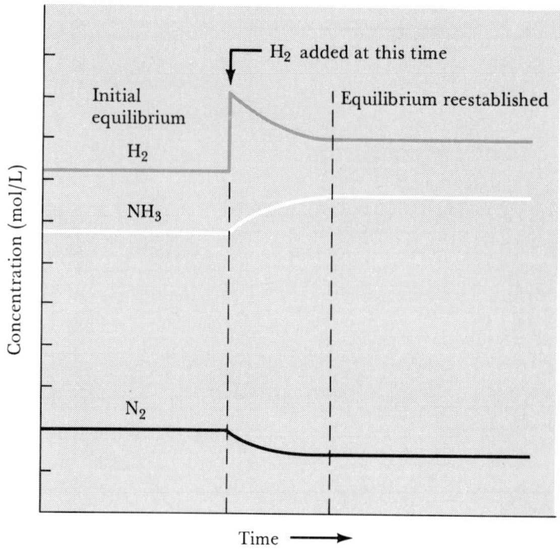

FIGURE 14.4 When H_2 is added to an equilibrium mixture of N_2, H_2, and NH_3, a portion of the H_2 reacts with N_2 to form NH_3, thereby establishing a new equilibrium position.

We can reach the same conclusions by considering the effect that adding or removing a substance has on the reaction quotient (Section 14.4). For example, removal of NH_3 from an equilibrium mixture gives

$$\frac{[NH_3]^2}{[N_2][H_2]^3} = Q < K$$

Because $Q < K$, the reaction shifts from left to right, forming more NH_3 and decreasing $[N_2]$ and $[H_2]$, to restore a new equilibrium that is still governed by K.

If the products of a reaction can be removed continuously, the reacting system can be continuously shifted to form more products. The yield of NH_3 in the Haber process can be increased dramatically by liquefying the NH_3; the liquid NH_3 is removed and the N_2 and H_2 are recycled to form more NH_3. The general way this is accomplished is shown in Figure 14.5. If a reaction is operated so that equilibrium cannot be achieved because of escape of products, or if the equilibrium constant is very large, the reaction will proceed essentially to completion. In such instances the chemical equation for the reaction is usually given with a single arrow: reactants $\longrightarrow$ products.

Effect of Pressure and Volume Changes

If a system is at equilibrium and the total pressure is increased by reducing the volume, the system responds by shifting in the direction that occupies the smaller volume. In practice, this means that the shift is in the direction that decreases the number of moles of gas. Conversely, decreasing the pressure by increasing the volume causes a shift in the direction that produces more gas molecules. For the reaction

$$N_2(g) + 3H_2(g) \rightleftharpoons 2NH_3(g)$$

there is 2 mol of gas on the right side of the chemical equation ($2NH_3$) and 4 mol of gas on the left ($1N_2 + 3H_2$). Consequently, an increase in

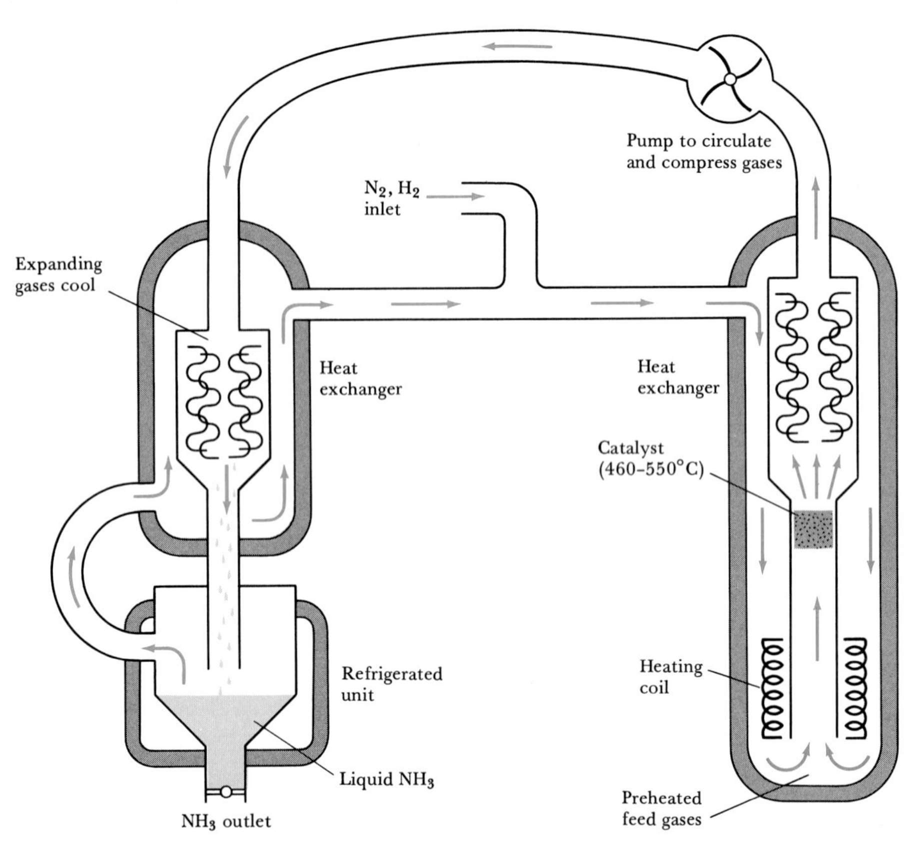

FIGURE 14.5 Schematic diagram summarizing the industrial production of ammonia. Incoming N_2 and H_2 gases are heated to approximately 500°C and passed over a catalyst. The resultant gas mixture is allowed to expand and cool, causing NH_3 to liquefy. Unreacted N_2 and H_2 gases are recycled.

pressure (decrease in volume) leads to the formation of NH_3; the reaction shifts toward the side with fewer gas molecules. In the case of the reaction

$$H_2(g) + I_2(g) \rightleftharpoons 2HI(g)$$

changing the pressure will not influence the position of the equilibrium. In this case there are the same number of gaseous product molecules as there are gaseous reactants.

It is important to keep in mind that pressure-volume changes do not change the value of K, as long as the temperature remains constant. Rather, they change the concentrations of the gaseous substances. In Sample Exercise 14.3, we calculated K_c for an equilibrium mixture at 472°C that contained $[H_2] = 0.1207\,M$, $[N_2] = 0.0402\,M$, and $[NH_3] = 0.00272\,M$. The value of K_c is 0.105. Consider what happens if we double the pressure of this equilibrium mixture by reducing the volume by half. If there were no shift in equilibrium, this volume change would cause the concentrations of all substances to double, giving new concentrations of $[H_2] = 0.2414\,M$, $[N_2] = 0.0804\,M$, and $[NH_3] =$

0.00544 M. The reaction quotient would then no longer equal the equilibrium constant:

$$Q = \frac{[NH_3]^2}{[N_2][H_2]^3} = \frac{(0.00544)^2}{(0.0804)(0.2414)^3} = 2.62 \times 10^{-2}$$

Because $Q < K_c$, the system is no longer at equilibrium. Equilibrium will be reestablished by increasing $[NH_3]$ and decreasing $[N_2]$ and $[H_2]$ until $Q = K_c = 0.105$. Therefore, the equilibrium shifts to the right as Le Châtelier's principle predicted.

It is possible to change the pressure of a system without changing its volume. Suppose that the N_2–H_2–NH_3 system is held in a constant-volume container. The total pressure of the system will increase if an inert gas such as argon is added. However, this addition will not alter the partial pressures of N_2, H_2, or NH_3. Consequently, the presence of the inert gas will not affect the equilibrium.

Effect of Temperature Changes

Changes in concentrations or total pressure can cause shifts in equilibrium without changing the equilibrium constant. In contrast, almost every equilibrium constant changes in value with change in temperature. Consider the decomposition of carbon dioxide into carbon monoxide and oxygen:

$$2CO_2(g) \rightleftharpoons 2CO(g) + O_2(g) \qquad [14.11]$$

TABLE 14.4 Percent dissociation of CO_2 into CO and O_2 as a function of temperature

Temperature (K)	Percent dissociation
1500	0.048
2000	2.05
2500	17.6
3000	54.8

From a knowledge of the heats of formation of $CO_2(g)$ and $CO(g)$, we can conclude that the forward reaction is highly endothermic; using the values of ΔH_f° from Appendix D, ΔH° for the overall reaction is calculated to be 566 kJ. At room temperature and thereabouts, CO_2 has no observable tendency to dissociate into CO and O_2. However, at high temperatures the equilibrium shifts to the right as shown in Table 14.4. Clearly, the equilibrium constant for reaction Equation 14.11 is very dependent on temperature and increases with increasing temperature.

The equilibrium constants for exothermic reactions, that is, those in which heat is evolved, decrease with increase in temperature. By contrast, equilibrium constants for endothermic reactions increase with increase in temperature.

Following the line of reasoning employed in Section 11.3, we can deduce the rules for the temperature dependence of the equilibrium constant by applying Le Châtelier's principle. When heat is added to a system by increasing the temperature, the equilibrium should shift in such a direction as to undo partially the effect of the added heat. It shifts, therefore, in the direction in which heat is absorbed. If a reaction is exothermic in the forward direction, it must be endothermic in the reverse direction. Thus, when heat is added to an equilibrium system that is exothermic in the forward direction, the equilibrium shifts in the reverse direction, in the direction of reactants. In summary, the rule is that *when heat is added at constant pressure to an equilibrium system, the equilibrium shifts in the direction that absorbs heat.* Conversely, if heat is removed from an equilibrium system, the equilibrium shifts in the direction that evolves heat.

SAMPLE EXERCISE 14.11

Consider the following reaction:

$$N_2O_4(g) \rightleftharpoons 2NO_2(g) \qquad \Delta H^\circ = 58.0 \text{ kJ}$$

In what direction will the equilibrium shift when each of the following changes is made to a system at equilibrium: (a) add N_2O_4; (b) remove NO_2; (c) increase pressure; (d) increase volume; (e) decrease temperature?

Solution: Le Châtelier's principle can be applied to determine the effects of each of these changes.

(a) The system will adjust so as to decrease the concentration of the added N_2O_4; the reaction consequently shifts toward the formation of more products (shifts toward the right side of the equation).

(b) The system will adjust to this change by forming more NO_2; the equilibrium shifts toward the product side of the equation, to the right.

(c) The system will establish a new equilibrium that has a smaller volume (fewer gas molecules); consequently, the reaction shifts to the left.

(d) The system will shift in the direction that occupies a larger volume (more gas molecules); it moves to the right.

(e) The system will adjust to a new equilibrium position by shifting in the direction that produces heat. The reaction is endothermic in the forward direction (left to right). It therefore shifts to the left, with the formation of more N_2O_4 in order to produce heat. Note that only this last change effects the numerical value of the equilibrium constant, K.

SAMPLE EXERCISE 14.12

Using the standard heat of formation data in Appendix D, determine the enthalpy change for the reaction

$$N_2(g) + 3H_2(g) \rightleftharpoons 2NH_3(g)$$

From this determine how the equilibrium constant for the reaction should change with temperature.

Solution: Recall that the standard enthalpy change for a reaction is given by the standard molar enthalpies of the products, each multiplied by its coefficient in the balanced chemical equation, less the same quantities for the reactants. ΔH_f° for $NH_3(g)$ at 25°C is −46.19 kJ/mol. The ΔH_f° values for $H_2(g)$ and $N_2(g)$ are zero by definition, because the enthalpies of formation of the elements in their normal states at 25°C are defined as zero (Section 4.5). Because 2 mol of NH_3 is formed, the total enthalpy change is

$$2 \text{ mol}(-46.19 \text{ kJ/mol}) - 0 = -92.38 \text{ kJ}$$

The reaction in the forward direction is exothermic. An increase in temperature should therefore cause the reaction to shift in the *reverse* direction, that is, in the direction of less NH_3 and more N_2 and H_2. This is what occurs, as reflected in the values for K_p presented in Table 14.5. Notice that K_p changes very markedly with change in temperature, and that it is larger at lower temperatures. This is a matter of great practical importance. To form ammonia at a reasonable rate, higher temperatures are required. Yet at higher temperatures, the equilibrium constant is smaller, so the percentage conversion to ammonia is smaller. To compensate for this, higher pressures are needed, because high pressure favors ammonia formation.

14.6 THE RELATIONSHIP BETWEEN CHEMICAL EQUILIBRIUM AND CHEMICAL KINETICS

It is very important to be clear on the distinction between a system at equilibrium and the rate at which the system approaches equilibrium in the first place. Imagine that we have a system initially containing only reactant molecules and no products. As the system begins to change, the only reaction occurring is formation of products. As products accumulate, however, the reverse reaction also begins to occur. In many systems,

TABLE 14.5 Variation in K_p for the equilibrium $N_2 + 3H_2 \rightleftharpoons 2NH_3$ as a function of temperature

Temperature (°C)	K_p
300	4.34×10^{-3}
400	1.64×10^{-4}
450	4.51×10^{-5}
500	1.45×10^{-5}
550	5.38×10^{-6}
600	2.25×10^{-6}

this reverse reaction is so slow that it is completely unimportant. The system then just keeps on changing until essentially all the reactants are converted to products. Reactions of this sort are said to proceed to completion. Even though the reverse reaction, conversion of products to reactants, is possible in principle, it is not observed.

In many other reactions, on the other hand, the reverse reaction occurs more rapidly. After a time, reactant molecules are being formed just as rapidly as they are themselves reacting to form products. When the rates of the two opposing processes are equal, the overall rate of change in the system is zero, and we have chemical equilibrium. A true chemical equilibrium always involves a balancing of equal and opposite rate processes. This is true regardless of the particular mechanism or pathway by which the reaction proceeds. Figure 14.6 shows an energy profile for the single-step, bimolecular reaction between reactants A and B and products C and D:

$$A + B \rightleftharpoons C + D \qquad [14.12]$$

At equilibrium, the rate at which reactants cross the energy barrier to form products equals the rate at which products cross the energy barrier in the reverse direction to form reactants. Because we have chosen the situation where the reaction occurs in a single, bimolecular step in each direction, the rate law for each direction is second order. Thus

$$r_f = \text{rate of forward reaction} = k_f[A][B]$$
$$r_r = \text{rate of reverse reaction} = k_r[C][D]$$

(Remember that we use a lowercase k to represent rate constants and a capital K to represent equilibrium constants.) At equilibrium these two rate processes must be equal:

$$k_f[A][B] = k_r[C][D]$$

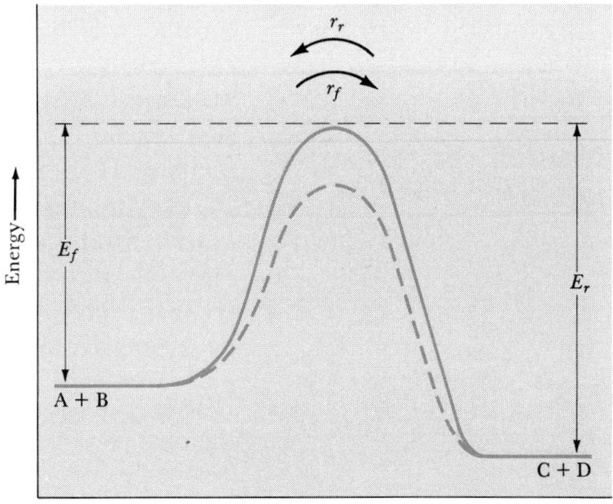

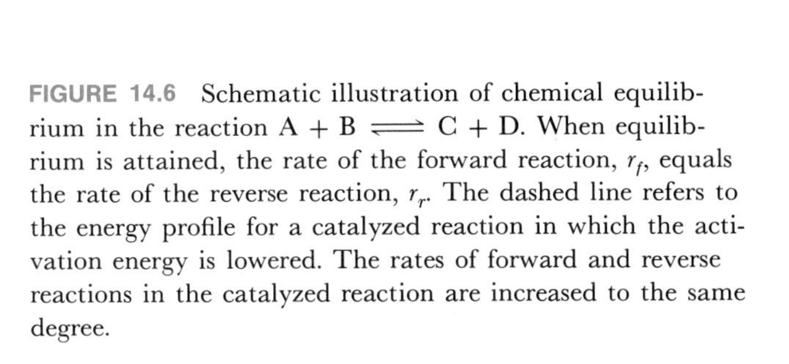

FIGURE 14.6 Schematic illustration of chemical equilibrium in the reaction $A + B \rightleftharpoons C + D$. When equilibrium is attained, the rate of the forward reaction, r_f, equals the rate of the reverse reaction, r_r. The dashed line refers to the energy profile for a catalyzed reaction in which the activation energy is lowered. The rates of forward and reverse reactions in the catalyzed reaction are increased to the same degree.

If we rearrange this equation we obtain

$$\frac{k_f}{k_r} = \frac{[C][D]}{[A][B]} = K \quad [14.13]$$

Thus the equilibrium constant is just the ratio of rate constants for the forward and reverse reactions.* Recall that the rate of a reaction decreases as the height of the energy barrier increases (Section 13.4). The barrier for the forward reaction (E_f) in Figure 14.6 is lower than for the reverse reaction (E_r). Thus k_f must be larger than k_r, so that K should be a large number. This is in keeping with the fact that the energies of the products are lower than the energies of the reactants.

The Effects of Catalysts

Suppose that we add a catalyst to the reaction system described in Equation 14.12 so that the energy barrier to reaction is lowered, as shown by the dashed line in Figure 14.6. The rates of *both* the forward and reverse reactions are increased by the presence of a catalyst. In fact, the two rate constants are affected to precisely the same degree. In other words, it is impossible for a catalyst to lower the barrier for just the forward reaction and not the reverse reaction. Because the forward and reverse reaction rate constants are affected to exactly the same degree, their ratio is unchanged. We therefore have the rule that *a catalyst may change the rate of approach to equilibrium, but it does not change the value of the equilibrium constant.*

The rate at which a reaction approaches equilibrium is an important practical consideration. As an example, let us again consider the synthesis of ammonia from N_2 and H_2. In designing a process for ammonia synthesis, Haber had to deal with a rather serious problem. He wished to synthesize ammonia at the lowest temperature possible, consistent with a reasonable reaction rate. But in the absence of a catalyst, hydrogen and nitrogen do not react with one another at a significant rate either at room temperature or even at much higher temperatures. On the other hand, Haber had to cope with a rapid decrease in equilibrium constant with increasing temperature as shown in Table 14.5. At temperatures sufficiently high to give a satisfactory reaction rate, the amount of ammonia formed was too small. The solution to this dilemma was to develop a catalyst that would produce a reasonably rapid approach to equilibrium, at a sufficiently low temperature so that the equilibrium constant was still reasonably large. The development of a catalyst thus became the focus of Haber's research efforts.

After trying different substances to see which would be most effective, Haber finally settled on iron mixed with metal oxides. Variants of the original catalyst formulations are still employed. With these catalysts it is possible to obtain a reasonably rapid approach to equilibrium at temperatures around 400 to 500°C, and with gas pressures of 200 to 600 atm.

*The relationship between rate constants and the equilibrium constant is somewhat more complicated if the reaction occurs by a multistep mechanism. However, even in that case the equilibrium constant can be related to a ratio of rate constants. Regardless of the mechanism, the equilibrium-constant expression can be written directly from the balanced equation for the overall reaction. By contrast, rate laws must be determined by experiment.

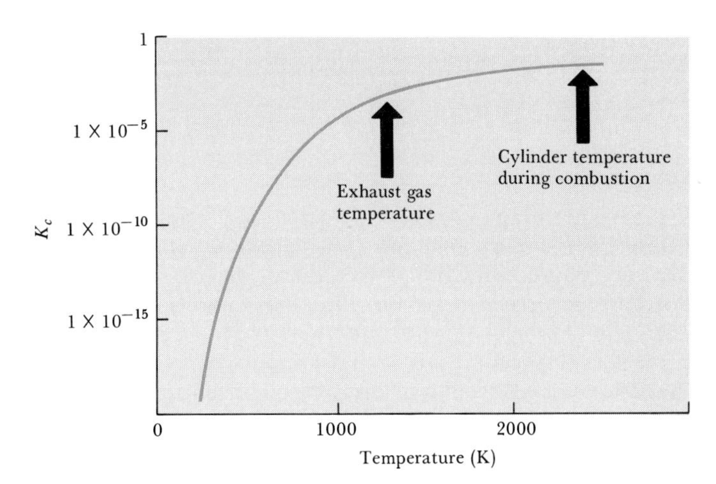

FIGURE 14.7 Variation of the equilibrium constant for the reaction $\frac{1}{2}N_2(g) + \frac{1}{2}O_2(g) \rightleftharpoons NO(g)$ as a function of temperature.

The high pressures are needed to obtain a satisfactory degree of conversion at equilibrium. You can see from Table 14.3 that if an improved catalyst could be found, one that would lead to sufficiently rapid reaction at temperatures lower than 400 to 500°C, it would be possible to obtain the same degree of equilibrium conversion at much lower pressures. This would result in a great savings in the cost of equipment for ammonia synthesis. In view of the growing need for nitrogen as fertilizer, the fixation of nitrogen is a process of ever-increasing importance, worthy of additional research effort.

The formation of NO from N_2 and O_2 provides another interesting example of the practical importance of changes in equilibrium constant and reaction rate with temperature. The balanced equation and the standard enthalpy change for the reaction are

$$\frac{1}{2}N_2(g) + \frac{1}{2}O_2(g) \rightleftharpoons NO(g) \qquad \Delta H^\circ = 90.4 \text{ kJ} \qquad [14.14]$$

The reaction is endothermic, that is, heat is absorbed when NO is formed from the elements. By applying Le Châtelier's principle, we deduce that an increase in temperature will shift the equilibrium in the direction of more NO. The equilibrium constant, K_c, for formation of 1 mol of NO from the elements at 300 K is only about 10^{-15}. On the other hand, at a much higher temperature, about 2400 K, the equilibrium constant is much larger, about 0.05. The manner in which K_c for reaction Equation 14.14 varies with temperature is shown in Figure 14.7.*

This graph helps to explain why NO is a pollution problem. In the cylinder of a modern high-compression auto engine, the temperatures during the fuel-burning part of the cycle may be on the order of 2400 K. Because there is a fairly large excess of air in the cylinder, these conditions provide an opportunity for the formation of some NO. After the combustion, however, the gases are quickly cooled. As the temperature drops the equilibrium of Equation 14.14 shifts strongly to the left, that is, in the direction of N_2 and O_2. But the lower temperatures also mean that the rate of the reaction also is decreased. The NO formed at high temperatures is essentially "frozen" in that form as the gas cools.

The gases exhausting from the cylinder are still quite hot, perhaps 1200 K. At this temperature, as shown in Figure 14.7, the equilibrium constant for formation of NO is much smaller. However, the rate of conversion of NO to N_2 and O_2 is too slow to permit much loss of NO before the gases are cooled still further. Getting the NO out of the exhaust gases, described in Chapter 12, depends on finding a catalyst that will work at the temperatures of the exhaust gases, and that will cause conversion of NO into something harmless. If a catalyst could be found which would convert the NO back into N_2 and O_2 the equilibrium would be sufficiently favorable. It has not proved possible to find a catalyst capable of withstanding the grueling conditions found in automobile exhaust systems that can catalyze the conversion of NO into N_2 and O_2. Instead, the catalysts used are designed to catalyze reaction of NO with H_2 or CO.

*It is necessary to use a log scale for K_c in Figure 14.7 because the values of K_c vary over such a huge range.

FOR REVIEW

Summary

If the reactants and products of a reaction are kept in contact, a chemical reaction can achieve a state of dynamic balance in which forward and reverse reactions are occurring at equal rates. This condition is known as **chemical equilibrium.** A system at equilibrium does not change with time. For such a system, the ratio of products to reactants, each concentration raised to the power corresponding to the coefficient in the balanced equation, is called the **equilibrium constant, K:** K = [products]/[reactants]. The equilibrium constant changes with temperature but is not affected by changes in relative concentrations of any reacting substances or by pressure or the presence of catalysts. In **heterogeneous equilibria,** the concentrations of pure solids or liquids are absent from the equilibrium-constant expression.

If concentrations are expressed in molarity, we label the equilibrium constant K_c; if the concentration units are atmospheres, we use K_p. K_c and K_p are related by the following equation: $K_p = K_c(RT)^{\Delta n}$. A large value for K_c or K_p indicates that the equilibrium mixture contains more products than reactants. A small value for the equilibrium constant means the equilibrium lies toward the reactant side.

The **reaction quotient, Q,** is found by substituting reactant and product concentrations into the equilibrium-constant expression. If the system is at equilibrium, $Q = K$. However, if $Q \neq K$, nonequilibrium conditions apply; if $Q < K$, the reaction will move toward equilibrium by forming more products (the reaction moves from left to right); if $Q > K$, the reaction will proceed from right to left. Knowledge of the value of K_c or K_p permits calculation of the equilibrium concentrations of reactants and products.

Le Châtelier's principle indicates that if we disturb a system that is at equilibrium, the equilibrium will shift to minimize the disturbing influence. The effects of adding (or removing) reactants or products, and of changing pressure, volume, or temperature can be predicted using this principle. The equilibrium constant changes value with changes in temperature. Catalysts affect the speed at which equilibrium is reached but do not affect K (do not affect the position of equilibrium).

Learning goals

Having read and studied this chapter, you should be able to:

1. Write the equilibrium-constant expression for a balanced chemical equation, whether heterogeneous or homogeneous.
2. Numerically evaluate K_c (or K_p) from a knowledge of the equilibrium concentrations (or pressures) of reactants or products or from the initial concentrations and the equilibrium concentration of at least one substance.
3. Interconvert K_c and K_p.
4. Calculate the reaction quotient, Q, and by comparison with the value of K_c or K_p, determine whether a reaction is at equilibrium. If it is not at equilibrium, you should be able to predict in which direction it will shift to reach equilibrium.
5. Use the equilibrium constant to calculate equilibrium concentrations. (You will also need to know either the equilibrium concentrations of all but one substance or the initial concentrations together with the equilibrium concentration of one substance.)
6. Explain how the relative equilibrium quantities of reactants and products are shifted by changes in temperature, pressure, or the concentrations of substances involved in the equilibrium reaction.
7. Explain how the change in equilibrium constant with change in temperature is related to the heat change in the reaction.
8. Describe the effect of a catalyst on a system as it approaches equilibrium.

Key terms

Among the more important terms and expressions used for the first time in this chapter are the following:

Chemical equilibrium (Section 14.2) is a state of a chemical system in which the rate of formation of products equals the rate of formation of reactants from products.

The **Haber process** (Section 14.1) refers to the catalyst system and conditions of temperature and pressure developed by Fritz Haber and co-workers for the formation of NH_3 from H_2 and N_2.

Heterogeneous equilibrium (Section 14.3) refers to the equilibrium state established between substances in two or more different phases, for example, between a gas and solid or between a solid and liquid.

Homogeneous equilibrium (Section 14.3) is the state of equilibrium established between reactant and product substances all in the same phase, for example, all gases, or all dissolved in solution.

The **law of mass action** (Section 14.2) provides the rules according to which the equilibrium constant is expressed in terms of the concentrations of reactants and products, in accordance with the balanced chemical equation for the reaction.

Le Châtelier's principle (Section 14.5) tells us that when we bring to bear some disturbing influence on a system at chemical equilibrium, the relative concentrations of reactants and products will shift so as to undo partially the effects of the disturbance.

The **reaction quotient, Q** (Section 14.4), is the value that is obtained when concentrations of reactants and products are inserted into the equilibrium-constant expression. If the concentrations are equilibrium concentrations, $Q = K$; otherwise, $Q \neq K$.

EXERCISES

Equilibrium-constant expressions

14.1 Write the equilibrium-constant expression for each of the following reactions:

(a) $2SO_3(g) \rightleftharpoons 2SO_2(g) + O_2(g)$
(b) $NH_4NO_2(s) \rightleftharpoons N_2(g) + 2H_2O(g)$
(c) $N_2O_4(g) \rightleftharpoons 2NO_2(g)$
(d) $Fe_3O_4(s) + H_2(g) \rightleftharpoons 3FeO(s) + H_2O(g)$
(e) $4NH_3(g) + 3O_2(g) \rightleftharpoons 2N_2(g) + 6H_2O(g)$

14.2 Write the equilibrium-constant expression for each of the following reactions. In each case indicate whether the reaction is homogeneous or heterogeneous.

(a) $2H_2(g) + O_2(g) \rightleftharpoons 2H_2O(g)$
(b) $CH_3OH(l) \rightleftharpoons CO(g) + 2H_2(g)$
(c) $4HCl(g) + O_2(g) \rightleftharpoons 2H_2O(l) + 2Cl_2(g)$
(d) $PbO(s) + CO_2(g) \rightleftharpoons PbCO_3(s)$
(e) $3O_2(g) \rightleftharpoons 2O_3(g)$

Calculation of K_c and K_p

14.3 A sample of chlorine gas is placed in a vessel and heated to 1400 K. The chlorine dissociates: $Cl_2(g) \rightleftharpoons 2Cl(g)$. When equilibrium is reached at this temperature it is found that $P_{Cl_2} = 1.00$ atm and $P_{Cl} = 2.97 \times 10^{-2}$ atm. What is the K_p value at 1400 K?

14.4 Gaseous hydrogen iodide is introduced into a container at 425°C, where it partly decomposes to hydrogen and iodine: $2HI(g) \rightleftharpoons H_2(g) + I_2(g)$. At equilibrium the resultant mixture is analyzed, and it is found that $[H_2] = 4.79 \times 10^{-4} M$, $[I_2] = 4.79 \times 10^{-4} M$, and $[HI] = 3.53 \times 10^{-3} M$. What is the value of K_c at this temperature?

14.5 A mixture of CH_4 and H_2O is passed over a nickel catalyst at 1000 K. The emerging gas is collected in a 5.00-L flask and found to contain 8.62 g of CO, 2.60 g of H_2, 43.0 g of CH_4, and 48.4 g of H_2O. Assuming that equilibrium has been reached, calculate K_c for the reaction

$CH_4(g) + H_2O(g) \rightleftharpoons CO(g) + 3H_2(g)$

14.6 The equilibrium constant for the reaction

$SO_2(g) + \frac{1}{2}O_2(g) \rightleftharpoons SO_3(g)$

is $K_c = 20.4$ at 700°C. (a) What is the value of K_c for

$SO_3(g) \rightleftharpoons SO_2(g) + \frac{1}{2}O_2(g)$

(b) What is the value of K_c for

$2SO_2(g) + O_2(g) \rightleftharpoons 2SO_3(g)$

(c) What is the value of K_p for

$2SO_2(g) + O_2(g) \rightleftharpoons 2SO_3(g)$

14.7 The equilibrium constant for the reaction

$2NO(g) + O_2(g) \rightleftharpoons 2NO_2(g)$

is $K_p = 1.48 \times 10^4$ at 184°C. (a) What is the value of K_p for

$2NO_2(g) \rightleftharpoons 2NO(g) + O_2(g)$

(b) What is the value of K_p for

$NO(g) + \frac{1}{2}O_2(g) \rightleftharpoons NO_2(g)$

(c) What is the value of K_c for

$2NO(g) + O_2(g) \rightleftharpoons 2NO_2(g)$

14.8 (a) At 700 K, $K_c = 0.11$ for the following reaction:

$CO_2(g) + H_2(g) \rightleftharpoons CO(g) + H_2O(g)$

What is the value of K_p? (b) At 1000 K, $K_c = 278$ for the reaction

$2SO_2(g) + O_2(g) \rightleftharpoons 2SO_3(g)$

What is the value of K_p? (c) For the equilibrium

$C(s) + CO_2(g) \rightleftharpoons 2CO(g)$

$K_p = 167.5$ at 1000°C. What is the value of K_c?

14.9 At temperatures near 800°C, steam passed over hot coke (a form of carbon obtained from coal) reacts to form CO and H_2:

$C(s) + H_2O(g) \rightleftharpoons CO(g) + H_2(g)$

The mixture of gases that results is an important industrial fuel called water gas. When equilibrium is achieved at 800°C, $[H_2] = 4.0 \times 10^{-2}\,M$, $[CO] = 4.0 \times 10^{-2}\,M$, and $[H_2O] = 1.0 \times 10^{-2}\,M$. Calculate K_c and K_p at this temperature.

14.10 A mixture of 0.100 mol of NO, 0.050 mol of H_2, and 0.100 mol of H_2O is placed in a 1.00-L vessel. The following equilibrium is established:

$$2NO(g) + 2H_2(g) \rightleftharpoons N_2(g) + 2H_2O(g)$$

At equilibrium, $[NO] = 0.070\,M$. (a) Calculate the equilibrium concentrations of H_2, N_2, and H_2O. (b) Calculate K_c.

14.11 A mixture of 1.000 mol of SO_2 and 1.000 mol of O_2 is placed in a 1.000-L vessel and kept at 1000 K until equilibrium is reached. At equilibrium the vessel is found to contain 0.919 mol of SO_3. Calculate K_c for the reaction

$$2SO_2(g) + O_2(g) \rightleftharpoons 2SO_3(g)$$

14.12 A sample of nitrosyl bromide, NOBr, decomposes according to the following equation:

$$2NOBr(g) \rightleftharpoons 2NO(g) + Br_2(g)$$

An equilibrium mixture in a 5.00-L vessel at 100°C contains 3.22 g of NOBr, 3.08 g of NO, and 4.19 g of Br_2. (a) Calculate K_c. (b) Calculate K_p. (c) What is the total pressure exerted by the mixture of gases?

14.13 A mixture of 1.374 g of H_2 and 70.31 g of Br_2 is heated together in a 2.00-L vessel at 700 K. These substances react as follows:

$$H_2(g) + Br_2(g) \rightleftharpoons 2HBr(g)$$

At equilibrium, the vessel is found to contain 0.566 g of H_2. What is the numerical value for K_c at 700 K?

Reaction quotient

14.14 As shown in Table 14.4, K_p for the equilibrium

$$N_2(g) + 3H_2(g) \rightleftharpoons 2NH_3(g)$$

is 4.51×10^{-5} at 450°C. Each of the mixtures listed below may or may not be at equilibrium at 450°C. Indicate in each case whether the mixture is at equilibrium; if it is not at equilibrium, indicate the direction (toward product or toward reactants) in which the mixture must shift to achieve equilibrium: (a) 100 atm NH_3, 30 atm N_2, 500 atm H_2; (b) 30 atm NH_3, 600 atm H_2, no N_2; (c) 26 atm NH_3, 42 atm H_2, 202 atm N_2; (d) 100 atm NH_3, 60 atm H_2, 5 atm N_2.

14.15 At 100°C the equilibrium

$$COCl_2(g) \rightleftharpoons CO(g) + Cl_2(g)$$

has an equilibrium constant, K_c, equal to 2.19×10^{-10}. Are the following mixtures of CO, Cl_2, and $COCl_2$ at equilibrium? If they are not at equilibrium, indicate the direction the reaction must proceed to reach equilibrium. (a) $[CO] = 1.0 \times 10^{-3}\,M$, $[Cl_2] = 1.0 \times 10^{-3}\,M$, $[COCl_2] = 2.19 \times 10^{-1}\,M$; (b) $[CO] = 3.31 \times 10^{-6}\,M$, $[Cl_2] = 3.31 \times 10^{-6}\,M$; $[COCl_2] = 5.00 \times 10^{-2}\,M$; (c) $[CO] = 4.50 \times 10^{-7}\,M$; $[Cl_2] = 5.73 \times 10^{-6}\,M$; $[COCl_2] = 8.57 \times 10^{-2}\,M$.

Equilibrium concentrations

14.16 At 1495°C, K_c for the equilibrium

$$H_2(g) + Br_2(g) \rightleftharpoons 2HBr(g)$$

is 3.5×10^4. If equilibrium concentrations at this temperature are 0.10 M H_2 and 0.20 M Br_2, what is the concentration of HBr?

14.17 The equilibrium constant, K_c, is 4.1×10^{-4} at 2000°C for the reaction

$$N_2(g) + O_2(g) \rightleftharpoons 2NO(g)$$

If 1.4 g of N_2 and 1.0 g of NO are in equilibrium with O_2 in a 0.500-L vessel at this temperature, how many grams of O_2 are present?

14.18 At 1558 K the equilibrium constant, K_c, for the reaction

$$Br_2(g) \rightleftharpoons 2Br(g)$$

is 1.04×10^{-3}. If a 0.200-L vessel contains 4.53×10^{-2} mol of Br_2 at equilibrium, how many moles of Br are present?

14.19 For the equilibrium

$$C(s) + CO_2(g) \rightleftharpoons 2CO(g)$$

$K_p = 167.5$ at 1000°C. What is the partial pressure of CO_2 that is in equilibrium with CO whose partial pressure is 0.500 atm?

14.20 The equilibrium constant for the reaction

$$I_2(g) + Br_2(g) \rightleftharpoons 2IBr(g)$$

has a value of 280 at 150°C. Suppose that a quantity of IBr is placed in a closed reaction vessel and the system allowed to come to equilibrium. When equilibrium is attained, the pressure of IBr is 0.20 atm. What are the pressures of $I_2(g)$ and $Br_2(g)$ at this point?

14.21 The following reaction can occur when N_2 and O_2 come in contact at high temperatures:

$$N_2(g) + O_2(g) \rightleftharpoons 2NO(g)$$

At 2400 K, the equilibrium constant, K_c, for this reaction is 2.5×10^{-3}. What are the equilibrium concentrations of N_2, O_2, and NO (in moles per liter) if equilibrium is obtained at 2400 K starting with a reaction mixture of 0.20 mol of N_2 and 0.20 mol of O_2 in a 5.00-L vessel?

14.22 At 21.8°C, the equilibrium constant, K_c, is 1.2×10^{-4} for the following reaction:

$$NH_4HS(s) \rightleftharpoons NH_3(g) + H_2S(g)$$

Calculate the equilibrium concentrations of NH_3 and H_2S if a sample of solid NH_4HS is placed in a closed vessel and allowed to decompose until equilibrium is reached at 21.8°C.

[14.23] The equilibrium constant for the reaction

$$H_2(g) + I_2(g) \rightleftharpoons 2HI(g)$$

has a value of 54.6 at 699 K. Suppose that 38.4 g of HI, 5.00 g of H_2, and 26.8 g of I_2 are placed together in a 2.00-L container at this temperature. What concentration of each substance will be present when the mixture reaches equilibrium?

Le Châtelier's principle

14.24 Consider the following equilibrium system:

$$C(s) + CO_2(g) \rightleftharpoons 2CO(g) \quad \Delta H° = 119.8 \text{ kJ}$$

If the reaction is at equilibrium, what would be the effect of (a) adding $CO_2(g)$; (b) adding $C(s)$; (c) adding heat; (d) increasing the pressure on the system by decreasing the volume; (e) adding a catalyst; (f) removing $CO(g)$?

14.25 In the reaction

$$6CO_2(g) + 6H_2O(l) \rightleftharpoons C_6H_{12}O_6(s) + 6O_2(g) \quad \Delta H° = 2816 \text{ kJ}$$

how is the equilibrium yield of $C_6H_{12}O_6$ affected by (a) increasing P_{CO_2}; (b) increasing temperature; (c) removing CO_2; (d) increasing the total pressure; (e) removing part of the $C_6H_{12}O_6$; (f) adding a catalyst?

14.26 How do each of the following changes affect the numerical value of the equilibrium constant for an exothermic reaction: (a) remove a reactant or product; (b) increase the total pressure; (c) decrease the temperature; (d) add a catalyst?

14.27 For the equilibrium

$$CO_2(g) + H_2(g) \rightleftharpoons CO(g) + H_2O(g)$$

K_c is 0.08 at 400°C and 0.41 at 600°C. Is the reaction as written exothermic or endothermic? Explain.

14.28 Using the standard heat of formation data in Appendix D, determine the enthalpy change for the reaction

$$2NOCl(g) \rightleftharpoons 2NO(g) + Cl_2(g)$$

Using this information, determine how the equilibrium constant for the reaction should change with temperature.

Chemical equilibrium and kinetics

14.29 Consider the following reaction, which occurs in a single step and is reversible:

$$CO(g) + Cl_2(g) \rightleftharpoons COCl(g) + Cl(g)$$

Kinetic studies indicate that the rate constant for the forward reaction, k_f, is $1.38 \times 10^{-28}/M$-s; for the reverse reaction, $k_r = 9.3 \times 10^{10}/M$-s (both rate constants at 25°C). What is the value for the equilibrium constant for this reaction at 25°C?

14.30 For the reaction

$$2SO_2(g) + O_2(g) \longrightarrow 2SO_3(g)$$

the standard enthalpy change is

$$\Delta H° = -196.6 \text{ kJ}$$

The activation energy for the uncatalyzed reaction is about 160 kJ/mol. Sketch the energy profile for this reaction, as in Figure 14.6. Sulfur dioxide is produced in the cylinder of an auto engine by oxidation of the small quantity of sulfur present in gas. The catalytic mufflers installed in cars since 1975 result in conversion of a large portion of this SO_2 to SO_3 (Section 13.6). Using a dotted line, sketch on your figure an energy profile which might apply for SO_2 oxidation in a catalytic muffler.

14.31 In the reaction

$$NO(g) + O_3(g) \rightleftharpoons NO_2(g) + O_2(g)$$

The rate law for the forward reaction is

$$r_f = k_f[NO][O_3]$$

and for the reverse reaction

$$r_r = k_r[NO_2][O_2]$$

Using these expressions, write the equilibrium condition in terms of opposing rates, and show how K_c relates to k_f and k_r.

14.32 Are any of the following statements false? For those that are, discuss the sense in which they are incorrect. (a) If a catalyst increases the rate of a forward reaction by a factor of 1000 over the uncatalyzed rate, it increases the rate of the reverse reaction by a factor of 1000 also. (b) A catalyst can promote the formation of product in some reactions by inhibiting the reverse reaction in an equilibrium. (c) Although heterogeneous catalysts must affect the rates of both forward and reverse reactions to an equal extent, homogeneous catalysts can be made to affect the rate of just the forward or just the reverse step. (d) All fast reactions have large equilibrium constants.

Additional exercises

14.33 Write the equilibrium-constant expression for each of the following reactions and indicate which are homogeneous and which are heterogeneous:

(a) $ZnSO_3(s) \rightleftharpoons ZnO(s) + SO_2(g)$
(b) $H_2(g) + CO_2(g) \rightleftharpoons CO(g) + H_2O(g)$
(c) $2NO_2(g) \rightleftharpoons 2NO(g) + O_2(g)$
(d) $SO_2Cl_2(g) \rightleftharpoons SO_2(g) + Cl_2(g)$
(e) $S(s) + O_2(g) \rightleftharpoons SO_2(g)$

14.34 Throughout this chapter we have used gas-phase reactions to illustrate the concept of equilibrium. However, solution reactions could also be used. Write the equilibrium-constant expressions for each of the following aqueous reactions:

(a) $Ag^+(aq) + 2NH_3(aq) \rightleftharpoons Ag(NH_3)_2{}^+(aq)$
(b) $Ag_2CrO_4(s) \rightleftharpoons 2Ag^+(aq) + CrO_4{}^{2-}(aq)$
(c) $HNO_2(aq) \rightleftharpoons H^+(aq) + NO_2{}^-(aq)$
(d) $Zn(s) + Cu^{2+}(aq) \rightleftharpoons Zn^{2+}(aq) + Cu(s)$
(e) $NH_3(aq) + H_2O(l) \rightleftharpoons NH_4{}^+(aq) + OH^-(aq)$

14.35 For the reaction $N_2O_4(g) \rightleftharpoons 2NO_2(g)$, an equilibrium mixture is found to contain 4.27×10^{-2} mol/L of N_2O_4 and 1.41×10^{-2} mol/L of NO_2 at 25°C. What is the value of K_c for this temperature?

14.36 A mixture of 3.0 mol of SO_2, 4.0 mol of NO_2, 1.0 mol of SO_3, and 4.0 mol of NO is placed in a 2.0-L vessel. The following reaction takes place:

$$SO_2(g) + NO_2(g) \rightleftharpoons SO_3(g) + NO(g)$$

When equilibrium is reached at 700°C, the vessel is found to contain 1.0 mol of SO_2. (a) Calculate the equilibrium concentrations of SO_2, NO_2, SO_3, and NO. (b) Calculate the value of K_c for this reaction at 700°C.

14.37 A 1.00-g sample of PCl_5 is introduced into a 250-mL flask, which is sealed and then heated to 250°C. The PCl_5 dissociates as follows:

$$PCl_5(g) \rightleftharpoons PCl_3(g) + Cl_2(g)$$

If 0.250 g of Cl_2 is present at equilibrium, what is the numerical value of K_c at 250°C?

[14.38] A 0.831-g sample of SO_3 is placed in a 1.00-L container and heated to 1100 K. The SO_3 undergoes decomposition to SO_2 and O_2:

$$2SO_3(g) \rightleftharpoons 2SO_2(g) + O_2(g)$$

At equilibrium, the total pressure in the container is 1.300 atm. Find the values of K_p and K_c for this reaction at 1100 K.

[14.39] PCl_5 is placed in a 2.00-L flask at 250°C. The following reaction occurs:

$$PCl_5(g) \rightleftharpoons PCl_3(g) + Cl_2(g)$$

At equilibrium, the total pressure of the mixture is 2.00 atm. The partial pressure of PCl_5 at equilibrium is 0.37 atm. Calculate K_p at 250°C.

14.40 A mixture of 1.000 mol of N_2 and 3.000 mol of H_2 is placed in a 1.00-L vessel at 600°C. At equilibrium it is found that the mixture contains 0.371 mol of NH_3. Calculate K_c at 600°C for the reaction

$$N_2(g) + 3H_2(g) \rightleftharpoons 2NH_3(g)$$

14.41 Nitric oxide, NO, rapidly oxidizes to nitrogen dioxide, NO_2, even at room temperature. At 1000 K, 0.0400 mol of NO and 0.0600 mol of O_2 are placed in a 2.00-L vessel. At equilibrium the concentration of NO_2 was 2.2×10^{-3} *M*. (a) Calculate the equilibrium concentrations of NO and O_2. (b) Calculate the equilibrium constant, K_c, for the reaction

$$2NO(g) + O_2(g) \rightleftharpoons 2NO_2(g)$$

14.42 Calculate K_p for the reaction

$$2SO_2(g) + O_2(g) \rightleftharpoons 2SO_3(g)$$

if at a particular temperature and a total pressure of 88.0 atm the equilibrium mixture consists of 56.6 mole percent SO_2, 10.6 mole percent O_2, and 32.8 mole percent SO_3.

[14.43] Write the equilibrium-constant expression for the equilibrium

$$C(s) + CO_2(g) \rightleftharpoons 2CO(g)$$

The table below shows the relative mole percentages of $CO_2(g)$ and $CO(g)$ at a total pressure of 1 atm for several temperatures. Calculate the value of K_c at each temperature. Is the reaction exothermic or endothermic? Explain.

Temperature (°C)	CO_2 (%)	CO (%)
850	6.23	93.77
950	1.32	98.68
1050	0.37	99.63
1200	0.06	99.94

14.44 Nitric oxide, NO, reacts readily with chlorine gas as follows:

$$2NO(g) + Cl_2(g) \rightleftharpoons 2NOCl(g)$$

At 700 K, the equilibrium constant, K_p, for this reaction is 0.26. Predict the behavior of each of the following mixtures at this temperature: (a) $P_{NO} = 0.15$ atm, $P_{Cl_2} = 0.31$ atm, and $P_{NOCl} = 0.11$ atm; (b) $P_{NO} = 0.12$ atm, $P_{Cl_2} = 0.10$ atm, and $P_{NOCl} = 0.050$ atm; (c) $P_{NO} = 0.15$ atm, $P_{Cl_2} = 0.20$ atm, and $P_{NOCl} = 5.10 \times 10^{-3}$ atm.

14.45 Consider the reaction

$$2CO(g) + O_2(g) \rightleftharpoons 2CO_2(g) \qquad \Delta H^\circ = -514.2 \text{ kJ}$$

In which direction will the equilibrium move if (a) CO_2 is added; (b) CO_2 is removed; (c) the volume is increased; (d) the pressure is increased; (e) the temperature is increased?

14.46 For the reaction shown in Exercise 14.45, what effect does increasing temperature have on the magnitude of the equilibrium constant?

14.47 The evaporation of any liquid requires energy added as heat. (a) How is the liquid-vapor equilibrium—for instance, $H_2O(l) \rightleftharpoons H_2O(g)$—affected by increasing temperature? (b) How does the equilibrium constant vary as the quantity of the liquid is increased?

14.48 NiO is to be reduced to nickel metal in an industrial process by use of the reaction

$$NiO(s) + CO(g) \rightleftharpoons Ni(s) + CO_2(g)$$

At 1600 K the equilibrium constant for the reaction is 600. If a CO pressure of 150 mm Hg is to be employed in the furnace, and total pressure never exceeds 760 mm Hg, will reduction occur?

14.49 Consider the reaction

$$CO(g) + 2H_2(g) \rightleftharpoons CH_3OH(l)$$

Using the thermochemical data in Appendix D, determine whether the equilibrium constant for this reaction increases or decreases with increasing temperature. Assuming equal pressures of CO and H_2, how would the extent of conversion of the gas mixture to methanol (CH_3OH) vary with total pressure?

[14.50] Suppose that there is a region in outer space where initially the hydrogen molecule concentration is 10^2 molecules/cm^3, the N_2 concentration is 1 molecule/cm^3, and the temperature is 100 K. At this temperature, K_p for the reaction

$$N_2(g) + 3H_2(g) \rightleftharpoons 2NH_3(g)$$

is approximately 6×10^{37}. Assuming that equilibrium is attained, is a significant fraction of the N_2 converted to NH_3?

14.51 A 0.0120-mol sample of N_2O_4 is allowed to dissociate and come to equilibrium with $NO_2(g)$ in a 0.186-L vessel at 25°C:

$$N_2O_4(g) \rightleftharpoons 2NO_2(g)$$

If $K_c = 4.61 \times 10^{-3}$ at this temperature, what is the percent dissociation of the N_2O_4?

[14.52] At 1558 K the equilibrium constant, K_c, for the reaction

$$Br_2(g) \rightleftharpoons 2Br(g)$$

is 1.04×10^{-3}. (a) Calculate the equilibrium concentration of Br atoms if the initial concentration of Br_2 is 1.00 M. (b) Calculate the fraction of the initial concentration of Br_2 that is dissociated into atoms.

14.53 At 400 K, the equilibrium constant, K_p, for the following reaction is 6.0×10^{-9}:

$$NH_4Cl(s) \rightleftharpoons NH_3(g) + HCl(g)$$

What are the equilibrium vapor pressures of NH_3 and HCl that are produced by the decomposition of solid NH_4Cl at 400 K?

[14.54] An equilibrium mixture of H_2, I_2, and HI at 458°C contains 2.24×10^{-2} M H_2, 2.24×10^{-2} M I_2, and 0.155 M HI in a 5.00-L vessel. What are the equilibrium concentrations when equilibrium is reestablished following the addition of 0.100 mol of HI?

14.55 Suppose that you worked at the U.S. Patent Office and a patent application came across your desk in which it was claimed that a newly developed catalyst was much superior to the Haber catalyst for ammonia synthesis, because the catalyst led to much greater equilibrium conversion of N_2 and H_2 into NH_3 than the Haber catalyst under the same conditions. What would be your response?

14.56 At 1200 K, the approximate temperature of automobile exhaust gases (Figure 14.6), the equilibrium constant for the reaction

$$2CO_2(g) \rightleftharpoons 2CO(g) + O_2(g)$$

is about 1×10^{-13} atm. Assuming that the exhaust gas (total pressure 1 atm) contains 0.2 percent CO by volume, 12 percent CO_2, and 3 percent O_2, is the system at equilibrium with respect to the above reaction? Based on your conclusion, would the CO concentration in the exhaust be lowered or increased by a catalyst that speeded up the reaction above?

14.57 For the single-step reaction

$$NO(g) + O_3(g) \underset{k_r}{\overset{k_f}{\rightleftharpoons}} NO_2(g) + O_2(g)$$

$K_p = 1.32 \times 10^{10}$ at 1000 K. If $k_f = 6.26 \times 10^8/M\text{-s}$ at this temperature, calculate k_r.

15 Aqueous Equilibria: Acids and Bases

One of the most important classifications of substances is in terms of acid and base properties. From the earliest days of experimental chemistry it was recognized that certain substances, called acids, possess a sour taste and are capable of dissolving active metals such as zinc. Acids also cause vegetable dyes to turn a characteristic color; for example, litmus, which is obtained from certain lichens (a composite plant made up of an alga and a fungus), turns red on contact with acids. Like acids, bases possess a set of characteristic properties that can be used to identify them. Whereas acids have a sour taste (the sour taste of lemons is due to the presence of citric acid in lemon juice), bases have a characteristic bitter taste. Bases also feel slippery to the touch. Whereas acids cause litmus to turn red, bases cause it to turn blue. Bases also react with many dissolved metal salts to form precipitates.

The fact that all acids and all bases show certain characteristic chemical properties suggests that there must be an essential feature common to all the members of each class. Lavoisier proposed that acids were oxygen-containing substances. In fact, Lavoisier derived the name *oxygen* from the Greek word for "acid former." However, careful studies by a number of other scientists showed that hydrochloric acid contains no oxygen. By 1830 it became evident that hydrogen was the one element present in all acids. It was subsequently shown that aqueous solutions of both acids and bases conduct electrical currents. In the 1880s, the Swedish chemist Svante Arrhenius (1859–1927) suggested the existence of ions to explain this electrical conduction. Shortly after, he proposed that acids are substances that form H^+ ions in water solutions and that bases produce OH^- ions. These definitions of acids and bases were the ones that were presented in Section 3.3 and that we have used in subsequent discussions.

In this chapter we shall take a closer look at acids and bases. We shall see how the properties of these substances can be understood in terms of their structures and bonding. The concept of equilibrium introduced in

Chapter 14 will figure prominently in these discussions. We shall see also that the properties of acids and bases, as we generally encounter them, are very much dependent on the fact that water is the solvent in which those properties are observed. To appreciate how really remarkable an aqueous (water) acid solution is, let's begin by considering a very common chemical reagent, hydrochloric acid.

15.1 WATER AND ACIDIC SOLUTIONS

Figure 15.1 is a reproduction of the label from a bottle of reagent-grade* concentrated hydrochloric acid. Although this label might at first seem to be rather dull subject matter, it contains fascinating and even amazing data. Consider first some properties of HCl. It is a gaseous substance, formed commercially by the controlled reaction of hydrogen and chlorine:

$$H_2(g) + Cl_2(g) \longrightarrow 2HCl(g)$$

Hydrogen chloride liquefies under 1 atm pressure at $-84°C$, and freezes at $-112°C$. Liquid HCl is a difficult and most unpleasant substance to work with, but some enterprising and persistent persons have learned that the liquid is a very poor conductor of electricity. This suggests that there are very few ions present in the liquid. Dry HCl gas is soluble to only a limited extent in a dry organic solvent such as benzene. The solutions do not conduct electric current. Furthermore, these solutions lack other properties that we associate with hydrochloric acid. For example, addition of an active metal such as zinc to them produces no evident chemical reaction. In contrast, zinc reacts readily with aqueous solutions of HCl to produce hydrogen gas:

$$Zn(s) + 2HCl(aq) \longrightarrow H_2(g) + ZnCl_2(aq) \qquad [15.1]$$

*The term "reagent grade" denotes that a substance is of high purity.

KEEP FROM CHILDREN
DANGER!
CAUSES BURNS.
BEFORE USING, READ MANUFACTURING CHEMISTS' ASSOCIATION, INC., SAFETY DATA SHEET SD-39
Do not get in eyes, on skin, on clothing. Avoid breathing vapor. Keep container closed. Use with adequate ventilation. Wash thoroughly after handling.
FIRST AID: In case of contact, immediately flush eyes or skin with plenty of water for at least 15 minutes while removing contaminated clothing and shoes. Call a physician. Wash clothing before re-use.
SPILL OR LEAK: Flush away by flooding with water applied quickly to entire spill. Neutralize washings with lime or soda ash.

POISON
ANTIDOTE
If taken Internally: Call Physician Immediately. Give at once large draughts of water containing Milk of Magnesia, or milk, soap and water, magnesium oxide, white of eggs beaten up with water, or olive oil. Avoid Carbonates. Give No Emetics.

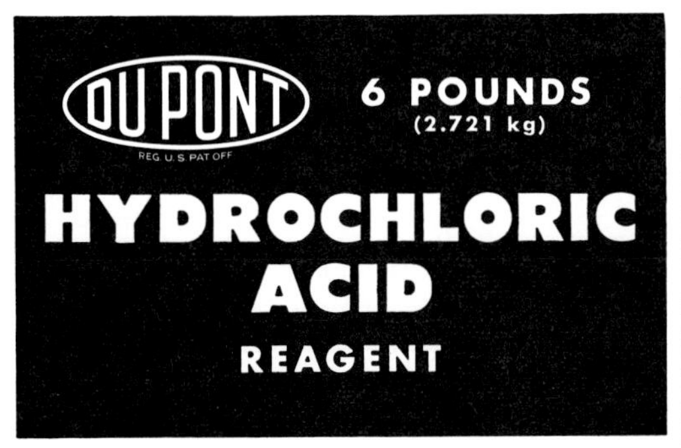

KEEP THIS BOTTLE IN A COOL PLACE AND REMOVE CAP CAREFULLY TO AVOID SPURTING

E. I. DU PONT DE NEMOURS & COMPANY (INC.)
INDUSTRIAL CHEMICALS DEPARTMENT
Wilmington, Del., U. S. A.

MEETS A. C. S. SPECIFICATIONS
HCl........ Min. 37.0% / Max. 38.0%
Sp. Gr. 60°/60°F, Min.. 1.19
Maximum of Impurities
Free Cl 0.00005%
Sulfites (SO_3) .. 0.00008%
Sulfates (SO_4) . 0.00008%
Hy. Met. (as Pb) 0.00002%
Residue after Ignition.... 0.0004%
Fe 0.00001%
As 0.0000005%
NH_4 .... 0.0003%
Br 0.005%
Cu 0.00005%
Ni 0.00005%
Color, APHA . 10
Extractable Organic Substances Passes A.C.S. Test (Approx. 0.0005%)

IC-20902 1-72 PRINTED IN U.S.A.

FIGURE 15.1 Reproduction of a label from a bottle of reagent-grade concentrated hydrochloric acid. (*Courtesy E. I. du Pont de Nemours & Co.*)

With these facts in mind let us now examine the label shown in Figure 15.1. We note that concentrated aqueous hydrochloric acid consists of 37 percent by weight of hydrogen chloride; HCl is obviously very soluble in water. At 15°C, 1 L of water dissolves up to 450 L of dry HCl gas at 1 atm pressure! This high solubility is nicely shown by the hydrogen chloride fountain, a popular lecture demonstration, illustrated in Figure 15.2. The molarity of concentrated hydrochloric acid solution is about 12 *M*, as shown in Sample Exercise 15.1. To have some appreciation for just how remarkable this solubility is, you should know that argon, which has essentially the same molecular weight as HCl, is soluble in water at 15°C and 1 atm pressure to the extent of only 0.002 *M*.

SAMPLE EXERCISE 15.1

From the data shown in Figure 15.1, calculate the molarity of a concentrated hydrochloric acid solution.

Solution: We see that the specific gravity, or density, of the solution is 1.19 g/cm³. Using this and the stated weight percentage, we have

$$\left(\frac{1.19 \text{ g soln}}{1 \text{ cm}^3}\right)\left(\frac{1000 \text{ cm}^3}{1 \text{ L}}\right)\left(\frac{0.37 \text{ g HCl}}{1 \text{ g soln}}\right) \times \left(\frac{1 \text{ mol HCl}}{36.5 \text{ g HCl}}\right) = \frac{12 \text{ mol HCl}}{1 \text{ L soln}} = 12\,M$$

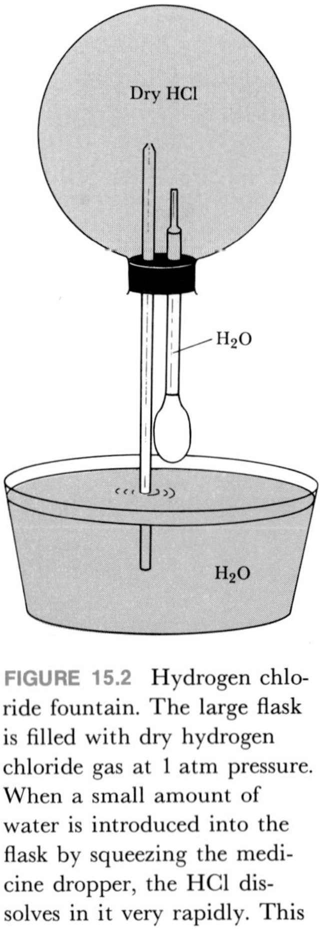

FIGURE 15.2 Hydrogen chloride fountain. The large flask is filled with dry hydrogen chloride gas at 1 atm pressure. When a small amount of water is introduced into the flask by squeezing the medicine dropper, the HCl dissolves in it very rapidly. This causes the pressure in the flask to decrease, and water is forced up the glass tube. The spray of water continues until the flask is almost completely filled.

In contrast with the solutions of HCl in dry benzene, aqueous solutions of HCl, that is, hydrochloric acid solutions, strongly conduct an electrical current. This behavior indicates that hydrochloric acid contains ions. As we have noted, this observation led Arrhenius to define an acid as any substance capable of producing an excess of H^+ ions in water. Arrhenius's definition of an acid is quite reasonable, but it doesn't tell us much about what actually happens when an acid dissolves in water. To appreciate the nature of acidic solutions, and thus understand more clearly why acids behave as they do, we must look more closely into the question of how and why a substance such as HCl reacts with water to form ions.

Nature of the Hydrated Proton

What is it about water that causes a molecule such as HCl to come apart, or dissociate, into H^+ and Cl^- ions? The first and most obvious point to make is that water is a polar liquid (Section 10.4). Consequently, the oxygen atom, which bears a partial negative charge, is attracted to the partially positive hydrogen end of the polar HCl molecule. Second, the water molecule has unshared electron pairs on the oxygen atom; these are capable of forming a covalent bond to the hydrogen ion in the following manner:

$$\text{Cl—H}\cdots\text{:}\underset{\displaystyle\text{H}}{\underset{|}{\ddot{\text{O}}}}\text{—H} \longrightarrow \text{Cl}^- + \left[\text{H—}\underset{\displaystyle\text{H}}{\underset{|}{\ddot{\text{O}}}}\text{—H}\right]^+ \qquad [15.2]$$

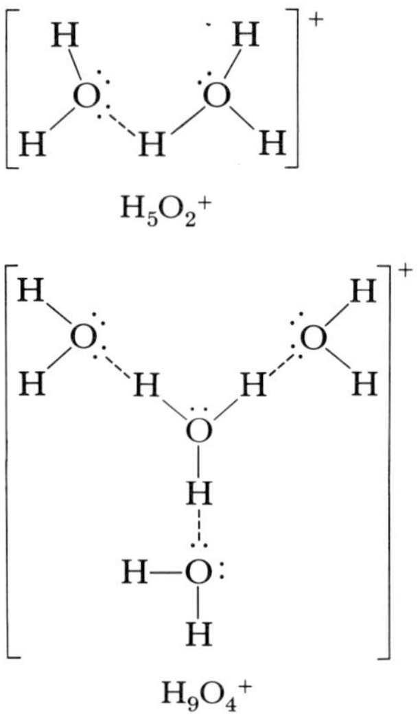

$H_5O_2^+$

$H_9O_4^+$

FIGURE 15.3 Two possible forms for the proton in water, in addition to H_3O^+. Experimental evidence indicates the existence of both these species.

The reaction between HCl and H_2O shown in Equation 15.2 involves the transfer of a proton (an H^+ ion) from HCl to the water molecule. This reaction produces the H_3O^+ ion, called the *hydronium ion.*

The situation in water is actually much more complex than Equation 15.2 suggests. We've learned (Section 10.3) that hydrogen bonds exist throughout liquid water. The existence of this hydrogen-bond network is responsible for many of the special properties of water, for example, its high polarity and high melting and boiling points. Much research has been devoted to learning how H^+ ions fit into the complex structure of liquid water. Experimental studies show that, in part, the H^+ ions must exist as hydronium ions. In fact, it is possible to isolate salts of the form $H_3O^+Cl^-$, $H_3O^+ClO_4^-$, and others, in which there is clearly an H_3O^+ ion in the solid lattice. But just as water molecules are strongly hydrogen bonded to one another, the H_3O^+ ion in solution is hydrogen bonded to other water molecules. Thus ions such as the two shown in Figure 15.3 are possible and have been shown to form.

We must conclude from these observations that no single species can adequately represent the proton in solution. Both $H^+(aq)$ and $H_3O^+(aq)$ are used to represent the hydrated, or aquated, hydrogen ion, that is, the hydrogen ion surrounded by solvent water. Thus we may write the reaction of HCl with water to form hydrochloric acid in either of the following ways:

$$HCl(aq) + H_2O(l) \longrightarrow H_3O^+(aq) + Cl^-(aq) \quad [15.3]$$

$$HCl(aq) \longrightarrow H^+(aq) + Cl^-(aq) \quad [15.4]$$

The hydrated proton will be represented through the remainder of this chapter as $H^+(aq)$. However, regardless of whether $H^+(aq)$ or $H_3O^+(aq)$ is used to represent the hydrated proton, you should keep in mind that *acidic solutions are formed by a chemical reaction in which an acid transfers a proton to water.*

15.2 BRØNSTED-LOWRY THEORY OF ACIDS AND BASES

The nature of the reaction between an acid and water as we have just described it was first appreciated by the Danish chemist Johannes Brønsted (1879–1947) and the English chemist Thomas M. Lowry (1874–1936). Brønsted and Lowry recognized that acid-base behavior could be described in terms of the ability of substances to transfer protons. In 1923, Brønsted and Lowry independently proposed that *acids be defined as substances that are capable of donating a proton, and bases as substances capable of accepting a proton.* In these terms, when HCl dissolves in water, it acts as an acid in donating a proton to the solvent. The solvent, H_2O, then acts as a base in accepting the proton (see Equation 15.2).

In earlier discussions, we have applied the term base to substances that produce an excess of OH^- ions in aqueous solution. Notice, however, that the OH^- ion is an acceptor of protons; it reacts readily with the hydrated proton to form water:

$$H^+(aq) + OH^-(aq) \rightleftharpoons H_2O(l) \quad [15.5]$$

Similarly, we have seen (Section 3.3) that aqueous solutions of ammonia are basic because NH_3 reacts with H_2O to form NH_4^+ and OH^-:

$$H_2O(l) + NH_3(aq) \rightleftharpoons NH_4^+(aq) + OH^-(aq) \qquad [15.6]$$

In this reaction the H_2O gives a proton to NH_3; H_2O is the acid, NH_3 is the base.

The reactions that we have cited above as examples of proton-transfer reactions are reversible. For example, when NH_4^+ (as from NH_4Cl) and OH^- (as from NaOH) are mixed they form H_2O and NH_3:

$$NH_4^+(aq) + OH^-(aq) \rightleftharpoons NH_3(aq) + H_2O(l) \qquad [15.7]$$

The net ionic equation is just the reverse of the reaction between NH_3 and H_2O (Equation 15.6). In this reverse reaction (Equation 15.7), NH_4^+ acts as the proton donor and OH^- as the proton acceptor. Thus we see that as the reaction proceeds in one direction, H_2O is the acid, and NH_3 is the base. In the other direction, NH_4^+ is the acid, and OH^- is the base. This example illustrates that every acid has associated with it a **conjugate base** * formed from the acid by the loss of a proton. For example, the conjugate base of NH_4^+ is NH_3; the conjugate base of H_2O is OH^-. Similarly, every base has a **conjugate acid** formed from the base by addition of a proton. H_2O is the conjugate acid of OH^-. An acid and a base, such as H_2O and OH^-, that differ only by the presence or absence of a proton are called a **conjugate acid-base pair.**

The more readily a substance gives up a proton, the less readily its conjugate base will accept a proton. In other words, *the stronger an acid, the weaker its conjugate base; the weaker an acid, the stronger its conjugate base.* For example, HCl is a good proton donor because its conjugate base, Cl^-, has less attraction for protons than water does. The proton is therefore transferred to H_2O to form $H^+(aq)$.

Figure 15.4 displays some common acids and their conjugate bases. Notice that the strongest acids have the weakest conjugate bases, and the weakest acids have the strongest conjugate bases. $H^+(aq)$ is the strongest proton donor that can exist at equilibrium in aqueous solution. Thus acids listed above $H^+(aq)$ in Figure 15.4 completely transfer protons to water to form $H^+(aq)$. Likewise, $OH^-(aq)$ is the strongest base that can exist at equilibrium in aqueous solution. Any stronger proton acceptor will completely react with water, removing a proton to form OH^- ions.

*The word "conjugate" means joined together as a pair, or coupled.

SAMPLE EXERCISE 15.2

Hydrocyanic acid, HCN, is ionized in water by the reaction

$$HCN(aq) \rightleftharpoons H^+(aq) + CN^-(aq)$$

to a lesser extent than is an HF solution of the same concentration. What is the conjugate base of HCN? Is it a stronger or weaker base than F^-?

Solution: The conjugate base of HCN is CN^-, the ion that remains after a proton has been lost to the solvent. Because HCN dissociates to a lesser extent than does HF, this means that the tendency of the reverse reaction to occur is greater. In the reverse reaction the conjugate base, CN^-, accepts a proton from the solvent. In other words, CN^- is a stronger conjugate base than is F^-.

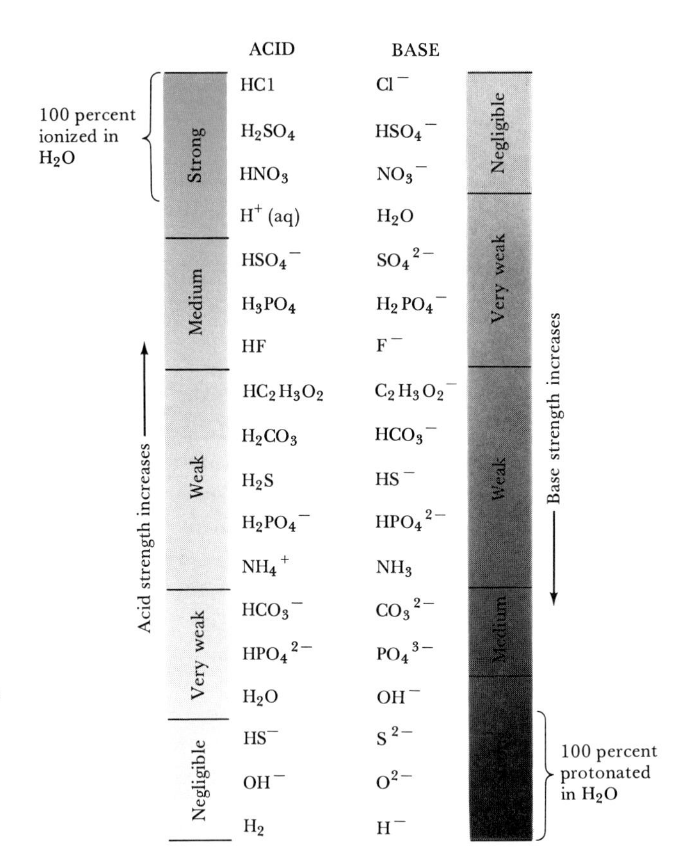

FIGURE 15.4 Relative strengths of some common conjugate acid-base pairs, which are listed opposite one another in the two columns. The conjugate bases of strong acids have negligible basicity; the conjugate acids of strong bases have negligible acidity. The strongest acid that can exist in water is the aquated proton, $H^+(aq)$. Any substance that gives up a proton more readily than the solvent simply loses that proton to the solvent, forming $H^+(aq)$. Such stong acids, for example, HNO_3, are essentially completely ionized in water. Similarly, any base that is a stronger proton acceptor than OH^- removes protons from the water to become essentially completely protonated.

15.3 THE DISSOCIATION OF WATER AND THE pH SCALE

In the presence of an acid such as HCl, water acts as a proton acceptor; in the presence of a base such as NH_3, it acts as a proton donor. An important feature of water is that it is also capable of acting as a proton donor and proton acceptor toward itself. The process by which this occurs is called autoionization:

$$\mathrm{H{-}\ddot{\underset{\substack{|\\H}}{O}}{:}} + \mathrm{H{-}\ddot{\underset{\substack{|\\H}}{O}}{:}} \rightleftharpoons \left[\mathrm{H{-}\ddot{\underset{\substack{|\\H}}{O}}{-}H}\right]^+ + \mathrm{{:}\ddot{\underset{..}{O}}{-}H^-} \qquad [15.8]$$

This reaction amounts to a spontaneous ionization of the solvent. It occurs only to a very small extent. At room temperature, only about one out of every 10^8 molecules is in the ionic form at any one instant. We know that water is a strongly hydrogen-bonded liquid. Thus the hydrogen of one molecule may be attracted to an unshared pair of electrons on the oxygen of an adjacent molecule. Occasionally, the hydrogen will transfer to the other molecule. Perhaps simultaneously there will be a transfer of H^+ from some other molecule to the oxygen that is losing a hydrogen ion. As a result of these ready transfers of H^+ from one molecule to another, there is, on the average, a certain very small fraction of molecules in the ionized form. No one molecule remains in that condition for long; the equilibria are extremely rapid. It has been found that, on the average, a proton transfers from one molecule to another in water at the rate of about 1000 times per second.

By expressing the hydrated proton as $H^+(aq)$ rather than $H_3O^+(aq)$, we can rewrite Equation 15.8 as

$$H_2O(l) \rightleftharpoons H^+(aq) + OH^-(aq) \qquad [15.9]$$

The equilibrium expression for this autoionization reaction can be written as

$$K = \frac{[H^+][OH^-]}{[H_2O]}$$

The concentration of water in aqueous solutions is typically very large, about 55 M, and remains essentially constant for dilute solutions. It is therefore customary to exclude the concentration of water from equilibrium-constant expressions for aqueous solutions, just as we exclude the concentrations of pure solids and liquids from the equilibrium-constant expressions for heterogeneous reactions (Section 14.3). Thus we can write the equilibrium-constant expression for the autoionization of water as

$$K[H_2O] = K_w = [H^+][OH^-]$$

The product of two constants, K and $[H_2O]$, defines a new constant, K_w. This important equilibrium constant is called the **ion-product constant** for water. K_w has the value of 1.0×10^{-14} at 25°C. This is an important equilibrium constant. You should memorize this expression:

$$K_w = [H^+][OH^-] = 1.0 \times 10^{-14} \qquad [15.10]$$

Equation 15.10 is valid for aqueous solutions as well as for pure water. A solution for which $[H^+] = [OH^-]$ is said to be *neutral*. In most solutions H^+ and OH^- concentrations are not equal. As the concentration of one of these ions increases, the concentration of the other must decrease so that the ion-product equals 1.0×10^{-14}. In acidic solutions, $[H^+]$ exceeds $[OH^-]$; in basic solutions, the reverse is true, $[OH^-]$ exceeds $[H^+]$.

SAMPLE EXERCISE 15.3

Calculate the values of $[H^+]$ and $[OH^-]$ in a neutral solution at 25°C.

Solution: By definition, in a neutral solution, $[H^+]$ equals $[OH^-]$. Let us call the concentration of each of these species in neutral solution x. Using Equation 15.10, we have

$$[H^+][OH^-] = (x)(x) = 1.0 \times 10^{-14}$$
$$x^2 = 1.0 \times 10^{-14}$$
$$x = 1.0 \times 10^{-7} = [H^+] = [OH^-]$$

In an acid solution, $[H^+]$ is greater than $1.0 \times 10^{-7}\,M$; in a basic solution it is less than $1.0 \times 10^{-7}\,M$.

SAMPLE EXERCISE 15.4

Calculate the concentration of $H^+(aq)$ in (a) a solution in which $[OH^-]$ is 0.010 M; (b) a solution in which $[OH^-]$ is $2.0 \times 10^{-9}\,M$.

Solution: (a) Using Equation 15.10, we have

$$[H^+][OH^-] = 1.0 \times 10^{-14}$$

$$[H^+] = \frac{1.0 \times 10^{-14}}{[OH^-]} = \frac{1.0 \times 10^{-14}}{0.010}$$
$$= 1.0 \times 10^{-12}\,M$$

This solution is basic because $[H^+] < [OH^-]$.

(b) In this instance

$$[H^+] = \frac{1.0 \times 10^{-14}}{[OH^-]} = \frac{1.0 \times 10^{-14}}{2.0 \times 10^{-9}}$$
$$= 5.0 \times 10^{-6}\,M$$

This solution is acidic because $[H^+] > [OH^-]$.

pH

In almost every area of pure and applied chemistry the acid-base properties of water are of importance. As examples, the fate of pollutant chemicals in a water body, the rapidity with which a metal object immersed in water corrodes, and the suitability of an aquatic environment for support of fish and plant life are all critically dependent on the acidity or basicity of the water. The concentration of $H^+(aq)$ in such solutions is often expressed in terms of **pH**. The pH is defined as the negative log in base 10, of the hydrogen-ion concentration.*

$$pH = -\log [H^+] = \log \left(\frac{1}{[H^+]}\right) \qquad [15.11]$$

Notice that a change in $[H^+]$ by a factor of 10 results in a unit change in pH. (If you need a review of exponential notation and of the use of logs, see Appendix A.) As an example of the use of Equation 15.11, let us calculate the pH of a neutral solution, that is, one in which $[H^+] = [OH^-] = 1.0 \times 10^{-7}$ (Sample Exercise 15.3). The pH is given by

$$pH = -\log [H^+] = -\log (1.0 \times 10^{-7}) = 7.00$$

Thus the pH of a neutral solution is 7.00.

Because pH is simply another means of expressing $[H^+]$, acidic and basic solutions can be distinguished on the basis of their pH values:

$$pH < 7 \text{ in acidic solutions}$$
$$pH > 7 \text{ in basic solutions}$$
$$pH = 7 \text{ in neutral solutions}$$

You should keep in mind that the pH is a measure only of the equilibrium concentration of dissociated hydrogen ion present as $H^+(aq)$. The pH values characteristic of several familiar solutions are shown in Figure 15.5.

With a log table (Appendix B) or log function on a calculator, it is a simple matter to convert from concentration of $H^+(aq)$ to pH, and vice versa, as outlined in Sample Exercises 15.5 and 15.6.

*Usually, you will see pH defined as $-\log [H^+]$, occasionally as $-\log [H_3O^+]$. As discussed in Section 15.1, the same species is involved in all cases.

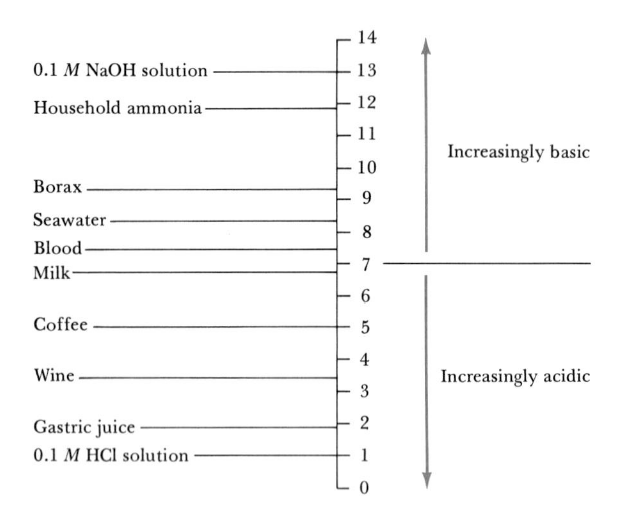

FIGURE 15.5 Values of pH for some more commonly encountered solutions. The pH scale in this figure is shown to extend from 0 to 14, because nearly all solutions commonly encountered have pH values in that range. In principle, however, the pH values for strongly acidic solutions can be less than 0, and for strongly basic solutions can be greater than 14.

SAMPLE EXERCISE 15.5

Calculate the pH values for the two solutions described in Sample Exercise 15.4.

Solution: In the first instance we found $[H^+]$ to be 1.0×10^{-12}. The pH of this solution is given by

$$\text{pH} = -\log(1.0 \times 10^{-12}) = -(-12.00) = 12.00$$

The pH of the second solution is given by

$$\begin{aligned}\text{pH} &= -\log(5.0 \times 10^{-6})\\ &= -(\log 5 + \log 10^{-6})\\ &= -(0.699 - 6.00) = 5.30\end{aligned}$$

SAMPLE EXERCISE 15.6

A sample of freshly pressed apple juice has a pH of 3.76. Calculate $[H^+]$.

Solution: From the equation defining pH (Equation 15.11), we have $-\log[H^+] = 3.76$. Thus $\log[H^+] = -3.76$. To find $[H^+]$ we need to find the antilog of -3.76. That is, we want the number whose log is -3.76. Some calculators have an antilog function (usually labeled INV log or $\log^{-1}$), which makes the calculation quite simple. Other calculators rely on 10^x or y^x functions to find antilogs: antilog $(-3.76) = 10^{-3.76}$. If you are relying on a log table, the simplest way to take the antilog is to write the log as a sum of an integer and a positive decimal fraction: $-3.76 = -4.00 + 0.24$. The antilog of -4 is 1×10^{-4}. From a log table we find that the antilog of 0.24 is approximately 1.7. Thus $[H^+] = 1.7 \times 10^{-4}\,M$.

Other pX Scales

The negative log is a convenient way of expressing the magnitudes of numbers that are generally very small. We use the convention that the negative log of a quantity is labeled p(quantity). For example, one can express the concentration of OH^- as pOH:

$$\text{pOH} = -\log[OH^-]$$

By taking the log of both sides of Equation 15.10 and multiplying through by -1, we can obtain

$$\text{pH} + \text{pOH} = -\log K_w = 14.00 \qquad [15.12]$$

This expression is often convenient to use. We will see later (Section 15.7) than the pX notation is also useful in dealing with equilibrium constants.

Indicators

Various means are available for quantitatively estimating pH. The simplest is the use of an indicator. An **indicator** is a colored substance, usually derived from plant material, that can exist in either an acid or base form. The two forms are differently colored. By adding a small amount of an indicator to a solution and noting its color, it is possible to determine whether it is in the acid or base form. If one knows the pH at which the indicator turns from one form to the other, one can then determine from the observed color whether the solution has a higher or lower pH than this value. For example, litmus, one of the most common indicators, changes color in the vicinity of pH 7. However, the color change is not very sharp. Red litmus indicates a pH of about 5 or lower, and blue litmus, a pH of about 8.2 or higher. Many other indicators change color at various pH values between 1 and 14. Some of the more commonly used are listed in Table 15.1. We see from this table that methyl orange, for example, changes color over the pH interval from 2.9 to 4.0. Below pH 2.9 it is in the acid form, which is red. In the interval from pH 2.9 to 4.0 it is gradually converted to its basic form, which has a yellow color. By pH 4.0 the conversion is complete, and the solution is yellow. Paper tape that is impregnated with various indicators and that comes complete with a comparator color scale is widely used for approximate determinations of pH.

TABLE 15.1 Some of the more common acid-base indicators

Name	pH interval for color change	Acid color	Base color
Methyl violet	0–2	Yellow	Violet
Methyl yellow	1.2–2.3	Red	Yellow
Methyl orange	2.9–4.0	Red	Yellow
Methyl red	4.2–6.3	Red	Yellow
Bromthymol blue	6.0–7.6	Yellow	Blue
Thymol blue	8.0–9.6	Yellow	Blue
Phenolphthalein	8.3–10	Colorless	Pink
Alizarin yellow G	10.1–12.0	Yellow	Red

The pH meter is a widely used, simple instrument for rapid and accurate determination of pH. It is so common that if you go on to further study in chemistry or in an applied science you are almost certain to encounter one. A complete understanding of how a pH meter works requires a knowledge of electrochemistry, a subject we take up in Chapter 18. However, we can say at this point that a pH meter consists of a pair of electrodes that are placed in the solution to be measured and a sensitive meter for measuring small voltages, on the order of millivolts. A typical pH meter with a pair of electrodes is shown in Figure 15.6. When the electrodes are placed in the solution, they form an electrochemical cell (something like a battery) that has a voltage. The voltage of the cell is dependent on $[H^+]$; thus, by measuring the voltage we

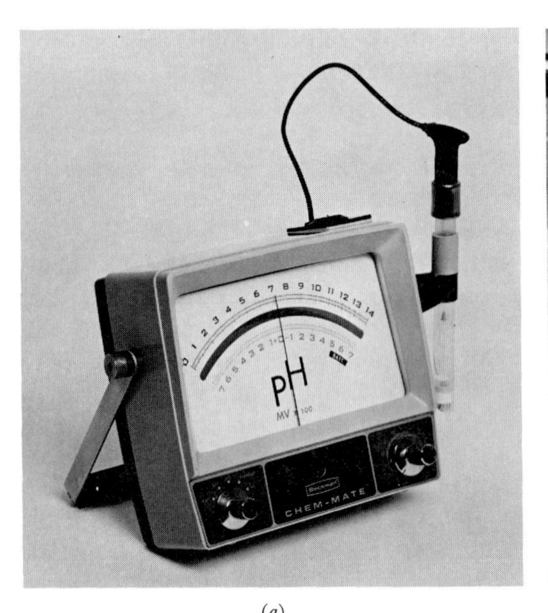

(a)

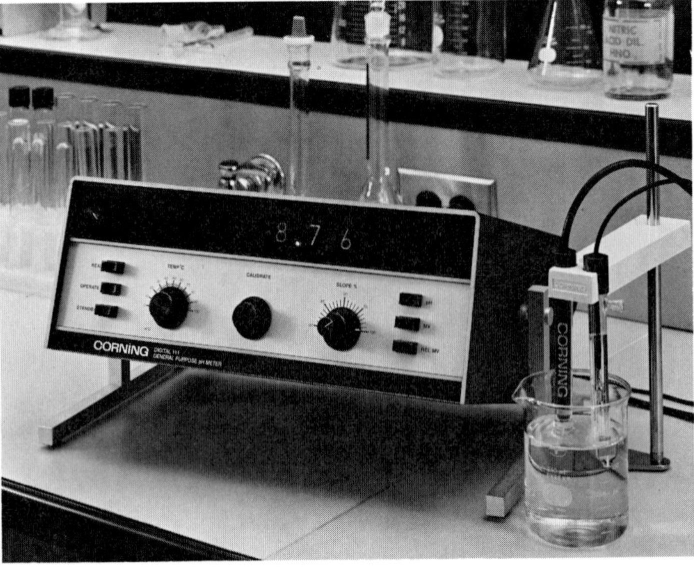

(b)

FIGURE 15.6 (*a*) A pH meter of the type normally used for student work. (*b*) A research-type pH meter with digital display. (*a, courtesy Beckman Instruments, Inc.; b, courtesy Corning Glass Works*)

obtain a measure of [H^+]. Electrodes that can be used with pH meters come in all shapes and sizes, depending on their intended use, but fundamentally they are nearly all the same. One of the electrodes is a reference electrode. The one that is actually sensitive to $H^+(aq)$ is almost always a so-called glass electrode. The wire in the inner compartment of the electrode is in contact with a solution of known and fixed $H^+(aq)$ concentration. The wall of the compartment is formed of a special thin glass that is permeable to $H^+(aq)$. As a result, the voltage that this electrode, together with a reference electrode, generates when placed in a solution depends on [H^+] in the solution.

To extend the range of possible pH measurements, much research has gone into the development of electrodes that can be used with very small quantities of solution. It is now possible to insert electrodes into single living cells in order to monitor pH of the cell medium. The pH meter is also widely used outside the laboratory. Pocket-sized models are available for use in environmental studies, monitoring of industrial effluents, and in agricultural work.

15.4 STRONG ACIDS AND STRONG BASES

The pH of an aqueous solution depends on the ability of the solute to give protons to, or remove them from, water. In this section and the following ones we will examine the extent to which these proton-transfer reactions occur.

In terms of the Brønsted-Lowry concept, a strong aqueous acid is any substance that reacts completely with water to form $H^+(aq)$; a weak acid is a substance that only partly reacts in this fashion. The number of strong acids is not very large; the six most important ones are listed in Table 15.2. We recommend that you commit them to memory if you have not already done so. In the case of H_2SO_4, only the first proton is completely ionized in aqueous solution.

We can consider aqueous solutions of all strong acids to consist entirely of ions, with no significant concentration of neutral solute molecules remaining; such acids are said to be completely ionized or dissociated. For example, a 0.10 *M* aqueous solution of HNO_3, nitric acid,

TABLE 15.2 Common strong acids and bases

Acids	Bases
HCl (hydrochloric acid)	Hydroxides and oxides of 1A metals
HBr (hydrobromic acid)	Hydroxides and oxides of 2A metals (except Be)
HI (hydroiodic acid)	
HNO_3 (nitric acid)	
$HClO_4$ (perchloric acid)	
H_2SO_4 (sulfuric acid)	

contains $[H^+] = 0.10\,M$, and $[NO_3^-] = 0.10\,M$; the concentration of HNO_3 is virtually zero.

Strong bases are strong proton acceptors. The most common strong bases are NaOH and KOH, which are ionic compounds containing OH^- ions in the solid. They are strong electrolytes, dissolving in water as would any other ionic substance (Section 11.4). For example, a $0.10\,M$ aqueous solution of NaOH contains $0.10\,M$ $Na^+(aq)$ and $0.10\,M$ $OH^-(aq)$ with no undissociated NaOH:

$$NaOH(s) \xrightarrow{H_2O} Na^+(aq) + OH^-(aq) \quad [15.13]$$

All of the hydroxides of the alkali metals (family 1A) are strong electrolytes, but the compounds of Li, Rb, and Cs are too expensive to be encountered commonly in the laboratory. The hydroxides of all of the alkaline earths (family 2A) except Be are also strong electrolytes. However, they have limited solubilities and are consequently used only when high solubility is not critical. $Mg(OH)_2$ has an especially low solubility (9×10^{-3} g/L of water at 25 °C). The least expensive and most common of the alkaline earth hydroxides is $Ca(OH)_2$, whose solubility at 25 °C is 0.97 g/L.

Basic solutions are also created when substances react with water to form $OH^-(aq)$. The most common of these is the oxide ion. Ionic metal oxides, especially Na_2O and CaO, are often used in industry when a strong base is needed. Each mole of O^{2-} reacts with water to form 2 mol of OH^-, leaving virtually no O^{2-} remaining in the solution:

$$O^{2-}(aq) + H_2O(l) \longrightarrow 2OH^-(aq) \quad [15.14]$$

Similarly, ionic hydrides and nitrides, such as $NaH(s)$ and $Mg_3N_2(s)$, react to form basic solutions:

$$NaH(s) + H_2O(l) \longrightarrow Na^+(aq) + H_2(g) + OH^-(aq) \quad [15.15]$$

$$Mg_3N_2(s) + 6H_2O(l) \longrightarrow 3Mg^{2+}(aq) + 2NH_3(aq) + 6OH^-(aq) \quad [15.16]$$

In these examples the anions, H^- and N^{3-}, are stronger bases than $OH^-(aq)$. They therefore remove a proton from H_2O, which is the conjugate acid of OH^-.

SAMPLE EXERCISE 15.7

What is the pH of a solution of (a) 0.010 M HCl; (b) 0.010 M $Ca(OH)_2$?

Solution: (a) HCl is a strong acid. Consequently, $[H^+] = 0.010\,M$ and $pH = -\log(0.010) = 2.00$.

(b) $Ca(OH)_2$ is a strong base; each $Ca(OH)_2$ forms $2OH^-$ ions. Consequently, $[OH^-] = 0.020\,M$ and

$$[H^+] = \frac{1.00 \times 10^{-14}}{[OH^-]} = \frac{1.00 \times 10^{-14}}{0.020}$$
$$= 5.0 \times 10^{-13}$$
$$pH = -\log(5.0 \times 10^{-13}) = 12.30$$

15.5 WEAK ACIDS

Most substances that are acidic in water are actually weak acids. The extent to which an acid ionizes in an aqueous medium can be expressed by the equilibrium constant for the ionization reaction. In general, we can represent any acid by the symbol HX, where X^- is the formula for the conjugate base that remains when the proton ionizes. The ionization equilibrium is then given by Equation 15.17:

$$HX(aq) \rightleftharpoons H^+(aq) + X^-(aq) \qquad [15.17]$$

The corresponding equilibrium-constant expression is

$$K_a = \frac{[H^+][X^-]}{[HX]} \qquad [15.18]$$

The equilibrium constant is often given the symbol K_a and is called the **acid-dissociation constant.**

Table 15.3 shows the names, structures, and values of K_a for several weak acids. A more complete table is given in Appendix E. Note that many weak acids are compounds composed largely of carbon and hydrogen. Generally speaking, hydrogen atoms bound to carbon are not ionized in an aqueous medium. The ionizable hydrogens are in most instances bound to oxygen. The smaller the value for K_a, the weaker the acid. For example, phenol is the weakest acid listed in Table 15.3.

SAMPLE EXERCISE 15.8

A student prepared a 0.10 M solution of formic acid, $HCHO_2$, and measured its pH using a pH meter of the type illustrated in Figure 15.6(b). The pH at 25°C was found to be 2.38. (a) Calculate K_a for formic acid at this temperature. (b) What percentage of the acid is dissociated in this 0.10 M solution?

Solution: (a) The first step in solving any equilibrium problem is to write the equation for the equilibrium reaction. The ionization equilibrium for formic acid can be written as follows:

$$HCHO_2(aq) \rightleftharpoons H^+(aq) + CHO_2^-(aq)$$

The equilibrium-constant expression for this equilibrium is

$$K_a = \frac{[H^+][CHO_2^-]}{[HCHO_2]}$$

From the measured pH we can calculate $[H^+]$:

$$pH = -\log[H^+] = 2.38$$
$$\log[H^+] = -2.38$$
$$[H^+] = 4.2 \times 10^{-3}\,M$$

(If you are using a calculator you would have used the INV log, or $\log^{-1}$ function, as appropriate, to obtain the antilog of -2.38.)

The Lewis structure for formic acid is as follows:

$$\mathrm{H{-}\overset{\overset{\displaystyle :O:}{\|}}{C}{-}\ddot{\underset{\cdot\cdot}{O}}{-}H}$$

The proton that ionizes is the one bound to oxygen, shown in color.

We can do a little accounting to determine the concentrations of the species involved in the equilibrium. For each H^+ produced in solution, there is also formed one CHO_2^- anion, and there is a loss of one $HCHO_2$ molecule. Let's write under the equilibrium expression the initial and equilibrium concentrations of the species involved:

	$HCHO_2$	$\rightleftharpoons$	H^+	+	CHO_2^-
Initial:	$0.10\ M$		$0\ M$		$0\ M$
Equilibrium:	$(0.10 - 4.2 \times 10^{-3})\ M$		$4.2 \times 10^{-3}\ M$		$4.2 \times 10^{-3}\ M$

Notice that the amount of $HCHO_2$ that dissociates is very small in comparison with the initial concentration of the acid. To the number of significant figures we are using, the subtraction yields just 0.10.

We can now insert the equilibrium concentrations into the expression for K_a:

$$K_a = \frac{(4.2 \times 10^{-3})(4.2 \times 10^{-3})}{0.10}$$
$$= 1.8 \times 10^{-5}$$

(b) The percentage of acid that dissociates is given by the concentration of H^+ or CHO_2^- at equilibrium, divided by the initial acid concentration, times 100:

$$\text{Percent dissociation} = \frac{[H^+] \times 100}{[HCHO_2]}$$
$$= \frac{(4.2 \times 10^{-3}) \times 100}{0.10}$$
$$= 4.2 \text{ percent}$$

TABLE 15.3 Some weak acids in water at 25°C[a]

Acid	Molecular formula	Structural formula[a]	Conjugate base	K_a
Hydrofluoric	HF	H—F	F^-	6.8×10^{-4}
Hydrocyanic	HCN	H—C≡N	$C{\equiv}N^-$	4.9×10^{-10}
Acetic	$HC_2H_3O_2$	H—O—C(=O)—CH₃	$C_2H_3O_2^-$	1.8×10^{-5}
Benzoic	$HC_7H_5O_2$	H—O—C(=O)—C₆H₅	$C_7H_5O_2^-$	6.5×10^{-5}
Nitrous	HNO_2	H—O—N=O	NO_2^-	4.5×10^{-4}
Phenol	HOC_6H_5	H—O—C₆H₅	$C_6H_5O^-$	1.3×10^{-10}
Ascorbic (vitamin C)	$HC_6H_7O_6$	[ring structure: H—O, O—H, C=C, H, O=C, O, C, H, H, C—C—OH, OH, H]	$C_6H_7O_6^-$	8.0×10^{-5}

[a]The proton that ionizes is shown in color.

From the value for K_a it is possible to calculate the concentration of $H^+(aq)$ in a solution of a weak acid. For example, consider acetic acid, $HC_2H_3O_2$, the substance that gives the characteristic odor and acidic properties to vinegar. Let us calculate the concentration of $H^+(aq)$ in a 0.10 M solution of acetic acid.

Our first step is to write the ionization equilibrium for acetic acid:

$$HC_2H_3O_2(aq) \rightleftharpoons H^+(aq) + C_2H_3O_2^-(aq) \qquad [15.19]$$

Note from the Lewis structure for acetic acid shown in Table 15.3 that the hydrogen that ionizes is attached to the oxygen atom. We write this hydrogen separate from the others in the formula to emphasize that this one hydrogen is readily ionized.

The second step is to write the equilibrium-constant expression and the value for the equilibrium constant, if that is known. From Table 15.3 we have $K_a = 1.8 \times 10^{-5}$. Thus we can write the following:

$$K_a = \frac{[H^+][C_2H_3O_2^-]}{[HC_2H_3O_2]} = 1.8 \times 10^{-5} \qquad [15.20]$$

As the third step, we need to express the concentrations that make up the equilibrium-constant expression. This can be done with a little accounting, as described in Sample Exercise 15.8:

	$HC_2H_3O_2(aq)$	$\rightleftharpoons$	$H^+(aq)$	+	$C_2H_3O_2^-(aq)$
Initial:	$0.10\ M$		$0\ M$		$0\ M$
Equilibrium:	$(0.10 - x)\ M$		$x\ M$		$x\ M$

Because we seek to find the equilibrium value for $[H^+]$, let us call this quantity x. The concentration of acetic acid before any of it dissociates is $0.10\ M$. The equation for the equilibrium tells us that for each molecule of $HC_2H_3O_2$ that dissociates, one $H^+(aq)$ and one $C_2H_3O_2^-(aq)$ are formed. Thus, if x moles per liter of $H^+(aq)$ are formed at equilibrium, x moles per liter of $C_2H_3O_2^-(aq)$ must also have formed, and x moles per liter of $HC_2H_3O_2$ must have been dissociated. This gives rise to the equilibrium concentrations shown above.

As the fourth step of the problem, we need to substitute the equilibrium concentrations into the equilibrium-constant expression. The substitution gives the following equation:

$$K_a = \frac{[H^+][C_2H_3O_2^-]}{[HC_2H_3O_2]} = \frac{(x)(x)}{0.10 - x} = 1.8 \times 10^{-5} \qquad [15.21]$$

Because this equation has only one unknown it can be solved using algebra. However, the solution is a little tedious, because it requires use of the quadratic formula (Appendix A.3). By taking account of what is actually occurring in the solution, we can make things simpler for ourselves. Because the value of K_a is small, we might guess that x will be quite small. (In other words, perhaps only a small fraction of the $HC_2H_3O_2$ is actually ionized.) Indeed, if we solve the problem using the quadratic formula we find that $x = 1.3 \times 10^{-3}\ M$. Now you know that if a small number is subtracted from a much larger one, the result is approximately equal to the larger number. In our example we have

$$0.10 - x = 0.10 - 0.0013 \simeq 0.10$$

We can therefore make the approximation of ignoring x relative to 0.10 in the denominator of Equation 15.21. This leads us to the following simplified expression:

$$K_a = \frac{(x)(x)}{0.10} = 1.8 \times 10^{-5}$$

Solving for x, we have

$$x^2 = (0.10)(1.8 \times 10^{-5}) = 1.8 \times 10^{-6}$$
$$x = \sqrt{1.8 \times 10^{-6}} = 1.3 \times 10^{-3}\,M = [H^+]$$

From the value calculated for x we see that our simplifying approximation is quite reasonable. This type of approximation can be used whenever conditions in solution are such that only a small fraction of acid ionizes. As a general rule, if the quantity x, which is subtracted from the initial concentration of the acid, is more than about 5 percent of the initial value, it is best to use the quadratic formula. In cases of doubt, assume that the approximation is valid and solve for x in the simplified equation. Compare this approximate value of x with the initial concentration of acid. If it is more than about 5 percent as large, the problem should be reworked using the quadratic formula. For example, if the initial concentration of acid were 0.050 M, and x turned out in a given case to be 0.0016 M, then

$$\left(\frac{0.0016\,M}{0.050\,M}\right)(100) = 3.2\%$$

SAMPLE EXERCISE 15.9

Calculate the pH of a 0.20 M solution of HCN (refer to Table 15.3 or Appendix E for value of K_a).

Solution: Preceeding as in the example worked out above, we write

$$HCN(aq) \rightleftharpoons H^+(aq) + CN^-(aq)$$

$$K_a = \frac{[H^+][CN^-]}{[HCN]} = 4.9 \times 10^{-10}$$

Let $x = [H^+]$ at equilibrium. Then we have the following concentrations:

	$HCN(aq)$	$\rightleftharpoons$ $H^+(aq)$	+ $CN^-(aq)$
Initial:	0.20 M	0 M	0 M
Equilibrium:	$(0.20 - x)\,M$	$x\,M$	$x\,M$

Substituting into the equilibrium constant expression yields

$$K_a = \frac{(x)(x)}{0.20 - x} = 4.9 \times 10^{-10}$$

We next make the simplifying approximation that x, the amount of acid that dissociates, is small in comparison with the initial concentration of acid; that is,

$$0.20 - x \simeq 0.20$$

Thus

$$\frac{x^2}{0.20} = 4.9 \times 10^{-10}$$

Solving for x, we have

$$x^2 = (0.20)(4.9 \times 10^{-10})$$
$$= 0.98 \times 10^{-10}$$
$$x = \sqrt{0.98 \times 10^{-10}}$$
$$= 0.99 \times 10^{-5} = 9.9 \times 10^{-6} = [H^+]$$
$$pH = -\log[H^+] = -\log(9.9 \times 10^{-6})$$
$$= 5.00$$

The result obtained in Sample Exercise 15.9 is typical of the behavior of weak acids; the concentration of $H^+(aq)$ is only a small fraction of the

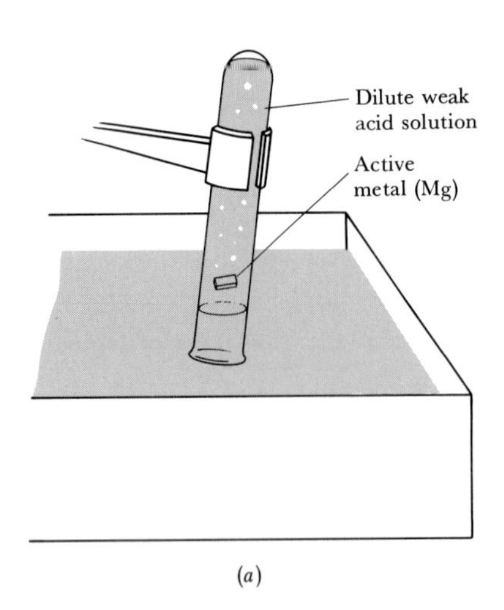

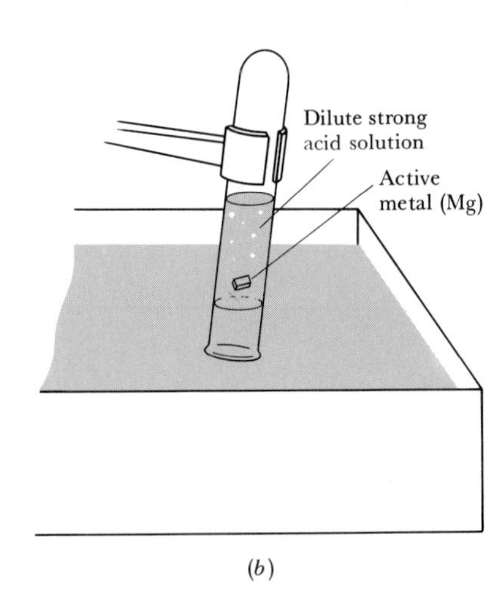

FIGURE 15.7 Demonstration of the relative rates of reaction of two acid solutions of the same concentration with an active metal. The solution in (*a*) is that of a weak acid, in (*b*) that of a strong acid. Reaction produces $H_2(g)$, which collects in the tube. From the relative amounts of gas collected in the two tubes after a period of time, it is evident that reaction is faster in (*b*). This indicates that even though the concentrations of acid are the same in the two tubes, the concentration of $H^+(aq)$ is much greater in (*b*).

concentration of the acid in solution. Thus those properties of the acid solution that relate directly to the concentration of $H^+(aq)$, such as electrical conductivity or rate of reaction with an active metal, are much less in evidence for a solution of a weak acid than for a solution of a strong acid. Figure 15.7 illustrates an experiment often carried out in the chemistry laboratory to demonstrate the difference in concentration of $H^+(aq)$ in weak and strong acid solutions of the same concentration. The rate of reaction with the active metal is much faster for the solution of a strong acid. Reactions in which the rate depends on $H^+(aq)$ are common.

Figure 15.8 illustrates an experiment in which the electrical conductivity of an HCl solution is compared with the conductivity of an HF solution. The conductivity of the solution of the strong acid increases

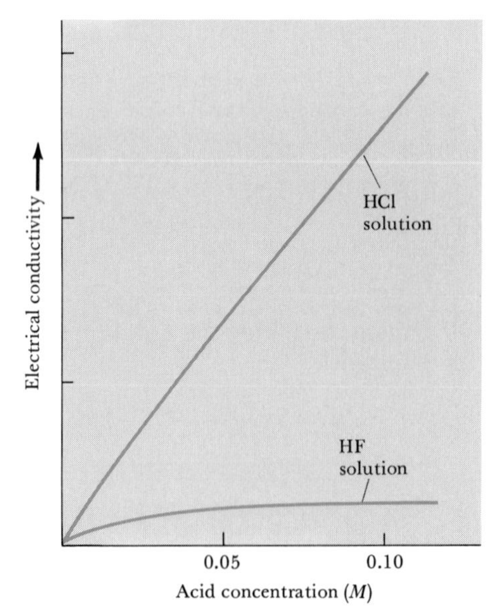

FIGURE 15.8 Electrical conductivity versus concentration for solutions of HCl, a strong acid, and HF, a weak acid. The conductivity of the HCl solution is not completely linear with concentration because of attractive forces between the ions at higher concentrations (Section 11.5). The conductivity for the HF solution is quite nonlinear with concentration and much lower than for HCl because only a fraction of the HF molecules ionize. It is nonlinear with concentration because the fraction of molecules ionizing changes with concentration.

approximately in proportion to the concentration. This is what one would expect; because all the acid molecules ionize, the concentration of ions in solution is directly proportional to the concentration of acid. The conductivity of the solution of the weak acid is very much less than that for a strong acid and does not vary linearly with the acid concentration. The nonlinearity of the graph arises from the fact that the percentage of acid ionized varies with the acid concentration. This is illustrated in Sample Exercise 15.10.

SAMPLE EXERCISE 15.10

Calculate the percentage of HF molecules ionized in a 0.10 *M* HF solution; in a 0.010 *M* HF solution.

Solution: The equilibrium reaction and equilibrium concentrations can be written as follows:

	$HF(aq)$	$\rightleftharpoons$	$H^+(aq)$ +	$F^-(aq)$
Initial:	0.10 *M*		0 *M*	0 *M*
Equilibrium:	$(0.10 - x)$ *M*		x *M*	x *M*

The equilibrium-constant expression is as follows:

$$K_a = \frac{[H^+][F^-]}{[HF]} = \frac{(x)(x)}{0.10 - x} = 6.8 \times 10^{-4}$$

We might be tempted to try solving this equation using the same approximation used in earlier examples, that is, by neglecting the concentration of acid that ionizes in comparison with the initial concentration (by neglecting x in comparison with 0.10). However K_a is large enough in this case to make that a poor approximation. We must therefore rearrange our equation, and write it in standard quadratic form:

$$x^2 = (0.10 - x)(6.8 \times 10^{-4})$$
$$= 6.8 \times 10^{-5} - (6.8 \times 10^{-4})x$$
$$x^2 + (6.8 \times 10^{-4})x - 6.8 \times 10^{-5} = 0$$

Solution of this equation by use of the standard quadratic formula,

$$x = \frac{-b \pm (b^2 - 4ac)^{1/2}}{2a}$$

gives

$$x = \frac{-6.8 \times 10^{-4} \pm [(6.8 \times 10^{-4})^2 + 4(6.8 \times 10^{-5})]^{1/2}}{2}$$

$$= \frac{-6.8 \times 10^{-4} \pm 1.65 \times 10^{-2}}{2}$$

Of the two solutions, only the one that gives a positive value for x is physically reasonable. Thus

$$x = 7.9 \times 10^{-3}$$

(You might also see what answer you would get by making the simplifying approximation of neglecting x with respect to 0.10.)

From our result we can calculate the percent of molecules ionized:

$$\text{Percent ionized} = \left(\frac{\text{concentration ionized}}{\text{original concentration}}\right)(100)$$
$$= \left(\frac{7.9 \times 10^{-3}\,M}{0.10\,M}\right)(100) = 7.9\%$$

Proceeding similarly for the 0.010 *M* solution, we have

$$\frac{x^2}{0.010 - x} = 6.8 \times 10^{-4}$$

Solving the resultant quadratic expression, we obtain

$$x = [H^+] = [F^-] = 2.3 \times 10^{-3}\,M$$

The percentage of molecules ionized is

$$\left(\frac{0.0023}{0.010}\right)(100) = 23\%$$

Notice that in diluting the solution by a factor of 10, the percentage of molecules ionized increases by a factor of 3. We could have arrived at this conclusion qualitatively by applying Le Châtelier's principle (Section 14.5) to the equilibrium. There are more "particles" or reaction components on the right side of the equation than on the left. Dilution causes the reaction to shift in the direction of the larger number of particles, because this counters the effect of the decreasing concentration of particles.

Polyprotic Acids

Many substances are capable of furnishing more than one proton to water. Substances of this type are called polyprotic acids. As an example, sulfurous acid, H_2SO_3, may react with water in two successive steps:

$$H_2SO_3(aq) \rightleftharpoons H^+(aq) + HSO_3^-(aq) \qquad K_a = 1.7 \times 10^{-2} \qquad [15.22]$$

$$HSO_3^-(aq) \rightleftharpoons H^+(aq) + SO_3^{2-}(aq) \qquad K_a = 6.4 \times 10^{-8} \qquad [15.23]$$

The values for K_a in each case show that the reactions are incomplete. Notice that loss of the second proton occurs much less readily than the first, as shown by the smaller value for K_a in the second reaction. This trend is intuitively reasonable; on the basis of electrostatic attractions we would expect the positively charged proton to be lost more readily from the neutral H_2SO_3 molecule than from the negatively charged HSO_3^- ion.

The successive acid-dissociation constants of polyprotic acids are sometimes labeled K_{a1}, K_{a2}, and so forth. This notation is often simplified to K_1, K_2, and so forth. For example, the equilibrium constant for the loss of a proton from HSO_3^- (Equation 15.23), can be labeled as K_{a2} or K_2, because this proton is the second one removed from the neutral acid, H_2SO_3. The acid-dissociation constants for a few common polyprotic acids are given in Table 15.4; a more complete list is found in Appendix E. Notice that the K_a values for successive losses of protons from these acids differ by at least a factor of 10^3.

Because K_{a1} is so much larger than subsequent dissociation constants for these polyprotic acids, almost all the $H^+(aq)$ in the solution comes from the first ionization reaction. As long as successive K_a values differ by a factor of 10^3 or more, it is possible to obtain a satisfactory estimate of the pH of polyprotic acid solutions by considering only K_{a1}.

SAMPLE EXERCISE 15.11

The solubility of CO_2 in pure water at 25°C and 0.1 atm pressure is 0.0037 *M*. The common practice is to assume that all of the dissolved CO_2 is in the form of H_2CO_3, which is produced by reaction between the CO_2 and H_2O:

$$CO_2(aq) + H_2O(l) \rightleftharpoons H_2CO_3(aq)$$

What is the pH of a 0.0037 *M* solution of H_2CO_3?

Solution: H_2CO_3 is a polyprotic acid; the two acid dissociation constants, K_{a1} and K_{a2} (Table 15.4), differ by more than a factor of 10^3. Consequently, the pH can be determined by considering only K_{a1}, thereby treating the acid as if it were a monoprotic acid. Proceeding as in Sample Exercises 15.9 and 15.10, we can write the equilibrium reaction and equilibrium concentrations as follows:

	$H_2CO_3(aq)$ $\rightleftharpoons$	$H^+(aq)$ +	$HCO_3^-(aq)$
Initial:	0.0037 *M*	0 *M*	0 *M*
Equilibrium:	$(0.0037 - x)$ *M*	x *M*	x *M*

The equilibrium-constant expression is as follows:

$$K_{a1} = \frac{[H^+][HCO_3^-]}{[H_2CO_3]} = \frac{(x)(x)}{0.0037 - x} = 4.3 \times 10^{-7}$$

Because K_{a1} is small, we make the simplifying approximation that x is small so that $0.0037 - x \simeq 0.0037$. Thus

$$\frac{(x)(x)}{0.0037} = 4.3 \times 10^{-7}$$

Solving for x, we have

$$x^2 = (0.0037)(4.3 \times 10^{-7}) = 1.6 \times 10^{-9}$$
$$x = \sqrt{1.6 \times 10^{-9}} = 4.0 \times 10^{-5} M = [H^+] = [HCO_3^-]$$

The small value of x indicates that our simplifying assumption was justified. The pH is therefore

$$pH = -\log [H^+] = -\log (4.0 \times 10^{-5}) = 4.40$$

If we had been asked to solve for $[CO_3^{2-}]$, we would need to use K_{a2}. Let's illustrate that calculation. Using the values of $[HCO_3^-]$ and $[H^+]$ calculated above, and setting $[CO_3^{2-}] = y$, we have the following initial and equilibrium concentration values:

	$HCO_3^-(aq)$	$\rightleftharpoons$	$H^+(aq)$	+	$CO_3^{2-}(aq)$
Initial:	$4.0 \times 10^{-5} M$		$4.0 \times 10^{-5} M$		$0 M$
Equilibrium:	$(4.0 \times 10^{-5} - y) M$		$(4.0 \times 10^{-5} + y) M$		$y M$

Assuming that y is small compared to 4.0×10^{-5}, we have

$$K_{a2} = \frac{[H^+][CO_3^{2-}]}{[HCO_3^-]} = \frac{(4.0 \times 10^{-5})(y)}{4.0 \times 10^{-5}} = 5.6 \times 10^{-11}$$
$$y = 5.6 \times 10^{-11} M = [CO_3^{2-}]$$

The value calculated for y is indeed very small in comparison with 4.0×10^{-5}, showing that our assumption was justified. It also shows that the ionization of the HCO_3^- is negligible in comparison with that of H_2CO_3 as far as production of H^+ is concerned. However, it is the *only* source of CO_3^{2-}, which has a very low concentration in the solution.

Our calculations thus tell us that in a solution of carbon dioxide in water most of the CO_2 is in the form of CO_2 or H_2CO_3, a small fraction ionizes to form H^+ and HCO_3^-, and an even smaller fraction ionizes to give CO_3^{2-}.

TABLE 15.4 Acid-dissociation constants of some common polyprotic acids at 25°C

Name	Formula	K_{a1}	K_{a2}	K_{a3}
Carbonic	H_2CO_3	4.3×10^{-7}	5.6×10^{-11}	
Oxalic	$H_2C_2O_4$	5.9×10^{-2}	6.4×10^{-5}	
Phosphoric	H_3PO_4	7.5×10^{-3}	6.2×10^{-8}	4.2×10^{-13}
Sulfurous	H_2SO_3	1.7×10^{-2}	6.4×10^{-8}	

15.6 WEAK BASES

Many substances behave as weak bases in water. Such substances react with water, removing protons from H_2O, therby forming the conjugate acid of the base and OH^- ions:

$$\text{Weak base} + H_2O \rightleftharpoons \text{conjugate acid} + OH^- \quad [15.24]$$

The most commonly encountered weak base is ammonia:

$$NH_3(aq) + H_2O(l) \rightleftharpoons NH_4^+(aq) + OH^-(aq) \quad [15.25]$$

The equilibrium-constant expression for this reaction can be written as

$$K = \frac{[NH_4^+][OH^-]}{[NH_3][H_2O]} \quad [15.26]$$

Because the concentration of water is essentially constant even when moderate concentrations of other substances are present, the $[H_2O]$ term is incorporated into the equilibrium constant, giving

$$K[H_2O] = K_b = \frac{[NH_4^+][OH^-]}{[NH_3]} \quad [15.27]$$

The constant K_b is called the **base-dissociation constant,** by analogy with the acid-dissociation constant, K_a, for weak acids. However, as we can see from Equations 15.24 and 15.25, the term "dissociation" in the case of weak bases has a slightly different meaning. It refers to the dissociation of *water* as a result of reaction with the base. Table 15.5 lists the names, formulas, Lewis structures, equilibrium reactions, and values of K_b for several weak bases in water. Appendix E includes a more extensive list. Notice that these bases contain one or more unshared pairs of electrons. An unshared pair is necessary to form the bond with H^+.

TABLE 15.5 Weak bases and their aqueous solution equilibria

Base	Lewis structure	Conjugate acid	Equilibrium reaction	K_b
Ammonia (NH_3)	$H-\ddot{N}-H$ with H below N	NH_4^+	$NH_3 + H_2O \rightleftharpoons NH_4^+ + OH^-$	1.8×10^{-5}
Pyridine (C_5H_5N)	ring with $N:$	$C_5H_5NH^+$	$C_5H_5N + H_2O \rightleftharpoons C_5H_5NH^+ + OH^-$	1.7×10^{-9}
Hydroxylamine (H_2NOH)	$H-\ddot{N}-OH$ with H below N	H_3NOH^+	$H_2NOH + H_2O \rightleftharpoons H_3NOH^+ + OH^-$	1.1×10^{-8}
Methylamine (NH_2CH_3)	$H-\ddot{N}-CH_3$ with H below N	$NH_3CH_3^+$	$NH_2CH_3 + H_2O \rightleftharpoons NH_3CH_3^+ + OH^-$	4.4×10^{-4}
Nicotine ($C_{10}H_{14}N_2$)	H_2C—CH_2, HC, CH_2, $\ddot{N}$, CH_3; pyridine ring with N	$HC_{10}H_{14}N_2^+$	$C_{10}H_{14}N_2 + H_2O \rightleftharpoons C_{10}H_{14}N_2H^+ + OH^-$	7×10^{-7} 1.4×10^{-11}
Hydrosulfide ion (HS^-)	$[H-\ddot{S}:]^-$	H_2S	$HS^- + H_2O \rightleftharpoons H_2S + OH^-$	1.8×10^{-7}
Carbonate ion (CO_3^{2-})	$[:\ddot{O}:$ above C, with two O$]^{2-}$	HCO_3^-	$CO_3^{2-} + H_2O \rightleftharpoons HCO_3^- + OH^-$	1.8×10^{-4}
Hypochlorite (ClO^-)	$[:\ddot{Cl}-\ddot{O}:]^-$	HClO	$ClO^- + H_2O \rightleftharpoons HClO + OH^-$	3.3×10^{-7}

SAMPLE EXERCISE 15.12

Calculate the concentration of OH^- in a 0.15 *M* solution of NH_3.

Solution: We use essentially the same procedure here as used in solving problems involving the dissociation of acids. The first step is to write the equilibrium expression and the corresponding equilibrium-constant expression:

$$NH_3(aq) + H_2O(l) \rightleftharpoons NH_4^+(aq) + OH^-(aq)$$

$$K_b = \frac{[NH_4^+][OH^-]}{[NH_3]} = 1.8 \times 10^{-5}$$

We then tabulate the equilibrium concentrations involved in the equilibrium:

$$\begin{array}{llll} & NH_3(aq) + H_2O(l) & \rightleftharpoons NH_4^+(aq) + & OH^-(aq) \\ \text{Initial:} & 0.15\,M & 0\,M & 0\,M \\ \text{Equilibrium:} & (0.15 - x)\,M & x\,M & x\,M \end{array}$$

(Notice that we ignore the concentration of H_2O, because this is not involved in the equilibrium-constant expression.) Inserting these quantities into the equilibrium-constant expression gives the following:

$$K_b = \frac{[NH_4^+][OH^-]}{[NH_3]} = \frac{(x)(x)}{0.15 - x} = 1.8 \times 10^{-5}$$

Because K_b is small we can neglect the small amount of NH_3 that reacts with water, as compared with the total NH_3 concentration; that is, we can neglect x in comparison with 0.15 M. Then we have

$$\frac{x^2}{0.15} = 1.8 \times 10^{-5}$$

$$x^2 = (0.15)(1.8 \times 10^{-5}) = 0.27 \times 10^{-5}$$

$$x = \sqrt{2.7 \times 10^{-6}} = 1.6 \times 10^{-3}\,M = [OH^-]$$

Notice that the value obtained for x is only about 1 percent of the NH_3 concentration, 0.15 M. Therefore, our neglect of x in comparison with 0.15 is justified.

Amines

The weak nitrogen bases listed in Table 15.5 belong to a family known as **amines**. These compounds can be thought of as being formed by replacing one or more of the N—H bonds in NH_3 with N—C bonds. (In hydroxylamine, H_2NOH, one of the N—H bonds of NH_3 has been replaced by a N—OH bond.) Like ammonia, such amines are able to extract a proton from the water molecule by forming a N—H bond. The following equation illustrates this behavior:

$$H{-}\ddot{N}(H){-}CH_3(aq) + H_2O(l) \rightleftharpoons \left[H{-}N(H)(H){-}CH_3\right]^+(aq) + OH^-(aq) \quad [15.28]$$

Many amines with low molecular weights have unpleasant, often "fishy" odors. Amines as well as NH_3 are produced by anaerobic (absence of O_2) decomposition of dead animal or plant matter, thereby contributing to their odor. One such amine, $H_2N(CH_2)_5NH_2$, is known as cadaverine.

Many drugs, including quinine, codeine, caffeine, and amphetamine (Benzedrine), are amines. Like other amines, these substances are weak bases; the amine nitrogen is readily protonated by treatment with an acid. The resulting products are called acid salts. If we use A as abbreviation for an amine, the acid salt formed by reaction with hydrochloric acid would be written as AH^+Cl^-. (It is sometimes written as $A \cdot HCl$ and referred to as a hydrochloride.) For example, amphetamine hydrochloride is the acid salt formed by treating amphetamine with HCl:

$$C_6H_5{-}CH_2{-}CH(CH_3){-}\ddot{N}H_2(aq) + HCl(aq) \longrightarrow$$

Amphetamine

$$C_6H_5{-}CH_2{-}CH(CH_3){-}NH_3^+Cl^-(aq)$$

Amphetamine hydrochloride

Such acid salts are less volatile, more stable, and generally more water soluble than the corresponding neutral amines. Many drugs that are amines are sold and administered as acid salts.

Anions of Weak Acids

A second common class of weak bases is composed of the anions of weak acids. Consider, for example, an aqueous solution of sodium acetate, $NaC_2H_3O_2$. This salt dissolves in water to give Na^+ and $C_2H_3O_2^-$ ions. The Na^+ ion is always a spectator ion in acid-base reactions. However, the $C_2H_3O_2^-$ ion is the conjugate base of a weak acid, acetic acid. Consequently, the $C_2H_3O_2^-$ ion is basic and reacts to a slight extent with water ($K_b = 5.6 \times 10^{-10}$):

$$C_2H_3O_2^-(aq) + H_2O(l) \rightleftharpoons HC_2H_3O_2(aq) + OH^-(aq) \qquad [15.29]$$

SAMPLE EXERCISE 15.13

Calculate the pH of a 0.010 *M* solution of sodium hypochlorite, NaClO.

Solution: NaClO is an ionic compound consisting of Na^+ and ClO^- ions. As such it is a strong electrolyte. The hypochlorite ion, ClO^-, supplied by this salt is a weak base. The base dissociation constant for ClO^- is given in Table 15.5, $K_b = 3.3 \times 10^{-7}$. We can write the reaction between ClO^- and water, and the equilibrium concentrations present in this solution as follows:

	$ClO^-(aq) + H_2O(l)$	$\rightleftharpoons$	$HClO(aq)$ +	$OH^-(aq)$
Initial:	0.010 *M*		0 *M*	0 *M*
Equilibrium:	(0.010 − *x*) *M*		*x M*	*x M*

Because K_b is small, we anticipate that *x* will be small, so that $(0.010 - x) \simeq 0.010$. Using this approximation, we have

$$K_b = \frac{[HClO][OH^-]}{[ClO^-]} = \frac{(x)(x)}{0.010} = 3.3 \times 10^{-7}$$

Solving for *x* yields

$$x^2 = (0.010)(3.3 \times 10^{-7}) = 3.3 \times 10^{-9}$$
$$x = [OH^-] = \sqrt{3.3 \times 10^{-9}} = 5.7 \times 10^{-5}\,M$$

$[H^+]$ can be obtained using the ion-product constant for water:

$$[H^+] = \frac{1.0 \times 10^{-14}}{[OH^-]} = \frac{1.0 \times 10^{-14}}{5.7 \times 10^{-5}}$$
$$= 1.8 \times 10^{-10}\,M$$
$$pH = -\log [H^+] = -\log (1.8 \times 10^{-10}) = 9.75$$

Thus we see that this solution of NaClO is slightly basic.

15.7 RELATION BETWEEN K_a AND K_b

We've seen in a qualitative way that the stronger acids have the weaker conjugate bases. The fact that this qualitative relationship exists suggests that we might be able to find a quantitative relationship. Let's explore this matter by considering the NH_4^+ and NH_3 conjugate acid-base pair. Each of these species reacts with water as follows:

$$NH_4^+(aq) \rightleftharpoons NH_3(aq) + H^+(aq) \qquad [15.30]$$
$$NH_3(aq) + H_2O(l) \rightleftharpoons NH_4^+(aq) + OH^-(aq) \qquad [15.31]$$

Each of these equilibria is expressed by a characteristic dissociation constant:

$$K_a = \frac{[NH_3][H^+]}{[NH_4^+]} \qquad K_b = \frac{[NH_4^+][OH^-]}{[NH_3]}$$

Now notice something very interesting and important. When Equations 15.30 and 15.31 are added together, the NH_4^+ and NH_3 species cancel, and we are left with just the autoionization of water:

$$\begin{array}{rcl} NH_4^+(aq) & \rightleftharpoons & NH_3(aq) + H^+(aq) \\ NH_3(aq) + H_2O(l) & \rightleftharpoons & NH_4^+(aq) + OH^-(aq) \\ \hline H_2O(l) & \rightleftharpoons & H^+(aq) + OH^-(aq) \end{array}$$

To determine what we should do about the equilibrium constants for the added reactions, we make use of a rule that can be derived from the general principles governing chemical equilibria: *When two reactions are added to give a third reaction, the equilibrium constant for the third reaction is given by the product of the equilibrium constants for the two added reactions.* Thus in general,

$$\begin{gathered} \text{If reaction 1 + reaction 2 = reaction 3} \\ \text{then } K_1 \times K_2 = K_3 \end{gathered}$$

Applying this to our present example, if we multiply K_a and K_b, we obtain the following result:

$$\begin{aligned} K_a \times K_b &= \left(\frac{[\cancel{NH_3}][H^+]}{[\cancel{NH_4^+}]}\right)\left(\frac{[\cancel{NH_4^+}][OH^-]}{[\cancel{NH_3}]}\right) \\ &= [H^+][OH^-] = K_w \end{aligned}$$

Thus the result of multiplying K_a times K_b is just the ion-product constant, K_w (Equation 15.10). This is, of course, just what we would expect, because addition of Equations 15.29 and 15.30 gave us just the autoionization equilibrium for water, for which the equilibrium constant is K_w.

The relationship we have just found is so important that it should be emphasized and called to special attention: *The product of the acid-dissociation constant for an acid and the base-dissociation constant for its conjugate base is the ion-product constant for water:*

$$K_a \times K_b = K_w \qquad [15.32]$$

As the strength of an acid increases (larger K_a), the strength of its conjugate base must decrease (smaller K_b), so that the product $K_a \times K_b$ remains equal to 1.0×10^{-14}.

Because of Equation 15.32 we can calculate K_a for any weak acid if we know K_b for its conjugate base. Similarly, we can calculate K_b for a weak base if we know K_a for its conjugate acid. As a practical consequence, ionization constants are often listed for only one member of a conjugate acid-base pair. For example, Appendix E does not contain K_b values for the anions of weak acids because these can be readily calculated from the tabulated K_a values for their conjugate acids.

SAMPLE EXERCISE 15.14

Calculate (a) the base-dissociation constant, K_b, for the fluoride ion, F^-; (b) the acid-dissociation constant, K_a, for the ammonium ion, NH_4^+.

Solution: (a) K_b for F^- is not included in Table 15.5 or in Appendix E. However, K_a for its conjugate acid, HF, is given in Table 15.3 and Appendix

E as $K_a = 6.8 \times 10^{-4}$. We can therefore use Equation 15.32 to calculate K_b:

$$K_b = \frac{K_w}{K_a} = \frac{1.0 \times 10^{-14}}{6.8 \times 10^{-4}} = 1.5 \times 10^{-11}$$

(b) K_b for NH_3 is listed in Table 15.5 and in Appendix E as $K_b = 1.8 \times 10^{-5}$. Using Equation 15.32, we can calculate K_a for the conjugate acid, NH_4^+:

$$K_a = \frac{K_w}{K_b} = \frac{1.0 \times 10^{-14}}{1.8 \times 10^{-5}} = 5.6 \times 10^{-10}$$

If you have the occasion to look up the values for acid or base dissociation constants in a chemistry handbook, you may find them expressed as pK_a or pK_b, that is, as $-\log K_a$ or $-\log K_b$ (Section 15.3). Equation 15.32 can be put in terms of pK_a and pK_b by taking the negative log of both sides:

$$pK_a + pK_b = pK_w = 14.00 \qquad [15.33]$$

This form is particularly useful when the tabulated pK value is that for the conjugate acid or base of the substance of interest. It is not uncommon to find the dissociation constants for bases tabulated as pK_a values for the corresponding conjugate acids. As an example, morphine, a nitrogen-containing base, is listed as the protonated cation, with $pK_a = 7.87$. This means that for the reaction

$$C_{17}H_{19}O_3NH^+(aq) \rightleftharpoons C_{17}H_{19}O_3N(aq) + H^+(aq)$$

the equilibrium constant, K_a, has the value $K_a = \text{antilog}(-7.87) = 10^{-7.87} = 1.3 \times 10^{-8}$. The reaction

$$C_{17}H_{19}O_3N(aq) + H_2O(l) \rightleftharpoons C_{17}H_{19}O_3NH^+(aq) + OH^-(aq)$$

is described by equilibrium constant K_b. Using Equation 15.33 and $pK_a = 7.87$, we have

$$pK_b = 14.00 - pK_a = 14.00 - 7.87 = 6.13$$

Thus $K_b = \text{antilog}(-6.13) = 10^{-6.13} = 7.4 \times 10^{-7}$.

15.8 ACID-BASE PROPERTIES OF SALT SOLUTIONS

Even before you began this chapter you were undoubtably aware of many substances that are acidic such as HNO_3, HCl, and H_2SO_4, and others that are basic, such as NaOH and NH_3. However, our recent discussions have indicated that ions can also exhibit acidic or basic properties. For example, we calculated K_a for NH_4^+ and K_b for F^- in Sample Exercise 15.14. Such behavior implies that salt solutions can be acidic or basic. Before proceeding with further discussions of acids and bases, let's summarize some features of salts that should bring their acid and base properties into sharper focus.

We can assume that when salts dissolve in water they are completely ionized; nearly all salts are strong electrolytes. Consequently the acid-base properties of salt solutions are due to the behavior of the cations and anions. Many ions are able to react with water to generate $H^+(aq)$ or $OH^-(aq)$. This type of reaction is often called **hydrolysis.**

The anions of weak acids, HX, are basic, and consequently they react with water to produce OH^- ions:

$$X^-(aq) + H_2O(l) \rightleftharpoons HX(aq) + OH^-(aq) \qquad [15.34]$$

In contrast, the anions of strong acids, such as the NO_3^- ion, exhibit no significant basicity; these ions do not hydrolyze, and consequently do not influence pH.

Anions of polyprotic acids such as HCO_3^- that still have ionizable protons are capable of acting as either proton donors or proton acceptors (that is, either acids or bases). Their behavior toward water will be determined by the relative magnitudes of K_a and K_b for the ion, as shown in Sample Exercise 5.15.

SAMPLE EXERCISE 15.15

Predict whether the salt Na_2HPO_4 will form an acidic or basic solution on dissolving in water.

Solution: The two possible reactions that HPO_4^{2-} may undergo on addition to water are

$$HPO_4^{2-}(aq) \rightleftharpoons H^+(aq) + PO_4^{3-}(aq) \qquad [15.35]$$

$$HPO_4^{2-}(aq) + H_2O \rightleftharpoons H_2PO_4^-(aq) + OH^-(aq) \qquad [15.36]$$

Depending on which of these has the larger equilibrium constant, the ion will cause the solution to be acidic or basic.The value of K_a for reaction Equation 15.35, as shown in Table 15.4, is 4.2×10^{-13}. We must calculate the value of K_b for reaction Equation 15.36 from the value of K_a for the conjugate acid formed, $H_2PO_4^-$. We make use of the relationship shown in Equation 15.32:

$$K_a \times K_b = K_w$$

We want to know K_b for the base HPO_4^{2-}, knowing the value of K_a for the conjugate acid $H_2PO_4^-$:

$$K_b(HPO_4^{2-}) \times K_a(H_2PO_4^-) = K_w = 1.0 \times 10^{-14}$$

Because K_a for $H_2PO_4^-$ is 6.2×10^{-8} (Table 15.4), we calculate K_b for HPO_4^{2-} to be 1.6×10^{-7}. This is considerably larger than K_a for HPO_4^{2-}; thus the reaction shown in Equation 15.36 predominates over that in Equation 15.35, and the solution is basic.

All cations except those of the alkali metals and the heavier alkaline earths (Ca^{2+}, Sr^{2+}, and Ba^{2+}) act as weak acids in water solution. Because the alkali metal and alkaline earth cations do not hydrolyze in water, the presence of any of these ions in solution does not influence pH. It may surprise you that metal ions such as Al^{3+} and the transition metal ions form weakly acidic solutions. We can take this observation for now as a point of fact. The reasons for this behavior are discussed in Section 15.10.

Among the cations that produce an acidic solution is, of course, NH_4^+, which is the conjugate acid of the base NH_3. The NH_4^+ ion dissociates in water as follows:

$$NH_4^+(aq) \rightleftharpoons H^+(aq) + NH_3(aq) \qquad [15.37]$$

The pH of a solution of a salt can be qualitatively predicted by considering the cation and anion of which the salt is composed. A convenient way to do this is to consider the relative strengths of the acids and bases from which the salt is derived:*

1. *Salt derived from a strong base and a strong acid.* Examples are NaCl and $Ca(NO_3)_2$, which are derived from NaOH and HCl and from $Ca(OH)_2$ and HNO_3, respectively. Neither cation nor anion hydrolyzes. The solution has a pH of 7.
2. *Salt derived from a strong base and a weak acid.* In this case the anion is a relatively strong conjugate base. Examples are NaClO and $Ba(C_2H_3O_2)_2$. The anion hydrolyzes to produce $OH^-(aq)$ ions. The solution has a pH above 7.

*These rules apply to what can be called normal salts. These salts are ones that contain no ionizable protons on the anion. The pH of an acid salt (such as $NaHCO_3$ and NaH_2PO_4) is affected not only by the hydrolysis of the anion but also by its acid dissociation as well, as shown in Sample Exercise 15.15.

3. *Salt derived from a weak base and a strong acid.* In this case the cation is a relatively strong conjugate acid. Examples are NH_4Cl and $Al(NO_3)_3$. The cation hydrolyzes to produce $H^+(aq)$ ions. The solution has a pH below 7.
4. *Salt derived from a weak base and a weak acid.* Examples are $NH_4C_2H_3O_2$, NH_4CN, and $FeCO_3$. Both cation and anion hydrolyze. The pH of the solution depends upon the extent to which each ion hydrolyzes. The pH of a solution of NH_4CN is greater than 7 because CN^- ($K_b = 2.0 \times 10^{-5}$) is more basic than NH_4^+ ($K_a = 5.6 \times 10^{-10}$) is acidic. Consequently, CN^- hydrolyzes to a greater extent than NH_4^+ does.

SAMPLE EXERCISE 15.16

List the following solutions in the order of increasing pH: (a) 0.1 M $Co(ClO_4)_3$; (b) 0.1 M RbCN; (c) 0.1 M $Sr(NO_3)_2$; (d) 0.1 M $KC_2H_3O_2$.

Solution: The most acidic solution will be (a), which has a metal ion that undergoes hydrolysis, and an anion derived from a strong acid. Solution (c) should have pH of about 7, because it is derived from an alkaline earth cation and the anion of a strong acid. Solutions (b) and (d) are both derived from an alkali metal ion, which does not undergo hydrolysis, and an anion of a weak acid. Anion hydrolysis should lead to a basic solution in both cases, but solution (b) will be more strongly basic because CN^- is a stronger base than is $C_2H_3O_2^-$. Thus the order of pH is 0.1 *M* $Co(ClO_4)_3$ < 0.1 *M* $Sr(NO_3)_2$ < 0.1 *M* $KC_2H_3O_2$ < 0.1 *M* RbCN.

15.9 ACID-BASE CHARACTER AND CHEMICAL STRUCTURE

From our discussion to this point, we have seen that when any substance is dissolved in water one of three things can happen. The $H^+(aq)$ concentration might increase, in which case the substance behaves as an acid; it might decrease (with corresponding increase in OH^-), in which case the substance is acting as a base; or, there might be no change in $[H^+]$, an indication that the substance possesses neither acid nor base character. It would be very helpful to have further guidelines as to how acid or base characteristics relate to chemical structure, so that we might be better able to predict how a compound will behave on dissolving in water. However, we must expect that any simple rules we might formulate won't always work. Many different factors contribute to ionization in a polar solvent such as water. The best we can hope for are a few rules that are *almost always* obeyed.

Effects of Bond Polarity and Bond Strength

When a substance HX transfers a proton to the solvent, an ionic rupture of the H—X bond occurs. Such a reaction will occur most readily when the H—X bond is already polarized in the following sense:

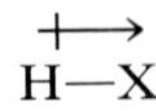

For example, compare NH_4^+ and CH_4. These two species have the same electronic structure; that is, they are isoelectronic. Both consist of a central atom with an octet of electrons bonding four hydrogens. The difference is in the nuclear charge of the central atom. Because the nuclear charge of N is one greater than for C, the electron pairs shared with the hydrogens are more closely attracted to N in NH_4^+ than to C in CH_4. That is, the N—H bonds are more polarized than the C—H bonds. Correspondingly, ammonium ion is an acid in water, whereas methane is not:

$$NH_4^+(aq) \rightleftharpoons H^+(aq) + NH_3(aq) \qquad K_a = 5.6 \times 10^{-10} \qquad [15.38]$$

$$CH_4(aq) \rightleftharpoons H^+(aq) + CH_3^-(aq) \qquad \text{no reaction} \qquad [15.39]$$

One other factor of major importance in determining whether a substance acts as an acid is the strength of the H—X bond. Very strong bonds are less easily ionized than weaker ones. This factor is of importance in the case of the hydrogen halides. The H—F bond is the most polar of any H—X bond. One might therefore expect that HF would be a very strong acid, if the first rule were all that mattered. However, the energy required to dissociate HF into H and F atoms is much higher than for the other hydrogen halides, as shown in Table 7.3. As a result, HF is a weak acid, whereas all the other hydrogen halides are strong acids in water.

The factors we have just considered can be used to relate the acid-base properties of the hydride of an element to its position in the periodic table. In any horizontal row of the table, the most basic hydrides are on the left, the most acidic hydrides are on the right. For example, in the second row of the table, NaH is a basic hydride. On addition to water it reacts as described by Equation 15.15, with formation of $OH^-(aq)$. On the right-hand side of the row, the acidity increases in the order $PH_3 < H_2S < HCl$. This general trend is related to the increasing electronegativity of the element as we move from left to right in any horizontal row. In general, metal hydrides are either basic or show no pronounced acid-base properties in water, whereas nonmetal hydrides range from acidic to showing no pronounced acid-base properties.

In any vertical row of nonmetallic elements there is a tendency toward increasing acidity with increasing atomic number. For example, among the group 6B elements the acid dissociation constants vary in the order $H_2O < H_2S < H_2Se < H_2Te$.

This order is due primarily to the fact that the bond strengths steadily decrease in this series as the central atom grows larger and overlaps of atomic orbitals grow smaller.

Hydroxides and Oxyacids

Many of the acids commonly encountered involve one or more O—H bonds. For example, H_2SO_4 contains two such bonds:

```
        ..
       :O:
   ..   |   ..
H—O—S—O—H
   ..   |   ..
       :O:
        ..
```

Such substances in which OH groups and possibly additional oxygen atoms are bound to a central atom are referred to as **oxyacids.** Let's consider, then, an OH group bound to some other atom Y, which might in turn have other groups attached to it:

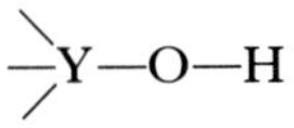

$$\rangle Y—O—H$$

At one extreme, Y might be a metal such as Na, K, or Mg. The pair of electrons shared between Y and O is then completely transferred to oxygen, and an ionic compound involving OH^- is formed. Because of the charge that surrounds it, the oxygen of the OH^- ion does not strongly attract to itself the electron pair it shares with hydrogen. That is, the O—H bond in OH^- is not strongly polarized. There is therefore no tendency for the hydrogen of OH^- to be transferred to the solvent as $H^+(aq)$. Such compounds therefore behave as bases.

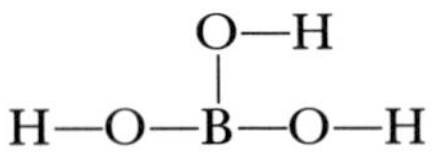

```
    O—H
    |
H—O—B—O—H
```

Orthoboric acid

I—O—H

Hypoiodous acid

CH_3—O—H

Methanol

When Y is an element of intermediate electronegativity, around 2.0, the bond to O is more covalent in character, and the substance does not readily lose OH^-. Elements with electronegativities in this range include B, C, P, As, and I (Figure 7.6). Examples of acids of such elements include orthoboric acid, hypoiodous acid, and methanol, the structures of which are shown at left.

Such substances might behave as acids in water, depending on the ease with which the proton is lost from oxygen. As a general rule, the more strongly the group Y attracts the electron pair it shares with the oxygen, the more polar the OH bond will be, and the more acidic the substance. In the three examples just given, the central atom does not strongly attract the electron pair it shares with oxygen. The acid-dissociation constant for orthoboric acid is 6.5×10^{-10}; for hypoiodous acid, 2.3×10^{-11}; no acidic or basic character is observed for methyl alcohol in water.

SAMPLE EXERCISE 15.17

Draw the Lewis structure for orthosilicic acid, $Si(OH)_4$. What do you predict for the acid-base properties of this substance?

Solution: The Lewis structure for $Si(OH)_4$ is as follows:

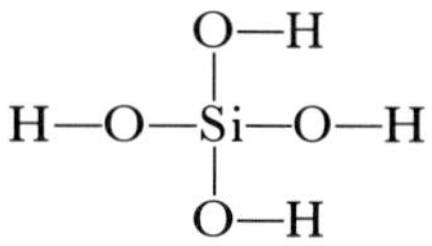

```
     O—H
     |
H—O—Si—O—H
     |
     O—H
```

Because silicon is an element of intermediate electronegativity (Figure 7.6), we would expect that $Si(OH)_4$ would not be strongly acidic or basic. By analogy with orthoboric acid, we might guess that it would be weakly acidic, as in fact it is:

$$K_a = 2 \times 10^{-10}$$

As the electronegativity of Y increases, or as groups with greater electron-attracting ability are placed on Y, the acidic properties of the substance increase. It is possible to relate the acid strengths of oxyacids both to the electronegativity of Y and to the number of groups attached to it. *For acids that have the same structure, but differ in the electronegativity of the central atom, Y, acid strength increases with increasing electronegativity.* Examples are shown in Table 15.6.

TABLE 15.6 Acid-dissociation constants (K_a) of oxyacids in comparison with electronegativity values (EN) of atom Y

H—O—Y	K_a	EN of Y	H—O—Y(=O)—O—H	K_{a1}	EN of Y
HClO	3×10^{-8}	3.2	H_2SO_3	1.7×10^{-2}	2.6
HBrO	2×10^{-9}	3.0	H_2SeO_3	3.5×10^{-3}	2.6
HIO	2×10^{-11}	2.7	H_2CO_3	4.3×10^{-7}	2.5
$HOCH_3$	~0	2.5[a]			

[a]This value is the electronegativity for carbon.

In a series of acids that have the same central atom, Y, but differing numbers of attached groups, the acid strength increases with increasing oxidation number of the central atom. For example, in the series of oxyacids of chlorine extending from hypochlorous to perchloric acid, acid strength steadily increases:

Acid	H—Ö—Cl:	H—Ö—Cl—Ö:	H—Ö—Cl(—:O:)—Ö:	H—Ö—Cl(—:O:)(—:O:)—Ö:
Chlorine oxidation numbers	Hypochlorous +1	Chlorous +3	Chloric +5	Perchloric +7

Increasing acid strength →

In this series the ability of chlorine to withdraw electrons from the OH group, and thus make the O—H bond even more polar, increases as electron-withdrawing oxygen atoms are added to the chlorine.

15.10 THE LEWIS THEORY OF ACIDS AND BASES

For a substance to be a proton acceptor (that is, a base in the Brønsted-Lowry sense), that substance must possess an unshared pair of electrons for binding the proton. For example, we have seen that NH_3 acts as a proton acceptor. Using Lewis structures we can write the reaction between H^+ and NH_3 as follows:

$$H^+ + :NH_3 \longrightarrow [NH_4]^+ \qquad [15.40]$$

G. N. Lewis was the first to notice this aspect of acid-base reactions. He proposed a definition of acid and base that emphasizes the shared electron pair: *An acid is defined as an electron-pair acceptor and a base as an electron-pair donor.*

Every base that we have discussed thus far, whether it be OH^-, H_2O, an amine, or an anion, is an electron-pair donor. Everything that is a base in the Brønsted-Lowry sense (a proton acceptor) is also a base in the Lewis sense (an electron-pair donor). However, in the Lewis theory, a base can donate its electron pair to something other than H^+. The Lewis definition therefore greatly increases the number of species that can be considered as acids; H^+ is a Lewis acid, but not the only one. For example, consider the reaction between NH_3 and BF_3. This reaction occurs because BF_3 has a vacant orbital in its valence shell (Section 7.7). It therefore acts as an electron-pair acceptor (a Lewis acid) toward NH_3, which donates the electron pair:

$$\underset{\text{Base}}{H_3N:} + \underset{\text{Acid}}{BF_3} \longrightarrow H_3N{-}BF_3 \qquad [15.41]$$

Our emphasis throughout this chapter has been on water as the solvent and on the proton as the source of acidic properties. In such cases we find the Brønsted-Lowry definition of acids and bases to be the most useful one to use. In fact, when we speak of a substance as being acidic or basic, we are usually thinking of aqueous solutions and using these terms in the Arrhenius or Brønsted-Lowry sense. The advantage of the Lewis theory is that it allows us to treat a wider variety of reactions, including ones that do not involve proton transfer, as acid-base reactions. To avoid confusion, a substance like BF_3 is rarely called an acid unless it is clear from the context that we are using the term in the sense of the Lewis definition. Instead, such substances which function as electron-pair acceptors are referred to explicitly as "Lewis acids."

Lewis acids include molecules like BF_3 that have an incomplete octet of electrons. In addition, many simple cations can function as Lewis acids. For example, Fe^{3+} interacts strongly with cyanide ions to form the ferricyanide ion, $Fe(CN)_6{}^{3-}$:

$$Fe^{3+} + 6\,:C{\equiv}N:^- \longrightarrow [Fe(C{\equiv}N)_6]^{3-}$$

Some compounds with multiple bonds can behave as Lewis acids. For example, the reaction of carbon dioxide with water to form carbonic acid, H_2CO_3, can be pictured as an attack by a water molecule on CO_2, in which the water acts as an electron-pair donor, and the CO_2 as an electron-pair acceptor:

$$H_2\ddot{O}: \curvearrowright O{=}C{=}O$$

The electron pair of one of the carbon-oxygen π bonds is moved onto the oxygen to leave a vacant orbital on the carbon, which can act as electron-pair acceptor. We have shown the shift of these electrons with arrows. After forming the initial "adduct," a proton moves from one oxygen to another, thereby forming carbonic acid:

$$H{-}\underset{H}{O}{-}C(=O){-}O:^{-} \longrightarrow H{-}O{-}C(=O){-}O{-}H$$

A similar kind of Lewis acid-base reaction takes place when any oxide of a nonmetal dissolves in water to form an acidic solution.

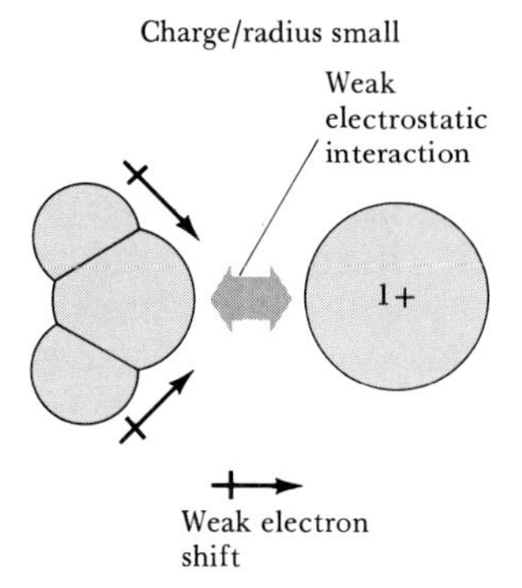

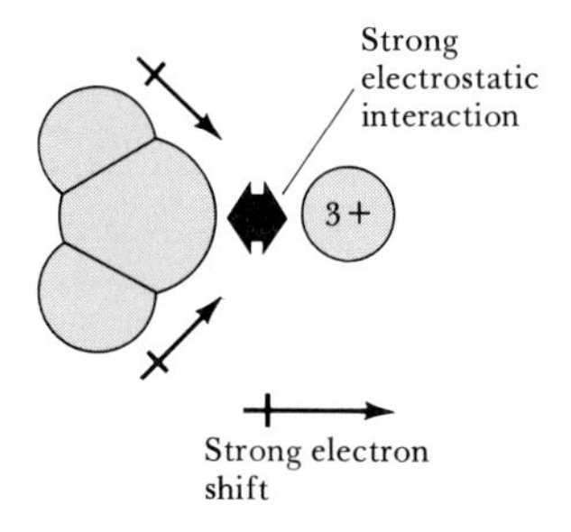

FIGURE 15.9 Interaction of a water molecule with a cation of 1+ charge or 3+ charge. The interaction is much stronger with the smaller ion of higher charge.

Hydrolysis of Metal Ions

The Lewis theory is also helpful in explaining why solutions of many metal ions show acidic properties (Section 15.8). For example, a solution of a salt such as $Cr(NO_3)_3$ is quite acidic. A solution of $ZnCl_2$ is also acidic, though to a lesser extent. To understand why this is so, we must examine the interaction between a metal ion and water molecules.

Because metal ions are positively charged, they attract the unshared electron pairs of water molecules. It is primarily this interaction, referred to as **hydration**, that causes salts to dissolve in water, as explained in Section 11.2. The strength of attraction increases with the charge of the ion and is strongest for the smallest ions. The ratio of ionic charge to ionic radius provides a good measure of the extent of hydration. This ratio is listed in Table 15.7 for a selection of metal ions. The process of hydration is a Lewis acid-base interaction, in which the metal ion acts as a Lewis acid, and the water molecules as Lewis bases. When the water molecule interacts with the positively charged metal ion, electron density is drawn from the oxygen, as illustrated in Figure 15.9. This flow of electron density causes the O—H bond to become more polarized; as a result, water molecules bound to the metal ion, M, are more acidic than those in the bulk solvent. The hydrated metal ion thus acts as a source of protons:

$$M(H_2O)_n^{z+} \rightleftharpoons M(H_2O)_{(n-1)}(OH)^{(z-1)+} + H^+(aq) \qquad [15.42]$$

In this equation z is the charge on the metal ion, and n is the number of hydrating water molecules. In this case of the 3+ metal ions listed in Table 15.7, n is 6; for the other ions it is probably closer to 4, although the

TABLE 15.7 Ionic charge/ionic radius ratio and acid hydrolysis constant for metal ions of various charges

Metal ion	Ionic charge / Ionic radius	Acid hydrolysis constant, K_h
Na^+	1.0	Negligible
Li^+	1.5	2×10^{-14}
Ca^{2+}	2.1	2×10^{-13}
Mg^{2+}	3.1	4×10^{-12}
Zn^{2+}	2.7	1×10^{-9}
Cu^{2+}	2.8	1×10^{-8}
Al^{3+}	6.7	1×10^{-5}
Cr^{3+}	4.8	1×10^{-4}
Fe^{3+}	4.7	2×10^{-3}

exact number is difficult to determine. The hydrolysis reaction shown in Equation 15.42 represents the behavior of an acid in just the same way as Equation 15.17, which applies to HX. The "acid" in Equation 15.42 is not just a single molecule, but a collection of molecules. For example, it might be $Fe(H_2O)_6^{3+}$, which we usually represent merely as $Fe^{3+}(aq)$. The acid hydrolysis constants (that is, the equilibrium constants for Equation 15.42) are listed in Table 15.7 for several ions whose charge/ionic radius ratios are also given. Notice that there is a general trend toward larger acid hydrolysis constants as the ionic charge/ionic radius ratio increases. The tendency to transfer a proton to water is greatest for the smallest and most highly charged ions. The effect of ionic size and charge is illustrated in Figure 15.9.

FOR REVIEW

Summary

In this chapter we have considered the general properties of acidic and basic solutions, with emphasis on water as the solvent. We have seen that an acid solution is created when a substance reacts with water in such a way as to increase the concentration of solvated hydrogen ions, which are represented as $H^+(aq)$ or $H_3O^+(aq)$. The concentration of $H^+(aq)$ is often expressed on the **pH scale:** $pH = -\log [H^+]$. Solutions of pH less than 7 are acidic; those with pH greater than 7 are basic.

Water spontaneously ionizes to a slight degree **(autoionization)**, forming $H^+(aq)$ and $OH^-(aq)$. The extent of ionization is expressed by the **ion-product constant** for water:

$$K_w = [H^+][OH^-] = 1.0 \times 10^{-14}$$

This relationship describes not only pure water, but aqueous solutions as well. Because the concentration of water is effectively constant in dilute solutions, $[H_2O]$ is omitted from this equilibrium-constant expression as well as from others associated with reactions in aqueous solution.

Through most of this chapter we have relied on the **Brønsted-Lowry theory** of acids and bases. According to this theory, an acid is a proton (H^+) donor, a base is a proton acceptor. Reaction of an acid with water results in the formation of $H^+(aq)$ and the **conjugate base** of the acid. **Strong acids** have conjugate bases that are weaker proton acceptors than H_2O. Such acids are strong electrolytes, ionizing completely in solution. The common strong acids are HCl, HBr, HI, HNO_3, $HClO_4$, and H_2SO_4. **Weak acids** are substances for which the reaction with water is incomplete, and an equilibrium is established. The extent to which the reaction proceeds is expressed by the **acid-dissociation constant,** K_a. **Polyprotic acids** are acids such as H_2SO_3 that have more than one ionizable proton. These acids have more than one acid-dissociation constant: K_{a1}, K_{a2}, and so forth, which decrease in magnitude in the order $K_{a1} > K_{a2} > K_{a3}$.

Aside from the ionic hydroxides such as NaOH, bases produce an increase of OH^- by reaction with water. **Strong bases** have **conjugate acids** that are no stronger than H_2O. The common strong bases are the hydroxides and oxides of the alkali metals and alkaline earths. **Weak bases** include H_2O, NH_3, amines, and the anions of weak acids. The extent to which a weak base reacts with water to generate OH^- and the conjugate acid of the base is measured by the **base-dissociation constant,** K_b.

The stronger an acid, the weaker its conjugate base; the weaker an acid, the stronger its conjugate base. This qualitative observation is expressed quantitatively by the expression $K_a \times K_b = K_w$ (where K_a and K_b are dissociation constants for conjugate acid-base pairs).

The acid-base properties of salts can be ascribed to the behavior of their respective cations and anions. The reaction of ions with water with a resultant change in pH is called **hydrolysis.** The cations of strong bases (the alkali metal ions and alkaline earth metal ions) and the anions of strong acids do not undergo hydrolysis.

The tendency of a substance to show acidic or basic characteristics in water can be correlated reasonably well with chemical structure. Acid character requires the presence of a highly polar H—X bond, promoting loss of hydrogen as H^+ on reaction with water. Basic character, on the other hand, requires the presence of an available pair of electrons. By considering the effects of changes in structure, it is possible to predict how a given structural change is likely to alter the acidity or basicity.

In the Lewis theory of acids and bases, the emphasis is on the shared electron pair rather than on the proton. An acid is defined as an electron-pair acceptor, a base as an electron-pair donor. The Lewis theory is more general than the Brønsted-Lowry model, because it applies to cases in which the proton is the acid, and to others as well.

Learning goals

Having read and studied this chapter, you should be able to:

1 Explain the process that occurs when an acid dissolves in water.

2 Describe the forms in which the proton exists in water.

3 Define an acid, base, conjugate acid, and conjugate base in terms of the Brønsted-Lowry theory of acids and bases.

4 Explain what is meant by the autoionization of water and write the ion-product-constant expression.

5 Explain what is meant by pH and calculate pH from a knowledge of $[H^+]$ or $[OH^-]$; also be able to perform the reverse operations.

6 Calculate $[OH^-]$ from pOH and K from pK, and be able to perform the reverse operations.

7 Identify the common strong acids and bases.

8 Write the acid-dissociation-constant expression for any weak acid in water.

9 Calculate $[H^+]$ for a weak acid solution in water, knowing acid concentration and K_a.

10 Write the base-dissociation constant expression for a weak base in water.

11 Calculate $[H^+]$ for any weak base solution in water, knowing the base concentration and K_b.

12 Calculate the percent dissociation for an acid or base, knowing its concentration in solution and the acid or base dissociation constant.

13 Explain the relationship between an acid and its conjugate base or between a base and its conjugate acid and calculate K_b from a knowledge of K_a, or vice versa.

14 Predict whether a particular salt solution will be acidic, basic, or neutral.

15 Explain how acid strength relates in a general way to the nature of the H—X bond.

16 Predict the relative acid strengths of oxyacids and oxyanions.

17 Define an acid or base in terms of the Lewis acid-base theory.

18 Predict the relative acidities of solutions of metal salts from a knowledge of metal-ion charges and ionic radii.

Key terms

Among the more important terms and expressions used for the first time in this chapter are the following:

An acid-base indicator (Section 15.3) is a substance whose color changes in passing from an acidic to a basic form, or vice versa.

The acid-dissociation constant, K_a (Section 15.5), is an equilibrium constant that expresses the extent to which an acid transfers a proton to solvent water.

Autoionization of water (Section 15.3) is the process whereby water spontaneously forms low concentrations of $H^+(aq)$ and $OH^-(aq)$ ions by proton transfer from one water molecule to another.

The base-dissociation constant, K_b (Section 15.6), is an equilibrium constant that expresses the extent to which a base reacts with solvent water, accepting a proton and forming $OH^-(aq)$.

A Brønsted acid (Section 15.2) is any substance capable of acting as a source of protons.

A Brønsted base (Section 15.2) is any substance capable of acting as a proton acceptor.

A conjugate acid (Section 15.2) is a substance formed by addition of a proton to a Brønsted base.

A conjugate base (Section 15.2) is a substance formed by loss of a proton from a Brønsted acid.

Hydrolysis (Section 15.8) is a process in which a cation or anion reacts with water so as to change the pH.

The ion-product constant (Section 15.3) for water, K_w, is the product of the aquated hydrogen ion and hydroxide ion concentrations: $[H^+][OH^-] = K_w = 1.0 \times 10^{-14}$.

A Lewis acid (Section 15.10) is defined as an electron-pair acceptor.

A Lewis base (Section 15.10) is defined as an electron-pair donor.

An oxyacid (Section 15.9) is a compound in which one or more OH groups, and possibly additional oxygen atoms, are bonded to a central atom.

The term pH (Section 15.3) is defined as the negative log in base 10 of the aquated hydrogen-ion concentration: $pH = -\log [H^+]$.

A polyprotic acid (Section 15.5) is a substance capable of dissociating more than one proton in water; H_2SO_4 is an example.

EXERCISES

Aqueous acids

15.1 Why does an aqueous solution of HCl conduct an electrical current? What species actually move through the solution to carry the current? Why does a benzene solution of HCl *not* carry an electrical current?

15.2 Nitric acid forms a monohydrate, $HNO_3 \cdot H_2O$, that melts at 235 K. Suggest a possible ionic structure for the monohydrate. Draw the Lewis structure of the cation and anion.

15.3 Gaseous hydrogen iodide is very soluble in water. A solution containing 52.4 percent HI by weight has a density of 1.60 g/cm^3. What is the ratio of moles of H_2O to HI in this solution? Suggest an explanation for the high solubility of HI in water. Suggest one or more experiments that would provide a test of your hypothesis.

15.4 The electrical conductivities of aqueous acid solutions are higher than those of other aqueous solutions of electrolytes. Using Lewis structures, indicate a mechanism by which the proton could appear to move through solution by a jumping process.

Brønsted acids and bases

15.5 Give the conjugate acid of each of the following bases: (a) NH_3; (b) Br^-; (c) NH_2^-; (d) $H_2PO_4^-$; (e) OH^-.

15.6 Give the conjugate base of each of the following proton sources: (a) H_3PO_4; (b) HBr; (c) $H_2C_2O_4$; (d) HS^-; (e) NH_4^+; (f) PH_3.

15.7 The $H_2PO_4^-$ ion is capable of acting as either a Brønsted acid or base. Illustrate these two behaviors by writing balanced equations for its reactions with CN^- and H_2SO_4 in aqueous solution.

15.8 Identify the acid and base in each of the following reactions:

(a) $NH_2^-(aq) + H_2O(l) \rightleftharpoons NH_3(aq) + OH^-(aq)$

(b) $H_2C_2O_4(aq) + H_2O(l) \rightleftharpoons HC_2O_4^-(aq) + H_3O^+(aq)$

(c) $H^+(aq) + HPO_4^{2-}(aq) \rightleftharpoons H_2PO_4^-(aq)$

(d) $HC_2O_4^-(aq) + CO_3^{2-}(aq) \rightleftharpoons C_2O_4^{2-}(aq) + HCO_3^-(aq)$

(e) $PH_4^+(aq) + H_2O(aq) \rightleftharpoons PH_3(aq) + H_3O^+(aq)$

15.9 In each of the following pairs, which is the stronger Brønsted base in water: (a) NH_3 or NH_2^-; (b) CN^- or Br^-; (c) H^- or H^+; (d) OH^- or NH_3?

15.10 Using Figure 15.4, predict whether the following reactions proceed to the right to any appreciable extent ($K > 1$):

(a) $HSO_4^-(aq) + NH_3(aq) \rightleftharpoons SO_4^{2-}(aq) + NH_4^+(aq)$

(b) $H_2PO_4^-(aq) + HCO_3^-(aq) \rightleftharpoons H_3PO_4(aq) + CO_3^{2-}(aq)$

(c) $HNO_3(aq) + HS^-(aq) \rightleftharpoons NO_3^-(aq) + H_2S(aq)$

(d) $HCN(aq) + NO_3^-(aq) \rightleftharpoons CN^-(aq) + HNO_3(aq)$

Autoionization of water: pH

15.11 Indicate whether the following solutions are acidic, basic, or neutral at 25°C: (a) $3.0 \times 10^{-5}\,M$ H^+; (b) $4.0 \times 10^{-10}\,M$ H^+; (c) $1.0 \times 10^{-7}\,M$ OH^-; (d) $8 \times 10^{-2}\,M$ H^+; (e) $5 \times 10^{-3}\,M$ OH^-; (f) $8 \times 10^{-9}\,M$ OH^-; (g) $6 \times 10^{-7}\,M$ OH^-.

15.12 Calculate the pH corresponding to each of the following concentrations of H^+ or OH^-: (a) $3.0 \times 10^{-5}\,M$ H^+; (b) $7.65 \times 10^{-4}\,M$ H^+; (c) $6.2 \times 10^{-12}\,M$ OH^-; (d) $8.3 \times 10^{-4}\,M$ OH^-. Indicate in each case whether the solution is acidic, basic, or neutral.

15.13 If $K_w = 5.47 \times 10^{-14}$ at 50°C, what is the value of $[H^+]$, and what is the pH, for a neutral solution at this temperature?

15.14 Calculate $[H^+]$ for solutions with the following pH values: (a) 3.54; (b) 5.66; (c) 8.80; (d) 11.33; (e) 0.00. Inicate in each case whether the solution is acidic or basic.

15.15 (a) The $[OH^-]$ in a given solution is $4.35 \times 10^{-3}\,M$. What is pOH? What is pH? (b) The pH of a solution is 8.20; what is pOH? What is $[OH^-]$? (c) A substance has an acid dissociation constant of 4.0×10^{-6}. What is pK_a? (d) Refer to Exercise 15.13. What is the value of the sum pH + pOH at 50°C?

15.16 A sample of lake water thought to be affected by acid rain is expected to have a pH in the range of 4.3. Which indicators from among those listed in Table 15.1 would you employ to obtain the best estimate of pH? Explain how you would carry out the experiment.

15.17 By what factor does $[H^+]$ change for a pH change of (a) 1.00 unit; (b) 0.3 unit; (c) 2.00 units.

15.18 A pH meter of the type shown in Figure 15.6(*a*) has a rated uncertainty of ± 0.05 pH unit. A solution is measured to have a pH of 4.65. What are the maximum and minimum values of pH the solution could have, based on the uncertainty in the pH meter reading? What is the percentage precision with which $[H^+]$ is determined?

Strong acids and bases

15.19 Calculate the mass of KOH necessary to form 2.00 L of solution of pH 9.00.

15.20 What is the pH of each of the following solutions: (a) $5.0 \times 10^{-3}\,M$ HBr; (b) $3.00 \times 10^{-4}\,M$ $Ba(OH)_2$; (c) $4.4 \times 10^{-2}\,M$ KOH; (d) $1 \times 10^{-3}\,M$ HCl + 0.1 M HNO_3?

15.21 Calculate the pH of a solution that contains (a) 1.600 g of HNO_3 in 300 mL of solution; (b) 3.00 g of NaOH in 0.600 L of solution; (c) 0.100 g of $Ca(OH)_2$ in 400 mL of solution; (d) 0.340 g of $HClO_4$ in 600 mL of solution.

[15.22] What is the pH of a 1×10^{-9} M solution of HBr?

Weak acids

15.23 Group the following compounds according to whether they produce an acidic or a basic solution: HF, KCN, CaO, NaH, $HClO_4$, $H_2C_2O_4$, KOH, H_3PO_4.

15.24 Using the data of Table 15.3, indicate which acid in each of the following pairs is the stronger in aqueous solution: (a) HNO_2 or $HC_2H_3O_2$; (b) HOC_6H_5 or HCN; (c) HCN or $HC_7H_5O_2$.

15.25 Write the acid-dissociation expression for each of the following acids: (a) chloric acid, $HClO_3$; (b) hydrogen sulfate anion, HSO_4^-.

15.26 Lactic acid, $HC_3H_5O_3$, has one dissociable hydrogen. A 0.10 M solution of lactic acid has a pH of 2.44. Calculate K_a.

15.27 A 0.050 M solution of $KHCrO_4$ has a pH of 3.80. Calculate K_a for $HCrO_4^-$.

15.28 Calculate the concentration of $H^+(aq)$ in each of the following solutions (K_a values are given in Appendix E): (a) 0.10 M hypochlorous acid, HClO; (b) 0.060 M hydroazoic acid, HN_3; (c) 0.070 M phenol, HOC_6H_5; (d) 0.040 M benzoic acid, $HC_7H_5O_2$.

15.29 In jelly making, the pH must be below 4.0. How many grams of benzoic acid, $HC_7H_5O_2$, must be added to attain this pH in 4.20 L of a solution that is to be cooled to form jelly, assuming that the benzoic acid is the only ingredient that determines the pH?

15.30 Ascorbic acid, better known as vitamin C, has the formula $HC_6H_7O_6$ (see Table 15.3). What is the pH of a solution formed by dissolving a 1.00-g tablet of this substance in enough water to form 0.350 L of solution?

15.31 What is the percent ionization of hydroazoic acid (K_a in Appendix E) in solutions of the following concentrations: (a) 0.500 M; (b) 0.100 M; (c) 0.0500 M?

15.32 Show that for a weak acid the percent ionization should vary as the inverse square root of the acid concentration.

15.33 A 0.100 M solution of a weak acid HX is 11.2 percent ionized. Using this information, calculate $[H^+]$, $[X^-]$, [HX], and K_a for HX.

15.34 Citric acid, $H_3C_6H_5O_7$, is present in citrus fruits. It is a triprotic acid (that is, one that can supply three protons per molecule), with acid-dissociation-constant values $K_{a1} = 7.4 \times 10^{-4}$, $K_{a2} = 1.7 \times 10^{-5}$, and $K_{a3} = 4.0 \times 10^{-7}$. Write the balanced equations for the three acid-dissociation equilibria. Calculate the pH of a 0.050 M solution of citric acid. Explain any approximations or assumptions that you make in your calculation.

15.35 Notice from Table 15.4 that the second acid-dissociation constant of oxalic acid is much smaller than the first. Draw the Lewis structures of both oxalic acid, and the anion that results from removal of one proton from the acid. Explain in terms of electrostatic arguments why the second acid dissociation should be smaller than the first.

Weak bases; K_a-K_b relationship

15.36 Write balanced net ionic equations for the reactions of each of the following substances with water. Write the base-dissociation-constant expression for each substance: (a) hydrazine, H_2NNH_2; (b) azide ion, N_3^-; (c) formate ion, CHO_2^-.

15.37 A 0.100 M solution of NH_3 at 50°C has a pH of 10.40; the pK_w for water at 50°C is 13.26. Calculate K_b for ammonia at this temperature.

15.38 Calculate $[OH^-]$ and pH for each of the following solutions: (a) 0.050 M pyridine; (b) 0.010 M hydroxylamine; (c) 3.0×10^{-3} M NH_3.

15.39 Dimethylamine is used in the insecticide Sevin. What is the pH of a 0.18 M solution of dimethylamine if this substance has a pK_b of 3.267?

15.40 Calculate the percentage of methylamine that forms $NH_3CH_3^+$ in a 0.030 M solution of the amine.

15.41 Using values of K_a from Appendix E, calculate the base-dissociation constants for each of the following species: (a) nitrite ion, NO_2^-; (b) azide ion, N_3^-; (c) hydrogen phosphate ion, HPO_4^{2-}; (d) formate ion, CHO_2^-.

15.42 Using the values of K_b from Appendix E, calculate K_a for each of the following species: (a) dimethylammonium ion, $(CH_3)_2NH_2^+$; (b) hydrazinium ion, $H_3NNH_2^+$.

15.43 The pK_a values for H_2SeO_3 at 25°C are 2.64 and 8.27. Write the base-hydrolysis-equilibrium equations for the $SeO_3^{2-}(aq)$ and $HSeO_3^-(aq)$ ions, and calculate the equilibrium-constant values for each equilibrium.

15.44 Using the results from Exercise 15.43, calculate the pH of a 0.010 M solution of K_2SeO_3.

15.45 Calculate the pH of each of the following solutions: (a) 0.020 M potassium azide, KN_3; (b) 0.050 M methylamine solution; (c) 0.100 M sodium oxalate, $Na_2C_2O_4$; (d) 0.050 M calcium acetate, $Ca(C_2H_3O_2)_2$.

15.46 Sodium carbonate, Na_2CO_3, sodium ascorbate, $NaC_6H_7O_6$, and sodium bicarbonate, $NaHCO_3$, are all used in antacid tablets. For equal molar amounts, which of these substances will produce the most basic solution on dissolving in water?

Salt solutions

15.47 Indicate whether each of the following substances would form an acidic, basic, or neutral solution in water: (a) $KC_2H_3O_2$; (b) CH_3NH_3Br; (c) $NaHC_2O_4$; (d) $Sr(NO_3)_2$; (e) $FeCl_3$; (f) KNO_2.

15.48 Arrange the following ions in the order of increasing base strength: N_3^-; NO_3^-; HPO_4^{2-}; H_3O^+.

15.49 Calculate the pH of each of the following solutions: (a) 0.100 M Na_2O; (b) 0.030 M NaOCl; (c) 2.5×10^{-3} M Na_2CO_3.

Acid-base character and chemical structure

15.50 List the following hydrides in the order of increasing acidity in aqueous solution: PH_3; HCl; MgH_2; HI.

15.51 How does the acidity of the hydride of an element vary as a function of the electronegativity of the element?

15.52 Which member of each of the following pairs is the stronger acid: (a) NH_2^- or NH_4^+; (b) H_2SO_4 or H_2SO_3; (c) HIO or HClO; (d) H_2SO_3 or H_2CO_3; (e) H_2Se or HBr?

[15.53] Whereas iodic acid, HIO_3, is a moderately strong acid, $K_a = 1.7 \times 10^{-1}$, periodic acid, H_5IO_6, is weaker, $K_{a1} = 2.8 \times 10^{-2}$, despite the fact that iodine is in a higher oxidation state in the latter compound. Use Lewis structures for these two compounds to account for the lower acidity of periodic acid.

15.54 Draw the Lewis structures for sulfurous acid, H_2SO_3, and sulfuric acid, H_2SO_4. Which is the stronger acid in water? Explain.

[15.55] The Lewis structure for acetic acid is shown in Table 15.3. Replacement of hydrogen atoms on the carbon by chlorine atoms causes an increase in acidity, as follows:

Acid	Formula	K_a(25°C)
Acetic	CH_3COOH	1.8×10^{-5}
Chloroacetic	$CH_2ClCOOH$	1.4×10^{-3}
Dichloroacetic	$CHCl_2COOH$	3.3×10^{-2}
Trichloroacetic	CCl_3COOH	2×10^{-1}

Using Lewis structures as the basis of your discussion, explain the observed trend in acidities in the series. Calculate the pH of a 0.10 *M* solution of each acid.

Lewis acids and bases

15.56 Prepare a table in which you compare the definitions of acids and bases according to the Lewis, Brønsted, and Arrhenius theories. Which is the most general; that is, which includes the others within its scope? Explain.

15.57 Identify the Lewis acid and Lewis base in each of the following reactions:

(a) $HNO_2(aq) + OH^-(aq) \rightleftharpoons NO_2^-(aq) + H_2O(l)$
(b) $FeBr_3(s) + Br^-(aq) \rightleftharpoons FeBr_4^-(aq)$
(c) $Zn^{2+}(aq) + 4NH_3(aq) \rightleftharpoons Zn(NH_3)_4^{2+}(aq)$
(d) $SO_2(g) + H_2O(l) \rightleftharpoons H_2SO_3(aq)$

15.58 Write the acid-dissociation process that occurs for $Fe(NO_3)_3$ dissolved in water. Using the data of Table 15.7, calculate the pH of a 0.010 *M* solution of $Fe(NO_3)_3$.

15.59 Which member of each of the following pair would you expect to produce the more acidic solution: (a) LiI or ZnI_2; (b) $FeCl_3$ or $CaCl_2$; (c) $CoCl_2$ or $CoCl_3$; (d) $Fe(NO_3)_3$ or $Fe(C_2H_3O_2)_3$?

Additional exercises

15.60 Calculate [H⁺] for each of the following solutions: (a) urine, pH 6.1; (b) lemon juice, pH 2.1; (c) gastric juice, pH 1.4; (d) household ammonia, pH 11.9. Indicate in each case whether the solution is acidic, basic, or neutral.

15.61 Arrange the following 0.10 *M* solutions in the order of decreasing pH: KNO_2, $KClO_4$, $HClO_4$, KCN, NH_4Br.

15.62 The decomposition of nitramide in aqueous solution,

$$NH_2NO_2 \longrightarrow N_2O + H_2O$$

proceeds at different rates in the presence of a fixed concentration of various anions. The rate decreases in the order hydroxide > acetate > benzoate > formate. Relate these results to the base strengths of the anions. What can you conclude about the reaction?

15.63 What is the pH of each of the following solutions: (a) 0.030 *M* NaOH; (b) 2.8×10^{-2} *M* $HClO_4$; (c) 0.0400 *M* HIO; (d) 0.150 *M* KCNO; (e) 0.020 *M* CH_3NH_2; (f) 0.0400 *M* CH_3NH_3Cl.

15.64 Calculate the number of moles of each of the following substances that must be present in 200 mL of solution to form a solution with pH 3.25: (a) HCl; (b) $HC_7H_5O_2$; (c) HF.

15.65 Calculate the percent ionization of each of the following substances in the solutions indicated: (a) 0.050 *M* HNO_2; (b) 0.050 *M* CH_3NH_2.

15.66 Liquid ammonia undergoes autoionization analogous to that of water:

$$2NH_3(l) \rightleftharpoons NH_4^+(lq) + NH_2^-(lq)$$

(Here we use *lq* as we would *aq*; it means that the ions are present in the solvent, which in this case is liquid ammonia.) At − 50°C, the ion product for this reaction, $[NH_4^+][NH_2^-]$, is 1.0×10^{-33}. What is the concentration of NH_4^+ in a liquid ammonia solution at −50°C (a) if 2.0×10^{-5} mol of NH_4Cl is dissolved to form 500 mL of solution; (b) if 1.2×10^{-5} mol of KNH_2 is dissolved to form 500 mL of solution?

15.67 The dye bromthymol blue is a weak acid whose ionization can be represented as

$$HBb(aq) \rightleftharpoons H^+(aq) + Bb^-(aq)$$

Which way will this equilibrium shift when NaOH is added? The acid form of the dye is yellow, whereas its conjugate base is blue. What color is the NaOH solution containing this dye?

15.68 Predict the effect that each of the following added substances would have on the pH of an aqueous solution of HNO_2: (a) KNO_2; (b) $HClO_4$; (c) NaCN; (d) KOH.

15.69 Hemoglobin is involved in a series of equilibria involving protonation-deprotonation and oxygenation-deoxygenation. The overall reaction is approximately as follows:

$$HbH^+(aq) + O_2(aq) \rightleftharpoons HbO_2(aq) + H^+(aq)$$

(where Hb stands for hemoglobin and HbO_2 for oxyhemoglobin). (a) $[O_2]$ is higher in the lungs and lower in the tissues. What effect does high $[O_2]$ have on the position of this equilibrium? (b) The normal pH of blood is 7.4. What is [H⁺] in normal blood? Is the blood acidic, basic, or neutral? (c) If the blood pH is lowered by the presence of large amounts of acidic metabolism products, a condition known as acidosis results. What effect does lowering blood pH have on the ability of hemoglobin to transport O_2?

[15.70] Although we think of NH_3 as a base, it is capable of donating a proton when a sufficiently strong base is present. (a) Could such a reaction occur in aqueous solution? Explain. (b) Calculate the pH of a solution obtained by adding 0.30 g of $NaNH_2$ to sufficient water to form 0.400 L of solution.

[15.71] Pure sulfuric acid, H_2SO_4, is a colorless liquid that melts at 10°C and boils at 338°C. This pure substance undergoes autoionization to a much larger extent than does water. (a) Write the equilibrium expression for the autoionization of sulfuric acid, and identify the acid, base, conjugate acid, and conjugate base. (b) Write the Lewis structures for the conjugate acid and conjugate base formed in the autoionization. (c) Write expressions for the reactions you would expect to occur when a small amount of each of the following substances is added to pure sulfuric acid: H_2O; $HClO_4$ (a stronger acid than H_2SO_4); K_2SO_4.

15.72 Saccharin, a sugar substitute, is a weak acid with $pK_a = 11.68$ at 25°C. It ionizes in aqueous solution as follows:

$$HNC_7H_4SO_3(aq) \rightleftharpoons H^+(aq) + NC_7H_4SO_3^-(aq)$$

What is the pH of a 0.10 *M* solution of this substance?

[15.73] What are the concentrations of H^+, HSO_4^-, and SO_4^{2-} in a 0.025 *M* solution of H_2SO_4?

[15.74] What are the concentrations of H^+, $H_2PO_4^-$, HPO_4^{2-}, and PO_4^{3-} in a 0.10 *M* solution of H_3PO_4?

15.75 Ephedrine, a central nervous system stimulant, is used in nasal sprays as a decongestant. This compound is a weak organic base:

$$C_{10}H_{15}ON(aq) + H_2O(l) \rightleftharpoons C_{10}H_{15}ONH^+(aq) + OH^-(aq)$$

K_b has the value 1.4×10^{-4}. What pH would you expect for a 0.035 *M* solution of ephedrine, assuming that no other substances are present? What is the value of pK_a for the conjugate acid, ephedrine hydrochloride?

15.76 Morphine, $C_{17}H_{19}NO_3$, is a weak base containing a nitrogen atom, with $pK_b = 6.1$. (a) What is the pH of a 0.050 *M* solution of morphine? (b) What is the value of pK_a for the conjugate acid, morphine hydrochloride? What is the pH of a 0.050 *M* solution of this substance?

15.77 Codeine, $C_{18}H_{21}NO_3$, is a weak organic base. A 5.0×10^{-3} *M* solution of codeine has a pH of 9.95; calculate the value of K_b for this substance.

[15.78] Many moderately large organic molecules containing basic nitrogen atoms are not very soluble in water as the netural molecule, but are frequently much more soluble as the acid salt. Assuming that the pH in the stomach is 2.5, indicate whether each of the following compounds would be present in the stomach as the neutral base or in the protonated form: nicotine, $K_b = 7 \times 10^{-7}$; caffeine, $K_b = 4 \times 10^{-14}$; strychnine, $K_b = 1 \times 10^{-6}$; quinine, $K_b = 1.1 \times 10^{-6}$.

[15.79] Amino acids contain an amino group, $-NH_2$, located on the carbon atom that also contains a carboxylic acid group, $-COOH$. Glycine, the simplest amino acid, could exist in water in either form I or II below:

$$\underset{\text{I}}{H_2N{-}CH_2{-}\overset{\overset{\displaystyle O}{\|}}{C}{-}OH} \qquad \underset{\text{II}}{{}^+H_3N{-}CH_2{-}\overset{\overset{\displaystyle O}{\|}}{C}{-}O^-}$$

K_a for the carboxylic acid group of glycine is 4.3×10^{-3}, and K_b for the amino group is 6.0×10^{-5}. (a) What is the pH of a 0.10 *M* aqueous solution of glycine? (b) In what forms, other than I and II, can glycine exist in solution, depending on pH? (c) What form of glycine would you expect to be present in a solution with pH 10; with pH 2?

[15.80] A 1.00 *m* solution of HF freezes at −1.90°C. Based on the extent of freezing-point lowering (Section 11.5), calculate the fraction of HF dissociated at this temperature. What is the value of K_a for HF at −1.90°C?

[15.81] Calculate the pH values for 0.100 *M* solutions of each of the following acids (K_a values in Appendix E): phenol, hydrogen chromate, acetic, and formic. Make a graph of pH for these solutions versus pK_a for the corresponding acid. If the relationship is linear for some or all of the data, determine its slope. Derive a general expression that accounts for the result you obtain.

[15.82] In an aqueous solution containing only a weak diprotic acid H_2X, the concentration of X^{2-} is numerically equal to K_{a2}. Show why this is so.

16 Aqueous Equilibria: Further Considerations

Water is the solvent of preeminent importance on our planet. It is also, in a sense, the solvent of life. It is difficult to imagine how living matter in all its complexity could exist with any liquid other than water as solvent. Water occupies its position of importance not only because of its abundance, but also because of its exceptional ability to dissolve a wide variety of substances. Aqueous solutions encountered in nature, such as biological fluids or seawater, contain many solutes. Consequently, many equilibria can take place simultaneously in these solutions. In Chapter 15 we discussed weak acid and base equilibria. However, we restricted our discussions to solutions containing a single solute. In this chapter we shall consider acid-base equilibria in aqueous solutions that contain two or more solutes. Furthermore, we shall broaden our treatment of aqueous equilibria to include other kinds of reactions, especially those involving slightly soluble salts.

16.1 THE COMMON-ION EFFECT

When $NaC_2H_3O_2$ is added to a solution of $HC_2H_3O_2$, the pH of the solution increases ($[H^+]$ is reduced). This result isn't surprising because $C_2H_3O_2^-$ is a weak base; like any other base it should increase the pH. However, it is instructive to view this effect from the perspective of Le Châtelier's principle. Like most salts, $NaC_2H_3O_2$ is a strong electrolyte. Consequently, it completely ionizes in aqueous solution to form Na^+ ions and $C_2H_3O_2^-$ ions. $HC_2H_3O_2$ is a weak electrolyte that dissociates as follows:

$$HC_2H_3O_2(aq) \rightleftharpoons H^+(aq) + C_2H_3O_2^-(aq) \qquad [16.1]$$

The addition of $C_2H_3O_2^-$, from $NaC_2H_3O_2$, causes this equilibrium, Equation 16.1, to shift to the left, thereby decreasing the equilibrium concentration of $H^+(aq)$. The $C_2H_3O_2^-$ is said to decrease or repress the dissociation of $HC_2H_3O_2$. In general, the dissociation of a weak electrolyte is decreased by adding to the solution a strong electrolyte that has an ion in common with the weak electrolyte. This shift in equilibrium position, which occurs when we add an ion that is common to an equilibrium reaction, is called the **common-ion effect.** Sample Exercises 16.1 and 16.2 illustrate how equilibrium concentrations may be calculated when a solution contains a mixture of a weak electrolyte and a strong electrolyte that share a common ion.

SAMPLE EXERCISE 16.1

Suppose that we add 8.20 g, or 0.100 mol, of sodium acetate, $NaC_2H_3O_2$, to 1 L of a 0.100 *M* solution of acetic acid, $HC_2H_3O_2$. What is the pH of the resultant solution?

Solution: Because $NaC_2H_3O_2$ is a strong electrolyte, the only equilibrium that we need to consider is that for dissociation of acetic acid. This equilibrium reaction and the concentrations of the species involved are summarized below:

	$HC_2H_3O_2(aq) \rightleftharpoons$	$H^+(aq)$ +	$C_2H_3O_2^-(aq)$
Initial:	0.100 *M*	0 *M*	0.100 *M*
Equilibrium:	$(0.100 - x)\,M$	$x\,M$	$(0.100 + x)\,M$

Notice that the equilibrium concentration of $C_2H_3O_2^-$ is the initial amount coming from the added $NaC_2H_3O_2$ (0.100 *M*), plus the amount (x) formed by dissociation of acetic acid. The Na^+ ion from $NaC_2H_3O_2$ is merely a spectator ion; it has no influence on pH (Section 15.8).

The equilibrium constant expression is

$$K_a = 1.8 \times 10^{-5} = \frac{[H^+][C_2H_3O_2^-]}{[HC_2H_3O_2]}$$

(The value of the equilibrium constant is taken from Appendix E.) Addition of the acetate salt does not change the value of the equilibrium constant. Substituting the equilibrium concentrations in this equilibrium-constant equation, we obtain

$$\frac{x(0.100 + x)}{0.100 - x} = 1.8 \times 10^{-5}$$

We can simplify this equation by ignoring x relative to 0.100. This simplification gives us

$$\frac{x(0.100)}{0.100} = 1.8 \times 10^{-5}$$

$$x = 1.8 \times 10^{-5}\,M = [H^+]$$

$$pH = -\log(1.8 \times 10^{-5}) = 4.74$$

Earlier (Section 15.5) we calculated that in a 0.10 *M* solution of $HC_2H_3O_2$, $[H^+]$ is $1.3 \times 10^{-3}\,M$, corresponding to a pH value of 2.89.

SAMPLE EXERCISE 16.2

Calculate the fluoride concentration and pH of a solution containing 0.10 mol of HCl and 0.20 mol of HF in a liter of solution.

Solution: This solution consists of a strong electrolyte (HCl) and a weak electrolyte (HF) which share a common ion (H^+). Consequently, the exercise is of the same general sort as Sample Exercise 16.1, except that the common ion is a cation rather than an anion.

Let's define the equilibrium concentration of F^- as x. The equilibrium concentration of $H^+(aq)$ is then the amount supplied by the strong acid, HCl (0.10 *M*), plus the amount formed by the dissociation of HF (x). The equilibrium process involving the weak acid and the concentrations of the species involved in that equilibrium are summarized as follows:

	$HF(aq)$	$\rightleftharpoons$	$H^+(aq)$ +	$F^-(aq)$
Initial:	0.20 *M*		0.10 *M*	0
Equilibrium:	$(0.20 - x)\,M$		$(0.10 + x)\,M$	$x\,M$

The equilibrium constant, from Appendix E, is 6.8×10^{-4}. Thus we have

$$K_a = 6.8 \times 10^{-4} = \frac{(0.10 + x)x}{0.20 - x}$$

If we assume x is small relative to 0.10 or 0.20 M, this expression simplifies to give

$$\frac{(0.10)x}{0.20} = 6.8 \times 10^{-4}$$

$$x = \frac{0.20}{0.10}(6.8 \times 10^{-4})$$

$$= 1.4 \times 10^{-3}\,M = [F^-]$$

The concentration of $H^+(aq)$ is then

$$[H^+] = (0.10 + x)\,M = 0.10\,M$$

Thus pH = 1.00. Notice that $[H^+]$ is for practical purposes due entirely to the HCl; the HF makes a negligible contribution by comparison.

Sample Exercises 16.1 and 16.2 both involve weak acids. However, weak bases could have been used just as easily to illustrate the common-ion effect. For example, a mixture of aqueous NH_3 and NH_4Cl have the NH_4^+ ion in common; a mixture of aqueous NH_3 and NaOH have the OH^- ion in common. In each case the pertinent equilibrium is dissociation of the weak base, NH_3:

$$NH_3(aq) + H_2O(l) \rightleftharpoons NH_4^+(aq) + OH^-(aq) \qquad [16.2]$$

Common Ions Generated by Acid-Base Reactions

Consider a solution formed by mixing a weak acid such as $HC_2H_3O_2$ and a strong base such as NaOH. The net ionic equation for the resultant proton transfer is

$$HC_2H_3O_2(aq) + OH^-(aq) \rightleftharpoons C_2H_3O_2^-(aq) + H_2O(l) \qquad [16.3]$$

The Na^+ ion is a spectator ion in this process (Section 11.4). Because this reaction is just the reverse of the base-dissociation reaction for $C_2H_3O_2^-$, the equilibrium constant is the reciprocal of K_b for $C_2H_3O_2^-$:

$$K = \frac{1}{K_b \text{ for } C_2H_3O_2^-} = \frac{K_a \text{ for } HC_2H_3O_2}{K_w}$$

$$= \frac{1.8 \times 10^{-5}}{1.0 \times 10^{-14}} = 1.8 \times 10^{9}$$

The large equilibrium constant indicates that the reaction proceeds to a large extent toward products. Reactions between strong acids and strong bases, strong acids and weak bases, and weak acids and strong bases (as in the present example) proceed virtually to completion.

If the number of moles of $HC_2H_3O_2$ exceeds the number of moles of NaOH, the reaction between them produces $C_2H_3O_2^-$ ions, leaving the excess $HC_2H_3O_2$ unreacted. For example, consider a 1-L solution that initially contains 0.20 mol of $HC_2H_3O_2$ and 0.10 mol of NaOH. The reaction (Equation 16.3) consumes 0.10 mol of $HC_2H_3O_2$, producing 0.10 mol of $C_2H_3O_2^-$ and leaving 0.10 mol of $HC_2H_3O_2$ unreacted. The resultant solution, which is 0.10 M in $C_2H_3O_2^-$ and 0.10 M in $HC_2H_3O_2$, is identical to a solution produced by mixing 0.10 mol of $NaC_2H_3O_2$ and

0.10 mol of $HC_2H_3O_2$ to form a liter of solution. As shown in Sample Exercise 16.1, such a solution has a pH of 4.74.

SAMPLE EXERCISE 16.3

Calculate the pH of a solution produced by mixing 0.60 L of 0.10 *M* NH_4Cl with 0.40 L of 0.10 *M* NaOH.

Solution: NH_4Cl is a strong electrolyte supplying NH_4^+ and Cl^- ions. NaOH is a strong electrolyte supplying Na^+ and OH^- ions. The molar quantities of these ions present in the solution before any reaction occurs are

$$\text{Moles } NH_4^+ = \text{moles } Cl^- = M_{NH_4Cl} \times V_{NH_4Cl}$$
$$= \left(0.10\ \frac{\text{mol}}{\text{L}}\right)(0.60\ \text{L}) = 0.060\ \text{mol}$$
$$\text{Moles } Na^+ = \text{moles } OH^- = M_{NaOH} \times V_{NaOH}$$
$$= \left(0.10\ \frac{\text{mol}}{\text{L}}\right)(0.40\ \text{L}) = 0.040\ \text{mol}$$

The weakly acidic NH_4^+ and the basic OH^- react as follows:

$$NH_4^+(aq) + OH^-(aq) \rightleftharpoons NH_3(aq) + H_2O(l) \qquad [16.4]$$

The Na^+ ion and Cl^- ion are not acidic or basic and are merely spectator ions. The reaction shown in Equation 16.4, involving a weak acid and a strong base, is virtually complete. Thus the 0.040 mol of OH^- consumes 0.040 mol of NH_4^+, producing 0.040 mol of NH_3. The remaining NH_4^+ is the original amount minus the amount consumed: 0.060 mol − 0.040 mol = 0.020 mol. The total volume of the solution is the sum of the two original solutions: 0.60 L + 0.40 L = 1.00 L. Thus the concentrations of NH_3 and NH_4^+ are

$$[NH_3] = \frac{0.040\ \text{mol}}{1.00\ \text{L}} = 0.040\ M$$
$$[NH_4^+] = \frac{0.020\ \text{mol}}{1.00\ \text{L}} = 0.020\ M$$

The pH of a solution containing NH_4^+ and NH_3 can be calculated by considering the common-ion effect of NH_4^+ on the dissociation equilibrium of the weakly basic NH_3. It can also be calculated by considering the effect of NH_3 on the dissociation equilibrium of the weakly acidic NH_4^+. The same result will be obtained in either case. Using the reverse reaction of that shown in Equation 16.4, the pertinent species are summarized below with *x* defined as the equilibrium concentration of OH^-:

	$NH_3(aq) + H_2O(l) \rightleftharpoons$	$NH_4^+(aq)$	$+\ OH^-(aq)$
Initial:	0.040 *M*	0.020 *M*	0
Equilibrium:	$(0.040 - x)\ M$	$(0.020 + x)\ M$	xM

Substituting the equilibrium concentrations in the equilibrium-constant expression (K_b for NH_3 is taken from Appendix E) and assuming *x* to be small compared with 0.020 and 0.040 gives

$$K_b = 1.8 \times 10^{-5} = \frac{(0.020)x}{0.040}$$
$$x = \frac{0.040}{0.020}(1.8 \times 10^{-5}) = 3.6 \times 10^{-5} = [OH^-]$$
$$\text{pOH} = -\log(3.6 \times 10^{-5}) = 4.44$$
$$\text{pH} = 14.00 - \text{pOH} = 14.00 - 4.44 = 9.56$$

16.2 BUFFER SOLUTIONS

Many aqueous solutions resist a change in pH upon addition of small amounts of acid or base. Such solutions are called buffer solutions, and are said to be buffered. Human blood, for example, is a complex aqueous medium with a pH buffered at about 7.4. Any significant variation of the pH from this value results in a severe pathological response and, eventually, death. As another example, the chemical behavior of seawater is determined in very important respects by its pH, buffered at about 8.1 to 8.3 near the surface. Addition of a small amount of an acid or base to either blood or seawater does not result in a large change in pH. Compare, for example, the behavior of a liter of seawater and a liter of pure water upon addition of 0.1 mL of 1 *M* HCl solution (1×10^{-4} mol of HCl), as shown in Figure 16.1. The pH of pure water changes by 3 units, from 7 to 4. The pH of seawater changes by only 0.6 pH unit, from 8.2 to 7.6. Substances already dissolved in seawater limit the change in $[H^+]$ on addition of HCl.

Buffers ordinarily require two species, an acidic one to react with added OH^- and a basic one to react with added H^+. It is, of course, necessary that these acidic and basic species not consume each other through a neutralization reaction. These requirements are fulfilled by an acid-base conjugate pair such as $HC_2H_3O_2$–$C_2H_3O_2^-$ or NH_4^+–NH_3. The $HC_2H_3O_2$–$C_2H_3O_2^-$ buffer mixture can be prepared by adding sodium acetate, $NaC_2H_3O_2$, to a solution of acetic acid, $HC_2H_3O_2$. The NH_4^+–NH_3 buffer mixture can be prepared by adding ammonium chloride, NH_4Cl, to a solution of ammonia, NH_3. In general, a buffer mixture consists of an aqueous solution of an acid-base conjugate pair prepared by mixing a weak acid or base with a salt of that acid or base.

To understand how a buffer works, let's consider a solution of $HC_2H_3O_2$ and $NaC_2H_3O_2$. Ionization of $HC_2H_3O_2$ is governed by the following equilibrium reaction:

$$HC_2H_3O_2(aq) \rightleftharpoons H^+(aq) + C_2H_3O_2^-(aq) \qquad [16.5]$$

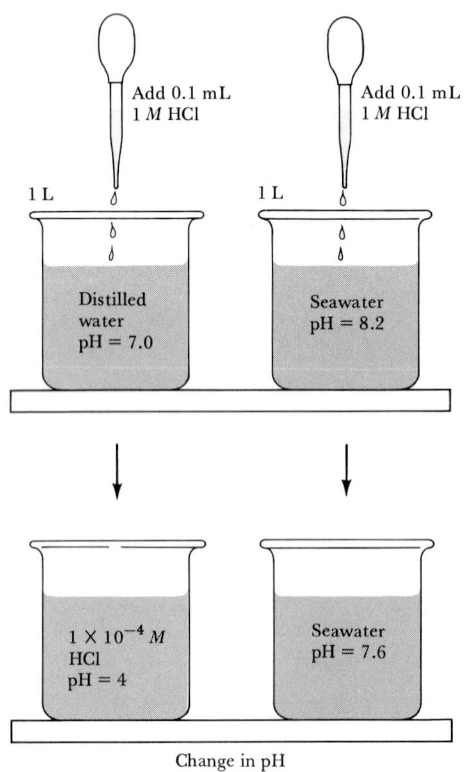

FIGURE 16.1 Comparison of the effect of added acid on the pH of distilled water as compared with the same quantity of seawater.

The $C_2H_3O_2^-$ in this equilibrium comes from both $HC_2H_3O_2$ and $NaC_2H_3O_2$. This mixture can either react with surplus H^+ ions or release them, according to the circumstances. For example, if a small quantity of acid is added to the solution, the equilibrium shifts to the left; acetate ion reacts with the added H^+. The solution thereby limits pH change due to added acid. On the other hand, if a small quantity of a base is added, it reacts with H^+. This reaction causes the equilibrium of Equation 16.5 to shift to the right; $HC_2H_3O_2$ dissociates to form more H^+. The solution thereby also resists change in pH due to added base.

Two important characteristics of a buffer are buffering capacity and pH. Buffering capacity is the amount of acid or base the buffer can neutralize before the pH begins to change to an appreciable degree. This capacity depends on the amount of acid and base from which the buffer is made. The pH of the buffer depends on K_a for the acid and on the relative concentrations of the acid and base that comprise the buffer.

For a better understanding of these characteristics, let's consider a buffer mixture consisting of a weak acid HX and a corresponding salt MX, where M could be Na^+, K^+, and so forth. The acid dissociation equilibrium is

$$HX(aq) \rightleftharpoons H^+(aq) + X^-(aq) \qquad [16.6]$$

and the corresponding acid dissociation constant expression is

$$K_a = \frac{[H^+][X^-]}{[HX]} \qquad [16.7]$$

Let us now take the logarithm in base 10 of both sides of the equation:

$$\log K_a = \log [H^+] + \log \frac{[X^-]}{[HX]} \qquad [16.8]$$

Multiplying through on both sides by -1, we have

$$-\log K_a = -\log [H^+] - \log \frac{[X^-]}{[HX]} \qquad [16.9]$$

Because $-\log K_a = pK_a$ and $-\log [H^+] = pH$, we have

$$pK_a = pH - \log \frac{[X^-]}{[HX]} \qquad [16.10]$$

$$pH = pK_a + \log \frac{[X^-]}{[HX]} \qquad [16.11]$$

In general,

$$pH = pK_a + \log \frac{[\text{base}]}{[\text{acid}]} \qquad [16.12]$$

This relationship, which is called the **Henderson-Hasselbalch equation,** is very useful in dealing with buffers. Notice that $pH = pK_a$ when the concentration of the acid and its conjugate base are equal. The pH range of most buffers is limited to the vicinity of the pK_a of the acid. For this reason, one usually tries to select a buffer whose acid has a pK_a close to the desired pH.

Although the pH of a particular buffer mixture depends on the relative concentrations of the acid and its conjugate base, its buffering capacity depends on the amounts of the acid and base. A 1-L solution that is 1 M in $HC_2H_3O_2$ and 1 M in $NaC_2H_3O_2$ will have the same pH as a 1-L solution that is 0.1 M in $HC_2H_3O_2$ and 0.1 M in $NaC_2H_3O_2$. However, the first solution has a greater buffering capacity because it contains more $HC_2H_3O_2$ and $C_2H_3O_2^-$.

SAMPLE EXERCISE 16.4

What is the pH of a buffer mixture composed of equal concentrations of NH_4Cl and NH_3?

Solution: Using Equation 16.12, we have

$$pH = pK_a + \log\frac{[NH_3]}{[NH_4^+]}$$

Because $[NH_3] = [NH_4^+]$,

$$\frac{[NH_3]}{[NH_4^+]} = 1 \quad \text{and} \quad \log\frac{[NH_3]}{[NH_4^+]} = 0$$

Thus

$$pH = pK_a + 0$$

The pK_a for NH_4^+ is related to K_b for NH_3 through the relationship $K_a \times K_b = K_w$ (Section 15.7). From Appendix E we have $K_b = 1.8 \times 10^{-5}$.

$$K_a = \frac{K_w}{K_b} = \frac{1.0 \times 10^{-14}}{1.8 \times 10^{-5}} = 5.6 \times 10^{-10}$$

Thus

$$pH = pK_a = -\log(5.6 \times 10^{-10}) = 9.25$$

SAMPLE EXERCISE 16.5

(a) How is the pH of an aqueous solution of NH_3 affected by the addition of NH_4Cl? (b) What concentration of NH_4Cl must be added to a 0.10 M solution of NH_3 to adjust the pH to 9.00?

Solution: (a) NH_4Cl is a strong electrolyte supplying NH_4^+ and Cl^- ions. NH_3 is a weak base that reacts with water as follows:

$$NH_3(aq) + H_2O(l) \rightleftharpoons NH_4^+(aq) + OH^-(aq)$$

The presence of NH_4^+, which is a weak acid, shifts the ionization equilibrium to the left, lowering $[OH^-]$ and thereby lowering pH. We have here simply another example of the common-ion effect; aqueous solutions of NH_3 and NH_4Cl have the NH_4^+ ion in common.

(b) A pH of 9.00 corresponds to $[H^+] = 1.0 \times 10^{-9}$. As shown in Sample Exercise 16.4, the pK_a for NH_4^+ is 9.25. The NH_4^+ and NH_3 are a conjugate acid-base pair. Thus we have

$$pH = pK_a + \log\frac{[\text{base}]}{[\text{acid}]}$$

$$9.00 = 9.25 + \log\frac{[NH_3]}{[NH_4^+]}$$

$$\log\frac{[NH_3]}{[NH_4^+]} = 9.00 - 9.25 = -0.25$$

$$\frac{[NH_3]}{[NH_4^+]} = 10^{-0.25} = 0.56$$

To calculate NH_4^+ from this ratio, we can assume that the concentration of NH_3 is 0.10 M. From the examples worked out in Chapter 15 (see, for example, Sample Exercise 15.11), we know that only a small fraction of the NH_3 reacts with water; by Le Châtelier's principle, addition of NH_4^+ to the solution will reduce the extent of reaction even more. Thus all the NH_4^+ present can be considered to be that which is added. Therefore, we have

$$[NH_4^+] = \frac{[NH_3]}{0.56} = \frac{0.10\,M}{0.56} = 0.18\,M$$

Addition of Acids or Bases to Buffers

Let us now consider in a more quantitative way the response of a buffer solution to addition of an acid or base.

SAMPLE EXERCISE 16.6

A liter of solution containing 0.100 mol of $HC_2H_3O_2$ and 0.100 mol of $NaC_2H_3O_2$ is being prepared to provide a buffer of pH 4.74. Calculate the pH of this solution (a) after 0.020 mol of NaOH is added (neglect any volume changes); (b) after 0.020 mol of HCl is added (again, neglect volume changes).

Solution: (a) The OH^- provided by the strong base NaOH reacts nearly completely with the weak acid $HC_2H_3O_2$:

$$HC_2H_3O_2(aq) + OH^-(aq) \longrightarrow H_2O(l) + C_2H_3O_2^-(aq)$$

The 0.020 mol of added OH^- therefore reacts with 0.020 mol of $HC_2H_3O_2$, producing 0.020 mol of $C_2H_3O_2^-$. Thus the final solution contains 0.080 mol of $HC_2H_3O_2$ (the original 0.100 mol minus that which reacts, 0.020 mol), and 0.120 mol of $C_2H_3O_2^-$ (the original 0.100 mol plus that formed in the reaction, 0.020 mol). In summary,

	$HC_2H_3O_2(aq)$ +	$OH^-(aq)$	$\longrightarrow H_2O(l) + C_2H_3O_2^-(aq)$
Initial:	0.100 *M*	0.020 *M*	0.100 *M*
Change:	−0.020	−0.020	+0.020
After rxn:	0.080 *M*	0.0 *M*	0.120 *M*

Using Equation 16.12, we have

$$\text{pH} = \text{p}K_a + \log\frac{[C_2H_3O_2^-]}{[HC_2H_3O_2]}$$

$$= 4.74 + \log\left(\frac{0.120}{0.080}\right) = 4.74 + 0.18 = 4.92$$

(b) The H^+ provided by the added HCl reacts essentially completely with the weak base $C_2H_3O_2^-$, to form $HC_2H_3O_2$:

$$C_2H_3O_2^-(aq) + H^+(aq) \longrightarrow HC_2H_3O_2(aq)$$

The 0.020 mol of H^+ therefore reacts with 0.020 mol of $C_2H_3O_2^-$ to form 0.020 mol of $HC_2H_3O_2$. The final solution then contains 0.120 mol of $HC_2H_3O_2$ and 0.080 mol of $C_2H_3O_2^-$. In summary,

	$C_2H_3O_2^-(aq)$ +	$H^+(aq)$	$\longrightarrow HC_2H_3O_2(aq)$
Initial:	0.100 *M*	0.020 *M*	0.100 *M*
Change:	−0.020	−0.020	+0.020
After rxn:	0.080 *M*	0.0 *M*	0.120 *M*

Using Equation 16.12, we obtain

$$\text{pH} = \text{p}K_a + \log\frac{[C_2H_3O_2^-]}{[HC_2H_3O_2]}$$

$$= 4.74 + \log\left(\frac{0.080}{0.120}\right) = 4.74 - 0.18 = 4.56$$

The pH decreases, as expected for addition of a strong acid to a solution. However, the magnitude of the decrease is small because the solution is a buffer.

To appreciate more fully the buffer action of the solution of acetic acid and sodium acetate that we've just considered, let's compare its behavior with the action of a solution that is not a buffer. We saw in Sample

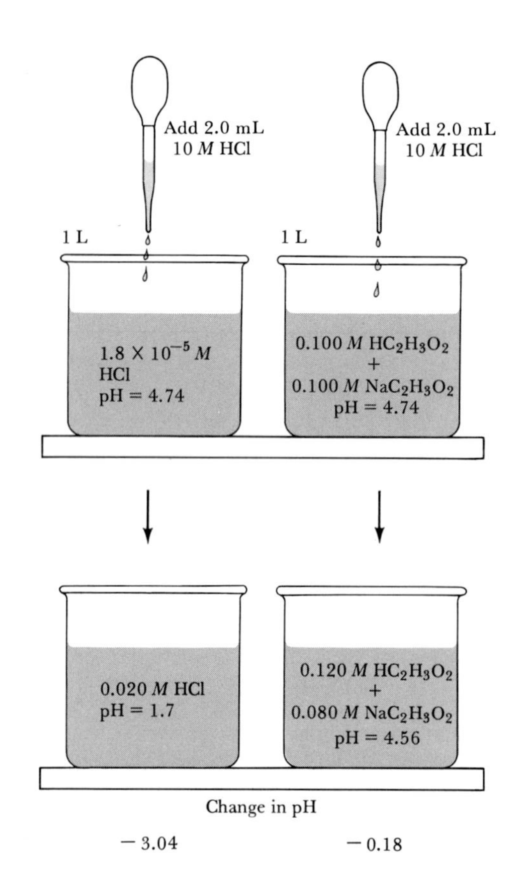

FIGURE 16.2 Comparison of the effect of added acid on a buffer solution of pH 4.74 compared with an HCl solution of pH 4.74.

Exercise 16.1 that the pH of a solution that is 0.100 *M* in acetic acid and 0.100 *M* in sodium acetate is 4.74. A solution of this same pH is obtained by addition of 1.8×10^{-5} mol of HCl to a liter of water. Because HCl is a strong acid, $1.8 \times 10^{-5}\,M$ HCl has a $H^+(aq)$ concentration of $1.8 \times 10^{-5}\,M$ and thus a pH of 4.74. Now suppose that 2.0 mL of 10 *M* HCl solution, that is, 0.020 mol of HCl, is added to a liter of each of these two solutions of pH 4.74.

Figure 16.2 illustrates the effect of the added HCl. As we've seen in Sample Exercise 16.6(b), the added HCl reacts with $C_2H_3O_2^-$ to form $HC_2H_3O_2$. From the calculations carried out there we found that the pH of the buffer solution decreased by 0.18 pH unit. Note, however, that addition of the 2.0 mL of HCl to a liter of HCl solution that has a pH of 4.74 causes the pH to drop to 1.7. In this case, the added HCl provides a much larger total amount of acid than is present in the very dilute HCl solution.

We might have chosen to add 0.02 mol of hydroxide to the solutions illustrated in Figure 16.2. This addition would have caused the pH of the dilute HCl solution to go from 4.74 to 12.3 (you should be able to explain why this is so), whereas the pH of the buffer solution would have increased by only 0.18 pH unit. Thus a buffer solution responds to approximately the same degree, but in opposite direction, to acid or base addition.

Blood is an important example of a buffered solution. Human blood is slightly basic with a pH of about 7.39 to 7.45. In a healthy person the pH never departs more than perhaps 0.2 pH unit from the average value. Whenever pH falls below about 7.4, the condition is called *acidosis;* when pH rises above 7.4, the condition is called *alkalosis.* Death may result if the pH falls below 6.8 or rises above 7.8. Acidosis is the more common tendency, because ordinary metabolism produces several acids.

The body uses three primary methods to control blood pH: (1) The blood contains several buffers, including H_2CO_3–HCO_3^- and $H_2PO_4^-$–HPO_4^{2-} pairs, and hemoglobin-containing conjugate acid-base pairs. (2) The kidneys serve to absorb or release $H^+(aq)$. The pH of urine is normally about 5.0 to 7.0. Acidosis is accompanied by increased loss of body fluids as the kidneys work to reduce $H^+(aq)$. (3) The concentration of $H^+(aq)$ is also altered by the rate at which CO_2 is removed from the lungs. The pertinent equilibria are

$$H^+(aq) + HCO_3^-(aq) \rightleftharpoons H_2CO_3(aq) \rightleftharpoons H_2O(l) + CO_2(g)$$

Removal of CO_2 shifts these equilibria to the right, thereby reducing $H^+(aq)$.

Acidosis or alkalosis disrupts the mechanism by which hemoglobin transports oxygen in blood. Hemoglobin (Hb) is involved in a series of equilibria whose overall result is approximately

$$HbH^+(aq) + O_2(aq) \rightleftharpoons HbO_2(aq) + H^+(aq)$$

In acidosis, this equilibrium is shifted to the left and the ability of hemoglobin to form oxyhemoglobin, HbO_2, is decreased. The lesser amount of O_2 thereby available to cells in the body causes fatigue and headaches; if great enough, it also triggers air hunger (the feeling of being "out of breath" that causes deep breathing).

Temporary acidosis occurs during strenuous exercise, when energy demands exceed the oxygen available for complete oxidation of glucose to CO_2. In this case the glucose is converted to an acidic metabolism product, lactic acid, $CH_3CHOHCOOH$. Acidosis also occurs when glucose is unavailable to the cells. This situation can arise, for example, during starvation or as a result of diabetes. In the case of diabetes, glucose is unable to enter the cells because of inadequate insulin, the substance responsible for passage of glucose from the bloodstream to the interior of cells. When glucose is unavailable, the body relies for energy on stored fats, which produce acidic metabolism products.

16.3 TITRATION CURVES

Many acid-base reactions are used in chemical analyses. For example, the carbonate content in a sample can be determined by a procedure that involves titration with a strong acid such as HCl. Titrations were described briefly in Chapter 3 (Section 3.11). In that earlier discussion we noted that acid-base indicators can be used to signal the equivalence point (that is, the point at which stoichiometrically equivalent quantities of acid and base have been brought together). But given the variety of indicators, changing colors at different pH values, which indicator is best for a particular titration? This question can be answered by examining a graph of pH changes during a titration. A graph of pH as a function of the volume of added titrant is called a **titration curve.**

Strong Acid-Strong Base

The titration curve produced when a strong base is added to a strong acid has the general shape shown in Figure 16.3. This curve depicts the pH change that occurs as 0.100 *M* NaOH is added to 50.0 mL of 0.100 *M* HCl. The pH can be calculated at various stages in the course of the titration. The pH starts out low; pH = 1.00 for 0.100 *M* HCl. As NaOH is added, the pH increases slowly at first and then rapidly in the vicinity of the equivalence point. The pH at the equivalence point in any acid-base titration is the pH of the resultant salt solution. In this example,

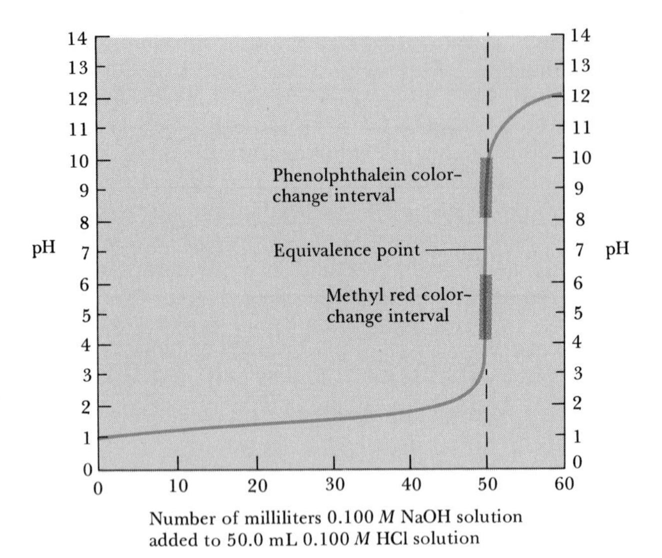

FIGURE 16.3 pH curve for titration of a solution of a strong acid with a solution of a strong base, in this case HCl and NaOH.

NaCl is formed, which does not hydrolyze (Section 15.8); the equivalence point therefore occurs at pH 7. Because the pH change is very large near the equivalence point, the indicator for the titration need not change color precisely at 7.0. Most strong acid-strong base titrations are carried out using phenolphthalein as an indicator because its color change is dramatic. From Table 15.1 we see that this indicator changes color in the pH range 8.3 to 10. Thus a slight excess of NaOH must be present to cause the observed color change. However, it requires such a tiny excess of base to make the color change occur that no serious error is introduced. Similarly, methyl red, which changes color in the slightly acid range, could also be used. The pH intervals of color change for these two indicators are shown in Figure 16.3.

Titration of a solution of a strong base with a solution of a strong acid would yield an entirely analogous curve of pH versus added acid. In this case, however, the pH would be high at the outset of the titration, and low at its completion.

SAMPLE EXERCISE 16.7

Calculate the pH when the following quantities of 0.100 M NaOH solution have been added to 50.00 mL of 0.100 M HCl solution: (a) 49.00 mL; (b) 49.90 mL; (c) 50.10 mL; (d) 51.00 mL.

Solution: (a) The number of moles of OH^- in 49.00 mL of 0.100 M NaOH is

$$0.04900 \text{ L soln} \left(\frac{0.100 \text{ mol OH}^-}{1 \text{ L soln}}\right) = 4.90 \times 10^{-3} \text{ mol OH}^-$$

The number of moles of H^+ in the original solution is 5.00×10^{-3}. Thus there remains

$$(5.00 \times 10^{-3}) - (4.90 \times 10^{-3}) = 1.0 \times 10^{-4} \text{ mol H}^+(aq)$$

in 0.0990 L of solution. The concentration of $H^+(aq)$ is thus

$$\frac{1.0 \times 10^{-4} \text{ mol}}{0.0990 \text{ L soln}} = 1.0 \times 10^{-3}\, M$$

The corresponding pH is 3.0.

To answer parts (b), (c), and (d) of the exercise we proceed in the same way. In each case, we calculate the amount of added OH^-, compare that with the amount of H^+ originally present, and determine what amount of H^+ or OH^- exists in excess over the other. The concentration is then calculated from a knowledge of the total volume of the solution. By proceeding in this way we can construct a table of pH versus volume of added NaOH solution, as shown in Table 16.1. By graphing the pH as a function of the volume of added NaOH, we obtain the titration curve shown in Figure 16.3.

TABLE 16.1 Titration of 50.00 mL of 0.100 *M* HCl solution with 0.100 *M* NaOH solution

Volume of HCl	Volume of NaOH	Total volume	Moles H^+	Moles OH^-	Molarity of excess ion	pH
50.00	0.00	50.00	5.00×10^{-3}	0.00	0.100	1.00
50.00	49.00	99.00	5.00×10^{-3}	4.90×10^{-3}	$1.00 \times 10^{-3}(H^+)$	3.00
50.00	49.90	99.90	5.00×10^{-3}	4.99×10^{-3}	$1.00 \times 10^{-4}(H^+)$	4.00
50.00	50.10	100.10	5.00×10^{-3}	5.01×10^{-3}	$1.00 \times 10^{-4}(OH^-)$	10.00
50.00	51.00	101.00	5.00×10^{-3}	5.10×10^{-3}	$1.00 \times 10^{-3}(OH^-)$	11.00
50.00	60.00	110.00	5.00×10^{-3}	6.00×10^{-3}	$9.09 \times 10^{-3}(OH^-)$	11.96

Titrations Involving a Weak Acid or Base

Titration of a weak acid by a strong base results in pH curves that look similar to those for strong acid-strong base titrations. Figure 16.4 shows the pH curve produced when 0.100 *M* NaOH is added to 50.0 mL of 0.100 *M* acetic acid. There are three noteworthy differences between this curve and that for a strong acid-strong base titration: (1) The weak acid will have a higher initial pH. The pH of 0.100 *M* $HC_2H_3O_2$ is 2.89 (see Section 15.5, where $[H^+]$ is calculated; $[H^+] = 1.3 \times 10^{-3}\,M$). (2) The pH will rise more rapidly in the early part of the titration, but more slowly near the equivalence point. The weaker the acid, the less marked the pH change near the equivalence point (see Figure 16.5). (3) The pH

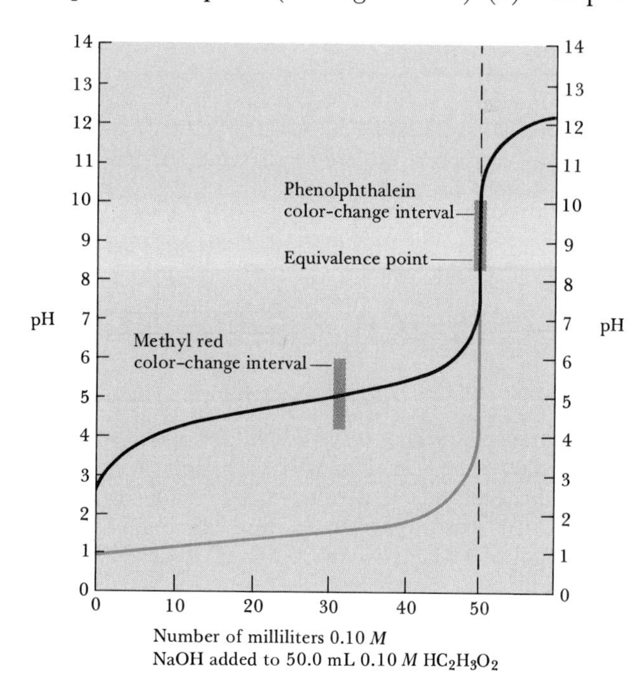

FIGURE 16.4 The black line shows the graph of pH versus added 0.10 *M* NaOH solution in the titration of 0.10 *M* acetic acid solution. The colored line segment shows the graph of pH versus added base for the titration of 0.10 *M* HCl.

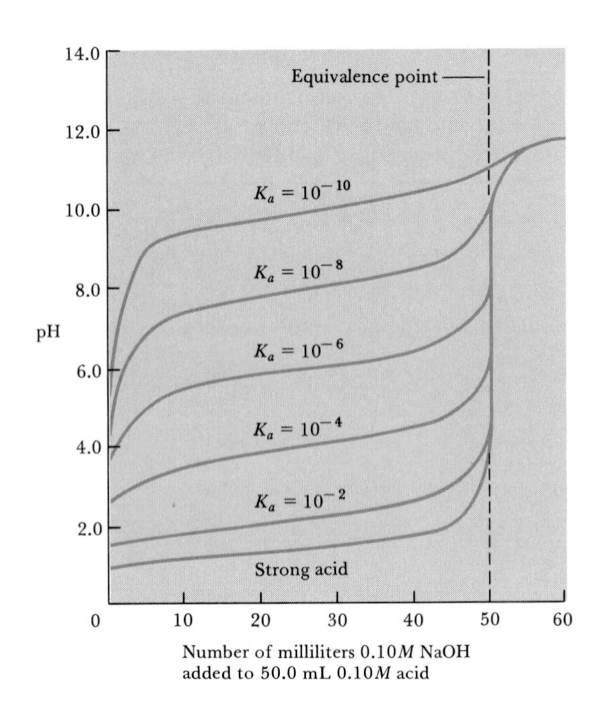

FIGURE 16.5 Influence of acid strength upon titration curves. Each curve represents titration of 50.0 mL of 0.10 *M* acid with 0.10 *M* NaOH.

at the equivalence point is *not* 7. In the present example the solution contains 0.050 *M* $NaC_2H_3O_2$ at the equivalence point (remember that the total volume of solution is doubled by addition of the NaOH solution to the $HC_2H_3O_2$ solution). The pH of this solution is 8.71 (this might be a good time to review how the pH of a $NaC_2H_3O_2$ solution can be calculated; an example of a similar calculation is given in Section 15.6, Sample Exercise 15.13).

SAMPLE EXERCISE 16.8

Calculate the pH in the titration of acetic acid by sodium hydroxide after 30.0 mL of 0.100 *M* NaOH solution has been added to 50.0 mL of 0.100 *M* acetic acid solution.

Solution: The total number of moles of $HC_2H_3O_2$ originally in solution is

$$\left(\frac{0.100 \text{ mol } HC_2H_3O_2}{1 \text{ L soln}}\right)(0.0500 \text{ L soln})$$
$$= 5.00 \times 10^{-3} \text{ mol } HC_2H_3O_2$$

Similarly, 30.0 mL of 0.100 *M* NaOH solution contains 3.00×10^{-3} mol of OH^-. During the titration, this OH^- reacts with the acetic acid, forming 3.00×10^{-3} mol of $C_2H_3O_2^-$ and leaving 2.00×10^{-3} mol of $HC_2H_3O_2$, in 80 mL of solution:

$$OH^-(aq) + HC_2H_3O_2(aq) \longrightarrow H_2O(l) + C_2H_3O_2^-(aq)$$

The resulting molarities are thus

$$[HC_2H_3O_2] = \frac{2.00 \times 10^{-3} \text{ mol } HC_2H_3O_2}{0.0800 \text{ L}}$$
$$= 0.0250\ M$$

$$[C_2H_3O_2^-] = \frac{3.00 \times 10^{-3} \text{ mol } C_2H_3O_2^-}{0.0800 \text{ L}}$$
$$= 0.0375\ M$$

The value for pK_a of acetic acid is $-\log(1.8 \times 10^{-5}) = 4.74$. Inserting these quantities into Equation 16.12, we find

$$pH = 4.74 + \log\left(\frac{0.0375}{0.0250}\right)$$
$$= 4.91$$

By proceeding in similar fashion, other points on the titration curve shown in Figure 16.4 could be calculated. Incidentally, you might note that at the halfway point in the titration, when $[C_2H_3O_2^-]$ equals $[HC_2H_3O_2]$, the pH equals pK_a, which in this case is 4.74.

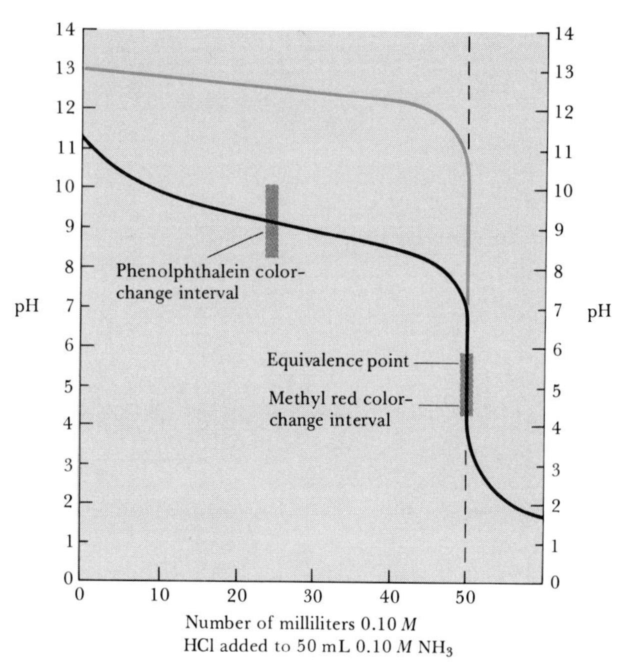

FIGURE 16.6 The black line shows the graph of pH versus added 0.10 M HCl solution in the titration of 0.10 M ammonia. The colored line segment shows the graph of pH versus added acid for the titration of 0.10 M NaOH.

The fact that the pH at the equivalence point in the titration of the weak acid is considerably higher than it is in the titration of the strong acid is important. We saw earlier that in titrating 0.10 M HCl with 0.10 M NaOH, either phenolphthalein or methyl red could be used as indicator. Although the pH of color change does not correspond precisely to the equivalence point for either indicator, both were close enough that no significant error would be introduced by using either of them (Figure 16.3). In a titration of acetic acid with NaOH, phenolphthalein is an ideal indicator, because it changes color just at the pH of the equivalence point. However, methyl red is not a good choice. The pH at the midrange of its color change is 5.2. At this pH we are still far short of the equivalence point, as shown in Figure 16.4.

If the acid to be titrated were weaker than acetic acid, the titration curve would look similar to that shown in Figure 16.4. However, the pH at the beginning of the reaction would be higher, and the pH at the equivalence point would also be higher. This means that the pH change at the equivalence point would not be as sharp, and the end point would be more difficult to detect (see Figure 16.5). With very weak acids, for which K_a is less than about 1×10^{-8}, an acid-base titration is not really a good quantitative procedure when using indicators.

Titration of a weak base, for example, 0.10 M NH_3, with a strong acid such as 0.10 M HCl solution leads to the titration curve shown in Figure 16.6. In this particular example, the equivalence point occurs at pH 5.3. In this case methyl red would be an ideal indicator, but phenolphthalein would be a poor choice.

Titration of Polyprotic Acids

In the case of weak acids containing more than one ionizable proton, reaction with OH^- occurs in a series of steps. An important example of the kind of equilibria that occur is carbonic acid, H_2CO_3. We saw in

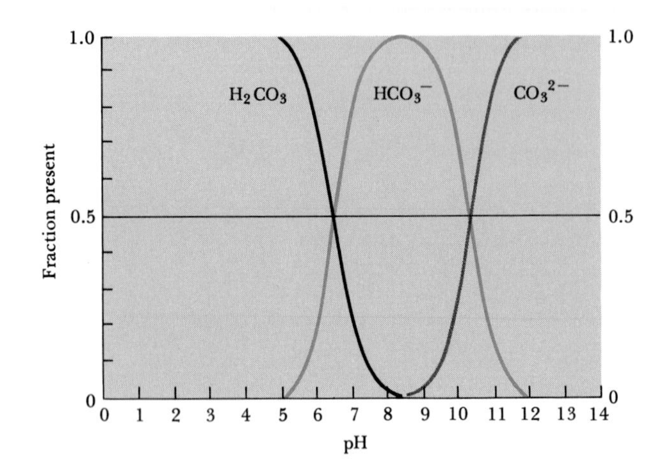

FIGURE 16.7 Relative abundances of the various species derived from dissolved CO_2 as a function of pH.

Sample Exercise 15.11 that in an aqueous solution of CO_2, the relative concentrations of the species involved are $[H_2CO_3] \gg [HCO_3^-] \gg [CO_3^{2-}]$. Neutralization of H_2CO_3 proceeds in two stages:

$$H_2CO_3(aq) + OH^-(aq) \rightleftharpoons H_2O(l) + HCO_3^-(aq) \quad [16.13]$$

$$HCO_3^-(aq) + OH^-(aq) \rightleftharpoons H_2O(l) + CO_3^{2-}(aq) \quad [16.14]$$

Figure 16.7 shows how the relative abundances of H_2CO_3, HCO_3^-, and CO_3^{2-} vary as a function of the solution pH. As OH^- is added to H_2CO_3 (as pH increases), $[H_2CO_3]$ decreases while $[HCO_3^-]$ increases. By pH 8.5 H_2CO_3 has been nearly completely converted to HCO_3^-. At this stage Figure 16.7 shows $[H_2CO_3] = 0$, and $[CO_3^{2-}] = 0$; the concentrations of these species are too small to register on the figure. As pH increases beyond 8.5, $[HCO_3^-]$ decreases while $[CO_3^{2-}]$ increases. Thus it is only after the first reaction goes nearly to completion that the second reaction starts.

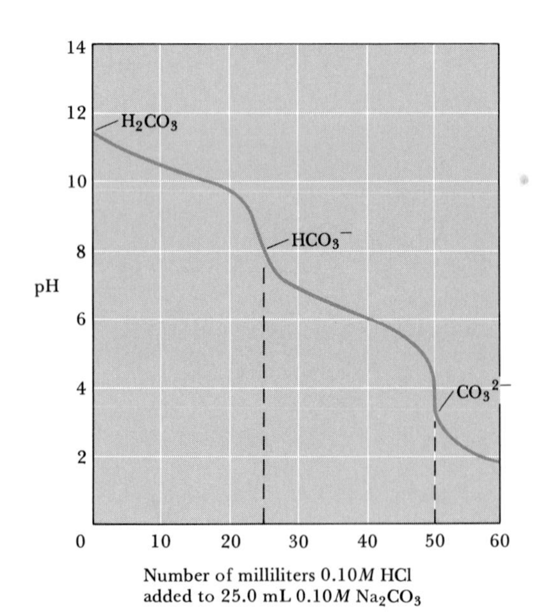

FIGURE 16.8 Titration curve for the titration of 25.0 mL of 0.10 M Na_2CO_3 with 0.10 M HCl.

When the neutralization steps of a polyprotic acid or polybasic base are sufficiently separated, the substance exhibits a titration curve with multiple equivalence points. Figure 16.8 shows the titration curve for the H_2CO_3–HCO_3^-–CO_3^{2-} system. Notice that there are two distinct equivalence points along the titration curve.

SAMPLE EXERCISE 16.9

In the titration of 50.0 mL of a 0.020 *M* solution of $NaHCO_3$ with 0.020 *M* HCl solution, what is the pH when 25.0 mL of acid solution has been added?

Solution: The number of moles of added HCl is

$$\left(\frac{0.020 \text{ mol HCl}}{1 \text{ L soln}}\right)(0.0250 \text{ L soln}) = 5.0 \times 10^{-4} \text{ mol HCl}$$

The total number of moles of HCO_3^- present originally is

$$\left(\frac{0.020 \text{ mol } HCO_3^-}{1 \text{ L soln}}\right)(0.0500 \text{ L soln}) = 1.0 \times 10^{-3} \text{ mol } HCO_3^-$$

H^+ reacts with HCO_3^- to form H_2CO_3:

$$H^+(aq) + HCO_3^-(aq) \rightleftharpoons H_2CO_3(aq)$$

After the addition of HCl solution we have $(1.0 \times 10^{-3}) - (5.0 \times 10^{-4}) = 5.0 \times 10^{-4}$ mol of H_2CO_3 and 5.0×10^{-4} mol of HCO_3^- in 75.0 mL of solution. Thus each of these species has a concentration of 6.67×10^{-3} *M*. In terms of Figure 16.7 we are at the crossing point for the H_2CO_3 and HCO_3^- curves. Because the solution contains a H_2CO_3–HCO_3^- conjugate acid-base pair, it is simplest to employ the acid dissociation equilibrium for H_2CO_3 to calculate pH:

$$H_2CO_3(aq) \rightleftharpoons HCO_3^-(aq) + H^+(aq)$$

Using Equation 16.12, we obtain

$$\text{pH} = \text{p}K_a + \log\frac{[HCO_3^-]}{[H_2CO_3]}$$

$$= -\log[4.3 \times 10^{-7}] + \log\left(\frac{6.67 \times 10^{-3}}{6.67 \times 10^{-3}}\right)$$

$$= 6.37$$

16.4 SOLUBILITY EQUILIBRIA

The equilibria that we have considered thus far in this chapter have involved acids and bases. Furthermore, they have been homogeneous; that is, all the species involved in an equilibrium have been in the same phase. In this section we will consider the equilibria involved in another important type of solution reaction, the dissolution or precipitation of slightly soluble salts. Such reactions are heterogeneous.

For an equilibrium to exist between a solid substance and its solution, the solution must be saturated and in contact with undissolved solid. As an example, consider a saturated solution of $BaSO_4$ that is in contact with solid $BaSO_4$. The chemical equation for the relevant equilibrium can be written as

$$BaSO_4(s) \rightleftharpoons Ba^{2+}(aq) + SO_4^{2-}(aq) \qquad [16.15]$$

The solid is an ionic compound (as are others that we will discuss in this section). Such compounds are almost invariably strong electrolytes; to the extent that they dissolve, these compounds are present in solutions as ions. We observed in Section 14.3 that for heterogeneous reactions the concentration of the solid is constant. It is thus incorporated into the equilibrium constant. Thus the equilibrium-constant expression for the dissolution of $BaSO_4$ can be written as

$$K = [Ba^{2+}][SO_4^{2-}]$$

When concentration is expressed in molarity, the equilibrium constant is called the **solubility-product constant** and designated as K_{sp}:

$$K_{sp} = [Ba^{2+}][SO_4^{2-}] \quad [16.16]$$

The rules for writing the solubility-product expression are the same as those for the writing of any equilibrium-constant expression: *The solubility product is equal to the product of the concentrations of the ions involved in the equilibrium, each raised to the power of its coefficient in the equilibrium equation.*

SAMPLE EXERCISE 16.10

Write the expression for the solubility-product constant for $Ca_3(PO_4)_2$.

Solution: We first write the equation for the solubility equilibrium:

$$Ca_3(PO_4)_2(s) \rightleftharpoons 3Ca^{2+}(aq) + 2PO_4^{3-}(aq)$$

Following the rule stated above, the power to which the Ca^{2+} concentration is raised is 3, the power to which the PO_4^{3-} concentration is raised is 2. The resulting expression for K_{sp} is

$$K_{sp} = [Ca^{2+}]^3[PO_4^{3-}]^2$$

It is important to distinguish carefully between solubility and solubility product. The solubility is the quantity of substance that dissolves in a given quantity of water. It is often expressed as grams of solute per 100 g of water, or in terms of molarity. The solubility of a substance can be changed by adding some other substance to the solution. For example, the addition of Na_2SO_4 lowers the solubility of $BaSO_4$; the SO_4^{2-} supplied by the Na_2SO_4 shifts the solubility equilibrium, Equation 16.15, to the left. This behavior is another example of the common-ion effect. However, the presence of Na_2SO_4 does not affect K_{sp} for $BaSO_4$.* K_{sp} is a true equilibrium constant; its value at a particular temperature is the same, regardless of whether other substances are present in solution. Nevertheless, K_{sp} and solubility are related and one can be calculated from the other.

*This is strictly true only for very dilute solutions. The values of equilibrium constants are somewhat altered when the total concentration of ionic substances in water is increased. However, we shall ignore these effects, which are taken into consideration only for very accurate work.

SAMPLE EXERCISE 16.11

Solid barium sulfate is shaken in contact with pure water at 25°C for several days. Each day a sample is withdrawn and analyzed for its barium concentration. After several days, the value of $[Ba^{2+}]$ is constant, indicating that equilibrium has been reached. The concentration of Ba^{2+} is $1.04 \times 10^{-5}\,M$. What is K_{sp} for $BaSO_4$?

Solution: The analysis provides values for both $[Ba^{2+}]$ and $[SO_4^{2-}]$. Because all of the Ba^{2+} and SO_4^{2-} ions come from $BaSO_4$, there must be one Ba^{2+} for each SO_4^{2-}. Consequently, $[Ba^{2+}] = [SO_4^{2-}] = 1.04 \times 10^{-5}\,M$. Thus we have

$$\begin{aligned} K_{sp} &= [Ba^{2+}][SO_4^{2-}] \\ &= (1.04 \times 10^{-5})(1.04 \times 10^{-5}) \\ &= 1.08 \times 10^{-10} \end{aligned}$$

SAMPLE EXERCISE 16.12

The K_{sp} for CaF_2 is 3.9×10^{-11}. What is the solubility of CaF_2 in water in grams per liter?

Solution: The solubility equilibrium involved is

$$CaF_2(s) \rightleftharpoons Ca^{2+}(aq) + 2F^-(aq)$$

For each mole of CaF_2 that dissolves, 1 mol of Ca^{2+} and 2 mol of F^- enter the solution. Thus if we let x equal the solubility of CaF_2 in moles per liter, the molar concentrations of Ca^{2+} and F^- will be $[Ca^{2+}] = x$ and $[F^-] = 2x$.

The K_{sp} expression for this reaction is

$$K_{sp} = [Ca^{2+}][F^-]^2 = 3.9 \times 10^{-11}$$

Substituting $[Ca^{2+}] = x$ and $[F^-] = 2x$ and solving for x, we have

$$x(2x)^2 = 3.9 \times 10^{-11}$$
$$4x^3 = 3.9 \times 10^{-11}$$
$$x = 2.1 \times 10^{-4}\ M$$

Thus the molar solubility of CaF_2 is 2.1×10^{-4} mol/L. The mass of CaF_2 that dissolves in a liter of solution is

$$\left(\frac{2.1 \times 10^{-4} \text{ mol } CaF_2}{1 \text{ L soln}}\right)\left(\frac{78.1 \text{ g } CaF_2}{1 \text{ mol } CaF_2}\right) = 1.6 \times 10^{-2} \text{ g } CaF_2/\text{L soln}$$

Appendix E contains K_{sp} values for a variety of salts at 25°C. From these equilibrium constants we can calculate solubilities under a variety of conditions. If a single solute is dissolved in water, the relative concentrations of the ions are determined by the formula of the solute. In the case of CaF_2, the concentration of F^- must be twice the concentration of Ca^{2+}.

$$CaF_2(s) \rightleftharpoons Ca^{2+}(aq) + 2F^-(aq) \qquad [16.17]$$

However, it is possible to alter the relative concentrations of Ca^{2+} and F^- by adding a soluble salt containing one of these ions. For example, we might add either $Ca(NO_3)_2$ or NaF. Either will lower the solubility of the CaF_2. If $Ca(NO_3)_2$ is added, the equilibrium concentration of Ca^{2+} is the sum of the Ca^{2+} concentrations from the two sources. This is another example of the common-ion effect introduced in Section 16.1.

SAMPLE EXERCISE 16.13

What is the molar solubility of CaF_2 in a solution containing 0.010 M NaF?

Solution: As in Sample Exercise 16.12, we have

$$K_{sp} = [Ca^{2+}][F^-]^2 = 3.9 \times 10^{-11}$$

The value of K_{sp} is unchanged by the fact that the solution initially contains 0.010 M NaF. In Sample Exercise 16.12, the relative concentrations of Ca^{2+} and F^- were determined entirely by the solubility of CaF_2. In the present exercise we must take account of the fact that there is a second source of F^-. Let's again let the molar solubility of CaF_2 be x. Then the concentration of Ca^{2+} is x and the concentration of F^- derived from CaF_2 is $2x$. But there is in addition a 0.010 M contribution to the F^- concentration from the dissolved NaF. We thus have

$$[Ca^{2+}] = x \qquad [F^-] = 0.010 + 2x$$
$$K_{sp} = [Ca^{2+}][F^-]^2 = (x)(0.010 + 2x)^2 = 3.9 \times 10^{-11}$$

This would be a messy problem to solve exactly, but fortunately it is possible to greatly simplify matters. Even without the common-ion effect of 0.010 M F^-, the solubility of CaF_2 in water is not very great, as illustrated by the small value of x obtained in Sample Exercise 16.12. We know from application of Le Châtelier's principle that the solubility of CaF_2 will be even smaller in the presence of 0.010 M NaF. We can therefore safely assume that the 0.010 M F^- concentration from the NaF is much greater than the small additional contribution resulting from the solubility of CaF_2. That is, we can neglect $2x$ in comparison with 0.010. We then have

$$3.9 \times 10^{-11} = x(0.010)^2$$
$$x = 3.9 \times 10^{-7}\,M = [Ca^{2+}]$$

This value for x represents the solubility of CaF_2 in a solution that is 0.010 M in NaF. Note that it is much smaller than $2.1 \times 10^{-4}\,M$, the solubility of CaF_2 in pure water.

16.5 CRITERIA FOR PRECIPITATION OR DISSOLUTION

Equilibrium can be achieved starting with the substances on either side of the chemical equation. The equilibrium between $BaSO_4(s)$, $Ba^{2+}(aq)$, and $SO_4^{2-}(aq)$ (Equation 16.15) can be achieved starting with solid $BaSO_4$. It can also be achieved starting with solutions of salts containing Ba^{2+} and SO_4^{2-}, say $BaCl_2$ and Na_2SO_4. When these two solutions are mixed, a precipitate of $BaSO_4$ will form if the product of ion concentrations $Q = [Ba^{2+}][SO_4^{2-}]$ is greater than K_{sp}.*

The possible relationships between Q and K_{sp} are summarized as follows:

if $Q > K_{sp}$ precipitation occurs until $Q = K_{sp}$

if $Q = K_{sp}$ equilibrium exists (saturated solution)

if $Q < K_{sp}$ solid dissolves until $Q = K_{sp}$

*The use of the reaction quotient, Q, to determine the direction a reaction must proceed to reach equilibrium was discussed earlier, in Section 14.4. In the present case the equilibrium expression contains no denominator and therefore is really not a quotient. Thus Q is often referred to simply as the ion product.

SAMPLE EXERCISE 16.14

Will a precipitate form when 0.100 L of $3.0 \times 10^{-3}\,M$ $Pb(NO_3)_2$ is added to 0.400 L of $5.0 \times 10^{-3}\,M$ Na_2SO_4?

Solution: The possible reaction products are $PbSO_4$ and $NaNO_3$. Sodium salts are quite soluble; however, $PbSO_4$ has a K_{sp} of 1.6×10^{-8} (Appendix E). To determine whether the $PbSO_4$ precipitates we must calculate $Q = [Pb^{2+}][SO_4^{2-}]$, and compare it with K_{sp}.

When the two solutions are mixed, the total volume becomes $0.100\text{ L} + 0.400\,L = 0.500\text{ L}$. The number of moles of Pb^{2+} in 0.100 L of $3.0 \times 10^{-3}\,M$ $Pb(NO_3)_2$ is

$$(0.100\text{ L})\left(3.0 \times 10^{-3}\,\frac{\text{mol}}{\text{L}}\right) = 3.0 \times 10^{-4}\text{ mol}$$

The concentration of Pb^{2+} in the 0.500-L mixture is therefore

$$[Pb^{2+}] = \frac{3.0 \times 10^{-4}\text{ mol}}{0.500\text{ L}} = 6.0 \times 10^{-4}\,M$$

The number of moles of SO_4^{2-} is

$$(0.400\text{ L})\left(5.0 \times 10^{-3}\,\frac{\text{mol}}{\text{L}}\right) = 2.0 \times 10^{-3}\text{ mol}$$

Therefore, $[SO_4^{2-}]$ in the 0.500-L mixture is

$$[SO_4^{2-}] = \frac{2.0 \times 10^{-3}\text{ mol}}{0.500\text{ L}} = 4.0 \times 10^{-3}\,M$$

We then have

$$Q = [Pb^{2+}][SO_4^{2-}] = (6.0 \times 10^{-4})(4.0 \times 10^{-3}) = 2.4 \times 10^{-6}$$

Because $Q > K_{sp}$, precipitation of $PbSO_4$ will occur.

SAMPLE EXERCISE 16.15

What concentration of OH^- must be exceeded in a 0.010 *M* solution of $Ni(NO_3)_2$ in order to precipitate $Ni(OH)_2$? (Assume that the added OH^- does not change the concentration of Ni^{2+}.)

Solution: From Appendix E we have $K_{sp} = 1.6 \times 10^{-14}$ for $Ni(OH)_2$. Any OH^- in excess of that in a saturated solution of $Ni(OH)_2$ will cause some $Ni(OH)_2$ to precipitate. For a saturated solution we have

$$K_{sp} = [Ni^{2+}][OH^-]^2 = 1.6 \times 10^{-14}$$

Thus if $Q = [Ni^{2+}][OH^-]^2 > 1.6 \times 10^{-14}$ precipitation will occur. Letting $[OH^-] = x$ and using $[Ni^{2+}] = 0.010\ M$, we have

$$(0.010)x^2 > 1.6 \times 10^{-14}$$

$$x^2 > \frac{1.6 \times 10^{-14}}{0.010} = 1.6 \times 10^{-12}$$

$$x > \sqrt{1.6 \times 10^{-12}} = 1.3 \times 10^{-6}\ M$$

This concentration of OH^- corresponds to a solution pH of 8.11 (pH = 14.00 − pOH = 14.00 − 5.89 = 8.11). Thus $Ni(OH)_2$ will precipitate when the solution pH is 8.11 or higher.

Solubility and pH

The solubility of any substance whose anion is basic will be affected to some extent by the pH of the solution. For example, consider $Mg(OH)_2$, for which the solubility equilibrium is

$$Mg(OH)_2(s) \rightleftharpoons Mg^{2+}(aq) + 2OH^-(aq) \qquad [16.18]$$

The value of K_{sp} for $Mg(OH)_2$ is 1.8×10^{-11}. Suppose that solid $Mg(OH)_2$ is equilibrated with a solution buffered at a pH of 9.0. Then pOH is 5.0, that is, $[OH^-] = 1.0 \times 10^{-5}$. Inserting this value for $[OH^-]$ into the solubility-product expression, we have

$$K_{sp} = [Mg^{2+}][OH^-]^2 = 1.8 \times 10^{-11}$$

$$[Mg^{2+}][1.0 \times 10^{-5}]^2 = 1.8 \times 10^{-11}$$

$$[Mg^{2+}] = 0.18\ M$$

Thus $Mg(OH)_2$ is quite soluble in a buffered, slightly basic medium. If the solution were made more acidic, the solubility of $Mg(OH)_2$ would increase, because the OH^- concentration decreases with increasing acidity. The Mg^{2+} concentration would thus increase to maintain the equilibrium condition.

The solubility of almost any salt is affected if the solution is made sufficiently acidic or basic. The effects are very noticeable, however, only when one or both ions involved is moderately acidic or basic. The metal hydroxides we've just discussed are good examples of compounds involving a strong base, the hydroxide ion. As an additional example, the fluoride ion of CaF_2 is a weak base; it is the conjugate base of the weak acid HF. As a result, CaF_2 is more soluble in acidic solutions than in neutral or basic ones, because of the reaction of F^- with H^+ to form HF. The solution process can be considered as two consecutive reactions:

$$CaF_2(s) \rightleftharpoons Ca^{2+}(aq) + 2F^-(aq) \qquad [16.19]$$

$$F^-(aq) + H^+(aq) \rightleftharpoons HF(aq) \qquad [16.20]$$

The equation for the overall process is

$$CaF_2(s) + 2H^+(aq) \rightleftharpoons Ca^{2+}(aq) + 2HF(aq) \qquad [16.21]$$

Qualitatively, we can understand what occurs in terms of Le Châtelier's principle: The solubility equilibrium is driven to the right because the free F^- concentration is reduced by reaction with H^+. The reduction of $[F^-]$ causes Q to be reduced so that it becomes smaller than K_{sp}. Thus more CaF_2 dissolves.

These examples illustrate a general rule: *The solubility of slightly soluble salts containing basic anions increases as* $[H^+]$ *increases (as pH is lowered).* Salts with anions of negligible basicity (the anions of strong acids) are largely unaffected by pH.

SAMPLE EXERCISE 16.16

Which of the following substances will be more soluble in acidic solution than in basic solution? (a) $Ni(OH)_2(s)$; (b) $CaCO_3(s)$; (c) $BaSO_4(s)$; (d) $AgCl(s)$.

Solution: (a) $Ni(OH)_2(s)$ will be more soluble in acidic solution, because of reaction of H^+ with the OH^- ion, forming water:

$$Ni(OH)_2(s) \rightleftharpoons Ni^{2+}(aq) + 2OH^-(aq)$$
$$2OH^-(aq) + 2H^+(aq) \rightleftharpoons 2H_2O(l)$$

$$\text{Overall: } Ni(OH)_2(s) + 2H^+(aq) \rightleftharpoons Ni^{2+}(aq) + 2H_2O(l)$$

(b) Similarly, $CaCO_3(s)$ reacts with acid, liberating gaseous CO_2:

$$CaCO_3(s) \rightleftharpoons Ca^{2+}(aq) + CO_3^{2-}(aq)$$
$$CO_3^{2-}(aq) + 2H^+(aq) \rightleftharpoons H_2CO_3(aq) \longrightarrow CO_2(g) + H_2O(l)$$

$$\text{Overall: } CaCO_3(s) + 2H^+(aq) \longrightarrow Ca^{2+}(aq) + CO_2(g) + H_2O(l)$$

(c) The solubility of $BaSO_4$ is largely unaffected by changes in solution pH, because SO_4^{2-} is a rather weak base and thus has little tendency to combine with a proton. However, $BaSO_4$ is slightly more soluble in strongly acidic solutions.

(d) The solubility of AgCl is unaffected by changes in pH, because Cl^- is the anion of a strong acid.

Precipitations of Metal Sulfides

Ions can be separated from each other on the basis of their solubilities. One widely used procedure for separating metal ions is based on the relative solubilities of their sulfides. Metals that form insoluble metal salts will not precipitate unless $[S^{2-}]$ is sufficiently high for Q to exceed K_{sp}. The sulfide concentration can be adjusted by regulating the pH of the solution. For example, CuS can be precipitated from a mixture of Cu^{2+} and Zn^{2+} by bubbling H_2S gas into a properly acidified solution of these ions. CuS ($K_{sp} = 6.3 \times 10^{-36}$) is less soluble than ZnS ($K_{sp} = 1.1 \times 10^{-21}$). Consequently, $[S^{2-}]$ can be regulated to cause CuS to precipitate while not exceeding K_{sp} for ZnS.

When H_2S is bubbled through an aqueous solution at 25°C and 1 atm, a saturated solution of H_2S forms that is approximately 0.1 *M* in H_2S. H_2S is a weak diprotic acid:

$$H_2S(aq) \rightleftharpoons H^+(aq) + HS^-(aq) \qquad K_{a1} = 5.7 \times 10^{-8} \qquad [16.22]$$

$$HS^-(aq) \rightleftharpoons H^+(aq) + S^{2-}(aq) \qquad K_{a2} = 1.3 \times 10^{-13} \qquad [16.23]$$

These equations can be combined to give an equation for the overall dissociation of H_2S into S^{2-}:

$$H_2S(aq) \rightleftharpoons 2H^+(aq) + S^{2-}(aq) \qquad [16.24]$$

The equilibrium-constant expression for this overall dissociation process is

$$K = \frac{[H^+]^2[S^{2-}]}{[H_2S]} = K_{a1} \times K_{a2} = 7.4 \times 10^{-21} \qquad [16.25]$$

Substituting the solubility of H_2S, 0.1 M, into this expression gives

$$\frac{[H^+]^2[S^{2-}]}{(0.1)} = 7.4 \times 10^{-21}$$
$$[H^+]^2[S^{2-}] = (0.1)(7.4 \times 10^{-21})$$
$$= 7 \times 10^{-22} \qquad [16.26]$$

Equation 16.26 can be used to calculate the concentration of S^{2-} in saturated solutions of H_2S at various pH values.

Now consider how we can calculate the pH that will allow us to prevent precipitation of ZnS while precipitating CuS from a solution that is 0.10 M in Zn^{2+}, 0.10 M in Cu^{2+}, and saturated with H_2S. The maximum concentration of S^{2-} that can be present in the solution before precipitation of ZnS occurs can be calculated from the solubility-product expression for this substance:

$$[Zn^{2+}][S^{2-}] = K_{sp} = 1.1 \times 10^{-21}$$
$$[S^{2-}] = \frac{1.1 \times 10^{-21}}{[Zn^{2+}]} = \frac{1.1 \times 10^{-21}}{0.10}$$
$$= 1.1 \times 10^{-20}\,M$$

The concentration of hydrogen ions necessary to give $[S^{2-}] = 1.1 \times 10^{-20}\,M$ can be calculated using Equation 16.26:

$$[H^+]^2[S^{2-}] = 7 \times 10^{-22}$$
$$[H^+]^2 = \frac{7 \times 10^{-22}}{[S^{2-}]} = \frac{7 \times 10^{-22}}{1.1 \times 10^{-20}} = 6 \times 10^{-2}$$
$$[H^+] = \sqrt{6 \times 10^{-2}} = 0.24\,M$$
$$\text{pH} = -\log(2.4 \times 10^{-1}) = 0.6$$

Thus ZnS will not precipitate if the pH is 0.6 or lower. However, at pH 0.6, where $[S^{2-}] = 1.1 \times 10^{-20}$, the ion product for a 0.10 M Cu^{2+} solution would be

$$Q = [Cu^{2+}][S^{2-}] = (0.10)(1.1 \times 10^{-20}) = 1.1 \times 10^{-21}$$

Because Q exceeds the K_{sp} of CuS (that is, 6.3×10^{-36}), CuS will precipitate under these conditions. Indeed, $[S^{2-}]$ must be less than 6.3×10^{-35} M to prevent the precipitation of CuS from this solution. Even under strongly acidic conditions $[S^{2-}]$ will be high enough to cause precipitation of CuS. Thus a fairly broad range of sulfide concentration will allow for effective separation of Cu^{2+} (as CuS) from Zn^{2+}.

Effect of Complex Formation on Solubility

It is a characteristic property of metal ions that they are able to act as Lewis acids, or electron-pair acceptors, toward water molecules, which act as Lewis bases, or electron-pair donors (Section 15.10). Lewis bases other than water can also interact with metal ions, particularly with transition metal ions. Such interactions can have a dramatic effect on the solubility of a metal salt. For example, AgCl, whose $K_{sp} = 1.82 \times 10^{-10}$, will dissolve in the presence of aqueous ammonia because of the interaction between Ag^+ and the Lewis base NH_3. This process can be viewed as the sum of two reactions, the solubility equilibrium of AgCl and the Lewis acid-base interaction between Ag^+ and NH_3:

$$AgCl(s) \rightleftharpoons Ag^+(aq) + Cl^-(aq) \quad [16.27]$$

$$Ag^+(aq) + 2NH_3(aq) \rightleftharpoons Ag(NH_3)_2{}^+(aq) \quad [16.28]$$

$$\text{Overall:} \quad AgCl(s) + 2NH_3(aq) \rightleftharpoons Ag(NH_3)_2{}^+(aq) + Cl^-(aq) \quad [16.29]$$

The presence of NH_3 drives the top reaction, the solubility equilibrium of AgCl, to the right as $Ag^+(aq)$ is removed to form $Ag(NH_3)_2{}^+$.

For a Lewis base such as NH_3 to increase the solubility of a metal salt, it must be able to interact more strongly with the metal ion than water does. The NH_3 must displace solvating H_2O molecules (Sections 11.2 and 15.10) in order to form $Ag(NH_3)_2{}^+$:

$$Ag^+(aq) + 2NH_3(aq) \rightleftharpoons Ag(NH_3)_2{}^+(aq) \quad [16.30]$$

An assembly of a metal ion and the Lewis bases bonded to it, such as $Ag(NH_3)_2{}^+$, is called a **complex ion.** The stability of a complex ion in aqueous solution can be judged by the magnitude of the equilibrium constant for the formation of the complex ion from the hydrated metal ion. For example, the equilibrium constant for formation of $Ag(NH_3)_2{}^+$ (Equation 16.30) is 1.7×10^7:

$$K_f = \frac{[Ag(NH_3)_2{}^+]}{[Ag^+][NH_3]^2} = 1.7 \times 10^7 \quad [16.31]$$

Such equilibrium constants are called **formation constants,** K_f. The formation constants for a few complex ions are shown in Table 16.2.

SAMPLE EXERCISE 16.17

Calculate the concentration of Ag^+ present in solution at equilibrium when concentrated ammonia is added to a 0.010 M solution of $AgNO_3$ to give an equilibrium concentration of $[NH_3] = 0.20$ M. Neglect the small volume change that occurs on addition of NH_3.

Solution: Because K_f is quite large we begin with the assumption that essentially all of the Ag^+ is converted to $Ag(NH_3)_2^+$, in accordance with Equation 16.30. Thus $[Ag^+]$ will be small at equilibrium. If $[Ag^+]$ was 0.010 M initially, then $[Ag(NH_3)_2^+]$ will be 0.010 M following addition of the NH_3. Let the concentration of Ag^+ at equilibrium be x. Then

$$\begin{array}{ccc} Ag^+(aq) + 2NH_3(aq) & \rightleftharpoons & Ag(NH_3)_2^+(aq) \\ x\,M \qquad 0.20\,M & & (0.010 - x)\,M \end{array}$$

Because the concentration of Ag^+ is very small, we can ignore x in comparison with 0.010. Thus $0.010 - x \simeq 0.010\,M$. Substituting these values into the equilibrium-constant expression, Equation 16.31, we obtain

$$\frac{[Ag(NH_3)_2^+]}{[Ag^+][NH_3]^2} = \frac{0.010}{x(0.20)^2} = 1.7 \times 10^7$$

Solving for x, we obtain $x = 1.4 \times 10^{-8}\,M = [Ag^+]$. It is evident that formation of the $Ag(NH_3)_2^+$ complex drastically reduces the concentration of free Ag^+ ion in solution.

TABLE 16.2 Formation constants for some metal complex ions in water at 25°C

Complex ion	K_f	Equilibrium equation
$Ag(NH_3)_2^+$	1.7×10^7	$Ag^+(aq) + 2NH_3(aq) \rightleftharpoons Ag(NH_3)_2^+(aq)$
$Ag(CN)_2^-$	1×10^{21}	$Ag^+(aq) + 2CN^-(aq) \rightleftharpoons Ag(CN)_2^-(aq)$
$Ag(S_2O_3)_2^{3-}$	2.9×10^{13}	$Ag^+(aq) + 2S_2O_3^{2-}(aq) \rightleftharpoons Ag(S_2O_3)_2^{3-}(aq)$
$CdBr_4^{2-}$	5×10^3	$Cd^{2+}(aq) + 4Br^-(aq) \rightleftharpoons CdBr_4^{2-}(aq)$
$Cr(OH)_4^-$	8×10^{29}	$Cr^{3+}(aq) + 4OH^-(aq) \rightleftharpoons Cr(OH)_4^-(aq)$
$Co(SCN)_4^{2-}$	1×10^3	$Co^{2+}(aq) + 4SCN^-(aq) \rightleftharpoons Co(SCN)_4^{2-}(aq)$
$Cu(NH_3)_4^{2+}$	5×10^{12}	$Cu^{2+}(aq) + 4NH_3(aq) \rightleftharpoons Cu(NH_3)_4^{2+}(aq)$
$Cu(CN)_4^{2-}$	1×10^{25}	$Cu^{2+}(aq) + 4CN^-(aq) \rightleftharpoons Cu(CN)_4^{2-}(aq)$
$Ni(NH_3)_6^{2+}$	5.5×10^8	$Ni^{2+}(aq) + 6NH_3(aq) \rightleftharpoons Ni(NH_3)_6^{2+}(aq)$
$Fe(CN)_6^{4-}$	1×10^{35}	$Fe^{2+}(aq) + 6CN^-(aq) \rightleftharpoons Fe(CN)_6^{4-}(aq)$
$Fe(CN)_6^{3-}$	1×10^{42}	$Fe^{3+}(aq) + 6CN^-(aq) \rightleftharpoons Fe(CN)_6^{3-}(aq)$

The general rule is that metal salts will dissolve in the presence of a suitable Lewis base such as NH_3 if the metal forms a sufficiently stable complex with the base. The ability of metal ions to form complexes is an extremely important aspect of their chemistry. In Chapter 24 we will take a much closer look at complex ions. In that chapter and others we shall see applications of complex ions to areas such as biochemistry, metallurgy, and photography.

Amphoterism

Many metal hydroxides and oxides that are relatively insoluble in neutral water dissolve in *either* a strongly acidic or a strongly basic medium. Such substances are said to be amphoteric (from the Greek word *amphoteros*, which means both). Examples include the hydroxides and oxides of Al^{3+}, Cr^{3+}, Zn^{2+}, and Sn^{2+}.

The dissolution of these species in acidic solutions should be anticipated based on the earlier discussions in this section. We have seen that acids promote the dissolving of compounds with basic anions. What makes amphoteric oxides and hydroxides special is that they dissolve in strongly basic solutions. This behavior results from the formation of complex anions containing several (typically four) hydroxides bound to the metal ion:

$$Al(OH)_3(s) + OH^-(aq) \rightleftharpoons Al(OH)_4^-(aq) \qquad [16.32]$$

Amphoterism is often interpreted in terms of the behavior of the water molecules that surround the metal ion and that are bonded to it by Lewis acid-base interactions (Section 15.10). For example, $Al^{3+}(aq)$ is more accurately represented as $Al(H_2O)_6^{3+}(aq)$; six water molecules are bonded to the Al^{3+} in aqueous solution. As discussed in Section 15.10, this hydrated ion is a weak acid. As a strong base is added, the $Al(H_2O)_6^{3+}$ loses protons in a stepwise fashion, eventually forming the neutral and water-insoluble $Al(H_2O)_3(OH)_3$. This substance then dissolves upon removal of an additional proton to form the anion $Al(H_2O)_2(OH)_4^-$. The reactions that occur are as follows:

$$Al(H_2O)_6^{3+}(aq) + OH^-(aq) \rightleftharpoons Al(H_2O)_5(OH)^{2+}(aq) + H_2O(l)$$
$$Al(H_2O)_5(OH)^{2+}(aq) + OH^-(aq) \rightleftharpoons Al(H_2O)_4(OH)_2^+(aq) + H_2O(l)$$
$$Al(H_2O)_4(OH)_2^+(aq) + OH^-(aq) \rightleftharpoons Al(H_2O)_3(OH)_3(s) + H_2O(l)$$
$$Al(H_2O)_3(OH)_3(s) + OH^-(aq) \rightleftharpoons Al(H_2O)_2(OH)_4^-(aq) + H_2O(l)$$

Further proton removals are possible, but each successive reaction occurs less readily than the one before. As the charge on the ion becomes more negative, it becomes increasingly difficult to remove a positively charged proton. Addition of an acid reverses these reactions. The proton adds in a stepwise fashion to convert the OH^- groups to H_2O, eventually reforming $Al(H_2O)_6^{3+}$. The common practice is to simplify the equations for these reactions by excluding the bound H_2O molecules. Thus we usually write Al^{3+} instead of $Al(H_2O)_6^{3+}$, $Al(OH)_3$ instead of $Al(H_2O)_3(OH)_3$, $Al(OH)_4^-$ instead of $Al(H_2O)_2(OH)_4^-$, and so forth.

The extent to which an insoluble metal hydroxide reacts with either acid or base varies with the particular metal ion involved. Many metal hydroxides, for example, $Ca(OH)_2$, $Fe(OH)_2$, and $Fe(OH)_3$, are capable of dissolving in acidic solution but do not react with excess base. These hydroxides are not amphoteric.

In Sample Exercise 16.18 we work through an example of the kinds of calculations that are involved in considering such equilibria from a quantitative point of view.

SAMPLE EXERCISE 16.18

The solubility product for zinc hydroxide, $Zn(OH)_2$, is 1.2×10^{-17}. The formation constant for the hydroxo complex, $Zn(OH)_4^{2-}$, is 4.6×10^{17}. What concentration of OH^- is required to dissolve 0.010 mol of $Zn(OH)_2$ in a liter of solution?

Solution: Let us first write the equilibria with which we will be concerned:

$$Zn(OH)_2(s) \rightleftharpoons Zn^{2+}(aq) + 2OH^-(aq)$$
$$Zn^{2+}(aq) + 4OH^-(aq) \rightleftharpoons Zn(OH)_4^{2-}(aq)$$

We know that if 0.010 mol of $Zn(OH)_2$ dissolves in the liter of solution, we will have a total of 0.010 mol of Zn(II) present *in some form.* Because the $Zn(OH)_2$ is considerably more soluble in the OH^- solution than it would be in pure water, we can assume that essentially all of the Zn^{2+} is present as $Zn(OH)_4^{2-}$. Thus $[Zn(OH)_4^{2-}] = 0.010\ M$. Let us now set up the equilibrium-constant expression for complex formation:

$$K_f = \frac{[Zn(OH)_4^{2-}]}{[Zn^{2+}][OH^-]^4} = 4.6 \times 10^{17} \qquad [16.33]$$

As we've already indicated, we can assume that $[Zn(OH)_4^{2-}] = 0.010\ M$. However, it would appear that we still have two unknowns in the expression, $[Zn^{2+}]$ and $[OH^-]$. We can eliminate one of these by making use of the solubility-product expression. No matter how much additional OH^- is added to the solution to dissolve the $Zn(OH)_2$, it is still true that

$$[Zn^{2+}][OH^-]^2 = 1.2 \times 10^{-17}$$

Let's solve this expression for $[Zn^{2+}]$ and substitute into Equation 16.33:

$$[Zn^{2+}] = \frac{1.2 \times 10^{-17}}{[OH^-]^2}$$

$$4.6 \times 10^{-17} = \frac{[Zn(OH)_4^{2-}]}{\left(\dfrac{1.2 \times 10^{-17}}{[OH^-]^2}\right)[OH^-]^4}$$

Substituting in $[Zn(OH)_4^{2-}] = 0.010$ and solving for $[OH^-]$, we obtain

$$[OH^-]^2 = \frac{0.010}{(1.2 \times 10^{-17})(4.6 \times 10^{17})}$$

$$[OH^-] = 0.042\ M$$

That is, at a hydroxide ion concentration of $0.042\ M$ or larger, 0.010 mol of $Zn(OH)_2$ will completely dissolve in a liter of solution, because of formation of the $Zn(OH)_4^{2-}$ complex.

The purification of aluminum ore in the manufacture of aluminum metal provides an interesting application of the property of amphoterism. As we have seen, $Al(OH)_3$ is amphoteric, whereas $Fe(OH)_3$ is not. Aluminum occurs in large quantities as the ore **bauxite**, which is essentially Al_2O_3 with additional water molecules. The ore is contaminated with Fe_2O_3 as an impurity. When bauxite is added to a strongly basic solution, the Al_2O_3 dissolves, because the aluminum forms complex ions such as $Al(OH)_4^-$. The Fe_2O_3 impurity, however, is not amphoteric and remains as a solid. The solution is filtered, getting rid of the iron impurity. Aluminum hydroxide is then precipitated by addition of acid. The purified hydroxide receives further treatment and eventually yields aluminum metal.

When writing reactions of amphoteric substances, we often show them as the oxides rather than the hydrated hydroxides. Although the various ways of showing the reactions are confusing at first, the only real differences are in the number of water molecules shown. Sample Exercise 16.19 shows how we can view the reaction of a metal oxide with base.

SAMPLE EXERCISE 16.19

Write the balanced net-ionic equation for the reaction between $Al_2O_3(s)$ and aqueous solutions of NaOH that causes the $Al_2O_3(s)$ to dissolve. (H_2O is also involved as a reactant.)

Solution: NaOH is a strong electrolyte that provides the necessary OH^- ions. The reaction product can be taken to be $Al(OH)_4^-$, an assembly of Al^{3+} and four OH^- ions. The unbalanced equation is

$$Al_2O_3(s) + OH^-(aq) \rightleftharpoons Al(OH)_4^-(aq)$$

Two Al are needed among the products:

$$Al_2O_3(s) + OH^-(aq) \rightleftharpoons 2Al(OH)_4^-(aq)$$

To balance the charge on both sides of the equation requires $2OH^-$. Sufficient H_2O is then added to balance the O and H counts. The balanced equation is

$$Al_2O_3(s) + 2OH^-(aq) + 3H_2O(l) \rightleftharpoons 2Al(OH)_4^-(aq)$$

16.6 QUALITATIVE ANALYSES FOR METALLIC ELEMENTS

In this chapter we have seen several examples of equilibria involving metal ions in aqueous solution. In this final section we look very briefly at how solubility equilibria and complex formation can be used to detect the presence of particular metal ions in solution. Before the development of modern analytical instrumentation, it was necessary to analyze mixtures of metals in a sample by so-called "wet" chemical methods. For example, a metallic sample that might contain several metallic elements was dissolved in a concentrated acid solution. This solution was then tested in a systematic way for the presence of various metallic ions.

Qualitative analysis involves simply determining the presence or absence of a particular metal ion. It should be distinguished from **quantitative analysis**, which involves determining how much of a given substance is present. Wet methods of qualitative analysis are no longer so important as a means of analysis. However, they are frequently used in freshman chemistry laboratory programs as a means of illustrating equilibria, teaching the properties of common metal ions in solution, and developing laboratory skills. Typically, such analyses proceed in three stages: (1) The ions are separated into broad groups on the basis of solubility properties. (2) The individual ions within each group are then separated by selectively dissolving members in the group. (3) The ions are then identified by means of specific tests.

A scheme in common use divides the common cations into five groups as shown in Figure 16.9. The order of addition of reagents is important.

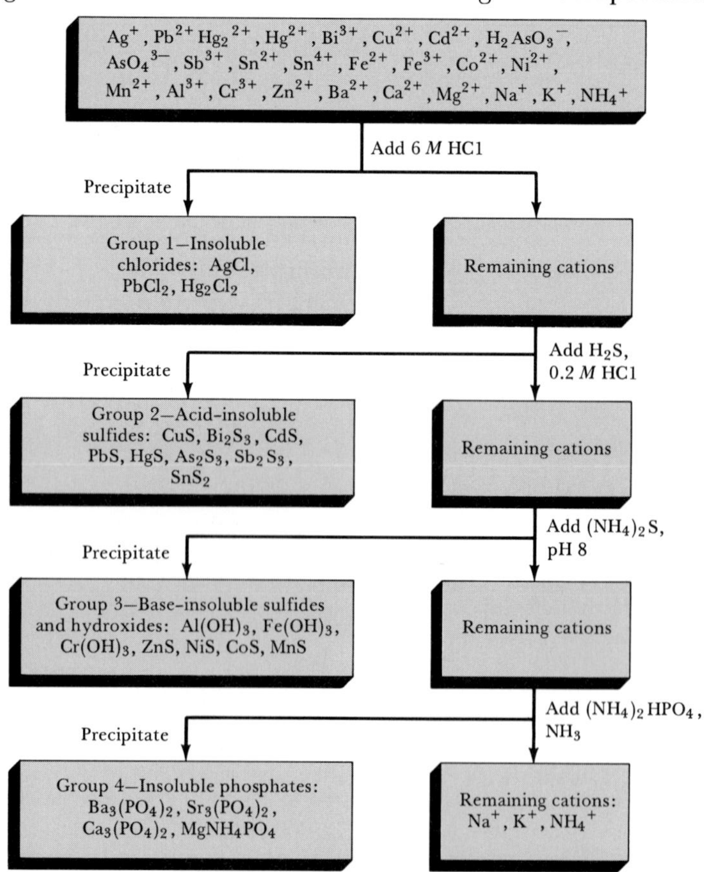

FIGURE 16.9 Qualitative analysis scheme for separating cations into groups.

The most selective separations—that is, those that involve the smallest number of ions—are carried out first. The reactions that are used must proceed so far toward completion that any concentration of cations remaining in the solution is too small to interfere with subsequent tests.

Let's take a closer look at each of these five groups of cations, examining briefly the logic used in this qualitative analysis scheme:

1. *Insoluble chlorides:* Of the common metal ions only Ag^+, Hg_2^{2+}, and Pb^{2+} form insoluble chlorides. Thus when dilute HCl is added to a mixture of cations, only AgCl, Hg_2Cl_2, and $PbCl_2$ will precipitate, leaving the other cations in solution. The absence of a precipitate indicates that the starting solution contains no Ag^+, Hg_2^{2+}, or Pb^{2+}.
2. *Acid-insoluble sulfides:* After any insoluble chlorides have been removed, the remaining, now acidic solution is treated with H_2S. As we saw in Section 16.5, the dissociation of H_2S is repressed in acidic solutions so that the concentration of free S^{2-} is very low. Consequently, only the most insoluble metal sulfides, CuS, Bi_2S_3, CdS, PbS, HgS, As_2S_3, Sb_2S_3, and SnS_2, can precipitate. (Note the very small values of K_{sp} for some of these sulfides in Appendix E.) Those metal ions whose sulfides are somewhat more soluble, for example ZnS or NiS, remain in solution.
3. *Base-insoluble sulfides:* After the solution is filtered to remove any acid-insoluble sulfides, the remaining solution is made slightly basic, and more H_2S is added. In basic solutions the concentration of S^{2-} is higher than in acidic solutions. Thus the ion products for many of the more soluble sulfides are caused to exceed their K_{sp} values and precipitation occurs. The metal ions precipitated at this stage are Al^{3+}, Cr^{3+}, Fe^{3+}, Zn^{2+}, Ni^{2+}, Co^{2+}, and Mn^{2+}. (Actually, the Al^{3+}, Fe^{3+}, and Cr^{3+} ions do not form insoluble sulfides, but instead are precipitated as the insoluble hydroxides at the same time.)
4. *Insoluble phosphates:* At this point the solution contains only metal ions from periodic table groups 1A and 2A. Addition of $(NH_4)_2HPO_4$ to a basic solution causes precipitation of the group 2A elements Mg^{2+}, Ca^{2+}, Sr^{2+}, and Ba^{2+}, because these metals form insoluble phosphates.
5. *The alkali metal ions and* NH_4^+: The ions that remain after removal of the insoluble carbonates form a small group in which each ion can be tested for by an individual test. For example, the flame test is useful to show the presence of K^+, because the flame turns a characteristic violet color if K^+ is present.

Additional separation and testing is necessary to determine which ions are present within each of the groups. As an example, consider the ions of the insoluble chloride group. The precipitate containing the metal chlorides is boiled in water. It happens that $PbCl_2$ is relatively soluble in hot water, whereas AgCl and Hg_2Cl_2 are not. The hot solution is filtered and a solution of Na_2CrO_4 added to the filtrate. If Pb^{2+} is present, a yellow precipitate of $PbCrO_4$ forms. The test for Ag^+ consists of treating the metal chloride precipitate with dilute ammonia. Only Ag^+ forms an ammonia complex. If AgCl is present in the precipitate it will dissolve in the ammonia solution:

$$AgCl(s) + 2NH_3(aq) \rightleftharpoons Ag(NH_3)_2^+(aq) + Cl^-(aq) \qquad [16.34]$$

After treatment with ammonia, the solution is filtered, and the filtrate made acidic by adding nitric acid. The nitric acid removes ammonia from solution by forming NH_4^+, thus releasing Ag^+, which should reform the AgCl precipitate:

$$Ag(NH_3)_2^+(aq) + Cl^-(aq) + 2H^+(aq) \rightleftharpoons AgCl(s) + 2NH_4^+(aq) \qquad [16.35]$$

The analyses for individual ions in the acid-insoluble and base-insoluble sulfides are a bit more complex, but the same general principles are involved. The detailed procedures for carrying out such analyses are given in laboratory manuals.

FOR REVIEW

Summary

In this chapter we've considered several types of important equilibria occurring in aqueous solution. Our primary emphasis has been on acid-base equilibria in solutions containing two or more solutes and on solubility equilibria. We observed that the dissociation of a weak acid or base is repressed by the presence of a strong electrolyte that provides an ion common to the equilibrium. This phenomenon is an example of the **common-ion effect.**

A particularly important type of acid-base mixture is one involving a weak conjugate acid-base pair. Such mixtures function as **buffers.** Addition of small amounts of additional acid or base to a buffered solution causes only small changes in pH, because the buffer reacts with the added acid or base. (Recall that strong acid-strong base, strong acid-weak base, and weak acid-strong base reactions proceed essentially to completion.) Buffer solutions are usually prepared from a weak acid and a salt of that acid or from a weak base and a salt of that base. Two important characteristics of a buffer solution are its buffering capacity and its pH.

The plot of the pH of an acid (or base) as a function of the volume of added base (or acid) is called a **titration curve.** Titration curves aid in selecting a proper pH indicator for an acid-base titration. The titration curve of a strong acid-strong base titration exhibits a large change in pH in the immediate vicinity of the equivalence point; at the equivalence point for this titration, pH = 7. For strong acid-weak base or weak acid-strong base titrations, the pH change in the vicinity of the equivalence point is not as large. Furthermore, the pH at the equivalence point is not 7 in either of these cases.

The equilibrium between a solid salt and its ions in solution provides an example of heterogeneous equilibrium. The **solubility-product constant,** K_{sp}, is an equilibrium constant that expresses quantitatively the extent to which the salt dissolves. Addition to the solution of an ion common to a solubility equilibrium causes the solubility of the salt to decrease. This phenomenon is another example of the common-ion effect.

Comparison of the ion product, Q, with the value of K_{sp} can be used to judge whether a precipitate will form when solutions are mixed or whether a slightly soluble salt will dissolve under various conditions. Solubility is affected by the common-ion effect, by pH, and by the presence of certain Lewis bases that react with metal ions to form stable **complex ions.** Solubility is affected by pH when one or more of the ions involved in the solubility equilibrium is an acid or base. For example, the solubility of MnS is increased on addition of acid, because S^{2-} is basic. **Amphoteric metal hydroxides** are those slightly soluble metal hydroxides that dissolve on addition of either acid or base. The reactions that give rise to the amphoterism are acid-base reactions involving the OH^- or H_2O groups bound to the metal ion. Complex-ion formation in aqueous solution involves the displacement by Lewis bases such as NH_3 or CN^- of water molecules attached to the metal ion. The extent to which such complex formation occurs is expressed quantitatively by the **formation constant** for the complex ion.

The fact that the ions of different metallic elements vary a great deal in the solubilities of their salts, in their acid-base behavior, and in their tendencies to form complexes can be used to separate and detect the presence of metal ions in mixtures. Qualitative analysis refers to determination of the presence or absence of a metal ion in a mixture of metal ions in solution. The analysis usually proceeds by separation of the ions into groups on the basis of precipitation reactions, and then by analyses for individual metal ions within each group.

Learning goals

Having read and studied this chapter, you should be able to:

1 Predict qualitatively and calculate quantitatively the effect of an added common ion on the pH of an aqueous solution of a weak acid or base.
2 Calculate the concentrations of each species present in a solution formed by mixing an acid and a base.
3 Describe how a buffer solution of a particular pH is made and how it operates to control pH.
4 Calculate the change in pH of a simple buffer solution of known composition caused by adding a small amount of strong acid or base.
5 Describe the form of the titration curves for titration of a strong acid by a strong base, a weak acid by a strong base, or a weak base by a strong acid.
6 Calculate the pH at any point, including the equivalence point, in an acid-base titration.
7 Set up the expression for the solubility-product constant for a salt.
8 Calculate K_{sp} from solubility data and solubility from the value for K_{sp}.
9 Calculate the effect of an added common ion on the solubility of a slightly soluble salt.
10 Predict whether a precipitate will form when two solutions are mixed, given appropriate K_{sp} values.
11 Explain the effect of pH on a solubility equilibrium involving a basic or acidic ion.
12 Formulate the equilibrium between a metal ion and a Lewis base to form a complex ion of a metal.
13 Describe how complex formation can affect the solubility of a slightly soluble salt.
14 Calculate the concentration of metal ion in equilibrium with a Lewis base with which it forms a soluble complex ion, from a knowledge of initial concentrations and K_f.
15 Explain the origin of amphoteric behavior and write equations describing the dissolution of an amphoteric metal hydroxide in either an acidic or basic medium.
16 Explain the general principles that apply to the groupings of metal ions in the qualitative analysis of an aqueous mixture.

Key terms

Among the more important terms and expressions used for the first time in this chapter are the following:

Amphoterism (Section 16.5) is a term used to describe the ability of certain slightly soluble metal hydroxides to dissolve in either an acidic or basic medium. The solubility results from formation of a complex ion via an acid-base reaction.

A buffer solution (Section 16.2) is one that undergoes a limited change in pH upon addition of a small amount of acid or base.

The common-ion effect (Section 16.1) refers to the effect of an ion common to an equilibrium in shifting the equilibrium. For example, added Na_2SO_4 decreases the solubility of the slightly soluble salt $BaSO_4$, or added $NaC_2H_3O_2$ decreases the percent ionization of $HC_2H_3O_2$.

The formation constant (Section 16.5) for a metal ion complex is the equilibrium constant for formation of the complex from the metal ion and base species present in solution. It is a measure of the tendency of the complex to form.

A metal ion complex (Section 16.5) consists of a metal ion and a well-defined group of ions or neutral molecules bound to the ion via a Lewis acid-base interaction.

Qualitative analysis (Section 16.6) refers to determining the presence or absence of a particular substance that might be present in a mixture.

The solubility-product constant (Section 16.4) is an equilibrium constant related to the equilibrium between a solid salt and its ions in solution. It provides a quantitative measure of the solubility of a slightly soluble salt.

EXERCISES

Common-ion effect; buffers

16.1 Describe the effect on pH (increase, decrease, or no change) that results from each of the following additions: (a) sodium formate, $NaCHO_2$, to a solution of formic acid, $HCHO_2$; (b) ammonium perchlorate, NH_4ClO_4, to a solution of ammonia, NH_3; (c) potassium bromide to a solution of potassium nitrite, KNO_2; (d) hydrochloric acid, HCl, to a solution of sodium acetate, $NaC_2H_3O_2$.

16.2 What is the pH of a solution that is (a) 0.050 *M* in sodium formate, $NaCHO_2$, and 0.100 *M* in formic acid, $HCHO_2$; (b) 0.060 *M* in sodium benzoate, $NaC_7H_5O_2$, and 0.090 *M* in benzoic acid, $HC_7H_5O_2$?

16.3 Calculate the pH of each of the following solutions: (a) 0.25 *M* hydroazoic acid, HN_3, and 0.125 *M* sodium azide, NaN_3; (b) 0.20 *M* benzoic acid, $HC_7H_5O_2$, and 0.20 *M* sodium benzoate, $NaC_7H_5O_2$; (c) 0.15 *M* pyridine, C_5H_5N, and 0.20 *M* pyridinium chloride, $C_5H_5NH^+Cl^-$; (d) 0.10 *M* NaOH and 0.20 *M* propionic acid, $HC_3H_5O_2$.

16.4 Assume that 5.0×10^{-3} mol of NaOH is added to each of the following solutions. Calculate the change in pH in each case: (a) 1 L of 0.050 *M* acetic acid, $HC_2H_3O_2$; (b) 1 L of 0.050 *M* hydrochloric acid, HCl; (c) 1 L of 0.050 *M* sodium acetate, $NaC_2H_3O_2$.

16.5 A certain organic compound that is used as an indicator for acid-base reactions exists in aqueous solution as equal concentrations of the acid form, HB, and the base form, B^-, at a pH of 7.80. What is pK_a for the acid form of this indicator, HB?

16.6 (a) Calculate the pH of a 0.060 *M* solution of hydrofluoric acid, HF. (b) Calculate the pH of the solution after sufficient sodium fluoride has been added to make the solution 0.120 *M* in fluoride ion, F^-. (c) If a solution of HF and NaF is 0.100 *M* in HF and has a pH of 3.35, what is the concentration of fluoride ion, F^-?

16.7 (a) Write the net ionic equation for the reaction that occurs when a solution of perchloric acid, $HClO_4$, is mixed with a solution of sodium benzoate, $NaC_7H_5O_2$. (b) Calculate the equilibrium constant for this reaction. (c) Calculate the concentrations of Na^+, ClO_4^-, H^+, $C_7H_5O_2^-$, and $HC_7H_5O_2$ when 0.50 L of 0.20 *M* $HClO_4$ is mixed with 0.50 L of 0.20 *M* $NaC_7H_5O_2$.

16.8 What is the pH of a solution formed by mixing 0.030 mol of anilinium hydrochloride, $C_6H_5NH_3^+Cl^-$, with 0.060 mol of aniline, $C_6H_5NH_2$, to form 1.0 L of solution?

16.9 Suppose that you have two solutions of pH 9.0, one a buffer solution, the other a solution of KOH. Explain how you would test a small sample from either of these solutions to determine which is the buffer solution.

16.10 Explain what is meant in speaking of the buffering *capacity* of a buffer solution.

16.11 Indicate whether each of the following statements is true or false, and explain your reasoning: (a) In a solution of equimolar weak base and strong base, the pOH is determined by the strong base alone. (b) Optimal buffer action is achieved when the conjugate acid-base pair are present in equal concentrations. (c) The capacity of a buffer solution to resist changes in pH is related to the ratio of acid and conjugate base (or base and conjugate acid), and independent of the total concentration of these reagents.

16.12 (a) What is the ratio of HCO_3^- to H_2CO_3 in blood of pH 7.4? (b) What is the ratio of $HCO^-{}_3$ to H_2CO_3 in an exhausted marathon runner whose blood pH is 7.1?

16.13 When asked as an examination question to give an example of an alkaline buffer solution, a student listed a solution of $NaHCO_3$. If you were the instructor, would you give no credit, partial credit, or full credit for this answer? Explain.

[16.14] You have 1.0 *M* solutions of H_3PO_4 and NaOH. Describe how you would prepare a buffer solution of pH 7.20 with the highest possible buffering capacity from these reagents. Describe the composition of the buffer solution.

16.15 Calculate the pH of each of the following solutions: (a) 0.250 *M* HClO–0.125 *M* NaClO; (b) 0.300 *M* NaH_2PO_4–0.100 *M* Na_2HPO_4; (c) 0.20 *M* NH_4Br–0.15 *M* NH_3; (d) 0.200 *M* Na_2CO_3–0.10 *M* $NaHCO_3$.

16.16 Consider 1 L of a buffer mixture that is 0.080 *M* in formic acid, $HCHO_2$, and 0.080 *M* in sodium formate, $NaCHO_2$. Calculate the change in pH that occurs when the following quantities of reagent are added: (a) 5.0×10^{-4} mol of HCl; (b) 5.0×10^{-3} mol of HCl; (c) 5.0×10^{-4} mol of NaOH; (d) 5.0×10^{-3} mol of NaOH.

16.17 One liter of a buffer solution contains 0.15 mol of acetic acid and 0.10 mol of sodium acetate. (a) What is the pH of this buffer? (b) What is the pH after addition of 0.010 mol of HNO_3? (c) What is the pH after addition of 0.010 mol of NaOH?

Titration curves

16.18 How does titration of a strong acid with a strong base differ from titration of a weak acid with a strong base with respect to the following points: (a) quantity of base required to reach the equivalence point; (b) pH at the beginning of the titration; (c) pH at the equivalence point; (d) pH after addition of a slight excess of base; (e) choice of indicator for determining the equivalence point?

16.19 How many mL of 0.048 *M* NaOH are required to reach the equivalence point in titrating each of the following solutions: (a) 30.0 mL of 0.038 *M* HBr; (b) 28.0 mL of 0.018 *M* H_2SO_4; (c) 32.0 mL of 0.034 *M* $HC_2H_3O_2$.

16.20 Four 20.0-mL samples of different HBr solutions were titrated with 0.100 *M* NaOH solution. The volumes of base required to reach the equivalence point in each case were (a) 27.5 mL; (b) 21.8 mL; (c) 48.9 mL; (d) 25.5 mL. Calculate the concentrations of the four HBr solutions.

16.21 A 20.00-mL sample of 0.200 *M* HBr solution is titrated with 0.200 *M* NaOH solution. Calculate the pH of the solution after the following volumes of base have been added: (a) 5.00 mL; (b) 15.00 mL; (c) 19.9 ml; (d) 20.0 ml; (e) 20.1 mL; (f) 35.0 mL.

16.22 Calculate the pH at the equivalence point for titrating 0.200 *M* solutions of each of the following bases with 0.200 *M* HBr: (a) sodium hydroxide, NaOH; (b) hydroxylamine, NH_2OH; (c) aniline, $C_6H_5NH_2$.

16.23 Suppose that you wished to titrate a 0.200 M solution of Na_2SO_3 with a 0.200 M solution of perchloric acid, $HClO_4$, a strong acid. Which indicator from among those listed in Table 15.1 would you use to observe the first equivalence point in this titration; which to observe the second?

[16.24] Sketch the titration curve for titration of 50.00 mL of 0.0100 M sodium phenolate, NaC_6H_5O, by 0.0100 M HCl.

Solubility equilibria

16.25 Write the expression for the solubility product constant for the solubility equilibrium of each of the following compounds: (a) CdS; (b) MgC_2O_4; (c) CeF_3; (d) $Fe_3(AsO_4)_2$; (e) Cu_2S.

16.26 If the molar solubility of CaF_2 at 35°C is 1.24×10^{-3} mol/L, what is its K_{sp} at this temperature?

16.27 For each salt listed below we have given the solubility at 25°C. From these data calculate the value for K_{sp} in each case: (a) $AgIO_3$, 0.0283 g/L; (b) $Cd(CN)_2$, 0.22 g/L; (c) Ag_2CO_3, 3.5×10^{-2} g/L; (d) CrF_3, 0.14 g/L.

16.28 The K_{sp} for $SrCO_3$ is 1.1×10^{-10}. What is the molar solubility (that is, moles per liter) for this substance in water?

16.29 Using the K_{sp} values listed in Appendix E, calculate the solubility in grams per liter of solution for each of the following substances: (a) $MnCO_3$; (b) $Mg(OH)_2$; (c) CeF_3.

16.30 Calcium hydroxide is widely used in industry because it is the cheapest strong base available. (a) Using K_{sp} from Appendix E, calculate the molar solubility of $Ca(OH)_2$ in pure water. (b) What is the pH of a saturated solution of $Ca(OH)_2$?

16.31 Calculate molar solubility of AgI in (a) pure water; (b) 3.0×10^{-3} M NaI solution; (c) 3.0×10^{-3} M $AgNO_3$ solution.

16.32 Calculate the molar solubility of Ag_2SO_4 in (a) pure water; (b) 0.10 M Na_2SO_4; (c) 0.10 M $AgNO_3$.

16.33 The K_{sp} for $AgIO_3$ is 3.0×10^{-8}. Should a precipitate of $AgIO_3$ form when 100 mL of 0.010 M $AgNO_3$ solution is mixed with 10 mL of 0.010 M $NaIO_3$ solution?

16.34 Calculate the pH at which $Mn(OH)_2$ just begins to precipitate when concentrated NaOH solution is added to a 0.010 M solution of $MnCl_2$.

16.35 Will $Mn(OH)_2$ precipitate from solution if the pH of a 0.050 M solution of $MnCl_2$ is adjusted to 8.0?

16.36 What is the ratio of $[Ca^{2+}]$ to $[Fe^{2+}]$ in a lake in which the water is in equilibrium with deposits of both $CaCO_3$ and $FeCO_3$?

Controlled precipitation; dissolution

16.37 Which of the following salts will be substantially more soluble in acid solution than in pure water: (a) $AgCO_3$; (b) CeF_3; (c) Ag_2SO_4; (d) PbI_2; (e) $Cd(OH)_2$?

16.38 Calculate the molar solubility of $Cd(OH)_2$ (a) at pH 8.0; (b) at pH 10.0.

16.39 Calculate the S^{2-} concentration in a saturated H_2S solution (solubility 0.1 M) at (a) pH 4.8; (b) pH 8.0.

16.40 A solution that is 0.10 M in Mn^{2+} and 0.050 M in H^+ is saturated with H_2S (solubility 0.1 M). Will MnS precipitate?

16.41 Calculate the solubility of ZnS (in moles per liter) in a 0.1 M solution of H_2S if the pH of the solution is 2.0.

16.42 (a) The K_{sp} for cerium iodate, $Ce(IO_3)_3$, is 3.2×10^{-10}. Calculate the molar solubility of $Ce(IO_3)_3$ in pure water. (b) What concentration of $NaIO_3$ in solution would be necessary to reduce the Ce^{3+} concentration in a saturated solution of $Ce(IO_3)_3$ by a factor of 10 below that calculated in part (a)?

16.43 Milk of magnesia is a suspension of $Mg(OH)_2$. Suppose that a person swallows a quantity of milk of magnesia that contains 12 g of $Mg(OH)_2$. If this is mixed in the stomach with 350 mL of solution buffered at pH 3.8, will the $Mg(OH)_2$ dissolve?

[16.44] The solubility of $Mg(OH)_2$ in water varies in water as a function of pH. If we let the molar solubility of $Mg(OH)_2$ be S, the functional relationship between solubility and pH is of the form $\log S = a + b\text{pH}$. What are the values for a and b?

16.45 Complete the following equations:

(a) $Al(OH)_3(s) + OH^-(aq) \longrightarrow$
(b) $Zn(OH)_2(s) + 2H^+(aq) \longrightarrow$
(c) $Cr(OH)_4^-(aq) + H^+(aq) \longrightarrow$
(d) $Zn^{2+}(aq) + 2OH^-(aq) \longrightarrow$
(e) $Cr^{3+}(aq) + 4OH^-(aq) \longrightarrow$

16.46 Compare the reactions

$$H^+(aq) + OH^-(aq) \longrightarrow H_2O(l)$$
$$Zn^{2+}(aq) + 4OH^-(aq) \longrightarrow Zn(OH)_4^{2-}(aq)$$
$$Cu^{2+}(aq) + 4CN^-(aq) \longrightarrow Cu(CN)_4^{2-}(aq)$$

as examples of Lewis acid-base reactions. In each case identify the Lewis acid and Lewis base.

16.47 From the value for K_f listed in Table 16.2, calculate the concentration of Cu^{2+} in 1 L of a solution that contains a total of 1×10^{-3} mol of copper(II) ion and that is 0.10 M in NH_3.

16.48 What concentration of CN^- is required to reduce the free silver ion concentration to 1.0×10^{-6} M in a solution that originally was 0.010 M in Ag^+?

[16.49] To what final concentration of NH_3 must a solution be adjusted to just dissolve 0.020 mol of NiC_2O_4 in a liter of solution? (*Hint:* You can neglect the hydrolysis of $C_2O_4^{2-}$, because the solution will be quite basic.)

16.50 Using the value of K_{sp} for AgCl and K_f for $Ag(CN)_2^-$, calculate the equilibrium constant for the following reaction:

$$AgCl(s) + 2CN^-(aq) \longrightarrow Ag(CN)_2^-(aq) + Cl^-(aq)$$

16.51 Using the value of K_{sp} for Ag_2S, K_{a1} and K_{a2} for H_2S, and $K_f = 1.1 \times 10^5$ for $AgCl_2^-$, calculate the equilibrium constant for the following reaction:

$$Ag_2S(s) + 4Cl^-(aq) + 2H^+(aq) \longrightarrow 2AgCl_2^-(aq) + H_2S(aq)$$

16.52 A total of 0.010 mol of $Cd(NO_3)_2$ is dissolved in 1 L of solution that is 1.0 M in NaSCN. K_f for $Cd(SCN)_4^{2-}$ is 4×10^3. What is the ratio $[Cd^{2+}]/[Cd(SCN)_4^{2-}]$ at equilibrium?

Qualitative analysis

16.53 A solution containing an unknown number of metal ions is treated with dilute HCl; no precipitate forms. The pH is adjusted to about 1, and H_2S is bubbled through. Again no precipitate forms. The pH of the solution is then adjusted to about 8. Again H_2S is bubbled through. This time a precipitate forms. The filtrate from this solution is treated with $(NH_4)_2HPO_4$. No precipitate forms. Which metal ions discussed in Section 16.6 are possibly present? Which are definitely absent within the limits of these tests?

16.54 A student who is in a great hurry to finish his laboratory work decides that his qualitative analysis unknown contains a metal ion from the insoluble phosphate group, group 4. He therefore tests his sample directly with $(NH_4)_2HPO_4$, skipping earlier tests for the metal ions in groups 1–3. He observes a precipitate and concludes that a metal ion from group 4 is indeed present. Why is this possibly an erroneous conclusion?

[16.55] What concentration of Pb^{2+} remains in a solution after $PbCl_2$ has been precipitated from a solution that is 0.1 M in Cl^- at 25°C? Will PbS form if the filtrate from the $PbCl_2$ precipitate is saturated with H_2S at pH 1?

16.56 In the course of various qualitative analysis procedures the following mixtures are encountered: (a) Zn^{2+} and Cd^{2+}; (b) $Cr(OH)_3$ and $Fe(OH)_3$; (c) Mg^{2+} and K^+; (d) Ag^+ and Mn^{2+}. Suggest how each mixture might be separated.

16.57 (a) Precipitation of the group 4 cations requires a basic medium. Why is this so? (b) What is the most significant difference between the sulfides precipitated in group 2 and those precipitated in group 3? (c) Suggest a procedure that would serve to redissolve all of the group 3 cations following their precipitation and separation.

Additional exercises

16.58 Which of the following solutions has the greatest buffering capacity, and which has the least: (a) 0.10 M $HC_2H_3O_2$ and 0.10 M $NaC_2H_3O_2$, pH 4.74; (b) 1.8×10^{-5} M HBr, pH 4.74; (c) 0.01 M $HC_2H_3O_2$ and 0.01 M $NaC_2H_3O_2$, pH 4.74? Explain.

16.59 Write balanced net-ionic equations for the reactions between (a) sodium hydroxide and benzoic acid; (b) sodium hydroxide and hydrobromic acid; (c) sodium formate and ammonium chloride; (d) nitrous acid and potassium hydroxide.

16.60 A hypothetical weak acid, HA, was combined with NaOH in the following proportions: 0.20 mol of HA, 0.080 mol of NaOH. The mixture was diluted to total volume of 1 L, and the pH measured. (a) If pH = 4.80, what is the pK_a of the acid? (b) How many additional moles of NaOH would need to be added to the solution to increase the pH to 5.00?

16.61 Calculate the pH of a solution formed by mixing 0.060 mol of NH_4Cl with 0.025 mol of NH_3 in 0.500 L of solution.

16.62 A solution containing 0.050 mol of a weak acid, HX, and 0.030 mol of a salt, KX, is diluted to a total volume of 0.500 L. The pH of the solution is 4.86. What is the acid-dissociation constant K_a?

16.63 Calculate the pH of the solution formed by mixing (a) 100 mL of 0.10 M benzoic acid, $HC_7H_5O_2$, and 50 mL of 0.10 M sodium benzoate, $NaC_7H_5O_2$; (b) 100 mL of 0.10 M HNO_3 and 100 mL of 0.10 M $HC_2H_3O_2$; (c) 50 mL of 0.10 M $NaC_2H_3O_2$ and 50 mL of 0.10 M NaOH; (d) 25.0 mL of 0.100 M NaOH and 75.0 mL of 0.100 M $HC_2H_3O_2$.

16.64 A 0.30 M solution of dimethylamine, $(CH_3)_2NH$, containing an unknown concentration of dimethylammonium chloride, $(CH_3)_2NH_2^+Cl^-$, has a pH of 10.40. What is the concentration of dimethylammonium ion in the solution?

16.65 The acid-base indicator bromcresol green is a weak acid. The yellow acid and blue base forms of the indicator are present in equal concentrations in a solution when the pH is 4.68. What is pK_a for bromcresol green?

16.66 What is the pH of the solution after 0.015 mol of NaBrO has been added to 0.500 L of a 0.100 M solution of hypobromous acid, HBrO?

16.67 How many moles of sodium hypobromite, NaBrO, should be added to 1.00 L of 0.200 M hypobromous acid, HBrO, to form a buffer solution of pH 8.80? Assume that no volume change occurs when the NaBrO is added.

16.68 Which of the following conjugate acid-base pairs is the best choice for preparing a buffer solution of pH 3.5? (a) HPO_4^{2-} and PO_4^{3-}; (b) HCNO and CNO^-; (c) NH_4^+ and NH_3; (d) $HC_7H_5O_2$ and $C_7H_5O_2^-$.

16.69 Would the solubility of $MnCO_3$ in pure water differ from that in a buffered solution of pH 8.0? Explain.

16.70 Phosphate buffer, consisting of $H_2PO_4^-$ and HPO_4^{2-}, helps control the pH of blood. Many carbonated soft drinks also use this buffer system. What is the pH of a soft drink in which the major buffer ingredients are 6.5 g of NaH_2PO_4 and 8.0 g of Na_2HPO_4 in 355 mL of solution?

[16.71] A biochemist needs 750 mL of an acetic acid-sodium acetate buffer with pH 4.50. Solid sodium acetate, $NaC_2H_3O_2$, and glacial acetic acid, $HC_2H_3O_2$, are available. Glacial acetic acid is 99 percent $HC_2H_3O_2$ by weight and has a density of 1.05 g/mL. If the buffer is to be 0.30 M in $HC_2H_3O_2$, how many grams of $NaC_2H_3O_2$ and how many milliliters of glacial acetic acid must be used?

16.72 What is the pH at the equivalence point in the following titrations: (a) 0.100 M HCl with 0.100 M methylamine; (b) 0.100 M NaOH with 0.100 M ascorbic acid; (c) 0.100 M HBr with 0.100 M KOH?

16.73 If 50.00 mL of 0.100 M Na_2SO_3 is titrated with 0.100 M HCl, calculate (a) the pH at the start of the titration; (b) the volume of HCl required to reach the first equivalence point. What is the predominant species present? (c) the volume of HCl required to reach the second equivalence point, and the pH at the second equivalence point.

[16.74] Equivalent quantities of 0.10 M solutions of an acid HA, and a base, B, are mixed. The pH of the resulting solution is 8.8. (a) Write the equilibrium equation and equilibrium-constant expression for the reaction between HA and B. (b) If K_a for HA is 5.0×10^{-6}, what is the value of the equilibrium constant for the reaction between HA and B?

[16.75] What should be the pH of a buffer solution that will result in a Mg^{2+} concentration of $3.0 \times 10^{-2} M$ in equilibrium with solid magnesium oxalate?

[16.76] Calculate the solubility of $CuCO_3$ in a strongly buffered solution of pH 7.5.

16.77 Write the net ionic equation for the reaction that occurs when (a) hydrogen sulfide gas is bubbled through a solution of mercury(II) chloride; (b) solid magnesium sulfite is added to a solution of hydrochloric acid; (c) aqueous ammonia is added to a solution of nickel(II) nitrate; (d) sodium carbonate solution is added to a solution of manganese(II) sulfate; (e) aluminum fluoride dissolves in a solution containing excess sodium fluoride (pK_f for the AlF_6^{3-} ion is 19.8).

16.78 Germanic acid, H_2GeO_3, has pK_{a1} and pK_{a2} values of 9.0 and 12.4, respectively. (a) Draw the Lewis structure for this acid. (b) What is the pH of a 0.010 M solution of the acid? (c) Would titration of this acid with NaOH solution, using phenolphthalein as indicator (Table 15.1), be a good quantitative procedure? Explain.

16.79 What is the concentration of HS^- in a saturated (0.1 M) solution of H_2S if the pH is (a) 3.0; (b) 5.5?

[16.80] The value of K_{sp} for $Mg_3(AsO_4)_2$ is 2.1×10^{-20}. The AsO_4^{3-} ion is, of course, derived from the weak acid H_3AsO_4, $pK_{a1} = 2.22$, $pK_{a2} = 6.98$, $pK_{a3} = 11.50$. When asked to calculate the molar solubility of $Mg_3(AsO_4)_2$ in water, a student employed the K_{sp} expression, and assumed that $[Mg^{2+}] = 1.5[AsO_4^{3-}]$. Why was this a mistake?

[16.81] Cadmium sulfide, CdS, is used commercially as a yellow paint pigment known as cadmium yellow. It is prepared by saturating a slightly acidic solution of Cd^{2+} with H_2S gas. The presence of Fe^{2+}, a common impurity, can result in the precipitate being contaminated with black FeS, thereby ruining the color of the pigment. (a) If a solution containing 0.10 M Cd^{2+} and $1 \times 10^{-4} M$ Fe^{2+} is saturated with H_2S ($[H_2S] = 0.10 M$), what pH is needed to keep the Fe^{2+} from precipitating? (b) If a $HC_2H_3O_2$-$C_2H_3O_2$-buffer is used to control pH, what ratio of $HC_2H_3O_2$ to $C_2H_3O_2^-$ is required?

16.82 The pH of a solution that is 0.020 M in Hg^{2+}, 0.020 M in Ni^{2+}, and 0.020 M in Pb^{2+} is adjusted to 2.0 and then saturated with H_2S gas. Which metal ions, if any, will precipitate?

[16.83] The solubility of manganese oxalate dihydrate, $MnC_2O_4 \cdot 2H_2O$, is 0.035 g per 100 mL of solution. The acid dissociation constants of oxalic acid are listed in Appendix E. Calculate the pH of a saturated solution of $MnC_2O_4 \cdot 2H_2O$.

[16.84] Calculate the molar solubility of $Fe(OH)_3$ in water. (This problem is not as simple as it might first appear; consider carefully the concentration of OH^-.)

[16.85] What pH range will permit PbS to precipitate while leaving Mn^{2+} in solution if the solution is 0.010 M in Pb^{2+} and 0.010 M in Mn^{2+} prior to saturation with H_2S?

[16.86] Using the K_f value for $Ag(NH_3)_2^+$ that is listed in Table 16.1, calculate the concentration of Ag^+ present in solution at equilibrium when concentrated NH_3 is added to a 0.010 M solution of $AgNO_3$ until the ammonia concentration is 0.20 M. (Neglect the small concentration change in Ag^+ that accompanies the change in the volume of the solution upon addition of NH_3).

17 Chemical Thermodynamics

Two of the major questions concerning any chemical reaction are: How far toward completion does the reaction proceed? How rapidly does the reaction approach equilibrium? We get the answer to the first question from a knowledge of the equilibrium constant. We learn about the second from a study of the reaction rate.

If we want to carry out a reaction that has a favorable equilibrium constant, that is, a reaction that proceeds with conversion of a sizable fraction of reactants into products, we need only worry about getting the reaction rate into a convenient range. This may not be easy, but hard work and ingenious research might eventually lead to a catalyst that can speed up a slow reaction or to some means of controlling a reaction that is too rapid. Haber's discovery of a suitable catalyst for ammonia synthesis (Chapter 14) provides an excellent example of successful research of this type. However, if Haber had attempted to fix nitrogen by reacting N_2 and O_2 at 400 to 500°C, instead of reacting N_2 and H_2, he would never have developed a successful process. The equilibrium constant for the reaction of N_2 with O_2 to form NO is so small in this temperature range that the amount of NO present at equilibrium would have been too small for practical purposes. Even if one had a catalyst that would enable rapid reaction of N_2 with O_2 to form NO, the reaction would still have no practical consequences. There is therefore a need for some way to predict in advance whether a reaction can proceed to any significant extent before coming to equilibrium.

When we first introduced chemical equilibrium, in Chapter 14, we defined it from a kinetic point of view: Equilibrium occurs when opposing reactions occur at equal rates. However, equilibrium also has a basis in thermodynamics, the area of science dealing with energy relationships. In this chapter we shall see that it is possible to predict the position of an equilibrium using certain principles and concepts from thermodynamics.

17.1 SPONTANEOUS PROCESSES

Thermodynamics is based on several fundamental laws that summarize our experience with energy changes. The first law of thermodynamics, which we discussed at length in Chapter 4, states that energy is conserved. By this we mean that energy is neither created nor destroyed in any process, such as the falling of a brick, the melting of an ice cube, or a chemical reaction. Energy flows from one part of nature to another, or is converted from one form to another, but the total remains constant. We saw in Chapter 4 that we could express the first law in the form $\Delta E = q - w$, where ΔE represents the change in energy of a system in a process, q is the heat absorbed by the system from its surroundings in the process, and w is the work done by the system on its surroundings during the process.

Once we specify a particular process or change, the first law helps us to balance the books, so to speak, on the heat released, work done, and so forth. However, it says nothing about whether the process or change we specify can in fact occur. That question is encompassed in the second law of thermodynamics.

The **second law of thermodynamics** expresses the notion that there is an inherent direction in which any system not at equilibrium moves. To reverse that direction requires addition of energy to the system. For example, if you hold a brick in your hand and let it go, the brick falls to the floor. Water placed in a freezer compartment converts to ice. A shiny nail left outdoors turns to rust. Every one of these processes proceeds without needing to be driven by an outside source of energy; such processes are said to be **spontaneous.** For every spontaneous process, we can imagine a reverse process occurring. For example, we can imagine a brick moving from the floor into your hand, ice cubes melting at $-10°C$, or a rusty iron nail being transformed into a shiny one. It is inconceivable that any of these occurrences is spontaneous. If we saw a film in which these things happened, we would conclude that the film was being run backward. Our years of observing nature at work have impressed us with a simple rule: *Processes that are spontaneous in one direction are not spontaneous in the reverse direction.*

Consider a reaction about which we had much to say in Chapter 14:

$$N_2(g) + 3H_2(g) \rightleftharpoons 2NH_3(g) \qquad [17.1]$$

When we mix N_2 and H_2 at any temperature, say 472 K, the reaction proceeds in the forward direction; this process is spontaneous. However, if we place a mixture of 1.00 mol of N_2, 3.00 mol of H_2, and 1.00 mol of NH_3 in a 1-L container at 472 K, it is not immediately obvious whether or not the formation of more NH_3 should be spontaneous. Nevertheless, if we have the equilibrium constant, which at this temperature happens to be $K_c = 0.105$, we can predict the direction in which the reaction proceeds to reach equilibrium. In this case

$$Q = \frac{[NH_3]^2}{[N_2][H_2]^3} = \frac{(1.00)^2}{(1.00)(3.00)^3} = 0.0370$$

Because the reaction quotient is smaller than K_c, the system moves spontaneously toward equilibrium by formation of NH_3 (Section 14.4). The opposite process, conversion of NH_3 into N_2 and H_2, is not spontaneous for this particular reaction mixture at 472 K.

The process by which a system reaches equilibrium is a spontaneous change. The process may be fast, or it may be slow; thermodynamics has nothing to say about how rapidly such a process occurs. For example, a mixture of $H_2(g)$ and $Cl_2(g)$ will react at 25°C to form $HCl(g)$. The equilibrium constant for this reaction, K_p, is very large, 5.2×10^{16}. When $H_2(g)$ and $Cl_2(g)$ are mixed, the reaction between them is spontaneous, yet this mixture may show no apparent reaction even over a period of centuries. The reaction is extremely slow unless initiated by a suitable catalyst, a spark, or a ray of light. Once initiated in one of these ways, this reaction proceeds with explosive speed nearly to completion.

What factors make a process spontaneous? We considered that question briefly in Section 11.2. In that earlier discussion we noted that two basic principles are involved: Processes in which the energy content of the system decreases (exothermic processes) tend to occur spontaneously. Furthermore, spontaneity characterizes processes in which the randomness or disorder of the system increases. In the next section we consider these factors, especially the matter of disorder, in greater detail.

17.2 SPONTANEITY, ENTHALPY, AND ENTROPY

The spontaneous motion of a brick released from your hand is toward the ground. As the brick falls, it loses potential energy. This potential energy is first converted into kinetic energy, the energy of motion of the brick. When the brick hits the floor, its kinetic energy is converted into heat. The overall result of the brick's fall is thus a conversion of the potential energy of the brick into heat in its surroundings. Our experience with other simple mechanical systems is similar: objects fall, clocks run down, stretched rubber bands contract. All of these phenomena can be summarized by saying that such systems seek a resting place of minimum energy.

It is clear that the tendency for a system to achieve the lowest possible energy is one of the driving forces that determines the behavior of molecular systems. For example, just as a brick possesses potential energy because of its position relative to the floor, so also a chemical substance possesses potential energy relative to other substances because of the arrangements of nuclei and electrons. When these arrangements change, energy may be released. For example, the combustion of propane (bottled gas), which is clearly a spontaneous process, is strongly exothermic:

$$C_3H_8(g) + 5O_2(g) \longrightarrow 3CO_2(g) + 4H_2O(l) \qquad \Delta H° = -2202 \text{ kJ} \qquad [17.2]$$

The rearrangements in space of nuclei and electrons in going from propane and oxygen to carbon dioxide and water lead to a lower chemical potential energy, so heat is evolved. Reactions that are exothermic are generally spontaneous. However, it is clear that the tendency toward minimum enthalpy cannot be the only factor that determines spontaneity in molecular processes. It is instructive to consider some spontaneous processes that are not exothermic.

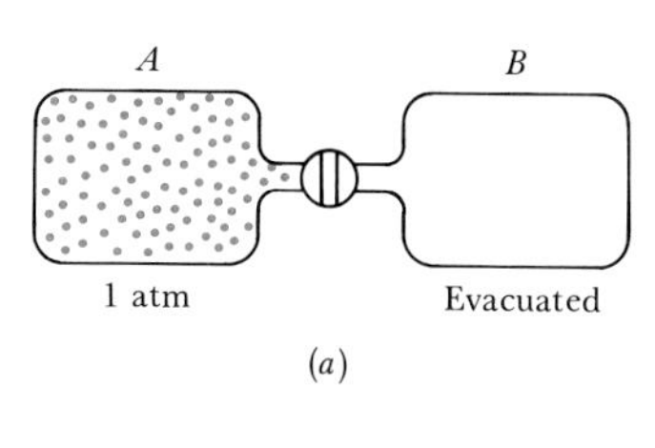

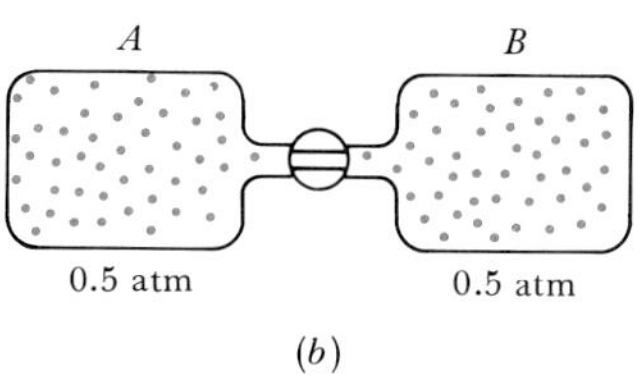

FIGURE 17.1 Expansion of an ideal gas into an evacuated space. In (*a*), flask *A* holds an ideal gas at 1 atm pressure whereas flask *B* is evacuated. In (*b*), the stopcock connecting the flasks has been opened. The ideal gas expands to occupy both flasks *A* and *B* at a pressure of 0.5 atm.

Spontaneity and Entropy Change

A bit of thinking brings to mind several processes that are spontaneous even though they are not exothermic. For example, consider an ideal gas confined at 1 atm pressure to a 1-L flask, as shown in Figure 17.1. The flask is connected via a closed stopcock to another 1-L flask that is evacuated. Now suppose the stopcock is opened. Is there any doubt about what would happen? We intuitively recognize that the gas would expand into the second flask until the pressure is equally distributed in both flasks, at 0.5 atm. In the course of expanding from the 1-L flask into the larger volume, the ideal gas neither absorbs nor emits heat. Nevertheless, the process is spontaneous. The reverse process, in which the gas that is evenly distributed between the two flasks suddenly moves entirely into one of the flasks, leaving the other vacant, is inconceivable. Yet this is also a process that would involve no emission or absorption of heat. It is evident that some other factor than heat emitted or absorbed is important in making the process of gas expansion spontaneous.

As another example, consider the melting of ice cubes at room temperature. The process

$$H_2O(s) \longrightarrow H_2O(l) \qquad [17.3]$$

at 27°C is highly spontaneous, as we all know. Yet this is an endothermic change. The melting of ice above 0°C thus represents an example of a spontaneous, endothermic process.

A similar type of process, discussed in Chapter 11, is the endothermic dissolving of many salts in water. If we add solid potassium chloride, KCl, to a glass of water at room temperature and stir, we can feel the solution growing colder as the salt dissolves. Thus the process, which we can write as

$$KCl(s) \xrightarrow{H_2O} KCl(aq) \qquad [17.4]$$

is endothermic, and yet spontaneous.

The three processes just described have something in common that accounts for the fact that they are spontaneous. In each instance, the products of the process are in a more random or disordered state than the reactants. Let's consider each case in turn.

When we have a gas confined to a 1-L volume as in Figure 17.1(*a*), we can specify the location of each and every gas molecule as being in that liter of space. After the gas has expanded, we can't be sure which gas molecules are at any one instant in the original volume, and which are on the other side. We must therefore say that the location of each and every gas molecule is specified as being in the entire 2-L space. In other words, the gas molecules, because they can be anywhere within a 2-L space, are more randomized than when they are confined to a 1-L space.

The molecules of water that make up an ice crystal are held rigidly in place in the ice crystal lattice (Figure 17.2). When the ice melts, the water molecules are free to move about with respect to one another and to turn over. Thus in liquid water the individual water molecules are more randomly distributed than in the solid. The highly ordered solid structure is replaced by the highly disordered liquid structure.

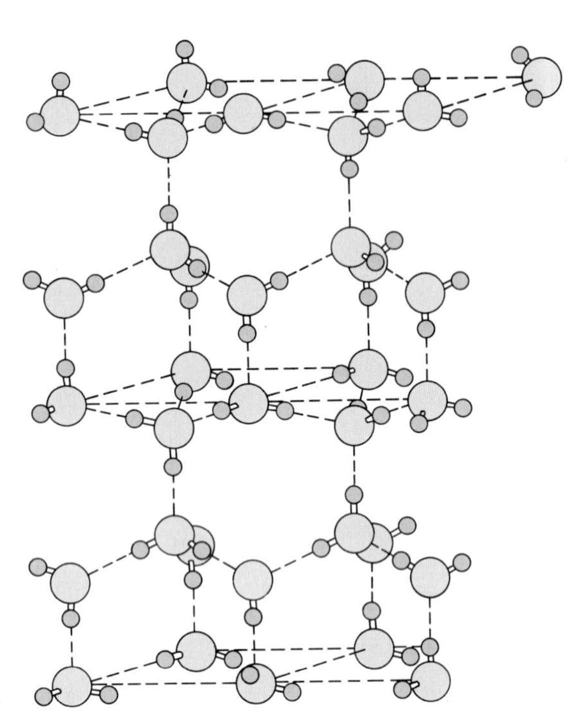

FIGURE 17.2 Structure of ice.

A similar situation applies when KCl dissolves in water, although here we must be a little careful not to take too much for granted. In solid KCl, the K^+ and Cl^- are in a highly ordered, crystalline state. When the solid dissolves, the ions are free to move about in the water. They are obviously in a much more random and disordered state than before. At the same time, though, water molecules are held around the ions, as water of hydration (Section 11.3), as illustrated in Figure 17.3. These water molecules, are in a *more* ordered state than before, because they are confined to the immediate environment of the ions. Thus the dissolving of a salt involves both ordering and disordering processes. It happens that the disordering processes are dominant, so the overall effect is an increase in disorder upon dissolving a salt in water.

SAMPLE EXERCISE 17.1

Indicate in each of the following cases whether the process shown results in an increase or decrease in randomness or disorder.

(a) $4Fe(s) + 3O_2(g) \longrightarrow 2Fe_2O_3(s)$
(b) $Ag^+(aq) + Cl^-(aq) \longrightarrow AgCl(s)$
(c) $H_2O(l) \longrightarrow H_2O(g)$

Solution: Process (a) results in a decrease in randomness, because a gas is converted into part of a solid lattice. The units of the solid oxide lattice are much more highly ordered and confined to specific locations than are the molecules of a gas. (Note that this reaction is spontaneous even though there is an overall decrease in randomness. This is so because the reaction is highly exothermic. The combined effects of enthalpy change and change in randomness are discussed in Section 17.5.)

Process (b) also represents a decrease in randomness, because the ions that are free to move about the volume of the solution form a solid lattice in which they are confined to highly regular locations.

Process (c) occurs with an increase in randomness or disorder, because the gaseous water molecules are distributed throughout a much larger volume than in the liquid state.

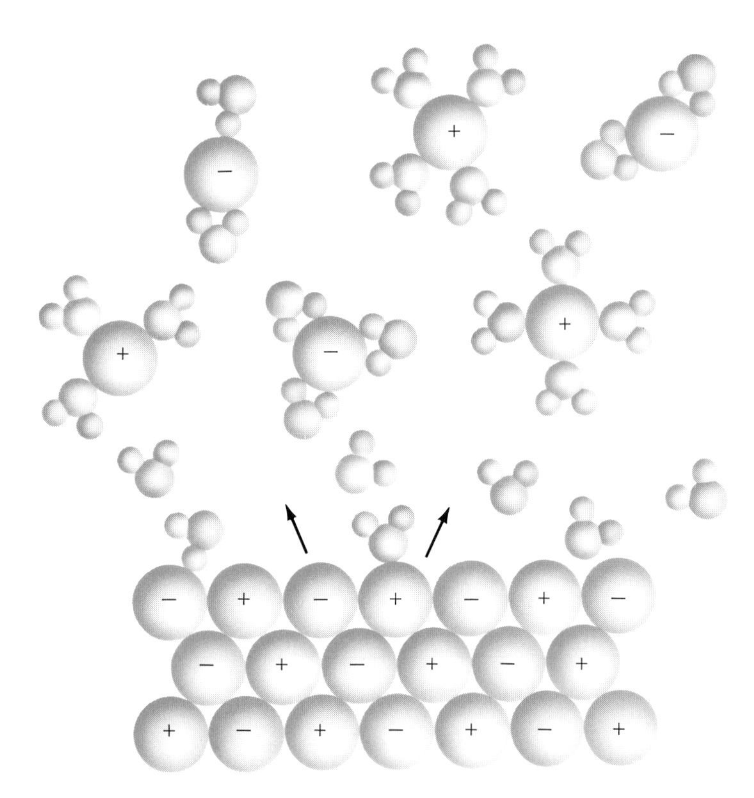

FIGURE 17.3 Changes in degree of order in the ions and solvent molecules on dissolving of an ionic solid in water. The ions become more randomized and the water molecules that hydrate the ions become less randomized.

As these examples illustrate, spontaneity is associated with an increase in randomness or disorder of a system. The randomness is expressed by a thermodynamic quantity called entropy, given the symbol S. The more random a system, the larger its entropy. Like enthalpy, entropy is a state function (Section 4.3). The change in entropy for a process, $\Delta S = S_{\text{final}} - S_{\text{initial}}$, depends only on the initial and final states of the system and not on the particular pathway by which it changes from one state to another.

The Second Law of Thermodynamics

Our introduction of the concept of entropy allows us to reexamine the second law of thermodynamics and its implications. In the discussion of spontaneity in Section 17.1, it was noted that the second law has to do with the direction in which processes move; it is associated with the idea that processes that are spontaneous in one direction are not spontaneous in the opposite direction. This idea applies not only to chemical changes but to other processes as well.

We all know that heat flows spontaneously from a hot object to a cold one. We also know that to cause heat to flow in the reverse direction, from a cold object to a hotter one or from a system at some temperature to surroundings at a higher temperature, requires an input of energy. For example, it requires electrical energy to maintain a refrigerator at a lower temperature than the surrounding kitchen.

A related but less obvious point is that heat cannot be completely converted into work. There is always some heat that is transferred to the surroundings. For example, in a steam turbine, the heat energy contained in high-temperature steam is converted into electrical energy; the kinetic energy of the steam molecules is converted into the kinetic energy of the moving turbine blades, and eventually into electrical energy. But not all the kinetic energy of the steam molecules can be converted into the kinetic energy of the turbine. Some energy is lost to the surroundings as heat. Every electrical power plant is a source of waste heat; the laws of thermodynamics tell us that this must be so. Indeed, the fact that heat cannot be completely converted into work is one of the earliest statements of the second law.

There are many ways to state the second law. In chemical contexts it's usually expressed in terms of entropy. To develop such a statement, let's think in terms of an **isolated system,** one that doesn't exchange energy or matter with its surroundings. When a process occurs spontaneously in an isolated system, the system always ends up in a more random state. For example, when a gas expands under the conditions shown in Figure 17.1, there is no exchange of heat, work, or matter with the surroundings; this is an isolated system. The spontaneous expansion corresponds to an increase in entropy.

In the real world we rarely deal with isolated systems. We are usually concerned with systems that exchange energy with their surroundings in the form of heat or work. When such a system changes spontaneously, it may undergo either an increase or decrease in its entropy. However, the second law tells us that *in any spontaneous process there is always an increase in the entropy of the universe.* As an example, consider the oxidation of iron to $Fe_2O_3(s)$:

$$4Fe(s) + 3O_2(g) \longrightarrow 2Fe_2O_3(s) \qquad [17.5]$$

As discussed in Sample Exercise 17.1, this chemical process results in a decrease in the degree of randomness; that is, ΔS for the process is negative. But when the process occurs, some change also occurs in the surroundings. For example, the reaction is exothermic; heat is therefore evolved and absorbed by the surroundings. In fact, the change that occurs in the surroundings causes an increase in the entropy of the surroundings that is larger than the decrease that occurs in the system itself. For any spontaneous process the sum of the entropy change in the system and that in the surroundings (which is the entropy change in the universe caused by that process) must be positive:

$$\Delta S_{\text{universe}} = \Delta S_{\text{system}} + \Delta S_{\text{surroundings}} > 0 \qquad [17.6]$$

No process that produces order (lower entropy) in a system can proceed without producing an even larger disorder (higher entropy) in its surroundings. That is, the disorder thereby introduced into the surroundings will always exceed the order achieved in the system. Thus while energy is conserved (the first law), entropy continues to increase (the second law).

The consequences of the statement of the second law that is given above are quite profound. We humans, for example, are very complex, highly organized, and well-ordered systems. We have a very low entropy content as compared with the same amount of carbon dioxide, water, and several other simple chemicals into which our bodies might be decomposed. But all of the thousands of chemical reactions necessary to produce one adult human have caused a very large increase in entropy of the rest of the universe. Thus the overall entropy change necessary to form and maintain a human, or for that matter any other living system, is positive.

In a similar way, the human activities that produce such an impressive ordering of the world around us—formation of copper metal from a widely dispersed copper ore; production from sand of silicon used in transistors; production of the paper on which this book is printed from trees—have along the way used a great deal of energy that has been converted, in a sense, to disorder—coal and oil burned to form CO_2 and H_2O; a sulfide ore roasted to form SO_2 that pollutes the atmosphere; various waste products scattered in the environment. Modern human society is, in effect, using up its limited storehouse of energy-rich materials in its headlong rush to exploit technology.

In recent years a few social scientists have begun to appreciate the importance of thermodynamic considerations in the economic laws that must eventually rule human activities. A leading economic theorist, Nicholas Georgescu-Roegen, published in 1971 a book entitled *The Entropy Law and the Economic Process*. He argues that the human race must eventually learn to live within the bounds of the energy supply that reaches earth daily from the sun, because it will soon have exhausted the supply of readily available energy of other sorts.

17.3 A MOLECULAR INTERPRETATION OF ENTROPY

It is useful to develop a qualitative sense of how entropy changes in a system depend on changes in structure, physical state, and so forth. In Section 17.2 some examples of spontaneous, endothermic processes were considered. We saw, for example, that the increase in volume that occurs when a gas expands results in an increase in the randomness of the system (positive ΔS). In a similar fashion, the distribution of a liquid or solid solute in a solution is accompanied by an increase in entropy. For example, ΔS is positive when KCl dissolves in water. However, the dissolving of a gas, such as CO_2 in H_2O, causes the gas molecules to move in a much smaller volume; consequently, for this process the entropy of the system decreases (negative ΔS). Similarly, a decrease in the number of gaseous particles as the result of a reaction causes a decrease in entropy (negative ΔS). For example, ΔS for the following reaction is negative:

$$2NO(g) + O_2(g) \longrightarrow 2NO_2(g) \quad [17.7]$$

Entropy changes can also be associated with molecular motions within a substance. A molecule consisting of more than one atom can engage in several types of motion. The entire molecule can move in one direction or another as in the movements of gas molecules. We call such movement **translational motion.** The atoms within a molecule may also undergo **vibrational motion** in which they move periodically toward and away from each other, much as a tuning fork vibrates about its equilibrium shape. Figure 17.4 shows the vibrational motions possible for the water molecule. In addition, molecules may possess **rotational motion,** as though they were spinning like a top. The rotational motion of the water molecule is also illustrated in Figure 17.4. These forms of motion are ways that the molecule has of storing energy. As the temperature of a system increases, the amounts of energy stored in these forms of motion increase.

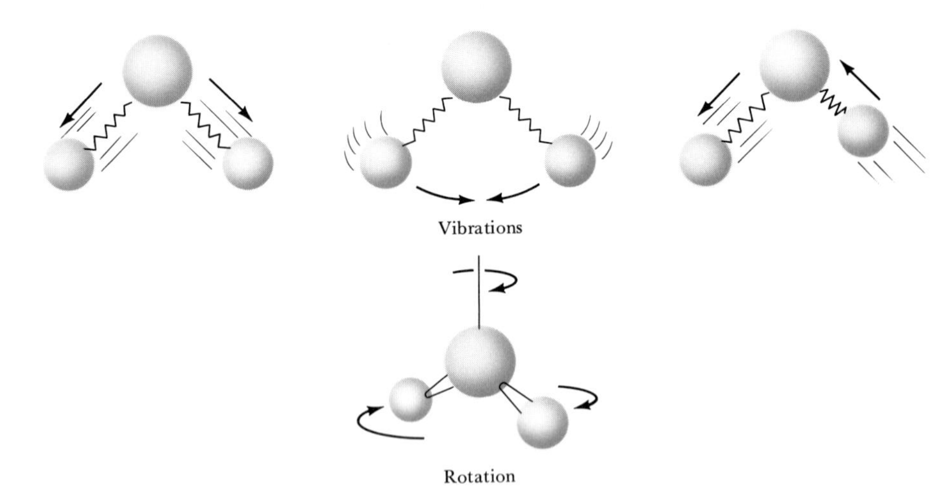

FIGURE 17.4 Examples of vibrational and rotational motion, as illustrated for the water molecule. Vibrational motions involve periodic displacements of the atoms with respect to one another, a phenomenon similar to the vibrations of the arms of a tuning fork. Rotational motions involve the spinning of a molecule about an axis.

To see what this has to do with entropy, let's imagine that we begin with a pure substance that forms a perfect crystalline lattice at the lowest temperature possible, absolute zero. At this stage, none of the kinds of motion that we have been talking about are present. The individual atoms and molecules are as well defined in position and in terms of energy as they can ever be. The **third law of thermodynamics** states that *the entropy of a pure crystalline substance at absolute zero is zero:* $S°(0\ K) = 0$. As the temperature is raised, the units of the solid lattice begin to acquire energy. In a crystalline solid, the molecules or atoms that occupy the lattice places are constrained to remain more or less in place. Nevertheless, they may store energy in the form of vibrational motion about their lattice positions. Instead of all the molecules necessarily being in the lowest possible energy state, there is a kind of expansion in the number of possible energies that the lattice atoms or molecules may have. This increase in possible energy states is not unlike the expansion of the gas illustrated in Figure 17.1. The entropy of the gas increases on expansion because the volume through which the gas molecules move is larger. The entropy of the lattice increases with temperature because the number of possible energy states in which the molecules or atoms are distributed is larger.

It is instructive to follow what happens to the entropy of our substance as we continue to heat it. Let's suppose that at some temperature a phase change occurs, converting the substance from one solid form to another. This means that the arrangement of the lattice units changes in some way, possibly so that the lattice is less regular.* This type of phase change occurs sharply at one temperature, just as do other types of phase changes, for example, from a solid to a liquid. When the change occurs,

*As an example of a solid-state phase change, gray tin converts at 13°C to another solid form called white tin. White tin is stable above the transition temperature, gray tin below it. White tin has a higher entropy than that of gray tin.

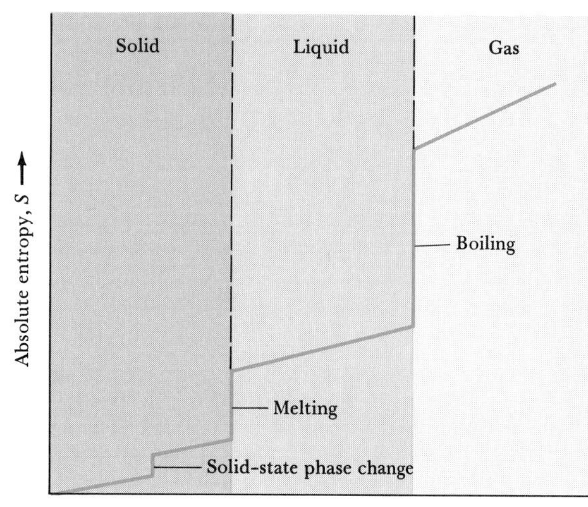

FIGURE 17.5 Entropy changes that occur as the temperature of a substance rises from absolute zero.

there is a change in entropy, because the two lattice arrangements do not have precisely the same degrees of randomness.

Figure 17.5 shows the variation in entropy with temperature for our sample. Note that the change in S with temperature is gradual up to the solid-state phase change and that there is then a sharp increase in S at that temperature. At temperatures above the phase change, the entropy increases with increasing temperature, up to the melting point of the solid.

When the solid melts, the units of the lattice are no longer confined to specific locations relative to other units, but are free to move about the entire volume of the unit. This added freedom of motion for the individual molecules adds greatly to the entropy content of the substance. At the temperature of melting, we therefore see a large increase in entropy content. After all the solid has melted, the temperature again increases, and with it the entropy.

SAMPLE EXERCISE 17.2

Figure 17.5 shows that the entropy of a liquid increases as its temperature increases. What factors are responsible for the increase in entropy?

Solution: The average kinetic energy of the molecules in a liquid increases with temperature. As temperature increases, more molecules at any given instant possess higher energies. This "expansion" in the energies that the molecules possess is measured by the increase in entropy. The increase in S with increasing temperature results from increased energy of motion of all kinds within the liquid.

At the boiling point of the liquid there is again a big increase in entropy. The increase in this case results largely from the increased volume in which the molecules may be found. This is intuitively in line with our earlier ideas about entropy, because an increase in volume means an increase in randomness. It is less likely that a given molecule will be found in a given volume element when there is a great expansion from the liquid to the gaseous state.

As the gas is heated, the entropy increases steadily, because more and more energy is being stored in the gas molecules. The distribution of molecular speeds is spread out toward higher values, as illustrated in Figure 9.9. Again, the idea of an expansion in the range of energies in which molecules may be found helps us remember that increased average energy means increased entropy.

SAMPLE EXERCISE 17.3

For each of the following pairs of substances, indicate which has the higher entropy and suggest a possible reason: (a) 1 mol of $NaCl(s)$ and 1 mol of $HCl(g)$ at 25°C; (b) 2 mol of $HCl(g)$ and 1 mol of $HCl(g)$ at 25°C; (c) 1 mol of $HCl(g)$ and 1 mol of $Ar(g)$ at 25°C; (d) 1 mol of $N_2(s)$ at 24 K and 1 mol of $N_2(g)$ at 298 K.

Solution: (a) Gaseous HCl has the higher entropy per mole, because it has acquired a high degree of randomness as a result of being in the gaseous state. (b) The sample containing 2 mol of HCl has twice the entropy of the sample containing 1 mol. (c) The HCl sample has the higher entropy, because the HCl molecule is capable of storing energy in more ways than is Ar. It may rotate, or the H—Cl distance may change periodically in a vibrational motion. (d) The gaseous N_2 sample has the higher entropy, because the entropy increases resulting from melting and then boiling of N_2 are included in its total entropy content.

SAMPLE EXERCISE 17.4

Predict whether the entropy change of the system in each of the following reactions is positive or negative.

(a) $H_2O(l) \longrightarrow H_2O(g)$ (at 25°C)

(b) $CaCO_3(s) \longrightarrow CaO(s) + CO_2(g)$

(c) $N_2(g) + 3H_2(g) \longrightarrow 2NH_3(g)$

(d) $N_2(g) + O_2(g) \longrightarrow 2NO(g)$

(e) $Ag^+(aq) + Cl^-(aq) \longrightarrow AgCl(s)$

Solution: (a) The entropy change in this process is positive, because the single substance involved is going from the liquid to the gaseous state. We have already seen that the phase change from liquid to gas results in an increase in entropy (Figure 17.5).

(b) The entropy change here is positive, because a solid is converted into a solid and a gas. Gaseous substances generally possess more entropy than solids, so whenever the products contain more moles of gas than the reactants, the entropy change is probably positive.

(c) The entropy change in formation of ammonia from nitrogen and hydrogen is negative, because there are fewer moles of gas in the product than in the reactants.

(d) This represents a case in which the entropy change will be small, because the same number of moles of gas is involved in the reactants and the product. The sign of ΔS is impossible to predict based on our discussions thus far, but we can predict that ΔS will be small.

(e) The entropy change in the precipitation of a salt from solution is negative. The ions in solution are free to move about the entire volume of the liquid, whereas in the solid lattice they are more confined.

17.4 CALCULATION OF ENTROPY CHANGES

The enthalpy change in a chemical reaction is often easily measured in a calorimeter, as described in Section 4.6. There is no comparable easy means for measuring the change in entropy. Nevertheless, various types of measurements have made it possible to determine the **absolute entropy** at any temperature for a great number of substances. These entropies are based on the fact that the entropies of pure solids are zero at absolute zero. A table of absolute entropy values (usually written $S°$) is included in Appendix D. These entropies are expressed in units of joules per degree Kelvin per mole: J/K-mol.

The entropy change in a chemical reaction is given by the sum of the entropies of the products less the sum of entropies of reactants. Thus, in the overall reaction

$$aA + bB + \cdots \rightleftharpoons pP + qQ + \cdots \qquad [17.8]$$

The total entropy change is given by

$$\Delta S° = [pS°(P) + qS°(Q) + \cdots] - [aS°(A) + bS°(B) + \cdots] \qquad [17.9]$$

In other words we sum the absolute entropies of all the products, multiplying each by the coefficient of the product in the balanced equation, and then subtract the same sort of sum of entropies of the reactants.

SAMPLE EXERCISE 17.5

Calculate $\Delta S°$ for the synthesis of ammonia from $N_2(g)$ and $H_2(g)$:

$$N_2(g) + 3H_2(g) \longrightarrow 2NH_3(g)$$

Solution: Using Equation 18.9, we have

$$\Delta S° = 2S°(NH_3) - [S°(N_2) + 3S°(H_2)]$$

Substituting the appropriate $S°$ values from Appendix D yields

$$\Delta S° = (2\text{ mol})\left(192.5 \frac{\text{J}}{\text{K-mol}}\right) - \left[(1\text{ mol})\left(191.5 \frac{\text{J}}{\text{K-mol}}\right) + (3\text{ mol})\left(130.58 \frac{\text{J}}{\text{K-mol}}\right)\right]$$

$$= -198.2\text{ J/K}$$

The value for $\Delta S°$ is negative, as we predicted in Sample Exercise 17.4(c).

17.5 THE FREE-ENERGY FUNCTION

We still haven't attempted to use thermodynamics to predict whether a given reaction will be spontaneous. We have seen that spontaneity involves two thermodynamic concepts, entropy and enthalpy. Before we can make the predictions we would like, we must introduce a third function that interrelates entropy and enthalpy. That function is called **free energy,** or Gibbs free energy after the American mathematician J. Willard Gibbs (1839–1903), who first proposed it (Figure 17.6). The free energy, G, is related to enthalpy and entropy by the expression

$$G = H - TS \qquad [17.10]$$

where T is the absolute temperature. Free energy, like the enthalpy and entropy functions to which it is related, is a state function.

For a process occurring at constant temperature and pressure, the change in free energy is given by the expression

$$\Delta G = \Delta H - T\Delta S \qquad [17.11]$$

A process that is driven spontaneously toward equilibrium both by decreasing energy (negative ΔH) and increasing randomness (positive ΔS) will have a negative ΔG. Indeed, there is a simple relationship between

FIGURE 17.6 Josiah Willard Gibbs (1839–1903) was the first person to be awarded a Ph.D. in science from an American university (Yale, 1863). He went on to become one of the foremost mathematical scientists of his day. From 1871 until his death he held the chair of mathematical physics at Yale University. He made significant contributions to thermodynamics and is credited with laying much of the theoretical foundation that led to the development of chemical thermodynamics. (*Culver Pictures*)

the sign of ΔG for a reaction and the spontaneity of that reaction operated at constant temperature and pressure:

1 If ΔG is negative, the reaction is spontaneous in the forward direction.
2 If ΔG is zero, the reaction is at equilibrium; there is no driving force tending to make the reaction go in either direction.
3 If ΔG is positive, the reaction in the forward direction is nonspontaneous; work must be supplied from the surroundings to make it occur. However, the reverse reaction will be spontaneous.

An analogy is often drawn between the free-energy change in a spontaneous reaction and the potential energy change in a boulder rolling down a hill. Potential energy in a gravitational field "drives" the boulder until it reaches a state of minimum potential energy in the valley [Figure 17.7(*a*)]. Similarly, the free energy of a chemical system decreases (negative ΔG) until it reaches a minimum value [Figure 17.7(*b*)]. When this minimum is reached, a state of equilibrium exists. In any spontaneous process at constant temperature and pressure, the free energy always decreases. As shown in Figure 17.7(*b*), the equilibrium condition can be approached by a spontaneous change from either direction, from the product side or the reactant side.

As an example of these ideas, let's return to the synthesis of ammonia from nitrogen and hydrogen. Imagine that we have a certain number of moles of nitrogen and three times that number of moles of hydrogen in a reaction vessel that permits us to maintain a constant temperature and pressure. We know from our earlier discussions that the formation of ammonia will not be complete; an equilibrium will be reached in which the reaction vessel contains some mixture of N_2, H_2, and NH_3. The free energy of the system decreases (negative ΔG) until this equilibrium is attained. Once the equilibrium has been reached, there is no further spontaneous formation of NH_3. The equilibrium condition is the minimum free energy available to the system at this temperature and pres-

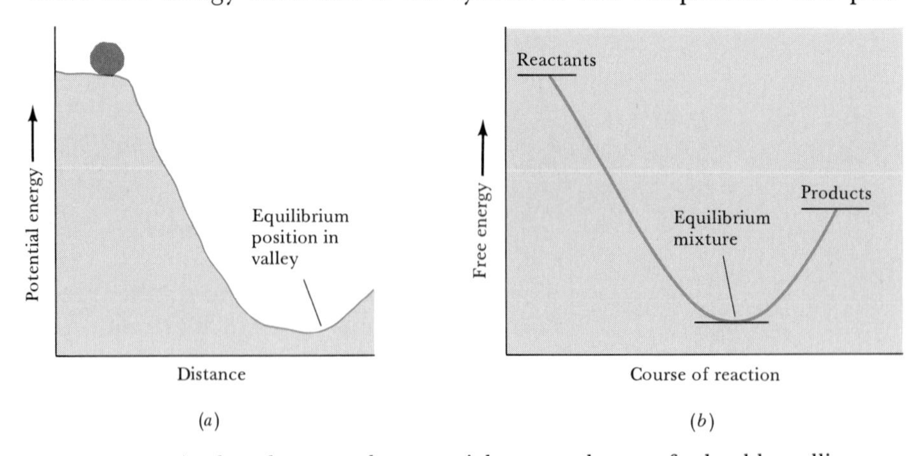

FIGURE 17.7 Analogy between the potential energy change of a boulder rolling down a hill (*a*) and the free-energy change in a spontaneous reaction (*b*). The equilibrium position in (*a*) is given by the minimum potential energy available to the system. The equilibrium position in (*b*) is given by the minimum free energy available to the system.

sure. To form more NH_3 from N_2 and H_2 once equilibrium has been reached requires an increase in free energy (positive ΔG).

It is not necessary that we reach equilibrium beginning only with N_2 and H_2. We could reach the same equilibrium by beginning with an appropriate amount of NH_3. Ammonia held at a constant temperature and pressure will decompose to form N_2 and H_2 until an equilibrium is attained. This process is also spontaneous, involving a decrease in free energy as the system approaches equilibrium. The equilibrium can be approached from either the reactant side or product side, as indicated in a general fashion in Figure 17.7(*b*).

Calculation of $\Delta G°$

We have noted that free energy is a state function. This means that it is possible to tabulate standard free energies of formation for substances, just as it is possible to tabulate standard enthalpies of formation. It is important to remember that standard values for these functions imply a particular set of conditions, or standard states (Section 4.5). The standard state for gaseous substances is 1 atm pressure. For solid substances the standard state is the pure solid; for liquids, the pure liquid. For substances in solution, the standard state is normally a concentration about 1 *M*; in accurate work it may be necessary to make certain corrections, but we need not worry about these. The temperature usually chosen for purposes of tabulating data is 25°C. Just as for the standard heats of formation, the free energies of elements in their standard states are arbitrarily set to zero. This arbitrary choice of reference point has no effect on the quantity in which we are really interested, namely the *difference* in free energy between reactants and products. The rules about standard states are summarized in Table 17.1. A table of standard free energies of formation is to be found in Appendix D.

The standard free energies of formation for substances are useful in calculating the standard free-energy change in a chemical process. For the general reaction

TABLE 17.1 Conventions used in establishing standard free-energy values

State of matter	Standard state
Solid	Pure solid
Liquid	Pure liquid
Gas	1 atm pressure[a]
Solution	Usually 1 *M*[b]
Elements	Standard free energy of formation of the element in its normal state is defined as zero

[a]Neglecting nonideal gas behavior.
[b]Neglecting nonideality of solutions.

$$aA + bB + \cdots \longrightarrow pP + qQ + \cdots \qquad [17.12]$$

the standard free-energy change is

$$\Delta G° = [p\,\Delta G_f°(P) + q\,\Delta G_f°(Q) + \cdots] - [a\,\Delta G_f°(A) + b\,\Delta G_f°(B) + \cdots] \qquad [17.13]$$

In this expression $\Delta G_f°(P)$ represents the standard free energy of formation of product P, and all the other $\Delta G°$ have similar meanings. Stated verbally, the standard free-energy change for a reaction equals the sum of the standard free-energy values per mole of each product, each multiplied by the corresponding coefficient in the balanced equation, less the same sort of sum for the reactants.

What use can be made of this standard free-energy change for a chemical reaction? The quantity $\Delta G°$ tells us whether a mixture of reactants and products, each present under standard conditions, would spontaneously react in the forward direction to produce more products ($\Delta G°$

negative) or in the reverse direction to form more reactants ($\Delta G°$ positive). Because standard free-energy values are readily available for a large number of substances, the standard free-energy change is easy to calculate for any reaction system of interest.

SAMPLE EXERCISE 17.6

Determine the standard free-energy change for the following reaction at 298 K:

$$N_2(g) + 3H_2(g) \rightleftharpoons 2NH_3(g)$$

Solution: Using Appendix D, we find that the standard free energies for the three substances of interest are as follows: $N_2(g)$, $\Delta G_f^\circ = 0.0$; $H_2(g)$, $\Delta G_f^\circ = 0.0$; $NH_3(g)$, $\Delta G_f^\circ = -16.66$ kJ/mol.

The standard free-energy change for the reaction of interest is

$$\Delta G° = 2\,\Delta G_f^\circ(NH_3) - [3\,\Delta G_f^\circ(H_2) + \Delta G_f^\circ(N_2)]$$

Inserting numerical quantities we obtain

$$\Delta G° = -33.32 \text{ kJ}$$

The fact that $\Delta G°$ is negative tells us that a mixture of H_2, N_2, and NH_3 at 25°C, each present at a pressure of 1 atm, would react spontaneously to form more ammonia. (Remember, however, that this says nothing about the rate at which the reaction occurs.)

Free Energy and Temperature

It is worthwhile to examine Equation 17.11 closely to see how the free-energy function depends on both the enthalpy and entropy changes for a given process. If it were not for entropy effects, all exothermic reactions, those in which ΔH is negative, would be spontaneous. The entropy contribution, represented by the quantity $-T\,\Delta S$, may increase or decrease the tendency of the reaction to proceed spontaneously. When ΔS is positive, meaning that the final state is more random or disordered than the initial state, the term $-T\,\Delta S$ makes a negative contribution to ΔG; that is, it increases the tendency of the reaction to occur spontaneously. When ΔS is negative, however, the term $-T\,\Delta S$ decreases the tendency of the reaction to occur spontaneously.

When ΔH and $-T\,\Delta S$ are of opposite sign, the relative importance of the two terms determines whether ΔG is negative or positive. In these instances, temperature is an important consideration. Both ΔH and ΔS are in principle capable of changing with temperature. In practice, how-

TABLE 17.2 Effect of temperature on reaction spontaneity

ΔH	ΔS	ΔG	Reaction characteristics	Examples
−	+	Always negative	Reaction is spontaneous at all temperatures: reverse reaction is always nonspontaneous	$2O_3(g) \longrightarrow 3O_2(g)$
+	−	Always positive	Reaction is nonspontaneous at all temperatures; reverse reaction occurs	$3O_2(g) \longrightarrow 2O_3(g)$
−	−	Negative at low temperatures; positive at high temperatures	Reaction is spontaneous at low temperatures; reverse reaction is nonspontaneous at low temperatures	$CaO(s) + CO_2(g) \longrightarrow CaCO_3(s)$
+	+	Positive at low temperatures; negative at high temperatures	Reaction is nonspontaneous at low temperatures but becomes spontaneous as temperature is raised	$CaCO_3(s) \longrightarrow CaO(s) + CO_2(g)$

ever, the changes that occur are not very large unless very large temperature changes are involved. The only quantity in the equation

$$\Delta G = \Delta H - T\Delta S$$

that changes markedly with temperature is therefore $-T\Delta S$. At high temperatures the entropy term becomes relatively more important.

Various possible situations for the relative signs of ΔH and ΔS are shown in Table 17.2, together with examples of each. By applying the concepts we have developed for predicting entropy changes it is often possible to predict how ΔG will change with change in temperature.

SAMPLE EXERCISE 17.7

(a) Predict the direction in which $\Delta G°$ for the equilibrium

$$N_2(g) + 3H_2(g) \rightleftharpoons 2NH_3(g)$$

will change with increase in temperature. (b) Calculate $\Delta G°$ at 500°C assuming that $\Delta H°$ and $\Delta S°$ do not change with temperature.

Solution: (a) In Sample Exercise 17.5 we saw that the change in $\Delta S°$ for the equilibrium of interest is negative. This means that the term $-T\Delta S°$ is positive and grows larger with increasing temperature. The standard free-energy change, $\Delta G°$, is the sum of the negative quantity $\Delta H°$ and the positive quantity, $-T\Delta S°$. Because only the latter grows larger with increasing temperature, $\Delta G°$ grows less negative.

(b) The $\Delta H_f°$ and $S°$ values necessary to calculate $\Delta H°$ and $\Delta S°$ for this reaction can be taken from Appendix D. We have previously performed these calculations in Sample Exercise 14.12 (Section 14.5) and in Sample Exercise 17.5, giving $\Delta H° = -92.38$ kJ and $\Delta S° = -198.2$ J/K. To calculate $\Delta G°$ given $\Delta H°$ and $\Delta S°$, we use the following relationship:

$$\Delta G° = \Delta H° - T\Delta S°$$

(Recall that the superscript ° indicates that the process is operated under standard-state conditions.) Assuming that $\Delta H°$ and $\Delta S°$ do not change with temperature, and using $T = 500 + 273 = 773$ K, we have

$$\Delta G° = -92.38 \text{ kJ} + (773 \text{ K})\left(-198.2 \frac{\text{J}}{\text{K}}\right)\left(\frac{1 \text{ kJ}}{10^3 \text{ J}}\right)$$

$$= -92.38 \text{ kJ} + 153.21 \text{ kJ}$$

$$= 60.83 \text{ kJ}$$

Notice that we changed $T\Delta S°$ from units of joules to kilojoules so it could be added to $\Delta H°$, which is in units of kilojoules.

In Sample Exercise 17.6 we calculated $\Delta G°$ for this reaction at 298 K: $\Delta G°_{298} = -33.32$ kJ. Thus we see that increasing the temperature from 298 K to 773 K changes $\Delta G°$ from -33.32 kJ to $+60.83$ kJ. Of course, the result at 773 K is not as accurate as that at 298 K, because $\Delta H°$ and $\Delta S°$ do change slightly with temperature. Nevertheless, the result should be a reasonable approximation. The positive increase in $\Delta G°$ with increasing temperature is in agreement with the qualitative prediction made in part (a) of this exercise. Our result indicates that a mixture of $N_2(g)$, $H_2(g)$, and $NH_3(g)$, each at 1 atm pressure (standard-state conditions), will react spontaneously at 298 K to form more $NH_3(g)$; however, this same reaction is not spontaneous at 773 K. In fact, at 773 K, the reaction will proceed in the opposite direction, to form more $N_2(g)$ and $H_2(g)$.

17.6 FREE ENERGY AND THE EQUILIBRIUM CONSTANT

Although it is very valuable to have a ready means of determining $\Delta G°$ for a reaction from tabulated values, we usually want to know about the direction of spontaneous change for systems that are not at standard conditions. For any chemical process the general relationship between the free-energy change under standard conditions, $\Delta G°$, and the free-energy change under any other conditions, ΔG, is given by the following expression:

$$\Delta G = \Delta G^\circ + 2.303RT \log Q \qquad [17.14]$$

R in this expression is the ideal-gas-equation constant, 8.314 J/K-mol; T is the absolute temperature; and Q is the reaction quotient (Section 14.4) that corresponds to the chemical reaction and particular reaction mixture of interest.

SAMPLE EXERCISE 17.8

Calculate ΔG at 298 K for the following reaction if the reaction mixture consists of 1.0 atm N_2, 3.0 atm H_2, and 1.0 atm NH_3:

$$N_2(g) + 3H_2(g) \longrightarrow 2NH_3(g)$$

Solution: For the balanced equation and set of concentrations given, the reaction quotient Q is

$$Q = \frac{P^2_{NH_3}}{P_{N_2}P^3_{H_2}} = \frac{(1.0)^2}{(1.0)(3.0)^3} = 3.7 \times 10^{-2}$$

ΔG°_{298} was calculated in Sample Exercise 17.6: $\Delta G^\circ_{298} = -33.32$ kJ. Thus we have

$$\Delta G = \Delta G^\circ + 2.303\, RT \log Q$$

$$= (-33.32 \text{ kJ}) + 2.303\left(8.314 \frac{\text{J}}{\text{K}}\right)(298 \text{ K}) \times \left(\frac{1 \text{ kJ}}{10^3 \text{ J}}\right) \log (3.7 \times 10^{-2})$$

$$= -33.32 \text{ kJ} + (-8.17 \text{ kJ})$$

$$= -41.49 \text{ kJ}$$

The free-energy change becomes more negative, changing from -33.32 kJ to -41.49 kJ, as the pressures of N_2, H_2, and NH_3 are changed from 1.0 atm each (standard-state conditions, ΔG°) to 1.0 atm, 3.0 atm, and 1.0 atm, respectively. The larger negative value for ΔG when the pressure of H_2 is increased from 1.0 atm to 3.0 atm indicates a larger "driving force" to produce NH_3. This result bears out the prediction of Le Châtelier's principle, which indicates that increasing P_{H_2} should shift the reaction more to the product side, thereby forming more NH_3.

When a system is at equilibrium, ΔG must be zero, and the reaction quotient Q must by definition equal K (Section 14.4). Thus for a system at equilibrium (when $\Delta G = 0$ and $Q = K$), Equation 17.14 transforms as follows:

$$\Delta G = \Delta G^\circ + 2.303RT \log Q \qquad [17.14]$$

$$0 = \Delta G^\circ + 2.303RT \log K$$

$$\Delta G^\circ = -2.303RT \log K \qquad [17.15]$$

From Equation 17.15 we can readily see that if ΔG° is negative, log K must be positive. A positive value for log K means that $K > 1$. On the other hand if ΔG° is positive, log K is negative, which means that $K < 1$. To summarize:

ΔG° negative	$K > 1$
ΔG° zero	$K = 1$
ΔG° positive	$K < 1$

It is possible from a knowledge of $\Delta G°$ for a reaction to calculate the value for the equilibrium constant, using Equation 17.15. Some care is necessary, however, in the matter of units. When dealing with gases, the concentrations of reactants should be expressed in units of atmospheres. The concentrations of solids are 1 if they are pure solids; the concentrations of pure liquids are 1 if they are pure liquids; if a mixture of liquids is involved, the concentration of each is expressed as mole fraction. For substances in solution, concentrations in moles per liter are appropriate.

SAMPLE EXERCISE 17.9

From standard free energies of formation, calculate the equilibrium constant for the reaction

$$N_2(g) + 3H_2(g) \rightleftharpoons 2NH_3(g)$$

at 25°C.

Solution: The equilibrium constant for this reaction is written as

$$K = \frac{P^2_{NH_3}}{P_{N_2}P^3_{H_2}}$$

where the gas concentrations are expressed in atmospheres pressure. The standard free-energy change for the reaction was determined in Sample Exercise 17.6 to be -33.32 kJ. Inserting this into Equation 17.15, we obtain

$$-33{,}320 \text{ J} = -2.303(8.314 \text{ J/K-mol})(298 \text{ K}) \log K_p$$
$$\log K_p = 5.85$$

Taking the antilog, we have

$$K_p = 7.0 \times 10^5$$

This is a large equilibrium constant. Compare its magnitude with the equilibrium constants at higher temperature, as listed in Table 14.5. If a catalyst could be found that would permit reasonably rapid reaction of N_2 with H_2 at room temperature, high pressures would not be required to force the equilibrium toward NH_3.

17.7 FREE ENERGY AND WORK

The free energy has one further interesting property that comes from the fact that it is related to the degree of spontaneity of a process. Any process that occurs spontaneously can be utilized for the performance of useful work, at least in principle. For example, the falling of water in a waterfall is certainly a spontaneous process; it is also one from which it is possible to extract work, by causing the water to turn the blades of a turbine as it falls. Similarly, the burning of gasoline in the cylinders of a car leads to production of useful work in the motion of the car. How much work is extracted from a particular process depends on how it is carried out. For example, we might burn a liter of gasoline in an open container and extract no useful work at all. In an automobile engine, the overall efficiency for production of work is low, perhaps 20 percent. If the gasoline were caused to react with oxygen under other, more favorable conditions, the amount of work we could extract would be higher. In practice, we never achieve the maximum amount of work possible from a theoretical point of view. However, it is useful to know what the maximum possible work derivable from a process is, so that we might have a measure of our success in extracting work from processes of practical importance. Thermodynamics tells us that *the maximum possible work that can be derived from a spontaneous process occurring at constant temperature and pressure is equal to the free-energy change.*

For processes that are not spontaneous ($\Delta G > 0$) the free-energy change is a measure of the *minimum* amount of work that must be done to cause the process to occur. In actual cases, we always need to do more than this theoretical minimum amount, because of inefficiencies in the way the change is caused to occur.

Free-energy considerations are very important in thinking about many nonspontaneous reactions that we might wish to carry out for our own purposes, or that occur in nature. For example, we might wish to extract a metal from an ore. If we look at a reaction such as

$$\begin{array}{c} Cu_2S(s) \longrightarrow 2Cu(s) + S(s) \\ \Delta G^\circ = +86.2 \text{ kJ}; \Delta H^\circ = +79.5 \text{ kJ} \end{array} \qquad [17.16]$$

we find that it is endothermic and highly nonspontaneous. Clearly, then, we cannot hope to obtain copper metal from Cu_2S merely by trying to catalyze the reaction shown in Equation 17.16. Instead, we must "do work" on the reaction in some way, in order to force it to occur as we wish. We might do this by coupling the reaction we've written with another reaction, so that we arrive at an overall reaction that *is* spontaneous. Consider, for example, the reaction

$$\begin{array}{c} S(s) + O_2(g) \longrightarrow SO_2(g) \\ \Delta G^\circ = -300.1 \text{ kJ}; \Delta H^\circ = -296.9 \text{ kJ} \end{array} \qquad [17.17]$$

This is an exothermic, spontaneous reaction. Adding reaction Equations 17.16 and 17.17, we obtain

$$\begin{array}{rcll} Cu_2S(s) & \longrightarrow & 2Cu(s) + S(s) & \Delta G^\circ = +86.2 \text{ kJ}; \Delta H^\circ = +79.5 \text{ kJ} \\ S(s) + O_2(g) & \longrightarrow & SO_2(g) & \Delta G^\circ = -300.1 \text{ kJ}; \Delta H^\circ = -296.9 \text{ kJ} \\ \hline Cu_2S(s) + O_2(g) & \longrightarrow & 2Cu(s) + SO_2(g) & \\ & & & \Delta G^\circ = -213.9 \text{ kJ}; \Delta H^\circ = -217.4 \text{ kJ} \end{array}$$

The free-energy change for the overall reaction is the sum of the free-energy changes of the two reactions, and similarly for the overall enthalpy change. Because the negative free-energy change for the second reaction is larger than the positive free-energy change for the first, the overall reaction has a large and negative standard free-energy change.

The coupling of two or more reactions together to cause a nonspontaneous chemical process to occur is very important in biochemical systems. Many of the reactions that are absolutely essential to the maintenance of life do not occur within the human body spontaneously. These necessary reactions are made to occur, however, by coupling them with reactions that are spontaneous and energy releasing. The energy releases that accompany the metabolism of foodstuffs provide the primary source of necessary free energy. For example, a compound such as glucose, $C_6H_{12}O_6$, is oxidized in the body, and a substantial amount of energy is released.

$$\begin{array}{c} C_6H_{12}O_6(s) + 6O_2(g) \longrightarrow 6CO_2(g) + 6H_2O(l) \\ \Delta G^\circ = -2880 \text{ kJ}; \Delta H^\circ = -2800 \text{ kJ} \end{array} \qquad [17.18]$$

This is energy is employed to "do work" in the body. However, some means is necessary to couple, or connect, the energy released by glucose

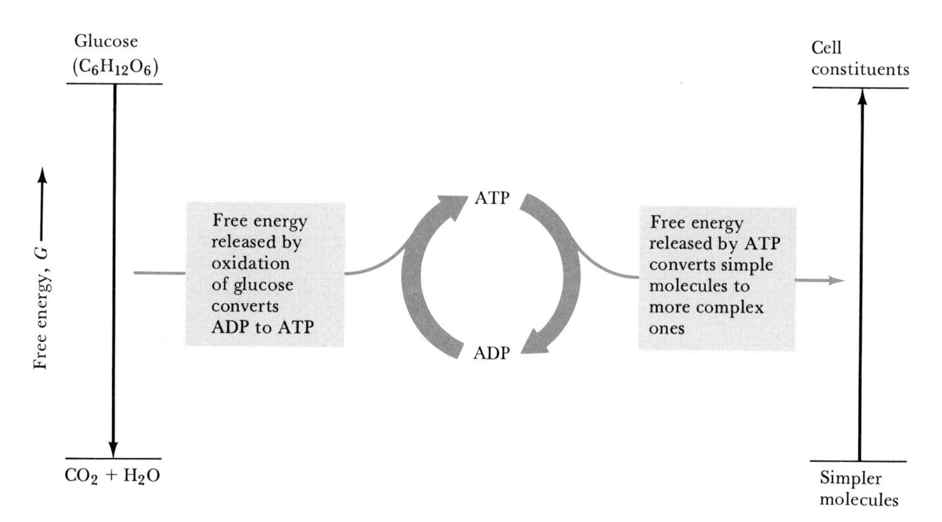

FIGURE 17.8 Schematic representation of a part of the free-energy changes that occur in cell metabolism. The oxidation of glucose to CO_2 and H_2O produces free energy. This released free energy is used to convert ADP into the more energetic ATP. The ATP is then used, as needed, as an energy source to convert simple molecules into more complex cell constituents. When ATP releases its free energy, it is converted to ADP.

oxidation to the reactions that require energy. One means of accomplishing this is shown graphically in Figure 17.8. Adenosine triphosphate (ATP) is a high-energy molecule. When ATP is converted to a lower-energy molecule, adenosine diphosphate (ADP), energy to drive other chemical reactions becomes available. The energy released in glucose oxidation is used in part to reconvert ADP back to ATP. Thus the ATP-ADP interconversions act as a means of storing energy and releasing it to drive needed reactions. The coupling of reactions so that the free energy released in one may be utilized by another requires particular enzymes as catalysts. In Chapter 26, which deals with the biosphere, we shall have occasion to examine the energy relationships in living systems in more detail.

FOR REVIEW

Summary

In this chapter we have examined the concept of equilibrium from a thermodynamic point of view. The enthalpy change of a system, ΔH, is a measure of the potential energy change in a process. Exothermic processes ($\Delta H < 0$) tend to occur spontaneously. The spontaneous character of a reaction is also determined by the change in randomness or disorder of the system, measured by the entropy, S. Processes that produce an increase in randomness or disorder of the system ($\Delta S > 0$) tend to occur spontaneously.

Entropy changes in a system are associated with an increase in the number of ways the particles of the system can be distributed among possible energy states or spatial arrangements. For example, an increase in volume, in translational energy, or in number of particles all lead to an increase in entropy. In any process the total entropy change is the sum of the entropy change of the system and the surroundings. The second law of thermodynamics tells us that in any spontaneous process the entropy of the universe increases. That is,

$$\Delta S_{\text{system}} + \Delta S_{\text{surroundings}} > 0$$

The standard entropy change in a system, $\Delta S°$, can be calculated from tabulated absolute entropy values, $S°$. Absolute entropy values are determined with the aid of the third law of thermodynamics, which states that the entropy of a pure crystalline solid at 0 K is zero.

The free energy, G, is a thermodynamic state function that combines the two state functions enthalpy and entropy: $G = H - TS$. For processes that occur at constant temperature and pressure, $\Delta G = \Delta H - T\Delta S$. The free-energy change for a process occurring at constant temperature and pressure relates directly to reaction spontaneity. ΔG is negative for all spontaneous processes, whereas a positive value for ΔG indicates a nonspontaneous process (one that is spontaneous in the reverse direction). ΔG is zero at equilibrium. The free energy also has the important property that it is a measure of the maximum useful work that can be performed by a system in a spontaneous process. In practice this amount of useful work is never realized, because processes in the real world have inherent inefficiencies. If a process is nonspontaneous, the value of ΔG is a measure of the minimum work that must be done on the system to cause the process to occur. Again, in practice we always need to do more than this minimum.

The standard free-energy change, $\Delta G°$, for any process can be calculated from tabulated standard free energies of formation, $\Delta G_f°$; it can also be calculated from standard enthalpy and entropy changes: $\Delta G° = \Delta H° - T\Delta S°$. Temperature changes will change the value of ΔG and can also change its sign.

The free-energy change under nonstandard conditions is related to the standard free-energy change: $\Delta G = \Delta G° + 2.303RT \log Q$. At equilibrium ($\Delta G = 0$, $Q = K$), $\Delta G° = -2.303RT \log K$. Thus the standard free-energy change is related to the equilibrium constant. Consequently, we can understand the position of a chemical equilibrium in terms of the $\Delta H°$ and $T\Delta S°$ functions of which $\Delta G°$ is composed.

Learning goals

Having read and studied this chapter, you should be able to:

1 Define the term *spontaneity* and apply it in identifying spontaneous processes.

2 Describe how entropy is related to randomness or disorder.

3 State the second law of thermodynamics.

4 Predict whether the entropy change in a given process is positive, negative, or near zero.

5 State the third law of thermodynamics.

6 Describe how and why the entropy of a substance changes with increasing temperature or when a phase change occurs, starting with the substance as a pure solid at 0 K.

7 Calculate $\Delta S°$ for any reaction from tabulated absolute entropy values, $S°$.

8 Define free energy in terms of enthalpy and entropy.

9 Explain the relationship between the sign of the free-energy change, ΔG, and whether a process is spontaneous in the forward direction.

10 Calculate the standard free-energy change at constant temperature and pressure, $\Delta G°$, for any process from tabulated values for the standard free energies of reactants and products.

11 List the usual conventions regarding standard states in setting the values for standard free energies.

12 Predict how ΔG will change with temperature, given the signs for ΔH and ΔS.

13 Estimate $\Delta G°$ at any temperature given $\Delta S°_{298}$ and $\Delta H°_{298}$.

14 Calculate the free-energy change under nonstandard conditions, ΔG, given $\Delta G°$, temperature, and the data needed to calculate the reaction quotient.

15 Calculate $\Delta G°$ from K and perform the reverse operation.

16 Describe the relationship between ΔG and the maximum work that can be derived from a spontaneous process, or the minimum work required to accomplish a nonspontaneous process.

Key terms

Among the more important terms and expressions used for the first time in this chapter are the following:

Absolute entropy, $S°$ (Section 17.4), refers to the entropy of a particular substance at some temperature and in some particular state, referenced to a zero entropy for the pure solid substance at a temperature of absolute zero.

Entropy (Section 17.2) is a thermodynamic function associated with the number of different, equivalent energy states or spatial arrangements in which a system may be found. It is a thermodynamic state function, which means that once we specify the conditions for a system, that is, the temperature, pressure, and so on, the entropy is defined.

Free energy (Section 17.5) is a thermodynamic state function that combines enthalpy and entropy, in the form $G = H - TS$. For a change occurring at constant temperature and pressure, the change in free energy is $\Delta G = \Delta H - T\Delta S$.

An isolated system (Section 17.2) is one that does not exchange heat or work with its surroundings.

The second law of thermodynamics (Section 17.1) is a statement of our experience that there is a direction to the way events occur in nature: When a process occurs spontaneously in one direction, it is nonspontaneous in the reverse direction. It is possible to state the second law in many different forms, but they all relate back to the same idea about spontaneity. One of the most common statements found in chemical contexts is that in any spontaneous process the entropy of the universe increases.

A spontaneous process (Section 17.1) is one that is capable of proceeding in a given direction, as written or described, without needing to be driven by an outside source of energy. A process may be spontaneous even though it is very slow. ΔG is negative for all spontaneous processes.

The third law of thermodynamics (Section 17.3) states that the entropy of a pure, crystalline solid at absolute zero temperature is zero: $S°(0\ K) = 0$.

Vibrational and rotational energies (Section 17.3) refer to the storing of energies in molecules in the form of vibrational motions between the atoms of the molecules or rotational motions of the molecule as a whole.

EXERCISES

Spontaneous Processes

17.1 Which of the following processes are spontaneous: (a) The melting of an ice cube at $-5°C$, 1 atm pressure; (b) formation of CO_2 and H_2O from CH_4 and O_2 at 300 K; (c) dissolution of sugar in water at 40°C; (d) $2N(g) \longrightarrow N_2(g)$; (e) rusting of iron; (f) alignment of iron filings in a magnetic field?

17.2 For each of the following pairs, indicate which substance you would expect to possess the larger absolute entropy: (a) 1 mol of $H_2(g)$, 298 K, 1 atm pressure, or 1 mol of $H_2(g)$, 298 K, 10 atm pressure; (b) 1 mol of $H_2O(s)$ at 5°C or 1 mol of $H_2O(l)$ at 5°C; (c) 1 mol of $Br_2(l)$ at 1 atm, 58.8°C, or 1 mol of $Br_2(g)$ at 1 atm, 58.8°C; (d) 1 mol of $KNO_3(s)$, or 1 mol of $KNO_3(aq)$ at 30°C; (e) 1 mol of $O(g)$ at 300 K or 1 mol of $O_2(g)$ at 300 K.

17.3 Indicate whether each of the following statements is true or false. If it is false, correct it. (a) The feasibility of manufacturing NH_3 from N_2 and H_2 depends entirely on the value of ΔH for the process $N_2(g) + 3H_2(g) \longrightarrow 2NH_3(g)$. (b) The reaction of $H_2(g)$ with $Cl_2(g)$ to form $HCl(g)$ is an example of a spontaneous process. (c) A spontaneous process is one that occurs rapidly. (d) A process that is nonspontaneous in one direction is spontaneous in the opposite direction. (e) Spontaneous processes are those that are exothermic and that lead to a higher degree of order in the system.

17.4 In each of the following, indicate whether the process leads to a higher or lower degree of ordering in the system: (a) $I_2(s) \longrightarrow I_2(g)$;
(b) $KNO_3(s) \longrightarrow KNO_3(aq)$;
(c) $2Na(s) + Cl_2(g) \longrightarrow 2NaCl(s)$; (d) 50 mL of $H_2O(l)$ + 50 mL of $C_2H_5OH(l)$ (ethyl alcohol) $\longrightarrow$ 100 mL of solution.

17.5 A nineteenth-century chemist, Marcellin Berthelot, suggested that all chemical processes that proceed spontaneously are exothermic. Is this correct? If you think not, offer some counterexamples.

17.6 Using Appendix D, compare the absolute entropy values at 298 K for the substances in each of the following pairs: (a) $Hg(g)$ and $Hg(l)$; (b) $F(g)$ and $F_2(g)$; (c) $FeO(s)$ and $Fe_2O_3(s)$; (d) $H_2O(l)$ and $H_2O_2(l)$. Explain the origin of the difference in $S°$ values in each case.

17.7 Predict the sign of ΔS for the system in each of the following processes: (a) freezing of 1 mol of $H_2O(l)$; (b) evaporation of 1 mol of $Br_2(l)$; (c) precipitation of $BaSO_4$ upon mixing $Ba(NO_3)_2(aq)$ and $H_2SO_4(aq)$; (d) oxidation of magnesium metal: $2Mg(s) + O_2(g) \longrightarrow 2MgO(s)$.

17.8 Predict whether the entropy change in the system is positive or negative for each of the following processes:

(a) $2C(s) + O_2(g) \longrightarrow 2CO(g)$
(b) $2K(s) + Br_2(l) \longrightarrow 2KBr(s)$
(c) $2MnO_2(s) \longrightarrow MnO(s) + O_2(g)$
(d) $O(g) + O_2(g) \longrightarrow O_3(g)$

17.9 For each of the following processes, indicate whether the sign of ΔS and ΔH is expected to be positive, negative, or about zero. (a) A solid sublimes. (b) The temperature of a solid is lowered by 25°C. (c) Ethyl alcohol evaporates from a beaker. (d) A diatomic molecule dissociates into atoms. (e) A piece of charcoal is combusted to form $CO_2(g)$ and $H_2O(g)$.

17.10 Using tabulated $S°$ values from Appendix D, calculate $\Delta S°$ for each of the following reactions:

(a) $2HBr(g) + F_2(g) \longrightarrow 2HF(g) + Br_2(g)$
(b) $2NO(g) + O_2(g) \longrightarrow 2NO_2(g)$
(c) $2CH_3OH(g) + 3O_2(g) \longrightarrow 2CO_2(g) + 4H_2O(g)$
(d) $2FeO(s) + 2O_2(g) \longrightarrow Fe_2O_3(s)$

In each case, account for the sign of $\Delta S°$.

17.11 When the insecticide DDT is applied it is sprayed over large areas. It later moves in the environment through soil into plants and water supplies. In lakes it concentrates in the fatty tissue of fish. If we think of DDT as the "system," describe the entropy change associated with each of the processes above. In terms of cleaning up

DDT in the environment, what entropy change to the DDT system is required? How can such an entropy change be brought about? Is the process feasible?

17.12 The reaction $Mg(s) + O_2(g) \longrightarrow 2MgO(s)$ has associated with it a negative value for $\Delta S°$, and it is a highly spontaneous process. The second law of thermodynamics states that in any spontaneous process there is always an increase in the entropy of the universe. Is there an inconsistency between the second law and what we know about the oxidation of magnesium? Explain.

17.13 The melting and boiling points of HCl are $-115°C$ and $-84°C$, respectively. Using this information and the value of $S°$ listed in Appendix D, draw a rough sketch of $S°$ for HCl from absolute zero to 298 K.

17.14 Gold and silver form a solid solution in which the Ag and Au atoms are randomly dispersed throughout the lattice. Would you expect the absolute entropy at 0 K of a solid solution of the two metals in 1:1 proportion to be zero, greater than zero, or less than zero? Explain.

17.15 What factors are responsible for the major portion of the entropy change in each of the following processes? (a) A solid melts. (b) A liquid evaporates. (c) The temperature of $SO_2(g)$ is increased by 25°C. (d) A solid salt dissolves in water.

Free energy

17.16 (a) Express the free-energy change in a process in terms of the changes that occur in the enthalpy and entropy of the system: (b) Using this expression, indicate the sign of the free-energy change for a process with $\Delta H = 0$ and ΔS positive. (c) What is the significance of a value of $\Delta G = 0$ for a system? (d) What is the relationship between ΔG for a process and the speed at which it occurs?

17.17 What is the meaning of the *standard* free-energy change in a process, $\Delta G°$, as contrasted with simply the free-energy change, ΔG?

17.18 Using the data from Appendix D, calculate the standard free-energy change for each of the following processes. In each case indicate whether the reaction is spontaneous under standard conditions.

(a) $MnO_2(s) + 2CO(g) \longrightarrow Mn(s) + 2CO_2(g)$
(b) $H_2(g) + Br_2(g) \longrightarrow 2HBr(g)$
(c) $6Cl_2(g) + 2Fe_2O_3(g) \longrightarrow 4FeCl_3(s) + 3O_2(g)$
(d) $CaO(s) + H_2O(l) \longrightarrow Ca(OH)_2(s)$

17.19 Calculate $\Delta G°_{298}$ for

$H_2O_2(g) \longrightarrow H_2O(g) + \frac{1}{2}O_2(g)$

given that $\Delta H°_{298} = -106$ KJ and $\Delta S°_{298} = +58$ J/K for this process. Would you expect $H_2O_2(g)$ to be very stable at 298 K? Explain briefly.

17.20 Using the data in Appendix D, calculate $\Delta H°$, $\Delta S°$, and $\Delta G°$ for each of the following reactions. In each case show that $\Delta G° = \Delta H° - T\Delta S°$.

(a) $BaO(s) + CO_2(g) \longrightarrow BaCO_3(s)$
(b) $2KClO_3(s) \longrightarrow 2KCl(s) + 3O_2(g)$
(c) $2CH_3OH(l) + 3O_2(g) \longrightarrow 2CO_2(g) + 4H_2O(g)$
(d) $NOCl(g) + Cl(g) \longrightarrow NO(g) + Cl_2(g)$

17.21 Classify each of the following reactions as belonging to one of the four possible types summarized in Table 17.2:

(a) $N_2(g) + 3F_2(g) \longrightarrow 2NF_3(g)$
$\Delta H° = -249$ kJ; $\Delta S° = -278$ J/K
(b) $N_2(g) + 3Cl_2(g) \longrightarrow 2NCl_3(g)$
$\Delta H° = +460$ kJ; $\Delta S° = -275$ J/K
(c) $N_2F_4(g) \longrightarrow 2NF_2(g)$
$\Delta H° = 85$ kJ; $\Delta S° = 198$ J/K
(d) $2H_2O(l) \longrightarrow 2H_2(g) + O_2(g)$
$\Delta H° = 572$ kJ; $\Delta S° = 329$ J/K

17.22 (a) Using the data of Appendix D, predict how $\Delta G°$ for the following process will change with increasing temperature:

$P_4(g) \longrightarrow 2P_2(g)$

(b) Calculate $\Delta G°$ at 900 K, assuming that $\Delta H°$ and $\Delta S°$ do not change with temperature.

17.23 (a) Using the data of Appendix D, predict how $\Delta G°$ for the following process will change with increasing temperature:

$SO_3(g) + H_2(g) \longrightarrow SO_2(g) + H_2O(g)$

(b) Calculate $\Delta G°$ at 600 K, assuming that $\Delta H°$ and $\Delta S°$ do not change with variation in temperature.

17.24 (a) Using the data of Appendix D, predict how $\Delta G°$ for the following process will change with increasing temperature:

$CaO(s) + SO_3(g) \longrightarrow CaSO_4(s)$

(b) Calculate $\Delta G°$ at 800 K, assuming that $\Delta H°$ and $\Delta S°$ do not change with increasing temperature.

Free energy and equilibrium

17.25 Explain qualitatively how ΔG changes for each of the following reactions as the partial pressure of N_2 is increased:

(a) $N_2H_4(g) \longrightarrow N_2(g) + 2H_2(g)$
(b) $N_2(g) + 3F_2(g) \longrightarrow 2NF_3(g)$
(c) $NH_4NO_2(s) \longrightarrow N_2(g) + 2H_2O(g)$

17.26 Using the data of Appendix D, calculate ΔG at 298 K for each of the following reactions, given the pressures listed: (a) $N_2(g) + 3H_2(g) \longrightarrow 2NH_3(g)$; $P_{N_2} = 12.0$ atm, $P_{H_2} = 2.0$ atm, $P_{NH_3} = 4.0$ atm; (b) $N_2H_4(g) \longrightarrow N_2(g) + 2H_2(g)$; $P_{N_2H_4} = 6.0$ atm, $P_{N_2} = 1 \times 10^{-3}$ atm, $P_{H_2} = 2 \times 10^{-4}$ atm.

17.27 Calculate the standard free-energy change for the reaction

$PbCO_3(s) \longrightarrow PbO(s) + CO_2(g)$

at 25°C. At what pressure of CO_2 gas does $\Delta G = 0$ at 25°C?

17.28 Write the equilibrium-constant expression and calculate the magnitude of the equilibrium constant at 298 K for each of the following reactions, using data from Appendix D:

(a) $2HBr(g) + Cl_2(g) \rightleftharpoons 2HCl(g) + Br_2(g)$
(b) $CaSO_4(s) + CO_2(g) \rightleftharpoons CaCO_3(s) + SO_3(g)$

17.29 Consider the weak acid $HY(aq)$. If a 0.10 M solution of HY has a pH of 2.86 at 25°C, what is $\Delta G°$ for the acid dissociation?

[17.30] The reaction

$$SO_2(g) + 2H_2S(g) \rightleftharpoons 3S(s) + 2H_2O(g)$$

is the basis of a suggested method for removal of SO_2 from power-plant stack gases. The standard free energy for each substance is given in Appendix D. (a) What is the equilibrium constant at 298 K for the reaction as written? (b) Is this reaction at least in principle a feasible method of removing SO_2? (c) Assuming that the H_2O vapor pressure is 25 mm Hg and adjusting the conditions so that P_{SO_2} equals P_{H_2S}, calculate the equilibrium SO_2 pressure in this system. (d) The reaction as written is exothermic. Would you expect the process to be more or less effective at higher temperatures?

Free energy and work

17.31 Give examples of work that can be realized from each of the following spontaneous processes; (a) combustion of 100 g of ethyl alcohol; (b) movement of large air masses in response to pressure differences in the atmosphere; (c) metabolism of 100 g of potato chips in a human; (d) irradiation of the planet by solar energy.

17.32 Calculate $\Delta G°$, $\Delta H°$ and $\Delta S°$ for the reactions, $2FeO(s) \longrightarrow 2Fe(s) + O_2(g)$ and $C(s) + O_2(g) \longrightarrow CO_2(g)$. Can recovery of Fe from FeO be carried out by coupling the first reaction with the second at 1200 K? (Use graphite carbon for your calculations.)

[17.33] Consider the apparatus shown in Figure 17.9, which is much like that shown in Figure 17.1. $N_2(g)$ is initially confined in flask A, of 1 L volume, at 2.0 atm pressure, and chamber B, of volume 1 L, is evacuated. When the stopcock is opened, N_2 expands to both sides. (a) Assuming that N_2 behaves as an ideal gas, what is ΔH for this process? (b) Describe the process by which the gas can be returned to its original condition. What change in ΔH occurs in the system in this process? (c) What are the changes in ΔG and ΔH of the system for the overall processes of expansion and compression? (d) In this overall process is work done by the system on the surroundings, or by the surroundings on the system? Explain. (e) Describe how this example illustrates the second law of thermodynamics.

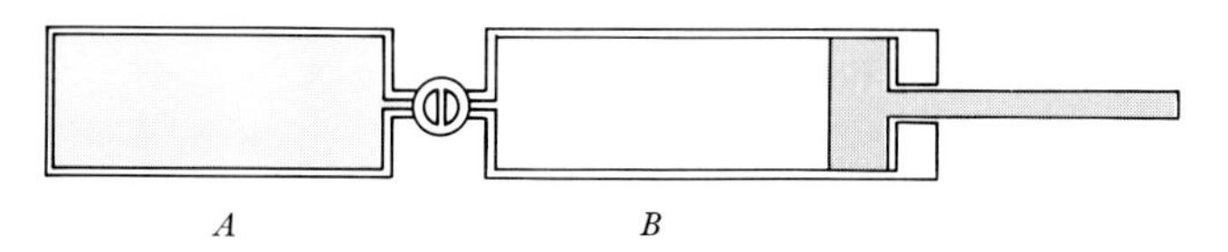

FIGURE 17.9

Additional exercises

17.34 Provide a brief definition of each of the following terms: (a) isolated system; (b) second law of thermodynamics; (c) state function; (d) spontaneous process; (e) entropy.

17.35 Proteins are long-chain biomolecules made from amino acids. The protein chain is held in specific arrangements by hydrogen bonds between N—H and >C=O groups along the chain (Section 10.4). When a protein is denatured, as when an egg is boiled, the increased thermal energy causes rupture of hydrogen bonds, and the regular arrangement along the protein chain is lost. What are the signs of ΔH and ΔS in the protein system associated with denaturation?

17.36 Explain briefly why entropy increases in each of the following processes. (a) Ice melts. (b) Carbon tetrachloride evaporates. (c) 5 g of dioxin becomes distributed over a 5-km^2 area in Times Beach, Missouri. (d) Solid niobium is warmed from 2.4 K to 300 K.

17.37 When a rubber band is stretched, the long-chain molecules of the rubber become increasingly aligned along the direction of the stretch. What are the signs of ΔH and ΔS in the rubber-band system associated with stretching of a rubber band?

17.38 Predict the signs of ΔH, ΔS, and ΔG for the following processes: (a) $KI(l)$ solidifies at 681°C, the normal melting point of KI under 1 atm pressure; (b) 0.1 mol of $KNO_3(s)$ dissolves in 1 L of water at 25°C (see Figure 11.10); (c) ice melts at 25°C, 1 atm pressure.

17.39 (a) Calculate $\Delta H°$, $\Delta S°$, and $\Delta G°$ for each of the following processes at 25°C, using the data of Appendix D:

(i) $4Cr(s) + 3O_2(g) \longrightarrow 2Cr_2O_3(s)$
(ii) $2F_2(g) + 2CaO(s) \longrightarrow 2CaF_2(g) + O_2(g)$
(iii) $C_2H_2(g) + 4Cl_2(g) \longrightarrow 2CCl_4(g) + H_2(g)$

(b) For each of these processes, predict the manner in which the free-energy change varies with an increase in temperature.

[17.40] (a) Below are given the normal boiling point (Section 10.3) for several substances, and their enthalpies of vaporization at those temperatures. From these data calculate ΔS of vaporization for each substance. (*Hint:* The gaseous and liquid states are in equilibrium; what does this imply about ΔG?)

Substance	Normal boiling point (°C)	Enthalpy of vaporization (kJ/mol)
Acetone, $(CH_3)_2O$	56.2	30.3
Benzene, C_6H_6	80.1	30.7
Ammonia, NH_3	−33.4	23.4
Water, H_2O	100.0	40.6

(b) What significance can be attached to the variations in ΔS among these compounds? Explain.

17.41 On the basis of the $\Delta H°$ and $\Delta S°$ values given for each reaction below, indicate (a) which are spontaneous under standard conditions at 25°C and (b) which can be expected to be spontaneous at high temperatures.

(a) $KClO_3(s) \longrightarrow KCl(s) + \frac{3}{2}O_2(g)$
$\Delta H^\circ = -44.7$ kJ; $\Delta S^\circ = +59.1$ J/K

(b) $2Al(s) + 3Cl_2(g) \longrightarrow 2AlCl_3(s)$
$\Delta H^\circ = -332$ kJ; $\Delta S^\circ = -93.4$ J/K

(c) $NOCl(g) \longrightarrow NO(g) + \frac{1}{2}Cl_2(g)$
$\Delta H^\circ = +37.6$ kJ; $\Delta S^\circ = +58.5$ J/K

17.42 (a) Given K_a for benzoic acid at 298 K (Appendix E), calculate ΔG° for the dissociation of this substance in aqueous solution. (b) What is the value of ΔG at equilibrium? (c) What is the value of ΔG when $[H^+] = 3.0 \times 10^{-3}$ *M*, $[C_7H_5O_2^-] = 2.0 \times 10^{-5}$ *M*, and $[HC_7H_5O_2] = 0.10$ *M*?

17.43 (a) The standard free energy of formation of ethylene oxide, C_2H_4O, at 25°C is -11.7 kJ/mol. What is the equilibrium constant for the following reaction at 25°C?

$$2C_2H_4(g) + O_2(g) \rightleftharpoons 2C_2H_4O(g)$$

(b) Ethylene oxide is an important industrial chemical. Would reaction of ethylene, C_2H_4, with ozone, $O_3(g)$, possess a larger or smaller equilibrium constant? Use thermodynamic quantities to support your answer.

17.44 At what temperature will K_p for the following reaction equal unity, assuming that ΔH° and ΔS° do not change with temperature:

$$2NOCl(g) \rightleftharpoons 2NO(g) + Cl_2(g)$$

17.45 Evaporating NH_3 and HCl from aqueous ammonia ("ammonium hydroxide") and hydrochloric acid, respectively, produce the white haze often seen on laboratory windows and glassware:

$$NH_3(g) + HCl(g) \longrightarrow NH_4Cl(s)$$

Calculate the equilibrium constant, K_p, for this reaction at 25°C.

17.46 The oxidation of glucose, $C_6H_{12}O_6$, in body tissue produces CO_2 and H_2O. In contrast, anaerobic decomposition, which occurs during fermentation, produces ethyl alcohol, C_2H_5OH. (a) Compare the equilibrium constants for the following reactions:

$$C_6H_{12}O_6(s) + 6O_2(g) \rightleftharpoons 6CO_2(g) + 6H_2O(l)$$
$$C_6H_{12}O_6(s) \rightleftharpoons 2C_2H_5OH(l) + 2CO_2(g)$$

The standard free energy of formation of $C_6H_{12}O_6(s)$ is -912 kJ/mol. (b) Compare the maximum amounts of work that can be obtained from these two processes under standard conditions.

17.47 Calculate ΔG at 298 K for the following reaction if the reaction mixture consists of 6.0 atm $CO(g)$, 3×10^3 atm $O_2(g)$, and 1.0 atm $CO_2(g)$:

$$2CO(g) + O_2(g) \rightleftharpoons 2CO_2(g)$$

17.48 (a) In each of the following reactions, predict the sign of ΔH° and ΔS°, and discuss briefly how these factors determine the magnitude of *K*. (b) Based on your general chemical knowledge, predict which of these reactions will have $K > 1$. (c) In each case, indicate whether *K* should increase or decrease with increasing temperature.

(i) $2Mg(s) + O_2(g) \rightleftharpoons 2MgO(s)$
(ii) $2KI(l) \rightleftharpoons 2K(g) + I_2(g)$
(iii) $Na_2(g) \rightleftharpoons 2Na(g)$
(iv) $V_2O_5(s) \rightleftharpoons 2V(s) + \frac{5}{2}O_2(g)$

17.49 For each of the following reactions at 25°C, calculate the values of ΔH°, ΔS°, and ΔG°, using the data in Appendix D.

$$4FeO(s) + O_2(g) \longrightarrow 2Fe_2O_3(s)$$
$$2Fe_3O_4(s) + \frac{1}{2}O_2(g) \longrightarrow 3Fe_2O_3(s)$$
$$3FeO(s) + \frac{1}{2}O_2(g) \longrightarrow Fe_3O_4(s)$$

Based on your calculations, indicate which form of iron oxide is most stable in the presence of 1 atm O_2 at 25°C. Explain.

[17.50] The relationship between the temperature of a reaction, its standard enthalpy change, and the equilibrium constant at that temperature can be expressed in terms of the following linear equation:

$$\log K = \frac{-\Delta H^\circ}{2.30RT} + \text{constant}$$

(a) Explain how this equation can be used to determine ΔH° experimentally from the equilibrium constants at several different temperatures. (b) Derive the equation above using relationships given in this chapter. What is the constant equal to?

[17.51] The potassium-ion concentration in blood plasma is about 5.0×10^{-3} *M* while the concentration in muscle-cell fluid is much greater, 0.15 *M*. The plasma and intracellular fluid are separated by the cell membrane, which we assume is permeable only to K^+. (a) What is ΔG for the transfer of 1 mol of K^+ from blood plasma to the cellular fluid at body temperature, 37°C? (b) What is the minimum amount of work that must be used to transfer this K^+?

[17.52] The standard free-energy change at absolute temperature *T* (K) for the equilibrium

$$CaCO_3(s) \rightleftharpoons CaO(s) + CO_2(g)$$

is given by the expression

$$\Delta G_T^\circ = 177.1 \text{ kJ/mol} - (158 \times T) \text{ J/mol-K}$$

(a) Account for the negative sign before the temperature-dependent term of this expression. (b) At what temperature does the equilibrium pressure of $CO_2(g)$ in the system equal 0.100 atm?

18 Electrochemistry

Many important chemical processes utilize electricity, whereas others can be used to produce it. Because of the importance of electricity in modern society, it is useful for us to examine the subject area of electrochemistry, which deals with the relationships that exist between electricity and chemical reactions. As we shall see, our discussions of electrochemistry will provide insight into such diverse topics as the construction and operation of batteries, spontaneity of chemical reactions, electroplating, and the corrosion of metals. Because electricity involves the flow of electrons, electrochemistry focuses on reactions in which electrons are transferred from one substance to another. Such reactions are known as oxidation-reduction or "redox" reactions.

18.1 OXIDATION-REDUCTION REACTIONS

Oxidation-reduction reactions were briefly discussed in Section 7.10. At that time we defined oxidation as an increase in oxidation number (loss of electrons) and reduction as a decrease in oxidation number (gain of electrons). If one substance gains electrons and is thereby reduced, another substance must lose electrons and be thereby oxidized. Oxidation and reduction must occur simultaneously; there cannot be one without the other. For example, consider the reaction between iron and hydrogen chloride, Equation 18.1:

$$\overset{0}{Fe}(s) + 2\overset{+1}{H}\overset{-1}{Cl}(g) \longrightarrow \overset{+2}{Fe}\overset{-1}{Cl_2}(s) + \overset{0}{H_2}(g) \qquad [18.1]$$

The oxidation number of each element is given above the symbol for the element. If you have forgotten the rules for finding oxidation numbers, it would be useful to refer back to Section 7.10 and review them. By looking at the oxidation states in Equation 18.1, we see that iron is oxidized while HCl is simultaneously reduced.

In discussing oxidation-reduction reactions, it is often useful to refer to the substance causing oxidation as the oxidizing agent or oxidant. Oxidizing agents possess an affinity for electrons and cause other substances to be oxidized by abstracting electrons from them. Because the oxidizing agent gains electrons, it is reduced. Similarly, a substance that causes reduction is called a reducing agent or reductant. In Equation 18.1, HCl is the oxidizing agent, whereas Fe is the reducing agent. The substance reduced in a reaction is always the oxidizing agent, whereas the substance oxidized is always the reducing agent.

Half-Reactions

Although oxidation and reduction must take place simultaneously, it is often convenient to consider them as separate processes. For example, the oxidation of Sn^{2+} by Fe^{3+},

$$Sn^{2+}(aq) + 2Fe^{3+}(aq) \longrightarrow Sn^{4+}(aq) + 2Fe^{2+}(aq) \qquad [18.2]$$

can be considered to consist of two processes: (1) the oxidation of Sn^{2+} (Equation 18.3), and (2) the reduction of Fe^{3+} (Equation 18.4).

$$\text{Oxidation:} \quad Sn^{2+}(aq) \longrightarrow Sn^{4+}(aq) + 2e^- \qquad [18.3]$$

$$\text{Reduction:} \quad 2Fe^{3+}(aq) + 2e^- \longrightarrow 2Fe^{2+}(aq) \qquad [18.4]$$

Such equations, which show either oxidation or reduction alone, are known as half-reactions. As shown in Equations 18.3 and 18.4, the number of electrons lost in an oxidation half-reaction must equal the number gained in a reduction half-reaction. When this condition is met and the half-reactions are balanced, they can be added to give the balanced total oxidation-reduction equation.

Balancing Equations by the Method of Half-Reactions

As we have suggested above, oxidation-reduction equations can be balanced using half-reactions. As an example, let's consider the reaction that occurs between permanganate ion, MnO_4^-, and oxalate ion, $C_2O_4^{2-}$, in acidic water solutions. When MnO_4^- is added to an acidified solution of $C_2O_4^{2-}$, the deep purple color of the MnO_4^- ion fades. Bubbles of CO_2 form, and the solution takes on the pale pink color of Mn^{2+}. We can therefore write the rough, unbalanced equation as follows:

$$MnO_4^-(aq) + C_2O_4^{2-}(aq) \longrightarrow Mn^{2+}(aq) + CO_2(g) \qquad [18.5]$$

Experiments would also show that H^+ is consumed and H_2O produced in the reaction. We shall see that this fact can be deduced in the course of balancing the equation.

To complete and balance Equation 18.5 by the method of half-reactions, we proceed through the following three steps. *Step 1* is to write two incomplete half-reactions, one involving the oxidant and the other involving the reductant.

$$MnO_4^-(aq) \longrightarrow Mn^{2+}(aq)$$

$$C_2O_4^{2-}(aq) \longrightarrow CO_2(g)$$

In *step 2,* the half-reactions are completed and balanced separately. First the atoms undergoing oxidation or reduction are balanced, then the remaining elements, and finally the charge. If the reaction occurs in acidic water solution, H^+ and H_2O can be added to either reactants or products to balance hydrogen and oxygen. Similarly, in basic solution, the equation can be completed using OH^- and H_2O. These species are in large supply in the respective solutions, and their formation as products or their utilization as reactants can easily go undetected experimentally. In the permanganate half-reaction, we already have one manganese atom on each side of the equation. However, we have four oxygens on the left and none on the right side; four H_2O molecules are needed among the products to balance the four oxygen atoms in MnO_4^-:

$$MnO_4^-(aq) \longrightarrow Mn^{2+}(aq) + 4H_2O(l)$$

The eight hydrogen atoms that this introduces among the products can then be balanced by adding $8H^+$ to the reactants:

$$8H^+(aq) + MnO_4^-(aq) \longrightarrow Mn^{2+}(aq) + 4H_2O(l)$$

At this stage there are equal numbers of each type of atom on both sides of the equation, but the charge still needs to be balanced. The total charge of the reactants is $+8 - 1 = +7$, while that of the products is $+2 + 4(0) = +2$. To balance the charge, five electrons are added to the reactant side.*

$$5e^- + 8H^+(aq) + MnO_4^-(aq) \longrightarrow Mn^{2+}(aq) + 4H_2O(l) \qquad [18.6]$$

Proceeding similarly with the oxalate half-reaction, we have

$$C_2O_4^{2-}(aq) \longrightarrow 2CO_2(g)$$

Charge is balanced by adding two electrons among the products:

$$C_2O_4^{2-}(aq) \longrightarrow 2CO_2(g) + 2e^- \qquad [18.7]$$

In *step 3,* we multiply each equation by an appropriate factor so that the number of electrons gained in one half-reaction equals the number of electrons lost in the other. The half-reactions are then added to give the overall balanced equation. In our example, the MnO_4^- half-reaction must be multiplied by 2 and the $C_2O_4^{2-}$ half-reaction must be multiplied by 5:

$$10e^- + 16H^+(aq) + 2MnO_4^-(aq) \longrightarrow 2Mn^{2+}(aq) + 8H_2O(l)$$

$$5C_2O_4^{2-}(aq) \longrightarrow 10CO_2(g) + 10e^-$$

$$16H^+(aq) + 2MnO_4^-(aq) + 5C_2O_4^{2-}(aq) \longrightarrow 2Mn^{2+}(aq) + 8H_2O(l) + 10CO_2(g) \qquad [18.8]$$

*Although the oxidation numbers of the elements need not be used in balancing a half-reaction by this method, oxidation numbers can be used as a check. In this example MnO_4^- contains manganese in a +7 oxidation state. Because manganese changes from a +7 to a +2 oxidation state, it must gain five electrons just as we have already concluded.

The balanced equation is the sum of the balanced half-reactions.

The equations for reactions that occur in basic solution can be balanced initially as if they occurred in acidic solution. The H^+ ions can then be "neutralized" by adding an equal number of OH^- ions to both sides of the equation. This procedure is shown in Sample Exercise 18.1.

SAMPLE EXERCISE 18.1

Complete and balance the following equations:

(a) $Cr_2O_7^{2-}(aq) + Cl^-(aq) \longrightarrow Cr^{3+}(aq) + Cl_2(g)$ (acidic solution)

(b) $CN^-(aq) + MnO_4^-(aq) \longrightarrow CNO^-(aq) + MnO_2(s)$ (basic solution)

Solution: (a) The incomplete and unbalanced half-reactions are

$$Cr_2O_7^{2-}(aq) \longrightarrow Cr^{3+}(aq)$$
$$Cl^-(aq) \longrightarrow Cl_2(g)$$

The half-reactions are first balanced with respect to the elements. In the first half-reaction the presence of $Cr_2O_7^{2-}$ among the reactants requires two Cr^{3+} among the products. The 7 oxygen atoms in $Cr_2O_7^{2-}$ are balanced by adding $7H_2O$ to the products. The 14 hydrogen atoms in $7H_2O$ are then balanced by adding $14H^+$ among the reactants:

$$14H^+(aq) + Cr_2O_7^{2-}(aq) \longrightarrow 2Cr^{3+}(aq) + 7H_2O(l)$$

Charge is balanced by adding electrons to the left side of the equation so that the total charge is the same on both sides:

$$6e^- + 14H^+(aq) + Cr_2O_7^{2-}(aq) \longrightarrow 2Cr^{3+}(aq) + 7H_2O(l)$$

In the second half-reaction, two Cl^- are required to balance one Cl_2:

$$2Cl^-(aq) \longrightarrow Cl_2(g)$$

We add two electrons to the right side to attain charge balance:

$$2Cl^-(aq) \longrightarrow Cl_2(g) + 2e^-$$

This second half-reaction must be multiplied by 3 to equalize electron loss and gain in the two half-reactions. The two half-reactions are then added to give the balanced equation:

$$14H^+(aq) + Cr_2O_7^{2-}(aq) + 6Cl^-(aq) \longrightarrow 2Cr^{3+}(aq) + 7H_2O(l) + 3Cl_2(g)$$

(b) The incomplete and unbalanced half-reactions are

$$CN^-(aq) \longrightarrow CNO^-(aq)$$
$$MnO_4^-(aq) \longrightarrow MnO_2(s)$$

The equations may be balanced initially as if they took place in acidic solution. The resultant balanced half-reactions are

$$CN^-(aq) + H_2O(l) \longrightarrow CNO^-(aq) + 2H^+(aq) + 2e^-$$
$$3e^- + 4H^+(aq) + MnO_4^-(aq) \longrightarrow MnO_2(s) + 2H_2O(l)$$

Because H^+ cannot exist in any appreciable concentration in basic solution, it is removed from the equations by the addition of an appropriate amount of $OH^-(aq)$. In the CN^- half-reaction, $2OH^-(aq)$ is added to both sides of the equation to "neutralize" the $2H^+(aq)$. The $2OH^-(aq)$ and $2H^+(aq)$ form $2H_2O(l)$:

$$2OH^-(aq) + H_2O(l) + CN^-(aq) \longrightarrow CNO^-(aq) + 2H_2O(l) + 2e^-$$

The half-reaction can be simplified because H_2O occurs on both sides of the equation. The simplified equation is

$$2OH^-(aq) + CN^-(aq) \longrightarrow CNO^-(aq) + H_2O(l) + 2e^-$$

For the MnO_4^- half-reaction, $4OH^-(aq)$ is added to both sides of the equation:

$$3e^- + 4H_2O(l) + MnO_4^-(aq) \longrightarrow MnO_2(s) + 2H_2O(l) + 4OH^-(aq)$$

Simplifying gives

$$3e^- + 2H_2O(l) + MnO_4^-(aq) \longrightarrow MnO_2(s) + 4OH^-(aq)$$

The top equation is multiplied by 3 and the bottom one by 2 to equalize electron loss and gain in the two half-reactions. The half-reactions are then added:

$$6OH^-(aq) + 3CN^-(aq) \longrightarrow 3CNO^-(aq) + 3H_2O(l) + 6e^-$$

$$6e^- + 4H_2O(l) + 2MnO_4^-(aq) \longrightarrow 2MnO_2(s) + 8OH^-(aq)$$

$$6OH^-(aq) + 3CN^-(aq) + 4H_2O(l) + 2MnO_4^-(aq) \longrightarrow 3CNO^-(aq) + 3H_2O(l) + 2MnO_2(s) + 8OH^-(aq)$$

The overall equation can be simplified, because H_2O and OH^- occur on both sides. The simplified equation is

$$3CN^-(aq) + H_2O(l) + 2MnO_4^-(aq) \longrightarrow 3CNO^-(aq) + 2MnO_2(s) + 2OH^-(aq)$$

The method of half-reactions is not the only method of balancing oxidation-reduction reactions. We have stressed this method because it is useful in studying other material in this chapter. Another popular method, the oxidation-number method, is considered below. To illustrate this approach, consider the following equation, which was balanced earlier using half-reactions:

$$MnO_4^-(aq) + H^+(aq) + C_2O_4^{2-}(aq) \longrightarrow Mn^{2+}(aq) + H_2O(l) + CO_2(g)$$

To balance this equation by the oxidation-number method, we proceed as follows:

1. Determine the oxidation number for each element on both sides of the equation, thereby determining the elements that have undergone oxidation and reduction. In the present example, hydrogen is +1 in both reactants and products while oxygen is −2; neither is oxidized or reduced. However, Mn changes from +7 in MnO_4^- to +2 in Mn^{2+}; C changes from +3 in $C_2O_4^{2-}$ to +4 in CO_2.

2. Determine the change in oxidation number for each element undergoing oxidation and reduction. These changes can be conveniently shown above arrows connecting the pertinent elements in the reactants and products:

$$\overset{(+7)}{Mn}O_4^- + H^+ + \underset{(+3)}{C_2}O_4^{2-} \longrightarrow \overset{(+2)}{Mn^{2+}} + H_2O + \underset{(+4)}{C}O_2$$

(Mn: −5; C: +2)

In this example the oxidation number of Mn changes by five units while that for each C changes by one unit. However, there are two carbons in $C_2O_4^{2-}$, so that the total change for each $C_2O_4^{2-}$ is two, one for each carbon.

3. Using the oxidation-number changes found in step 2, determine the simplest ratio of moles of oxidant and reductant that will lead to equal gains and losses in oxidation numbers. The procedure is equivalent to requiring that the number of electrons lost by $C_2O_4^{2-}$ equals the number gained by MnO_4^-. We must have $5C_2O_4^{2-}$ for each $2MnO_4^-$ so that each substance is involved in a total change in oxidation numbers of 10: 2×5 for Mn and $5 \times 2 \times 1$ for C.

4. Having determined coefficients for species being oxidized and reduced, balance the number of atoms of the remaining elements in the usual manner. In this example, the $5C_2O_4^{2-}$ contain 10 carbon atoms, requiring $10CO_2$ among the products. There are 28 oxygen atoms among the reactants (8 in $2MnO_4^-$ and 20 in $5C_2O_4^{2-}$), which require there to be $8H_2O$ among the products; there are then 28 oxygen atoms in the products (8 in $8H_2O$ and 20 in $10CO_2$). Finally, $8H_2O$ requires the presence of $16H^+$ among the reactants:

$$2MnO_4^-(aq) + 16H^+(aq) + 5C_2O_4^{2-}(aq) \longrightarrow 2Mn^{2+}(aq) + 8H_2O(l) + 10CO_2(g)$$

In cases where an incomplete oxidation-reduction equation is given, H^+ and H_2O (in acidic solutions) or OH^- and H_2O (in basic solutions) can be added to complete the balancing.

18.2 VOLTAIC CELLS

In principle, the energy released in any spontaneous redox reaction can be directly harnessed to perform electrical work. This task is accomplished through a voltaic (or galvanic) cell, which is merely a device in which electron transfer is forced to take place through an external pathway rather than directly between reactants.

One such spontaneous reaction occurs when a piece of zinc is placed in contact with a solution containing Cu^{2+}. As the reaction proceeds, the blue color that is characteristic of $Cu^{2+}(aq)$ ions fades, and copper metal begins to deposit on the zinc. At the same time, the zinc begins to dis-

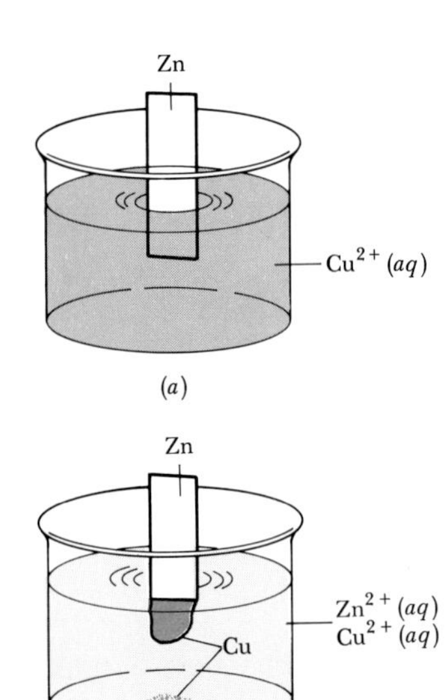

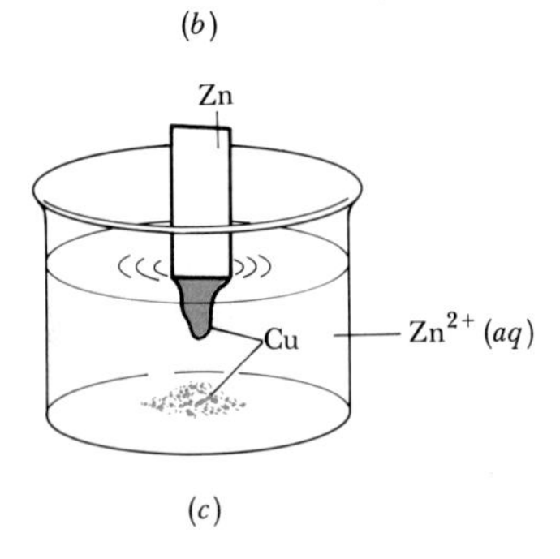

FIGURE 18.1 (*a*) A strip of zinc is placed in a colored solution containing Cu^{2+} ions. (*b*) Electrons are transferred from zinc to Cu^{2+}, and the zinc metal dissolves and copper metal deposits. During this process, the color characteristic of Cu^{2+} ions in the solution fades. (*c*) After reaction has gone to completion, zinc is in excess, and the color characteristic of Cu^{2+} has disappeared.

solve. These transformations are shown in Figure 18.1 and are summarized by Equation 18.9:

$$Zn(s) + Cu^{2+}(aq) \longrightarrow Zn^{2+}(aq) + Cu(s) \qquad [18.9]$$

Figure 18.2 shows a voltaic cell that utilizes the oxidation-reduction reaction between Zn and Cu^{2+} expressed in Equation 18.9. Although the experimental design shown in Figure 18.2 is more complex than that in Figure 18.1, it is important to recognize that the chemical reaction involved is the same in both cases. The major difference between the two arrangements is that in Figure 18.2 the zinc metal and $Cu^{2+}(aq)$ are no longer in direct contact. Consequently, reduction of the Cu^{2+} can occur only by a flow of electrons through the wire that connects Zn and Cu (the external circuit).

The two solid metals that are connected by the external circuit are called electrodes. By definition, the electrode at which oxidation occurs is called the **anode;** the electrode at which reduction occurs is called the **cathode.*** In our present example, Zn is the anode and Cu is the cathode:

$$\begin{array}{lrcl} \text{Oxidation; anode:} & Zn(s) & \longrightarrow & Zn^{2+}(aq) + 2e^- \\ \text{Reductions; cathode:} & Cu^{2+}(aq) + 2e^- & \longrightarrow & Cu(s) \end{array}$$

The voltaic cell may be regarded as composed of two half-cells, one corresponding to the oxidation process and one corresponding to the reduction process. Electrons become available as zinc metal is oxidized at the anode. They flow through the external circuit to the cathode, where they are consumed as $Cu^{2+}(aq)$ is reduced.

We must be careful about the signs we attach to the electrodes in a voltaic cell. We have seen that electrons are released at the anode, as the zinc is oxidized. Electrons are thus flowing out of the anode and into the external circuit. Because the electrons are negatively charged, we thus assign a negative sign to the anode. Conversely, electrons flow into the cathode, where they are consumed in the reduction of copper. A positive sign is thus assigned to the cathode, because it appears to attract the negative electrons.

As the cell pictured in Figure 18.2 operates, oxidation of Zn introduces additional Zn^{2+} ions into the anode compartment. Unless a means is provided to neutralize this positive charge, no further oxidation can take place. Similarly, the reduction of Cu^{2+} at the cathode leaves an excess of negative charge in solution in that compartment. Electrical neutrality is maintained by a migration of ions through a "salt bridge," as illustrated in Figure 18.2. A salt bridge consists of a U-shaped tube that contains an electrolyte solution such as $NaNO_3(aq)$ whose ions will not react with other ions in the cell or with the electrode materials. The ends of the U-tube may be loosely plugged with glass wool or the electrolyte incorporated into a gel so that the electrolyte solution does not run out when the

*To help remember these definitions, it is useful to note that anode and oxidation both begin with a vowel, whereas cathode and reduction both begin with a consonant.

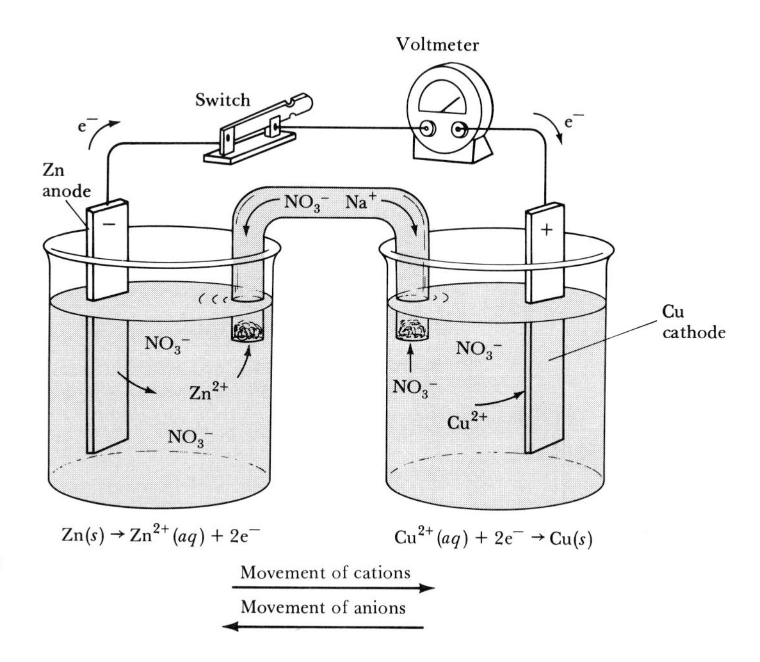

FIGURE 18.2 Complete and functioning voltaic cell using a salt bridge to complete the electrical circuit.

U-tube is inverted. As oxidation and reduction proceed at the electrodes, ions from the salt bridge migrate to neutralize charge in the anode and cathode compartments. Anions migrate toward the anode and cations toward the cathode. In fact, no measurable electron flow will occur through the external circuit unless a means is provided for ions to migrate through the solution from one electrode compartment to another, thereby completing the circuit.

SAMPLE EXERCISE 18.2

The following oxidation-reduction reaction is spontaneous in the direction indicated:

$$Cr_2O_7^{2-}(aq) + 14H^+(aq) + 6I^-(aq) \longrightarrow 2Cr^{3+}(aq) + 3I_2(s) + 7H_2O(l)$$

A solution containing $K_2Cr_2O_7$ and H_2SO_4 is poured into one beaker and a solution of KI is poured into another. A salt bridge is used to join the beakers. A metallic conductor that will not react with either solution (such as platinum foil) is suspended in each solution, and the two conductors are connected with wires through a voltmeter or some other device to detect an electric current. The resultant voltaic cell generates an electric current. Indicate the reaction occurring at the anode, the reaction at the cathode, the direction of electron and ion migrations, and the signs of the electrodes.

Solution: The half-reactions for the given chemical equation are

$$Cr_2O_7^{2-}(aq) + 14H^+(aq) + 6e^- \longrightarrow 2Cr^{3+}(aq) + 7H_2O(l)$$

$$6I^-(aq) \longrightarrow 3I_2(s) + 6e^-$$

The first half-reaction is the reduction process, which by definition occurs at the cathode. The second half-reaction is an oxidation, which occurs at the anode. The I^- ions are the source of electrons and the $Cr_2O_7^{2-}$ ions are the receptors. Consequently, the electrons flow through the external circuit from the electrode immersed in the KI solution (the anode) to the electrode immersed in the $K_2Cr_2O_7$–H_2SO_4 solution (the cathode). The electrodes themselves do not react in any way; they merely provide a means of transferring electrons from or to the solutions. The cations move through the solutions toward the cathode, while the anions move toward the anode. The anode is the negative electrode and the cathode the positive one.

18.3 CELL EMF

A voltaic cell may be thought to possess a "driving force" or "electrical pressure" that pushes electrons through the external circuit. This driving force is called the electromotive force (abbreviated emf); emf is measured in units of volts and is also referred to as the voltage or *cell potential.* One volt (V) is the emf required to impart 1 J of energy to a charge of 1 C:

$$1\ \text{V} = 1\,\frac{\text{J}}{\text{C}} \qquad [18.10]$$

The accurate determination of cell emf requires the use of special apparatus. The measurement must be made in such a manner that only a negligible current flows. If a significant current is allowed to flow, the apparent voltage of the cell is lowered as a result of the internal resistance of the cell, and because of changes in concentrations around the electrodes.

The voltaic cell pictured in Figure 18.2 generates an emf of 1.10 V when operated under standard-state conditions. You will recall from Section 17.5 (see Table 17.1) that standard-state conditions include 1 *M* concentrations for reactants and products that are in solution and 1 atm pressure for those that are gases. In the present example, the cell would be operated with $[Cu^{2+}]$ and $[Zn^{2+}]$ both at 1 *M*.* The emf generated by a cell is given the symbol E. When operated under standard-state conditions it is called the standard emf, $E°$ (sometimes called *standard cell potential*):

$$Zn(s) + Cu^{2+}(aq) \longrightarrow Zn^{2+}(aq) + Cu(s) \qquad E° = 1.10\ \text{V} \qquad [18.11]$$

The emf of any voltaic cell depends on the nature of the chemical reactions taking place in the cell, the concentrations of reactants and products, and the temperature of the cell, which we shall take to be 25°C unless otherwise noted.

Standard Electrode Potentials

Just as an overall cell reaction can be thought of as the sum of two half-reactions, the emf of a cell can be thought of as the sum of two half-cell potentials: that due to the loss of electrons at the anode (the oxidation potential, E_{ox}) and that due to electron gain at the cathode (the reduction potential, E_{red}):

$$E_{cell} = E_{ox} + E_{red} \qquad [18.12]$$

It is impossible directly to measure an isolated oxidation or reduction potential. However, if one half-reaction is arbitrarily assigned a standard half-cell potential, standard potentials of other half-reactions can be determined relative to that reference. The half-reaction that corresponds to

*Actually, solutions of 1 *M* represent the standard state only if they behave ideally. Salts dissolved in an aqueous medium at concentrations on the order of 1 *M* do not in general behave ideally; substantial corrections need to be made to allow for nonideal behavior. However, in this introduction to electrochemistry we will not concern ourselves with these corrections.

reduction of H^+ to form H_2 has been chosen as the reference and assigned a standard reduction potential of exactly 0 volts:

$$2H^+(1\ M) + 2e^- \longrightarrow H_2(1\ \text{atm}) \qquad E^\circ_{red} = 0\ V \qquad [18.13]$$

Figure 18.3 illustrates a voltaic cell constructed to use the following redox reaction between Zn and H^+:

$$Zn(s) + 2H^+(aq) \longrightarrow Zn^{2+}(aq) + H_2(g) \qquad [18.14]$$

Oxidation of zinc occurs in the anode compartment, whereas reduction of H^+ occurs in the cathode compartment. The standard hydrogen electrode, which is designed to operate under standard-state conditions ($[H^+] = 1\ M$ and $P_{H_2} = 1$ atm), consists of a platinum wire and a piece of platinum foil covered with finely divided platinum that serves as an inert surface for the cathode reaction. This electrode is encased in a glass tube so that hydrogen gas can bubble over the platinum. The voltaic cell generates a standard emf (E°_{cell}) of 0.76 V. By using the defined standard reduction potential of H^+ ($E^\circ_{red} = 0$), it is possible to calculate the standard oxidation potential of Zn:

$$E^\circ_{cell} = E^\circ_{ox} + E^\circ_{red}$$
$$0.76\ V = E^\circ_{ox} + 0$$

Thus a standard oxidation potential of 0.76 V can be assigned to Zn:

$$Zn(s) \longrightarrow Zn^{2+}(aq) + 2e^- \qquad E^\circ_{ox} = 0.76\ V \qquad [18.15]$$

The standard potentials for other half-reactions can be established from other cell emfs in a similar fashion.

The half-cell potential for any oxidation is equal in magnitude but opposite in sign to that of the reverse reduction. For example,

$$Zn^{2+}(aq) + 2e^- \longrightarrow Zn(s) \qquad E^\circ_{red} = -0.76\ V \qquad [18.16]$$

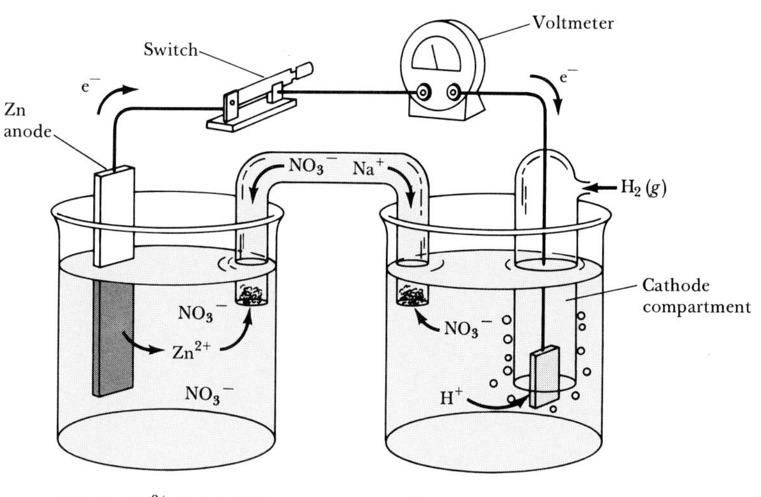

FIGURE 18.3 Voltaic cell using a standard hydrogen electrode.

TABLE 18.1 Standard electrode potentials in water at 25°C

Standard potential (V)	Reduction half-reaction
2.87	$F_2(g) + 2e^- \longrightarrow 2F^-(aq)$
1.51	$MnO_4^-(aq) + 8H^+(aq) + 5e^- \longrightarrow Mn^{2+}(aq) + 4H_2O(l)$
1.36	$Cl_2(g) + 2e^- \longrightarrow 2Cl^-(aq)$
1.33	$Cr_2O_7^{2-}(aq) + 14H^+(aq) + 6e^- \longrightarrow 2Cr^{3+}(aq) + 7H_2O(l)$
1.23	$O_2(g) + 4H^+(aq) + 4e^- \longrightarrow 2H_2O(l)$
1.06	$Br_2(l) + 2e^- \longrightarrow 2Br^-(aq)$
0.96	$NO_3^-(aq) + 4H^+(aq) + 3e^- \longrightarrow NO(g) + 2H_2O(l)$
0.80	$Ag^+(aq) + e^- \longrightarrow Ag(s)$
0.77	$Fe^{3+}(aq) + e^- \longrightarrow Fe^{2+}(aq)$
0.68	$O_2(g) + 2H^+(aq) + 2e^- \longrightarrow H_2O_2(aq)$
0.59	$MnO_4^-(aq) + 2H_2O(l) + 3e^- \longrightarrow MnO_2(s) + 4OH^-(aq)$
0.54	$I_2(s) + 2e^- \longrightarrow 2I^-(aq)$
0.40	$O_2(g) + 2H_2O(l) + 4e^- \longrightarrow 4OH^-(aq)$
0.34	$Cu^{2+}(aq) + 2e^- \longrightarrow Cu(s)$
0	$2H^+(aq) + 2e^- \longrightarrow H_2(g)$
−0.28	$Ni^{2+}(aq) + 2e^- \longrightarrow Ni(s)$
−0.44	$Fe^{2+}(aq) + 2e^- \longrightarrow Fe(s)$
−0.76	$Zn^{2+}(aq) + 2e^- \longrightarrow Zn(s)$
−0.83	$2H_2O(l) + 2e^- \longrightarrow H_2(g) + 2OH^-(aq)$
−1.66	$Al^{3+}(aq) + 3e^- \longrightarrow Al(s)$
−2.71	$Na^+(aq) + e^- \longrightarrow Na(s)$
−3.05	$Li^+(aq) + e^- \longrightarrow Li(s)$

By convention, half-cell potentials are tabulated as standard reduction potentials, also referred to as merely as **electrode potentials.** Table 18.1 contains a selection of electrode potentials; a more complete list is found in Appendix F. These electrode potentials may be combined to calculate the standard emfs of a large variety of voltaic cells.

SAMPLE EXERCISE 18.3

Given that E°_{cell} is 1.10 V for the Zn–Cu^{2+} cell shown in Figure 18.2, and that E°_{ox} is 0.76 V for the oxidation of zinc (Equation 18.15), calculate E°_{red} for the reduction of Cu^{2+} to Cu:

$$Cu^{2+}(aq) + 2e^- \longrightarrow Cu(s)$$

Solution:

$$E^\circ_{cell} = E^\circ_{ox} + E^\circ_{red}$$
$$1.10\ V = 0.76\ V + E^\circ_{red}$$
$$E^\circ_{red} = 1.10\ V - 0.76\ V = 0.34\ V$$

SAMPLE EXERCISE 18.4

Using standard electrode potentials tabulated in Table 18.1, calculate the standard emf for the cell described in Sample Exercise 18.2:

$$Cr_2O_7^{2-}(aq) + 14H^+(aq) + 6I^-(aq) \longrightarrow 2Cr^{3+}(aq) + 3I_2(s) + 7H_2O(l)$$

Solution: The standard emf of the cell, E°_{cell}, is the sum of the standard oxidation and reduction potentials for the appropriate half-reactions:

$$E^\circ_{cell} = E^\circ_{ox} + E^\circ_{red}$$

The half-reactions and their potentials are as follows:

$$Cr_2O_7^{2-}(aq) + 14H^+(aq) + 6e^- \longrightarrow 2Cr^{3+}(aq) + 7H_2O(l) \quad E^\circ_{red} = 1.33\ V$$

$$6I^-(aq) \longrightarrow 3I_2(s) + 6e^- \quad E^\circ_{ox} = -0.54\ V$$

$$Cr_2O_7^{2-}(aq) + 14H^+(aq) + 6I^-(aq) \longrightarrow 2Cr^{3+}(aq) + 7H_2O(l) + 3I_2(s) \quad E^\circ_{cell} = 0.79\ V$$

Notice that E°_{ox} for I^- has the opposite sign from the reduction potential of I_2 listed in Table 18.1. Also notice that even though the iodide half-reaction must be multiplied by 3 in order to obtain the balanced equation for the reaction, the half-cell potential is *not* multiplied by 3. The standard potential is an intensive property; it does not depend on the quantity of reactants and products, but only on their concentration. Thus it does not matter whether there are 6 mol of I^- or 1 mol as long as the concentration of iodide is 1 *M*.

Just as the sum of the half-reactions gives the chemical equation for the overall cell reaction, the sum of the corresponding oxidation and reduction potentials gives the cell emf. In this case $E^\circ_{cell} = 0.79$ V; a positive emf is characteristic of all voltaic cells.

Oxidizing and Reducing Agents

Half-cell potentials indicate the ease with which a species is oxidized or reduced. *The more positive the E° value for a half-reaction, the greater the tendency for that reaction to occur as written.* A negative reduction potential indicates that the species is more difficult to reduce than $H^+(aq)$, whereas a negative oxidation potential indicates that the species is more difficult to oxidize than H_2. Examination of the half-reactions in Table 18.1 shows that F_2 is the most easily reduced species and consequently the strongest oxidizing agent listed:

$$F_2(g) + 2e^- \longrightarrow 2F^-(aq) \quad E^\circ_{red} = 2.87\ V \qquad [18.17]$$

Lithium ion, Li^+, is the most difficult to reduce and therefore the poorest oxidizing agent:

$$Li^+(aq) + e^- \longrightarrow Li(s) \quad E^\circ_{red} = -3.05\ V \qquad [18.18]$$

The most frequently encountered oxidizing agents are the halogens, oxygen, and oxyanions such as MnO_4^-, $Cr_2O_7^{2-}$, and NO_3^- whose central atoms have high positive oxidation states. Metal ions in high positive oxidation states such as Ce^{4+}, which is readily reduced to Ce^{3+}, are also employed as oxidizing agents.

Among the substances listed in Table 18.1, lithium is the most easily oxidized and consequently the strongest reducing agent:

$$Li(s) \longrightarrow Li^+(aq) + e^- \quad E^\circ_{ox} = 3.05\ V \qquad [18.19]$$

Fluoride ion, F^-, is the most difficult to oxidize and therefore the poorest reducing agent:

$$2F^-(aq) \longrightarrow F_2(g) + 2e^- \quad E^\circ_{ox} = -2.87\ V \qquad [18.20]$$

H_2 and a variety of metals having positive oxidation potentials (such as Zn and Fe) are employed as reducing agents. Some metal ions in low oxidation states, such as Sn^{2+}, which is oxidized to Sn^{4+}, also function as reducing agents. Solutions of reducing agents are difficult to store for extended periods because of the ubiquitous presence of O_2, a good oxidizing agent. For example, developer solutions used in photography are mild reducing agents; consequently, they have only a limited shelf life because they are readily oxidized by O_2 from the air.

SAMPLE EXERCISE 18.5

Using Table 18.1, determine which of the following species is the strongest oxidizing agent: MnO_4^- (in acid solution), $I_2(s)$, $Zn^{2+}(aq)$.

Solution: The strongest oxidizing agent will be the species that is most readily reduced. Therefore, we should compare reduction potentials. From Table 19.1 we have

$$MnO_4^-(aq) + 8H^+(aq) + 5e^- \longrightarrow Mn^{2+}(aq) + 4H_2O(l) \quad E^\circ_{red} = 1.51\ V$$
$$I_2(s) + 2e^- \longrightarrow 2I^-(aq) \quad E^\circ_{red} = 0.54\ V$$
$$Zn^{2+}(aq) + 2e^- \longrightarrow Zn(s) \quad E^\circ_{red} = -0.76\ V$$

Because the reduction of MnO_4^- has the highest positive potential, MnO_4^- is the strongest oxidizing agent of the three.

18.4 SPONTANEITY AND EXTENT OF REDOX REACTIONS

We have observed that voltaic cells utilize redox reactions that proceed spontaneously. Conversely, any reaction that can occur in a voltaic cell to produce a positive emf must be spontaneous. Consequently, it is possible to decide whether a redox reaction will be spontaneous by using half-cell potentials to calculate the emf associated with it: *A positive emf indicates a spontaneous process, whereas a negative emf indicates a nonspontaneous one.*

SAMPLE EXERCISE 18.6

Using the standard electrode potentials listed in Table 18.1, determine whether the following reactions are spontaneous under standard conditions:

(a) $Cu(s) + 2H^+(aq) \longrightarrow Cu^{2+}(aq) + H_2(g)$

(b) $Cl_2(g) + 2I^-(aq) \longrightarrow 2Cl^-(aq) + I_2(s)$

Solution: (a) We utilize Table 18.1 to obtain the necessary half-reaction potentials. Because the overall reaction converts Cu to Cu^{2+}, we use the standard oxidation potential for Cu. Because the overall reaction converts H^+ to H_2 we use the reduction potential for H^+. Adding these, we obtain the standard emf for the overall reaction:

$$Cu(s) \longrightarrow Cu^{2+}(aq) + 2e^- \quad E^\circ_{ox} = -0.34\ V$$
$$2e^- + 2H^+(aq) \longrightarrow H_2(g) \quad E^\circ_{red} = 0\ V$$
$$\overline{Cu(s) + 2H^+(aq) \longrightarrow Cu^{2+}(aq) + H_2(g) \quad E^\circ = -0.34\ V}$$

Because the standard emf is negative, the reaction is not spontaneous in the direction written. Copper does not react with acids in this fashion. However, the reverse reaction is spontaneous: Cu^{2+} can be reduced by H_2.

(b) We write equations to obtain the standard emf for the overall reaction:

$$2e^- + Cl_2(g) \longrightarrow 2Cl^-(aq) \quad E^\circ_{red} = 1.36\ V$$
$$2I^-(aq) \longrightarrow I_2(s) + 2e^- \quad E^\circ_{ox} = -0.54\ V$$
$$\overline{Cl_2(g) + 2I^-(aq) \longrightarrow 2Cl^-(aq) + I_2(s) \quad E^\circ = 0.82\ V}$$

This reaction is spontaneous and could be used to build a voltaic cell. It is often used as a qualitative test for the presence of I^- in aqueous solution. The solution is treated with a solution of Cl_2. If I^- is present I_2 forms. If CCl_4 is added, the I_2 dissolves in the CCl_4, imparting a characteristic purple color to the solution.

Emf and Free-Energy Change

We've seen that the free-energy change, ΔG, accompanying a chemical process is a measure of its spontaneity (Chapter 17). Because the cell emf indicates whether a redox reaction is spontaneous, we might expect some

FIGURE 18.4 Michael Faraday (1791–1867) was born in England, one of ten children of a poor blacksmith. At the age of 14 he was apprenticed to a bookbinder who, with uncommon leniency, gave the young man time to read and even to attend lectures. In 1812 he became an assistant in Humphry Davy's laboratory in the Royal Institution. He eventually succeeded Davy as the most famous and influential scientist in England. During his scientific career he made an amazing number of important discoveries in chemistry and physics. He developed methods for liquefying gases, discovered benzene, and formulated the quantitative relationships between electrical current and the extent of chemical reaction in electrochemical cells that either produce or use electricity. In addition, he worked out the design of the first electric generator and laid the theoretical foundations for the development of our modern theory of electricity. (*Culver Pictures*)

relationship to exist between the cell emf and the free-energy change. Indeed, this is the case; the emf, E, and the free-energy change, ΔG, are related by Equation 18.21:

$$\Delta G = -n\mathfrak{F}E \qquad [18.21]$$

In this equation n is the number of moles of electrons transferred in the reaction, and $\mathfrak{F}$ is Faraday's constant, named after Michael Faraday (Figure 18.4). Faraday's constant is the electrical charge on 1 mol of electrons; consequently, this quantity of charge is referred to as a **faraday**:

$$1\ \mathfrak{F} = 96{,}500\ \frac{\text{C}}{\text{mol e}^-} = 96{,}500\ \frac{\text{J}}{\text{V-mol e}^-} \qquad [18.22]$$

For the situation in which reactants and products are in their standard states, Equation 18.21 is modified to relate $\Delta G°$ and $E°$:

$$\Delta G° = -n\mathfrak{F}E° \qquad [18.23]$$

SAMPLE EXERCISE 18.7

Use the standard electrode potentials given in Table 18.1 to calculate the standard free-energy change, $\Delta G°$, for the following reaction:

$$2Br^-(aq) + F_2(g) \longrightarrow Br_2(l) + 2F^-(aq)$$

Solution:

$$\begin{array}{llr} 2Br^-(aq) \longrightarrow Br_2(l) + 2e^- & E°_{ox} = & -1.06\ \text{V} \\ F_2(g) + 2e^- \longrightarrow 2F^-(aq) & E°_{red} = & 2.87\ \text{V} \\ \hline 2Br^-(aq) + F_2(g) \longrightarrow Br_2(l) + 2F^-(aq) & & \\ & E° = & 1.81\ \text{V} \end{array}$$

Notice that two electrons are transferred in the reaction, so $n = 2$.

$$\Delta G^\circ = -n\mathfrak{F}E^\circ$$
$$= -(2 \text{ mol e}^-)\left(96{,}500 \frac{\text{J}}{\text{V-mol e}^-}\right)(1.81 \text{ V})$$
$$= -3.49 \times 10^5 \text{ J} = -349 \text{ kJ}$$

Notice also that whereas a negative sign for ΔG° indicates spontaneity under standard conditions, a positive sign for E° indicates the same.

Emf and Equilibrium Constant

In Chapter 17 we discussed the significance of ΔG for a chemical system. Among other things, we saw that the standard free-energy change is related to the equilibrium constant, K, through Equation 18.24:

$$\Delta G^\circ = -2.30RT \log K \qquad [18.24]$$

This relationship indicates that E° should also be related to the equilibrium constant. Substituting the relationship $\Delta G^\circ = -n\mathfrak{F}E^\circ$ (Equation 18.23) into Equation 18.24 gives:

$$-n\mathfrak{F}E^\circ = -2.30RT \log K$$
$$E^\circ = \frac{+2.30RT}{n\mathfrak{F}} \log K \qquad [18.25]$$

When $T = 298$ K, this equation can be simplified by substituting the numerical values for R and $\mathfrak{F}$:

$$E^\circ = \frac{+2.30(8.314 \text{ J/K-mol})(298 \text{ K})}{n(96{,}500 \text{ J/V-mol})} \log K$$
$$= \frac{0.0591 \text{ V}}{n} \log K \qquad [18.26]$$

Thus the standard emf generated by a cell increases as the equilibrium constant for the cell reaction increases.

SAMPLE EXERCISE 18.8

Using standard electrode potentials, calculate the equilibrium constant at 25°C for the reaction

$$O_2(g) + 4H^+(aq) + 4Fe^{2+}(aq) \longrightarrow 4Fe^{3+}(aq) + 2H_2O(l)$$

Solution:

$$O_2(g) + 4H^+(aq) + 4e^- \longrightarrow 2H_2O(l) \qquad E^\circ_{red} = 1.23 \text{ V}$$
$$4Fe^{2+}(aq) \longrightarrow 4Fe^{3+}(aq) + 4e^- \qquad E^\circ_{ox} = -0.77 \text{ V}$$
$$O_2(g) + 4H^+(aq) + 4Fe^{2+}(aq) \longrightarrow 4Fe^{3+}(aq) + 2H_2O(l) \qquad E^\circ = 0.46 \text{ V}$$

From the half-reactions given above we see that $n = 4$. Using Equation 18.26, we have

$$\log K = \frac{nE^\circ}{0.0591 \text{ V}}$$
$$= \frac{4(0.46 \text{ V})}{0.0591 \text{ V}}$$
$$= 31.1$$
$$K = 1 \times 10^{31}$$

Thus Fe^{2+} ions are stable in acidic solutions only in the absence of O_2 (unless a suitable reducing agent is present).

Emf and Concentration

In practice, voltaic cells are unlikely to be operated under standard-state conditions. However, the emf generated under nonstandard conditions can be calculated from $E°$, temperature, and the concentrations of reactants and products in the cell. The equation that permits this calculation can be derived from the relationship between ΔG and $\Delta G°$, given earlier in Section 17.6:

$$\Delta G = \Delta G° + 2.30RT \log Q \qquad [18.27]$$

Because $\Delta G = -n\mathfrak{F}E$ (Equation 18.21), we can write

$$-n\mathfrak{F}E = -n\mathfrak{F}E° + 2.30RT \log Q$$

Solving this equation for E gives

$$E = E° - \frac{2.30RT}{n\mathfrak{F}} \log Q \qquad [18.28]$$

This relationship is known as the Nernst equation after Walther Hermann Nernst (1864–1941), a German chemist who established a good part of the theoretical foundations of electrochemistry. At 298 K, the quantity $2.30RT/\mathfrak{F}$ is equal to 0.0591 V-mol, so that the Nernst equation can be written in the following simplified form:

$$E = E° - \frac{0.0591}{n} \log Q \qquad [18.29]$$

As an example of how Equation 18.29 might be used, consider the following reaction:

$$Zn(s) + Cu^{2+}(aq) \longrightarrow Zn^{2+}(aq) + Cu(s) \qquad E° = 1.10 \text{ V}$$

In this case $n = 2$ and the Nernst equation gives

$$E = 1.10 \text{ V} - \frac{0.0591 \text{ V}}{2} \log \frac{[Zn^{2+}]}{[Cu^{2+}]} \qquad [18.30]$$

Recall that Q includes expressions for the species in solution, but not for solids. Experimentally it is found that the emf generated by a cell is independent of the size or shape of the solid electrodes used. From Equation 18.30 it is evident that the emf of a cell based on this chemical reaction increases as $[Cu^{2+}]$ increases and as $[Zn^{2+}]$ decreases. For example, when $[Cu^{2+}]$ is 5.0 M and $[Zn^{2+}]$ is 0.050 M, we have

$$\begin{aligned} E &= 1.10 \text{ V} - \frac{0.0591 \text{ V}}{2} \log \left(\frac{(0.050)}{(5.0)}\right) \\ &= 1.10 \text{ V} - \frac{0.0591 \text{ V}}{2}(-2.00) = 1.16 \text{ V} \end{aligned}$$

We could have anticipated this result by applying Le Châtelier's principle (Section 14.5). If the concentrations of reactants increase relative to the concentrations of products, the cell reaction becomes more highly spontaneous and the emf increases. Conversely, if the concentrations of products increase relative to reactants, emf decreases. As a cell operates, reactants are consumed and products form. Resultant decreases in reactant concentrations and increases in product concentrations cause the emf to decrease.

SAMPLE EXERCISE 18.9

Calculate the emf generated by the cell described in Sample Exercise 18.2 when $[Cr_2O_7^{2-}] = 2.0\,M$, $[H^+] = 1.0\,M$, $[I^-] = 1.0\,M$, and $[Cr^{3+}] = 1.0 \times 10^{-5}\,M$:

$$Cr_2O_7^{2-}(aq) + 14H^+(aq) + 6I^-(aq) \longrightarrow 2Cr^{3+}(aq) + 3I_2(s) + 7H_2O(l)$$

Solution: The standard emf for this reaction was calculated in Sample Exercise 18.4: $E° = 0.79$ V. As you will see if you refer back to that exercise, n is 6. The reaction quotient, Q, is

$$Q = \frac{[Cr^{3+}]^2}{[Cr_2O_7^{2-}][H^+]^{14}[I^-]^6} = \frac{(1.0 \times 10^{-5})^2}{(2.0)(1.0)^{14}(1.0)^6}$$
$$= 5.0 \times 10^{-11}$$

Using Equation 18.29, we have

$$E = 0.79\text{ V} - \frac{0.0591\text{ V}}{6}\log(5.0 \times 10^{-11})$$
$$= 0.79\text{ V} - \frac{0.0591\text{ V}}{6}(-10.30)$$
$$= 0.79\text{ V} + 0.10\text{ V}$$
$$= 0.89\text{ V}$$

This result is qualitatively what we expect: Because the concentration of $Cr_2O_7^{2-}$ (a reactant) is above 1 M, whereas the concentration of Cr^{3+} (a product) is below 1 M, the emf is greater than $E°$.

SAMPLE EXERCISE 18.10

If the measured voltage in a $Zn–H^+$ cell such as that shown in Figure 18.3 is 0.45 V at 25°C when $[Zn^{2+}]$ is 1 M and P_{H_2} is 1 atm, what is the concentration of H^+?

Solution: The cell reaction is

$$Zn(s) + 2H^+(aq) \longrightarrow Zn^{2+}(aq) + H_2(g)$$

The standard emf is $E° = 0.76$ V. Applying Equation 18.29 with $n = 2$ gives

$$0.45 = 0.76 - \frac{0.0591}{2}\log\frac{[Zn^{2+}]P_{H_2}}{[H^+]^2}$$
$$= 0.76 - \frac{0.0591}{2}\log\frac{1}{[H^+]^2}$$

Since

$$\log\frac{1}{x^2} = -\log x^2 = -2\log x$$

we can write

$$0.45 = 0.76 - \frac{0.0591}{2}(-2\log[H^+])$$

Solving for $\log[H^+]$ gives us

$$\log[H^+] = \frac{0.45 - 0.76}{0.0591} = -5.25$$
$$[H^+] = 10^{-5.25} = 5.7 \times 10^{-6}\,M$$

This example shows how a voltaic cell whose cell reaction involves H^+ can be used to measure $[H^+]$ or pH. A pH meter (Section 15.3) is merely a specially designed voltaic cell with a voltmeter calibrated to read pH directly.

18.5 SOME COMMERCIAL VOLTAIC CELLS

Voltaic cells that receive wide use are convenient energy sources whose primary virtue is portability. Although any spontaneous redox reaction can be used as the basis of a voltaic cell, making a commercial cell that utilizes a particular redox reaction can require considerable ingenuity. Salt-bridge cells such as those that we have been discussing provide considerable insight into the operation of voltaic cells. However, they are generally unsuitable for commercial use, because they have high internal resistances. As a result, if we attempt to draw a large current from them, their voltage drops sharply. Furthermore, the cells that we have pictured so far lack the compactness and ruggedness required for portability.

Voltaic cells cannot yet compete with other common energy sources on the basis of cost alone. The cost of electricity from a common flashlight battery is on the order of $80 per kilowatt-hour. By comparison, electrical energy from power plants normally costs the consumer less than 10¢ per kilowatt-hour.

In this section we shall consider some common batteries. A battery merely consists of one or more voltaic cells. When the cells are connected in series (that is, with the positive terminal of one attached to the negative terminal of another), the battery produces an emf that is the sum of the emfs of the individual cells.

Lead Storage Battery

One of the most common batteries is the lead storage battery used in automobiles. A 12-V lead storage battery consists of six cells, each producing 2 V. The anode of each cell is composed of lead; the cathode is composed of lead dioxide, PbO_2, packed on a metal grid. Both electrodes are immersed in sulfuric acid. The electrode reactions that occur during discharge are as follows:

Anode: $$Pb(s) + SO_4^{2-}(aq) \longrightarrow PbSO_4(s) + 2e^- \qquad E° = +0.356\ V$$

Cathode: $$PbO_2(s) + SO_4^{2-}(aq) + 4H^+(aq) + 2e^- \longrightarrow PbSO_4(s) + 2H_2O(l) \qquad E° = +1.685\ V$$

$$Pb(s) + PbO_2(s) + 4H^+(aq) + 2SO_4^{2-}(aq) \longrightarrow 2PbSO_4(s) + 2H_2O(l) \qquad E° = +2.041\ V \qquad [18.31]$$

The reactants Pb and PbO_2, between which electron transfer occurs, serve as the electrodes. Because the reactants are solids, there is no need to separate the cell into anode and cathode compartments; there is no way the Pb and PbO_2 can come into direct physical contact unless one electrode plate touches another. To keep the electrodes from touching, wood or glass-fiber spacers are placed between them. To increase the current output, each cell contains a number of anode and cathode plates, as shown in Figure 18.5.

The cell emf of a lead storage battery varies during use, because the concentration of H_2SO_4 varies with the extent of cell discharge. As Equation 18.31 indicates, H_2SO_4 is used up during the discharge of a lead storage battery. Because sulfuric acid has a high density, the density

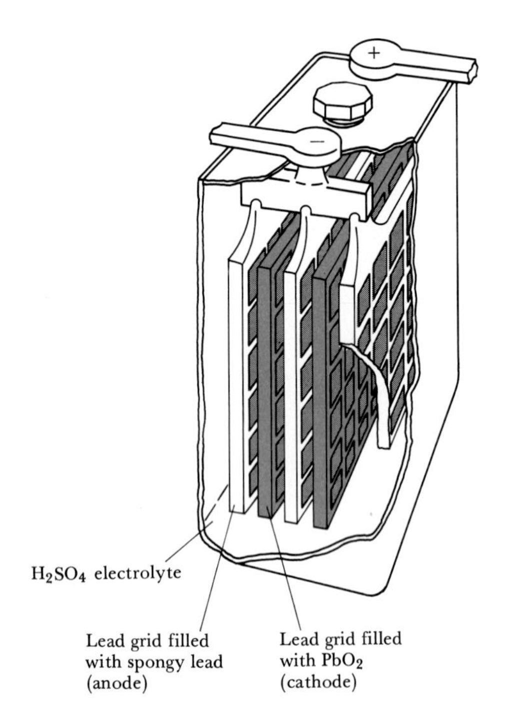

FIGURE 18.5 Lead storage cell.

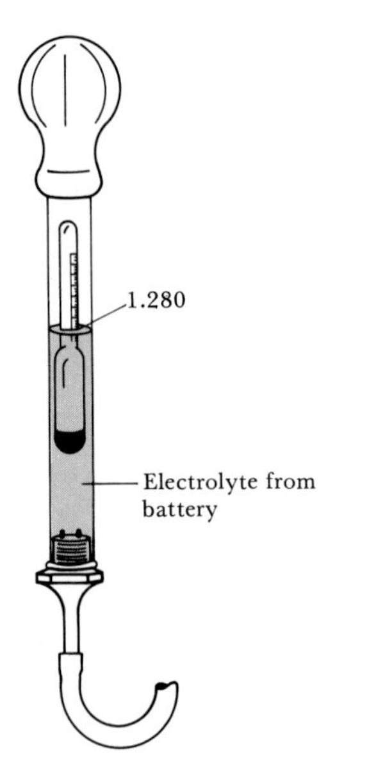

FIGURE 18.6 Hydrometer of the type used to measure the density of the electrolyte in a lead storage battery.

of the electrolyte in the cell decreases during discharge. When fully charged the electrolyte has a density of 1.25 to 1.30 g/cm^3. If the density falls below 1.20 g/cm^3, the battery needs recharging. A service-station attendant can test the density using a hydrometer. This device, shown in Figure 18.6, has a float that sinks to a depth that is a function of the density of the liquid in which it is immersed.

One advantage of the lead storage battery is that it can be recharged. During recharging, an external source of energy is used to reverse the direction of the spontaneous redox reaction, Equation 18.31. Thus the overall process during charging is as follows:

$$2PbSO_4(s) + 2H_2O(l) \longrightarrow Pb(s) + PbO_2(s) + 4H^+(aq) + 2SO_4^-(aq) \quad [18.32]$$

The energy necessary for recharging the battery is provided in an automobile by a generator that is driven by the engine. The recharging is made possible by the fact that $PbSO_4$ formed during discharge adheres to the electrodes. Thus as the external source forces electrons from one electrode to another, the $PbSO_4$ is converted to Pb at one electrode and to PbO_2 at the other; these, of course, are the materials of a fully charged cell. If the battery is charged too rapidly, water may be decomposed to form H_2 and O_2. Besides the explosive potential of H_2–O_2 mixtures, this secondary reaction can shorten the lifetime of the battery. The evolution of these gases can dislodge Pb, PbO_2, or $PbSO_4$ from the plates. Solids accumulate as a sludge at the bottom of the battery. In time they may form a short circuit that renders the cell useless.

It has been found that this problem can be substantially reduced by adding calcium to the extent of about 0.07 percent by weight to the lead in forming the electrodes. The presence of the calcium reduces the extent to which water is decomposed during the charging cycle. In the newer batteries, electrolysis of water loss is sufficiently low so that the battery is "sealed"; that is, no provision is made for addition of water and escape of gases, as was necessary in the older designs.

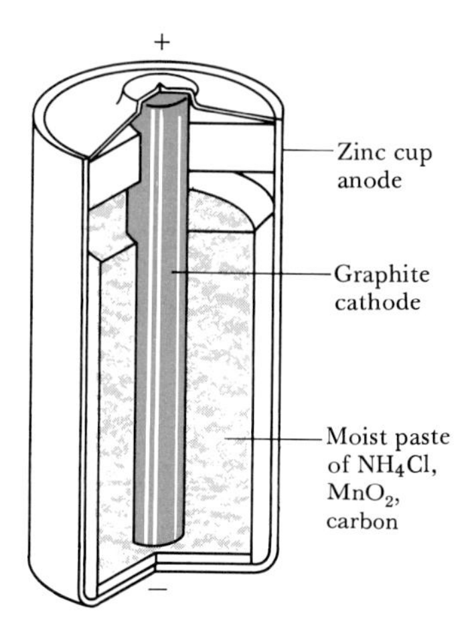

FIGURE 18.7 Cutaway view of a dry cell.

Dry Cell

The common dry cell is familiar because of its wide use in flashlights and portable radios. Because of this use, it is often referred to as a flashlight battery. It is also known as the Leclanché cell after its inventor, who patented it in 1866. In the acid version, the anode consists of a zinc can that is in contact with a paste of MnO_2, NH_4Cl, and carbon. An inert cathode, consisting of a graphite rod, is immersed in the center of the paste as shown in Figure 18.7. The cell has an exterior layer of cardboard or metal to seal the cell against the atmosphere. The electrode reactions are complex, and the cathode reaction appears to vary with the rate of discharge. The reactions at the electrodes are generally represented as shown below:

Anode: $$Zn(s) \longrightarrow Zn^{2+}(aq) + 2e^- \quad [18.33]$$

Cathode: $$2NH_4^+(aq) + 2MnO_2(s) + 2e^- \longrightarrow Mn_2O_3(s) + 2NH_3(aq) + H_2O(l) \quad [18.34]$$

Only a fraction of the cathode material, that near the electrode, is electrochemically active because of the limited mobility of the chemicals in the cell.

In the alkaline version of the dry cell, NH_4Cl is replaced by KOH. The anode reaction still involves oxidation of Zn, but the zinc is present as a powder, mixed with the electrolyte in a gel formulation. As in the common dry cell, the cathode reaction involves reduction of MnO_2. Figure 18.8 shows a cutaway view of a miniature alkaline cell, of the type used in camera exposure controls, calculators, and some watches. Although more costly than the common dry cell, alkaline cells provide improved performance. They maintain usable voltage over a longer fraction of consumption of anode and cathode materials, and provide up to 50 percent more total energy than a common dry cell of the same size.

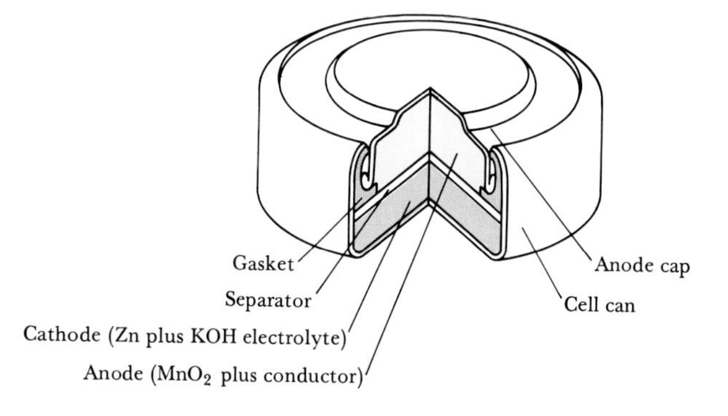

FIGURE 18.8 Cutaway view of a miniature alkaline $Zn–MnO_2$ dry cell. (*Courtesy Union Carbide Corporation*)

Nickel-Cadmium Batteries

Because dry cells are not rechargeable, they have to be replaced frequently. Thus a rechargeable cell, the nickel-cadmium battery, has become increasingly popular, especially for use in battery-operated tools and calculators. Cadmium metal acts as the anode. $NiO_2(s)$ serves as cathode; it is reduced to $Ni(OH)_2(s)$. The electrode reactions occurring within this cell during discharge are as follows:

$$\text{Anode:} \quad Cd(s) + 2OH^-(aq) \longrightarrow Cd(OH)_2(s) + 2e^- \qquad [18.35]$$

$$\text{Cathode:} \quad NiO_2(s) + 2H_2O(l) + 2e^- \longrightarrow Ni(OH)_2(s) + 2OH^-(aq) \qquad [18.36]$$

As in the lead storage cell, the reaction products adhere to the electrodes. This permits the reactions to be readily reversed during charging. Because no gases are produced during either charging or discharging, the battery can be sealed.

Fuel Cells

Many substances, for example H_2 or CH_4, are used as fuels. The heat released in their reaction with oxygen is the source of energy. The thermal energy released by combustion is often subsequently converted to electrical energy. Because combustion reactions are oxidation-reduction reactions, it would be possible to use such reactions to generate an electrical current directly if suitable voltaic cells could be designed. It is, in fact, advantageous to perform such a direct conversion to electrical energy. In producing electricity via combustion, heat is used to convert water to steam. The steam is used to drive a turbine that drives a generator. Some energy is lost as waste heat whenever energy is converted from one form to another or from one material to another. Typically, a maximum of only 40 percent of the energy from combustion is converted to electricity; the remainder is lost as heat. Direct production of electricity from fuels via a voltaic cell should yield a higher rate of conversion of the chemical energy of the fuels into electricity. Voltaic cells that utilize conventional fuels are known as **fuel cells.**

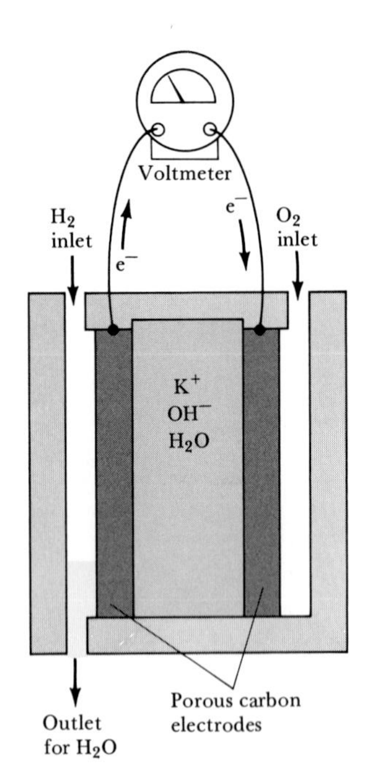

FIGURE 18.9 Cross section of a H_2-O_2 fuel cell.

A great deal of research has gone into attempts to develop practical fuel cells. One of the problems encountered is the high operating temperatures of most cells, which not only siphons off energy but also accelerates corrosion of cell parts. A low-temperature cell has been developed that uses H_2, but the present cost of the cell makes it too expensive for large-scale use. However, it has been used in special situations, such as in space vehicles. For example, a H_2–O_2 fuel cell was used as the primary source of electrical energy on the *Apollo* moon flights. The weight of the fuel cell sufficient for 11 days in space was approximately 500 lb. This may be compared to the several tons that would have been required for an engine-generator set.

The electrode reactions in the H_2–O_2 fuel cell are as follows:

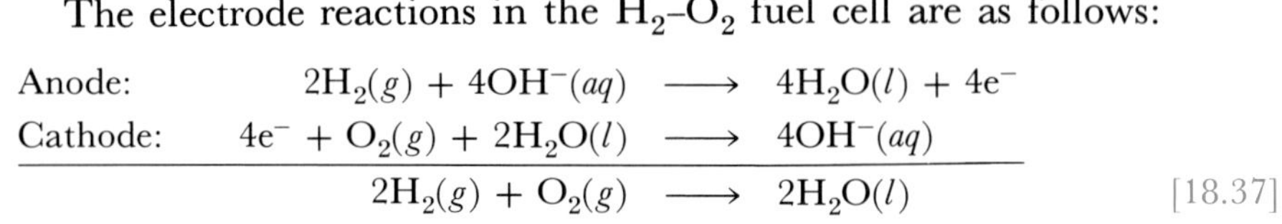

$$\text{Anode:} \quad 2H_2(g) + 4OH^-(aq) \longrightarrow 4H_2O(l) + 4e^-$$

$$\text{Cathode:} \quad 4e^- + O_2(g) + 2H_2O(l) \longrightarrow 4OH^-(aq)$$

$$2H_2(g) + O_2(g) \longrightarrow 2H_2O(l) \qquad [18.37]$$

The cell is illustrated in Figure 18.9. The electrodes are composed of hollow tubes of porous, compressed carbon impregnated with catalyst; the electrolyte is KOH. Because the reactants are supplied continuously, a fuel cell does not "go dead."

18.6 ELECTROLYSIS AND ELECTROLYTIC CELLS

We have seen that spontaneous oxidation-reduction reactions are used to make voltaic cells, electrochemical devices that generate electricity. Conversely, it is possible to use electrical energy to cause nonspontaneous oxidation-reduction reactions to occur. For example, electricity can be used to decompose molten sodium chloride into its component elements:

$$2NaCl(l) \longrightarrow 2Na(l) + Cl_2(g)$$

Such processes, which are driven by an outside source of electrical energy, are referred to as **electrolysis** reactions and are performed in **electrolytic cells.**

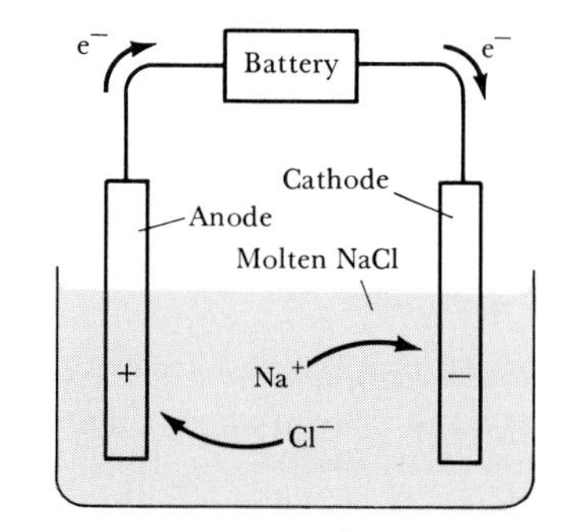

FIGURE 18.10 Electrolysis of molten sodium chloride.

An electrolytic cell consists of two electrodes in a molten salt or aqueous solution. The cell is driven by a battery or some other source of direct electrical current. The battery acts as an electron pump, pushing electrons into one electrode and pulling them from the other. Withdrawing electrons from an electrode gives it a positive charge, whereas adding electrons to an electrode makes it negative. In the electrolysis of molten NaCl, shown in Figure 18.10, Na^+ ions pick up electrons at the negative electrode and are thereby reduced. As the Na^+ ions in the vicinity of this electrode are depleted, additional Na^+ ions migrate in. In a related fashion, there is a net movement of Cl^- ions to the positive electrode, where they give up electrons and are thereby oxidized. Just as in voltaic cells, the electrode at which reduction occurs is called the cathode, and the electrode at which oxidation occurs is called the anode.

Anode:	$2Cl^-(l)$	$\longrightarrow$	$Cl_2(g) + 2e^-$
Cathode:	$2Na^+(l) + 2e^-$	$\longrightarrow$	$2Na(l)$
	$2Na^+(l) + 2Cl^-(l)$	$\longrightarrow$	$2Na(l) + Cl_2(g)$ [18.38]

Notice that the sign convention for the electrodes in an electrolysis cell is just the opposite of that for a voltaic cell. The cathode in the electrolysis cell is negative, because electrons are being forced onto it by the external voltage source. The anode is positive because electrons are being withdrawn by the external source.

Electrolyses of molten salts and molten salt solutions for the production of active metals such as sodium and aluminum are important industrial processes. We shall have more to say about them in Chapter 22, when processes for obtaining metals are discussed.

Electrolysis of Aqueous Solutions

Sodium cannot be prepared by electrolysis of aqueous solutions of NaCl, because water is more easily reduced than $Na^+(aq)$:

$$2H_2O(l) + 2e^- \longrightarrow H_2(g) + 2OH^-(aq) \qquad E^\circ_{red} = -0.83 \text{ V}$$
$$Na^+(aq) + e^- \longrightarrow Na(s) \qquad E^\circ_{red} = -2.71 \text{ V}$$

Consequently, H_2 is produced at the cathode.

The possible anode reactions are the oxidation of Cl^- and of H_2O:

$$2Cl^-(aq) \longrightarrow Cl_2(g) + 2e^- \qquad E^\circ_{ox} = -1.36 \text{ V}$$
$$2H_2O(l) \longrightarrow 4H^+(aq) + O_2(g) + 4e^- \qquad E^\circ_{ox} = -1.23 \text{ V}$$

These standard oxidation potentials are not greatly different, but they do suggest that H_2O should be oxidized more readily than Cl^-. However, the voltage required for a reaction is sometimes much greater than that indicated by the electrode potentials. The additional voltage required to cause electrolysis is called the **overvoltage**. The overvoltage is believed to be caused by slow reaction rates at the electrodes. Overvoltages for the deposition of metals are low, but those required for the liberation of hydrogen gas or oxygen gas are usually high. In the present case the overvoltage for O_2 formation is sufficiently high to permit oxidation of Cl^- rather than H_2O. Consequently, electrolysis of aqueous solutions of NaCl, known as brines, produces H_2 and Cl_2 unless the concentration of Cl^- is quite low:

$$\begin{array}{llcl} \text{Anode:} & 2Cl^-(aq) & \longrightarrow & Cl_2(g) + 2e^- \\ \text{Cathode:} & 2H_2O(l) + 2e^- & \longrightarrow & H_2(g) + 2OH^-(aq) \\ \hline & 2Cl^-(aq) + 2H_2O(l) & \longrightarrow & Cl_2(g) + H_2(g) + 2OH^-(aq) \end{array} \qquad [18.39]$$

The Na^+ ion is merely a spectator ion (Section 11.6) in the electrolysis. This process is used commercially because all of the products (H_2, Cl_2, and NaOH) are commercially important chemicals.

Electrode potentials can be used to determine the minimum potential required for an electrolysis. In the case of the formation of H_2 and Cl_2 from a brine solution under standard conditions, a minimum voltage of 2.19 V is required.

$$\begin{aligned} E^\circ &= E^\circ_{ox}(Cl^-) + E^\circ_{red}(H_2O) \\ &= -1.36 \text{ V} + (-0.83 \text{ V}) = -2.19 \text{ V} \end{aligned}$$

The emf calculated above is negative, reminding us that the process is not spontaneous but must be driven by an outside source of energy. Higher voltages than those calculated are invariably needed. One reason is the resistance of the cell; another is the overvoltage phenomenon discussed above.

SAMPLE EXERCISE 18.11

Explain why the electrolysis of an aqueous solution of $CuCl_2$ produces Cu(*s*) and $Cl_2(g)$. What is the minimum emf required for this process under standard conditions?

Solution: At the cathode we can envision either reduction of Cu^{2+} or of H_2O:

$$Cu^{2+}(aq) + 2e^- \longrightarrow Cu(s) \qquad E^\circ_{red} = 0.34 \text{ V}$$
$$2H_2O(l) + 2e^- \longrightarrow H_2(g) + 2OH^-(aq) \qquad E^\circ_{red} = -0.83 \text{ V}$$

The electrode potentials indicate that reduction of Cu^{2+} occurs more readily. Reduction of H_2O is made even more difficult because of the overvoltage for H_2 formation.

At the anode, we can envision either the oxidation of Cl^- or H_2O. As in the case of NaCl solutions, Cl_2 is usually produced because of the overvoltage for O_2 formation.

The minimum emf required for this electrolysis under standard conditions is 1.02 V:

$$\begin{aligned} E^\circ &= E^\circ_{red}(Cu^{2+}) + E^\circ_{ox}(Cl^-) \\ &= 0.34\ V + (-1.36\ V) = -1.02\ V \end{aligned}$$

Electrolysis with Active Electrodes

In our discussion of the electrolysis of molten NaCl and of NaCl solutions, we considered the electrodes to be inert. Consequently, the electrodes did not undergo reaction but merely served as the surface at which oxidation and reduction occurred. However, it often happens that the electrodes themselves participate in the electrolysis process.

When aqueous solutions are electrolyzed using metal electrodes, the electrode will be oxidized if its oxidation potential is greater than that for water. For example, nickel is oxidized more readily than water:

$$Ni(s) \longrightarrow Ni^{2+}(aq) + 2e^- \qquad E^\circ_{ox} = +0.28\ V$$

$$2H_2O(l) \longrightarrow 4H^+(aq) + O_2(g) + 4e^- \qquad E^\circ_{ox} = -1.23\ V$$

If nickel is made the anode in an electrolysis cell, nickel metal is oxidized as the anode reaction. If $Ni^{2+}(aq)$ is present in the solution, it is reduced at the cathode in preference to reduction of water. An electrolytic cell of this kind is illustrated in Figure 18.11. As current flows, nickel dissolves from the anode and deposits on the cathode:

$$\text{Anode:} \qquad Ni(s) \longrightarrow Ni^{2+}(aq) + 2e^- \qquad [18.40]$$

$$\text{Cathode:} \qquad Ni^{2+}(aq) + 2e^- \longrightarrow Ni(s) \qquad [18.41]$$

Electrolytic processes involving active metal electrodes, that is, in which the metal electrodes participate in the cell reaction, have several important applications. We will see in Chapter 22 that electrolysis af-

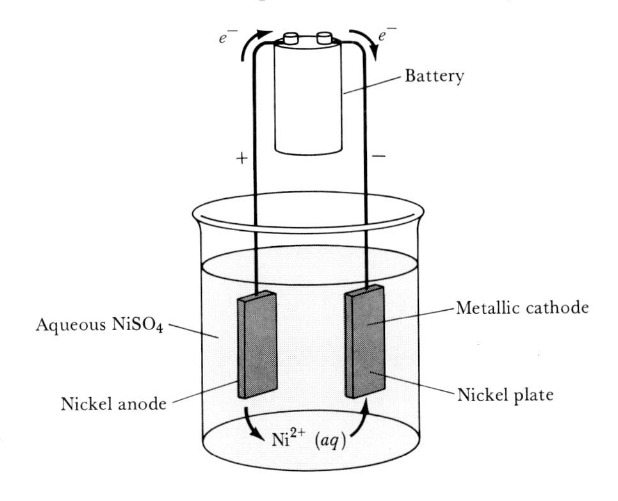

FIGURE 18.11 Electrolytic cell with an active metal electrode. Nickel dissolves from the anode to form $Ni^{2+}(aq)$. At the cathode $Ni^{2+}(aq)$ is reduced and forms a nickel "plate."

fords a means of purifying crude metals such as copper, zinc, cobalt, and nickel. An additional important application is in **electroplating,** in which one metal is "plated," or deposited, on another. Electroplating is used to protect objects against corrosion and to improve their appearance. For example, car bumpers and other exposed metal parts are protected against corrosion by electroplating the steel with a coat of nickel, and then a final coat of chromium. The object to be electroplated is made the cathode of the electrolytic cell. The cathode in the cell illustrated in Figure 18.11 might, for example, be a car bumper or some other metallic object. As another example, the deposition of a silver plate on a spoon is shown in Figure 18.12.

18.7 QUANTITATIVE ASPECTS OF ELECTROLYSIS

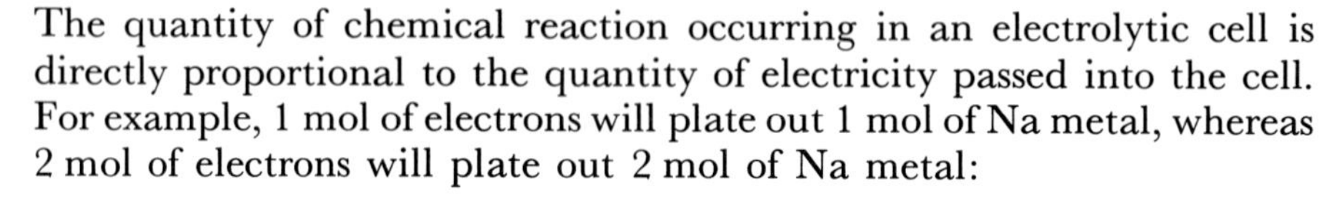

The quantity of chemical reaction occurring in an electrolytic cell is directly proportional to the quantity of electricity passed into the cell. For example, 1 mol of electrons will plate out 1 mol of Na metal, whereas 2 mol of electrons will plate out 2 mol of Na metal:

$$Na^+ + e^- \longrightarrow Na$$

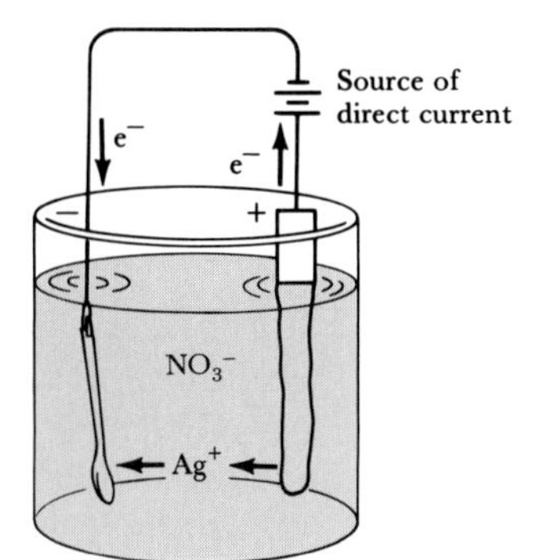

FIGURE 18.12 Electroplating of an object with silver. The object to be plated is made the cathode, whereas the plating metal is made the anode.

Similarly, it requires 2 mol of electrons to produce 1 mol of copper from Cu^{2+} and 3 mol of electrons to produce 1 mol of aluminum from Al^{3+}:

$$Cu^{2+} + 2e^- \longrightarrow Cu$$
$$Al^{3+} + 3e^- \longrightarrow Al$$

The quantity of charge passing through an electrical circuit, such as that in an electrolytic cell, is generally measured in **coulombs.** As noted in Section 18.4, there are 96,500 C in a faraday:

$$1\ \mathfrak{F} = 96{,}500\ C = \text{charge of 1 mol of electrons} \qquad [18.42]$$

In terms of other, perhaps more familiar electrical units, a coulomb is the quantity of electrical charge passing a point in a circuit in 1 s when the current is 1 ampere (A).* Therefore, the number of coulombs passing through a cell can be obtained by multiplying the amperage and the elapsed time in seconds:

$$\text{Coulombs} = \text{amperes} \times \text{seconds} \qquad [18.43]$$

These ideas are applied in Sample Exercises 18.12 and 18.13. Although these exercises involve electrolytic cells, the same relationships can be applied to voltaic cells.

*Conversely, current is the rate of flow of electricity. An ampere is the current associated with the flow of 1 C past a point each second.

SAMPLE EXERCISE 18.12

Calculate the amount of aluminum produced in 1.00 h by the electrolysis of molten $AlCl_3$ if the current is 10.0 A.

Solution: Using Equation 18.43, we can write

$$\text{Coulombs} = (10.0\ \text{A})(1.00\ \text{h}) \times \left(\frac{3600\ \text{s}}{1\ \text{h}}\right)\left(\frac{1\ \text{C}}{1\ \text{A-s}}\right) = 3.60 \times 10^4\ \text{C}$$

The half-reaction for the reduction of Al^{3+} is

$$Al^{3+} + 3e^- \longrightarrow Al$$

The amount of aluminum produced depends on the number of available electrons: 1 mol Al $\simeq$ 3 $\mathfrak{F}$. We can therefore write

$$\text{Grams Al} = (3.60 \times 10^4\ \text{C})\left(\frac{1\ \mathfrak{F}}{96{,}500\ \text{C}}\right) \times \left(\frac{1\ \text{mol Al}}{3\ \mathfrak{F}}\right)\left(\frac{27.0\ \text{g Al}}{1\ \text{mol Al}}\right) = 3.36\ \text{g}$$

SAMPLE EXERCISE 18.13

A constant current was passed through a solution of Cu^{2+} for a period of 5.00 min. During this time the cathode increased in mass by 1.24 g. How many amperes of current were used?

Solution: In this case the half-reaction that we are focusing on is

$$Cu^{2+} + 2e^- \longrightarrow Cu$$

Therefore, 1 mol Cu $\simeq$ 2 $\mathfrak{F}$. Using this information, we can calculate the amperes as follows:

$$\text{Coulombs} = (1.24\ \text{g Cu})\left(\frac{1\ \text{mol Cu}}{63.5\ \text{g Cu}}\right) \times \left(\frac{2\ \mathfrak{F}}{1\ \text{mol Cu}}\right)\left(\frac{96{,}500\ \text{C}}{1\ \mathfrak{F}}\right) = 3.77 \times 10^3\ \text{C}$$

$$\text{Amperes} = \frac{\text{Coulombs}}{\text{seconds}} = \left(\frac{3.77 \times 10^3\ \text{C}}{5.00\ \text{min}}\right)\left(\frac{1\ \text{min}}{60\ \text{s}}\right)\left(\frac{1\ \text{A-s}}{1\ \text{C}}\right) = 12.6\ \text{A}$$

Electrical Work

This is a good point at which to consider the relationship between electrochemical processes and work. We have already seen that a positive value for E is associated with a negative value for the free-energy change, and thus with a spontaneous process. We also know that for any spontaneous process ΔG is a measure of the maximum useful work, w_{max}, that can be extracted from the process: $-\Delta G = w_{max}$. Since $\Delta G = -n\mathfrak{F}E$, the maximum useful electrical work obtainable from a voltaic cell should be simply

$$w_{max} = n\mathfrak{F}E \qquad [18.44]$$

Keep in mind that w_{max} is the maximum useful work done *by* the system on its sourroundings. A positive value for w_{max} requires a spontaneous process (positive E) for its accomplishment.

Notice that the maximum work obtainable is proportional to the cell potential E. We can think of E as a kind of pressure, a measure of the driving force for the process. Recall from Section 18.3 that the units of E are J/C; w_{max} is also proportional to the number of coulombs that flow, measured by $n\mathfrak{F}$. When we put these quantities together we can see how the units cancel to leave us with units of energy:

$$\begin{aligned} w_{max} &= n \times \mathfrak{F} \times E \\ (\text{J}) &= (\text{mol}) \times \left(\frac{\text{C}}{\text{mol}}\right) \times \left(\frac{\text{J}}{\text{C}}\right) \end{aligned}$$

In an electrolysis cell we employ an external source of energy to cause a nonspontaneous electrochemical process to occur. In this case, ΔG for the cell process is positive and the cell potential is negative. The *minimum* amount of work required to cause the nonspontaneous cell reaction to occur, w_{min}, is given by

$$w_{min} = n\mathfrak{F}E \qquad [18.45]$$

where E is the calculated cell potential. In practice, we always need to expend more than this minimum amount of work, because of inefficiencies in the process. In addition, a higher potential E' may be required to cause the cell reaction to occur, because of overvoltage. In this case the minimum work required is given by

$$w_{min} = n\mathfrak{F}E' \qquad [18.46]$$

Electrical work is usually expressed in energy units of watts times time. The **watt** is a unit of electrical power, that is, the rate of energy expenditure:

$$1 \text{ watt (W)} = \frac{1 \text{ J}}{\text{s}} \qquad [18.47]$$

Thus a watt-second is a joule. The unit employed by electric utilities is the kilowatt-hour, which works out to be 3.6×10^6 J:

$$1 \text{ kWh} = (1000 \text{ W})(1 \text{ h})\left(\frac{3600 \text{ s}}{1 \text{ h}}\right)\left(\frac{1 \text{ J/s}}{1 \text{ W}}\right) = 3.6 \times 10^6 \text{ J} \qquad [18.48]$$

Using these considerations, we can calculate the maximum work obtainable from voltaic cells and the minimum work required to bring about desired electrolysis reactions.

SAMPLE EXERCISE 18.14

Calculate the minimum number of kilowatt-hours of electricity required to produce 1000 kg of aluminum by electrolysis of Al^{3+} if the required emf is 4.5 V.

Solution: We need to employ Equation 18.46 to calculate w_{min} for an applied potential of 4.5 V. First we need to calculate $n\mathfrak{F}$, the number of coulombs required.

$$\text{Coulombs} = (1000 \text{ kg Al})\left(\frac{1000 \text{ g Al}}{1 \text{ kg Al}}\right) \times \left(\frac{1 \text{ mol Al}}{27.0 \text{ g Al}}\right)\left(\frac{3\ \mathfrak{F}}{1 \text{ mol Al}}\right) \times \left(\frac{96{,}500 \text{ C}}{1\ \mathfrak{F}}\right) = 1.07 \times 10^{10} \text{ C}$$

We can now employ Equation 18.46 to calculate w_{min}. In doing so we must apply the unit conversion factor of Equation 18.48:

$$\text{Kilowatt-hours} = (1.07 \times 10^{10} \text{ C})(4.5 \text{ V}) \times \left(\frac{1 \text{ J}}{1 \text{ C-V}}\right)\left(\frac{1 \text{ kWh}}{3.6 \times 10^6}\right) = 1.33 \times 10^4 \text{ kWh}$$

This quantity of energy does not include the energy used to mine, transport, and process the aluminum ore, and to keep the electrolysis bath molten during electrolysis. A typical electrolytic cell used to reduce aluminum is only 40 percent efficient, 60 percent of the electrical energy being dissipated as heat. It therefore requires on the order of 33 kWh of electricity to produce 1 kg of aluminum.

The aluminum industry consumes about 2 percent of the electrical energy generated in the United States. Because this is used mainly for reduction of aluminum, recycling this metal saves large quantities of energy. Interestingly, the United States uses 5 percent of the world's production of aluminum to make food and beverage cans.

SAMPLE EXERCISE 18.15

A 12-V lead storage battery contains 410 g of lead in its anode plates and a stoichiometrically equivalent amount of PbO_2 in the cathodes. (a) What is the maximum number of coulombs of electrical charge it can deliver without being recharged? (b) For how many hours could the battery deliver a steady current of 1.0 A assuming that the current does not fall during discharge? (c) What is the maximum electrical work that the battery can accomplish in kilowatt-hours?

Solution: (a) The lead anode undergoes a two-electron oxidation:

$$Pb \longrightarrow Pb^{2+} + 2e^-$$

Consequently, $2\ \mathfrak{F} \simeq 1$ mol Pb. Using this relationship, we have

$$\text{Coulombs} = (410\text{ g Pb})\left(\frac{1\text{ mol Pb}}{207\text{ g Pb}}\right) \times \left(\frac{2\ \mathfrak{F}}{1\text{ mol Pb}}\right)\left(\frac{96{,}000\text{ C}}{1\ \mathfrak{F}}\right) = 3.8 \times 10^5\text{ C}$$

Although the emf of a cell is independent of the masses of the solid reactants involved in the cell, the total electrical charge the cell can deliver does depend on these quantities. The size and surface area further affects the *rate* at which electrical charge can be delivered.

(b) We calculate the number of hours of operation at a current level of 1.0 A by recalling that a coulomb corresponds to a current of 1 A flowing for 1 s:

$$\text{Hours} = (3.8 \times 10^5\text{ C})\left(\frac{1\text{ A-s}}{1\text{ C}}\right) \times \left(\frac{1\text{ h}}{3600\text{ s}}\right)\left(\frac{1}{1.0\text{ A}}\right) = 1.1 \times 10^2\text{ h}$$

This battery might be described as a 110 amp-hour battery.

(c) The maximum work is given by the product $n\mathfrak{F}E$, Equation 18.44:

$$\text{Kilowatt-hours} = (3.8 \times 10^5\text{ C})(12\text{ V}) \times \left(\frac{1\text{ J}}{1\text{ C-V}}\right)\left(\frac{1\text{ kWh}}{3.6 \times 10^6\text{ J}}\right) = 1.3\text{ kWh}$$

18.8 CORROSION

Before we close our discussion of electrochemistry, let's apply some of what we have learned to a very important problem, the **corrosion** of metals. Corrosion reactions are redox reactions in which a metal is attacked by some substance in its environment and converted to an unwanted compound.

One of the most familiar corrosion processes is the rusting of iron. From an economic standpoint this is a significant process. It is estimated that up to 20 percent of the iron produced annually in this country is used to replace iron objects that have been discarded because of rust damage.

The rusting of iron is known to involve oxygen; iron does not rust in water unless O_2 is present. Rusting also involves water; iron does not rust in oil, even if it contains O_2, unless H_2O is also present. Other factors such as the pH of the solution, the presence of salts, contact with metals more difficult to oxidize than iron, and stress on the iron can accelerate rusting.

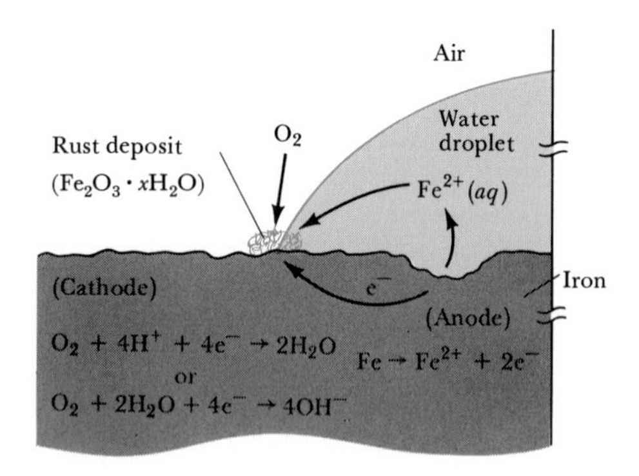

FIGURE 18.13 Corrosion of iron in contact with water.

The corrosion of iron is generally believed to be electrochemical in nature. A region on the surface of the iron serves as an anode at which the iron undergoes oxidation:

$$Fe(s) \longrightarrow Fe^{2+}(aq) + 2e^- \qquad E^\circ_{ox} = 0.44\ V \qquad [18.49]$$

The electrons so produced migrate through the metal to another portion of the surface that serves as the cathode. Here oxygen can be reduced:

$$O_2(g) + 4H^+(aq) + 4e^- \longrightarrow 2H_2O(l) \qquad E^\circ_{red} = 1.23\ V \qquad [18.50]$$

Notice that H^+ is involved in the reduction of O_2. As the concentration of H^+ is lowered (that is, as pH is increased), the reduction of O_2 becomes less favorable. It is observed that iron in contact with a solution whose pH is above 9 to 10 does not corrode. In the course of the corrosion, the Fe^{2+} formed at the anode is further oxidized to Fe^{3+}. The Fe^{3+} forms the hydrated iron(III) oxide known as rust:*

$$4Fe^{2+}(aq) + O_2(g) + 4H_2O(l) + 2xH_2O(l) \longrightarrow 2Fe_2O_3 \cdot xH_2O(s) + 8H^+(aq) \qquad [18.51]$$

Because the cathode is generally the area having the largest supply of O_2, the rust often deposits there. If you look closely at a shovel after it has stood outside in the moist air with wet dirt adhered to its blade, you may notice that pitting has occurred under the dirt but that rust has formed elsewhere, where O_2 is more readily available. The corrosion process is summarized by way of illustration in Figure 18.13.

The enhanced corrosion caused by the presence of salts is usually quite evident on autos in areas where there is heavy salting of roads during winter. The effect of salts is readily explained by the voltaic mechanism: The ions of a salt provide the electrolyte necessary for completion of the electrical circuit.

The presence of anodic and cathodic sites on the iron requires two different chemical environments on the surface. These can occur through

*Frequently, metal compounds that are obtained from aqueous solution have water associated with them. For example, copper(II) sulfate crystallizes from water with 5 mol of water per mole of $CuSO_4$. We represent this formula as $CuSO_4 \cdot 5H_2O$. Such compounds are called hydrates. Rust is a hydrate of iron(III) oxide with a variable amount of water of hydration. We represent the variable water content by writing the formula as $Fe_2O_3 \cdot xH_2O$.

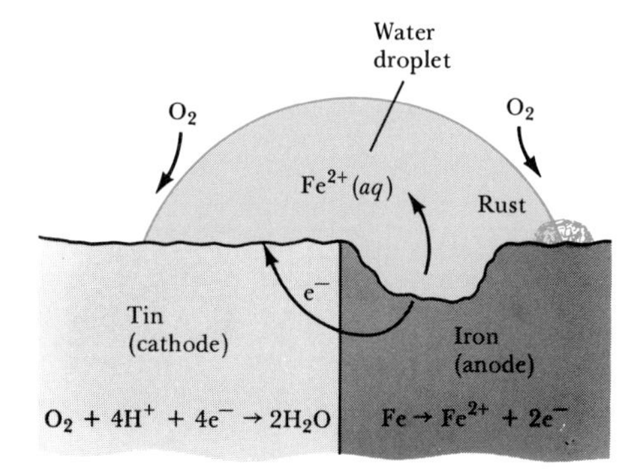

FIGURE 18.14 Corrosion of iron in contact with tin.

the presence of impurities or lattice defects (perhaps introduced by strain on the metal). At the sites of such impurities or defects, the atomic-level environment around the iron atom may permit the metal to be either more or less easily oxidized than at normal lattice sites. Thus these sites may serve as either anodes or cathodes. Ultrapure iron, prepared in such a way as to minimize lattice defects, is far less susceptible to corrosion than is ordinary iron.

Iron is often covered with a coat of paint or another metal such as tin, zinc, or chromium to protect its surface against corrosion. "Tin cans" are produced by applying a thin layer of tin over steel. The tin protects the iron only as long as the protective layer remains intact. Once it is broken and the iron exposed to air and water, tin actually promotes the corrosion of the iron. It does so by serving as the cathode in the electrochemical corrosion. As shown by the following half-cell potentials, iron is more readily oxidized than tin:

$$Fe(s) \longrightarrow Fe^{2+}(aq) + 2e^- \qquad E^\circ_{ox} = 0.44\ V \qquad [18.52]$$

$$Sn(s) \longrightarrow Sn^{2+}(aq) + 2e^- \qquad E^\circ_{ox} = 0.14\ V \qquad [18.53]$$

The iron therefore serves as the anode and is oxidized as shown in Figure 18.14.

"Galvanized iron" is produced by coating iron with a thin layer of zinc. The zinc protects the iron against corrosion even after the surface coat is broken. In this case the iron serves as the cathode in the electrochemical corrosion because zinc is oxidized more easily than iron:

$$Zn(s) \longrightarrow Zn^{2+}(aq) + 2e^- \qquad E^\circ_{ox} = 0.76\ V \qquad [18.54]$$

The zinc therefore serves as the anode and is corroded instead of the iron, as shown in Figure 18.15. Such protection of a metal by making it the

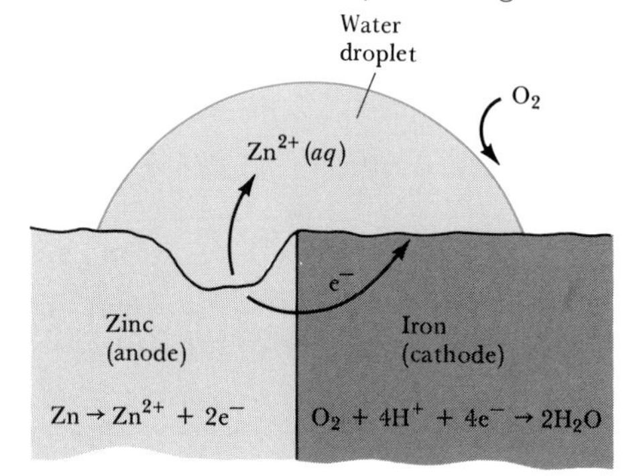

FIGURE 18.15 Cathodic protection of iron in contact with zinc.

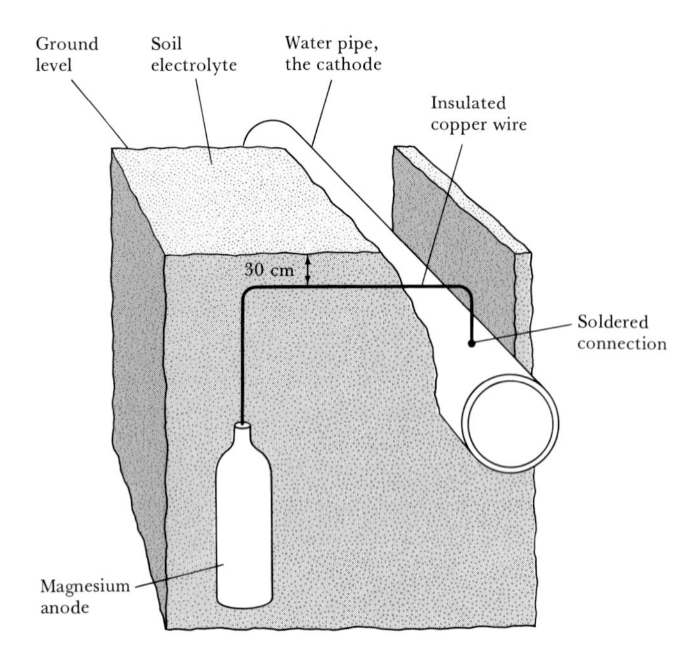

FIGURE 18.16 Cathodic protection of an iron water pipe. The magnesium anode is surrounded by a mixture of gypsum, sodium sulfate, and clay to promote conductivity of ions. The pip, in effect, is the cathode of a voltaic cell.

cathode in an electrochemical cell is known as **cathodic protection.** Underground pipelines are often protected against corrosion by making the pipeline the cathode of a voltaic cell. Pieces of an active metal such as magnesium are buried along the pipeline and connected to it by wire as shown in Figure 18.16. In moist soil, where corrosion can occur, the active metal serves as the anode and the pipe experiences cathodic protection.

Although our discussions have centered on iron, this is not the only metal subject to corrosion. One thing that may be surprising in light of our discussions is the fact that an aluminum can, disposed of carelessly in the environment, corrodes so much more slowly than a steel can. On the basis of the standard oxidation potentials of aluminum ($E^\circ_{ox} = 1.66$ V) and iron ($E^\circ_{ox} = 0.44$ V), we would expect the aluminum to be much more readily corroded. The slow corrosion of aluminum is explained by the formation of a thin, compact oxide coating that forms on its surface. This protects the underlying metal from further corrosion. Magnesium, which also has a large oxidation potential, is similarly protected. The oxide coat on iron is too porous to offer similar protection. However, when iron is alloyed with chromium, a protective oxide coating does form. Such alloys are called stainless steels.

SAMPLE EXERCISE 18.16

Predict the nature of the corrosion that would take place if an iron gutter were nailed to a house using aluminum nails.

Solution: A voltaic cell can be formed at the point of contact of the two metals. The metal that is most easily oxidized will serve as the anode, whereas the other metal serves as the cathode. By comparing standard oxidation potentials of Al and Fe we see that Al will be the anode:

$$\mathrm{Al}(s) \longrightarrow \mathrm{Al}^{3+}(aq) + 3e^- \qquad E^\circ_{ox} = 1.66\ \mathrm{V}$$
$$\mathrm{Fe}(s) \longrightarrow \mathrm{Fe}^{2+}(aq) + 2e^- \qquad E^\circ_{ox} = 0.44\ \mathrm{V}$$

The gutter will thus be protected against corrosion in the vicinity of the nail, because the iron serves as the cathode. However, the nail would quickly corrode, leaving the gutter on the ground.

What do you think would happen if aluminum siding were nailed to a house using iron nails?

FOR REVIEW

Summary

We have seen that oxidation-reduction (or redox) reactions can be considered to consist of two half-reactions, one due to **oxidation** (loss of electrons) and the other due to **reduction** (gain of electrons). The substance oxidized is called the **reducing agent,** or **reductant,** whereas the substance reduced is called the **oxidizing agent,** or **oxidant.** If the half-reactions are balanced, they can be added to obtain the balanced redox equation. The number of electrons lost by the reducing agent must equal the number gained by the oxidizing agent. Consequently, different types of atoms and total charges must each balance.

Spontaneous redox reactions can be used to generate electricity in **voltaic cells.** Conversely, electricity can be used to bring about nonspontaneous reactions in **electrolytic cells.** In either type of cell, the electrode at which oxidation occurs is called the **anode,** whereas the electrode at which reduction occurs is called the **cathode.**

A voltaic cell may be thought to possess a "driving force" that moves the electrons through the external circuit, from anode to cathode. This driving force is called the **electromotive force (emf),** and is measured in volts. The emf of a cell can be regarded as being composed of two parts: that due to oxidation at the anode and that due to reduction at the cathode: $E_{cell} = E_{ox} + E_{red}$.

Oxidation potentials (E_{ox}) and reduction potentials (E_{red}) can be assigned to half-reactions by defining the standard hydrogen electrode as a reference:

$$2\mathrm{H}^+(1\ M) + 2e^- \longrightarrow \mathrm{H}_2(1\ \mathrm{atm}) \qquad E^\circ = 0$$

Standard reduction potentials are referred to merely as **standard electrode potentials** and are tabulated for a great variety of reduction half-reactions. The oxidation potential for an oxidation half-reaction will be of the same magnitude as, but opposite in sign to, the electrode potential for the reverse reduction process. The more positive the potential associated with a half-reaction, the greater the tendency for that reaction to occur as written. Electrode potentials can be used to determine the maximum voltages generated by voltaic cells or the minimum voltages required in electrolytic cells. They can also be used to predict whether certain redox reactions are spontaneous (positive E). The emf is related to free-energy changes: $\Delta G = -n\mathfrak{F}E$, where $\mathfrak{F}$ is **Faraday's constant,** 96,500 J/V-mol = 96,500 C/mol.

The emf of a cell varies in magnitude with temperature and with the concentrations of reactants and products. The Nernst equation relates emf under nonstandard conditions to the standard emf:

$$E = E^\circ - \frac{2.30RT}{n\mathfrak{F}}\log Q$$

At equilibrium

$$E = 0 \qquad \text{and} \qquad E^\circ = \frac{2.30RT}{n\mathfrak{F}}\log K$$

Thus standard emfs are related to equilibrium constants. The maximum electrical work that can be obtained from a voltaic cell is the product of the total charge it delivers, $n\mathfrak{F}$, and its emf, E: $w_{max} = n\mathfrak{F}E$. To illustrate the principles involved in voltaic cells, simple cells utilizing salt bridges were discussed. Commercial cells need to be more rugged. Four common batteries were discussed: the lead storage battery, the nickel-cadmium battery, the common dry cell, and the alkaline dry cell. The first two are rechargeable, whereas the latter two are not.

In an electrolytic cell an external source of electricity is used to "pump" electrons from the anode to the cathode. The current-carrying medium within the cell may be either a molten salt or an aqueous electrolyte solution. The products produced during electrolysis can generally be predicted by comparing electrode potentials associated with possible oxidation and reduction processes. However, because of the **overvoltage** phenomenon, some reactions (such as those that generate H_2 and O_2) occur less readily than electrode potentials would suggest.

The extent of chemical reaction in either an electrolytic or voltaic cell can be related to the quantity of electricity that passes through the external circuit. The amount of electrical charge possessed by a mole of electrons is known as a faraday (abbreviated $\mathfrak{F}$), 96,500 C; 1 coulomb equals 1 ampere-second: 1 C = 1 A-s.

Our knowledge of electrochemistry allows us to design batteries and to bring about desirable redox reactions such as those involved in electroplating and in the reduction and refining of metals. Electrochemical principles also help us to understand and combat corrosion. Corrosion of a metal such as iron can be shown to be electrochemical in origin. A metal can be protected against corrosion by putting it in contact with another metal that more readily undergoes oxidation. This process is known as cathodic protection.

Learning goals

Having read and studied this chapter, you should be able to:

1. Recognize redox reactions and identify the reductant and oxidant.
2. Complete and balance redox equations using the method of half-reactions.
3. Diagram simple voltaic and electrolytic cells, labeling anode, cathode, the directions of ion and electron movements, and the signs of the electrodes.
4. Calculate the emf generated by a voltaic cell or the minimum emf required to cause an electrolytic cell reaction to proceed, having been given appropriate electrode potentials.
5. Use electrode potentials to predict whether a reaction will be spontaneous.
6. Interconvert $E°$, $\Delta G°$, and K for redox reactions.
7. Use the Nernst equation to calculate emfs under nonstandard conditions.
8. Use the Nernst equation to calculate the concentration of an ion, given E, $E°$, and the concentrations of the remaining ions.
9. Describe the lead storage battery, the dry cell, and the nickel-cadmium cell.
10. Calculate the third quantity, having been given any two of the following: time, current, amount of substance produced or consumed in an electrolysis reaction.
11. Calculate the maximum electrical work performed by a voltaic cell (or the minimum electrical work to cause a reaction in an electrolytic cell), having been given emf and electrical charge or information from which they can be determined.
12. Describe the phenomenon of corrosion in terms of electrochemical principles, and explain the principle that underlies cathodic protection.

Key terms

Among the more important terms and expressions used for the first time in this chapter are the following:

An anode (Section 18.2) is an electrode at which oxidation occurs.

A cathode (Section 18.2) is an electrode at which reduction occurs.

Cathodic protection (Section 18.8) is a means of protecting a metal against corrosion by making it the cathode in a voltaic cell. This can be done by attaching a more active metal.

Corrosion (Section 18.8) is the process by which a metal is oxidized by substances in its environment.

An electrolytic cell (Section 18.6) is a device in which a nonspontaneous reaction is caused to occur by passage of current under a sufficient external electrical potential.

The electromotive force, emf (Section 18.3), is a measure of the driving force or "electrical pressure" for completion of a chemical reaction via passage of electrons through an external circuit. It is measured in volts, which is equivalent to joules per coulomb.

A faraday (Section 18.4) is the total charge of a mole of electrons, 96,500 C.

A half-reaction (Section 18.1) is an equation for either oxidation or reduction that explicitly shows the electrons involved [for example, $2H^+(aq) + 2e^- \longrightarrow H_2(g)$].

The Nernst equation (Section 18.4) relates the cell emf, E, to the standard cell emf, $E°$, and the reaction quotient, Q:

$$E = E° - \frac{2.30RT}{n\mathfrak{F}} \log Q$$

An oxidant, or oxidizing agent (Section 18.1), is a substance that is reduced and thereby causes the oxidation of some other substance.

A reductant, or reducing agent (Section 18.1), is a substance that is oxidized and thereby causes the reduction of some other substance.

A standard electrode potential, $E°$ (Section 18.3), is the reduction potential of a half-reaction with all solution species at 1 M concentration and all gaseous species at 1 atm, measured relative to the hydrogen electrode, for which $E°$ is exactly 0 V.

A voltaic cell (Section 18.2) is a device in which a spontaneous chemical reaction is made to occur via passage of electrons through an external circuit.

EXERCISES

Oxidation-reduction reactions

18.1 Complete and balance the following reaction or half-reactions, indicating in each case whether oxidation or reduction is occurring:

(a) $Al(OH)_3(s) \longrightarrow Al(OH)_4^-(aq)$ (basic solution)
(b) $Ni(s) \longrightarrow Ni^{2+}(aq)$ (acidic solution)
(c) $ClO^-(aq) \longrightarrow ClO_3^-(aq)$ (basic solution)
(d) $H_2S(aq) \longrightarrow S(s)$ (acidic solution)
(e) $CrO_4^{2-}(aq) \longrightarrow Cr_2O_7^{2-}(aq)$ (acidic solution)

18.2 Indicate whether the following reactions involve oxidation-reduction. If they do, identify the substance oxidized and the substance reduced.

(a) $2H_2SO_4(l) + 2NaBr(s) \longrightarrow Br_2(g) + SO_2(g) + Na_2SO_4(s) + 2H_2O(g)$
(b) $PBr_3(l) + 3H_2O(l) \longrightarrow H_3PO_3(l) + 3HBr(g)$
(c) $SiO_2(s) + 6HF(aq) \longrightarrow H_2SiF_6(s) + 2H_2O(l)$
(d) $2KClO_3(s) \longrightarrow 2KCl(s) + 3O_2(g)$
(e) $2Hg^{2+}(aq) + N_2H_4(aq) \longrightarrow 2Hg(l) + N_2(g) + 4H^+(aq)$

18.3 Balance the following equations, identifying in each case the substance oxidized and the substance reduced:

(a) $F_2(g) + Br_2(g) \longrightarrow BrF_3(g)$
(b) $ZnS(s) + O_2(g) \longrightarrow ZnO(s) + SO_2(g)$
(c) $MnO_4^-(aq) + I^-(aq) + H_2O(l) \longrightarrow MnO_2(s) + I_2(s) + OH^-(aq)$
(d) $Fe(s) + NO_3^-(aq) + H^+(aq) \longrightarrow Fe^{2+}(aq) + NO(g) + H_2O(l)$
(e) $H_2O_2(aq) + HClO(aq) + H^+(aq) \longrightarrow O_2(g) + Cl^-(aq) + H_2O(l)$

18.4 Complete and balance each of the following equations:

(a) $Cr_2O_7^{2-}(aq) + I^-(aq) \longrightarrow Cr^{3+}(aq) + IO_3^-(aq)$ (acidic solution)
(b) $MnO_4^-(aq) + CH_3OH(aq) \longrightarrow Mn^{2+}(aq) + HCO_2H(aq)$ (acidic solution)
(c) $As(s) + ClO_3^-(aq) \longrightarrow H_3AsO_3(aq) + HClO(aq)$ (acidic solution)
(d) $As_2O_3(s) + NO_3^-(aq) \longrightarrow H_3AsO_4(aq) + N_2O_3(aq)$ (acidic solution)
(e) $MnO_4^-(aq) + Br^-(aq) \longrightarrow MnO_2(s) + BrO_3^-(aq)$ (basic solution)
(f) $H_2O_2(aq) + Cl_2O_7(aq) \longrightarrow ClO_2^-(aq) + O_2(g)$ (basic solution)
(g) $Pb(OH)_4^{2-}(aq) + ClO^-(aq) \longrightarrow PbO_2(s) + Cl^-(aq)$ (basic solution)
(h) $TlOH(s) + NH_2OH(aq) \longrightarrow Tl_2O_3(s) + N_2(g)$ (basic solution)

18.5 Hydrazine (N_2H_4) and dinitrogen tetraoxide (N_2O_4) form a self-igniting mixture that has been employed as a rocket propellant. The reaction products are N_2 and H_2O. (a) Write a balanced chemical equation for this reaction. (b) Which substance serves as reducing agent and which as oxidizing agent?

[18.6] Oxalic acid, $H_2C_2O_4$, occurs in certain vegetables, for example, rhubarb and spinach. This acid is toxic but occurs below toxic limits in foods. However, it may be concentrated during certain types of food processing. The concentration of $H_2C_2O_4$ in a sample may be determined by titration with permanganate ion (MnO_4^-) in an acid solution forming $CO_2(g)$ and $Mn^{2+}(aq)$. (a) Write a balanced net ionic equation for this reaction. (b) If a 50.0-g sample requires 22.0 mL of 0.0500 M MnO_4^- solution to reach an end point in the titration, what is the weight percentage of $H_2C_2O_4$ in the sample?

Voltaic cells; emf; spontaneity

18.7 Indicate whether each of the following statements regarding voltaic cells is true or false. Correct any false statements. (a) The anode of a voltaic cell is assigned a negative sign. (b) All voltaic cells require inert electrodes. (c) The cathode is the electrode at which oxidation occurs. (d) A salt bridge serves to maintain electrical neutrality in the anode and cathode compartments

18.8 Given the following half-cell reactions and associated standard half-cell potentials:

$$AuBr_4^-(aq) + 3e^- \longrightarrow Au(s) + 4Br^-(aq) \quad E° = -0.858\ V$$
$$Eu^{3+}(aq) + e^- \longrightarrow Eu^{2+}(aq) \quad E° = -0.43\ V$$
$$IO^-(aq) + H_2O(l) + 2e^- \longrightarrow I^-(aq) + 2OH^-(aq) \quad E° = 0.49\ V$$
$$Sn^{2+}(aq) + 2e^- \longrightarrow Sn(s) \quad E° = -0.14\ V$$

(a) Write the cell reaction for the combination of these half-cell reactions that leads to the largest cell emf, and calculate the value. (b) Write the cell reaction for the combination of half-cell reactions that leads to the smallest cell emf and calculate that value.

18.9 A 1 M solution of $Cu(NO_3)_2$ is placed in a beaker with a strip of Cu metal. A 1 M solution of $SnSO_4$ is placed in a second beaker with a strip of Sn metal. The two beakers are connected by a salt bridge, and the two metal electrodes are linked by wires to a voltmeter. (a) Which electrode serves as anode, and which as cathode? (b) Which electrode gains mass and which loses mass as the cell reaction proceeds? (c) Indicate the sign of each electrode. (d) What is the emf generated by the cell under standard conditions?

18.10 A voltaic cell that utilizes the reaction

$$Tl^{3+}(aq) + 2Cr^{2+}(aq) \longrightarrow Tl^{+}(aq) + 2Cr^{3+}(aq)$$

has a measured emf of 1.19 V under standard conditions. (a) What is $E°$ for the half-cell reaction: $Tl^{3+}(aq) + 2e^- \longrightarrow Tl^{+}(aq)$? (b) What is meant by the expression "under standard conditions"? (c) Sketch the voltaic cell; label the anode and cathode, and indicate the direction of electron flow.

18.11 Sketch voltaic cells based on the following reactions:

(a) $Ni(s) + Cu^{2+}(aq) \longrightarrow Ni^{2+}(aq) + Cu(s)$

(b) $Sn(s) + ClO^-(aq) + H_2O(l) \longrightarrow Sn^{2+}(aq) + Cl^- + 2OH^-(aq)$

In each case label the anode and cathode and identify the positive and the negative terminal. Also indicate the nature of the electrolytic solution in each compartment and the nature of the electrodes. Show the directions of ion and electron movements. Calculate the emf generated by each cell under standard conditions.

18.12 The following oxidation-reduction reaction is spontaneous in the direction indicated:

$$5Fe^{2+}(aq) + MnO_4^-(aq) + 8H^+(aq) \longrightarrow 5Fe^{3+}(aq) + Mn^{2+}(aq) + 4H_2O(l)$$

A solution containing $KMnO_4$ and H_2SO_4 is poured into one beaker while a solution of $FeSO_4$ is poured into another. A salt bridge is used to join the beakers. A platinum foil is placed in each solution, and the two solutions connected by a wire that passes through a voltmeter. (a) Indicate the reactions occurring at the anode and at the cathode, the direction of electron movement through the external circuit, the direction of ion migrations through the solutions, and the signs of the electrodes. (b) Calculate the emf of the cell under standard conditions.

18.13 Which of the following does not ordinarily function as a reducing agent (that is, cannot be readily oxidized): (a) F^-; (b) Zn; (c) $Cr_2O_7^{2-}$; (d) I_2; (e) NO?

18.14 (a) Arrange the following species in order of increasing strength as oxidizing agents: $Cr_2O_7^{2-}$, H_2O_2, Cu^{2+}, Cl_2, O_2. (b) Arrange the following species in order of increasing strength as reducing agents: Zn, I^-, Sn^{2+}, H_2O_2, Al.

18.15 Based on the data in Appendix F, which of the following is the strongest oxidizing agent, and which the weakest: $Ce^{4+}(aq)$, $Br_2(l)$, $H_2O_2(aq$ acid), $Zn(s)$?

18.16 Using Table 18.1, suggest one or more reducing agents capable of reducing $Eu^{3+}(aq)$ to $Eu^{+2}(aq)$ ($E° = -0.43$ V).

18.17 Which of the following processes are spontaneous under standard conditions?

(a) $H_2S(aq) + Cl_2(aq) \longrightarrow S(s) + 2Cl^-(aq) + 2H^+(aq)$

(b) $5H_2O_2(l) + Cl_2(aq) \longrightarrow 4H_2O(l) + 2H^+(aq) + 2ClO_3^-(aq)$

(c) $2Cl^-(aq) + Cu^{2+}(aq) \longrightarrow Cu(s) + Cl_2(g)$

(d) $10NO_3^-(aq) + 3I_2(s) + 4H^+(aq) \longrightarrow 10NO(g) + 6IO_3^-(aq) + 2H_2O(l)$

Relationships between $E°$, $\Delta G°$, and K

18.18 Indicate whether each of the following statements is true or false. Correct those that are false. (a) For a given set of concentrations, the cell emf is independent of the number of moles of reacting substances present. (b) The free-energy change for a cell reaction is also independent of the number of moles of substance reacting. (c) $E°$ and $\Delta G°$ have the same units. (d) $E°$ is directly proportional to the log of the equilibrium constant for the cell reaction.

18.19 Show that for two cell reactions that involve the same number of moles of electrons, n, the difference in standard emf values is related to the log of the ratio of equilibrium constants.

18.20 From standard potentials calculate $\Delta G°$ for each of the reactions listed in Exercise 18.17.

18.21 Given the following half-cell potentials:

$$Fe^{2+}(aq) \longrightarrow Fe^{3+}(aq) + e^- \quad E° = -0.771 \text{ V}$$

$$S_2O_6^{2-}(aq) + 4H^+(aq) + 2e^- \longrightarrow 2H_2SO_3(aq) \quad E° = 0.60 \text{ V}$$

$$N_2O(aq) + 2H^+(aq) + 2e \longrightarrow N_2(g) + H_2O(l) \quad E° = -1.77 \text{ V}$$

(a) Write balanced chemical equations for oxidation of $Fe^{2+}(aq)$ by the other two reagents for which the half-cell reaction is given. (b) Calculate $\Delta G°$ and the equilibrium constant K for each reaction.

18.22 Assuming that $n = 2$, what magnitude of equilibrium constant is associated with a standard cell emf of (a) 1.00 V; (b) 0.10 V; (c) -0.20 V?

18.23 Using the standard half-cell potentials listed in Appendix F, calculate the equilibrium constant for each of the following reactions:

(a) $Zn(s) + Sn^{2+}(aq) \longrightarrow Zn^{2+}(aq) + Sn(s)$

(b) $Co(s) + 2H^+(aq) \longrightarrow Co^{2+}(aq) + H_2(g)$

(c) $10Br^-(aq) + 2MnO_4^-(aq) + 16H^+(aq) \longrightarrow 2Mn^{2+}(aq) + 8H_2O(l) + 5Br_2(l)$

(d) $2Cr(s) + 3Ni^{2+}(aq) \longrightarrow 2Cr^{3+}(aq) + 3Ni(s)$

18.24 Cytochrome, a complicated molecule that we shall represent as $CyFe^{2+}$, reacts with the air we breathe to supply energy required to synthesize adenosine triphosphate, ATP. The ATP is employed by the body as an energy source to drive other reactions (Section 17.7). At pH 7 the following electrode potentials pertain to this oxidation of $CyFe^{2+}$:

$$O_2(g) + 4H^+(aq) + 4e^- \longrightarrow 2H_2O(l) \quad E = 0.82 \text{ V}$$

$$CyFe^{3+}(aq) + e^- \longrightarrow CyFe^{2+}(aq) \quad E = 0.22 \text{ V}$$

(a) What is ΔG for the oxidation of $CyFe^{2+}$ by air? (b) If the synthesis of 1 mol of ATP from adenosine diphosphate, ADP, requires a ΔG of 37.7 kJ, how many moles of ATP are synthesized per mole of O_2?

Nernst equation

18.25 In a cell that utilizes the reaction

$$Sn(s) + Br_2(aq) \longrightarrow Sn^{2+}(aq) + 2Br^-(aq)$$

what is the effect on cell emf of each of the following changes? (a) NaBr is dissolved in the cathode compartment. (b) $SnSO_4$ is dissolved in the anode compartment. (c) The electrode in the cathode compartment is changed from platinum to graphite. (d) The area of the anode is doubled.

18.26 Calculate the emf of the cell described in Exercise 18.25 under the following conditions: $[Br^-] = 0.10\ M$; $[Sn^{2+}] = 0.050\ M$.

18.27 A voltaic cell is constructed that is based on the following reaction:

$$Sn^{2+}(aq) + Pb(s) \longrightarrow Pb^{2+}(aq) + Sn(s)$$

If the concentration of Sn^{2+} in the cathode compartment is 1.00 M and the cell generates an emf of 0.22 V, what is the concentration of Pb^{2+} in the anode compartment? If the anode compartment contains $[SO_4^{2-}] = 1.00\ M$, what is the K_{sp} of the $PbSO_4$?

18.28 What is the effect on the emf of the cell shown in Figure 18.3 of each of the following changes? (a) The H_2 gas is diluted with an equal volume of Ar. (b) The area of the anode is doubled. (c) Sulfuric acid is added to the cathode compartment. (d) Sodium nitrate is added to the anode compartment.

18.29 The cell of Figure 18.3 could be used to provide a measure of the pH in the cathode compartment. Calculate the pH of the cathode compartment solution if the cell emf is measured to be 0.720 V when $[Zn^{2+}] = 0.1\ M$, $P_{H_2} = 1$ atm.

18.30 Calculate the ratio Fe^{2+}/Fe^{3+} at equilibrium in the chemical system.

$$Ag(s) + Fe^{3+}(aq) \rightleftharpoons Ag^+(aq) + Fe^{2+}(aq)$$

if $[Ag^+]$ at equilibrium is 0.010 M.

[18.31] From the potentials for the half-cell reactions

$$Cu^+(aq) + e^- \longrightarrow Cu(s) \qquad E° = +0.521\ V$$

$$Cu(s) + Br^-(aq) \longrightarrow CuBr(s) + e^- \qquad E° = -0.030\ V$$

calculate K_{sp} for $CuBr(s)$, using the Nernst equation.

Commercial voltaic cells

18.32 Explain why it is possible to assess the state of charge or discharge of a lead-acid battery by measuring the density of the battery fluid.

18.33 (a) Write the reactions for discharge and charge of the nickel-cadmium rechargeable cell. (b) Given the following half-cell potentials, calculate the standard emf of the cell:

$$Cd(OH)_2(s) + 2e^- \longrightarrow Cd(s) + 2OH^-(aq)^- \qquad E° = -0.76\ V$$

$$NiO_2(s) + 2H_2O(l) + 2e^- \longrightarrow Ni(OH)_2(s) + 2OH^-(aq) \qquad E° = +0.49\ V$$

18.34 If 120 g of zinc is employed in the casing of a particular Leclanché dry cell, and assuming that all of this is consumed in the cell reaction, how many grams of MnO_2 undergo reaction?

18.35 Suppose that an alkaline dry cell were manufactured using cadmium metal rather than zinc. What effect would this have on the cell emf?

Electrolysis

18.36 Why are different products obtained when molten $MgCl_2$ and aqueous $MgCl_2$ are electrolyzed with inert electrodes? Predict the products in each case.

18.37 Sketch a cell for electrolysis of a $CoCl_2$ solution using inert electrodes. Indicate the directions in which ions and electrons move. Give the electrode reactions and label the anode and cathode, indicating which is positive and which is negative.

18.38 Sketch a cell for the electrolysis of aqueous HCl using copper electrodes. Give the electrode reactions, labeling the anode and cathode. Calculate the minimum applied voltage required to cause electrolysis to occur, assuming standard conditions.

18.39 (a) Calculate the minimum applied voltage required to cause the following electrolysis reaction to occur, assuming that the anode is platinum and the cathode is nickel:

$$Ni^{2+}(aq) + 2Br^-(aq) \longrightarrow Ni(s) + Br_2(l)$$

(b) In practice, a larger voltage than this calculated minimum is required to produce the electrode reactions. Why is this so?

18.40 Determine the value of the faraday from the mass of iodine, 4.8285 g, released by the passage of 3671.3 C of electricity through an aqueous solution of HI. Compare this result with the accepted value.

18.41 How many faradays are required to produce 1 mol of each of the following in a voltaic or electrolytic cell from the indicated starting material: (a) $Cr(OH)_3(s)$ from $CrO_4^{2-}(aq)$; (b) $Fe^{3+}(aq)$ from $Fe(s)$; (c) $IO_3^-(aq)$ from $I^-(aq)$; (d) $Zn(OH)_4^{2-}(aq)$ from $Zn(s)$?

18.42 A $Zn^{2+}(aq)$ solution is electrolyzed using a current of 0.600 A. What mass of Zn is plated out after 300 min?

18.43 In the cell described in Exercise 18.25, the cell is operated until 3.00 g of Sn have dissolved from the anode. How many coulombs has the cell delivered at this point?

18.44 (a) In the electrolysis of aqueous NaCl, how many liters of $Cl_2(g)$ (measured at STP) are generated by a current of 5.50 A for a period of 100 min? (b) How many moles of NaOH(aq) have formed in the solution in this period?

18.45 If 0.500 L of a 0.600 M $SnSO_4$ solution is electrolyzed for a period of 30.0 min using a current of 4.60 A, and using inert electrodes, what are the final concentrations of each ion remaining in the solution? (Assume that the volume of the solution does not change.)

[18.46] Making stamping masters for phonograph records involves placing a fine coating of silver on a plastic record to give it a surface that conducts an electrical current. The record is then electroplated with nickel. (a) Should the record be made the anode or cathode in the electrolytic cell? (b) How much time is required to provide a 0.0010-cm coating of nickel on both faces of the record using a current of 0.20 A if the record has a diameter of 30.0 cm? (The density of nickel is 8.90 g/cm^3.) Ignore the nickel deposited on the edges.

Electrical work

18.47 Indicate whether each of the following statements is true or false. Correct those that are false. (a) The theoretical maximum work obtainable from a voltaic cell is exactly equal to the theoretical minimum work required to drive the cell reaction in the opposite direction under the same conditions. (b) For a given cell reaction the maximum work obtainable is proportional to the total mass of reactants that pass into products. (c) The minimum work required to cause an electrolytic cell reaction to proceed is often less than the theoretical value, because of overvoltage.

18.48 Provide a simple, direct argument to support the statement that the maximum work obtainable from a given voltaic cell is a state function; that is, it depends only on the initial and final states of the cell. Is this true also of the *actual* work obtainable from the cell? Give examples to illustrate your answer.

18.49 What is the maximum electrical work, in joules, that the cell described in Exercise 18.25 can accomplish under standard conditions if 0.200 mol of Sn is consumed?

18.50 (a) Calculate the mass of magnesium produced in a large facility by electrolysis of molten $MgCl_2$, by a current of 90,000 A flowing for 16 h. Assume that the cell is 50 percent efficient. (b) What is the total energy requirement for this electrolysis if the applied emf is 4.20 V?

Corrosion

18.51 An iron object is plated with a coating of cobalt to protect against corrosion. Does the cobalt protect iron by cathodic protection?

18.52 When an iron object is plated with tin, does the tin act as a sacrificial anode in protecting against corrosion? Explain.

18.53 Amines are compounds related to ammonia. One of their characteristics is their ability to function as Brønsted bases (Section 15.6). Suggest how they protect against corrosion when added to antifreeze as corrosion inhibitors.

18.54 Considering the following standard half-cell potentials:

$$Ti(s) \longrightarrow Ti^{2+}(aq) + 2e^- \qquad E^\circ = +1.63\ V$$

$$Ti(s) + 2H_2O(l) \longrightarrow TiO_2(s) + 4H^+ + 4e^- \qquad E^\circ = +0.86\ V$$

How do you account for the fact that titanium metal is quite corrosion resistant?

Additional exercises

18.55 Complete and balance the following equations:

(a) $HNO_2(aq) + MnO_4^-(aq) \longrightarrow NO_3^-(aq) + Mn^{2+}(aq)$ (acidic solution)

(b) $UO^{2+}(aq) + ClO_3^-(aq) \longrightarrow UO_2^{2+}(aq) + Cl^-(aq)$ (basic solution)

(c) $SCN^-(aq) + BrO_3^-(aq) \longrightarrow CN^-(aq) + Br^-(aq) + SO_4^{2-}(aq)$ (basic solution)

(d) $H_5IO_6(aq) + I^-(aq) \longrightarrow I_2(s)$ (acidic solution)

(e) $Br_2(l) \longrightarrow BrO_3^-(aq) + Br^-(aq)$ (basic solution)

(f) $CrO_4^{2-}(aq) + HSnO_2^-(aq) \longrightarrow HSnO_3^-(aq) + CrO_2^-(aq)$ (basic solution)

(g) $H_2SeO_3(aq) + H_2S(aq) \longrightarrow Se(s) + HSO_4^-(aq)$ (acidic solution)

(h) $Te(s) + ClO_3^-(aq) \longrightarrow H_2TeO_3(aq) + Cl^-(aq)$ (acidic solution)

(i) $HCN(aq) + Cu^{2+}(aq) \longrightarrow (CN)_2(aq) + CuCN(s)$ (acidic solution)

18.56 Hydrogen peroxide is unusual in its ability to act sometimes as an oxidizing agent, sometimes as a reducing agent. Write balanced equations for the following reactions involving hydrogen peroxide: (a) Oxidation of $ZnS(s)$ to $ZnSO_4(aq)$. (b) Reduction of $MnO_4^-(aq)$ to $MnO_2(s)$ in basic solution. (c) Reduction of $BrO_3^-(aq)$ to $Br^-(aq)$ in basic solution. (d) Use of hydrogen peroxide as an oxidizing agent has the advantage of forming only H_2O as a by-product. In World War II, the German V2 rockets used pure H_2O_2 as the oxidizing agent for a mixture of methanol (CH_3OH) and hydrazine (N_2H_4). The methanol is oxidized to CO_2 and H_2O, while the hydrazine is oxidized to N_2 and H_2O. Write balanced equations for both oxidations.

18.57 Iodic acid, HIO_3, can be prepared by oxidizing iodine with concentrated nitric acid. The HIO_3 settles out of the reaction mixture as a white solid. The unbalanced equation is

$$I_2(s) + H^+(aq) + NO_3^-(aq) \longrightarrow HIO_3(s) + NO_2(g)$$

Calculate the quantity of NO_2 produced in the formation of 10.0 g of HIO_3.

18.58 (a) Complete and balance the following reaction:

$$XeO_3(s) + I^-(aq) \longrightarrow Xe(g) + I_3^-(aq)$$

(b) What volume of dry Xe(g) at 305K, 500 mm Hg pressure, would be formed from reaction of 3.40 g of $XeO_3(s)$ with excess I^- solution? (c) How many moles of $I_3^-(aq)$ would be formed?

18.59 A common shorthand way of representing a voltaic cell is to list the reactants and products from left to right in the following form:

anode | anode solution || cathode solution | cathode

A double vertical line represents a salt bridge or porous barrier. A single vertical line represents a change of phase, such as from solid to solution. (a) Write the half-reactions and overall cell reaction represented by $Fe|Fe^{2+}||Ag^+|Ag$; sketch the cell. (b) Write the half-reactions and overall cell reaction represented by $Zn|Zn^{2+}||H^+|H_2$; sketch the cell. (c) Using the notation just described, represent a cell based on the following reaction:

$$ClO_3^-(aq) + Cu(s) \longrightarrow Cl^-(aq) + Cu^{2+}(aq)$$

Sketch the cell.

18.60 The zinc-silver oxide cell used in hearing aids and electrical watches is based on the following half-reactions:

$$Zn^{2+}(aq) + 2e^- \longrightarrow Zn(s) \quad E° = -0.763 \text{ V}$$

$$Ag_2O(s) + H_2O(l) + 2e^- \longrightarrow 2Ag(s) + 2OH^-(aq) \quad E° = 0.344 \text{ V}$$

(a) What substance is oxidized and what is reduced in the cell during discharge? (b) What is the positive electrode and what is the negative electrode? (c) What emf does this cell generate under standard conditions?

18.61 A voltaic cell consists of Co and a 0.1 *M* solution of $Co^{2+}(aq)$ in one compartment, and Cu and a 0.1 *M* solution of $Cu^{2+}(aq)$ in the other. (a) Sketch the cell and label the anode and cathode. (b) What is the emf of the cell? (c) What would be the effect on the emf of adding phosphate ion to the cobalt compartment [K_{sp} for $Co_3(PO_4)_2$ is 2×10^{-35}]?

18.62 Predict whether the following reactions will be spontaneous in acidic solution under standard conditions: (a) oxidation of Sn to Sn^{2+} by I_2 (to form I^-); (b) reduction of Ni^{2+} to Ni by I^- (to form I_2); (c) reduction of Ce^{4+} to Ce^{3+} by Br^- (to form Br_2); (d) reduction of Ag^+ to Ag by H_2O_2.

18.63 For a cell based on the cell reaction $2Cr(s) + 3Br_2(l) \longrightarrow 2Cr^{3+}(aq) + 6Br^-(aq)$, (a) sketch the cell, labeling anode and cathode; (b) calculate the standard cell emf; (c) calculate the cell emf for the case in which $[Cr^{3+}] = 0.200\ M$ and $[Br^-] = 0.0100\ M$.

18.64 Concentration cells involve the same reactants and products in anode and cathode compartments, but at different concentrations. Calculate the emf of a cell containing 0.040 *M* Cr^{3+} in one compartment and 1.0 *M* Cr^{3+} in the other if Cr electrodes are used in both. Which is the anode compartment?

[18.65] (a) Write the cell reaction for the cell shown in Figure 18.17. (b) Write the full Nernst equation expression for this cell, and rearrange it so that the pH of the solution is given as a function of the other cell variables. (c) If the cell emf is 0.660 V, the pressure of H_2 is 1.0 atm, and $[Cl^-]$ is $1.0 \times 10^{-3}\ M$, what is the pH of the solution?

18.66 From the following half-cell reactions and their potentials, construct a cell with the largest possible standard emf; calculate the equilibrium constant for the cell reaction.

$$PbO_2(s) + H_2O(l) + 2e^- \longrightarrow PbO(s) + 2OH^-(aq) \quad E° = +0.28 \text{ V}$$

$$IO_3^-(aq) + 2H_2O(l) + 4e^- \longrightarrow IO^-(aq) + 4OH^-(aq) \quad E° = +0.56 \text{ V}$$

$$PO_4^{3-}(aq) + 2H_2O(l) + 2e^- \longrightarrow HPO_3^{2-} + 3OH^-(aq) \quad E° = -1.05 \text{ V}$$

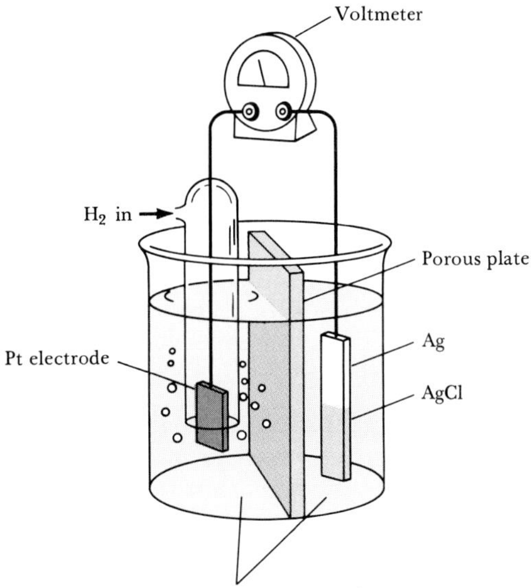

FIGURE 18.17

[18.67] The standard potential for the reduction of AgSCN is 0.0895 V:

$$AgSCN(s) + e^- \longrightarrow Ag(s) + SCN^-(aq)$$

Using this value together with another electrode potential, calculate K_{sp} for AgSCN.

18.68 Calculate the electrode potential for reduction of H_2O at pH 4.00. Assume that $P_{H_2} = 1.00$ atm.

18.69 (a) How many coulombs are required to plate a layer of chromium metal 0.23 mm thick on an auto bumper with a total area of 0.32 m^2 from a solution containing CrO_4^{2-}? The density of chromium metal is 7.20 g/cm^3. (b) What current flow is required for this electroplate if the bumper is to be plated in 6.0 s?

18.70 The element indium is to be obtained by electrolysis of a molten halide of the element. Passage of a current of 3.20 A for a period of 40.0 min results in formation of 4.57 g of In. What is the oxidation state of indium in the halide melt?

18.71 Explain why electrolysis of an aqueous solution of Na_2SO_4 containing litmus develops a blue color at the cathode and a red color at the anode.

18.72 Peroxyborate bleaches, such as found in Borateem, have replaced older "chlorine" bleaches in many bleaching agents. Sodium peroxyborate, $NaBO_3$, can be prepared by electrolytic oxidation of borax ($Na_2B_4O_7$) solutions:

$$Na_2B_4O_7(aq) + 10NaOH(aq) \longrightarrow 4NaBO_3(aq) + 5H_2O(l) + 8Na^+(aq) + 8e^-$$

How many grams of $NaBO_3$ can be prepared in 24.0 h if the current is 20.0 A?

18.73 How long could an alkaline dry cell be used to power a hearing aid if a current of 3.0×10^{-3} A is required, and if the cell life is limited by the amount of Zn that can react, 1.68 g?

18.74 If charging efficiency is 90 percent, how much time is required to convert 30.0 g of $PbSO_4$ into PbO_2, using a current of 6.0 A?

18.75 (a) What is the maximum amount of work that a 6-V golf-cart lead storage battery can accomplish if it is rated at 300 A-h? (b) List some of the reasons why this amount of work is never realized.

18.76 The type of lead storage cell used in automobiles does not tolerate "deep discharge" (in which the cell reaction is allowed to proceed to near completion) very well. Typically, the battery fails after 20 to 30 such cycles. (a) Why does deep discharge lead to battery failure? (b) How is deep discharge avoided during normal auto use?

18.77 Write the reaction that might be expected to occur if a block of Mg is attached to a steel ship hull. Explain how the Mg acts to protect the ship against corrosion.

18.78 If you were going to apply a small potential to a steel ship resting in the water, as a means of inhibiting corrosion, would you apply a negative or a positive charge? Explain.

18.79 Describe the role of atmospheric oxygen in the corrosion of iron.

[18.80] A family owns an antique set of silverware that has a fine, dark coating of Ag_2S in the crevices of the pattern that adds to its beauty. The set is placed in a galvanized container together with soap and water in order to be cleaned. The Ag_2S disappears as the set sits in the container, leaving the silverware with the appearance of a new rather than an antique set. Explain the electrochemical processes that have occurred.

[18.81] Several years ago, a unique proposal was made to raise the *Titanic*. The plan involved placing pontoons within the ship using a surface-controlled submarine-type vessel. The pontoons would contain cathodes and would be filled with hydrogen gas formed by the electrolysis of water. It has been estimated that it would require about 7×10^8 mol of H_2 to provide the bouyancy to lift the ship (*Journal of Chemical Education*, vol. 50, p. 61, 1973). (a) How many coulombs of electrical charge would be required? (b) What is the minimum voltage required to generate H_2 and O_2 if the pressure on the gases at the depth of the wreckage (2 mi) is 300 atm? (c) What is the minimum electrical energy required to raise the *Titanic* by electrolysis? (d) What is the minimum cost of the electrical energy required to generate the necessary H_2 if the electricity costs 23¢ per kilowatt-hour?

[18.82] Edison's invention of the light bulb and its public demonstration in December 1879 generated considerable demand for the distribution of electricity to homes. One problem was how to measure the amount of electricity consumed by each household. Edison invented a coulometer (described in the *Journal of Chemical Education*, vol. 49, p. 627, 1972) that could be used with alternating current. Zinc plated out at the cathode of the coulometer. Every month the cathode was removed and weighed to determine the quantity of electricity used. If the cathode increased in mass by 1.62 g and the coulometer drew 0.35 percent of the current entering the home, how many coulombs of electricity were used in that month?

19 Nuclear Chemistry

As we have progressed through this book, our focus has been on chemical reactions, specifically reactions in which electrons play a dominant role. In this chapter, we shall consider nuclear reactions, changes in matter whose origin is the nucleus of the atom. Some experts predict that we shall have to depend more and more on nuclear energy to replace our dwindling supplies of fossil fuels and to meet our rising energy demands. Thus our consideration of nuclear chemistry continues a minor theme of energy generation started in Chapter 18. Even before we begin, you should have some awareness of the controversy surrounding nuclear energy—how do you feel about having a nuclear power plant in your town? Because the topic of nuclear energy evokes such emotional reaction, it is difficult to sift fact from opinion and begin to weigh pros and cons rationally. It is therefore important for any educated person of our time to have some understanding of nuclear reactions and the use of radioactive substances.

However, before we get too deeply involved in our discussions, it is useful to review and extend slightly some ideas introduced in Section 2.4. First, we should recall that there are two subatomic particles that reside in the nucleus, the **proton** and the **neutron.** We shall refer to these particles as **nucleons.** Recall also that all atoms of a given element have the same number of protons; this number is known as the element's **atomic number.** However, the atoms of a given element can have different numbers of neutrons and therefore different **mass numbers;** the mass number is the total number of nucleons in the nucleus. Atoms with the same atomic number but different mass numbers are known as **isotopes.** The different isotopes of an element are distinguished by citing their mass numbers. For example, the three naturally occurring isotopes of uranium are identified as uranium-233, uranium-235, and uranium-238, where the numbers given are the mass numbers. These isotopes are also labeled, using chemical symbols, as ${}^{233}_{92}U$, ${}^{235}_{92}U$, and ${}^{238}_{92}U$. The superscript is the mass number, the subscript is the atomic number. Different isotopes have different natural abundances. For example, 99.3 percent of the naturally occurring uranium is uranium-238, whereas 0.7 percent is uranium-235, and only a trace is uranium-233. One reason that it now becomes important to distinguish between different isotopes is that the nuclear properties of an atom depend on the number of both protons and neutrons in its nucleus. In contrast, we have found that an atom's chemical properties are unaffected by the number of neutrons in the nucleus. Now let's begin to discuss the reactions that a nucleus can undergo.

19.1 NUCLEAR REACTIONS: AN OVERVIEW

There are several ways in which a nucleus can undergo a reaction and thereby change its identity. Some nuclei are unstable and spontaneously emit particles and electromagnetic radiation. Such spontaneous emission from the nucleus of the atom is known as **radioactivity**. The discovery of this phenomenon by Henri Becquerel in 1896 was described in Section 2.4. Those isotopes that are radioactive are known as **radioisotopes**. An example is uranium-238, which spontaneously emits **alpha rays**; these rays consist of a stream of helium-4 nuclei known as **alpha particles**. When a uranium-238 nucleus loses an alpha particle, the remaining fragment has an atomic number of 90 and a mass number of 234. It is therefore a thorium-234 nucleus. We can represent this reaction by the following nuclear equation:

$$^{238}_{92}U \longrightarrow ^{234}_{90}Th + ^{4}_{2}He \qquad [19.1]$$

When a nucleus spontaneously decomposes in this way, it is said to have decayed, or undergone **radioactive decay**.

Notice in Equation 19.1 that the sum of the mass numbers is the same on both sides of the equation (238 = 234 + 4). Likewise, the sum of the atomic numbers of nuclear charges on both sides of the equation is equal (92 = 90 + 2). Mass numbers and atomic numbers are similarly balanced in writing other nuclear equations. In writing nuclear equations, we are not concerned with the chemical form of the atom in which the nucleus resides. The radioactive properties of the nucleus are essentially independent of the state of chemical combination of the atom. It makes no difference whether we are dealing with the atom in the form of an element or of one of its compounds.

Another way a nucleus can change identity is to be struck by a neutron or by another nucleus. Nuclear reactions that are induced in this way are known as **nuclear transmutations**. Such a transmutation occurs when the chlorine-35 nucleus is struck by a neutron ($^{1}_{0}n$); this collision produces a sulfur-35 nucleus and a proton ($^{1}_{1}p$ or $^{1}_{1}H$). The nuclear equation for this reaction is shown in Equation 19.2:

$$^{35}_{17}Cl + ^{1}_{0}n \longrightarrow ^{35}_{16}S + ^{1}_{1}H \qquad [19.2]$$

Notice again that the sum of mass numbers and of atomic numbers is the same on both sides of the equation. By bombarding nuclei with various particles, it is possible to prepare nuclei not found in nature. The sulfur-35 produced in Equation 19.2 is such an isotope.

SAMPLE EXERCISE 19.1

Write a balanced nuclear equation for the nuclear transmutation in which an aluminum-27 nucleus is struck by a helium-4 nucleus, producing a phosphorus-30 nucleus and a neutron.

Solution: By referring to a periodic table or a list of elements, we find that aluminum has an atomic number of 13; its chemical symbol is therefore $^{27}_{13}Al$. The atomic number of phosphorus is 15; its chemical symbol is therefore $^{30}_{15}P$. The balanced equation is

$$^{27}_{13}Al + ^{4}_{2}He \longrightarrow ^{30}_{15}P + ^{1}_{0}n$$

We shall take a closer look at radioactive decay in sections that follow. We shall also discuss the preparation of nuclei via nuclear transmutation. In Sections 19.7 and 19.8, we shall discuss two other types of nuclear reactions. One is known as nuclear fission and the other as nuclear fusion. Fission involves the fragmentation of a large nucleus into two roughly equal-sized nuclei. Fusion involves the combination of two small nuclei to form a larger nucleus.

19.2 RADIOACTIVITY

As we were discussing radioactivity in the preceding section, two general questions might have occurred to you: Which nuclei are radioactive, and what types of radiation do they emit? These are important questions, and we shall now examine them. Let us first consider the type of radiation involved in the phenomenon of radioactivity.

Types of Radioactive Decay

Emission of radiation is one of the ways by which an unstable nucleus is transformed into a stable one with less energy. The emitted radiation is the carrier of the excess energy. In Section 2.4, we discussed the three most common types of radiation emitted by radioactive substances: alpha (α), beta (β), and gamma (γ) rays.

As we noted in Section 19.1, alpha rays consist of streams of helium-4 nuclei known as **alpha particles.** Equation 19.3 gives another example of this type of radioactive decay:

$$^{222}_{86}Rn \longrightarrow ^{218}_{84}Po + ^{4}_{2}He \quad [19.3]$$

Beta rays consist of streams of electrons. Because the **beta particles** are electrons, they are represented as $^{0}_{-1}e$. The superscript zero reflects the exceedingly small mass of the electron by comparison to the mass of a nucleon. The subscript -1 indicates the negative charge of the particle, which is opposite that of the proton. Iodine-131 is an example of an isotope that undergoes decay by beta emission. This reaction is summarized by Equation 19.4:

$$^{131}_{53}I \longrightarrow ^{131}_{54}Xe + ^{0}_{-1}e \quad [19.4]$$

Emission of a beta particle has the effect of converting a neutron within the nucleus into a proton, thereby increasing the atomic number of the nucleus by one:

$$^{1}_{0}\text{n} \longrightarrow {}^{1}_{1}\text{p} + {}^{0}_{-1}\text{e} \qquad [19.5]$$

However, just because electrons are ejected from the nucleus, we need not think that the nucleus is composed of these particles, any more than we consider a match to be composed of sparks simply because it gives them off when struck. The electrons come into being when the nucleus is disrupted.

Gamma rays consist of electromagnetic radiation of very short wavelength (that is, high-energy photons). The position of gamma rays in the electromagnetic spectrum is shown in Figure 5.3. Gamma rays can be represented as $^{0}_{0}\gamma$. Such radiation changes neither the atomic number nor the mass number of a nucleus. It almost always accompanies other radioactive emission, because it represents the energy lost when the remaining nucleons reorganize into more stable arrangements. Generally, we shall not show the gamma rays when writing nuclear equations.

Two other types of radioactive decay that occur are positron emission and electron capture. A **positron** is a particle that has the same mass as an electron but an opposite charge.* The positron is represented as $^{0}_{1}\text{e}$. Carbon-11 is an example of an isotope that decays by positron emission:

$$^{11}_{6}\text{C} \longrightarrow {}^{11}_{5}\text{B} + {}^{0}_{1}\text{e} \qquad [19.6]$$

Emission of a positron can be thought of as converting a proton into a neutron as shown in Equation 19.7. The atomic number of the nucleus is thereby decreased by one:

$$^{1}_{1}\text{p} \longrightarrow {}^{1}_{0}\text{n} + {}^{0}_{1}\text{e} \qquad [19.7]$$

Electron capture involves capture of an electron from the electron cloud surrounding the nucleus. Rubidium-81 undergoes decay in this fashion as shown in Equation 19.8:

$$^{81}_{37}\text{Rb} + {}^{0}_{-1}\text{e (orbital electron)} \longrightarrow {}^{81}_{36}\text{Kr} \qquad [19.8]$$

Electron capture has the effect of converting a proton within the nucleus into a neutron as shown in Equation 19.9:

$$^{1}_{1}\text{p} + {}^{0}_{-1}\text{e} \longrightarrow {}^{1}_{0}\text{n} \qquad [19.9]$$

The symbols used to represent the various elementary particles involved in radioactive decay and nuclear transformations are summarized in Table 19.1.

TABLE 19.1 Particles commonly involved in radioactive decay and nuclear transformations

Particle	Symbol
Neutron	$^{1}_{0}\text{n}$
Proton	$^{1}_{1}\text{p}$ or $^{1}_{1}\text{H}$
Electron	$^{0}_{-1}\text{e}$
Alpha particle	$^{4}_{2}\alpha$ or $^{4}_{2}\text{He}$
Beta particle	$^{0}_{-1}\beta$ or $^{0}_{-1}\text{e}$
Positron	$^{0}_{1}\text{e}$

*The positron has a very short life, because it is annihilated when it collides with an electron: $^{0}_{1}\text{e} + {}^{0}_{-1}\text{e} \longrightarrow 2{}^{0}_{0}\gamma$.

SAMPLE EXERCISE 19.2

Write balanced nuclear equations for the following reactions: (a) thorium-230 undergoes alpha decay; (b) thorium-231 undergoes decay to form protactinium-231.

Solution: (a) The information given in the problem can be summarized as

$$^{230}_{90}Th \longrightarrow ^{4}_{2}He + X$$

The remaining product, X, must be deduced. Because mass numbers must have the same sum on both sides of the equation, we deduce that X has a mass number of 226. Similarly, the atomic number of X must be 88. Element number 88 is radium (refer to periodic table or list of elements). The equation is therefore as follows:

$$^{230}_{90}Th \longrightarrow ^{4}_{2}He + ^{226}_{88}Ra$$

(b) In this case we must determine what type of particle is emitted in the course of the radioactive decay. We can write the following equation:

$$^{231}_{90}Th \longrightarrow ^{231}_{91}Pa + X$$

The atomic numbers are obtained from a list of elements such as that given on the inside front cover. In order for the mass numbers to balance, X must have a mass number of 0. Its atomic number must be -1. The particle with these characteristics is the beta particle (electron). We therefore write the following:

$$^{231}_{90}Th \longrightarrow ^{231}_{91}Pa + ^{0}_{-1}e$$

Nuclear Stability

There is no single rule that will allow us to predict whether a particular nucleus is radioactive and how it might decay. However, we can list some empirical observations that are helpful in making predictions:

1. All nuclei with 84 or more protons are unstable. For example, all isotopes of uranium, atomic number 92, are radioactive.
2. Nuclei with a total of 2, 8, 20, 50, 82, or 126 protons or neutrons are generally more stable than nuclei found near them in the periodic table. For example, there are three stable nuclei with an atomic number of 18, two with 19, five with 20, and one with 21; there are three stable nuclei with 18 neutrons, none with 19, four with 20, and none with 21. Thus there are more stable nuclei with 20 protons or 20 neutrons than with 18, 19, or 21. The numbers 2, 8, 20, 50, 82, and 126 are called **magic numbers.** Just as enhanced chemical stability is associated with the presence of 2, 10, 18, 36, 54, or 86 electrons, the noble gas configurations, enhanced nuclear stability is associated with the magic number of nucleons.
3. Nuclei with even numbers of both protons and neutrons are generally more stable than those with odd numbers of nucleons, as shown in Table 19.2.
4. The stability of a nucleus can be correlated to a certain degree with its neutron-to-proton ratio. All nuclei with two or more protons contain neutrons. Neutrons are apparently involved in some way in holding protons together within the nucleus. As shown in Figure 19.1, the number of neutrons necessary to create a stable nucleus increases rapidly as the number of protons increases; the neutron-to-proton ratios of stable nuclei increase with increasing atomic number. The area within which all stable nuclei are found is known as the **belt of stability.**

TABLE 19.2 The number of stable isotopes with even and odd numbers of protons and neutrons

Number of stable isotopes	Protons	Neutrons
157	Even	Even
52	Even	Odd
50	Odd	Even
5	Odd	Odd

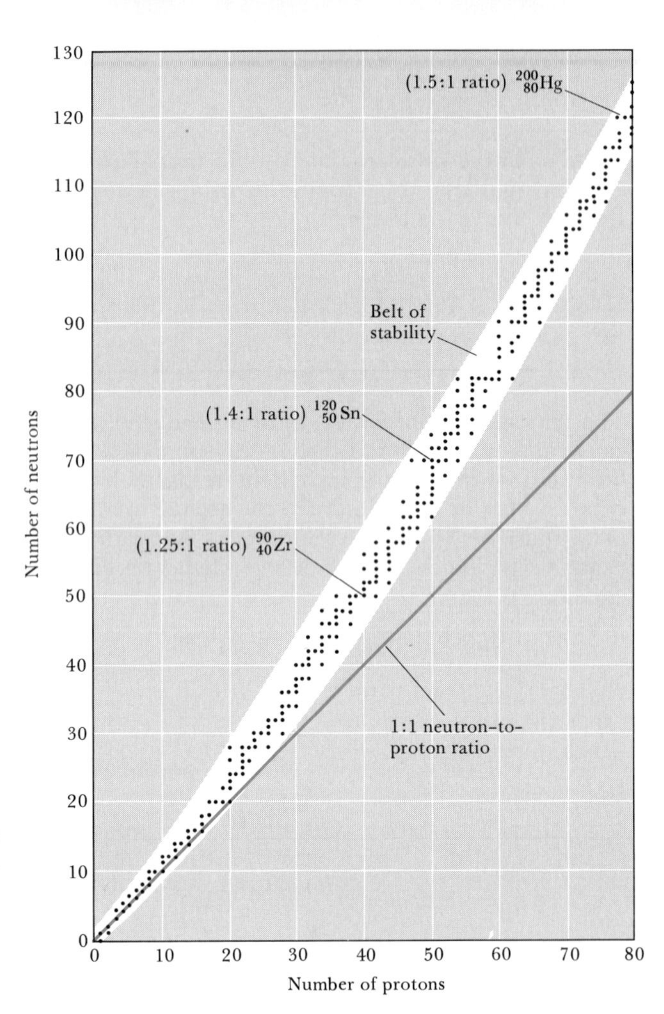

FIGURE 19.1 Plot of the number of neutrons versus the number of protons in stable nuclei. As the atomic number increases, the neutron-to-proton ratio of the stable nuclei increases. The stable nuclei are located in an area of the graph known as the belt of stability. The majority of radioactive nuclei occur outside this belt.

SAMPLE EXERCISE 19.3

Would you expect the following nuclei to be radioactive: $^{4}_{2}He$, $^{39}_{20}Ca$, $^{210}_{85}At$?

Solution: Helium-4 has a magic number of both protons and neutrons (two each). We would therefore expect $^{4}_{2}He$ to be stable.

Calcium-39 has an even number of protons (20) and an odd number of neutrons (19); 20 is one of the magic numbers. Nevertheless, we should suspect that this nuclide is radioactive because the neutron-to-proton ratio is less than 1. This would place it below the belt of stability.

Astatine-210 is radioactive. Recall that there are no stable nuclei beyond atomic number 83.

The type of radioactive decay that a particular radioisotope will undergo depends to a large extent on its neutron-to-proton ratio compared to those of nearby nuclei that are within the belt of stability. Consider a nucleus whose high neutron-to-proton ratio places it above the belt of stability. This nucleus can lower its ratio and move toward the belt of stability by emitting a beta particle. Beta emission decreases the number of neutrons and increases the number of protons in a nucleus as shown in Equation 19.5.

Nuclei that have low neutron-to-proton ratios and that therefore lie below the belt of stability either emit positrons or undergo electron capture. Both modes of decay decrease the number of protons and increase the number of neutrons in the nucleus as shown in Equations 19.7 and 19.9. Positron emission is more common than electron capture among

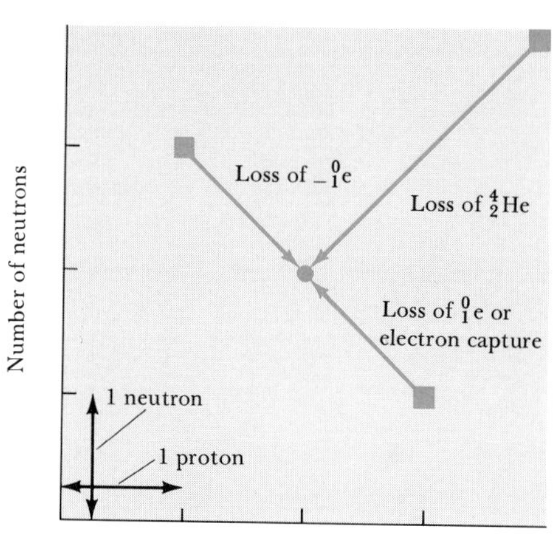

FIGURE 19.2 Result of alpha emission ($_{2}^{4}He$), beta emission ($_{-1}^{0}e$), positron emission ($_{1}^{0}e$), and electron capture on the number of protons and neutrons in a nucleus. The squares represent unstable nuclei, and the circle represents a stable one. Moving from right to left or from bottom to top, each tick mark represents an additional proton or neutron, respectively. Moving in the reverse direction indicates the loss of a proton or neutron.

the lighter nuclei; however, electron capture becomes increasingly common as nuclear charge increases.

Alpha emission is found primarily among nuclei whose atomic numbers are greater than 83. These nuclei would lie beyond the upper right edge of Figure 19.1, outside the belt of stability. Emission of an alpha particle moves the nucleus diagonally toward the belt of stability by decreasing both the number of protons and the number of neutrons by two. The result of each type of radioactive decay relative to a stable nucleus is shown in Figure 19.2.

SAMPLE EXERCISE 19.4

By referring to Figure 19.1, predict the mode of radioactive decay of the following nuclei: (a) $_{11}^{20}Na$; (b) $_{40}^{97}Zr$; (c) $_{92}^{235}U$.

Solution: To answer this question we use the guidelines given above in the text.

(a) This nucleus has a neutron-to-proton ratio below 1. It therefore lies below the belt of stability. It can gain stability either by positron emission or electron capture. Because the atomic number is small we might predict that the nucleus undergoes positron emission. If we refer to a standard reference like the *Handbook of Chemistry and Physics,* we find that this prediction is correct. The nuclear reaction is

$$_{11}^{20}Na \longrightarrow {}_{1}^{0}e + {}_{10}^{20}Ne$$

(b) In referring to Figure 19.1, we find that this nucleus has a neutron-to-proton ratio that is too high. We would therefore predict that it undergoes beta decay. Again the prediction is correct. The nuclear reaction is

$$_{40}^{97}Zr \longrightarrow {}_{-1}^{0}e + {}_{41}^{97}Nb$$

(c) This nucleus lies outside the belt of stability to the upper right. We might therefore predict that it would undergo alpha emission. Again the prediction is correct. The nuclear equation is

$$_{92}^{235}U \longrightarrow {}_{2}^{4}He + {}_{90}^{231}Th$$

At this point we should note that our guidelines don't always work. For example, thorium-233, $_{90}^{233}Th$, which we might expect to undergo alpha decay, undergoes beta decay instead. Furthermore, a few radioactive nuclei actually lie within the belt of stability. For example, both $_{60}^{146}Nd$ and $_{60}^{148}Nd$ are stable and lie in the belt of stability; however, $_{60}^{147}Nd$, which lies between them, is radioactive.

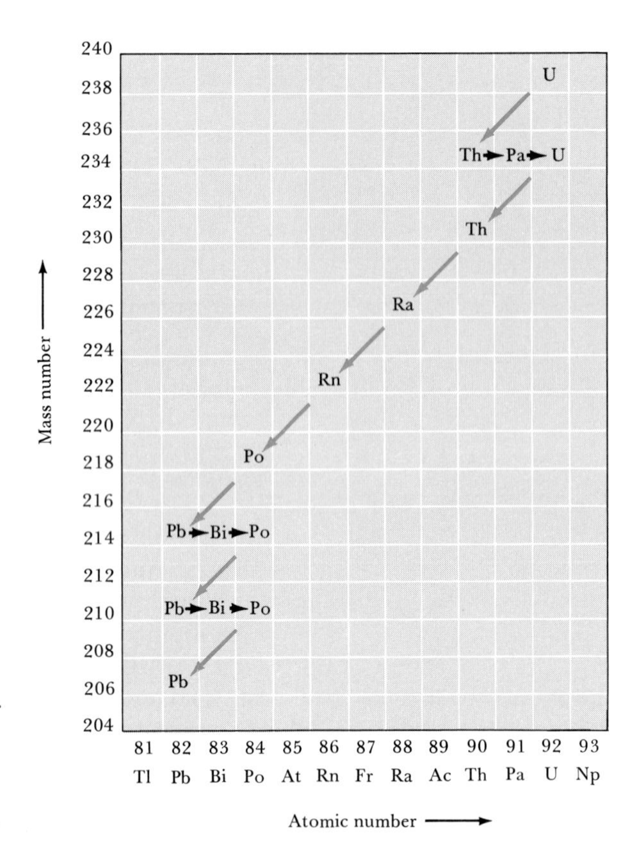

FIGURE 19.3 Nuclear disintegration series for uranium-238. The $^{238}_{92}U$ nucleus decays to $^{234}_{90}Th$. Subsequent decay processes eventually form the stable $^{206}_{82}Pb$ nucleus. Each of the colored arrows corresponds to the loss of an alpha particle. Each black arrow corresponds to the loss of a beta particle.

Radioactive Series

Some nuclei, like uranium-238, cannot gain stability by a single emission. Consequently, a series of successive emissions occur. As shown in Figure 19.3, uranium-238 decays to thorium-234, which is radioactive and decays to protactinium-234. This nucleus is also unstable and subsequently decays. Such successive reactions continue until a stable nucleus, lead-206, is formed. A series of nuclear reactions that begins with an unstable nucleus and terminates with a stable one is known as a **radioactive series** or a **nuclear disintegration series.** Altogether there are three such series found in nature. In addition to the series that begins with uranium-238 and terminates with lead-206, there is one that begins with uranium-235 and ends with lead-207. The third series begins with thorium-232 and ends with lead-208.

19.3 PREPARATION OF NEW NUCLEI

In 1919, Rutherford performed the first artificial conversion of one nucleus into another. He succeeded in converting nitrogen-14 into oxygen-17 using the high-velocity alpha (α) particles emitted by radium. The reaction is shown in Equation 19.10:

$$^{14}_{7}N + ^{4}_{2}He \longrightarrow ^{17}_{8}O + ^{1}_{1}H \qquad [19.10]$$

This reaction demonstrated that nuclear reactions can be induced by striking nuclei with particles such as alpha particles. Such reactions have permitted synthesis of hundreds of radioisotopes in the laboratory. As noted in Section 19.1, these conversions of one nucleus into another are called nuclear transmutations. It is common to represent such conversions by listing, in order, the target nucleus, the bombarding particle, the ejected particle, and the product nucleus. Written in this fashion, Equation 19.10 is $^{14}_{7}N(\alpha, p)^{17}_{8}O$. The alpha particle, proton, and neutron are abbreviated as α, p, and n, respectively.

SAMPLE EXERCISE 19.5

Write the balanced nuclear equation for the process summarized as $^{27}_{13}Al(n, \alpha)^{24}_{11}Na$.

Solution: The n is the abbreviation for a neutron, whereas α represents an alpha particle. The neutron is the bombarding particle, and the alpha particle is a product. Therefore, the nuclear equation is

$$^{27}_{13}Al + ^{1}_{0}n \longrightarrow ^{4}_{2}He + ^{24}_{11}Na$$

Charged particles such as alpha particles must be moving very fast in order to overcome the electrostatic repulsion between them and the target nucleus. The higher the nuclear charge on either the projectile or the target, the faster the projectile must be moving to bring about a nuclear reaction. Therefore, many methods have been devised to accelerate charged particles using strong magnetic and electrostatic fields. These **particle accelerators** bear such names as the **cyclotron** and **synchrotron.** The cyclotron is illustrated in Figure 19.4. The hollow D-shaped electrodes are called "dees." The projectile particles are introduced into a vacuum chamber within the cyclotron. The particles are then accelerated by making the dees alternately positively and negatively charged. Magnets placed above and below the dees keep the particles moving in a spiral path until they are finally deflected out of the cyclotron and emerge to strike a target substance. Particle accelerators have been used mainly to probe the secrets of nuclear structure and to synthesize heavy elements.

Most synthetic isotopes used in quantity in medicine and scientific research are made using neutrons as projectiles. Because neutrons are neutral, they are not repelled by the nucleus; consequently, they do not need to be accelerated, as do the charged particles, in order to cause

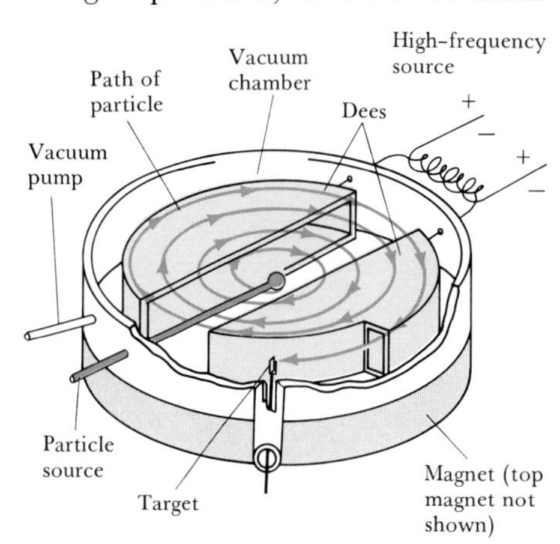

FIGURE 19.4 Schematic representation of a cyclotron.

nuclear reactions (indeed, they cannot be so accelerated). The necessary neutrons are produced by the reactions that occur in nuclear reactors (Section 19.7). Cobalt-60, used in radiation therapy for cancer, is produced by neutron capture. Iron-58 is placed in a nuclear reactor, where it is bombarded by neutrons. The sequence of reactions shown in Equations 19.11 through 19.13 takes place:

$$^{58}_{26}\text{Fe} + ^{1}_{0}\text{n} \longrightarrow ^{59}_{26}\text{Fe} \qquad [19.11]$$

$$^{59}_{26}\text{Fe} \longrightarrow ^{59}_{27}\text{Co} + ^{0}_{-1}\text{e} \qquad [19.12]$$

$$^{59}_{27}\text{Co} + ^{1}_{0}\text{n} \longrightarrow ^{60}_{27}\text{Co} \qquad [19.13]$$

Transuranium Elements

Artificial transmutations have been used to produce the elements from atomic number 93 to 105. These are known as the **transuranium elements,** because they occur immediately following uranium in the periodic table. Elements 93 (neptunium) and 94 (plutonium) were first discovered in 1940. They were produced by bombarding uranium-238 with neutrons as shown in Equations 19.14 and 19.15.

$$^{238}_{92}\text{U} + ^{1}_{0}\text{n} \longrightarrow ^{239}_{92}\text{U} \longrightarrow ^{239}_{93}\text{Np} + ^{0}_{-1}\text{e} \qquad [19.14]$$

$$^{239}_{93}\text{Np} \longrightarrow ^{239}_{94}\text{Pu} + ^{0}_{-1}\text{e} \qquad [19.15]$$

Elements with larger atomic numbers are normally formed in small quantities in particle accelerators. For example, curium-242 is formed when a plutonium-239 target is struck with accelerated alpha particles:

$$^{239}_{94}\text{Pu} + ^{4}_{2}\text{He} \longrightarrow ^{242}_{96}\text{Cm} + ^{1}_{0}\text{n} \qquad [19.16]$$

19.4 HALF-LIFE

Again our discussions may have raised some questions in your mind. For example, why is it that some radioisotopes, like uranium-238, are found in nature, whereas others are not and must be synthesized? The key to this question is the fact that different nuclei undergo decay at different speeds. Uranium-238 undergoes decay very slowly, whereas many other nuclei such as sulfur-35 decay rapidly. To understand the phenomenon of radioactivity, it is important to consider the rates of radioactive decay.

Radioactive decay is a first-order process. As shown in Section 13.3, first-order processes have characteristic half-lives. The **half-life** is the time required for half of any given quantity of a substance to react. The rates of decay of nuclei are commonly discussed in terms of their half-lives.

Each isotope has its own characteristic half-life. For example, the half-life of strontium-90 is 29 yr. If we started with 10.0 g of strontium-90, only 5.0 g of that isotope would remain after 29 yr. The other half of the strontium-90 would have been converted to yttrium-90 as shown in Equation 19.17:

$$^{90}_{38}\text{Sr} \longrightarrow ^{90}_{39}\text{Y} + ^{0}_{-1}\text{e} \qquad [19.17]$$

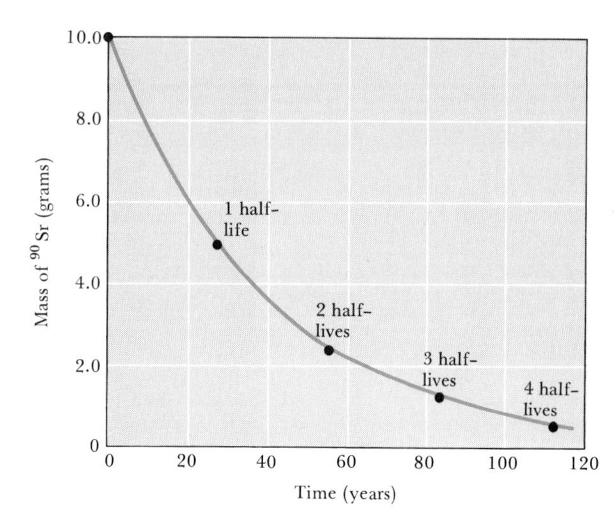

FIGURE 19.5 Decay of a 10.0-g sample of $^{90}_{38}Sr$ ($t_{1/2}$ = 28.8 yr).

After another 29-yr period half of the remaining 5.0 g of strontium-90 would likewise decay. The loss of strontium-90 as a function of time is shown in Figure 19.5.

Half-lives as short as millionths of a second and as long as billions of years have been observed. The half-lives of some radioisotopes are listed in Table 19.3. One important feature of half-lives for nuclear decay is that they are unaffected by external conditions such as temperature, pressure, or state of chemical combination. Therefore, unlike chemical toxins, radioactive atoms cannot be rendered harmless by chemical reaction or by any other practical treatment. As far as we know now our only choice is merely to allow these nuclei to lose activity at their own characteristic rates. In the meantime, of course, we must take precautions to isolate the radioisotopes as much as possible because of the damage radiation can cause (see Section 19.9).

SAMPLE EXERCISE 19.6

The half-life of cobalt-60 is 5.3 yr. How much of a 1.000-mg sample of cobalt-60 is left after a 15.9-yr period?

Solution: A period of 15.9 yr is three half-lives for cobalt-60. At the end of one half-life, 0.500 mg of cobalt-60 remains; 0.250 mg remains at the end of two half-lives, and 0.125 mg at the end of three half-lives.

TABLE 19.3 Some radioactive isotopes and their half-lives and type of decay

	Isotope	Half-life (yr)	Type of decay
Natural radioisotopes	$^{238}_{92}U$	4.5×10^{9}	Alpha
	$^{235}_{92}U$	7.1×10^{8}	Alpha
	$^{232}_{90}Th$	1.4×10^{10}	Alpha
	$^{40}_{19}K$	1.3×10^{9}	Beta
	$^{14}_{6}C$	5700	Beta
Synthetic radioisotopes	$^{239}_{94}Pu$	24,000	Alpha
	$^{137}_{55}Cs$	30	Beta
	$^{90}_{38}Sr$	28.8	Beta
	$^{131}_{53}I$	0.022	Beta

FIGURE 19.6 Ancient manuscripts and other cultural artifacts made of organic materials are often dated by means of carbon-14 analysis.

Dating

Because the half-life of any particular nuclide is so constant, it can serve as a molecular clock and can be used to determine the ages of different objects. For example, carbon-14 has been used to determine the age of organic materials (see Figure 19.6). The procedure is based on the formation of carbon-14 by neutron capture in the upper atmosphere:

$$^{14}_{7}N + ^{1}_{0}n \longrightarrow ^{14}_{6}C + ^{1}_{1}H \qquad [19.18]$$

This reaction provides a small but reasonably constant source of carbon-14. The carbon-14 is radioactive, undergoing beta decay with a half-life of 5700 yr:

$$^{14}_{6}C \longrightarrow ^{14}_{7}N + ^{0}_{-1}e \qquad [19.19]$$

In using radiocarbon dating, it is generally assumed that the ratio of carbon-14 to carbon-12 in the atmosphere has been constant for at least 50,000 yr. The carbon-14 is incorporated into carbon dioxide, which is in turn incorporated, through photosynthesis, into more complex carbon-containing molecules within plants. The plants are eaten by animals, and the carbon-14 thereby becomes incorporated within them as well. Because a living plant or animal has a constant intake of carbon compounds, it is able to maintain a ratio of carbon-14 to carbon-12 that is identical with that of the atmosphere. However, once the organism dies, it no longer ingests carbon compounds to replenish the carbon-14 that is lost through radioactive decay. The ratio of carbon-14 to carbon-12 therefore decreases. If the ratio diminishes to half that of the atmosphere, we can conclude that the object is one half-life, or 5700 yr, old. The method cannot be used to date objects that are over about 20,000 to 50,000 yr old. After this length of time, the radioactivity is too low to be measured accurately. The radiocarbon-dating technique has been checked by comparing the ages of trees determined by counting their rings and by radiocarbon analysis. As the tree grows, it adds a ring each year. The old growth no longer replenishes the supply of carbon. Thus the carbon-14 decays while the concentration of carbon-12 remains constant. Most of the wood used in these tests was from California bristlecone pines, which reach ages up to 2000 yr. By using trees that died at a known time thousands of years ago, it is possible to make comparisons back to about 5000 B.C. The two dating methods agree to within about 10 percent.

Other isotopes can be similarly used to date other types of objects. For example, it takes 4.5×10^9 yr for half of a sample of uranium-238 to decay to a stable product, lead-206. The age of rocks containing uranium can be determined by measuring the ratio of lead-206 to uranium-238. If the lead-206 had somehow become incorporated into the rock by normal chemical processes instead of by radioactive decay, the rock would contain large amounts of the more abundant isotope lead-208. In the absence of large amounts of this "geonormal" isotope of lead, it is assumed that all of the lead-206 was at one time uranium-238.

The oldest rocks found on the earth are approximately 3×10^9 yr old. This age indicates that the crust of the earth has been solid for at least this length of time. Prior to the crystallization of rock, lead-206 and uranium-238 could separate. It is estimated that it required $1\text{–}1.5 \times 10^9$ yr for the earth to cool and its surface to become solid. This places the age of earth at $4.0\text{–}4.5 \times 10^9$ yr.

Calculations Based on Half-Life

So far our discussion has been mainly qualitative. We now consider the topic of half-lives from a more quantitative point of view. This permits us to answer questions of the following types: How do we determine the half-life of uranium-238? We certainly don't sit around for 4.5×10^9 yr waiting for half of it to decay! Similarly, how do we quantitatively determine the age of an object that can be dated by radiometric means?

The rate of radioactive decay of any radioisotope is first order. It can therefore be described by Equation 19.20,

$$\text{Rate} = kN \qquad [19.20]$$

where N is the number of nuclei of a particular radioisotope, and k is the first-order rate constant. This equation can be transformed into Equation 19.21. (Note that the equations that follow are the same as those we encountered in Section 13.3, in dealing with first-order chemical processes.)

$$\frac{-kt}{2.30} = \log \frac{N_t}{N_0} \qquad [19.21]$$

where t is the time interval during which decay is measured, k is the rate constant, N_0 is the initial number of nuclei (at zero time), and N_t is the number remaining after the time interval. The relationship between the rate constant, k, and half-life, $t_{1/2}$, is given by Equation 19.22:

$$k = \frac{0.693}{t_{1/2}} \qquad [19.22]$$

SAMPLE EXERCISE 19.7

A rock contains 0.257 mg of lead-206 for every milligram of uranium-238. How old is the rock if the half-life for decay of uranium-238 to lead-206 is 4.5×10^9 yr?

Solution: The amount of uranium-238 in the rock when it was first formed is assumed to be the 1.000 mg present at the time of the analysis plus the quantity that decayed to lead-206:

$$\text{Original } ^{238}_{92}\text{U} = 1.000 \text{ mg} + \frac{238}{206}(0.257 \text{ mg})$$
$$= 1.297 \text{ mg}$$

Using Equation 19.22, we can calculate the rate constant for the process from its half-life:

$$k = \frac{0.693}{4.5 \times 10^9 \text{ yr}} = 1.5 \times 10^{-10}/\text{yr}$$

Rearranging Equation 19.21 to solve for time, t, and substituting known quantities gives

$$t = \frac{-2.30}{k} \log \frac{N_t}{N_0}$$
$$= \frac{-2.30}{1.5 \times 10^{-10}/\text{yr}} \log \frac{1.000}{1.297} = 1.7 \times 10^9 \text{ yr}$$

SAMPLE EXERCISE 19.8

If we start with 1.000 g of strontium-90, 0.953 g will still remain after 2.00 yr. (a) What is the half-life of strontium-90? (b) How much strontium-90 would remain after 5.00 yr?

Solution: (a) Equation 19.21 can be solved for the rate constant, k, and then Equation 19.22 used to calculate half-life, $t_{1/2}$:

$$k = \frac{-2.30}{t} \log \frac{N_t}{N_0} = \frac{-2.30}{2.00 \text{ yr}} \log \frac{0.953 \text{ g}}{1.000 \text{ g}}$$
$$= \frac{-2.30}{2.00 \text{ yr}}(-0.0209) = 0.0240/\text{yr}$$
$$t_{1/2} = \frac{0.693}{k} = \frac{0.693}{0.0240/\text{yr}} = 28.8 \text{ yr}$$

(b) Again using Equation 19.21, with $k = 0.0240/\text{yr}$, we have

$$\log \frac{N_t}{N_0} = \frac{-kt}{2.30}$$
$$= \frac{-(0.0240/\text{yr})(5.00 \text{ yr})}{2.30} = -0.0522$$

Calculation of N_t/N_0 from $\log (N_t/N_0) = -0.0522$ is readily accomplished using the INV log or 10^x function of a calculator:

$$\frac{N_t}{N_0} = 10^{-0.0522} = 0.887$$

Because $N_0 = 1.000$ g, we have

$$N_t = 0.887N_0 = 0.887(1.000 \text{ g}) = 0.887 \text{ g}$$

19.5 DETECTION OF RADIOACTIVITY

A variety of methods have been devised to detect emissions from radioactive substances. Becquerel discovered radioactivity because of the effect of radiation on photographic plates. Photographic plates and film have long been used to detect radioactivity. The radiation affects photographic film in the same way as ordinary light. With care, film can be used to give a quantitative measure of activity. The greater the extent of exposure to radiation, the darker the area of the developed negative.

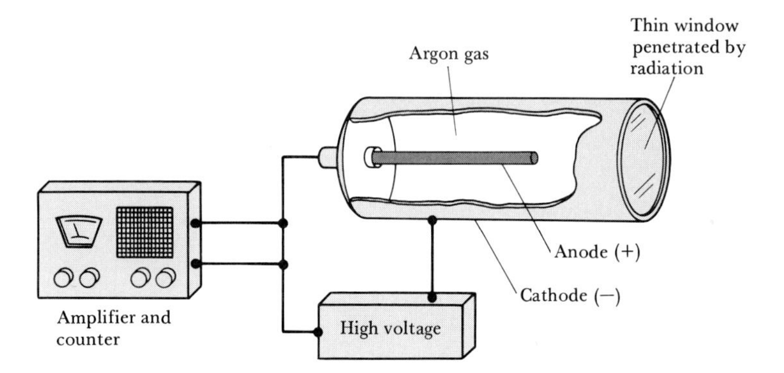

FIGURE 19.7 Schematic representation of a Geiger counter.

People who work with radioactive substances carry film badges to record the extent of their exposure to radiation.

Radioactivity can also be detected and measured using a device known as a Geiger counter. The operation of the Geiger counter is based on the ionization of matter caused by radiation (Section 19.9). The ions and electrons produced by the ionizing radiation permit conduction of an electrical current. The basic design of a Geiger counter is shown in Figure 19.7. It consists of a metal tube filled with gas. The cylinder has a "window" made of material that can be penetrated by alpha, beta, or gamma rays. In the center of the tube is a wire. The wire is connected to one terminal of a source of direct current and the metal cylinder is attached to the other terminal. Current flows between the wire and metal cylinder whenever ions are produced by entering radiation. The current pulse created when radiation enters the tube is amplified; each pulse is counted as a measure of the amount of radiation.

Certain substances that are electronically excited by radiation can also be used as means for detecting and measuring radiation. Some substances excited by radiation give off light (fluoresce) as electrons return to their lower-energy states. For example, dials of luminous watches are painted with a mixture of ZnS and a tiny quantity of $RaSO_4$. The zinc sulfide fluoresces when struck by the radioactive emissions from the radium. An instrument known as a scintillation counter can be used to detect and measure fluorescence and thereby the radiation that causes it.

Radiotracers

Because radioisotopes can be detected so readily, they can be used to follow an element through its chemical reactions. For example, the incorporation of carbon atoms from CO_2 into glucose in photosynthesis, has been studied using CO_2 containing carbon-14:

$$6CO_2 + 6H_2O \xrightarrow[\text{chlorophyll}]{\text{sunlight}} C_6H_{12}O_6 + 6O_2 \qquad [19.23]$$

The CO_2 is said to be labeled with the carbon-14. Detection devices such as Geiger counters can then be used to follow the carbon-14 as it moves from the CO_2 through the various intermediate compounds to glucose.

Such use of radioisotopes is made possible by the fact that all isotopes of an element have essentially identical chemical properties. If a small quantity of a radioisotope is mixed with the naturally occurring stable isotopes of the same element, all of the isotopes will go through the same reactions together. Where the element goes is then revealed by the radioactivity of the radioisotope. Because the radioisotope can be used to trace the path of the element, it is called a **radiotracer.**

Radiotracers have found wide use as a diagnostic tool in medicine. For example, iodine-131 has been used to test the activity of the thyroid gland. This gland is the only important user of iodine in the body. A solution of NaI containing a small quantity of iodine-131 is drunk by the patient. The ability of the thyroid to take up the iodine is determined using a Geiger tube placed close to the thyroid, in the neck region. A normal thyroid will absorb about 12 percent of the iodine within a few hours. When radioisotopes are used in this manner, only a very small amount is used so that the patient does not receive a harmful dose of radioactivity.

In 1977, Rosalyn S. Yalow (Figure 19.8) received the Nobel Prize in medicine for her pioneering work in developing the technique of radioimmunoassay (RIA). This technique is an extraordinarily sensitive method involving radiotracers for detecting the presence of minute amounts of drugs, hormones, peptides, antibiotics, and a host of other substances. RIA methods can be used, for example, to provide an early indication of pregnancy or to detect substances that signal the early stages of a disease. The tests are carried out on a small sample of tissue, blood, or other fluid taken from the subject.

19.6 MASS-ENERGY CONVERSIONS

So far we have said little about the energies associated with nuclear reactions. These energies can be considered with the aid of Einstein's famous equation relating mass and energy:

$$E = mc^2 \qquad [19.24]$$

FIGURE 19.8 Rosalyn S. Yalow was born in New York City in 1921. She is a graduate of Hunter College in New York and received her Ph.D. in physics from the University of Illinois (Urbana) in 1945. For the past several years she has carried out research in medical physics and is chief of the Nuclear Medicine Service at the Veterans Administration Hospital in Bronx, New York. Among her many awards and recognitions is the Nobel Prize in medicine and physiology, which she was awarded in 1977.

In this equation E stands for energy, m for mass, and c for the speed of light, 3.00×10^8 m/s. This equation states that the mass and energy of an object are proportional. The greater an object's mass, the greater its energy. Because the proportionality constant in the equation, c^2, is such a large number, small changes in mass are accompanied by large changes in energy.

The mass changes accompanying chemical reactions are too small to detect. For this reason, it is possible to speak of the conservation of mass in chemical reactions. The calculation of the loss of mass that accompanies the combustion of a mole of CH_4 illustrates this point. This calculation is shown in Sample Exercise 19.9.

SAMPLE EXERCISE 19.9

Calculate the mass that disappears from the system when a mole of CH_4 is combusted:

$$CH_4(g) + 2O_2(g) \longrightarrow CO_2(g) + 2H_2O(l)$$

The system loses 890 kJ of energy in this process ($\Delta E = -890$ kJ).

Solution: From Equation 19.24 we can write that the change in mass, Δm, in the system is proportional to its change in energy, ΔE:

$$\Delta E = c^2 \, \Delta m$$

Rearranging to solve for Δm, we have

$$\Delta m = \frac{\Delta E}{c^2}$$

Substituting the value given for ΔE and remembering that a negative sign is associated with an exothermic process, we have

$$\Delta m = \frac{-890 \text{ kJ}}{(3.00 \times 10^8 \text{ m/s})^2}\left(\frac{1000 \text{ J}}{1 \text{ kJ}}\right) \times \left(\frac{1 \text{ kg-m}^2/\text{s}^2}{1 \text{ J}}\right) = -9.89 \times 10^{-12} \text{ kg}$$

The negative sign indicates mass loss. This mass change is far too small to detect.

The mass changes and the associated energy changes in nuclear reactions are much greater than in chemical reactions. The energy released through the nuclear fission of only about a pound of uranium (Section 19.7) is equivalent to that released by combustion of 1500 tons of coal.

Nuclear Binding Energies

It was discovered in the 1930s that the masses of nuclei are always less than the masses of the individual nucleons of which they are composed. For example, the helium-4 nucleus has a mass of 4.00150 amu. The mass of a proton is 1.00728 amu, while that of a neutron is 1.00867 amu. Consequently, two protons and two neutrons have a total mass of 4.03190 amu:

$$\begin{aligned}
\text{Mass of two protons} &= 2(1.00728 \text{ amu}) = 2.01456 \text{ amu} \\
\text{Mass of two neutrons} &= 2(1.00867 \text{ amu}) = \underline{2.01734 \text{ amu}} \\
\text{Total mass} &= 4.03190 \text{ amu}
\end{aligned}$$

The individual nucleons weigh 0.03040 amu more than the helium-4 nucleus:

$$\begin{aligned} \text{Mass of two protons and two neutrons} &= 4.03190 \text{ amu} \\ \text{Mass of } ^{4}_{2}\text{He nucleus} &= \underline{4.00150 \text{ amu}} \\ \text{Mass difference} &= 0.03040 \text{ amu} \end{aligned}$$

The mass difference between a nucleus and its constituent nucleons is called the **mass defect.** The origin of the mass defect is readily understood if we consider that energy must be added to a nucleus in order to break it into separated protons and neutrons:

$$\text{Energy} + {}^{4}_{2}\text{He} \longrightarrow 2\,{}^{1}_{1}\text{p} + 2\,{}^{1}_{0}\text{n} \qquad [19.25]$$

According to Einstein's mass-energy equivalence relationship (Equation 19.24), the addition of energy to a system must be accompanied by a proportional increase of mass. The mass change, Δm, is defined as the total mass of the products minus the total mass of the reactants. The mass change for the conversion of helium-4 into separated nucleons is $\Delta m = 0.03040$ amu, as shown in the calculations above. The associated energy change, therefore, is readily calculated:

$$\begin{aligned} \Delta E &= c^2\,\Delta m \\ &= (3.00 \times 10^{8}\text{ m/s})^2(0.0304\text{ amu}) \\ &\quad \times \left(\frac{1.00\text{ g}}{6.02 \times 10^{23}\text{ amu}}\right)\left(\frac{1\text{ kg}}{1000\text{ g}}\right) \\ &= 4.52 \times 10^{-12}\,\frac{\text{kg-m}^2}{\text{s}^2} = 4.52 \times 10^{-12}\text{ J} \end{aligned}$$

Decomposition of a mole of helium-4 in this fashion would require a tremendous quantity of energy:

$$\left(6.02 \times 10^{23}\,\frac{\text{nuclei}}{\text{mol}}\right)\left(4.52 \times 10^{-12}\,\frac{\text{J}}{\text{nucleus}}\right) = 2.72 \times 10^{12}\text{ J/mol}$$

The energy change calculated from the mass defect of a nucleus is called the **binding energy** of the nucleus. It is the energy required to decompose the nucleus into separated protons and neutrons. Thus the larger the binding energy, the more stable the nucleus is toward such decomposition. The mass defects and binding energies of three nuclei (helium-4, iron-56, and uranium-238) are compared in Table 19.4. The binding energies per nucleon (that is, the binding energy of each nucleus divided by the total number of nucleons in that nucleus) are also compared in the table. Similar calculations for other nuclei indicate that the

TABLE 19.4 Mass differences and binding energies for three nuclei

Nucleus	Mass of nucleus (amu)	Mass of individual nucleons (amu)	Mass difference (amu)	Binding energy (J)	Binding energy per nucleon (J)
${}^{4}_{2}\text{He}$	4.00150	4.03190	0.0304	4.52×10^{-12}	1.13×10^{-12}
${}^{56}_{26}\text{Fe}$	55.92066	56.44938	0.52872	7.90×10^{-11}	1.41×10^{-12}
${}^{238}_{92}\text{U}$	238.0003	239.9356	1.9353	2.89×10^{-10}	1.22×10^{-12}

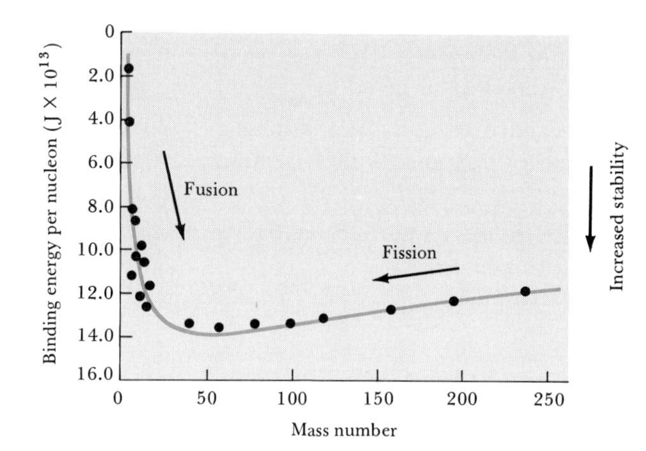

FIGURE 19.9 The average binding energy per nucleon increases to a maximum at a mass number of 50 to 60 and decreases slowly thereafter. As a result of these trends, fusion of light nuclei and fission of heavy nuclei are exothermic processes.

binding energy per nucleon increases in magnitude to about 1.4×10^{-12} J with nuclei whose mass numbers are in the vicinity of iron-56. It then decreases slowly to about 1.2×10^{-12} J for very heavy nuclei. This trend is shown in Figure 19.9. These results indicate that heavy nuclei will gain stability, and therefore give off energy, if they are fragmented. This process, known as **fission,** occurs in the atomic bomb and in nuclear power plants. Figure 19.9 also indicates that even greater amounts of energy should be available if very light nuclei are fused together. Such **fusion reactions** take place in the hydrogen bomb and are the essential energy-producing reactions in the sun. We shall look more closely at fission and fusion in Sections 19.7 and 19.8.

In calculating energy changes for nuclear reactions thus far, we have used nuclear masses. However, the masses of nuclides are normally expressed in tables as atomic masses (that is, the masses of nuclei plus electrons). It is a simple matter to calculate nuclear masses from atomic masses. For example, the mass of the $^{56}_{26}Fe$ atom is given in a table of nuclides as 55.93492 amu. This includes the mass of 26 electrons. When the mass of 26 electrons (26 electrons $\times$ 5.486×10^{-4} amu/electron = 0.01426 amu) is subtracted from the atomic mass, we obtain the mass of the nucleus listed in Table 19.4, 55.92066 amu. In calculating the mass change in a nuclear reaction, it is usually acceptable to employ the masses of the atoms containing the nuclides of interest, because the number of electrons in the reactants and products is usually the same. Thus the difference in atomic masses is usually the same as the difference in nuclear masses. If you have any doubt about whether you can use atomic masses like this in a particular situation, it is never wrong to subtract the mass of electrons from the atomic masses and use the resultant nuclear masses.

SAMPLE EXERCISE 19.10

How much energy is lost or gained when a mole of cobalt-60 undergoes beta decay: $^{60}_{27}Co \longrightarrow {}^{0}_{-1}e + {}^{60}_{28}Ni$? The mass of the $^{60}_{27}Co$ atom is 59.9338 amu, that of a $^{60}_{28}Ni$ atom is 59.9308 amu, and that of an electron is 0.000549 amu.

Solution: The products of the nuclear reaction are $^{60}_{28}Ni^{+}$ (the 27 electrons of cobalt are carried along in the reaction) and an electron (the beta particle). The mass of these products is just the mass of the neutral $^{60}_{28}Ni$ atom. The mass change in the reaction, therefore, is given by

$$\Delta m = \text{mass of } ^{60}_{28}\text{Ni atom} - \text{mass of } ^{60}_{27}\text{Co atom}$$
$$= 59.9308 \text{ amu} - 59.9338 \text{ amu} = -0.0030 \text{ amu}$$

Thus for a mol of cobalt-60, $\Delta m = -0.0030$ g. The energy produced by the reaction can be calculated from this mass:

$$\Delta E = c^2 \Delta m$$
$$= (3.00 \times 10^8 \text{ m/s})^2(-0.0030 \text{ g})\left(\frac{1 \text{ kg}}{1000 \text{ g}}\right)$$
$$= -2.7 \times 10^{11} \frac{\text{kg-m}^2}{\text{s}^2} = -2.7 \times 10^{11} \text{ J}$$

By comparison, it takes only 9×10^5 J to break all the chemical bonds in a mole of water.

19.7 NUCLEAR FISSION

Our discussion of the energy changes in nuclear reactions (Section 19.6) revealed an important observation: Both the splitting of heavy nuclei (fission) and the union of light nuclei (fusion) are exothermic. Commercial nuclear power plants and the most common forms of nuclear weaponry depend for their operation on the process of nuclear fission. The first nuclear fission to be discovered was that of uranium-235. This nucleus, as well as those of uranium-233 and plutonium-239, undergoes fission when struck by slow-moving neutrons.* The fission process is illustrated in Figure 19.10. A heavy nucleus can split in many different ways, just as can a piece of glass. Two different ways that the uranium-235 nucleus splits are shown in Equations 19.26 and 19.27:

$$^{1}_{0}n + ^{235}_{92}U \longrightarrow ^{137}_{52}Te + ^{97}_{40}Zr + 2^{1}_{0}n \qquad [19.26]$$

$$^{1}_{0}n + ^{235}_{92}U \longrightarrow ^{142}_{56}Ba + ^{91}_{36}Kr + 3^{1}_{0}n \qquad [19.27]$$

Over 200 different isotopes of 35 different elements have been found among the fission products of uranium-235. In general, these are radioactive.

On the average, 2.4 neutrons are produced by every fission of uranium-235. If one fission produces two neutrons, these two neutrons can cause two fissions. The four neutrons thereby released produce four fissions, and so forth, as shown in Figure 19.11. The number of fissions and their associated energies quickly escalate, and, if the process is unchecked, the result is a violent explosion. Reactions that multiply in this fashion are called **branching chain reactions.**

In order for a chain fission reaction to occur, the sample of fissionable

*There are other heavy nuclei that can be induced to undergo fission. However, these three are the only ones of practical importance.

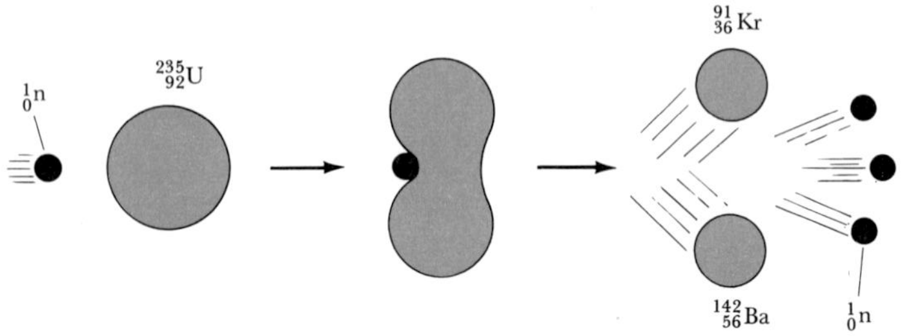

$^{1}_{0}n$ $^{235}_{92}U$ $^{91}_{36}Kr$ $^{142}_{56}Ba$ $^{1}_{0}n$

FIGURE 19.10 Schematic representation of the fission of uranium-235 showing one of its many fission patterns. In this process, 3.5×10^{-11} J of energy is produced.

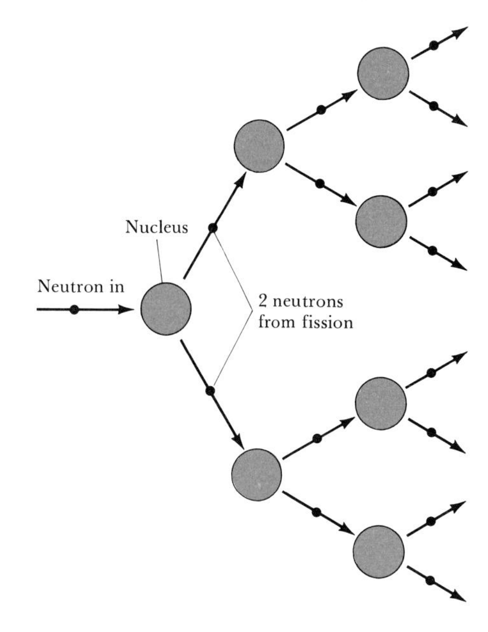

FIGURE 19.11 Chain fission reaction in which each fission produces two neutrons. The process leads to an accelerating rate of fission, with the number of fissions doubling at each stage.

material must have a minimum size. Otherwise, neutrons escape from the sample before they have an opportunity to strike a nucleus and cause fission. The chain stops if enough neutrons are lost. The reaction is then said to be subcritical. If the mass is large enough to maintain the chain reaction with a constant rate of fission, the reaction is said to be critical. This situation results if only one neutron from each fission is subsequently effective in producing another fission. If the mass is larger still, few of the neutrons produced are able to escape. The chain reaction then multiplies the number of fissions and the reaction is said to be supercritical. The effect of mass size on whether a reaction is subcritical, critical, or supercritical is illustrated in Figure 19.12. One of the ways that an atomic bomb is triggered to produce a supercritical mass is shown in Figure 19.13. As shown, two subcritical masses are brought together by use of chemical explosives to form a supercritical mass. We have seen

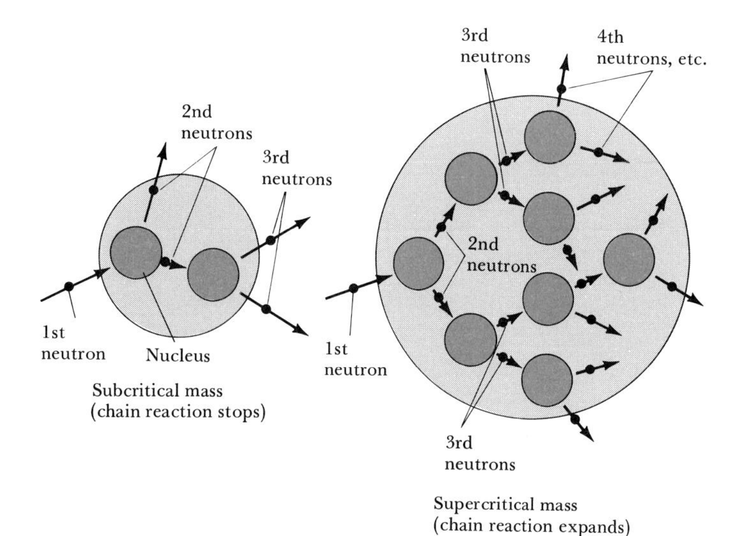

FIGURE 19.12 The chain reaction in a subcritical mass soon stops because neutrons are lost from the mass without causing fission. As the size of the mass increases, fewer neutrons are able to escape, and the chain reaction is able to expand.

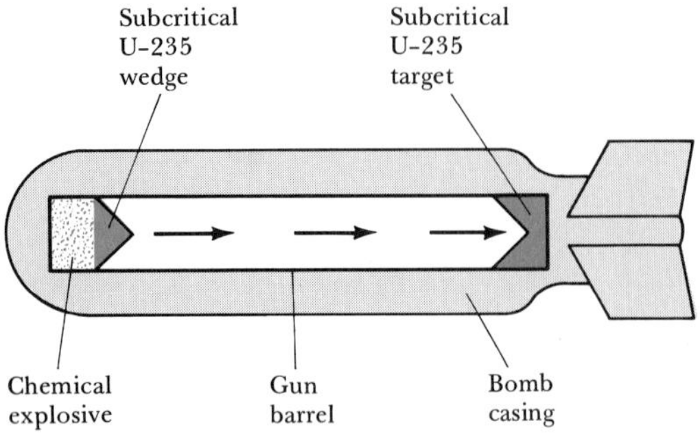

Subcritical U-235 wedge

Subcritical U-235 target

Chemical explosive

Gun barrel

Bomb casing

FIGURE 19.13 One design used in atomic bombs. A conventional explosive is used to bring two subcritical masses together to form a supercritical mass.

that the basic design of such a bomb is quite simple. The fissionable materials are potentially available to any nation with a nuclear reactor. This simplicity has already resulted in the proliferation of atomic weapons. It is possible that such weapons could come into the hands of terrorist groups or groups intent on using the weapons to extort money from corporations or nations. Such events would add a frightening aspect to the nuclear age in which we live.

The fission of uranium-235 was first achieved in the late 1930s by Enrico Fermi and his colleagues in Rome and shortly thereafter by Otto Hahn and his co-workers in Berlin. Both groups were trying to produce transuranium elements. In 1938, Hahn identified barium among his reaction products. He was puzzled by this observation and questioned the identification because the presence of barium was so unexpected. He sent a detailed letter describing his experiments to Lise Meitner, a former co-worker. Meitner had been forced to leave Germany because of the anti-Semitism of the Third Reich, and she had settled in Sweden. She surmised that Hahn's experiment indicated that a new nuclear process was occurring in which the uranium-235 split. She called this nuclear fission. Meitner passed word of this discovery to her nephew, Otto Frisch, a physicist working at Niels Bohr's institute in Copenhagen. He repeated the experiment, verifying Hahn's observations and finding that tremendous energies were involved. In January 1939, Meitner and Frisch published a short article describing this new reaction. In March 1939, Leo Szilard and Walter Zinn at Columbia University discovered that more neutrons are produced than were used in each fission. As we have seen, this allows a branching chain reaction. News of these discoveries and an awareness of their potential use in explosive devices spread rapidly within the scientific community. Several scientists finally persuaded Albert Einstein, the most famous physicist of the time, to write a letter to President Roosevelt outlining the implications of these discoveries. Einstein's letter, written in August 1939, outlined the possible military applications of nuclear fission and emphasized the danger that weapons based on fission would pose if they were to be developed by the Nazis. Roosevelt judged it imperative that the United States investigate the possibility of such weapons. Late in 1941 the decision was made to build a bomb based on the fission reaction. An enormous research project, known as the "Manhattan Project," was initiated. This project led to the development of the atomic bomb and the dawning of the nuclear age.

Nuclear Reactors

Nuclear fission produces the energy generated by nuclear power plants. The "fuel" of the nuclear reactor is therefore a fissionable substance such as uranium-235. Typically, uranium is enriched to about 3 percent uranium-235 and then used in the form of UO_2 pellets. These enriched uranium pellets are encased in zirconium or stainless-steel tubes. Rods composed of materials such as cadmium or boron are used to control the fission process by absorbing neutrons. These **control rods** permit a sufficient flux of neutrons to keep the reaction chain self-sustaining but yet

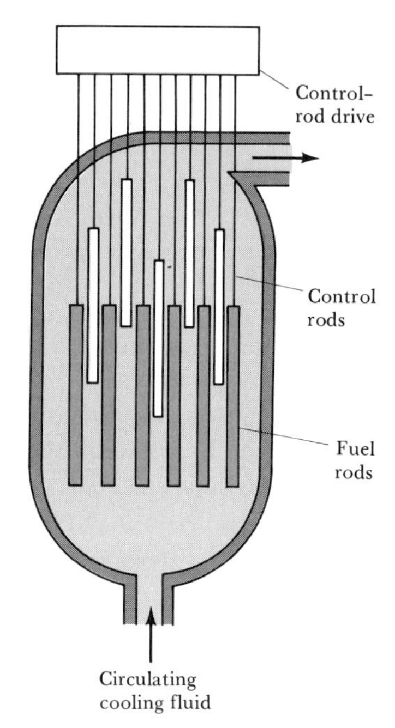

FIGURE 19.14 Reactor core showing fuel elements, control rods, and cooling fluid.

prevent the reactor core from overheating.* The reactor is started by using a neutron source; it is stopped by inserting the control rods more deeply into the reactor core, the site of the fission (Figure 19.14). The reactor core also contains a **moderator** that acts to slow down neutrons, so that they can be captured more readily by the fuel. Finally a **cooling liquid** circulates through the reactor core to carry off the heat generated by the nuclear fission. The cooling liquid can also serve as the neutron moderator.

The design of the power plant is basically the same as that of a power plant that burns fossil fuel (except that the burner is replaced by the reactor core). In both instances, steam is used to drive a turbine that is connected to an electrical generator. Because the steam must be condensed, additional cooling water is needed. This water is generally obtained from a large source such as a river or lake. The water is returned to its source at a higher temperature than when it was removed. Power plants are therefore a significant source of thermal pollution. The nuclear power plant design shown in Figure 19.15 is the one that is currently most popular. The primary coolant, which passes through the core, is in a closed system. Subsequent coolants never pass through the reactor core at all. This lessens the chance that radioactive products could escape the core. Additionally, the reactor is surrounded by a concrete shell to shield personnel and nearby residents from radiation.

Fission products accumulate as the reactor operates. These products lessen the efficiency of the reactor by capturing neutrons. Therefore, the reactor must be stopped periodically, so that the nuclear fuel can be reprocessed. When the fuel rods are removed from the reactor they are initially very radioactive. It was originally intended that they be stored

*The reactor core cannot reach supercritical levels and explode with the violence of an atomic bomb because the concentration of uranium-235 is too low. However, if the core overheats, sufficient damage might be done to release materials into the environment.

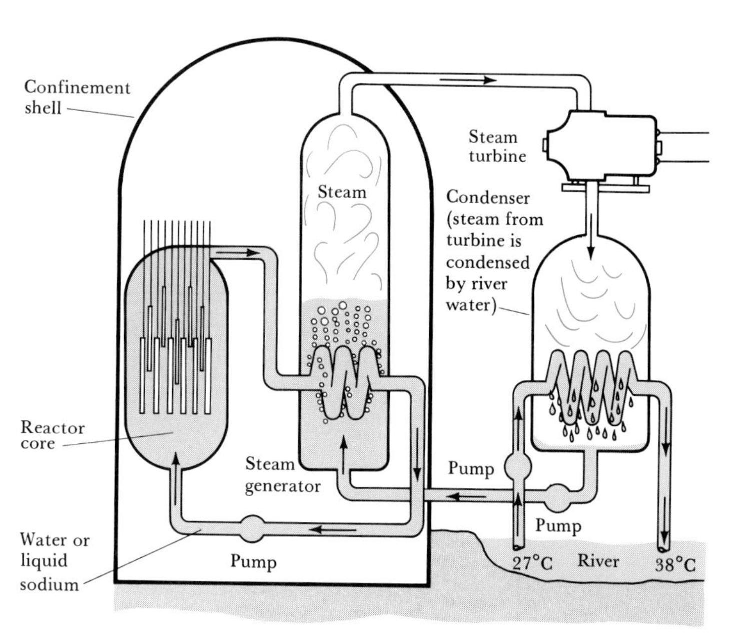

FIGURE 19.15 Basic design of a nuclear power plant. Heat produced by the reactor core is carried by a cooling fluid such as water or liquid sodium to a steam generator. The steam so produced is used to drive an electrical generator.

for several months in pools at the reactor site to allow decay of short-lived radioactive nuclei. They were then to be transported in shielded containers to reprocessing plants, where the fuel would be separated from the fission products. However, reprocessing plants have been plagued with operational difficulties, and there is intense opposition to the transport of nuclear wastes on the nation's highways. Even if the transportation difficulties could be overcome, the high radioactivity of the spent fuel makes reprocessing a hazardous operation. At present the spent fuel rods are simply being kept in storage at reactor sites.

In reprocessing, the uranium is separated from nuclear waste products and repackaged into new fuel rods. The problem then becomes one of disposal of the leftover fission products. Storage poses a major problem because the fission products are extremely radioactive. It is estimated that 20 half-lives are required for their radioactivity to reach levels acceptable for biological exposure. Based on the 28.8-yr half-life of strontium-90, one of the longer-lived and most dangerous of the products, the wastes must be stored for 600 yr. If plutonium-239 is not removed, storage must be for longer periods; plutonium-239 has a half-life of 24,000 yr. It is advantageous, however, to remove plutonium-239 because it can be used as a fissionable fuel.

There has been and will continue to be a considerable amount of research devoted to disposal of radioactive wastes. At present the most attractive possibilities appear to be formation of a glass, ceramic, or synthetic rock from the wastes, as a means of immobilizing them. The solid materials could then be buried deep underground. Because the radioactivity will persist for a long time, there must be assurances that the solids will not crack from the heat generated by nuclear decay, possibly allowing radioactivity to find its way into underground water supplies. At present we simply do not have clear-cut answers to all the difficult and important questions relating to nuclear waste disposal.

Although the precise figures are kept secret, experts estimate that from 1000 to 3000 nuclear weapons are added to the U.S. nuclear stockpile each year. The radioactive waste from this military production in a single year far exceeds the total of all the wastes yet produced from operation of nuclear power plants. Production continues despite the fact that the United States already possesses about 30,000 nuclear bombs.

Breeder Reactors

Some scientists have estimated that it will take 40 yr or less to use all of the relatively low-cost uranium-235 available in this country. Because uranium-235 is very rare and because its supply can be exhausted so readily, methods of generating other fissionable materials are being actively investigated. Fissionable plutonium-239 and uranium-233 can be produced in nuclear reactors from nuclides that are far more abundant than uranium-235. The reactions involved are as follows:

$$^{238}_{92}\mathrm{U} + ^{1}_{0}\mathrm{n} \longrightarrow ^{239}_{92}\mathrm{U} \qquad [19.28]$$

$$^{239}_{92}\mathrm{U} \longrightarrow ^{239}_{93}\mathrm{Np} + ^{0}_{-1}\mathrm{e} \qquad (t_{1/2} = 24\ \mathrm{min}) \qquad [19.29]$$

$$^{239}_{93}\mathrm{Np} \longrightarrow ^{239}_{94}\mathrm{Pu} + ^{0}_{-1}\mathrm{e} \qquad (t_{1/2} = 2.3\ \mathrm{days}) \qquad [19.30]$$

$$ {}^{232}_{90}\mathrm{Th} + {}^{1}_{0}\mathrm{n} \longrightarrow {}^{233}_{90}\mathrm{Th} \qquad [19.31] $$

$$ {}^{233}_{90}\mathrm{Th} \longrightarrow {}^{233}_{91}\mathrm{Pa} + {}^{0}_{-1}\mathrm{e} \qquad (t_{1/2} = 22 \text{ min}) \qquad [19.32] $$

$$ {}^{233}_{91}\mathrm{Pa} \longrightarrow {}^{233}_{92}\mathrm{U} + {}^{0}_{-1}\mathrm{e} \qquad (t_{1/2} = 27 \text{ days}) \qquad [19.33] $$

It is theoretically possible to build a reactor that both produces energy and converts uranium-238 or thorium-232 into fissionable fuel. We can envision a fission of a uranium-235 nucleus producing two neutrons, one causing another fission, the second initiating the change of uranium-238 into plutonium-239. Plutonium-239 is produced in ordinary reactors because of the presence of uranium-238. However, the hope is to be able to produce more fuel than the reactor uses. Reactors able to do this are still under development; they are called **breeder reactors.**

The design of breeder reactors poses many difficult technical problems. In addition, the breeder reactor program has become the subject of intense political debate. Because both breeder reactors and nuclear fuel reprocessing plants produce plutonium-239, any nation that acquired a breeder reactor or reprocessing technology would have in hand the raw materials for atomic weaponry. Fear of nuclear proliferation or the possibility of thefts of nuclear materials by terrorist groups has resulted in uncertainty as to whether the United States should proceed with breeder reactor development and fuel reprocessing. Development of breeder reactors is, however, progressing in several other countries.

19.8 NUCLEAR FUSION

As shown in Section 19.6, energy is produced when light nuclei are fused to form heavier ones. Reactions of this type are responsible for the energy produced by the sun. Spectroscopic studies indicate that the sun is composed of 73 percent H, 26 percent He, and only 1 percent of all other elements, by mass. Among the several fusion processes that are believed to occur are the following:

$$ {}^{1}_{1}\mathrm{H} + {}^{1}_{1}\mathrm{H} \longrightarrow {}^{2}_{1}\mathrm{H} + {}^{0}_{1}\mathrm{e} \qquad [19.34] $$

$$ {}^{1}_{1}\mathrm{H} + {}^{2}_{1}\mathrm{H} \longrightarrow {}^{3}_{2}\mathrm{He} \qquad [19.35] $$

$$ {}^{3}_{2}\mathrm{He} + {}^{3}_{2}\mathrm{He} \longrightarrow {}^{4}_{2}\mathrm{He} + 2{}^{1}_{1}\mathrm{H} \qquad [19.36] $$

$$ {}^{3}_{2}\mathrm{H} + {}^{1}_{1}\mathrm{H} \longrightarrow {}^{4}_{2}\mathrm{He} + {}^{0}_{1}\mathrm{e} \qquad [19.37] $$

Theories have been proposed for generation of other elements via fusion processes.

Fusion is appealing as an energy source because of the availability of light isotopes and because fusion products are generally not radioactive. It is therefore potentially a cleaner process than is fission. The problem is that high energies are required to overcome the repulsion between nuclei. The required energies are achieved by high temperatures. Fusion reactions are therefore also known as **thermonuclear reactions.** The lowest temperature required for any fusion is that needed to fuse ${}^{2}_{1}\mathrm{H}$ and ${}^{3}_{1}\mathrm{H}$, as shown in Equation 19.38. This reaction requires a temperature of 40,000,000 K:

$$ {}^{2}_{1}\mathrm{H} + {}^{3}_{1}\mathrm{H} \longrightarrow {}^{4}_{2}\mathrm{He} + {}^{1}_{0}\mathrm{n} \qquad [19.38] $$

Such high temperatures have been achieved by using an atomic bomb to initiate the fusion process. This is done in the thermonuclear, or hydrogen, bomb. Clearly, this approach is unacceptable for controlled power generation.

Numerous problems must be overcome before fusion becomes a practical energy source. Besides the high temperatures necessary to initiate the reaction, there is the problem of confining the reaction. No known structural material is able to withstand the enormous temperatures necessary for fusion. Research has centered on the use of strong magnetic fields to contain the reaction. Use of powerful lasers to generate the temperatures required for fusion has also been the subject of much recent research. Although there is reason for some degree of optimism, it is impossible to tell if and when the tremendous technical difficulties involving nuclear fusion will be overcome. It is therefore not yet clear whether fusion will ever be a practical source of energy for humankind.

19.9 BIOLOGICAL EFFECTS OF RADIATION

The increased pace of synthesis and use of radioisotopes has led to increased concern over the effects of radiation on matter, particularly in biological systems. We therefore close this chapter by examining the health hazards associated with radioisotopes.

Alpha, beta, and gamma rays (as well as X rays) possess energies far in excess of ordinary bond energies and ionization energies. Consequently, these forms of radiation are able to fragment and ionize molecules, generating unstable, highly reactive particles as they pass through matter. For example, gamma rays are able to ionize water molecules, forming unstable H_2O^+ ions. An H_2O^+ ion can react with another water molecule to form an H_3O^+ ion and a neutral OH molecule:

$$H_2O^+ + H_2O \longrightarrow H_3O^+ + OH \qquad [19.39]$$

The unstable and highly reactive OH molecule is an example of a **free radical,** a substance with one or more unpaired electrons. In a biological system such particles can attack a host of other compounds to produce new free radicals, which in turn attack yet other compounds. Thus the formation of a single free radical can instigate a large number of chemical reactions that are ultimately able to disrupt the normal operations of cells.

The resultant radiation damage to living systems can be classified as either **somatic** or **genetic.** Somatic damage is that which affects the organism during its own lifetime. Genetic damage is that which has a genetic effect; it harms offspring through damage to genes and chromosomes, the body's reproductive material. It is more difficult to study genetic effects than somatic ones because it may take several generations for genetic damage to become apparent. Somatic damage includes "burns," molecular disruptions similar to those produced by high temperatures. It also includes cancer. Cancer is brought about by damage to the growth-regulation mechanism of cells, which causes them to reproduce in an uncontrolled manner. In general, the tissues that show the greatest damage from radiation are those that reproduce at a rapid rate, such as bone marrow, blood-forming tissues, and lymph nodes. Leukemia is probably the major cancer problem associated with radiation.

The clinical symptoms of acute (short-term) exposure to radiation include a decrease in the number of white blood cells, fatigue, nausea, and diarrhea. Sufficient exposure can result in death from blood disorders, gastrointestinal failure, and damage to the central nervous system. In light of these effects, it is important to determine whether there are any safe levels of exposure to radiation. What are the maximum levels of radiation that we should permit from various human activities? Unfortunately, we are hampered in our attempts to set realistic standards by our lack of understanding of the effects of chronic (long-term) exposure to radiation. Most scientists presently believe that the effects of radiation are proportional to exposure, even down to low exposures. This means that *any* amount of radiation causes some finite risk of injury. The extent of the risk depends on the type of radiation, its location, its energy, and the magnitude of the radiation dose.

Radiation Doses

A gram of radium undergoes 3.7×10^{10} nuclear disintegrations in 1 s. This number of disintegrations is known as a **curie** (Ci). The curie is the unit used in measuring nuclear activity. For example, a 5.0-millicurie sample of cobalt-60 undergoes $(5.0 \times 10^{-3})(3.7 \times 10^{10}) = 1.8 \times 10^{8}$ disintegrations per second.

The damage produced by radiation outside the body depends not only on the nuclear activity, but also on the energy and penetrating power of the radiation. Gamma rays are particularly dangerous because they penetrate human tissue very effectively, just as do X rays. Consequently, their damage is not limited to the skin. In contrast, most alpha rays are stopped by skin, and beta rays are able to penetrate only about 1 cm beyond the surface of the skin. Hence, neither is as dangerous as gamma rays unless the radiation source somehow enters the body. Within the body, alpha rays are particularly dangerous because they leave a very dense trail of damaged molecules as they move through matter.

Two units, the rad and rem, are commonly used to measure radiation doses. (A third unit, the roentgen, is essentially the same as the rad.) A **rad** (*r*adiation *a*bsorbed *d*ose) is the amount of radiation that deposits 1×10^{-2} J of energy per kilogram of tissue. A rad of alpha rays can produce more damage than a rad of beta rays. Consequently, the rad is often multiplied by a factor that measures the relative biological damage caused by the radiation. This factor is known as the relative biological effectiveness of the radiation, abbreviated **RBE.** The RBE is approximately 1 for beta and gamma rays and 10 for alpha rays. The exact value for the RBE varies with dose rate, total dose, and type of tissue affected. The product of the number of rads and the RBE of the radiation gives the effective dosage in **rems** (*r*oentgen *e*quivalent for *m*an):

$$\text{Number of rems} = (\text{number of rads})(\text{RBE}) \qquad [19.40]$$

The effects of some short-time exposures to radiation are given in Table 19.5. For most persons the effects of long-term exposure to low levels of radiation are more important. Each of us receives an average

TABLE 19.5 Effects of short-time exposures to radiation

Dose (rem)	Effect
0–25	No detectable clinical effects
25–50	Slight, temporary decrease in white blood cell counts
100–200	Nausea; marked decrease in white blood cells
500	Death of half the exposed population within 30 days after exposure

exposure of 0.1 to 0.2 rem each year as background radiation from natural sources, such as naturally occurring radioisotopes, and cosmic rays coming in from outer space. This amount may vary widely from one person to the next, depending on living situation. For example, persons who live in stone or brick houses normally receive more exposure than those who live in wooden houses. Those who live at higher elevations receive more cosmic ray exposure, because the earth's atmosphere absorbs cosmic rays to some extent.

In addition to natural sources, we all receive some exposure to radiation resulting from human activity. Table 19.6 shows radiation levels and the percent contributions of the more important sources to the average exposure of the U.S. population. It must be emphasized that these are merely averages; the relative contributions of the various sources will vary widely from one person to another. For example, a chest X ray involves a radiation dose of 20 to 40 millirem. Some people have never been X-rayed, others have had many exposures. Uranium miners clearly receive larger exposures than the general populace; similarly, those who live near a nuclear power plant may receive more exposure from this source than the average population.

Radiation exposure limits are presently set by the Environmental Protection Agency at 500 millirem each year for the general population and 5 rem for occupational exposure, exclusive of background radiation. As more scientists have become convinced that even low radiation dosage poses some risk, there has been increasing pressure to set tougher standards for radiation exposure. As with many other aspects of human activity, the question at issue is one of risk versus benefit. Before we can make intelligent decisions we must have a deeper understanding of the risks than we do at present.

TABLE 19.6 Contributions to the average exposure of the U.S. population to ionizing radiation

Source	Percent contribution	Radiation level (millirem)
Natural background	52	100
X rays, other medical	43	80
Uranium mining, other technology-related sources	2	4
Fallout from nuclear weapons testing	3	6
Nuclear power plants	0.14	0.3
Consumer products—watch dials, color TV, etc.	0.02	0.04

FOR REVIEW

Summary

Certain nuclei are radioactive. Most of these nuclei gain stability by emitting alpha particles ($^4_2$He), beta particles ($_{-1}^{0}$e), and/or gamma radiation ($^0_0\gamma$). Some nuclei undergo decay by positron ($^0_1$e) emission or by electron capture. The neutron-to-proton ratio is one factor determining nuclear stability. The presence of magic numbers of nucleons and an even number of protons and neutrons are also important. Nuclear transmutations can be induced by bombarding nuclei with charged particles using particle accelerators or with neutrons in a nuclear reactor.

Radioisotopes have characteristic rates of decay. These rates are generally expressed in terms of half-lives. The constant half-lives of nuclides permit their use in dating objects. The ease of detection of radioisotopes also permits their use as tracers, to follow elements through their reactions. Three methods of detection were discussed: use of photographic film, scintillation counters, and Geiger counters.

The energy produced in nuclear reactions is accompanied by measurable losses of mass in accordance with Einstein's relationship, $\Delta E = c^2 \Delta m$. The difference in mass between nuclei and the nucleons of which they are composed is known as the mass defect. The mass defect of a nuclide allows calculation of its nuclear binding energy, the energy required to separate the nucleus into individual nucleons. Examination of binding energies per nucleon reveals that energy is produced when heavy nuclei split (fission) and when light nuclei fuse (fusion).

Uranium-235, uranium-233, and plutonium-239 undergo fission when they capture a neutron. The resulting nuclear reaction is a chain reaction. If the reaction maintains a constant rate it is said to be critical. If it slows it is said to be subcritical. In the atomic bomb, subcritical masses are brought together to form a mass that is supercritical. In nuclear reactors the fission is controlled to generate a constant power. The reactor core consists of fissionable fuel, control rods, moderator, and cooling fluid. The nuclear power plant resembles a conventional power plant except that the core replaces the fuel burner. In breeder reactors more nuclear fuel is produced than is used to generate energy. There is concern regarding the safety with which nuclear power plants can be operated. Beyond this, the reprocessing of spent fuel rods and disposal of highly radioactive nuclear wastes are unsolved problems.

Nuclear fusion requires high temperatures because nuclei must have large kinetic energies to overcome their mutual repulsions. It has not yet been possible to generate a controlled fusion process.

The radioactivity of a sample is measured in curies. One curie corresponds to 3.7×10^{10} disintegrations per second. The amount of energy deposited in biological tissue by radiation is measured in terms of the rad; one rad corresponds to 1×10^{-2} J per kilogram of tissue. The rem is a more useful measure of the biological damage created by the deposited energy. We receive radiation from both naturally occurring and human sources in roughly equal amounts for the general population. The effects of long-term exposure to low levels of radiation are not completely understood, but the most commonly accepted hypothesis is that the extent of biological damage varies in direct proportion to the level of exposure.

Learning goals

Having read and studied this chapter, you should be able to:

1. Write the nuclear symbols for protons, neutrons, electrons, alpha particles, and positrons.
2. Determine the effect of different types of decay on the proton-neutron ratio and predict the type of decay that a nucleus will undergo based on its position relative to the belt of stability.
3. Complete and balance nuclear equations, having been given all but one of the particles involved.
4. Use the half-life of a substance to predict the amount of radioisotope present after a given period of time.
5. Calculate half-life, age of an object, or the remaining amount of radioisotope, having been given any two of these pieces of information.
6. Explain how radioisotopes can be used in dating objects and as radiotracers.
7. Explain how radioactivity is detected, including a description of the basic design of a Geiger counter.
8. Use Einstein's relation, $\Delta E = c^2 \Delta m$, to calculate the energy change or the mass change of a reaction, having been given one of these quantities.
9. Calculate the binding energies of nuclei, having been given their masses and the masses of protons and neutrons.
10. Explain what fission and fusion are and what types of nuclei produce energy when undergoing these processes.

11 Describe the design of a nuclear power plant, including an explanation of the role of fuel elements, control rods, moderator, and cooling fluid.

12 Explain the role played by the mode of radioactivity of a radioisotope in determining its ability to damage biological systems.

13 Define the units used to describe the level of radioactivity (curie) and to measure the effects of radiation on biological systems (rem and rad).

14 Describe various sources of radiation to which the general population is exposed and indicate the relative contributions of each.

Key terms

Among the more important terms and expressions used for the first time in this chapter are the following:

Alpha particles (Section 19.1) are identical to helium-4 nuclei, consisting of two protons and two neutrons, symbol ^4_2He.

Beta particles (Section 19.2) are energetic electrons emitted from the nucleus, symbol $_{-1}^{0}e$.

The binding energy (Section 19.6) of a nucleus is the energy required to decompose that nucleus into nucleons; it is usually calculated from the mass defect of the nucleus.

A breeder reactor (Section 19.7) is a nuclear-fission reactor that produces more fissionable fuel than it consumes.

A chain reaction (Section 19.7) is a series of reactions in which one reaction initiates the next.

A critical mass (Section 19.7) is the amount of fissionable material necessary to maintain a chain reaction. Smaller masses are said to be subcritical, whereas larger ones are supercritical.

A curie (Section 19.9) is a measure of radioactivity: 1 curie $= 3.7 \times 10^{10}$ nuclear disintegrations per second.

Electron capture (Section 19.2) is a mode of radioactive decay in which an inner-shell orbital electron is captured by the nucleus.

Fission (Section 19.7) is the splitting of a large nucleus into two intermediate-sized ones.

A free radical (Section 19.9) is a substance with one or more unpaired electrons.

Fusion (Section 19.8) is the joining of small nuclei to form larger ones.

Gamma rays (Section 19.2) are energetic electromagnetic radiation emanating from the nucleus of a radioactive atom.

Half-life (Section 19.3) is the time required for half of a sample of a particular radioisotope to decay.

The mass defect (Section 19.6) of a nucleus is the difference between the mass of the nucleus and the total masses of the individual nucleons that it contains.

A nuclear transmutation (Section 19.1) is a conversion of one kind of atom to another.

A positron (Section 19.2) is a particle with the same mass as an electron but with a positive charge, symbol ^0_1e.

A rad (Section 19.9) is a measure of the energy absorbed from radiation by tissue or other biological material; 1 rad = transfer of 1×10^{-2} J of energy per kilogram of material.

A radioisotope (Section 19.1) is an isotope that is radioactive; that is, it is undergoing nuclear changes with emission of nuclear radiation.

RBE (Section 19.9) (relative biological effectiveness) is an adjustment factor used to convert rads to rems; it accounts for differences in biological effects of different particles having the same energy.

A rem (Section 19.9) is a measure of the biological damage caused by radiation; rems = rads $\times$ RBE.

EXERCISES

Nuclear reactions

19.1 Provide a short description or definition of each of the following terms: (a) nucleon; (b) mass number; (c) nuclear transmutation; (d) alpha ray; (e) positron; (f) gamma ray.

19.2 Indicate the number of protons and neutrons in each of the following nuclei: (a) oxygen-17; (b) $^{99}_{42}Mo$; (c) cesium-136; (d) ^{115}Ag.

19.3 Indicate the number of protons, neutrons, and nucleons in each of the following: (a) cerium-137; (b) ^{234}Pu; (c) cadmium-113; (d) $^{37}_{17}Cl$.

19.4 Write balanced nuclear equations for the following transformations. (a) ^{181}Hf undergoes beta decay. (b) Radium-226 decays to a radon isotope. (c) Lead-205 undergoes positron emission. (d) Tungsten-179 undergoes orbital electron capture.

19.5 Write balanced nuclear equations for each of the following nuclear transformations. (a) Zirconium-93 undergoes beta decay. (b) Neptunium-233 undergoes alpha decay. (c) Francium-218 is formed by decay of an actinium nuclide. (d) Einsteinium-246 undergoes orbital electron capture.

19.6 Complete and balance the following nuclear equations by supplying the missing particle:

(a) ${}^{32}_{16}S + {}^{1}_{0}n \longrightarrow {}^{1}_{1}H + ?$
(b) ${}^{7}_{4}Be + {}^{0}_{-1}e$ (orbital electron) $\longrightarrow ?$
(c) $? \longrightarrow {}^{187}_{76}Os + {}^{0}_{-1}e$
(d) ${}^{98}_{42}Mo + {}^{2}_{1}H \longrightarrow {}^{1}_{0}n + ?$
(e) ${}^{235}_{92}U + {}^{1}_{0}n \longrightarrow {}^{135}_{54}Xe + ? + 2{}^{1}_{0}n$

19.7 Complete and balance the following nuclear equations by supplying the missing particle:

(a) ${}^{252}_{98}Cf + {}^{10}_{5}B \longrightarrow 3{}^{1}_{0}n + ?$
(b) ${}^{2}_{1}H + {}^{3}_{2}He \longrightarrow {}^{4}_{2}He + ?$
(c) ${}^{1}_{1}H + {}^{11}_{5}B \longrightarrow 3?$
(d) ${}^{122}_{53}I \longrightarrow {}^{122}_{54}Xe + ?$
(e) ${}^{59}_{26}Fe \longrightarrow {}^{0}_{-1}e + ?$

19.8 Fermium-250, formed in a nuclear synthesis experiment, decays rather rapidly. The decay chain proceeds through several steps to form ${}^{234}_{92}U$, which has a very long half-life. Write the several reactions that lead from ${}^{250}_{100}Fm$ to ${}^{234}_{92}U$.

19.9 Write balanced equations for the following nuclear reactions: (a) ${}^{238}_{92}U(n,\gamma){}^{239}_{92}U$; (b) ${}^{14}_{7}N(p,\alpha){}^{11}_{6}C$; (c) ${}^{18}_{8}O(n,\beta){}^{19}_{9}F$; (d) ${}^{59}_{26}Fe(\alpha,\beta){}^{63}_{29}Cu$.

19.10 The naturally occurring radioactive decay series that begins with ${}^{235}_{92}U$ stops with formation of the stable ${}^{207}_{82}Pb$. The decays proceed through a series of alpha particle and beta particle emissions. How many of each type of emission are involved in this series?

Nuclear stability

19.11 Which of the following nuclides would you expect to be radioactive: (a) ${}^{17}_{8}O$; (b) ${}^{176}_{74}W$; (c) ${}^{108}_{50}Sn$; (d) ${}^{92}_{40}Zr$; (e) ${}^{238}_{94}Pu$? Justify your choices.

19.12 In each of the following pairs, which nuclide would you expect to be the more abundant in nature: (a) ${}^{19}_{9}F$ or ${}^{18}_{9}F$; (b) ${}^{80}_{34}Se$ or ${}^{81}_{34}Se$; (c) ${}^{56}_{26}Fe$ or ${}^{57}_{26}Fe$; (d) ${}^{118}_{50}Sn$ or ${}^{118}_{51}Sb$? Justify your choice.

19.13 Which of the following nuclides is most likely to be a positron emitter: (a) ${}^{53}_{24}Cr$; (b) ${}^{51}_{25}Mn$; (c) ${}^{59}_{26}Fe$? Explain.

19.14 Indicate whether each of the following nuclides lie within the belt of stability in Figure 19.1: (a) ${}^{108}_{49}In$; (b) ${}^{102}_{47}Ag$; (c) ${}^{17}_{7}N$; (d) ${}^{210}_{86}Rn$. If they do not, describe a nuclear decay process that would alter the neutron-to-proton ratio in the direction of increased stability.

19.15 All of the following nuclides are radioactive and undergo either beta or positron emission: (a) ${}^{66}_{32}Ge$; (b) ${}^{105}_{45}Rh$; (c) ${}^{137}_{53}I$; (d) ${}^{133}_{58}Ce$. Indicate which nuclei undergo each of these types of emission.

19.16 We have seen that one possible mode of nuclear transformation is orbital electron capture. Which electron in a many-electron atom is most likely to be captured in such a process? How would you expect the likelihood of orbital electron capture to change with atomic number?

19.17 The neutron-to-proton ratio for maximum stability for nuclei with mass numbers in the range 205 to 220 is 1.52. Based on this fact, account for the following observations: (a) Thallium-210 undergoes beta decay. (b) Actinium-204 undergoes orbital electron capture. (c) Bismuth-197 undergoes alpha particle emission.

19.18 Sulfur-35 is radioactive and undergoes beta decay. What differences would you expect in the chemical behavior of atoms containing sulfur-35 as compared with those containing sulfur-32, which is nonradioactive? Explain.

Half-life; dating

19.19 Germanium-66 decays by positron emission, with a half-life of 2.5 h. Write the equation for the nuclear reaction. How much ${}^{66}Ge$ remains from a 25.0-mg sample after 10.0 h?

19.20 The half-life of tritium (hydrogen-3) is 12.3 yr. If 48.0 mg of tritium is released from a nuclear power plant during the course of an accident, what mass of this nuclide remains after 12.3 yr? After 49.2 yr?

19.21 A sample of the synthetic nuclide curium-243 was prepared. After 1 yr the radioactivity of the sample had declined from 3012 disintegrations per second to 2921 disintegrations per second. What is the half-life of the decay process?

19.22 The half-life for decay of ${}^{139}Ba$ is 85 min. What mass of ${}^{139}Ba$ remains after 16 h from a sample that originally contained 24.0 μg of the nuclide?

19.23 The half-life of ${}^{239}Pu$ is 24,000 yr. What fraction of the ${}^{239}Pu$ present in nuclear wastes generated today will be present in the year 3000?

19.24 An experiment was designed to determine whether an aquatic plant absorbed iodide ion from water. Iodine-131 ($t_{1/2} = 8.1$ days) was added as a tracer in the form of iodide ion to a tank containing the plants. The initial activity of a 1.00-μL sample of the water was determined to be 89 counts per minute. After 32 days the level of activity in a 1.00μL sample was determined to be 5.7 counts per minute. Did the plants absorb iodide from the water?

19.25 A sample of strontium-89 has an initial activity of 4600 counts per minute on a device that measures the level of radioactivity. After exactly 30 days the activity has declined to 3130 counts per minute. What is the half-life for decay of strontium-89?

19.26 The half-life for the process ${}^{238}U \longrightarrow {}^{206}Pb$ is 4.5×10^{9} yr. A mineral sample contains 50.0 mg of ${}^{238}U$ and 14.0 mg of ${}^{206}Pb$. What is the age of the mineral?

19.27 A wooden artifact from a Chinese temple has a ${}^{14}C$ activity of 25.8 counts per minute as compared with an activity of 31.7 counts per minute for a standard of zero age. From the half-life for ${}^{14}C$ decay, 5.7×10^{3} yr, determine the age of the artifact.

19.28 Potassium-40 decays to argon-40 with a half-life of 1.27×10^{9} yr. What is the age of a rock in which the weight ratio of ${}^{40}Ar$ to ${}^{40}K$ is 3.6?

[19.29] The synthetic ratioisotope technetium-99, which decays by beta emission, is the most widely used isotope in nuclear medicine. The following data were collected on a sample of ^{99}Tc:

Disintegrations per minute	Time (h)
180	0
130	2.5
104	5.0
77	7.5
59	10.0
46	12.5
24	17.5

Make a graph of these data similar to Figure 19.5 and determine the half-life. (You may wish to make a graph of the log of the disintegration rate versus time; a little rearranging of Equation 19.21 will produce an equation for a linear relation between log N_t and t; from the slope you can obtain k.)

[19.30] A 26.00-g sample of water containing tritium, $^{3}_{1}H$, emits 1.50×10^3 beta particles per second. Tritium is a weak beta emitter, with a half-life of 12.26 yr. What fraction of all the hydrogen in the water sample is tritium? (Hint: Use Equations 19.20 and 19.22.)

Mass-energy relationships

19.31 Calculate the binding energy per nucleon for the following nuclei: (a) $^{12}_{6}C$ (atomic mass, 12.00000 amu); (b) $^{61}_{28}Ni$ (atomic mass, 60.93106 amu); (c) $^{206}_{82}Pb$ (atomic mass, 205.97447 amu).

19.32 Calculate the total binding energy and the binding energy per nucleon for each of the following nuclei: (a) $^{64}_{30}Zn$ (atomic mass, 63.92914); (b) $^{37}_{17}Cl$ (atomic mass, 36.96590 amu); (c) $^{4}_{2}He$ (atomic mass, 4.00260 amu).

19.33 How much energy must be supplied to break a single $^{6}_{3}Li$ nucleus into separated protons and neutrons if the nucleus has a mass of 6.01347 amu? What does this energy correspond to for 1 mol of ^{6}Li nuclei?

19.34 What mass change occurs when 1 kg of crystalline silicon is converted to pure silica, $SiO_2(s)$? ΔH of the reaction is − 910.8 kJ/mol of Si.

19.35 The solar radiation falling on earth amounts to 1.07×10^{16} kJ/min. What is the mass equivalence of the solar energy falling on earth in a 24-h period? If the energy released in the reaction

$$^{235}_{92}U + ^{1}_{0}n \longrightarrow ^{141}_{56}Ba + ^{92}_{36}K + 3^{1}_{0}n$$

(^{235}U atomic mass, 235.0439 amu; ^{141}Ba atomic mass, 140.9140 amu; ^{92}Kr atomic mass, 91.9218 amu)

is taken as typical of that occurring in a nuclear reactor, what mass of uranium-235 is required to equal 0.10 percent of the solar energy that falls on earth in 1 day?

19.36 Based on the following atomic mass values—($^{1}_{1}H$, 1.00782; $^{2}_{1}H$, 2.01410 amu; $^{3}_{1}H$, 3.01605 amu; $^{3}_{2}He$, 3.01603 amu; $^{4}_{2}He$, 4.00260 amu)—and the mass of the neutron given in the text, calculate the energy released in each of the following nuclear reactions, all of which are possibilities for a controlled fusion process:

(a) $^{2}_{1}H + ^{3}_{1}H \longrightarrow ^{4}_{2}He + ^{1}_{0}n$
(b) $^{2}_{1}H + ^{2}_{1}H \longrightarrow ^{3}_{2}H + ^{1}_{0}n$
(c) $^{2}_{1}H + ^{3}_{2}He \longrightarrow ^{4}_{2}He + ^{1}_{1}H$

19.37 It has been proposed that a nuclear reactor be built that derives its energy from fission of calcium-48. Is this a feasible project? Explain.

19.38 Which of the following nuclei is likely to have the largest mass defect per nucleon: (a) $^{59}_{27}Co$; (b) $^{11}_{5}B$; (c) $^{118}_{50}Sn$; (d) $^{243}_{96}Cm$? Explain your answer.

Effects and uses of radioisotopes

19.39 Complete and balance the nuclear equations for the following fission reactions:

(a) $^{235}_{92}U + ^{1}_{0}n \longrightarrow ^{160}_{62}Sm + ^{72}_{30}Zn + \_\_\_\_ ^{1}_{0}n$
(b) $^{239}_{94}Pu + ^{1}_{0}n \longrightarrow ^{144}_{58}Ce + \_\_\_\_ + 2^{1}_{0}n$
(c) $^{233}_{92}U + ^{1}_{0}n \longrightarrow ^{133}_{51}Sb + ^{98}_{41}Nb + \_\_\_\_ ^{1}_{0}n$

19.40 Why is it not possible for the fuel in a nuclear reactor to blow up, as in an atomic bomb, when the reactor goes critical?

19.41 The stresses placed on materials in nuclear reactor cores are very stringent, especially in breeder reactors. In terms of what we have learned in this chapter, what sorts of damage would you expect materials in the reactor core to undergo?

19.42 Why does it require such enormously high temperatures to initiate controlled fusion, whereas no such requirement exists for initiating the nuclear reaction in a fission reactor?

19.43 Describe or define the following: (a) rem; (b) curie; (c) moderator; (d) breeder reactor; (e) critical mass.

19.44 Compare the characteristics of alpha, beta, and gamma rays as they pertain to the health hazards associated with radioactivity.

19.45 Plutonium-239 emits alpha particles that possess energies of about 5×10^8 kJ/mol. The half-life for the decay is about 24,000 yr. It has been said that the main danger from plutonium is inhalation of plutonium-containing dust. Why is such inhalation, rather than simply exposure to plutonium in the environment, the major concern?

19.46 Explain how one might use radioactive ^{59}Fe (a beta emitter with $t_{1/2}$ = 46 days) to determine the extent to which rabbits are able to convert a particular iron compound in their diet into blood hemoglobin, which contains an iron atom.

19.47 Explain how you might use radiotracer techniques to determine the extent to which nitrogen-containing fertilizer applied to farm soil finds its way into water supplies. Look in *Lange's Handbook of Chemistry*, or any other handbook containing a table of nuclear properties, to determine whether there is a nitrogen nuclide that would be suitable as a radiotracer in such a study.

19.48 Chlorine-36 is a convenient radiotracer. It is a weak beta emitter, with $t_{1/2} = 3 \times 10^5$ yr. Describe how you would use this radiotracer to carry out each of the following experiments. (a) Determine whether trichloroacetic acid, CCl_3COOH, undergoes any ionization of its chlorines as chloride ion in aqueous solution. (b) Demonstrate that the equilibrium between dissolved $BaCl_2$ and solid $BaCl_2$ in a saturated solution is a dynamic process. (c) Determine the effects of soil pH on the uptake of chloride ion from the soil by soybeans.

[19.49] Tests on human subjects in Boston in 1965 and 1966, following the era of atomic bomb testing, revealed average quantities of about 2 picocuries of plutonium radioactivity in the average person. How many disintegrations per second does this level of activity imply? If each alpha particle deposits 8×10^{-13} J of energy and if the average person weighs 75 kg, calculate the number of rads of radiation dose in 1 yr from such a level of plutonium, and also calculate the number of rems.

Additional exercises

19.50 Distinguish between the terms in each of the following pairs: (a) electron and positron; (b) binding energy and mass defect; (c) curie and rem; (d) gamma rays and beta rays.

19.51 Figure 19.3 shows the stepwise decay of uranium-238 to form the stable lead-206 nucleus. Write balanced nuclear equations for each step in this sequence.

19.52 Harmful chemicals are often destroyed by chemical treatment. For example, an acid can be neutralized by a base. Why can't chemical treatment be applied to destroy the fission products produced in a nuclear reactor?

19.53 A sample of cobalt-60 was purchased in 1968 for use as a source of beta rays in some biomedical experiments. The half-life for decay of ^{60}Co is 5.25 yr. What fraction of the activity present in the original sample remains in 1985?

[19.54] According to current regulations, the maximum permissible dose of strontium-90 in the body of an adult is 1 microcurie (1×10^{-6} Ci). Using the relationship.

$$\text{Rate} = kN$$

calculate the number of atoms of strontium-90 to which this corresponds. To what mass of strontium-90 does this correspond ($t_{1/2}$ for strontium-90 is 27.6 yr)?

19.55 During the past 30 yr nuclear scientists have synthesized approximately 1600 nuclei not known in nature. Many more might be discovered by using heavy-ion bombardment, which is possible only if high-energy instruments are used to accelerate the ions. Complete and balance the following reactions, which involve heavy-ion bombardments:

(a) $^{6}_{3}Li + ^{63}_{28}Ni \longrightarrow ?$
(b) $^{48}_{20}Ca + ^{248}_{96}Cm \longrightarrow ?$
(c) $^{88}_{38}Sr + ^{84}_{36}Kr \longrightarrow ^{116}_{46}Pd + ?$
(d) $^{48}_{20}Ca + ^{238}_{92}U \longrightarrow ^{70}_{20}Ca + 4^{1}_{0}n + 2?$

19.56 Suppose that the strontium-90 in the soil near the site of a nuclear bomb explosion gives rise to a radioactivity level of 7000 counts per second. How long will it be before the level drops to 1000 counts per second? ($t_{1/2}$ for strontium-90 is 28.8 yr.)

19.57 A portion of the sun's energy comes from the reaction

$$4^{1}_{1}H \longrightarrow ^{4}_{2}He + 2^{0}_{+1}e$$

Calculate the energy change in this reaction (see Problem 19.36 and the text for needed masses). This reaction requires a temperature of about 10^6 to 10^7 K. Why is such a high temperature required?

19.58 The 13 known nuclides of zinc range from $^{60}_{30}Zn$ to $^{72}_{30}Zn$. The naturally occurring nuclides have mass numbers 64, 66, 67, 68, and 70. What mode or modes of decay would you expect for the least massive radioactive nuclides of zinc? What mode for the most massive nuclides?

19.59 The sun radiates energy into space at the rate of 3.9×10^{26} J/s. Calculate the rate of mass loss from the sun.

19.60 An ancient wooden object is found to have an activity of 9.6 disintegrations per minute per gram of carbon. By contrast, the carbon in a living tree undergoes 18.4 disintegrations per minute per gram of carbon. Based on the activity of carbon-14 in the object, calculate its age.

19.61 It has been suggested that strontium-90 deposited in the hot desert from nuclear testing will undergo radioactive decay more rapidly because it will be exposed to much higher average temperatures. Is this a reasonable suggestion?

[19.62] A radioactive decay series that begins with $^{237}_{93}Np$ ends with formation of the stable nuclide $^{209}_{83}Bi$. How many alpha particle emissions and how many beta particle emissions are involved in the sequence of radioactive decays?

[19.63] The half-life for decay of $^{230}_{90}Th$ is 8.0×10^4 yr. Because one cannot take the time to collect data for a graph such as that in Figure 19.5, how might one determine the half-life for such a long-lived isotope?

19.64 Rutherford was able to carry out the first nuclear transmutation reactions by bombarding nitrogen-14 nuclei with alpha particles. However, in the famous experiment on scattering of alpha particles by gold foil (Section 2.6), a nuclear transmutation reaction did not occur. What is the difference in the two experiments? What would one need to do to carry out a successful nuclear transmutation reaction involving gold nuclei and alpha particles?

19.65 Determine the product nucleus in the following cases: (a) $^{75}_{33}As(\alpha,n)$______; (b) $^{7}_{3}Li(p,n)$______; (c) $^{31}_{15}P(^{2}_{1}H,p)$______.

[19.66] Fusion energy is generally advanced as potentially a cleaner form of nuclear power than fission. It is anticipated, however, that the materials used to construct the inner components of a fusion reactor will become intensely radioactive and will need frequent replacement. What is the origin of this problem?

[19.67] Suppose you had a detection device that could count every decay from a radioactive sample of plutonium-239 ($t_{1/2}$ is 24,000 yr). How many counts per second would you obtain from a sample that contained 0.500 g of plutonium-239? (Hint: Look at Equations 19.20 and 19.22.)

[19.68] When a positron is annihilated by combination with an electron, two photons of equal energy result. What is the wavelength of these photons? Are they gamma-ray photons?

20 Chemistry of Hydrogen, Oxygen, Nitrogen, and Carbon

For the most part, the previous chapters of this book have involved chemical principles, such as rules for bonding, the laws of thermodynamics, the factors influencing reaction rates, and so forth. In the course of explaining these principles, we have described the chemical and physical properties of many substances. However, we have not systematically examined the chemical elements and the compounds they form. This aspect of chemistry, often referred to as **descriptive chemistry**, is the subject of the next several chapters.

In this chapter we examine four important nonmetals: hydrogen, oxygen, nitrogen, and carbon. These nonmetals form many commercially important compounds and are the primary elements in biological systems. In Chapter 21 we will examine the remaining nonmetals on a group-by-group basis.

In studying descriptive chemistry, it is important to look for trends and general types of behavior, rather than trying to memorize all the facts presented. The periodic table is, of course, an invaluable tool in this task. Before we begin our examination of particular nonmetals, it is useful to review briefly some general periodic trends.

20.1 PERIODIC TRENDS

Figure 20.1 summarizes the way in which several important properties of elements vary in relation to the periodic chart. One of the most useful features shown is the division of elements into the broad categories of metals and nonmetals. The nonmetals, of course, are located in the upper right-hand corner of the chart. Some of the distinguishing properties of metals and nonmetals are listed in Table 20.1. Metals in the bulk form have a characteristic luster. All except mercury are solids at room temperature. Strongly metallic elements also exhibit good electrical and thermal conductivity and are malleable (can be flattened) and ductile (can be drawn into wire). By contrast, nonmetallic elements are not lustrous and are generally poor conductors of heat and electricity. Seven

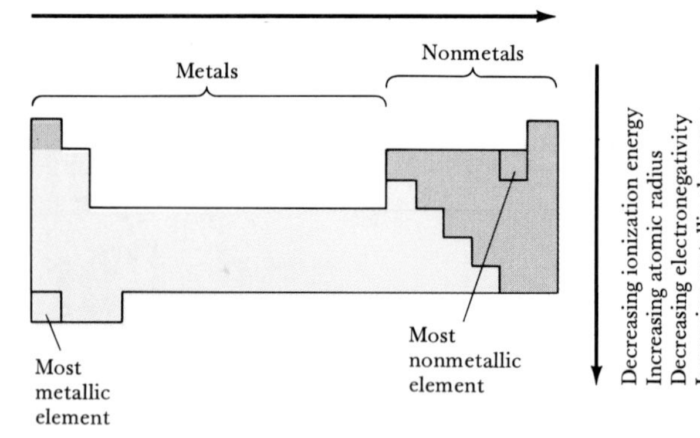

FIGURE 20.1 Trends in key properties of the elements as a function of position in the periodic table.

of the nonmetals exist under ordinary conditions as diatomic molecules. Included in this list are five gases (hydrogen, nitrogen, oxygen, fluorine, and chlorine), one liquid (bromine), and one volatile solid (iodine). The remaining nonmetals are solids that may be hard like diamond or soft like sulfur. These variations in properties are accounted for in terms of the bonding in the element, as discussed in Section 8.7.

Elements that border on the division between metals and nonmetals may have *some* characteristic metallic properties and not others. For example, antimony *looks* like a metal, but it is brittle rather than malleable and is a poor conductor of heat and electricity. Elements such as antimony that exhibit some properties of metals and some of nonmetals are called **semimetals** or **metalloids.** These elements are shown in the colored boxes in Figure 20.2.

Another important trend shown in Figure 20.1 involves electronegativity. Recall that electronegativity decreases as we move down a given family or group and increases from left to right across the table. As a result, nonmetals have higher electronegativities than do metals. Consequently, compounds formed between strongly metallic and strongly nonmetallic elements tend to be ionic (for example, metal fluorides and metal oxides). These substances are solids at room temperature. In contrast, compounds formed between nonmetals are molecular substances.

TABLE 20.1 Characteristic properties of metallic and nonmetallic elements

Metallic elements	Nonmetallic elements
Distinguishing luster	Nonlustrous; various colors
Malleable and ductile as solids	Solids are usually brittle, may be hard or soft
Good thermal and electrical conductivity	Poor conductors of heat and electricity
Most metallic oxides are ionic solids; dissolve in water to form basic solutions	Most nonmetallic oxides are covalent compounds; dissolve in water to form acidic solutions
Exist in aqueous solution mainly as cations	Exist in aqueous solution mainly as anions or oxyanions

1A	2A	3B	4B	5B	6B	7B	8B			1B	2B	3A	4A	5A	6A	7A	8A
H																	He
Li	Be											B	C	N	O	F	Ne
Na	Mg											Al	Si	P	S	Cl	Ar
K	Ca	Sc	Ti	V	Cr	Mn	Fe	Co	Ni	Cu	Zn	Ga	Ge	As	Se	Br	Kr
Rb	Sr	Y	Zr	Nb	Mo	Tc	Ru	Rh	Pd	Ag	Cd	In	Sn	Sb	Te	I	Xe
Cs	Ba	La	Hf	Ta	W	Re	Os	Ir	Pt	Au	Hg	Tl	Pb	Bi	Po	At	Rn

FIGURE 20.2 Portion of the periodic table. Colored boxes indicate the semimetals.

Molecular substances with low molecular weight tend to be gases, liquids, or volatile solids at room temperature.

Among the nonmetals, the chemistry of the first member of each family often differs in several important ways from that of subsequent members. The differences are due in part to the smaller size and greater electronegativity of the first member. In addition, the first member is restricted to forming a maximum of four bonds because it has only the 2*s* and the three 2*p* orbitals available for bonding. Subsequent members of the family are able to use *d* orbitals in bonding in addition to *s* and *p* orbitals; they can therefore form more than four bonds. As an example, consider the chlorides of nitrogen and phosphorus, the first two members of group 5A. Nitrogen forms a maximum of three bonds with chlorine, NCl_3. Although phosphorus can form the trichloride compound, PCl_3, it is also able to form five bonds with chlorine, PCl_5. These compounds are shown in Figure 20.3. Because hydrogen has only the 1*s* orbital available for bonding, it is restricted to forming one bond.

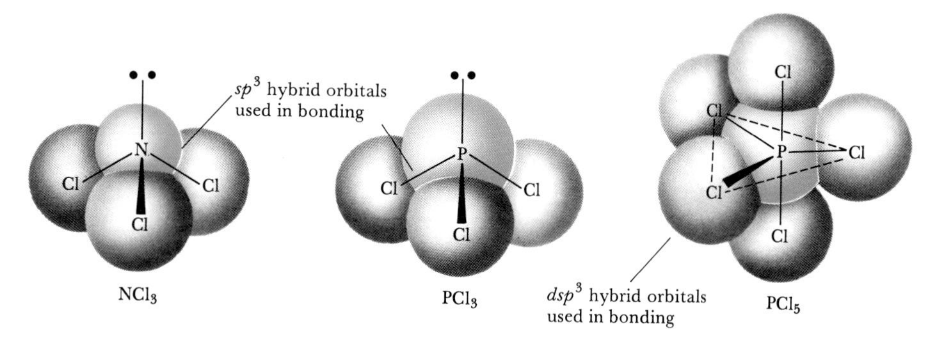

FIGURE 20.3 Comparison of NCl_3, PCl_3, and PCl_5. Nitrogen is unable to form NCl_5 because it does not have *d* orbitals available for binding.

SAMPLE EXERCISE 20.1

Sulfur forms a fluoride compound, SF_6, that has six sulfur-fluorine bonds. In contrast, oxygen is able to form at most two bonds to fluorine, OF_2. (a) Rationalize this difference. (b) Predict the molecular geometry of each compound and describe the hybridization employed by S and O.

Solution: (a) The electron configuration of oxygen is $[He]2s^22p^4$. The next-highest-energy orbitals available for bonding are in the third shell. Thus oxygen uses only the 2*s* and 2*p* orbitals in bonding. These orbitals can accommodate only eight electrons. In OF_2, oxygen already has eight electrons, six of its own and one from each fluorine atom:

F—Ö—F (with a lone pair above and below the O)

Consequently, it is unable to accommodate additional fluorine atoms.

The electron configuration of sulfur is $[Ar]3s^23p^4$. In this case the 3*d* orbitals are available for bonding. Thus sulfur is able to expand its octet to accommodate more than eight electrons:

SF_6 structure: S bonded to six F atoms (F—S—F)

(b) According to the VSEPR theory (Section 8.1), the four electron pairs around oxygen are disposed at the corners of a tetrahedron. The four electron pairs are accommodated in sp^3 hybrid orbitals. The geometric arrangement of atoms associated with two bonded pairs and two nonbonded pairs is nonlinear.

In the case of SF_6, the six electron pairs around sulfur are disposed at the corners of an octahedron. The associated hybridization is d^2sp^3. Because all six electrons are bonded pairs, the geometric arrangement of atoms is also octahedral.

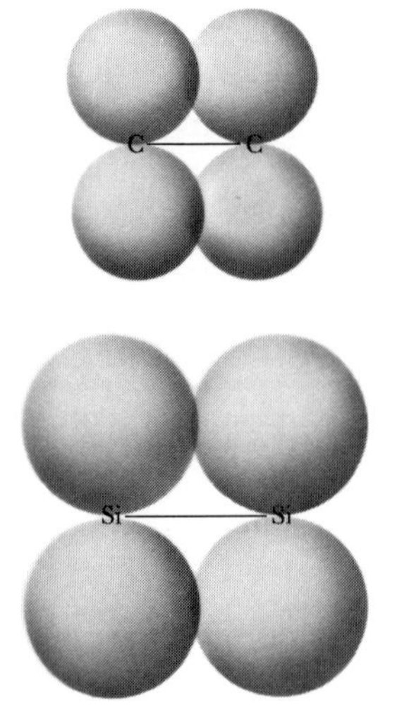

FIGURE 20.4 Comparison of π-bond formation by sideways overlap of *p* orbitals between two carbon atoms and between two silicon atoms. The distance between nuclei increases as we move from carbon to silicon because of the larger size of the silicon atom. The *p* orbitals do not overlap as effectively between two silicon atoms because of this greater separation.

Another difference between the first member of any family and subsequent members of the same family is the greater ability of the former to form π bonds. We can understand this, in part, in terms of atomic size. As atoms increase in size, the sideways overlap of *p* orbitals, which form the strongest type of π bond, becomes less effective. This is shown in Figure 20.4. As an illustration of this effect, consider two differences in the chemistry of carbon and silicon, the first two members of group 4A. Carbon has two crystalline allotropes, diamond and graphite. In diamond there are σ bonds between carbon atoms but no π bonds. In graphite, π bonds result from sideways overlap of *p* orbitals (Section 8.7). Silicon occurs only in the diamondlike crystal form. Silicon does not exhibit a graphitelike structure because of the low stability of π bonds between silicon atoms.

We see the same type of difference in the dioxides of these elements (Figure 20.5). In CO_2, carbon forms double bonds to oxygen, thereby achieving its valence by π bonding. In contrast, SiO_2 contains no double

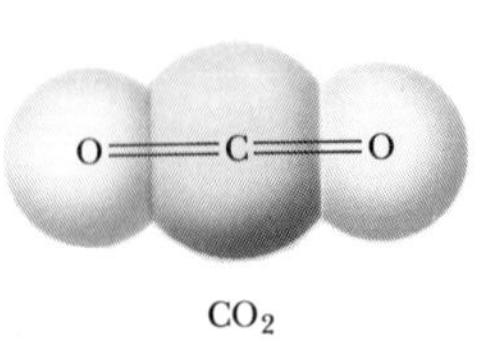

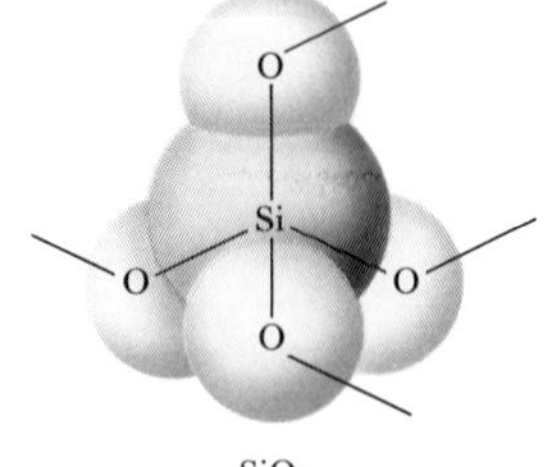

FIGURE 20.5 Comparison of the structures of CO_2 and SiO_2; CO_2 has double bonds, whereas SiO_2 has only single bonds.

bonds. Instead, four oxygens are bonded to each silicon, forming an extended structure reminiscent of diamond.*

20.2 CHEMICAL REACTIONS

Throughout this chapter and subsequent ones, you will find a large number of chemical reactions presented. To remember all of them is an overwhelming task. However, you will probably need to remember some, especially those of a more general nature. In earlier discussions we have encountered several general categories of reactions: combustion reactions (Section 3.3), metathesis reactions (Sections 3.3 and 11.6), Brønsted acid-base (proton-transfer) reactions (Section 15.2), Lewis acid-base reactions (Section 15.10), and redox reactions (Section 18.1). In this chapter about a third of the reactions discussed involve oxidation by O_2 (combustion) or are proton-transfer reactions.

In oxidation reactions involving O_2, hydrogen-containing compounds produce H_2O. Carbon-containing ones produce CO_2 (unless O_2 is in short supply, in which case CO or even C can form). Nitrogen-containing compounds tend to form N_2, although NO can form in special cases. The following reactions are illustrative of these generalizations:

$$2CH_3OH(l) + 3O_2(g) \longrightarrow 2CO_2(g) + 4H_2O(l) \quad [20.1]$$

$$4CH_3NH_2(g) + 9O_2(g) \longrightarrow 4CO_2(g) + 10H_2O(l) + 2N_2(g) \quad [20.2]$$

The formation of H_2O, CO_2, and N_2 reflects the high thermodynamic stabilities of these compounds, which are indicated by the large bond energies for the O—H, C=O, and N≡N bonds that they contain (463, 799, and 941 kJ/mol, respectively). The formation of the stable H_2O and CO_2 molecules also occurs when H_2 and C are used to reduce metal oxides:

$$NiO(s) + H_2(g) \longrightarrow Ni(s) + H_2O(l) \quad [20.3]$$

$$2CuO(s) + C(s) \longrightarrow 2Cu(s) + CO_2(g) \quad [20.4]$$

In dealing with proton-transfer reactions, you should remember that the weaker a Brønsted acid, the stronger its conjugate base. For example, H_2, OH^-, NH_3, CH_4, and C_2H_2, which we encounter in this chapter, are exceedingly weak proton donors. In fact, they have *no* tendency to act as acids in water. Thus species formed from them by removing one or more protons (such as H^-, O^{2-}, NH_2^-, N^{3-}, CH_3^-, C^{4-}, and C_2^{2-}) are extremely strong bases. All react readily with water, removing protons from the H_2O to form OH^-. The following reactions are illustrative:

$$CH_3^-(aq) + H_2O(l) \longrightarrow CH_4(g) + OH^-(aq) \quad [20.5]$$

$$N^{3-}(aq) + 3H_2O(l) \longrightarrow NH_3(aq) + 3OH^-(aq) \quad [20.6]$$

*The formula SiO_2 is consistent with this structure, because each oxygen is shared by two silicon atoms (not shown in Figure 20.5). For bookkeeping purposes, we may therefore count a half of each of the four oxygens that are bound to a given silicon atom as belonging to that silicon. We shall consider silicon-oxygen compounds in some detail in Chapter 21.

Of course, substances that are stronger proton donors than H_2O, such as HCl, H_2SO_4, $HC_2H_3O_2$, and other acids, also react readily with basic anions.

SAMPLE EXERCISE 20.2

Predict the products formed in each of the following reactions and write a balanced chemical equation:

(a) $CH_3NHNH_2(g) + O_2(g) \longrightarrow$

(b) $Mg_3P_2(s) + H_2O(l) \longrightarrow$

(c) $NaCN(s) + HCl(aq) \longrightarrow$

Solution: (a) This combustion reaction should produce CO_2, H_2O, and N_2:

$$2CH_3NHNH_2(g) + 5O_2(g) \longrightarrow 2CO_2(g) + 6H_2O(l) + 2N_2(g)$$

(b) The Mg_3P_2 consists of Mg^{2+} and P^{3-} ions. The P^{3-} ion, like N^{3-}, has a strong affinity for protons and thus reacts with H_2O to form OH^- and PH_3 (PH^{2-}, PH_2^-, and PH_3 are all exceedingly weak proton donors):

$$Mg_3P_2(s) + 6H_2O(l) \longrightarrow 2PH_3(g) + 3Mg(OH)_2(aq)$$

The $Mg(OH)_2$ is of low solubility in water and may precipitate.

(c) The NaCN consists of Na^+ and CN^- ions. The CN^- ion is basic (HCN is a weak acid). Thus it reacts with protons to form the conjugate acid:

$$NaCN(s) + HCl(aq) \longrightarrow HCN(aq) + NaCl(aq)$$

or

$$NaCN(s) + H^+(aq) \longrightarrow HCN(aq) + Na^+(aq)$$

The HCN has limited solubility in water and readily escapes as a gas.

20.3 HYDROGEN

Formation of elemental hydrogen was first recorded in the sixteenth century by the alchemist Paracelsus (1493–1541), who observed the formation of an "air" (gas) produced by the action of acids on iron. However, it was the English chemist Henry Cavendish (1731–1810) who first isolated pure hydrogen and distinguished it from other gases. The element was subsequently named by the French chemist Lavoisier and means "water producer" after the fact that water is formed when the gas burns in air (Greek: *hydor,* water; *gennao,* to produce).

Hydrogen is the most abundant element in the universe. It is the nuclear fuel consumed by our sun and other stars to produce energy (Section 19.8). Although about 70 percent of the universe is composed of hydrogen, it accounts for only 0.87 percent of the earth's mass. Presumably, considerable hydrogen escaped from the earth during its early history. Unlike larger planets such as Saturn and Jupiter, the earth has a gravitational field too weak to hold the light molecules of the gaseous element. What remains is most commonly found in combination with oxygen. Water, which is 11 percent hydrogen by weight, is the most abundant hydrogen compound. Because water covers about 70 percent of the earth's surface, hydrogen is readily available. Hydrogen is also an important part of petroleum, cellulose, starch, fats, alcohols, acids, and a wide variety of other materials.

Isotopes of Hydrogen

The most common isotope of hydrogen, ${}^1_1H$, has a nucleus consisting of a single proton. This isotope, sometimes referred to as **protium,** comprises 99.9844 percent of naturally occurring hydrogen.

Two other isotopes are known: ${}^{2}_{1}H$, whose nucleus contains a proton and a neutron, and ${}^{3}_{1}H$, whose nucleus contains a proton and two neutrons. The ${}^{2}_{1}H$ isotope is referred to as **deuterium.** Deuterium comprises 0.0156 percent of naturally occurring hydrogen. It is not radioactive. In writing the chemical formulas of compounds containing deuterium, that isotope is often given the symbol D, as in D_2O. D_2O, which is called deuterium oxide or **heavy water,** can be obtained by electrolysis of ordinary water. The heavier D_2O undergoes electrolysis at a slower rate than does the lighter H_2O and is thus concentrated during the electrolysis. Typically, electrolysis of 2400 L of water will produce 83 mL of 99 percent D_2O. D_2O, which is presently available in ton quantities, is used as a moderator and coolant in certain nuclear reactors. Some of the physical properties of H_2O and D_2O are compared in Table 20.2. Notice the small but discernible difference in properties.

The third isotope, ${}^{3}_{1}H$, is known as **tritium** and is often given the symbol T. It is radioactive, with a half-life of 12.3 yr:

$$ {}^{3}_{1}H \longrightarrow {}^{3}_{2}He + {}^{0}_{-1}e \qquad t_{1/2} = 12.3 \text{ yr} \qquad [20.7] $$

It is formed continuously in the upper atmosphere in nuclear reactions induced by cosmic rays; however, because of its short half-life, only trace quantities exist naturally. The isotope can be synthesized in nuclear reactors. The preferred method of production involves neutron bombardment of lithium-6:

$$ {}^{6}_{3}Li + {}^{1}_{0}n \longrightarrow {}^{3}_{1}H + {}^{4}_{2}He \qquad [20.8] $$

Each isotope of hydrogen contains a single electron and consequently undergoes identical chemical reactions. However, the heavier deuterium and tritium generally undergo reactions at a somewhat slower rate than protium. This rate effect is greater for the isotopes of hydrogen than for other elements because these isotopes show the greatest relative mass differences.

Deuterium and tritium have proved valuable in studying the reactions of compounds containing hydrogen. A compound can be "labeled" by replacing one or more ordinary hydrogen atoms at specific locations within a molecule with deuterium or tritium. By comparing the location of the heavy hydrogen label in the reactants with that in the products, the mechanism of the reaction can often be inferred. As an example of the chemical insight that can be gained by using deuterium, consider

TABLE 20.2 Comparison of properties of H_2O and D_2O

Property	H_2O	D_2O
Melting point (°C)	0.00	3.81
Boiling point (°C)	100.00	101.42
Density at 25°C (g/mL)	0.997	1.104
Heat of fusion (kJ/mol) at m.p.	6.008	6.276
Heat of vaporization (kJ/mol) at b.p.	40.67	41.61
Ion product, (K_w) at 25°C	1.01×10^{-14}	1.95×10^{-15}

what happens when methyl alcohol, CH_3OH, is placed in D_2O. The H atom of the O—H bond exchanges rapidly with the D atoms in D_2O, forming CH_3OD. The H atoms of the CH_3 group do not exchange. This experiment demonstrates the kinetic stability of C—H bonds and reveals the speed at which the O—H bond in the molecule breaks and reforms.

Properties of Hydrogen

Hydrogen is the only element that is not a member of any family in the periodic table. Because of its $1s^1$ electron configuration, it is generally placed above lithium in the periodic table. However, it is definitely not an alkali metal. It forms a positive ion much less readily than any alkali metal; the ionization energy of the hydrogen atom is 1310 kJ/mol, whereas that of lithium is 517 kJ/mol. Furthermore, the hydrogen atom has no electrons below its valence shell; the H^+ ion is just a bare proton. The simple H^+ ion is not known to exist in any compound. Its small size gives it a strong attraction for electrons; it either strips electrons from surrounding matter (forming hydrogen atoms that combine to form H_2), or it shares electron pairs, forming covalent bonds with other atoms. While we may represent the aquated hydrogen ion as $H^+(aq)$, that proton is bonded to one or more water molecules and is thus often represented as $H_3O^+(aq)$ (Section 15.1).

Hydrogen is also sometimes placed above the halogens in the periodic table because the hydrogen atom can also pick up one electron to form the **hydride ion,** H^-. However, the electron affinity of hydrogen, $\Delta H_{EA} = -71$ kJ/mol, is not as large as that of any halogen; the electron affinity of fluorine is -332 kJ/mol, whereas that of iodine is -295 kJ/mol. In general, hydrogen shows no closer resemblance to the halogens than it does to the alkali metals.

In its elemental form, hydrogen exists at room temperature as a colorless, odorless, tasteless gas composed of diatomic molecules, H_2. We can call H_2 dihydrogen, but it is more commonly referred to as molecular hydrogen or merely hydrogen. Because H_2 is nonpolar and has only two electrons, attractive forces between molecules are extremely weak. Consequently, the melting point (-259°C) and boiling point (-253°C) of H_2 are very low.

The H—H bond-dissociation energy (436 kJ/mol) is high for a single bond (see Table 7.3). By comparison, the Cl—Cl bond-dissociation energy is only 242 kJ/mol. As a result of its strong bond, most reactions of H_2 are slow at room temperature. However, the molecule is readily activated by heating, irradiation, or catalysis. The activation process generally produces hydrogen atoms, which are very reactive. The activation of H_2 by finely divided nickel, palladium, and platinum was considered briefly in our earlier discussions of heterogeneous catalysis (Section 13.6). Once H_2 is activated, it reacts rapidly and exothermally with a wide variety of substances.

Hydrogen forms strong covalent bonds with many elements, including oxygen; the H—O bond-dissociation energy is 464 kJ/mol. The strong H—O bond makes hydrogen an effective reducing agent for many metal oxides. For example, when H_2 is passed over heated CuO, copper is produced:

FIGURE 20.6 Burning of the airship *Hindenburg* while landing at Lakehurst, New Jersey, on May 6, 1937. This picture was taken only 22 seconds after the first explosion occurred. (*UPI*)

$$CuO(s) + H_2(g) \longrightarrow Cu(s) + H_2O(g) \quad [20.9]$$

When H_2 is ignited in air, a vigorous reaction occurs, forming H_2O:

$$2H_2(g) + O_2(g) \longrightarrow 2H_2O(l) \qquad \Delta H = -571.7 \text{ kJ} \quad [20.10]$$

Air containing as little as 4 percent H_2 (by volume) is potentially explosive. The disastrous burning of the hydrogen-filled airship the *Hindenberg* in 1937 (Figure 20.6) dramatically demonstrated the high flammability of H_2.

Preparation of Hydrogen

When a small quantity of H_2 is needed in the laboratory, it is usually obtained by the reaction between an active metal such as zinc and a dilute acid such as HCl or H_2SO_4:

$$Zn(s) + 2H^+(aq) \longrightarrow Zn^{2+}(aq) + H_2(g) \quad [20.11]$$

Because H_2 has an extremely low solubility in water, it can be collected by displacement of water, as shown in Figure 20.7.

When commercial quantities of H_2 are needed, the raw materials are usually hydrocarbons (from either natural gas or petroleum) or water.

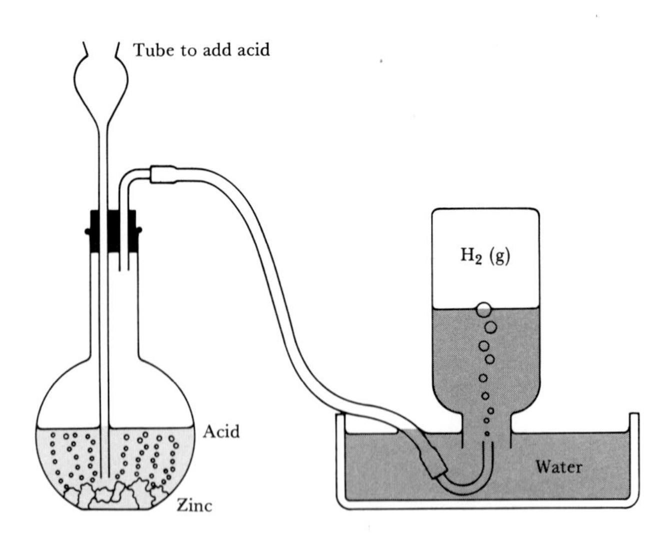

FIGURE 20.7 Apparatus commonly used in the laboratory for preparation of hydrogen.

Hydrocarbons are substances like CH_4 and C_8H_{18} that consist of carbon and hydrogen. Much hydrogen is presently obtained in the course of refining petroleum. In the refining process, large hydrocarbons are catalytically broken into smaller molecules with the accompanying production of H_2 as a by-product.

Hydrogen is also produced by the reaction of methane, CH_4, the principal component of natural gas, with steam at 1100°C:

$$CH_4(g) + H_2O(g) \longrightarrow CO(g) + 3H_2(g) \quad [20.12]$$

$$CO(g) + H_2O(g) \longrightarrow CO_2(g) + H_2(g) \quad [20.13]$$

When heated to about 1000°C, carbon also reacts with steam to produce a mixture of H_2 and CO gases:

$$C(s) + H_2O(g) \longrightarrow H_2(g) + CO(g) \quad [20.14]$$

Because both CO and H_2 burn in air to produce heat, this mixture, known as **water gas**, is used as an industrial fuel.

Simple electrolysis of water consumes too much energy and is consequently too costly a process to be used commercially to produce H_2. However, hydrogen is produced as a by-product in the electrolysis of brine (NaCl) solutions in the course of Cl_2 and NaOH manufacture:

$$2NaCl(aq) + 2H_2O(l) \xrightarrow{\text{electrolysis}} H_2(g) + Cl_2(g) + 2NaOH(aq) \quad [20.15]$$

Uses of Hydrogen

Hydrogen is a commercially important substance; about 2×10^8 kg (200,000 tons) is produced annually in the United States. Over two-thirds of the annual production is consumed in the synthesis of ammonia by the Haber process (Section 14.1). Hydrogen is also used to manufacture methanol, CH_3OH. As shown in Equation 20.16, the synthesis involves the catalytic combination of carbon monoxide and hydrogen under high pressures and temperatures:

$$CO(g) + 2H_2(g) \xrightarrow[\substack{200\text{–}300 \text{ atm} \\ \text{catalyst}}]{300\text{–}400^\circ\text{C}} CH_3OH(l) \qquad [20.16]$$

Hydrogenation of vegetable oils in the manufacture of margarine and vegetable shortening is another important use. In this process H_2 is added to carbon-carbon double bonds in the oil. A simple example of a hydrogenation occurs in the conversion of ethylene to ethane:

$$\underset{\text{Ethylene}}{H_2C{=}CH_2}(g) + H_2(g) \xrightarrow{\text{catalyst}} \underset{\text{Ethane}}{H{-}CH_2{-}CH_2{-}H}\,(g) \qquad [20.17]$$

Because organic compounds with double bonds have the ability to add additional hydrogen atoms, they are said to be unsaturated. The term "polyunsaturated," which often appears in food advertisements, refers to molecules that have several (poly) double bonds between carbon atoms. The following molecule is a polyunsaturated hydrocarbon:

$$H_3C{-}CH{=}CH{-}CH_2{-}CH_2{-}CH{=}CH_2$$

The polyunsaturated molecules that occur in vegetable oils are much more complex (Section 26.5).

Binary Hydrogen Compounds

Hydrogen reacts with other elements to form compounds of three general types: (1) ionic hydrides, (2) metallic hydrides, and (3) molecular hydrides.

The ionic hydrides are formed by the alkali metals and by the heavier alkaline earths (Ca, Sr, and Ba). These active metals are much less electronegative than hydrogen. Consequently, hydrogen acquires electrons from them to form hydride ions, H^-, as shown in Equations 20.18 and 20.19:

$$2Li(s) + H_2(g) \longrightarrow 2LiH(s) \qquad [20.18]$$

$$Ca(s) + H_2(g) \longrightarrow CaH_2(s) \qquad [20.19]$$

The resultant ionic hydrides are solids with high melting points (LiH melts at 680°C).

The hydride ion is very basic and reacts readily with compounds having even weakly acidic protons to form H_2. For example, H^- reacts readily with H_2O:

$$H^-(aq) + H_2O(l) \longrightarrow H_2(g) + OH^-(aq) \qquad [20.20]$$

Thus ionic hydrides can be used as convenient (although expensive) sources of H_2. Calcium hydride, CaH_2, is sold in commercial quantities and used for inflation of life rafts, weather balloons, and other like uses where a simple, compact means of H_2 generation is desired. CaH_2 is also used to remove H_2O from organic liquids.

The reaction between H^- and H_2O (Equation 20.20) is not only an acid-base reaction, but a redox reaction as well. The H^- ion can be viewed not only as a good base, but also as a good reducing agent. In fact, hydrides are able to reduce O_2 to H_2O:

$$2NaH(s) + O_2(g) \longrightarrow Na_2O(s) + H_2O(l) \qquad [20.21]$$

Thus hydrides are normally stored in an environment that is free of both moisture and air.

Metallic hydrides are formed when hydrogen reacts with transition metals. These compounds are so named because they retain their metallic conductivity and other metallic properties. In many metallic hydrides the ratio of metal atoms to hydrogen atoms is not a ratio of small whole numbers, nor is it a fixed ratio. The composition can vary within a range, depending on the conditions of synthesis. For example, although TiH_2 can be prepared, usual preparations yield substances with about 10 percent less hydrogen than this, $TiH_{1.8}$. These nonstoichiometric metallic hydrides are sometimes called **interstitial hydrides.** They may be considered to be solutions of hydrogen atoms in the metal, with the hydrogen atoms occupying the holes or interstices between metal atoms in the solid lattice. However, this description is an oversimplification; there is evidence for chemical interaction between metal and hydrogen.

The ready absorption of H_2 by palladium metal has been used to separate H_2 from other gases and in purifying H_2 on an industrial scale. At 300 to 400 K H_2 dissociates into atomic hydrogen on the Pd surface. The H atoms dissolve in the Pd, and under H_2 pressure they diffuse through, recombining to form H_2 on the opposite surface. Because no other molecules exhibit this property, absolutely pure H_2 results.

Research is presently in progress investigating metallic hydrides as storage media for hydrogen. In the case of many metals, hydride formation occurs directly upon contact between the metal and H_2. Furthermore, the reaction is often readily reversible so that H_2 can be obtained from the hydride by reducing the pressure of the hydrogen gas above the metal. Metals accommodate an extremely high density of hydrogen because the hydrogen atoms are closely packed into the interstitial sites between metal atoms. Indeed, the number of hydrogen atoms per unit volume is greater in some metallic hydrides than in liquid H_2. In some hydrides the metal can accommodate two or three times as many hydrogen atoms as there are metal atoms; that is, the stoichiometry approaches MH_3 (where M is the metal).

The absorption of hydrogen by metals also has detrimental effects. The hydrides tend to be more brittle than the metal. Thus absorption of hydrogen can weaken and embrittle steel and other structural metals (Figure 20.8).

The **molecular hydrides**, formed by nonmetals and semimetals, are either gases or liquids under standard conditions. The simple molecular hydrides are listed in Figure 20.9 together with their standard free energies of formation, ΔG_f°. In each family, the thermal stability (measured by ΔG_f°) decreases as we move down the family. (Recall that the more stable a compound with respect to its elements under standard conditions, the more negative ΔG_f° is.) We will discuss the molecular hydrides further in the course of examining the other nonmetallic elements.

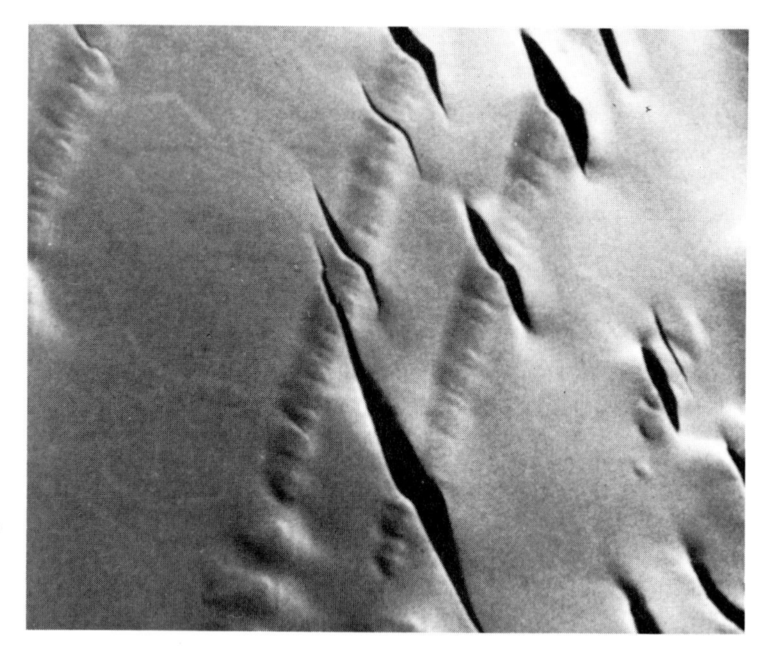

FIGURE 20.8 Cracking in niobium metal due to hydride formation when the metal is stressed under a 1 percent H_2 atmosphere. The raised lines running nearly vertically are due to formation of NbH at certain planes within the metal. The cracks begin in these regions because NbH is very brittle. (*Photo courtesy of Professor Howard Birnbaum, University of Illinois; electron microscope photo, about 200× magnification. Reprinted with permission from* Acta Metallurgica, *vol. 25, M. L. Grossbeck and H. K. Birnbaum, "Low Temperature Hydrogen Embrittlement of Niobium II—Microscopic Observations," copyright 1977, Pergamon Press Ltd.*)

4A	5A	6A	7A
$CH_4(g)$ −50.8	$NH_3(g)$ −16.7	$H_2O(l)$ −237	$HF(g)$ −271
$SiH_4(g)$ +56.9	$PH_3(g)$ +18.2	$H_2S(g)$ −33.0	$HCl(g)$ −95.3
$GeH_4(g)$ +117	$AsH_3(g)$ +111	$H_2Se(g)$ +71	$HBr(g)$ −53.2
	$SbH_3(g)$ +187	$H_2Te(g)$ +138	$HI(g)$ +1.30

FIGURE 20.9 Standard free energies of formation (kJ/mol) of molecular hydrides.

20.4 OXYGEN

By the middle of the seventeenth century it was recognized that air contained a component associated with burning and breathing. It wasn't until 1774, however, that oxygen was discovered by Joseph Priestley (Figure 20.10). The name "oxygen" was subsequently given by Lavoisier.

Oxygen plays an important role in the chemistries of most other elements and is found in combination with other elements in a great variety of compounds. Indeed, oxygen is the most abundant element both in the earth's crust and in the human body. It constitutes 89 percent of water by mass and 20.9 percent of air by volume (23 percent by mass). It also comprises 50 percent by mass of the sand, clay, limestone, and igneous rocks that make up the bulk of the earth's crust.

Properties of Oxygen

Oxygen has two allotropes, O_2 and O_3. When we speak of elemental or molecular oxygen, it is usually understood that we are speaking of O_2, the normal form of the element; O_3 is called **ozone.**

Molecular oxygen, O_2, exists at room temperature as a colorless, odorless, and tasteless gas. It melts at $-218°C$ and has a normal boiling point of $-183°C$. It is only slightly soluble in water, but its presence in water is essential to marine life.

The electron configuration of the oxygen atom is $[He]2s^22p^4$. Thus oxygen can complete its octet of electrons by either picking up two electrons to form the oxide ion, O^{2-}, or by sharing two electrons. In its covalent compounds it tends to form two bonds, either two single bonds as in H_2O or a double bond as in formaldehyde, $H_2C{=}O$. The O_2 molecule itself contains a double bond.

The bond in O_2 is very strong (the bond-dissociation energy is 495 kJ/mol). Oxygen also forms strong bonds with many other elements. Consequently, many oxygen-containing compounds are thermodynamically stable compared to O_2. However, in the absence of a catalyst most reactions of O_2 have high activation energies and thus require high temperatures to proceed at a suitable rate. Once a sufficiently exothermic reaction begins, it may accelerate rapidly, producing a reaction of explosive violence.

Preparation of Oxygen

Oxygen can be obtained either from air or from certain oxygen-containing compounds. Nearly all commercial oxygen is obtained by fractional distillation of liquefied air. The normal boiling point of O_2 is $-183°C$, whereas that of N_2, the other principal component of air, is $-196°C$. Thus when liquefied air is warmed, the N_2 boils off, leaving liquid O_2 contaminated mainly with small amounts of N_2 and argon.

The common laboratory preparation of O_2 involves the thermal decomposition of potassium chlorate, $KClO_3$, with manganese dioxide, MnO_2, added as a catalyst:

$$2KClO_3(s) \longrightarrow 2KCl(s) + 3O_2(g) \qquad [20.22]$$

Like H_2, O_2 can be collected by displacement of water because of its relatively low solubility.

Uses of Oxygen

Oxygen is one of the most widely used industrial chemicals. In 1982, it ranked behind only sulfuric acid, H_2SO_4, nitrogen, N_2, and ammonia, NH_3. About 1.4×10^{10} kg (15 million tons) of O_2 is used annually in the United States. It is shipped and stored either as the liquid, or in steel containers as compressed gas; however, about 70 percent of O_2 output is generated at the site where it is needed.

Oxygen is by far the most widely used oxidizing agent. Over half of the O_2 produced is used in the steel industry, mainly to remove impurities from steel (Section 22.3). It is also used to bleach pulp and paper. (Oxidation of intensely colored compounds often gives colorless prod-

FIGURE 20.10 Joseph Priestley (1733–1804); Priestley became interested in chemistry at the age of 39, perhaps through his personal acquaintance with Benjamin Franklin. Because he lived next door to a brewery where he could obtain carbon dioxide, his initial studies involved this gas and were later extended to other gases. Because he was suspected of sympathizing with the American and French Revolutions, his church, home, and laboratory in Birmingham were burned by a mob in 1791. Priestley had to flee in disguise. He eventually emigrated to the United States in 1794, where he lived his remaining years in relative seclusion in Pennsylvania. Although his discovery of "dephlogisticated air" (oxygen) eventually led to the downfall of the phlogiston theory, Priestley stubbornly continued to support this theory even after strong evidence had brought it into serious question. Priestley was a scientific conservative, but he was very liberal in his religious and political views. (*National Portrait Gallery*)

ucts.) Other applications include its use in medicine to ease breathing difficulties. It is also used together with acetylene, C_2H_2, in oxyacetylene welding (Figure 20.11). The reaction between C_2H_2 and O_2 is highly exothermic, producing temperatures in excess of 3000°C:

$$2C_2H_2(g) + 5O_2(g) \longrightarrow 4CO_2(g) + 2H_2O(g) \qquad \Delta H^\circ = -2510 \text{ kJ} \qquad [20.23]$$

FIGURE 20.11 Oxyacetylene welding. (*Courtesy Bethlehem Steel Corporation*)

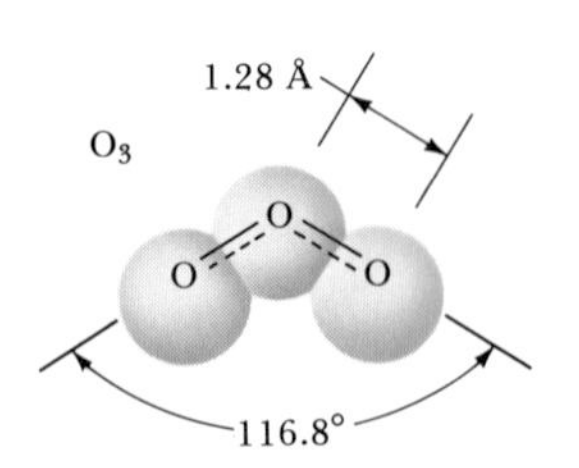

FIGURE 20.12 Structure of the ozone molecule.

Ozone

Ozone, O_3, is a pale blue gas with a sharp, irritating odor. It is poisonous, but no human deaths have been attributed to it. Most people can detect about 0.01 ppm in air. Exposure to 0.1 to 1 ppm produces headaches, burning of the eyes, and irritation to the respiratory passages.

The structure of the O_3 molecule is shown in Figure 20.12. The molecule possesses a π bond that is delocalized over the three oxygen atoms (Section 8.7). The molecule dissociates readily, forming reactive oxygen atoms:

$$O_3(g) \longrightarrow O_2(g) + O(g) \qquad \Delta H^\circ = 107 \text{ kJ/mol} \qquad [20.24]$$

Not surprisingly, ozone is a stronger oxidizing agent than dioxygen. One measure of this oxidizing power is the high reduction potential of O_3, compared to that of O_2:

$$O_3(g) + 2H^+(aq) + 2e^- \longrightarrow O_2(g) + H_2O(l) \qquad E^\circ = 2.07 \text{ V} \qquad [20.25]$$

$$O_2(g) + 4H^+(aq) + 4e^- \longrightarrow 2H_2O(l) \qquad E^\circ = 1.23 \text{ V} \qquad [20.26]$$

Ozone forms oxides with many elements under conditions where O_2 will not react; indeed, it oxidizes all of the common metals except gold and platinum.

Ozone can be prepared by passing electricity through dry O_2:

$$3O_2(g) \xrightarrow{\text{electricity}} 2O_3(g) \qquad \Delta H = 287 \text{ kJ} \qquad [20.27]$$

The pungent odor of ozone gas can sometimes be detected where there is a spark jump, and in the atmosphere during lightning storms. The preparation of O_3 may be accomplished in an apparatus such as that shown in Figure 20.13. The gas cannot be stored for long except at low temperature because it readily decomposes to O_2. The decomposition is catalyzed by certain metals, such as Ag, Pt, and Pd, and by many transition metal oxides.

SAMPLE EXERCISE 20.3

Using ΔG_f° for ozone from Appendix D, calculate the equilibrium constant K_p, for Equation 20.27 at 298.0 K.

Solution: From Appendix D we have $\Delta G_f^\circ(O_3) = 163.4$ kJ/mol. Thus for Equation 20.27, $\Delta G^\circ = (2 \text{ mol } O_3)(163.4 \text{ kJ/mol } O_3) = 326.8$ kJ. From Chapter 17, Equation 17.15, we have $\Delta G^\circ = -2.303RT \log K$. Thus

$$\log K = \frac{-\Delta G^\circ}{2.303RT}$$

$$= \frac{-326.8 \times 10^3 \text{ J}}{(2.303)(8.314 \text{ J/K-mol})(298.0 \text{ K})} = -57.27$$

$$K = 5.3 \times 10^{-58}$$

In spite of the unfavorable equilibrium constant, ozone can be prepared from O_2 as described in the text above. The unfavorable free energy of formation is overcome by energy from the electrical discharge, and O_3 is removed before the reverse reaction can occur, so a nonequilibrium mixture results.

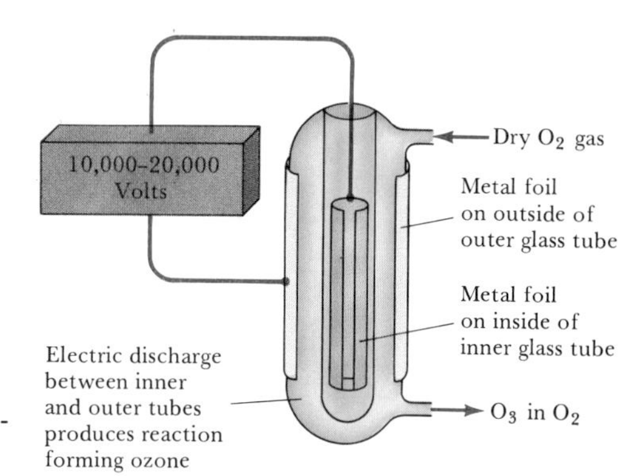

FIGURE 20.13 Apparatus for producing ozone from O_2.

At the present time, uses of ozone as an industrial chemical are relatively small. It is sometimes used in treatment of domestic water in place of chlorine. Like Cl_2, it serves to kill bacteria and oxidize organic compounds. The largest use of ozone, however, is in the preparation of pharmaceuticals, synthetic lubricants, and other commercially useful organic compounds, where O_3 is used to sever carbon-carbon double bonds.

Ozone is an important component of the upper atmosphere, where it serves to screen out ultraviolet radiation (Section 12.3). In this way ozone protects the earth from the effects of these high-energy rays. However, in the lower atmosphere ozone is considered an air pollutant. It is a major constituent of smog. Because of its oxidizing power, it causes damage to living systems and structural materials, especially rubber.

Oxides

The electronegativity of oxygen is second only to that of fluorine. Consequently, oxygen exhibits negative oxidation states in all compounds except those with fluorine, OF_2 and O_2F_2. The -2 oxidation state is by far the most common. As we have seen, compounds in this oxidation state are called *oxides*. Compounds containing O—O bonds and oxygen in an oxidation state of -1 are called **peroxides**. Oxygen has an oxidation state of $-\frac{1}{2}$ in O_2^-, which is called the **superoxide** ion.

Nonmetals form covalent oxides (Section 7.11). Most of these oxides are simple molecules with low melting and boiling points. However, SiO_2 and B_2O_3 have polymeric structures. Most nonmetal oxides combine with water to give oxyacids. For example, sulfur dioxide, SO_2, dissolves in water to give sulfurous acid, H_2SO_3:

$$SO_2(g) + H_2O(l) \longrightarrow H_2SO_3(aq) \qquad [20.28]$$

Such oxides are called **acidic anhydrides** (anhydride means "without water") or **acidic oxides.**

Most metal oxides are ionic compounds. Those ionic oxides that dissolve in water form hydroxides and are consequently referred to as **basic anhydrides** or **basic oxides.** For example, barium oxide, BaO, dissolves in water to form barium hydroxide, $Ba(OH)_2$:

$$BaO(s) + H_2O(l) \longrightarrow Ba(OH)_2(aq) \qquad [20.29]$$

Such reactions can be attributed to the high basicity of the O^{2-} ion and consequently its virtually complete hydrolysis in water:

$$O^{2-}(aq) + H_2O(l) \longrightarrow 2OH^-(aq) \quad [20.30]$$

Even those ionic oxides that are water insoluble tend to dissolve in acids. Iron(III) oxide, for example, dissolves in acids:

$$Fe_2O_3(s) + 6H^+(aq) \longrightarrow 2Fe^{3+}(aq) + 3H_2O(l) \quad [20.31]$$

This reaction is used to remove rust ($Fe_2O_3 \cdot nH_2O$) from iron or steel prior to application of a protective coat of zinc or tin.

Oxides that are borderline in acidic and basic character are said to be *amphoteric* (Section 16.5). If a metal forms more than one oxide, the basic character of the oxide decreases as the oxidation state of the metal increases:

Compound	Oxidation state of Cr	Nature of oxide
CrO	+2	Basic
Cr_2O_3	+3	Amphoteric
CrO_3	+6	Acidic

Peroxides and Superoxides

The most active metals (Cs, Rb, and K) react with O_2 to give superoxides (CsO_2, RbO_2, and KO_2). Their active neighbors in the periodic table (Na, Ca, Sr, and Ba) react with O_2, producing peroxides (Na_2O_2, CaO_2, SrO_2, and BaO_2). Less active metals and nonmetals produce normal oxides.

When superoxides dissolve in water, O_2 is produced:

$$2KO_2(s) + 2H_2O(l) \longrightarrow 2K^+(aq) + 2OH^-(aq) + O_2(g) + H_2O_2(aq) \quad [20.32]$$

Because of this reaction, potassium superoxide, KO_2, is used as an oxygen source in masks worn for rescue work. Moisture in the breath causes the compound to decompose to form O_2 and KOH. The KOH so formed serves to remove CO_2 from the exhaled breath:

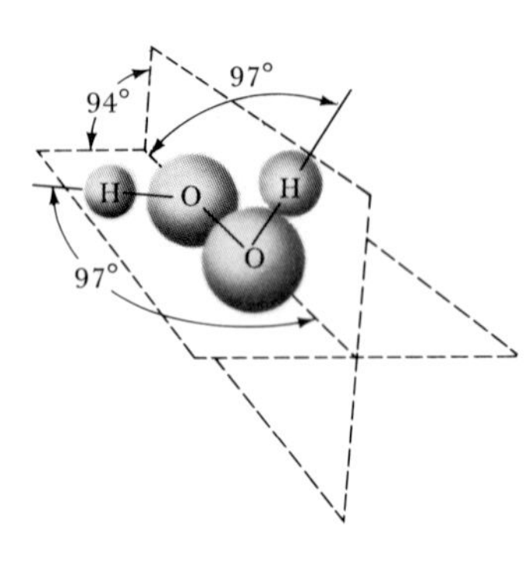

FIGURE 20.14 Structure of the hydrogen peroxide molecule.

$$2OH^-(aq) + CO_2(g) \longrightarrow H_2O(l) + CO_3^{2-}(aq) \quad [20.33]$$

Sodium peroxide, Na_2O_2, is used commercially as an oxidizing agent. When dissolved in water it produces hydrogen peroxide:

$$Na_2O_2(s) + 2H_2O(l) \longrightarrow 2Na^+(aq) + 2OH^-(aq) + H_2O_2(aq) \quad [20.34]$$

Hydrogen peroxide, H_2O_2, is the most familiar and commercially important peroxide. The structure of H_2O_2 is shown in Figure 20.14. The peroxide linkage (—O—O—) exists in species other than H_2O_2 and O_2^{2-}

ion. For example, the peroxydisulfate ion, $S_2O_8^{2-}$ or $O_3SOOSO_3^{2-}$, contains this linkage. The $S_2O_8^{2-}$ ion is produced by electrolysis of a 50 percent aqueous solution of sulfuric acid. The reaction of this ion with water produces hydrogen peroxide:

$$2H_2O(l) + S_2O_8^{2-}(aq) \longrightarrow H_2O_2(aq) + 2H^+(aq) + 2SO_4^{2-}(aq) \qquad [20.35]$$

Pure hydrogen peroxide is a clear syrupy liquid, density 1.47 g/cm^3 at 0°C. It melts at −0.4°C, and its normal boiling point is 151°C. These properties are characteristic of a highly polar, strongly hydrogen-bonded liquid such as water. Concentrated hydrogen peroxide is a dangerously reactive substance, because the decomposition to form water and oxygen gas is very exothermic:

$$2H_2O_2(l) \longrightarrow 2H_2O(l) + O_2(g) \qquad \Delta H^\circ = -196.0 \text{ kJ} \qquad [20.36]$$

The decomposition can occur with explosive violence if highly concentrated hydrogen peroxide comes in contact with substances that can catalyze the reaction. Hydrogen peroxide is marketed as a chemical reagent in aqueous solutions of up to about 30 percent by weight. A solution containing about 3 percent by weight H_2O_2 is commonly used as a mild antiseptic; somewhat more concentrated solutions are employed to bleach fabrics such as cotton, wool, or silk.

The peroxide ion is a by-product of metabolism that results from the reduction of molecular oxygen, O_2. The body disposes of this reactive species with enzymes with such names as peroxidase and catalase.

The fizzing that occurs when a dilute H_2O_2 solution is applied to an open wound is due to decomposition of the H_2O_2 into O_2 and H_2O, a reaction catalyzed by the aforementioned enzymes.

Hydrogen peroxide is capable of acting as either an oxidizing or reducing agent. Equations 20.37 and 20.38 show the half-reactions for reaction in acid solution.

$$2H^+(aq) + H_2O_2(aq) + 2e^- \longrightarrow 2H_2O(l) \qquad E^\circ = 1.77 \text{ V} \qquad [20.37]$$

$$H_2O_2(aq) \longrightarrow O_2(g) + 2H^+(aq) + 2e^- \qquad E^\circ = -0.67 \text{ V} \qquad [20.38]$$

In basic solution, the corresponding standard electrode potentials are 0.87 V for reduction of H_2O_2 and 0.08 V for its oxidation.

In many old oil paints the white lead carbonate pigments have become discolored due to formation of black lead sulfide. Hydrogen peroxide has been used in restoration work, to convert the black sulfide to white lead sulfate. Both salts are insoluble in water. The reaction is as shown in Equation 20.39.

$$PbS(s) + 4H_2O_2(aq) \longrightarrow PbSO_4(s) + 4H_2O(l) \qquad [20.39]$$

The Oxygen Cycle

Oxygen accounts for about one-fourth of the atoms in living matter. Because the number of oxygen atoms is fixed, as O_2 is removed from air through respiration and other processes, it needs to be replenished. The

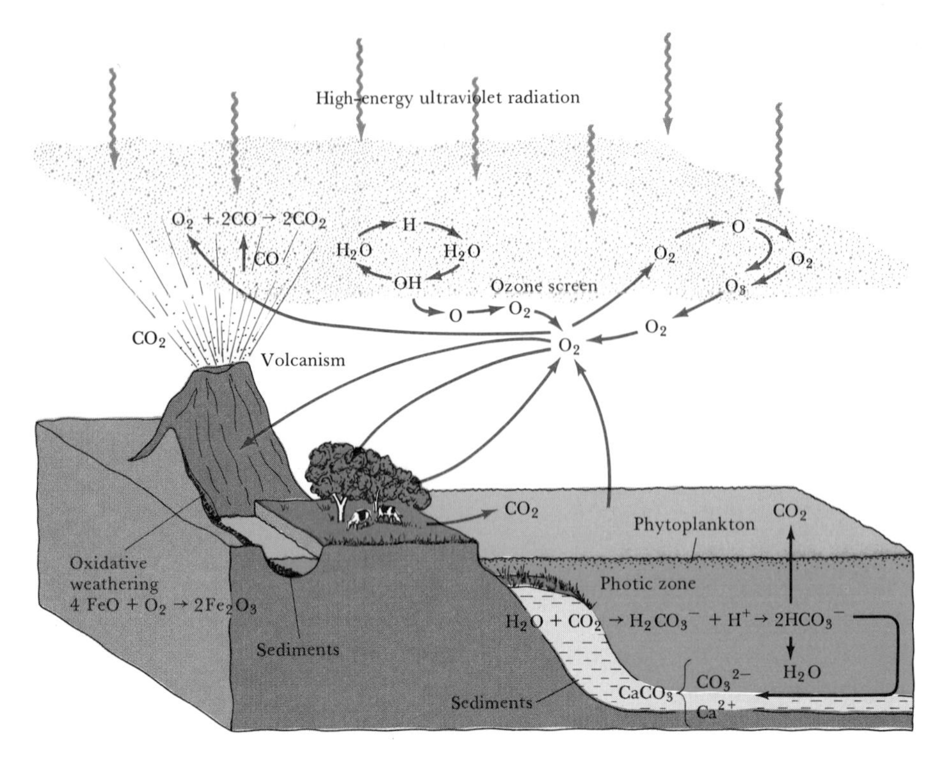

FIGURE 20.15 Simplified view of the oxygen cycle, showing some of the primary reactions involving oxygen in nature. The atmosphere, which contains O_2, is one of the primary sources of the element. Some O_2 is produced by radiation-induced dissociation of H_2O in the upper atmosphere. Some O_2 is produced by green plants from H_2O and CO_2 in the course of photosynthesis. Atmospheric CO_2, in turn, results from combustion reactions, animal respiration, and the dissociation of bicarbonate in water. The O_2 is used to produce ozone in the upper atmosphere in oxidative weathering of rocks, in animal respiration, and in combustion reactions.

major nonliving sources of oxygen other than O_2 are CO_2 and H_2O. A simplified picture of the movement of oxygen in our environment is shown in Figure 20.15. This figure points out both how O_2 is removed from the atmosphere and how it is replenished. Oxygen, O_2, is reformed mainly from CO_2 through the process of photosynthesis. Energy is produced when O_2 is converted to CO_2; energy must therefore be supplied to reform O_2 from CO_2. This energy is provided by the sun. Thus life on earth depends on chemical recycling made possible by solar energy. The situation is represented schematically in Figure 20.16.

20.5 NITROGEN

Nitrogen was discovered in 1772 by the Scottish botanist Daniel Rutherford. He found that when a mouse was enclosed in a sealed jar, the animal quickly consumed the life-sustaining component of air and died. When the "fixed air" (CO_2) in the container was removed, a "noxious air" remained that would not sustain combustion or life. That gas is known to us now as nitrogen.

Nitrogen constitutes 78 percent by volume of the earth's atmosphere, where it occurs as N_2 molecules. Although nitrogen is a key element in living creatures, compounds of nitrogen are not abundant in the earth's crust. The major natural deposits of nitrogen compounds are those of KNO_3 (saltpeter) in India, and $NaNO_3$ (Chile saltpeter) in Chile and other desert regions of South America.

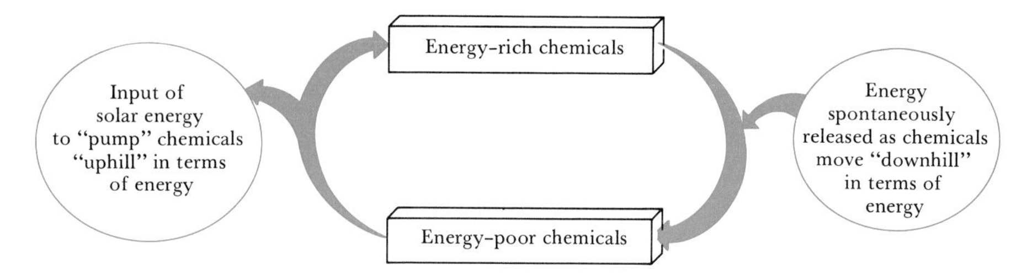

FIGURE 20.16 Schematic representation of the role of solar energy in the cycling of elements in nature.

Properties of Nitrogen

Nitrogen is a colorless, odorless, and tasteless gas composed of N_2 molecules. Its melting point is $-210°C$ and its normal boiling point is $-196°C$.

The N_2 molecule is very unreactive because of the strong triple bond between nitrogen atoms (the $N{\equiv}N$ bond-dissociation energy is 941 kJ/mol, nearly twice that for the bond in O_2; see Table 7.3). When substances burn in air they normally react with O_2 but not with N_2. However, when magnesium burns in air, reaction with N_2 also occurs to form magnesium nitride, Mg_3N_2. A similar reaction occurs with lithium:

$$3Mg(s) + N_2(g) \longrightarrow Mg_3N_2(s) \qquad [20.40]$$

$$6Li(s) + N_2(g) \longrightarrow 2Li_3N(s) \qquad [20.41]$$

The nitride ion is a strong Brønsted base. It reacts with water to form ammonia, NH_3:

$$Mg_3N_2(s) + 6H_2O(l) \longrightarrow 2NH_3(aq) + 3Mg(OH)_2(s) \qquad [20.42]$$

The electron configuration of the nitrogen atom is $[He]2s^22p^3$. The element exhibits all formal oxidation states from $+5$ to -3, as shown in Table 20.3. Because nitrogen is the third most electronegative element after fluorine and oxygen, it exhibits positive oxidation states only in combination with those two elements.

Figure 20.17 summarizes the standard electrode potentials for interconversion of several common nitrogen species. The fact that the potentials in the diagram are large and positive is an indication that the nitrogen oxides and oxyanions shown are strong oxidizing agents.

TABLE 20.3 Oxidation states of nitrogen

Oxidation state	Examples
+5	N_2O_5, HNO_3, NO_3^-
+4	NO_2, N_2O_4
+3	HNO_2, NO_2^-, NF_3
+2	NO
+1	N_2O, $H_2N_2O_2$, $N_2O_2^{2-}$, HNF_2
0	N_2
−1	NH_2OH, H_2NF
−2	N_2H_4
−3	NH_3, NH_4^+, NH_2^-

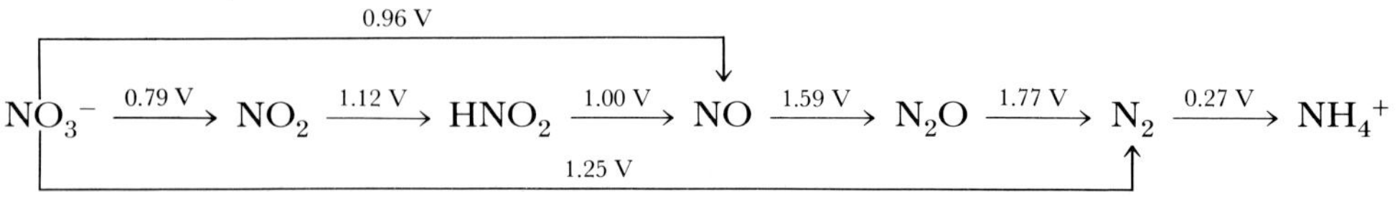

0.96 V (NO_3^- to NO)

$$NO_3^- \xrightarrow{0.79\ V} NO_2 \xrightarrow{1.12\ V} HNO_2 \xrightarrow{1.00\ V} NO \xrightarrow{1.59\ V} N_2O \xrightarrow{1.77\ V} N_2 \xrightarrow{0.27\ V} NH_4^+$$

1.25 V (NO_3^- to N_2)

FIGURE 20.17 Standard reduction potentials in acid solution for nitrogen-containing compounds in various oxidation states.

Notice in Figure 20.17 that NO_3^- is written as an anion, whereas HNO_2 is written as an acid. This convention follows from the fact that HNO_2 is a weak acid, whereas HNO_3 is a strong acid. Thus HNO_3 can be considered to be completely ionized even in acidic solution. The voltages listed on the lines connecting various nitrogen species are the standard reduction potentials for the half-reactions that convert one species to any other. For example, conversion of NO_3^- to NO_2 in acid solution has a standard reduction potential of 0.79 V (leftmost entry). The complete balanced half-reaction is written as follows:

$$NO_3^-(aq) + 2H^+(aq) + e^- \longrightarrow NO_2(g) + H_2O(l) + H_2O(l) \qquad E° = 0.79\ V \qquad [20.43]$$

You should be able to write a similar, complete balanced half-reaction for other changes shown in Figure 20.17, using $H^+(aq)$ and $H_2O(l)$ to achieve a balanced equation, as described in Section 18.1.

By using the standard reduction potentials in Figure 20.17, it is possible to calculate the standard reduction potentials for other half-reactions for which values are not given. For example, we can calculate the standard potential for the reduction of N_2O to NH_4^+ using the N_2O to N_2 and N_2 to NH_4^+ potentials given. The balanced equations for these processes are as follows:

$$2e^- + N_2O(g) + 2H^+(aq) \longrightarrow N_2(g) + H_2O(l) \qquad E° = 1.77\ V \qquad [20.44]$$

$$6e^- + 8H^+(aq) + N_2(g) \longrightarrow 2NH_4^+(aq) \qquad E° = 0.27\ V \qquad [20.45]$$

$$8e^- + N_2O(g) + 10H^+(aq) \longrightarrow 2NH_4^+(aq) + H_2O(l) \qquad [20.46]$$

As shown above, adding the two half-reactions results in the desired half-reaction for conversion of N_2O to NH_4^+. However, the desired electrode potential cannot be obtained by merely adding the given $E°$ values.

To understand why this is so, we must recall (Section 18.6) that when we add reactions (or half-reactions) to obtain a new reaction (or half-reaction), the free-energy change, ΔG, for the new reaction is the sum of the free-energy changes for the reactions that are added. We also know that the standard potential is related to the free-energy change by the equation $\Delta G° = -n\mathfrak{F}E°$. Thus we have that

$$\begin{aligned} \Delta G°\ (\text{rxn } 20.46) &= \Delta G°\ (\text{rxn } 20.44) + \Delta G°\ (\text{rxn } 20.45) \\ -n\mathfrak{F}E°\ (\text{rxn } 20.46) &= -n\mathfrak{F}E°\ (\text{rxn } 20.44) - n\mathfrak{F}E°\ (\text{rxn } 20.45) \\ -8\mathfrak{F}E°\ (\text{rxn } 20.46) &= -2\mathfrak{F}E°\ (\text{rxn } 20.44) - 6\mathfrak{F}E°\ (\text{rxn } 20.45) \\ &= -2\mathfrak{F}(1.77\ V) - 6\mathfrak{F}(0.27\ V) \\ E°\ (\text{rxn } 20.46) &= \frac{-2\mathfrak{F}(1.77\ V) - 6\mathfrak{F}(0.27\ V)}{-8\mathfrak{F}} \\ &= \frac{5.16\ V}{-8} \\ &= 0.64\ V \end{aligned}$$

You may wonder why this procedure was not employed in Chapter 18, in which we added half-reactions and their corresponding $E°$ values. In Chapter 18 we added two half-reactions to give a balanced equation, in which the number of electrons gained in one half-reaction is just balanced by the number of electrons lost in the other half-reaction. When the half-reactions are added to give a complete balanced equation, n, the number of electrons, is the same for all reactions. It thus cancels out, and the result is the same whether we directly add $E°$ values or use the procedure described above. The important point to remember is that when half-reactions are added to give a new half-reaction in which there is a net gain or loss of electrons, the procedure described above must be used to obtain the correct half-reaction potential.

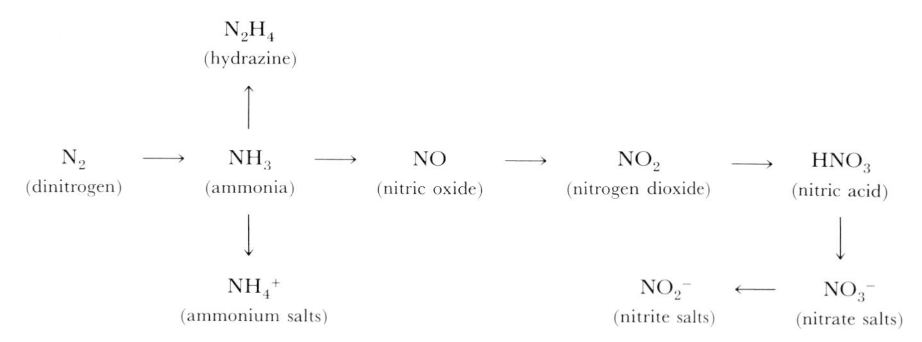

FIGURE 20.18 Conversion of N_2 into common nitrogen compounds.

Preparation and Uses

Nitrogen is obtained in commercial quantities by fractional distillation of liquid air. About 1.6×10^{10} kg (18 million tons) of N_2 is produced annually in the United States.

Because of its low reactivity, large quantities of N_2 are used as an inert gaseous blanket to exclude O_2 during the processing and packaging of foods, the manufacture of chemicals, the fabrication of metals, and the production of electronic devices. Liquid N_2 is employed as a coolant to freeze foods rapidly.

The largest use of nitrogen is in the manufacture of nitrogen-containing chemicals. The formation of nitrogen compounds from N_2 is known as nitrogen fixation. The demand for fixed nitrogen is high because the element is required in maintaining soil fertility. Although we are immersed in an ocean of air that contains abundant N_2, our supply of food is limited more by the availability of fixed nitrogen than by that of any other plant nutrient. Thus N_2 is used primarily for the manufacture of nitrogen-containing fertilizers. It is also used to manufacture explosives, plastics, and many important chemicals.

The chain of conversion of N_2 into a variety of useful, simple nitrogen-containing species is given in Figure 20.18. The processes shown are discussed in more detail in later portions of this section.

Hydrogen Compounds of Nitrogen

Ammonia is one of the most important compounds of nitrogen. It is a colorless, toxic gas that has a characteristic irritating odor. As we have noted in previous discussions (Section 15.6), The NH_3 molecule is basic ($K_b = 1.8 \times 10^{-5}$).

In the laboratory, NH_3 is prepared by the action of NaOH on an ammonium salt. The NH_4^+ ion, which is the conjugate acid of NH_3, loses a proton to OH^-. The resultant NH_3 is volatile and is driven from the solution by mild heating:

$$NH_4Cl(aq) + NaOH(aq) \longrightarrow NH_3(g) + H_2O(l) + NaCl(aq) \qquad [20.47]$$

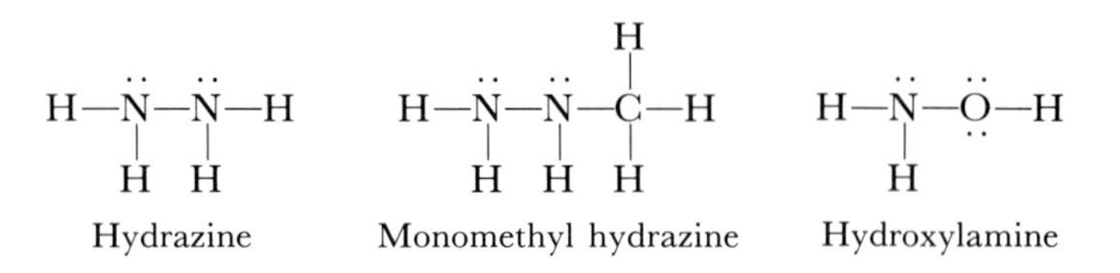

FIGURE 20.19 Lewis structures of hydrazine, N_2H_4, monomethylhydrazine, CH_3NHNH_2, and hydroxylamine, NH_2OH.

Commercial production of NH_3 is achieved by the Haber process (Section 14.1), in which N_2 and H_2 are catalytically combined at high pressure and high temperature:

$$N_2(g) + 3H_2(g) \longrightarrow 2NH_3(g) \qquad [20.48]$$

About 75 percent of the ammonia produced in this country is used for fertilizer.

Hydrazine, N_2H_4, bears the same relationship to ammonia that hydrogen peroxide does to water. As shown in Figure 20.19, the hydrazine molecule contains an N—N single bond. Hydrazine is quite poisonous. It can be prepared by the reaction of ammonia with hypochlorite ion, OCl^-, in aqueous solution:

$$2NH_3(aq) + OCl^-(aq) \longrightarrow N_2H_4(aq) + Cl^-(aq) + H_2O(l) \qquad [20.49]$$

The possible formation of N_2H_4 from household ammonia and chlorine bleach, which contains OCl^-, is one reason for the oft-cited warning not to mix household cleaning agents.

Pure hydrazine is an oily, colorless liquid with a melting point of 1.5°C and a boiling point of 113°C. The pure substance explodes on heating and is a highly reactive reducing agent. N_2H_4 is normally em-

FIGURE 20.20 Launch of the first space shuttle, April 12, 1981. Monomethyl hydrazine is one of the fuels used in the space shuttle. (*Courtesy NASA*)

ployed in aqueous solution, where it can be handled safely. The substance is weakly basic, and salts of $N_2H_5^+$ can be formed:

$$N_2H_4(aq) + H_2O(l) \rightleftharpoons N_2H_5^+(aq) + OH^-(aq) \qquad K_b = 1.3 \times 10^{-6} \qquad [20.50]$$

The combustion of hydrazine is highly exothermic:

$$N_2H_4(l) + O_2(g) \longrightarrow N_2(g) + 2H_2O(g) \qquad \Delta H^\circ = -534 \text{ kJ} \qquad [20.51]$$

Hydrazine and compounds derived from it, such as monomethyl hydrazine (Figure 20.16), are used as rocket fuels. Monomethyl hydrazine is one of the fuels used in the space shuttle Columbia (Figure 20.20).

Hydroxylamine, NH_2OH (Figure 20.16) can be thought of as being derived from NH_3 by replacing a hydrogen with an OH group. Like N_2H_4 it is highly reactive and rather unstable as a pure substance. It normally acts as a reducing agent with formation of N_2. However, it may be oxidized by some strong oxidants to N_2O or even NO_3^-.

SAMPLE EXERCISE 20.4

Hydroxylamine reduces copper(II) to the free metal in acid solutions. Write a balanced equation for the reaction assuming that N_2 is the oxidation product.

Solution: The unbalanced and incomplete half-reactions are

$$Cu^{2+}(aq) \longrightarrow Cu(s)$$
$$NH_2OH(aq) \longrightarrow N_2(g)$$

Balancing these equations as described in Section 18.1 gives

$$Cu^{2+}(aq) + 2e^- \longrightarrow Cu(s)$$
$$2NH_2OH(aq) \longrightarrow N_2(g) + 2H_2O(l) + 2H^+(aq) + 2e^-$$

Adding these half-reactions gives the balanced equation:

$$Cu^{2+}(aq) + 2NH_2OH(aq) \longrightarrow Cu(s) + N_2(g) + 2H_2O(l) + 2H^+(aq)$$

Hydrogen azide, HN_3 (Figure 20.21), is the parent compound of a number of covalent and ionic azides. It can be prepared by reaction of sodium azide, NaN_3, with sulfuric acid. Hydrogen azide is a highly dangerous liquid that has a boiling point of 36°C and decomposes explosively to the free elements. In water it is a weak acid ($K_a = 1.9 \times 10^{-5}$) and its aqueous solutions are called *hydroazoic acid.* Its salts, called azides, are unstable. When azide salts of heavy metals are heated or struck, they decompose explosively to form N_2 and the free metal. Lead azide, $Pb(N_3)_2$, is used as a primer in ammunition.

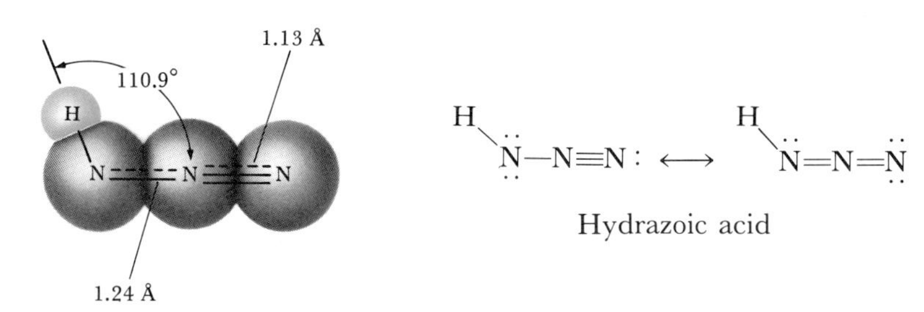

FIGURE 20.21 Resonance structures and molecular structure of hydrogen azide, HN_3.

Oxides and Oxyacids of Nitrogen

Nitrogen forms three common oxides: N_2O (nitrous oxide), NO (nitric oxide), and NO_2 (nitrogen dioxide). It also forms two unstable oxides that we will not discuss, N_2O_3 (dinitrogen trioxide) and N_2O_5 (dinitrogen pentoxide).

Nitrous oxide, N_2O, is also known as laughing gas because a person becomes somewhat giddy after inhaling only a small amount of it. This colorless gas was the first substance used as a general anesthetic. It is used as the compressed gas propellant in several aerosols and foams, such as in whipped cream. It can be prepared in the laboratory by carefully heating ammonium nitrate to about 200°C:

$$NH_4NO_3(s) \xrightarrow{\Delta} N_2O(g) + 2H_2O(g) \qquad [20.52]$$

Nitric oxide, NO, is also a colorless gas, but unlike N_2O, it is slightly toxic. It can be prepared in the laboratory by reduction of dilute nitric acid, using copper or iron as a reducing agent:

$$3Cu(s) + 2NO_3^-(aq) + 8H^+(aq) \longrightarrow 3Cu^{2+}(aq) + 2NO(g) + 4H_2O(l) \qquad [20.53]$$

It is also produced by direct combination of N_2 and O_2 at elevated temperatures. As we saw in Section 12.4, this reaction is a significant source of nitrogen oxide air pollutants, which form during combustion reactions in air. However, the direct combination of N_2 and O_2 is not presently used for commercial production of NO because the yield is low; the equilibrium constant K_c at 2400 K is only 0.05.

The commercial route to NO (and hence to other oxygen-containing compounds of nitrogen) is by means of the catalytic oxidation of NH_3 (Figure 20.22):

$$4NH_3(g) + 5O_2(g) \xrightarrow[800°C]{\text{Pt catalyst}} 4NO(g) + 6H_2O(g) \qquad [20.54]$$

In the absence of the platinum catalyst, NH_3 is converted to N_2 instead of NO:

$$4NH_3(g) + 3O_2(g) \xrightarrow{1000°C} 2N_2(g) + 6H_2O(g) \qquad [20.55]$$

The catalytic conversion of NH_3 to NO is the first step in a three-step process known as the Ostwald process, by which NH_3 is converted commercially into nitric acid, HNO_3. Nitric oxide reacts readily with O_2, forming NO_2 when exposed to air:

$$2NO(g) + O_2(g) \longrightarrow 2NO_2(g) \qquad [20.56]$$

When dissolved in water, NO_2 forms nitric acid:

$$3NO_2(g) + H_2O(l) \longrightarrow 2H^+(aq) + 2NO_3^-(aq) + NO(g) \qquad [20.57]$$

FIGURE 20.22 The oxidation of ammonia during the production of nitric acid is one of the most important chemical processes undertaken on a large industrial scale. The reaction is carried out in the presence of a platinum-rhodium catalyst, which is usually in the form of a woven gauze. Here new catalyst gauzes are being installed in an ammonia oxidation plant. (*Courtesy Johnson-Matthey Metals Limited*)

Note that nitrogen is both oxidized and reduced in this reaction. The NO_2 is said to have undergone disproportionation. The reduction product NO can be converted back into NO_2 by exposure to air and thereafter dissolved in water to prepare more HNO_3.

Nitrogen dioxide is a yellow-brown gas. It is poisonous and has a choking odor. At lower temperatures two NO_2 molecules combine to form the colorless N_2O_4:

$$2NO_2(g) \longrightarrow N_2O_4(g) \qquad \Delta H^\circ = -58\ \text{kJ} \qquad [20.58]$$

The two common oxyacids of nitrogen are nitric acid, HNO_3, and nitrous acid, HNO_2 (Figure 20.23). Nitric acid is a colorless, corrosive

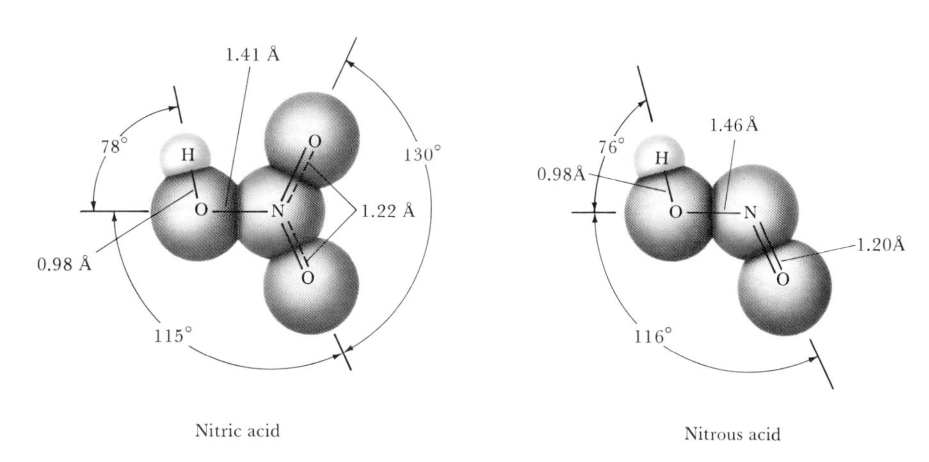

FIGURE 20.23 Structures of nitric acid and nitrous acid.

liquid. It is both a strong acid and a good oxidizing agent. It is used in the production of plastics, drugs, nitrate fertilizers, and explosives. The development of the Haber and Ostwald processes in Germany just prior to World War I permitted Germany to make munitions even though naval blockades prevented access to traditional sources of nitrates. Among the explosives made from nitric acid are nitroglycerin, trinitrotoluene (TNT), and nitrocellulose. The reaction of nitric acid with glycerin to form nitroglycerin is shown in Equation 20.59:

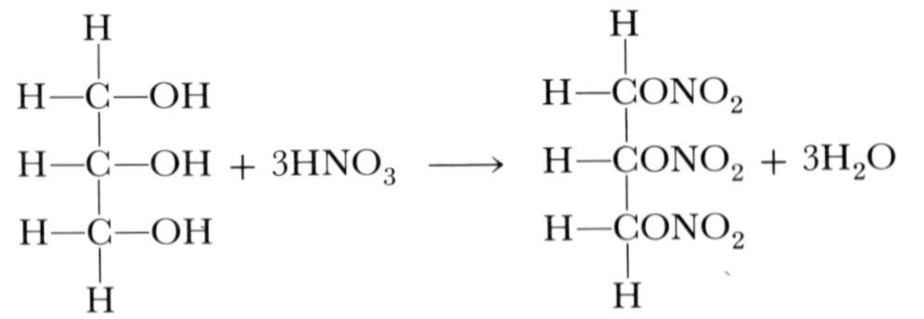

$$\begin{array}{c} \text{H} \\ | \\ \text{H—C—OH} \\ | \\ \text{H—C—OH} \\ | \\ \text{H—C—OH} \\ | \\ \text{H} \end{array} + 3HNO_3 \longrightarrow \begin{array}{c} \text{H} \\ | \\ \text{H—CONO}_2 \\ | \\ \text{H—CONO}_2 \\ | \\ \text{H—CONO}_2 \\ | \\ \text{H} \end{array} + 3H_2O \qquad [20.59]$$

When nitroglycerin explodes, the reaction summarized in Equation 20.60 occurs:

$$4C_3H_5N_3O_9(l) \longrightarrow 6N_2(g) + 12CO_2(g) + 10H_2O(g) + O_2(g) \qquad [20.60]$$

A considerable amount of gaseous products form from the liquid. The sudden formation of these gases, together with their expansion resulting from the heat generated by the reaction, produces the explosion.

Nitrous acid, HNO_2, is considerably less stable than HNO_3 and tends to disproportionate into NO and HNO_3. It is normally made by action of a strong acid such as H_2SO_4 on a cold solution of a nitrite salt such as $NaNO_2$. Nitrous acid is a weak acid ($K_a = 4.5 \times 10^{-4}$).

Sodium nitrite is used as a food additive in cured meats such as hot dogs, ham, and cold cuts. The NO_2^- serves two functions: It inhibits growth of bacteria, especially *Clostridium botulinum*, which produces the potentially fatal food poisoning known as botulism. The nitrite also preserves the red color of the meat and thereby its appetizing appearance. Debate over the continued use of nitrites in cured meat products centers on the fact that HNO_2 can react with certain organic compounds to form compounds known as nitrosoamines. These are compounds of the type given in Figure 20.24. These organic compounds have been shown to produce cancer in laboratory animals. Although no human cancer has been linked to nitrites in meat, that potential has caused the U.S. Food and Drug Administration to reduce the limits of allowable concentrations of NO_2^- in foods.

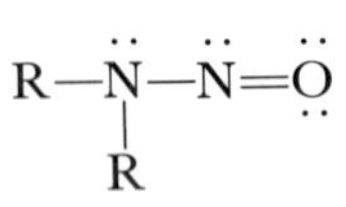

$$\begin{array}{l} R—\ddot{N}—\ddot{N}=\ddot{O} \\ \quad\;\; | \\ \quad\;\; R \end{array}$$

Nitrosoamine

FIGURE 20.24 General formula for a nitrosoamine (R = groups such as CH_3).

The Nitrogen Cycle in Nature

There are two primary routes for nitrogen fixation in nature: Lightning causes the formation of NO from N_2 and O_2 in air. In addition, the root nodules of certain leguminous plants, such as peas, beans, peanuts, and alfalfa, contain nitrogen-fixing bacteria. It is well known that both iron and molybdenum are involved in the enzyme system responsible for the nitrogen fixation in these root nodules. It is of interest to understand this process and to develop catalysts that, like enzymes, fix nitrogen at ambient pressure and temperature (and hence potentially at an energy savings over the Haber process).

Nitrogen is found in many compounds that are vital to life, including proteins, enzymes, nucleic acids, vitamins, and hormones. Plants employ very simple compounds as starting materials from which such complex, biologically necessary compounds are formed. Plants are able to use several forms of nitrogen, especially NH_3, NH_4^+, and NO_3^-. Liquid ammonia, ammonium nitrate, NH_4NO_3, and urea, $(NH_2)_2CO$, are among the most commonly applied fertilizers. Urea is made by the reaction of ammonia and carbon dioxide:

$$2NH_3(aq) + CO_2(aq) \rightleftharpoons H_2N\overset{\overset{\displaystyle O}{\|}}{C}NH_2(aq) + H_2O(l) \qquad [20.61]$$

NH_3 is slowly released as the urea reacts with water in the soil.

Animals are unable to synthesize the complex nitrogen compounds they require from the simple substances used by plants. Instead, they rely on more complicated precursors present in foods. Those nitrogen compounds not needed by the animal are excreted as nitrogenous waste. Certain microorganisms are able to convert this waste back into N_2. Nitrogen is recycled in this fashion. As in the case of the cycling of oxygen, energy from the sun is required. A simplified picture of the nitrogen cycle is shown in Figure 20.25.

Large-scale cultivation of nitrogen-fixing legumes and industrial fixation have increased the quantity of fixed nitrogen in the biosphere. One effect of this intrusion into the nitrogen cycle has been increased water pollution. Much fixed nitrogen ends up as nitrates in the soil. These compounds are highly water soluble. They are therefore readily washed from the soil into water bodies. Only a portion of the added fertilizer ends up being used by the plants for which it was intended. Once in a lake, nitrates stimulate plant growth, encouraging, for example, rapid growth of algae. When these plants die, their decay consumes O_2 in the water, thereby killing fish and other oxygen-dependent organisms. The resultant anaerobic environment leads to the foul odors that we associate with highly polluted bodies of water.

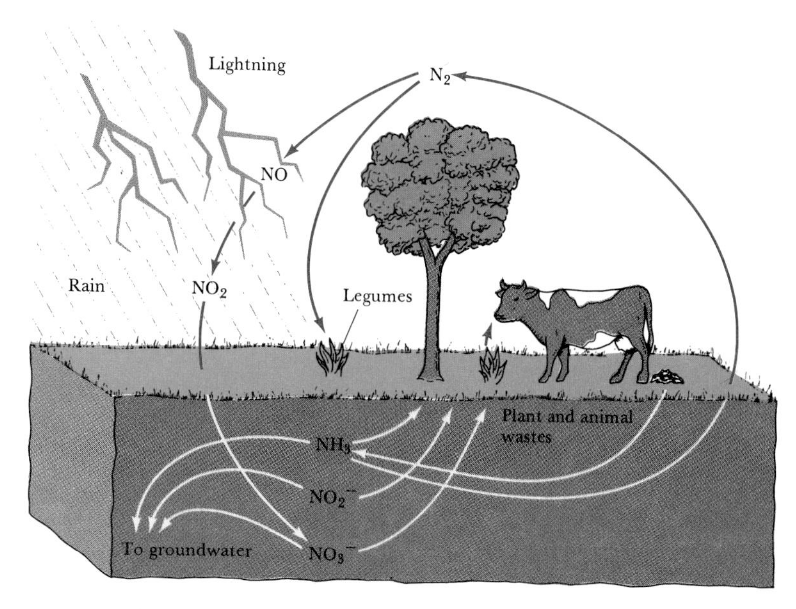

FIGURE 20.25 Simplified picture of the nitrogen cycle, showing some of the primary reactions involved in the utilization and formation of nitrogen in nature. The main reservoir of nitrogen is the atmosphere, which contains N_2. This nitrogen is fixed through the action of lightning and leguminous plants. The compounds of nitrogen reside in the soil as NH_3 (and NH_4^+), NO_2^-, and NO_3^-. All are water soluble and can be washed out of the soil by ground water. These nitrogen compounds are utilized by plants in their growth and are incorporated into animals that eat the plants. Animal waste and dead plants and animals are attacked by certain bacteria that free N_2, which escapes into the atmosphere, thereby completing the cycle.

20.6 CARBON

Carbon is not an abundant element; it constitutes only 0.027 percent of the earth's crust. Although some occurs in elemental form as graphite and diamond, most is found in combined form. Over half occurs in carbonate compounds such as $CaCO_3$. Carbon is also found in coal, petroleum, and natural gas. The importance of the element stems in large part from its occurrence in all living organisms; life as we know it is based on carbon compounds. About 150 years ago scientists believed that these life-sustaining compounds could be made only within living systems. They were consequently called **organic compounds.** It is now known that organic compounds can be synthesized in the laboratory from simple inorganic (nonorganic) substances. Although the name "organic chemistry" persists, it is now used to describe the portion of chemistry that focuses on hydrocarbons and compounds derived from them by substituting some hydrogen atoms with other atoms. In this section we will take a brief look at carbon and its most common inorganic compounds. Organic chemistry and its application to biological systems are considered in Chapters 25 and 26.

Elemental Forms of Carbon

We have seen from earlier discussions (Section 8.7) that carbon exists in two allotropic forms, graphite and diamond. Charocal, carbon black, and coke are microcrystalline or amorphous forms of graphite.

Graphite is mined in several parts of the world, but it can also be produced synthetically from amorphous carbon. The word "graphite" comes from the Greek and means "to write." The so-called lead of pencils contains graphite together with a binder, usually clay, to make it harder. Graphite is a soft, black, slippery solid that has a metallic luster and conducts electricity. Not surprisingly, graphite is used commercially as a lubricant and also to construct electrodes. The relative softness and electrical conductivity of graphite can be related to its structure. Crystalline graphite consists of parallel sheets of carbon atoms, each sheet containing hexagonal arrays of carbon atoms (Figure 8.30). Each atom exhibits sp^2 hybridization and is involved in delocalized π bonding with other atoms in the sheet. This delocalized π system is responsible for the electrical conductivity of the graphite; the interaction of π electrons with light is responsible for the black color. The softness and lubricity arise from the weak binding, by London-dispersion forces, that exists between sheets.

The properties of graphite are anisotropic (that is, they differ in different directions through the solid). Along the carbon planes graphite possesses great strength because of the number and strength of the carbon-carbon bonds in this direction. In contrast, we have seen that the bonds between planes are relatively weak, making graphite weak in that direction.

Fibers of graphite can be prepared in which the carbon planes are aligned to varying extents parallel to the fiber axis. These fibers are also lightweight (density of about 2 g/cm^3) and chemically quite unreactive. The oriented fibers are made by first slowly pyrolyzing organic fibers to about 150 to 300°C. These fibers are then heated to about 2500°C to graphitize them (that is to convert amorphous carbon to graphite). Stretching the fiber during pyrolysis helps orient the graphite planes parallel to the fiber axis. More amorphous carbon fibers are formed by pyrolysis of organic fibers at 1200 to 1400°C. These amorphous materials, commonly referred to as carbon fibers, are the type most commonly employed in commercial materials.

Composite materials that take advantage of the strength, stability, and low density of carbon fibers are widely used. Composites are combinations of two or more materials; these materials are present as separate phases and are combined to form structures that take advantage of certain desirable properties of each component. In carbon composites, the graphite fibers are often woven into a fabric that is embedded in a matrix that binds them into a solid structure. The fibers transmit loads evenly thoughout the matrix. The finished composite is thus stronger than any one of its components. Epoxy systems are useful matrices because of their excellent adherence, but they are costly and limited to service temperatures below 150°C. More heat-resistent resins are required for many aerospace applications where carbon composites find wide use. The fuselage and wings of the aircraft shown in Figure 20.26(a) are manufactured from such carbon composites. Figure 20.26(b) shows a photomicrograph of a section through a carbon-epoxy composite.

Diamonds are found in many parts of the world, but most come from South Africa. The conversion of graphite to diamond is endothermic:

$$\text{C(graphite)} \longrightarrow \text{C(diamond)} \qquad \Delta H^\circ = 1.87 \text{ kJ} \qquad [20.62]$$

Furthermore, diamond is the more compact allotrope ($d = 2.25 \text{ g/cm}^3$ for graphite; $d = 3.51 \text{ g/cm}^3$ for diamond). Thus its formation from graphite is favored by high temperatures and high pressures. Such conditions are used in synthesizing industrial-grade diamonds (Figure 20.27) employed in cutting, grinding, and polishing tools. Diamond is the hardest known substance. Because of this property, diamonds are used as styluses in high-quality record players. The hardness of diamond can be attributed to its structure (Fig. 8.29). Each carbon atom has four nearest neighbors to which it is bonded by σ bonds. The bond angles are all 109°, typical of sp^3 hybridization. The resultant interlocking network of covalent bonds makes the structure very rigid. The fact that all valence electrons are involved in σ bonding makes diamond a nonconductor.

Carbon black is formed when hydrocarbons are heated in a very limited supply of oxygen:

(*a*)

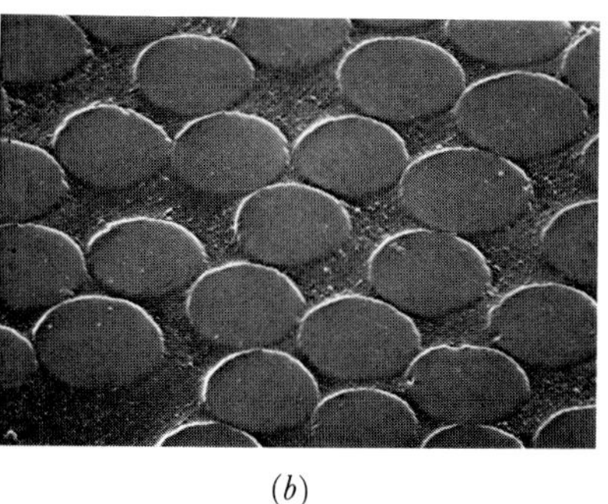

(*b*)

FIGURE 20.26 (*a*) The fuselage and wings of this aircraft are manufactured from graphite composites. (*Courtesy Lear Fan Ltd.*) (*b*) Section through a carbon-epoxy composite. Magnification 1000 times. (*Courtesy American Cyanamid Co.*)

FIGURE 20.27 Graphite and synthetic diamond prepared from graphite. Most synthetic diamonds lack the size, color, and clarity of natural diamonds and are therefore not used in jewelry. The conversion of graphite into diamond is usually carried out in the presence of catalysts at about 2000°C and at pressures exceeding 40,000 atm. (*General Electric Research and Development Center*)

$$CH_4(g) + O_2(g) \longrightarrow C(s) + 2H_2O(g) \quad [20.63]$$

It is used as a pigment in black inks; large amounts are also used in making automobile tires.

Charcoal is formed when wood is heated strongly in the absence of air. Charcoal has a very open structure, giving it an enormous surface area per unit mass. Activated charcoal, a pulverized form whose surface is cleaned by heating with steam, is widely used to adsorb molecules. It is used in filters to remove offensive odors from air and colored or bad-tasting impurities from water.

Coke is an impure form of carbon formed when coal is heated strongly in the absence of air. It is widely used as a reducing agent in metallurgical operations (Chapter 22).

Oxides of Carbon

Carbon forms two principal oxides, carbon monoxide, CO, and carbon dioxide, CO_2. Carbon monoxide is formed when carbon or hydrocarbons are burned in a limited supply of oxygen:

$$2C(s) + O_2(g) \longrightarrow 2CO(g) \quad [20.64]$$

It is a colorless, odorless, and tasteless gas (m.p. = −199°C; b.p. = −192°C). It is toxic because of its ability to bind to hemoglobin and thus interfere with oxygen transport (Section 12.4). Low-level poisoning results in headaches and drowsiness; high-level poisoning, of course, can cause death.

CO is an unusual carbon compound because it has a lone pair of electrons on carbon: $:C\equiv O:$. One might imagine that CO would be unreactive, like the isoelectronic N_2 molecule. Both substances have high bond energies (1072 kJ/mol for $C\equiv O$ and 941 kJ/mol for $N\equiv N$). However, because of the lower nuclear charge on carbon (compared with either N or O), the lone pair on carbon is not held as strongly as that on N or O. Consequently, CO is better able to function as an electron-pair donor (Lewis base) than is N_2. CO is able to form a wide variety of covalent compounds, known as metal carbonyls, with transition metals. An example of such a compound is $Ni(CO)_4$, a volatile, very toxic compound that is formed by simply warming metallic nickel in the presence of CO. The formation of such metal carbonyls is the first step in the transition metal catalysis of a variety of reactions of CO.

Carbon monoxide has several commercial uses. Because it burns readily, forming CO_2, it is employed as a fuel:

$$2CO(g) + O_2(g) \longrightarrow 2CO_2(g) \qquad \Delta H = -566\ \text{kJ} \qquad [20.65]$$

It is also an important reducing agent, widely employed in metallurgical operations to reduce metal oxides. For example, it is the most important reducing agent in the blast furnace reduction of iron(III) oxide:

$$Fe_2O_3(s) + 3CO(g) \longrightarrow 2Fe(s) + 3CO_2(g) \qquad [20.66]$$

This reaction is discussed in greater detail in Chapter 22. Carbon monoxide is also used in the preparation of several organic compounds. In Section 20.3 we saw that it can be combined catalytically with H_2 to manufacture methanol, CH_3OH (Equation 20.16).

Use of carbon monoxide as a basic starting material for the synthesis of hydrocarbons and other organic chemicals is presently the subject of considerable research. A major impetus for these studies is the belief that increasing costs or decreasing availability of petroleum could eventually cause the raw material base for commercial organic chemicals to shift from petroleum to coal. Coal can be converted to a mixture of carbon monoxide and hydrogen by treatment with steam at elevated temperatures (Equation 20.13). This gaseous mixture could be used as a feedstock for a variety of chemical syntheses.

Interest in CO as a raw material goes back to German efforts to produce synthetic fuels in the 1930s and 1940s. One process developed in that period is the Fischer-Tropsch synthesis, in which CO is catalytically hydrogenated to form hydrocarbons and CO_2. The overall process is given by Equation 20.67, where $-(CH_2)-$ represents a portion of a hydrocarbon:

$$2CO(g) + H_2(g) \longrightarrow -(CH_2)- + CO_2(g) \qquad [20.67]$$

A drawback of this process is that it produces a wide mixture of hydrocarbons. About 100 different hydrocarbons were formed in the original process; about half of these were in the C_3 to C_8 range. The mechanism is believed to involve H_2 and CO dissociation on the catalyst surface. The resultant O atoms are attacked by H atoms, forming H_2O. The carbon atoms are attacked to form CH_2 units bound to the catalyst, $M{=}CH_2$. These CH_2 units combine, one carbon at a time, to form hydrocarbon chains. Much present research is directed to find catalysts that will increase the selectivity of the reaction, producing a narrow range of hydrocarbons as products.

Carbon dioxide is produced when carbon-containing substances are burned in excess oxygen:

$$C(s) + O_2(g) \longrightarrow CO_2(g) \qquad [20.68]$$

$$CH_4(g) + 2O_2(g) \longrightarrow CO_2(g) + 2H_2O(l) \quad [20.69]$$

$$C_2H_5OH(l) + 3O_2(g) \longrightarrow 2CO_2(g) + 3H_2O(l) \quad [20.70]$$

It is also produced when many carbonates are heated:

$$CaCO_3(s) \longrightarrow CaO(s) + CO_2(g) \quad [20.71]$$

Large quantities are also obtained as a by-product of the fermentation of sugar during the production of alcohol:

$$\underset{\text{Glucose}}{C_6H_{12}O_6(aq)} \xrightarrow{\text{yeast}} \underset{\text{Ethanol}}{2C_2H_5OH(aq)} + 2CO_2(g) \quad [20.72]$$

In the laboratory CO_2 is normally produced by the action of acids on carbonates:

$$CO_3^{2-}(aq) + 2H^+(aq) \longrightarrow CO_2(g) + H_2O(l) \quad [20.73]$$

Carbon dioxide is a colorless and odorless gas. It is not toxic, but high concentrations increase respiration rate and can cause suffocation. It is readily liquefied by compression. However, when cooled at atmospheric pressure it condenses as a solid rather than as a liquid. The solid sublimes at atmospheric pressure (at $-78°C$). This property makes solid CO_2 valuable as a refrigerant that is always free of the liquid form; solid CO_2 is thus known as **dry ice.** About half of the CO_2 consumed annually is used for such refrigeration. The other major use is in the production of carbonated beverages. Large quantities are also used to manufacture **washing soda,** $Na_2CO_3 \cdot 10H_2O$, and **baking soda,** $NaHCO_3$. Baking soda is so named because the following reaction occurs in baking:

$$NaHCO_3(s) + H^+(aq) \longrightarrow Na^+(aq) + CO_2(g) + H_2O(l) \quad [20.74]$$

The $H^+(aq)$ is provided by vinegar, sour milk, or by hydrolysis of certain salts. The bubbles of CO_2 that form are trapped in the dough, causing it to raise. Washing soda is used to precipitate metal ions that interfere with the cleansing action of soap (Section 12.6).

Carbonic Acid and Carbonates

Carbon dioxide is moderately soluble in H_2O at atmospheric pressure. The resultant solutions are moderately acidic as a result of the formation of carbonic acid, H_2CO_3:

$$CO_2(aq) + H_2O(l) \rightleftharpoons H_2CO_3(aq) \quad [20.75]$$

Carbonic acid is a weak diprotic acid. Its acidic character causes carbonated beverages to have a sharp, slightly acidic taste.

Although carbonic acid cannot be isolated as a pure compound, two types of salts can be obtained by neutralization of carbonic acid solutions: hydrogen carbonates (bicarbonates) and carbonates. Partial neu-

FIGURE 20.28 Sinkhole filled with water. (© *1980 Kent and Donna Dannen; Photo Researchers, Inc.*)

tralization produces HCO_3^-, whereas complete neutralization gives CO_3^{2-}.

The HCO_3^- ion is a stronger base than acid ($K_a = 5.6 \times 10^{-11}$; $K_b = 2.3 \times 10^{-8}$). Consequently, aqueous solutions of HCO_3^- are weakly alkaline:

$$HCO_3^-(aq) + H_2O(l) \rightleftharpoons H_2CO_3(aq) + OH^-(aq) \qquad [20.76]$$

The carbonate ion is much more strongly basic ($K_b = 1.8 \times 10^{-4}$):

$$CO_3^{2-}(aq) + H_2O(l) \rightleftharpoons HCO_3^-(aq) + OH^-(aq) \qquad [20.77]$$

Minerals containing the carbonate ion are plentiful.* The principal carbonate minerals are calcite ($CaCO_3$), magnesite ($MgCO_3$), dolomite [$MgCa(CO_3)_2$], and siderite ($FeCO_3$). Calcite is the principal mineral in limestone rock, large deposits of which occur in many parts of the world. It is also the main constituent of marble, chalk, pearls, coral reefs, and the shells of marine animals such as clams and oysters. Although $CaCO_3$ has low solubility in pure water, it dissolves readily in acidic solutions, with evolution of CO_2:

$$CaCO_3(s) + 2H^+(aq) \longrightarrow Ca^{2+}(aq) + H_2O(l) + CO_2(g) \qquad [20.78]$$

Because water containing CO_2 is slightly acidic (Equation 20.75), $CaCO_3$ dissolves slowly in this medium:

**Minerals* are solid substances that occur in nature. They are usually known by their common names rather than by their chemical names. What we know as *rock* is merely an aggregate of different kinds of minerals.

$$CaCO_3(s) + H_2O(l) + CO_2(g) \rightleftharpoons Ca^{2+}(aq) + 2HCO_3^-(aq) \quad [20.79]$$

This reaction occurs when surface waters move underground through limestone deposits. It is the principal way that Ca^{2+} enters groundwater, producing "hard water" (Section 12.6). If the dissolving limestone underlies a comparatively thin layer of earth, sinkholes such as that shown in Figure 20.28 are produced. If the limestone deposit is deep enough underground, the dissolution of the limestone produces a cave; two well-known limestone caves are the Mammoth Cave in Kentucky and the Carlsbad Caverns in New Mexico (Figure 20.29).

The chemical reaction between $CaCO_3$ and acidic solutions formed by CO_2 is responsible for the erosion of marble and limestone monuments. The greatly increased concentrations of SO_2 in rainfall resulting from industrial processes has accelerated this corrosion process. To slow the erosion of monuments some have been treated with a mixture of $Ba(OH)_2$ and urea, $(NH_2)_2CO$. These two substances react with each other to form a layer of $BaCO_3$ on the monument's surface:

$$H_2N\overset{\overset{\displaystyle O}{\|}}{C}NH_2(aq) + H_2O(l) \longrightarrow 2NH_3(aq) + CO_2(aq) \quad [20.80]$$

$$CO_2(aq) + Ba^{2+}(aq) + 2OH^-(aq) \longrightarrow BaCO_3(s) + H_2O(l) \quad [20.81]$$

Barium carbonate ($K_{sp} = 5.1 \times 10^{-9}$) is about as insoluble as calcium carbonate ($K_{sp} = 2.8 \times 10^{-9}$). However, when it reacts with SO_2, it forms an even more insoluble compound, $BaSO_4$, as shown in Equation 20.82:

$$2BaCO_3(s) + 2SO_2(aq) + O_2(g) \longrightarrow 2BaSO_4(s) + 2CO_2(g) \quad [20.82]$$

The solubility product of $BaSO_4$ is 1.1×10^{-10}, whereas that of $CaSO_4$ is only 9.1×10^{-6}.

One of the most important reactions of $CaCO_3$ is its decomposition into CaO and CO_2 at elevated temperatures, given earlier in Equation 20.71. Over 1.3×10^{10} kg (15 million tons) of calcium oxide, known as lime or quicklime, is used in the United States each year. Because calcium oxide reacts with water to form $Ca(OH)_2$, it is an important com-

FIGURE 20.29 The Queen's Chamber, Carlsbad Caverns, New Mexico. (*NPS, Peter Sanchez*)

mercial base. It is also important in making mortar, a mixture of sand, water, and CaO used in construction to bind bricks, blocks, and rocks together. The CaO reacts with water and CO_2 to form $CaCO_3$, which binds the sand in the mortar:

$$CaO(s) + H_2O(l) \rightleftharpoons Ca^{2+}(aq) + 2OH^-(aq) \qquad [20.83]$$

$$Ca^{2+}(aq) + 2OH^-(aq) + CO_2(aq) \longrightarrow CaCO_3(s) + H_2O(l) \qquad [20.84]$$

Carbides

The binary compounds of carbon with metals, metalloids, and certain nonmetals are called carbides. There are three types: ionic, interstitial, and covalent. The ionic, or saltlike, carbides are formed by the more active metals. The most common ionic carbides contain C_2^{2-} ions: $:C{\equiv}C:^{2-}$. This ion hydrolyzes to form acetylene:

$$CaC_2(s) + 2H_2O(l) \longrightarrow Ca(OH)_2(aq) + C_2H_2(g) \qquad [20.85]$$

Carbides containing C_2^{2-} are thus referred to as **acetylides.** The most important ionic carbide is calcium carbide, CaC_2, which is produced industrially by the reduction of CaO with carbon at high temperature:

$$2CaO(s) + 5C(s) \longrightarrow 2CaC_2(s) + CO_2(g) \qquad [20.86]$$

The CaC_2 is used to prepare acetylene, which is used in welding.

Interstitial carbides are formed by many transition metals. The carbon atoms occupy open spaces (interstices) between metal atoms, in a manner analogous to the interstitial hydrides (Section 20.3). An example is tungsten carbide, WC, which is very hard and heat resistant and consequently is used in making cutting tools.

Covalent carbides are formed by boron and silicon. Silicon carbide, SiC, is known as carborundum. It is made by heating SiO_2 (sand) and carbon to high temperatures:

$$SiO_2(s) + 3C(s) \longrightarrow SiC(s) + 2CO(g) \qquad [20.87]$$

SiC has the same structure as diamond except that silicon atoms alternate with carbon atoms. Like industrial diamond, carborundum is very hard and thus used as an abrasive and in cutting tools.

Other Inorganic Compounds of Carbon

Hydrogen cyanide, HCN (Figure 20.30), is an extremely toxic gas that has the odor of bitter almonds. It is formed commercially by passing a mixture of methane, ammonia, and air over a catalyst at 800°C:

$$2CH_4(g) + 2NH_3(g) + 3O_2(g) \longrightarrow 2HCN(g) + 6H_2O(g) \qquad [20.88]$$

In the laboratory it is made by heating a cyanide salt such as NaCN with an acid:

$$CN^-(aq) + H^+(aq) \longrightarrow HCN(g) \qquad [20.89]$$

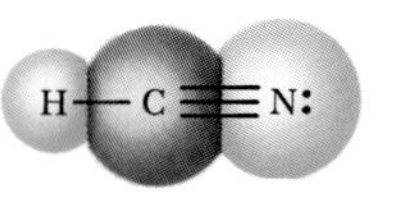

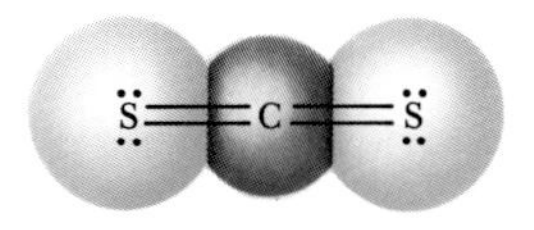

FIGURE 20.30 Structures of hydrogen cyanide and carbon disulfide.

Aqueous solutions of HCN are known as hydrocyanic acid. Neutralization with a base such as NaOH produces cyanide salts such as NaCN. Cyanides find use in the manufacture of several well-known plastics, including nylon and Orlon. The CN^- ion forms very stable complexes with most transition metals (Sections 22.4 and 24.1). The toxic action of CN^- is caused by its combination with iron(III) in cytochrome oxidase, a key enzyme that promotes respiration.

Carbon disulfide, CS_2, is an important industrial solvent for waxes, greases, celluloses, and other nonpolar substances. It is a colorless, volatile liquid (b.p. 46.3°C). The vapor is very poisonous and highly flammable. The compound is formed by direct reaction of carbon and sulfur at high temperature.

FOR REVIEW

Summary

Metallic and nonmetallic elements are distinguished by several physical and chemical properties. Nonmetals lack luster, are not malleable or ductile, and are not good conductors of heat or electricity. They possess higher ionization energies and electronegativities than do metallic elements.

Among elements of a given family, size increases with increasing atomic number. Correspondingly, electronegativity and ionization energy decrease. Metallic character parallels electronegativity trends. Among the nonmetallic elements, the first member of each family differs dramatically from the other members; it forms a maximum of four bonds to other atoms (that is, it is confined to an octet of valence-shell electrons). Second, it exhibits a much greater tendency to form π bonds than do the heavier elements in that family.

Hydrogen has three isotopes: **protium** (^1_1H), **deuterium** (^2_1H), and **tritium** (^3_1H). Hydrogen is not a member of any periodic family, although it is usually placed above lithium. The hydrogen atom can either lose an electron, forming H^+, or gain one, forming H^-, the **hydride** ion. Because the H—H bond is relatively strong, H_2 is fairly unreactive unless activated by heat, irradiation, or catalysis. Industrially, H_2O or hydrocarbons are used as sources of hydrogen; in the lab H_2O is usually obtained by the action of acids on active metals such as Zn. The binary compounds of hydrogen are of three general types: ionic hydrides (formed by active metals), metallic hydrides (formed by transition metals), and molecular hydrides (formed by nonmetals).

Oxygen exhibits two allotropes, O_2 and O_3 (ozone). O_2 is separated from air; small amounts can be obtained by heating $KClO_3$. Ozone is obtained by passing an electrical discharge through O_2. Both O_2 and O_3 are good oxidizing agents, but O_3 is the stronger. Oxygen normally exhibits an oxidation state of -2 in compounds (oxides). The soluble oxides of nonmetals generally produce acidic aqueous solutions; they are called **acidic anhydrides.** By contrast, soluble metal oxides produce basic solutions and are referred to as **basic anhydrides. Peroxides** contain O—O bonds and oxygen in a -1 oxidation state. In superoxides, which contain the O_2^- ion, oxygen has an oxidation state of $-\frac{1}{2}$.

The primary source of nitrogen is the atmosphere, where it occurs as N_2 molecules. Molecular nitrogen is chemically very stable because of the strong N≡N bond. In its compounds nitrogen exhibits oxidation states ranging from -3 to $+5$. The most important process for converting N_2 into compounds is the **Haber process,** used to prepare ammonia. Another commercially important process is the **Ostwald process,** which involves preparation of HNO_3 beginning with the catalytic oxidation of NH_3. Nitrogen has three common oxides: N_2O **(nitrous oxide),** NO **(nitric oxide),** and NO_2 **(nitrogen dioxide).** It has two common oxyacids, HNO_3 **(nitrous acid)** and HNO_3

(nitric acid). Nitric acid is both a strong acid and a good oxidizing agent. Other important nitrogen compounds include hydrazine, N_2H_4, hydroxylamine, NH_2OH, and hydrogen azide, HN_3. Nitrogen compounds are important fertilizers.

Carbon exhibits two allotropes, diamond and graphite. Amorphous forms of graphite include charcoal, carbon black, and coke. Carbon exhibits two common oxides, CO and CO_2. Aqueous solutions of CO_2 produce carbonic acid, H_2CO_3, which is the parent acid of carbonate salts. Binary compounds of carbon are called carbides. Carbides may be ionic, interstitial, or covalent. Calcium carbide, CaC_2, is used to prepare acetylene.

Learning goals

Having read and studied this chapter, you should be able to:

1 Identify an element as a metal, semimetal, or nonmetal, based on its position in the periodic table or its properties.

2 Give examples of how the first member in each family of nonmetallic elements differs from the other elements of the same family and account for these differences.

3 Predict the relative electronegativities and metallic character of any two members of a periodic family or a horizontal row of the periodic table.

4 Cite the most common occurrences of each element discussed in this chapter and how each is obtained in its elemental form.

5 Cite at least two uses for each element discussed in this chapter.

6 Cite the instances where allotropy is observed among the elements discussed in this chapter, and cite the most common molecular form of each element.

7 Describe and name the three isotopes of hydrogen.

8 Distinguish between ionic, metallic, and molecular hydrides.

9 Explain what is meant by the terms *basic anhydride* and *acidic anhydride* and give examples of each.

10 Distinguish between oxides, peroxides, and superoxides.

11 Describe the chemical and physical properties of hydrogen peroxide and its method of preparation.

12 Write balanced chemical equations for formation of nitric acid via the Ostwald process starting from NH_3.

13 Cite at least one example each of a nitrogen compound in which nitrogen is in each of the oxidation states from -3 to $+5$ and be able to name them.

14 Account for the differences in physical properties of diamond and graphite based on their structures.

15 Be able to identify and name the important inorganic compounds of carbon discussed in this chapter, and describe their salient chemical and physical properties.

16 Distinguish between ionic, interstitial, and covalent carbides.

Key Terms

Among the more important terms and expressions used for the first time in this chapter are the following:

An acidic anhydride (Section 20.4) is an oxide that forms an acid when added to water; soluble nonmetal oxides are acidic anhydrides.

A basic anhydride (Section 20.4) is an oxide that forms a base when added to water; soluble metal oxides are basic anhydrides.

A disproportionation reaction (Section 20.5) is one in which a species undergoes simultaneous oxidation and reduction [as in $N_2O_3(g) \longrightarrow NO(g) + NO_2(g)$].

The Ostwald process (Section 20.5) is used to make nitric acid from ammonia. The NH_3 is catalytically oxidized by O_2 to form NO; NO in air is oxidized to NO_2; HNO_3 is formed in a disproportionation reaction when NO_2 dissolves in water.

EXERCISES

Periodic trends

20.1 State two chemical and two physical characteristics that distinguish a nonmetallic element from a metallic one.

20.2 State two important ways in which the first member of each family of nonmetals differs from subsequent members.

20.3 Identify each of the following elements as a metal, nonmetal, or semimetal: (a) germanium; (b) rubidium; (c) xenon; (d) selenium; (e) antimony; (f) zirconium.

20.4 Which element of group 5A would you expect to have the most metallic character? Describe two properties of the elements in the group that should relate to this degree of metallic character.

20.5 Consider the following list of elements: Li, K, N, P, Ne, Ar. From this list select the element that (a) is most electronegative; (b) has the greatest metallic character; (c) most readily forms a positive ion; (d) has the smallest atomic radius; (e) forms π bonds most readily.

20.6 In each of the following pairs of substances, select the one that has the more polar bonds: (a) NF_3 or IF_3; (b) N_2O_3 or B_2O_3; (c) BN or BF_3; (d) HF or H_2Te; (e) CO_2 or SnO_2. Explain your choice in each case.

20.7 The highest fluoride compound formed by nitrogen is NF_3, whereas phosphorus and arsenic readily form PF_5 and AsF_5, respectively. Account for this difference.

20.8 Explain the following observations. (a) Although CO is a well-known compound, SiO doesn't exist under ordinary conditions. (b) AsH_3 is a stronger reducing agent than NH_3. (c) HNO_3 is a stronger oxidizing agent than H_3PO_4. (d) Cl_2 is capable of oxidizing sulfur in H_2S. (e) Silicon is able to form a compound with six fluorides, SiF_6, whereas carbon is able to bond to a maximum of four, CF_4.

[20.9] Using this chapter and material elsewhere in the text (use the index), compile a list of physical and chemical properties of the elements oxygen and zinc. How do these properties justify the classification of one as a nonmetal and the other as a metal?

[20.10] The electronegativity of a given element can be regarded as varying with its oxidation state. How would you expect the electronegativity to change as a function of oxidation state? Although manganese (Mn) is completely different from chlorine in properties as an element, the characteristics of MnO_4^- rather closely resemble those of ClO_4^-. Cite at least two instances of this similarity (you may need to look in a handbook; think in terms of acid-base properties, oxidation-reduction behavior, solubility, and so on). Discuss your observations in terms of the oxidation states, electron configurations, and electronegativities of the central atoms.

20.11 Complete and balance the following equations:

(a) $NiO(s) + C(s) \longrightarrow$
(b) $C_3H_7OH(l) + O_2(g) \longrightarrow$
(c) $NaOCH_3(s) + H_2O(l) \longrightarrow$
(d) $Na_2O(s) + HC_2H_3O_2(aq) \longrightarrow$
(e) $WO_3(s) + H_2(g) \longrightarrow$
(f) $Al_4C_3(s) + H_2O(l) \longrightarrow$
(g) $Na_2S(s) + HCl(aq) \longrightarrow$
(h) $NH_2OH(l) + O_2(g) \longrightarrow$
(i) $AlP(s) + H_2O(aq) \longrightarrow$

Hydrogen

20.12 Why are the properties of hydrogen different from those of either the group 1A or 7A elements?

20.13 Give the names and chemical symbols for the three isotopes of hydrogen. Which isotope is radioactive? Write a nuclear equation for the process that occurs when this isotope undergoes radioactive decay.

20.14 Give a balanced chemical equation for the preparation of hydrogen using (a) Mg; (b) H_2O; (c) CH_4; (d) C.

20.15 List three industrial uses for hydrogen.

20.16 Complete and balance the following equations:

(a) $NaH(s) + H_2O(l) \longrightarrow$
(b) $Fe(s) + H_2SO_4(aq) \longrightarrow$
(c) $H_2(g) + Br_2(g) \longrightarrow$
(d) $Na(l) + H_2(g) \longrightarrow$
(e) $PbO(s) + H_2(g) \longrightarrow$

20.17 Write balanced chemical equations for each of the following reactions (some of these are analogous to reactions shown in the chapter). (a) Aluminum metal reacts with acids to form hydrogen gas. (b) Steam reacts with magnesium metal to give magnesium oxide and hydrogen (magnesium fires cannot be put out with water). (c) Manganese(IV) oxide is reduced to manganese(II) oxide by hydrogen gas. (d) Calcium hydride reacts with water to generate hydrogen gas.

20.18 Identify the following hydrides as ionic, metallic, or molecular: (a) BaH_2; (b) H_2Te; (c) $TiH_{1.7}$; (d) B_2H_6; (e) RbH.

20.19 About 1.1×10^{10} kg of ethylene, C_2H_4, is produced annually in the United States by the pyrolysis of ethane, C_2H_6: $C_2H_6(g) \longrightarrow C_2H_4(g) + H_2(g)$. How many kilograms of H_2 are produced as a by-product?

20.20 Hydrogen gas has a higher fuel value than natural gas on a weight basis but not on a volume basis. Thus hydrogen is not competitive with natural gas as a fuel transported long distance through pipelines. Calculate the heat of combustion of H_2 and of CH_4 (the principal component of natural gas) (a) per mole of each; (b) per gram of each; (c) per cubic meter of each at STP. Assume $H_2O(l)$ as a product.

Oxygen

20.21 What is the oxidation state of O in the following compounds: (a) MgO; (b) BaO_2; (c) OF_2; (d) SO_2; (e) H_2O_2?

20.22 List three industrial uses for oxygen.

20.23 Complete and balance the following equations:

(a) $CaO(s) + H_2O(l) \longrightarrow$
(b) $Al_2O_3(s) + H^+(aq) \longrightarrow$
(c) $Na_2O_2(s) + H_2O(l) \longrightarrow$
(d) $N_2O_3(g) + H_2O(l) \longrightarrow$
(e) $KO_2(s) + H_2O(l) \longrightarrow$
(f) $NO(g) + O_3(g) \longrightarrow$

20.24 Write the balanced chemical equations for each of the following reactions. (a) When mercury(II) oxide is heated, it decomposes to form O_2 and mercury metal. (b) When copper(II) nitrate is heated strongly, it decomposes to form copper(II) oxide, nitrogen dioxide, and oxygen. (c) Lead(II) sulfide, PbS(s), reacts with ozone to form

$PbSO_4(s)$ and $O_2(g)$. (d) When heated in air, $ZnS(s)$ is converted to ZnO. (e) Potassium peroxide reacts with $CO_2(g)$ to give the carbonate ion and O_2. (f) Although silver does not react with oxygen at room temperature, it reacts with ozone, forming Ag_2O.

20.25 Write balanced chemical equations showing the amphoterism of ZnO.

20.26 In acidic solution the ferrocyanide ion, $Fe(CN)_6^{4-}$, is oxidized by H_2O_2 to the ferricyanide ion, $Fe(CN)_6^{3-}$. In basic solution, the ferricyanide ion is reduced by H_2O_2 to the ferrocyanide ion. Write balanced net ionic equations for both reactions.

Nitrogen

20.27 Write the formulas for each of the following compounds and indicate the oxidation state of nitrogen in each: (a) nitrous acid; (b) hydrazine; (c) potassium azide; (d) sodium cyanide; (e) ammonium chloride; (f) nitrous oxide; (g) ammonia; (h) hydroxylamine; (i) nitric oxide; (j) nitric acid; (k) dinitrogen pentoxide; (l) lithium nitride.

20.28 List three industrial uses for nitrogen.

20.29 Write the Lewis structures for each of the following species: (a) NH_4^+; (b) $N_2H_5^+$; (c) HNO_3; (d) NO_2; (e) NO_2^-; (f) N_2O.

20.30 Complete and balance the following equations:

(a) $Mg_3N_2(s) + H_2O(l) \longrightarrow$
(b) $NO(g) + O_2(g) \longrightarrow$
(c) $NH_3(g) + O_2(g) \xrightarrow{\Delta}$ (no catalyst)
(d) $NaNH_2(s) + H_2O(l) \longrightarrow$
(e) $N_2O_5(g) + H_2O(l) \longrightarrow$
(f) $NH_3(aq) + H^+(aq) \longrightarrow$
(g) $N_2H_4(aq) + H^+(aq) \longrightarrow$

20.31 Using chemical equations, outline the preparation of nitric acid starting with only air and water as starting materials.

20.32 Write balanced net ionic equations for each of the following reactions. (a) Dilute nitric acid reacts with zinc metal with formation of nitrous oxide. (b) Concentrated nitric acid reacts with sulfur with formation of nitrogen dioxide. (c) Concentrated nitric acid oxidizes sulfur dioxide with formation of nitric oxide. (d) Urea reacts with water to form ammonia and carbon dioxide.

20.33 Write balanced net ionic equations for each of the following reactions. (a) Hydrazine is burned in excess fluorine gas, forming NF_3. (b) Hydrazine reduces CrO_4^{2-} to $Cr(OH)_4^-$ (hydrazine is oxidized to N_2). (c) Hydroxylamine is oxidized to N_2 by Cu^{2+} in aqueous solution (Cu^{2+} is reduced to copper metal). (d) Aqueous azide ion reacts with Cl_2, forming N_2.

20.34 Hydrazine has been employed as a reducing agent for metals. Using standard electrode potentials, predict whether the following metals can be reduced to the metallic state by hydrazine under standard conditions in acidic solution: (a) Fe^{2+}; (b) Sn^{2+}; (c) Cu^{2+}; (d) Ag^+; (e) Cr^{3+}.

20.35 Write complete balanced half-reactions for (a) reduction of nitrate ion to N_2 in acidic solution; (b) oxidation of NH_4^+ to N_2 in acidic solution. What is the standard electrode potential in each case? (See Figure 20.17.)

[20.36] Write the complete balanced half-reaction for (a) reduction of NO_2 to NO in acidic solution; (b) oxidation of NH_4^+ to NO_3^- in acidic solution. What is the standard potential in each case? (See Figure 20.17.)

20.37 Each of the following compounds is used as a nitrogen fertilizer: $(NH_2)_2CO$ (urea), NH_3, $(NH_4)_2SO_4$, and $NaNO_3$. Calculate the percentage of nitrogen in each compound.

20.38 A traditional laboratory preparation of nitrogen gas involves the decomposition of ammonium nitrite. Write a balanced equation for this reaction. Calculate the volume of nitrogen gas collected over water at 22°C and 1 atm pressure from decomposition of 3.26 g of NH_4NO_2.

[20.39] Thermodynamic data for $HNO_3(aq)$ at 25°C are as follows: $\Delta H_f^\circ = -207.3$ kJ/mol; $\Delta G_f^\circ = -111.3$ kJ/mol. Calculate ΔH° and ΔG° for the reaction

$$\tfrac{1}{2}N_2(g) + \tfrac{1}{2}H_2O(g) + \tfrac{5}{4}O_2(g) \longrightarrow HNO_3(aq)$$

It has been suggested that if a suitable catalyst were present, the atmospheric gases could react to make the oceans a dilute nitric acid solution. In thermodynamic terms, is this a possible process?

Carbon

20.40 Write the Lewis structures of each of the following species: (a) CN^-; (b) CO; (c) C_2^{2-}; (d) CS_2; (e) CO_2; (f) CO_3^{2-}.

20.41 Give the chemical formula for (a) calcium acetylide; (b) carborundum; (c) sodium cyanide; (d) calcium carbonate; (e) potassium hydrogen carbonate.

20.42 Give three industrial uses for (a) carbon dioxide; (b) carbon monoxide.

20.43 What aspects of the chemical behavior of the following compounds makes them hazardous to work with: (a) CS_2; (b) NaCN; (c) CO?

20.44 Calculate the heat of combustion of diamond and of graphite (forming CO_2).

20.45 Rationalize the general trend in Lewis basicities in the following isoelectronic series: $N_2 < CO < CN^-$.

20.46 Complete and balance the following equations:

(a) $ZnCO_3(s) \xrightarrow{\Delta}$
(b) $BaC_2(s) + H_2O(l) \longrightarrow$
(c) $C_2H_4(g) + O_2(g) \longrightarrow$
(d) $CH_3OH(l) + O_2(g) \longrightarrow$
(e) $NaCN(s) + HCl(aq) \longrightarrow$
(f) $CO_2(g) + OH^-(aq) \longrightarrow$
(g) $NaHCO_3(s) + H^+(aq) \longrightarrow$
(h) $CaO(s) + C(s) \xrightarrow{\Delta}$
(i) $C(s) + H_2O(g) \xrightarrow{\Delta}$
(j) $CuO(s) + CO(g) \longrightarrow$

20.47 Write balanced chemical equations for each of the following reactions. (a) Burning magnesium metal in a carbon dioxide atmosphere reduces the CO_2 to carbon. (b) In photosynthesis, solar energy is used to produce glucose, $C_6H_{12}O_6$, and O_2 out of carbon dioxide and water. (c) When carbonate salts dissolve in water, they produce basic solutions.

20.48 Indicate the geometry and the types of hybrid orbitals used by each carbon atom in the following species: (a) $CH_3C{\equiv}CH$; (b) NaCN; (c) CS_2; (d) C_2H_6.

20.49 What volume of dry acetylene, measured at 27°C and 720 mm Hg, is formed when 10.0 g of calcium carbide is placed in water?

Additional exercises

20.50 Account for the fact that H_2 is a more costly chemical than O_2.

20.51 Starting with D_2O, suggest a preparation for (a) ND_3; (b) D_2SO_4; (c) NaOD; (d) DNO_3; (e) C_2D_2; (f) DCN.

20.52 The annual production of H_2, N_2, and O_2 in the United States is generally reported in cubic feet, measured at STP. If 2.0×10^8 kg of H_2, 1.1×10^{10} kg of N_2, and 1.4×10^{10} kg of O_2 are produced in a particular year, how many cubic feet are produced of (a) H_2; (b) N_2; (c) O_2?

20.53 Write a balanced chemical equation for the reaction of each of the following compounds with water: (a) $SO_2(g)$; (b) $Cl_2O(g)$; (c) $Na_2O(s)$; (d) $BaC_2(s)$; (e) $RbO_2(s)$; (f) $Mg_3N_2(s)$; (g) $Na_2O_2(s)$; (h) $NaH(s)$.

20.54 Select the more acidic member of each of the following pairs: (a) Mn_2O_7 and MnO_2; (b) SnO and SnO_2; (c) SO_2 and SO_3; (d) SiO_2 and SO_2; (e) Ga_2O_3 and In_2O_3; (f) SO_2 and SeO_2.

20.55 What is the anhydride of each of the following acids: (a) H_2SO_3; (b) HNO_3; (c) H_3PO_4; (d) $HClO_4$?

20.56 Write a series of chemical equations that describes how Na_2CO_3 could be prepared starting with only H_2O, NaCl, and $CaCO_3$ as raw materials.

20.57 Hydrogen peroxide is capable of oxidizing (a) K_2S to S; (b) SO_2 to SO_4^{2-}; (c) NO_2^- to NO_3^-; (d) As_2O_3 to AsO_4^{3-}; (e) Fe^{2+} to Fe^{3+}. Write balanced net ionic equations for each of these oxidations.

20.58 Hydrogen peroxide reduces (a) MnO_4^- to Mn^{2+}; (b) Cl_2 to Cl^-; (c) Ce^{4+} to Ce^{3+}; (d) O_3 to H_2O. Write balanced net ionic equations for each of these reductions.

20.59 Hydrogen peroxide can be prepared in the laboratory by air oxidation of barium metal followed by treatment of the resultant product with dilute sulfuric acid. Write balanced chemical equations for the two reactions that take place.

20.60 Hydrazine has long been used to help control corrosion in boilers. It does so by oxidizing Fe to Fe_3O_4 and by reducing Fe_2O_3 to Fe_3O_4. The Fe_3O_4 forms a protective film on the boiler walls. Write balanced chemical equations for the oxidation of Fe and reduction of Fe_2O_3 by hydrazine.

20.61 Both dimethylhydrazine, $(CH_3)_2NNH_2$, and monomethylhydrazine, CH_3HNNH_2, have been used as rocket fuels. When dinitrogen tetraoxide, N_2O_4, is used as the oxidizer, the products are H_2O, CO_2, and N_2. If the thrust of the rocket depends on the volume of the products produced, which of the substituted hydrazines produces a greater thrust per gram total weight of oxidizer plus fuel. [Assume that both fuels generate the same temperature and that $H_2O(g)$ is formed.]

20.62 Suggest why each of the following compounds is either unstable or does not exist: (a) NCl_5; (b) H_3NO_4; (c) $(CH_3)_2Si{=}O$; (d) P_2O; (e) H_3.

20.63 Explain why N_2O is able to support combustion, though not as completely as pure oxygen.

20.64 A commerical lawn fertilizer contains 18 percent nitrogen by weight. All of this nitrogen is present as urea. What is the weight percentage of urea in the fertilizer?

20.65 Annual production of nitric acid in the United States is about 7.3×10^9 kg. What volume of ammonia, measured at 45 atm and 25°C, is required for production of this much HNO_3 via the Ostwald process, assuming an overall 92 percent conversion efficiency?

20.66 It is estimated that 95 percent of the ammonia production in the United States uses H_2 obtained from natural gas (Equations 20.12 and 20.13). If 50 percent of the CH_4 used to manufacture H_2 is burned to maintain proper temperature for ammonia synthesis, how many kilograms of CH_4 are consumed in supplying the 1.5×10^{10} kg of NH_3 produced annually?

20.67 From the thermodynamic data in Appendix D, calculate $\Delta H°$ and $\Delta G°$ at 25°C for the oxidation of NH_3 in the following reactions:

$$4NH_3(g) + 5O_2(g) \longrightarrow 4NO(g) + 6H_2O(g)$$
$$4NH_3(g) + 3O_2(g) \longrightarrow 2N_2(g) + 6H_2O(g)$$

20.68 K_2C_2 is an acetylide, whereas Mg_2C is a methanide (it contains C^{4-}). Write balanced equations for the hydrolysis of each compound.

20.69 A certain carbide of magnesium is reacted with water to form $Mg^{2+}(aq)$ and a volatile hydrocarbon. Hydrolysis of 0.3052 g of the carbide produces 0.1443 g of the hydrocarbon. The hydrocarbon consists of 90.0 percent C and 10.0 percent H. The density of the hydrocarbon gas at 25°C and 742 mm Hg is 1.60 g/L. What is the formula for the carbide, and what is the molecular formula for the hydrocarbon?

[20.70] Account for the fact that an acidic solution of hydrogen peroxide oxidizes $Fe(CN)_6^{4-}$ to $Fe(CN)_6^{3-}$, whereas in basic solution the $Fe(CN)_6^{3-}$ oxidizes H_2O_2. Write balanced equations and calculate the equilibrium constant for each reaction from $E°$ values.

20.71 Peroxosulfuric acid, $HO_3S{-}O{-}O{-}SO_3H$, can be made by reaction of pure H_2O_2 with chlorosulfonic acid, $ClSO_3H$. Write a balanced chemical equation for (a) the hydrolysis of $H_2S_2O_8$ by water; (b) reaction of $H_2S_2O_8$ with aqueous KI to form I_2.

20.72 Complete and balance the following equations:

(a) $Li_3N(s) + H_2O(l) \longrightarrow$

(b) $NH_3(aq) + H_2O(l) \longrightarrow$

(c) $NO_2(g) + H_2O(l) \longrightarrow$

(d) $2NO_2(g) \longrightarrow$

(e) $NH_3(g) + O_2(g) \xrightarrow{\text{catalyst}}$

(f) $CO(g) + O_2(g) \longrightarrow$

(g) $H_2CO_3(aq) \xrightarrow{\Delta}$

(h) $Ni(s) + CO(g) \longrightarrow$

(i) $CS_2 + O_2(g) \longrightarrow$

(j) $CaO(s) + SO_2(g) \longrightarrow$

(k) $Na(s) + H_2O(l) \longrightarrow$

(l) $CH_4(g) + H_2O(g) \xrightarrow{\Delta}$

(m) $LiH(s) + H_2O(l) \longrightarrow$

(n) $Fe_2O_3(s) + 3H_2(g) \longrightarrow$

20.73 Complete and balance the following equations:

(a) $MnO_4^-(aq) + H_2O_2(aq) + H^+(aq) \longrightarrow$

(b) $Fe^{2+}(aq) + H_2O_2(aq) \longrightarrow$

(c) $I^-(aq) + H_2O_2(aq) + H^+(aq) \longrightarrow$

(d) $MnO_2(s) + H_2O_2(aq) + H^+(aq) \longrightarrow$

(e) $I^-(aq) + O_3(g) \longrightarrow I_2(s) + O_2(g) + OH^-(aq)$

20.74 (a) How many grams of H_2 can be stored in 1.00 kg of the alloy FeTi if the hydride $FeTiH_2$ is formed? (b) What volume does this quantity of H_2 occupy at STP?

21 Further Chemistry of the Nonmetallic Elements

In Chapter 20 we considered some differences between metals and nonmetals and then selected four nonmetals (hydrogen, oxygen, nitrogen, and carbon) for close examination. In this chapter we take a more panoramic view of nonmetals; we will begin with the noble gases and then move, group by group, from right to left through the periodic table. We will, of course, have little to say about those elements discussed in Chapter 20. Of the remaining elements, those that are more important will require a more complete discussion. Thus we will spend most of our time considering the halogens, sulfur, phosphorus, and silicon.

21.1 THE NOBLE GAS ELEMENTS

We have referred at several points in the text to the fact that the elements of group 8A are chemically unreactive. Indeed, most of our references to these elements have been in relation to their physical properties, as when we discussed intermolecular forces in Section 10.4. According to the Lewis theory of chemical bonding, the relative inertness of these elements is due to the formation of a completed octet of valence-shell electrons. The stability of such an arrangement is reflected in the high ionization energies of the group 8A elements (Section 6.5).

The group 8A elements are all gases at room temperature. All are components of the earth's atmosphere, except for radon, which exists only as a short-lived radioisotope. Only argon is relatively abundant (Table 12.1). Argon and the heavier noble gases are recovered from liquid air by fractional distillation. Argon is used as a blanketing atmosphere in electric light bulbs. The gas conducts heat away from the filament but does not react with it. It is also used as a protective atmosphere to prevent oxidation in welding and certain high-temperature metallurgical processes. Neon is used in electric signs; the gas is caused to radiate by passing an electric discharge through the tube.

Helium is, in many ways, the most important of the noble gases.

Helium boils at 4.2 K under 1 atm pressure, the lowest boiling point of any substance. Liquid helium is used as a coolant to conduct experiments at very low temperatures. Because helium has such a low abundance in the atmosphere and boils at such a low temperature, recovery of the gas from the atmosphere would require an immense expenditure of energy. Helium is found in relatively high concentrations in many natural-gas wells. Some of this helium is separated to meet current demands, and a little is kept for later use. Unfortunately, however, most of the helium is allowed to escape.

Because the noble gases are exceedingly stable, it is reasonable to expect that they will undergo reaction only under rather rigorous conditions. Furthermore, we might expect that the heavier noble gases would be most likely to undergo chemical transformation, because the ionization energies are lower for the heavier elements, as illustrated in Figure 6.8. A lower ionization energy would suggest the possibility of losing an electron in formation of an ionic bond. In addition, since the group 8A elements already contain eight electrons in their valence shell (except, of course, for helium, which contains just two), formation of covalent bonds requires an expanded valence shell. We have seen (Section 7.7) that valence-shell expansion occurs most readily with larger atoms.

The first noble gas compound was prepared by Neil Bartlett in 1962. His work caused a sensation, because it undercut the belief that the noble gas elements were truly chemically inert. Bartlett's initial work involved xenon in combination with fluorine, the element we would expect to be most reactive. Since then several xenon compounds of fluorine and oxygen have been prepared. Some properties of these substances are listed in Table 21.1. The three fluorides XeF_2, XeF_4, and XeF_6 are made by direct reaction of the elements. By varying the ratio of reactants and altering reaction conditions, one or the other of the three compounds can be obtained. The oxygen-containing compounds are formed when the fluorides are reacted with water, as in Equations 21.1 and 21.2:

$$XeF_6(s) + H_2O(l) \longrightarrow XeOF_4(l) + 2HF(g) \qquad [21.1]$$

$$XeF_6(s) + 3H_2O(l) \longrightarrow XeO_3(aq) + 6HF(aq) \qquad [21.2]$$

TABLE 21.1 Properties of xenon compounds

Compound	Oxidation state of Xe	Melting point (°C)	ΔH_f° (kJ/mol)[a]
XeF_2	+2	129	−109(*g*)
XeF_4	+4	117	−218(*g*)
XeF_6	+6	49	−298(*g*)
$XeOF_4$	+6	−41 to −28	+146(*l*)
XeO_3	+6	—[b]	+402(*s*)
XeO_2F_2	+6	31	+145(*s*)
XeO_4	+8	—[c]	—

[a]At 25°C, for the compound in the state indicated.
[b]A solid; decomposes at 40°C.
[c]A solid; decomposes at −40°C.

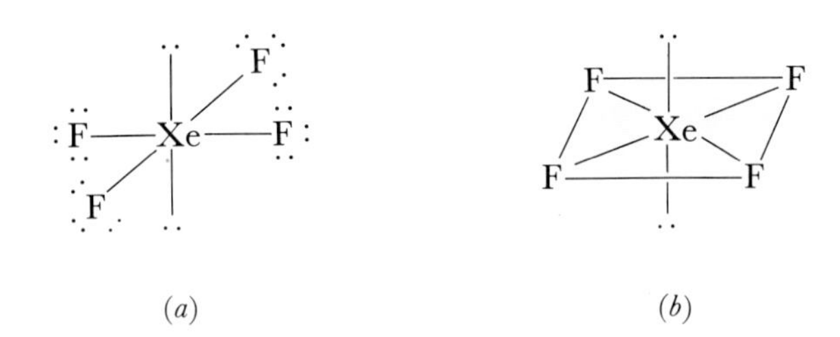

FIGURE 21.1 Lewis and geometrical structures of XeF_4.

SAMPLE EXERCISE 21.1

Predict the structure of XeF_4.

Solution: To predict the structure, we must first write the Lewis structure for the molecule. The total number of valence-shell electrons involved is 36 (8 from xenon and 7 from each of the four fluorines). This leads to the Lewis structure shown in Figure 21.1(*a*). We see that Xe has 12 electrons in its valence shell. We thus expect an octahedral disposition of six electron pairs. Two of these are nonbonded pairs. Because nonbonded pairs have a larger volume requirement than do bonded pairs (Section 8.1), it is reasonable to expect these nonbonded pairs to be opposite one another. The expected structure is square planar, as shown in Figure 21.1(*b*). The experimentally determined structure agrees with this prediction.

The fact that the enthalpies of formation of the xenon fluorides are negative suggests that these compounds should be reasonably stable, and this is indeed found to be the case. They are, however, powerful fluorinating agents and must be handled in containers that do not readily react to form fluorides. Notice that the enthalpies of formation of the oxyfluorides and oxides of xenon are positive (Table 21.1); these compounds are quite unstable.

As we might expect, formation of compounds by the other noble gas elements occurs much less readily than in the case of xenon. Only one binary krypton compound, KrF_2, is known with certainty; it decomposes to the elements at $-10°C$.

21.2 THE HALOGENS

The elements of group 7A consist of fluorine, chlorine, bromine, iodine, and astatine. These elements are known as the halogens, after the Greek words *halos* and *gennao,* meaning "salt-formers." The halogens have played an important part in the development of chemistry as a science. Chlorine was first prepared by the Swedish chemist Karl Wilhelm Scheele in 1774, but it was not until 1810 that the English chemist Humphry Davy identified it as an element. The discovery of iodine followed shortly thereafter, in 1811, and bromine was discovered in 1825. Compounds of fluorine were known for a long time, but it was not until 1886 that the French chemist Henri Moissan succeeded in preparing the very reactive free element.

The general valence-electron configuration of the halogens is ns^2np^5, where n may have values ranging from 2 through 6. When a halogen atom shares all seven of its electrons with a more electronegative atom, it is assigned the $+7$ oxidation state. The halogens also achieve a noble gas configuration by gaining an electron, which results in a -1 oxidation state. Thus we expect that the halogens might occur in oxidation states ranging from $+7$ at one extreme to -1 at the other. Oxidation states over this entire range are observed for all the halogens except fluorine; this most electronegative of all the elements is observed only in the 0 or −1 oxidation states.

Occurrences

Table 21.2 summarizes the occurrences of the halogens in nature. Both fluorine and chlorine are fairly abundant, but they are quite differently distributed. This happens because the salts of chlorine are generally quite soluble, whereas some of those of fluorine are not. Bromine is comparatively much less abundant than chlorine or fluorine, and iodine is much rarer still.

TABLE 21.2 Occurrences of the halogens

Element	Occurrences
Fluorine	Fluorspar, CaF_2; fluorapatite, $Ca_5(PO_4)_3F$; cryolite, Na_3AlF_6 Biologically: teeth, bones
Chlorine	Seawater (0.55 *M*, chiefly as NaCl); salt beds, salt lakes (for example, Dead Sea, Great Salt Lake); NaCl deposits Biologically: gastric juice [as HCl(*aq*)], tissue fluids
Bromine	Seawater (8.3×10^{-4} *M*); underground brines; salt beds, salt lakes Biologically: minor concentrations of bromide along with chloride
Iodine	Seawater (4×10^{-7} *M*); seaweeds; $NaIO_3$ in minor amounts in nitrate deposits; oil-well brines Biologically: in human thyroid gland

Chlorine, bromine, and iodine occur as the halides in seawater and in salt deposits. The concentration of iodine in these sources is generally very small. However, iodine is concentrated by certain seaweeds, which are harvested, dried, and burned. Iodine is then extracted from the ashes. The element is also extracted commercially from oil-well brines in California. Fluorine occurs in the minerals fluorspar (CaF_2), cryolite (Na_3AlF_6), and fluorapatite [$Ca_5(PO_4)_3F$]. Only the first of these is an important commercial source of fluorine for the chemical industry. All isotopes of astatine are radioactive. The longest-lived isotope is astatine-210, which has a half-life of 8.3 h and decays mainly by electron capture. Astatine was first synthesized by bombarding bismuth-209 with high-energy alpha particles, as shown in Equation 21.3:

$$^{209}_{83}Bi + ^{4}_{2}He \longrightarrow ^{211}_{85}At + 2^{1}_{0}n \qquad [21.3]$$

The fact that a cyclotron must be used to synthesize astatine makes it very expensive and limits its application and study.

Properties and Preparation

Before we begin a discussion of the chemical characteristics of the halogens, it will be useful to review key properties of the elements, summarized in Table 21.3. The symbol X represents any one of the halogens. Notice that all of the properties listed except the last two refer to the halogen atom. The last two refer to the diatomic molecules, X_2, with a

TABLE 21.3 Some properties of the halogen atoms

Property	F	Cl	Br	I
Atomic radius (Å)	0.72	1.00	1.15	1.40
Ionic radius, X^- (Å)	1.33	1.81	1.96	2.17
First ionization energy (kJ/mol)	1.68×10^3	1.25×10^3	1.14×10^3	1.01×10^3
Electron affinity (kJ/mol)	-332	-349	-325	-295
Electronegativity	4.0	3.2	3.0	2.7
X—X single-bond energy (kJ/mol)	155	242	193	151
Reduction potential: $\frac{1}{2}X_2(aq) + e^- \longrightarrow X^-(aq)$	2.87	1.36	1.07	0.54

single bond joining the atoms. You will recall from Section 8.7 that this is the stable form of the halogens as free elements.

Most of the properties listed in Table 21.3 vary in a regular fashion as a function of atomic number. Within each horizontal row of the periodic table each halogen has a high ionization energy, second only to the noble gas element adjacent to it in the table. Similarly, each halogen has the highest electronegativity of all the elements in its horizontal row. Within the halogen family, atomic and ionic radii increase with increasing atomic number. Correspondingly, the ionization energy and electronegativity steadily decrease as we go down the family, from fluorine to iodine.

Under ordinary conditions the halogens exist as diatomic molecules. The molecules are held together in the solid and liquid states by London dispersion forces (Section 10.4). Because I_2 is the largest and most polarizable of the halogen molecules, it is not surprising that the intermolecular forces between I_2 molecules are the strongest. Thus I_2 has the highest melting point and boiling point. At room temperature and 1 atm pressure, I_2 is a solid, Br_2 is a liquid, and Cl_2 and F_2 are gases.

The comparatively low bond energy in F_2 (155 kJ/mol) accounts in part for the extreme reactivity of elemental fluorine. Because of its high reactivity, F_2 is very difficult to work with. Certain metals, such as copper and nickel, can be used to contain F_2 because their surfaces form a protective coating of metal fluoride. Chlorine must also be handled with care. Because chlorine liquefies upon compression at room temperature, it is normally stored and handled in liquid form in steel containers. Chlorine and the heavier halogens are reactive, although less so than fluorine. They combine directly with most elements except the rare gases.

Because of the high electronegativities of the halogens compared with other elements, the halogens tend to gain electrons from other substances and thereby serve as oxidizing agents. The oxidizing ability of the halogens, which is indicated by their reduction potentials, decreases going down the group. As a result, we find that a given halogen is able to oxidize the anions of the halogens below it in the family. For example, Cl_2 will oxidize Br^- and I^-, but not F^-.

SAMPLE EXERCISE 21.2

Write the balanced chemical equation for the reaction, if any, that occurs between (a) $I^-(aq)$ and $Br_2(l)$; (b) $Cl^-(aq)$ and $I_2(s)$.

Solution: (a) Br_2 is able to oxidize (remove electrons) from the anions of the halogens below it in the periodic table. Thus it will oxidize I^-:

$$2I^-(aq) + Br_2(l) \longrightarrow I_2(s) + 2Br^-(aq)$$

(b) Cl^- is the anion of a halogen above iodine in the periodic table. Thus I_2 cannot oxidize Cl^-; there is no reaction.

Notice from Table 21.3 that the reduction potential of F_2 is exceptionally high. Fluorine gas readily oxidizes water according to the reaction

$$F_2(aq) + H_2O(l) \longrightarrow 2HF(aq) + \tfrac{1}{2}O_2(g) \qquad E° = +1.64\ V \qquad [21.4]$$

Fluorine cannot be prepared by electrolytic oxidation in water, because water itself is oxidized more readily than F^-, with formation of $O_2(g)$. In practice the element is formed by electrolytic oxidation of anhydrous HF. Because HF is not by itself a good conductor of electricity, a solution of KF in anhydrous HF is used. The KF reacts with HF to form a new salt, $K^+HF_2^-$, which acts as the current carrier in the liquid. (The HF_2^- ion is stable because of the very strong hydrogen bond between the two fluoride ions, as described in Section 10.4.) The overall cell reaction is

$$2KHF_2(l) \longrightarrow H_2(g) + F_2(g) + 2KF(l) \qquad [21.5]$$

Chlorine is produced mainly by electrolysis of either molten or aqueous sodium chloride, as described in Sections 18.6 and 22.5. Both bromine and iodine are obtained commercially from brines containing the halide ions by oxidation with Cl_2. The reaction between I^- and Cl_2 is given in Equation 21.6:

$$2I^-(aq) + Cl_2(g) \longrightarrow 2Cl^-(aq) + I_2(s) \qquad [21.6]$$

Uses of the Halogens

Fluorine has become an important industrial chemical. It is used, for example, to prepare fluorocarbons, very stable carbon-fluorine compounds. An example is CF_2Cl_2, known as Freon 12, which is used as a refrigerant and as a propellant for aerosol cans. As we noted in Section 12.3, the effects of these substances on the ozone layer have been under recent investigation. Fluorocarbons are also used as lubricants and in plastics; Teflon (Figure 21.2) is a polymeric fluorocarbon noted for its high thermal stability and lack of chemical reactivity.

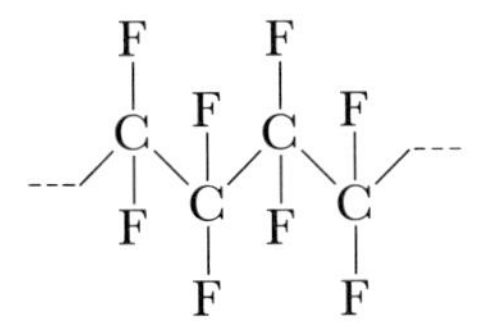

FIGURE 21.2 Structure of Teflon, a fluorocarbon polymer.

Chlorine is by far the most important halogen commercially. About 8.3×10^9 kg (9.1 million tons) of Cl_2 is produced in the United States each year. In addition, hydrogen chloride production is about 2.3×10^9 kg (2.5 million tons) annually. About half of this inorganic chlorine finds its way eventually into vinyl chloride, C_2H_3Cl, used in polyvinyl chloride (PVC) plastics manufacture; ethylene dichloride, $C_2H_2Cl_2$, an organic solvent; and other chlorine-containing organic compounds. Much of the remainder of the chlorine is used as a bleach in the paper and textile industries. When Cl_2 dissolves in cold dilute base it forms Cl^- and the hypochlorite ion, ClO^-:

$$2OH^-(aq) + Cl_2(aq) \longrightarrow Cl^-(aq) + ClO^-(aq) + H_2O(l) \qquad [21.7]$$

Sodium hypochlorite, NaClO, is the active ingredient in some liquid bleaches. Many solid bleaching powders employ Ca(ClO)Cl as their active ingredient. This substance is obtained by reaction of Cl_2 with a suspension of $Ca(OH)_2$:

$$Ca(OH)_2(s) + Cl_2(aq) \longrightarrow Ca^{2+}(aq) + ClO^-(aq) + Cl^-(aq) + H_2O(l) \qquad [21.8]$$

Ca(ClO)Cl is recovered as crystals upon evaporation of the solution. Chlorine is also used in water treatment to oxidize and thereby destroy bacteria (Section 12.6).

Annual U.S. production of bromine is about 2×10^8 kg (220,000 tons). Bromine is used in the production of silver bromide used in photographic film. It is also used to prepare dibromoethane, $C_2H_4Br_2$, a gasoline additive used in gasoline that contains tetraethyl lead, $Pb(C_2H_5)_4$. The dibromoethane prevents formation of lead deposits in the engine by forming volatile $PbBr_2$, which is emitted in exhaust. Bromine-containing compounds also are used as fire retardants in clothing.

Iodine has not found as wide use as the other halogens. One familiar use is its addition, as KI, to salt to form iodized salt. Table salt contains about 0.02 percent potassium iodide by weight. Iodized salt is able to provide the small amount of iodine necessary in our diets; it is essential for the formation of thyroxin, a hormone secreted by the thyroid gland. Lack of iodine in the diet results in an enlarged thyroid gland, a condition called goiter. Iodine is also familiar as tincture of iodine, a solution of I_2 in alcohol that is used as an antiseptic.

The Hydrogen Halides

All of the halogens form stable diatomic molecules with hydrogen. These are very important compounds, in part because aqueous solutions of the hydrogen halides other than HF are strongly acidic. Table 21.4 lists some of the more important properties of the hydrogen halides.

Notice that the melting and boiling points of HF are abnormally high as compared with the other hydrogen halides. The cause of this unusual behavior is the strong hydrogen bonding that exists between HF molecules in the liquid state, as described in Section 10.4. Because HF is a small molecule and is capable of strong hydrogen-bonding interactions, it is completely miscible with water. The other hydrogen halides, though not capable of strong hydrogen-bonding interactions, are also very soluble in water. We learned in Section 15.1 that this very high solubility is the result of reaction of the hydrogen halide with water, producing halide anions and the solvated proton, $H^+(aq)$. Thus the aqueous solutions

TABLE 21.4 Properties of the hydrogen halides

Property	HF	HCl	HBr	HI
Molecular weight	20.01	36.46	80.92	127.91
Melting point (°C)	−83	−114	−87	−51
Boiling point (°C)	19.9	−85	−67	−35
Bond-dissociation energy (kJ/mol)	565	431	364	297
H—X distance (Å)	0.92	1.27	1.41	1.61
Solubility in H_2O (g/100 g H_2O, 10°C)	∞	78	210	234

of the hydrogen halides are acidic; HF is a weak acid in water, whereas the other hydrogen halides are strong acids.

The hydrogen halides can be formed by direct reaction of the elements. The reaction can be explosive in character, as in the reaction of H_2 with F_2 or when reaction between a mixture of H_2 and Cl_2 is initiated by a spark or a photon of sufficient energy. On the other hand, the reaction of either Br_2 or I_2 with hydrogen is much less vigorous, and the reaction mixtures must be heated to cause the system to approach equilibrium.

The most important means of formation of the hydrogen halides is through reaction of a salt of the halide with a strong, nonvolatile acid. Hydrogen fluoride and hydrogen chloride are prepared in this manner by reaction of a cheap, readily available salt with concentrated sulfuric acid:

$$CaF_2(s) + H_2SO_4(l) \longrightarrow 2HF(g) + CaSO_4(s) \qquad [21.9]$$

$$NaCl(s) + H_2SO_4(l) \longrightarrow HCl(g) + NaHSO_4(s) \qquad [21.10]$$

Because the hydrogen halide is the only volatile component in the mixture, it can be easily removed. The hydrogen halide is usually absorbed in water and marketed as the corresponding acid. In both cases the reaction mixtures must be heated to cause reaction to occur. The $NaHSO_4$ formed in producing HCl can be forced to react still one stage further by more extensive heating and the addition of more NaCl:

$$NaCl(s) + NaHSO_4(s) \longrightarrow HCl(g) + Na_2SO_4(s) \qquad [21.11]$$

This reaction provides an illustration of the acidic properties of thc hydrogen sulfate ion, HSO_4^-.

Neither hydrogen bromide nor hydrogen iodide can be prepared by analogous reactions of salts with H_2SO_4, because HBr and HI undergo oxidation by H_2SO_4 at the higher temperatures of the reaction. The overall reactions are described by Equations 21.12 and 21.13:

$$2NaBr(s) + 2H_2SO_4(l) \longrightarrow Br_2(g) + SO_2(g) + Na_2SO_4(s) + 2H_2O(g) \qquad [21.12]$$

$$8NaI(s) + 9H_2SO_4(l) \longrightarrow 8NaHSO_4(s) + H_2S(g) + 4I_2(g) + 4H_2O(g) \qquad [21.13]$$

Notice that in the case of the bromide, part of the H_2SO_4 is reduced to SO_2, in which sulfur is in the $+4$ oxidation state. In the reaction with iodide, sulfur is reduced all the way to H_2S, in which sulfur is in the -2 oxidation state. This difference in products reflects the greater ease of oxidation of the iodide. The difficulties associated with use of H_2SO_4 can be avoided by using a nonvolatile acid that is a poorer oxidizing agent than H_2SO_4; concentrated phosphoric acid serves well.

SAMPLE EXERCISE 21.3

Write balanced equations for formations of HBr and HI from reaction of the appropriate sodium salt with phosphoric acid.

Solution: The formula for phosphoric acid is H_3PO_4. Let us assume that only one of the dissociable hydrogens of this acid undergoes reaction. (The actual number that reacts depends on reaction conditions.) The balanced equations are then:

$$NaBr(s) + H_3PO_4(l) \xrightarrow{\Delta} NaH_2PO_4(s) + HBr(g)$$

$$NaI(s) + H_3PO_4(l) \xrightarrow{\Delta} NaH_2PO_4(s) + HI(g)$$

The hydrogen halides are also formed when certain molecular (covalent) halides are hydrolyzed, as in the following examples:

$$PBr_3(l) + 3H_2O(l) \longrightarrow H_3PO_3(l) + 3HBr(g) \qquad [21.14]$$

$$SeCl_4(s) + 3H_2O(l) \longrightarrow H_2SeO_3(s) + 4HCl(g) \qquad [21.15]$$

SAMPLE EXERCISE 21.4

Write the balanced chemical equation for the reaction between $SiCl_4$ and H_2O.

Solution: By analogy to the reactions given above, we would expect the formation of HCl together with the oxyacid of silicon, which we may write as $Si(OH)_4$. Thus the equation is

$$SiCl_4(l) + 4H_2O(l) \longrightarrow Si(OH)_4(s) + 4HCl(g)$$

Because $Si(OH)_4$ readily undergoes loss of water to form SiO_2, the reaction is also written in the following way:

$$SiCl_4(l) + 2H_2O(l) \longrightarrow SiO_2(s) + 4HCl(g)$$

Some nonmetal halides, such as NF_3, CCl_4, and SF_6, are quite unreactive toward water. The lack of reactivity of these halides is not due to thermodynamic factors, but rather to the kinetic features of the reaction. For example, $\Delta G°$ for the hydrolysis of CCl_4 is highly negative, -377 kJ/mol CCl_4, which implies that the reaction should proceed very nearly to completion (Section 17.6). However, no low-energy reaction mechanism is available; the hydrolysis reaction has a very high activation barrier.

Generally, a molecular halide is unreactive when the central atom is unable to expand further the number of electrons in its valence shell (that is, the central atom has its maximum stable coordination number). For example, carbon can accommodate a maximum of eight electrons in its valence shell. Silicon, on the other hand, can accommodate a greater number of electrons by using *d* orbitals in bonding. The hydrolysis of $SiCl_4$ is believed to occur by attack of the silicon atom by the H_2O molecule, followed by loss of H^+ and Cl^- ions, as shown in Figure 21.3. This reaction pathway is not available to CCl_4.

Because the hydrogen halides form acidic solutions in water, they exhibit the characteristic properties of acids in their reactions in water. For example, they may react with an active metal to produce hydrogen gas or with a base in a neutralization reaction. In such reactions the

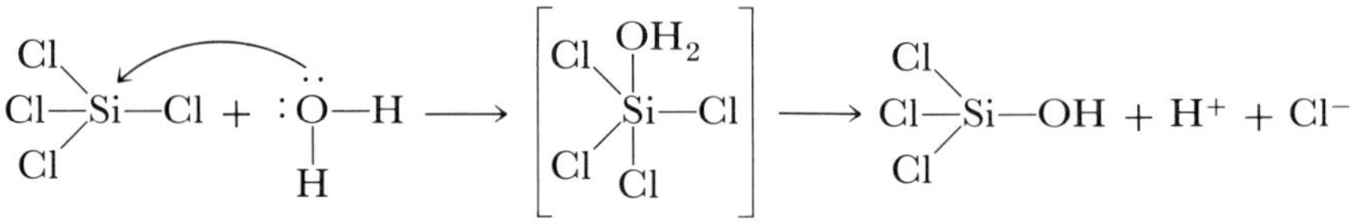

$$\mathrm{Cl_3Si{-}Cl} + \mathrm{:\ddot{O}H_2} \longrightarrow \left[\mathrm{Cl_4Si(OH_2)}\right] \longrightarrow \mathrm{Cl_3Si{-}OH} + H^+ + Cl^-$$

FIGURE 21.3 Proposed mechanism for the first stage of the reaction between $SiCl_4$ and H_2O. After the H_2O forms an adduct with $SiCl_4$, that adduct loses H^+ and Cl^- ions. The process of H_2O attack is then repeated three additional times, with ultimate formation of $Si(OH)_4$.

halide ion is merely a spectator ion; it does not directly participate in the reaction. On the other hand, in oxidation-reduction reactions, the halide ion may be oxidized, as in the following example:

$$14H^+(aq) + 6Cl^-(aq) + Cr_2O_7^{2-}(aq) \longrightarrow 2Cr^{3+}(aq) + 3Cl_2(g) + 7H_2O(l) \quad [21.16]$$

The halide ion also plays a role in precipitation reactions, in which an insoluble metal halide is formed:

$$Pb^{2+}(aq) + 2Br^-(aq) \longrightarrow PbBr_2(s) \quad [21.17]$$

In this reaction the solvated proton is a spectator ion.

Hydrofluoric acid is unusual among the aqueous hydrogen halides in that it reacts readily with silica, SiO_2, and with various silicates to form hexafluorosilicic acid, as in these examples:

$$SiO_2(s) + 6HF(aq) \longrightarrow H_2SiF_6(aq) + 2H_2O(l) \quad [21.18]$$

$$CaSiO_3(s) + 8HF(aq) \longrightarrow H_2SiF_6(aq) + CaF_2(aq) + 3H_2O(l) \quad [21.19]$$

Hydrofluoric acid must therefore be stored in wax or plastic bottles, because it reacts with ordinary glasses, which consist mostly of silicate structures (Section 21.5).

Interhalogen Compounds

Because the halogens form diatomic molecules in their most stable state at ordinary temperatures and pressures, it is not surprising to discover that diatomic molecules consisting of two different halogen atoms exist. These compounds are the simplest example of **interhalogens**, that is, compounds formed between two different halogen elements. Some properties of the diatomic interhalogens are listed in Table 21.5. The table is incomplete because certain of the interhalogens are not stable; they undergo decomposition to form the diatomic halogen elements or to form more complex interhalogen compounds.

The higher interhalogen compounds have formulas of the form XX'_3, XX'_5, or XX'_7, where X is chlorine, bromine, or iodine, and X′ is fluorine (the one exception is ICl_3, in which X′ is chlorine). Using the VSEPR model (Section 8.1), we can predict the geometrical structures of these compounds. We can also describe the bonding about the central atom in terms of a hybrid orbital description (Section 8.2).

TABLE 21.5 Properties of interhalogen compounds, XX′

Compound	ClF	BrF	BrCl	IF	ICl	IBr
Molecular weight	54.6	98.9	115.4	145.9	162.4	206.8
Melting point (°C)	−155	−33	−54	—	27	42
Boiling point (°C)	−90	20	5	—	98	116
Bond distance (Å)	1.63	1.76	2.16	1.91	2.32	—
Dipole moment (D)	0.9	1.3	0.6	—	0.5	—
Bond-dissociation energy (kJ/mol)	253	237	218	278	208	175

SAMPLE EXERCISE 21.5

Account for the valence-shell electron distribution and geometrical structure in BrF_3. What hybrid orbital description is most suitable for the central atom in this molecule?

Solution: Bromine has seven valence-shell electrons. If the Br atom is singly bonded to three fluorine atoms, there are an additional three electrons from this source. According to the VSEPR model, these ten electrons are disposed as five electron pairs about the central atom at the vertices of a trigonal bipyramid (Table 8.3). Three of the electron pairs are used in bonding to fluorine, the other two are unshared electron pairs. These unshared pairs require a larger space, so they are placed in the equatorial plane of the trigonal bipyramid:

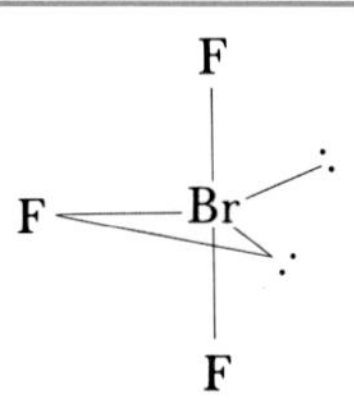

Because the unshared pairs push the bonding pairs back a little, the molecule should have the shape of a bent T. In terms of a hybridization description, we need to employ one of the bromine valence-shell *d* orbitals (the 4*d*) in addition to the 4*s* and three 4*p* orbitals to provide the five atomic orbitals to contain the five electron pairs in the valence shell. Thus the appropriate hybrid orbital description is sp^3d, which results in orbitals directed toward the vertices of a trigonal bipyramid.

Notice that the central atom in these higher interhalogen compounds has valence-shell *d* orbitals available for bonding. Thus the valence shell can be expanded beyond the octet. Second, the central atom in all cases is relatively large compared to the atoms grouped around it. Because fluorine is small and forms very strong bonds, it is ideally suited as the X′ atom. Only when the central atom is very large, as in the case of iodine, can the larger chlorine atom form an interhalogen, in ICl_3. The importance of size is also seen in the fact that iodine is capable of forming IF_7, whereas with bromine a maximum of five fluorines can be fitted around the central atom, in BrF_5. Chlorine and fluorine can form the ClF_5 molecule, but only with great difficulty.

The interhalogen compounds are exceedingly reactive. They attack

TABLE 21.6 The oxyacids of the halogens

Oxidation state of halogen	Formula of acid			Type of name
	Cl	Br	I	
+1	$HClO$	$HBrO$	HIO	*Hypo*hal*ous* acid
+3	$HClO_2$	—	—	Hal*ous* acid
+5	$HClO_3$	$HBrO_3$	HIO_3	Hal*ic* acid
+7	$HClO_4$	$HBrO_4$	HIO_4, H_5IO_6	*Per*hal*ic* acid

glass very readily and must be placed in special metal containers. They are very active fluorinating agents, as in these examples:

$$2CoCl_2(s) + 2ClF_3(g) \longrightarrow 2CoF_3(s) + 3Cl_2(g) \quad [21.20]$$

$$Se(s) + 3BrF_5(l) \longrightarrow SeF_6(l) + 3BrF_3(l) \quad [21.21]$$

The *polyhalide ions* are closely related to the interhalogens. Many of these ions are stable as salts of the alkali metal ions—for example, KI_3, $CsIBr_2$, $KICl_4$, and $KBrF_4$. Some of them, notably I_3^-, are also stable in aqueous solution.

Oxyacids and Oxyanions

Table 21.6 summarizes the formulas of the known oxyacids of the halogens and the way they are named.* The oxyacids are rather unstable; they generally decompose (sometimes explosively) when one attempts to isolate them. All of the oxyacids are strong oxidizing agents. The oxyanions, formed on removal of the proton from the oxyacids, are generally more stable than the oxyacids themselves. (Review the nomenclature of the oxyanions, Section 2.7.) Hypochlorite salts are used as bleaches and disinfectants because of the powerful oxidizing capabilities of the hypochlorite ion. Sodium chlorite, which can be isolated as the trihydrate, $NaClO_2 \cdot 3H_2O$, is used as a bleaching agent. Chlorite salts form potentially explosive mixtures with organic materials. Chlorate salts are similarly very reactive. For example, a mixture of potassium chlorate and sulfur may explode when struck. Potassium chlorate is used in making matches and fireworks. Perchloric acid and its salts are the most stable of the oxyacids and oxyanions. Dilute perchloric acid solutions are quite safe, and most perchlorate salts are stable except when heated with organic materials.

There are two oxyacids that have iodine in the +7 oxidation state. These periodic acids are HIO_4 (called metaperiodic acid), and H_5IO_6 (called paraperiodic acid). The two forms exist in equilibrium in aqueous solution:

$$H_5IO_6(aq) \rightleftharpoons H^+(aq) + IO_4^-(aq) + 2H_2O(l) \qquad K = 0.015 \quad [21.22]$$

*Fluorine forms one oxyacid, HFO. Because the electronegativity of fluorine is greater than that of oxygen, we must consider fluorine to be a −1 oxidation state and oxygen to be in the 0 oxidation state.

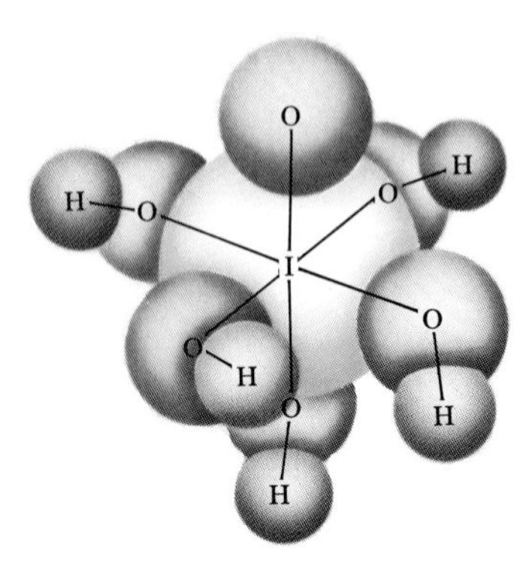

FIGURE 21.4 Paraperiodic acid, H_5IO_6.

HIO_4 is a strong acid, whereas H_5IO_6 is a weak one; the first two acid-dissociation constants for H_5IO_6 are $K_{a1} = 2.8 \times 10^{-2}$ and $K_{a2} = 4.9 \times 10^{-9}$. Crystalline H_5IO_6 is obtained when periodic acid solutions are evaporated at low temperatures. Mild heating of H_5IO_6 under vacuum produces HIO_4. The structure of H_5IO_6 is given in Figure 21.4. The fact that a molecule forms in which iodine is surrounded by six oxygen atoms can be attributed to the large size of the iodine. The smaller halogens do not form acids of this type.

The acid strengths of the oxyacids increase with increasing oxidation state of the central atom; the origin of this trend is discussed in Section 15.9. The stability of the oxyacids and of the corresponding oxyanions toward reduction increases with increasing oxidation state of the central halogen atom. Because the halogens are relatively electronegative elements, we might expect that compounds in which the halogen has an increasingly positive oxidation number would be *less* stable. The origins of this unexpected trend are rather complicated and beyond the scope of our survey. However, the result is important, because it affects the chemical behavior of the oxyacids and oxyanions. The stabilities of the oxyacids and oxyanions toward reduction in acid solution are summarized in Figure 21.5. The notation in this figure is the same as that discussed in reference to Figure 20.17 (Section 20.5). Notice in Figure 21.5 that ClO_4^- and ClO_3^- are written as anions, whereas $HClO_2$ and $HClO$ are written as acids. This difference follows from the fact that $HClO_4$ and $HClO_3$ are strong acids and thus can be considered to be completely ionized even in acid solution. $HClO_2$ and $HClO$, on the other hand, are weak acids.

Notice that all the reduction potentials given in Figure 21.5 are positive. This is in keeping with the strongly oxidizing character of the oxychlorides. Notice also that the potentials are greatest for reduction of the lower-oxidation-state species, $HClO_2$ and $HClO$. These acids are the most strongly oxidizing agents among the oxychlorine compounds.

$$ClO_4^- \xrightarrow{1.19\ V} ClO_3^- \xrightarrow{1.21\ V} HClO_2 \xrightarrow{1.64\ V} HClO \xrightarrow{1.63\ V} Cl_2 \xrightarrow{1.36\ V} Cl^-$$

(1.38 V: $ClO_3^- \rightarrow Cl^-$; 1.47 V: $ClO_3^- \rightarrow Cl_2$)

FIGURE 21.5 Standard electrode (reduction) potentials in acid solution for oxychlorine species. Note that Cl^-, ClO_4^-, and ClO_3^- are written as anions because HCl, $HClO_4$, and $HClO_3$ are strong acids and are thus largely dissociated, even in acidic solution.

21.3 THE GROUP 6A ELEMENTS

The elements of group 6A are oxygen, sulfur, selenium, tellurium, and polonium. We have already discussed oxygen in Section 20.4. In this section we will examine the group as a whole and then look at sulfur, selenium, and tellurium, focusing mainly on sulfur. We will not have much to say about polonium, an element produced by radioactive decay of radium. There are no stable isotopes of this element. As a result, it is found only in trace quantities in radium-containing minerals.

General Characteristics of Group 6A

The group 6A elements possess the general outer electron configuration ns^2np^4, where n may have values ranging from 2 through 6. These elements thus may attain a noble gas electron configuration by the addition of two electrons, which results in a -2 oxidation state. Because the group 6A elements are nonmetals, this is a common oxidation state. Except for oxygen, however, the group 6A elements are also commonly found in positive oxidation states up to $+6$, which corresponds to sharing of all six valence-shell electrons with atoms of a more electronegative element. Sulfur, selenium, and tellurium also differ from oxygen in being able to use d orbitals in bonding. Thus compounds with expanded valence shells such as SF_6, SeF_6, and TeF_6 occur.

The group 6A elements vary considerably in their abundances. Oxygen is everywhere about us, in the atmosphere and in the earth's crust. Sulfur is generally abundant in the earth's crust, in a variety of forms as we shall see, but mainly in the form of sulfide ores. Selenium is rather scarce; it generally occurs as a minor constituent in sulfur-containing minerals. Tellurium is among the rarest of the elements and is less abundant than gold or platinum.

Some of the more important properties of the atoms of the group 6A elements are summarized in Table 21.7. The energy of the X—X single bond is estimated from data for the elements, except for oxygen. In this case, because the O—O bond in O_2 is not a single bond (Sections 8.6 and 8.7), the estimated O—O bond energy in hydrogen peroxide is employed. The reduction potential listed in the last line of the table refers to the reduction of the element in its standard state to form $H_2X(aq)$ in acidic solution. In most of the properties listed in Table 21.7 we again see a regular variation as a function of atomic number. Atomic and ionic radii increase; correspondingly, the ionization energy decreases, as we expect from the discussion in Section 6.5.

The electron affinities listed apply to the process shown in Equation 21.23:

$$X(g) + e^- \longrightarrow X^-(g) \qquad [21.23]$$

This, of course, does not produce the commonly observed stable ionic form of these elements, X^{2-}, but it is the first step in its formation. It is

TABLE 21.7 Some properties of the atoms of the group 6A elements

Property	O	S	Se	Te
Atomic radius (Å)	0.73	1.02	1.17	1.35
X^{2-} ionic radius (Å)	1.45	1.90	2.02	2.22
First ionization energy (kJ/mol)	1312	1004	946	870
Electron affinity (kJ/mol)	−141	−201	−195	−186
Electronegativity	3.4	2.6	2.6	2.1
X—X single-bond energy (kJ/mol)	142[a]	266	172	126
Reduction potential to H_2X in acidic solution (V)	1.23	0.14	−0.40	−0.72

[a] Based on O—O bond energy in H_2O_2.

interesting to note that addition of an electron to oxygen is less exothermic than addition of an electron to one of the other group 6A elements. The origin of this effect has to do with the relatively small size of the oxygen atom. Addition of a single extra electron to the smaller atom results in larger electron-electron repulsions, offsetting the gain in stability that results from the closer approach of the electron to the nucleus.

The decrease in ionization energy is the factor mainly responsible for the decrease in electronegativity in the series. In this connection it is interesting to note that sulfur and selenium are closely similar in many respects, whereas tellurium is substantially less electronegative. Note that the ease of reduction of the free element to form H_2X varies greatly throughout the series. Whereas oxygen is very readily reduced to the -2 oxidation state, the potential for reduction of tellurium is actually quite strongly negative. These observations indicate an increasingly metallic character in the group 6A elements with increasing atomic number. The physical properties of the elements show a corresponding trend. At one extreme we have oxygen, a diatomic molecule, and sulfur, a yellow, nonconducting solid that melts at 114°C. At the other extreme, the stable form of tellurium has a bright luster and low electrical conductivity and melts at 452°C.

Several interesting comparisons can be made between the data listed in Table 21.7 and the corresponding values for the halogens listed in Table 21.3. Notice that the ionization energies and electron affinities of the halogens are generally higher. Correspondingly, the atomic radii of the halogens are smaller and their electronegativities are higher. The potentials for reduction of the free elements to the stable negative oxidation state are greater for the halogens, as expected. The X—X single-bond energies for the corresponding members of the two families are not greatly different. For example, the S—S bond in S_8 has a bond energy of 226 kJ/mol as compared with 243 kJ/mol for the Cl—Cl bond in Cl_2. It is interesting that in both families of elements the X—X bond energy for the first member in each series is unusually low. With this background, let us now consider the occurrences and preparation of sulfur, selenium, and tellurium.

Occurrences and Preparation of Sulfur, Selenium, and Tellurium

Sulfur occurs in the elemental state in large underground deposits that serve as the principal source of the element. The Frasch process, illustrated in Figure 21.6, is used to obtain the element from these deposits. The method is based on the low melting point and low density of sulfur. Superheated water is forced into the deposit, where it melts the sulfur. Compressed air then forces the molten sulfur up a pipe that is concentric with the ones that introduce the hot water and compressed air into the deposit. The method is particularly useful for removing sulfur from beds that lie under water or quicksand. Figure 21.7 shows an offshore sulfur mine.

Sulfur also occurs widely as sulfide and sulfate minerals and as a minor component of coal and petroleum. The presence of sulfur in coal and petroleum poses a major problem. As we saw in Section 12.4, combustion of these “unclean” fuels leads to serious sulfur oxide pollution.

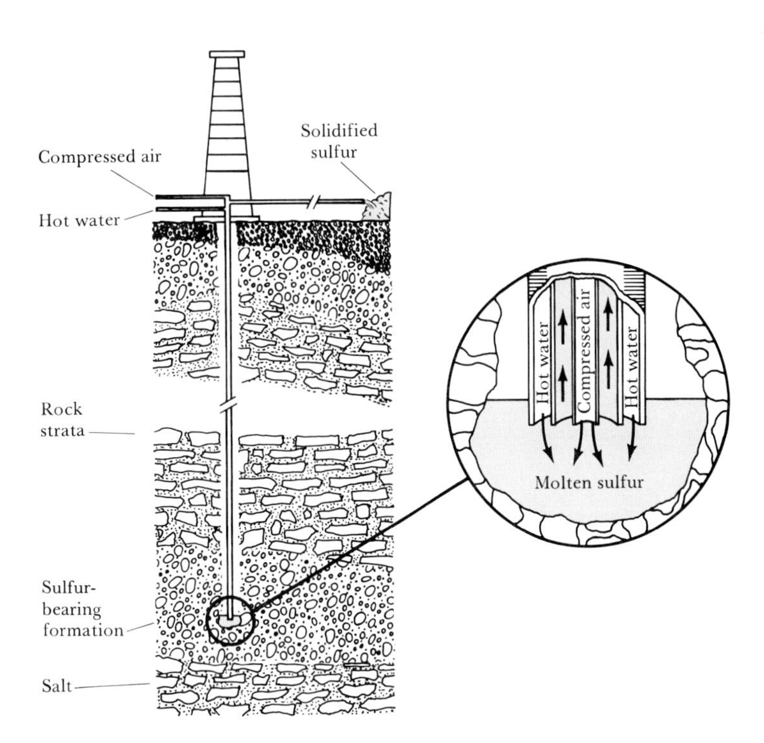

FIGURE 21.6 Mining of sulfur by the Frasch process. The process is named after Herman Frasch, who invented the process in the early 1890s. The process is particularly useful for recovering sulfur from deposits located under quicksand or water.

Also, operations that use sulfide minerals as sources of metals liberate sulfur oxides. Much effort has therefore been directed at removing this sulfur, and these efforts have increased the availability of sulfur. The sale of this sulfur helps partially to offset the costs of the desulfurizing pro-

FIGURE 21.7 Production platform of the world's first offshore sulfur mine. The mine stands in 50 ft of water 7 mi off Grand Isle, Louisiana. (*Exxon Corp.*)

cesses and equipment. However, sulfur obtained from sulfur deposits by the Frasch process is about 99.5 percent pure and can be used for most commercial processes without purification. It is therefore relatively cheap. Consequently, sulfur from desulfurizing processes must be sold at prices below its cost in order to compete on the market. Nevertheless, about half the sulfur used in the United States each year is produced by means other than the Frasch process.

Selenium and tellurium occur in rare minerals as Cu_2Se, $PbSe$, Ag_2Se, Cu_2Te, $PbTe$, Ag_2Te, and Au_2Te. They also occur as minor constituents in sulfide ores of copper, iron, nickel, and lead. From a commercial point of view, the most important sources of these elements are the copper ores. When these are treated to form copper metal, the selenium and tellurium are retained to a large extent in the copper as the elements. When the copper is purified by electrolysis (Section 22.5), the impurities such as selenium and tellurium along with precious metals such as gold and silver collect beneath the anode. If the sludge beneath the anode is treated with concentrated sulfuric acid at about 400°C, selenium is oxidized to selenium dioxide, which sublimes from the reaction mixture:

$$Se(s) + 2H_2SO_4(l) \xrightarrow{\Delta} SeO_2(g) + 2SO_2(g) + 2H_2O(g) \qquad [21.24]$$

The dioxide is recovered as a white powder from the upper furnace walls. It is then dissolved in dilute hydrochloric acid solution, in which it is quite soluble. Selenous acid, a weak acid, is formed:

$$SeO_2(s) + H_2O(l) \longrightarrow H_2SeO_3(aq) \qquad [21.25]$$

Sulfur dioxide is then bubbled into the solution, and an oxidation-reduction reaction occurs:

$$H_2O(l) + H_2SeO_3(aq) + 2SO_2(g) \longrightarrow Se(s) + 2H^+(aq) + 2HSO_4^-(aq) \qquad [21.26]$$

Tellurium is recovered from the anode sludge by other procedures.

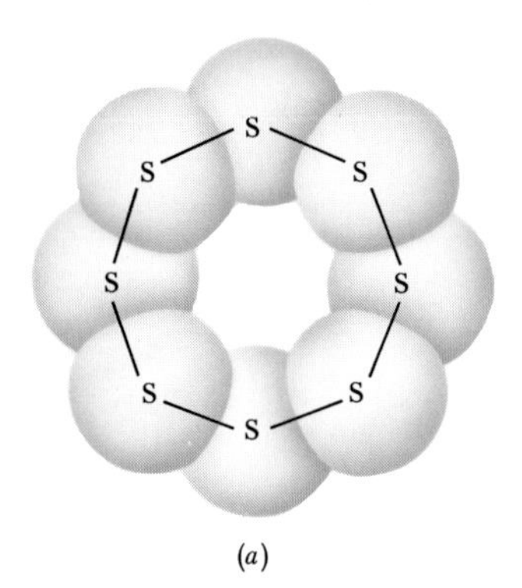

(a)

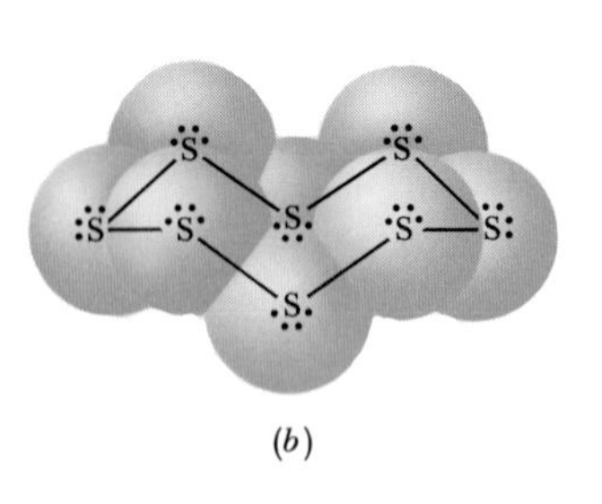

(b)

FIGURE 21.8 Top view (*a*) and side view (*b*) of the sulfur molecule in rhombic sulfur.

Properties and Uses

As we normally encounter it, sulfur is yellow, tasteless, and nearly odorless. It is insoluble in water and exists in several allotropic forms. The thermodynamically stable form at room temperature is rhombic sulfur, which consists of puckered S_8 rings, as shown in Figure 21.8. When heated above its melting point, at 113°C, sulfur undergoes a variety of changes. The molten sulfur first contains S_8 molecules and is fluid because the rings readily slip over each other. Further heating of this straw-colored liquid causes rings to break, the fragments joining to form very long molecules that can become entangled. The sulfur consequently becomes highly viscous. This change is marked by a color change to dark reddish brown. Further heating breaks the chains and the viscosity again decreases.

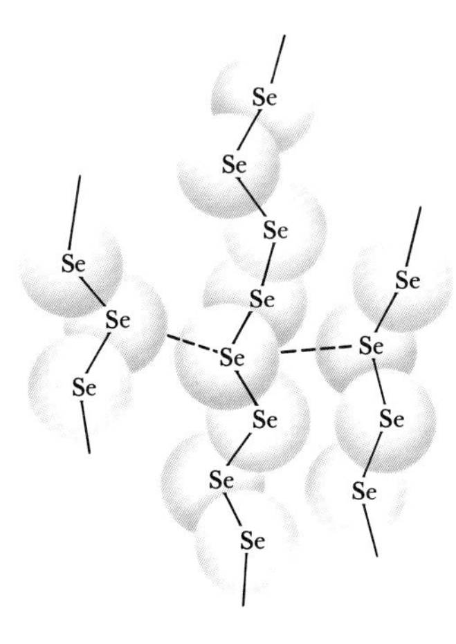

FIGURE 21.9 Portion of the structure of crystalline seleneium. The dashed lines represent weak bonding interactions between atoms in adjacent chains. Tellurium has the same structure.

Most of the 1.1×10^{10} kg (12 million tons) of sulfur produced in the United States each year is used in the manufacture of sulfuric acid. Sulfur is also used in vulcanizing rubber, a process that toughens rubber by introducing cross-linking between polymer chains (Section 10.5).

The most stable allotropes of both selenium and tellurium are crystalline substances containing helical chains of atoms, as illustrated in Figure 21.9. Each atom of the chain, however, is close to atoms in adjacent chains, and it appears that there is some sharing of electron pairs between these atoms, as indicated by the dotted lines in Figure 21.9. The electrical conductivity of selenium is very small in the dark but increases greatly upon exposure to light. This property of the element is used in photoelectric cells and in light meters such as those used in cameras. Xerox copiers also depend on the photoconductivity of selenium for their operation.

Sulfur is widely distributed in biological systems. It is present in most proteins, as a component of the amino acids cysteine and methionine (Section 26.2). In contrast, selenium is rare in biological systems. It has only recently been established that there is a human nutritional requirement for the element. Selenium is present in trace quantities in most vegetables, especially spinach. The amount of selenium required for adequate nutrition is very small. In quanitities much larger than this small nutritional requirement, the element is toxic. Tellurium does not have a known role in human nutrition. Its compounds are poisonous and, if they are volatile, usually have highly offensive odors.

Oxides, Oxyacids, and Oxyanions

Sulfur dioxide was first discovered by Joseph Priestley in 1774, when he heated mercury with concentrated sulfuric acid. The reaction that occurs is shown in Equation 21.27:

$$Hg(l) + 2H_2SO_4(l) \longrightarrow HgSO_4(s) + SO_2(g) + 2H_2O(l) \qquad [21.27]$$

In the laboratory, SO_2 is prepared by the action of aqueous acid on a sulfite salt, as shown in Equation 21.28:

$$2H^+(aq) + SO_3^{2-}(aq) \longrightarrow SO_2(g) + H_2O(l) \qquad [21.28]$$

Sulfur dioxide is formed when sulfur is combusted in air; it has a choking odor and is poisonous. The gas is particularly toxic to lower organisms such as fungi and is consequently used for sterilizing dried fruit. At 1 atm pressure and room temperature, SO_2 dissolves in water to the extent of 45 volumes of gas per volume of water, to produce a solution of about 1.6 M concentration. The solution is acidic, and we describe it as $H_2SO_3(aq)$. Actually, there is evidence that much of the sulfur dioxide exists in solution as hydrated SO_2. When the saturated solution is cooled, crystals of the hydrate, $SO_2 \cdot 6H_2O$, can be recovered. It is convenient, however, to assume that all of the SO_2 that dissolves is in the form of the weak acid, H_2SO_3, which ionizes according to Equations 21.29 and 21.30:

$$H_2SO_3(aq) \rightleftharpoons H^+(aq) + HSO_3^-(aq) \quad K_{a1} = 1.7 \times 10^{-2}(25°C) \qquad [21.29]$$

$$HSO_3^-(aq) \rightleftharpoons H^+(aq) + SO_3^{2-}(aq) \quad K_{a2} = 6.4 \times 10^{-8}(25°C) \qquad [21.30]$$

Aqueous sulfur dioxide can act as a reducing agent, for example, in the reaction with iodate ion in acidic solution:

$$2IO_3^-(aq) + 5H_2SO_3(aq) \longrightarrow I_2(s) + 5HSO_4^-(aq) + H_2O(l) + 3H^+(aq) \qquad [21.31]$$

As we have seen, combustion of sulfur in air produces mainly SO_2, but small amounts of SO_3 are also formed. The reaction produces mainly SO_2 because the activation-energy barrier for further oxidation to SO_3 is very high unless the reaction is catalyzed. Sulfur trioxide is of great commercial importance because it is the anhydride of sulfuric acid. In the manufacture of sulfuric acid, SO_2 is first obtained by burning sulfur. The SO_2 is then oxidized to SO_3 using a catalyst such as V_2O_5 or platinum. The SO_3 is dissolved in H_2SO_4 because it does not dissolve quickly in water. The reaction is shown in Equation 21.32. The $H_2S_2O_7$ formed in this reaction, called pyrosulfuric acid, is then added to water to form H_2SO_4, as shown in Equation 21.33:

$$SO_3(g) + H_2SO_4(l) \longrightarrow H_2S_2O_7(l) \qquad [21.32]$$

$$H_2S_2O_7(l) + H_2O(l) \longrightarrow 2H_2SO_4(l) \qquad [21.33]$$

Commercial sulfuric acid is generally 98 percent H_2SO_4 and boils at 340°C. It is a colorless, oily liquid. Sulfuric acid has many useful properties. Most important, it is a strong acid, a good dehydrating agent,* and a moderately good oxidizing agent.

Year after year, the output of sulfuric acid has been the largest of any chemical produced in the United States. About 3.0×10^{10} kg (33 million tons) is produced annually in this country. Sulfuric acid is employed in some way in almost all manufacturing. Consequently, its consumption is considered a standard measure of industrial activity. The primary uses for sulfuric acid are shown in Figure 21.10.

Only the first proton in sulfuric acid is completely ionized in aqueous solution. The second proton ionizes only partially:

$$H_2SO_4(aq) \longrightarrow H^+(aq) + HSO_4^-(aq)$$

$$HSO_4^-(aq) \rightleftharpoons H^+(aq) + SO_4^{2-}(aq) \qquad K_a = 1.1 \times 10^{-2}$$

Consequently, sulfuric acid forms two series of compounds: sulfates and bisulfates (or hydrogen sulfates).

Related to the sulfate ion is the thiosulfate ion, $S_2O_3^{2-}$, formed by boiling an alkaline solution of SO_3^{2-} with elemental sulfur, as shown in Equation 21.34:

$$8SO_3^{2-}(aq) + S_8(s) \longrightarrow 8S_2O_3^{2-}(aq) \qquad [21.34]$$

*Considerable heat is given off when sulfuric acid is diluted with water. Consequently, dilution must always be done carefully by pouring the acid into water to distribute the heat as uniformly as possible and to avoid spattering of the acid.

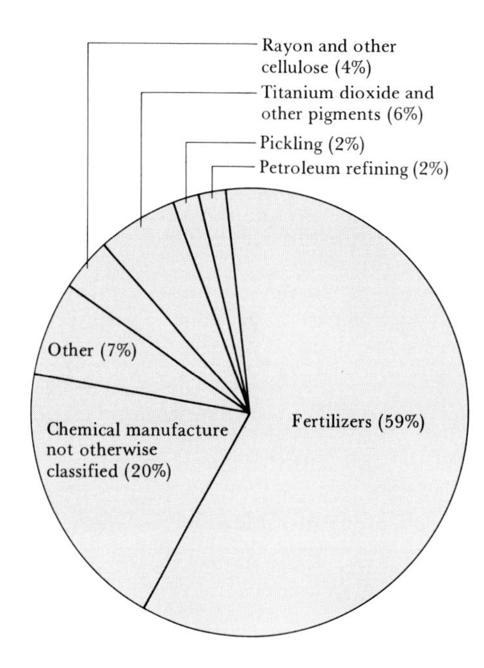

FIGURE 21.10 Sulfuric acid use in the United States.

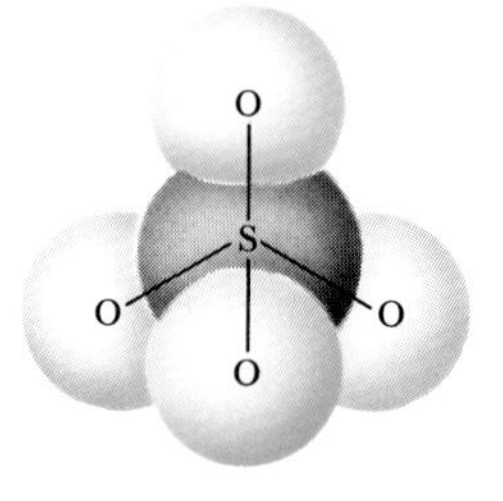

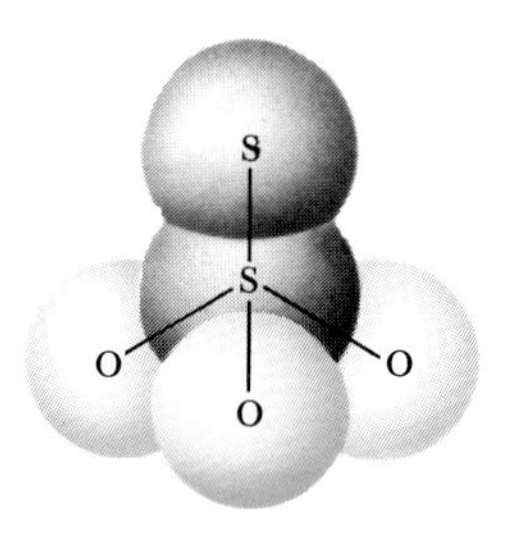

FIGURE 21.11 Comparison of the structures of the sulfate, SO_4^{2-}, and thiosulfate, $S_2O_3^{2-}$, ions.

The term "thio" indicates substitution of sulfur for oxygen. The structures of the sulfate and thiosulfate ions are compared in Figure 21.11. When acidified, the thiosulfate ion decomposes to form sulfur and H_2SO_3. The pentahydrated salt of sodium thiosulfate, $Na_2S_2O_3 \cdot 5H_2O$, is known as "hypo." It is used in photography to furnish thiosulfate ion as a complexing agent for silver ion. Photographic film consists of a suspension of microcrystals of AgBr in gelatin. In those portions of the film that are exposed to light the AgBr is "activated." This means that when the film is treated with a mild reducing agent (called the developer) the exposed AgBr is reduced. Thus black, metallic silver is formed in the film in concentrations proportional to the intensity of light exposure. The film is then treated with sodium thiosulfate solution to remove the remaining silver bromide, by forming the soluble silver thiosulfate complex:

$$AgBr(s) + 2S_2O_3^{2-}(aq) \rightleftharpoons Ag(S_2O_3)_2^{3-}(aq) + Br^-(aq) \qquad [21.35]$$

This step in the process is called "fixing." Thiosulfate ion is also used in quantitative analysis as a reducing agent for iodine:

$$2S_2O_3^{2-}(aq) + I_2(s) \longrightarrow 2I^-(aq) + S_4O_6^{2-}(aq) \qquad [21.36]$$

Both selenium and tellurium form dioxides upon burning the element in air or oxygen. Selenium dioxide is very soluble in water; it forms an acidic solution from which crystals of selenous acid, H_2SeO_3, can be isolated. Neutralization of selenous acid solutions yields salts of hydrogen selenite or selenite ion, for example, $NaHSeO_3$ or Na_2SeO_3. Tellurium dioxide is insoluble in water; thus the aqueous acid H_2TeO_3 is unknown. However, it is possible to prepare tellurite salts by dissolving TeO_2 in an aqueous base.

SAMPLE EXERCISE 21.6

In the vapor phase, SeO_2 exists as monomeric molecules. In the solid state it consists of zigzag chains involving Se—O—Se—O bonds. Write the Lewis structures of each form of SeO_2 and predict the bond angles about the selenium in each case.

Solution: In the gas phase, SeO_2 should have the Lewis structures

$$:\!\ddot{O}\!-\!\ddot{Se}\!=\!\ddot{O} \longleftrightarrow \ddot{O}\!=\!\ddot{Se}\!-\!\ddot{O}\!:$$

Using the VSEPR model (Section 8.1) we predict that the molecule will be bent, with an O—Se—O bond angle of somewhat less than 120°.

The Lewis structure for the solid-state form is as follows:

$$\diagdown\ddot{Se}(-:\!\ddot{O}\!:)\!-\!\ddot{O}\!-\!\ddot{Se}(-:\!\ddot{O}\!:)\!-\!\ddot{O}\!\diagdown$$

According to the VSEPR model, the four valence-shell electron pairs about Se are arranged approximately at the vertices of a tetrahedron. The three Se—O bonds thus form a trigonal-pyramidal arrangement; the O—Se—O bond angles should be somewhat less than 109°. (It is found that the angles are not all the same; the O—Se—O angle involving the oxygens in the chain is 98°, the other two O—Se—O angles are 90°.)

Both selenium trioxide and tellurium trioxide are known. These are the anhydrides of selenic and telluric acids, but the acids are formed other than by dissolving the trioxides in water. Selenic acid, H_2SeO_4, can be formed by oxidation of aqueous SeO_2 with 30 percent hydrogen peroxide, as shown in Equation 21.37:

$$H_2SeO_3(aq) + H_2O_2(aq) \longrightarrow HSeO_4^-(aq) + H_2O(l) + H^+(aq) \qquad [21.37]$$

Notice that we represent $H_2SeO_3(aq)$ in the nonionized form; the acid-dissociation constants for H_2SeO_3 at 25°C are $K_{a1} = 2.3 \times 10^{-3}$ and $K_{a2} = 5.3 \times 10^{-9}$. The nonionized acid is thus probably most representative of the species present in solution. On the other hand, selenic acid, H_2SeO_4, is a strong acid in its first dissociation constant; the second dissociation has an equilibrium constant $K_{a2} = 2.2 \times 10^{-2}$ at 25°C. Either SeO_4^{2-} or $HSeO_4^-$ might therefore be used to represent the most abundant species in acidic solution.

Telluric acid is formed when $TeO_2(s)$ is oxidized by various aqueous oxidizing agents. We might expect the formula for telluric acid to be H_2TeO_4, by analogy with the corresponding acid of sulfur or selenium. However, the acid which is recovered from solution as a white crystalline material is H_6TeO_6. [The formula can also be written as $Te(OH)_6$.] This substance is called orthotelluric acid. Notice the close similarity to iodine, which forms paraperiodic acid, H_5IO_6. In both cases, the large size of the central atom permits bonding with six surrounding oxygens. Orthotelluric acid is a weak acid, capable of dissociating up to two protons (at 25°C, $K_{a1} = 2.4 \times 10^{-8}$, $K_{a2} = 1.0 \times 10^{-11}$).

Sulfides, Selenides, and Tellurides

Sulfur forms compounds by direct combination with many elements. When the element is less electronegative than sulfur, sulfides, which contain S^{2-}, form. For example, iron(II) sulfide, FeS, forms by direct combination of iron and sulfur. Many metallic elements are found in the form of sulfide ores, for example, PbS (galena) and HgS (cinnabar). A series of related ores, containing the S_2^{2-} ion (analogous to the peroxide ion) are known as *pyrites*. Iron pyrite, FeS_2, occurs as golden-yellow cubic crystals. Because it has been on occasion mistaken for gold by overeager gold miners, in some locales it is called "fool's gold."

One of the most important sulfides is hydrogen sulfide, H_2S. This substance is not normally produced by direct union of the elements because it is unstable at elevated temperature and decomposes into the elements. It is normally prepared by action of dilute sulfuric acid on iron(II) sulfide, as shown in Equation 21.38:

$$FeS(s) + 2H^+(aq) \longrightarrow H_2S(aq) + Fe^{2+}(aq) \qquad [21.38]$$

A common laboratory source of H_2S is the reaction of thioacetamide with water, as shown in Equation 21.39:

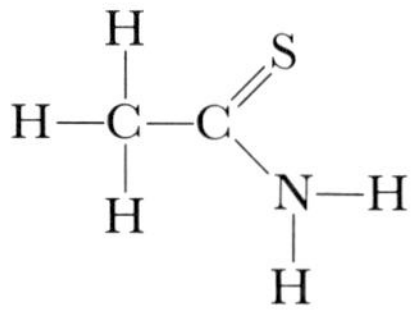

$$CH_3C(=S)NH_2(aq) + H_2O(l) \longrightarrow$$

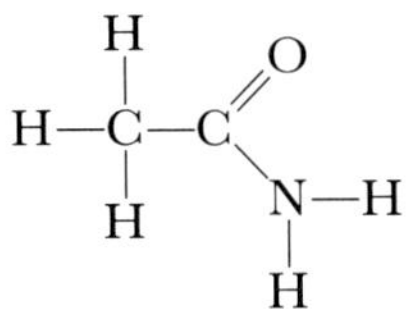

Thioacetamide

$$CH_3C(=O)NH_2(aq) + H_2S(aq) \qquad [21.39]$$

Acetamide

Hydrogen sulfide is often used in the laboratory for qualitative analysis of certain metal ions (Section 16.6).

One of hydrogen sulfide's most readily recognized properties is its odor; H_2S is largely responsible for the offensive odor of rotten eggs. Hydrogen sulfide is actually quite toxic; it has about the same level of toxicity as hydrogen cyanide, the gas that was used in California's gas chambers. The volatile hydrides H_2Se and H_2Te are similar to H_2S in many respects. Both compounds possess very offensive, lingering odors and are toxic. In aqueous solutions the acid strength increases in the order $H_2S < H_2Se < H_2Te$.

21.4 THE GROUP 5A ELEMENTS

The elements of group 5A are nitrogen, phosphorus, arsenic, antimony, and bismuth. We have already discussed the chemistry of nitrogen (Section 20.5). In our present discussion we will find it convenient to discuss the general characteristics of the group, then to consider phosphorus, and finally to comment briefly on the heavier elements of the group.

General Characteristics of the Group 5A Elements

The group 5A elements possess the outer electron configuration ns^2np^3, where n may have values ranging from 2 to 6. A noble-gas configuration results from addition of three electrons to form the -3 oxidation state. Ionic compounds containing X^{3-} ions are not common, however, except for salts of the more active metals, for example, Na_3N. More commonly the group 5A element acquires an octet of electrons via covalent bonding. The oxidation number may range from -3 to $+5$, depending on the nature and number of the atoms to which it is bound.

Because of its lower electronegativity, phosphorus is found more frequently in positive oxidation states than is nitrogen. Furthermore, compounds in which phosphorus has the $+5$ oxidation state are not strong oxidizing agents, as are corresponding compounds of nitrogen. Conversely, compounds in which phosphorus has a -3 oxidation state are much stronger reducing agents than are corresponding compounds of nitrogen.

Both nitrogen and phosphorus are abundant and important components of our environment. Nitrogen is, of course, the major component of the earth's atmosphere and is present in substantial quantities in biological systems. Phosphorus is found in minerals as phosphate; this element is also an important component of biological systems. Arsenic, antimony, and bismuth are much less abundant but are readily obtainable from accessible mineral sources. Bismuth holds an interesting place in chemistry. The only naturally occurring nuclide of this element, ^{209}Bi, has the highest atomic number of any nuclide that is stable with respect to radioactive decay. (This nuclide probably does decompose, but its half-life is estimated to be in excess of 10^{18} years.)

Some of the important properties of the atoms of the group 5A elements are listed in Table 21.8. The general pattern that emerges from these data is similar to what we have seen before; size and metallic character increase as atomic number increases within the group. Note also that in comparison with the corresponding elements of groups 6A and 7A, the atomic radii are larger, and ionization energies and electronegativities are lower.

The variation in properties among the elements of group 5A is more striking than that seen in the case of groups 6A and 7A. Nitrogen at the

TABLE 21.8 Properties of the atoms of group 5A elements

Property	N	P	As	Sb	Bi
Atomic radius (Å)	0.75	1.10	1.22	1.43	—
First ionization energy (kJ/mol)	1402	1012	947	834	703
Electron affinity (kJ/mol)	+6.8	−72	−77	−101	−106
Electronegativity	3.0	2.2	2.2	2.0	2.0
X—X single-bond[a] energy (kJ/mol)	163	200	150	120	—
X≡X triple-bond energy (kJ/mol)	941	480	380	295	192

[a] Approximate values only.

one extreme exists as a gaseous diatomic molecule; it is clearly nonmetallic in character. At the other extreme, bismuth is a reddish-white, metallic-looking substance that has most of the characteristics of a metal.

The electron affinities of the group 5A elements are especially interesting. In the ground state, the atoms of group 5A elements possess the electron configuration $ns^2np^1np^1np^1$. That is, according to Hund's rule, the valence-level *p* orbitals are exactly half-filled with electrons, with spins aligned. Addition of an electron to this rather stable arrangement is not favored, and the electron affinity of nitrogen is in fact zero or slightly positive (Section 6.6). The electron affinities of the other group 5A elements are exothermic, but addition of an electron to any of the group 5A elements releases considerably less energy than the analogous process for the group 6A or 7A elements. The existence of a stable, half-filled valence shell is also responsible for the relatively high ionization energies of the group 5A elements. This is especially evident in the case of nitrogen, which has a higher ionization energy than oxygen.

The values listed for X—X single-bond energies are not very reliable, because it is difficult to obtain such data from thermochemical experiments. However, there is no doubt about the general trend: an abnormally low value for the N—N single bond, a sharp increase at phosphorus, then a gradual decline among the other elements of the series. From observations of the group 5A elements in the gas phase (for all except N_2 this requires high temperatures), it is possible to estimate the X≡X triple-bond energy, as listed in Table 21.8. Here we see a trend that is very much different than that for the X—X single bond. Nitrogen forms a much stronger bond than do the other elements, and there is a steady decline in the triple-bond energy down through the group. These data help us to appreciate why nitrogen alone of the group 5A elements exists as a diatomic molecule in its stable state at 25 °C. All the other elements exist in structural forms involving single bonds between the atoms.

Occurrence, Isolation, and Properties of Phosphorus

Phosphorus occurs mainly in the form of phosphate minerals. The principal source of phosphorus is phosphate rock, which contains phosphate mainly in the form of $Ca_3(PO_4)_2$. Deposits of phosphate rock occur mainly in Florida, the western United States, North Africa, and parts of the USSR. The element is produced commercially by reduction of phosphate with coke* in the presence of SiO_2, as shown in Equation 21.40:

$$2Ca_3(PO_4)_2(s) + 6SiO_2(s) + 10C(s) \xrightarrow{1500\,^\circ C} P_4(g) + 6CaSiO_3(l) + 10CO(g) \qquad [21.40]$$

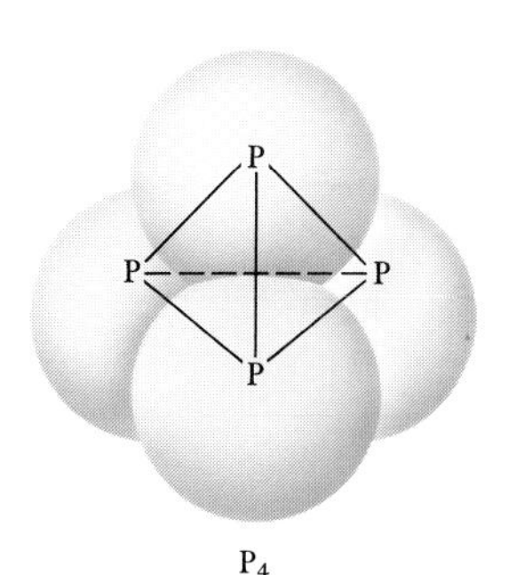

FIGURE 21.12 Structure of the P_4 molecule of white phosphorus.

The phosphorus produced in this fashion is the allotrope known as white phosphorus. This form distills from the reaction mixture as the reaction proceeds.

White phosphorus consists of P_4 tetrahedra as shown in Figure 21.12. As we noted in Section 8.7, the 60° bond angles in P_4 are unusually small

*Coke is a form of carbon made by heating coal in the absence of air.

for molecules. There must consequently be much strain in the bonding, a fact that is consistent with the high reactivity of white phosphorus. This allotrope bursts spontaneously into flames if exposed to air. It is a white, waxlike solid that melts at 44.2°C and boils at 280°C. When heated in the absence of air to about 400°C, it is converted to a more stable allotrope known as red phosphorus. This form does not ignite on contact with air. It is also considerably less poisonous than the white form.

Phosphorus Halides

Phosphorus forms a wide range of compounds with the halogens. A few properties of the more important phosphorus-halogen compounds are listed in Table 21.9.

We have already discussed the Lewis structures and geometries of the phosphorus halide molecules at other places in the text (Sections 7.5 and 7.6). The dipole moments (Section 8.2) listed in Table 21.9 reflect these geometries. The PX_3 compounds have a pyramidal shape (Figure 20.3), and the overall molecular dipole moment reflects the P—X bond polarity. We conclude that the bond polarities decrease in the order P—F > P—Cl > P—Br > P—I. This is the order expected on the basis of the electronegativity difference between phosphorus and the halogen. The PX_5 molecules possess trigonal-bipyramidal structures (Figure 20.3) with five electron pairs shared with the five X groups. The bond dipole moments of the five P—X bonds cancel to yield a net zero dipole moment in all cases.

Phosphorus trifluoride is prepared by reaction with another fluoride, as in Equation 21.41:

$$PCl_3(l) + AsF_3(l) \longrightarrow PF_3(g) + AsCl_3(l) \qquad [21.41]$$

The driving force for this reaction is the greater strength of the phosphorus-fluorine bond as compared with the other bond energies involved (490 kJ/mol for P—F as compared with 406 kJ/mol for As—F, 326 kJ/mol for P—Cl, and 322 kJmol for As—Cl). Phosphorus pentafluoride can be prepared by direct combination of the elements, by reaction of PCl_5 with AsF_3, or by heating P_4O_{10} in a sealed tube with CaF_2:

$$4P_4O_{10}(l) + 15CaF_2(s) \longrightarrow 6PF_5(l) + 5Ca_3(PO_4)_2(s) \qquad [21.42]$$

Phosphorus trichloride is formed by passing a stream of dry chlorine gas over white or red phosphorus. In an excess of Cl_2, the PCl_3 reacts further

TABLE 21.9 Properties of phosphorus-halogen compounds

Property	PF_3	PF_5	PCl_3	PCl_5	PBr_3	PBr_5	PI_3
Melting point (°C)	−160	−94	−112	167	−40	—	61
Boiling point (°C)	−95	−85	76	—[a]	173	106	120[b]
Dipole moment (D)	1.0	0	0.78	0	0,5	0	0

[a] Sublimes at 163°C at 1 atm pressure.
[b] At 15 mm Hg pressure.

to form PCl_5. Phosphorus tribromide is formed by reaction of liquid bromine on red phosphorus. Addition of still more bromine results in formation of PBr_5. However, this compound readily dissociates in the vapor state:

$$PBr_5(g) \rightleftharpoons PBr_3(g) + Br_2(g) \qquad [21.43]$$

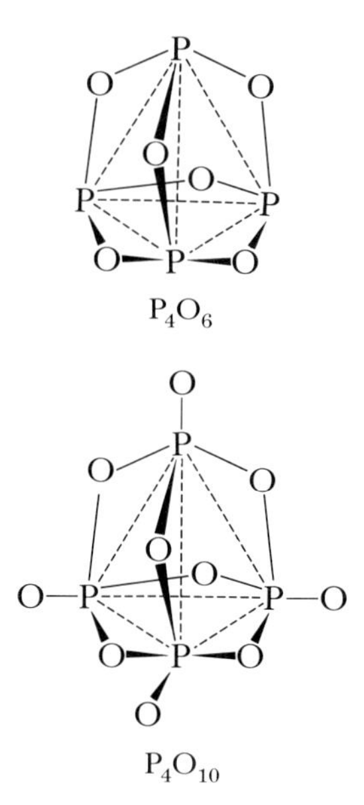

FIGURE 21.13 Structures of P_4O_6 and P_4O_{10}.

Phosphorus triiodide can be formed by reacting white phosphorus and iodine in an appropriate solvent.

The phosphorus halides hydrolyze on contact with water. The reactions occur readily, and most of the phosphorus halides fume in air as a result of reaction with water vapor. In the presence of excess water, the products are the corresponding phosphorus oxyacid and hydrogen halide, as in the examples shown in Equations 21.44 and 21.45:

$$PF_3(g) + 3H_2O(l) \longrightarrow H_3PO_3(aq) + 3HF(aq) \qquad [21.44]$$

$$PCl_5(l) + 4H_2O(l) \longrightarrow H_3PO_4(aq) + 5HCl(aq) \qquad [21.45]$$

When only a little water is used, the hydrogen halide can be obtained as a gas; these reactions thus provide convenient laboratory preparations of the hydrogen halides.

Oxy Compounds of Phosphorus

Probably the most significant compounds of phosphorus are those in which the element is combined in some way with oxygen. Phosphorus(III) oxide, P_4O_6, is obtained by allowing white phosphorus to oxidize in a limited supply of oxygen. When oxidation takes place in the presence of excess oxygen, phosphorus(V) oxide, P_4O_{10}, forms. This compound is also readily formed by oxidation of P_4O_6. Phosphorus(III) oxide is often called phosphorus trioxide after its empirical formula; similarly, phosphorus(V) oxide is often called phosphorus pentoxide, and written as the empirical formula P_2O_5. These two oxides represent the two most common oxidation states for phosphorus, +3 and +5. The structural relationship between P_4O_6 and P_4O_{10} is shown in Figure 21.13. Notice the resemblance these molecules have to the P_4 molecule, Figure 21.12.

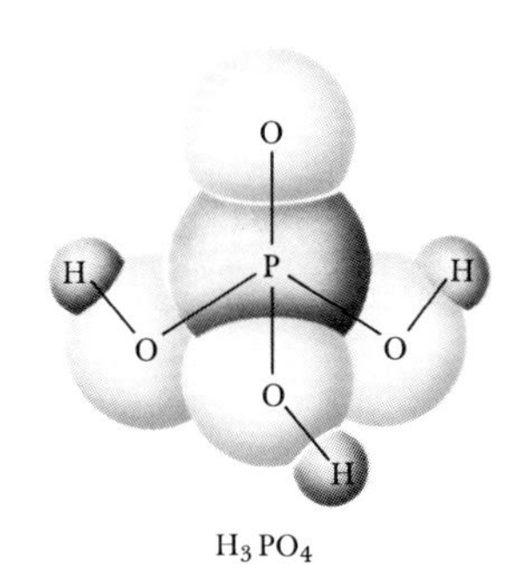

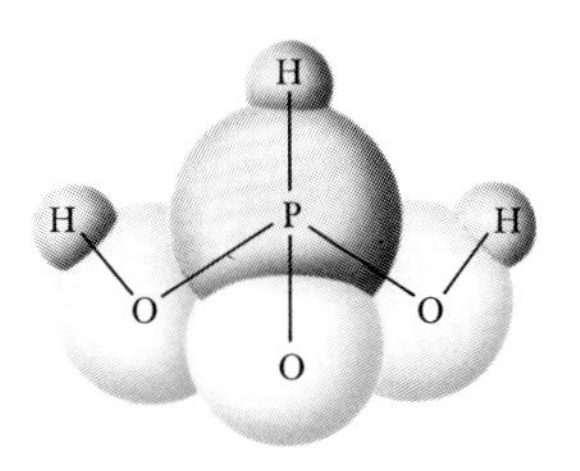

FIGURE 21.14 Structures of H_3PO_4 and H_3PO_3.

Phosphorus(V) oxide is the anhydride of phosphoric acid, H_3PO_4, a weak triprotic acid. In fact, P_4O_{10} has a very high affinity for water and is consequently used as a drying agent. Phosphorus(III) oxide is the anhydride of phosphorous acid, H_3PO_3, a weak diprotic acid. The structures of H_3PO_4 and H_3PO_3, are shown in Figure 21.14. The hydrogen atom that is attached directly to phosphorus in H_3PO_3 is not acidic.

One characteristic of phosphoric and phosphorous acids is their tendency to undergo condensation reactions when heated. A **condensation reaction** is one in which two or more molecules combine to form a larger molecule by eliminating a small molecule such as H_2O. Such a reaction in which two H_3PO_4 molecules are joined by the elimination of one H_2O molecule is shown in Equation 21.46:

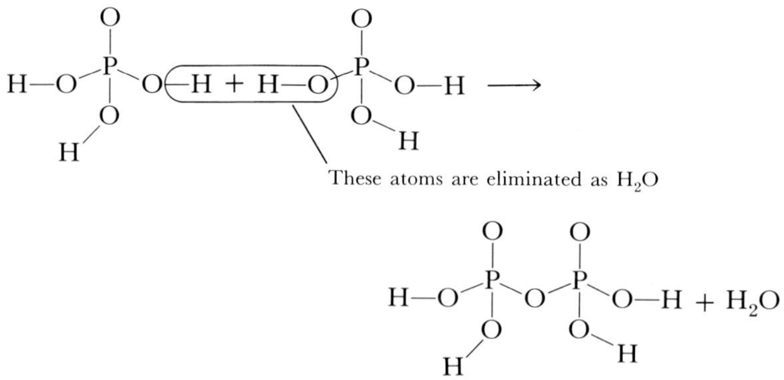

$$H-O-P(=O)(-O-H)-O-H + H-O-P(=O)(-O-H)-O-H \longrightarrow$$

These atoms are eliminated as H_2O

$$H-O-P(=O)(-O-H)-O-P(=O)(-O-H)-O-H + H_2O \qquad [21.46]$$

$$2H_3PO_4 \longrightarrow H_4P_2O_7 + H_2O$$

Further condensation produces phosphates having an empirical formula of HPO_3:

$$nH_3PO_4 \longrightarrow (HPO_3)_n + nH_2O \qquad [21.47]$$

Two phosphates having this empirical formula, one cyclic and the other polymeric, are shown in Figure 21.15. The three acids H_3PO_4, $H_4P_2O_7$, and $(HPO_3)_n$ all contain phosphorus in its +5 oxidation state, and all are therefore called phosphoric acids. To differentiate them, the prefixes *ortho-*, *pyro-*, and *meta-* are used: H_3PO_4 is orthophosphoric acid, $H_4P_2O_7$ is pyrophosphoric acid, and HPO_3 is metaphosphoric acid.

Phosphoric acid and its salts find their most important uses in detergents and fertilizers. The phosphates in detergents are often in the form of sodium tripolyphosphate, $Na_5P_3O_{10}$ (see Section 24.2). A typical detergent formulation contains 47 percent phosphate, 16 percent bleaches, perfumes, and abrasives, and 37 percent linear alkylsulfonate (LAS) surfactant, such as that shown below:

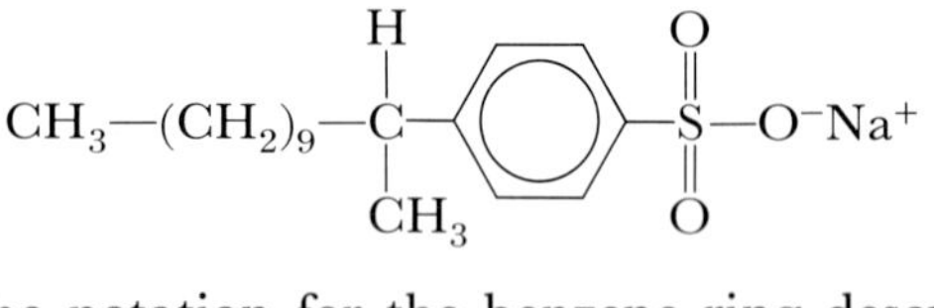

$$CH_3-(CH_2)_9-CH(CH_3)-C_6H_4-SO_2-O^-Na^+$$

(We have used the notation for the benzene ring described in Section 8.4.) The detergent action of such molecules has been described in Sec-

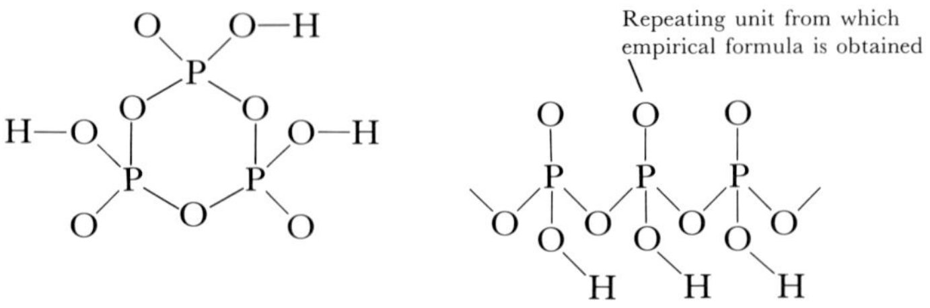

Repeating unit from which empirical formula is obtained

$(HPO_3)_3$
Trimetaphosphoric acid

$(HPO_3)_n$
Polymetaphosphoric acid

FIGURE 21.15 Structures of trimetaphosphoric acid and polymetaphosphoric acid.

tion 11.7. The phosphate ions form bonds with metal ions that contribute to the hardness of water. This keeps these ions from interfering with the action of the surfactant. The phosphates also keep the pH above 7 and thus prevent the surfactant molecules from becoming protonated (gaining an H^+ ion).

Most phosphate rock mined is converted to fertilizers. The mined phosphate rock contains large amounts of sand and clay. At a treatment plant, sand, clay, and organic materials are removed from the raw ore. The resultant concentrate is shipped to plants that convert it to phosphoric acid or water-soluble phosphate fertilizers. The $Ca_3(PO_4)_2$ in phosphate rock is insoluble ($K_{sp} = 2.0 \times 10^{-29}$). It is converted to a soluble form for use in fertilizers. This can be accomplished by treating the concentrated phosphate rock with sulfuric or phosphoric acid:

$$Ca_3(PO_4)_2(s) + 4H^+(aq) + 3SO_4{}^{2-}(aq) \longrightarrow 3CaSO_4(s) + 2H_2PO_4{}^-(aq) \qquad [21.48]$$

$$Ca_3(PO_4)_2(s) + 4H^+(aq) \longrightarrow 3Ca^{2+}(aq) + 2H_2PO_4{}^-(aq) \qquad [21.49]$$

The mixture formed when ground phosphate rock is treated with sulfuric acid and then dried and pulverized is known as superphosphate. The $CaSO_4$ formed in this process is of little use in soil except when deficiencies in calcium or sulfur exist. It also dilutes the phosphorus, which is the nutrient of interest. If the phosphate rock is treated with phosphoric acid, the product contains no $CaSO_4$ and has a higher percentage of phosphorus. This product is known as triple superphosphate. Although the solubility of $Ca(H_2PO_4)_2$ allows it to be assimilated by plants, it also allows it to be washed from the soil and into water bodies, thereby contributing to eutrophication.

Phosphorus compounds are important in biological systems. The element occurs, for example, in phosphate groups in RNA and DNA, the molecules responsible for control of protein biosynthesis and transmission of genetic information. It also occurs in adenosine triphosphate (ATP), which stores energy within biological cells:

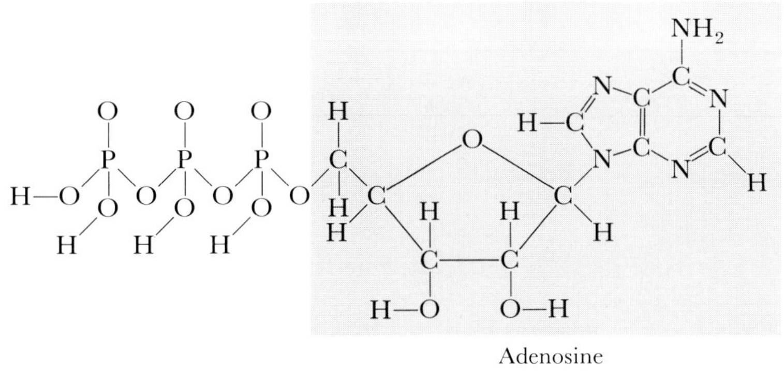

Adenosine

The P—O—P bond of the end phosphate group is broken by hydrolysis with water, forming adenosine diphosphate (ADP). This reaction produces 33 kJ of energy:

$$H_2O + \underset{\text{ATP}}{H{-}O{-}\overset{\overset{\displaystyle O}{|}}{\underset{\underset{\displaystyle O{-}H}{|}}{P}}{-}O{-}\overset{\overset{\displaystyle O}{|}}{\underset{\underset{\displaystyle O{-}H}{|}}{P}}{-}O{-}\overset{\overset{\displaystyle O}{|}}{\underset{\underset{\displaystyle O{-}H}{|}}{P}}{-}O{-}\text{adenosine}} \longrightarrow$$

$$H{-}O{-}\overset{\overset{\displaystyle O}{|}}{\underset{\underset{\displaystyle O{-}H}{|}}{P}}{-}O{-}H + \underset{\text{ADP}}{H{-}O{-}\overset{\overset{\displaystyle O}{|}}{\underset{\underset{\displaystyle O{-}H}{|}}{P}}{-}O{-}\overset{\overset{\displaystyle O}{|}}{\underset{\underset{\displaystyle O{-}H}{|}}{P}}{-}O{-}\text{adenosine}} \qquad [21.50]$$

This energy is used to perform the mechanical work of muscle contraction and in many other biochemical reactions (see Section 17.6 and Figure 17.8). There is more on the biological properties of phosphorus-containing substances in Chapter 26.

Arsenic, Antimony, and Bismuth

Arsenic, antimony, and bismuth occur in nature in the form of sulfide minerals such as As_2S_3, Sb_2S_3, and Bi_2S_3. In addition, the elements are found as minor components in ores of various metals, such as Cu, Pb, Ag, and Hg. Arsenic and antimony exhibit allotropy similar to that of phosphorus. Both elements can be prepared as soft, yellow, nonmetallic solids by quickly cooling the high-temperature vapor of the element. In this allotropic form, analogous to the white allotrope of phosphorus, the elements are present as As_4 or Sb_4 tetrahedra. Heating or the action of light converts the substances into the gray, more metallic forms containing sheets of atoms. In its common form bismuth has a reddish-white, rather metallic appearance. The element forms alloys with many metallic elements. Alloys of lead, bismuth, and tin are used in the construction of plugs in sprinkler systems. Water is discharged when the fusible metal plug melts.

Arsenic and antimony resemble phosphorus in much of their chemical behavior. For example, both elements form halides of formula MX_3 and MX_5 with structures and chemical properties similar to the halides of phosphorus. The oxy compounds of these two elements are also rather similar to those of phosphorus, except that the higher oxidation state is not so easily attained. Thus the product of burning arsenic in oxygen is As_4O_6, not As_4O_{10}. The higher oxide can be obtained by oxidation of As_4O_6 with a strong oxidizing agent such as nitric acid:

$$As_4O_6(s) + 4NO_3^-(aq) + 6H_2O(l) + 4H^+(aq) \longrightarrow 4H_3AsO_4(aq) + 4HNO_2(aq) \qquad [21.51]$$

In terms of the criteria discussed at the beginning of Chapter 20, bismuth is more logically considered a metal rather than a nonmetal. Bismuth usually appears in the +3 oxidation state; there is little tendency to attain the higher +5 oxidation state that is so common for phosphorus. The common oxide of bismuth is Bi_2O_3. This substance is insoluble in water or basic solution but is soluble in acidic solution. It thus is classified as a basic anhydride. As we have seen, the oxides of metals characteristically behave as basic anhydrides.

21.5 THE GROUP 4A ELEMENTS

The elements of group 4A are carbon, silicon, germanium, tin, and lead. The general trend from nonmetallic to metallic as we go down a family is strikingly evident in group 4A. Carbon is strictly nonmetallic; silicon is essentially nonmetallic, although it does exhibit some characteristics of a metalloid, particularly in its electrical and physical properties; germanium is a metalloid; tin and lead are both metallic. The chemistry of carbon was discussed in Section 20.5. In the present discussion we will consider a few general characteristics of group 4A, and then look more thoroughly at silicon.

General Characteristics of the Group 4A Elements

Some properties of the group 4A elements are given in Table 21.10. The elements possess the outer electron configuration ns^2np^4. The electronegativities of the elements are generally low; carbides that formally contain C^{4-} ions are observed only in the case of a very few compounds of carbon with very active metals. Formation of +4 ions by electron loss is not observed for any of these elements; the ionization energies are too high. However, the +2 state is found in the chemistry of germanium, tin, and lead; it is the principal oxidation state for lead. The vast majority of the compounds of the group 4A elements are covalently bonded. Carbon forms a maximum of four bonds. The other members of the family are able to form higher coordination numbers because of the availability of *d* orbitals for bonding.

Carbon differs from the other group 4A elements in its pronounced ability to form multiple bonds both with itself and with other nonmetals, especially N, O, and S. The origin of this behavior was considered earlier, in Section 20.1.

Table 21.10 shows that the strength of a bond between two atoms of a given element decreases as we go down group 4A. Carbon-carbon bonds are quite strong. As a consequence, carbon has a striking ability to form compounds in which carbon atoms are bonded to each other. This property, called **catenation**, permits the formation of extended chains and rings of carbon atoms and accounts for the large number of organic compounds that exist. Catenation is also exhibited by other elements,

TABLE 21.10 Some Properties of the Group 4A Elements

Property	C	Si	Ge	Sn	Pb
Atomic radius (Å)	0.77	1.17	1.22	1.41	1.54
First ionization energy (kJ/mol)	1090	780	782	704	714
Electronegativity	2.5	1.9	2.0	1.8	2.3
X—X single-bond energy (kJ/mol)	348	226	188	151	—

especially ones in the vicinity of carbon in the periodic table, such as boron, nitrogen, phosphorus, oxygen, sulfur, silicon, and germanium. However, such self-linkage is far less important in the chemistries of these other elements. For example, the Si—Si bond strength (226 kJ/mol) is far smaller than the Si—O bond strength (368 kJ/mol). As a result, the chemistry of silicon is dominated by the formation of Si—O bonds, and Si—Si bonds play a rather minor role.

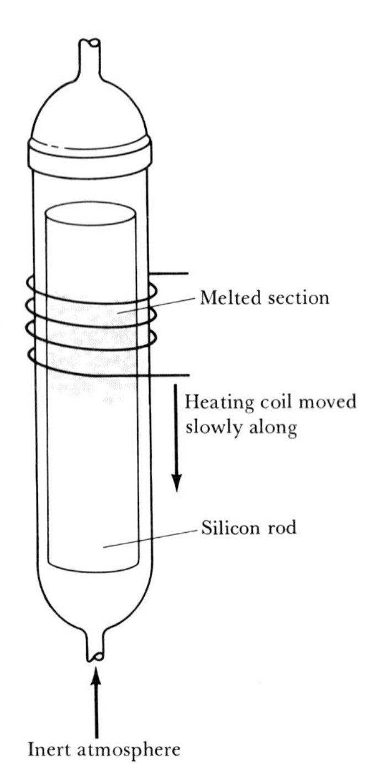

FIGURE 21.16 Zone-refining apparatus.

Occurrence and Preparation of Silicon

Silicon is the second most abundant element, after oxygen, in the earth's crust. It occurs in SiO_2 and in an enormous variety of silicate minerals. The element is obtained by the reduction of silicon dioxide with carbon at high temperature:

$$SiO_2(l) + 2C(s) \longrightarrow Si(l) + 2CO(g) \qquad [21.52]$$

Elemental silicon has a diamond type of structure; a graphitelike allotrope is not known, presumably because of the weakness of π bonds between silicon atoms. Crystalline silicon is a gray, metallic-looking solid that melts at 1410°C. The element is a semiconductor and is thus used in making transistors and solar cells. In order to be used as a semiconductor, the element must be extremely pure. One method of purification involves treatment of the element with Cl_2 to form $SiCl_4$. The $SiCl_4$ is a volatile liquid that is purified by fractional distillation and then reduced by H_2 to elemental silicon:

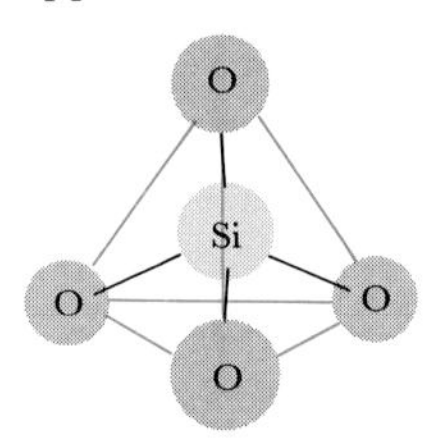

FIGURE 21.17 Structure of the SiO_4 tetrahedron of the SiO_4^{4-} ion. This ion is found in several minerals such as zircon, $ZrSiO_4$.

$$SiCl_4(g) + 2H_2(g) \longrightarrow Si(s) + 4HCl(g) \qquad [21.53]$$

The element can be further purified by the process of zone refining. In the zone-refining process a heated coil is passed slowly along a silicon rod as shown in Figure 21.16. A narrow band of the element is thereby melted. As the molten area is swept slowly along the length of the rod the impurities concentrate in the molten region, following it to the end of the rod. The end in which the impurities are collected is cut off and recycled through the purification process starting with formation of $SiCl_4$. The purified top portion of the rod is retained for manufacture of electronic devices.

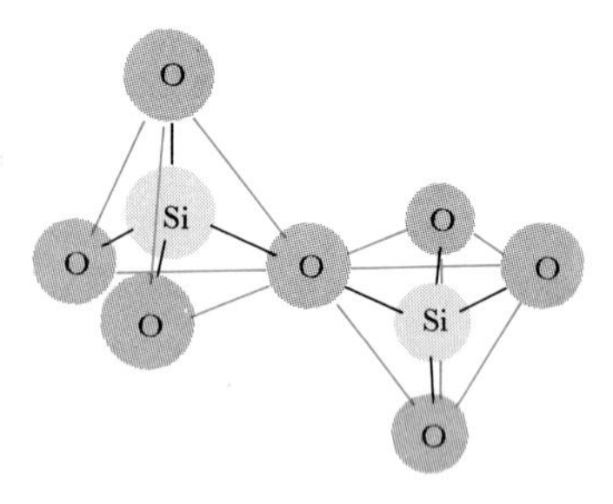

FIGURE 21.18 Geometrical structure of the $Si_2O_7^{6-}$ ion, which is formed by the sharing of an oxygen atom by two silicon atoms. This ion occurs in several minerals such as hardystonite, $Ca_2Zn(Si_2O_7)$.

Silicates

It is estimated that over 90 percent of the earth's crust consists of silicates, if SiO_2 is included. The basic structural unit of these silicates consists of a silicon atom surrounded in a tetrahedral fashion by four oxygens, as shown in Figure 21.17. Because the silicon has an oxidation state of +4 and oxygen −2, a simple ion of this composition has a charge of $+4 + 4(-2) = -4$. The simple SiO_4^{4-} ion, which is known as the orthosilicate ion, is found in very few silicate minerals. Usually, silicate tetrahedra share oxygen atoms to build up more complex structures containing Si—O—Si linkages. When two tetrahedra share a single oxygen, the $Si_2O_7^{6-}$ ion shown in Figure 21.18 results.

SAMPLE EXERCISE 21.7

Draw the Lewis structures, showing all valence electrons, for the SiO_4^{4-} and $Si_2O_7^{6-}$ ions. Indicate how the $Si_2O_7^{6-}$ ion is related to SiO_4^{4-}.

Solution: The total number of valence electrons in SiO_4^{4-} is 32 (remember, we must count the 4 electrons that make the ionic charge -4). The total number of valence electrons in $Si_2O_7^{6-}$ is 56. When these electrons are placed about the atoms connected as shown in Figures 21.17 and 21.18, we obtain the following Lewis structures:

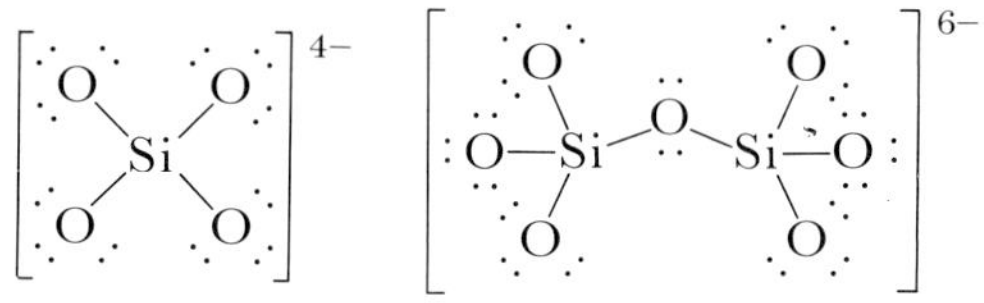

Notice that the bridging oxygen in $Si_2O_7^{6-}$ contains two rather than three unshared electron pairs.

It is possible to form a chain of SiO_4 tetrahedra by sharing oxygens at two corners of each tetrahedron, as shown in Figure 21.19(*a*). The simplest formula associated with this single-chain silicate anion is SiO_3^{2-}. This anion occurs, for example, in the mineral enstatite, $MgSiO_3$, which consists of silicate chains (hence the SiO_3^{2-} portion of the mineral's simplest formula) with Mg^{2+} ions between strands to balance charge.

The other structures shown in Figure 21.19 are also found in silicates. For example, it is possible for each tetrahedron to share three corners, giving rise to sheet structures. Such an arrangement results in a simplest formula $Si_2O_5^{2-}$ as in talc, $Mg_3(Si_2O_5)_2(OH)_2$. When all four oxygens of each SiO_4 unit are involved in Si—O—Si linkages, the structure extends in all three dimensions in space, forming quartz, SiO_2. Because the structure is locked together in a three-dimensional array much like diamond, quartz is harder than the strand or sheet-type silicates.

Asbestos is a general term applied to a group of fibrous silicate minerals. These minerals possess chainlike arrangements of the silicate tetrahedra, or sheet structures in which the sheets are formed into rolls. The result is that the minerals have a fibrous character, as shown in Figure 21.20. Asbestos has been widely used as thermal insulation material, especially in high-temperature applications where the high chemical stability of the silicate structure serves well. However, in recent years there has been a growing awareness that certain forms of asbestos pose a health hazard. Tiny fibers readily penetrate the tissues of the lungs and digestive tract, and remain there over a long period of time. Eventually, lesions, including cancer, may result.

Aluminosilicates

In many silicate minerals Si^{4+} ions are replaced by Al^{3+} ions within the silicate tetrahedra. This replacement produces **aluminosilicates**. In order to maintain charge balance, an extra cation such as K^+ must accompany each of these substitutions. Muscovite, $KAl_2(AlSi_3O_{10})(OH)_2$,* a mica mineral, is an aluminosilicate.

Replacement of a quarter of the silicon atoms in a sheet silicate, $Si_4O_{10}^{4-}$, with aluminum produces the $AlSi_3O_{10}^{5-}$ sheets found in this mineral. In both the silicate-sheet and aluminosilicate-sheet minerals,

*Aluminum is found in this mineral in two different environments. The first two Al^{3+} ions are located between the aluminosilicate sheets. The aluminum that is shown in parentheses is located within the sheets. It has replaced a silicon and is therefore located in an AlO_4 tetrahedron.

(*a*) Single-strand chain:

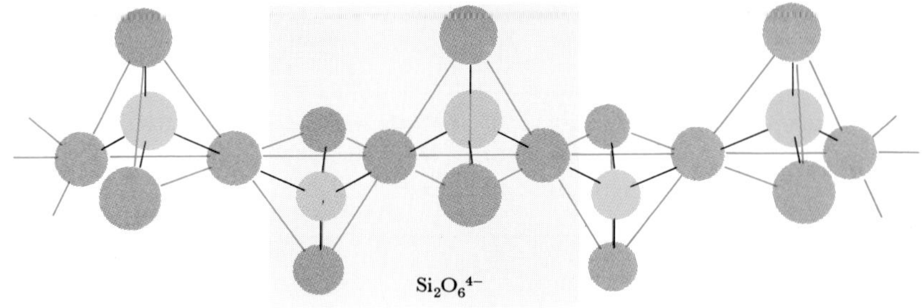

(*b*) Double-strand chain:

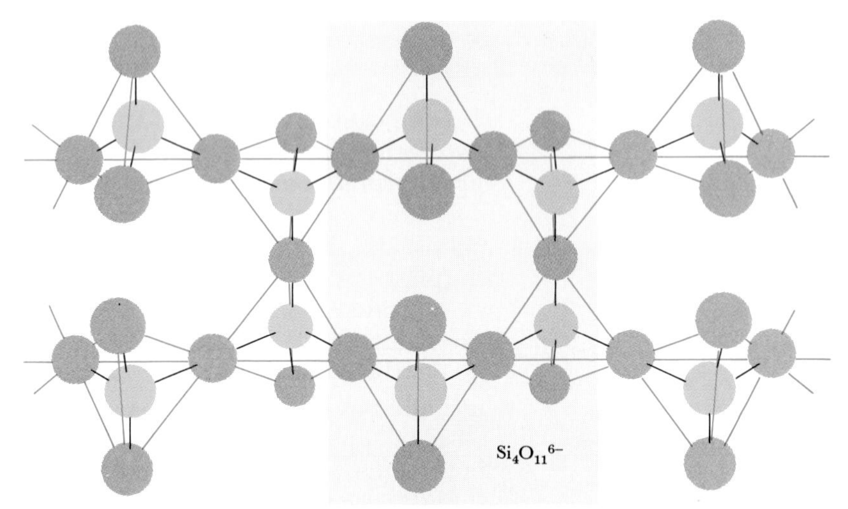

(*c*) Sheet (or layer) structure:

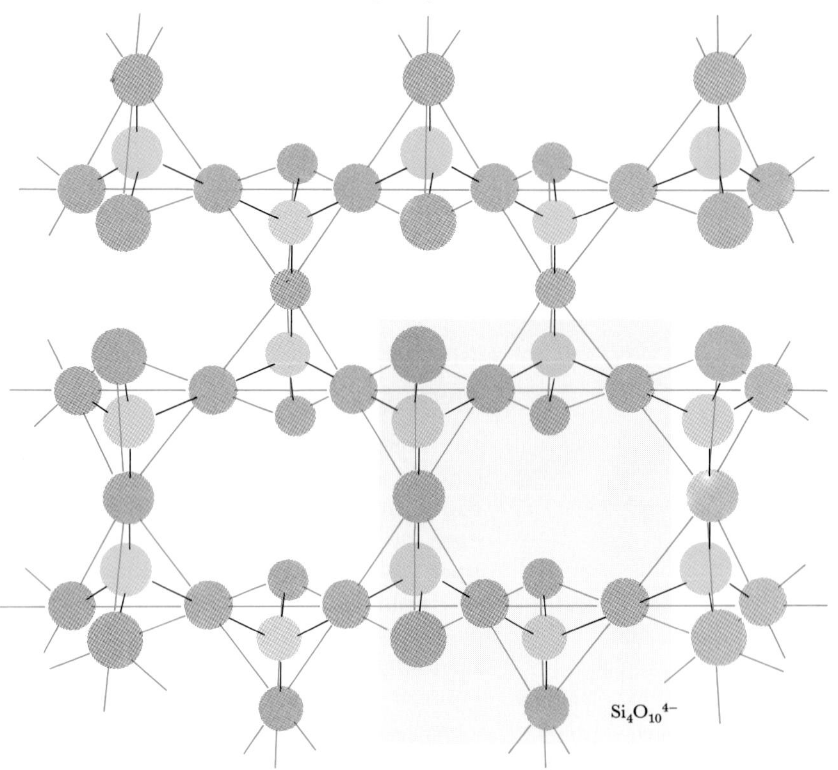

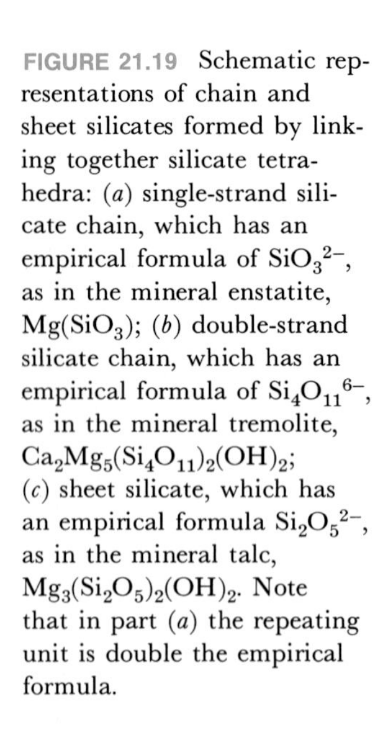

FIGURE 21.19 Schematic representations of chain and sheet silicates formed by linking together silicate tetrahedra: (*a*) single-strand silicate chain, which has an empirical formula of SiO_3^{2-}, as in the mineral enstatite, $Mg(SiO_3)$; (*b*) double-strand silicate chain, which has an empirical formula of $Si_4O_{11}^{6-}$, as in the mineral tremolite, $Ca_2Mg_5(Si_4O_{11})_2(OH)_2$; (*c*) sheet silicate, which has an empirical formula $Si_2O_5^{2-}$, as in the mineral talc, $Mg_3(Si_2O_5)_2(OH)_2$. Note that in part (*a*) the repeating unit is double the empirical formula.

FIGURE 21.20 Serpentine asbestos; note the fibrous character of this mineral.

cations are located between sheets to balance the charge. The electrostatic attraction between these cations and the charges in the sheets is greater for the aluminosilicate than for the corresponding silicate, because there are higher charges in the aluminosilicates. Thus, whereas the sheets in the silicate mineral talc slide readily over each other, the sheets in the aluminosilicate mica do not. Nevertheless, mica does cleave readily into sheets, as illustrated in Figure 21.21.

When aluminum replaces up to half of the silicon atoms in SiO_2, the *feldspar* minerals result. For example, orthoclase, $KAlSi_3O_8$, is a common feldspar mineral. The cations that compensate the extra negative charge

FIGURE 21.21 Piece of mica, showing its cleavage into thin sheets.

accompanying this replacement are usually Na^+, K^+, or Ca^{2+} ions. The feldspars are the most abundant rock-forming silicates, comprising about 50 percent of the minerals in the earth's crust.

SAMPLE EXERCISE 21.8

The mineral anorthite is a feldspar mineral formed by replacing half of the silicon atoms in SiO_2 with aluminum and maintaining charge balance with Ca^{2+} ions. What is the simplest formula for this mineral?

Solution: If we write SiO_2 as Si_2O_4, the replacement of half of the Si^{4+} with Al^{3+} produces $AlSiO_4^-$. One Ca^{2+} therefore requires two $AlSiO_4^-$ units to maintain charge balance and the empirical formula of the mineral is $CaAl_2Si_2O_8$.

The **clay minerals** are hydrated aluminosilicates having sheet-type structures. For example, kaolinite, $Al_2Si_2O_5(OH)_4$, is a clay mineral. These minerals have small particle size and correspondingly large surface areas. They have the ability to adsorb cations on their surfaces. Often the metal ions displace H^+ ions from the OH groups on the surfaces of the clay particle:

$$M^+(aq) + H\text{—}O\text{—}clay \rightleftharpoons H^+(aq) + M\text{—}O\text{—}clay \qquad [21.54]$$

This situation gives rise to pH-dependent equilibria. Notice that the higher the concentration of $H^+(aq)$, the more the equilibrium is shifted to the left. If the soil is basic, the equilibrium lies to the right, and $M^+(aq)$ is not available to plants. Thus the pH of the soil plays an important role in determining the soil's fertility (that is, its ability to supply plants with essential nutrients).

Glass

Quartz melts at approximately 1600°C, forming a tacky liquid. In the course of melting, many silicon-oxygen bonds are broken. When the liquid is rapidly cooled, silicon-oxygen bonds are reformed before the atoms are able to arrange themselves in a regular fashion. An amorphous solid, known as quartz glass or silica glass, therefore results (see Figure 10.30). Many different substances can be added to SiO_2 to cause it to melt at a lower temperature. The common glass used in windows and bottles is known as soda-lime glass. It contains CaO and Na_2O in addition to SiO_2 from sand. The CaO and Na_2O are produced by heating two inexpensive chemicals, limestone, $CaCO_3$, and soda ash, Na_2CO_3. These carbonates decompose at elevated temperatures, as shown in Equations 21.55 and 21.56:

$$CaCO_3(s) \longrightarrow CaO(s) + CO_2(g) \qquad [21.55]$$

$$Na_2CO_3(s) \longrightarrow Na_2O(s) + CO_2(g) \qquad [21.56]$$

Other substances can be added to soda-lime glass to produce color or to change the properties of the glass in various ways. For example, addi-

TABLE 21.11 Compositions, properties, and uses of various types of glass

Type of glass	Composition by weight	Properties and uses
Soda-lime	12% Na_2O, 12% CaO, 76% SiO_2	Window glass, bottles
Aluminosilicate	5% B_2O_3, 10% MgO, 10% CaO, 20% Al_2O_3, 55% SiO_2	High melting—used in cooking ware
Lead alkali	10% Na_2O, 20% PbO, 70% SiO_2	High refractive index—used in lenses, decorative glass
Borosilicate	5% Na_2O, 3% CaO, 16% B_2O_3, 76% SiO_2	Low coefficient of thermal expansion—used in laboratory ware, cooking utensils
Bioglass	24% Na_2O, 24% CaO, 6% P_2O_5, 46% SiO_2	Compatible with bone—used as coating on surgical implants

tion of CoO produces the deep blue color of "cobalt glass." Replacement of Na_2O by K_2O results in a harder glass that has a higher melting point. Replacement of CaO by PbO results in a denser glass with a higher refractive index. Addition of nonmetal oxides such as B_2O_3 or P_2O_5, which form network structures related to the silicates, also causes a change in properties of the glass. For example, addition of B_2O_3 results in a glass with a higher melting point and a lower coefficient of thermal expansion. Such glasses, sold commercially under trade names such as Pyrex or Kimax, are used in applications where resistance to thermal shock is important, for example, in laboratory glassware or coffee makers. Table 21.11 lists the formulations and properties of several representative types of glass.

A recent popular development has been that of "photochromic" glasses that darken in the sun and return to their clear state in the dark. This glass contains a dispersion of AgCl or AgBr. These substances are photosensitive, in that they decompose to silver and halogen atoms in the presence of light. The finely divided silver is black. The silver and halogen atoms are kept in close proximity by the glass matrix and reform AgCl or AgBr in the dark:

$$AgCl \underset{\text{dark reaction}}{\overset{\text{light reaction, } h\nu}{\rightleftharpoons}} Ag + Cl \qquad [21.57]$$

Optical fibers represent a relatively new, high-technology application of glassmaking. An optical fiber is a glass thread that conducts light, just as a copper wire conducts electrons. A beam of light that enters one end of an optical fiber travels along the length of the fiber and emerges at the other end. Optical fibers are of interest because they can be used to transmit information, using the light waves that pass through the fiber. Optical fibers are capable of carrying much more information for a given cross section of fiber than can be transmitted via the conventional coaxial cable, as illustrated by the comparison shown in Figure 21.22.

The key to long-distance transmission of signals via optical fibers is glass purity. Impurity ions such as Fe^{2+} cause absorption of the light waves, with resulting signal loss. Today it is possible to obtain optical fibers that undergo a loss of only about 1 percent of signal over a distance of 1 km. To achieve that level of performance, impurities must be reduced to the level of 1 part per billion. This ultrahigh purity is achieved by distilling high-purity liquid glass, to separate it from any remaining impurities.

FIGURE 21.22 Optical fiber cable for transmitting communications information is held on the right. The optical fiber has the same information-carrying capacity as the much larger copper cable with its many individual wires on the left. (*Courtesy Corning Glass Works*)

21.6 BORON

At this point there is only one additional element to consider in our survey of nonmetallic elements: boron. Boron is the only element of group 3A that can be considered to be nonmetallic. The element has an extended network structure. The melting point of boron, 2300°C, is intermediate between that for carbon, 3550°C, and that for silicon, 1410°C. The electronic configuration of boron is $[He]2s^22p^1$. The element exhibits a valence of 3 in all its common chemical compounds. We have seen (Section 7.7) that the electron configuration about boron in the boron halides provides an exception to the octet rule, in that there are but six electrons in the boron valence shell. The boron halides are thus strong Lewis acids (Section 15.10).

Salts of the borohydride ion, BH_4^-, are widely used as reducing agents. This ion can be thought of as an isoelectronic analog of CH_4 or NH_4^+. The lower charge of the central atom in BH_4^- means that the hydrogens of the BH_4^- are "hydridic," that is, they carry a partial negative charge. Thus it is not surprising that borohydrides are good reducing agents, as illustrated by the reaction in Equation 21.58:

$$3BH_4^-(aq) + 4IO_3^-(aq) \longrightarrow 4I^-(aq) + 3H_2BO_3^-(aq) + 3H_2O(l) \qquad [21.58]$$

The only important oxide of boron is boric oxide, B_2O_3. This substance is the anhydride of boric acid, which we may write as H_3BO_3 or $B(OH)_3$. Boric acid is a weak acid, so much so that solutions of boric acid are used as an eyewash. Upon warming to 100°C, orthoboric acid loses water by a condensation reaction similar to that described in Section 21.4. The product of the reaction, as shown in Equation 21.59, is metaboric acid, a polymeric substance of formula HBO_2.

$$H_3BO_3(s) \longrightarrow HBO_2(s) + H_2O(g) \quad [21.59]$$

Heating of metaboric acid results in still more loss of water, as in Equations 21.60 and 21.61:

$$4HBO_2(s) \longrightarrow H_2B_4O_7(s) + H_2O(l) \quad [21.60]$$

$$H_2B_4O_7(s) \longrightarrow 2B_2O_3(s) + H_2O(g) \quad [21.61]$$

The acid $H_2B_4O_7$ is called tetraboric acid. The sodium salt, $Na_2B_4O_7 \cdot 10H_2O$, called borax, occurs in dry lake deposits in California and can also be readily prepared from other borate minerals. Solutions of borax are alkaline; the substance is used in various laundry and cleaning products.

FOR REVIEW

Summary

The noble gas elements exhibit a very limited chemical behavior because of the exceptional stability of their electronic configuration. The xenon fluorides and oxides and KrF_2 are the only established examples of chemical reactivity among these elements.

The halogens occur as diatomic molecules. These elements possess the highest electronegativities of the elements in each row of the periodic table. All except fluorine exhibit oxidation states varying from -1 to $+7$. Fluorine, being the most electronegative element, is restricted to the oxidation states 0 and -1. The tendency to form the -1 oxidation state from the free element (that is, the oxidizing power of the element) decreases with increasing atomic number in the family. The halogens form compounds with one another, called interhalogens. In the higher interhalogens, XX'_n, the element X may be Cl, Br, or I, and X′ is nearly always F; *n* may have values 3, 5, or 7.

The group 6A elements range from the very abundant and strongly nonmetallic oxygen to the rare and rather metallic tellurium. Sulfur occurs widely in the form of sulfide ores and in elemental sulfur beds. The element has several allotropic forms; the most stable one consists of S_8 rings. The most important compound of the element is sulfuric acid, a strong acid that is a good dehydrating agent and has a high boiling point. Selenium and tellurium are chemically rather similar to sulfur, especially with respect to formation of oxides and oxyanions.

The group 5A elements exhibit a wide range of behavior, from strongly nonmetallic in the case of nitrogen to distinctly metallic in the case of bismuth. Phosphorus occurs in nature in certain phosphate minerals. The element exhibits several allotropes, including one known as white phosphorus, a reactive form consisting of P_4 tetrahedra. Phosphorus forms compounds of formula PX_3 and PX_5 with the halogens. These undergo hydrolysis in water to produce the corresponding oxyacid of phosphorus and HX. Phosphorus forms two oxides, P_4O_6 and P_4O_{10}. Their corresponding acids, phosphorous acid and phosphoric acid, show a strong tendency to undergo condensation reactions when heated. Compounds of phosphorus are important components of fertilizers.

The group 4A elements show great diversity in physical and chemical properties. Carbon excels in being able to form multiple bonds and in undergoing catenation. Silicon is noteworthy as a semiconductor and for its tendency to form Si—O bonds. Silicon is the second most abundant element and it occurs in a wide variety of silicates. Silicates are composed of SiO_4 tetrahedra, which, through sharing oxygen atoms, are able to link together to form chains, sheets, and three-dimensional arrays. In many minerals, Si^{4+} ions are replaced by Al^{3+} ions, thus forming aluminosilicates. Silicates are important components of glass.

Boron commonly exhibits a valence of +3, as in boric oxide, B_2O_3, and boric acid, H_3BO_3. The acid readily undergoes condensation reactions.

Learning goals

Having read and studied this chapter, you should be able to:

1 Cite the most common occurrences of each nonmetallic element discussed in this chapter (for example, fluorine is found in CaF_2).

2 Cite the most common molecular form of each nonmetallic element discussed in the chapter (for example, selenium is found as chains of Se atoms).

3 Account for the fact that xenon forms several compounds with fluorine and oxygen, krypton forms only KrF_2, and no chemical reactivity is known for the lighter noble gas elements.

4 Write the formulas of the known fluorides, oxyfluorides, and oxides of xenon, and describe the relative stabilities of the oxides as compared with the fluorides.

5 Describe the electronic and geometrical structures of the known compounds of xenon.

6 Write balanced chemical equations describing at least one means of preparation of each halogen from naturally occurring sources.

7 Describe at least one important use of each halogen element.

8 Write a balanced chemical equation describing the preparation of each of the hydrogen halides.

9 Give examples of diatomic and higher interhalogen compounds and describe their electronic and geometrical structures.

10 Name the oxyacids or oxyanions of the halogens, given their formula, or vice versa.

11 Describe the variation in acid strength and oxidizing strength of the oxyacids of chlorine.

12 Indicate the formulas of the common oxides of sulfur and the properties of their aqueous solutions.

13 Write balanced chemical equations for formation of sulfuric acid from sulfur and describe the important properties of the acid.

14 Compare the chemical behaviors of selenium and tellurium with that of sulfur, with respect to common oxidation states and formulas of oxides and oxyacids.

15 Describe the preparation of elemental phosphorus from its ores, using balanced chemical equations.

16 Describe the formulas and structures of the stable halides and oxides of phosphorus.

17 Write balanced chemical equations for the reactions of the halides and oxides of phosphorus with water.

18 Describe a condensation reaction and give examples involving compounds of phosphorus or boron.

19 Describe the structures possible for silicates and their empirical formulas (for example, silicate tetrahedra can combine through bridging oxygens to form a single-string silicate chain whose empirical formula is SiO_3^{2-}).

20 Correlate the physical properties of certain silicate minerals, such as asbestos, with their structures.

21 Explain the changes in composition and properties that accompany substitution of Al^{3+} for Si^{4+} in a silicate.

22 Describe what is meant by a clay mineral.

23 Describe the role of clay minerals in soil fertility and the effects of soil pH.

24 Describe the composition of soda-lime glass.

25 Describe the manufacture of glass.

Key terms

Among the more important terms and expressions used for the first time in this chapter are the following:

Aluminosilicates (Section 21.5) are compounds that are structurally related to silicates in which some Si^{4+} ions are replaced by Al^{3+}.

Catenation (Section 21.5) refers to the linking of like atoms to form chains or rings.

Clay minerals (Section 21.5) are hydrated aluminosilicates.

A condensation reaction (Section 21.4) is one in which two or more molecules combine to form larger ones by elimination of small molecules such as H_2O.

Glass (Section 21.5) is an amorphous solid formed by fusion of SiO_2, CaO, and Na_2O. Other oxides may also be used to form glasses with differing characteristics.

An interhalogen compound (Section 21.2) is one formed between two different halogen elements. Examples include IBr and BrF_3.

Silicates (Section 21.5) are compounds containing silicon and oxygen and based on SiO_4 tetrahedra.

EXERCISES

The noble gases and halogens

21.1 Write the chemical formula for each of the following compounds and indicate the oxidation state of the halogen or noble gas atom in each: (a) iodate ion; (b) bromic acid; (c) bromine trifluoride; (d) sodium hypochlorite; (e) iodous acid; (f) potassium triiodide; (g) calcium bromide; (h) xenon trioxide; (i) xenon oxytetrafluoride.

21.2 Name the following compounds: (a) $KClO_2$; (b) $Ca(IO_3)_2$; (c) $AlCl_3$; (d) $HBrO_2$; (e) H_5IO_6; (f) $Fe(ClO_4)_2$; (g) $HClO$; (h) XeF_4; (i) IF_5; (j) XeO_3.

21.3 Why were the noble gases among the last elements to be discovered?

21.4 What are the major factors responsible for the fact that xenon forms stable compounds with fluorine, whereas argon does not?

21.5 List the halogens in order of increasing X—X halogen bond energies. Suggest a reason for the low F—F bond energy.

21.6 Explain each of the following observations. (a) At room temperature, I_2 is a solid, Br_2 is a liquid, and Cl_2 and F_2 are both gases. (b) F_2 cannot be prepared by electrolytic oxidation of aqueous F^- solutions. (c) The boiling point of HF is much higher than those of the other hydrogen halides. (d) The halogens decrease in oxidizing power in the order $F_2 > Cl_2 > Br_2 > I_2$. (e) For a given oxidation state the acid strength of the oxyacid in aqueous solution decreases in the order chlorine > bromine > iodine. (f) Hydrofluoric acid cannot be stored in glass bottles. (g) HI cannot be prepared by treating NaI with sulfuric acid. (h) The interhalogen ICl_3 is known but $BrCl_3$ is not. (i) I_2 is more soluble in aqueous solutions of I^- than in pure water.

21.7 Predict the geometrical structures of the following: (a) I_3^-; (b) ICl_4^-; (c) ClO_3^-; (d) H_5IO_6; (e) XeF_4.

21.8 The interhalogen compound $BrF_3(l)$ reacts with antimony(V) fluoride to form the salt $(BrF_2)(SbF_6)$. Write the Lewis structure for both the cation and anion in this substance, and describe the likely geometrical structure of each.

21.9 Write a balanced chemical equation showing the commercial preparation of each halogen element.

21.10 Write balanced net ionic equations for the reaction of each of the following substances with water: (a) PBr_5; (b) IF_5; (c) $SiBr_4$; (d) F_2; (e) ClO_2 (chloric acid is a product); (f) $HI(g)$.

21.11 Write a balanced chemical equation that describes a suitable means of preparing each of the following substances: (a) HF; (b) I_2; (c) XeF_4; (d) $Ca(OCl)Cl$; (e) SF_6, (f) NaClO.

21.12 Write balanced chemical equations for each of the following reactions (some of which are analogous but not identical to reactions shown in this chapter). (a) Bromine forms hypobromite ion on addition to aqueous base. (b) Chlorine reacts with an aqueous solution of sodium bromide. (c) Bromine reacts with an aqueous solution of hydrogen peroxide, liberating, O_2. (d) Hydrogen bromide is produced upon heating calcium bromide with phosphoric acid. (e) Hydrogen bromide is formed upon hydrolysis of aluminum bromide. (f) Aqueous hydrogen fluoride reacts with solid calcium carbonate, forming water-insoluble calcium fluoride.

21.13 (a) Write the balanced net-ionic equation for the reduction of ClO_3^- to Cl_2 by Fe^{2+} in acidic aqueous solution. (b) Calculate the standard emf for this reaction.

21.14 Chloride ion is oxidized in aqueous solution to $Cl_2(aq)$ by each of the following reagents: (a) $MnO_2(s)$; (b) $MnO_4^-(aq)$; (c) $Cr_2O_7^{2-}(aq)$. In each case, write a complete, balanced net ionic equation.

[21.15] Write the balanced half-reaction for (a) reduction of $HClO_2$ to Cl_2; (b) reduction of $HClO_2$ to Cl^-. In each case calculate the standard electrode potential for the half-reaction using information summarized in Figure 21.5.

[21.16] The redox behavior of xenon and some of its oxy species in aqueous alkaline solution is summarized as follows:

$$HXeO_6^{3-} \xrightarrow{0.94\ V} HXeO_4^- \xrightarrow{1.26\ V} Xe$$

Calculate the electrode potential for conversion of $HXeO_6^{3-}$ into Xe.

[21.17] Using the thermochemical data of Table 21.1 and Appendix D, calculate the average Xe—F bond energies in XeF_2, XeF_4, and XeF_6, respectively. What is the significance of the trend in these quantities?

[21.18] The solubility of Cl_2 in 100 g of water at STP is 310 cm^3. Assume that this quantity of Cl_2 is dissolved and equilibrated as follows:

$$Cl_2(aq) + H_2O(l) \rightleftharpoons Cl^-(aq) + HClO(aq) + H^+(aq)$$

If the equilibrium constant for this reaction is 4.7×10^{-4}, calculate the equilibrium concentration of HOCl formed.

Group 6A

21.19 Write the chemical formula for each of the following compounds and indicate the oxidation state of the group 6A element in each: (a) selenous acid; (b) potassium hydrogen sulfite; (c) hydrogen telluride; (d) carbon disulfide; (e) selenium trioxide; (f) orthotelluric acid; (g) selenate ion.

21.20 Name each of the following compounds: (a) $K_2S_2O_3$; (b) Al_2S_3; (c) $NaHSeO_3$; (d) SeF_6; (e) H_2Se; (f) FeS_2; (g) $NaHSO_4$; (h) Na_2SeO_4.

21.21 Sulfur and the group 6A elements below it in the periodic chart exhibit oxidation states ranging from -2 to $+6$. What factors control the lowest and highest oxidation states? Explain.

21.22 Write the Lewis structures for each of the following species and indicate the geometrical structure of each: (a) SeO_4^{2-}; (b) H_6TeO_6; (c) $TeO_2(g)$; (d) S_2Cl_2; (e) chlorosulfonic acid, HSO_3Cl (chlorine is bonded to sulfur).

21.23 The SF_5^- ion is formed when $SF_4(g)$ reacts with fluoride salts containing large cations such as $CsF(s)$. Draw the Lewis structure for the SF_5^- ion.

21.24 Draw the Lewis structures for the following species and predict the relative S—O bond lengths in each: SO_2, SO_3, SO_4^{2-}.

21.25 Write a balanced chemical equation for each of the following reactions. (a) Selenium dioxide dissolves in water. (b) Solid zinc sulfide reacts with hydrochloric acid. (c) Elemental sulfur reacts with sulfite ion to form thiosulfate. (d) Elemental selenium is heated with sulfuric acid. (e) Sulfur trioxide is dissolved in sulfuric acid. (f) Selenous acid is oxidized by hydrogen peroxide.

21.26 Write balanced chemical equations for each of the following reactions. (You may have to guess at one or more of the reaction products, but you should be able to make a reasonable guess based on your study of this chapter.) (a) Selenous acid is reduced by hydrazine in aqueous solution to yield elemental selenium (see Chapter 20 for a discussion of hydrazine). (b) Heating orthotelluric acid to temperatures over 200°C yields the acid anhydride. (c) Hydrogen selenide can be prepared by reaction of aqueous acid solution on aluminum selenide. (d) Sodium thiosulfate is used to remove excess Cl_2 from chlorine-bleached fabrics. The thiosulfate ion forms SO_4^{2-} and elemental sulfur while Cl_2 is reduced to Cl^-.

21.27 An aqueous solution of SO_2 acts as a reducing agent to reduce (a) aqueous $KMnO_4$ to $MnSO_4(s)$; (b) acidic aqueous $K_2Cr_2O_7$ to aqueous Cr^{3+}; (c) aqueous $Hg_2(NO_3)_2$ to mercury metal. Write balanced chemical equations for these reactions.

21.28 In aqueous solution, hydrogen sulfide reduces (a) Fe^{3+} to Fe^{2+}; (b) Br_2 to Br^-; (c) MnO_4^- to Mn^{2+}; (d) HNO_3 to NO_2. In all cases, under appropriate conditions, the product is elemental sulfur. Write a balanced net ionic equation for each reaction.

[21.29] The maximum allowable concentration of $H_2S(g)$ in air is 20 mg per kilogram of air (20 ppm by weight). How many grams of FeS would be required to react with hydrochloric acid to produce this concentration in an average room measuring 2.7 m × 4.3 m × 4.3 m?

[21.30] Suggest an explanation for the fact that telluric acid is a weak acid, whereas sulfuric acid and selenic acids are strong.

[21.31] SF_4 is very reactive; for example, it reacts readily with water to produce HF and SO_2. By contrast, SF_6 is very stable. It has even been used, with O_2, for X-ray examination of the lungs. (a) What features of the molecules make attack of sulfur by Lewis bases much more likely for SF_4 than for SF_6? (b) What feature of SF_4 makes attack of sulfur by Lewis acids possible? (c) Why is SF_4 subject to oxidation, whereas SF_6 is not?

Group 5A

21.32 Write formulas for each of the following compounds and indicate the oxidation state of the group 5A element in each: (a) potassium phosphite; (b) phosphorus(III) oxide; (c) calcium dihydrogen arsenate; (d) arsenous acid; (e) antimony(III) sulfide.

21.33 Name each of the following compounds: (a) Na_3P; (b) H_3PO_3; (c) AsF_3; (d) P_4O_{10}; (e) H_3AsO_4.

21.34 Sodium trimetaphosphate ($Na_3P_3O_9$) and sodium tetrametaphosphate ($Na_4P_4O_{12}$) are used as water-softening agents. They contain cyclic $P_3O_9^{3-}$ and $P_4O_{12}^{4-}$ ions, respectively. Propose reasonable structures for these ions.

21.35 Account for the following observations. (a) H_3PO_3 is a diprotic acid. (b) Nitric acid is a strong acid, whereas phosphoric acid is weak. (c) Phosphate rock is not effective as a phosphate fertilizer. (d) Phosphorus does *not* exist at room temperature as a diatomic molecule, P_2, but nitrogen does. (e) Phosphonium salts such as PH_4Cl can be formed under anhydrous conditions, but they can't be made in aqueous solution.

21.36 Phosphorus pentachloride exists in one form in the solid state as an ionic lattice of PCl_4^+ and PCl_6^- ions. (a) Draw the Lewis structures of these ions and predict their geometries. (b) What set of hybrid orbitals is employed by phosphorus in each case? (c) Why should PCl_5 exist as an ionic substance in the solid state, whereas it is stable as the neutral molecule in the gas phase?

21.37 Write balanced chemical equations for each of the following reactions: (a) preparation of white phosphorus from calcium phosphate; (b) hydrolysis of PCl_3; (c) preparation of PCl_3 from P_4; (d) preparation of PF_5 from P_4O_{10}; (e) reaction of P_4O_{10} with water; (f) reaction of As_4O_6 with water; (g) dehydration of orthophosphoric acid to form pyrophosphoric acid.

21.38 Whereas PCl_3 hydrolyzes readily in water to form H_3PO_3, $SbCl_3$ hydrolyzes only in part, forming SbOCl. Explain this difference.

[21.39] The redox behavior of several phosphorus compounds in basic solution is summarized in the following diagram:

$$PH_3 \xrightarrow{0.89\text{ V}} P_4 \xrightarrow{2.05\text{ V}} H_2PO_2^- \xrightarrow{1.57\text{ V}} HPO_3^{2-}$$

Calculate the electrode potential for oxidation of (a) PH_3 to $H_2PO_2^-$; (b) P_4 to HPO_3^{2-}.

[21.40] (a) Calculate the P-to-P distance in both P_4O_6 and P_4O_{10} from the following data: the P—O—P bond angle for P_4O_6 is 127.5°, while that for P_4O_{10} is 124.5°. The P—O distance (to bridging oxygens) is 0.165 nm in P_4O_6 and 0.160 nm in P_4O_{10}. (b) Rationalize the relative P-to-P distances in the two compounds.

Group 4A and boron

21.41 Write the formulas for each of the following compounds and indicate the oxidation state of the group 4A element or of boron in each: (a) silicon dioxide;

(b) germanium tetrachloride; (c) sodium borohydride; (d) boric acid; (e) stannous chloride.

21.42 Pure silicon is a semiconductor, whereas titanium, which also possesses four valence-shell electrons is a metallic conductor. Account for this difference.

21.43 Covalent silicon-hydrogen compounds are called silanes. The silane Si_2H_6, known as disilane, exists, but no Si_2H_4 and Si_2H_2 compounds are known. In contrast, carbon forms C_2H_6, C_2H_4, and C_2H_2. Explain why silicon doesn't form Si_2H_4 and Si_2H_2.

21.44 Suggest some reasons why carbon is more suitable than silicon as the major structural element in living systems.

21.45 Select the member of group 4A that best fits each of the following descriptions: (a) forms the most acidic oxide; (b) is most commonly found as a +2 ion; (c) has the lowest ionization energy; (d) catenates to the greatest extent; (e) forms the strongest bond with hydrogen; (f) a component of sand.

21.46 Both $GeCl_4$ and $SiCl_4$ fume in moist air because of hydrolysis to GeO_2 and SiO_2. Write balanced equations for these reactions.

21.47 Germanium differs markedly from silicon in that the +2 halides are fairly stable. These halides can be prepared by the reduction of the tetrahalide with germanium metal. (a) Write the balanced chemical equation for the formation of $GeCl_2$. (b) Predict the geometrical structure of the gaseous $GeCl_2$ molecule.

21.48 Ultrapure germanium, like silicon, is used in semiconductors. Germanium of "ordinary" purity is prepared by the high-temperature reduction of GeO_2 with carbon. The Ge is converted to $GeCl_4$ by treatment with Cl_2 and then purified by distillation; $GeCl_4$ is then hydrolyzed in water to GeO_2 and reduced to the elemental form with H_2. The element is then zone-refined. Write a balanced chemical equation for each of the chemical transformations in the course of forming ultrapure Ge from GeO_2.

21.49 Describe the fundamental structural unit present in all silicate minerals. How is this unit modified in aluminosilicates?

21.50 Provide a brief explanation for each of the following observations. (a) The Ba^{2+} ion does not readily substitute for Mg^{2+} in minerals. (b) In basic soils, metallic ions are not readily released for use by plants. (c) Acidic soils may be treated with lime, CaO, a process called "sweetening the soil." (d) The darkening of photochromic glasses that occurs in bright light is reversible.

21.51 Which of the following elements would you expect to be able to substitute for Al^{3+} in an aluminosilicate mineral: (a) iron; (b) lithium; (c) nickel; (d) chromium, (e) lanthanum? Indicate in each case the expected oxidation state and the reason for your answer.

21.52 The mineral orthoclase is a feldspar mineral formed by replacing a quarter of the Si^{4+} ions in SiO_2 with Al^{3+} and maintaining charge balance with K^+ ions. What is the empirical formula for this mineral?

21.53 How is the mica mineral $KMg_3(AlSi_3O_{10})(OH)_2$ related structurally to the mica mineral muscovite mentioned in the text?

21.54 Draw the Lewis structure for the cyclic $Si_3O_9^{6-}$ ion.

21.55 What empirical formula and unit charge are associated with each of the following structural types: (a) isolated SiO_4 tetrahedra; (b) a chain structure of SiO_4 tetrahedra joined at corners to adjacent units; (c) a structure consisting of tetrahedra joined at corners to form a six-membered ring of alternating Si and O atoms?

21.56 Propose a reasonable description of the structure of each of the following minerals: (a) albite, $NaAlSi_3O_8$; (b) leucite, $KAlSi_2O_6$; (c) zircon, $ZrSiO_4$; (d) sphene, $CaTiSiO_5$.

[21.57] Considering the chemical formulas and structures of clay minerals, suggest an explanation for why they can be molded and then become hard and brittle when heated.

Additional exercises

21.58 Name each of the following compounds: (a) $H_2B_4O_7$; (b) SiC; (c) HPO_3; (d) XeF_2; (e) Na_2S; (f) $KClO_3$.

21.59 Explain each of the following observations. (a) H_2S is a better reducing agent than H_2O. (b) H_2SO_4 is a stronger acid than H_2SeO_4. (c) Astatine is generally not discussed in any detail in discussion of halogens. (d) White phosphorus is quite volatile, whereas red phosphorus is not. (c) Xenon hexafluoride is a stable compound, whereas krypton hexafluoride is unknown. (f) Addition of SF_4 to water results in an acidic solution. (g) Silicate-sheet minerals are softer than aluminosilicate-sheet ones.

21.60 Write balanced chemical equations to account for the following observations. (There may not be closely similar reactions shown in the chapter; however, you should be able to make reasonable guesses at the likely products.) (a) When burning sodium metal is immersed in a pure HCl atmosphere, it continues to burn. (b) Bubbling SO_2 gas through liquid bromine that is covered with a layer of water results in formation of a strongly acidic solution; upon distillation, an aqueous HBr solution is collected. The remaining liquid is still strongly acidic. (c) When bromine is added to a basic solution containing potassium hypochlorite, insoluble potassium bromate is formed. (d) When bromic acid is reacted with SO_2, $Br_2(aq)$ is formed. (e) Uranium(VI) fluoride is formed by the action of ClF_3 on uranium (IV) chloride.

21.61 Show, by chemical equation, the preparation of (a) $XeOF_4$ from XeF_6; (b) I_2 from IO_3^-; (c) Se from H_2SeO_3.

21.62 The cyano group behaves in some ways like a halogen. Thus cyanogen gas, $(CN)_2$, has been called a pseudohalogen and the cyanide ion, CN^-, a pseudohalide. Cyanogen reacts with an aqueous solution of NaOH in a fashion analogous to Cl_2. (a) Write a balanced chemical equation for this reaction. (b) Write the Lewis structure for $(CN)_2$, and describe its geometrical structure.

21.63 Xenon trioxide disproportionates in strongly alkaline solution to form the thermally stable perxenate ion, XeO_6^{4-}. Predict the geometry of this ion. Describe the bonding in terms of the hybridization of xenon valence-shell orbitals.

21.64 Elemental sulfur is capable of reacting under suitable conditions with Fe, F_2, O_2, or H_2. Write balanced chemical equations to describe the reaction in each case. In which reactions is sulfur acting as a reducing agent and in which as an oxidizing agent?

21.65 Compare the first ionization energies of phosphorus and sulfur and account for the relative magnitudes in terms of the electronic structures of the atoms.

21.66 Calculate $\Delta H°$ for the reaction $P_4(g) \longrightarrow 2P_2(g)$ at 298 K.

21.67 Calculate the molarity of an aqueous hydrofluoric acid solution that is 50.0 percent HF by weight and has a density of 1.155 g/cm^3.

21.68 What pressure of gas is formed when 0.654 g of XeO_3 decomposes completely to the free elements at 48°C in a 0.452-L volume?

21.69 Calculate the mass of sodium iodide that must react with excess phosphoric acid to produce the quantity of HI required to form 2.50 L of 4.80 *M* solution.

21.70 One method proposed for removal of SO_2 from the flue gases of power plants involves reaction with aqueous H_2S. Elemental sulfur is the product. (a) Write a balanced chemical equation for the reaction. (b) What volume of H_2S at 27°C and 740 mm Hg would be required to remove the SO_2 formed by burning 1.0 ton of coal containing 3.5 percent S by weight? (c) What mass of elemental sulfur is produced? Assume that all reactions are 100 percent efficient.

21.71 Although H_2Se is toxic, no deaths have been attributed to it. One reason is its vile odor, which serves as a sensitive warning of its presence. In addition, H_2Se is readily oxidized by O_2 in air to nontoxic elemental red selenium before harmful amounts of H_2Se can enter the body. (a) Write a balanced chemical equation for this oxidation. (b) Calculate $\Delta G°$ and the equilibrium constant (at 298 K) for the reaction.

[21.72] The standard heats of formation of $H_2O(g)$, $H_2S(g)$, $H_2Se(g)$, and $H_2Te(g)$ are −241.8, −20.17, +29.7, and 99.6 kJ/mol, respectively. The enthalpies necessary to convert the elements in their standard states to one mole of gaseous atoms are 248, 277, 227, and 197 kJ/mol of atoms for O, S, Se, and Te, respectively. The enthalpy for dissociation of H_2 is 436 kJ/mol. Calculate the average H—O, H—S, H—Se, and H—Te bond energies and comment on their trend.

[21.73] The N—X bond distances in the nitrosyl halides, NOX, are 1.52, 1.98, and 2.14 Å for NOF, NOCl, and NOBr, respectively. Compare the distances with the atomic radii for the halogens (Table 21.3). Is the variation in N—X distance what you expect from these covalent radii? If not, account for the deviations.

[21.74] Boron nitride has a graphitelike structure with B—N bond distances of 1.45 Å within sheets and a separation of 3.30 Å between sheets. At high temperatures the BN assumes a diamondlike form that is harder than diamond. Rationalize the similarity between BN and elemental carbon.

22 Metallurgy, Metallic Structure, and Alloys

In this chapter we will be concerned with the metallic elements. We will examine the chemical forms in which the metals occur in nature and the means by which we obtain the metals from these sources. We will then examine the characteristics of the metallic state and see how these properties can be accounted for in terms of the electronic structures of metals. We will also see how metals and mixtures of metals, called alloys, are employed in modern technology. Finally, we will look at some of the properties and uses of an important group of substances called intermetallic compounds. Much of this subject matter comes under the heading of **metallurgy**, which is traditionally thought of as the science of extracting metals from their natural sources. Modern metallurgy, however, is concerned also with the properties and structures of metals and alloys.

Let us begin, then, with a look at the sources from which we obtain the metals that are so essential to the workings of modern society.

22.1 OCCURRENCE AND DISTRIBUTION OF METALS

The portion of our environment that is the solid earth beneath our feet is called the **lithosphere.** The lithosphere provides us with most of the materials with which we feed, clothe, shelter, support, and entertain ourselves. Although the bulk of the earth is solid, we have access to only a small region near the surface. The deepest well ever drilled is only 7.7 km deep, and the deepest mine extends only 3.4 km into the earth. In comparison, the earth has a radius of 6370 km. Many of the substances most useful to us are not especially abundant in that portion of the lithosphere to which we have ready access. Furthermore, most of the metals of interest to us occur in chemical forms that are not useful. Deposits that contain the metal in economically exploitable quantities are known as *ores.* Usually, the compounds or elements that we desire must be separated from a large quantity of undesirable material and then chemically processed to render them useful. By consuming large quantities of substances located close to the surface we have literally changed the surface

FIGURE 22.1 Large open-pit mining operation.

of the planet (see Figure 22.1). Experts estimate that about 2.3×10^4 kg (25 tons) of materials is extracted from the lithosphere and processed annually to support each person in our society. Because the richest sources of many substances are becoming exhausted, it will be necessary in the future to process larger volumes of lower-quality raw materials. Thus the extraction of the compounds and elements we need will cost more in terms of energy and environmental impact.

Although we have not penetrated very far into the earth, we have been able to construct a general picture of its structure and composition from indirect evidence. The outer portion, or *crust,* constitutes only about 0.4 percent of the total mass of the earth. It has an average thickness of 17 km, ranging in depth from 4 to 70 km. The term "crust" is a carryover from the time when the entire interior of the planet was thought to be molten. Eighty-eight elements occur in the earth's crust. Twelve of these, listed in Table 22.1, make up 99.5 percent of the crust by weight. However, if you were to analyze a shovelful of dirt from your backyard, you would not find it to have the composition listed in Table 22.1. Elements are not distributed uniformly throughout the crust. It is interesting to compare the order of abundance of the elements with the order of estimated world consumption shown in Table 22.2. The ranking of elements in the two lists shows little correlation; some elements that are not very abundant are widely used. Consequently, the occurrence and distribution of *concentrated* deposits of certain important elements in the lithosphere are important and often play a role in international politics.

TABLE 22.1 The twelve most abundant elements in the earth's crust

Element	Percent by weight
Oxygen	50
Silicon	26
Aluminum	7.5
Iron	4.7
Calcium	3.4
Sodium	2.6
Potassium	2.4
Magnesium	1.9
Hydrogen	0.9
Titanium	0.6
Chlorine	0.2
Phosphorus	0.1

Most elements occur in nature in combination with other elements,

TABLE 22.2 Estimated annual world consumption of elements

Element	Annual consumption (kg)
C	10^{12}–10^{13}
Na, Fe	10^{11}–10^{12}
N, O, S, K, Ca	10^{10}–10^{11}
H, F, Mg, Al, P, Cl, Cr, Mn, Cu, Zn, Ba, Pb	10^{9}–10^{10}
B, Ti, Ni, Zr, Sn	10^{8}–10^{9}
Ar, Co, As, Mo, Sb, W, U	10^{7}–10^{8}
Li, V, Se, Sr, Nb, Ag, Cd, I, rare earths, Au, Hg, Bi	10^{6}–10^{7}
He, Be, Te, Ta	10^{5}–10^{6}

TABLE 22.3 Some common minerals

Mineral name	Chemical formula
Calcite	$CaCO_3$
Chalcopyrite	$CuFeS_2$
Cinnabar	HgS
Corundum	Al_2O_3
Fluorite	CaF_2
Galena	PbS
Gypsum	$CaSO_4 \cdot 2H_2O$
Halite	$NaCl$
Hematite	Fe_2O_3
Malachite	$Cu_2(CO_3)(OH)_2$
Perovskite	$CaTiO_3$
Pyrite	FeS_2
Quartz	SiO_2
Talc	$Mg_3(Si_4O_{10})(OH)_2$
Turquoise	$CuAl_6(PO_4)_4(OH)_2 \cdot 4H_2O$
Wulfenite	$PbMoO_4$

that is, in compounds. Solid inorganic substances occurring in nature are referred to as **minerals.** Table 22.3 lists several common minerals. Some of these are pictured in Figure 22.2. Minerals are usually known by their common names rather than by their chemical names. They are often named by the person who first described them. Although there is no systematic method of nomenclature, the name frequently ends in -ite. Minerals often have variable compositions owing to the substitution of one element for another within the solid. For example, in the mineral *olivine,* Fe_2SiO_4 and Mg_2SiO_4 are present together in variable relative amounts. One can view the substance as Mg_2SiO_4 in which some of the Mg^{2+} ions have been replaced by Fe^{2+}. Such a mineral nevertheless has a well-defined crystal structure.

Minerals can be classified into three groups: native elements, silicate minerals, and nonsilicate minerals. Examples are listed in Table 22.4. Metals found in the elemental form include silver, gold, palladium, platinum, ruthenium, rhodium, osmium, and iridium. These metals, from groups 8B and 1B of the periodic table, are known as *noble metals* because of their lack of reactivity. All of these metals have very high electrode potentials and are thus difficult to oxidize. For example:

(*a*)

(*b*)

(*c*)

FIGURE 22.2 Three common minerals: (*a*) calcite; (*b*) fluorite; (*c*) wulfenite. Note the variety of crystal shapes.

TABLE 22.4 Types of minerals

General category	Examples
Native elements	Cu, Ag, Au, Bi, Pt, Pd, S
Silicate minerals	$ZrSiO_4$, $Be_3AlSi_6O_{18}$
Nonsilicate minerals	
Oxides	Al_2O_3, Fe_2O_3, Fe_3O_4, Cu_2O, TiO_2
Hydroxides	$Mg(OH)_2$
Carbonates	$CaCO_3$, $MgCO_3$, $PbCO_3$, $ZnCO_3$
Sulfates	$BaSO_4$, $PbSO_4$, $CaSO_4$
Sulfides	Ag_2S, Cu_2S, HgS, ZnS
Halides	NaCl, $MgCl_2$
Phosphates	$Ca_3(PO_4)_2$

$$Ag^+(aq) + e^- \longrightarrow Ag(s) \qquad E° = +0.799\ V \qquad [22.1]$$

$$Os^{2+}(aq) + 2e^- \longrightarrow Os(s) \qquad E° = +0.85\ V \qquad [22.2]$$

$$Ir^{3+}(aq) + 3e^- \longrightarrow Ir(s) \qquad E° = +1.15\ V \qquad [22.3]$$

Nonsilicate minerals are the principal sources of metals. The most abundant and commercially important minerals are oxides, sulfides, and carbonates.

22.2 EXTRACTIVE METALLURGY

Extractive metallurgy is concerned with all the steps involved in obtaining a metal from its ore. A great many different processes, both physical and chemical, may be involved. The first steps are generally physical processes that increase the concentration of the desired mineral and prepare it for subsequent operations. The concentrated ore is then subjected to chemical processes that may further concentrate it and that eventually lead to reduction to the free metal.

Physical Concentration Processes

The initial steps in treating an ore involve crushing and grinding to produce a material that can be further processed. Then the desired mineral is sorted from the undesired material, called gangue (pronounced "gang"), by taking advantage of some difference in a property of the mineral as compared with the gangue. For example, advantage might be taken of the fact that the ore has a higher density. The denser material would settle more rapidly when the finely ground ore is stirred in water or some other fluid medium. As a further example, certain minerals are magnetic; that is, particles of the mineral would be strongly attracted to the poles of a magnet. The magnetic material can be separated from the nonmagnetic gangue by moving finely ground ore on a conveyor belt past a series of magnets.

Flotation is a very important method for concentrating ores, particularly sulfides. A suspension of ground ore in water is agitated while air is blown through, as shown in Figure 22.3. Certain chemicals are added so that a froth or foam is created. Particles of the desired mineral become attached to the air bubbles and are floated out with the froth, which is

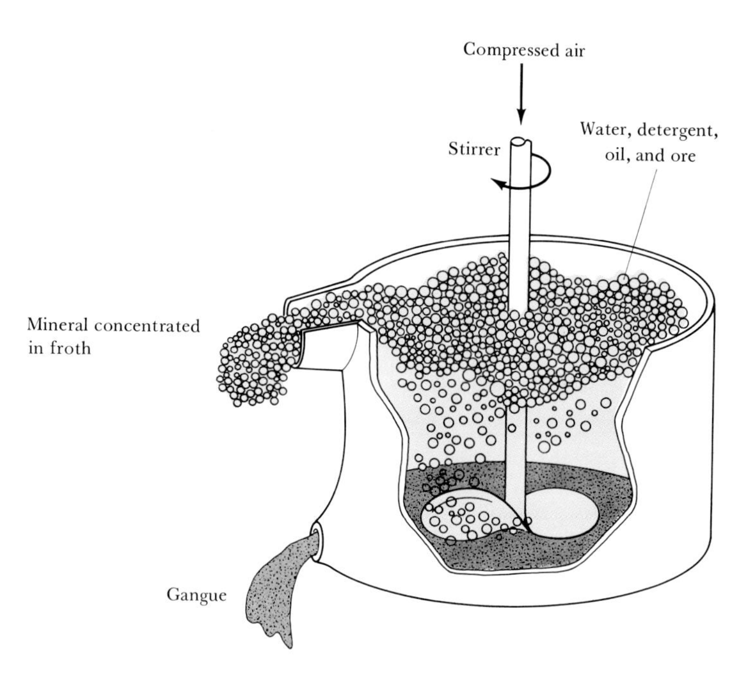

FIGURE 22.3 Schematic diagram of froth flotation, by means of which a mineral can be separated from a larger amount of gangue.

skimmed off the top. At the same time, the gangue settles to the bottom of the tank. Flotation methods depend on the fact that the surfaces of the desired mineral particles are hydrophobic (not wetted by water, or can be made so by addition of a flotation agent. Chemicals are usually added that selectively adsorb on the surfaces of the desired mineral particles to reduce wetting by water. The molecules of such substances usually contain a polar end that binds to the mineral surface and a nonpolar end that protrudes into the water. The mode of action of flotation agents is similar to that described in Section 11.7 for soaps and detergents. One of the most successful flotation agents is potassium ethyl xanthate (pronounced "zanthate"):

$$S{=}C\begin{matrix} \diagup O{-}C_2H_5 \\ \diagdown S^- \quad K^+ \end{matrix}$$

The negative end of this molecule attaches to the surface of a sulfide mineral such as Cu_2S, while the nonpolar organic end of the molecule extends away from the surface to inhibit the approach of water molecules.

Ores that have been enriched in the desired minerals by one or more of the processes we have just described may be too finely divided for use in later stages of treatment. The ore must then be pressed into pellets or briquets, or heated to cause sintering into larger pieces.

Once all the appropriate physical methods have been employed, the ore is ready for further enrichment in the desired metal by chemical means. These chemical processes may conveniently be considered under three headings: pyrometallurgy, hydrometallurgy, and electrometallurgy.

22.3 PYROMETALLURGY

Pyrometallurgy involves the use of heat to convert the desired mineral in an ore from one chemical to another, and eventually to the free metal. We can group the chemical reactions involved according to the type of products formed. Calcination involves heating an ore to bring about its decomposition and the elimination of a volatile product. The volatile product could, for example, be water or CO_2. Carbonate ores are frequently calcined to drive off CO_2, forming the metal oxide, for example:

$$PbCO_3(s) \longrightarrow PbO(s) + CO_2(g) \qquad [22.4]$$

Most carbonates decompose reasonably rapidly at temperatures in the range 400 to 500°C, although $CaCO_3$ requires a higher temperature, about 1000°C. Similarly, most hydrated minerals lose water at temperatures on the order of 100 to 300°C.

Roasting is a thermal treatment that causes chemical reactions involving the furnace atmosphere. Roasting may produce oxidation or reduction and may be accompanied by calcination. One of the most important types of roasting process involves oxidation of sulfide ores. When sulfide ores are roasted, the sulfide is oxidized, and the metal is partly or entirely converted to the oxide, as in these examples:

$$2ZnS(s) + 3O_2(g) \longrightarrow 2ZnO(s) + 2SO_2(g) \qquad [22.5]$$

$$2MoS_2(s) + 7O_2(g) \longrightarrow 2MoO_3(s) + 4SO_2(g) \qquad [22.6]$$

In such processes it often happens that the metal oxide formed is relatively volatile at the temperature of the roast and sublimes from the furnace.

Depending on the conditions of the roasting process, the metal may form a sulfate rather than an oxide. For example:

$$CoS(s) + 2O_2(g) \longrightarrow CoSO_4(s) \qquad [22.7]$$

Formation of the sulfate can sometimes be used to advantage; because sulfates are generally water soluble, the material containing the desired metal can be leached away from insoluble residue.

The ore of a less active metal such as mercury can be roasted to the free metal:

$$HgS(s) + O_2(g) \longrightarrow Hg(g) + SO_2(g) \qquad [22.8]$$

The free metal can also be produced when a reducing atmosphere is present during the roast. Carbon monoxide, CO, provides such an atmosphere:

$$PbO(s) + CO(g) \longrightarrow Pb(l) + CO_2(g) \qquad [22.9]$$

Roasting is not always a feasible method of reduction. Some metals that are difficult to obtain as the free metal by roasting are best converted to the metal halide, which can then be reduced. To obtain the metal halide, the metal oxide or another compound, such as metal carbide, is roasted with an atmosphere of the halogen, usually chlorine. For example,

$$TiC(s) + 4Cl_2(g) \longrightarrow TiCl_4(g) + CCl_4(g) \quad [22.10]$$

Smelting is a melting process in which the materials formed in the course of chemical reactions separate into two or more layers. Smelting often involves a roasting stage in the same furnace. Two of the important types of layers formed in smelters are molten metal and slag. The molten metal may consist almost entirely of a single metal, or may be a solution of two or more metals.

Slag consists mainly of molten silicate minerals, with aluminates, phosphates, fluorides, and other ionic compounds as constituents. A slag is formed when a metal oxide such as CaO reacts with molten silica, SiO_2. Slags are classified as acidic or basic, according to the acidity or basicity of the oxides added to the silica to form the slag. In this way of looking at such reactions the most basic oxides are those of the alkali and alkaline earth metal ions. The most acidic oxides are those of the nonmetals, for example, SiO_2 and P_2O_5. Somewhat less acidic are the oxides of the more highly charged metal ions, such as TiO_2, Fe_2O_3, and Al_2O_3. Thus a slag containing mainly CaO is basic in character. On the other hand, a slag containing mainly SiO_2, or SiO_2 with a nonmetallic oxide such as P_2O_5, is acidic in character. The reaction of a basic oxide such as CaO with an acidic oxide such as SiO_2 can be thought of as a reaction of an acid with a base to form a "salt" as a reaction product:

$$CaO(l) + SiO_2(l) \longrightarrow CaSiO_3(l) \quad [22.11]$$

We can also think of such reactions as occurring between ions in the molten mixture:

$$Ca^{2+}(l) + O^{2-}(l) + SiO_2(l) \longrightarrow Ca^{2+}(l) + SiO_3^{2-}(l)$$

We will see several applications of these ideas in the sections on metallurgy that follow.

SAMPLE EXERCISE 22.1

Predict the product of the reaction between Na_2O and TiO_2. Write a balanced chemical equation for the reaction. Which oxide is acting as base? Which as acid?

Solution: The salt that is formed in this reaction is likely to be Na_2TiO_3, in which there are present Na^+ and TiO_3^{2-} ions. Another possibility would be Na_4TiO_4; indeed, some compounds containing the TiO_4 unit are known, for example, Mg_2TiO_4 and Zn_2TiO_4.

The balanced equation for the reaction is

$$Na_2O(l) + TiO_2(s) \longrightarrow Na_2TiO_3(l)$$

In this reaction Na_2O acts as the base, and TiO_2 acts as the acid. The equation could be written in ionic form as

$$2Na^+(l) + O^{2-}(l) + TiO_2(l) \longrightarrow 2Na^+(l) + TiO_3^{2-}(l)$$

The chemical changes that occur when oxides are melted together can best be understood as acid-base reactions. However, since there is no water or other proton-donating solvent present, we cannot think of these reactions in terms of the Brønsted acid-base theory. Rather, we must view them in terms of the Lewis acid-base theory. Recall that in the Lewis theory, discussed in Section 15.10, a base is an electron-pair donor, and an acid is an electron-pair acceptor. In the molten oxides the electron-pair donor is the oxide ion, O^{2-}. A base is a substance that furnishes oxide ions to the medium. In contrast, an acid is a substance that can react with an oxide ion by forming a bond with one of its unshared electron pairs. The oxides of highly charged metal or nonmetal ions are acidic because the central ion is capable of forming a polar covalent bond with the oxide ion. In the oxides of the alkali metal or alkaline earth metals, the oxide ion is readily available to donate an electron pair because the bond to the metal ion is essentially ionic in character; thus these oxides are basic.

Refining involves treatment of a crude, relatively impure metal product from a metallurigical process to improve its purity and to better define its composition. Sometimes the goal of the refining process is the metal itself in as pure a form as possible. However, it may also be the production of a mixture with well-defined composition, as in the production of steels from crude iron.

To make some of these ideas and definitions more concrete, let us now consider the pyrometallurgy of iron.

The Pyrometallurgy of Iron

The major sources of iron are ores rich in either hematite, Fe_2O_3, or magnetite, Fe_3O_4. As the richest hematite ores of the vast Mesabi range in northern Minnesota have been depleted, attention has been turned to ores of lower iron content. The most actively mined ores today are the taconites. In a typical taconite ore of good quality the iron is present mainly as Fe_3O_4, with an overall 30 to 40 percent iron content. To enrich the ore it is ground to a fine particle size so that the Fe_3O_4 can be separated from the gangue by a magnetic separation process. The concentrate from this separation step has an iron content of about 60 to 65 percent, and contains 2 to 8 percent silica, SiO_2. After preliminary concentration the ore must be formed into pellets 6 to 25 mm in diameter, a size suitable as feed for the blast furnace. Reduction of the iron ore occurs in a blast furnace, illustrated in Figure 22.4. A blast furnace is essentially a huge chemical reactor, capable of continuous operation. The largest furnaces are over 60 m high and 14 m wide. When operating at full capacity they produce up to 10,000 tons of iron per day. The blast furnace is charged at the top with a mixture of iron ore, coke, and limestone. Coke is coal that has been heated in the absence of air to drive off volatile components. It is about 85 to 90 percent carbon. Coke serves as the fuel, producing heat as it is burned in the lower part of the furnace. It is also the source of the reducing gases, CO and H_2. Limestone, $CaCO_3$, serves as the source of basic oxide in slag formation. Air, which enters the blast furnace at the bottom after preheating, is also an important raw material; it is required for combustion of the coke. Production of 1 kg of pig iron requires about 2 kg of ore, 1 kg of coke, 0.3 kg of limestone, and 1.5 kg of air.

Coke is burned mainly in the lower part of the furnace where the temperatures are highest. In the hottest part of the furnace the tempera-

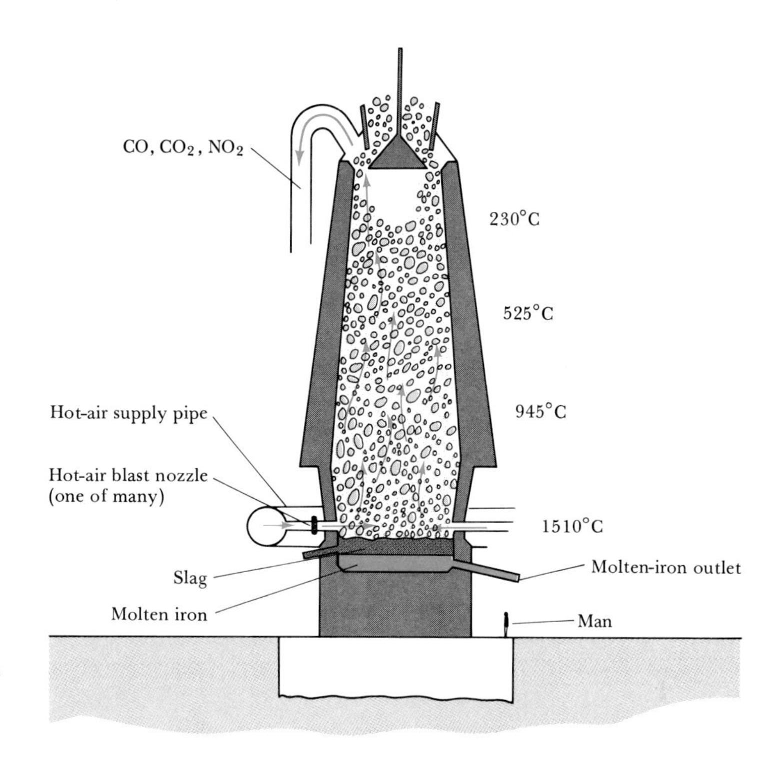

FIGURE 22.4 Blast furnace used for reduction of iron ore. Notice the approximate temperatures in the various regions of the furnace.

tures reach in excess of 1900°C. At this temperature CO_2 is not stable in the presence of excess carbon; it reacts with the coke to form CO:

$$C(s) + CO_2(g) \longrightarrow 2CO(g) \qquad [22.12]$$

Thus the overall reaction in the combustion of coke is

$$2C(s) + O_2(g) \longrightarrow 2CO(g) \qquad \Delta H = -110 \text{ kJ} \qquad [22.13]$$

Water vapor present in the air also reacts with carbon:

$$C(s) + H_2O(g) \longrightarrow CO(g) + H_2(g) \qquad \Delta H = +131 \text{ kJ} \qquad [22.14]$$

Note that whereas the reaction with oxygen is exothermic and provides heat for furnace operation, the reaction with water vapor is endothermic. Addition of water to the air thus provides a means of controlling furnace temperature.

In the upper part of the furnace limestone is calcined (Equation 20.72). Here also the iron oxides are reduced by CO and H_2. For example, the important reactions for Fe_3O_4 are

$$Fe_3O_4(s) + 4CO(g) \longrightarrow 3Fe(s) + 4CO_2(g) \qquad \Delta H = -19 \text{ kJ} \qquad [22.15]$$

$$Fe_3O_4(s) + 4H_2(g) \longrightarrow 3Fe(s) + 4H_2O(g) \qquad \Delta H = 149 \text{ kJ} \qquad [22.16]$$

Reduction of other elements present in the ore also occurs in the hottest parts of the furnace, where carbon is the major reducing agent.

Among the most important of these other elements are manganese, silicon, and phosphorus:

$$MnO(s) + C(s) \longrightarrow Mn(l) + CO(g) \qquad [22.17]$$

$$SiO_2(l) + 2C(s) \longrightarrow Si(l) + 2CO(g) \qquad [22.18]$$

$$P_2O_5(l) + 5C(s) \longrightarrow 2P(l) + 5CO(g) \qquad [22.19]$$

Molten iron collects at the base of the furnace, as shown in Figure 22.4. It is overlaid with a layer of molten slag, formed by the reaction of CaO with the silica present in the ore, as described by Equation 22.11. The layer of slag over the molten iron helps to protect it from reaction with the incoming air. Periodically, the furnace is tapped to drain off slag and molten iron. The iron produced in the furnace, called *pig iron,* may be cast into solid ingots. However, most is used directly in the manufacture of steel. For this purpose it is transported, while still liquid, to the steelmaking shop.

Refining of Iron; Formation of Steel

Iron is refined in a vessel called a **converter**, which has a capacity of about 200 tons. In a typical process the converter is charged with about 75 tons of scrap steel and CaO to form a basic slag, then filled with molten iron fresh from the blast furnace. This iron from the blast furnace contains 0.6 to 1.2 percent silicon, about 0.2 percent phosphorus, 0.4 to 2.0 percent manganese, and about 0.03 percent sulfur. In addition, there will be considerable dissolved carbon. All of these impurity elements are removed by oxidation. In modern steelmaking the oxiding agent is pure O_2, or O_2 diluted with argon. The presence of N_2 is avoided, because the incorporation of nitrogen in steel causes brittleness.

A cross-sectional view of one design of converter is shown in Figure 22.5. In this design O_2, diluted with argon, is blown directly into the molten metal. The oxygen reacts exothermally with carbon, silicon, and impurity metals, reducing the concentrations of these elements in the iron. Carbon and sulfur are expelled as CO or SO_2, respectively. Silicon is oxidized to SiO_2 and adds to whatever slag may have been present initially in the melt. Metal oxides react with the SiO_2 to form silicates. The presence of a basic slag is important for removal of phosphorus:

$$3CaO(l) + P_2O_5(l) \longrightarrow Ca_3(PO_4)_2(l) \qquad [22.20]$$

Nearly all of the O_2 blown into the converter is consumed in the oxidation reactions. By monitoring the O_2 concentration in the gas coming from the converter it is possible to tell when the oxidation is essentially complete. It normally requires only about 20 minutes for oxidation of the impurities present in the iron. At this point a sample of the molten metal is withdrawn and subjected to analysis for all of the elements of importance in determining the quality of the steel produced: carbon, manganese, phosphorus, sulfur, silicon, nickel, chromium, copper, and other metallic elements that might be present. Depending on the results, additional oxygen may be blown in or additional ore might be added.

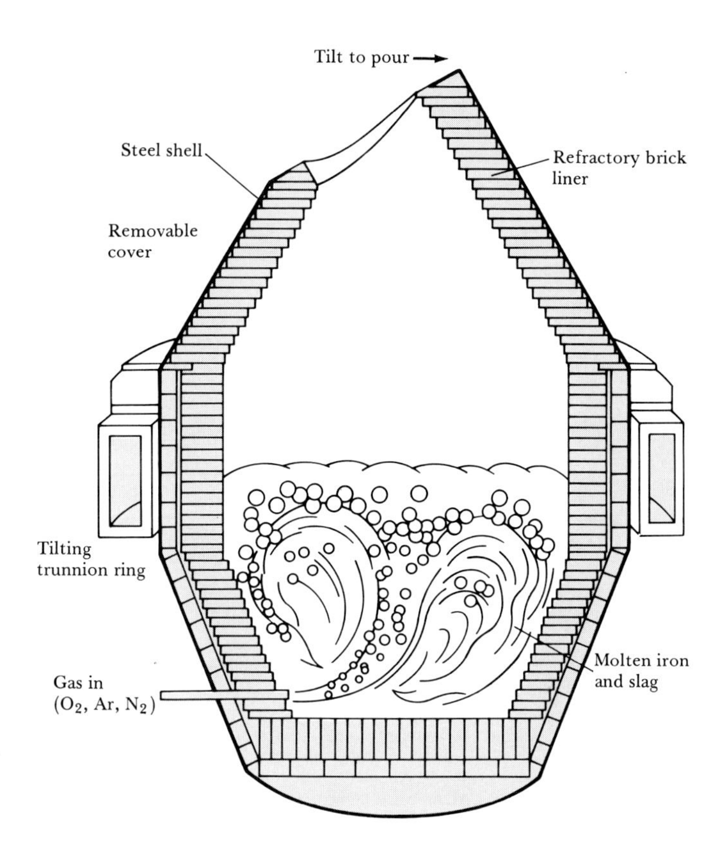

FIGURE 22.5 Converter for refining of iron. A mixture of oxygen and inert gas is blown through the molten iron and slag. The heat of oxidation of impurities maintains the mixture in a molten state. When the desired composition is attained, the converter is tilted to pour.

When the desired composition is attained, the contents of the converter are dumped into a large ladle. As the ladle is being filled, alloying elements are added. These added elements produce steels with various kinds of properties. (We will have more to say about alloys in Section 22.7.) The still-molten mixture is then poured into molds for solidification.

Now that we have seen examples of reductions in pyrometallurgical operations, let us examine the thermodynamic basis for such reductions. Figure 22.6 shows the free energies of formation, ΔG_f°, of several oxides as a function of temperature. As the figure shows, metal oxides generally become less stable (ΔG_f° more positive) as temperature increases. This trend follows from the fact that $\Delta G_f^\circ = \Delta H_f^\circ - T\,\Delta S_f^\circ$ (Section 17.5). The formation of an oxide from a metal and O_2,

$$2M(s) + O_2(g) \longrightarrow 2MO(s) \qquad [22.21]$$

generally represents a decrease in disorder, so that ΔS_f° is negative. The factor $-T\;\Delta S_f^\circ$ therefore becomes increasingly positive with increasing temperature. In Figure 22.6 we notice that for CuO, ΔG_f° becomes positive at approximately 900°C. Consequently, CuO spontaneously decomposes to metal and O_2 above this temperature. The negative slope of the CO line is noteworthy. For the formation of CO from carbon and oxygen (Equation 22.13), ΔS_f° is positive, because the number of moles of gas in the products is larger than in the reactants. For a reduction of the type illustrated in Equation 22.22,

$$MO(s) + C(s) \longrightarrow M(l) + CO(g) \qquad [22.22]$$

ΔG° is negative, and the reaction is therefore spontaneous in all cases where the CO line is below the line for the metal oxide. To illustrate, we can use Figure 22.6 to estimate ΔG° at 1500°C for the reaction

$$FeO(s) + C(s) \longrightarrow Fe(l) + CO(g)$$

For this reaction we have

$$\Delta G^\circ = \Delta G_f^\circ(CO) - \Delta G_f^\circ(FeO)$$

From Figure 22.6 we obtain

$$\Delta G_f^\circ(CO) = (-550/2)\ \text{kJ/mol}$$

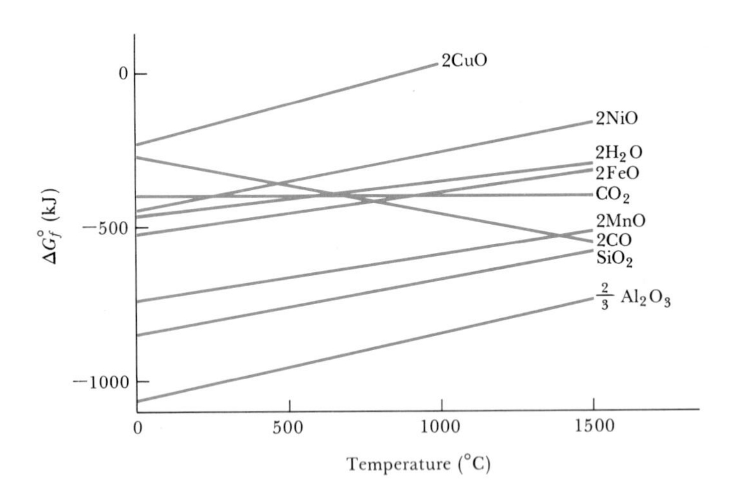

FIGURE 22.6 Standard free energies of formation per mole of O_2 for several oxides.

and

$$\Delta G_f^\circ(\text{FeO}) = (-300/2) \text{ kJ/mol}$$

Therefore,

$$\Delta G_f^\circ = -275 \text{ kJ/mol} + 150 \text{ kJ/mol}$$
$$= -120 \text{ kJ/mol} \quad \text{(rounded to two significant figures)}$$

Notice that at 1500°C the stability line for CO in the figure is below the stability line for FeO. Correspondingly, ΔG_f° is negative, and the reaction is therefore spontaneous.

22.4 HYDROMETALLURGY

Until now we have considered only high-temperature metallurgical reactions that characterize pyrometallurgy. However, many metals of great economic importance are produced in large quantities by hydrometallurgical techniques. Hydrometallurgy involves a selective separation of a mineral or group of minerals from the rest of the ore by an aqueous chemical process. The most important hydrometallurgical process is leaching, in which the desired mineral is selectively dissolved. To illustrate how hydrometallurgical processes work, let us consider the recovery of several elements, beginning with gold.

Hydrometallurgy of Gold

In the recovery of gold from low-grade ores, the ores are finely ground, then treated with an aerated, aqueous solution of NaCN. Because of formation of the stable, water-soluble cyano complex, gold is oxidized:

$$4Au(s) + 8CN^-(aq) + O_2(aq) + 2H_2O(l) \longrightarrow 4Au(CN)_2^-(aq) + 4OH^-(aq) \quad [22.23]$$

Following this leaching operation the solution is treated with CaO to cause precipitation as hydroxides of several metallic species that might be present, and to promote settling of finely divided material. The solution is filtered, then deaerated by vacuum pumping to remove O_2. Next Zn dust is added to precipitate Au:

$$2Au(CN)_2^-(aq) + Zn(s) \longrightarrow Zn(CN)_4^{2-}(aq) + 2Au(s) \quad [22.24]$$

A slimy mixture of Zn and Au is recovered by filtration. Zinc is removed from this mixture by heating in air; ZnO forms and is sublimed away. The crude gold that remains is smelted with a charge of borax ($Na_4B_4O_7 \cdot 10H_2O$) and silica (SiO_2). The slag that forms helps to scavenge out any remaining metal oxides.

This example illustrates how a mineral, in this case the native metal, is leached from the gangue by taking advantage of the fact that a very stable complex is formed [K_f for $Au(CN)_2^-$ is 2×10^{38}]. Equilibria involving metal complexes play a very important role in hydrometallurgy. The second reaction in the process involves an oxidation-reduction reaction in which the more active and less valuable zinc replaces gold. Hydrometallurgical reactions are characterized by this two-reaction sequence: dissolution, then reprecipitation. The dissolution step might be an oxidation, as in the example we have just considered, but other types of reaction, such as metathesis or complex formation, might also be employed. Similarly, the reprecipitation process might be a reduction, but formation of an insoluble salt is also a common process. We can best illustrate these points by considering the hydrometallurgy of some additional metals.

Hydrometallurgy of Aluminum

Among metals, aluminum is second only to iron in the extent to which it is used commercially. World production of the metal is about 14×10^{10} kg (15 million tons) per year. The most useful ore of aluminum is *bauxite,* in which Al is present as hydrated oxides, $Al_2O_3 \cdot xH_2O$. The value of x varies, depending on the particular mineral present. Large bauxite deposits are found in Guiana, Australia, Jamaica, Mexico, and Brazil. Because bauxite deposits in the United States are very limited, the nation imports most of the ore used in production of aluminum.

The major impurities found in bauxite are silica, SiO_2, and iron(III) oxide, Fe_2O_3. It is essential to separate alumina, Al_2O_3, from these impurities before the metal is recovered by electrochemical reduction, as described in Section 22.7. The process used to purify bauxite, called the **Bayer process,** is a hydrometallurgical procedure. The ore is first crushed and ground, then digested in a concentrated aqueous NaOH solution, about 30 percent NaOH by weight, at a temperature in the range 150 to 230°C. Sufficient pressure, between 4 and 30 atm, is maintained to prevent boiling. The trihydrate, $Al_2O_3 \cdot 3H_2O$, dissolves more readily than the monohydrate, $Al_2O_3 \cdot H_2O$. The reaction in either case leads to a complex aluminate anion of the form $Al(H_2O)_2(OH)_4^-$ (see Section 16.5). For example:

$$Al_2O_3 \cdot H_2O(s) + 6H_2O(l) + 2OH^- \longrightarrow 2Al(H_2O)_2(OH)_4^-(aq) \qquad [22.25]$$

Silica dissolves in the strongly basic medium, but then forms insoluble aluminosilicate salts with the aluminate ion. These settle out with a red "mud" consisting mostly of iron(III) oxides, which are not soluble in the strong base. Notice that in this procedure advantage is taken of the fact that Al^{3+} is amphoteric, whereas Fe^{3+} is not. After filtration, the solution is diluted to reduce the concentration of hydroxide. In effect, this causes a partial reversal of the equilibrium shown in Equation 22.25, although the product is not the original mineral, but a highly hydrated aluminum hydroxide.

After the aluminum hydroxide precipitate has been filtered, it is calcined in preparation for the electroreduction of the ore to the metal. The solution recovered from the filtration must be reconcentrated so that it can be used again. This is accomplished by heating to evaporate water from the solution. The energy requirements of this evaporative stage are high, and it is the most costly part of the process.

Hydrometallurgy of Copper

Most copper is obtained from chalcopyrite, $CuFeS_2$, using pyrometallurgical methods. However, the pyrometallurgical procedures do have distinct disadvantages. The necessity for conducting reactions at high temperatures means that the energy costs for the process are high. Second, roasting and smelting of sulfide ores generates substantial air pollution, because of the large quantities of waste slag and SO_2 produced. For each kilogram of Cu produced, about 1.5 kg of iron slag and 2 kg of SO_2 are also formed. To absorb all this SO_2 in the form of $CaSO_3$ or $CaSO_4$, in the manner described in Section 12.3, is impractical because of the enormous quantities of material involved. Hydrometallurgical procedures offer the possibility of avoiding some of these waste material problems.

One promising hydrometallurgical approach to recovery of copper involves the aqueous oxidation of the chalcopyrite ore following concentration by froth flotation methods. When a slurry of the ore in aqueous sulfuric acid is agitated with oxygen, oxidation of the sulfide occurs, with dissolution of copper:

$$2CuFeS_2(s) + 2H^+(aq) + 4O_2(g) \longrightarrow 2Cu^{2+}(aq) + SO_4^{2-}(aq) + Fe_2O_3(s) + 3S(s) + H_2O(l) \qquad [22.26]$$

The resulting solution can be electrolyzed to recover the copper. The solution that remains from the electrolysis is simply aqueous sulfuric acid; it is recycled back to the leaching step with fresh ore.

22.5 ELECTROMETALLURGY

The term **electrometallurgy** refers to the use of electrolysis methods to obtain free metal from one of its compounds or to purify a crude form of the metal. The principles involved in electrolysis were discussed in Section 18.6. You should be sure that you understand this material, by reviewing it at this time if necessary. Sample Exercise 22.2 is illustrative of what you should know.

SAMPLE EXERCISE 22.2

Write the electrode reactions for electrolysis of the solution that results from the process described in Equation 22.26.

Solution: After filtering to remove solids formed in the reaction, the solution contains just aqueous copper sulfate, with any excess acid that may be present. Upon electrolysis, Cu^{2+} ion is preferentially reduced at the cathode, because $E°$ for reduction of Cu^{2+} is +0.34 V. At the anode, water is oxidized. The two half-reactions are:

$$Cu^{2+}(aq) + 2e^- \longrightarrow Cu(s)$$
$$H_2O(l) \longrightarrow 2H^+(aq) + \tfrac{1}{2}O_2(g) + 2e^-$$

$$Cu^{2+}(aq) + H_2O(l) \longrightarrow Cu(s) + 2H^+(aq) + \tfrac{1}{2}O_2(g)$$

The overall effect is that copper is replaced by H^+, thus forming sulfuric acid.

Electrometallurgical procedures can be broadly differentiated according to whether they involve electrolysis of a molten salt or of an aqueous solution. Electrolytic methods are very important as a means of obtaining the more active metals, such as sodium, magnesium, and aluminum, in the free state. Metals such as these could not be reduced at a cathode in aqueous medium, because the solvent itself is reduced at a lower voltage. The standard potentials for reduction of water under acidic and basic conditions are both more positive than the standard potentials for reductions of such active metals as Na ($E° = -2.71$ V), Mg ($E° = -2.37$ V), or Al ($E° = -1.66$ V):

$$2H^+(aq) + 2e^- \longrightarrow H_2(g) \qquad E° = 0.00\ V \qquad [22.27]$$

$$H_2O(l) + 2e^- \longrightarrow H_2(g) + 2OH^-(aq) \qquad E° = -0.83\ V \qquad [22.28]$$

To form such metals by electrochemical reduction, therefore, we must employ a molten salt medium, in which the metal ion of interest is the most readily reduced species.

Electrometallurgy of Sodium

Metallic sodium is employed in the manufacture of several compounds of sodium that are of importance in the chemical industry, including sodium amide, $NaNH_2$, and sodium hydride, NaH. It is also used in the manufacture of tetraethyl lead and related compounds.

In the commercial preparation of sodium, molten NaCl is electrolyzed in a specially designed cell, called the Downs cell, illustrated in Figure 22.7. The electrolyte medium through which current flows is molten NaCl. Calcium chloride, $CaCl_2$, is added to lower the melting point of the cell medium from the normal boiling point of NaCl, 804°C, to around 600°C. The $Na(l)$ and $Cl_2(g)$ produced in the electrolysis are kept from coming in contact and reforming NaCl. In addition, the Na must be kept from contact with oxygen, since the metal would quickly oxidize under the high-temperature conditions of the cell reaction.

Electrometallurgy of Aluminum

In Section 22.4 we discussed the Bayer process, in which bauxite or other ore of aluminum is concentrated to produce a relatively pure hydrous aluminum hydroxide. When this concentrate is calcined at several hundred degrees Celsius, a partially hydrated aluminum oxide, $Al_2O_3 \cdot xH_2O$, called alumina, is formed. At temperatures in excess of

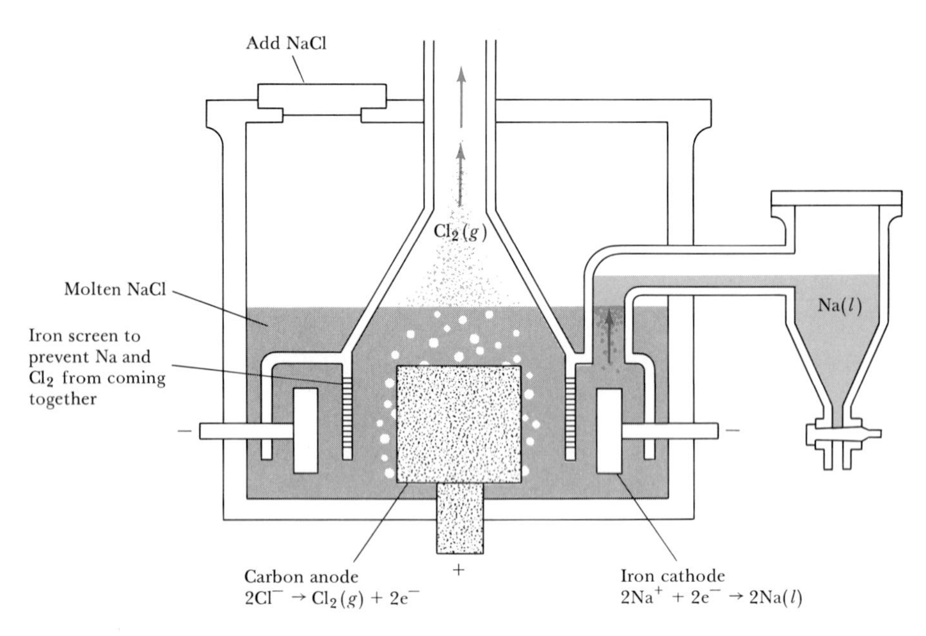

FIGURE 22.7 Downs cell used in the commercial production of sodium.

1000°C, anhydrous aluminum oxide is formed. Anhydrous aluminum oxide melts at over 2000°C. This is too high to permit its use as a molten medium for electrolytic formation of free aluminum. The electrolytic process used commercially to produce aluminum is known as the **Hall process,** named after its inventor, Charles M. Hall (Figure 22.8). The purified Al_2O_3 is dissolved in molten cryolite, Na_3AlF_6, which has a melting point of 1012°C and is an effective conductor of electric current. A schematic diagram of the electrolysis cell is shown in Figure 22.9. Graphite rods are employed as anodes and are consumed in the electrolysis process. The electrode reactions are:

$$\text{Anode:} \quad C(s) + 2O^{2-}(l) \longrightarrow CO_2(g) + 4e^- \qquad [22.29]$$

$$\text{Cathode:} \quad 3e^- + Al^{3+}(l) \longrightarrow Al(l) \qquad [22.30]$$

Charles M. Hall began work on the problem of reducing aluminum in about 1885, after he had learned from a professor of the difficulty of reducing ores of very active metals. Prior to the development of his electrolytic process, aluminum was obtained by a chemical reduction using sodium or potassium as reducing agent. The procedure was very costly; as late as 1852 the cost of aluminum was $545 per pound. During the Paris exposition in 1855, aluminum was exhibited as a rare metal in spite of the fact that it is the third most abundant element in the earth's crust. Hall, who was 21 years old when he began his research, utilized handmade and borrowed equipment in his studies, and his laboratory was a woodshed near his home. In about a year's time he was able to solve the problem. The solution consisted of finding an ionic compound that could be melted to form a conducting medium that would dissolve Al_2O_3 and that would not interfere in the electrolysis reactions. The relatively rare mineral cryolite, Na_3AlF_6, found in Greenland, met these criteria. Ironically, Paul Héroult, who was the same age as Hall, made the same discovery in France at about the same time. As a result of the discovery of Hall and Héroult, large-scale production of aluminum became commercially feasible, and aluminum became a common and familiar metal.

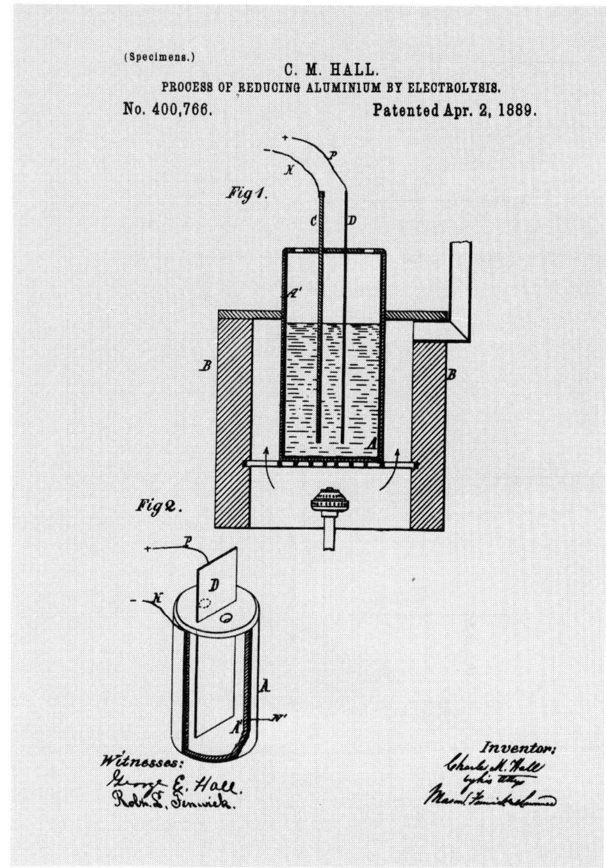

FIGURE 22.8 Charles M. Hall (1863–1914) as a young man, and the patent diagram of Hall's device for reducing aluminum. (*Courtesy of ALCOA*)

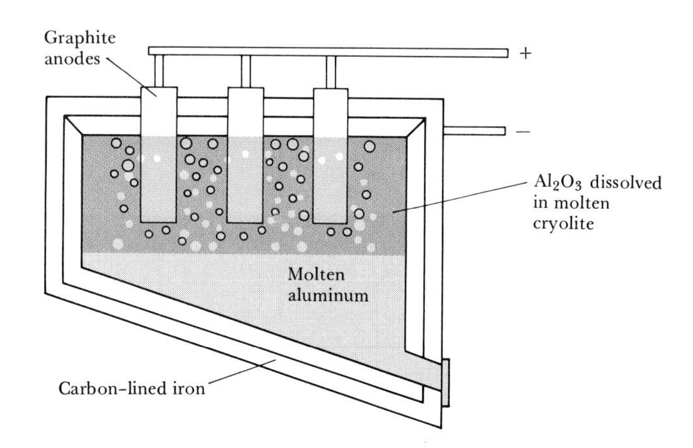

FIGURE 22.9 Typical Hall-process electrolysis cell used to reduce aluminum. Because molten aluminum is more dense than the molten mixture of Na_3AlF_6 and Al_2O_3, the metal collects at the bottom of the cell.

Electrometallurgy of Copper

The crude copper produced from $CuFeS_2$ or Cu_2S by pyrometallurgical methods is not sufficiently free of various impurities to serve in electrical applications. Further purification is achieved by an electrochemical

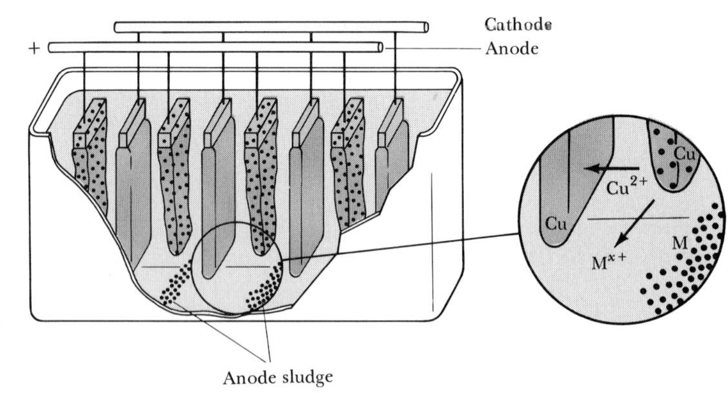

FIGURE 22.10 Electrolysis cell for refining of copper. Notice that as the anodes dissolve away, the cathodes on which the pure metal is deposited grow in size.

method, as illustrated in Figure 22.10. Crude copper is made the anode in a cell in which an acidic solution of copper sulfate is the electrolyte medium. Pure copper sheets are used as cathodes. Application of a suitable voltage across such a cell results in oxidation of copper metal at the anode and reduction of Cu^{2+} at the cathode to form copper metal. These reactions occur because copper is oxidized more readily than water:

$$Cu(s) \longrightarrow Cu^{2+}(aq) + 2e^- \qquad E° = -0.34 \text{ V} \qquad [22.31]$$

$$2H_2O(l) \longrightarrow 4H^+(aq) + O_2(g) + 4e^- \qquad E° = -1.23 \text{ V} \qquad [22.32]$$

Similarly, Cu^{2+} is reduced at the cathode in preference to H^+. The crude copper employed as anode material in the electrolysis cells contains many impurities, including lead, zinc, nickel, arsenic, selenium, tellurium, and several precious metals, such as gold, silver, and platinum. Metallic impurities that are more active than copper are readily oxidized at the anode but do not plate out at the cathode, because their reduction potentials are more negative than for copper. On the other hand, the less active metals are not oxidized at the anode. Instead, they collect below the anode as a sludge that is collected and processed to recover the valuable metals. The anode sludges from copper refining cells provide one-fourth of U.S. silver production and about one-eighth of U.S. gold production.

SAMPLE EXERCISE 22.3

Nickel is one of the chief impurities in the crude copper that is subjected to electrorefining. What happens to this nickel in the course of the electrolytic process?

Solution: The standard potential for oxidation of nickel is more positive than that for copper:

$$Ni(s) \longrightarrow Ni^{2+}(aq) + 2e^- \qquad E° = 0.28 \text{ V}$$

$$Cu(s) \longrightarrow Cu^{2+}(aq) + 2e^- \qquad E° = -0.34 \text{ V}$$

Nickel will thus be more readily oxidized than copper at the anode, assuming standard conditions. Of course, we do not have standard conditions in the electrolytic cell. The crude copper anode is nearly all copper, so we can assume that the activity, or effective concentration, of copper in the anode is essentially the same as that for pure metal. However, the nickel is present as a highly dilute impurity in the copper. Thus the activity, or effective concentration, of nickel in the anode is much lower than it would be if we had a pure nickel anode. Nonetheless, the nickel is sufficiently more electropositive than copper to preferentially undergo oxidation at the anode. The reduction of Ni^{2+}, which is just the reverse of the oxidation process, occurs less readily than the reduction of Cu^{2+}. The Ni^{2+} thus accumulates in the electrolyte solution while the Cu^{2+} is reduced at the cathode. After a time it is necessary to recycle the electrolyte solution to remove the accumulated metal ion impurities such as Ni^{2+} and Zn^{2+}.

Electrochemical methods such as those described for copper are also employed in the purification of commercially important metals such as zinc, cobalt, and nickel. It is interesting that this can be done even though the electrode potentials for reduction of these metals in aqueous solution are negative. We might expect that reduction of hydrogen ions would occur at the cathode in preference to reduction of the metal ion, especially when the solution is acidic. In practice, very little hydrogen is formed at the cathode, because of the phenomenon of *overvoltage,* described in Section 18.6. The overvoltage for formation of H_2 at the cathode when the electrode is the purified metal is generally on the order of several tenths of a volt. Thus formation of H_2 requires a voltage several tenths of a volt above the calculated value. By contrast, the overvoltage for deposition of the metallic elements is very low. By careful control of the voltage applied to the cell, it is possible to minimize gas formation. It is important that this be done, because the hydrogen produced at the cathode where the pure metal is being deposited could adversely affect the quality of the metal, and because any reduction or oxidation of solvent represents wasted energy.

22.6 THE ELECTRONIC STRUCTURES OF METALS

In our discussions of metallurgy we have so far confined ourselves to discussing the methods employed for obtaining metals in pure form. Metallurgy is also concerned with understanding the properties of metals and alloys, and with the development of useful new materials. As with any branch of science and engineering, our ability to make advances is coupled to our understanding of the fundamental properties of the systems with which we work. At several places in the text we have referred to the differences between metals and nonmetals with regard to both physical and chemical behavior. Let us now consider the distinctive properties of metals and then relate those properties to a model for metallic bonding.

Properties of Metals

You have no doubt at some time held in your hand a length of copper wire or an iron bolt. Perhaps you have had occasion to observe the surface of a freshly cut piece of sodium metal. These substances, although distinct from one another, share certain similarities that make it easy to classify them as metallic. A fresh metal surface has a characteristic luster. In addition, we know that those metals that can be handled with bare hands have a characteristic cold, metallic feeling, related to their high heat conductivity. Metals also have high electrical conductivities; when a voltage is applied across a length of metal, current flows. The current flow occurs without any displacement of atoms within the metal structure and is due to the flow of electrons within the metal. Current flow in metals is different from current flow in an aqueous ionic solution or molten salt, in which ions rather than electrons move under the influence of the electrical potential. Heat and electrical conductivities vary from one metal to another in the same way. For example, silver and copper, which possess the highest electrical conductivities, also possess the highest heat conductivities. This observation suggests that the two types of conductivity have the same origin in metals.

The lustrous appearance of fresh metal surfaces is due to the fact that light of all wavelengths is reflected. The color in metals such as copper or gold is due to absorption of light from the blue, high-energy region of the visible spectrum.

Most metals are *malleable,* which means that they can be hammered into thin sheets, and *ductile,* which means that they can be drawn into wires. These properties indicate that the atoms of the metallic lattice are capable of slipping with respect to one another. Ionic solids or crystals of most covalent compounds do not exhibit such behavior. These types of solids are typically brittle and fracture easily along certain planes. Consider, for example, the difference between dropping an ice cube and a block of aluminum metal on a concrete floor.

The metals form solid structures in which the atoms are arranged as close-packed spheres or in a similar packing arrangement. For example, copper possesses a close-packing arrangement called cubic close packing (Section 10.5), in which each copper atom is in contact with 12 other copper atoms. The number of valence-shell electrons available for bond formation is insufficient for a copper atom to form a localized electron-pair bond to each of its neighbors. As another example, magnesium has only two valence electrons, yet each Mg atom in the metal is surrounded by 12 other Mg atoms. If each atom is to share its bonding electrons with all of its neighbors, these electrons must be able to move from one bonding region to another.

Electron-Sea Model for Metallic Bonding

One very simple model that accounts for some of the most important characteristics of metals is referred to as the **electron-sea model.** The metal is pictured as a lattice of metal ions in close-packed array, with the valence electrons completely free to move about the solid, as illustrated in Figure 22.11. The electrons are confined to the metal by electrostatic attraction to the cations, and they are uniformly distributed throughout the structure. No individual electron, however, is confined to any particular metal ion. Thus under the influence of an electrical field, the electrons migrate toward the positive end of the metal, while at the negative end electrons flow into the metal from the external source. The high heat conductivity of the metal is accounted for by supposing that the electrons readily transfer kinetic energy throughout the solid. In such a model a deformation of the lattice that caused a change in the relative

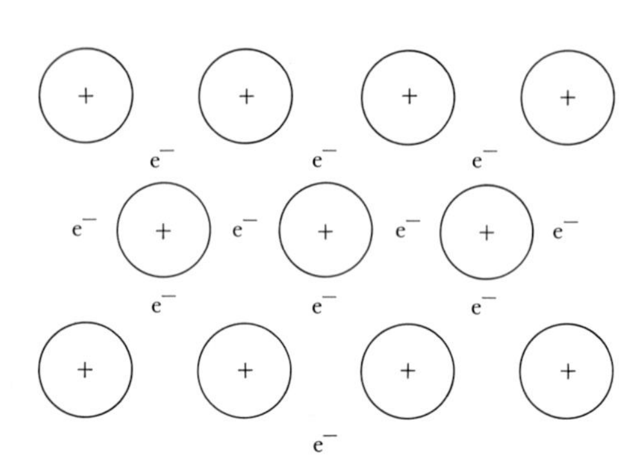

FIGURE 22.11 Schematic illustration of the electron-sea model for the electronic structure of metals.

positions of the cations would be compensated for by a redistribution of the electrons. Thus the malleability and ductility of metal is also accounted for.

The electron-sea model for metallic structure is qualitative and simple. Although it helps to account for conductivity, luster, malleability, and ductility, other properties of metals, such as melting points, heats of atomization, and hardness, are not explainable by the model. To understand such properties we must consider the number of electrons involved in metallic bonding. In transition metals the valence-shell s, p, and d metal orbitals may overlap, and electrons from all these orbitals might be involved in metallic bonding to some extent. Figure 22.12 shows the melting points of the metals through the transition series. Notice that the melting-point maximum in each period occurs for groups 5B and 6B. These elements possess electron configurations $(n-1)d^3ns^2$, $(n-1)d^4ns^2$, or $(n-1)d^5ns^1$. That is, there are five or six electrons beyond the noble gas arrangement. Curves for heats of fusion, hardness and densities of the solid metals, and for the boiling points and heats of vaporization of the liquid metals have very similar appearances to that of Figure 22.12.

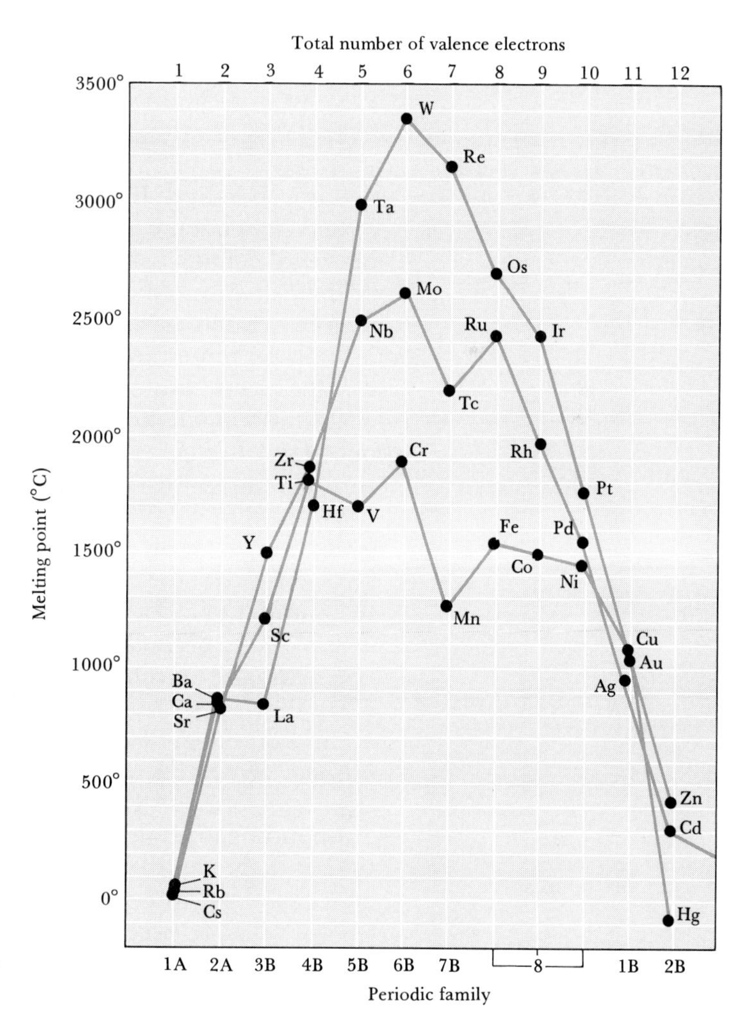

FIGURE 22.12 Melting points of the metals as a function of their location in the periodic table and their number of valence electrons.

It is reasonable to assume that the variation in melting point shown in Figure 22.12, and in other related physical properties, is due to variations in the number of electrons per metal atom involved in bonding. We can qualitatively understand the shape of the curve in Figure 22.12 if we assume that only the outer *s* and *d* electrons are involved in metallic bonding. Because there are one *s* and five *d* orbitals of a given quantum number, each metal atom can contribute a total of six atomic orbitals to metallic bond formation. As we add electrons to these orbitals, in proceeding from group 1A up to group 6B, the average number of bonds per metal atom should increase. However, once we have reached six electrons, addition of further electrons results in pairing of electrons in metal orbitals. The paired electrons cannot be used in bond formation to neighbors. Thus the average number of metallic bonds per metal atom decreases as we proceed from group 6B to the end of each transition series. This analysis accounts rather nicely for the way in which the physical properties related to bond order vary with electronic configurations of the elements. You should keep in mind, however, that other factors, such as atomic radius, nuclear charge, and the particular packing structure adopted by the metal, also play a role.

The Molecular-Orbital Model for Metals

We now have an idea of how the number of electrons per atom involved in metallic bonding depends on electron configuration. Let us now see whether we can improve somewhat on the simple electron-sea picture for metallic bonding, by applying the concepts of molecular orbitals, discussed in Section 8.5. In considering the structures of molecules such as benzene (Section 8.4), we saw that electrons can in some cases be delocalized, or distributed over several nuclear centers. This happens when the atomic orbitals on one nuclear center are able to interact with atomic orbitals on more than one other neighboring atom. In graphite (Section 8.7) the electrons are delocalized over an entire plane. It is useful to think of the bonding in metals in a similar way. The valence atomic orbitals on one metal atom overlap with those on several nearest neighbors. These in turn overlap with the atomic orbitals of still other atoms.

We saw in Section 8.5 that overlap of atomic orbitals can lead to formation of molecular orbitals; the number of molecular orbitals formed is equal to the number of atomic orbitals that overlap. In a metal the number of atomic orbitals involved in a particular molecular orbital is very large, because the atomic orbitals each overlap with several others. Thus the number of molecular orbitals is also very large. Figure 22.13 shows schematically what happens as increasing numbers of metal atoms come together to form molecular orbitals. As overlap of atomic orbitals occurs, bonding and antibonding molecular-orbital combinations are formed. The difference in energy between the bonding and antibonding orbitals is no greater than in ordinary covalent-bond formation, but the number of molecular orbitals is very much larger. The energies of these molecular orbitals lie at closely spaced intervals in the energy range between the highest- and lowest-energy orbitals. Thus the interaction of all the valence atomic orbitals of the metal atoms with the orbitals of adjacent metal atoms gives rise to a huge number of very

FIGURE 22.13 Schematic illustration of the interactions of atomic metal orbitals to form the delocalized orbitals of the metal lattice. The two atomic orbitals on each metal atom in this example might represent an s and p orbital. The main point is the formation of a very large number of molecular orbitals with very closely spaced energy levels. The number of electrons available does not completely fill these orbitals.

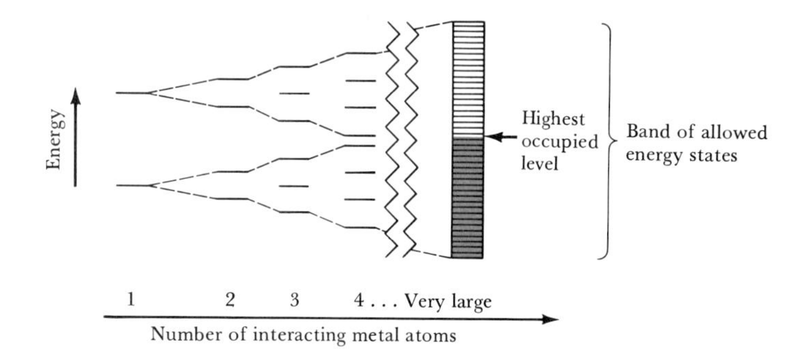

closely spaced molecular orbitals that extend over the entire metal lattice. The energy separations between these metal orbitals are so tiny that for all practical purposes it is possible to think of them as forming a continuous *band* of allowed energy states, as shown in Figure 22.13. The electrons available do not completely fill this set of allowed orbitals; one can think of them as a partially filled container for electrons. The incomplete filling of the energy band gives rise to characteristic metallic properties. The electrons in orbitals near the top of the occupied levels require very little energy input to be "promoted" to a still higher energy orbital that is unoccupied. Under the influence of any source of excitation, such as an applied electrical potential or an input of thermal energy, electrons move into previously vacant levels and are thus freed to move through the lattice, giving rise to electrical or thermal conductivity.

You can see that this molecular-orbital model is not so different in some respects from the electron-sea model. In both models the electrons are free to move about in the solid. The molecular-orbital model is more quantitative than is the simple electron-sea model. Although we cannot go into the details here, many properties of metals can be accounted for by quantum-mechanical calculations using molecular-orbital theory.

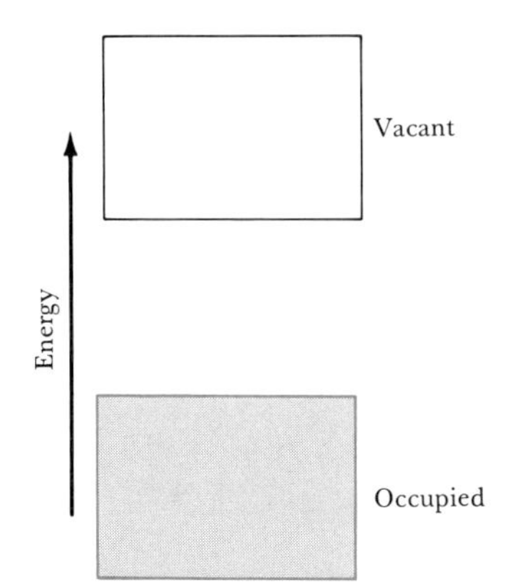

FIGURE 22.14 Bands of allowed energies of the molecular orbitals in diamond. The molecular orbitals are formed from the carbon $2s$ and $2p$ atomic orbitals. The particular structure of diamond, with just four nearest neighbors for each carbon atom, causes the molecular orbitals in the solid to fall within two widely separated bands of allowed energies. The number of valence electrons is just sufficient to completely occupy the lower allowed set. As a result, diamond is an insulator.

One final point should be made about the molecular-orbital, or band, theory as it is applied to solids. The metallic character of the solid depends on the fact that there are more molecular orbitals in the band than are needed to accommodate all the electron pairs provided by the atoms that make up the structure. In some solids, however, the number of electrons available is just the right number to completely fill the allowed levels of a band. Such a situation applies, for example, to diamond. When we apply molecular-orbital theory to this substance, for which the solid structure is shown in Figure 8.30, we find that the bands of allowed energies are as shown in Figure 22.14. The carbon $2s$ and $2p$ atomic orbitals combine to form two bands of allowed energies. One of these is completely filled with the paired valence electrons. There is a large gap in energy between this occupied set of orbitals and the second set of allowed energies. Because all the orbitals that are readily accessible to the valence electrons are filled, the characteristic metallic properties are absent. For example, there is no readily available vacant orbital into which the highest-energy electrons can move under the influence of an applied electrical potential. Thus diamond has none of the characteristic properties of metals. A solid substance that has an electronic structure such as that for diamond, that is, in which the valence band is completely filled, is called an *insulator*.

SAMPLE EXERCISE 22.4

What type of electrical behavior is expected for each of the following substances: (a) Ge; (b) Zn; (c) I_2; (d) NaCl(*aq*)?

Solution (a) Germanium, an element of group 4, has the same structure as diamond (Section 8.7). We expect that, like diamond, it will have a very low electrical conductivity as a pure substance. (b) Zinc, a metal should have high electrical conductivity of the kind characteristic of metals, in which electrons are the current carriers in the metal. (c) I_2 is a molecular solid consisting of I_2 molecules. Because all the electrons in each molecule are paired in stable orbitals, there should be no electrical conductivity. (d) In aqueous NaCl, the Na^+ and Cl^- ions are free to move through the solution. NaCl(*aq*) is thus capable of conducting an electrical current (Section 11.4).

A **semiconductor** possesses a conductance much lower than that of metals. A semiconductor can be formed from an insulator by adding different kinds of atoms into the structure. This process, called *doping,* results in too few or too many electrons to just fill the allowed energy band. To see how this works, consider silicon, which has the same structure as diamond. Pure silicon is not a good conductor; the valence-level electrons just fill the allowed energy band, as illustrated in Figure 22.15(a). Now suppose that a small amount of phosphorus is added. The phosphorus atoms substitute for silicon atoms at random sites in the structure. Phosphorus, however, possesses five valence-shell electrons per atom, as compared with four for silicon. There is no room for these extra electrons in the allowed energy band (often called the *valence band*). They must therefore occupy the higher-energy band, as illustrated in Figure 22.15(*b*). These higher-energy electrons have access to many vacant orbitals within the energy band they occupy, and serve as current carriers. Silicon doped with phosphorus in this manner is called an ***n*-type** semiconductor, because the current carrier is negatively charged.

If the silicon is doped instead with a group 3 element such as gallium, the atoms that substitute for silicon have one electron too few to meet their bonding requirements to the four neighboring silicon

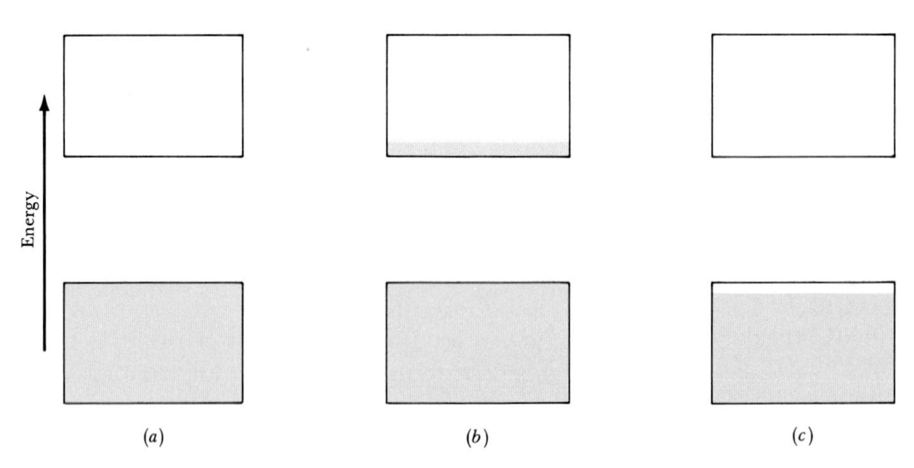

FIGURE 22.15 Effect of doping on the occupancy of the allowed energy levels in silicon. (*a*) Pure silicon. The valence-shell electrons just fill the lower-energy allowed energy levels. (*b*) Silicon doped with phosphorus. Excess electrons occupy the lowest-energy orbitals in the higher-energy band of allowed energies. These electrons are capable of conducting current. (*c*) Silicon doped with gallium. There are not quite enough electrons to fully occupy the orbitals of the lower-energy allowed band. The presence of vacant orbitals in this band permits current flow.

atoms. The valence band is thus not completely filled, as illustrated in Figure 22.15(*c*). Under the influence of an applied field, electrons can move from occupied molecular orbitals to one of the few orbitals that are vacant in the allowed energy band. A semiconductor that is formed by doping silicon with a group 3 element is called a ***p*-type** semiconductor, because it appears that the current is being carried by a positive hole created by the vacancy. The modern electronics industry is based on integrated circuitry formed from silicon doped with various elements to create the desired electronic characteristics.

22.7 ALLOYS

An alloy is a material that contains more than one element and has the characteristic properties of metals. The alloying of metals is of great importance because it is one of the primary ways of modifying the properties of the pure metallic elements. For example, nearly all the uses we make of iron involve alloy compositions of one sort or another. As another example, pure gold is too soft to be used in jewelry, whereas alloys of gold and copper are quite hard. Pure gold is termed 24 karat; the common alloy used in jewelry is 14 karat, meaning that it is 58 percent gold ($\frac{14}{24} \times 100$). A gold alloy of this composition has suitable hardness to be used in jewelry. The alloy can be either yellow or white, depending on the elements added. Some further examples of alloys are given in Table 22.5.

Alloys can be classified as solution alloys, heterogeneous alloys, and intermetallic compounds. Solution alloys are homogeneous mixtures with the components dispersed randomly and uniformly. Atoms of the solute can take positions normally occupied by a solvent atom, thereby forming a substitutional alloy, or they can occupy interstitial positions, thereby forming an interstitial alloy. These types are diagrammed in Figure 22.16.

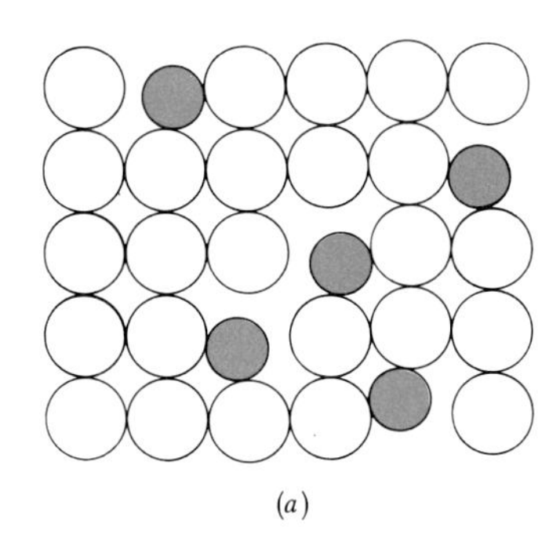

(*a*)

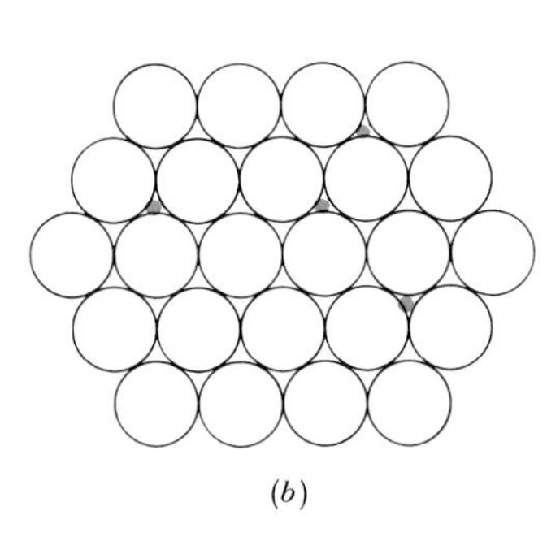

(*b*)

FIGURE 22.16 (*a*) Substitutional and (*b*) interstitial alloys. The open circles are the host metal; the colored circles, the other component of the alloy.

Substitutional alloys are formed when the two metallic components have similar atomic radii and chemical-bonding characteristics. For example, silver and gold form such an alloy over the entire range of possible compositions. When the two metals differ in radii by more than about 15 percent, they have more limited solubility behavior. For an interstitial alloy to form, the component present in the interstitial positions between the solvent atoms must have a relatively much smaller covalent radius. Typically, the interstitial element is a nonmetal that participates in bonding to neighboring atoms. The presence of the extra bonds provided by the interstitial component causes the metal lattice to become harder, stronger, and less ductile. For example, iron containing less than 3 percent carbon is much harder than pure iron and has a much higher tensile strength and other desirable physical properties. *Mild steels* contain less than 0.2 percent carbon; they are malleable and ductile and are used to make cables, nails, and chains. *Medium steels* contain 0.2 to 0.6 percent carbon; they are tougher than the mild steels and are used to make girders and rails. *High-carbon steel,* used in cutlery, tools, and springs, contains 0.6 to 1.5 percent carbon. In all these cases other elements may be added to form *alloy steels.* For example, vanadium and chromium may be added to impart strength and resistance to fatigue as well as corrosion resistance. For example, a rail steel used in Sweden on lines bearing heavy ore carriers contains 0.7 percent carbon, 1 percent chromium, and 0.1 percent vanadium.

One of the most important iron alloys is stainless steel, which contains

TABLE 22.5 Some common alloys

Primary element	Name of alloy	Composition by weight	Properties	Uses
Bismuth	Wood's metal	50% Bi, 25% Pb, 2.5% Sn, 12.5% Cd	Low melting point (70°C)	Fuse plugs, automatic sprinklers
Copper	Yellow brass	67% Cu, 33% Zn	Ductile, takes polish	Hardware items
Iron	Stainless steel	80.6% Fe, 0.4% C, 18% Cr, 1% Ni	Resists corrosion	Tableware
Lead	Plumber's solder	67% Pb, 33% Sn	Low melting point (275°C)	Soldering joints
Silver	Sterling silver	92.5% Ag, 7.5% Cu	Bright surface	Tableware
	Dental amalgam	70% Ag, 18% Sn, 10% Cu, 2% Hg	Easily worked	Dental fillings
Tin	Pewter	85% Sn, 6.8% Cu, 6% Bi, 1.7% Sb		Utensils

0.4 percent carbon, 18 percent chromium, and 1 percent nickel. The chromium is obtained by carbon reduction of chromite, $FeCr_2O_4$, in an electric furnace. The product of the reduction is ferrochrome, $FeCr_2$, which is then added in appropriate amount to the molten iron that comes from the converter to achieve the desired steel composition. Alloys of the type we have been discussing differ from ordinary chemical compounds in that the composition is not fixed. The ratio of elements present may vary over a wide range, imparting a variety of specific physical and chemical properties to the materials.

In **heterogeneous alloys,** the components are not dispersed uniformly. For example, in one form of steel known as pearlite, two distinct phases, essentially pure iron and the compound Fe_3C, known as cementite, are present in alternating layers. In general, the properties of heterogeneous alloys depend not only on the composition but also on the manner in which the solid is formed from the molten mixture. Rapid quenching leads to distinctly different properties than does slow cooling.

Intermetallic Compounds

Intermetallic compounds are homogeneous alloys that have definite properties and composition. For example, copper and aluminum form a compound, $CuAl_2$, known as duralumin. Intermetallic compounds made their appearance very early in the history of technology. For example, when a copper or bronze object is exposed to arsenic or an appropriate arsenic compound, the object becomes coated with Cu_3As, which can be polished to a beautiful silvery color. Examples of this technique are found in objects from Anatolia dated from 2100 B.C. As another example, a recipe for forming a dental amalgam from the elements silver, tin, and mercury is found in the Chinese literature of the seventh century A.D. The prescribed mixture results in formation of the two intermetallic compounds Ag_2Hg_3 and Sn_8Hg.

These examples of early intermetallic compounds illustrate some rather strange and unfamiliar ratios of combining elements. Nothing that we have yet discussed in this text would lead us to predict such compositions. The fact that intermetallic compounds possess unfamiliar and apparently bizarre relative proportions is further complicated by the fact that any two elements may form several intermetallic compounds. For example, potassium and mercury form KHg, KHg_2, KHg_3, K_2Hg_9, and KHg_{10}. When we further consider that one cannot readily isolate and purify intermetallic compounds by many of the techniques used frequently in working with ordinary ionic and covalent substances, it is perhaps not surprising that there was for a long time considerable doubt whether intermetallic compositions ever amounted to pure compounds. However, by the early part of this century it was clear that many intermetallic compounds did indeed exist, and it was left for future research to understand their compositions. Although much progress has been made since then, that research still goes on today. At the same time, intermetallic compounds have many very important applications in current high technology.

Although intermetallic compounds seem to have a bewildering array of empirical formulas, chemists and metallurgists have made substantial progress in understanding why certain types of formulas arise. One of the most important breakthroughs in this area came in 1926, when the British scientist William Hume-Rothery proposed that certain compositions represented favorable ratios of electrons to atoms. He had noticed that many intermetallic compounds could be placed into one of three groups, which he labeled as β (beta), γ (gamma), and ϵ (epsilon) phases. For example, the zinc, aluminum, and tin alloys of copper could be so grouped, as shown in Table 22.6. Hume-Rothery pointed out that each phase corresponded to a particular ratio of total valence electrons to atoms. For the β phases the ratio is three electrons to two atoms (for example, in Cu_5Sn, each Cu atom contributes 1 electron, the Sn atom contributes 4, for a total of 9 electrons per 6 atoms, or a 3:2 ratio); for the γ phases the ratio is 21:13, and for the ϵ phases, 21:12. (You can most easily remember these fractions if they are all put in terms of the same numerator: β 21:14, γ 21:13, and ϵ 21:12.)

Hume-Rothery's discovery prompted a great deal of study to determine why these particular ratios of electrons to atoms are favorable for the formation of stable intermetallic compounds. An account of the theory that has evolved is beyond the scope of this text. The most important point is that there are rules for bonding arrangements in intermetallic compounds, just as in other substances. It turns out that the details of the geometrical structures are important, as well as the number of electrons per atom. Attempts to understand Hume-Rothery's rules have also led to a better understanding of the factors involved in stabilizing intermetallic compounds with other compositions.

Modern Applications of Intermetallic Compounds

TABLE 22.6 Compositions of zinc, aluminum, and tin alloys of copper

β phase	γ phase	ϵ phase
CuZn	Cu_5Zn_8	$CuZn_3$
Cu_3Al	Cu_9Al_4	Cu_5Al_3
Cu_5Sn	$Cu_{31}Sn_8$	Cu_3Sn

We are literally surrounded with examples of intermetallic compounds that play important roles in modern society because of their special properties. High-performance aircraft are composed of a wide range of alloy materials, few of which contain a significant amount of iron, the traditional structural material. The intermetallic compound Ni_3Al is a major component of jet aircraft engines, chosen because of its light weight and strength. The coating on the new, very hard razor blades that are so widely advertised is an intermetallic compound, Cr_3Pt.

Materials that can be made into permanent magnets have many important uses. Intermetallic compounds of formula Co_5M, where M is a rare-earth element (atomic number between 58 and 71), possess unusual magnetic properties. The most useful of these compounds is Co_5Sm, from which extraordinarily strong magnets per unit weight of metal can be formed. The lightweight headsets that are used with portable stereo systems employ Co_5Sm permanent magnets (Figure 22.17).

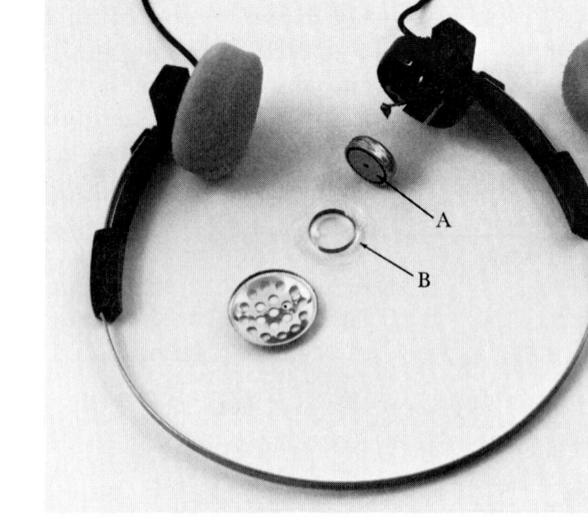

FIGURE 22.17 Interior of a lightweight audio headset. The permanent Co_5Sm alloy magnet, A, is surrounded by the copper coil, B, that fits in the annular space around the magnet. When electric current corresponding to the audio signal passes through the coil, the magnet is moved in and out in response to current variations. This movement in turn drives the speaker diaphragm. The assembly can be made small and lightweight because of the very strong magnetism of the alloy.

Intermetallic compounds have played a very important role in the development of superconducting materials. *Superconductivity* refers to the ability of a substance to conduct current with no apparent resistance whatever. Substances become superconducting only at very low temperatures, near absolute zero. The first superconducting alloy, Au_2Bi, was discovered in 1929. Since then many superconducting intermetallic com-

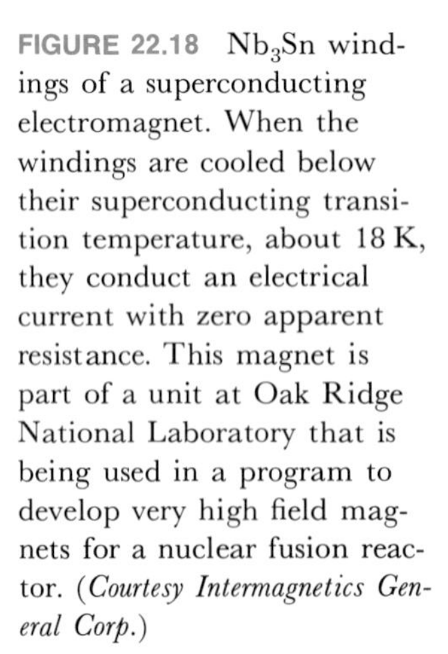
FIGURE 22.18 Nb_3Sn windings of a superconducting electromagnet. When the windings are cooled below their superconducting transition temperature, about 18 K, they conduct an electrical current with zero apparent resistance. This magnet is part of a unit at Oak Ridge National Laboratory that is being used in a program to develop very high field magnets for a nuclear fusion reactor. (*Courtesy Intermagnetics General Corp.*)

pounds have been discovered, with relatively high transition temperatures (that is, the temperature at which superconductivity sets in). One of the most important of these is Nb_3Sn, which has a transition temperature of 18 K, and which can be made into filaments to form the coil of a magnet. By passing a very large current through the superconducting material (which can be done because the resistance of the metal is essentially zero), a very high magnetic field can be developed. An example of a superconducting magnet with Nb_3Sn windings is shown in Figure 22.18. Such magnets are important in many areas of basic research. It requires considerable energy to keep the currently known superconducting materials below their transition temperatures, all below 20 K. There is therefore great interest in finding materials that will be superconducting at higher temperatures, perhaps even at room temperature.

FOR REVIEW

Summary

In this chapter our focus has been on the metallic elements: their occurrences, the means by which they can be obtained from their ores, the nature of metallic bonding and its relationship to the properties of metals, and the characteristics of alloys.

The metallic elements occur in nature as **minerals**, inorganic substances found in various kinds of deposits. **Metallurgy** is concerned with how to obtain the desired metals from these natural sources, and with the properties and structures of metallic elements. **Extractive metallurgy** is concerned with the processes that must be carried out to obtain the metal from the crude ore. These processes include physical concentration steps such as froth flotation, or magnetic separation. The chemical processes for obtaining the metal from the ore, once it has been enriched by physical concentration processes, can be grouped according to the type of chemical process involved. **Pyrometallurgy** involves the use of heat to bring about chemical reactions that convert an ore from one chemical form to another. One such process is **calcination**, in which an ore is heated to drive off a volatile substance. For example, a carbonate ore might be heated to drive off CO_2. In **roasting** the ore is heated under conditions that bring about reaction with the furnace atmosphere. For example, a sulfide ore such as ZnS might be roasted to cause oxidation of the sulfur to SO_2. In the course of such pyrometallurgical operations it often happens that two or more layers of mutually insoluble materials form in the furnace. Processes in which this happens are referred to as **smelting**. One of the layers often formed in this way, called **slag**, is composed of molten silicate minerals, with other ionic materials such as fluorides or phosphates as ionic constituents.

Iron is a most important metal in modern society. The metal is obtained from oxide ores, Fe_2O_3 and Fe_3O_4, by reduction in a **blast furnace**. The reducing agent is carbon, in the form of coke. Limestone, $CaCO_3$, is added to react with the silica present in the crude ore to form a slag. The raw iron from the blast furnace, called **pig iron**, is usually taken directly to a **converter**, in which it is refined to form various kinds of steel. In the converter, the crude molten iron is reacted with pure oxygen to oxidize impurity elements such as manganese, sulfur, phosphorus, and others.

Hydrometallurgy involves use of chemical processes occurring in aqueous solution to effect separation of a mineral or a particular element from others. In **leaching** an ore is treated with an aqueous reagent to dissolve selectively one desired component. For example, gold metal can be leached from raw ore by treatment with aqueous cyanide solution in air. In the **Bayer process** aluminum is selectively dissolved from bauxite as the $Al(H_2O)_2(OH)_4^-$ ion by treatment with concentrated NaOH solution. **Electrometallurgy** involves use of electrolytic methods for preparation or purification of a metallic element. Sodium is prepared by electrolysis of molten NaCl solution, in a **Downs cell**. Aluminum is obtained in the **Hall process** by electrolysis of Al_2O_3 in molten cryolite, Na_3AlF_6. Copper is purified by electrolysis of aqueous copper sulfate solution, using crude copper anodes.

The properties of metals can be accounted for in general by a model in which the electrons are free to move throughout the metal atom structure. The chemical bonds between metal atoms are delocalized over the entire solid. In the molecular-orbital model for metallic bonding, the available electrons do not completely fill the band of closely spaced allowed energy orbitals. As a result, promotion of electrons to higher-energy orbitals requires very little energy, giving rise to high electrical and thermal conductivity, as well as other characteristic metallic properties. By contrast, in an insulator all the orbitals of the band of allowed orbitals are completely filled.

Alloys are materials composed of more than one element, and possessing characteristic metallic properties. Usually, one or more metallic elements are major components. Solution alloys are homogeneous alloys, with the components distributed uniformly throughout. In heterogeneous alloys the components are not distributed uniformly; instead, two or more distinct phases with characteristic compositions are present. Intermetallic compounds are homogeneous alloys that have definite properties and compositions. For example, the alloy duralumin is an intermetallic compound of composition $CuAl_2$. The intermetallic compounds possess a wide variety of interesting properties that have important applications in modern technology.

Learning goals

Having read and studied this chapter, you should be able to:

1 Describe what is meant by the term *mineral,* and provide a few examples of common minerals.

2 Define various terms employed in discussions of metallurgy, notably *gangue, calcination, roasting, smelting, refining,* and *leaching.*

3 Describe the process of froth flotation, and explain the physical principles on which it is based.

4 Distinguish between pyrometallurgy, hydrometallurgy, and electrometallurgy, and provide examples of each type of metallurgical process.

5 Describe the nature of slag, and indicate the means by which it is formed in pyrometallurgical operations.

6 Describe the pyrometallurgy of iron. You should know which ores are employed, the general design of a blast furnace, the ingredients in the blast furnace reactions, and the chemical reactions of major importance. You should know how a converter is used to refine the crude pig iron that is the product of the blast furnace operation.

7 Describe the hydrometallurgy of gold, including the chemical reactions of major importance.

8 Describe the Bayer process for purification of bauxite, including balanced chemical equations.

9 Describe how copper may be recovered from chalcopyrite by means of a hydrometallurgical process.

10 Describe the process by which sodium metal is obtained from NaCl, including balanced chemical equations for electrode processes.

11 Describe the Hall process for obtaining aluminum, including balanced chemical equations for electrode processes.

12 Describe the electrometallurgical purification of copper, including balanced chemical equations for electrode processes.

13 Describe how the extent of metallic bonding among the first transition elements is related to their $3d$ electron configurations.

14 Discuss the simple electron-sea model for metals, and indicate how it accounts for certain important properties of metals.

15 Describe the molecular-orbital model for metals, including the idea of bands of allowed energy levels. You should also be able to distinguish between metals and insulators in terms of this model.

16 Name the important types of alloys, and distinguish between them.

17 Describe the nature of intermetallic compounds, and provide examples of such compounds and their uses.

Key terms

Among the more important terms and expressions used for the first time in this chapter are the following:

An alloy (Section 22.7) has the characteristic properties of a metal, and contains more than one element. Often there is one principal metallic component, with one or more other elements present in smaller amounts. Alloys may be homogeneous or heterogeneous in nature.

The Bayer process (Section 22.4) is a hydrometallurgical procedure for purification of bauxite in recovery of aluminum from bauxite-containing ores.

Calcination (Section 22.3) involves heating an ore to bring about its decomposition, and the elimination of a volatile product. For example, a carbonate ore might be calcined to drive off CO_2.

A converter (Section 22.3) is a container in which crude iron from the blast furnace is refined.

The Downs cell (Section 22.5) is used to obtain sodium metal by electrolysis of molten NaCl.

Electrometallurgy (Section 22.5) refers to the use of electrolysis to obtain free metal from one of its compounds, or to purify a crude form of the metal.

In the electron-sea model (Section 22.6) for metals, the metallic structure is pictured as consisting of metal ions at the locations of the metallic atoms, with the electrons moving freely about in the spaces between the ions.

Extractive metallurgy (Section 22.2) is concerned with the steps involved in obtaining a metal in as pure a form as possible, from raw ore as starting material.

Flotation (Section 22.2) is a technique for concentrating raw, finely ground ore by a process of physical separation of the desired ore from gangue through selective wetting with the use of flotation agents.

Gangue (Section 22.2) is material of little or no value that accompanies the desired mineral in most raw ores.

The **Hall process** (Section 22.5) is employed to obtain aluminum by electrolysis of Al_2O_3 dissolved in molten cryolite, Na_3AlF_6.

Hydrometallurgy (Section 22.4) refers to aqueous chemical processes for recovery of a metal from an ore.

An **intermetallic compound** (Section 22.7) is a homogeneous alloy with definite properties and composition. Intermetallic compounds are stoichiometric compounds in the full sense, but their compositions are not readily understood in terms of ordinary chemical bonding theory.

Leaching (Section 22.4) refers to the selective dissolution of a desired mineral by passing an aqueous reagent solution through an ore.

The **lithosphere** (Section 22.1) is that portion of our environment consisting of the solid earth.

Metallurgy (Introduction) is the science of extracting metals from their natural sources, by a combination of chemical and physical processes. It is also concerned with the properties and structures of metals and alloys.

A **mineral** (Section 22.1) is a solid inorganic substance occurring in nature, such as calcium carbonate, which occurs as limestone.

Pig iron (Section 22.3) is the product of reduction of iron ore in a blast furnace.

Pyrometallurgy (Section 22.3) involves the use of heat to convert a mineral in an ore from one chemical form to another, and eventually to the free metal.

Refining (Section 22.3) is the process of converting an impure form of a metal into a more usable substance of well-defined composition. For example, crude pig iron from the blast furnace is refined in a converter to produce steels of desired compositions.

Roasting (Section 22.3) is thermal treatment of an ore to bring about chemical reactions involving the furnace atmosphere. For example, a sulfide ore might be roasted to form a metal oxide and SO_2.

Slag (Section 22.3) is a mixture of molten silicate minerals. Slags may be acidic or basic, according to the acidity or basicity of the oxide added to silica.

Smelting (Section 22.3) is a melting process in which the materials formed in the course of the chemical reactions that occur separate into two or more layers. For example, the layers might be slag and molten metal.

EXERCISES*

Minerals; metallurgy

22.1 Two of the most heavily utilized elements listed in Table 22.2 are Al and Fe. Indicate the most important natural sources of these elements, and the oxidation state of the metal in each case.

22.2 What conditions would you expect to be required in order that one metallic element be able to substitute for another in a mineral, as in the example of olivine described on page 725?

22.3 Provide a brief definition of each of the following terms: (a) calcination; (b) leaching; (c) gangue; (d) slag; (e) flotation.

22.4 Using the concepts presented in Chapter 10, describe the forces that operate in the binding of potassium ethyl xanthate to Cu_2S in froth flotation, and indicate how this binding leads to flotation of the mineral.

22.5 Complete and balance each of the following equations:

(a) $ZnCO_3(s) \xrightarrow{\Delta}$

(b) $Cu_2S(s) + O_2(g) \longrightarrow$

(c) $ZnS(s) + O_2(aq) \longrightarrow SO_4^{2-}(aq) +$

(d) $Al(OH)_3(s) + OH^-(aq) \longrightarrow$

(e) $UO_2(s) + CO_3^{2-}(aq) + O_2(g) \longrightarrow [UO_2(CO_3)_3]^{4-} + OH^-(aq)$

(f) $Cu^{2+}(aq) + Cl^-(aq) + H_2SO_3(aq) \longrightarrow CuCl(s) + SO_4^{2-}(aq)$

22.6 In an electrolytic process nickel sulfide is oxidized in a two-step reaction:

$$Ni_3S_2(s) \longrightarrow Ni^{2+}(aq) + 2NiS(s) + 2e^-$$
$$NiS(s) \longrightarrow Ni^{2+}(aq) + S(s) + 2e^-$$

What mass of Ni^{2+} is produced in solution by passage of a current of 66 A for a period of 8.0 h, assuming the cell to be 100 percent efficient?

22.7 What mass of copper is deposited in an electrolytic refining cell by a passage of 240 A current for a period of 10 h, assuming 80 percent current efficiency?

22.8 Complete and balance each of the following equations:

(a) $H_2SeO_3(aq) + SO_2(aq) \longrightarrow Se(s) + SO_4^{2-}(aq)$
(b) $CaO(l) + P_2O_5(l) \longrightarrow$
(c) $K(l) + TiCl_4(g) \longrightarrow$
(d) $Pb(s) + NaOH(l) + NaNO_3(l) \longrightarrow Na_2PbO_3(l) + N_2(g) + H_2O(g)$
(This is a fused salt reaction.)
(e) $Cu_2O(l) + CH_4(g) \longrightarrow Cu(l)$
(f) $WO_3(s) + H_2(g) \longrightarrow$

22.9 Write balanced chemical equations for molten state reactions between (a) CaO and Al_2O_3; (b) FeO and SiO_2; (c) MgO and PbO_2; (d) Na_2O and P_2O_5.

22.10 Place the following metal oxides in a list with the most basic oxide first and the most acidic oxide last: MgO, Fe_2O_3, Re_2O_7, Li_2O, TiO_2.

[22.11] In steelmaking, when the slag present is "acidic," containing SiO_2, FeO, and MnO, but little CaO, removal of phosphorus from iron through oxidation of the phosphorus does not occur, whereas it does occur when the slag is "basic." Suggest an explanation.

22.12 Describe how electrometallurgy could be employed to purify crude cobalt metal. Describe the compositions of the electrodes and electrolyte, and write out all electrode reactions.

22.13 Write balanced chemical equations for the reductions of FeO and Fe_2O_3 by H_2 or CO in a blast furnace.

22.14 What is the major reducing agent in the reduction of iron ore in a blast furnace? Write a balanced chemical equation for the reduction process.

22.15 A charge of 2.0×10^4 kg of material containing 32 percent Cu_2S and 7 percent FeS is added to a converter and oxidized. What mass of $SO_2(g)$ is formed?

22.16 A charge of 1.2×10^4 kg of concentrate from a partial roast of a chalcopyrite ore, containing 26 percent FeO, is added to a furnace. SiO_2 is added to form a slag with FeO. Write the balanced equation for the reaction leading to slag formation and calculate the mass of SiO_2 required to react with the FeO to form slag.

22.17 In terms of the concepts discussed in Chapter 11, indicate why the molten metal and slag phases formed in the blast furnace shown in Figure 22.4 are immiscible.

[22.18] Why do the free energies of formation of the metal oxides grow increasingly positive with increasing temperature, as illustrated in Figure 22.6? Why, by contrast, is the free energy of formation of CO_2 nearly independent of temperature?

22.19 Using Figure 22.6, estimate the free-energy change for each of the following reactions at 1200°C.

(a) $NiO(s) + CO(g) \longrightarrow Ni(s) + CO_2(g)$
(b) $Si(s) + 2MnO(s) \longrightarrow SiO_2(s) + 2Mn(s)$
(c) $FeO(s) + H_2(g) \longrightarrow Fe(s) + H_2O(g)$

[22.20] In the electrolytic purification of nickel some iron may be present in the crude metal. It is found that as iron accumulates in the electrolyte solution the current efficiency of the electrolysis cell decreases when the contents of the cathode cell compartment are not isolated from the air. Write reactions involving iron species that could account for the loss in current efficiency. (Hint: Recall that iron can exist in solution in more than one oxidation state.)

22.21 The element tin is generally recovered from deposits of the ore cassiterite, SnO_2. The oxide is reduced with carbon, and the crude metal is purified by electrolysis. Write balanced chemical equations for the reduction process and the electrode reactions in the elecrolysis, assuming that an acidic solution of $SnSO_4$ is employed as electrolyte.

22.22 In the electrolytic purification of tin, it is possible to employ as the electrolyte medium a basic solution in which tin is present as the $HSnO_2^-$ ion. Write the electrode reactions for both anode and cathode reactions in this medium.

22.23 What role does each of the following materials play in the chemical processes that occur in the blast furnace: (a) air; (b) limestone; (c) coke; (d) water? Write balanced chemical equations to illustrate your answers.

[22.24] In terms of thermodynamic functions, particularly free energies of formation, what conditions must be operative for refining iron in the converter shown in Figure 22.15 to be a workable process?

Metals and alloys

22.25 Suppose that you have samples of magnesium, iron, sulfur, high-carbon steel, and calcium sulfate, all formed into solid, pencil-shaped rods. Compare these substances with respect to appearance, flexibility, melting point, and thermal conductivity. Account for the differences you would expect to observe.

22.26 Sodium is a highly malleable substance, whereas sodium chloride is not. Explain this difference in properties.

22.27 The enthalpies of sublimation of copper metal, copper(I) iodide, and iodine at 25°C are 338, 326, and 62 kJ/mol, respectively. Account for this variation in terms of the attractive forces that operate in each solid.

22.28 The interatomic distance in silver metal, 2.88 Å, is much larger than the calculated Ag—Ag single-bond distance, 1.34 Å. Nevertheless, the melting point of silver, 906°C, and its high boiling point, 2164°C, suggest that there is strong bonding between silver atoms in the metal. Describe a model for the structure of silver metal that is consistent with these observed properties.

22.29 How does the electron-sea model account for the high electrical conductivity of copper metal?

[22.30] The melting point of neodymium, Nd, is

1019°C. Considering its place in the periodic table and the melting points of other elements as shown in Figure 22.12, what number of electrons would you estimate to be involved in metallic bonding in this element? What do your conclusions suggest about the role of $4f$ electrons in the bonding?

22.31 The densities of the elements K, Ca, Sc, and Ti are 0.86, 1.5, 3.2, and 4.5 g/cm^3, respectively. What factors are likely to be of major importance in determining this variation?

[22.32] Indicate whether each of the following is likely to be an insulator, a metallic conductor, an n-type semiconductor, or a p-type semiconductor: (a) germanium doped with Ga; (b) pure $CuAl_2$; (c) pure boron; (d) silicon doped with As; (e) pure Nb; (f) pure MgO.

Additional exercises

22.33 Distinguish between (a) a substitutional and interstitial alloy; (b) an intermetallic compound and a heterogeneous alloy; (c) a metal and an insulator; (d) the bonding in $Na_2(g)$ and in $Na(s)$.

22.34 Complete and balance each of the following equations:

(a) $MoO_3(g) + MnO(l)$

(b) $FeCl_2(s) + H_2(g) \xrightarrow{650°C}$

(c) $Mn_3O_4(s) + Al(l) \longrightarrow$

(d) $FeS(l) + Cu_2O(l) \longrightarrow FeO +$

22.35 The crude copper that is subjected to electrorefining contains selenium and tellurium as impurities. Describe the probable fate of these elements during electrorefining, and relate your answer to the positions of these elements in the periodic table.

22.36 Magnesium is obtained by electrolysis of molten $MgCl_2$. Several cells are connected in parallel by very large copper buses that convey current to the cells. Assuming that the cells are 96 percent efficient in producing the desired products in electrolysis, what mass of Mg is formed by passage of a current of 97,000 A for a period of 24 h?

22.37 Write a balanced chemical equation to illustrate the use of each of the following reducing agents in metallurgy: (a) carbon; (b) H_2; (c) Na.

22.38 Write balanced chemical equations to correspond to each of the following verbal descriptions. (a) $NiO(s)$ can be solubilized by leaching with aqueous sulfuric acid. (b) After concentration, an ore containing the mineral carrolite, $CuCo_2S_4$, is leached with aqueous sulfuric acid to produce a solution containing copper and cobalt ions. (c) Titanium dioxide is treated with chlorine in the presence of carbon as reducing agent to form $TiCl_4$. (d) Under oxygen pressure, $ZnS(s)$ reacts at 150°C with aqueous sulfuric acid to form soluble zinc sulfate, with deposition of elemental sulfur.

22.39 Zinc can be purified by electrolytic refining. A current of 75 A is used at an applied voltage of 3.5 V. Assuming that the electrolytic process is 94 percent efficient, how much electrical energy (kilowatt hours) is required to produce 1.0×10^8 kg of zinc metal?

22.40 In the early days of iron mining in the United States, the iron oxide mined in the Lake Superior region was reduced to pig iron in blast furnaces located near the mines. The furnaces used charcoal made from hardwood as the reducing agent (transportation savings prompted this procedure). What mass of pig iron containing 2.5 percent carbon was produced from each kilogram of iron ore, assuming that it is mined as magnetite, Fe_3O_4, of 70 percent purity?

22.41 In the weathering of rocks many metallic elements (for example, Ca^{2+}, Fe^{2+}, Mn^{2+}) pass into solution as the bicarbonate salts. Explain why this is so.

22.42 Pure silicon is a very poor conductor of electricity. Titanium, which also possesses four valence-shell electrons, is a metallic conductor. Explain the difference.

22.43 Treatment of iron surfaces at high temperatures with ammonia leads to "nitriding," in which nitrogen atoms are incorporated into the metal structure. What kind of alloy formation occurs in this case? How would you expect nitriding to alter the properties of the metal?

22.44 Tin exists in two allotropic forms; gray tin has a diamond structure and white tin has a close-packed structure. Which of these allotropic forms would you expect to be more metallic in character? Explain why the electrical conductivity of white tin is much greater than that of gray tin. Which form would you expect to have the longer Sn—Sn bond distance?

[22.45] Suppose that a rich deposit of the mineral absolite, $CoO \cdot 2MnO_2 \cdot 4H_2O$, were found in the western United States. Sketch out a scheme for recovery of cobalt metal from the ore containing this mineral.

[22.46] Write balanced chemical equations to correspond to the steps in the following brief account of the metallurgy of molybdenum. "Molybdenum occurs primarily as the sulfide, MoS_2. On boiling with concentrated nitric acid, a white residue of MoO_3 is obtained. This is an acidic oxide; from a solution of hot, excess concentrated ammonia, ammonium molybdate crystallizes on cooling. On heating ammonium molybdate, white MoO_3 is obtained. On heating to 1200°C in hydrogen, a gray powder of metallic molybdenum is obtained."

[22.47] To satisfy the Hume-Rothery rules, how many valence electrons per atom must be assumed for each of the three elements that make up the intermetallic compounds listed in Table 22.6?

[22.48] Which of the following compounds satisfy the Hume-Rothery rules: (a) Na_2Sn; (b) Ag_5Cd_8; (c) Au_5Ge; (d) Zr_3Al? For those that do, indicate whether they correspond to a β, γ, or ϵ phase. (Where more than one valence for the metal is possible, you should consider all the reasonable choices.)

23 Chemistry of the Transition Elements

In Chapter 22 we considered the distribution of metallic elements in nature, the means by which they can be obtained from their ores, and the important characteristics of metals and alloys. Let's now turn our attention to the chemical characteristics of the metals. In Chapter 6 we had a brief look at the rather simple chemistry associated with the group 1 and 2 elements. In this chapter we will focus on the transition metals. Recall that the transition elements comprise three series of 10 elements each in the middle part of the periodic table, as depicted on the inside front cover. For these elements the *d* orbitals are being filled (Figure 6.4). The transition elements include some of the most familiar and important elements, such as iron, copper, nickel, cobalt, and silver. They also include many elements that are likely to be unfamiliar to you, but which are nonetheless important in modern technology. Even the least abundant of the transition elements has found uses as chemists and other scientists search for new materials to meet new performance requirements. As an illustration of this point, Figure 23.1 shows the approximate composition of a high-performance jet engine. Note that nearly the entire mass of the engine consists of transition elements. Note also that iron, long considered the workhorse element of technology, is not even present to a significant extent.

The transition elements are numerous, and each has distinctive chemical properties. There is no point in our attempting to master the individual chemical properties of these elements. Rather, we will approach the subject by considering general aspects of their chemical behavior, such as trends in characteristic oxidation states. If we can understand patterns of chemical behavior in terms of electron configurations, nuclear charge, and other key properties, we have the basis for making sense of a mass of descriptive facts. To show how these general considerations can be used to understand a body of data, we will consider some selected aqueous solution behavior of the transition elements, and the properties of solid transition metal oxides. We will consider next the way in which the

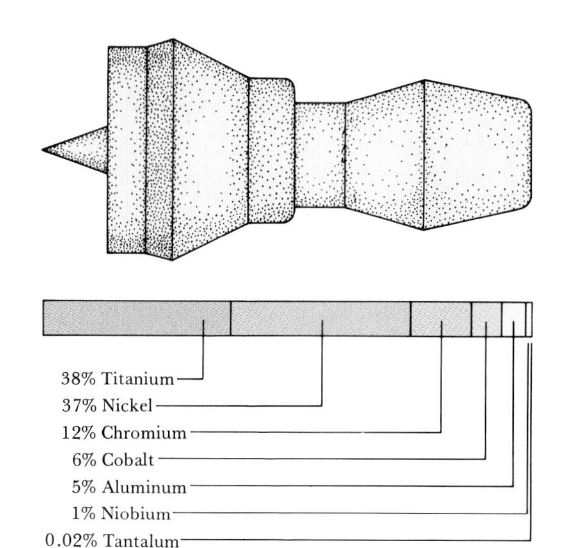

FIGURE 23.1 Metallic elements employed in construction of an aircraft jet engine.

presence of unpaired electrons in the transition elements and their compounds gives rise to interesting and useful magnetic properties. Finally, we will give brief attention to metals as strategic materials.

23.1 PHYSICAL PROPERTIES OF THE TRANSITION ELEMENTS

To understand many of the applications of transition elements in technology, we must understand their electronic structures. The electronic configurations of the transition elements are shown in Table 23.1. As we proceed across a given transition series, electrons are being placed into the $3d$, $4d$, or $5d$ orbitals. You should recall from Section 6.3 that the s orbital with major quantum number one higher than the d orbitals is very close in energy to those d orbitals. Whereas the s orbital normally contains two electrons, there are occasions when an electron moves from the s orbital to a d orbital. Several examples are to be found in Table 23.1. You should recall also that the lanthanides, or inner transition elements, with atomic numbers from 58 through 71, occur at the front of the third transition series. We will not have much occasion to talk about the chemistry of these elements as such, but their existence does have interesting consequences for the properties of the elements of the third transition series, as we shall soon see.

TABLE 23.1 Electronic configurations of the transition elements

Sc	Ti	V	Cr	Mn	Fe	Co	Ni	Cu	Zn
$3d^{1}4s^{2}$	$3d^{2}4s^{2}$	$3d^{3}4s^{2}$	$3d^{5}4s^{1}$	$3d^{5}4s^{2}$	$3d^{6}4s^{2}$	$3d^{7}4s^{2}$	$3d^{8}4s^{2}$	$3d^{10}4s^{1}$	$3d^{10}4s^{2}$
Y	Zr	Nb	Mo	Tc	Ru	Rh	Pd	Ag	Cd
$4d^{1}5s^{2}$	$4d^{2}5s^{2}$	$4d^{4}5s^{1}$	$4d^{5}5s^{1}$	$4d^{5}5s^{2}$	$4d^{7}5s^{1}$	$4d^{8}5s^{1}$	$4d^{10}$	$4d^{10}5s^{1}$	$4d^{10}5s^{2}$
La	Hf	Ta	W	Re	Os	Ir	Pt	Au	Hg
$5d^{1}6s^{2}$	$4f^{14}5d^{2}6s^{2}$	$4f^{14}5d^{3}6s^{2}$	$4f^{14}5d^{4}6s^{2}$	$4f^{14}5d^{5}6s^{2}$	$4f^{14}5d^{6}6s^{2}$	$4f^{14}5d^{7}6s^{2}$	$4f^{14}5d^{9}6s^{1}$	$4f^{14}5d^{10}6s^{1}$	$4f^{14}5d^{10}6s^{2}$

As we examine the properties of the transition elements, it is important to distinguish those that are characteristic of the solid or liquid metal from those characteristic of isolated atoms in the gas phase. You should not expect these two types of properties to vary in the same way in a given transition element series. For example, Figure 22.16 of the preceding chapter shows how the melting points of the transition elements vary in the three series. The melting points of the solids are related to the extent of metallic bonding that holds the solid together. In a similar way, the heats of sublimation of the transition elements, shown in Figure 23.2, also reflect the extent of metallic bonding in the solid; in the atomization process $M(s) \longrightarrow M(g)$, metal-metal bonds are broken. On the other hand, a property such as ionization energy or atomic radius is characteristic of the *isolated* atoms of each element. To illustrate this distinction further, some properties of the elements of the first transition series are listed in Table 23.2. In contrast to those properties characteristic of the bulk metal, the ionization energy and atomic radius, properties characteristic of the individual atoms, show a relatively smooth and rather

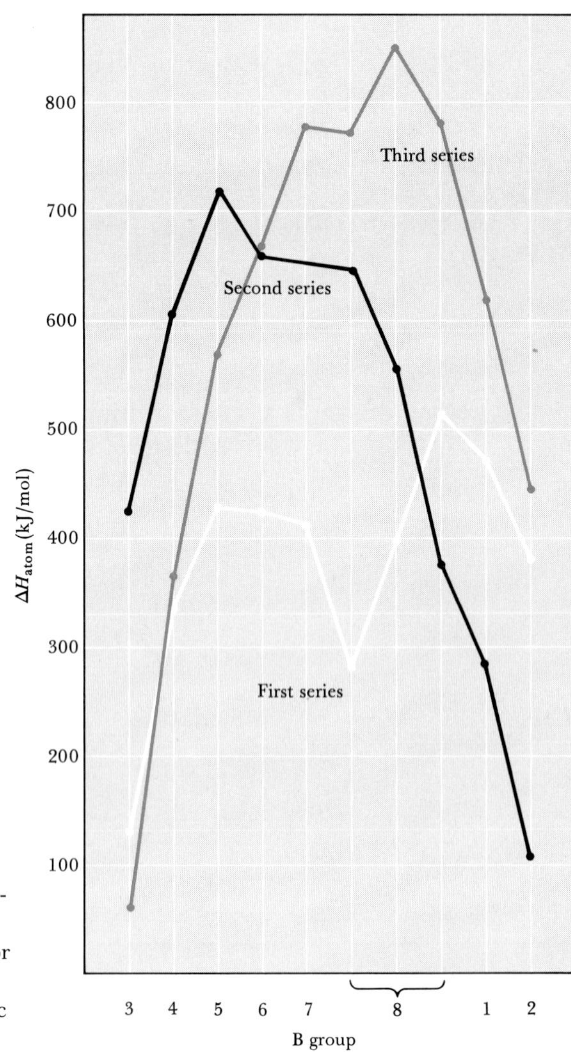

FIGURE 23.2 Heats of atomization of the transition elements. These are the heats for the process, $M(s) \rightarrow M(g)$, where $M(g)$ is the monatomic gas species.

TABLE 23.2 Properties of the first-series transition elements

Group:	3B	4B	5B	6B	7B	8B			1B	2B
Element:	Sc	Ti	V	Cr	Mn	Fe	Co	Ni	Cu	Zn
Electron configuration	$3d^14s^2$	$3d^24s^2$	$3d^34s^2$	$3d^54s^1$	$3d^54s^2$	$3d^64s^2$	$3d^74s^2$	$3d^84s^2$	$3d^{10}4s^1$	$3d^{10}4s^2$
First ionization energy (kJ/mol)	631	658	650	653	717	759	758	737	745	906
Atomic radius (Å)	1.44	1.32	1.22	1.17	1.17	1.16	1.16	1.15	1.17	1.24
Density (g/cm^3)	3.0	4.5	6.1	7.9	7.2	7.9	8.7	8.9	8.9	7.1
ΔH_{atom} (kJ/mol)	377	470	514	398	283	415	428	430	338	130
Melting point (°C)	1538	1660	1917	1857	1244	1537	1494	1455	1084	420

small variation across the series. The observed trend in these properties can be understood in terms of the concept of increasing effective nuclear charge, as described in Chapter 6.

The incomplete screening of the nuclear charge by added electrons produces a very interesting and important effect in the third transition series. In the lanthanide series, electrons are added to the $4f$ orbitals. These orbitals extend radially fairly far from the nucleus. The result is that the $4f$ electrons do not completely shield the nucleus from the $5d$ electrons of the third series. You can see what effect this has by looking at the atomic covalent radii for the transition elements, graphed in Figure 23.3. We would expect that in any one group the radius would increase as the major quantum number increases. This trend is seen in group 3B where the covalent atomic radius increases in the order Sc < Y < La. In

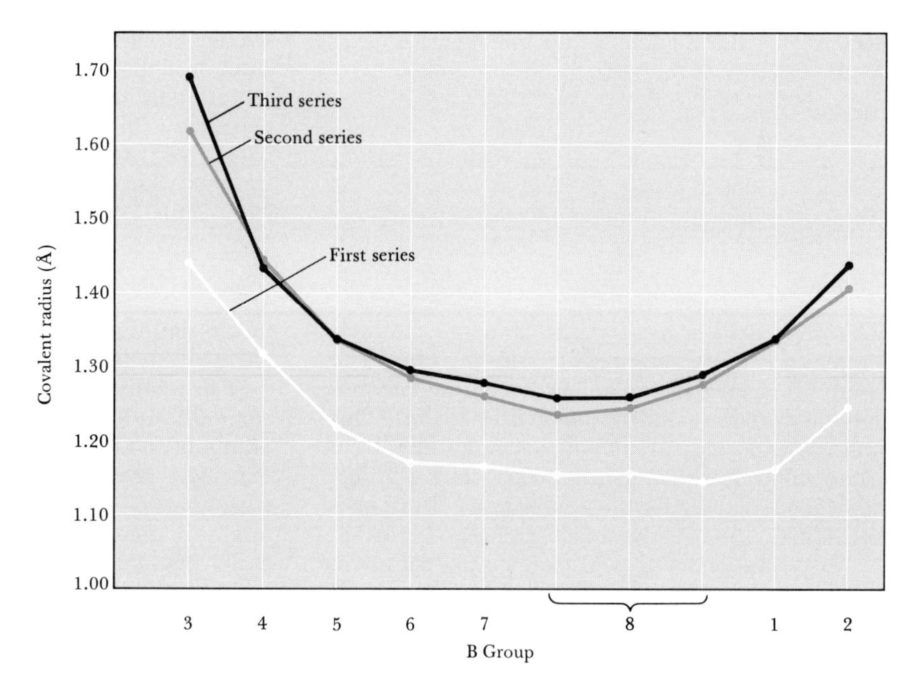

FIGURE 23.3 Variation in covalent radius of the transition elements as a function of the periodic table group number.

group 4B we would similarly expect the radius to increase in the order Ti < Zr < Hf. We find the expected increase in going from Ti to Zr, but Hf has the *same* radius as Zr. This effect has its origin in the lanthanide series that precedes the third transition series; note the difference in electronic configurations of Hf and Zr (Table 23.1). In the lanthanide series, as electrons are added to the 4*f* orbitals, the atomic radius grows gradually smaller because of an increasing effective nuclear charge. The contraction in radius in the lanthanide series, often called the **lanthanide contraction,** is of just about the right magnitude to offset the increase in radius that comes from the increase of one in major quantum number. Thus Hf and Zr have about the same radius. You can see from Figure 23.3 that second- and third-row transition elements of the same group have about the same radii all the way across the series. One consequence of this effect is that the second- and third-series transition elements are more similar to one another than they are to the first-series element of the same group.

SAMPLE EXERCISE 23.1

The first ionization energies, I_1, of the group 3B and 6B elements are as follows:

Group 3B		Group 6B	
Sc	631 kJ/mol	Cr	653 kJ/mol
Y	616 kJ/mol	Mo	685 kJ/mol
La	538 kJ/mol	W	769 kJ/mol

Account for the relative values among these elements in terms of their electronic structures.

Solution: The decrease in I_1 with increasing atomic number in group 3B is what we expect. As the major quantum number increases, the outermost electrons are farther from the nucleus. If they experience about the same effective nuclear charge, as we might expect if the electron configurations are comparable in the series, the energy required for removal of the electron should decrease. The group 6B elements have higher ionization energies than do the corresponding group 3B elements. This again is expected; in moving from group 3B to group 6B the nuclear charge increases by 3; because the screening by the additional *d* electrons is incomplete, the effective nuclear charge increases.

The variation in I_1 with increasing period in group 6B is not as easy to explain. The covalent atomic radius of molybdenum is larger than that of chromium. Thus we might expect a lower value for I_1 for Mo than for Cr, just as I_1 is lower for Y than for Sc. The electronic configurations of Cr and Mo are entirely comparable (Table 23.1). The fact that I_1 is larger for Mo tells us that the effective nuclear charge experienced by the 5*s* electron in Mo is larger than that experienced by the 4*s* electron in Cr. The 5*s* electron in Mo must penetrate farther inside the electron distribution of the 4*d* orbitals than does the 4*s* electron inside the 3*d* electron distribution of Cr. The 5*s* electron in Mo therefore experiences a higher effective nuclear charge, with a resulting increase in I_1. Note that the electronic configuration for W is different from that for Cr or Mo (Table 23.1). The much higher ionization energy for W as compared with Mo, in spite of comparable covalent atomic radii, indicates that the outermost electrons in this element penetrate closer inside the other electrons, with a resultant higher effective nuclear charge.

The relative values of I_1 among the group 6B elements could not have been predicted on the basis of any of the ideas we have presented so far in this text. The example illustrates that in large atoms with many electrons the factors that influence energies and chemical behavior are complex and not easily predicted.

23.2 OXIDATION STATES

Recall from our discussion in Section 7.2 that in the oxidations of the transition elements to form metal ions, the electrons that are lost first are the outer *s* electrons. Loss of additional electrons occurs from the *d* subshell that lies just below the *s* subshell in energy. Very often, the ions of the transition elements possess partially occupied *d* subshells. The existence of these *d* electrons is primarily responsible for several characteristics that the transition elements share with one another:

1. They often exhibit more than one stable oxidation state.
2. Many compounds of the transition elements are colored. We will discuss the origin of these colors in Chapter 24.
3. Many compounds of the transition elements exhibit interesting magnetic properties. We will discuss this topic in Section 23.5.

The maximum possible oxidation state for any given element cannot exceed the number of electrons beyond the nearest rare-gas configuration. However, this oxidation state may not be a particularly stable one. Further, the most common oxidation states in solution are not always the same as those found in solid compounds. For example, manganese is commonly found in aqueous solution as the Mn^{2+} ion. The +4 oxidation state is not common in solution; nevertheless, MnO_2 is a quite stable solid. Figure 23.4 shows the oxidation states most often encountered for the transition elements of all three series. This table is not a complete accounting of all the oxidation states observed. First, there is a large class of compounds of the transition elements, called organometallic compounds, in which the metals are in low positive oxidation states (such as +1), or even in zero or negative oxidation states. We will not be able to discuss these compounds at length and have thus omitted references to the oxidation states that are observed in them. Second, for many of the transition elements, there may be one or more exotic compounds in which the element is in one of the oxidation states not shown in Figure 23.4. The oxidation states shown are those for which several well-characterized examples exist. The filled circles in the figure represent the oxidation states most commonly observed, or most important in the chemistry of the element, either in solution or in solid compounds. To return to Mn as our example, the oxidation states most important for solution are the +2 (Mn^{2+}) and +7 (MnO_4^-); in the solid state the +4 oxidation state (MnO_2) is common. The +3, +5, and +6 oxidation states are observed as relatively less stable species in either the solid state or in solution.

SAMPLE EXERCISE 23.2

The manganate ion, MnO_4^{2-}, is stable only in strongly basic solution. In acidic solution it reacts to form the permanganate ion and solid MnO_2. Write a balanced chemical equation to show this reaction.

Solution: In a disproportionation reaction the same species is both the oxidizing and the reducing agent. In this case we have manganese going from the +6 to the +7 and +4 oxidation states. Because there is a change in oxidation state of +1 for the oxidation and −2 for the reduction, there must be twice as much MnO_4^- produced as MnO_2. The two half-reactions and overall reaction are:

$$2[MnO_4^{2-}(aq) \longrightarrow MnO_4^-(aq) + e^-]$$

$$MnO_4^{2-}(aq) + 2e^- + 4H^+(aq) \longrightarrow MnO_2(s) + 2H_2O(l)$$

$$3MnO_4^{2-}(aq) + 4H^+(aq) \longrightarrow 2MnO_4^-(aq) + MnO_2(s) + 2H_2O(l)$$

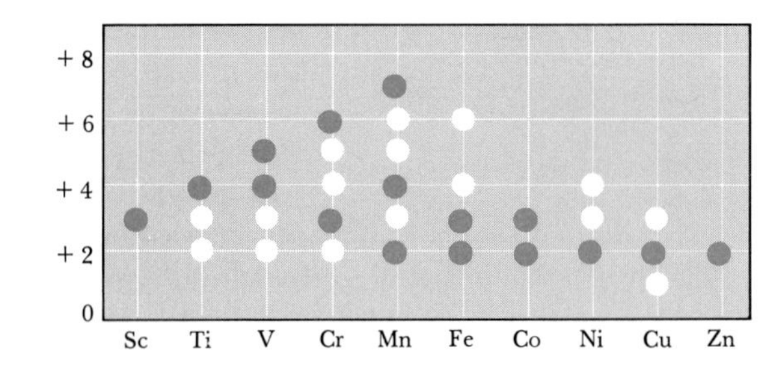

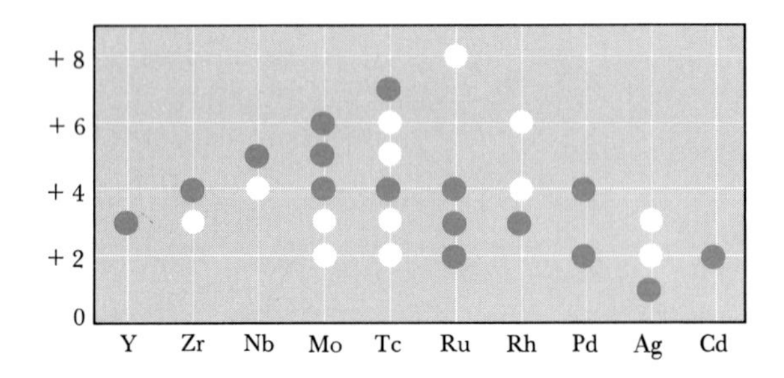

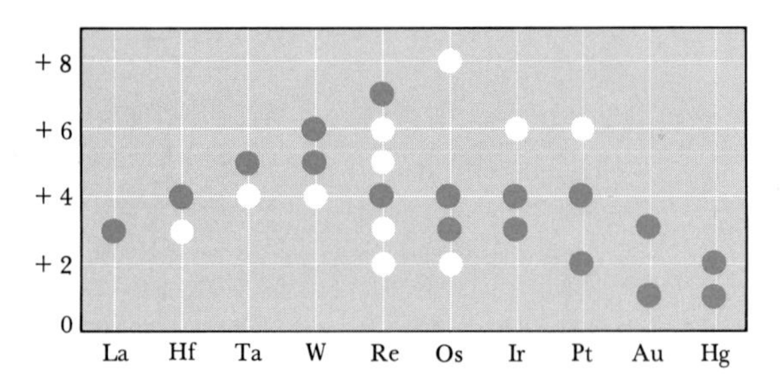

FIGURE 23.4 Oxidation states of the transition elements. The most common oxidation states are indicated by filled circles.

Inspection of Figure 23.4 reveals some interesting trends. Notice that the highest observed oxidation state increases with the number of valence-shell electrons up through group 7B. The +8 oxidation state is known for at least two elements of the next family (RuO_4 and OsO_4), but as we move to the right beyond this, the maximum observed oxidation state decreases. We can understand this general trend by recalling that as we move to the right in each transition element series the effective nuclear charge increases. It thus becomes increasingly difficult to remove electrons from the metal. Thus the element palladium forms no compounds in which it has an oxidation number higher than +4. Platinum, in the same family, exhibits the +6 oxidation state in only one compound, PtF_6, in which it is combined with the most electronegative of all the elements. In general, the highest oxidation states are to be found only when the metals are combined with the most electronegative elements: O, F, or possibly Cl. When the metals are in high oxidation states in such compounds the bonds to the other elements are not entirely ionic. Thus in PtF_6, for example, the Pt—F bonds are best thought of as polar covalent.

We often encounter the transition elements as ions in aqueous solution. This is especially true for the elements of the first transition series. Table 23.3 summarizes the common oxidation states and related electrode potentials for the first-series transition elements. This table summarizes a great deal of important chemical information about the first-series transition elements. In considering these data it is important to keep several factors in mind and to note certain important trends:

1 Except for copper, all the elements of the first transition series exhibit a negative standard electrode (reduction) potential for the +2 oxidation state. It is thus not surprising that, with rare exceptions, all these elements are found in nature in the combined state.

2 The metal ions are present in water as the aquated ions (Section 15.10). The metal ion acts as a Lewis acid toward the surrounding water molecules. When ions or molecules other than water are also present in solution, the metal ion may form a metal complex (Section 16.5). We will have a great deal more to say about these complexes in Chapter 24. For the moment you should just keep in mind that the electrode potentials shown in Table 23.3 can be greatly modified by complex formation. In general, the formation of a very stable metal complex may be a key factor in stabilizing a particular oxidation state

TABLE 23.3 Oxidation states in aqueous solution for the first-series transition elements

Element	Oxidation state	Reduction reaction	Standard electrode potential, E^0_{298}(V)
Sc	+3	$Sc^{3+}(aq) + 3e^- \longrightarrow Sc(s)$	−2.1
Ti	+2	$Ti^{2+}(aq) + 2e^- \longrightarrow Ti(s)$	−1.63
	+3	$Ti^{3+}(aq) + 3e^- \longrightarrow Ti(s)$	−0.37
V	+2	$V^{2+}(aq) + 2e^- \longrightarrow V(s)$	−1.2
	+3	$V^{3+}(aq) + e^- \longrightarrow V^{2+}(aq)$	−0.27
Cr	+2	$Cr^{2+}(aq) + 2e^- \longrightarrow Cr(s)$	−0.91
	+3	$Cr^{3+}(aq) + 3e^- \longrightarrow Cr(s)$	−0.74
	+6	$CrO_4^{2-}(aq) + 4H_2O(l) + 3e^- \longrightarrow Cr(OH)_3(s) + 5OH^-(aq)$	−0.13
	+6	$Cr_2O_7^{2-}(aq) + 14H^+(aq) + 6e^- \longrightarrow 2Cr^{3+}(aq) + 7H_2O(l)$	+1.33
Mn	+2	$Mn^{2+}(aq) + 2e^- \longrightarrow Mn(s)$	−1.18
	+7	$MnO_4^-(aq) + 8H^+(aq) + 5e^- \longrightarrow Mn^{2+}(aq) + 4H_2O(l)$	+1.51
	+7	$MnO_4^-(aq) + 2H_2O(l) + 3e^- \longrightarrow MnO_2(s) + 4OH^-(aq)$	+0.59
Fe	+2	$Fe^{2+}(aq) + 2e^- \longrightarrow Fe(s)$	−0.440
	+3	$Fe^{3+}(aq) + e^- \longrightarrow Fe^{2+}(aq)$	+0.771
Co	+2	$Co^{2+}(aq) + 2e^- \longrightarrow Co(s)$	−0.277
	+3	$Co^{3+}(aq) + e^- \longrightarrow Co^{2+}(aq)$	+1.842
Ni	+2	$Ni^{2+}(aq) + 2e^- \longrightarrow Ni(s)$	−0.28
Cu	+1	$Cu^+(aq) + e^- \longrightarrow Cu(s)$	+0.521
	+2	$Cu^{2+}(aq) + 2e^- \longrightarrow Cu(s)$	+0.337
Zn	+2	$Zn^{2+}(aq) + 2e^- \longrightarrow Zn(s)$	−0.763

for a metal ion. As an example, the Co^{3+} ion is not stable in water as the $Co(H_2O)_6{}^{3+}$ ion, but it is stable as the complex with ammonia, $Co(NH_3)_6{}^{3+}$:

$$Co(H_2O)_6{}^{3+}(aq) + e^- \longrightarrow Co(H_2O)_6{}^{2+}(aq) \qquad E° = +1.95\ V \qquad [23.1]$$

$$Co(NH_3)_6{}^{3+}(aq) + e^- \longrightarrow Co(NH_3)_6{}^{2+}(aq) \qquad E° = +0.1\ V \qquad [23.2]$$

3 For the simple aquated metal ions, the elements from the early part of the series tend to exhibit lower reduction potentials than do those at the other end of the series. For example, the standard potential for reduction of Ti^{3+} is negative. In fact, $Ti^{3+}(aq)$ is readily oxidized. [A standard potential for a titanium(IV) species is not shown in the table because the aquated form of this ion is not soluble.] By contrast, there is no comparably stable oxidation state of nickel higher than +2. The general shift toward lower oxidation states in aqueous solution as we move from left to right in a transition series is consistent with an increasing effective nuclear charge, as discussed earlier.
4 In very broad terms, the same trends in characteristic oxidation states apply as well to the elements of the second and third transition series. For example, rhenium, Re, forms a perrhenate anion, $ReO_4{}^-$, analogous to the permanganate anion, $MnO_4{}^-$, and cadmium exhibits the +2 oxidation state, as does zinc. However, there are many important differences in the chemical properties of the heavier metals, and only rather crude comparisons can be made.

Let's now consider briefly some of the chemistry of four elements from the first transition series. As you read this material you should look for evidence of trends that illustrate the generalizations outlined above.

23.3 SELECTED CHEMICAL PROPERTIES

Chromium

Chromium is obtained mainly from the ore *chromite,* $FeCr_2O_4$. You can think of this substance as a mixed oxide of formula $FeO \cdot Cr_2O_3$. About 60 percent of the chromium produced goes into forming chromium-based alloys, about 20 percent into chemical use (including electroplating), and most of the rest into furnace bricks and other refractory (high-temperature) materials.

Chromium dissolves slowly in dilute hydrochloric or sulfuric acid, liberating hydrogen and forming a blue solution of chromous ion:

$$Cr(s) + 2H^+(aq) \longrightarrow Cr^{2+}(aq) + H_2(g) \qquad [23.3]$$

Normally, however, you do not observe this blue color, because the chromium(II) ion is rapidly oxidized in air to the violet-colored chromium(III) ion:

$$4Cr^{2+}(aq) + O_2(g) + 4H^+(aq) \longrightarrow 4Cr^{3+}(aq) + 2H_2O(l) \qquad [23.4]$$

When hydrochloric acid solution is used, the solution appears green due to the formation of complex ions containing chloride coordinated to chromium, for example, $Cr(H_2O)_4Cl_2^+$.

Chromium is frequently encountered in aqueous solution in the +6 oxidation state. In basic solution the yellow chromate ion, CrO_4^{2-}, is the most stable. In acidic solution, the dichromate ion, $Cr_2O_7^{2-}$, is formed:

$$CrO_4^{2-}(aq) + H^+(aq) \rightleftharpoons HCrO_4^-(aq) \quad [23.5]$$

$$2HCrO_4^-(aq) \rightleftharpoons Cr_2O_7^{2-}(aq) + H_2O(l) \quad [23.6]$$

You may recognize the second of these reactions as a condensation reaction, in which water is split out from between two $HCrO_4^-$ ions. Similar reactions occur among the oxyanions of the nonmetallic and other metallic elements (Section 21.4). The equilibrium between the dichromate and chromate ions is readily observable, because CrO_4^{2-} is a bright yellow, and $Cr_2O_7^{2-}$ is a deep orange. Note from the electrode potential listed in Table 23.3 that dichromate ion in acidic solution is a strong oxidizing agent, as evidenced by the large, positive reduction potential. By contrast, chromate ion in basic solution is not a particularly strong oxidizing agent.

Chromium in trace quantities is an essential element in human nutrition. However, at the concentrations employed in laboratory work the element is highly toxic.

Iron

We have already discussed the metallurgy of iron in considerable detail in Section 22.3. Let's consider here some of the important aqueous solution chemistry of the metal. Iron exists in aqueous solution in either the +2 (ferrous) or +3 (ferric) oxidation state. It often appears in natural waters because of contact of these waters with deposits of $FeCO_3$ ($K_{sp} = 3.2 \times 10^{-11}$). Dissolved CO_2 in the water can lead to dissolution of the mineral:

$$FeCO_3(s) + CO_2(aq) + H_2O(l) \longrightarrow Fe^{2+}(aq) + 2HCO_3^-(aq) \quad [23.7]$$

The dissolved Fe^{2+}, together with Ca^{2+} and Mg^{2+}, contributes to water hardness (Section 12.6).

The standard reduction potentials given in Table 23.3 tell us much about the kind of chemical behavior we should expect to observe for iron. The potential for reduction from the +2 state to the metal is negative; on the other hand, the reduction from the +3 to the +2 state is positive. This tells us that iron should react with nonoxidizing acids such as dilute sulfuric acid or acetic acid to form $Fe^{2+}(aq)$, as indeed occurs. However, in the presence of air, $Fe^{2+}(aq)$ tends to oxidize to $Fe^{3+}(aq)$, as evidenced by the positive standard voltage for equation 23.8:

$$4Fe^{2+}(aq) + O_2(g) + 4H^+(aq) \longrightarrow 4Fe^{3+}(aq) + 2H_2O(l) \qquad E° = +0.46\ V \quad [23.8]$$

You may have seen instances in which water dripping from a faucet has left a brown stain in a sink. The brown color is due to insoluble iron(III) oxide, formed by oxidation of iron(II) present in the water:

$$4Fe^{2+}(aq) + 8HCO_3^-(aq) + O_2(g) \longrightarrow 2Fe_2O_3(s) + 8CO_2(g) + 4H_2O(l) \quad [23.9]$$

When iron metal reacts with an oxidizing acid such as warm dilute nitric acid, $Fe^{3+}(aq)$ is formed directly:

$$Fe(s) + NO_3^-(aq) + 4H^+(aq) \longrightarrow Fe^{3+}(aq) + NO(g) + 2H_2O(l) \quad [23.10]$$

In the +3 oxidation state, iron is soluble in acidic solution as the hydrated ion, $Fe(H_2O)_6^{3+}$. However, this ion hydrolyzes readily (Section 15.10):

$$Fe(H_2O)_6^{3+}(aq) \rightleftharpoons Fe(H_2O)_5(OH)^{2+}(aq) + H^+(aq) \quad [23.11]$$

As an acidic solution of iron(III) is made more basic, a gelatinous red-brown precipitate, most accurately described as a hydrous oxide, $Fe_2O_3 \cdot nH_2O$, is formed. In this formulation n represents an indefinite number of water molecules, depending on the precise conditions of the precipitation. Usually, the precipitate that forms is represented merely as $Fe(OH)_3$. The solubility of $Fe(OH)_3$ is very low ($K_{sp} = 4 \times 10^{-38}$). It dissolves in strongly acidic solution, but not in basic solution. The fact that it does *not* dissolve in basic solution is the basis of the Bayer process, in which aluminum is separated from impurities, primarily iron(III) (Sections 16.5 and 22.4).

Nickel

Nickel is not a particularly abundant element. It occurs chiefly in combination with arsenic, antimony, or sulfur. (You will note that both nickel and copper, which we will discuss next, are found mainly in combination with sulfur and related nonmetals, whereas the earlier transition elements such as chromium or iron are often found as oxide ores.) After separating nickel from other metallic elements that are usually present, the ore is roasted to NiO, which is then reduced with coke to form the crude metal. Crude nickel is normally refined by electrolysis, using nickel electrodes and an electrolyte containing both nickel sulfate and ammonium sulfate.

About 80 percent of the nickel produced goes into alloys. Among the more important of these is *stainless steel,* which also contains substantial amounts of chromium, and *Nichrome,* composed of 80 percent nickel and 20 percent chromium. Nichrome is used to make heating elements and in other applications that require a corrosion- and heat-resistant metal.

Nickel metal dissolves slowly in dilute mineral acids. Rather surprisingly, it does not dissolve in concentrated nitric acid. Attack of the strongly oxidizing acid on the metal surface results in formation of a protective oxide coat that stops further reaction. In its compounds nickel is almost always found in the +2 oxidation state. For example, anhydrous nickel(II) chloride, $NiCl_2$, is a yellow solid. It dissolves in water to form a green solution that contains the hydrated nickel(II) cation, $Ni(H_2O)_6^{2+}$. The Ni^{2+} ion forms complexes with many different anions and other neutral molecules such as NH_3; these are discussed in Chapter 24. Among the less soluble compounds of nickel are the carbonate, $NiCO_3$, the oxalate, NiC_2O_4, and the sulfide, NiS.

Copper

In Chapter 22 we discussed means by which copper can be produced from the ore chalcopyrite, $CuFeS_2$, and refined by electrolysis. In its aqueous solution chemistry, copper exhibits two oxidation states, +1 (cuprous) and +2 (cupric). Note that in the +1 oxidation state copper possesses a d^{10} electron configuration. Salts of Cu^+ are often water insoluble and are mostly white in color. In solution the Cu^+ ion readily disproportionates:

$$2Cu^+(aq) \longrightarrow Cu^{2+}(aq) + Cu(s) \qquad K = 1.2 \times 10^6 \qquad [23.12]$$

Because of this reaction, and because copper(I) is readily oxidized to copper(II) under most solution conditions, the +2 oxidation state is by far the more common.

Many salts of Cu^{2+} are water soluble, including $Cu(NO_3)_2$, $CuSO_4$, and $CuCl_2$. A widely used salt, copper sulfate pentahydrate, $CuSO_4 \cdot 5H_2O$, has four water molecules bound to the copper ion, and a fifth held to the SO_4^{2-} ion by hydrogen bonding. The salt is blue (it is called *blue vitriol*). Aqueous solutions of Cu^{2+}, in which the copper ion is coordinated by water molecules, are also blue. Among the insoluble compounds of copper(II) are $Cu(OH)_2$, which is formed when NaOH is added to an aqueous Cu^{2+} solution. This blue compound readily loses water on heating to form black copper(II) oxide:

$$Cu(OH)_2(s) \longrightarrow CuO(s) + H_2O(l) \qquad [23.13]$$

CuS is one of the least soluble copper(II) compounds ($K_{sp} = 6.3 \times 10^{-36}$). This black substance does not dissolve in NaOH, NH_3, or in nonoxidizing acids such as HCl. However, it does dissolve in HNO_3, which oxidizes the sulfide to sulfur.

SAMPLE EXERCISE 23.3

Write a balanced equation to represent the reaction of CuS with aqueous 3 *M* nitric acid. In the reaction sulfur is oxidized to the free element, and nitrate is reduced to NO.

Solution: The overall nonbalanced chemical equation is

$$CuS(s) + H^+(aq) + NO_3^-(aq) \longrightarrow Cu^{2+}(aq) + S(s) + NO(g) + H_2O(l)$$

Using the method of half-reactions (Section 19.1), we obtain

$$CuS(s) \longrightarrow Cu^{2+}(aq) + S(s) + 2e^-$$

$$NO_3^-(aq) + 3e^- + 4H^+(aq) \longrightarrow NO(g) + 2H_2O(l)$$

We balance in the usual way, by multiplying the first half-reaction by 3, the second by 2, and adding:

$$3CuS(s) + 2NO_3^-(aq) + 8H^+(aq) \longrightarrow 3Cu^{2+}(aq) + 3S(s) + 2NO(g) + 4H_2O(l)$$

$CuSO_4$ is often added to water to stop algae or fungal growth, and other copper preparations are used to spray or dust plants to protect them from lower organisms and insects. However, copper compounds are not generally toxic to human beings, except in massive quantities. Our daily diet normally includes from 2 to 5 mg of copper.

23.4 TRANSITION METAL OXIDES

The oxides of the transition elements are an interesting and very important group of compounds. That undesirable phenomenon known as corrosion is due primarily to formation of metal oxides. In the effort to inhibit corrosion, we are thus concerned with finding ways to inhibit oxide formation. On the other hand, many metal oxides have very useful properties and play important roles in chemical technology and electronics.

Table 23.4 shows the oxides formed by metals of the first transition series. By comparing the entries in this table with those in Figure 23.4, you can see that most of the oxidation states listed for the first-series transition elements are represented in the oxides. Notice in particular that for the elements through Mn, the highest oxidation state seen for each element is equal to the number of valence-shell electrons. The properties of the metal oxides reflect certain trends and regularities that it is useful to keep in mind:

1. For any particular element the oxide of highest oxidation state will act in general as an oxidizing agent. However, the oxidizing power increases steadily as the oxidation state of the metal increases. Thus, whereas Sc_2O_3 or TiO_2 are not good oxidizing agents at all, CrO_3 and Mn_2O_7 are powerful ones.
2. The acid-base behavior of a metal oxide depends on the oxidation state of the metal. In general, oxides in which the metal is in a low oxidation state are basic in character; those in which it is in a high oxidation state are acidic in character. Metal oxides are often insoluble in water, but their acidic or basic properties are revealed by whether they dissolve on reaction with acid or base. For example, FeO is insoluble in water or aqueous base, but dissolves in acidic solution, thus exhibiting basic character.
3. The acid-base characteristics just described are related to the fact that there is a shift from ionic to covalent character in the metal-oxygen bond as the oxidation state of the metal increases. Thus, whereas the bonding in MnO is essentially ionic, the bonding in Mn_2O_7 is polar covalent. This shift in bond character arises because, as the metal charge increases, its attraction for the electrons on the oxygen increases.

TABLE 23.4 Oxides of the first-series transition elements

Element									
Sc	**Ti**	**V**	**Cr**	**Mn**	**Fe**	**Co**	**Ni**	**Cu**	**Zn**
Sc_2O_3	TiO	VO	Cr_2O_3	MnO	FeO	CoO	NiO	Cu_2O	ZnO
	Ti_2O_3	V_2O_3	CrO_2	Mn_2O_3	Fe_3O_4	Co_3O_4	NiO_2	CuO	
	TiO_2	VO_2	CrO_3	MnO_2	Fe_2O_3	Co_2O_3			
		V_2O_5		Mn_2O_7		CoO_2			

SAMPLE EXERCISE 23.4

Predict the products that will occur upon reaction of (a) V_2O_5 with H_2O; (b) VO with a strong mineral acid. (Note: Assume that no oxidation-reduction reaction occurs.)

Solution: (a) V_2O_5 should be the anhydride of a strong acid. We note that the group 5B elements, like the group 5A elements, have five electrons in the orbitals beyond the nearest rare-gas element. By analogy with the behavior of group 5A elements, we would thus expect the acid formed by V_2O_5 to be H_3VO_4:

$$V_2O_5(s) + 3H_2O(l) \longrightarrow 2H_3VO_4(aq)$$

Here we have assumed that H_3VO_4 is a weak acid, in analogy with H_3PO_4. In fact, K_{a1} for H_3VO_4 is 1.7×10^{-4}.

(b) The oxide VO should be basic; that is, its reaction with acid should be analogous to that of CaO or other basic oxides, leading to the +2 aquated metal ion:

$$VO(s) + 2H^+(aq) \longrightarrow V^{2+}(aq) + H_2O(l)$$

The colors of the transition metal oxides are of considerable importance. Many oxides are employed as paint pigments, because they are very stable compounds that are able to withstand exposure to sunlight and air without undergoing chemical change. We have noted that the colors in compounds of the transition elements are usually due to the presence of electrons in incompletely filled *d* orbitals. In Chapter 24 we will take up a discussion of the electronic transitions that give rise to these colors. As you might expect, the oxides that possess completely empty 3*d* subshells (for example TiO_2 and ZnO) are white. In fact, both of these oxides are employed as white-paint pigments.

Very finely divided metal oxides are employed in high-grade paint finishes used in the automobile industry. When the particle size of the metal oxide pigment is on the order of 0.1 μm or smaller, the particles no longer efficiently scatter visible light rays. Thus, while they impart color to the lacquer in which they are mixed, they are transparent. A special form of Fe_2O_3 is widely used in this application. Small oxide particles are formed by precipitating $Fe(OH)_2$, then oxidizing it in air to form a substance of composition FeOOH. This is then dehydrated at high temperature to form small particles of Fe_2O_3. The iron oxide particles help prevent degradation of the lacquer by sunlight, because they strongly absorb the visible and ultraviolet rays. The brilliant metallic finish seen on many automobiles is achieved by mixing aluminum flakes about 10 to 50 μm in size into the lacquer.

The Structures of Metal Oxides

The structures of the metal oxides can best be considered in terms of some extensions of the idea of close packing, discussed in Section 11.5. You may recall from our discussion of ionic radii in Section 7.3 that anions generally have larger radii than do cations. In metal oxides and other ionic compounds, the large anions in the lattice often assume a close-packed arrangement identical to one of those discussed in Section 11.4. A close-packed array of spheres, whether hexagonal or cubic close packed, occupies 74 percent of the total volume taken up by the structure as a whole. The remaining 26 percent consists of holes between the spheres. The smaller cations in the lattice occupy these holes. For example, Fe_2O_3 consists of a hexagonal close-packed arrangement of oxide ions, with Fe^{3+} ions in the spaces between the oxide ions.

Examination of the close-packed structures reveals that they possess

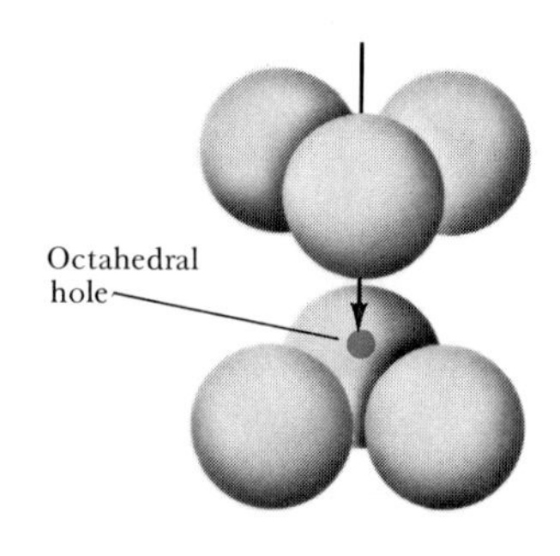

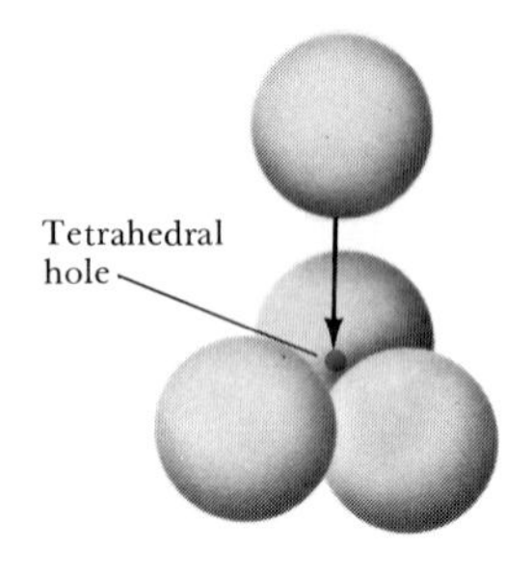

FIGURE 23.5 Views of the octahedral and tetrahedral holes that exist between two layers of close-packed spheres.

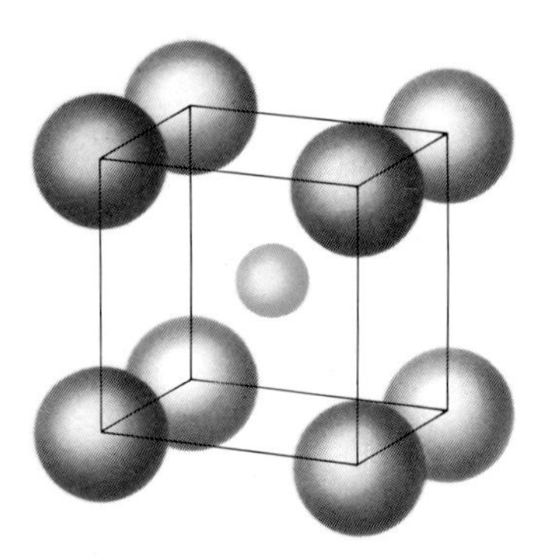

FIGURE 23.6 Eight-coordination of a cation in an anion lattice.

various types of holes. Two of these are shown in Figure 23.5. The first type is created by four large spheres arranged in a tetrahedral fashion. The second type is created by six large spheres arranged in an octahedral fashion. Using trigonometry, it can be shown that a **tetrahedral hole** is smaller than an **octahedral hole.** A small sphere whose radius is 0.225 times that of the larger sphere will fit perfectly into a tetrahedral hole. If the smaller sphere has a radius that is 0.414 times that of the larger spheres, it will just fit into an octahedral hole. The size of the cation relative to that of the anion helps to determine the type of hole that the cation occupies. The most stable arrangement is one that maximizes the number of cation-anion interactions, because these lead to electrostatic attractive forces. At the same time, the arrangement should be one that prevents direct anion-anion contacts, because these represent repulsive electrostatic forces. To illustrate, consider the case where a cation occupies a tetrahedral hole. If the ratio of cation radius to anion radius, r_c/r_a, is 0.225, the cation just touches all four anions, and they just touch one another. If the cation were a little larger, so that r_c/r_a were greater than 0.225, the cation would still be in touch with the four anions, but they would have been forced apart from one another, thus reducing the repulsive forces. In other words, this leads to a more stable arrangement. If we put that same cation in an octahedral hole, it would, in a sense, "rattle around" in the hole, because r_c/r_a is much less than 0.414. The anions would all be in contact, and a relatively less stable arrangement would result. For cations with radii such that $0.225 < r_c/r_a < 0.414$, therefore, a tetrahedral hole is normally the position of lowest energy.

When the ratio of radii, r_c/r_a, reaches a value of 0.414 or greater, the cation is no longer most stable in the tetrahedral hole. Instead, the most stable location is the octahedral hole, which permits the cation to contact a greater number of anions. When $r_c/r_a > 0.414$, the cation forces the anions apart, preventing direct anion-anion contact while maintaining the stabilizing cation-anion contacts. Thus for cation radii such that $0.414 < r_c/r_a < 0.732$, the most stable arrangement results when the cation is in an octahedral hole. When r_c/r_a exceeds 0.732, the anions are no longer most stable in a close-packed array. Instead, they assume a simple cubic arrangement, which permits the cation to be in contact with eight anions, as illustrated in Figure 23.6. The number of stabilizing cation-anion contacts is thus further increased. These radius ratio guidelines, which indicate the most likely environment for cations in an an-

TABLE 23.5 Radius ratios and cation location in an anion lattice

r_c/r_a[a]	Coordination number of cation	Arrangement of anions about the cation
0.225–0.414	4	Tetrahedral
0.414–0.732	6	Octahedral
0.732–1.000	8	Cubic

[a]If $r_c > r_a$, the ratio r_a/r_c gives the coordination number of the anion and the special arrangement of cations.

ionic lattice, are summarized in Table 23.5. The radius ratio rules are not always followed, because other factors, such as the electronic configurations of the metal ions and covalency in the bonding, also play a role. Nevertheless, they are useful guidelines to the structures that might be expected.

SAMPLE EXERCISE 23.5

The mineral hematite, Fe_2O_3, consists of a hexagonal close-packed array of oxide ions with Fe^{3+} ions occupying interstitial positions. Predict whether the Fe^{3+} ions are in octahedral or tetrahedral holes. The radius of Fe^{3+} is 0.65 Å; that of O^{2-} ion is 1.45 Å.

Solution: The radius ratio r_c/r_a is $0.65/1.45 = 0.45$. Based on the radius ratio, we would predict (Table 23.5) that the Fe^{3+} ions would be located in octahedral holes, as in fact they are.

One of the most common oxide structures is that adopted by *corundum*, a form of Al_2O_3. In this structure, the oxide anions form a hexagonal close-packed lattice, with the cation in octahedral holes. The oxides Ti_2O_3, V_2O_3, Cr_2O_3, Fe_2O_3, and Rh_2O_3 all adopt the corundum structure. Many other oxides adopt the rutile (TiO_2) structure, illustrated in Figure 7.2. The oxide ions in the rutile structure are not close packed. The metal ions are surrounded by six anions, but there is some distortion, so that two of the anions are located at a different distance than the other four. Among the transition metal oxides that adopt the rutile structure are MnO_2, ZrO_2, and RuO_2.

23.5 MAGNETISM IN TRANSITION ELEMENT COMPOUNDS

The magnetic properties of transition metal elements, which result from the presence of unpaired electrons in the valence orbitals of the metal atom or ion, are very important. Measurements of magnetic properties provide us with information about bonding in the metals and in their compounds. In addition, many important uses are made of magnetic properties in modern technology. In our earlier discussion of chemical bonding in Section 8.6, it was noted that substances that contain unpaired electons are **paramagnetic**; that is, they are drawn into a magnetic field. The electron is the source of magnetism. You will recall that the electron possesses a "spin" (Section 6.2). The presence of spin causes the electron to have a magnetic moment; that is, it behaves like a magnet on the microscopic scale. An atom or ion has a magnetic moment only when there are unpaired electrons in its valence-shell orbitals. For the transition and lanthanide elements, these unpaired electrons are in the *d* or *f* orbitals.

Magnetic behavior can be observed in isolated atoms or ions, as in the Stern-Gerlach experiment, illustrated in Figure 6.3. But what happens when atoms or ions that may individually be magnetic are present in a solid? The electrons that are unpaired in the isolated atom or ion may be involved in the bonding in the solid, and thus effectively paired up. If this occurs, the solid will not possess a magnetic moment. On the other hand, if there remain unpaired electrons on each atom or ion, the solid as a whole will exhibit some form of magnetism. It turns out that many different types of magnetic behavior are possible in solids. We will not be able to consider all the possible situations, some of which are very com-

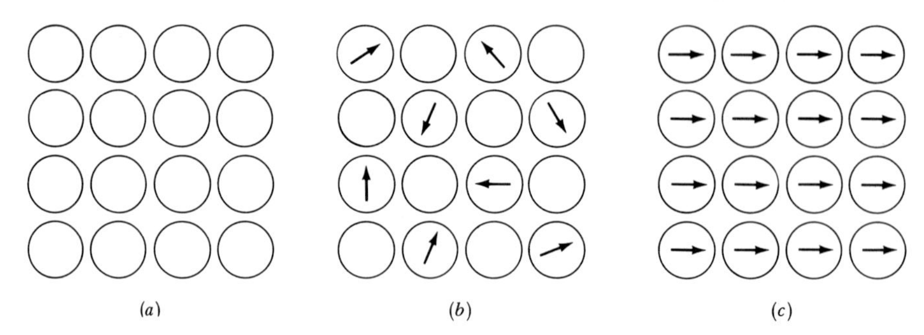

FIGURE 23.7 Types of magnetic behavior. (*a*) Diamagnetic; no centers with magnetic moments. (*b*) Simple paramagnetic; isolated centers with magnetic moments. (*c*) Ferromagnetic; coupled centers aligned in a common direction.

plex. However, the major classes of magnetic properties are readily undrstood on the basis of principles that we have already discussed in other contexts.

Diamagnetism

When all the electrons in a substance are paired, the substance exhibits diamagnetic behavior. In pairing, the magnetic moments of the individual electrons are effectively canceled. The motions of the electrons are perturbed by a magnetic field in such a way as to cause the substance to be repelled by a magnetic field. A diamagnetic lattice is illustrated schematically in Figure 23.7(*a*).

An experimental procedure for demonstrating diamagnetic behavior is illustrated in Figure 23.8. The sample to be studied is placed in a long, thin container and the mass of the assembly determined by weighing, as shown in Figure 23.8(a). Next a magnet is brought into place in such a manner that the sample will lie partly in the magnetic field. If the sample is diamagnetic, it will tend to move out of the field. Thus its apparent mass will decrease, as in Figure 23.8(*b*). Examples of diamagnetic substances include NaCl, SiO_2, copper metal, and most organic substances such as methane, CH_4, or benzene, C_6H_6.

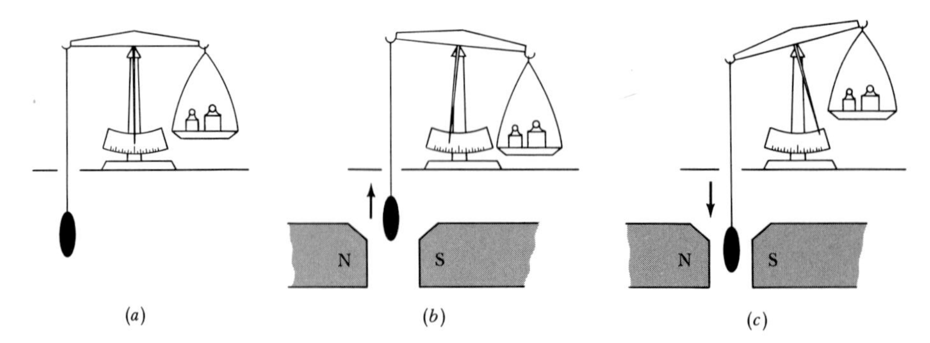

FIGURE 23.8 Experiment for determining the magnetic properties of a solid. The sample is first weighed in the absence of a magnetic field (*a*). When a field is applied, a diamagnetic sample tends to move out of the field (b), and thus appears to have a lower mass. A paramagnetic or ferromagnetic sample is drawn into the field (*c*), and thus appears to gain mass.

Simple Paramagnetism

If the solid contains transition metal ions in which there are unpaired electrons that do not feel the influence of the unpaired electrons at other like ions in the solid, the substance exhibits simple paramagnetism [Figure 23.7(b)]. A paramagnetic substance is drawn into a magnetic field, as shown in Figure 23 .8(*c*). As a result, when the magnetic field is applied, the mass of the sample appears to increase. A paramagnetic substance develops a magnetic moment in proportion to the applied field, and in proportion to the number of unpaired electrons. When the magnetic field is removed, the orientations of the magnetic moment on the individual metal ions again become randomly oriented with respect to one another. Thus the substance no longer possesses a magnetic moment.

In a substance that shows simple paramagnetic behavior the magnetic moment that develops in a magnetic field is proportional to the quantity $[j(j+2)]^{1/2}$, where j is the number of unpaired electrons at each paramagnetic center. We will encounter many examples of simple paramagnetic substances in Chapter 24 when we discuss the properties of coordination compounds of the transition elements.

Ferromagnetism

In a ferromagnetic solid the individual magnetic centers, which may be transition metal atoms or metal ions, interact with one another through the overlaps of atomic orbitals. By this means the electrons on a given center sense the orientations of electrons on neighboring centers. The most stable (lowest-energy) arrangement results when the spins of electrons on different centers are aligned in the same direction. When a ferromagnetic solid is placed in a magnetic field, as illustrated in Figure 23.8, the electrons tend to align strongly along the magnetic field. The increase in apparent mass that results may be as much as 1 million times larger than for a simple paramagnetic substance. When the external magnetic field is removed, the interactions between the electrons cause the solid as a whole to maintain a magnetic moment. Ferromagnetic behavior is illustrated schematically in Figure 23.7(*c*).

The most common examples of ferromagnetic materials are the elements Fe, Co, and Ni. You have perhaps done the experiment of inserting an iron nail into the core of a wire winding, and passing a direct current through for a time, as illustrated in Figure 23.9. When the current is turned off, the nail is seen to have developed a permanent magnetic moment, as evidenced by the manner in which it causes alignment of iron filings. We refer to it as a permanent magnet. Many alloy compositions exhibit ferromagnetism to a higher degree than for the pure metals themselves. Such alloys are employed in most of the applications for permanent magnets. Certain metal oxides (for example, CrO_2) also exhibit ferromagnetic behavior.

Several ferromagnetic metal oxides are used as recording media in magnetic recording tape. CrO_2, which is one of the most widely used oxides, is formed by reduction of CrO_3, or by a high-temperature reaction of two different chromium oxides with one another:

$$2CrO_3(s) \longrightarrow 2CrO_2(s) + O_2(g) \quad [23.14]$$

$$CrO_3(s) + Cr_2O_3(s) \longrightarrow 3CrO_2(s) \quad [23.15]$$

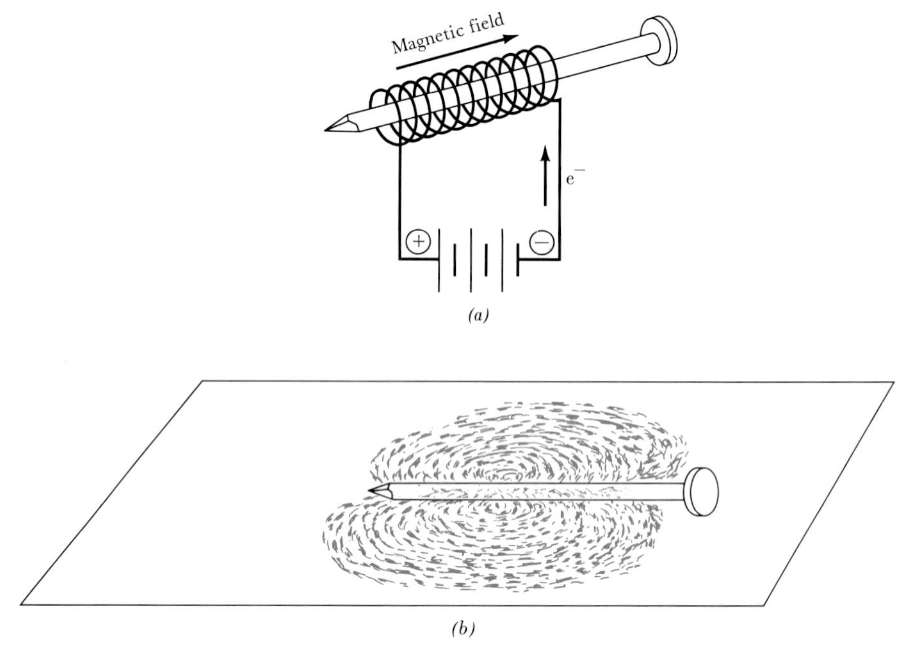

FIGURE 23.9 (*a*) Formation of a permanent magnet from an iron nail by the passage of direct current (*b*). The presence of permanent magnetism is seen in the alignment of iron filings when the nail is laid on a sheet of paper and the filings are allowed to align.

Other oxide materials used include a special crystallographic form of Fe_2O_3, and the mixed oxide, $CoFe_2O_4$. In magnetic recording it is important that the ferromagnetic material employed be in a very finely divided form, and that the particles be needle-shaped. A diagram of the recording process is shown in Figure 23.10. When an electrical current is passed through the coil, lines of magnetic field are established in the core. Where there is a gap, the field lines "leak" out and flow through the magnetic tape that passes in front of the gap. Variation in the magnetic field that is caused by modulation of the signal applied to the core causes a variation in the orientation and strength of magnetic moment imparted to the ferromagnetic particles in the tape.

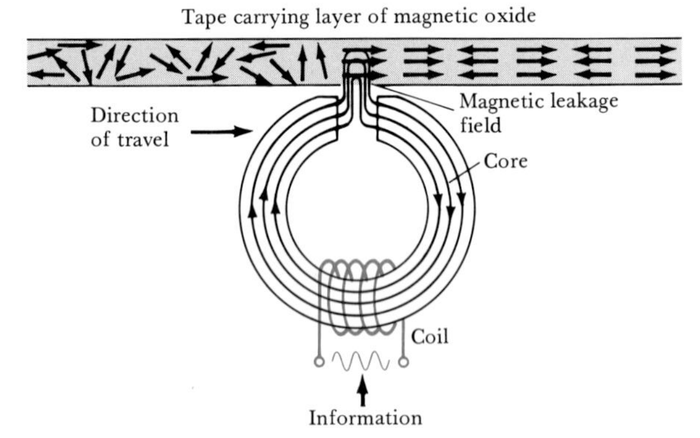

FIGURE 23.10 Schematic illustration of magnetic recording. As the tape is drawn past the gap in the core, magnetic field lines extending into the magnetic oxide in the tape cause an orientation of the magnetic moments within the domains of the oxide. The orientation and strength of the magnetic field varies as the information is supplied to the core in the form of a variable current.

23.6 METALS AS STRATEGIC MATERIALS

It is evident from the material we have discussed in this and the preceding chapter that a modern high-technology society depends critically on the availability of many metallic elements. It is important to the economic and military security of a highly industrialized nation that it have dependable access to the mineral sources from which these essential metals are derived. In recent years this has become a matter of deep concern

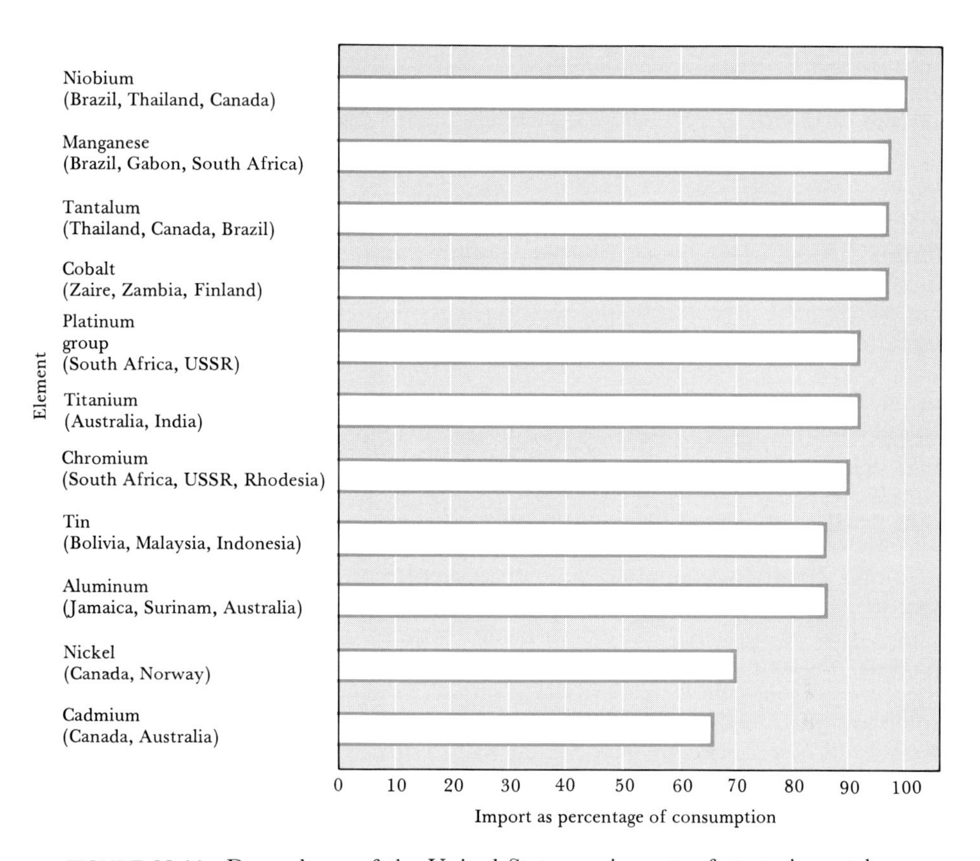

FIGURE 23.11 Dependence of the United States on imports of strategic metals.

in the United States, because in fact this country has no domestic sources of many of the most critical elements. Figure 23.11 shows the extent of our dependence on imports in terms of percentages of annual consumption that are imported. Note that all of the elements listed in Figure 23.1 as important components of a modern jet engine are on this list. Furthermore, the most important sources of some of the metals are not highly developed nations with which our diplomatic relations are secure.

In response to this situation the United States has developed a strategic minerals policy. A part of this policy involves establishment of stockpiles of the most critical metals, so that if all foreign sources were cut off the United States could function for an extended time by drawing on the stockpiled reserves. Second, more intensive efforts to develop new reserves within the United States, and to recycle old metal, have been initiated.

As part of the effort to locate new sources, increased attention has been paid to underseas resources. The **manganese nodules** that lie on the sea bottom should in time provide a plentiful source of manganese and cobalt. In addition, massive deposits of metallic sulfides have recently been discovered off the west coast of the United States and Mexico. The deposits occur at the site of an active rift in the earth's crust. The contact of ocean water with the molten magma produces high-temperature solutions containing dissolved metal salts and suspended materials. When these materials issue under pressure into the cooler ocean waters, they become supersaturated and deposition of solids results.

FOR REVIEW

Summary

In this chapter we have considered some of the more important general properties of the transition elements. We have also seen how those properties find useful application.

The transition elements are characterized by the incomplete filling of the *d* orbitals. As we proceed through a given series of transition elements (that is, through a horizontal row), the effective nuclear charge as seen by the valence electrons increases slowly. The result is that the later transition elements in a given row tend to adopt lower oxidation states and have slightly smaller ionic radii for a given oxidation state. In comparing the elements of a given group, that is, elements located vertically with respect to one another in the table, we find that the atomic and ionic radii increase in the second series as compared with the first, but that the elements of the second and third series are very similar with respect to these properties. The effect is due to the **lanthanide contraction.** The insertion of the lanthanide elements, atomic numbers 58 through 71, before the third transition series causes an increase in effective nuclear charge that compensates for the increase in major quantum number in the third series as compared with the second.

The presence of *d* electrons in the transition elements leads to the existence of multiple oxidation states. In the transition elements of groups 3B through 7B the maximum oxidation state observed corresponds to the total number of *s* and *d* electrons available in the valence-shell orbitals. Beyond group 7B the maximum observed oxidation state declines because of increasing effective nuclear charge.

Metal oxides are a very important class of transition metal compound. Because oxygen is a highly electronegative element, oxides of metals in high oxidation states are observed. Oxides of metals in high oxidation states, for example, V_2O_5 or Mn_2O_7, react with water to form acidic solutions. Oxides of metals in low oxidation states, for example, FeO, behave as bases in their reactions with water or with acidic solutions. The structures of solid metal oxides can be viewed as close-packed lattices of oxide ions, with the relatively smaller metal ions occupying the holes between oxide ions. The metal ion is most commonly found in either a tetrahedral hole, in which it has four oxide ion neighbors, or an octahedral hole, in which it has six. When the ratio of cation radius to anion radius, r_c/r_a, is in the range $0.225 < r_c/r_a < 0.414$, a tetrahedral hole is preferred on electrostatic energy grounds. When the ratio is in the range $0.414 < r_c/r_a < 0.732$, an octahedral hole is preferred. When r_c/r_a exceeds 0.732, an arrangement involving eight oxide ions about the metal ion is energetically preferred. This type of coordination requires a slight rearrangement of the oxide ions from the close-packed structure.

The presence of unpaired electrons in valence orbitals leads to interesting magnetic behavior in the transition elements and their compounds. Many substances can be classified in terms of one of the following magnetic types:

1. In a **diamagnetic** substance there are no unpaired electrons. The substance is repelled by a magnetic field.
2. In a simple **paramagnetic** substance, unpaired electrons on ions in a solid lattice or in solution are oriented by a magnetic field, so that the substance is drawn into the magnetic field. In a paramagnetic substance the unpaired spins on a given site do not interact significantly with those on other sites in the lattice or in solution.
3. In a **ferromagnetic** substance the unpaired spins on ions in the lattice are strongly affected by those on neighboring ions. In a magnetic field the orientations of spins in different domains become aligned along the magnetic field direction. When the magnetic field is removed, this common orientation remains, giving rise to a magnetic moment, as observed in permanent magnets.

Because so many uses are made of transition elements in modern industry, these elements are important for national security. For many of the elements, there are no significant deposits in the United States. For others, the deposits in other countries are more abundant, or provide the element in a more accessible form. To ensure an adequate supply of essential metals, the government has a strategic minerals policy, involving the establishment of stockpiles of critical metals and the search for alternative sources, particularly the discovery of new reserves within the United States.

Learning goals

Having read and studied this chapter, you should be able to:

1 Identify the transition elements on a periodic table.
2 Write the electronic configurations for the transition elements. (However, you need not memorize the special arrangements adopted by some of the elements that lack two electrons in the valence *s* orbital.)
3 Describe the manner in which atomic properties such as ionization energy and atomic radius vary within each transition element series, and within each vertical group of transition elements.
4 Explain why the atomic radii and other atomic properties of second- and third-series transition elements of the same group are closely similar. To put it another way, you should be able to describe the "lanthanide contraction" and explain its origin.
5 Describe the general manner in which the maximum observed oxidation state varies as a function of group number among the transition elements, and account for the observed variation.
6 Describe the general manner in which the most stable oxidation state or states in aqueous solution vary as a function of group number among the first-series transition elements.
7 Write balanced chemical equations corresponding to the simple aqueous solution chemistry of chromium, iron, nickel, and copper, and account for the variations in chemical properties observed among these elements in terms of the characteristics of the elements themselves.
8 Describe the manner in which the highest oxidation state of the first-series transition metal oxides varies as a function of atomic number, and account for the observed variation.
9 Describe and explain the manner in which oxidizing strength, acid-base characteristics, and metal-oxygen bond character vary with the oxidation state of the metal.
10 Describe the factors that determine the most stable coordination environment for a metal ion in an oxide lattice.
11 Explain how the most stable coordination number for a metal ion in an oxide lattice varies with the ratio of cation to anion radii, r_c/r_a.
12 Name and describe the types of magnetic behavior discussed in this chapter.
13 Provide an explanation of each type of magnetic behavior in terms of the arrangements in the lattice of metal ions with unpaired spins, and their interactions with one another.
14 Describe an experimental method for determining the magnetic moment of a solid sample.
15 Explain what is meant by the term *strategic metal.*

Key terms

Among the more important terms and expressions used for the first time in this chapter are the following:

The term **diamagnetism** (Section 23.5) refers to the behavior of a substance with no unpaired electrons when it is placed in a magnetic field. Because of the manner in which the magnetic field perturbs the motions of the electrons in orbitals, there is a repulsion of the magnetic field lines by a diamagnetic substance.

A substance exhibits **ferromagnetism** (Section 23.5) when it possesses unpaired electrons on metal atoms or ions in the solid lattice, and when the magnetic moments on each metal ion are aligned with one another.

The **lanthanide contraction** (Section 23.1) refers to the gradual decrease in atomic and ionic radii with increasing atomic number among the lanthanide elements, atomic numbers 58 through 71. The decrease arises because of a gradual increase in effective nuclear charge as the nuclear charge increases and electrons are concurrently added to the $4f$ orbitals. The $4f$ electrons do not completely shield the $6s$ or $5d$ electrons from the added nuclear charge.

An **octahedral hole** (Section 23.4) in a close-packed lattice is an opening in the lattice formed by six lattice units in contact with one another. The radius of the largest sphere that can be placed in this hole without forcing the larger spheres apart is 0.414 times as large as the radius of the larger spheres.

Simple **paramagnetism** (Section 23.5) is observed when a substance possesses unpaired electrons on metal ions in a lattice or in solution, and when there is no appreciable interaction between the magnetic moments on different ions. A simple paramagnetic substance is weakly attracted into a magnetic field, because of the alignment of the magnetic moments on the metal ions within the magnetic field.

A **tetrahedral hole** (Section 23.4) in a close-packed lattice is an opening in the lattice formed by four lattice units in contact with one another. The radius of the largest sphere that can be placed in this hole without forcing the larger lattice spheres apart is 0.225 as large as the radius of the larger spheres.

EXERCISES

Physical properties of transition elements

23.1 Using only the periodic table for guidance, write the electronic configurations for (a) Mn; (b) Zr; (c) Ag; (d) Re.

23.2 What can you deduce regarding the relative energies of the 5*s* and 4*d* orbitals by comparing the electron configurations of Zr, Nb, Mo, and Tc?

23.3 How would you expect the heat of fusion for vanadium to compare with the heat of vaporization? Explain.

23.4 Which of the following properties is most closely related to bonding in the metal, and which is characteristic of individual atoms: (a) heat of sublimation; (b) boiling point; (c) first ionization energy; (d) malleability; (e) electrode potential for $M(s) \longrightarrow M^{2+}(aq) + 2e^-$?

23.5 What is the order of energies of the following atomic orbitals in Re: (a) 4*f*; (b) 6*s*; (c) 5*d*; (d) 5*f*; (e) 5*p*? List the lowest-energy orbital first, the highest-energy orbital last.

[23.6] The atomic radii of the later elements in each transition series (Figure 23.3) are slightly larger than the lowest value observed in each series. This effect is related to the buildup of filled *d* orbitals on the metal. Explain how this might cause an increase in covalent radius.

Oxidation states

23.7 Write the formula for the chloride corresponding to the highest expected oxidation state for each of the following elements: (a) Nb; (b) W; (c) Co; (d) Cd; (e) Re; (f)Hf.

23.8 What accounts for the fact that chromium exhibits several oxidation states in its compounds, whereas aluminum exhibits only the +3 oxidation state?

23.9 Write the expected electron configuration for each of the following ions: (a) Ru^{4+}; (b) Hf^{3+}; (c) Ag^{3+}; (d) Co^{3+}; (e) V^{3+}; (f) Pt^{2+}.

23.10 Write balanced chemical equations for each of the following verbal descriptions. (a) Vanadium(V) oxide is formed by thermal decomposition of ammonium metavanadate, NH_4VO_3. (b) Vanadium(V) oxide is reduced to vanadium(IV) oxide by thermal reduction with gaseous SO_2. (c) Vanadium(V) oxide is reduced by magnesium in hydrochloric acid to give a green solution of vanadium(III). (d) Niobium(V) chloride reacts with water to yield crystals of niobic acid, $HNbO_3$.

[23.11] Graph the electrode potentials for the process $M^{2+}(aq) + 2e^- \longrightarrow M(s)$ for the first-series transition elements (Table 23.3). What trend is discernible in these data? How do you account for the observed variation?

23.12 (a) What effect does hydration of the metal ion have on the electrode potential for the process $M^{2+}(aq) + 2e^- \longrightarrow M(s)$? (b) How would formation of a very stable metal complex in solution affect the value of the potential for reduction of the metal ion to the metal?

Chemical properties

23.13 How does the presence of air affect the relative stabilities of Cr^{2+} and Cr^{3+} in aqueous solution? Write a balanced chemical reaction as part of your response.

23.14 When a solution of $Cu^{2+}(aq)$ is reacted with aqueous sodium iodide, CuI and elemental iodine are formed. Write a balanced chemical equation for the reaction, and explain why it proceeds as it does.

23.15 When H_2S is bubbled through a solution containing $Fe^{3+}(aq)$, the precipitate that eventually forms is a mixture of iron(II) sulfide and sulfur. Write a balanced chemical equation for the reaction, and explain why it proceeds as it does.

23.16 The hydrolysis constant for $Fe(H_2O)_6{}^{3+}$ is 8.9×10^{-4}. Calculate the pH of a 0.010 *M* solution of $Fe(NO_3)_3$.

23.17 Chromium(III) ion is amphoteric. Write equations to describe the reactions of Cr_2O_3 with sulfuric acid solution and with aqueous sodium hydroxide.

23.18 The vanadate ion, $VO_4{}^{3-}(aq)$, exists as such in strongly alkaline solution, pH > 13. As acid is added, a divanadate ion, and ions containing still more vanadium atoms, are formed. Write balanced chemical equations to describe the formation of the divanadate and trivanadate ions.

23.19 Which of the following compounds would be expected to yield the most acidic aqueous solution: (a) VCl_3; (b) $FeSO_4$; (c) CrO_3; (d) $Co(NO_3)_2$?

23.20 Write balanced half-reactions to represent the electrochemical refining of nickel.

Metal oxides

23.21 Predict whether each of the following oxides will exhibit predominantly acidic or basic properties: (a) NiO; (b) CrO_3; (c) Cu_2O; (d) Re_2O_7.

23.22 Write balanced chemical equations for each of the following descriptions. (a) MnO_2 is formed when $Mn(NO_3)_2$ is heated. (b) Mn_3O_4 is formed when Mn metal is heated in air. (c) Rhenium metal reacts with nitric acid to form an aqueous solution of the strong acid $HReO_4$. (d) Technetium(VII) oxide decomposes on heating to form technetium(IV) oxide. (e) Cobalt(II) oxide is converted to Co_3O_4 on heating in oxygen at 700 K. (f) Silver(I) oxide is oxidized by aqueous $S_2O_8{}^{2-}$ to AgO.

23.23 Predict the type of hole in an oxide lattice that would be occupied by each of the following cations (radii in parentheses): (a) Li^+ (0.68 Å); (b) Ti^{4+} (0.53 Å); (c) Ni^{2+} (0.70 Å); (d) Ca^{2+} (0.94 Å).

23.24 From the ionic radii of each of the following metal ions, would you expect that the metal oxide would possess the corundum structure, as is found to be the case: V^{3+}, 0.74 A; Ti^{3+}, 0.77 A; Rh^{3+}, 0.75 A? Explain.

Magnetic properties

23.25 What characteristics of a ferromagnetic material distinguish it from one that is paramagnetic? What type of interaction must occur in the solid to bring about ferromagnetic behavior?

[23.26] (a) What characteristics of CrO_2 render it useful as a magnetic tape recording medium? (b) Would a paramagnetic substance serve as a recording medium? Explain.

Additional exercises

[23.27] The metals cobalt, chromium, manganese, platinum, and titanium receive most attention as strategic metals. By using references other than this text (for example, *Advanced Inorganic Chemistry,* F. A. Cotton and G. Wilkinson, 4th ed., John Wiley & Sons, Inc., New York, 1980; *Encyclopedia of Chemical Technology* (Kirk-Othmer), 3rd ed. (also published by Wiley); various issues of *Science* and *Chemical and Engineering News* during the period 1980 to present), describe for any three of these metals the principal mineral sources and their locations, annual U.S. consumption, and major uses.

23.28 The radius of the sulfide ion is 1.90 Å. Assuming that the sulfides can be viewed as close-packed structures of S^{2-} ions, predict the type of hole that each of the following cations will occupy: (a) Zn^{2+} (0.74 Å); (b) V^{5+} (0.54 Å); (c) Tl^{3+}(0.88 Å).

23.29 If cryolite, Na_3AlF_6, is viewed as a close-packed structure of fluoride ions, what types of interstitial positions do you predict the Na^+ and Al^{3+} ions will occupy (radii: F^-, 1.33 Å, Na^+, 0.98 Å; Al^{3+}, 0.45 Å)?

23.30 Low-valent metal oxides generally dissolve in water only under conditions that lead to the aquated metal ion. High-valent metal oxides, on the other hand, generally dissolve to form oxygen-containing species, for example, VO_4^{3-} or CrO_4^{2-}. Account for this difference in behavior.

23.31 Which compound in each of the following pairs is likely to be the stronger oxidizing agent: (a) TiO_2 or Mn_2O_7; (b) Cr_2O_3 or CrO_3; (c) $FeCl_2$ or $FeCl_3$; (d) NiO_2 or HfO_2; (e) Cu_2O or Ag_2O? Explain in each case.

[23.32] Using the data in Table 23.3, determine the standard electrode potentials for each of the following half-reactions:

(a) $Cr^{3+}(aq) + e^- \longrightarrow Cr^{2+}(aq)$

(b) $Cr_2O_7^{2-}(aq) + 12\,e^- + 14H^+(aq) \longrightarrow 2Cr(s) + 7H_2O(l)$

(c) $Co^{3+}(aq) + 3e^- \longrightarrow Co(s)$

23.33 The statement has been made that the chemistries of zirconium and hafnium are more nearly identical than for any other two elements of a given group. What accounts for this similarity?

23.34 Explain why the heat of atomization of zinc is much smaller than the heat of atomization of vanadium (Table 23.2).

23.35 Using the data of Table 23.3, calculate the equilibrium constant for each of the following reactions:

(a) $Ti^{2+}(aq) + Co(s) \rightleftharpoons Ti(s) + Co^{2+}(aq)$

(b) $Fe(s) + 2Fe^{3+}(aq) \rightleftharpoons 3Fe^{2+}(aq)$

(c) $Cr_2O_7^{2-}(aq) + 14H^+(aq) + 6V^{2+}(aq) \rightleftharpoons 2Cr^{3+}(aq) + 6V^{3+}(aq) + 7H_2O(l)$

23.36 Discuss the trend in most stable oxidation state in solution as a function of atomic number for the four elements discussed in Section 23.3, and account for any trend observed.

24 Chemistry of Coordination Compounds

In earlier chapters we noted that metallic elements are characterized by a tendency to lose electrons in their chemical reactions. For this reason, positively charged metal ions play a primary role in the chemical behavior of metals. Of course, metal ions do not exist in isolation. In the first place they are accompanied by anions that serve to maintain charge balance. In addition, metal ions act as Lewis acids (Section 15.10). Neutral molecules or anions with unshared pairs of electrons may be bound to the metal center. On several occasions we have discussed compounds in which a metal ion is surrounded by a group of anions or neutral molecules. Examples include $[Au(CN)_2]^-$, discussed in connection with metallurgy in Section 22.4; hemoglobin, discussed in connection with the oxygen-carrying capacities of the blood in Section 10.5; $[Cu(CN)_4]^{2-}$ and $[Ag(NH_3)_2]^+$, encountered in our discussion of equilibria in Section 16.5. Such species are known as **complex ions** or merely **complexes.** Compounds containing them are called **coordination compounds.**

24.1 THE STRUCTURE OF COMPLEXES

The molecules or ions that surround a metal ion in a complex are known as **ligands** (from the Latin word *ligare,* meaning "to bind"). Ligands are normally either anions or polar molecules. Furthermore, they have at least one unshared pair of valence electrons, as illustrated in the following examples:

$$:\ddot{O}(H)—H \qquad :N(H)(H)—H \qquad :\ddot{Cl}:^- \qquad :C\equiv N:^-$$

We will see that for most purposes it is adequate to think of the bonding between a metal ion and its ligands as an electrostatic interaction between the positive cation and the surrounding negative ions or dipoles oriented with their negative ends toward the metal ion. The ability of metal ions to form complexes normally increases as the positive charge of the cation increases and its size decreases. The weakest complexes are formed by the alkali metal ions such as Na^+ and K^+. On the other hand, the +2 and +3 ions of the transition elements generally excel in complex formation. In fact, many of the transition metal ions form complexes more readily than their charge and size would suggest. For example, on the basis of size alone we might expect that Al^{3+} ($r = 0.45$ Å) would form complexes more readily than the larger Cr^{3+} ion ($r = 0.62$ Å). However, with most ligands Cr^{3+} forms much more stable complexes than does Al^{3+}. Thus the bonding in these complexes cannot be explained entirely on the basis of electrostatic attraction between metal ion and ligands. To account for some of the observed differences in complexes, we must assume that there is some degree of covalent character in the metal-ligand bond. Because metal ions have empty valence orbitals, they can act as Lewis acids (electron-pair acceptors). Because ligands have unshared pairs of electrons, they can function as Lewis bases (electron-pair donors). The bond between metal and ligand can thus be pictured as resulting from a sharing of a pair of electrons that was initially on the ligand:

$$Ag^+(aq) + 2:NH_3(aq) \longrightarrow [H_3N:Ag:NH_3]^+(aq) \qquad [24.1]$$

We shall examine the bonding in complexes more closely in Section 24.8.

In forming a complex, the ligands are said to coordinate to the metal or to complex the metal. The central metal and the ligands bound to it constitute the **coordination sphere.** In writing the chemical formula for a coordination compound, we use square brackets to set off the groups within the coordination sphere from other parts of the compound. For example, the formula $[Cu(NH_3)_4]SO_4$ represents a coordination compound consisting of the $[Cu(NH_3)_4]^{2+}$ ion and the SO_4^{2-} ion. The four ammonia groups in the compound are bound directly to the copper(II) ion.

As you might expect, a metal complex has different properties than either the metal ion or ligands from which it is derived. We are normally most interested in the effects of complex formation on the properties of the metal ion. The ease of oxidation or reduction of the metal ion may be drastically changed by complex formation. For example, Ag^+ is readily reduced in water:

$$Ag^+(aq) + e^- \longrightarrow Ag(s) \qquad E° = +0.799\ V \qquad [24.2]$$

By contrast, the $[Ag(CN)_2]^-$ ion is not at all readily reduced, because formation of the cyano complex stabilizes the higher oxidation state:

$$[Ag(CN)_2]^-(aq) + e^- \longrightarrow Ag(s) + 2CN^-(aq) \qquad E° = -0.31\ V \qquad [24.3]$$

Solubility properties may be dramatically different. For example, $CrBr_3$ is insoluble in water, but the complex $[Cr(H_2O)_6]Br_3$ is very soluble. Colors may be different; $CrBr_3$ is a deep green-black, whereas $[Cr(H_2O)_6]Br_3$ is violet. These differences in properties occur because the metal ion plus its surrounding ligands is a distinct species in its own right, and has characteristic physical and chemical properties.

The charge of a complex is the sum of the charges on the central metal and on its surrounding ligands. In $[Cu(NH_3)_4]SO_4$ we can deduce the charge on the complex if we first recognize SO_4 as being the sulfate ion and therefore having a -2 charge. Because the compound is neutral, the complex ion must have a $+2$ charge, $[Cu(NH_3)_4]^{2+}$. The oxidation number of the copper must be $+2$ because the NH_3 groups are neutral:

$$+2 + 4(0) = +2$$

$$[Cu(NH_3)_4]^{2+}$$

SAMPLE EXERCISE 24.1

What is the oxidation number of the central metal in $[Co(NH_3)_5Cl](NO_3)_2$?

Solution: The NO_3 group is the nitrate anion and has a -1 charge, NO_3^-. The NH_3 ligands are neutral; the Cl is a coordinated chloride and therefore has a -1 charge. The sum of all the charges must be zero:

$$x + 5(0) + (-1) + 2(-1) = 0$$

$$[Co(NH_3)_5Cl](NO_3)_2$$

The charge on the cobalt, x, must therefore be $+3$.

SAMPLE EXERCISE 24.2

Given that a complex ion contains a chromium(III) bound to four water molecules and two chloride ions, write its formula.

Solution: The metal has a $+3$ oxidation number, water is neutral, and chloride has a -1 charge:

$$+3 + 4(0) + 2(-1) = +1$$

$$Cr(H_2O)_4Cl_2$$

Therefore, the charge on the ion is $+1$, $[Cr(H_2O)_4Cl_2]^+$.

Coordination Numbers and Geometries

The ligand atom that is bound directly to the metal is known as the **donor atom.** For example, nitrogen is the donor atom in the $[Ag(NH_3)_2]^+$ complex shown in Equation 24.1. The number of donor atoms attached to a metal is known as its **coordination number.** In $[Ag(NH_3)_2]^+$, silver has a coordination number of 2; in $[Cr(H_2O)_4Cl_2]^+$, chromium has a coordination number of 6.

Some metal ions exhibit constant coordination numbers. For example, the coordination number of chromium(III) and cobalt(III) is invariably

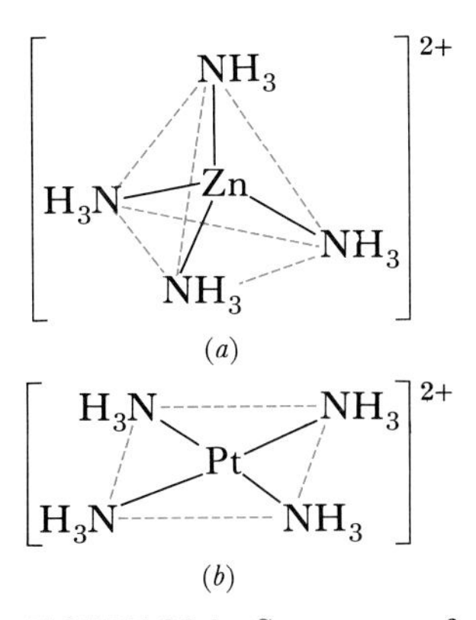

FIGURE 24.1 Structures of (a) $[Zn(NH_3)_4]^{2+}$ and (b) $[Pt(NH_3)_4]^{2+}$, illustrating the tetrahedral and square-planar geometries, respectively. These are the two common geometries for complexes in which the metal ion has a coordination number of four. The dashed lines shown in the figure are not bonds; they are included merely to assist in visualizing the shape of the metal complex.

6, whereas that of platinum(II) is always 4. However, the coordination numbers of most metal ions vary with the ligand. The most common coordination numbers are 4 and 6.

The coordination number of a metal ion is often influenced by the relative sizes of the metal ion and the surrounding ligands. As the ligand gets larger, fewer can coordinate to the metal. This explains why iron is able to coordinate to six fluorides in $[FeF_6]^{3-}$, but to only four chlorides in $[FeCl_4]^{-}$. Ligands that transfer substantial negative charge to the metal also produce reduced coordination numbers. For example, six neutral ammonia molecules can coordinate to nickel(II), forming $[Ni(NH_3)_6]^{2+}$; however, only four negatively charged cyanide ions coordinate, forming $[Ni(CN)_4]^{2-}$.

Four-coordinate complexes have two common geometries, tetrahedral and square planar, as shown in Figure 24.1. The tetrahedral geometry is the more common of the two and is especially common among the nontransition metals. The square-planar geometry is characteristic of transition metal ions with eight *d* electrons in the valence shell, for example, platinum(II) and gold(III); it is also found in some copper(II) complexes.

Six-coordinate complexes have an octahedral geometry, as shown in Figure 24.2(*a*). Notice that the octahedron can be represented as a planar square with ligands above and below the plane, as in Figure 24.2(*b*). In this representation, the ligands along the vertical axis are geometrically equivalent to those in the plane.

24.2 CHELATES

The ligands that we have discussed so far, such as NH_3 and Cl^-, are known as monodentate ligands (from the Latin meaning "one-toothed"). These ligands possess a single donor atom. Some ligands have two or more donor atoms situated so that they can simultaneously coordinate to a metal ion. They are called polydentate ligands ("many-toothed"); because they appear to grasp the metal between two or more donor atoms, they are also known as chelating agents (from the Greek word *chele,* "claw"). One such ligand is ethylenediamine:

$$\ddot{N}H_2{-}CH_2{-}CH_2{-}\ddot{N}H_2$$

This ligand, which is abbreviated en, has two nitrogen atoms (shown in color) that have unshared pairs of electrons. These donor atoms are sufficiently far apart that the ligand can wrap around a metal ion with the two nitrogen atoms simultaneously complexing to the metal in adjacent

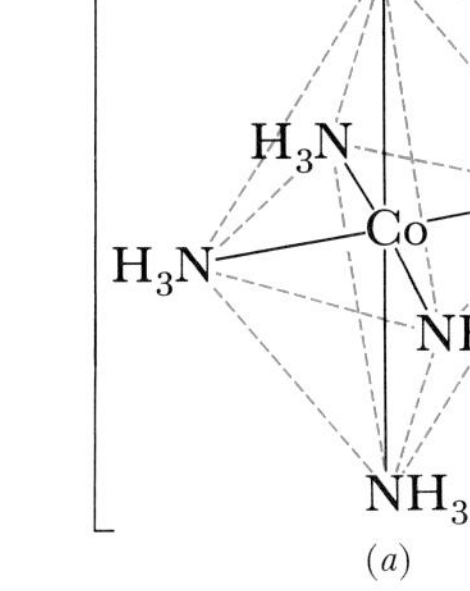

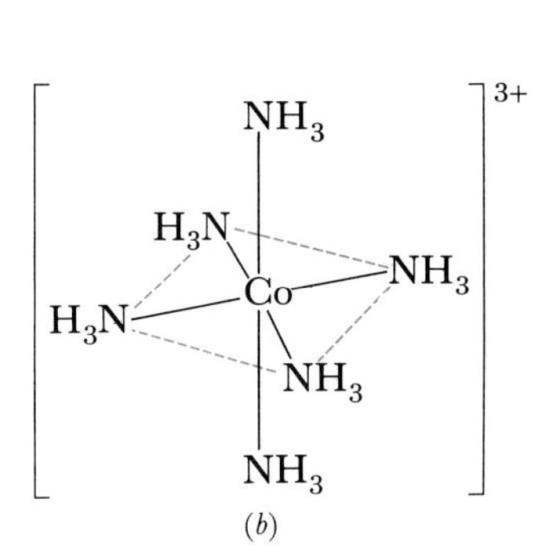

FIGURE 24.2 Two representations of an octahedral coordination sphere, the common geometric arrangement for complexes in which the metal ion has a coordination number of 6.

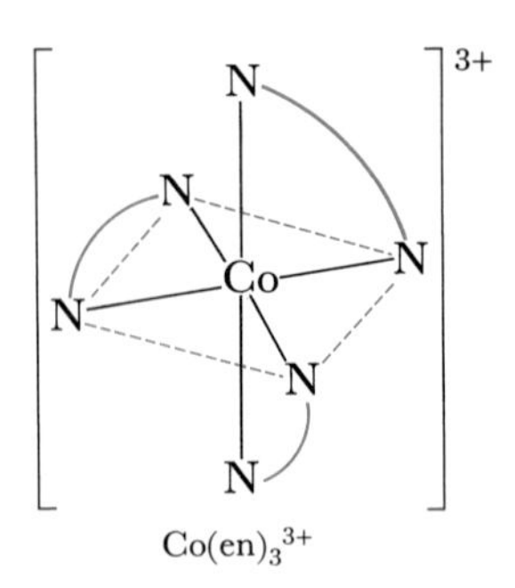

$Co(en)_3{}^{3+}$

FIGURE 24.3 $[Co(en)_3]^{3+}$ ion showing how each bidentate ethylenediamine ligand is able to occupy two positions in the coordination sphere.

positions. The $[Co(en)_3]^{3+}$ ion, which contains three ethylenediamine ligands bound to the octahedral coordination sphere of cobalt(III), is shown in Figure 24.3. Notice that the ethylenediamine has been written in a shorthand notation as two nitrogen atoms connected by a line.

Ethylenediamine is an example of a **bidentate ligand.** Other common bidentate ligands include oxalate, $C_2O_4{}^{2-}$, and carbonate, $CO_3{}^{2-}$ (the donor atoms are shown in color):

Another common polydentate ligand is the ethylenediaminetetraacetate ion:

$[EDTA]^{4-}$

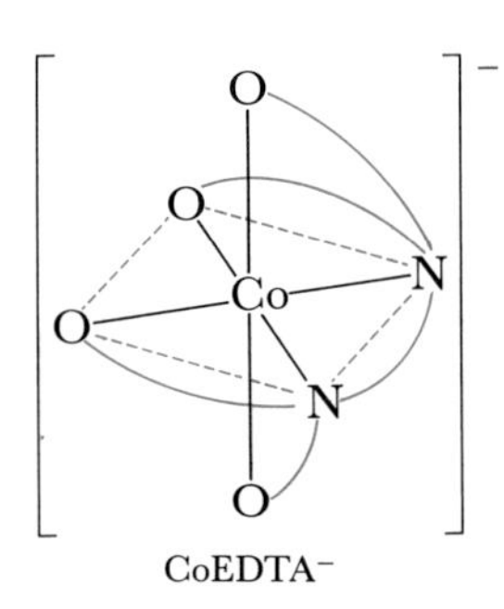

$CoEDTA^{-}$

FIGURE 24.4 $CoEDTA^{-}$ ion, showing how the ethylenediaminetetraacetate ion is able to wrap around a metal ion, occupying six positions in the coordination sphere.

This ion, abbreviated $EDTA^{4-}$, has six donor atoms. It can wrap around a metal ion using all six of these donor atoms as shown in Figure 24.4.

In general, chelating agents form more stable complexes than do related monodentate ligands. This is illustrated by the formation constants for $[Ni(NH_3)_6]^{2+}$ and $[Ni(en)_3]^{2+}$, shown in Equations 24.4 and 24.5.

$$[Ni(H_2O)_6]^{2+}(aq) + 6NH_3(aq) \rightleftharpoons [Ni(NH_3)_6]^{2+}(aq) + 6H_2O(l) \qquad K_f = 4 \times 10^8 \qquad [24.4]$$

$$[Ni(H_2O)_6]^{2+}(aq) + 3en(aq) \rightleftharpoons [Ni(en)_3]^{2+}(aq) + 6H_2O(l) \qquad K_f = 2 \times 10^{18} \qquad [24.5]$$

Although the donor atom is nitrogen in both instances, $[Ni(en)_3]^{2+}$ has a stability constant nearly 10^{10} times larger than $[Ni(NH_3)_6]^{2+}$.

The generally larger formation constants for polydentate ligands as compared with the corresponding monodentate ligands is known as the **chelate effect.** Thermochemical studies of complex formation in aqueous solution show that in nearly all cases *the chelate effect is due to a more favorable entropy change for complex formation involving polydentate ligands.* As an example, let's compare the thermodynamic data at 25°C for formation of two closely related complexes of Cd^{2+}:

$$Cd^{2+}(aq) + 4CH_3NH_2(aq) \rightleftharpoons [Cd(CH_3NH_2)_4]^{2+}(aq)$$
$$\Delta G° = -37.2\ kJ; \quad \Delta H° = -57.3\ kJ; \quad \Delta S° = -67.3\ J/K$$

$$Cd^{2+}(aq) + 2H_2NCH_2CH_2NH_2(aq) \rightleftharpoons [Cd(H_2NCH_2CH_2NH_2)_2]^{2+}(aq)$$
$$\Delta G° = -60.7\ kJ; \quad \Delta H° = -56.5\ kJ; \quad \Delta S° = +14.1\ J/K$$

Recall from Section 17.6 that a large, negative value for $\Delta G°$ corresponds to a large equilibrium constant for the forward reaction. The equilibrium constant for complex formation is thus much greater for the second reaction, involving en as ligand. We see that the enthalpy changes in the two reactions are nearly the same. On the other hand, the entropy change is much more positive for the second reaction. It is indeed that more positive entropy change that accounts for the larger negative value for $\Delta G°$ for the second reaction.

The more positive entropy change associated with reactions involving polydentate ligands is related to some of the ideas regarding entropy discussed in Section 17.3. Ligands that are bound to the metal ion are constrained to remain with that metal ion and its other ligands and are not free to move about the solution independently. There is therefore a negative entropy change associated with binding to the metal ion. In the first reaction above, four CH_3NH_2 molecules become bound to the metal ion, and four water molecules are freed to move about the solution. Although it might seem at first that this should lead to a net entropy change of zero, differences in hydrogen-bonding ability of H_2O as compared with the CH_3NH_2 molecules, and the tighter binding of CH_3NH_2 to the Cd^{2+} ion as compared with water, lead to a net negative entropy change. On the other hand, when one ethylenediamine molecule binds to Cd^{2+} it releases two H_2O molecules to move about freely in the solution. There is a net increase in the number of species that move about independently in the solution. The system has become more disordered, as it were, and the entropy change is accordingly more positive.

Complexing agents can often be added to a solution to prevent one or more of the customary reactions of a metal ion without actually removing it from the solution. For example, a metal ion that interferes with a chemical analysis can often be complexed and its interference thereby removed. In a sense the complexing agent hides the metal ion and is therefore referred to as a sequestering agent. Because chelates perform this role more effectively than do monodentate ligands, the term *sequestering agent* is usually reserved for chelates.

One of the most common applications of sequestering agents is in complexing cations in natural waters to keep them from interfering with the action of soap or detergent molecules. As noted in Section 12.6, Mg^{2+} and Ca^{2+} ions react with soap to form a precipitate commonly known as soap scum. Although these and other metal ions do not precipitate detergent molecules, they do complex with them thereby interfering with their cleansing action.* Phosphates are effective and cheap sequestering agents for these ions. The most important phosphate used for this purpose is sodium tripolyphosphate:

$$Na_5\left[\begin{array}{c} \quad\;\; O \quad\;\;\; O \quad\;\;\; O \\ \quad\;\; \| \quad\;\;\;\; \| \quad\;\;\;\; \| \\ O{-}P{-}O{-}P{-}O{-}P{-}O \\ \quad\;\; | \quad\;\;\;\; | \quad\;\;\;\; | \\ \quad\;\; O \quad\;\;\; O \quad\;\;\; O \end{array}\right]$$

Complexing agents also enhance the solubility of metal salts. For example, AgBr, the photosensitive material in photographic film, is insoluble in water, but dissolves in the presence of thiosulfate ion, $S_2O_3^{2-}$:

$$AgBr(s) \rightleftharpoons Ag^+(aq) + Br^-(aq) \qquad K_{sp} = 7.7 \times 10^{-13}$$

$$Ag^+(aq) + 2S_2O_3^{2-}(aq) \rightleftharpoons [Ag(S_2O_3)_2]^{3-}(aq) \qquad K_f = 1.6 \times 10^{13}$$

The thiosulfate can be visualized as shifting the first equilibrium to the right by complexing the Ag^+. Sodium thiosulfate decahydrate, $Na_2S_2O_3 \cdot 10H_2O$, known as hypo, is used in black-and-white photography to dissolve unexposed and undeveloped AgBr from the photographic film (Section 21.3).

*The action of soaps and detergents was discussed earlier, in Section 12.7; the formula of a typical detergent molecule is shown in Section 21.4.

1A	2A	3B	4B	5B	6B	7B	8B	8B	8B	1B	2B	3A	4A	5A	6A	7A	8A
H																	He
Li	Be											B	C	N	O	F	Ne
Na	Mg											Al	Si	P	S	Cl	Ar
K	Ca	Sc	Ti	V	Cr	Mn	Fe	Co	Ni	Cu	Zn	Ga	Ge	As	Se	Br	Kr
Rb	Sr	Y	Zr	Nb	Mo	Tc	Ru	Rh	Pd	Ag	Cd	In	Sn	Sb	Te	I	Xe
Cs	Ba	La	Hf	Ta	W	Re	Os	Ir	Pt	Au	Hg	Tl	Pb	Bi	Po	At	Rn

FIGURE 24.5 The elements that are essential for life are indicated by the shaded areas. The dark color indicates the four most abundant elements in living systems (hydrogen, carbon, nitrogen, and oxygen). The light color indicates the seven next most common elements. The gray shading indicates the elements needed in only trace amounts.

Metals and Chelates in Living Systems

Living systems consist mainly of hydrogen, oxygen, carbon, and nitrogen. In fact, more than 99 percent of the atoms required by living cells are one of these four elements. Nevertheless, many other elements are known to be essential for life. Those presently known to be essential are shown in Figure 24.5. Nine of the essential elements are transition metals—vanadium, chromium, iron, copper, zinc, manganese, cobalt, nickel, and molybdenum. These elements owe their roles in living systems mainly to their ability to form complexes with a variety of donor groups present in biological systems. Many enzymes, the body's catalysts, require metal ions to function. We shall take a close look at enzymes in Chapter 25.

Although our bodies require only small quantities of certain metals, a deficiency of the element can lead to serious illness. For example, it was recently discovered that a deficiency of manganese in the diet can lead to convulsive disorders. This is an especially important consideration during pregnancy. Some epilepsy patients have been helped by addition of manganese to their diets.

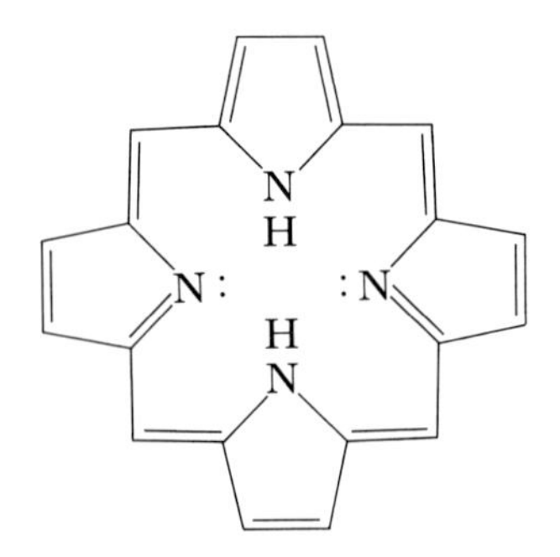

FIGURE 24.6 Structure of the porphine molecule. This molecule forms a tetradentate ligand with the loss of the two protons bound to nitrogen atoms. Porphine is the basic structure of porphyrins, compounds whose complexes play a variety of important roles in nature.

Among the most important chelating agents in nature are those derived from the porphine molecule shown in Figure 24.6. This molecule can coordinate to a metal using the four nitrogen atoms as donors. Upon coordination to a metal, the two H^+ shown bonded to nitrogen are displaced. Complexes derived from porphine are called **porphyrins.** Different porphyrins contain different metals and have different substituent groups attached to the carbon atoms at the ligand's periphery. Two of the most important porphyrins are heme, which contains iron(II), and chlorophyll, which contains magnesium(II). We discussed heme earlier in Section 10.5. Hemoglobin contains four heme subunits as shown in Figure 10.10. The iron is coordinated to the four nitrogen atoms of the porphyrin and also to a nitrogen atom from the protein that composes

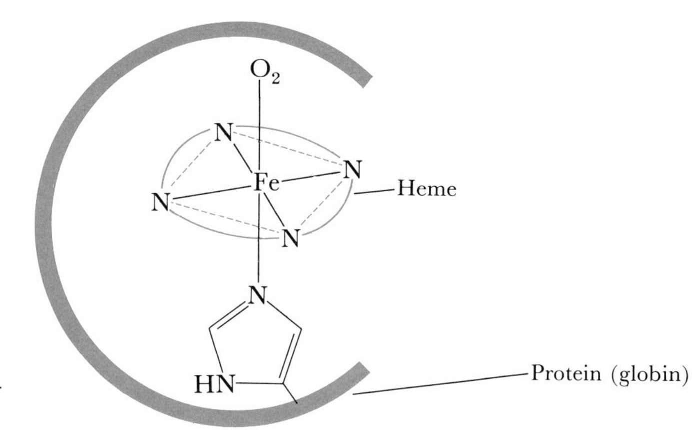

FIGURE 24.7 Schematic representation of oxyhemoglobin, showing one of the four heme units in the molecule. The iron is bound to four nitrogen atoms of the porphyrin, to a nitrogen from the surrounding protein, and to an O_2 molecule.

the bulk of the hemoglobin molecule. The sixth position around the iron is occupied either by O_2 (in oxyhemoglobin, the bright red form) or by water (in deoxyhemoglobin, the purplish-red form). Oxyhemoglobin is shown in Figure 24.7. As noted in Section 10.5, some groups such as CO act as poisons because of their ability to bind to iron more strongly than O_2 can.

When we have an insufficient quantity of iron in our diet, we suffer from iron-deficiency anemia. The lack of iron leads to a reduction in the amount of hemoglobin; we develop what advertisements have referred to as "iron-poor blood." Without hemoglobin to transport oxygen, our body's cells are unable to produce energy. Therefore, the symptoms of anemia include weakness and drowsiness.

Plants also need iron. Iron is part of a plant enzyme that participates in the production of chlorophyll, the green pigment essential to photosynthesis. Plants suffering from a deficiency in iron develop a condition known as iron chlorosis. The effect is usually first noticed in young leaves, which have a yellow coloration. Often the plant develops chlorosis because the soil conditions interfere with the availability of the iron. For example, iron may be present in the soil, but only in insoluble forms unavailable to the plant.

Iron may occur in soil either as iron(II) or iron(III), depending on the ability of oxygen to penetrate the soil and oxidize the iron. Although both Fe^{2+} and Fe^{3+} can be used by plants, the Fe^{3+} is much less soluble than Fe^{2+} in normal soils, which usually have pH's in the vicinity of 7. For example, in typical soil, Fe^{3+} precipitates as $Fe(OH)_3$ when the pH exceeds 3, whereas Fe^{2+} precipitates as $Fe(OH)_2$ when the pH exceeds 6. Therefore, plants growing in alkaline soils readily suffer from iron deficiency. Simple addition of iron salts to the soil does not correct this condition, because the iron is merely precipitated. Current practice in agriculture is to add the iron in a complexed form, such as $[FeEDTA]^{2-}$. The chelated iron will not precipitate and is therefore available to the plant; once in the plant, the iron is removed as needed.

Some plants, such as certain strains of soybeans, secrete their own chelates to solubilize the needed iron. Similarly, mosses and lichens growing on rocks generate chelating agents to extract the metal that they need.

All living cells need iron. One mechanism that our body uses to fight invading bacteria is to withhold iron needed by the bacteria. The bacteria use powerful chelating agents to obtain their required iron. Our body, in turn, keeps its iron tightly complexed. Thus there is a confrontation between the chelating agents of our body and those of the microbial invaders. It has been found that the ability of bacteria to synthesize their chelates is suppressed at elevated temperatures. Consequently, fever is part of the body's attempt to overcome the invading bacteria.

Chelates and chelating agents have also been used as drugs. For example, they have been used to destroy bacteria by depriving them of essential metals. In this fashion the drug mimics the body's natural defenses described above. Chelating agents are also used to remove metals such as Hg^{2+}, Pb^{2+}, and Cd^{2+}, which are detrimental to health. For example, one method of treating lead poisoning is to administer $Na_2[CaEDTA]$. The EDTA chelates the lead, allowing its removal from the body in urine.

24.3 NOMENCLATURE

Before we go too far in our discussions of complexes, it is useful to describe how these substances are named. When complexes were first discovered and few were known, they were named after the chemist who originally prepared them. A few of these names persist; for example, $NH_4[Cr(NH_3)_2(NCS)_4]$ is known as Reinecke's salt. As the number of known complexes grew, they began to be named by color. For example, $[Co(NH_3)_5Cl]Cl_2$, whose formula was then written $CoCl_3 \cdot 5NH_3$, was known as purpureocobaltic chloride after its purple color. Once the structures of complexes were more fully understood, it became possible to name them in a more systematic manner. Before we give the rules of naming complexes, let's consider two examples. We can thereby get a little better idea of how the nomenclature rules apply. The two complexes are:

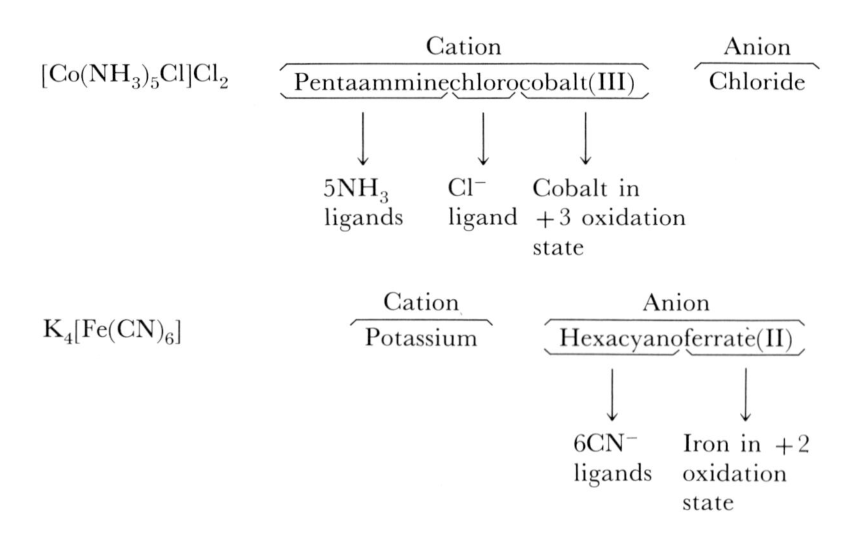

The rules of nomenclature are as follows:*

1 *In naming salts, the name of the cation is given before the name of the anion.* Thus in $[Co(NH_3)_5Cl]Cl_2$ we name the $[Co(NH_3)_5Cl]^{2+}$ and then Cl^-.

*The rules of nomenclature are approved by the International Union of Pure and Applied Chemistry and are subject to periodic revision. Some rules are clearly more important than others. For example, the order in which ligands are named (rule 2) is not as important as assigning each ligand its correct name.

TABLE 24.1 Some common ligands

Ligand	Ligand name
Azide, N_3^-	Azido
Bromide, Br^-	Bromo
Chloride, Cl^-	Chloro
Cyanide, CN^-	Cyano
Hydroxide, OH^-	Hydroxo
Carbonate, CO_3^{2-}	Carbonato
Oxalate, $C_2O_4^{2-}$	Oxalato
Ammonia, NH_3	Ammine
Ethylenediamine, en	Ethylenediamine
Water, H_2O	Aqua

2 *Within a complex ion or molecule, the ligands are named before the metal. Ligands are listed in alphabetical order, regardless of charge on the ligand. Prefixes that give the number of ligands are not considered part of the ligand name in determining alphabetical order.* Thus in the $[Co(NH_3)_5Cl]^{2+}$ ion, we name the ammonia ligands first, then the chloride, then the metal: pentaamminechlorocobalt(III). Note, however, that in writing the formula the metal is listed first.

3 *Anionic ligands end in the letter o, whereas neutral ones ordinarily bear the name of the molecule.* Some common ligands and their names are listed in Table 24.1. Special names are given to H_2O (aqua) and NH_3 (ammine). Thus the terms *chloro* and *ammine* occur in the name for $[Co(NH_3)_5Cl]Cl_2$.

4 *A Greek prefix (for example,* di-, tri-, tetra-, penta-, *and* hexa-) *is used to indicate the number of each kind of ligand when more than one is present.* Thus in the name for $[Co(NH_3)_5Cl]^{2+}$ we have pentaammine, indicating five NH_3 ligands. *If the name of the ligand itself contains a Greek prefix such as* mono- *or* di-, *the ligand is enclosed in parentheses and alternate prefixes* (bis-, tris-, tetrakis-, pentakis,- *and* hexakis-) *used.* For example, the name for $[Co(en)_3]Cl_3$ is tris(ethylenediamine)cobalt(III) chloride.

5 *If the complex is an anion, its name ends in* -ate. For example, in $K_4[Fe(CN)_6]$ the anion is called the hexacyanoferrate(II) ion. The suffix *-ate* is often added to the Latin stem as in this example.

6 *The oxidation number of the metal is given in parentheses in Roman numerals following the name of the metal.* For example, the Roman numeral III is used to indicate the $+3$ oxidation state of cobalt in $[Co(NH_3)_5Cl]^{2+}$.

Using these rules the compounds listed below on the left receive the names given in the second column:

$K[Pt(NH_3)(N_3)_5]$	potassium amminepentaazidoplatinate(IV)
$[Ni(C_5H_5N)_6]Br_2$	hexapyridinenickel(II) bromide
$[Co(NH_3)_4(H_2O)CN]Cl_2$	tetraammineaquacyanocobalt(III) chloride
$Na_2[MoOCl_4]$	sodium tetrachlorooxomolybdate(IV)
$Na[Al(OH)_4]$	sodium tetrahydroxoaluminate

In the last example, the oxidation state of the metal is not mentioned in the name because aluminum is nearly always in the $+3$ oxidation state.

SAMPLE EXERCISE 24.3

Give the name of the following compounds: (a) $[Cr(H_2O)_4Cl_2]Cl$; (b) $K_4[Ni(CN)_4]$.

Solution: We begin with the four waters, which are indicated as tetraaqua. Then there are two chloride ions, indicated as dichloro. The oxidation state of Cr is $+3$:

$$+3 + 4(0) + 2(-1) + (-1) = 0$$

$$[Cr(H_2O)_4Cl_2]Cl$$

Thus we have chromium(III). Finally, the anion is chloride. Putting these parts together we have the compound's name: tetraaquadichlorochromium(III) chloride.

(b) The complex has four CN^-, which we indicate as tetracyano. The oxidation state of the nickel is zero:

$$\begin{array}{c} 4(+1) + 0 + 4(-1) = 0 \\ \swarrow \quad \swarrow \quad \swarrow \\ K_4[Ni(CN)_4] \end{array}$$

Because the complex is an anion, the metal is indicated as nickelate(0). Putting these parts together and naming the cation first we have: potassium tetracyanonickelate(0).

Given the name of a coordination compound, you should be able to write out the formula. Remember that cations are listed before anions and that the metal and ligands of the coordination sphere are enclosed within brackets.

SAMPLE EXERCISE 24.4

Write the formula for bis(ethylenediamine)difluorocobalt(III) perchlorate.

Solution: The complex cation contains two fluorides, two ethylenediamines, and a cobalt with a +3 oxidation number. Knowing this, we can determine the charge on the complex:

$$\begin{array}{c} +3 + 2(0) + 2(-1) = +1 \\ \downarrow \quad \swarrow \quad \swarrow \\ [Co(en)_2F_2] \end{array}$$

The perchlorate anion has a single negative charge, ClO_4^-. Therefore, only one is needed to balance the charge on the complex cation. The formula is thus $[Co(en)_2F_2]ClO_4$.

24.4 ISOMERISM

It sometimes happens that two or more compounds have the same formula but differ in one or more physical or chemical properties, such as color, solubility, or rate of reaction with some reagent. Such compounds, which have the same collection of atoms arranged in different ways, are called **isomers.** Several types of isomerism are possible, as outlined in Figure 24.8.

Structural Isomerism

Structural isomers differ in the bonding arrangements of the atoms; that is, they are chemically distinct. Many different types of structural isomerism are known in coordination chemistry. The two listed in Figure 24.8 are given as examples. **Linkage isomerism** is a relatively rare but interesting type that arises when a particular ligand is capable of coordinating to a metal in two different ways. For example, the nitrite ion, NO_2^-, can coordinate through either a nitrogen or an oxygen atom, as shown in Figure 24.9. When it coordinates through the nitrogen atom the NO_2^- ligand is called *nitro:* when it coordinates through the oxygen atom, it is referred to as *nitrito,* and is generally written as ONO^-. The isomers shown in Figure 24.9 differ in their chemical and physical properties. For example, the N-bonded isomer is yellow, while the O-bonded isomer is red. Other ligands that are capable of coordinating through either of two donor atoms include thiocyanate, SCN^-, whose potential donor atoms are N and S.

Coordination-sphere isomers differ in the ligands that are directly bonded to the metal, as opposed to being outside the coordination sphere

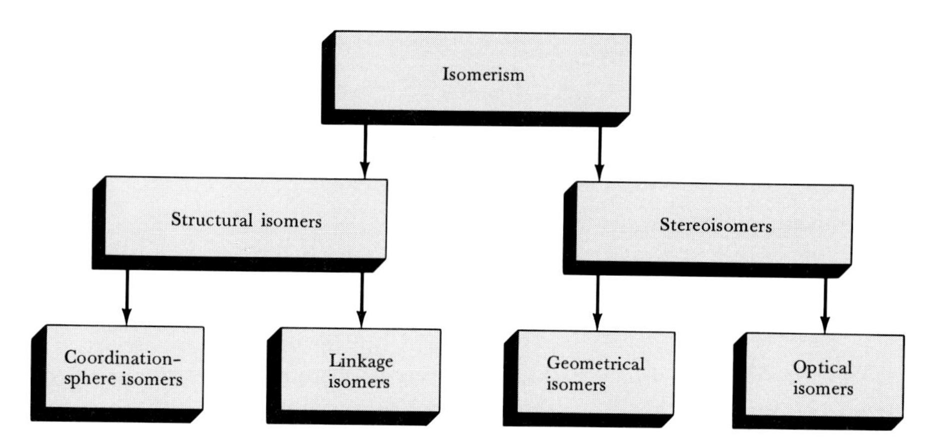

FIGURE 24.8 Forms of isomerism in coordination compounds.

in the solid lattice. For example, the compound $CrCl_3(H_2O)_6$ exists in three forms: $[Cr(H_2O)_6]Cl_3$, a violet compound, $[Cr(H_2O)_5Cl]Cl_2 \cdot H_2O$, a green compound, and $[Cr(H_2O)_4Cl_2]Cl \cdot 2H_2O$, also a green compound. In the second and third compounds the water has been displaced from the coordination sphere by chloride ions and occupies a site in the solid lattice.

Stereoisomerism

The most important form of isomerism is **stereoisomerism.** Stereoisomers have the same chemical bonds, but different spatial arrangements. For example, in $[Pt(NH_3)_2Cl_2]$, the chloro ligands can be either adjacent or opposite to one another, as illustrated in Figure 24.10. This particular form of isomerism, in which the arrangements of the constituent atoms is different though the same bonds are present, is called **geometric isomer-**

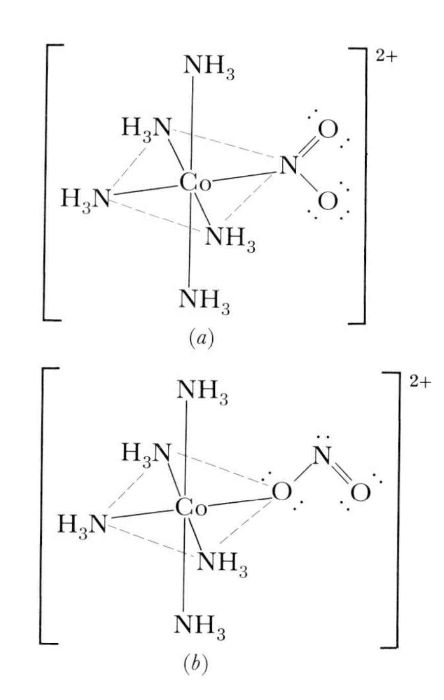

FIGURE 24.9 (*a*) Yellow N-bound and (*b*) red O-bound isomers of $[Co(NH_3)_5NO_2]^{2+}$.

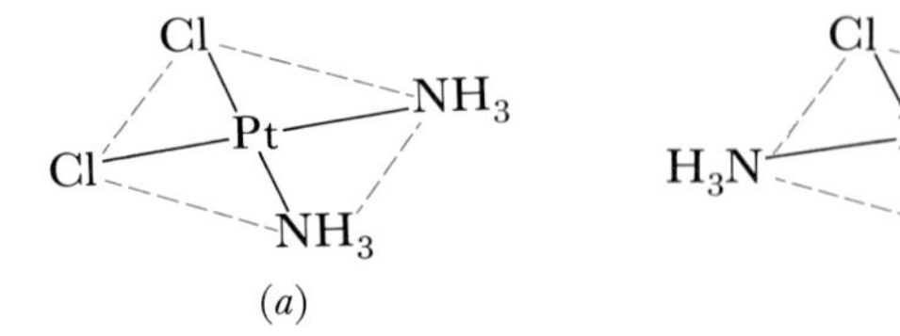

FIGURE 24.10 (*a*) *Cis* and (*b*) *trans* geometric isomers of the square-planar $[Pt(NH_3)_2Cl_2]$.

ism. Isomer (*a*), with like groups in adjacent positions, is called the *cis* isomer, whereas isomer (*b*), with like ligands across from one another, is called the *trans* isomer.

Geometric isomerism is possible also in octahedral complexes when there are two or more different ligands present. The *cis* and *trans* isomers of tetraamminedichlorocobalt(III) ion are shown in Figure 24.11. Note that these two isomers possess different colors. Their salts also possess different solubilities in water. In general, geometric isomers possess distinct physical and chemical properties.

Because all of the corners of a tetrahedron are adjacent to one another, *cis-trans* isomerism is not observed in tetrahedral complexes.

SAMPLE EXERCISE 24.5

How many geometric isomers are there for $[Cr(H_2O)_2Br_4]^-$?

Solution: This complex has a coordination number of six and therefore presumably an octahedral geometry. Like $[Co(NH_3)_4Cl_2]^+$ (Figure 24.11), it has four ligands of one type and two of another. Consequently it possesses two isomers: one with H_2O ligands across the metal from each other (the *trans* isomer) and one with H_2O ligands adjacent (the *cis* isomer).

In general, the number of isomers of a complex can be determined by making a series of drawings of the structure with ligands in different locations. It is easy to overestimate the number of geometric isomers; sometimes different orientations of a single isomer are incorrectly thought to be different isomers. Therefore, you should keep in mind that if two structures can be rotated so that they are equivalent, they are not isomers of one another. The problem of identifying isomers is compounded by the difficulty we often have in visualizing three-dimensional molecules from their two-dimensional representations. It is easier to determine the number of isomers if we are working with three-dimensional models.

A second type of stereoisomerism is known as optical isomerism. Optical isomers are nonsuperimposable mirror images of one another. They bear the same resemblance to one another that our left hand bears to our right hand. If you look at your left hand in a mirror, as shown in Figure 24.12, the image is identical to your right hand. Furthermore, your two hands are not superimposable on one another. A good example of a

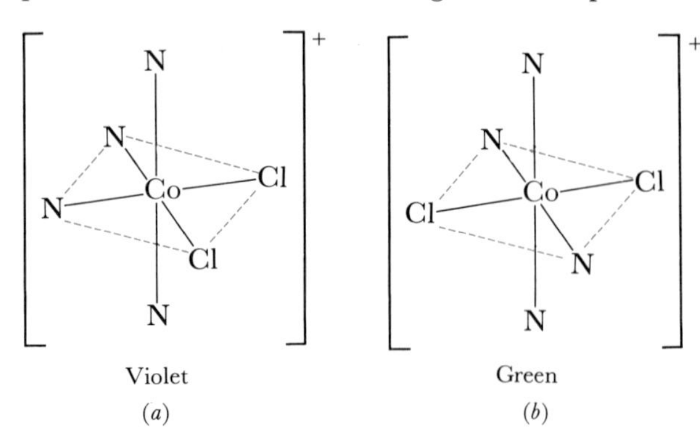

FIGURE 24.11 (*a*) *Cis* and (*b*) *trans* geometric isomers of the octahedral $[Co(NH_3)_4Cl_2]^+$ ion. (The symbol N represents the coordinated NH_3 group.)

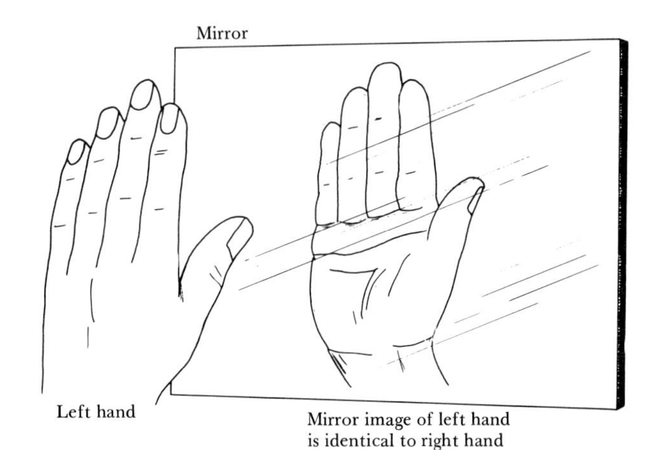

FIGURE 24.12 Our hands are nonsuperimposable mirror images of each other.

complex that exhibits this type of isomerism is the $[Co(en)_3]^{3+}$ ion. The two isomers of $[Co(en)_3]^{3+}$ and their mirror-image relationship to one another are shown in Figure 24.13. Just as there is no way that we can twist or turn our right hand to make it look identical to our left, so also there is no way to rotate one of these isomers to make it identical to the other. If we had models of each that we could handle, perhaps we could more easily satisfy ourselves of this fact. Molecules or ions that have nonsuperimposable mirror images are said to be **chiral** (pronounced KY-rul). Enzymes, the body's catalysts, are among the most highly chiral molecules known. As noted in Section 24.2, many enzymes contain complexed metal ions.

SAMPLE EXERCISE 24.6

Does either *cis*- or *trans*-$[Co(en)_2Cl_2]^+$ have optical isomers?

Solution: To answer this question you should draw out both the *cis* and *trans* isomers of $[Co(en)_2Cl_2]^+$, and then their mirror images. Note that the mirror image of the *trans* isomer is identical to the original. Consequently *trans*-$[Co(en)_2Cl_2]^+$ has no optical isomer. However, the mirror image of *cis*-$[Co(en)_2Cl_2]^+$ is not identical to the original. Consequently, there are optical isomers for this complex.

Most of the physical and chemical properties of optical isomers are identical. The properties of two optical isomers differ only if they are in a chiral environment—that is, one in which there is a sense of right- and

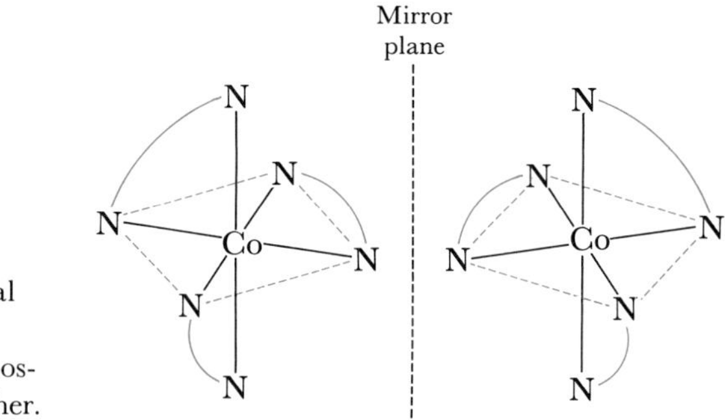

FIGURE 24.13 The two optical isomers of $[Co(en)_3]^{3+}$; notice that the ions are nonsuperimposable mirror images of each other.

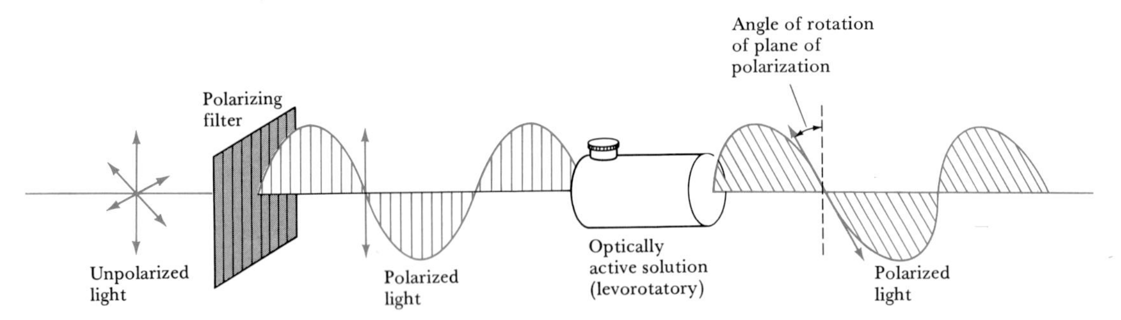

FIGURE 24.14 Effect of an optically active solution on the plane of polarization of plane-polarized light. The unpolarized light is passed through a polarizing filter. The resultant polarized light thereafter passes through a solution containing a levorotatory optical isomer. As a result, the plane of polarization of the light is rotated to the left relative to an observer looking toward the light source.

left-handedness. For example, in the presence of a chiral enzyme, the reaction of one optical isomer might be catalyzed, whereas the other isomer remains totally unreacted. Consequently, one optical isomer may produce a specific physiological effect within our body, whereas its mirror image produces a different effect or none at all.

Optical isomers are usually distinguished from each other by their interaction with plane-polarized light. If light is polarized, for example, by passage through a sheet of Polaroid film, the light waves are vibrating in a single plane, as shown in Figure 24.14. If the polarized light is passed through a solution containing one optical isomer, the plane of polarization is rotated either to the right (clockwise) or to the left (counterclockwise). The isomer that rotates the plane of polarization to the right is said to be **dextrorotatory;** it is labeled the dextro, or *d*, isomer (Latin *dexter,* "right"). Its mirror image will rotate the plane of polarization to the left; it is said to be **levorotatory** and is labeled the levo, or *l*, isomer (Latin *laevus,* "left"). Because of their effect on plane polarized light, chiral molecules are said to be **optically active.**

When a substance with optical isomers is prepared in the laboratory, the chemical environment during the synthesis is not usually chiral. Consequently, equal amounts of the two isomers are obtained; the mixture is said to be **racemic.** A racemic mixture will not rotate polarized light because the rotatory effects of the two isomers cancel each other. In order to separate the isomers from the racemic mixture, they must be placed in a chiral environment. For example, one optical isomer of the chiral tartrate anion,* $C_4H_4O_6^{2-}$, can be used to separate a racemic mixture of $[Co(en)_3]Cl_3$. If *d*-tartrate is added to an aqueous solution of the $[Co(en)_3]Cl_3$, *d*-$[Co(en)_3](d\text{-}C_4H_4O_6)Cl$ will precipitate, leaving *l*-$[Co(en)_3]^{3+}$ in solution.

*When sodium ammonium tartrate, $NaNH_4C_4H_4O_6$, is crystallized from solution, the two isomers form separate crystals whose shapes are mirror images of each other. In 1848, Louis Pasteur achieved the first separation of a racemic mixture into optical isomers; using a microscope he picked the "right-handed" crystals of this compound from the "left-handed" ones.

24.5 LIGAND EXCHANGE RATES

If we were to examine a number of different complexes in solution, we would observe that some exchange ligands at an extremely rapid rate, whereas others do so quite slowly. For example, addition of ammonia to an aqueous solution of $CuSO_4$ produces an essentially instantaneous color change as the pale blue $[Cu(H_2O)_4]^{2+}$ ion is converted to the deep blue $[Cu(NH_3)_4]^{2+}$ ion. When this solution is acidified, the pale blue color is regenerated again at a a rapid rate:

$$[Cu(NH_3)_4]^{2+}(aq) + 4H_2O(l) + 4H^+(aq) \longrightarrow [Cu(H_2O)_4]^{2+}(aq) + 4NH_4^+(aq) \qquad [24.6]$$

In contrast, $[Co(NH_3)_6]^{3+}$ is more difficult to prepare than is $[Cu(NH_3)_4]^{2+}$. However, once it has been formed and is then placed in an acidic solution, reaction to form NH_4^+ takes several days. This tells us that the coordinated NH_3 groups are not readily removed from the metal. Complexes like $[Cu(NH_3)_4]^{2+}$ that undergo rapid ligand exchange are said to be **labile**; those like $[Co(NH_3)_6]^{3+}$ that undergo slow ligand exchange are said to be **inert**. The distinction between labile and inert complexes applies to how rapidly equilibrium is attained and not to the position of the equilibrium. For example, although $[Co(NH_3)_6]^{3+}$ is inert in acidified aqueous solutions, the equilibrium constant indicates that the complex is not thermodynamically stable under these conditions:

$$[Co(NH_3)_6]^{3+}(aq) + 6H_2O(l) + 6H^+(aq) \rightleftharpoons [Co(H_2O)_6]^{3+}(aq) + 6NH_4^+(aq) \qquad K_f \simeq 10^{20} \qquad [24.7]$$

The kinetic inertness of $[Co(NH_3)_6]^{3+}$ can be attributed to a high activation energy for the reaction.

Cobalt(III) is one of few metal ions to consistently form inert complexes; others include chromium(III), platinum(IV), and platinum(II). Complexes of these ions maintain their identity in solution long enough to permit study of their structures and properties. They were therefore among the first complexes studied. Much of our understanding of structure and isomerism comes from studies of these complexes.

24.6 STRUCTURE AND ISOMERISM: A HISTORICAL PERSPECTIVE

Many early systematic studies of coordination compounds involved complexes of cobalt(III), chromium(III), platinum(II), and platinum(IV) with ammonia as a ligand. One of the earliest reports of the preparation of an ammine complex dates from 1798, when a chemist by the name of Tassaert accidentally prepared an ammonia complex of cobalt. The compound was found to have the empirical formula $CoN_6H_{18}Cl_3$. Tassaert wrote the formula as $CoCl_3 \cdot 6NH_3$, suggesting that the compound was analogous to hydrated salts like $CoCl_2 \cdot 6H_2O$.

By 1890, many ammine complexes had been prepared, and a great deal of information about them had been gathered by a number of different investigators. By this time chemists had begun to wonder about how the atoms in these complexes were connected to each other and about the possible effect of these arrangements on the properties of the complexes. Among the observations that any successful theory would have to account for were the electrical conductivity of the complexes in solution and their behavior toward $AgNO_3$. Recall from our previous discussions (Sections 11.4 and 18.6) that solutions of ionic substances conduct electrical current. The ease with which the solution conducts current is referred to as the conductivity. The conductivity of solutions of ionic substances increases with the total concentration of ions and is also greater for ions of higher charge. By comparing the conductivities of solutions of coordination compounds with those of simple salts, it was possible to determine the number and types of ions present in the solution. For example, the molar conductivity of an aqueous solution of $CoCl_3 \cdot 5NH_3$ is about the same as that of $CaCl_2$ and other 1:2 electrolytes. We can thus conclude that an aqueous solution of $CoCl_3 \cdot 5NH_3$ produces three ions, one that carries a +2 charge and two carrying negative charges. The number of free chloride ions present in the solution was determined by treating solutions with $AgNO_3$. When cold, freshly prepared solutions of $CoCl_3 \cdot 5NH_3$ were treated with $AgNO_3$, 2 mol of AgCl precipitated for each mole of complex; one chloride in the compound did not precipitate. These results are summarized in Table 24.2.

In 1893, a 26-year-old Swiss chemist, Alfred Werner, proposed a theory that successfully explained these facts and became the basis for our subsequent understanding of metal complexes. Werner's first basic postulate was that metals exhibit both primary and secondary valences. We now refer to these as the metal's oxidation state and coordination number, respectively. This postulate had no theoretical basis; it predated Lewis's theory of covalent bonding by 23 years. However, it allowed Werner to explain many experimental facts. Werner postulated a primary valence of three and a secondary valence of six for cobalt(III). He therefore wrote the formula for $CoCl_3 \cdot 5NH_3$ as $[Co(NH_3)_5Cl]Cl_2$. The ligands within the brackets satisfied cobalt's secondary valence of six; the three chlorides satisfied the primary valence of three. Werner proposed that the chlorides within the coordination sphere of cobalt(III) are bound so tightly that they are unavailable to contribute to the compound's conductivity or to react with $AgNO_3$. Thus $CoCl_3 \cdot 5NH_3$ consists of a $[Co(NH_3)_5Cl]^{2+}$ ion and two Cl^- ions.

Werner also sought to deduce the arrangement of ligands around the central metal. He postulated that cobalt(III) complexes exhibit an octahedral geometry and sought to verify this postulate by comparing the number of observed isomers with the number expected for various geometries. For example, $[Co(NH_3)_4Cl_2]^+$ should exhibit two geometric iso-

TABLE 24.2 Properties of some ammonia complexes of cobalt(III)

Original formulation	Color	Ions per formula unit	Cl^- ions in solution per formula unit	Modern formulation
$CoCl_3 \cdot 6NH_3$	Orange	4	3	$[Co(NH_3)_6]Cl_3$
$CoCl_3 \cdot 5NH_3$	Purple	3	2	$[Co(NH_3)_5Cl]Cl_2$
$CoCl_3 \cdot 4NH_3$	Green	2	1	*trans*-$[Co(NH_3)_4Cl_2]Cl$
$CoCl_3 \cdot 4NH_3$	Violet	2	1	*cis*-$[Co(NH_3)_4Cl_2]Cl$

mers if it was octahedral. When he first postulated an octahedral geometry for cobalt(III), only one isomer of $[Co(NH_3)_4Cl_2]^+$ was known, the green *trans* isomer. In 1907, after considerable effort, Werner succeeded in isolating the violet *cis* isomer. Even before that, however, he had succeeded in isolating other *cis* and *trans* isomers of cobalt(III) complexes. The occurrence of two isomers was consistent with his postulate of octahedral geometry. Another result consistent with octahedral geometry was the demonstration that $[Co(en)_3]^{3+}$ and certain other complexes were optically active. In 1913, Werner was awared the Nobel Prize in chemistry for his outstanding research work in the field of coordination chemistry.

SAMPLE EXERCISE 24.7

Suggest the structure for $CoCl_2 \cdot 6H_2O$.

Solution: By analogy to the ammonia complexes of cobalt(III), we might write the formula of this compound as $[Co(H_2O)_6]Cl_2$. Indeed, experimental evidence indicates that the water molecules are attached to the metal as ligands. Hydrated metal salts generally have water coordinated to the metal. However, water can also be hydrogen bonded to the anion, particularly to oxyanions. For example, the familiar $CuSO_4 \cdot 5H_2O$ has four water molecules coordinated to Cu^{2+}, and one to SO_4^{2-}.

24.7 COLOR AND MAGNETISM

Werner's theory helps us to understand many properties of complexes, including isomerism and conductivity. However, his theory must be extended before we can use it to explain other properties such as the colors and magnetic properties of transition metal complexes. Studies of these two properties have played an important role in the development of more modern models for metal-ligand bonding. We have discussed the various types of magnetic behavior in Section 23.5; we also discussed the interaction of radiant energy with matter in Section 5.1. Let's briefly examine the significance of these two properties for transition metal complexes before we try to develop a model for metal-ligand bonding.

Color

We have already made reference to the fact that those oxides of the transition elements in which the valence level *d* orbitals are partially occupied tend to be colored (Section 23.4). Similarly, the complexes of transition metal ions exhibit a variety of colors. For example, the colors of several cobalt(III) complexes are listed in Table 24.2. From the list you can see that the color of the complex changes as the ligands surrounding the metal ion change. In general, the colors also depend on the particular metal, and on its oxidation state.

Before we can attempt to explain the origin of these colors, we need to review our earlier discussion of light and introduce some new concepts. Recall first that visible light consists of electromagnetic radiation whose wavelength, λ, ranges from 400 nm to 700 nm (Figure 5.3). The energy of this radiation is inversely proportional to its wavelength, as discussed earlier, in Section 5.2:

$$E = h\nu = h(c/\lambda) \qquad [24.8]$$

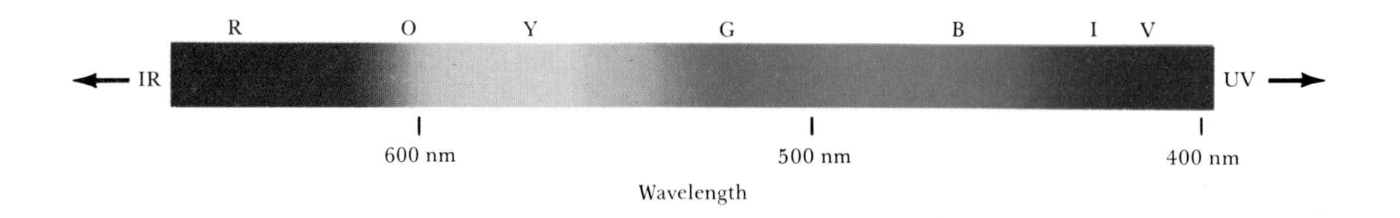

FIGURE 24.15 Visible spectrum showing the relation between color and wavelength.

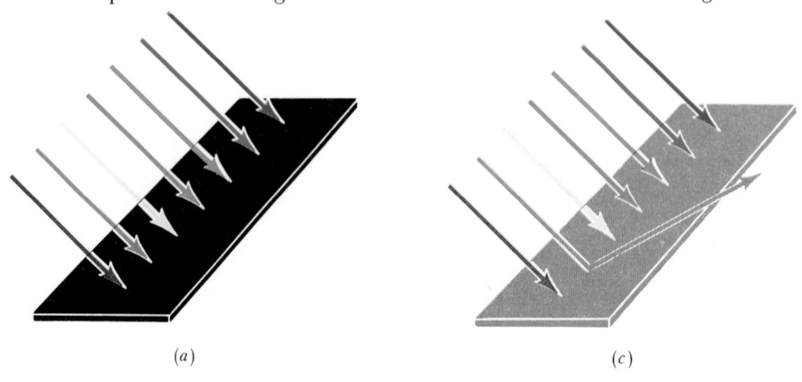

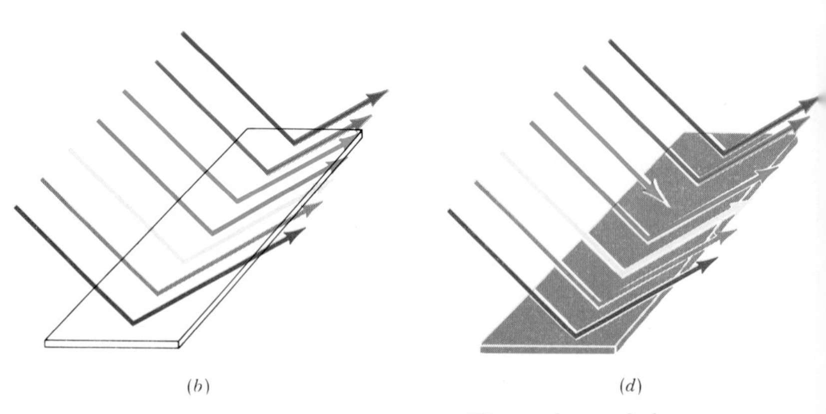

FIGURE 24.16 (*a*) An object is black if it absorbs all colors of light. (*b*) An object is white if it reflects all colors of light. (*c*) An object is orange if it reflects only this color and absorbs all others. (*d*) An object is also orange if it reflects all colors except blue, the complementary color of orange.

The visible spectrum is shown in Figure 24.15. The colors of the spectrum, indicated by their first letters, spell out what appears to be a man's name: Roy G. Biv.

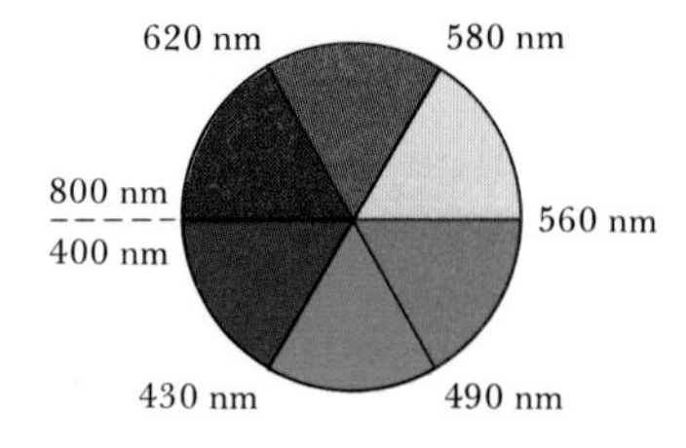

FIGURE 24.17 Artist's color wheel, showing the colors that are complementary to one another and the wavelength range of each color.

When a sample absorbs light, what we see is the sum of the remaining colors that strike our eyes. If a sample absorbs all wavelengths of visible light, none reaches our eyes from that sample. Consequently, it appears black. If the sample absorbs no light, it is white or colorless. If it absorbs all but orange, the sample appears orange. Each of these situations is shown in Figure 24.16. That figure shows one further situation; we also perceive an orange color when visible light of all colors except blue strikes our eyes. In a complementary fashion, if the sample absorbed only orange, it would appear blue; blue and orange are said to be **complementary colors.** Complementary colors can be determined with the aid of the artist's color wheel, shown in Figure 24.17. The colors that are complementary to one another, like orange and blue, are across the wheel from each other.

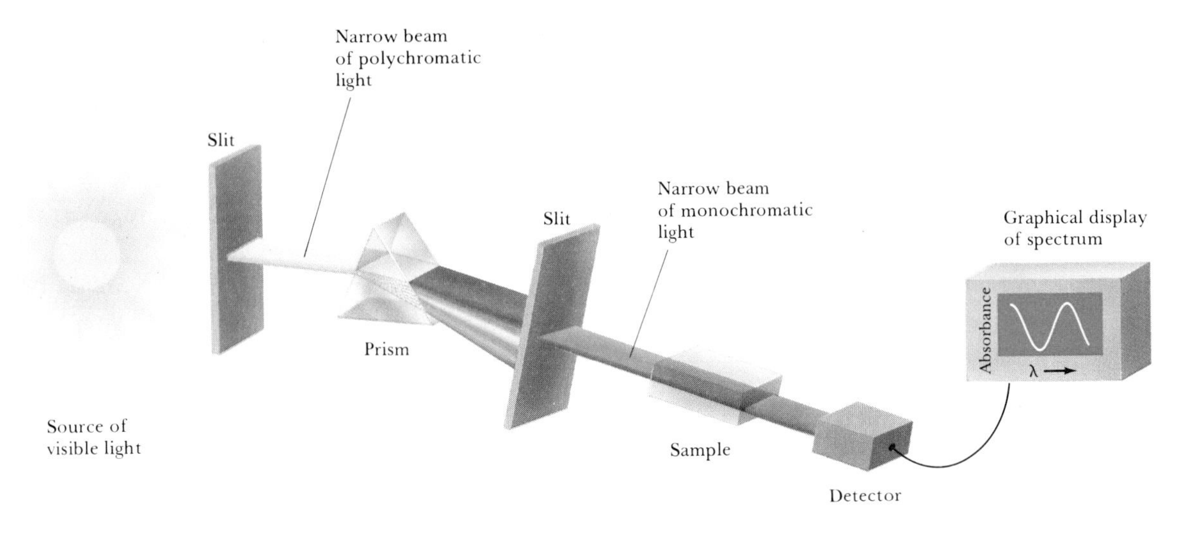

FIGURE 24.18 Experimental determination of the absorption spectrum of a solution. The prism is rotated so that different wavelengths of light pass through the sample. The detector measures the amount of light reaching it, and this information can be displayed as the absorption at each wavelength.

SAMPLE EXERCISE 24.8

The complex ion *trans*-$[Co(NH_3)_4Cl_2]^+$ absorbs light primarily in the red region of the visible spectrum (the most intense absorption is at 640 nm). What is the color of the complex?

Solution: Because the complex absorbs red light, its color will be the complementary color of red. From Figure 24.17, we see that this is green.

The amount of light absorbed by a sample as a function of wavelength is known as its **absorption spectrum.** The visible absorption spectrum of a transparent sample, such as a solution of *trans*-$[Co(NH_3)_4Cl_2]^+$, can be determined as shown in Figure 24.18. The spectrum of $[Ti(H_2O)_6]^{3+}$, which we shall discuss in the next section, is shown in Figure 24.19. The absorption maximum of $[Ti(H_2O)_6]^{3+}$ is at 510 nm. Because the sample absorbs most strongly in the green and yellow regions of the visible spectrum, it appears purple.

Magnetism

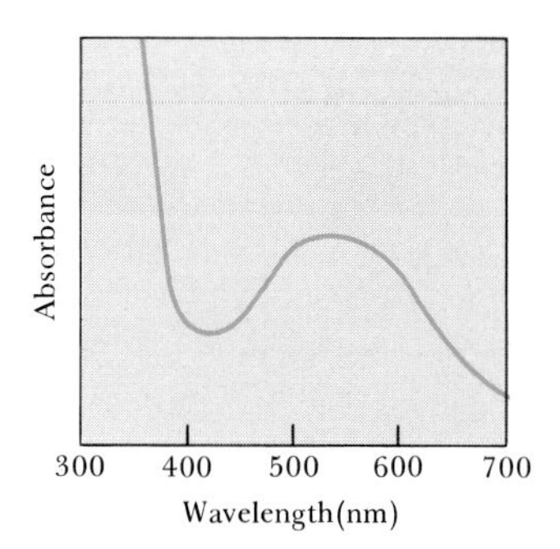

FIGURE 24.19 Visible absorption spectrum of the $[Ti(H_2O)_6]^{3+}$ ion.

Many transition metal complexes exhibit simple paramagnetism, as described in Section 23.5. In such compounds the individual metal ions possess some number of unpaired electrons that interact strongly with one another. However, the metal ions are usually so far apart that there is not a strong interaction between the magnetic moment on one metal ion with that on the others, so the magnetic behavior observed is simple paramagnetism. Using the experimental method illustrated in Figure 23.8, it is possible to determine quantitatively the effective magnetic moment of the sample, and from that to deduce the number of unpaired electrons per metal ion. The experiments reveal some interesting comparisons. For example, in compounds of the complex ion $[Co(NH_3)_6]^{3+}$ there are no unpaired electrons, but in compounds of the $[CoF_6]^{3-}$ ion there are four per metal ion. Clearly, there is a major difference in the way in which the electrons are arranged in the metal orbitals in these two cases, yet in both cases we are dealing with complexes of cobalt(III), with a $3d^6$ electron configuration. Any successful bonding theory must explain this and other related observations.

24.8 CRYSTAL-FIELD THEORY

Although the ability to form complexes is common to all metal ions, the most numerous and interesting complexes are formed by the transition elements. It has been recognized for a long time that the magnetic properties and colors of transition metal complexes are related to the presence of *d* electrons in metal orbitals. In this section we will consider a model for bonding in transition metal complexes, called the **crystal-field theory,** that accounts very well for the observed properties of these interesting substances.*

We have already noted that the ability of a metal ion to attract ligands such as water around itself can be viewed as a Lewis acid-base interaction (Section 15.10). The base, that is, the ligand, can be considered to donate a pair of electrons into a suitable empty hybrid orbital on the metal, as shown in Figure 24.20. However, we can assume that much of the attractive interaction between the metal ion and the surrounding ligands is due to the electrostatic forces between the positive charge on the metal and negative charges on the ligands. If the ligand is ionic, as in the case of Cl^- or SCN^-, the electrostatic interaction occurs between the positive charge on the metal center and the negative charge on the ligand. When the ligand is neutral, as in the case of H_2O or NH_3, the negative ends of these polar molecules, containing an unshared electron pair, are directed toward the metal. In this case the attractive interaction is of the ion-dipole type (Section 10.4). In either case, the result is the same; the ligands are attracted strongly toward the metal center. At the same time, however, the ligands repel one another, because they possess the same charge, or because the negative ends of the molecular dipoles are directed toward the same center. In any metal complex, then, there is a balance between the attractive forces between metal and ligands and the ligand-ligand repulsive interactions. The most stable complex re-

*The name "crystal field" arose because the theory was first developed to explain the properties of solid, crystalline materials such as ruby. The same theoretical model applies to complexes in solution.

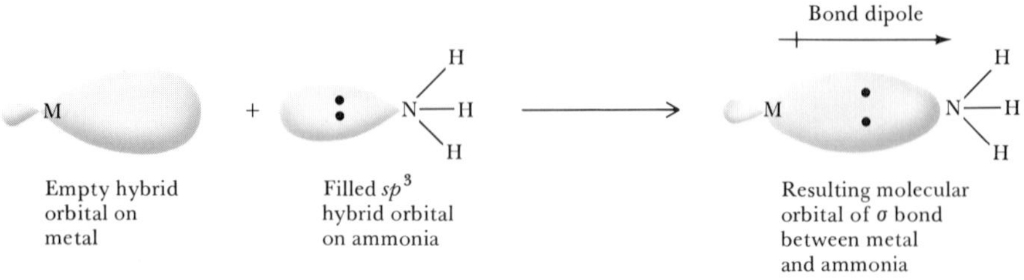

Bond dipole
M
+
N—H
H
H
M
N—H
H
H
Empty hybrid orbital on metal
Filled sp^3 hybrid orbital on ammonia
Resulting molecular orbital of σ bond between metal and ammonia

FIGURE 24.20 Representation of the metal-ligand bond in a complex as a Lewis acid-base interaction. The ligand, which acts as a Lewis base, donates charge to the metal via a metal hybrid orbital. The bond that results is strongly polar, with some covalent character. For many purposes it is sufficient to assume that the metal-ligand interaction is entirely electrostatic in character, as is done in the crystal-field model.

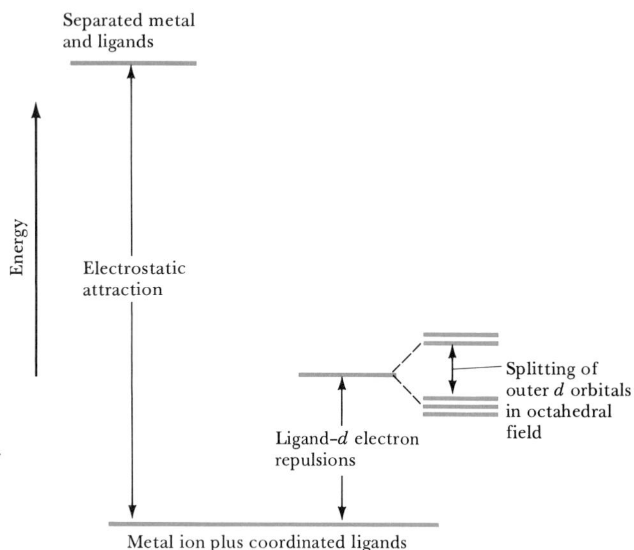

FIGURE 24.21 In the crystal-field model, the bonding between metal ion and donor atoms is considered to be largely electrostatic. The energy of the metal ion plus coordinated ligands is lower than that of the separated metal ion plus ligands because of the electrostatic attraction between metal ion and ligands. At the same time the energies of the metal d electrons are increased by the repulsive interaction between these electrons and the electrons of the ligands. However, if the arrangement of ligands is octahedral, the electrons that occupy the d_{z^2} and $d_{x^2-y^2}$ orbitals are more strongly repelled by the ligands than the electrons occupying the d_{xy}, d_{xz}, and d_{yz} orbitals. This difference in repulsive interactions gives rise to the splitting of the metal d-orbital energies shown on the right and is referred to as the crystal-field splitting.

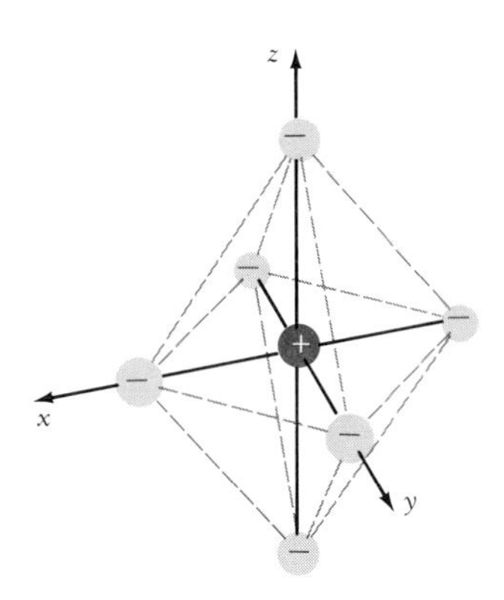

FIGURE 24.22 Octahedral array of negative charges surrounding a positive charge.

sults when the geometrical arrangement of the ligands around the metal is such as to maximize metal-ligand attractions while minimizing the ligand-ligand repulsions. The assembly of metal ion and ligands is lower in energy than the fully separated charges, as illustrated on the left in Figure 24.21.

In a six-coordinate complex the ligand-ligand repulsions cause the ligands to approach along the x, y, and z axes, as shown in Figure 24.22. With the physical arrangement of ligands and metal ion shown in this figure as our starting point, let us now consider what happens to the energies of electrons in the metal d orbitals as the ligands approach the metal ion. Keep in mind that the d electrons are the outermost electrons of the metal ion. We know that the overall energy of the metal ion plus ligands will be lower (that is, more stable) when the ligands are drawn toward the metal center. At the same time, however, there is a repulsive interaction between the outermost electrons on the metal and the negative charges on the ligands. This interaction is called the crystal field. As a result, if we consider just the energies of the d electrons on the metal ion, we find that these have *increased.* This increase in energy is shown in the center part of Figure 24.21. But the d orbitals of the metal ion do not all behave in the same way under the influence of the ligand field. To see why this is so, recall the shapes of the five d orbitals, illustrated in Figure 24.23. In the isolated metal ion, these five orbitals are equivalent in energy. However, as the ligands approach the metal ion, the $d_{x^2-y^2}$ and d_{z^2} orbitals, which are directed *along* the x, y, and z axes, are more strongly repelled by the ligands than the d_{xy}, d_{xz}, and d_{yz} orbitals. These latter are directed *between* the axes along which the ligands approach. Thus an energy separation, or splitting, occurs. The $d_{x^2-y^2}$ and d_{z^2} orbitals are raised in energy, and the d_{xz}, d_{yz}, and d_{xy} orbitals are lowered. This energy splitting is illustrated on the right side of Figure 24.21. In the material that follows we will concentrate on just the splitting of the d orbital energies by the ligand field. This is illustrated in a slightly different form in Figure 24.24.

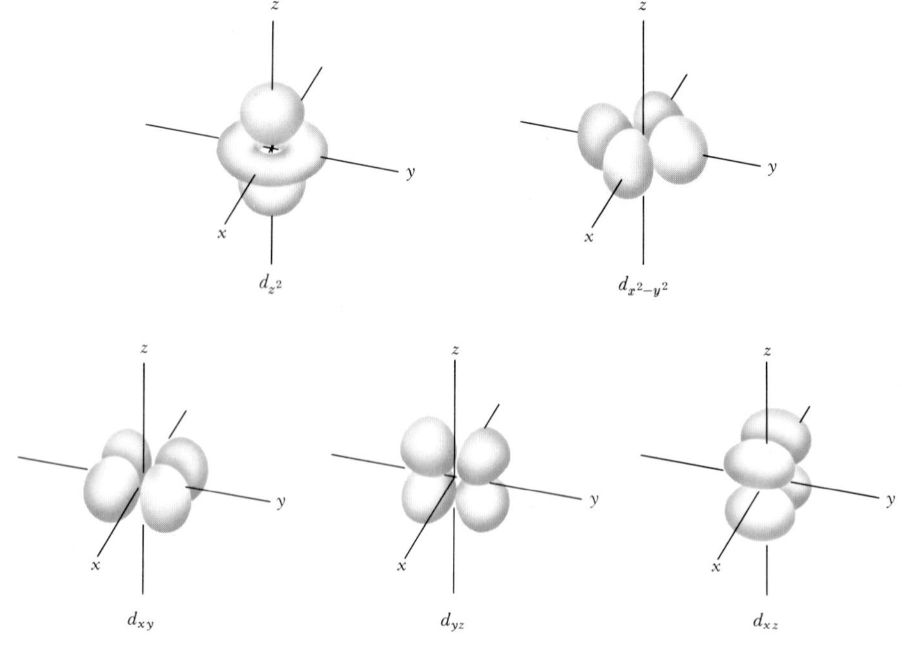

FIGURE 24.23 Shapes of the five *d* orbitals. Remember that the lobes represent regions in which the electrons occupying an orbital are most likely to be found.

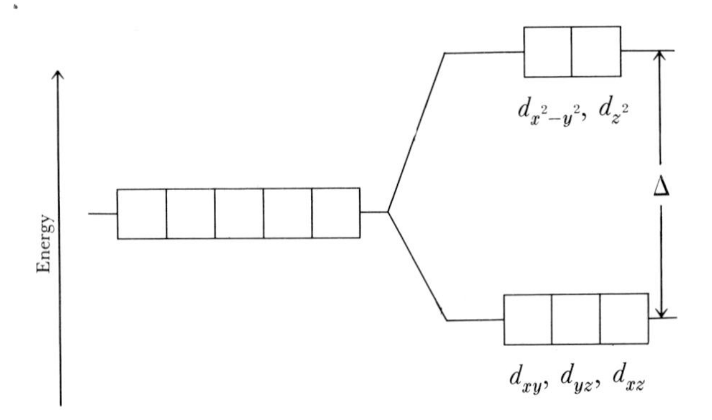

FIGURE 24.24 Energies of the *d* orbitals in an octahedral crystal field.

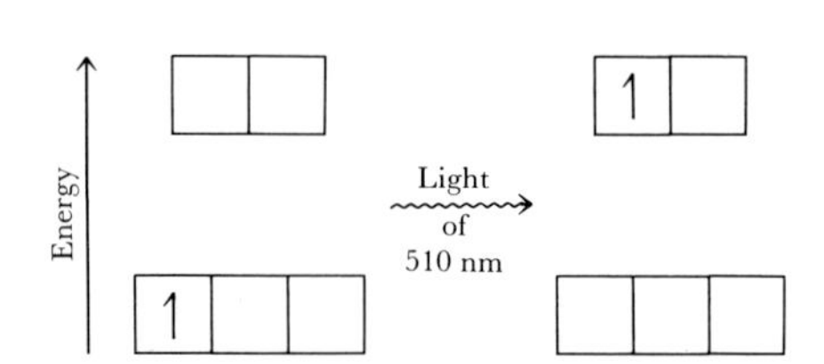

FIGURE 24.25 The 3*d* electron of $[Ti(H_2O)_6]^{3+}$ is excited from the lower-energy *d* orbitals to the higher-energy ones when irradiated with light of 510-nm wavelength.

Let's examine how the crystal-field model accounts for the observed colors in transition metal complexes. The energy gap between the *d* orbitals, labeled Δ, is of the same order of magnitude as the energy of a photon of visible light. (Δ is sometimes referred to as the crystal-field splitting energy.) It is therefore possible for a transition-metal complex to absorb visible light, which thereby excites an electron from the lower energy *d* orbitals into the higher-energy ones. The $[Ti(H_2O)_6]^{3+}$ ion provides a simple example, because titanium(III) has only one 3*d* electron. As shown in Figure 24.19, $[Ti(H_2O)_6]^{3+}$ has a single absorption peak in the visible region of the spectrum. The maximum absorption is at 510 nm (3.9×10^{-19} J/molecule). Light of this wavelength causes the *d* electron to move from the lower-energy set of *d* orbitals into the higher-energy set, as shown in Figure 24.25.

SAMPLE EXERCISE 24.9

Al^{3+}, Zn^{2+}, and Co^{2+} ions are placed in octahedral environments. Which can absorb visible light and thereby exhibit color?

Solution: The Al^{3+} ion has an electron configuration of [Ne]. Because it has no outer *d* electrons, it is colorless.

The Zn^{2+} ion has an electron configuration of $[Ar]3d^{10}$. In this case all of the 3*d* orbitals are filled. There is no room in the d_{z^2} and $d_{x^2-y^2}$ orbitals to accept an electron from a lower-energy d_{xy}, d_{yz}, or d_{xz} orbital. The complex is therefore colorless.

The Co^{2+} ion has an electron configuration of $[Ar]3d^7$. In this case there is room for movement of a *d* electron from the lower-energy d_{xy}, d_{yz}, and d_{xz} into the higher-energy d_{z^2} and $d_{x^2-y^2}$ orbitals. The complex is therefore colored.

Gemstones such as ruby and emerald owe their color to the presence of trace amounts of transition metal ions. For example, replacement of a fraction of the aluminum in the colorless mineral corundum, Al_2O_3, produces several different gems: chromium forms ruby, manganese forms amethyst, and iron forms topaz. Sapphire, which occurs in a variety of colors but is most often blue, contains titanium and cobalt. Several other gems are produced by replacing a trace of aluminum in the colorless mineral beryl, $Be_3Al_2Si_6O_{18}$, with transition metal ions. For example, emerald contains chromium, whereas aquamarine contains iron. See Figure 24.26.

The magnitude of the energy gap, Δ, and consequently the color of a complex, depend on both the metal and the surrounding ligands. For example, $[Fe(H_2O)_6]^{3+}$ is light violet, $[Cr(H_2O)_6]^{3+}$ is violet, and $[Cr(NH_3)_6]^{3+}$ is yellow. Ligands can be arranged in order of their abili-

FIGURE 24.26 Gemstones. (*a*) Uncut emerald crystal; (*b*) emerald; (*c*) ruby; (*d*) amethyst. [(*a*) © *American Museum of Natural History;* (*c*)–(*d*) © *Russ Kinne, 1979; Photo Researchers, Inc.*)

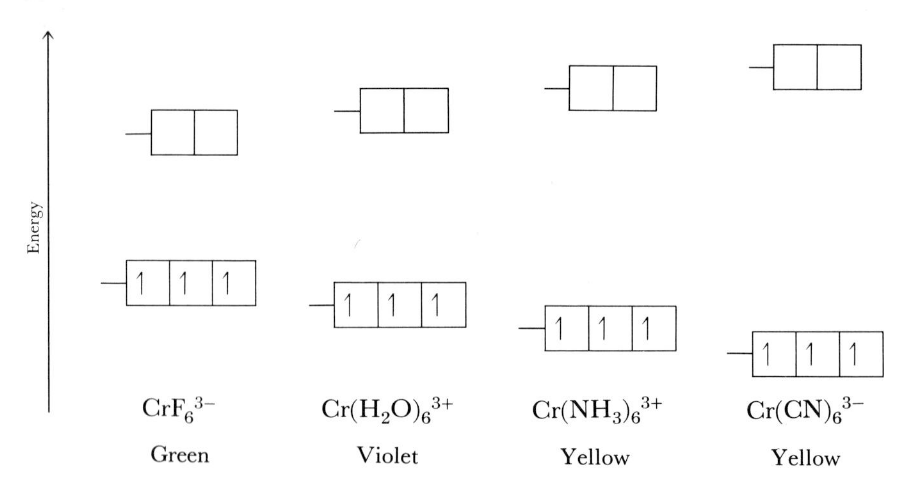

FIGURE 24.27 Crystal-field splitting in a series of chromium(III) complexes.

ties to increase the energy gap, Δ. The following is an abbreviated list of common ligands arranged in order of increasing Δ:

$$Cl^- < F^- < H_2O < NH_3 < \text{en} < NO_2^- \text{ (N-bonded)} < CN^-$$

This list is known as the **spectrochemical series.**

Ligands that lie on the low end of the spectrochemical series are termed weak-field ligands. Those that lie on the high end are termed strong-field ligands. Figure 24.27 shows schematically what happens to the crystal-field splitting when the ligand is varied in a series of chromium(III) complexes. (This is a good place to remind you that when a transition metal is ionized, the valence *s* electrons are removed first. Thus the outer electron configuration for chromium is $[Ar]4s^13d^5$; that for Cr^{3+} is $[Ar]3d^3$.) Notice that as the field exerted by the six surrounding ligands increases, the splitting of the metal *d* orbitals increases. Because the absorption spectrum is related to this energy separation, these complexes vary in color.

Electron Configurations in Octahedral Complexes

The crystal-field model helps us understand the magnetic properties and some important chemical properties of the transition metal ions. From our earlier discussion of electronic structure in atoms (Section 6.3) we expect that electrons will always occupy the lowest-energy vacant orbitals first, and that they will occupy a set of degenerate orbitals one at a time with their spins parallel (Hund's rule). Thus, if we have one, two, or three electrons to add to the *d* orbitals in a complex ion, the electrons will go into the lower-energy set of orbitals, with their spins parallel, as shown in Figure 24.28. However, when we come to add a fourth electron, a new problem arises. If the electron is added to the lower-energy orbital, an energy gain of magnitude Δ is realized, as compared with placing the electron in the higher-energy orbital. On the other hand, there is a penalty for doing this, because the electron must now be paired up with the electron already occupying the orbital. The energy required to do this as compared with putting it in another orbital with parallel spin is called

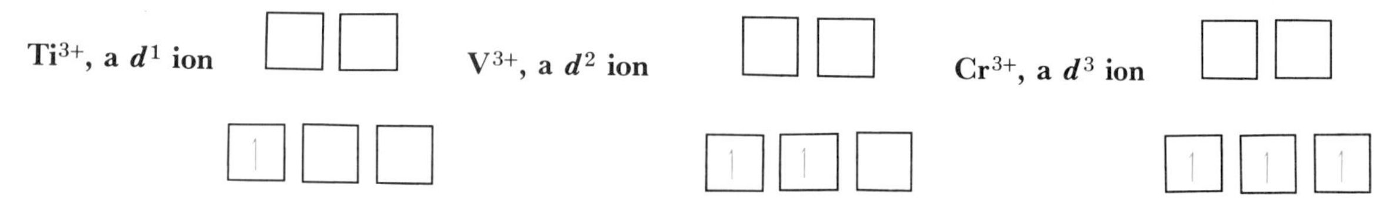

FIGURE 24.28 Electron configurations associated with one, two, and three electrons in the $3d$ orbitals in octahedral complexes.

the spin pairing energy. The spin pairing energy arises from the greater electrostatic repulsion of electrons that share an orbital as compared with two that are in different orbitals.

The ligands that surround the metal ion, and the charge on the metal, play major roles in determining which of the two electronic arrangements arises in many cases. To illustrate, consider the $[CoF_6]^{3-}$ and $[Co(CN)_6]^{3-}$ ions. In both cases the ligands have a -1 charge. However, the F^- ion, on the low end of the spectrochemical series, is a weak-field ligand. The CN^- ion, on the high end of the spectrochemical series, is a strong-field ligand. It produces a larger energy gap than does the F^- ion. The splitting of the d orbital energies in the complexes is compared in Figure 24.29. A count of the electrons in cobalt(III) tells us that we have six electrons to place in the $3d$ orbitals. Let us imagine that we add these electrons one at a time to the d orbitals of the $CoF_6{}^{3-}$ ion. The first three will, of course, go into the lower-energy orbitals with spins parallel. The fourth electron could go into a lower-energy orbital, pairing up with one of those already present. This would result in an energy gain of Δ over putting it in one of the higher-energy orbitals. However, it would cost energy in an amount equal to the spin pairing energy. Because F^- is a weak-field ligand, Δ is not so large, and it turns out that the most stable arrangement is one in which the electron is placed in the higher-energy orbital. Similarly, the fifth electron we add goes into a higher-energy orbital. With all of the orbitals containing at least one electron, the sixth must be paired up, and it goes into a lower-energy orbital. In the case of the $[Co(CN)_6]^{3-}$ complex, the crystal-field splitting is much larger. The spin pairing energy is smaller than Δ, so electrons are paired in the lower-energy orbitals, as illustrated in Figure 24.29.

The $[CoF_6]^{3-}$ complex is referred to as high spin; that is, the electrons are arranged so that they remain unpaired as much as possible. On the other hand, the electron arrangement found in the $[Co(CN)_6]^{3-}$ ion is referred to as low spin. These two different electronic arrangements can

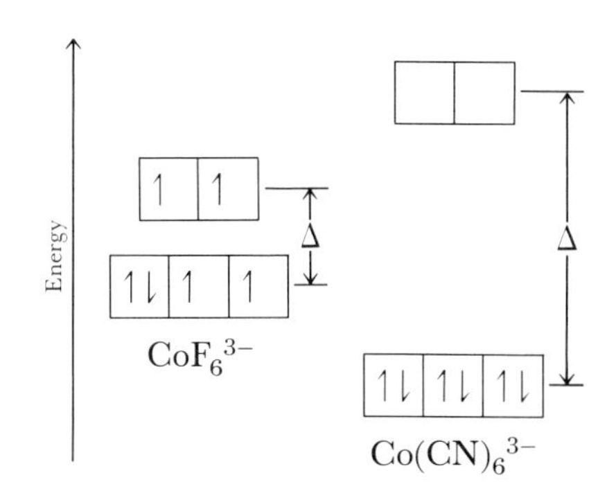

FIGURE 24.29 Population of d orbitals in the high-spin $[CoF_6]^{3-}$ ion (small Δ) and low-spin $[Co(CN)_6]^{3-}$ ion (large Δ).

be readily distinguished by measuring the magnetic properties of the complex, as described earlier. The absorption spectrum also shows characteristic features that indicate the electronic arrangement.

SAMPLE EXERCISE 24.10

Predict the number of unpaired electrons to be expected in six-coordinate high-spin and low-spin complexes of Fe^{3+}.

Solution: The Fe^{3+} ion possesses five $3d$ electrons. In a high-spin complex these are all unpaired. In a low-spin complex the electrons are confined to the lower-energy set of *d* orbitals, with the result that there is one unpaired electron:

High spin	Low spin
[↑][↑]	[ ][ ]
[↑][↑][↑]	[↑↓][↑↓][↑]

Crystal-Field Stabilization Energy

In discussing data relating to the energies of transition metal complexes, the **crystal-field stabilization energy,** CFSE, is a useful concept. Figure 24.30 shows a modification of Figure 24.24, in which the crystal-field splitting energy, Δ, is divided between the orbitals of lower energy and those of higher energy. An electron located in one of the lower-energy orbitals is stabilized relative to the average energy by an amount 0.4Δ. This is called the crystal-field stabilization energy, CFSE. On the other hand, an electron placed in one of the higher-energy orbitals is *destabilized* by an amount 0.6Δ. It thus has a negative CFSE. The total CFSE is just the sum of the CFSEs of all the *d* electrons.

The idea of a crystal-field stabilization energy, CFSE, can be tested by considering the enthalpies of hydration of the divalent metal ions of the transition elements. The **enthalpy of hydration,** ΔH_h, of a metal ion is the heat evolved in the process

$$M^{2+}(g) + \text{water} \longrightarrow [M(H_2O)_6]^{2+}(aq) \qquad [24.9]$$

The heat of such a process can be estimated quite well from various thermochemical data. The $3d$ electron configurations of the divalent ions are listed in Table 24.3. Because H_2O is a relatively weak field ligand, all the hexaaqua complexes of the metal ions listed are high-spin, octahedral complexes of the form $[M(H_2O)_6]^{2+}$. Thus the $3d$ electrons must occupy one or the other of the two sets of orbitals that are split in energy

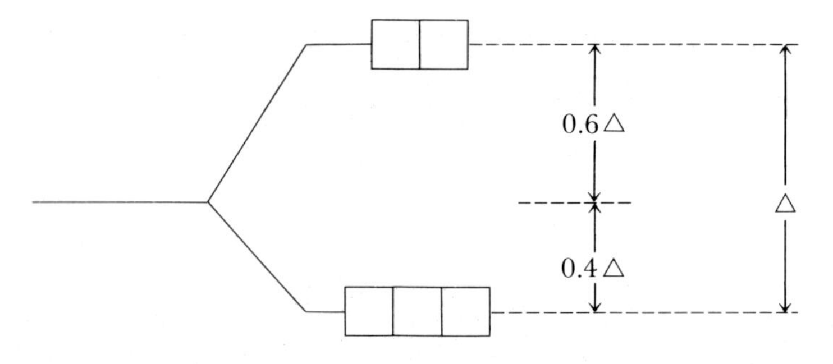

FIGURE 24.30 Crystal-field stabilization energy (CFSE) of 0.4Δ and destabilization energy of 0.6Δ for electrons in *d* orbitals split by an octahedral crystal field. Keep in mind that the actual magnitude of Δ depends on the metal ion, its charge, and on the ligands.

TABLE 24.3 Total crystal-field stabilization energies (CFSE) of the divalent metal ions in high-spin $M(H_2O)_6^{2+}$ complexes

Ion	Number of 3*d* electrons	Electrons in lower-energy orbitals	Electrons in higher-energy orbitals	Total CFSE
Ca^{2+}	0	0	0	0
Sc^{2+}	1	1	0	0.4Δ
Ti^{2+}	2	2	0	0.8Δ
V^{2+}	3	3	0	1.2Δ
Cr^{2+}	4	3	1	0.6Δ
Mn^{2+}	5	3	2	0
Fe^{2+}	6	4	2	0.4Δ
Co^{2+}	7	5	2	0.8Δ
Ni^{2+}	8	6	2	1.2Δ
Cu^{2+}	9	6	3	0.6Δ
Zn^{2+}	10	6	4	0

by the crystal field of the coordinated water molecules. Table 24.3 lists the number of electrons in the lower- and higher-energy sets for each metal ion. The total CFSE for the metal complex is the sum of the CFSE terms for each electron. For example, for the Cr^{2+} ion, with four 3*d* electrons, the total CFSE is the sum of $3 \times (0.4\Delta) - 0.6\Delta = 0.6\Delta$. A positive total CFSE represents a stabilization, that is, a lower energy, as compared with the average energy of the 3*d* orbitals in the field of the surrounding ligands.

If it were not for the influence of the CFSE, we would expect ΔH_h to vary in a smooth way with the metal-ion radius. The smaller the metal ion, the greater is the interaction between ligands and metal ion, and thus the more negative is ΔH_h. In passing from Ca^{2+} to Zn^{2+}, ΔH_h should become steadily more negative because of the steady decrease in ionic radius in proceeding across the transition-element series (Sections 6.7 and 7.3). Superimposed on this general trend, however, is the effect of the CFSE. Figure 24.31 shows the variation in ΔH_h as a function of atomic number for the divalent metal ions listed in Table 24.3. For the three ions that have no CFSE, Ca^{2+}, Mn^{2+}, and Zn^{2+}, the ΔH_h values lie near a smooth curve. For the other transition metal ions, the enthalpies are larger than expected from radius considerations alone. The extent of departure from the line is related to the magnitude of the total CFSE for the complex, as listed in Table 24.3. Notice that the discrepancies are quite large, on the order of ordinary bond energies. These data provide strong support for the crystal-field model. The idea of a crystal-field stabilization energy can be used to explain other properties of transition metal ions, such as the lattice energies of transition metal ionic solids and the relative stabilities of different oxidation states.

Tetrahedral and Square-planar Complexes

Thus far we have considered the crystal-field model only for complexes of octahedral geometry. When there are only four ligands about the metal, the geometry is tetrahedral, except for the special case of metal

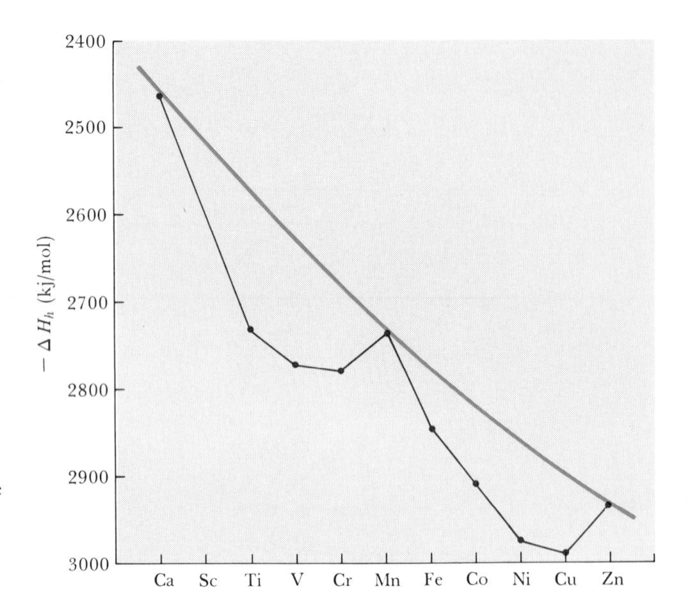

FIGURE 24.31 Enthalpies of hydration of the divalent transition metal ions. The smooth curve connecting the data for Ca^{2+}, Mn^{2+}, and Zn^{2+} represents the expected variation in the absence of crystal field stabilization effects. (Sc^{2+} is not known.)

ions with a d^8 electron configuration, which we will discuss in a moment. The crystal-field splitting of the metal *d* orbitals in tetrahedral complexes differs from that in octahedral complexes. Four equivalent ligands can interact with a central metal ion most effectively by approaching along the vertices of a tetrahedron. (Figure 23.5 offers a good comparison of the octahedral and tetrahedral geometries.) It turns out, and this is not easy to explain in just a few sentences, that the splitting of the metal *d* orbitals in a tetrahedral crystal is just the opposite of that for the octahedral case. That is, three of the metal *d* orbitals are raised in energy, and the other two are lowered, as illustrated in Figure 24.32. Because there are only four ligands instead of six as in the octahedral case, the crystal-field splitting is much smaller for tetrahedral complexes. Calculations show that for the same metal ion and ligand set, the crystal-field splitting for a tetrahedral complex is only $\frac{4}{9}$ as large as for the octahedral complex. For this reason, all tetrahedral complexes are characterized by high spin; the crystal field is never large enough to overcome the spin-pairing energies.

Square-planar complexes, in which four ligands are arranged about the metal ion in a plane, represent a common geometrical form. One can think of the square-planar complex as formed by removing two ligands from along the vertical *z* axis of the octahedral complex. As this happens the four ligands in the plane are drawn in more tightly. The changes that occur in the energy levels of the *d* orbitals are illustrated in Figure 24.33.

Square-planar complexes are characteristic of metal ions with a d^8 electron configuration. They are nearly always low spin; that is, the eight

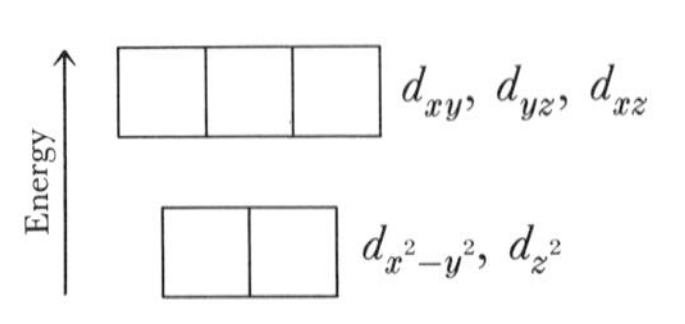

FIGURE 24.32 Energies of the *d* orbitals in a tetrahedral crystal field.

FIGURE 24.33 Effect on the relative energies of the d orbitals of removing the two negative charges from the z axis of an octahedral complex. When the charges are completely removed, the square-planar geometry results.

d electrons are spin-paired to form a diamagnetic complex. Such an electronic arrangement is particularly common among the heavier metals, such as Pd, Pt, Ir, and Au.

SAMPLE EXERCISE 24.11

Four-coordinate nickel(II) complexes exhibit both square-planar and tetrahedral geometries. The tetrahedral ones such as $[NiCl_4]^{2-}$ are paramagnetic; the square-planar ones such as $[Ni(CN)_4]^{2-}$ are diamagnetic. Show how the d electrons of nickel(II) populate the d orbitals in the appropriate crystal-field-splitting diagram in each of these cases.

Solution: Nickel(II) has an electron configuration of $[Ar]3d^8$. The population of the d electrons in the two geometries is given at right:

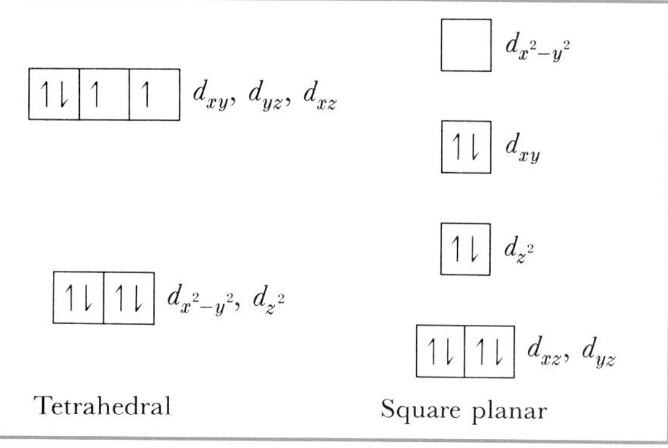

One of the simplest examples of a square-planar metal complex, *cis*-diamminedichloroplatinum(II), Figure 24.10(*a*), is a potent anticancer drug that is now used in treatment of cancer of the ovaries, testes, prostate, and other organs. When used along with other chemotherapeutic drugs, the platinum complex has led to complete, long-term remissions in 60 to 70 percent of cases treated. The anticancer activity of *cis*-$[Pt(NH_3)_2Cl_2]$ was discovered by accident by Barnett Rosenberg, of Michigan State University, and his co-workers, when they noted that bacterial cell division in growth media containing platinum electrodes was inhibited near the electrodes. The causative agent proved to be *cis*-$[Pt(NH_3)_2Cl_2]$.

We have seen that the crystal-field model provides a basis for explaining many features of transition metal complexes. In fact, it can be used to explain many observations in addition to those we have discussed. However, many lines of evidence show that the bonding between transition-metal ions and ligands must have some covalent character. Molecular-orbital theory (Sections 8.5 and 8.6) can also be used to describe the bonding in complexes. However, the application of molecular-orbital theory to coordination compounds is beyond the scope of our discussion. The crystal-field model, although not entirely accurate in all details, provides an adequate and useful description.

FOR REVIEW

Summary

Coordination compounds or complexes contain metal ions bonded to several surrounding anions or molecules known as ligands. The metal ion and its ligands comprise the coordination sphere of the complex. The atom of the ligand that bonds to the metal ion is known as the donor atom. The number of donor atoms attached to the metal ion is known as the coordination number of the metal ion. The common coordination numbers are four and six; the common coordination geometries are tetrahedral, square planar, and octahedral.

If a ligand has several donor atoms that can coordinate simultaneously to the metal, it is said to be polydentate and is referred to as a chelating agent. Two common examples are ethylenediamine (en), which is potentially bidentate, and ethylenediaminetetraacetate ($EDTA^{4-}$), which is potentially hexadentate. Many biologically important molecules such as the prophyrins are complexes of chelating agents.

Isomerism is common among coordination compounds. Structural isomerism involves differences in the bonding arrangements of the ligands. One simple form of structural isomerism, known as linkage isomerism, occurs when a ligand is capable of coordinating to a metal through either of two donor atoms. Coordination-sphere isomerism occurs when two compounds with the same overall formula contain different ligands in the coordination sphere. Stereoisomerism involves complexes with the same chemical bonding arrangements but with differing spatial arrangements of ligands. The most common forms are geometric and optical isomerism. Geometric isomers differ from one another in the relative locations of donor atoms in the coordination sphere; the most common are *cis-trans* isomers. Optical isomers differ from one another in that they are nonsuperimposable mirror images of one another. Isomers differ from one another in their chemical and physical properties; however, optical isomers differ only in the presence of a chiral environment. Most often optical isomers are distinguished from one another by their interactions with plane-polarized light; solutions of one isomer rotate the plane of polarization to the right, whereas solutions of its mirror image rotate the plane to the left. The isomer that rotates the plane of polarization to the right is said to be dextrorotatory, whereas its isomer is levorotatory. Chiral molecules are said to be optically active. A 50–50 mixture of two optical isomers does not rotate plane-polarized light and is said to be racemic.

Many of the early studies that served as a basis for our current understanding of complexes involved complexes of chromium(III), cobalt(III), platinum(II), and platinum(IV). Complexes of these metal ions are inert, in that they undergo ligand exchange at a slow rate. Complexes undergoing rapid exchange are said to be labile.

Studies of the magnetic properties and colors of transition metal complexes have played an important role in formulation of bonding theories for these compounds. The crystal-field theory successfully accounts for many properties of coordination compounds. In this model the interaction between metal ion and ligand is viewed as electrostatic. The ligands produce an electric field that causes a splitting in the energies of the metal *d* orbitals. In the spectrochemical series the ligands are listed in order of their ability to split the *d* orbital energies in octahedral complexes.

Electrons in the lower-energy *d* orbitals experience a stabilization relative to the average energy of the *d* orbitals, called the crystal-field stabilization energy. In strong-field complexes the splitting of *d* orbital energies is large enough to overcome spin-pairing energies, and *d* electrons preferentially pair up in the lower-energy orbitals. Such complexes are referred to as low spin. When the ligands exert a relatively weak crystal field the electrons occupy the higher-energy *d* orbitals in preference to pairing up in the lower-energy set, and the complexes are high spin.

The crystal-field model is applicable also to tetrahedral and square-planar complexes. However, the ordering of the *d* orbital energies is different than in octahedral complexes.

Learning goals

Having read and studied this chapter, you should be able to:

1 Determine either the charge of a complex ion, having been given the oxidation state of the metal, or the oxidation state, having been given the charge of the complex. (You will need to recognize the common ligands and their charges.)

2 Describe, with the aid of drawings, the common geometries of complexes. (You will need to recog-

nize whether the common ligands are functioning as monodentate or polydentate ligands.)

3 Name coordination compounds, having been given their formulas, or write their formulas, having been given their names.

4 Describe the common types of isomerism and distinguish between structural isomerism and stereoisomerism.

5 Determine the possible number of stereoisomers for a complex, having been given its composition.

6 Distinguish between inert and labile complexes.

7 Explain how the conductivity, precipitation reactions, and isomerism of complexes are used to infer their structures.

8 Explain how the magnetic properties of a compound can be measured and used to infer the number of unpaired electrons.

9 Explain how the colors of substances are related to their absorption and reflection of incident light.

10 Explain how the electrostatic interaction between ligands and metal *d* orbitals in an octahedral complex results in a splitting of energy levels.

11 Explain the significance of the spectrochemical series.

12 Account for the tendency of electrons to pair in strong-field, low-spin complexes.

13 Explain what is meant by the term *crystal-field stabilization energy,* and be able to calculate the total CFSE for a complex.

14 Account for the larger-than-expected enthalpies of hydration among the divalent metal ions of the transition metals.

15 Sketch a representation of the *d* orbital energy levels in a tetrahedral complex and explain the reason for a smaller crystal-field splitting in this geometry as compared with octahedral complexes.

16 Sketch the *d*-orbital energy levels in a square-planar complex.

Key terms

Among the more important terms and expressions used for the first time in this chapter are the following:

A **chelating agent** (Section 24.2) is a polydentate ligand that is capable of occupying two or more sites in the coordination sphere.

Chiral (Section 24.4) means having a nonsuperimposable mirror image; for example, we might refer to a molecule as being chiral.

A ***cis*** geometrical arrangement (Section 24.4) refers to one with like groups adjacent to each other.

The **coordination number** (Section 24.1) of an atom is the number of adjacent atoms to which it is directly bonded; in a complex, the coordination number of the metal ion is the number of donor atoms to which it is bonded.

The **coordination sphere** (Section 24.1) of a complex consists of the metal ion and its surrounding ligands.

Coordination-sphere isomers (Section 24.4) are structural isomers of coordination compounds in which the ligands within the coordination sphere differ.

Crystal-field stabilization energy, CFSE (Section 24.8) is the stabilization, as compared with the average *d* orbital energy, that results when an electron is placed in a lower-energy *d* orbital in an octahedral complex.

The **crystal-field theory** (Section 24.8) accounts for the colors and magnetic and other properties of transition metal complexes in terms of the splitting of the energies of metal-ion *d* orbitals by the electrostatic interaction with the ligands.

The term **dextrorotatory,** or merely **dextro** or ***d*** (Section 24.4), is used to label a chiral molecule that rotates the plane of polarization of plane-polarized light to the right (clockwise).

The **donor atom** (Section 24.1) of a ligand is the one that bonds to the metal.

Geometric isomers (Section 24.4) have different spatial arrangements of donor atoms around the coordination sphere.

A **high-spin** complex (Section 24.8) has the same number of unpaired electrons as does the isolated metal ion.

An **inert** complex (Section 24.5) exchanges ligands at a slow rate.

Isomers (Section 24.4) are compounds whose molecules have the same overall composition but different structures.

A **labile** complex (Section 24.5) exchanges ligands at a rapid rate.

The term **levorotatory,** or merely **levo** or ***l*** (Section 24.4), is used to label a chiral molecule that rotates the plane of polarization of plane-polarized light to the left (counterclockwise).

A **ligand** (Section 24.1) is an ion or molecule that coordinates to a single central metal atom or to an ion to form a complex.

Linkage isomers (Section 24.4) are structural isomers of coordination compounds in which a ligand differs in its mode of attachment to a metal ion.

A **low-spin** complex (Section 24.8) has fewer unpaired electrons than does the isolated metal ion.

A monodentate ligand (Section 24.2) is one that binds to the metal ion via a single donor atom. It occupies one position in the coordination sphere.

Optical isomers (Section 24.4) are stereoisomers in which the two forms of the compound are nonsuperimposable mirror images.

A polydentate ligand (Section 24.2) is one in which two or more donor atoms can coordinate to the same metal ion. In a bidentate ligand the number of coordinating atoms bound to the metal is two.

The spectrochemical series (Section 24.8) is a list of ligands arranged in order of their abilities to split the *d*-orbital energies (using the terminology of the crystal-field model).

Stereoisomers (Section 24.4) are compounds possessing the same formula and bonding arrangement but differing in the spatial arrangements of the atoms.

Structural isomers (Section 24.4) are compounds possessing the same formula but differing in the bonding arrangements of the atoms.

The *trans* geometrical arrangement (Section 24.4) refers to one with like groups opposite each other.

EXERCISES

Structure and Nomenclature

24.1 Indicate the coordination number about the metal in each of the following compounds: (a) $[Zn(en)_2]Br_2$; (b) $[Co(NH_3)_4Cl_2]Cl$; (c) $K[Co(C_2O_4)_2(NH_3)_2]$; (d) $K_2[MoOCl_4]$; (e) $K[Au(CN)_2]$.

24.2 Indicate the oxidation number of the central metal ion in each of the following coordination compounds: (a) $K_2[V(C_2O_4)_3]$; (b) $K_4[Fe(CN)_6]$; (c) $[Pd(NH_3)_2Cl_2]$; (d) $[Cr(en)_2F_2]NO_3$.

24.3 Sketch the structure of each of the following complexes: (a) $[Zn(NH_3)_4]^{2+}$; (b) $[PtCl_2(en)]$; (c) $[Ag(CN)_2]^-$; (d) *trans*-$[PtH(Br)(ONO)_2]^{2-}$.

24.4 Sketch the structure of each of the following complexes: (a) *cis*-$[CrCl_2(en)_2]^+$; (b) $[FeBr_4]^-$; (c) *trans*-$[Pt(NH_3)_4Cl_2]^{2+}$; (d) $[Cd(en)(SCN)_2]$.

24.5 Name each of the coordination compounds or complex ions listed in Exercises 24.3 and 24.4.

24.6 Name each of the following complexes: (a) $[Co(NO_2)_3(NH_3)_3]$; (b) $Cs[Cr(C_2O_4)_2Cl_2]$; (c) $K_2[NiCl_4]$; (d) $[Pd(en)_2][Cr(NH_3)_2Br_4]_2$; (e) $K_3[IrCl_5(S_2O_3)]$; (f) $[Co(en)(H_2O)F_3]$.

24.7 Using Br^- and NH_3 as ligands, (a) give the formula of a six-coordinate palladium(IV) complex that would be a nonelectrolyte in aqueous solution; (b) give the coordination compound of Pt(II) that has about the same electrical conductivity as RbBr; (c) give an octahedral complex of Fe(III) containing two NH_3 groups.

24.8 From the following list of metal complexes and simple salts, pair up each complex with a simple salt that is likely to have about the same electrical conductivity in solution: (a) $Pt(NH_3)_4Cl_4$; (b) $K_3Cr(C_2O_4)_3$; (c) $Co(NH_3)_4Cl_3$; (d) CsCl; (e) $Sr(NO_3)_2$; (f) $ScBr_3$. Write the formula for each complex with the brackets placed to show properly the species that lie within the coordination sphere.

24.9 Write the formulas for each of the following compounds, being sure to use brackets to indicate the coordination sphere: (a) hexaamminenickel(II) nitrate; (b) tetraamminediazidocobalt(III) fluoride; (c) potassium diaquatetranitritovanadate(III); (d) dichlorobis(ethylenediamine)cobalt(III) chloride; (e) bis(ethylenediamine) zinc(II) tetrabromomercurate(II); (f) pentaamminethiosulfatomanganese(III) sulfate.

24.10 Polydentate ligands can vary in the number of coordination positions they occupy. In each of the following indicate the most probable number of coordination positions occupied by the polydentate ligand present: (a) $[Co(NH_3)_4CO_3]Cl$; (b) $[Cr(C_2O_4)_2(H_2O)_2]Br$; (c) $[Cr(EDTA)(H_2O)]^-$; (d) $[Zn(en)_2](ClO_4)_2$.

Isomerism

24.11 By writing formulas or drawing structures related to any one of the following complexes, illustrate (a) geometrical isomerism; (b) linkage isomerism; (c) optical isomerism; (d) coordination-sphere isomerism. The complexes are: $[Co(NH_3)_4(C_2O_4)]Cl_2$; $[Pd(NH_3)_2(ONO)_2]$; and *cis*-$[V(en)_2Cl_2]^+$.

24.12 A palladium complex formed from a solution containing bromide ion and pyridine, C_5H_5N (a good donor toward metal ions), is found on elemental analysis to contain 37.6 percent bromine, 28.3 percent carbon, 6.60 percent nitrogen, and 2.37 percent hydrogen. The compound is slightly soluble in several organic solvents; its solutions in water or alcohol do not conduct electricity. It is found experimentally to have a zero dipole moment. Write the chemical formula and indicate its probable structure.

24.13 What types of isomerism are present in each of the following pairs of compounds: (a) $[CoCl(H_2O)(en)_2]Cl_2$ and $[CoCl_2(en)_2]Cl \cdot H_2O$; (b) $[PtCl_2(NH_3)_4]Br_2$ and $[PtBr_2(NH_3)_4]Cl_2$; (c) $[Co(NH_3)_5SCN](NO_3)_2$ and $[Co(NH_3)_5NCS](NO_3)_2$?

24.14 Which of the compounds listed in Exercise 24.13 are capable of stereoisomerism? In each case indicate the type of stereoisomerism involved, and draw the structures of the isomers.

24.15 Two forms of a platinum complex having identical chemical formulas have different colors and different solubilities in various solvents. Neither substance forms an electrically conducting solution when dissolved in water. On the basis of this limited information, what types of isomerism are possibly involved? Which types are definitely excluded by the data?

24.16 Sketch all possible isomeric structures (there might be only one) for each of the following complex ions or coordination compounds: (a) $[Cd(en)Cl_2]$; (b) $[Fe(C_2O_4)_2(NH_3)Cl]^{2-}$; (c) $[PdCl_2(SCN)_2]^{2-}$; (d) $[IrBr_3Cl(H_2O)_2]$.

[24.17] Diethylenetriamine, $H_2NCH_2CH_2NHCH_2CH_2NH_2$ (dien), can act as a tridentate ligand. Indicate all the possible isomeric forms and the types of isomerism involved for each of the following coordination compounds or complex ions: (a) $[Cr(dien)_2]^{3+}$; (b) $[Cd(dien)Br]^+$; (c) $[Co(dien)Br_3]$.

24.18 The compound $Co(NH_3)_5(SO_4)Br$ exists in two forms, one red and one violet. Both forms dissociate in solution to form two ions. Solutions of the red compound form a precipitate of AgBr on addition of $AgNO_3$ solution, but no precipitate of $BaSO_4$ on addition of $BaCl_2$ solution. For the violet compound, just the reverse occurs. From this evidence, indicate the structures of the complex ions in each case, and give the correct name of each compound.

24.19 Write the formulas for, and properly name, two possible coordination sphere isomers with the formula $Cr(H_2O)_4(OH)Br_2$.

[24.20] Write formulas showing the coordination spheres and give the correct names for two different substances with formula $PtCu(NH_3)_4Cl_4$.

Color; magnetism; crystal-field theory

24.21 What is the observed color of a coordination compound that absorbs radiation of wavelength about 580 nm?

24.22 Give the number of *d* electrons associated with the central metal in each of the following complexes: (a) $[Co(CN)_5]^{3-}$; (b) $[AuCl_4]^-$; (c) $[V(NH_3)_3Cl_3]^-$; (d) $[Ru(en)_3]^{2+}$; (e) $[Mn(CN)_6]^{3-}$.

24.23 Explain why many cyano complexes of divalent transition metal ions are yellow, whereas many aqua complexes of these ions are blue or green.

24.24 Assuming that the gemstones shown listed in Figure 24.26 have just one absorption in the visible region of the spectrum, predict the approximate wavelength of the absorption maximum.

24.25 What characteristics of the *d* orbitals are responsible for their splitting into two groups in an octahedral field of ligands?

24.26 For each of the following metals, write the electronic configuration of the metal atom and the +3 ion: (a) Rh; (b) Mn; (c) Pd. Draw the crystal-field energy-level diagram for the *d* orbitals of an octahedral complex and show the placement of the *d* electrons in each case, assuming a strong-field complex.

24.27 Classify the following complexes as either high spin or low spin: (a) $[Mn(H_2O)_6]^{3+}$ (two unpaired electrons); (b) $[Ru(CN)_6]^{3-}$ (one unpaired electron); (c) $[Cr(en)_3]^{2+}$ (four unpaired electrons).

24.28 The value of Δ for the $[CrF_6]^{3-}$ complex is 182 kJ/mol. Calculate the expected wavelength of the absorption corresponding to promotion of an electron from the lower-energy to the higher-energy *d*-orbital set in this complex. Should the complex absorb in the visible range? (You may need to review Sample Exercise 5.2; remember to divide by Avogadro's number.)

24.29 How would you go about determining the number of unpaired electrons in the compound $Na_2[CoCl_4]$? If the compound proved to have three unpaired electrons, how would you account for this in terms of crystal-field theory?

24.30 The compound $[Fe(bipy)_3](NO_3)_3$ (where bipy represents a bidentate nitrogen ligand) is paramagnetic. The magnitude of the paramagnetism indicates that there is one unpaired electron in the metal complex. Sketch the energy-level diagram for the complex and indicate the placement of electrons. Is the complex characterized by high spin or by low spin?

24.31 Draw the crystal-field energy-level diagrams and show the placement of electrons for the following complexes: (a) $[ZrCl_6]^{2-}$; (b) $[MnF_6]^{3-}$ (a high-spin complex); (c) $[Rh(NH_3)_6]^{3+}$ (a low-spin complex); (d) $[NiCl_4]^{2-}$; (e) $[PtBr_4]^{2-}$; (f) $[Cu(en)_3]^{2+}$.

24.32 Calculate the total CFSE for each of the following complexes: (a) $[HfCl_6]^{3-}$; (b) $[MnF_6]^{3-}$ (a high-spin complex); (c) $[IrCl_6]^{3-}$ (a low-spin complex); (d) $[V(NH_3)_6]^{3+}$.

[24.33] Both Fe^{2+} and Fe^{3+} complexes of Cl^- and Br^- are known, but only Fe^{2+} complexes of I^-. Suggest a reason for this.

[24.34] Suppose that a transition metal ion were located in a lattice in which it was in contact with just two nearby anions, located along the + and − axis directions. Diagram the splitting of the metal *d* orbitals that would result from such a crystal field. Assuming a strong field, how many unpaired electrons would you expect for a metal ion with six *d* electrons?

Additional exercises

24.35 Distinguish between the following terms: (a) a chelate and monodentate ligand; (b) coordination sphere and coordination number; (c) a labile and inert complex; (d) high-spin and low-spin complex; (e) coordination-sphere isomerism and linkage isomerism; (f) optical isomerism and geometrical isomerism.

24.36 Based on the molar conductance values listed below for the series of platinum(IV) complexes, write the formula for each complex so as to show which ligands are in the coordination sphere of the metal.

Complex	Molar conductance (Ω^{-1})[a] of 0.05 M solution
$Pt(NH_3)_6Cl_4$	523
$Pt(NH_3)_4Cl_4$	228
$Pt(NH_3)_3Cl_4$	97
$Pt(NH_3)_2Cl_4$	0
$KPt(NH_3)Cl_5$	108

[a]The ohm (Ω) is a unit of resistance; conductance is the inverse of resistance.

24.37 In Werner's early studies he observed that when the complex $[Co(NH_3)_4Br_2]Br$ was placed in water, the electrical conductivity of a 0.05 M solution changed from an initial value of 191 Ω^{-1} to a final value of 374 Ω^{-1} over a period of an hour or so. Suggest an explanation of the observed results. Write a balanced chemical equation to describe the reaction.

24.38 Solutions containing the $[Co(H_2O)_6]^{2+}$ ion absorb at about 520 nm; those containing the $[CoCl_4]^{2-}$ ion absorb at about 690 nm. What colors do you expect for the solutions? Why is the absorption maximum for the $[CoCl_4]^{2-}$ ion at longer wavelength than for the $[Co(H_2O)_6]^{2+}$ ion?

24.39 Draw out the *d*-orbital energy-level diagram for each of the following complexes, and indicate the most likely placement of *d* electrons: (a) $[Au(CN)_4]^-$; (b) $[Ni(C_2O_4)_3]^{2-}$; (c) $[PtF_6]$; (d) $[Rh(CN)_6]^{2-}$.

24.40 The complex $[Co(NH_3)_6]^{3+}$ contains no unpaired electrons, whereas the complex $[Mn(NH_3)_6]^{2+}$ contains five. Account for the difference in terms of the crystal-field model.

24.41 Acetylacetone forms very stable complexes with many metallic ions. It acts as a bidentate ligand, coordinating to the metal at two adjacent positions. Suppose that one of the CH_3 groups of the ligand is replaced by a CF_3 group, as shown:

Trifluoromethyl acetylacetonate (tfac)

$$\left[CF_3{-}\underset{\underset{:O:}{\|}}{C}{-}\overset{\overset{H}{|}}{C}{=}\underset{\underset{:\ddot{O}:}{|}}{C}{-}CH_3 \right]^-$$

Sketch all possible isomers for the tris complex of tfac with cobalt(III). (You can use the symbol ●‿○ to represent the ligand.)

24.42 What changes would you expect in the absorption spectrum of complexes of V(III) as the ligand is varied in the order F^-, NH_3, NO_2^-?

24.43 Indicate whether each of the following statements is true or false. If it is false, indicate how it would need to be modified to make it true. (a) Spin pairing energy is larger than Δ in low-spin complexes. (b) Δ is larger for complexes of Mn^{3+} with a given ligand than for complexes of Mn^{2+}. (c) $[NiCl_4]^{2-}$ is more likely to be square-planar than is $[Ni(CN)_4]^{2-}$.

24.44 The absorption maxima in the visible absorption spectra of several complexes are as follows: (a) $[Co(NH_3)_6]^{3+}$, 470 nm; (b) *trans*-$[Co(NH_3)_4(NO_2)_2]^+$, 440 nm; (c) *cis*-$[Co(NH_3)_4(H_2O)_2]^{3+}$, 510 nm; (d) *cis*-$[Co(en)_2Cl_2]^+$, 535 nm. Predict the color of each complex.

24.45 Many trace metal ions exist in the bloodstream as complexes with amino acids (Figure 26.3) or small peptides. The anion of the amino acid glycine,

$$H_2NCH_2\overset{\overset{O}{\|}}{C}{-}O^-$$

symbol gly, is capable of acting as a bidentate ligand, coordinating to the metal through nitrogen and $-O^-$ atoms. How many isomers are possible for (a) $[Zn(gly)_2]$ (tetrahedral); (b) $[Pt(gly)_2]$ (square planar); (c) $[Co(gly)_3]$ (octahedral). Sketch all possible isomers. Use N‿O to represent the ligand.

24.46 Write balanced chemical equations to represent the following observations. (In some instances, the complex involved has been discussed previously at some point in the text.) (a) Solid silver chloride dissolves in an excess of aqueous ammonia. (b) The green complex $[Cr(en)_2Cl_2]Cl$, on treatment with water over a long time converts to a brown-orange complex. Reaction of $AgNO_3$ with a molar solution of the product results in precipitation of 3 mol of AgCl. (Write *two* reactions.) (c) Insoluble zinc hydroxide dissolves in excess aqueous ammonia. (d) A pink solution of $Co(NO_3)_2$ turns deep blue on addition of concentrated hydrochloric acid.

24.47 From the following Δ values for each complex, calculate the total CFSE: (a) $[CrF_6]^{3-}$, 182 kJ/mol; (b) $[Cr(NH_3)_6]^{3+}$, 258 kJ/mol; (c) $[MoCl_6]^{3-}$, 230 kJ/mol; (d) $[Rh(CN)_6]^{3-}$, 545 kJ/mol.

24.48 In each of the following pairs of complexes, which would you expect to absorb at the longer wavelength: (a) $[CoF_6]^{4-}$, $[Co(CN)_6]^{4-}$; (b) $[V(H_2O)_6]^{2+}$, $[V(H_2O)_6]^{3+}$; (c) $[Mn(CN)_6]^{3-}$, $[MnCl_4]^-$? Explain your reasoning in each case.

24.49 Draw the geometrical structure of each of the following complexes: (a) $[Pt(en)_2]^{2+}$; (b) *cis*-dichloroethylenediamineoxalatoferrate(III) ion; (c) *trans*-diamminedibromopalladium(II); (d) *trans*-$[Co(en)_2Br_2]^+$; (e) tetrabromocobaltate(II) ion; (f) *trans*-$[Mo(NCS)_2(en)_2]^+$; (g) pentaamminebromovanadium(II) ion.

[24.50] The red color of ruby is due to the presence of Cr(III) ions in octahedral sites in the close-packed oxide lattice. Draw the crystal-field splitting diagram for Cr(III) in this environment. Suppose that the ruby crystal is subjected to high pressure. What do you predict for the variation in the wavelength of absorption of the ruby as a function of pressure? Explain.

[24.51] The d^3 and d^6 electronic configurations are favorable for octahedral coordination, but not for tetrahedral. Explain why this is so in terms of crystal-field theory.

25 Organic Chemistry

Organic chemistry deals mainly with compounds in which carbon is a principal element. There are no precise boundaries between organic chemistry and other areas of chemical science. The concept of a separate area of chemistry that could be thought of as organic developed out of the vitalist theory, which held that substances that make up living matter are fundamentally different from those that form inanimate matter. In 1828, Friedrich Wöhler, a German chemist, reacted potassium cyanate, KOCN, with ammonium chloride, NH_4Cl; much to his surprise he obtained urea, H_2NCONH_2, a well-known substance that had been isolated from the urine of mammals. Following Wöhler, many other organic substances were prepared from inorganic starting materials, and the vitalist theory gradually disappeared. Nevertheless organic chemistry continued to develop as a distinct area of chemistry. This may be explained partly by the fact that the raw materials for much of organic chemistry—oil, coal, wood, animal matter, and so forth—are of plant or animal origin.

In this chapter, we shall present a brief view of some of the elementary aspects of organic chemistry. We can do no more than hint at the magnitude of the subject. It has been estimated that there are now more than a million known organic substances. Each year several thousand new organic substances are discovered in nature or synthesized in the laboratory. These huge numbers might lead one to think that learning the subject of organic chemistry is hopelessly difficult. However, certain arrangements of atoms and groups of atoms, called **functional groups,** occur repeatedly in organic substances. These arrangements lead to particular chemical characteristics that are very similar among the compounds containing that functional group. Thus, by learning the characteristic chemical properties of these functional groups, you can understand the chemical characteristics of many organic substances.

25.1 THE HYDROCARBONS

Hydrocarbons contain only two elements, carbon and hydrogen. With such a limited range of composition, you might suppose that there would be little variety in the chemical properties of the hydrocarbons. However, such is not the case. The key structural feature of hydrocarbons, and for that matter of most other organic substances, is the presence of stable carbon-carbon bonds. Carbon alone among the elements is able to form stable, extended chains of atoms bonded through single, double, or triple bonds. No other element is capable of forming similar structures.

The hydrocarbons can be divided into four groups, the alkanes, alkenes, alkynes, and aromatic hydrocarbons. We have already encountered at least one member from each of these series. Figure 25.1 shows the name, molecular formula, and geometrical structure of the simplest member that contains a carbon-carbon bond in each series.

The alkanes consist of carbon atoms bonded either to hydrogen or to other carbon atoms by four single bonds. Depending on its place in the alkane structure, a carbon atom may be bonded to three hydrogens and one carbon, two hydrogens and two carbons, a hydrogen and three car-

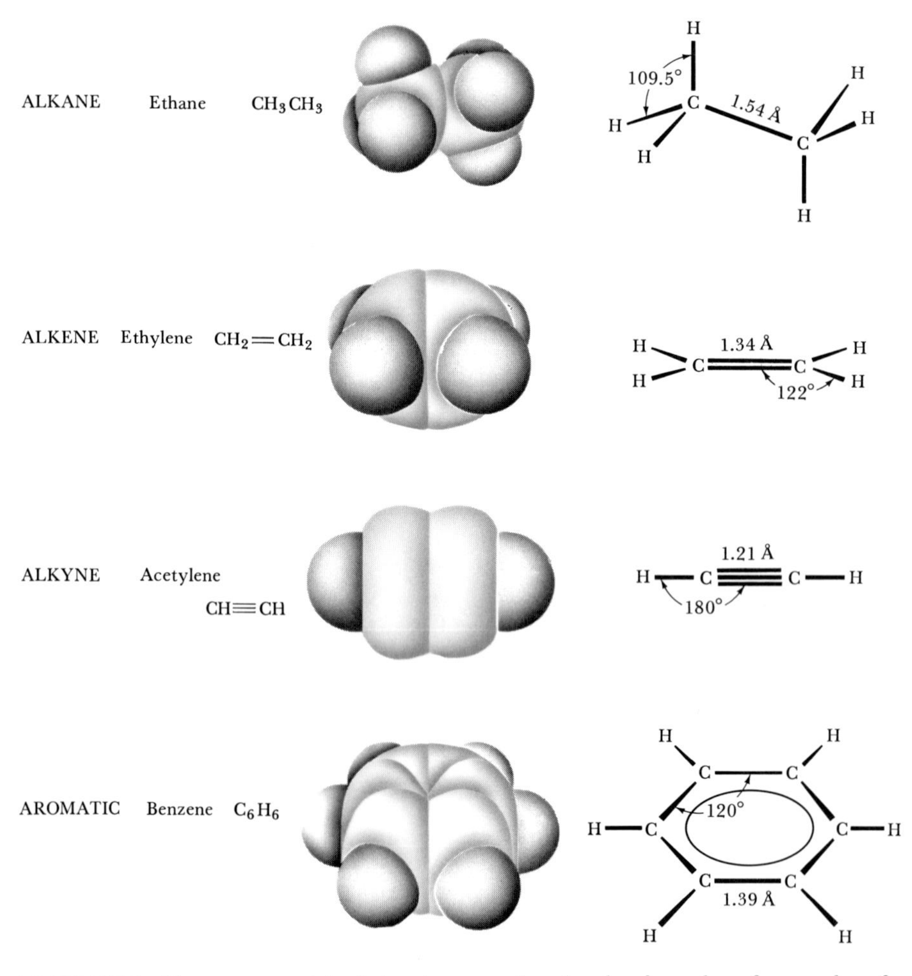

FIGURE 25.1 Names, geometrical structures, and molecular formulas of examples of each type of hydrocarbon.

bons, or four carbons. The alkenes are hydrocarbons with one or more carbon-carbon double bonds. The simplest member of the alkene series is ethylene; you might wish to review the electronic structure of ethylene, discussed in Section 8.4. In the alkynes, there is at least one carbon-carbon triple bond, as in the simplest member of the series, acetylene (Section 8.4). In the aromatic hydrocarbons the carbon atoms are connected in a planar ring structure, joined by both σ and π bonds between carbon atoms. Benzene is the best-known example of an aromatic hydrocarbon. Other examples are illustrated in Figure 8.14. The nonaromatic hydrocarbons, that is, the alkanes, alkenes, and alkynes, are referred to as **aliphatic** compounds to distinguish them from aromatic substances.

The members of the different series of hydrocarbons exhibit different chemical behaviors, as we shall see shortly. However, the hydrocarbons are very similar in many ways. Because carbon and hydrogen are not greatly different in electronegativity (2.5 for carbon, 2.2 for hydrogen), the C—H bond is not very polar. Hydrocarbons are formed entirely from C—H bonds and bonds between carbon atoms; this means that hydrocarbon molecules are relatively nonpolar. Thus they are very much unlike water; hydrocarbons are almost completely insoluble in water. Those that are liquids are good solvents toward nonpolar molecules, but poor solvents toward ionic substances, such as sodium chloride, or polar substances, such as NH_3.

The Alkanes

Table 25.1 lists several of the simplest alkanes. Many of these substances are familiar because of their widespread use. Methane is a major component of natural gas and is used for home heating and in gas stoves and hot-water heaters. Propane is the major component of bottled, or LP, gas used for home heating, cooking, and so forth in areas where natural gas is not available. Butane is used in the disposable lighters sold in drug stores and in the fuel cannisters for gas camping stoves and lanterns. Alkanes with from 5 to 12 carbon atoms per molecule are found in gasoline.

The formulas for the alkanes given in Table 25.1 are written in a notation called the condensed structural formula. This notation reveals the way in which atoms are bonded to one another, but does not require

TABLE 25.1 First several members of the straight-chain alkane series

Molecular formula	Condensed structural formula	Name	Boiling point (°C)
CH_4	CH_4	Methane	−161
C_2H_6	CH_3CH_3	Ethane	−89
C_3H_8	$CH_3CH_2CH_3$	Propane	−44
C_4H_{10}	$CH_3CH_2CH_2CH_3$	Butane	−0.5
C_5H_{12}	$CH_3CH_2CH_2CH_2CH_3$	Pentane	36
C_6H_{14}	$CH_3CH_2CH_2CH_2CH_2CH_3$	Hexane	68
C_7H_{16}	$CH_3CH_2CH_2CH_2CH_2CH_2CH_3$	Heptane	98
C_8H_{18}	$CH_3CH_2CH_2CH_2CH_2CH_2CH_2CH_3$	Octane	125
C_9H_{20}	$CH_3CH_2CH_2CH_2CH_2CH_2CH_2CH_2CH_3$	Nonane	151
$C_{10}H_{22}$	$CH_3CH_2CH_2CH_2CH_2CH_2CH_2CH_2CH_2CH_3$	Decane	174

drawing in all the bonds. For example, the Lewis structure and condensed structural formula for butane, C_4H_{10}, are:

```
     H   H   H   H
     |   |   |   |
 H—C—C—C—C—H
     |   |   |   |
     H   H   H   H
```

Lewis structure

$CH_3CH_2CH_2CH_3$

Condensed structural formula

We shall frequently use either Lewis structures or condensed structural formulas to represent organic compounds. You should practice drawing the structural formulas from the condensed ones. To aid you in this task, notice that each carbon atom in an alkane has four single bonds, while each hydrogen atom forms one single bond.

Notice that each succeeding compound in the series listed in Table 25.1 is related to the one before it by the addition of a CH_2 unit. A series such as that shown in Table 25.1 is known as a **homologous series.** The general formula for all the compounds listed in the table is C_nH_{2n+2}, where n is the number of carbon atoms. One of the characteristics of a homologous series is that all the compounds of the series can be described by the same general formula. We shall see several other examples of homologous series as we proceed.

The alkanes listed in Table 25.1 are called straight-chain alkanes, because all the carbon atoms are joined in a continuous chain. However, for alkanes consisting of four or more carbon atoms, other arrangements of the carbon atoms, consisting of branched chains, are possible. Figure 25.2 shows the Lewis structures and condensed structural formulas for all the possible structures of alkanes containing four or five carbon atoms. Notice that the two possible forms of butane have the same molecular formula, C_4H_{10}. Similarly, the three possible forms of pentane have the same molecular formula, C_5H_{12}. Compounds with the same molecular formula, but with different structures, are called **isomers.** The isomers of a given alkane differ slightly from one another in physical properties. By way of illustration, the melting and boiling points (°C) of the isomers of butane and pentane are given in Figure 25.2. The number of possible isomers increases rapidly with the number of carbon atoms in the alkane. For example, there are 18 possible isomers of octane, C_8H_{18}, and 75 possible isomers of decane, $C_{10}H_{22}$.

Nomenclature of Alkanes

The first names given to the structural isomers shown in Figure 25.2 are the so-called common names. The straight-chain isomer is referred to as the normal isomer, abbreviated by the prefix *n*-. The isomer in which one CH_3 group is branched off the major chain is labeled the iso-isomer; for example, isobutane. However, as the number of isomers grows, it becomes impossible to find a suitable prefix to denote each isomer. The need for a systematic means of naming organic compounds was recognized early in the history of organic chemistry. In 1892 an organization called the International Union of Chemistry met in Geneva, Switzerland, to formulate rules for systematic naming of organic substances.

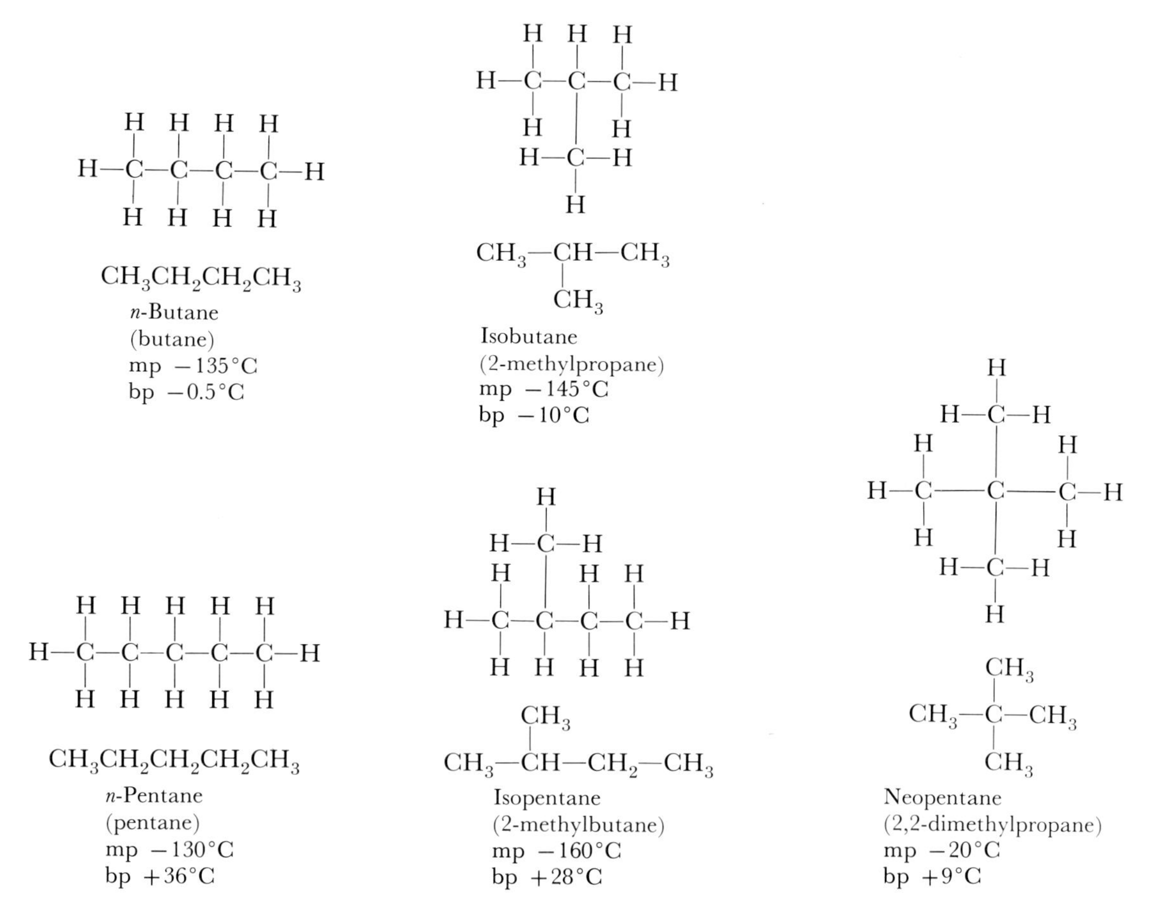

FIGURE 25.2 Possible structures, names, and melting and boiling points of alkanes of formula C_4H_{10} and C_5H_{12}.

Since that time the task of keeping the rules for naming compounds up to date has fallen to the International Union of Pure and Applied Chemistry (IUPAC). It is interesting to note that through two devastating world wars and major social upheavals, the work of IUPAC has continued. Chemists everywhere, regardless of their nationality or political affiliation, subscribe to a common system for naming compounds.

The IUPAC names for the isomers of butane and pentane are the second ones given for each compound in Figure 25.2. The following rules summarize the procedures used to arrive at these names. We shall see that a similar approach is taken to write the names for other organic compounds.

1 Each compound is named for the longest continuous chain of carbon atoms present. For example, the longest chain of carbon atoms in isobutane is three (Figure 25.2). Consequently, this compound is named as a derivative of propane, which has three carbon atoms; in the IUPAC system it is called 2-methylpropane.

2 A CH_3 group that branches off the main hydrocarbon chain is called a methyl group. In general, a group that is formed by removing a hydrogen atom from an alkane is called an **alkyl group.** The names for alkyl

TABLE 25.2 Names and condensed structural formulas for several alkyl groups

Group	Name
$CH_3—$	Methyl
$CH_3CH_2—$	Ethyl
$CH_3CH_2CH_2—$	*n*-Propyl
$CH_3CH_2CH_2CH_2—$	*n*-Butyl
CH_3 (above) / $HC—$ / CH_3 (below)	Isopropyl
CH_3 (above) / $CH_3—C—$ / CH_3 (below)	*t*-Butyl

groups are derived by dropping the *-ane* ending from the name of the parent alkane and adding *-yl*. For example, the methyl group, CH_3, is derived from methane, CH_4; likewise, the ethyl group, C_2H_5, is derived from ethane, C_2H_6. Table 25.2 lists several of the more common alkyl groups.

3 The location of an alkyl group along a carbon-atom chain is indicated by numbering the carbon atoms along the chain. Thus the name 2-methylpropane indicates the presence of a methyl (CH_3) group on the second carbon atom of a propane (three-carbon) chain. In general, the chain is numbered from the end that gives the lowest numbers for the alkyl positions.

4 If there is more than one substituent group of a certain type along the chain, the number of groups of that type is indicated by a prefix: *di-* (two), *tri-* (three), *tetra-* (four), *penta-* (five), and so forth. Thus the IUPAC name for neopentane (Figure 25.2) is 2,2-dimethylpropane. Dimethyl indicates the presence of two methyl groups; the 2,2- prefix indicates that both are on the second carbon atom of the propane chain.

SAMPLE EXERCISE 25.1

Name the following alkane:

```
CH3—CH—CH3
    |
CH3—CH—CH2
        |
        CH3
```

Solution: To name this compound properly, you must first find the longest continuous chain of carbon atoms. This chain, extending from the upper left CH_3 group to the lower right CH_3 group, is five carbon atoms long:

```
(1)CH3—(2)CH—CH3
          |
   CH3—(3)CH—(4)CH2
               |
            (5)CH3
```

The compound is thus named as a derivative of pentane. We could number the carbon atoms starting from either end. However, the IUPAC rules state that the numbering should be done so that the numbers of those carbons which bear side chains are as low as possible. This means that we should start numbering with the upper carbon. There is a methyl group on carbon number two, and one on carbon number three. The compound is thus called 2,3-dimethylpentane.

SAMPLE EXERCISE 25.2

Write the condensed structural formula for 2-methyl-3-ethylpentane.

Solution: The longest continuous chain of carbon atoms in this compound is five. We can therefore begin by writing out a string of five C atoms:

C—C—C—C—C

We next place a methyl group on the second carbon, and an ethyl group on the middle carbon atom of the chain. Hydrogens are then added to all the other carbon atoms to make their covalences equal to four:

```
     CH3
     |
CH3—CH—CH—CH2CH3
        |
        CH2CH3
```

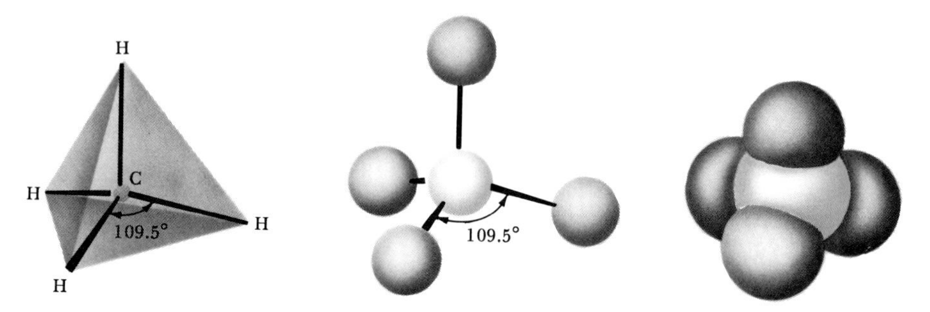

FIGURE 25.3 Representations of the three-dimensional arrangement of bonds about carbon in alkanes.

Structures of Alkanes

The Lewis structures or condensed structural formulas for alkanes do not tell us anything about the three-dimensional structures of these substances. As we would predict from the VSEPR model (Section 8.1), the geometry about each carbon atom in an alkane is tetrahedral; that is, the four groups attached to each carbon are located at the vertices of a tetrahedron. The three-dimensional structures can be represented as shown for methane in Figure 25.3. The bonding may be described as involving sp^3 hybridized orbitals on the carbon, as discussed earlier, in Section 8.2.

One of the characteristics of the carbon-carbon single bond is that rotation about this bond is relatively free. You might imagine grasping the top left methyl group in Figure 25.4, which shows the structure of propane, and twisting it relative to the rest of the structure. Motion of this sort occurs very rapidly in alkanes at room temperature. Thus a long-chain alkane is constantly undergoing motions that cause it to change its shape, something like a length of chain that is being shaken.

One possible structural form for alkanes is that in which the carbon chain forms a ring, or cycle. Alkanes with this form of structure are called **cycloalkanes.** A few examples of cycloalkanes are shown in Figure 25.5. The cycloalkane structures are sometimes drawn as simple polygons, as illustrated in Figure 25.5. In this shorthand notation, each corner of the polygon represents a CH_2 group. This method of representation is similar to that used for aromatic rings, as illustrated in Figure 8.14. In the case of the aromatic structures, each corner represents a CH group.

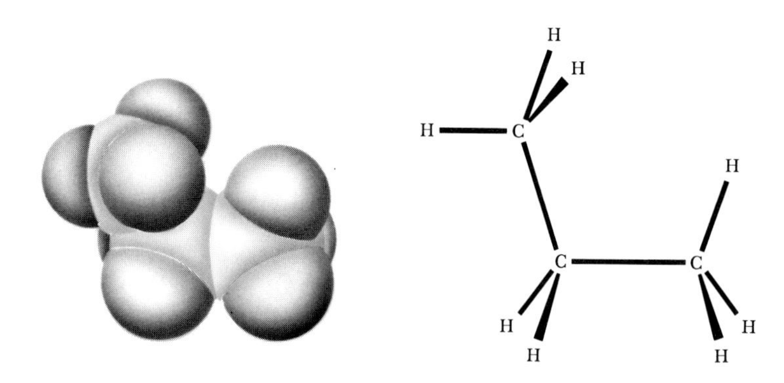

FIGURE 25.4 Three-dimensional models for propane, C_3H_8.

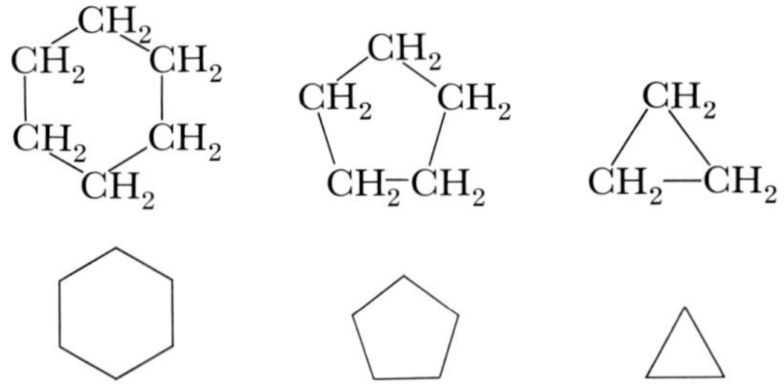

CH_2 CH_2 CH_2 CH_2 CH_2 CH_2 CH_2 CH_2 CH_2 CH_2-CH_2 CH_2 CH_2-CH_2

Cyclohexane	Cyclopentane	Cyclopropane

FIGURE 25.5 Condensed structural formulas of three cycloalkanes.

Carbon rings containing less than six carbon atoms are strained, because the C—C—C bond angle in the smaller rings must be less than the 109.5° tetrahedral angle. The amount of strain increases as the rings get smaller. In cyclopropane, which has the shape of an equilateral triangle, the angle is only 60°; this molecule is therefore much more reactive than either propane, its straight-chain analogue, or cyclohexane, which has no ring strain. Note that the empirical formula for the cycloalkanes is C_nH_{2n}, which differs from that for the straight-chain alkanes. The cycloalkanes thus form a separate homologous series.

Alkenes

The alkenes are close relatives of the alkanes. They differ in that there is at least one carbon-carbon double bond in the molecule. Alkenes are sometimes referred to as **olefins.** The presence of a double bond results in two fewer hydrogens than would be present in an alkane. Because the alkenes possess fewer hydrogens than are needed to form the alkane, they are said to be **unsaturated.** As we shall see a little later, the presence of the double bond confers considerably more chemical reactivity on the alkenes than is found in the alkanes. The simplest alkene is C_2H_4, called ethene, or ethylene. The next member of the series is $CH_3—CH{=}CH_2$, called propene or propylene. When there are more than three carbon atoms in the molecule, there are several possibilities for forming isomers. For example, the possible alkenes with four carbon atoms, and with molecular formula C_4H_8, are shown in Figure 25.6. The first compound shown has a branched chain; the other three all have continuous chains of four carbon atoms. In naming alkenes, the compound is named for the length of the longest continuous chain of carbon atoms that contains the double bond. Of course, the ending of the name listed in Table 25.1 is changed from *-ane* to *-ene*. The location of the double bond is indicated by a prefix number that designates the number of the carbon atom that is part of the double bond and is nearest an end of the chain. The chain is

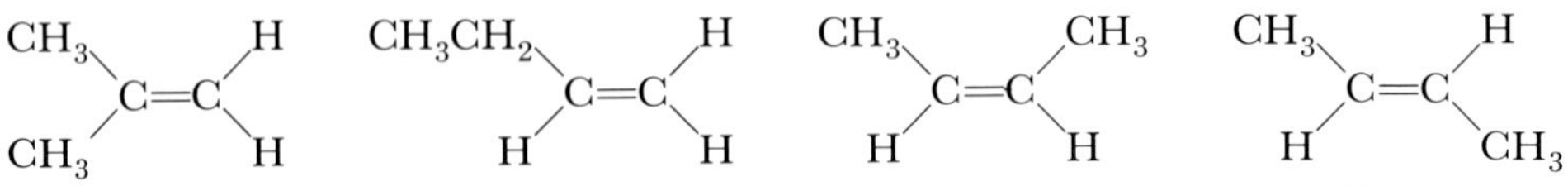

$(CH_3)(CH_3)C{=}C(H)(H)$	$(CH_3CH_2)(H)C{=}C(H)(H)$	$(CH_3)(H)C{=}C(CH_3)(H)$	$(CH_3)(H)C{=}C(H)(CH_3)$
2-Methylpropene	1-Butene	*cis*-2-Butene	*trans*-2-Butene
bp −7°C	bp −6°C	bp 4°C	bp 1°C

FIGURE 25.6 Structures, names, and boiling points of the alkenes with molecular formula C_4H_8.

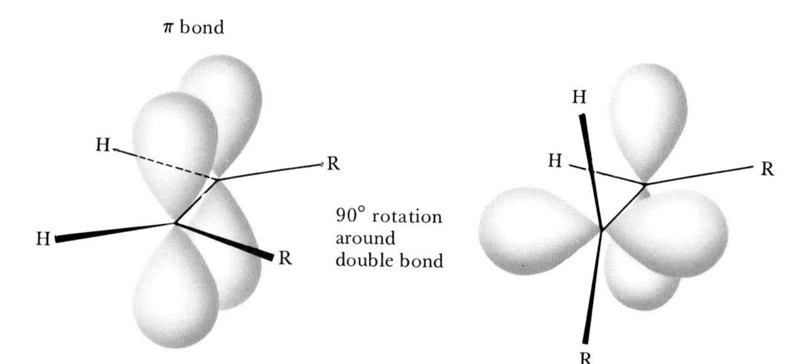

FIGURE 25.7 Schematic illustration of rotation about a carbon-carbon double bond in an alkene. The overlap of the *p* orbitals that form the π bond is lost in the rotation. For this reason, rotation about the carbon-carbon double bonds does not occur readily.

always numbered from the end that brings us to the double bond sooner and hence gives the smallest number prefix. These rules are used to name the isomers of butene that are listed in Figure 25.6. For example, the compound on the left has a three-carbon chain; thus the parent alkene is considered to be propene. The only possible location of the double bond in propene is between the first and second carbons; thus a prefix indicating its location is unnecessary. Numbering the carbon chain from the end closest to the double bond places a methyl group on the second carbon. Thus the name of the isomer is 2-methylpropene.

Cis- and *trans*-2-butene are geometrical isomers; that is, they are compounds that have the same molecular formula and the same groups bonded to one another, but that differ in the spatial arrangement of those groups. In the *cis* form, the two methyl groups are on the same side of the double bond; in the *trans* form they are on opposite sides. We have already met examples of geometrical isomerism, in the chemistry of transition metal coordination compounds (Section 24.4). Geometrical isomers possess distinct physical properties and may even differ significantly in their chemical behavior in certain circumstances.

Recall from our earlier discussion of the geometry about carbon (Section 8.4) that the double bond between two carbon atoms consists of a σ and a π part. Figure 25.7 shows the geometrical arrangement as in a *cis* alkene. The bonding arrangement around each carbon is planar; that is, the carbon-carbon bond axis and the bonds to the other two groups, either hydrogen or carbon, are all in a plane. It is easy to see from Figure 25.7 that rotation about the carbon-carbon double bond will not be easy. Such a rotation would cause the *p* orbitals which form the π bond to lose their overlap, and the π bond would be destroyed. Because rotation about the carbon-carbon bond is difficult the *cis* and *trans* isomers can be separated and studied. If the interconversion were easy, we would always have an equilibrium mixture of the two forms rather than the pure isomers.

SAMPLE EXERCISE 25.3

Name the following compound:

$$\begin{array}{l} \quad\quad\quad\quad\quad\quad\; CH_3 \\ \quad\quad\quad\quad\quad\quad\quad | \\ CH_3CH_2CH_2{-}CH \quad\quad CH_3 \\ \quad\quad\quad\quad\quad\quad\quad\; \diagdown \quad \diagup \\ \quad\quad\quad\quad\quad\quad\quad\;\; C{=}C \\ \quad\quad\quad\quad\quad\quad\quad\; \diagup \quad \diagdown \\ \quad\quad\quad\quad\quad\quad\;\; H \quad\quad\;\; H \end{array}$$

Solution: Because this compound possesses a double bond, it is an alkene. The longest continuous chain of carbons that contains the double bond is seven in length. Thus the parent compound is considered to be a heptene. The double bond begins at carbon number two (numbering from the end closest to the double bond); thus the parent hydrocarbon chain is named 2-heptene. Continuing the numbering along the chain, a methyl group is bound at carbon atom number four. Thus the compound is 4-methyl-2-heptene. Finally, we note that the geometrical configuration at the double bond is *cis;* that is, the alkyl groups are bonded to the double bond on the same side. Thus the full name is 4-methyl-*cis*-2-heptene.

In substances containing two alkene functional groups, each must be located by a number. The ending of the name is altered to identify the number of functional groups: diene (two double bonds), triene (three double bonds), and so forth. For example, 1,4-pentadiene is $CH_2{=}CH{-}CH_2{-}CH{=}CH_2$.

Alkynes

The alkynes are yet another series of unsaturated hydrocarbons, in which there are one or more carbon-carbon triple bonds between carbon atoms. The empirical formula for simple alkynes is C_nH_{2n-2}. The simplest alkyne, acetylene, is a highly reactive molecule. When acetylene is burned in a stream of oxygen in an oxyacetylene torch, the flame reaches a very high temperature, about 3200 K (Section 20.4). The oxyacetylene torch is widely used in welding, where high temperatures are required. Alkynes in general are highly reactive molecules. Because of their higher reactivity, they are not as widely distributed in nature as the alkenes; however, they are important intermediates in many industrial processes.

The alkynes are named by identifying the longest continuous chain in the molecule containing the triple bond, and modifying the ending of the name as listed in Table 25.1 from *-ane* to *-yne,* as shown in the following sample exercise.

SAMPLE EXERCISE 25.4

Name the following compounds:

(a) $CH_3CH_2CH_2{-}C{\equiv}C{-}CH_3$

(b) $CH_3CH_2CH_2\underset{\displaystyle CH_2CH_2CH_3}{\underset{|}{C}H}{-}C{\equiv}CH$

Solution: In (a) the longest chain of carbon atoms is six. There are no side chains. The triple bond begins at carbon atom number two (remember, we always arrange the numbering so that the smallest possible number is assigned to the carbon containing the multiple bond). Thus the name is 2-hexyne.

In (b) the longest continuous chain of carbon atoms is seven; but because this chain does not contain the triple bond we do not count it as derived from heptane. The longest chain containing the triple bond is six, so this compound is named as a derivative of hexyne, 3-propyl-1-hexyne.

Aromatic Hydrocarbons

The aromatic hydrocarbons are a large and important class of hydrocarbons. The simplest member of the series is benzene (see Figure 25.1), with molecular formula C_6H_6. As we have already noted, benzene is a planar, highly symmetrical molecule. The molecular formula for benzene suggests a high degree of unsaturation. One might thus expect that benzene would be highly reactive, and that it might resemble the unsaturated hydrocarbons in reactivity. In fact, however, benzene is not at all similar to alkenes or alkynes in chemical behavior. The great stability of benzene and the other aromatic hydrocarbons as compared with alkenes and alkynes is due to stabilization of the π electrons through delocalization in the π orbitals (Section 8.4).

We can obtain an estimate of the stabilization of the π electrons in benzene by comparing the energy involved in adding hydrogen to benzene to form a saturated compound, as compared with that involved in hydrogenating simple alkenes. The hydrogenation of benzene to form cyclohexane can be represented as

$$\text{C}_6\text{H}_6 + 3\text{H}_2 \longrightarrow \text{(s)} \qquad \Delta H^\circ = -208 \text{ kJ/mol} \qquad [25.1]$$

(The s in the ring on the right indicates that it is a cycloalkane, with CH_2 groups at each corner.) The enthalpy change in this reaction is -208 kJ/mol. The heat of hydrogenation of the cyclic alkene cyclohexene, is -120 kJ/mol:

Cyclohexene

$$+ \text{H}_2 \longrightarrow \text{(s)} \qquad \Delta H^\circ = -120 \text{ kJ/mol} \qquad [25.2]$$

Similarly, the heat released on hydrogenating 1,4-cyclohexadiene is -232 kJ/mol:

1,4-Cyclohexadiene

$$+ 2\text{H}_2 \longrightarrow \text{(s)} \qquad \Delta H^\circ = -232 \text{ kJ/mol} \qquad [25.3]$$

From these last two reactions it would appear that the heat of hydrogenating a double bond is about 116 kJ/mol for each bond. There is the equivalent of three double bonds in benzene. Thus we might expect that the heat of hydrogenating benzene would be about three times -116, or -348 kJ/mol, if benzene behaved as though it were "cyclohexatriene," that is, if it behaved as though it were three double bonds in a ring. Instead, the heat released is much less than this, indicating that benzene is more stable than would be expected for three double bonds. The difference of 140 kJ/mol between -348 kJ/mol and the observed heat of hydrogenation, -208 kJ/mol, can be ascribed to stabilization of the π electrons through delocalization in the π orbitals that extend around the ring.

There is no widely used systematic nomenclature for naming aromatic rings. Each ring system is given a common name; several aromatic compounds are shown in Figure 25.8. The aromatic rings are represented by

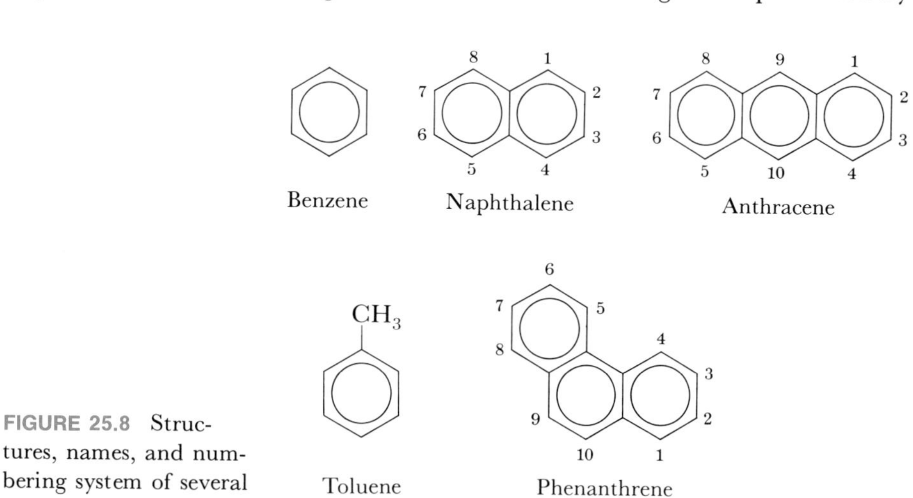

FIGURE 25.8 Structures, names, and numbering system of several aromatic compounds.

hexagons with a circle inscribed inside to denote aromatic character. Each corner represents a carbon atom. Each carbon is bound to three other atoms—either three carbons or two carbons and a hydrogen. The hydrogen atoms are not shown. In naming derivatives of the aromatic hydrocarbons, it is often necessary to indicate the position in the aromatic ring at which some side chain or other group is located. For this purpose the numbering shown in Figure 25.8 is used.

25.2 PETROLEUM

Petroleum is a complex mixture of organic compounds, mainly hydrocarbons, with smaller quantities of other organic compounds containing nitrogen, oxygen, or sulfur. Petroleum is formed over a period of millions of years by the decomposition of marine plants and animals.

The usual first step in the **refining,** or processing, of petroleum is separation of the crude oil into fractions on the basis of boiling points. The fractions commonly taken are shown in Table 25.3. As you might expect, the fractions that boil at higher temperatures are made up of molecules with larger numbers of carbon atoms per molecule. The fractions collected in the initial separation may require further processing to yield a usable product. For example, modifications must be made to the straight-run gasoline obtained from fractionation of petroleum to render it suitable for use as a fuel in automobile engines. Similarly, the fuel oil fraction may need additional processing to remove sulfur before it is suitable for use in an electrical power station or a home heating system. At present, the most commercially important single product from petroleum refining is gasoline.

Gasoline

Gasoline is a mixture of volatile hydrocarbons. Depending on the source of the crude oil, it may contain varying amounts of cyclic alkanes and aromatic hydrocarbons in addition to alkanes. Straight-run gasoline consists mainly of straight-chain hydrocarbons, which in general are not very suitable for use as fuel in an automobile engine. In an automobile engine, a mixture of air and gasoline vapor is ignited by the spark plug at the moment when the gas mixture inside the cylinder has been com-

TABLE 25.3 Hydrocarbon fractions from petroleum

Fraction	Size range of molecules	Boiling-point range (°C)	Uses
Gas	C_1–C_5	−160 to 30	Gaseous fuel, production of H_2
Straight-run gasoline	C_5–C_{12}	30 to 200	Motor fuel
Kerosene, fuel oil	C_{12}–C_{18}	180 to 400	Diesel fuel, furnace fuel, cracking
Lubricants	C_{16} and up	350 and up	Lubricants
Paraffins	C_{20} and up	Low-melting solids	Candles, matches
Asphalt	C_{36}	Gummy residues	Surfacing roads, fuel

TABLE 25.4 Octane numbers of some C_7 and C_8 hydrocarbons

Name	Condensed structural formula	Octane number
n-Heptane	$CH_3CH_2CH_2CH_2CH_2CH_2CH_3$	0
2-Methylhexane	$CH_3CH_2CH_2CH_2-CH(CH_3)-CH_3$	40
Methylcyclohexane	ring of five CH_2 groups and one CH bearing CH_3	75
2,3-Dimethylpentane	$CH_3CH_2-CH(CH_3)-CH(CH_3)-CH_3$	90
2,2,4-Trimethylpentane ("isooctane")	$CH_3-CH(CH_3)-CH_2-C(CH_3)_2-CH_3$	100

pressed by the piston. The burning of the gasoline should create a strong, smooth expansion of gas in the cylinder, forcing the piston outward and imparting force along the drive shaft of the engine. If the gas burns too rapidly, the piston receives a single hard slam rather than a strong, smooth push. The result is a "knocking" or pinging sound; the efficiency with which the energy of gasoline combustion is converted to power is reduced.

Gasolines are rated according to **octane number.** Gasolines with high octane numbers burn more slowly and smoothly and thus are more effective fuels, especially in engines in which the gas-air mixture is highly compressed. It happens that the more highly branched alkanes have higher octane numbers than the straight-chain compounds; some examples are shown in Table 25.4. The octane number of gasoline is obtained by comparing its knocking characteristics with those of "isooctane" (2,2,4-trimethylpentane) and heptane. Isooctane is assigned an octane number of 100, whereas heptane is assigned 0. Gasoline with the same knocking characteristics as a mixture of 95 percent isooctane and 5 percent heptane would be rated as 95 octane.

Because straight-run gasoline contains mostly straight-chain hydrocarbons, it has a low octane number. It is therefore subjected to a process called **cracking** to convert the straight-chain compounds into more desirable branched-chain molecules. Cracking is also used to convert some of the less volatile kerosene and fuel-oil fractions into compounds with lower molecular weights that are suitable for use as automobile fuel. In the cracking process, the hydrocarbons are mixed with a catalyst and heated to 400 to 500°C. The catalysts used are naturally occurring clay minerals, or synthetic Al_2O_3–SiO_2 mixtures. In addition to forming molecules more suitable for gasoline, cracking results in the formation of hydrocarbons of lower molecular weight, such as ethylene and propene. These are used in a variety of processes to form plastics and other chemicals.

The octane number of a given blend of hydrocarbons can be improved by adding an antiknock agent, a substance that helps control the burning rate of the gasoline. The most widely used substances for this purpose are tetraethyl lead, $(CH_3CH_2)_4Pb$, and tetramethyl lead, $(CH_3)_4Pb$. The premium gasolines contain 2 or 3 mL of one of these lead compounds per gallon, with a resultant increase of 10 to 15 in octane rating. Although alkyl lead compounds are undoubtedly effective in improving gasoline performance, their use in gasolines has been drastically curtailed because of the environmental hazards associated with lead. This metal is highly toxic, and there is good evidence that the lead released from automobile exhausts is a general health hazard. Although other substances have been tried as antiknock agents in gasolines, none of these has proved to be an effective and inexpensive antiknock agent that is environmentally safe. The 1975 and later model cars are designed to operate with unleaded gasolines. The gasolines blended for these cars are made up of more highly branched components and more aromatic components, because these have relatively high octane ratings.

25.3 REACTIONS OF HYDROCARBONS

The hydrocarbons are capable of undergoing a variety of reactions with other substances. Many of these reactions are of considerable importance in the chemical industry, because they lead to useful products or to substances that can in turn be converted to useful products.

Oxidation

The most common oxidation reactions of hydrocarbons are those that result in complete oxidation to form carbon dioxide and water:

$$CH_3CH_2CH_3 + 5O_2 \longrightarrow 3CO_2 + 4H_2O \quad [25.4]$$

$$CH_3CH{=}CH_2 + \tfrac{9}{2}O_2 \longrightarrow 3CO_2 + 3H_2O \quad [25.5]$$

$$CH_3C{\equiv}CH + 4O_2 \longrightarrow 3CO_2 + 2H_2O \quad [25.6]$$

$$C_6H_6 + \tfrac{15}{2}O_2 \longrightarrow 6CO_2 + 3H_2O \quad [25.7]$$

These reactions are all highly exothermic. Combustion of the hydrocarbons results in release of energy that can be used to propel an auto or an airplane, to generate steam in the boiler of an electric power plant, or to heat a building. The tremendous demand for petroleum to meet the world's energy needs has led to the tapping of oil wells in such forbidding places as the North Sea and the North Slope of the continent in Alaska, facing the Arctic Ocean (Figure 25.9).

Controlled oxidation of hydrocarbons yields organic substances that contain oxygen in addition to carbon and hydrogen. These products of controlled oxidations will be considered later when we discuss hydrocarbon derivatives.

Addition Reactions

Under appropriate conditions, it is possible to add atoms or groups of atoms to alkenes, alkynes, or aromatic hydrocarbons by disrupting the

FIGURE 25.9 Drilling for oil on the North Slope in Alaska (*Courtesy Exxon Corp.*)

carbon-carbon multiple bonds. For example, ethylene reacts with bromine to form 1,2-dibromoethane, as in Equation 25.8:

$$H_2C{=}CH_2 + Br_2 \longrightarrow \underset{\substack{| \\ Br}}{H_2C}{-}\underset{\substack{| \\ Br}}{CH_2} \qquad [25.8]$$

The pair of electrons that form the π bonds in ethylene are uncoupled and are used to form two new bonds to the two bromine atoms. The σ bond between the carbon atoms remains.

Addition of halogens to alkynes also occurs readily, as in the example shown in Equation 25.9:

$$CH_3C{\equiv}CH + 2Cl_2 \longrightarrow CH_3{-}\overset{\substack{Cl \\ |}}{\underset{\substack{| \\ Cl}}{C}}{-}\overset{\substack{Cl \\ |}}{\underset{\substack{| \\ Cl}}{CH}} \qquad [25.9]$$

1,1,2,2-Tetrachloropropane

Reaction of alkenes or alkynes with hydrogen halides also results in addition:

$$CH_3CH{=}CH_2 + HBr \longrightarrow CH_3\underset{\substack{| \\ Br}}{CH}{-}CH_3 \qquad [25.10]$$

Notice that this reaction might have proceeded to give another product, in which the bromine atom is on the end carbon. It turns out that when a hydrogen halide adds to an alkene, the more electronegative halogen atom always ends up on the carbon atom of the double bond that has the fewer hydrogen atoms. This rule, known as **Markovnikoff's rule**, was first formulated by the Russian chemist V. V. Markovnikoff.

SAMPLE EXERCISE 25.5

Predict the product of reaction of HCl with the following compound:

$$\begin{array}{c} \quad\quad\quad\quad CH_3 \\ \quad\quad\quad\quad | \\ CH_3CH_2C{=}CH \\ \quad\quad | \\ \quad\quad CH_3 \end{array}$$

Solution: The double-bonded carbon atom on the right has one hydrogen on it, the other has none. According to Markovnikoff's rule, the chloride of HCl will end up on the carbon atom on the left:

$$\begin{array}{c} \quad\quad\quad\quad Cl\ \ CH_3 \\ \quad\quad\quad\quad |\quad\ | \\ CH_3CH_2{-}C{-}CH_2 \\ \quad\quad\quad\quad | \\ \quad\quad\quad\quad CH_3 \end{array}$$

Markovnikoff's rule also applies to the addition of unsymmetrical reagents to alkynes. For example, reaction of HBr with 1-butyne yields 2,2-dibromobutane:

$$CH_3CH_2C{\equiv}CH + 2HBr \longrightarrow CH_3CH_2{-}\overset{\overset{\displaystyle Br}{|}}{\underset{\underset{\displaystyle Br}{|}}{C}}{-}CH_3 \qquad [25.11]$$

Using an acid such as H_2SO_4 as catalyst, it is possible to add H_2O to a double bond. The products of such reactions are **alcohols**; that is, compounds containing an OH group bonded to carbon. Markovnikoff's rule applies here as well; we consider the water molecule as being polarized as follows: $H^+{-}OH^-$. Thus the OH group adds to the carbon atom with fewer hydrogens:

$$\underset{\text{Propene}}{CH_3CH{=}CH_2} + HOH \xrightarrow{H_2SO_4} \underset{\text{2-Propanol}}{CH_3{-}\overset{\overset{\displaystyle OH}{|}}{C}H{-}CH_3} \qquad [25.12]$$

Addition of H_2 to an alkene converts it into an alkane. This reaction, referred to as hydrogenation, does not occur readily under ordinary temperature and pressure conditions. One of the reasons for the lack of reactivity of H_2 toward an alkene is the high bond energy of the H_2 bond. To promote the reaction it is necessary to use a catalyst that assists in the rupture of the H—H bond. The most widely used catalysts are heterogeneous and consist of finely divided metals on which H_2 is adsorbed. The action of these heterogeneous catalysts in reaction of H_2 with an alkene is described in detail in Section 13.6. Molecular hydrogen also reacts with alkynes in the presence of a catalyst to yield alkanes, as in the example of Equation 25.13.

$$CH_3C{\equiv}CH + 2H_2 \xrightarrow{Ni} CH_3CH_2CH_3 \qquad [25.13]$$

Addition reactions of aromatic hydrocarbons proceed much less readily than such reactions of alkenes and alkynes. For example, benzene does not react at all with Cl_2 or Br_2 under ordinary conditions. However, if conditions are sufficiently rigorous, such addition reactions may be forced to occur.

Substitution Reactions

In a substitution reaction of a hydrocarbon, one or more hydrogen atoms are replaced by other atoms or groups. Substitution reactions are difficult to carry out with aliphatic compounds. One of the most important substitution reactions of alkanes involves replacement of hydrogen by a halogen atom. The chlorination of an alkane is a photo-initiated reaction. That is, it requires the use of light, which causes dissociation of the Cl_2 molecule, forming chlorine atoms:

$$Cl_2 \longrightarrow 2Cl\cdot \qquad [25.14]$$

The chlorine atoms have only seven valence-shell electrons. Species with an odd number of electrons are called **free radicals** (or simply radicals). They are frequently represented in chemical equations by showing a dot next to their chemical formula, representing their unpaired electron. Chlorine atoms are highly reactive and attack the alkane, removing a hydrogen and forming an alkyl radical. For example, when the alkane is ethane we have

$$Cl\cdot + CH_3CH_3 \longrightarrow HCl + CH_3CH_2\cdot \qquad [25.15]$$

The ethyl radical, $CH_3CH_2\cdot$, is another reactive species. It combines with molecular chlorine to give ethyl chloride and yet another chlorine atom:

$$CH_3CH_2\cdot + Cl_2 \longrightarrow CH_3CH_2Cl + Cl\cdot \qquad [25.16]$$

The resultant chlorine atom reacts with more ethane, continuing the cycle. Thus for each quantum of light absorbed by a chlorine molecule, many molecules of ethyl chloride may be formed. This reaction provides an example of a **radical chain** process. One of the disadvantages of radical chain reactions such as this is that they are not very selective. As the concentration of ethyl chloridc builds up in the reaction, chlorine atoms may abstract hydrogen atoms from it, so that eventually dichloroethane and even more highly chlorinated molecules may be formed. Thus several products are formed in the reaction, and these must be carefully separated by distillation or some other separation procedure.

Substitution into alkenes or alkynes is difficult to carry out, because the presence of the reactive double or triple bond usually leads to addition reactions rather than substitution. By contrast, substitution into aromatic hydrocarbons is relatively easy. For example, when benzene is warmed in a mixture of nitric and sulfuric acid, hydrogen is replaced by the nitro group, NO_2:

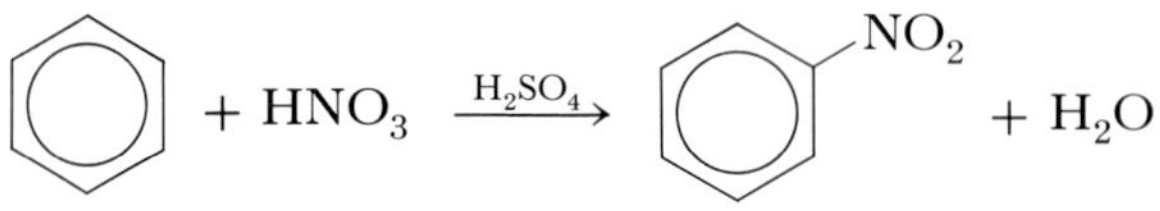

$$C_6H_6 + HNO_3 \xrightarrow{H_2SO_4} C_6H_5NO_2 + H_2O \qquad [25.17]$$

More vigorous treatment results in substitution of a second nitro group into the molecule:

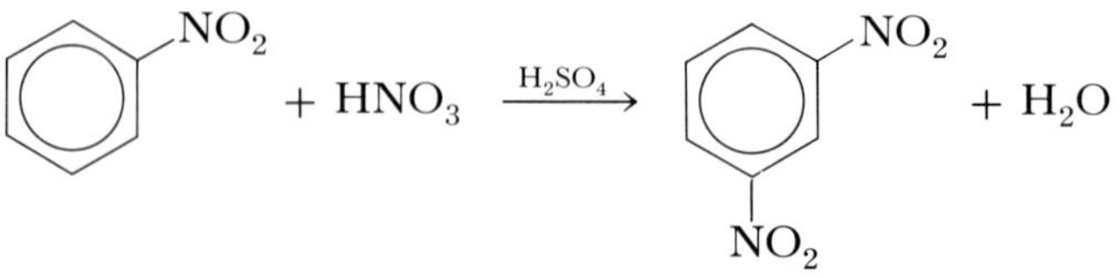

$$C_6H_5NO_2 + HNO_3 \xrightarrow{H_2SO_4} C_6H_4(NO_2)_2 + H_2O \qquad [25.18]$$

There are three possible isomers of benzene with two nitro groups attached. These three isomers are named *ortho-*, *meta-*, and *para*-dinitrobenzene:

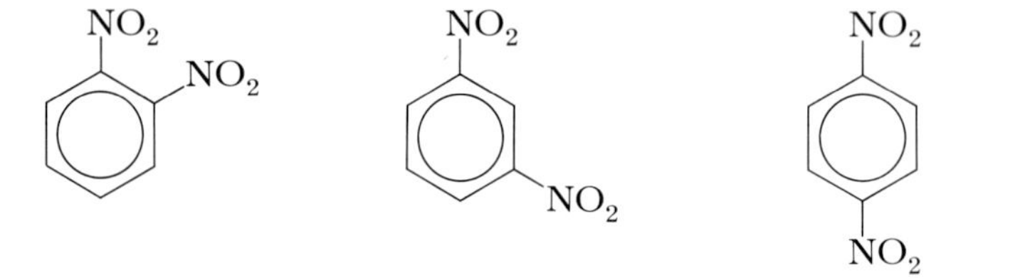

ortho-Dinitrobenzene
m.p. 118°C

meta-Dinitrobenzene
m.p. 90°C

para-Dinitrobenzene
m.p. 174°C

Only the *meta* isomer is formed in the reaction of nitric acid with nitrobenzene.

Bromination of benzene is carried out using $FeBr_3$ as a catalyst:

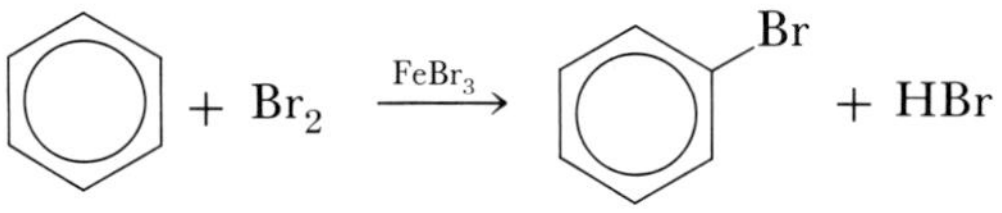

$$C_6H_6 + Br_2 \xrightarrow{FeBr_3} C_6H_5Br + HBr \qquad [25.19]$$

In a similar reaction, called the **Friedel-Crafts reaction,** alkyl groups can be substituted onto an aromatic ring by reaction of an alkyl halide with an aromatic compound in the presence of $AlCl_3$:

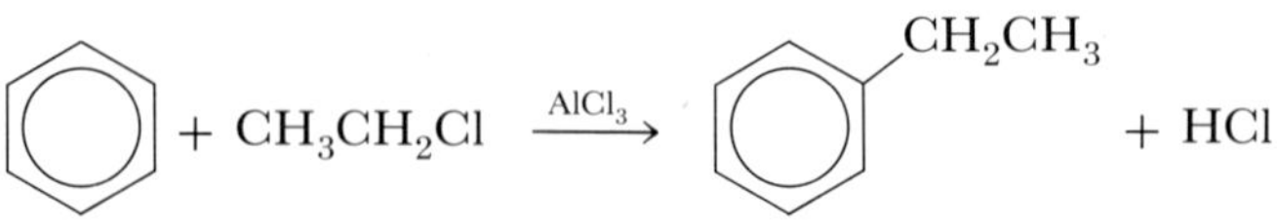

$$C_6H_6 + CH_3CH_2Cl \xrightarrow{AlCl_3} C_6H_5CH_2CH_3 + HCl \qquad [25.20]$$

The substitution reactions of aromatic compounds occur via attack of a positively charged reagent on the ring. Substances such as H_2SO_4, $FeCl_3$, or $AlCl_3$ serve as catalysts by generating the positively charged species. For example, the bromination of benzene proceeds via the following steps:

$$FeBr_3 + Br_2 \rightleftharpoons FeBr_4^- + Br^+ \quad [25.21]$$

$$C_6H_6 + Br^+ \longrightarrow C_6H_6Br^+ \text{ (carbocation: H and Br on same carbon, + charge on ring)} \quad [25.22]$$

$$C_6H_6Br^+ + FeBr_4^- \longrightarrow C_6H_5Br + FeBr_3 + HBr \quad [25.23]$$

The Br^+ ion has only six electrons in its valence shell. Using a pair of π electrons from the benzene, it forms a single bond to a carbon atom. In doing so, it leaves the adjacent carbon atom with only six electrons in its valence shell. Such a species is not very stable. Loss of a proton from the carbon atom to which the bromine is attached converts the molecule to a more stable one, bromobenzene. The overall result is that bromine has replaced hydrogen on the benzene ring.

25.4 HYDROCARBON DERIVATIVES

We noted in the introduction to this chapter that certain groups or arrangements of atoms impart characteristic behavior to an organic molecule. These groups are called functional groups. Two of these groups have already been discussed, the double and triple carbon-carbon bonds, both of which impart considerable chemical reactivity to a hydrocarbon. The other functional groups contain elements other than carbon and hydrogen, most often oxygen, nitrogen, or a halogen. Compounds containing these elements are generally considered to be hydrocarbon derivatives; they may be viewed as being derived from a hydrocarbon by replacing one or more of the hydrogens with functional groups. The compound can then be considered to consist of two parts: a hydrocarbon fragment such as an alkyl group (often designated R) and one or more functional groups.

Although a hydrocarbon derivative may consist largely of carbon and hydrogen, the hydrocarbon portion of the molecule may be essentially unchanged in various chemical reactions. The chemical properties are generally determined by the functional group. Now let's take a brief look at some of these functional groups and some of the properties they impart to organic compounds.

Alcohols

Alcohols are hydrocarbon derivatives in which one or more hydrogens of a parent hydrocarbon have been replaced by a hydroxyl or alcohol functional group, OH:

$$-\overset{|}{\underset{|}{C}}-O-H$$

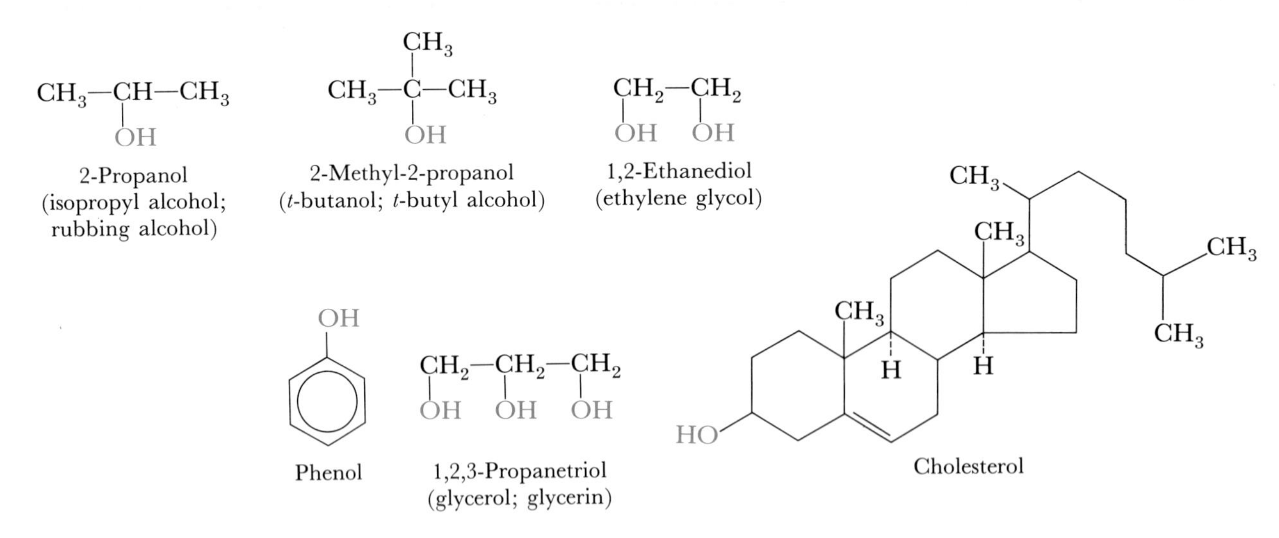

FIGURE 25.10 Structural formulas of several important alcohols.

Figure 25.10 shows the structural formulas and names of several alcohols. Note that the accepted name for an alcohol ends in *-ol.* The simple alcohols are named by changing the last letter in the name of the corresponding alkane to *-ol*—for example, ethan*e* becomes ethan*ol.* Where necessary, the location of the OH group is designated by an appropriate prefix numeral that indicates the number of the carbon atom bearing the OH group, as shown in the examples in Figure 25.10.

Aliphatic alcohols are classified according to the number of carbon groups bonded to the carbon that contains the OH group. If there is only one other carbon atom, as with ethanol or 1-propanol, the alcohol is termed a **primary alcohol.** If there are two other carbon groups attached, as in 2-propanol, the alcohol is **secondary.** If there are three other carbons, as in *t*-butanol, the alcohol is **tertiary.**

Because the OH group is quite polar, the presence of the hydroxyl group confers considerable polarity on the hydrocarbon molecule. Table 25.5 lists the solubilities of several alcohols in water at room temperature. In the alcohols having low molecular weights, the presence of the OH groups plays an important role in determining physical properties. However, as the length of the hydrocarbon chain increases, the OH group becomes less important in determining overall behavior (see also Section 11.3).

TABLE 25.5 Solubilities of several straight-chain alcohols in water

Alcohol	Boiling point (°C)	Solubility in water at 25°C (g/100 g H_2O)
Methanol	65	Miscible
Ethanol	78	Miscible
1-Propanol	97	Miscible
1-Butanol	117	9
Cyclohexanol	161	5.6
1-Hexanol	158	0.6

The simplest alcohol, methanol, has many important industrial uses, and it is produced on a large scale. Carbon monoxide and hydrogen are heated together under pressure in the presence of a metal oxide catalyst:

$$CO(g) + 2H_2(g) \xrightarrow[400^\circ C]{200\text{–}300 \text{ atm}} CH_3OH(g) \qquad [25.24]$$

Methanol has a very high octane rating of about 110 when it is blended with gasoline. There has been considerable discussion of using it as a fuel extender by mixing it with gasolines. Methanol boils at 65°C and freezes at −98°C. It is therefore suitable in terms of volatility for use as an automotive fuel. It could probably be manufactured for about twice the present cost of gasoline (on a per mile basis) if the manufacture were carried out on a larger scale than at present. To offset its higher cost, methanol has the advantage of requiring less pollution-control equipment. A disadvantage of methanol for fuel use is that it is miscible with water in all proportions; it thus might attract water into methanol-gasoline mixtures. Second, methanol is quite toxic. However, gasoline is also a toxic substance, and only moderate additional precautions would need to be taken if methanol were added to gasoline.

Ethanol, C_2H_5OH, is a product of the fermentation of carbohydrates such as sugar or starch. Bacterial cultures such as yeast work in the absence of air to convert carbohydrates into a mixture of ethanol and CO_2, as shown in Equation 25.25. In the process, they derive the energy necessary for growth:

$$C_6H_{12}O_6(aq) \xrightarrow{\text{yeast}} 2C_2H_5OH(aq) + 2CO_2(g) \qquad [25.25]$$

This naturally occurring reaction is carried out under carefully controlled conditions to produce beer, wine, and other beverages in which ethanol is the active ingredient. It is often said that ethanol is the least toxic of the straight-chain alcohols. Although this is true in the strictest sense, the combination of ethanol and automobiles produces far more human fatalities each year than any other chemical agent.

Many alcohols are formed industrially by addition of the elements of water to an olefin. For example, isopropyl alcohol is formed by hydration of propene using sulfuric acid as a catalyst:

$$CH_3CH{=}CH_2 + H_2O \xrightarrow{H_2SO_4} CH_3{-}\underset{\displaystyle OH}{\underset{|}{CH}}{-}CH_3 \qquad [25.26]$$

Ethylene glycol, the major ingredient in automobile antifreeze, is formed in a two-stage reaction. First ethylene is reacted with hypochlorous acid:

$$CH_2{=}CH_2 + HClO \longrightarrow \underset{\displaystyle OH}{\underset{|}{CH_2}}{-}\underset{\displaystyle Cl}{\underset{|}{CH_2}} \qquad [25.27]$$

Chlorine is removed from the product of this reaction by displacing the chloride ion with hydroxide ion:

$$\underset{\displaystyle OH}{\underset{|}{CH_2}}{-}\underset{\displaystyle Cl}{\underset{|}{CH_2}} + OH^- \longrightarrow \underset{\displaystyle OH}{\underset{|}{CH_2}}{-}\underset{\displaystyle OH}{\underset{|}{CH_2}} + Cl^- \qquad [25.28]$$

A reaction of this sort, in which hydroxide ion displaces another functional group is called base hydrolysis.

Phenol is the simplest example of a compound with an OH group attached to an aromatic ring. One of the most striking effects of the aromatic group is the greatly increased acidity of the proton. Phenol is about 1 million times more acidic in water than a typical aliphatic alcohol such as ethanol. Even so, it is not a very strong acid (K_a, 1.3×10^{-10}). Phenol is used industrially in the making of several kinds of plastics and in the preparation of dyes.

Cholesterol, shown in Figure 25.10, is an example of a biochemically important alcohol. Notice that the OH group forms only a small component of this rather large molecule. As a result, cholesterol is not very soluble in water (0.26 g per 100 mL of H_2O). Cholesterol is a normal component of our bodies. However, when present in excessive amounts it may precipitate from solution. It precipitates in the gallbladder to form crystalline lumps called gallstones. It may also precipitate against the walls of veins and arteries and thus contribute to high blood pressure and other cardiovascular problems. The amount of cholesterol in our blood is determined not only by how much cholesterol we eat, but by total dietary intake. There is evidence that excessive caloric intake leads the body to synthesize excessive cholesterol.

Ethers

Compounds in which two hydrocarbon groups are bonded to one oxygen are called ethers:

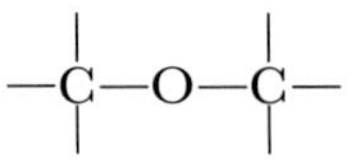

$$-\overset{|}{\underset{|}{C}}-O-\overset{|}{\underset{|}{C}}-$$

Ethers can be formed from two molecules of alcohol by splitting out a molecule of water. The reaction is thus a dehydration process; it is catalyzed by sulfuric acid, which takes up water to remove it from the system:

$$CH_3CH_2-OH + H-OCH_2CH_3 \xrightarrow{H_2SO_4} CH_3CH_2-O-CH_2CH_3 + H_2O \qquad [25.29]$$

A reaction in which water is split out from two substances is called a condensation reaction. An inorganic example of a condensation reaction was presented earlier (Section 21.4) in our discussion of the chemistry of phosphates. We shall encounter several additional examples in this chapter and in Chapter 26.

Ethers are used as solvents; both dimethyl ether and tetrahydrofuran are common solvents for organic reactions.

CH_3-O-CH_3

Dimethyl ether

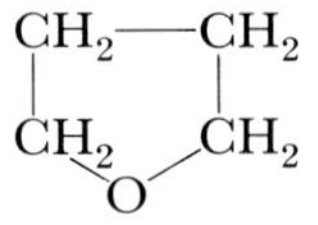

$$\begin{array}{ccc} CH_2 & - & CH_2 \\ | & & | \\ CH_2 & & CH_2 \\ & \diagdown O \diagup & \end{array}$$

Tetrahydrofuran (THF)

Aldehydes and Ketones

Several classes of organic compounds contain the carbonyl group

$$>C=O$$

In aldehydes the carbonyl group has at least one hydrogen atom attached, as in the following examples:

$$\underset{\text{Acetaldehyde}}{CH_3\overset{\overset{O}{\|}}{C}H} \qquad \underset{\text{Formaldehyde}}{H\overset{\overset{O}{\|}}{C}H}$$

In ketones, the carbonyl group occurs at the interior of a carbon chain and is therefore flanked by carbon atoms:

$$\underset{\text{Acetone}}{CH_3-\overset{\overset{O}{\|}}{C}-CH_3} \qquad \underset{\text{Methyl ethyl ketone}}{CH_3-\overset{\overset{O}{\|}}{C}-CH_2CH_3}$$

Aldehydes and ketones can be prepared by careful oxidation of alcohols. It is fairly easy to oxidize alcohols. Complete oxidation results in formation of CO_2 and H_2O, as in the burning of methanol:

$$CH_3OH(g) + \tfrac{3}{2}O_2(g) \longrightarrow CO_2(g) + 2H_2O(g) \qquad [25.30]$$

Controlled oxidation to form other organic substances is carried out by using various oxidizing agents such as air, hydrogen peroxide (H_2O_2), ozone (O_3), or potassium dichromate ($K_2Cr_2O_7$). We shall not concern ourselves with balancing most of these oxidation-reduction equations; instead, we shall simply show the source of oxygen as an O in parentheses.

Aldehydes are produced by oxidation of primary alcohols. For example, acetaldehyde is formed by oxidation of ethanol:

$$\underset{\text{Ethanol}}{CH_3CH_2OH} + (O) \longrightarrow \underset{\text{Acetaldehyde}}{CH_3\overset{\overset{O}{\|}}{C}H} + H_2O \qquad [25.31]$$

Ketones are produced by oxidation of secondary alcohols. For example, acetone is formed in large quantities by oxidation of isopropyl alcohol:

$$CH_3-\overset{\overset{OH}{|}}{C}H-CH_3 + (O) \longrightarrow \underset{\text{Acetone}}{CH_3-\overset{\overset{O}{\|}}{C}-CH_3} + H_2O \qquad [25.32]$$

Ketones are less reactive than are aldehydes because they do not possess the reactive C—H bond attached to the carbonyl carbon. Ketones are used extensively as solvents. Acetone, which boils at 56°C, is the most widely used. The carbonyl functional group imparts polarity to the solvent. Acetone is completely miscible with water, yet it dissolves a wide range of organic substances. Methyl ethyl ketone, $CH_3COCH_2CH_3$, which boils at 80°C, is also used industrially as a solvent.

Carboxylic Acids

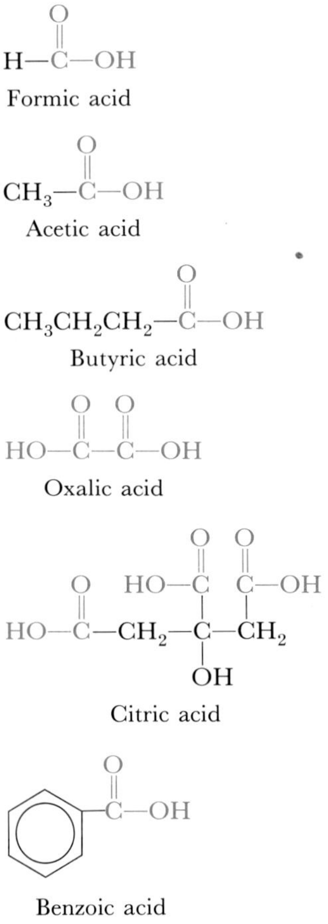

Oxidation of a primary alcohol may yield either an aldehyde or a carboxylic acid, depending on reaction conditions. For example, mild oxidation of ethanol produces acetaldehyde (Equation 25.31), which under more vigorous conditions may be further oxidized to acetic acid:

$$CH_3\overset{\overset{\displaystyle O}{\|}}{C}H + (O) \longrightarrow CH_3\overset{\overset{\displaystyle O}{\|}}{C}OH \qquad [25.33]$$

The oxidation of ethanol to acetic acid is responsible for causing wine to turn sour, producing vinegar.

As seen in this example, carboxylic acids contain the following functional group:

$$-\overset{\overset{\displaystyle O}{\|}}{C}-O-H$$

$$H-\overset{\overset{\displaystyle O}{\|}}{C}-OH$$

Formic acid

$$CH_3-\overset{\overset{\displaystyle O}{\|}}{C}-OH$$

Acetic acid

$$CH_3CH_2CH_2-\overset{\overset{\displaystyle O}{\|}}{C}-OH$$

Butyric acid

$$HO-\overset{\overset{\displaystyle O}{\|}}{C}-\overset{\overset{\displaystyle O}{\|}}{C}-OH$$

Oxalic acid

$$HO-\overset{\overset{\displaystyle O}{\|}}{C}-CH_2-\underset{\underset{\displaystyle OH}{|}}{\overset{\overset{\displaystyle HO-\overset{\overset{\displaystyle O}{\|}}{C}}{|}}{C}}-\overset{\overset{\displaystyle \overset{\overset{\displaystyle O}{\|}}{C}-OH}{|}}{CH_2}$$

Citric acid

$$C_6H_5-\overset{\overset{\displaystyle O}{\|}}{C}-OH$$

Benzoic acid

FIGURE 25.11 Structural formulas of several familiar carboxylic acids.

Carboxylic acids are widely distributed in nature and are important compounds in many industrial chemical processes. Figure 25.11 shows the structural formulas of several substances containing one or more carboxylic acid functional groups. Note that these substances are not named in a systematic way. The names of many acids are based on their historical origins. For example, formic acid was first prepared by extraction from ants; its name is therefore derived from the Latin word *formica*, meaning "ant." Oxalic acid was first identified as a constituent of the *oxalis* family of plants.

Carboxylic acids are important in the manufacture of polymers used to make fibers, films, and paints. Among the compounds having low molecular weights, acetic acid is an industrially important compound. A relatively new method for manufacture of acetic acid involves reaction of methanol with carbon monoxide, in the presence of a rhodium catalyst:

$$CH_3OH + CO \xrightarrow{\text{catalyst}} CH_3-\overset{\overset{\displaystyle O}{\|}}{C}-OH \qquad [25.34]$$

Notice that this reaction involves, in effect, the insertion of a carbon monoxide molecule between the CH_3 and OH groups. A reaction of this kind is called carbonylation.

Esters

Carboxylic acids may react with alcohols, with splitting out of water, to form esters:

$$CH_3-\overset{\overset{\displaystyle O}{\|}}{C}-OH + HO-CH_2CH_3 \longrightarrow$$

Acetic acid   Ethanol

$$CH_3-\overset{\overset{\displaystyle O}{\|}}{C}-O-CH_2CH_3 + H_2O \qquad [25.35]$$

Ethyl acetate

As seen in this example, esters are compounds in which the H atom of a carboxylic acid is replaced by a hydrocarbon group:

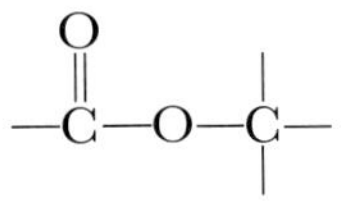

$$-\overset{\overset{\displaystyle O}{\|}}{C}-O-\overset{|}{\underset{|}{C}}-$$

The reaction between a carboxylic acid and an alcohol is another example of a condensation reaction. The reaction product, the ester, is named by using first the group from which the alcohol is derived and then the group from which the acid is derived.

SAMPLE EXERCISE 25.6

Name the following esters:

(a) 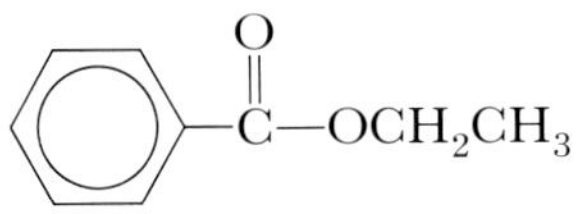 $C_6H_5-\overset{\overset{\displaystyle O}{\|}}{C}-OCH_2CH_3$

(b) 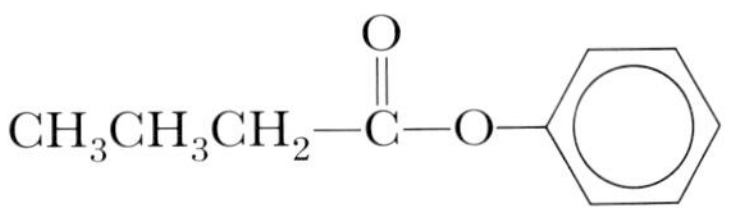 $CH_3CH_3CH_2-\overset{\overset{\displaystyle O}{\|}}{C}-O-C_6H_5$

Solution: In (a) the ester is derived from ethanol and benzoic acid. Its name is therefore ethyl benzoate. In (b) the ester is derived from phenol and butyric acid. The residue from the phenol, C_6H_5, is called the phenyl group. The ester is therefore named phenyl butyrate.

SAMPLE EXERCISE 25.7

Beginning with ethene as the only carbon-containing compound, suggest a series of reactions by which ethyl acetate might be prepared.

Solution: As shown in Equation 25.35, ethyl acetate can be prepared by a condensation reaction between acetic acid and ethanol. The ethanol can be prepared from ethene by H_2SO_4-catalyzed hydration:

$$CH_2{=}CH_2 + H_2O \xrightarrow{H_2SO_4} CH_3CH_2OH$$

This process is analogous to that given in Equation 25.26 for the hydration of propene. A portion of the ethanol thus produced can be oxidized to acetic acid, as shown in Equations 25.31 and 25.33.

Esters generally have very pleasant odors; they are largely responsible for the pleasant aromas of fruit. Ethyl butyrate, for example, is responsible for the odor of apples. When esters are treated with acid or base in aqueous solution, they are hydrolyzed; that is, the molecule is split into its alcohol and acid components:

$$\underset{\text{Methyl propionate}}{CH_3CH_2-\overset{\overset{\displaystyle O}{\|}}{C}-O-CH_3} + Na^+ + OH^- \longrightarrow$$

$$\underset{\text{Sodium propionate}}{CH_3CH_2-\overset{\overset{\displaystyle O}{\|}}{C}-O^-} + Na^+ + \underset{\text{Methanol}}{CH_3OH} \qquad [25.36]$$

In this example, the hydrolysis was carried out in basic medium. The products of the reaction are the sodium salt of the carboxylic acid and the alcohol.

Amines and Amides

Amines are organic bases (Section 15.6). They have the general formula R_3N, where R may be H or a hydrocarbon group as in the following examples:

$$\underset{\text{Ethylamine}}{CH_3CH_2NH_2} \qquad \underset{\text{Trimethylamine}}{(CH_3)_3N} \qquad \underset{\text{Aniline}}{C_6H_5-NH_2}$$

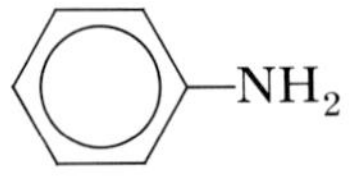

Amines containing a proton can undergo condensation reactions with carboxylic acids to form amides:

$$CH_3\overset{\overset{\displaystyle O}{\|}}{C}-OH + H-N(CH_3)_2 \longrightarrow CH_3\overset{\overset{\displaystyle O}{\|}}{C}-N(CH_3)_2 + H_2O \qquad [25.37]$$

The amide functional group may be considered to be derived from a carboxylic acid with an NR_2 group replacing the OH of the acid, as in these additional examples:

$$\underset{\text{Acetamide}}{CH_3\overset{\overset{\displaystyle O}{\|}}{C}-NH_2} \qquad \underset{\text{Benzamide}}{C_6H_5-\overset{\overset{\displaystyle O}{\|}}{C}-NH_2}$$

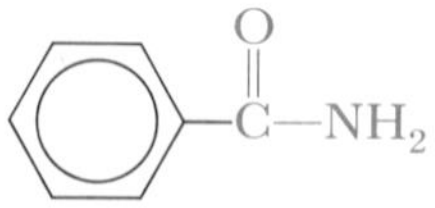

The amide linkage,

$$R-\overset{\overset{\displaystyle O}{\|}}{C}-\underset{\underset{\displaystyle H}{|}}{N}-R'$$

where R and R′ are organic groups, is the key functional group in the structures of proteins, about which we shall have more to say in Chapter 26.

25.5 POLYMERS

A polymer is a high-molecular-weight material formed from simple molecules called monomers. Polymers may be of either synthetic or natural origin. Those of natural origin include proteins, starch, and cellulose, substances that we will discuss in Chapter 26. Our focus in the present discussion is on synthetic organic polymers. Synthetic polymers may be formed by either addition reactions (addition polymerization) or condensation reactions (condensation polymerization).

Addition Polymerization

In addition polymerization, unsaturated hydrocarbons or hydrocarbon derivatives are caused to react with one another. For example, addition polymerization of ethylene produces polyethylene:

$$\cdots + \underset{\text{Ethylene}}{H_2C{=}CH_2} + \underset{\text{Ethylene}}{H_2C{=}CH_2} + \cdots \longrightarrow \underset{\text{Polyethylene}}{\cdots{-}CH_2{-}CH_2{-}CH_2{-}CH_2{-}\cdots} \qquad [25.38]$$

In such a polymerization reaction, π bonds are broken and the electrons used to form new σ bonds between the monomer units. In the preceding example, two ethylene units are shown to combine. In practice, more than a thousand monomer units may combine to form a polymer (see Section 10.5).

Polymerization reactions are often written in a simplified form to stress the repeating unit of the polymer chain. For example, the addition polymerization of 2-methylpropene (whose common name is isobutylene) to produce polysiobutylene can be represented as follows:

$$n\,\underset{\text{Isobutylene}}{HHC{=}C(CH_3)_2} \longrightarrow \underset{\text{Polyisobutylene}}{\left[{-}CH_2{-}C(CH_3)_2{-}\right]_n} \qquad [25.39]$$

Polyisobutylene is a rubbery, rather gummy material that can be further treated to form a wide variety of useful products. In its finished form, it is known as butyl rubber. Many other alkenes can be similarly polymerized to form the variety of useful materials summarized in Table 25.6.

The polymerization reactions of alkenes may be catalyzed by a number of different types of substances. The choice of catalyst is important because it influences the structure and molecular weight of the resulting polymer and thus its properties. For example, polyethylene can be a pliable material with a low melting point or a much stiffer substance with a higher melting point, depending on the nature of the catalyst. One of the most important types of catalyst is the free radical (Section 25.3). Organic peroxides, compounds with O—O bonds, are exceptionally good sources of free radicals. These compounds decompose on heating, with rupture of the O—O bond, to form a pair of free radicals:

TABLE 25.6 Alkenes that undergo addition polymerization

Monomer formula	Monomer name	Polymer name	Uses
$CH_2{=}CH_2$	Ethylene	Polyethylene	Coating for milk cartons, wire insulation, plastic bags
$CF_2{=}CF_2$	Tetrafluoroethylene	Teflon	Insulation, bearings, frying-pan surfaces
$CH_2{=}\underset{\underset{\displaystyle Cl}{\vert}}{C}H$	Vinyl chloride	Polyvinyl chloride (PVC)	Phonograph records, rainwear, piping
$CH_2{=}\underset{\underset{\displaystyle CN}{\vert}}{C}H$	Acrylonitrile	Polyacrylonitrile (Orlon)	Rug fibers
$CH_2{=}\underset{\underset{\displaystyle C_6H_5}{\vert}}{C}H$	Vinyl benzene (styrene)	Polystyrene	TV lead-in wire, combs, styrofoam insulation
$CH_2{=}\underset{\underset{\displaystyle CH_3}{\vert}}{C}H$	Propylene	Polypropylene	Fibers for fabrics and carpets

$$R{-}\underset{\cdot\cdot}{\ddot{O}}{-}\underset{\cdot\cdot}{\ddot{O}}{-}R \rightleftharpoons 2R{-}\underset{\cdot\cdot}{\ddot{O}}\cdot \qquad [25.40]$$

In this equation R could be any of a number of different organic groups. By placing a small amount of a peroxide in contact with an alkene and then heating it, radicals that can catalyze addition reactions are generated. For example, consider the polymerization of ethylene initiated (that is, started) by a radical formed from a peroxide:

$$R{-}\underset{\cdot\cdot}{\ddot{O}}{-}\underset{\cdot\cdot}{\ddot{O}}{-}R \rightleftharpoons 2R{-}\underset{\cdot\cdot}{\ddot{O}}\cdot$$

$$R{-}\underset{\cdot\cdot}{\ddot{O}}\cdot + H_2C{=}CH_2 \longrightarrow RO{-}\overset{\displaystyle H}{\underset{\displaystyle H}{\overset{|}{\underset{|}{C}}}}{-}\overset{\displaystyle H}{\underset{\displaystyle H}{\overset{|}{\underset{|}{C}}}}\cdot \qquad [25.41]$$

In Equation 25.41, the R—O radical reacts with a molecule of ethylene, forming a bond to one of the carbons. For this to occur, the pair of electrons forming the π bond in ethylene must have uncoupled. One of them is involved in the bond to the O—R group, the other is on the second carbon. Thus, the product of this reaction is itself a free radical. This species goes on to react with a second molecule of ethylene:

$$RO{-}\overset{\displaystyle H}{\underset{\displaystyle H}{\overset{|}{\underset{|}{C}}}}{-}\overset{\displaystyle H}{\underset{\displaystyle H}{\overset{|}{\underset{|}{C}}}}\cdot + H_2C{=}CH_2 \longrightarrow RO{-}\overset{\displaystyle H}{\underset{\displaystyle H}{\overset{|}{\underset{|}{C}}}}{-}\overset{\displaystyle H}{\underset{\displaystyle H}{\overset{|}{\underset{|}{C}}}}{-}\overset{\displaystyle H}{\underset{\displaystyle H}{\overset{|}{\underset{|}{C}}}}{-}\overset{\displaystyle H}{\underset{\displaystyle H}{\overset{|}{\underset{|}{C}}}}\cdot \qquad [25.42]$$

By means of such successive reactions, the polymer grows in length. A reaction of this type is known as a chain reaction, because it is self-propagating. However, chain-termination reactions might occur. If two radicals come into contact, their unpaired spins might couple, and no further reaction would occur. In forming polymers by such free-radical reactions, we must control conditions carefully so that a polymer chain with the desired average length results.

Condensation Polymerization

In condensation polymerization, molecules are joined together through condensation reactions. The most important condensation polymers are polyamides and polyesters. The nylons are polyamides. They can be formed by reaction between a substance with two carboxylic acid functional groups and a substance with two amine functional groups. For example, nylon 66 is formed by heating a six-carbon diacid with a six-carbon diamine:

$$\text{---COOH} + H_2N\text{—}(CH_2)_6\text{—}NH_2 + HOOC\text{—}(CH_2)_4\text{—}COOH + H_2N\text{—}(CH_2)_6\text{—}NH_2 + HOOC\text{—}(CH_2)_4\text{—}COOH + H_2N\text{---}$$

$$\big\downarrow \text{heat}$$

$$\text{---}\overset{O}{\overset{\|}{C}}\text{—NH—}(CH_2)_6\text{—NH—}\overset{O}{\overset{\|}{C}}\text{—}(CH_2)_4\text{—}\overset{O}{\overset{\|}{C}}\text{—NH—}(CH_2)_6\text{—NH—}\overset{O}{\overset{\|}{C}}\text{—}(CH_2)_4\text{—}\overset{O}{\overset{\|}{C}}\text{—NH---} \qquad [25.43]$$

The polymer is formed by loss of water from between each pair of reacting functional groups. The resulting polymer is labeled 66 because there are six carbon atoms in the sections between each NH group along the chain.

Condensation polymers can also result from formation of ester linkages, that is, by condensation of carboxylic acids with alcohols. In forming a polymer, it is important to use difunctional acids and alcohols. By varying the acid and alcohol functions, various kinds of polyester materials can be formed. The familiar polyester Dacron, from which so much of our clothing is produced, is a condensation polymer of ethylene glycol and terephthalic acid:

$$\underset{\text{Ethylene glycol}}{\text{---}HOCH_2CH_2OH} + \underset{\text{Terephthalic acid}}{HO\overset{O}{\overset{\|}{C}}\text{—}C_6H_4\text{—}\overset{O}{\overset{\|}{C}}OH} + HOCH_2CH_2OH + HO\overset{O}{\overset{\|}{C}}\text{—}C_6H_4\text{—}\overset{O}{\overset{\|}{C}}OH\text{---} \xrightarrow{\text{heat}}$$

$$\text{---}OCH_2CH_2O\overset{O}{\overset{\|}{C}}\text{—}C_6H_4\text{—}\overset{O}{\overset{\|}{C}}OCH_2CH_2O\overset{O}{\overset{\|}{C}}\text{—}C_6H_4\text{—}\overset{O}{\overset{\|}{C}}OCH_2CH_2O\overset{O}{\overset{\|}{C}}\text{—}C_6H_4\text{—}\overset{O}{\overset{\|}{C}}O\text{---} \qquad [25.44]$$

Dacron

The polymer formed under typical reaction conditions may have 80 to 100 units per molecule. This material can be melted without decomposition. The molten polymer is forced through tiny holes, and then it cools and solidifies in the form of fibers. As the tiny fibers are stretched, the long molecules are forced to lie in a more or less parallel arrangement. The drawn or stretched fibers are then spun into a yarn.

FOR REVIEW

Summary

In this chapter we have studied the chemical and physical characteristics of simple organic substances.

Hydrocarbons are composed of only carbon and hydrogen. There are four major classes of hydrocarbons. The alkanes are composed of only carbon-hydrogen and carbon-carbon single bonds. The alkenes contain one or more carbon-carbon double bonds. Alkynes contain one or more carbon-carbon triple bonds. Aromatic hydrocarbons contain cyclic arrangements of carbon atoms bonded through both σ and π bonds. Isomers are substances that possess the same molecular formula but that differ in some other respect. Isomers may differ in the bonding arrangements of atoms within the molecule or in the geometrical arrangements of groups. Geometrical isomerism is possible in the alkenes because of restricted rotation about the C=C double bond.

The major sources of hydrocarbons are fossil fuels such as coal and oil. Crude oil is separated by distillation into several fractions according to variations in volatility. The most important fraction is gasoline, used as automotive fuel. Less volatile fractions are cracked; that is, they are heated with catalysts to convert them to more volatile substances that have lower molecular weights and are suitable for gasoline. In the same process, straight-chain hydrocarbons are caused to rearrange to the more desirable branched-chain isomers, which have higher octane numbers.

Combustion of hydrocarbons is a highly exothermic process. The chief use of hydrocarbons is as a source of heat energy via combustion. The unsaturated hydrocarbons, the alkenes and alkynes, readily undergo addition reactions to the carbon-carbon multiple bonds. By contrast, addition reactions to aromatic hydrocarbons are difficult to carry out.

Substitution reactions are those in which one or more hydrogens of a hydrocarbon are replaced by some other atom or group. The substitution reactions of aliphatic (nonaromatic) hydrocarbons are not easily carried out. However, substitution reactions of aromatic hydrocarbons are easily carried out in the presence of acid catalysts.

The chemistry of hydrocarbon derivatives is often dominated by the nature of their functional groups. The functional groups we have considered are summarized here:

R—O—H (Alcohol)

R—C(=O)—H (Aldehyde)

>C=C< (or H) (Alkene)

—C≡C— (Alkyne)

R—C(=O)—N< (Amide)

R—N(—R′ (or H))—R″ (or H) (Amine)

R—C(=O)—O—H (Carboxylic acid)

R—C(=O)—O—R′ (Ester)

R—O—R′ (Ether)

R—C(=O)—R′ (Ketone)

(Remember that R, R′, and R″ represent some hydrocarbon group—for example, methyl, CH_3, or phenyl, C_6H_5.)

Alcohols are hydrocarbon derivatives containing one or more OH groups. Ethers are related compounds that are formed by a condensation reaction of two molecules of alcohol, with splitting out of water. Oxidation of primary alcohols can lead to formation of aldehydes; further oxidation of the aldehydes produces carboxylic acids. Oxidation of secondary alcohols leads to formation of ketones.

Carboxylic acids can form esters by a condensation reaction with alcohols, or they can form amides by condensation reaction with amines.

Polymers are large molecules formed by the combination of small, repeating units called monomers. Addition polymerization of alkenes is a source of many useful synthetic materials. These reactions can be catalyzed by many types of substances, including free radicals. Condensation polymerization of molecules with functional groups on both ends is also a source of important synthetic polymers, including polyamides and polyesters.

Learning goals

Having read and studied this chapter, you should be able to:

1. List the four groups of hydrocarbons and draw the structural formula of an example from each group.
2. List the names of the first 10 members of the alkane series of hydrocarbons.
3. Write the structural formula of an alkane, alkene, or alkyne, given its systematic (IUPAC) name.
4. Name an alkane, alkene, or alkyne, given its structural formula.
5. Give an example of geometrical isomerism in alkenes.
6. Explain why aromatic hydrocarbons do not readily undergo addition reactions as do the alkenes and alkynes.

7 List and describe the fractions obtained in petroleum refining.
8 Explain the general relationship between octane number and structure in alkanes and describe the methods used to increase the octane numbers of straight-run gasolines.
9 Give examples of addition reactions of alkenes and alkynes, showing the structural formulas of reactants and products.
10 Describe Markovnikoff's rule and apply it to the addition reactions of alkenes and alkynes.
11 Write the steps in the photo-initiated chlorination of an alkane.
12 Give two or three examples of substitution of an aromatic compound.
13 Write structural formulas of molecules that are examples of an alcohol, an ether, an aldehyde, a ketone, a carboxylic acid, an ester, an amine, and an amide.
14 Describe the general character of condensation reactions and given an example of the condensation of alcohols to form ethers, condensations of alcohols and carboxylic acids to form esters, and condensation of amines and carboxylic acids to form amides.
15 Give an example of addition polymerization of an alkene and of condensation polymerization to form a polyamide or polyester.

Key terms

Among the more important terms and expressions used for the first time in this chapter are the following:

Addition polymerization (Section 25.5) is a reaction in which alkenes add together end to end by opening of the carbon-carbon double bond, to form long polymeric chain molecules.

An **addition reaction** (Section 25.3) is one in which a reagent adds to the two carbon atoms of a carbon-carbon multiple bond.

Alcohols (Section 25.4) are hydrocarbons in which one or more hydrogens have been replaced by a hydroxyl group, OH.

Aliphatic hydrocarbons (Section 25.1) are those that are not aromatic; that is, the alkanes, cycloalkanes, alkenes, and alkynes.

Alkanes (Section 25.1) are compounds of carbon and hydrogen containing only single carbon-carbon bonds.

Alkenes (Section 25.1) are hydrocarbons containing one or more carbon-carbon double bonds.

Alkynes (Section 25.1) are hydrocarbons containing one or more carbon-carbon triple bonds.

An **antiknock agent** (Section 25.2) is a substance added to automotive fuel to increase its octane number. This agent slows the rate of burning of fuel in the cylinder.

Aromatic hydrocarbons (Section 25.1) are hydrocarbon compounds that contain a planar, cyclic arrangement of carbon atoms linked by both σ and π bonds.

Base hydrolysis (Section 25.4) is a chemical reaction involving the replacement of a negatively charged functional group by aqueous hydroxide ion.

The **carbonyl functional group** (Section 25.4), C=O, is a characteristic feature of several organic functional groups, such as ketones or carboxylic acids.

Carbonylation (Section 25.4) is a reaction in which the carbonyl functional group is introduced into a molecule.

A **condensation polymer** (Section 25.5) is a substance having high molecular weight that is formed by elimination of water from between molecules.

Cycloalkanes (Section 25.1) are saturated hydrocarbons of general formula C_nH_{2n}, in which the carbon atoms form a closed ring.

A **free radical** (Section 25.3) is a species with an odd number of electrons.

A **functional group** (introduction; Section 25.4) is an atom or group of atoms that has characteristic chemical properties.

Geometrical isomers (Section 25.1) have the same molecular formulas and functional groups, but differ in the geometrical arrangements of atoms. For example, *cis*- and *trans*-2-butene are geometrical isomers.

In **homologous series** (Section 25.1), compounds contain common structural elements, but differ in the number of atoms making up the molecule. Members of a homologous series have the same general formula. For example, the aliphatic saturated alcohols have the general formula $C_nH_{2n+2}O$.

A **hydrocarbon** (Section 25.1) is a compound composed of carbon and hydrogen. Substituted hydrocarbons may contain functional groups composed of atoms other than carbon and hydrogen.

Markovnikoff's rule (Section 25.3) states that when a reagent of general formula XY adds to an alkene, the more negatively charged species, X or Y, ends up on the carbon with the fewer attached hydrogen atoms. For example, on addition of HCl to propene, Cl ends up on the middle carbon atom.

An **olefin** (Section 25.1) is a compound containing one or more carbon-carbon double bonds.

In a **substitution reaction** (Section 25.3), one

atom or functional group bonded to a larger molecule is replaced by another.

Unsaturated hydrocarbons (Section 25.1) are nonaromatic compounds of carbon and hydrogen that contain one or more carbon-carbon multiple bonds.

EXERCISES

Hydrocarbon structures and nomenclature

25.1 Write the molecular formulas of an alkane, alkene, alkyne, and aromatic hydrocarbon that in each case contains six carbon atoms.

25.2 Write the structural formulas for all the alkanes with the molecular formula C_6H_{14}. Name each compound.

25.3 Write the structural formula for each of the following hydrocarbons: (a) $CH_3CH(CH_3)CH_2CH_3$; (b) $CH_3C{\equiv}CCH_3$; (c) $CH_2{=}CHCH_3$; (d) $CH_2{=}CHCH_2CH{=}CH_2$; (e) $(CH_3)_2CHCH_3$.

25.4 Write the condensed structural formulas of as many aliphatic compounds as you can think of that have the molecular formula (a) C_5H_8; (b) C_5H_{10}.

25.5 What are the characteristic bond angles (a) about carbon in an alkane; (b) about the carbon-carbon double bond in an alkene; (c) about the carbon-carbon triple bond in an alkyne?

25.6 Write the condensed structural formula for each of the following compounds: (a) 5-methyl-*trans*-2-heptene; (b) 3-chloropropyne; (c) *ortho*-dichlorobenzene; (d) 2,2,4,4-tetramethylpentane; (e) 2-methyl-3-ethylhexane; (f) 2-methyl-6-chloro-3-heptyne; (g) 1,5-dimethylnaphthalene; (h) methylcyclobutane.

25.7 Write the structural formula for each of the following compounds: (a) 2,2-dimethylpentane; (b) 2,3-dimethylhexane; (c) *cis*-2-hexene; (d) methylcyclopentane; (e) 2-chlorobutane; (f) 1,2-dibromobenzene.

25.8 Name each of the following compounds:

(a) CH_3CH_2 and CH_3 substituents on $CH_2CHCHCH_2CH_3$, with a further CH_3 below:

```
CH3CH2    CH3
   |       |
   CH2CHCHCH2CH3
       |
       CH3
```

(b)

```
                   CH3
                   |
CH3CH2        CH2CHCH2CH3
      \      /
       C=C
      /    \
     H      H
```

(c) para-dibromobenzene ring: Br on top and Br on bottom of a benzene ring

```
   Br
   |
 (benzene)
   |
   Br
```

(d)

```
           CH2CH3
           |
CH≡CCH2CCH3
           |
           CH3
```

(e)

```
CH3CHCH2CHCH3
   |     |
   Br    Br
```

(f) $CH_3CH{=}CHCH_2CH{=}CHCH_2CH_3$

(g)

```
       Cl       Cl
       |        |
CH3CHCHCHCH2CHCH3
     |
  (benzene)
```

25.9 Indicate whether each of the following molecules is capable of *cis-trans* geometric isomerism; for those that are, draw the structures of the isomers: (a) 2,3-dichlorobutane; (b) 2,3-dichloro-2-butene; (c) 1,3-dimethylbenzene; (d) 4,4-dimethyl-2-pentyne.

25.10 Draw the structural formulas for all possible geometric isomers for 2,6-octadiene.

25.11 Draw the structures of all isomers of (a) dichlorobenzene; (b) trichlorobenzene; (c) chlorobromomethylbenzene.

Petroleum

25.12 Indicate briefly the meaning of each of the following terms: (a) cracking; (b) isomerization; (c) antiknock agent; (d) octane number; (e) branched-chain alkane; (f) refining.

25.13 What is the octane number of a mixture of 20 percent *n*-heptane, 30 percent methylcyclohexane, and 50 percent isooctane? (Refer to Table 25.4.)

25.14 Describe two ways in which the octane number of a gasoline consisting of alkanes can be increased.

25.15 In places where the average temperature varies widely with the season, gasoline used during the winter contains a greater proportion of hydrocarbons with lower molecular weights than does gasoline used during the summer. Suggest a reason for this.

Reactions of hydrocarbons

25.16 Give an example, in the form of a balanced equation, of each of the following chemical reactions: (a) substitution reaction of an alkane; (b) oxidation of an alkene; (c) addition reaction of an alkyne; (d) substitution reaction of an aromatic hydrocarbon.

25.17 In each of the following pairs, indicate which molecule is the more reactive and give a reason for the greater reactivity: (a) butane and cyclobutane; (b) cyclohexane and cyclohexene; (c) benzene and 1-hexene; (d) 2-hexyne and 2-hexene.

25.18 Using condensed structural formulas, write a balanced chemical equation for each of the following reactions: (a) hydrogenation of 1-butene; (b) addition of H_2O to *cis*-2-butene using H_2SO_4 as a catalyst; (c) complete oxidation of cyclobutane; (d) reaction of benzene with 2-chloropropane in the presence of $AlCl_3$.

25.19 Using Markovnikoff's rule, predict the product formed when HCl is reacted with each of the following compounds:

(a) $CH_3CH{=}CH_2$

(b) 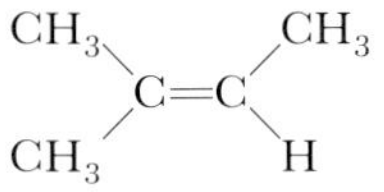 $(CH_3)_2C{=}C(CH_3)H$

$$\begin{matrix} CH_3 & & & & CH_3 \\ & \diagdown & & \diagup & \\ & & C{=}C & & \\ & \diagup & & \diagdown & \\ CH_3 & & & & H \end{matrix}$$

(c) $CH_3C{\equiv}CH$

25.20 Why do addition reactions occur more readily with alkenes and alkynes than with aromatic hydrocarbons?

25.21 Predict the product or products formed in each of the following reactions:

(a) *cis*-$CH_3CH{=}CHCH_3 + H_2 \xrightarrow{Ni}$

(b) $CH_3CH_2C{\equiv}CH + HI \longrightarrow$

(c) $CH_3\underset{\underset{CH_3}{|}}{C}{=\!=}\underset{\underset{CH_3}{|}}{C}CH_2CH_3 + H_2O \xrightarrow{H_2SO_4}$

(d) $CH_2{=}CHCH_3 + Br_2 \longrightarrow$

25.22 When cyclopropane is treated with HI, 1-iodopropane is formed. A similar type of reaction does not occur with cyclopentane or cyclohexane. How do you account for the reactivity of cyclopropane?

25.23 Using Appendix D, calculate the heat of combustion per mole for (a) ethane; (b) ethene; (c) ethyne. In each case assume gaseous products.

25.24 The heat of combustion of decahydronaphthalene, $C_{10}H_{18}$, is -6286 kJ/mol. The heat of combustion of naphthalene, $C_{10}H_8$, is -5157 kJ/mol. (In each case $CO_2(g)$ and $H_2O(l)$ are the products.) Using these data and data in Appendix D, calculate the heat of hydrogenation of naphthalene. Does this value provide any evidence for aromatic character in naphthalene?

25.25 Suggest a method of preparing ethylbenzene starting with benzene and ethylene as the only organic reagents.

Hydrocarbon derivatives

25.26 Identify the functional groups in each of the following compounds:

(a) $CH_3\underset{\underset{O}{\|}}{C}CH_2CH_3$ (b) $CH_3\underset{\underset{OH}{|}}{C}{=}O$

(c) $\underset{\underset{HO}{|}}{C}H_2CH_2CH_3$ (d) $CH_3O\underset{\underset{O}{\|}}{C}CH_2CH_3$

(e) $H\underset{\underset{O}{\|}}{C}CH_2CH_3$ (f) $CH_3\overset{\overset{CH_3}{|}}{C}{=}O$

(g) $NH_2\underset{\underset{O}{\|}}{C}CH_3$ (h) $CH_3CH_2NH(CH_3)$

25.27 What are the bond angles expected around each carbon atom in acetone, whose structural formula is shown below?

$$CH_3\overset{\overset{O}{\|}}{C}CH_3$$

25.28 Alcohols of low molecular weight are moderately soluble in water, whereas ethers of about the same molecular weight are not. Explain.

25.29 What products (if any) occur when each of the following compounds is mildly oxidized:

(a) $CH_3\underset{\underset{CH_3}{|}}{C}HCH_2OH$ (b) $CH_3CH_2CH_2OH$

(c) $CH_3\underset{\underset{CH_3}{|}}{\overset{\overset{CH_3}{|}}{C}}OH$ (d) $CH_3\overset{\overset{O}{\|}}{C}CH_3$

(e) $CH_3\overset{\overset{O}{\|}}{C}H$

25.30 Draw the structures of the esters formed from (a) acetic acid and 2-propanol; (b) acetic acid and 1-propanol; (c) formic acid and ethanol.

25.31 Name each of the compounds in Exercise 25.29.

25.32 Write structural formulas for each of the following compounds: (a) 2-butanol; (b) 1,2-ethanediol; (c) methylformate; (d) diethyl ketone; (e) diethyl ether.

25.33 The IUPAC name for a carboxylic acid is based on the name of the hydrocarbon with the same number of carbon atoms. The ending *-oic* is appended as in ethanoic acid, which is the IUPAC name for acetic acid,

$$CH_3\overset{\overset{O}{\|}}{C}OH$$

Give the IUPAC name for each of the following acids:

(a) $HCOH$ (with C=O; i.e. $HC(=O)OH$)

(b) $CH_3CH_2CH_2\overset{O}{\overset{\|}{C}}OH$

(c) $CH_3CH_2\underset{CH_3}{\underset{|}{C}}HCH_2\overset{O}{\overset{\|}{C}}OH$

25.34 Aldehydes and ketones can be named in a systematic way by counting the number of carbon atoms (including the carbonyl carbon) that they contain. The name of the aldehyde or ketone is based on the hydrocarbon with the same number of carbon atoms. The endings *-al,* for aldehyde, or *-one,* for ketone, are added as appropriate. Draw the structural formulas for the following aldehydes or ketones: (a) propanal; (b) 2-pentanone; (c) 3-methyl-2-butanone; (d) 2-methylbutanal.

25.35 Write balanced chemical equations, with organic substances given by their condensed formulas, describing the preparation of (a) acetone from 1-propene; (b) propanoic acid (a carboxylic acid with three carbon atoms) from 1-propanol; (c) sodium acetate from ethene; (d) methyl acetate from methyl alcohol and acetic acid; (e) methyl ethyl ketone from 2-butanol.

25.36 Write a balanced chemical equation to describe what happens when (a) methyl acetate is treated with sodium hydroxide; (b) methanol is treated with sulfuric acid.

25.37 Using condensed formulas, write a balanced chemical equation for the condensation reactions that occur between the following substances: (a) methanol with itself; (b) methanol with ethanol; (c) acetic acid with 2-propanol; (d) formic acid with 1-propanol; (e) diethylamine with benzoic acid (see Figure 25.11); (f) methylethylamine with acetic acid.

Polymers

25.38 Write a chemical equation for the addition polymerization of (a) vinyl chloride; (b) tetrafluoroethylene. (See Table 25.6.)

25.39 If two different monomers are involved in addition polymerization, the resultant polymer is said to be a copolymer of the two monomers. Draw the formula for the copolymer of vinyl chloride ($CH_2{=}CHCl$) and 1,1-dichloroethylene ($CH_2{=}CCl_2$) to form Saran, the polymer used to make food wrapping.

25.40 What is the mass, in grams, of 1 mol of polyethylene if the polymer is composed of 950 ethylene units. What is this weight in pounds?

25.41 Proteins are condensation polymers of amino acids, substances that contain both carboxylic acid and amine functional groups. What is the polymer formed by condensation polymerization of glycine, H_2NCH_2COOH?

25.42 A rather simple polymer can be made from ethylene glycol, $HOCH_2CH_2OH$, and oxalic acid, $HO{-}\underset{O}{\underset{\|}{C}}{-}\underset{O}{\underset{\|}{C}}{-}OH$. Sketch a portion of the polymer chain obtained from these monomers.

Additional exercises

25.43 Draw the structural formulas of two molecules with the formula C_4H_6.

25.44 What is the molecular formula of (a) an alkane with 20 carbon atoms; (b) an alkene with 18 carbon atoms; (c) an alkyne with 12 carbon atoms?

25.45 Draw the Lewis structures for the *cis* and *trans* isomers of 2-pentene. Does cyclopentane exhibit *cis-trans* isomerism? Explain.

25.46 Classify each of the following substances as alkane, alkene, or alkyne (assuming that none are cyclic hydrocarbons): (a) C_5H_{12}; (b) C_5H_8; (c) C_6H_{12}; (d) C_8H_{18}.

25.47 Give the IUPAC name for each of the following molecules:

(a) $CH_3\overset{OH}{\overset{|}{C}}HCH_3$ (b) CH_3OCH_3

(c) $CH_2{=}CHCH_2CH_2OH$

(d) $CH_3\overset{CH_2CH_3}{\overset{|}{C}}{=}CHCH_2CH_3$ (e) $CH_3C{\equiv}C{-}\underset{CH_3}{\underset{|}{\overset{CH_3}{\overset{|}{C}}}}H$

25.48 Describe how you would prepare (a) 2-bromo-2-chloropropane from propyne; (b) 1,2-dichloroethane from ethene; (c) a condensation polymer from $HOCH_2CH_2OH$ and

$HO\overset{O}{\overset{\|}{C}}CH_2CH_2\overset{O}{\overset{\|}{C}}OH$

(d) sodium benzoate from methylbenzoate; (e) polypropylene from propene; (f) 1-ethylnaphthalene from naphthalene and ethene.

25.49 Write the formulas for all of the structural isomers of C_3H_8O.

25.50 Explain the fact that *trans*-1,2-dichloroethene has no dipole moment while *cis*-1,2-dichloroethene has a dipole moment.

25.51 Would you expect cyclohexyne to be a stable compound? Explain.

25.52 Identify all of the functional groups in each of the following molecules:

(a) $CH_2{=}CH{-}O{-}CH{=}CH_2$ (an anesthetic)

(b) 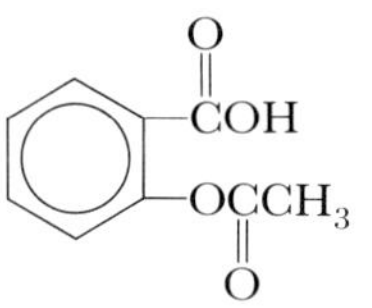 C_6H_4(COOH)(OCOCH$_3$) — COH, OCCH$_3$ (acetylsalicylic acid, aspirin)

(c) 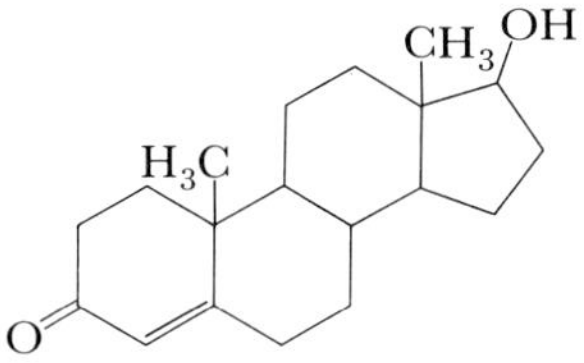 H_3C, CH_3, OH, O (testosterone, a male sex hormone)

25.53 Write the structural formulas for each of the following: (a) an ether with the formula C_3H_8O; (b) an aldehyde with the formula C_3H_6O; (c) a ketone with the formula C_3H_6O; (d) a secondary alkyl fluoride with the formula C_3H_7F; (e) a primary alcohol with the formula $C_4H_{10}O$; (f) a carboxylic acid with the formula $C_3H_6O_2$; (g) an ester with the formula $C_3H_6O_2$.

25.54 Give the condensed formulas for the carboxylic acid and the alcohol from which each of the following esters is formed:

(a) 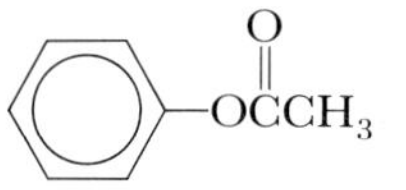 C_6H_5–OCCH$_3$ (with C=O) (b) 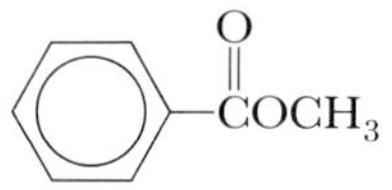 C_6H_5–COCH$_3$ (with C=O)

25.55 A 256-mg sample of an organic compound is combusted producing 512 mg of CO_2 and 209 mg of H_2O. At 127°C, 155 mg of this substance occupies a volume of 0.100 L with a pressure of 1.16 atm. What is the molecular formula of the compound? Suggest a possible structure for the compound and suggest some chemical test for the functional group.

[25.56] Suggest a process that would cause the breakup of a condensation polymer by essentially reversing the condensation reaction.

[25.57] An unknown organic compound is found on analysis to have the empirical formula $C_5H_{12}O$. It is slightly soluble in water. Upon careful oxidation it is converted into a compound of empirical formula $C_5H_{10}O$, which behaves chemically like a ketone. Indicate two or more reasonable structures for the unknown.

[25.58] Consider two hydrocarbons, A and B, whose molecular formulas are both C_4H_8. The first compound, A, is converted to $C_4H_{10}O$ when treated with water and sulfuric acid. Mild oxidation of this product produces C_4H_8O, which is an excellent solvent. Treatment with $K_2Cr_2O_7$ does not produce any further change in the C_4H_8O. In contrast, compound B gives a different $C_4H_{10}O$ substance upon treatment with sulfuric acid and water. Upon treating this $C_4H_{10}O$ substance with $K_2Cr_2O_7$, no oxidation occurs. Identify the two compounds, A and B, and trace all the reactions described that begin with these compounds, writing the names and structural formulas for all the organic compounds involved.

[25.59] An unknown substance is found to contain only carbon and hydrogen. It is a liquid that boils at 49°C at 1 atm pressure. Upon analysis it is found to contain 85.7 percent carbon and 14.3 percent hydrogen. At 100°C and 735 mm Hg, the vapor of the unknown has a density of 2.21 g/L. When it is dissolved in hexane solution and bromine water added, no reaction occurs. Suggest the identity of the unknown compound.

25.60 Bromination of butane requires irradiation. Write a series of reaction steps that can account for the bromination of butane under irradiation conditions.

26 Biochemistry

In earlier chapters of this text, we have considered various aspects of the physical world in which we live. We have discussed the chemistry of the atmosphere and of natural waters (Chapter 12), and of the solid earth (Chapter 22). In this final chapter we consider the **biosphere,** which is defined as that part of the earth in which living organisms are formed and live out their life cycles. The biosphere is not distinct from the atmosphere, natural waters, or the solid earth; rather, it is an integral part of it in those places where conditions permit life to exist. For living organisms to be sustained, there must be a supply of available energy, because the growth and maintenance of organisms requires energy. Second, there must be an adequate supply of water, because organisms are composed largely of water and use it for exchange of materials with their environment.

26.1 ENERGY REQUIREMENTS OF ORGANISMS

Living organisms require energy for their maintenance, growth, and reproduction. The ultimate source of this energy is the sun. However, as living matter proliferated on earth, and as organisms became more and more specialized, many of them developed the capacity for obtaining energy indirectly, by utilizing the energy stored in other organisms. For example, our bodies have essentially no capacity for directly utilizing solar energy. We consume animal and plant materials to obtain substances that our bodies can utilize as energy sources.

There are two distinct reasons why living systems need energy. First, organisms rely on substances readily available in their surroundings for synthesis of compounds needed for their existence. Most of these reactions are endothermic. To make such reactions proceed, energy must be supplied from an outside source. Second, living organisms are highly organized. The complexity of all the substances that make up even the simplest of single-cell organisms and the relationships between all the many chemical processes occurring are truly amazing. In thermodynamic terms, this means that living systems are very low in entropy as compared with the raw materials from which they are formed. The orderliness that is characteristic of living systems is attained by expenditure of energy.

Recall from the discussion in Section 17.5 that the free-energy change, ΔG, is related to both the enthalpy and the entropy changes that occur in a process:

$$\Delta G = \Delta H - T\Delta S$$

If the entropy change in a process that builds up a living organism is negative (in other words, if a more highly ordered state results), the contribution to ΔG is positive. This means that the process becomes less spontaneous. Thus both the enthalpy and the entropy changes that result in the formation, maintenance, and reproduction of living systems are such that the overall process is nonspontaneous. To overcome the positive values for ΔG associated with the essential processes, living systems must be coupled to some outside source of energy that can be converted into a form useful for driving the biochemical processes. The ultimate source of this needed energy is the sun.

The major means for conversion of solar energy into forms that can be utilized by living organisms is **photosynthesis.** The photosynthetic reaction that occurs in the leaves of plants is conversion of carbon dioxide and water to carbohydrate, with release of oxygen:

$$6CO_2 + 6H_2O \xrightarrow{48\ h\nu} C_6H_{12}O_6 + 6O_2 \qquad [26.1]$$

Note that formation of a mole of sugar, $C_6H_{12}O_6$, requires the absorption and utilization of 48 photons. This needed radiant energy comes from the visible region of the spectrum (Figure 5.3). The photons are absorbed by photosynthetic pigments in the leaves of plants. The key pigments are the **chlorophylls;** the structure of the most abundant chlorophyll, called chlorophyll *a*, is shown in Figure 26.1. Chlorophyll is a coordination

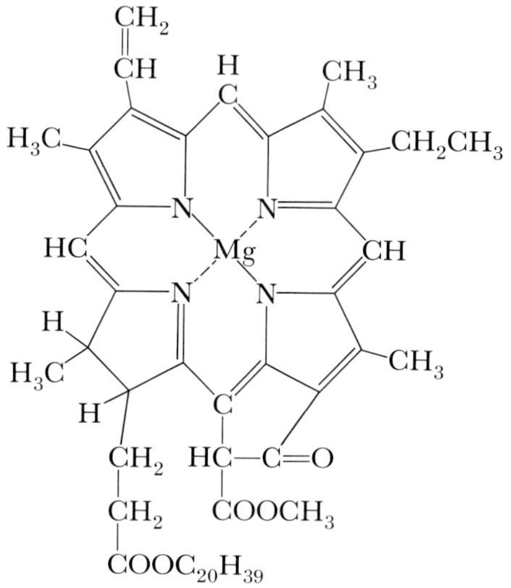

FIGURE 26.1 Structure of chlorophyll *a*. All chlorophyll molecules are essentially alike; they differ only in details of the side chains.

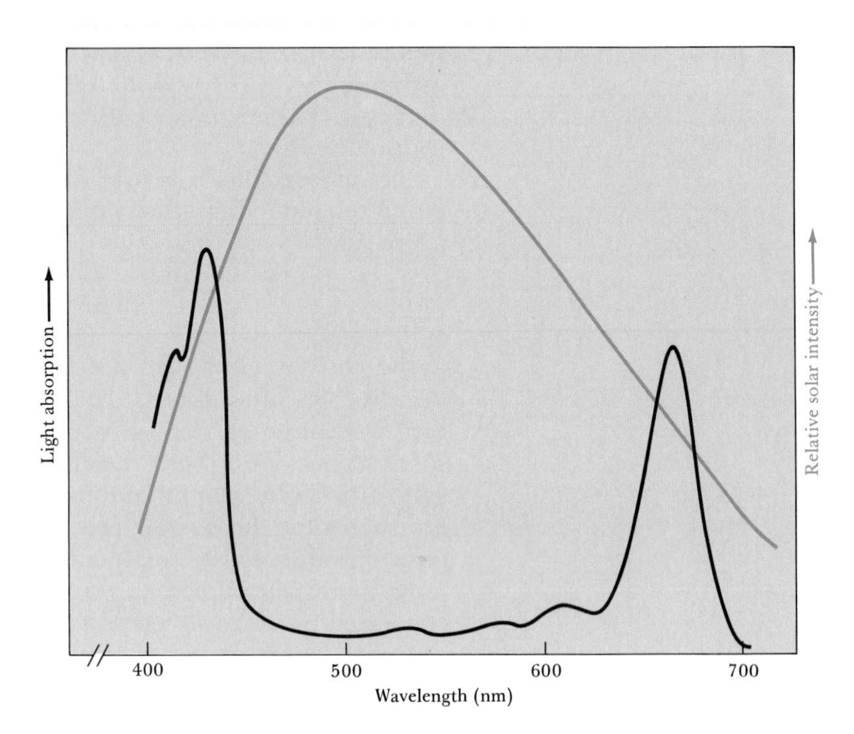

FIGURE 26.2 Absorption spectrum of chlorophyll (black curve), in comparison with the solar radiation at ground level (colored curve).

compound; it contains a Mg^{2+} ion bound to four nitrogen atoms arranged around the metal in a planar array. The nitrogen atoms are part of a porphyrin ring (Section 24.2). Notice that there is a series of alternating double bonds in the ring that surrounds the metal ion. This system of alternating, or conjugated, double bonds gives rise to the strong absorptions of chlorophyll in the visible region of the spectrum. Figure 26.2 shows the absorption spectrum of chlorophyll as compared with the distribution of solar energy at the earth's surface. Chlorophyll is green in color because it absorbs red light (maximum absorption at 655 nm) and blue light (maximum absorption at 430 nm) and transmits green light.

The solar energy absorbed by chlorophyll is converted in a complex series of steps into chemical energy. This stored energy is then used to drive the reaction in Equation 26.1 to the right, a direction in which it is highly endothermic. Thus, plant photosynthesis is nature's solar-energy-conversion machine; all living systems on earth are dependent on it for continued existence. A field of corn in the Midwest at the height of its growing season converts several percent of all the incident solar radiation into plant matter. It has been suggested that by cultivating about 6 percent of the land surface in the United States and using optimal growing conditions, we could obtain sufficient energy to satisfy all the energy needs of modern society.

Let us now turn our attention to the materials from which living organisms are formed. At some level of concentration, in some organism somewhere, almost every element seems to have a role to play in the biosphere. However, the elements of major importance in terms of their abundances in living systems are carbon, hydrogen, oxygen, nitrogen, phosphorus, and sulfur. Many other elements, including many metallic elements, are present in lesser quantities (refer back to Figure 24.5).

Much of the material present in living systems is in the form of **macro-**

molecules, polymers of high molecular weight. These biopolymers can be classified into three broad groups: proteins, carbohydrates, and nucleic acids. The proteins and carbohydrates, along with a class of molecules called fats and oils, are the major sources of energy in the food supply of animals. In addition, polymeric carbohydrates are the major construction material that gives form to plant systems, whereas proteins perform a similar role in animal systems. The nucleic acids are the storehouse of information that determines the form of a living system and that regulates its reproduction and development.

26.2 PROTEINS

Proteins are macromolecular substances present in all living cells. They serve as the major structural component in animal tissues; they are a major component of skin, cartilage, nails, and the skeletal muscles. Enzymes, the catalysts for the biochemical reactions occurring in all living systems, are proteins. Proteins transport vital substances within the body. For example, hemoglobin, which carries O_2 from the lungs to cells, is a protein. The antibodies that serve as a defense mechanism against undesirable substances within the organism are also composed of proteins.

The molecular weights of proteins vary from about 10,000 to over 50 million. Simple proteins are composed entirely of amino acids; other proteins, called conjugated proteins, are made up of simple proteins bonded to other kinds of biochemical structures. The most abundant elements present in proteins are carbon (50 to 55 percent), hydrogen (7 percent), oxygen (23 percent), and nitrogen (16 percent). Sulfur is present in most proteins to the extent of about 1 or 2 percent; phosphorus either is absent or is present to only a very slight extent.

Simple proteins are linear polymers of amino acids. The characteristic polymeric linkage in proteins is the amide linkage, introduced in Section 25.4:

$$\mathrm{R-\overset{\displaystyle O}{\overset{\|}{C}}-\underset{\displaystyle H}{\underset{|}{N}}-R'}$$

The particular functional group is called the peptide linkage when it is formed from amino acids. Proteins are sometimes referred to as polypeptides. Usually, however, the term *polypeptide* refers to a polymer of amino acids that has a molecular weight of less than 10,000. To see how the peptide linkage might serve to form a polymer we must investigate the structures of the amino acids, substances from which proteins are formed.

Amino Acids

Twenty amino acids are found to occur commonly in various proteins. All of these are α-amino acids; that is, the amino group is located on the carbon atom immediately adjacent to the carboxylic acid functional group. The general formula for an amino acid is as follows:

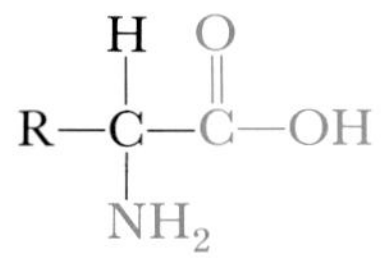

$$\mathrm{R{-}\overset{\displaystyle H}{\underset{\displaystyle NH_2}{C}}{-}\overset{\displaystyle O}{\overset{\|}{C}}{-}OH}$$

The amino acids differ in the nature of the R group. Figure 26.3 shows the structural formulas of the 20 amino acids found in most proteins. Although certain of the amino acids are more common than others, most large proteins contain most of the amino acids.

You can see from the condensed structural formula shown above that the α-carbon atom, which bears both the amino and carboxylic acid groups, has four different groups attached to it.* Any molecule containing a carbon with four different attached groups will be **chiral.** As we pointed out in Section 24.4, a chiral molecule is one that is not superimposable with its mirror image. Let us consider a particular amino acid, say alanine, and suppose that we have two such molecules that are mirror images of one another, as in Figure 26.4. This figure also shows two familiar objects that are mirror images of one another, a pair of hands. Your right and left hands are not superimposable. That is, if there were some way that you could hollow out your left hand and place it face down on a surface, there is no way that you could slide your right hand, also face down, into the left. (That is, of course, why you can't wear your right-hand glove on your left hand.) In the same way, one of the mirror-image forms of alanine cannot be superimposed on the other.

*The sole exception is glycine, for which R = H. For this amino acid, there are two H atoms on the α-carbon atom.

Amino acids with hydrocarbon side chains

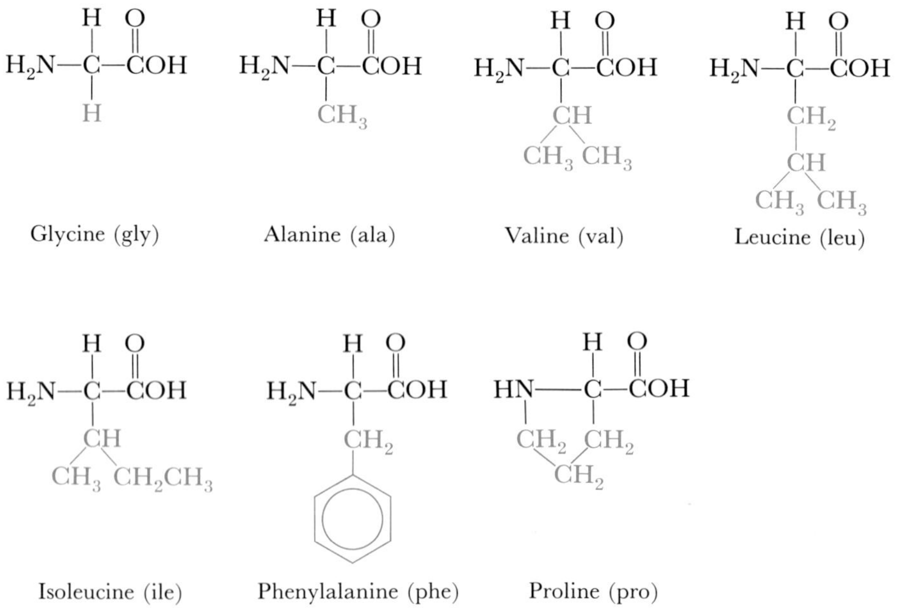

$$\mathrm{H_2N{-}\overset{\displaystyle H}{\underset{\displaystyle H}{C}}{-}\overset{\displaystyle O}{\overset{\|}{C}}OH}$$

Glycine (gly)

$$\mathrm{H_2N{-}\overset{\displaystyle H}{\underset{\displaystyle CH_3}{C}}{-}\overset{\displaystyle O}{\overset{\|}{C}}OH}$$

Alanine (ala)

$$\mathrm{H_2N{-}\overset{\displaystyle H}{\underset{\displaystyle CH(CH_3)_2}{C}}{-}\overset{\displaystyle O}{\overset{\|}{C}}OH}$$

Valine (val)

$$\mathrm{H_2N{-}\overset{\displaystyle H}{\underset{\displaystyle CH_2CH(CH_3)_2}{C}}{-}\overset{\displaystyle O}{\overset{\|}{C}}OH}$$

Leucine (leu)

$$\mathrm{H_2N{-}\overset{\displaystyle H}{\underset{\displaystyle CH(CH_3)CH_2CH_3}{C}}{-}\overset{\displaystyle O}{\overset{\|}{C}}OH}$$

Isoleucine (ile)

$$\mathrm{H_2N{-}\overset{\displaystyle H}{\underset{\displaystyle CH_2C_6H_5}{C}}{-}\overset{\displaystyle O}{\overset{\|}{C}}OH}$$

Phenylalanine (phe)

$$\mathrm{HN{-}\overset{\displaystyle H}{C}{-}\overset{\displaystyle O}{\overset{\|}{C}}OH \quad (ring: HN{-}CH_2{-}CH_2{-}CH_2{-}C)}$$

Proline (pro)

FIGURE 26.3 Condensed structural formulas of the amino acids, with the three-letter abbreviation for each acid.

Amino acids with polar, neutral side chains

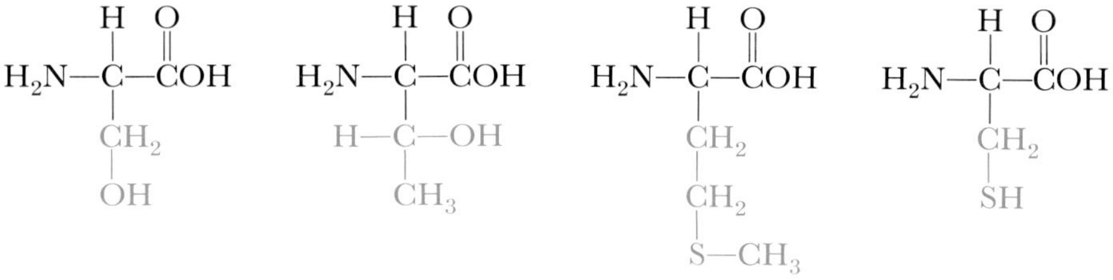

Serine (ser)

Threonine (thr)

Methionine (met)

Cysteine (cys)

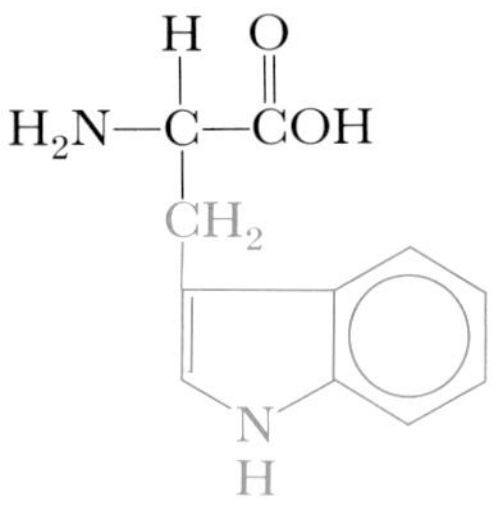

Tryptophan (trp)

Asparagine (asn)

Glutamine (gln)

Amino acids with acidic or basic side chain functional groups

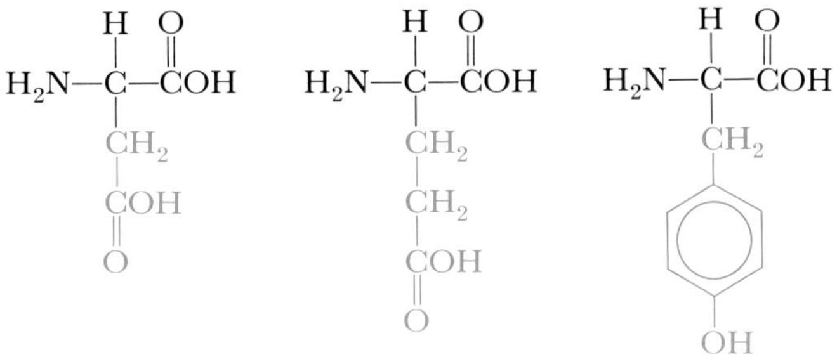

Aspartic acid (asp)

Glutamic acid (glu)

Tyrosine (tyr)

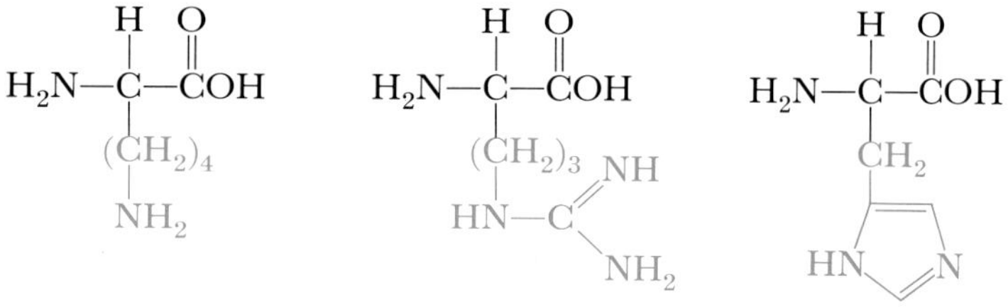

Lysine (lys)

Arginine (arg)

Histidine (his)

FIGURE 26.3 (*Continued.*)

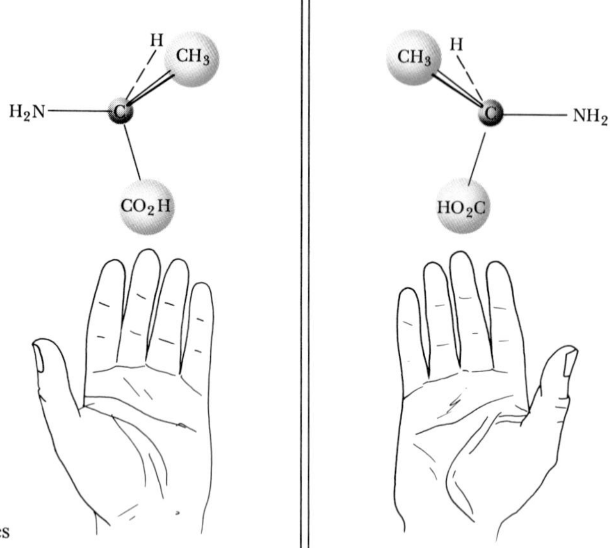

FIGURE 26.4 Illustration of chiral character. The pair of hands shown are mirror images of one another and are not superimposable. In the same way, the two alanine molecules that are mirror images are also not superimposable.

The two molecules of a chiral substance that are mirror images of one another are called **enantiomers.** Because two enantiomers are not exactly the same, they are isomers of one another. This type of isomerism is called **configurational isomerism,** or **optical isomerism.** The two enantiomers of a pair are sometimes labeled *R*- (from the Latin *rectus,* "right") or *S*- (from the Latin *sinister,* "left") to distinguish them. Another widely used notation for distinguishing the enantiomers uses the prefix labels D- (from the Latin *dexter,* "right") or L- (from the Latin *laevus,* "left"). The enantiomers of a chiral substance possess the same physical properties, such as solubility, melting point, and so forth. Their chemical behavior toward ordinary chemical reagents is also the same. However, they differ in chemical reactivity toward other chiral molecules. It is a striking fact that all the amino acids in nature are of the *S*-, or L-, configuration at the carbon center (except glycine, which is not chiral). Only amino acids of this specific configuration at the chiral carbon center are biologically effective in forming polypeptides and proteins in most organisms; peptide linkages are formed in cells under such specific conditions that enantiomeric molecules can be distinguished.

Chiral substances differ from one another in their effect on the plane of **polarized light.** We saw an example of this in Section 24.4, which dealt with the isomerism of coordination compounds. Suppose that a beam of polarized light is passed through a solution containing a chiral substance such as alanine, using the arrangement shown in Figure 24.14. A solution of one of the enantiomers causes the plane of polarization to be rotated in one direction; a solution of the other enantiomer has the opposite effect. In each case the amount of rotation is proportional to the concentration of the solution and the length of the cell containing it. A solution containing equal concentrations of the two enantiomers, called a **racemic mixture,** causes no net rotation. Enantiomers of a chiral substance are often called **optical isomers** because of their effect on polarized light.

The relative amounts of the various amino acids in a protein vary with the nature and function of the protein material. Amino acids that have hydrocarbon side chains predominate in the insoluble, fiberlike proteins such as silk, wool, and collagen (a tissue-supporting protein). The amino acids with polar side chains are relatively more abundant in the water-soluble proteins. In addition, the polar groups often play an important role in determining the overall structure of the protein molecule, about which we shall have more to say later. Amino acids with acid or base side chains also help to increase water solubility. In addition, the functional groups on the side chains can act as sources of acid or base character in reactions catalyzed by acids or bases.

Protein Structure

We have noted that the characteristic functional group in proteins is the amide, or peptide, linkage. This linkage is formed by a condensation reaction (Sections 25.4 and 25.5) between two amino acid molecules. As an example, alanine and glycine might form the dipeptide glycylalanine:

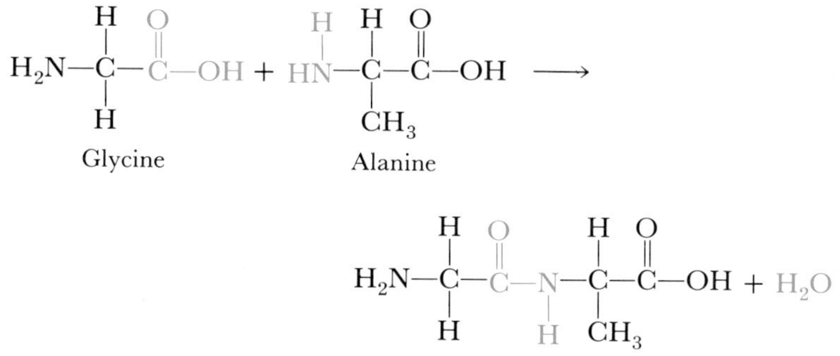

$$\begin{array}{ccccccccccccc} & & H & & O & & & & H & & H & & O & & \\ & & | & & \| & & & & | & & | & & \| & & \\ H_2N & - & C & - & C & - & OH & + & HN & - & C & - & C & - & OH & \longrightarrow \\ & & | & & & & & & & & | & & & & \\ & & H & & & & & & & & CH_3 & & & & \end{array}$$

Glycine Alanine

$$\begin{array}{ccccccccccccccc} & & H & & O & & & & H & & O & & & & \\ & & | & & \| & & & & | & & \| & & & & \\ H_2N & - & C & - & C & - & N & - & C & - & C & - & OH & + & H_2O \\ & & | & & & & | & & | & & & & & & \\ & & H & & & & H & & CH_3 & & & & & & \end{array}$$

Glycylalanine

Notice that the acid that furnishes the carboxyl group is named first, with a *-yl* ending; then the amino acid furnishing the amino group.

SAMPLE EXERCISE 26.1

Draw the structural formula for histidylglycine.

Solution: The name for this dipeptide tells us that the amino group of glycine and the carboxylic acid function of the histidine are involved in formation of the peptide bond. The structure is therefore

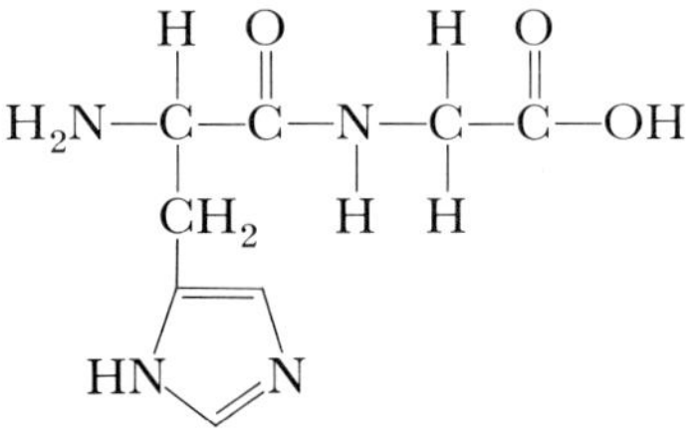

$$\begin{array}{ccccccccccccc} & & H & & O & & & & H & & O & & \\ & & | & & \| & & & & | & & \| & & \\ H_2N & - & C & - & C & - & N & - & C & - & C & - & OH \\ & & | & & & & | & & | & & & & \\ & & CH_2 & & & & H & & H & & & & \\ & & | & & & & & & & & & & \\ & & \text{(imidazole ring: HN, N)} & & & & & & & & & & \end{array}$$

Because 20 amino acids are commonly found in proteins and because the protein chain may consist of hundreds of amino acids, the number of possible arrangements of amino acids is huge beyond imagination. Nature makes use of only certain of these combinations, but even so the number of different proteins is very large.

The arrangement, or sequence, of amino acids along the chain determines the **primary structure** of a protein. This primary structure gives the protein its unique identity. A change of even one amino acid can alter the biochemical characteristics of the protein. For example, sickle-cell anemia is a genetic disorder resulting from a single misplacement in a protein chain in hemoglobin. The chain that is affected contains 146 amino acids. The first seven in the chain are valine, histidine, leucine, threonine, proline, glutamic acid, and glutamic acid. In a person suffering from sickle-cell anemia, the sixth amino acid is valine instead of glutamic acid. This one substitution of an amino acid with a hydrocarbon side chain for one that has an acidic functional group in the side chain alters the solubility properties of the hemoglobin and normal blood flow is impeded (see also Section 11.7).

Proteins in living organisms are not simply long, flexible chains with more or less random shape. Rather, the chain coils or stretches in particular ways that are essential to the proper functioning of the protein. This aspect of the protein structure is called the **secondary structure.** One of the most important and common secondary structure arrangements is the **α-helix,** first proposed by Linus Pauling and R. B. Corey. The helix arrangement is shown in schematic form in Figure 26.5. Imagine winding a long protein chain in a helical fashion around a long cylinder. The helix is held in position by hydrogen-bond interactions between an N—H bond and the oxygen of a carbonyl function located directly above or below. The pitch of the helix and the diameter of the cylinder must be such that (1) no bond angles are strained and (2) the N—H and C=O functional groups on adjacent turns are in proper position for hydrogen bonding. An arrangement of this kind is possible for some amino acids along the chain, but not for others. Large protein molecules may contain segments of chain that have the α-helical arrangement interspersed with sections in which the chain is in a random coil. The overall shape of the protein, determined by all the bends, kinks, and sections of a rodlike α-helical structure, is called the **tertiary structure.** You might suppose that a long, complex protein molecule could adopt almost any shape under a given set of conditions, but this is not the case. Certain foldings of the protein chain lead to lower-energy (more stable) arrangements than do other folding patterns. For example, a protein dissolved in aqueous solution folds in such a way that the nonpolar hydrocarbon portions are tucked within the molecule, whereas the more polar acidic and basic side-chain functional groups are projected into the solution.

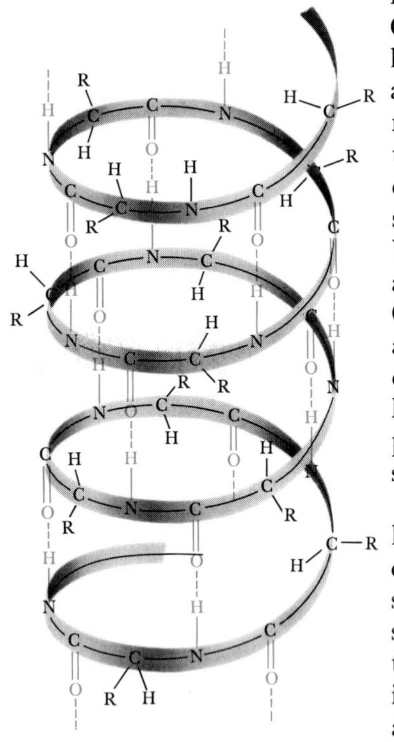

FIGURE 26.5 α-Helix structure for a protein. The symbol R represents any one of the several side chains shown in Figure 26.3.

When proteins are heated above the temperatures characteristic of living organisms, or when they are subjected to unusual acid or base conditions, they begin to lose their particular tertiary and secondary structure. As this happens the protein loses its biological activity; it is said to be **denatured.** When denaturation occurs under very mild conditions, it is often reversible. That is, a return to normal conditions results in a return of biological activity. On the other hand, denaturation may also be irreversible; chemical changes that permanently alter the character of the protein may occur. As an example, the protein material in the white of an egg is irreversibly denatured when the egg is placed for a time in boiling water.

With this brief introduction to proteins, let us now consider the characteristics of a very important group of proteins, the enzymes.

26.3 ENZYMES

The human body is characterized by an extremely complex system of interrelated chemical reactions. All of these reactions must occur at carefully controlled rates, so that thousands of individual chemical components are maintained at proper concentration levels and the system is able to respond as needed to demands made upon it. Every one of the many thousands of chemical reactions occurring in a biochemical system can be represented by an ordinary chemical equation. Indeed, many of the same reactions can be carried out under ordinary laboratory conditions. What is extraordinary about biochemical systems is that so many reactions occur with great rapidity at moderate temperatures. We cited in Chapter 13 the example of sugar, which is quite rapidly oxidized in the body at 37°C to carbon dioxide and water. In the laboratory, sugar does not react with oxygen at room temperature at a measurable rate. It must be heated to rather high temperature, for example, in a burner flame, before it burns.

The oxidation of sugar in the body occurs by means of a series of more than two dozen biochemically catalyzed steps. The catalyst in each step is called an **enzyme.** Enzymes are formed in living cells and are protein in nature. The molecular weights of enzymes range from a low of perhaps 10,000 to a high of about 1 million. The enzyme may consist of just a single protein chain or may involve several chains that are loosely held together.

The reaction that the enzyme catalyzes occurs at a specific site on the protein; this is called the **active site.** The substances that undergo reaction at this site are called **substrates.** Besides the substrate, other substances, called **cofactors** or **coenzymes,** may be required in the enzyme-catalyzed reaction. For example, the enzyme may require the presence of Mg^{2+} or some other metal ion, or it may require the presence and participation of a small organic molecule.

In many enzyme-catalyzed reactions, the coenzyme is one of the vitamins. An enzyme without its required coenzyme is called an **apoenzyme.** The combination of apoenzyme and coenzyme is called the **holoenzyme:**

$$\text{Apoenzyme} + \text{coenzyme} \longrightarrow \text{holoenzyme}$$

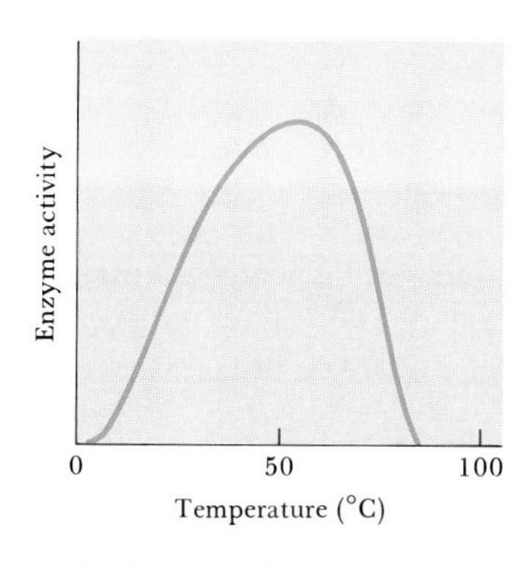

FIGURE 26.6 Enzyme activity as a function of temperature. At the higher temperature at which the activity of the enzyme drops to zero, the protein is denatured.

Enzymes possess certain characteristics not common to other types of catalysts. In the first place, they have a rather special sensitivity to temperature. Experimental studies have shown that the activity of a particular enzyme maximizes at around the normal temperature of the organism in which the enzyme is found. Figure 26.6 illustrates a typical activity versus temperature curve. Often it happens that when the temperature is raised above the usual operating temperature of the enzyme, the activity temporarily increases, but then subsequently decreases. The secondary and tertiary structures of the protein chain, on which the activity of the active site depends, are the result of many weak forces that induce the chain to take up that particular arrangement. Heating causes the protein chain to come undone, as it were; the enzyme is denatured and loses its activity completely.

A second respect in which enzymes are special is that they often show a sharp change in activity with change in the acidity or basicity of the solution. This suggests that acid-base reactions are important in the catalyzed reaction. Third, enzymes may be very specific in catalyzing exclusively a particular kind of reaction and no other, or even in catalyzing a particular reaction for only one compound and no other. The degree of specificity of enzymes varies widely. For example, enzymes called carboxypeptidases are rather general in their action.* They catalyze the hydrolysis of polypeptides into amino acids:

$$\begin{array}{ccccccccc} & \mathrm{H} & & \mathrm{O} & & \mathrm{H} & & \mathrm{O} & \\ & | & & \| & & | & & \| & \\ \cdots & \mathrm{C} & - & \mathrm{C} & - \mathrm{N} - & \mathrm{C} & - & \mathrm{C} & -\mathrm{OH} + \mathrm{H_2O} \\ & | & & & | & | & & & \\ & \mathrm{R'} & & & \mathrm{H} & \mathrm{R} & & & \end{array} \xrightarrow{\text{carboxypeptidase}}$$

$$\begin{array}{cccccccccc} & \mathrm{H} & & \mathrm{O} & & & \mathrm{H} & & \mathrm{O} & \\ & | & & \| & & & | & & \| & \\ \cdots & \mathrm{C} & - & \mathrm{C} & -\mathrm{OH} + & \mathrm{H_2N} - & \mathrm{C} & - & \mathrm{C} & -\mathrm{OH} \\ & | & & & & & | & & & \\ & \mathrm{R'} & & & & & \mathrm{R} & & & \end{array} \qquad [26.3]$$

This equation shows only the end member of a peptide chain; the carboxypeptidase attacks only the amide group at the end of the chain. However, it is active regardless of the nature of the particular side chains R and R′. Although carboxypeptidases catalyze the hydrolysis of peptides, they are not at all active in the hydrolysis of fats; an entirely separate set of enzymes is responsible for the latter reaction. The high degree of specificity that enzymes possess is necessary to maintain some degree of independence of all the reactions occurring in a complex organism.

Finally, many enzymes differ from ordinary nonbiochemical catalysts in that they are enormously more efficient. The number of individual, catalyzed-reaction events occurring per second at a particular active site, called the **turnover number,** is generally in the range 10^3 to 10^7 per second. Such large turnover numbers are indicative of reactions with very low activation energies.

SAMPLE EXERCISE 26.2

Solutions of dilute (about 3 percent) hydrogen peroxide are stable for long periods of time, especially if a small amount of a stabilizer is added. However, when such a solution is poured onto an open cut or wound, the hydrogen peroxide decomposes very rapidly with evolution of oxygen and formation of water:

$$2H_2O_2(aq) \longrightarrow O_2(g) + 2H_2O(l)$$

How can you account for this result?

Solution: Because the reaction occurs so rapidly on contact with the open wound, some chemical species present at the wound must catalyze the decomposition of hydrogen peroxide. That substance is an enzyme called peroxidase. Peroxidase is present to ensure that hydrogen peroxide does not accumulate in cells, because it could interfere with many other cell reactions. Peroxidase is an efficient enzyme, with a very high turnover number.

*The names of most enzymes end in *-ase*. The *-ase* ending is attached to the name of the substrate on which the enzyme acts, or to the name of the type of reaction that the enzyme catalyzes. Thus *peptidases* are enzymes that act on peptides; *lipase* is an enzyme that acts on lipids; a *transmethylase* is an enzyme that catalyzes transfer of a methyl group.

Biochemists have been trying for a long time to discover how and why enzymes are so fantastically efficient and selective. During the past 20 years, the development of new experimental techniques for studying the structures of molecules and for following the rates of very rapid reactions has led to a much better understanding of how enzymes work.

The high turnover numbers observed for enzymes suggest that the substrate molecules cannot be very tightly bound to the enzyme; if they were, they might block the active site. Reaction would then be slow, because the active site is not quickly cleared out. Most enzyme systems that have been studied behave as though there were an equilibrium between the substrate (S) and the active site (E); we can write this equilibrium as an equation:

$$E + S \underset{k_{-1}}{\overset{k_1}{\rightleftharpoons}} ES \qquad [26.4]$$

The symbol ES represents a species in which the substrate is attached in some way to the enzyme. This **enzyme-substrate complex,** as it is called, then reacts to give product (P) and the free active site:

$$ES \xrightarrow{k_2} E + P \qquad [26.5]$$

In this picture, it is assumed that the substrate molecules may come off and on the active site very rapidly compared with the rate at which they undergo reaction to form products P. This means that the equilibrium described by Equation 26.4 is rapidly established between enzyme and substrate. It is also supposed that the equilibrium is such that most of the enzyme sites are not occupied by S when the substrate is present at normal concentrations. Now suppose that we study and graph (Figure 26.7) the rate of the enzyme-catalyzed reaction as a function of increasing concentration of substrate S. When S is present in low concentration, most of the enzyme active sites are not in use. Increasing the concentration of S thus shifts the equilibrium in Equation 26.4 to the right, so that a larger number of enzyme-substrate complexes are formed. This in turn increases the overall rate of the reaction, because that rate depends on the concentration of ES, $[ES]$:

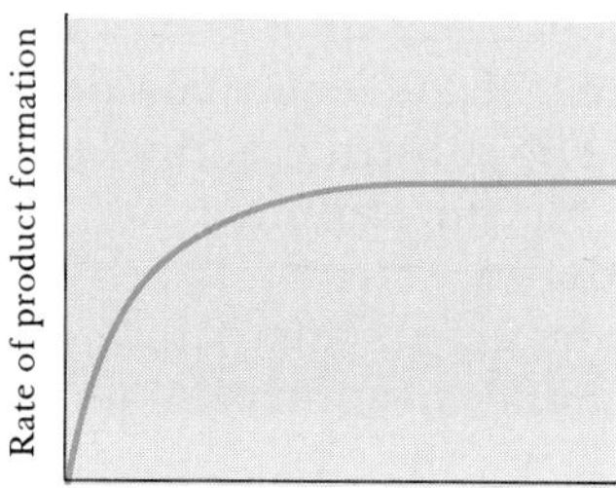

FIGURE 26.7 Rate of product formation as a function of concentration in a typical enzyme-catalyzed reaction. The rate of product formation is proportional to substrate concentration in the region of low concentration. At high substrate concentrations, the active sites on the enzyme are all complexed; further addition of substrate does not affect reaction rate.

$$\text{Rate} = \frac{\Delta[P]}{\Delta t} = k_2[ES] \qquad [26.6]$$

However, with still further increases in the concentration of S, a sizable fraction of the active sites are occupied. Adding still more S thus does not result in the same degree of increase in $[ES]$, and the rate does not increase as much. Eventually, with still higher concentration of S, *all* the active sites are effectively occupied. Further increases in S cannot result in faster reaction, because all available active sites are in use.

Many enzyme systems have been found to obey the sort of behavior illustrated in Figure 26.7. This strongly indicates that an enzyme-substrate complex is involved in the action of enzymes.

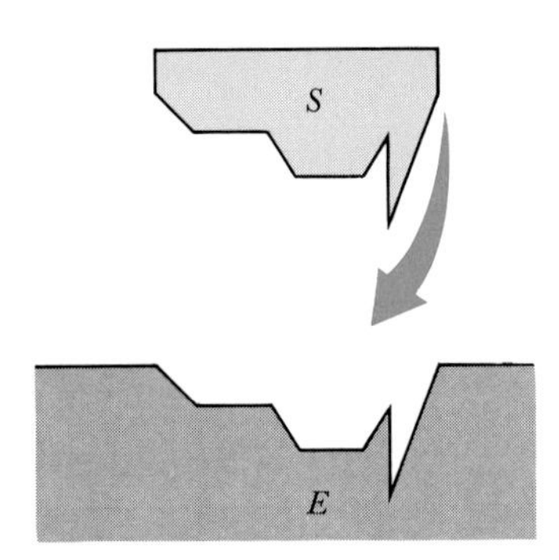

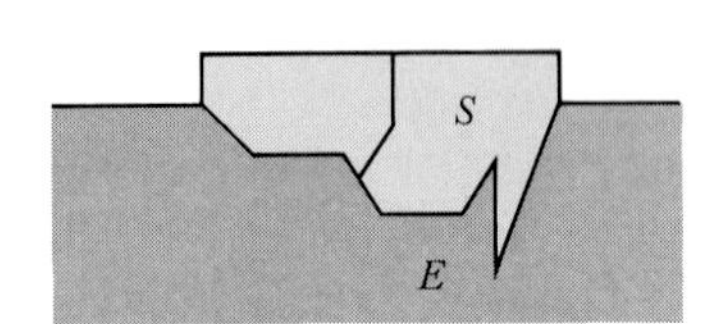

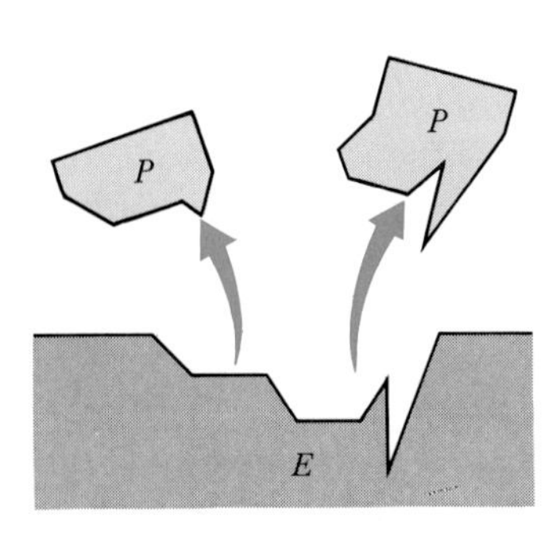

FIGURE 26.8 Lock-and-key model for formation of the enzyme-substrate complex.

The Lock-and-Key Model

One of the earliest models for the enzyme-substrate interaction was the **lock-and-key model** illustrated in Figure 26.8. The substrate is pictured as fitting neatly into a special place on the protein (the active site) tailored more or less specifically for that particular substrate. The catalyzed reaction occurs while the substrate is bound to the enzyme, and the reaction products then depart. Clearly, this picture of enzyme action has much in common with the models for heterogeneous catalysis that were discussed in Section 13.7. The difference is that an element of greater specificity is brought into the picture.

The lock-and-key model for enzyme action goes a long way toward explaining many aspects of enzyme action, but it cannot be the whole story. As the substrate molecules are drawn into the active site, they are somehow activated so that they are capable of extremely rapid reaction. This activation may result from the withdrawal or donation of electron density at a particular bond by the enzyme. In addition, in the process of fitting into the active site, the molecule may be distorted by the portions of the enzyme protein chain near the active site and thus made more reactive. If coenzymes are involved, the enzyme may promote reaction by binding both the coenzyme and substrate to the same site or to adjacent sites, thus keeping the reactants in close proximity. Enzymes, like all proteins, contain many acidic and basic sites along the chain, and these may be involved in acid-base reactions with the substrate. It is clear that the enzyme-coenzyme-substrate interactions can be very complex and are often difficult to unravel.

One of the more interesting corollaries of the lock-and-key model for enzyme action is that certain molecules should be capable of inhibiting the enzyme. Suppose that a certain molecule is capable of fitting the active site on the enzyme but is not reactive for one reason or another. If it is present in the solution along with substrate, this molecule competes with substrate for binding at the active sites. The substrate is thus prevented from forming the necessary enzyme-substrate complex, and the rate of product formation decreases. Metals that are highly toxic—for example, lead and mercury—probably operate as enzyme inhibitors. The heavy metal ions are especially strongly bound to sulfur-containing groups on protein side chains. By complexing very strongly to these sites on proteins, they interfere with the normal reactions of enzymes.

SAMPLE EXERCISE 26.3

Many bacteria employ *para*-aminobenzoic acid (PABA) in a vital enzyme-catalyzed metabolic reaction. Sulfa drugs such as sulfanilamide (*para*-aminobenzenesulfonamide) kill bacteria. Explain this in terms of the concept of the enzyme-substrate complex.

$$H_2N-C_6H_4-\overset{\overset{O}{\|}}{C}-OH \qquad H_2N-C_6H_4-\underset{\underset{O}{\|}}{\overset{\overset{O}{\|}}{S}}-NH_2$$

para-Aminobenzoic acid (PABA)

Sulfanilamide (*para*-aminobenzenesulfonamide)

Solution: The structural formulas reveal that PABA and sulfanilamide are very similar in shape and in the locations of polar functional groups. The sulfa drug acts as an inhibitor by binding at the active site in the enzyme and blocking access by PABA. Without this needed ingredient, the metabolic processes of the bacteria cease.

26.4 CARBOHYDRATES

The carbohydrates are a very important class of naturally occurring substances found in both plant and animal matter. The name carbohydrate (hydrate of carbon) comes from the empirical formulas for most substances in this class; they can be written as $C_x(H_2O)_y$. For example, glucose, the most abundant carbohydrate, has molecular formula $C_6H_{12}O_6$, or $C_6(H_2O)_6$. The carbohydrates are not really hydrates of carbon; rather, they are polyhydroxy aldehydes and ketones. The structural formula for glucose can be written as shown in Figure 26.9.

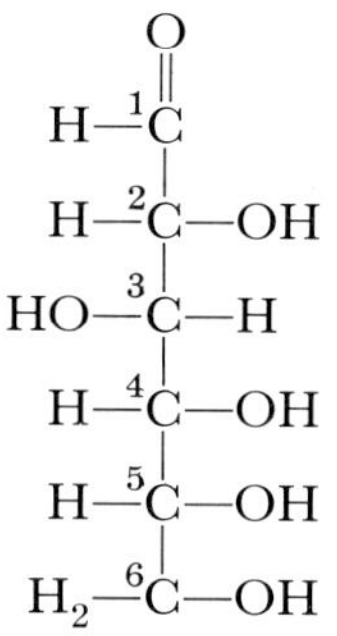

O
‖
H—^{1}C
H—^{2}C—OH
HO—^{3}C—H
H—^{4}C—OH
H—^{5}C—OH
H_2—^{6}C—OH

FIGURE 26.9 Structure of glucose.

Notice that four of the carbon atoms in this molecule, numbered 2, 3, 4, and 5, are chiral. Thus there are many configurational isomers of glucose. Several naturally occurring sugars differ from glucose only in the configuration at one of these four chiral carbon atoms. The fact that these sugars have different biological properties is another example of the extreme specificity of biochemical systems. Many sugars are optically active substances; their solutions cause the plane of polarization of polarized light to be rotated, as illustrated in Figure 24.14.

When glucose is placed in solution, the aldehyde functional group at one end of the molecule reacts with the hydroxy group at the other end to form a cyclic structure, called a hemiacetal:

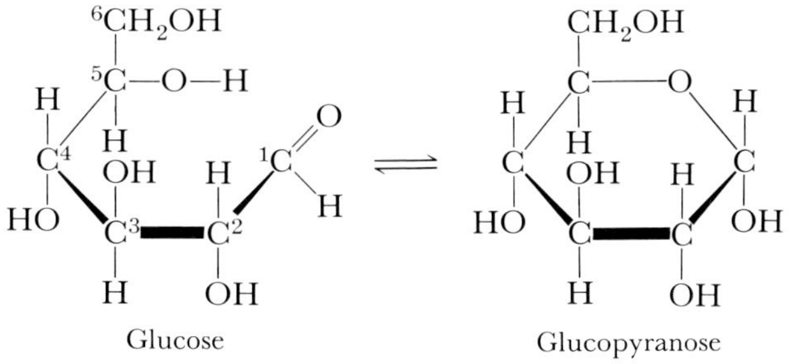

Glucose ⇌ Glucopyranose [26.7]

Note that in forming the ring structure in Equation 26.7, the OH group on carbon 1 can be on the same side of the ring as the OH group of carbon 2 (the α form) or on the opposite side (the β form). This seemingly minor difference is very important in distinguishing starch from cellulose, as we shall see shortly.

Six-membered cyclic hemiacetal structures are called pyranoses. The name *glucopyranose* indicates that the sugar glucose forms a six-membered ring. Sugars may also form five-membered rings called furanoses.

Some sugars are ketones rather than aldehydes. One important example is fructose, shown here in its straight-chain (linear) form as well as the five-membered cyclic form called fructofuranose:

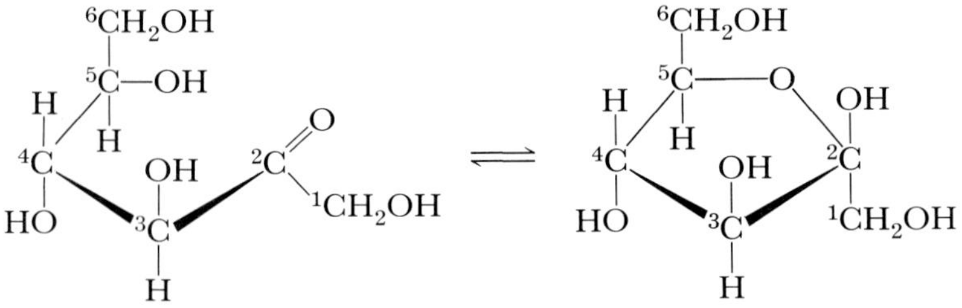

Fructose $\rightleftharpoons$ Fructofuranose [26.8]

Both glucose and fructose are examples of simple sugars, also called **monosaccharides.** Two or more such units can be joined together to form **polysaccharides.**

SAMPLE EXERCISE 26.4

Draw the structural formula for fructopyranose.

Solution: The name of the substance for which we are to draw the structure tells us that it involves a six-membered ring. Thus, beginning with fructose, for which the linear formula was given above, we complete a six-membered ring by closure with the OH group on the terminal carbon atom:

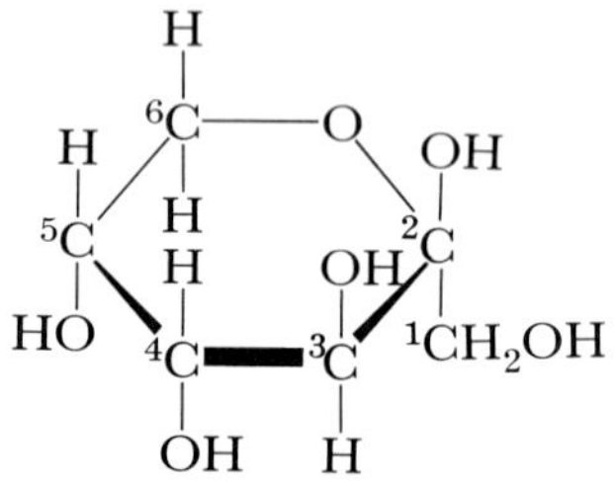

Notice that we might have drawn this structure so that the OH group of carbon atom 2 was on the opposite side of the ring as compared with the OH group of carbon 3. Thus there are two forms of fructopyranose.

Disaccharides

Disaccharides are formed by the joining together of two monosaccharide units. The monosaccharides are linked by a condensation reaction, in which a molecule of water is split out from between two hydroxyl groups, one on each sugar. Because there are several hydroxyl groups, there are several ways in which disaccharides can be formed. The structures of three common disaccharides, **sucrose** (table sugar), **maltose** (malt sugar), and **lactose** (milk sugar) are shown in Figure 26.10. The word "sugar" makes us think of sweetness; all sugars are sweet, but they differ in the degree of sweetness we perceive when we taste them. Sucrose is about six times sweeter than lactose, about three times sweeter than maltose, slightly sweeter than glucose, but only about half as sweet as fructose. Disaccharides can be hydrolyzed; that is, they can be reacted with water in the presence of an acid catalyst to form the monosaccharides. When sucrose is hydrolyzed, the mixture of glucose and fructose that forms, called *invert sugar,* * is sweeter to the taste than the original sucrose. The sweet syrup present in canned fruits and candies is largely invert sugar formed from hydrolysis of added sucrose.

*The term *invert sugar* comes from the fact that the rotation of the plane of polarized light by the glucose-fructose mixture is in the opposite direction, or inverted, from that of the sucrose solution.

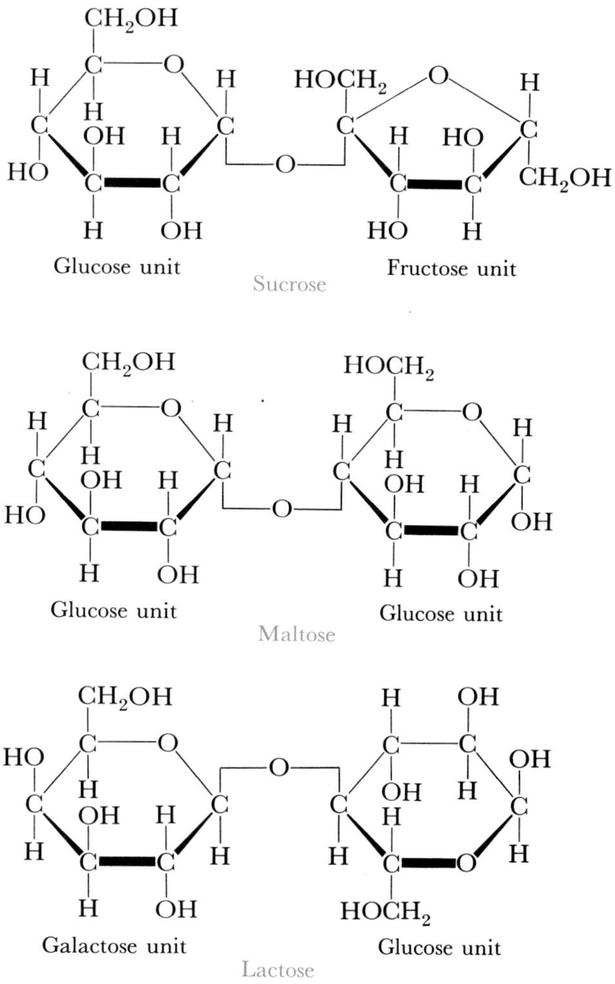

FIGURE 26.10 Structures of three disaccharides.

Polysaccharides

Polysaccharides are made up of several monosaccharide units joined together by a bonding arrangement similar to those shown for disaccharides in Figure 26.10. The most important polysaccharides are starch, glycogen, and cellulose, which are formed from repeating glucose units.

Starch is not a pure substance; the term refers to a group of polysaccharides found in plants. Starches serve as a major method of food storage in plant seeds and tubers. Corn, potatoes, wheat, and rice all contain substantial amounts of starch. These plant products serve as major sources of needed food energy for humans. Enzymes within the digestive system catalyze the hydrolysis of starch to glucose.

Some starch molecules are linear chains, whereas others are branched. All contain the type of chain structure illustrated in Figure 26.11. It is particularly important to note the geometrical arrangement in the joining of one ring to another; the glucose units are in the α form.

Glycogen is a starchlike substance synthesized in the body. Glycogen molecules vary in molecular weight from about 5000 to more than 5 million. Glycogen acts as a kind of energy bank in the body. It is concen-

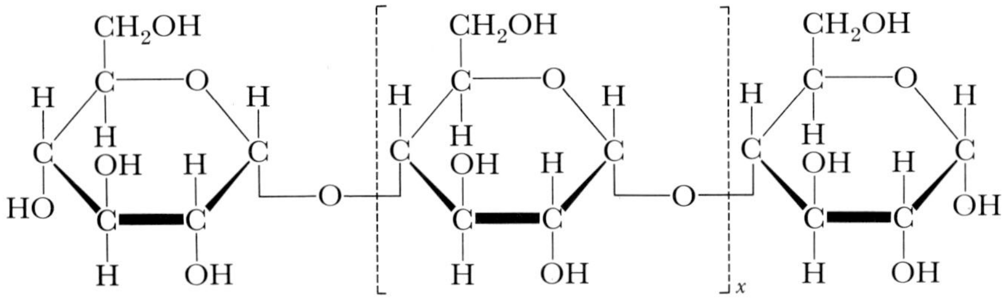

FIGURE 26.11 Structure of starch molecules.

trated in the muscles and liver. In muscles, it serves as an immediate source of energy; in the liver, it serves as a storage place for glucose and helps to maintain a constant glucose level in the blood.

Cellulose forms the major structural unit of plants. Wood is about 50 percent cellulose; cotton fibers are almost entirely cellulose. Cellulose consists of a straight chain of glucose units, with molecular weights averaging more than 500,000. The structure of cellulose is shown in Figure 26.12. At first glance, this structure looks very similar to that of starch. However, the differences in the arrangements of bonds that join the glucose units are important. Notice that glucose in cellulose is in the β form. The distinction is made clearer when we examine the structures of starch and cellulose in a more realistic three-dimensional representation, as shown in Figure 26.13. You can see that the individual glucose units have different relationships to one another in the two structures. Enzymes that readily hydrolyze starches do not hydrolyze cellulose. Thus you might chew up and swallow a pound of cellulose and receive no caloric value from it whatsoever, even though the heat of combustion per unit weight is essentially the same for both cellulose and starch. A pound of starch, on the other hand, would represent a substantial caloric intake. The difference is that the starch is hydrolyzed to glucose, which is eventually oxidized with release of energy. Cellulose, on the other hand, is not hydrolyzed by any enzymes present in the body, and so it passes through the body unchanged. Many bacteria contain enzymes, called cellulases, that hydrolyze cellulose. These bacteria are present in the digestive systems of grazing animals, such as cattle, that utilize cellulose for food.

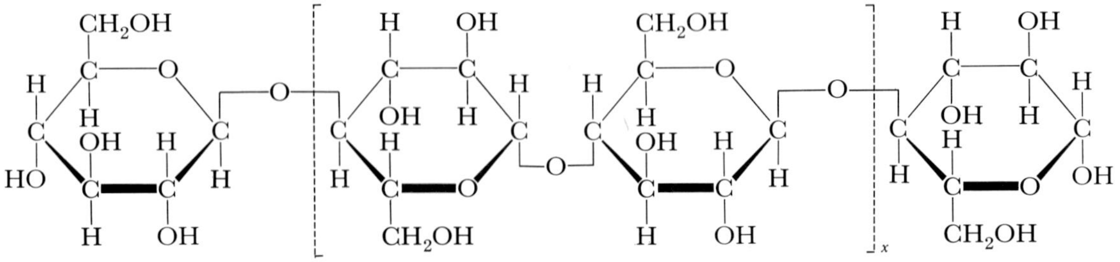

FIGURE 26.12 Structure of cellulose.

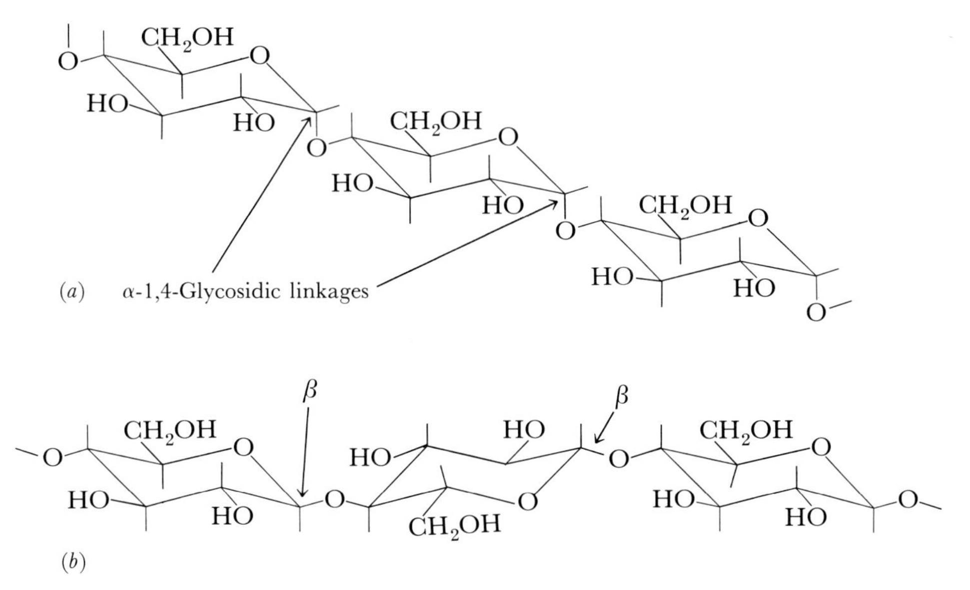

FIGURE 26.13 Structures of starch (*a*) and cellulose (*b*). These representations show the geometrical arrangements of bonds about each carbon atom. It is easy to see that the glucose rings are oriented differently with respect to one another in the two structures.

26.5 FATS AND OILS

$$\begin{array}{l} H_2C{-}O\overset{\overset{\large O}{\|}}{C}{-}R_1 \\ \;| \\ HC{-}O\overset{\overset{\large O}{\|}}{C}{-}R_2 \\ \;| \\ H_2C{-}O\overset{\overset{\large O}{\|}}{C}{-}R_3 \end{array}$$

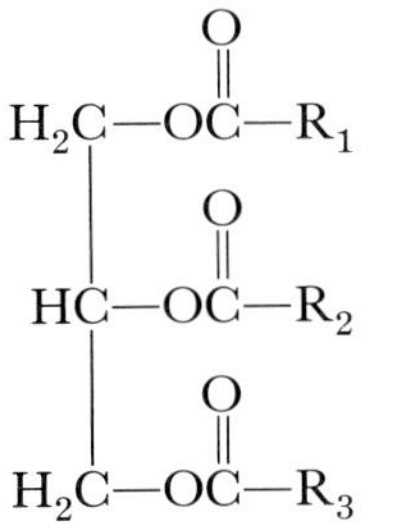

FIGURE 26.14 Structure of triglycerides. The groups R_1, R_2, and R_3 are hydrocarbon chains of from 3 to 21 carbons. The chains may be entirely saturated or may contain one or more *cis* olefin groups.

Both plant and animal systems need means of storing energy in various chemical forms so that this energy supply can be tapped as needed later on. In plant seeds, the stored energy is used to promote rapid growth after germination. In animals that hibernate during cold periods, the stored energy is needed when other sources of food are absent or scarce. One of the most important classes of compounds used for energy storage are the fats and oils. These substances are esters of long-chain carboxylic acids with 1,2,3-trihydroxypropane (glycerol) and are called triglycerides. Their general structural formula is shown in Figure 26.14. The groups R_1, R_2, and R_3 can be alike or different. The triglycerides are found as fat deposits in animals and as oils in nuts and seeds. Those that are liquid at room temperature are generally referred to as oils; those that are solid are called fats.

Hydrolysis of a fat or oil yields glycerol and the carboxylic acids of the R_1, R_2, and R_3 chains. The acids are commonly referred to as fatty acids. Most commonly, the fatty acids contain from 12 to 22 carbons. It is an interesting fact that nearly all fatty acids contain an even number of carbon atoms, including the carboxylic acid carbon. Several of the more common fatty acids are listed in Table 26.1. The oils, which are obtained mainly from plant products (corn, peanuts, and soybeans), are formed primarily from unsaturated fatty acids. On the other hand, animal fats (beef, butter, and pork) contain mainly saturated fatty acids.

Some of the oil extracted from cottonseed, corn, or soybean is utilized as liquid cooking oil. Other plant-derived oils are hydrogenated to convert the carbon-carbon double bonds in the acid chains to carbon-carbon single bonds. In the process, the oils are converted to solids; they are used to produce oleomargarine, peanut butter, vegetable shortening, and sim-

TABLE 26.1 Name, source, and structure of several fatty acids

Formula or structure[a]	Name	Source
$CH_3(CH_2)_{10}COOH$	Lauric acid	Coconuts
$CH_3(CH_2)_{12}COOH$	Myristic acid	Butter
$CH_3(CH_2)_{16}COOH$	Stearic acid	Animal fats
$CH_3(CH_2)_7CH{=}CH(CH_2)_7COOH$ (cis; H, H on the C=C)	Oleic acid	Corn oil
$CH_3(CH_2)_4CH{=}CHCH_2CH{=}CH(CH_2)_7COOH$ (cis)	Linoleic acid	Vegetable oils
$CH_3CH_2CH{=}CHCH_2CH{=}CHCH_2CH{=}CH(CH_2)_7COOH$ (cis)	Linolenic acid	Linseed oil

[a]Notice that the *cis* configuration is adopted at the double bonds in the unsaturated fatty acids.

ilar food products. As an example, trilinolein, a major constituent of cottonseed oil, is hydrogenated to form tristearin, which is solid at room temperature:

$$\begin{array}{l} CH_3(CH_2)_4{-}CH{=}CH{-}CH_2{-}CH{=}CH{-}(CH_2)_7\overset{O}{\overset{\|}{C}}OCH_2 \\ CH_3(CH_2)_4{-}CH{=}CH{-}CH_2{-}CH{=}CH{-}(CH_2)_7\overset{O}{\overset{\|}{C}}OCH \\ CH_3(CH_2)_4{-}CH{=}CH{-}CH_2{-}CH{=}CH{-}(CH_2)_7\overset{O}{\overset{\|}{C}}OCH_2 \\ \text{Trilinolein (liquid)} \end{array} \xrightarrow[Ni]{6H_2} \begin{array}{l} CH_3(CH_2)_{16}\overset{O}{\overset{\|}{C}}OCH_2 \\ CH_3(CH_2)_{16}\overset{O}{\overset{\|}{C}}OCH \\ CH_3(CH_2)_{16}\overset{O}{\overset{\|}{C}}OCH_2 \\ \text{Tristearin (mp 71°C)} \end{array} \qquad [26.9]$$

Fats and oils are an important source of energy in our food supply. In the body they undergo hydrolysis to form glycerol and carboxylic acids. The hydrolysis is promoted by enzymes called **lipases:**

$$\begin{array}{l} H_2C{-}O\overset{O}{\overset{\|}{C}}{-}R \\ HC{-}O\overset{O}{\overset{\|}{C}}{-}R \\ H_2C{-}O\overset{O}{\overset{\|}{C}}{-}R \end{array} + 3H_2O \xrightarrow{\text{lipase}} \begin{array}{l} H_2C{-}OH \\ HC{-}OH \\ H_2C{-}OH \end{array} + 3R{-}\overset{O}{\overset{\|}{C}}{-}OH \qquad [26.10]$$

This hydrolysis reaction, which takes place in aqueous solution, is hampered by the fact that fats and oils are essentially insoluble in water. As a result, not much hydrolysis occurs in the stomach. The gallbladder excretes compounds called bile salts into the small intestine to assist in the hydrolysis. The bile salts break up the larger droplets of fats into an emulsion (a suspension of very small droplets), so that hydrolysis can proceed more rapidly.

26.6 NUCLEIC ACIDS

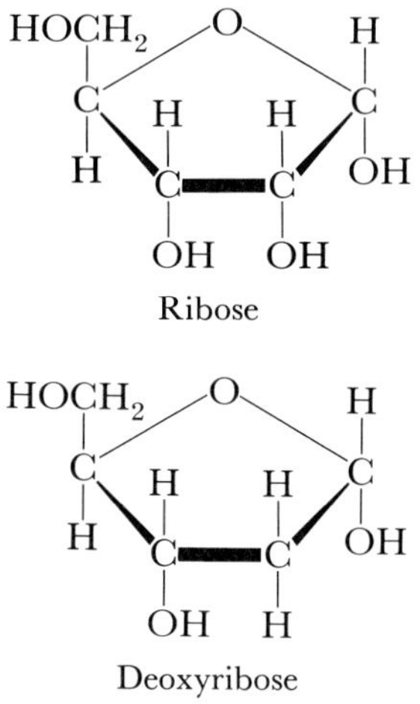

The nucleic acids are a class of biopolymers present in nearly all cells. They are classified into two groups—deoxyribonucleic acids (DNA) and ribonucleic acids (RNA). DNA molecules are very large; their molecular weights may range from 6 to 16 million. RNA molecules are much smaller, with molecular weights in the range 20,000 to 40,000. DNA is found primarily in the nucleus of the cell; RNA is found outside the nucleus, in the surrounding fluid called the cytoplasm.

Both DNA and RNA are polymers of a basic repeating unit called a nucleotide. To understand the structure of the polymer, we must first examine the structure of the nucleotide units. The nucleotide consists of three parts: (1) a phosphoric acid molecule, (2) a sugar in the furanose (five-membered ring) form, and (3) a nitrogen-containing organic base with a ring structure analogous to the ring structures of aromatic molecules. The sugar molecule found in RNA is ribose, shown in Figure 26.15. In DNA the sugar is deoxyribose, which differs from ribose only in the substitution of hydrogen for an —OH group on one carbon, as illustrated in Figure 26.15. In DNA the organic base may be any one of the four shown in Figure 26.16. An example of a complete nucleotide (deoxyadenylic acid) formed from these three components is illustrated in Figure 26.17. Substitution of one of the other bases for adenine in this structure produces one of the other three nucleotides that make up DNA.

Ribose

Deoxyribose

FIGURE 26.15 Ribose and deoxyribose, the two sugars found in RNA and DNA, respectively.

Just as a polypeptide is formed by a condensation reaction between amino acids, creating a peptide bond, a polynucleotide is formed by a condensation reaction creating a phosphate ester bond. The condensation reaction results in elimination of water from between an OH group of the phosphoric acid and an OH group of the sugar. The formation of the polymeric chain is shown schematically in Figure 26.18. The organic bases in this illustration are simply represented as B; they might be the same or different bases.

DNA molecules consist of two linear strands of the sort shown in Figure 26.18, which are wound together in the form of a double helix. It is very confusing to look at a model of double-stranded DNA in which all the atoms are represented. However, we can see from a more schematic

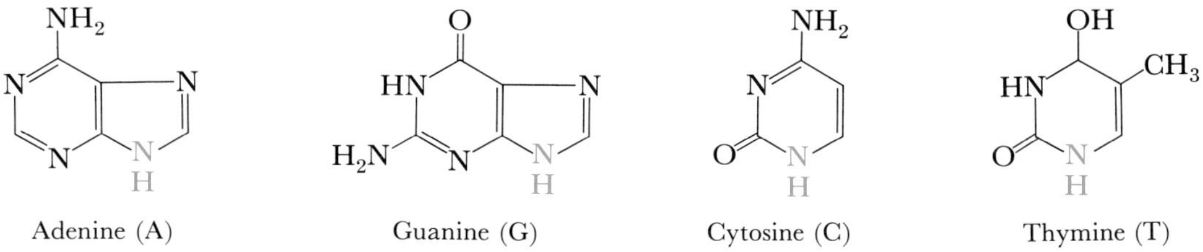

Adenine (A) Guanine (G) Cytosine (C) Thymine (T)

FIGURE 26.16 The four organic bases present in DNA. In DNA each base is attached to a ribose molecule through a bond from the nitrogen shown in color.

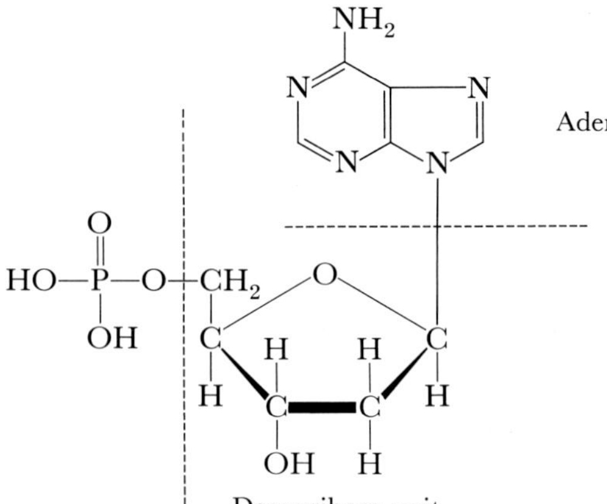

FIGURE 26.17 Structure of deoxyadenylic acid, a nucleotide formed from phosphoric acid, deoxyribose, and an organic base, adenine.

illustration, Figure 26.19, how the DNA molecule is constructed. Remember that the polymeric strand itself consists of alternating sugar and phosphate groups, represented as —S— and —P—, respectively. The various bases are attached to the polymeric strand at each sugar unit. The key to the double-helix structure for DNA is the formation of hydro-

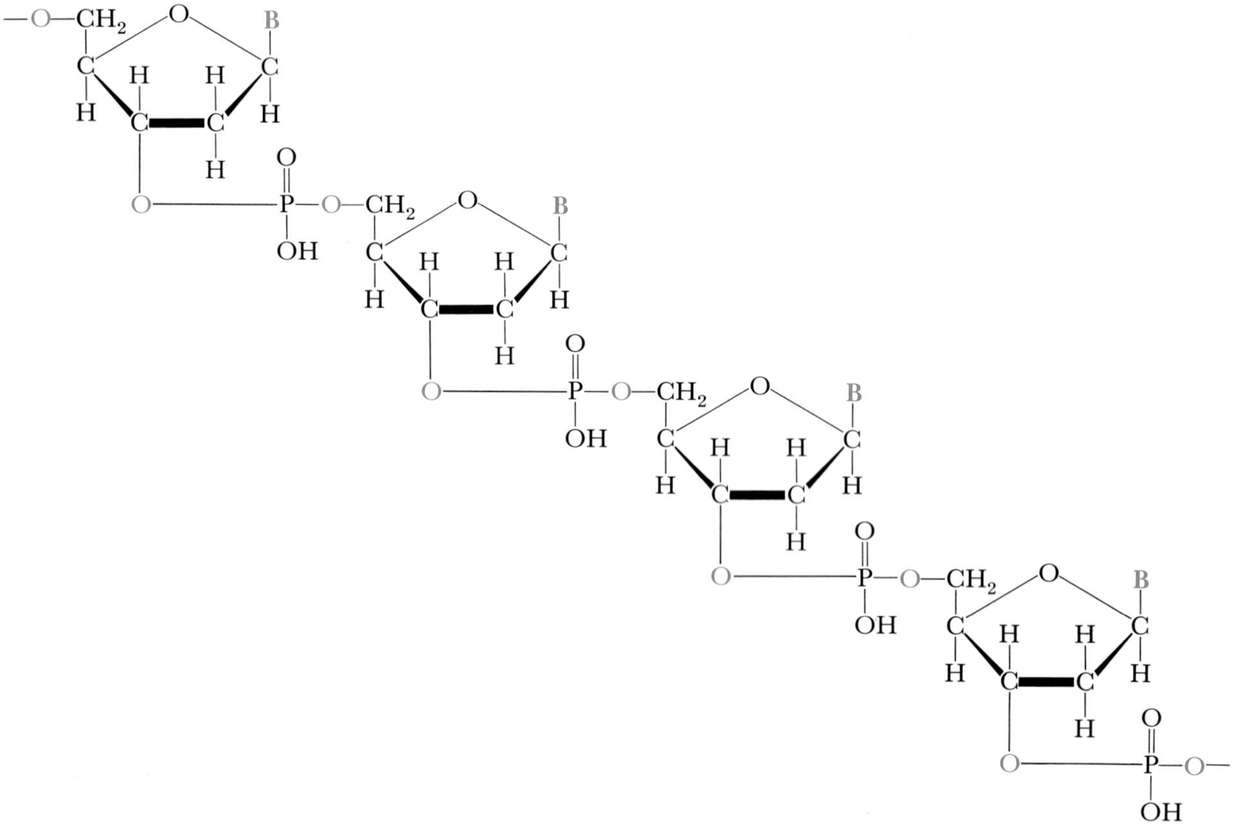

FIGURE 26.18 Structure of a polynucleotide. The symbol B represents any one of the four bases shown in Figure 26.16. Because the sugar in each nucleotide is deoxyribose, this polynucleotide is of the form found in DNA.

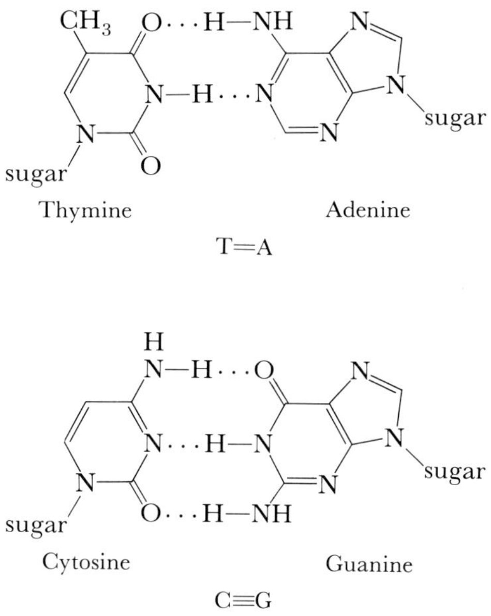

Thymine Adenine

T=A

Cytosine Guanine

C≡G

FIGURE 26.20 Hydrogen bonding between complementary base pairs. The hydrogen bonds shown here are responsible for formation of the double-stranded helical structure for DNA, as shown in Figure 26.19.

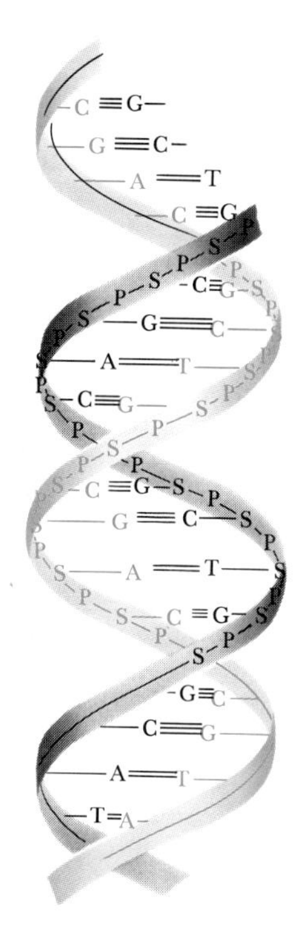

FIGURE 26.19 Schematic illustration of the double-stranded helical structure for DNA. The hydrogen-bond interactions between complementary base pairs is represented as illustrated in Figure 26.20.

gen bonds between bases on the two chains. When adenine and thymine are located opposite one another along the chain, they form a strong hydrogen bond, as shown in Figure 26.20. Similarly, guanine and cytosine form especially strong hydrogen bonds with one another. This means that the two strands of DNA are complementary; opposite each adenine on one chain there is always a thymine; opposite each guanine there is a cytosine. The double-helix structure for DNA, first proposed by James Watson and Francis Crick in 1953, has proved to be the key to understanding protein synthesis in cells, the means by which viral particles infect cells, the means by which genetic information is transmitted in reproduction of cells, and many other problems of central importance in molecular biology. Those themes are beyond the scope of this book; however, if you are interested in study in the area of life sciences, you will learn a good deal about such matters in other courses.

FOR REVIEW

Summary

In this chapter, we have taken a brief look at some of the major constituents of the biosphere—that part of the physical world in which organisms live out their life cycles. The maintenance of life requires a source of energy as well as appropriate environmental conditions. The ultimate source of the needed energy is the sun. Plants convert solar energy to chemical energy in the process called photosynthesis. Solar energy is absorbed by a plant pigment called chlorophyll and then utilized to form carbohydrate and O_2 from CO_2 and H_2O.

Proteins are polymers of amino acids. They are the major structural materials in animal systems. Enzymes, the catalysts of biochemical reactions, are protein in nature. All naturally occurring proteins are formed from some 20 amino acids. The amino acids are chiral substances; that is, they are capable of existing as nonsuperimposable mirror-image isomers called enantiomers. Usually, only one of the enantiomers is found to be biologically active. Protein structure is determined by the sequence of amino acids in the chain, the coiling or stretching of the chain, and the overall shape of the complete molecule. All these aspects of protein structure are important in determining its biological activity. Heating or other forms of treatment may inactivate, or denature, the protein.

In this chapter, we also have considered the action of enzymes—their specificity, a model for how they may operate, and the ways in which they are sometimes inhibited.

Carbohydrates, which are formed from polyhydroxy aldehydes and ketones, are the major structural constituent of plants and a source of energy in both plants and animals. The three most important groups of carbohydrates are starch, which is found in plants, glycogen, which is found in mammals, and cellulose, which is also found in plants. All are polysaccharides: that is, they are polymers of a simple sugar, glucose.

Fats and oils, along with proteins and carbohydrates, form the important sources of energy in our food supply. The fats and oils are esters of long-chain acids with glycerol, 1,2,3-trihydroxypropane; they are often called triglycerides. The major sources of these substances are animal fats and the oils of plant seeds, such as corn, peanuts, cottonseeds, and soybeans.

The nucleic acids are biopolymers of high molecular weight that carry the genetic information necessary for cell reproduction. In addition, the nucleic acids control cell development through control of protein synthesis. The nucleic acids consist of a polymeric backbone of alternating phosphate and ribose sugar groups, with organic bases attached to the sugar molecules. The DNA polymer is a double-stranded helix that is held together by hydrogen bonding between matching organic bases situated across from one another on the two strands.

Learning goals

Having read and studied this chapter, you should be able to:

1 Describe two distinct reasons why energy is required by living organisms.
2 List the several functions of proteins in living systems.
3 Define the terms *chiral* and *enantiomer* and draw the enantiomer of a given chiral molecule.
4 Write the reaction for formation of a peptide bond between two amino acids.
5 Explain the structures of proteins in terms of primary, secondary, and tertiary structure.
6 Explain the following characteristics of enzymes in terms of what is known about proteins: specificity; temperature dependence of enzyme activity, as shown in Figure 26.6; inhibition.
7 Explain the lock-and-key mechanism for enzyme action.
8 Distinguish the cyclic hemiacetal and linear forms of a sugar and describe the difference between the pyranose and furanose forms of sugar.
9 Describe the manner in which monosaccharides are joined together to form polysaccharides.
10 Enumerate the major groups of polysaccharides and indicate their sources and general functions.
11 Describe the structures of fats and oils and list the sources of these substances.
12 Draw the structures of any of the nucleotides that make up the polynucleotide DNA.
13 Describe the nature of the polymeric unit of polynucleotides.
14 Describe the double-stranded structure of DNA and explain the principle that determines the relationship between bases in the two strands.

Key terms

Among the more important terms and expressions used for the first time, or in a new context, in this chapter are the following:

The active site (Section 26.3) is the specific site in an enzyme at which the enzyme-catalyzed reaction occurs.

An amino acid (Section 26.2) is a carboxylic acid that contains an amino (—NH_2) group attached to the carbon atom adjacent to the carboxylic acid (CO_2H) functional group. Twenty different amino acids are important in living systems.

The biosphere (introduction) is that part of the earth in which living organisms can exist and live out their life cycles.

Carbohydrates (Section 26.4) are a class of substances formed from polyhydroxy aldehydes or ketones.

Cellulose (Section 26.4) is a polysaccharide of glucose; it is the major structural element in plant matter.

A chiral molecule (Section 26.2) is one that is not superimposable on its mirror image.

Chlorophyll (Section 26.1), found in plant leaves, is a plant pigment that plays a major role in conversion of solar energy to chemical energy in photosynthesis.

A coenzyme, or cofactor (Section 26.3), is a substance that is needed along with the enzyme if an enzyme-catalyzed reaction is to occur.

Configurational isomers (Section 26.2) are molecules that differ only by being nonsuperimposable mirror images of one another. Such isomers are often called optical isomers because their solutions affect the plane of polarized light differently.

Denaturation (Section 26.2) is the loss of biological activity in a protein because of disruption of its tertiary structure by heating, by the action of acids or bases, or by other influence.

Deoxyribonucleic acid, DNA (Section 26.6), is a polynucleotide in which the sugar component is deoxyribose (in the furanose-ring form).

The double-helix structure for DNA (Section 26.6) involves the winding of two DNA polynucleotide chains together in a helical arrangement. The two strands of the double helix are complementary, in that the organic bases on the two strands are paired for optimal hydrogen-bond interaction.

Enantiomers (Section 26.2) are two mirror-image molecules of a chiral substance. The enantiomers are nonsuperimposable.

Enzymes (Section 26.3) are proteins that act as catalysts in biochemical reactions.

Fats and oils (Section 26.5) are esters of long-chain carboxylic acids and the alcohol 1,2,3-trihydroxypropane (glycerol). These esters are also called **triglycerides.**

A furanose (Section 26.4) is a cyclic sugar molecule in the form of a five-membered ring.

Glucose (Section 26.4), a polyhydroxy aldehyde of formula $CH_2OH(CHOH)_4CHO$, is the most important of the monosaccharides.

Glycogen (Section 26.4) is the general name given to a group of polysaccharides of glucose that are synthesized in mammals and used as a means of carbohydrate energy storage.

In the α-helix structure (Section 26.2) for a protein, the protein is coiled in the form of a helix, with hydrogen bonds between C=O and N—H groups on adjacent turns.

A hemiacetal (Section 26.4) is a bonding arrangement formed by reaction of an aldehyde functional group with an alcohol. The hemiacetal linkage is formed when aldehyde sugars form the cyclic structure.

Lipases (Section 26.5) are enzymes that catalyze the hydrolysis of fats.

In the lock-and-key model for enzyme action (Section 26.3), the substrate molecule is pictured as fitting rather specifically into the active site on the enzyme. It is assumed that in being bound to the active site the substrate is somehow activated for reaction.

Macromolecules (Section 26.1) are polymeric molecules of high molecular weight. In referring to substances of biochemical origin, the term *biopolymers* is often used.

A monosaccharide (Section 26.4) is a simple sugar, most commonly containing six carbon atoms. The joining together of monosaccharide units by a condensation reaction results in formation of polysaccharides.

Nucleic acids (Section 26.6) are high-molecular-weight polymers of nucleotides.

A nucleotide (Section 26.6) is formed from a molecule of phosphoric acid, a sugar molecule, and an organic base. Nucleotides form linear polymers called DNA and RNA that are involved in protein synthesis and cell reproduction.

Photosynthesis (Section 26.1) is the process occurring in plant leaves by which light energy is used to convert CO_2 and water to carbohydrates and oxygen.

Polarized light (Section 26.2) is electromagnetic radiation in which the waves that form the light move in a single plane.

The primary structure of a protein (Section 26.2) refers to the sequence of amino acids along the protein chain.

Proteins (Section 26.2) are biopolymers formed from amino acids.

A pyranose (Section 26.4) is a cyclic sugar molecule in the form of a six-membered ring.

Ribonucleic acid, RNA (Section 26.6), is a polynucleotide in which ribose (in the furanose-ring form) is the sugar component.

The secondary structure of a protein (Section 26.2) refers to the manner in which the protein is coiled or stretched.

Starch (Section 26.4) is the general name given to a group of polysaccharides that act as energy-storage substances in plants.

The tertiary structure of a protein (Section 26.2) refers to the overall shape of the molecule—specifically, the manner in which sections of the chain fold back upon themselves or intertwine.

Turnover number (Section 26.3) refers to the number of individual reaction events per unit time at a particular active site in an enzyme.

EXERCISES

Energy requirements

26.1 Explain the relationship between entropy and the energy needs of organisms.

26.2 Name at least four processes occurring in an organism (such as yourself) that require energy. These may occur at the molecular level or may involve the complete organism.

[26.3] ΔG°_{298} for oxidation of glucose in solution is -2878 kJ:

$$C_6H_{12}O_6(aq) + 6O_2(g) \longrightarrow 6CO_2(g) + 6H_2O(l) \qquad \Delta G^\circ_{298} = -2878 \text{ kJ}$$

(a) What is ΔG°_{298} for photosynthesis? Is this reaction spontaneous under standard conditions? (b) What is ΔG_{298} for photosynthesis if $P_{O_2} = 0.02$ atm, $P_{CO_2} = 3.1 \times 10^{-4}$ atm, and $C_6H_{12}O_6 = 1.0 \times 10^{-3}\ M$ (assume that H_2O can be omitted from the reaction quotient; refer to Section 17.6, if necessary).

26.4 What is the role of pigments in photosynthesis? What characteristics would you expect their absorption spectra to have?

26.5 A tree might convert about 50 g of CO_2 per day into carbohydrate at its greatest rate of photosynthesis. How many grams of oxygen does the tree produce in 1 day at this rate? How many liters of O_2 (at STP) does this correspond to?

26.6 Assume that 1×10^{14} kg of carbon is photosynthetically fixed as glucose each year on the earth's surface. If the total solar energy falling on the earth's surface is 4×10^{21} kJ/yr, and 2878 kJ are required to produce a mole of glucose, calculate the percentage of the solar energy that is absorbed annually by photosynthetic processes.

Proteins

26.7 Name at least four distinct roles for proteins in animal organisms.

26.8 (a) What is an α-amino acid? (b) How do amino acids react to form proteins?

26.9 How do the side chains (R groups) of amino acids affect their behavior?

26.10 Draw the dipeptides formed by condensation reaction between glycine and valine.

26.11 Write a chemical equation for the formation of aspartylcysteine from the constituent amino acids.

26.12 Draw the structure of the tripeptide trp-gly-ser.

26.13 What amino acids would be obtained by hydrolysis of the following tripeptide?

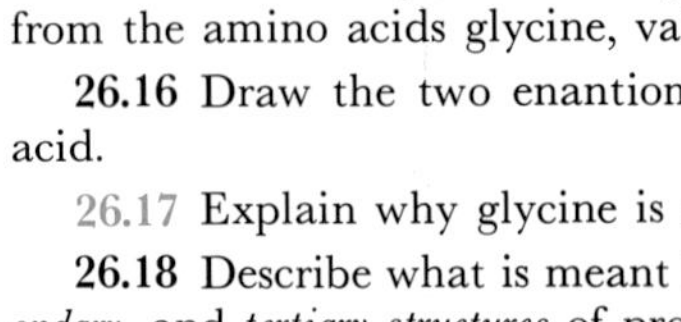

```
       O        O        O
       ||       ||       ||
H2NCHCNHCHCNHCHCOH
    |        |        |
   CH3     CH2     CH2
             |        |
            OH       SH
```

26.14 In what form would you expect glycine to exist in basic solution? In acidic solution?

26.15 How many different tripeptides could one make from the amino acids glycine, valine, and alanine?

26.16 Draw the two enantiometric forms of aspartic acid.

26.17 Explain why glycine is not optically active.

26.18 Describe what is meant by the term *primary, secondary,* and *tertiary structures* of proteins.

26.19 Describe the role of hydrogen bonding in determining the α-helix structure of a protein.

26.20 What is meant by the term *denaturation?* Describe how increased temperature could lead to denaturation.

Enzymes

26.21 Provide a definition of each of the following terms: (a) enzyme; (b) apoenzyme; (c) denaturation; (d) active site; (e) holoenzyme; (f) specificity; (g) peptidase; (h) turnover number.

26.22 In terms of the lock-and-key model, what characteristics must an enzyme inhibitor possess?

26.23 The names of most enzymes end in *-ase*. The *-ase* ending is attached to the name of the substrate on which the enzyme acts, or to the type of reaction it catalyzes. Match the following enzyme names and reactions:

1 esterase
2 decarboxylase
3 urease
4 transmethylase
5 peptidase

(a) removal of carboxyl groups from compounds
(b) hydrolysis of peptide linkages
(c) transfers a methyl group
(d) formation of ester linkages
(e) hydrolysis of urea

26.24 Normally, the rates of enzyme-catalyzed reactions increase linearly with increase in substrate concentration. What does this tell us about the position of equilibrium in Equation 26.4? Explain.

26.25 Protein enzymes are sold commercially for use as meat tenderizers and for stain removal (for example, blood stains) from cloth. What kind of enzymes do you expect that these are? Explain how they work.

26.26 One of the many remarkable enzymes in the human body is carbonic anhydrase, which catalyzes the release of dissolved carbon dioxide from the blood into the air of our lungs. If it were not for this enzyme, the body could not rid itself rapidly enough of the CO_2 accumulated by cell metabolism. The enzyme has a molecular weight of 30,000, contains one atom of zinc per protein molecule, and catalyzes the dehydration (the release to air) of up to 10^7 CO_2 molecules per second. Which components of this description correspond to the terms *holoenzyme, apoenzyme, cofactor,* and *turnover number?*

Carbohydrates

26.27 What is the difference between α-glucose and β-glucose? Show the condensation of two glucose molecules to form a disaccharide with α-linkages; with β-linkages.

26.28 Identify each of the following as an α and a β form of a hemiacetal:

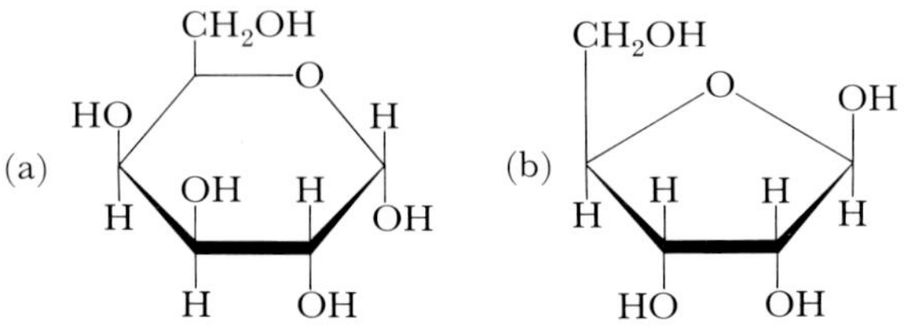

(a) [six-membered ring: CH_2OH, O, HO, H, OH, H, H, OH, H, OH]

(b) [five-membered ring: CH_2OH, O, OH, H, H, H, H, HO, OH]

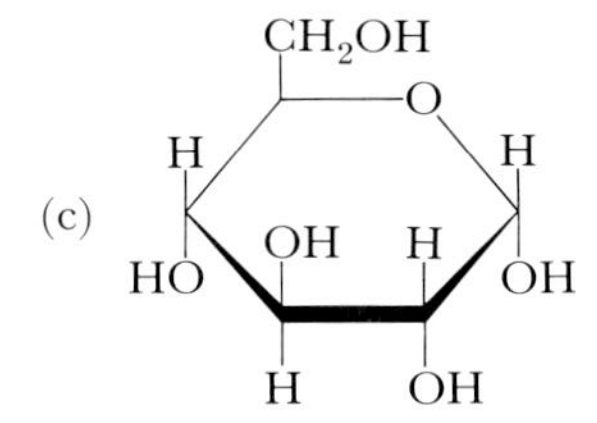

26.29 The structural formula for the linear form of galactose is shown here. Draw the structure of the pyranose form of this sugar.

```
      O
      ‖
      CH
      |
  H—C—OH
      |
 HO—C—H
      |
 HO—C—H
      |
  H—C—OH
      |
      CH2OH
```

26.30 Which carbon atoms in galactose (see Problem 26.29) are chiral?

26.31 Glucose is a hexose that can exist in solution as a pyranose; fructose, which is also a hexose, forms a furanose. Explain what is meant by the terms *hexose, pyranose,* and *furanose.* Why does glucose form a pyranose, whereas fructose forms a furanose?

Fats and oils

26.32 Draw the structural formula for the triglyceride of oleic acid.

26.33 The reaction between glycerol and a fatty acid is called esterification. Write the chemical equation for the esterification of a mole of glycerol with 3 mol of lauric acid.

26.34 The hydrolysis of fats and oils is catalyzed by bases. Many household cleaners (such as aqueous ammonia solutions and Drano, which contains NaOH) are basic. Write the chemical equation for the hydrolysis of the triglyceride of myristic acid.

26.35 Write a balanced chemical equation for complete hydrogenation of trilinolein. If partial hydrogenation were to occur, with uptake of only 4 mol of H_2 per mole of fat, draw the structural formulas of two possible products.

26.36 Fats and oils belong to a general class of compounds called lipids, which are distinguished by the fact that they are soluble in organic solvents. Discuss, in terms of their comparative structures, why fats belong to this class, whereas carbohydrates such as glucose and sucrose do not.

Nucleic acids

26.37 Describe a nucleotide. Draw the structural formula for deoxycytidine monophosphate in which cytosine is the organic base.

26.38 Write a balanced chemical equation for the condensation reaction between a mole of deoxyribose and a mole of phosphoric acid.

26.39 Imagine a single DNA strand containing a section with the following base sequence: A, C, T, C, G, A. What is the base sequence of the complementary strand?

26.40 When samples of double-stranded DNA are analyzed, the quantity of adenine present equals that of thymine. Similarly, the quantity of guanine equals that of cytosine. Explain the significance of these observations.

Additional exercises

26.41 In a temperate climate 1.0 m^2 of leaf area absorbs about 2.0×10^4 kJ of energy per day. About 1.2 percent of this energy is used in photosynthesis. (a) Calculate the leaf area required to convert 10,000 kJ per day into plant matter. This is the approximate energy requirement of a person doing an average quantity of work (Section 4.8). (b) In terms of what you know about the composition of plants, explain why more area than this would be required to provide a 10,000-kJ daily diet for a person.

26.42 In each of the following substances, locate the chiral carbon atoms, if any:

(a) $HOCH_2CH_2\overset{O}{\overset{\|}{C}}CH_2OH$ (b) $HOCH_2\overset{OH}{\overset{|}{C}H}\underset{O}{\underset{\|}{C}}CH_2OH$

(c) $HO\overset{O}{\overset{\|}{C}}\underset{NH_2}{\underset{|}{C}H}\overset{CH_3}{\overset{|}{C}H}C_2H_5$

26.43 Predict the products of the hydrolysis of each of the following compounds:

(a) $CH_3\underset{NH_2}{\underset{|}{C}H}\overset{O}{\overset{\|}{C}}NHCH_2\overset{O}{\overset{\|}{C}}OH$

(b)
```
         O
         ‖
C17H33COCH2
         O      |
         ‖      |
C17H33COCH
         O      |
         ‖      |
C17H33COCH2
```

(c)
```
     CH2OH                    CH2OH
        |——O                     |——O
  H   /      \   H         H   /      \   H
     /        \  |          |  /        \  |
HO  | OH    H  | └———O———┘ | OH    H  | OH
     \________/              \________/
     |        |              |        |
     H        OH             H        OH
```

(d)
```
                                OH
                                |
HO          O           CH2OPOH
  \      /     \       /        ‖
   |           |               O
 H  \ H     H  / H
     \_______/
     |       |
     OH      OH
```

26.44 Draw the condensed structural formula of each of the following tripeptides: (a) val-gly-thr; (b) pro-ser-ala.

[26.45] Explain why the peptide bond is planar.

26.46 Glutathione is a tripeptide found in most living cells. Partial hydrolysis yields cys-gly and glu-cys. What structures are possible for glutathione?

26.47 Phenylketonuria (PKU) is a disease caused by the lack in some individuals of an enzyme, phenylalanine hydroxylase. This enzyme catalyzes the conversion of phenylalanine to tyrosine (both amino acids). The disease can lead to severe mental retardation. (a) Write the condensed structural formulas for phenylalanine and tyrosine. (b) Suggest the origin for the name given to the enzyme.

26.48 (a) Describe in qualitative terms how an enzyme works. (b) What is meant by *enzyme substrate?* By *enzyme inhibition?*

[26.49] Ingestion of large amounts of alcohol by humans causes some enzymes to lose their active sites. Suggest a possible mechanism at the molecular level.

26.50 The popular flavor enhancer MSG (monosodium glutamate) is the monosodium salt of glutamic acid. (a) What is its condensed structural formula? (b) Only the L-isomer is effective. Is this surprising?

26.51 The standard free energy of formation of glycine(*s*) is -369 kJ/mol, whereas that of glycylglycine(*s*) is -488 kJ/mol. What is $\Delta G°$ for condensation of glycine to form glycylglycine?

26.52 Give the chain structural formulas for ribose and deoxyribose (see Figure 26.15).

[26.53] The enzyme *invertase* catalyzes the conversion of sucrose, a disaccharide, to invert sugar. When the concentration of invertase is 3×10^{-7} *M*, and the concentration of sucrose is 0.01 *M*, invert sugar is formed at the rate of 2×10^{-4} *M*/s. When the sucrose concentration is doubled, the rate of formation of invert sugar is doubled also. Assuming that the enzyme-substrate model is operative, is the fraction of enzyme tied up as complex large or small? Explain. Addition of innositol, another sugar, causes a decrease in rate of formation of invert sugar. Suggest a mechanism by which this occurs.

26.54 Give a specific example of each of the following: (a) a disaccharide; (b) a sugar present in nucleic acids; (c) a sugar present in human blood serum; (d) a polysaccharide.

[26.55] The standard free energy of formation for aqueous solutions of glucose is -917.2 kJ/mol, whereas that of glycogen is -662.3 kJ/mol of glucose units. Derive a general expression for $\Delta G°$ for the formation of a glycogen molecule that contains *n* units of glucose.

26.56 Define, in your own words, the terms *condensation* and *hydrolysis*. Why are such reactions so important in biochemistry?

26.57 Write a complementary nucleic acid strand for the following strand, using the concept of complementary base pairing: TATGCA.

26.58 The monoanion of adenosine monophosphate (AMP) is an intermediate in phosphate metabolism:

$$\text{A—O—}\overset{\displaystyle \text{O}^-}{\underset{\displaystyle \text{O}}{\overset{|}{\underset{\|}{\text{P}}}}}\text{—OH} = \text{AMP—OH}^-$$

where A = adenosine. If the pK_a for this anion is 7.21, what is the ratio of [AMP—OH$^-$] to [AMP—O^{2-}] in blood at pH = 7.40?

Appendices

APPENDIX A MATHEMATICAL OPERATIONS

A.1 Exponential Notation

The numbers used in chemistry are often either extremely large or extremely small. Such numbers are conveniently expressed in the form

$$N \times 10^n$$

where N is a number between 1 and 10, and n is the exponent. Some examples of this exponential notation are as follows:

1,200,000 is 1.2×10^6 (read "one point two times ten to the sixth power")
0.000604 is 6.04×10^{-4} (read "six point oh four times ten to the negative fourth power")

A positive exponent, as in our first example, tells us how many times a number must be multiplied by 10 to give the long form of the number:

$$1.2 \times 10^6 = 1.2 \times 10 \times 10 \times 10 \times 10 \times 10 \times 10 \quad \text{(six tens)}$$
$$= 1{,}200{,}000$$

It is also convenient to think of the positive exponent as the number of places the decimal point must be moved to the *left* to give a number greater than 1 and less than 10: if we begin with 3450 and move the decimal point three places to left, we end up with 3.45×10^3.

In a related fashion, a negative exponent tells us how many times we must divide a number by 10 to give the long form of the number:

$$6.04 \times 10^{-4} = \frac{6.04}{10 \times 10 \times 10 \times 10}$$
$$= 0.000604$$

It is convenient to think of the negative exponent as the number of places the decimal point must be moved to the *right* to give a number greater than 1 but less than 10: if we begin with 0.0048 and move the decimal point three places to right, we end up with 4.8×10^{-3}.

In the system of exponential notation, with each shift of the decimal point one place to the right, the exponent *decreases* by 1:

$$4.8 \times 10^{-3} = 48 \times 10^{-4}$$

Similarly, with each shift of the decimal point one place to the left, the exponent *increases* by 1:

$$4.8 \times 10^{-3} = 0.48 \times 10^{-2}$$

In working with exponents, it is important to know that $10^0 = 1$. The following rules are useful for carrying exponents through calculations:

1 **Addition and subtraction** In order to add or subtract numbers expressed in exponential notation, the powers of 10 must be the same:

$$\begin{aligned}(5.22 \times 10^4) + (3.21 \times 10^2) &= (522 \times 10^2) + (3.21 \times 10^2)\\ &= 525 \times 10^2\\ &= 5.25 \times 10^4\\ (6.25 \times 10^{-2}) - (5.77 \times 10^{-3}) &= (6.25 \times 10^{-2}) - (0.577 \times 10^{-2})\\ &= 5.67 \times 10^{-2}\end{aligned}$$

2 **Multiplication and division** When numbers expressed in exponential notation are multiplied, the exponents are added; when numbers expressed in exponential notation are divided, the exponent of the divisor is subtracted from the exponent of the dividend:

$$\begin{aligned}(5.4 \times 10^2)(2.1 \times 10^3) &= (5.4)(2.1) \times 10^{2+3}\\ &= 11 \times 10^5\\ &= 1.1 \times 10^6\\ (1.2 \times 10^5)(3.22 \times 10^{-3}) &= (1.2)(3.22) \times 10^{5-3}\\ &= 3.8 \times 10^2\end{aligned}$$

$$\begin{aligned}\frac{3.2 \times 10^5}{6.5 \times 10^2} = \frac{3.2}{6.5} \times 10^{5-2} &= 0.49 \times 10^3\\ &= 4.9 \times 10^2\end{aligned}$$

$$\begin{aligned}\frac{5.7 \times 10^7}{8.5 \times 10^{-2}} = \frac{5.7}{8.5} \times 10^{7-(-2)} &= 0.67 \times 10^9\\ &= 6.7 \times 10^8\end{aligned}$$

3 **Powers and roots** When numbers expressed in exponential notation are raised to a power, the exponents are multiplied by the power; when the roots of numbers expressed in exponential notation are taken, the exponents are divided by the root:

$$(1.2 \times 10^5)^3 = 1.2^3 \times 10^{5\times3}$$
$$= 1.7 \times 10^{15}$$
$$\sqrt[3]{2.5 \times 10^6} = \sqrt[3]{2.5} \times 10^{6/3}$$
$$= 1.3 \times 10^2$$

A.2 Logarithms

The common, or base 10, logarithm (abbreviated log) of any number is the power to which 10 must be raised to equal the number. For example, the common logarithm of 1000 (written log 1000) is 3, because raising 10 to the third power gives 1000: $10^3 = 1000$. Further examples are:

$$\log 10^5 = 5$$

$$\log 1 = 0$$

$$\log 10^{-2} = -2$$

In these examples, the logarithm can be obtained by simple inspection. However, it is not possible to obtain the logarithm of a number like 3.2 by inspection. Many electronic calculators have a log key that can be used to obtain logs. If you do not have such a calculator, you can easily use a log table such as that given in Table 1. To find the logs of numbers from 1 to 10, we locate the first digit of the number in the first vertical column. We then move horizontally to the column headed by the second digit of the number. In this way we find that the log of 3.2 is 505. Because decimals are omitted from the table, the log of 3.2 is actually 0.505. Further examples are:

$$\log 2.50 = 0.398$$

$$\log 2.55 = 0.406$$

In using Table 1 to find the log of 2.55, we must estimate, or interpolate, the value from the logs given for 2.5 and 2.6. Appendix B is a table of four-place logarithms from which the logarithm of 2.55 can be read directly as 0.4065. (In using Appendix B, you should mentally insert a decimal point between the two-digit numbers that form the first column.)

TABLE 1 Three-place common logarithms

	0	1	2	3	4	5	6	7	8	9
1	000	041	079	114	146	176	204	230	255	279
2	301	322	342	362	380	398	415	431	447	462
3	477	491	505	519	532	544	556	568	580	591
4	602	613	623	634	644	653	663	672	681	690
5	699	708	716	724	732	740	748	756	763	771
6	778	785	792	799	806	813	820	826	833	839
7	845	851	857	863	869	875	881	887	892	898
8	903	909	914	919	924	929	935	940	945	949
9	954	959	964	969	973	978	982	987	991	996

The logarithm of a number that is less than 1 or greater than 10 can be determined by writing the number first in standard exponential notation, as the following examples show:

$$\begin{aligned}\log 450 &= \log(4.50 \times 10^2)\\ &= \log 4.50 + \log 10^2\\ &= 0.653 + 2 = 2.653\\ \log 0.0673 &= \log(6.73 \times 10^{-2})\\ &= \log 6.73 + \log 10^{-2}\\ &= 0.828 - 2\\ &= -1.172\end{aligned}$$

Check these examples yourself, using Appendix B.

Because logarithms are exponents, mathematical operations involving logarithms follow the rules for the use of exponents:

1 Multiplication and division:

$$\log ab = \log a + \log b$$

$$\log \frac{a}{b} = \log a - \log b$$

2 Powers and roots:

$$\log a^n = n(\log a)$$

$$\log a^{1/n} = \frac{1}{n}(\log a)$$

Obtaining antilogarithms The process of finding a number given its logarithm is known as obtaining an antilogarithm. It is the reverse of taking a logarithm. Many electronic calculators employ a key labeled $\log^{-1}$ or INV log to obtain antilogs. Others employ the 10^x or y^x key ($10^{\log x} = x$). If you do not have a calculator that performs these operations, you can use a log table to obtain antilogs as the following examples illustrate:

1 Find the number whose logarithm is 5.322.

$$\begin{aligned}\text{antilog } 5.322 &= \text{antilog } 0.322 \times \text{antilog } 5\\ &= 2.10 \times 10^5\end{aligned}$$

2 Find the number whose logarithm is -2.133.

$$\begin{aligned}\text{antilog }(-2.133) &= \text{antilog } 0.867 \times \text{antilog }(-3)\\ &= 7.37 \times 10^{-3}\end{aligned}$$

pH problems In general chemistry, logarithms are used most frequently in working pH problems. The pH is defined as $-\log [H^+]$, as discussed in Section 15.3. The following sample exercise illustrates this application.

SAMPLE EXERCISE 1

(a) What is the pH of a solution whose hydrogen-ion concentration is 0.015 M? (b) If the pH of a solution is 3.80, what is its hydrogen-ion concentration?

Solution:

(a)
$$\begin{aligned} \text{pH} &= -\log [\text{H}^+] \\ &= -\log 0.015 \\ &= -\log (1.5 \times 10^{-2}) \\ &= -\log 1.5 - \log (10^{-2}) \\ &= -0.18 + 2 = 1.82 \end{aligned}$$

(b)
$$\begin{aligned} \text{pH} = -\log [\text{H}^+] &= 3.80 \\ \log [\text{H}^+] &= -3.80 \\ [\text{H}^+] &= \text{antilog}\,(-3.80) \\ &= \text{antilog}\, 0.20 \times \text{antilog}\,(-4) \\ &= 1.6 \times 10^{-4}\, M \end{aligned}$$

Natural logarithms Natural, or base e, logarithms (abbreviated ln) are the power to which e, which has the value 2.71828 . . . , must be raised to equal a number. The relation between common and natural logarithms is as follows:

$$\ln a = 2.303 \log a$$

A.3 Quadratic Equations

An algebraic equation of the form $ax^2 + bx + c = 0$ is called a quadratic equation. The two solutions to such an equation are given by the quadratic formula:

$$x = \frac{-b \pm \sqrt{b^2 - 4ac}}{2a}$$

SAMPLE EXERCISE 2

Find x if $2x^2 + 4x = 1$.

Solution: To solve the given equation for x, we must first put it in the form

$$ax^2 + bx + c = 0$$

and then use the quadratic formula. If

$$2x^2 + 4x = 1$$

then

$$2x^2 + 4x - 1 = 0$$

Using the quadratic formula, where $a = 2$, $b = 4$, and $c = -1$, we have

$$\begin{aligned} x &= \frac{-4 \pm \sqrt{(4)(4) - 4(2)(-1)}}{2(2)} \\ &= \frac{-4 \pm \sqrt{16 + 8}}{4} = \frac{-4 \pm \sqrt{24}}{4} \\ &= \frac{-4 \pm 4.899}{4} \end{aligned}$$

The two solutions are

$$x = \frac{0.899}{4} = 0.225$$

$$x = \frac{-8.899}{4} = -2.225$$

Often in chemical problems the negative solution has no physical meaning, and only the positive answer is used.

A.4 Graphs

TABLE 2 Interrelation between pressure and temperature

Temperature (°C)	Pressure (atm)
20.0	0.120
30.0	0.124
40.0	0.128
50.0	0.132

Often the clearest way to represent the interrelationship between two variables is to graph them. Usually, the variable that is being experimentally varied, called the independent variable, is shown along the horizontal axis (x-axis). The variable that responds to the change in the independent variable, called the dependent variable, is then shown along the vertical axis (y-axis). For example, consider an experiment in which we vary the temperature of an enclosed gas and measure its pressure. The independent variable is temperature, whereas the dependent variable is pressure. The data shown in Table 2 could be obtained by means of this experiment. These data are shown graphically in Figure 1. The relationship between temperature and pressure is linear. The equation for any straight-line graph has the form

$$y = ax + b$$

where a is the slope of the line and b is the intercept with the y-axis. In the case of Figure 1, we could say that the relationship between temperature and pressure takes the form

$$P = aT + b$$

where P is pressure in atm and T is temperature in °C. As shown on Figure 1, the slope is 4.10×10^{-4} atm/°C, and the intercept, the point where the line crosses the y-axis, is 0.112 atm. Therefore, the equation for the line is

$$P = \left(4.10 \times 10^{-4} \frac{\text{atm}}{°\text{C}}\right) T + 0.112 \text{ atm}$$

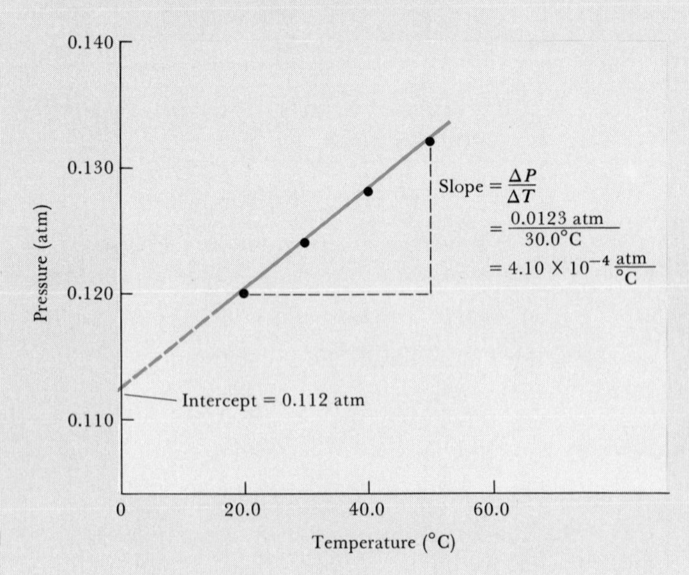

FIGURE 1

APPENDIX B TABLE OF FOUR-PLACE LOGARITHMS

	0	1	2	3	4	5	6	7	8	9
10	0000	0043	0086	0128	0170	0212	0253	0294	0334	0374
11	0414	0453	0492	0531	0569	0607	0645	0682	0719	0755
12	0792	0828	0864	0899	0934	0969	1004	1038	1072	1106
13	1139	1173	1206	1239	1271	1303	1335	1367	1399	1430
14	1461	1492	1523	1553	1584	1614	1644	1673	1703	1732
15	1761	1790	1818	1847	1875	1903	1931	1959	1987	2014
16	2041	2068	2095	2122	2148	2175	2201	2227	2253	2279
17	2304	2330	2355	2380	2405	2430	2455	2480	2504	2529
18	2553	2577	2601	2625	2648	2672	2695	2718	2742	2765
19	2788	2810	2833	2856	2878	2900	2923	2945	2967	2989
20	3010	3032	3054	3075	3096	3118	3139	3160	3181	3201
21	3222	3243	3263	3284	3304	3324	3345	3365	3385	3404
22	3424	3444	3464	3483	3502	3522	3541	3560	3579	3598
23	3617	3636	3655	3674	3692	3711	3729	3747	3766	3784
24	3802	3820	3838	3856	3874	3892	3909	3927	3945	3962
25	3979	3997	4014	4031	4048	4065	4082	4099	4116	4133
26	4150	4166	4183	4200	4216	4232	4249	4265	4281	4298
27	4314	4330	4346	4362	4378	4393	4409	4425	4440	4456
28	4472	4487	4502	4518	4533	4548	4564	4579	4594	4609
29	4624	4639	4654	4669	4683	4698	4713	4728	4742	4757
30	4771	4786	4800	4814	4829	4843	4857	4871	4886	4900
31	4914	4928	4942	4955	4969	4983	4997	5011	5024	5083
32	5051	5065	5079	5092	5105	5119	5132	5145	5159	5172
33	5185	5198	5211	5224	5237	5250	5263	5276	5289	5302
34	5315	5328	5340	5353	5366	5378	5391	5403	5416	5428
35	5441	5453	5465	5478	5490	5502	5514	5527	5539	5551
36	5563	5575	5587	5599	5611	5623	5635	5647	5658	5670
37	5682	5694	5705	5717	5729	5740	5752	5763	5775	5786
38	5798	5809	5821	5832	5843	5855	5866	5877	5888	5899
39	5911	5922	5933	5944	5955	5966	5977	5988	5999	6010
40	6021	6031	6042	6053	6064	6075	6085	6096	6107	6117
41	6128	6138	6149	6160	6170	6180	6191	6201	6212	6222
42	6232	6243	6253	6263	6274	6284	6294	6304	6314	6325
43	6335	6345	6355	6365	6375	6385	6395	6405	6415	6425
44	6435	6444	6454	6464	6474	6484	6493	6503	6513	6522
45	6532	6542	6551	6561	6571	6580	6590	6599	6609	6618
46	6628	6637	6646	6656	6665	6675	6684	6693	6702	6712
47	6721	6730	6739	6749	6758	6767	6776	6785	6794	6803
48	6812	6821	6830	6839	6848	6857	6866	6875	6884	6893
49	6902	6911	6920	6928	6937	6946	6955	6964	6972	6981
50	6990	6998	7007	7016	7024	7033	7042	7050	7059	7067
51	7076	7084	7093	7101	7110	7118	7126	7135	7143	7152
52	7160	7168	7177	7185	7193	7202	7210	7218	7226	7235
53	7243	7251	7259	7267	7275	7284	7292	7300	7308	7316
54	7324	7332	7340	7348	7356	7364	7372	7380	7388	7396
55	7404	7412	7419	7427	7435	7443	7451	7459	7466	7474
56	7482	7490	7497	7505	7513	7520	7528	7536	7543	7551
57	7559	7566	7574	7582	7589	7597	7604	7612	7619	7627
58	7634	7642	7649	7657	7664	7672	7679	7686	7694	7701
59	7709	7716	7723	7731	7738	7745	7752	7760	7767	7774

	0	1	2	3	4	5	6	7	8	9
60	7782	7789	7796	7803	7810	7818	7825	7832	7839	7846
61	7853	7860	7868	7875	7882	7889	7896	7903	7910	7917
62	7924	7931	7938	7945	7952	7959	7966	7973	7980	7987
63	7993	8000	8007	8014	8021	8028	8035	8041	8048	8055
64	8062	8069	8075	8082	8089	8096	8102	8109	8116	8122
65	8129	8136	8142	8149	8156	8162	8169	8176	8182	8189
66	8195	8202	8209	8215	8222	8228	8235	8241	8248	8254
67	8261	8267	8274	8280	8287	8293	8299	8306	8312	8319
68	8325	8331	8338	8344	8351	8357	8363	8370	8376	8382
69	8388	8395	8401	8407	8414	8420	8426	8432	8439	8445
70	8451	8457	8463	8470	8476	8482	8488	8494	8500	8506
71	8513	8519	8525	8531	8537	8543	8549	8555	8561	8567
72	8573	8579	8585	8591	8597	8603	8609	8615	8621	8627
73	8633	8639	8645	8651	8657	8663	8669	8675	8681	8686
74	8692	8698	8704	8710	8716	8722	8727	8733	8739	8745
75	8751	8756	8762	8768	8774	8779	8785	8791	8797	8802
76	8808	8814	8820	8825	8831	8837	8842	8848	8854	8859
77	8865	8871	8876	8882	8887	8893	8899	8904	8910	8915
78	8921	8927	8932	8938	8943	8949	8954	8960	8965	8971
79	8976	8982	8987	8993	8998	9004	9009	9015	9020	9025
80	9031	9036	9042	9047	9053	9058	9063	9069	9074	9079
81	9085	9090	9096	9101	9106	9112	9117	9122	9128	9133
82	9138	9143	9149	9154	9159	9165	9170	9175	9180	9186
83	9191	9196	9201	9206	9212	9217	9222	9227	9232	9238
84	9243	9248	9253	9258	9263	9269	9274	9279	9284	9289
85	9294	9299	9304	9309	9315	9320	9325	9330	9335	9340
86	9345	9350	9355	9360	9365	9370	9375	9380	9385	9390
87	9395	9400	9405	9410	9415	9420	9425	9430	9435	9440
88	9445	9450	9455	9460	9465	9469	9474	9479	9484	9489
89	9494	9499	9504	9509	9513	9518	9523	9628	9533	9538
90	9542	9547	9552	9557	9562	9566	9571	9576	9581	9586
91	9590	9595	9600	9605	9609	9614	9619	9624	9628	9633
92	9638	9643	9647	9652	9657	9661	9666	9671	9675	9680
93	9685	9689	9694	9699	9703	9708	9713	9717	9722	9727
94	9731	9736	9741	9745	9750	9754	9759	9763	9768	9773
95	9777	9782	9786	9791	9795	9800	9805	9809	9814	9818
96	9823	9827	9832	9836	9841	9845	9850	9854	9859	9863
97	9868	9872	9877	9881	9886	9890	9894	9899	9903	9908
98	9912	9917	9921	9926	9930	9934	9939	9943	9948	9952
99	9956	9961	9965	9969	9974	9978	9983	9987	9991	9996

APPENDIX C PROPERTIES OF WATER

Density:
0.99987 g/cm^3 at 0°C
1.00000 g/cm^3 at 4°C
0.99707 g/cm^3 at 25°C
0.95838 g/cm^3 at 100°C

Heat of fusion: 6.008 kJ/mol at 0°C

Heat of vaporization:
44.94 kJ/mol at 0°C
44.02 kJ/mol at 25°C
40.67 kJ/mol at 100°C

Specific heat:
Ice (−3°C)—2.092 J/°C-g
Water at 14.5°C—4.184 J/°C-g
Steam (100°C)—1.841 J/°C-g

Vapor pressure (mm Hg):

T (°C)	VP	T (°C)	VP	T (°C)	VP	T (°C)	VP
0	4.58	21	18.65	35	42.2	92	567.0
5	6.54	22	19.83	40	55.3	94	610.9
10	9.21	23	21.07	45	71.9	96	657.6
12	10.52	24	22.38	50	92.5	98	707.3
14	11.99	25	23.76	55	118.0	100	760.0
16	13.63	26	25.21	60	149.4	102	815.9
17	14.53	27	26.74	65	187.5	104	875.1
18	15.48	28	28.35	70	233.7	106	937.9
19	16.48	29	30.04	80	355.1	108	1004.4
20	17.54	30	31.82	90	525.8	110	1074.6

APPENDIX D THERMODYNAMIC QUANTITIES FOR SELECTED SUBSTANCES AT 25°C

Substance	ΔH_f° (kJ/mol)	ΔG_f° (kJ/mol)	S° (J/mol-K)
$Al(s)$	0.00	0.00	28.32
$Al_2O_3(s)$	−1669.8	−1576.5	51.00
$Ag^+(aq)$	105.90	77.11	73.93
$AgCl(s)$	−127.0	−109.70	96.11
$Ba(s)$	0.00	0.00	63.2
$BaCO_3(s)$	−1216.3	−1137.6	112.1
$BaO(s)$	−553.5	−525.1	70.42
$Br(g)$	111.8	82.38	174.9
$Br^-(aq)$	−120.9	−102.8	80.71
$Br_2(g)$	30.71	3.14	245.3
$Br_2(l)$	0.00	0.00	152.3
$C(g)$	718.4	672.9	158.0
C(diamond)	1.88	2.84	2.43
C(graphite)	0.00	0.00	5.69
$CCl_4(g)$	−106.7	−64.0	309.4
$CCl_4(l)$	−139.3	−68.6	214.4
$CF_4(g)$	−679.9	−635.1	262.3
$CH_4(g)$	−74.8	−50.8	186.3
$C_2H_2(g)$	226.7	209.2	200.8
$C_2H_4(g)$	52.30	68.11	219.4
$C_2H_6(g)$	−84.68	−32.89	229.5
$C_3H_8(g)$	−103.85	−23.47	269.9
$CH_3OH(g)$	−201.2	−161.9	237.6
$CH_3OH(l)$	−238.6	−166.23	126.8
$C_2H_5OH(l)$	−277.7	−174.76	160.7
$CH_3COOH(l)$	−487.0	−392.4	159.8
$C_6H_6(l)$	49.0	124.5	172.8
$C_6H_6(g)$	82.9	129.7	269.2
$CO(g)$	−110.5	−137.3	197.9
$CO_2(g)$	−393.5	−394.4	213.6
$Ca(s)$	0	0	41.4
$Ca(g)$	179.3	145.5	154.8
$CaCO_3$(calcite)	−1207.1	−1128.76	92.88
$CaF_2(s)$	−1219.6	−1167.3	68.87
$CaO(s)$	−635.5	−604.17	39.75
$Ca(OH)_2(s)$	−986.2	−898.5	83.4
$CaSO_4(s)$	−1434.0	−1321.8	106.7
$Cl(g)$	127.7	105.7	165.2
$Cl_2(g)$	0.00	0.00	222.96
$Co(s)$	0.00	0.00	28.4
$Co(g)$	439	393	179
$Cr(s)$	0.00	0.00	23.6
$Cr_2O_3(s)$	−1139.7	−1058.1	81.2
$Cr(g)$	397.5	352.6	174.2
$Cu(s)$	0.00	0.00	33.30
$Cu(g)$	338.4	298.6	166.3
$F(g)$	80.0	61.9	158.7
$F_2(g)$	0.00	0.00	202.7
$Fe(s)$	0.00	0.00	27.15
$Fe^{2+}(aq)$	−87.86	−84.93	113.4
$Fe^{3+}(aq)$	−47.69	−10.54	293.3
$FeCl_3(s)$	−400	−334	142.3
$FeO(s)$	−271.9	−255.2	60.75
$Fe_2O_3(s)$	−822.16	−740.98	89.96

Substance	ΔH_f° (kJ/mol)	ΔG_f° (kJ/mol)	S° (J/mol-K)
$Fe_3O_4(s)$	−1117.1	−1014.2	146.4
$H(g)$	217.94	203.26	114.60
$H^+(aq)$	0.00	0.00	0.00
$H_2(g)$	0.00	0.00	130.58
$HBr(g)$	−36.23	−53.22	198.49
$HCl(g)$	−92.30	−95.27	186.69
$HF(g)$	−268.61	−270.70	173.51
$HI(g)$	25.94	1.30	206.3
$H_2O(g)$	−241.8	−228.61	188.7
$H_2O(l)$	−285.85	−236.81	69.96
$H_2O_2(g)$	−136.10	−105.48	232.9
$H_2O_2(l)$	−187.8	−120.4	109.6
$H_2S(g)$	−20.17	−33.01	205.6
$H_2Se(g)$	29.7	15.9	219.0
$Hg(g)$	60.83	31.76	174.89
$Hg(l)$	0.00	0.00	77.40
$I(g)$	106.60	70.16	180.66
$I_2(s)$	0.00	0.00	116.73
$I_2(g)$	62.25	19.37	260.57
$K(g)$	89.99	61.17	160.2
$KCl(s)$	−435.9	−408.3	82.7
$KClO_3(s)$	−391.2	−289.9	143.0
$KClO_3(aq)$	−349.5	−284.9	265.7
$KNO_3(s)$	−492.70	−393.13	288.1
$Mg(s)$	0.00	0.00	32.51
$MgCl_2(s)$	−641.6	−592.1	89.6
$Mn(s)$	0	0	32.0
$Mn(g)$	280.7	238.5	173.6
$MnO(s)$	−385.2	−362.9	59.7
$MnO_2(s)$	−519.6	−464.8	53.14
$NH_3(g)$	−46.19	−16.66	192.5
$N_2H_4(g)$	95.40	159.4	238.5
$NH_4CN(s)$	0.00	—	—
$NH_4Cl(s)$	−314.4	−203.0	94.6
$NH_4NO_3(s)$	−365.6	−184.0	151
$NO(g)$	90.37	86.71	210.62
$NO_2(g)$	33.84	51.84	240.45
$NOCl(g)$	52.6	66.3	264
$N_2(g)$	0.00	0.00	191.50
$N_2O(g)$	81.6	103.59	220.0
$N_2O_4(g)$	9.66	98.28	304.3
$Na(g)$	107.7	77.3	51.5
$NaBr(aq)$	−360.6	−364.7	141
$NaCl(s)$	−410.9	−384.0	72.33
$NaCl(aq)$	−407.1	−393.0	115.5
$NaHCO_3(s)$	−947.7	−851.8	102.1
$Na_2CO_3(s)$	−1130.9	−1047.7	136.0
$NaNO_3(aq)$	−446.2	−372.4	207
$NaOH(s)$	−425.6	−379.5	64.46
$Ni(s)$	0	0	29.9
$Ni(g)$	429.7	384.5	182.1
$O(g)$	247.5	230.1	161.0
$O_2(g)$	0.00	0.00	205.0
$O_3(g)$	142.3	163.4	237.6
$OH^-(aq)$	−230.0	−157.3	−10.7
$P_2(g)$	144.3	103.7	218.1

Substance	ΔH_f° (kJ/mol)	ΔG_f° (kJ/mol)	S° (J/mol-K)
$P_4(g)$	58.9	24.4	280
$PCl_3(l)$	−319.6	−272.4	217
$PH_3(g)$	5.4	13.4	210.2
$POCl_3(g)$	−542.2	−502.5	325
$POCl_3(l)$	−597.0	−520.9	222
$P_4O_6(s)$	−1640.1	—	—
$P_4O_{10}(s$, hexagonal)	−2940.1	−2675.2	228.9
$PbBr_2(s)$	−277.4	−260.7	161
$PbCO_3(s)$	−699.1	−625.5	131.0
$Pb(NO_3)_2(s)$	−451.9	—	—
$Pb(NO_3)_2(aq)$	−421.3	—	—
$PbO(s)$	−217.3	−187.9	68.70
$Rb(g)$	85.8	55.8	170.0
$RbCl(s)$	−430.5	−412.0	92
$RbClO_3(s)$	−392.4	−292.0	152
$S(s$, rhombic)	0.00	0.00	31.88
$SO_2(g)$	−296.9	−300.4	248.5
$SO_3(g)$	−395.2	−370.4	256.2
$SOCl_2(l)$	−245.6	—	—
$Sc(s)$	0	0	34.6
$Sc(g)$	377.8	336.1	174.7
$Si(g)$	368.2	323.9	167.8
$SiCl_4(l)$	−640.1	−572.8	239.3
$SiO_2(s)$ (quartz)	−910.9	−856.5	41.84
$Ti(g)$	468	422	180.3
$V(s)$	0	0	28.9
$V(g)$	514.2	453.1	182.2
$Zn(s)$	0.00	0.00	41.63
$Zn(g)$	130.7	95.2	160.9
$ZnO(s)$	−348.0	−318.2	43.9

APPENDIX E AQUEOUS-EQUILIBRIUM CONSTANTS

E.1 DISSOCIATION CONSTANTS FOR ACIDS AT 25°C

Name	Formula	K_{a1}	K_{a2}	K_{a3}
Acetic	$HC_2H_3O_2$	1.8×10^{-5}		
Ascorbic	$HC_6H_7O_6$	8.0×10^{-5}		
Arsenic	H_3AsO_4	5.6×10^{-3}	1.0×10^{-7}	3.0×10^{-12}
Arsenous	H_3AsO_3	6×10^{-10}		
Benzoic	$HC_7H_5O_2$	6.5×10^{-5}		
Boric	H_3BO_3	5.8×10^{-10}		
Carbonic	H_2CO_3	4.3×10^{-7}	5.6×10^{-11}	
Chloroacetic	$HC_2H_2O_2Cl$	1.4×10^{-3}		
Cyanic	$HCNO$	3.5×10^{-4}		
Citric	$H_3C_6H_5O_7$	7.4×10^{-4}	1.7×10^{-5}	4.0×10^{-7}
Formic	$HCHO_2$	1.8×10^{-4}		
Hydroazoic	HN_3	1.9×10^{-5}		
Hydrocyanic	HCN	4.9×10^{-10}		
Hydrofluoric	HF	6.8×10^{-4}		
Hydrogen chromate ion	$HCrO_4^-$	3.0×10^{-7}		
Hydrogen peroxide	H_2O_2	2.4×10^{-12}		
Hydrogen selenate ion	$HSeO_4^-$	2.2×10^{-2}		
Hydrogen sulfate ion	HSO_4^-	1.2×10^{-2}		
Hydrogen sulfide	H_2S	5.7×10^{-8}	1.3×10^{-13}	
Hypobromous	$HBrO$	2×10^{-9}		
Hypochlorous	$HClO$	3.0×10^{-8}		
Hypoiodus	HIO	2×10^{-11}		
Iodic	HIO_3	1.7×10^{-1}		
Lactic	$HC_3H_5O_3$	1.4×10^{-4}		
Malonic	$H_2C_3H_2O_4$	1.5×10^{-3}	2.0×10^{-6}	
Nitrous	HNO_2	4.5×10^{-4}		
Oxalic	$H_2C_2O_4$	5.9×10^{-2}	6.4×10^{-5}	
Phenol	HC_6H_5O	1.3×10^{-10}		
Phosphoric	H_3PO_4	7.5×10^{-3}	6.2×10^{-8}	4.2×10^{-13}
Paraperiodic	H_5IO_6	2.8×10^{-2}	5.3×10^{-9}	
Propionic	$HC_3H_5O_2$	1.3×10^{-5}		
Pyrophosphoric	$H_4P_2O_7$	3.0×10^{-2}	4.4×10^{-3}	
Selenous	H_2SeO_3	2.3×10^{-3}	5.3×10^{-9}	
Sulfuric	H_2SO_4	strong acid	1.2×10^{-2}	
Sulfurous	H_2SO_3	1.7×10^{-2}	6.4×10^{-8}	
Tartaric	$H_2C_4H_4O_6$	1.0×10^{-3}	4.6×10^{-5}	

E.2 DISSOCIATION CONSTANTS FOR BASES AT 25°C

Name	Formula	K_b
Ammonia	NH_3	1.8×10^{-5}
Aniline	$C_6H_5NH_2$	4.3×10^{-10}
Dimethylamine	$(CH_3)_2NH$	5.4×10^{-4}
Ethylamine	$C_2H_5NH_2$	6.4×10^{-4}
Hydrazine	H_2NNH_2	1.3×10^{-6}
Hydroxylamine	$HONH_2$	1.1×10^{-8}
Methylamine	CH_3NH_2	4.4×10^{-4}
Pyridine	C_5H_5N	1.7×10^{-9}
Trimethylamine	$(CH_3)_3N$	6.4×10^{-5}

E.3 SOLUBILITY-PRODUCT CONSTANTS FOR COMPOUNDS AT 25°C

Name	Formula	K_{sp}
Barium carbonate	$BaCO_3$	5.1×10^{-9}
Barium chromate	$BaCrO_4$	1.2×10^{-10}
Barium fluoride	BaF_2	1.0×10^{-6}
Barium hydroxide	$Ba(OH)_2$	5×10^{-3}
Barium oxalate	BaC_2O_4	1.6×10^{-7}
Barium phosphate	$Ba_3(PO_4)_2$	3.4×10^{-23}
Barium sulfate	$BaSO_4$	1.1×10^{-10}
Cadmium carbonate	$CdCO_3$	5.2×10^{-12}
Cadmium hydroxide	$Cd(OH)_2$	2.5×10^{-14}
Cadmium sulfide	CdS	8.0×10^{-27}
Calcium carbonate	$CaCO_3$	2.8×10^{-9}
Calcium chromate	$CaCrO_4$	7.1×10^{-4}
Calcium fluoride	CaF_2	3.9×10^{-11}
Calcium hydroxide	$Ca(OH)_2$	5.5×10^{-6}
Calcium phosphate	$Ca_3(PO_4)_2$	2.0×10^{-29}
Calcium sulfate	$CaSO_4$	9.1×10^{-6}
Cerium(III) fluoride	CeF_3	8×10^{-16}
Chromium(III) fluoride	CrF_3	6.6×10^{-11}
Chromium(III) hydroxide	$Cr(OH)_3$	6.3×10^{-31}
Cobalt(II) carbonate	$CoCO_3$	1.4×10^{-13}
Cobalt(II) hydroxide	$Co(OH)_2$	1.6×10^{-15}
Cobalt(III) hydroxide	$Co(OH)_3$	1.6×10^{-44}
α-Cobalt(II) sulfide[a]	CoS	4.0×10^{-21}
Copper(I) bromide	$CuBr$	5.3×10^{-9}
Copper(I) chloride	$CuCl$	1.2×10^{-6}
Copper(I) sulfide	Cu_2S	2.5×10^{-48}
Copper(II) carbonate	$CuCO_3$	1.4×10^{-10}
Copper(II) chromate	$CuCrO_4$	3.6×10^{-6}
Copper(II) hydroxide	$Cu(OH)_2$	2.2×10^{-20}
Copper(II) phosphate	$Cu_3(PO_4)_2$	1.3×10^{-37}
Copper(II) sulfide	CuS	6.3×10^{-36}
Gold(I) chloride	$AuCl$	2.0×10^{-13}
Gold(III) chloride	$AuCl_3$	3.2×10^{-25}
Iron(II) carbonate	$FeCO_3$	3.2×10^{-11}
Iron(II) hydroxide	$Fe(OH)_2$	8.0×10^{-16}
Iron(II) sulfide	FeS	6.3×10^{-18}
Iron(III) hydroxide	$Fe(OH)_3$	4×10^{-38}
Lanthanum fluoride	LaF_3	7×10^{-17}
Lanthanum iodate	$La(IO_3)_3$	6.1×10^{-12}
Lead carbonate	$PbCO_3$	7.4×10^{-14}
Lead chloride	$PbCl_2$	1.6×10^{-5}
Lead chromate	$PbCrO_4$	2.8×10^{-13}
Lead fluoride	PbF_2	2.7×10^{-8}
Lead hydroxide	$Pb(OH)_2$	1.2×10^{-15}
Lead sulfate	$PbSO_4$	1.6×10^{-8}
Lead sulfide	PbS	8.0×10^{-28}
Magnesium hydroxide	$Mg(OH)_2$	1.8×10^{-11}
Magnesium oxalate	MgC_2O_4	8.6×10^{-5}
Manganese carbonate	$MnCO_3$	1.8×10^{-11}

[a] Some substances exist in more than one crystalline form; the prefix indicates the particular form for which K_{sp} is listed.

Name	Formula	K_{sp}
Manganese hydroxide	$Mn(OH)_2$	1.9×10^{-13}
Manganese(II) sulfide	MnS	1.0×10^{-13}
Mercury(I) chloride	Hg_2Cl_2	1.3×10^{-18}
Mercury(I) oxalate	$Hg_2C_2O_4$	2.0×10^{-13}
Mercury(I) sulfide	Hg_2S	1.0×10^{-47}
Mercury(II) hydroxide	$Hg(OH)_2$	3.0×10^{-26}
Mercury(II) sulfide	HgS	4×10^{-53}
Nickel carbonate	$NiCO_3$	6.6×10^{-9}
Nickel hydroxide	$Ni(OH)_2$	1.6×10^{-14}
Nickel oxalate	NiC_2O_4	4×10^{-10}
α-Nickel sulfide[a]	NiS	3.2×10^{-19}
Silver arsenate	Ag_3AsO_4	1.0×10^{-22}
Silver bromide	$AgBr$	5.0×10^{-13}
Silver carbonate	Ag_2CO_3	8.1×10^{-12}
Silver chloride	$AgCl$	1.8×10^{-10}
Silver chromate	Ag_2CrO_4	1.1×10^{-12}
Silver cyanide	$AgCN$	1.2×10^{-16}
Silver iodide	AgI	8.3×10^{-17}
Silver sulfate	Ag_2SO_4	1.4×10^{-5}
Silver sulfide	Ag_2S	6.3×10^{-50}
Strontium carbonate	$SrCO_3$	1.1×10^{-10}
Tin(II) hydroxide	$Sn(OH)_2$	1.4×10^{-28}
Tin(II) sulfide	SnS	1.0×10^{-25}
Zinc carbonate	$ZnCO_3$	1.4×10^{-11}
Zinc hydroxide	$Zn(OH)_2$	1.2×10^{-17}
Zinc oxalate	ZnC_2O_4	2.7×10^{-8}
α-Zinc sulfide[a]	ZnS	1.1×10^{-21}

[a] Some substances exist in more than one crystalline form; the prefix indicates the particular form for which K_{sp} is listed.

APPENDIX F STANDARD ELECTRODE POTENTIALS AT 25°C

Half-reaction	$E°$ (V)
$Ag^+(aq) + e^- \rightleftharpoons Ag(s)$	+0.799
$AgBr(s) + e^- \rightleftharpoons Ag(s) + Br^-(aq)$	+0.095
$AgCl(s) + e^- \rightleftharpoons Ag(s) + Cl^-(aq)$	+0.222
$Ag(CN)_2^-(aq) + e^- \rightleftharpoons Ag(s) + 2CN^-(aq)$	−0.31
$Ag_2CrO_4(s) + 2e^- \rightleftharpoons 2Ag(s) + CrO_4^{2-}(aq)$	+0.446
$AgI(s) + e^- \rightleftharpoons Ag(s) + I^-(aq)$	−0.151
$Ag(S_2O_3)_2^{3-} + e^- \rightleftharpoons Ag(s) + 2S_2O_3^{2-}(aq)$	+0.01
$Al^{3+}(aq) + 3e^- \rightleftharpoons Al(s)$	−1.66
$H_3AsO_4(aq) + 2H^+(aq) + 2e^- \rightleftharpoons H_3AsO_3(aq) + H_2O(l)$	+0.559
$Ba^{2+}(aq) + 2e^- \rightleftharpoons Ba(s)$	−2.90
$BiO^+(aq) + 2H^+(aq) + 3e^- \rightleftharpoons Bi(s) + H_2O(l)$	+0.32
$Br_2(l) + 2e^- \rightleftharpoons 2Br^-(aq)$	+1.065
$BrO_3^-(aq) + 6H^+(aq) + 5e^- \rightleftharpoons \frac{1}{2}Br_2(l) + 3H_2O(l)$	+1.52
$Ca^{2+}(aq) + 2e^- \rightleftharpoons Ca(s)$	−2.87
$2CO_2(g) + 2H^+(aq) + 2e^- \rightleftharpoons H_2C_2O_4(aq)$	−0.49
$Cd^{2+}(aq) + 2e^- \rightleftharpoons Cd(s)$	−0.403
$Ce^{4+}(aq) + e^- \rightleftharpoons Ce^{3+}(aq)$	+1.61
$Cl_2(g) + 2e^- \rightleftharpoons 2Cl^-(aq)$	+1.359
$HClO(aq) + H^+(aq) + e^- \rightleftharpoons \frac{1}{2}Cl_2(g) + H_2O(l)$	+1.63
$ClO^-(aq) + H_2O(l) + 2e^- \rightleftharpoons Cl^-(aq) + 2OH^-(aq)$	+0.89
$ClO_3^-(aq) + 6H^+(aq) + 5e^- \rightleftharpoons \frac{1}{2}Cl_2(g) + 3H_2O(l)$	+1.47
$Co^{2+}(aq) + 2e^- \rightleftharpoons Co(s)$	−0.277
$Co^{3+}(aq) + e^- \rightleftharpoons Co^{2+}(aq)$	+1.842
$Cr^{3+}(aq) + 3e^- \rightleftharpoons Cr(s)$	−0.74
$Cr^{3+}(aq) + e^- \rightleftharpoons Cr^{2+}(aq)$	−0.41
$Cr_2O_7^{2-}(aq) + 14H^+(aq) + 6e^- \rightleftharpoons 2Cr^{3+}(aq) + 7H_2O(l)$	+1.33
$CrO_4^{2-}(aq) + 4H_2O(l) + 3e^- \rightleftharpoons Cr(OH)_3(s) + 5OH^-(aq)$	−0.13
$Cu^{2+}(aq) + 2e^- \rightleftharpoons Cu(s)$	+0.337
$Cu^{2+}(aq) + e^- \rightleftharpoons Cu^+(aq)$	+0.153
$Cu^+(aq) + e^- \rightleftharpoons Cu(s)$	+0.521
$CuI(s) + e^- \rightleftharpoons Cu(s) + I^-(aq)$	−0.185
$F_2(g) + 2e^- \rightleftharpoons 2F^-(aq)$	+2.87
$Fe^{2+}(aq) + 2e^- \rightleftharpoons Fe(s)$	−0.440
$Fe^{3+}(aq) + e^- \rightleftharpoons Fe^{2+}(aq)$	+0.771
$Fe(CN)_6^{3-}(aq) + e^- \rightleftharpoons Fe(CN)_6^{4-}(aq)$	+0.36
$2H^+(aq) + 2e^- \rightleftharpoons H_2(g)$	0.000
$2H_2O(l) + 2e^- \rightleftharpoons H_2(g) + 2OH^-(aq)$	−0.83
$HO_2^-(aq) + H_2O(l) + 2e^- \rightleftharpoons 3OH^-(aq)$	+0.88
$H_2O_2(aq) + 2H^+(aq) + 2e^- \rightleftharpoons 2H_2O(l)$	+1.776
$Hg_2^{2+}(aq) + 2e^- \rightleftharpoons 2Hg(l)$	+0.789
$2Hg^{2+}(aq) + 2e^- \rightleftharpoons Hg_2^{2+}(aq)$	+0.920
$Hg^{2+}(aq) + 2e^- \rightleftharpoons Hg(l)$	+0.854
$I_2(s) + 2e^- \rightleftharpoons 2I^-(aq)$	+0.536
$IO_3^-(aq) + 6H^+(aq) + 5e^- \rightleftharpoons \frac{1}{2}I_2(s) + 3H_2O(l)$	+1.195
$K^+(aq) + e^- \rightleftharpoons K(s)$	−2.925
$Li^+(aq) + e^- \rightleftharpoons Li(s)$	−3.05
$Mg^{2+}(aq) + 2e^- \rightleftharpoons Mg(s)$	−2.37

Half-reaction	$E°$ (V)
$Mn^{2+}(aq) + 2e^- \rightleftharpoons Mn(s)$	−1.18
$MnO_2(s) + 4H^+(aq) + 2e^- \rightleftharpoons Mn^{2+}(aq) + 2H_2O(l)$	+1.23
$MnO_4^-(aq) + 8H^+(aq) + 5e^- \rightleftharpoons Mn^{2+}(aq) + 4H_2O(l)$	+1.51
$MnO_4^-(aq) + 2H_2O(l) + 3e^- \rightleftharpoons MnO_2(s) + 4OH^-(aq)$	+0.59
$HNO_2(aq) + H^+(aq) + e^- \rightleftharpoons NO(g) + H_2O(l)$	+1.00
$N_2(g) + 4H_2O(l) + 4e^- \rightleftharpoons 4OH^-(aq) + N_2H_4(aq)$	−1.16
$N_2(g) + 5H^+(aq) + 4e^- \rightleftharpoons N_2H_5^+(aq)$	−0.23
$NO_3^-(aq) + 4H^+(aq) + 3e^- \rightleftharpoons NO(g) + 2H_2O(l)$	+0.96
$Na^+(aq) + e^- \rightleftharpoons Na(s)$	−2.71
$Ni^{2+}(aq) + 2e^- \rightleftharpoons Ni(s)$	−0.28
$O_2(g) + 4H^+(aq) + 4e^- \rightleftharpoons 2H_2O(l)$	+1.23
$O_2(g) + 2H_2O(l) + 4e^- \rightleftharpoons 4OH^-(aq)$	+0.40
$O_2(g) + 2H^+(aq) + 2e^- \rightleftharpoons H_2O_2(aq)$	+0.68
$O_3(g) + 2H^+(aq) + 2e^- \rightleftharpoons O_2(g) + H_2O(l)$	+2.07
$Pb^{2+}(aq) + 2e^- \rightleftharpoons Pb(s)$	−0.126
$PbO_2(s) + HSO_4^-(aq) + 3H^+(aq) + 2e^- \rightleftharpoons PbSO_4(s) + 2H_2O(l)$	+1.685
$PbSO_4(s) + H^+(aq) + 2e^- \rightleftharpoons Pb(s) + HSO_4^-(aq)$	−0.356
$PtCl_4^{2-}(aq) + 2e^- \rightleftharpoons Pt(s) + 4Cl^-(aq)$	+0.73
$S(s) + 2H^+(aq) + 2e^- \rightleftharpoons H_2S(g)$	+0.141
$H_2SO_3(aq) + 4H^+(aq) + 4e^- \rightleftharpoons S(s) + 3H_2O(l)$	+0.45
$HSO_4^-(aq) + 3H^+(aq) + 2e^- \rightleftharpoons H_2SO_3(aq) + H_2O(l)$	+0.17
$Sn^{2+}(aq) + 2e^- \rightleftharpoons Sn(s)$	−0.136
$Sn^{4+}(aq) + 2e^- \rightleftharpoons Sn^{2+}(aq)$	+0.154
$Zn^{2+}(aq) + 2e^- \rightleftharpoons Zn(s)$	−0.763

Answers to Selected Exercises

CHAPTER 1

1.1 Air, dust, water, and steak are matter. 1.3 A *theory* is an explanation or model that accounts for a set of related observations. A *law* is a mathematical expression or verbal statement that summarizes a large body of observations. 1.6 (a) m; (b) m^3; (c) kg; (d) m^2; (e) s. 1.8 (a) μg; (b) ns; (c) mL; (d) pm. 1.10 (a) 4.53×10^6; (b) 6.50×10^2; (c) 2.57×10^{-3}; (d) $35 \times 10^6 = 3.5 \times 10^7$; (e) 2.53×10^{-2}. 1.13 (a) $4.0\ g/cm^3$. 1.17 (a) 77°F; (c) 248 K. 1.19 Exact numbers are (a), (c), (e) and (f). 1.20 (a) five; (b) four; (c) three; (d) three; (e) two; (f) two at minimum, possibly four. 1.22 (a) 4.568×10^6; (b) 6.338×10^3; (c) 2.389×10^{-3}; (d) 0.9876; (e) 3.226×10^{-4}. 1.24 (a) 0.55; (b) 3.5×10^3; (c) 5.74×10^{-9}; (d) 2×10^6; (e) 1.5; (f) 1.82×10^3. 1.26 (a) 5.24 L; (b) 1.0×10^2 km/h; (c) 2.20×10^3 kg; (d) 1.8 m. 1.29 $71\ cm^3 = 7.1 \times 10^{-5}\ m^3$. 1.32 15 g. 1.34 Color, density, and melting point are *intensive* properties. Mass and volume are *extensive* properties. 1.36 (a) 1.6 m, 49.9 kg–55.3 kg; (b) 1.75 m, 69.8 kg, 71 g protein. 1.40 $0.807\ g/cm^3$. 1.42 (a) 62.2 kg; (b) 4.0×10^1 kg 1.44 Mass of toluene is 29.295 g; $33.8\ cm^3$. 1.47 $0.82/kg. 1.51 463 mg; to one significant figure, 500 mg.

CHAPTER 2

2.1 Heterogeneous mixtures: wood, Bufferin or Anacin, milk. Pure substances: salt, gold. Solutions: wine. 2.3 Chemical changes: hard-boiling of an egg; rusting of iron; electrolysis of water. Physical changes: melting of ice; dissolving of salt in water; evaporation of water from a lake; pulverizing of rocks. 2.5 *Gases:* oxygen, hydrogen, chlorine, helium. *liquids:* mercury, alcohol, water. *solids:* iron, aluminum. 2.7 Compound A is a compound; gaseous product is also a compound because it is identical to the product of reaction between carbon and oxygen. Can't tell whether the white residue from heating of A is a compound or element. 2.9 (a) Lead; (b) cesium; (c) sulfur; (d) aluminum; (e) helium; (f) iron; (g) calcium. 2.11 Excess of one or another element can't alter the composition of the product, because the elements combine in a fixed ratio (Law of Constant Proportions). 2.13 Calculate the ratio mass O/mass N. A, 1.14; B, 2.28; C, 1.705. Dividing through by smallest, obtain A, 1.00, B, 2.00, C, 1.50. These data tell us that the ratios of O to N in the three compounds are related as small integers, 2, 4, 3. 2.15 There may be 1, 2, 3 or more excess electrons on the droplet. Assuming that the electronic charge is the lowest common factor in observed charges, obtain an average value of 1.59×10^{-19} C. 2.17 (a) Must contain the same number of protons, neutrons, and electrons; (b) Must contain the same number of protons and electrons; (c) must contain the same total number of protons plus neutrons. 2.19 (a) ^{6}Li 3 protons, 3 neutrons, 3 electrons; (b) ^{57}Fe 26 protons, 31 neutrons, 26 electrons; (c) ^{27}Al 13 protons, 14 neutrons, 13 electrons; (d) ^{19}F 9 protons, 10 neutrons, 9 electrons.

2.21 Symbol	^{79}Se	^{80}Br	^{137}Ba
Protons	34	35	56
Electrons	34	35	56
Neutrons	45	45	81
Mass number	79	80	137

2.22 (a) Mn (metal); (b) Br (nonmetal); (c) Cr (metal); (d) Ge (metalloid); (e) Hf (metal); (f) Se (nonmetal); (g) Ar (nonmetal). 2.24 (a) Calcium (Ca) and strontium (Sr) are in group 2A, and adjacent rows. Should possess closely similar chemical and physical properties. (b) Al and Ga are members of same family, thus should be similar in many ways. 2.25 (a) CH_2; (b) C_3H_8; (c) P_2O_5. 2.27 (a) $CaBr_2$; (b) $(NH_4)_2SO_4$. 2.32 (a) potassium chlorate; (b) sodium phosphate; (c) barium hydrogen sulfite (or barium bisulfite); (d) zinc hydroxide; (e) calcium cyanide. 2.33 (a) $CaCl_2$; (b) $Ba(OH)_2$; (c) KCN; (d) CuBr. 2.35 (a) HNO_2; (b) H_2SO_4; (c) HI. 2.36 (a) Sulfur dichloride; (b) xenon trioxide. 2.38 (a) $ZnCO_3$, ZnO, H_2O; (b) HF, SiO_2, SiF_4, H_2O; (c) SO_2, H_2O, H_2SO_3; (d) H_3P (or PH_3); (e) $HClO_4$, Cd, $Cd(ClO_4)_2$; (f) VBr_3. 2.40 (a) *Homogeneous* indicates the same throughout, as opposed to *heterogeneous,* which indicates variation throughout a material. (b) A *gas* expands so as to uniformly occupy the volume containing it. A *liquid* is of more fixed volume, but it flows to occupy the container. Gases are highly compressible, liquids are not. (c) *Atomic number* is a measure of the number of protons in the atomic nucleus; the *mass number* gives the number of protons plus neutrons. (d) A *chemical property* is a characteristic of a substance as it is converted into one or more other substances. A *physical property* is exhibited by a substance while it retains its chemical identity (e.g., a melting point). 2.42 (a) Ca_2S, $Ca(HS)_2$; (b) HBr, $HBrO_3$; (c) AlN, $Al(NO_2)_3$; (d) FeO, Fe_2O_3; (e) NH_3, NH_4^+; (f) K_2SO_3, $KHSO_3$; (g) Hg_2Cl_2, $HgCl_2$; (h) $HClO_3$, $HClO_4$. 2.47 (a) Be; (b) Mg; (c) Na.

2.49

Symbol	${}^{12}_{6}C$	${}^{17}_{8}O^{2-}$	${}^{25}_{12}Mg$	${}^{23}_{11}Na^+$	${}^{18}_{8}O^{2-}$
Protons	6	8	12	11	8
Neutrons	6	9	13	12	10
Electrons	6	10	12	10	10
Net charge	0	2−	0	1+	2−

2.56 (a) $CdCl_2$; (b) CdS; (c) ZnO. 2.58 *Elements:* I_2, S_8, P_4. *Compounds:* SO_2, HN_3, H_2O_2.

CHAPTER 3

3.1 (a) Consistent; (c) inconsistent. 3.2 (a) $2Al(s) + 3Cl_2(g) \longrightarrow 2AlCl_3(s)$; (b) $P_2O_3(s) + 3H_2O(l) \longrightarrow 2H_3PO_3(aq)$; (c) $Ca(OH)_2(aq) + 2HBr(aq) \longrightarrow CaBr_2(aq) + 2H_2O(l)$. 3.4 (a) $2PH_3(g) + 4O_2(g) \longrightarrow P_2O_5(s) + 3H_2O(g)$; (c) $B_2S_3(s) + 6H_2O(l) \longrightarrow 2H_3BO_3(aq) + 3H_2S(g)$. 3.7 (a) $2C_4H_{10}(g) + 13O_2(g) \longrightarrow 8CO_2(g) + 10H_2O(l)$; (c) $Al(OH)_3(s) + 3HNO_3(aq) \longrightarrow Al(NO_3)_3(aq) + 3H_2O(l)$; (e) $2AgNO_3(aq) + H_2SO_4(aq) \longrightarrow Ag_2SO_4(s) + 2HNO_3(aq)$. 3.10 (a) 44.10; (b) 64.06; (c) 104.1; (d) 132.1; (e) 32.04. 3.13 (a) H = 2.06%, S = 32.69%, O = 65.25%; (c) Mg = 16.39%, N = 18.89%, O = 64.72%. 3.16 Chlorine and fluorine are present in ratio of volumes of gases reacting. Two volumes of product gas tells us that compound must be ClF_3, assuming that chlorine and fluorine are diatomic. 3.17 Atomic weight = 24.319. 3.21 Berzelius value is 129.0 g/mol for Zn. High by a factor of two because Berzelius assumed formula for oxide is ZnO_2, not ZnO, as is the case. 3.23 (a) Molar wt. = 164.1; (b) 0.0152 mol $Cu(NO_3)_2$; (c) 53.3 g $Ca(NO_3)_2$; (d) 3.20×10^{22} Ca^{2+} ions. 3.25 (a) 132 g CO_2; (b) 2.37 g Ar; (c) 111 g HCl; (d) 0.22 g C_2H_4. 3.27 (a) 2.8×10^{-6} mol H_2O; (b) 5.1×10^{-4} g H_2O. 3.31 (a) CH_4; (b) 0.209 mol Fe, 0.313 mol O, thus Fe_2O_3. (c) NH_2 is empirical formula. 3.33 Empirical formula is CH_3O_2. 3.35 (a) Molecular formula is $C_2H_4Cl_2$; (b) $C_4H_8O_2$; (c) empirical and molecular formulas are $C_9H_8O_4$. 3.37 $Sr(OH)_2 \cdot 8H_2O$. 3.40 (a) 2.63 g NH_3; (b) 6.53 g $CaCO_3$. 3.42 13.2 kg PCl_5. 3.44 (a) 0.0515 mol SiH_4; (b) 3.71 g H_2O; (c) 5.54×10^{-4} mol SiH_4 from Mg_2Si, 3.74×10^{-4} mol SiH_4 from H_2O. Thus, H_2O is the limiting reagent; can form a maximum of 12.0 mg SiH_4. 3.46 KBr is the limiting reagent; 5.45 g AgBr formed in precipitation reaction. 3.48 (a) 60.3 g C_6H_5Br; (b) percentage yield = 94.0%. 3.50 (a) 0.134 *M*; (c) 5.90×10^{-6} *M*. 3.51 (a) 4.03 g CdF_2; (b) 1.42 g HNO_3; (c) 5.56 g NH_4CN; (d) 3.6×10^{10} g $Cl_4C_{12}H_6$. 3.52 (a) 0.171 L solution; (b) 0.496 L solution. 3.53 (a) 3.55 g Na_2SO_4; (b) 0.125 L solution. 3.55 (a) 0.133 *M*. 3.57 0.0912 L Na_2CrO_4 solution. 3.59 Weight percentage = 9.05×10^{-2}%. 3.62 (a) $Li_3N(s) + 3H_2O(l) \longrightarrow NH_3(g) + 3LiOH(aq)$; (c) $PBr_3(l) + 3H_2O(l) \longrightarrow H_3PO_3(aq) + 3HBr(aq)$; (e) $2CCl_4(g) + O_2(g) \longrightarrow 2CCl_2O(g) + 2Cl_2(g)$. 3.65 Weight percentages of P: (a) 18.4%; (b) 24.5%; (c) 23.4%. Weight percentage is highest in $Ca(H_2PO_4)_2 \cdot H_2O$. 3.69 C_2HF_3ClBr. 3.71 (a) $C_{10}H_{18}O$; (b) empirical formula is also molecular formula. 3.74 (a) $Al(OH)_3(s) + 3HCl(aq) \longrightarrow 3H_2O(l) + AlCl_3(aq)$; (b) 0.19 mol HCl. 3.76 Balanced equation for reaction is $2H_2(g) + O_2(g) \longrightarrow 2H_2O(g)$. Obtain 0.22 g H_2, 29.2 g H_2O upon completion of reaction. 3.81 (a) 3.47×10^2 kg $C_6H_{10}O_4$; (b) percent yield = 85.0%. 3.83 (a) 0.595 *M*; (b) 0.0231 mol HNO_3, 0.0462 *M*; (c) 0.382 *M*; (d) 0.041 mol KBr, 0.41 *M*. 3.85 (a) 2.2×10^{-9} *M*; (b) 1.3×10^{12} Na atoms or Na^+ ions/cm^3. 3.88 2.75 L solution.

CHAPTER 4

4.2 Water could turn a paddle wheel, pass through turbine blades in converting kinetic energy into work or electrical energy. 4.4 2.1 kJ; this kinetic energy is converted into work needed to deform slug upon impact, and into heat. 4.5 4.5×10^2 kJ, 1.1×10^2 kcal. 4.7 (a) $q = 0$; work done on surroundings is positive. Thus, $\Delta E = -(450\ J) = -450\ J$. 4.9 Additional heat needed to perform work of expanding the volume of the CO_2, working against atmospheric pressure. 4.10 $q_p = 660$ J. Work of expansion, $P\Delta V = 140$ J. $\Delta E = 660 - 140 = 520$ J. $\Delta H = q_p = 660$ J. 4.12 (a) Endothermic; (b) exothermic; (c) endothermic; (d) endothermic. 4.14 29.8 kJ evolved. 4.17 $\Delta H = +180.8$ kJ. 4.18 (a) $\Delta H = 2(-46.2) - 0 - 0 = -92.4$ kJ; (c) $\Delta H = 2(-425.6) + 0 - 0 - 2(-241.8) = -367.7$ kJ. 4.21 Nonzero values for ΔH_f° imply that the element is not in its standard state. Examples: Br(*g*), Co(*g*), $I_2(g)$, $P_2(g)$. 4.22 (b) $\Delta H = 30.71 - 0 = 30.71$ kJ/mol. 4.24 ΔH for reaction of three moles of MnO_2 is $2(-1699.8) + 3(0) - 3(-519.6) - 4(0) = -1780.8$ kJ. Heat per mol MnO_2 is -593 kJ. ΔH per g reactants is -4.829 kJ. 4.27 Heat capacity of 1000 gal H_2O is 1.58×10^4 kJ/°C. Need 1.0×10^4 bricks. 4.30 (a) 4220 kJ/mol caffeine; (b) overall uncertainty is approximately equal to the sum of the uncertainties due to each effect (somewhat less than this because unrelated uncertainties may partially cancel). Overall uncertainty about 44 kJ or less. 4.34 Total calorie content = 350 Cal. 4.35 94 Cal 4.37 7.3×10^6 tons. 4.39 (a) See list of key terms. 4.41 38 quad/yr. 4.44 For reaction $H(g) + Br(g) \longrightarrow HBr(g)$, $\Delta H = -366$ kJ. 4.45 (a) $\Delta H_{rxn} = 4(5.4) = +21.6$ kJ. 4.48 7.89°C. 4.50 2.86×10^{10} kJ/h from burning of coal; 1.1×10^7 g S/h; 1.0×10^8 kJ/h from combustion of S.

CHAPTER 5

5.1 Cosmic radiation; X rays; green light; radiation from heaters; FM station. 5.3 1.52×10^4 s; 4.22 h. 5.4 6.59×10^{14}/s; green. 5.8 (a) $E = 2.98 \times 10^{-19}$ J. 5.10 $E = 3.68 \times 10^{-19}$ J; 11 photons in the signal. 5.12 247 nm; an increase in intensity causes an increase in electrons emitted per unit time; however, the energies of the emitted electrons are unchanged. 5.15 (a) False; the speed of light in a vacuum is a constant, independent of frequency. 5.17 (a) Absorbed; (b) emitted; (c) absorbed. 5.19 Initial $n = 4$. 5.21 It should be larger; higher nuclear charge of Li^{2+}. 5.22 $E = 8.72 \times 10^{-18}$ J. 5.23 (a) Radius = 4.8 Å;

(b) radius = 0.18 Å. Smaller radius of lowest energy orbit due to greater nuclear-electron attraction in Li^{2+}. 5.26 $v = 4.51 \times 10^3$ m/s. 5.27 (a) False; it depends only on mass and speed. (c) True. 5.29 The three $3p$ orbitals differ only in their orientations in space. The $2p$ and $3p$ orbitals differ in their radial extensions. The $3p$ orbitals possess an additional small node near the nucleus; the major lobes are more extended in space, at a greater distance from the nucleus than for the $2p$. 5.31 $3f$ is incorrect; an f orbital corresponds to $l = 3$, and l must be no greater than $n - 1$. $2d$ is incorrect; a d orbital corresponds to $l = 2$, and l cannot be greater than 1 when $n = 2$. 5.34 (a) True. 5.36 Total energy = 945 J. 5.39 (a) 9.47×10^{12} km. 5.42 (a) $E = 3.46 \times 10^{-19}$ J; in molar units, 208 kJ/mol; (b) energy of 420 nm photon is 4.74×10^{-19} J. Excess energy over that required for photoemission is 1.28×10^{-19} J. Emitted electron could have up to this much kinetic energy. 5.45 (a) Five; (b) one; (c) essentially an infinite number; (d) one. 5.48 (a) $n = 3$, $l = 0$, $m_l = 0$. 5.50 Essentially no difference. 5.52 Presence of distinct sharp wavelengths of lower solar emission indicate absorptions of certain discrete wavelengths. Sharp lines indicate presence of states of well-defined energies, with transitions between them giving rise to absorption or emission.

CHAPTER 6

6.1 It is smaller for the $2s$ electron. 6.2 (a) $2s$. 6.4 Both n and l must have the same value. 6.6 The ionization energy varies as Z^2. Thus for He, the value is $4 \times 1312 = 5248$ kJ. 6.8 (b) 2. 6.12 (a) [Ar] $4s^2$; (c) [Ar] $4s^2 3d^{10} 4p^5$.

6.14 (b) [↑↓] [↑ ↑ ↑ ↑ ↑]
$4s$ $3d$

6.18 $3p$; $5s$; $4d$; $5p$; $6s$; $4f$; $6p$. 6.20 These lines are due for the most part to transitions from higher energy orbitals back to the $3s$ orbital. 6.21 (b) Cl (see Figure 6.10 and Table 6.5); (c) As. 6.23 (a) Cl. 6.25 (b) Al has larger ionization energy (barely so); Ga has larger size and probably the larger electron affinity, because it is larger and can more readily accommodate the added electron. 6.27 Nuclear charge is one higher in Al^{2+}. Note that I_3 for Al is higher than I_2 for Mg, Table 6.4. 6.30 Electron added to Cl experiences some nuclear charge because of incomplete shielding by the other $3p$ electrons. In addition, a stable octet is formed. In Ar the added electron must go into a $4s$ orbital, which lies outside the $3p$ electrons, so it is nearly completely shielded from the nucleus. 6.31 (a) $2K(s) + 2H_2O(l) \longrightarrow 2KOH(aq) + H_2(g)$; (c) $CaO(s) + H_2O(l) \longrightarrow Ca(OH)_2(aq)$. 6.33 The electron affinity is higher for Cu than for K (Figure 6.10) because the effective nuclear charge is higher in the group 1B element. We expect that the electron affinity for Zn should be a positive quantity, i.e., the process should be endothermic, because the element has a completed subshell of electrons. 6.36 Reversals occur at Ar 39.948, K 39.098; Co 58.93, Ni 58.69; Te 127.60, I 126.9045; Th 232.04, Pa 231.04; U 238.03, Np 237.048. The violations occur because of varying numbers of neutrons in the nuclei of the most abundant nuclides. 6.39 (a) $3s$; (c) $3d$. 6.43 The g orbitals correspond to $l = 4$. Thus, the lowest value which n can have would be 5. There are $2l + 1 = 9$ g orbitals of a given major quantum number. 6.45 (a) oxygen. 6.48 (a) Kr; (c) Discounting those elements that have completed d subshells, only O and S. Otherwise, all group 6A elements.

6.50 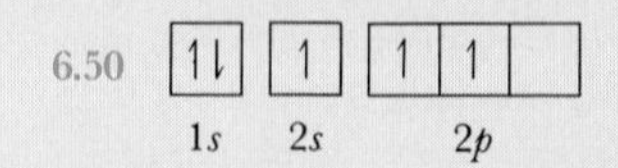

In this excited state one of the paired $2s$ electrons has been promoted to the $2p$ orbital.

CHAPTER 7

7.1 (b) Group 2A. 7.3 (a) $:\ddot{\underset{..}{S}}:^{2-}$; (d) $\cdot$Ca$\cdot$. 7.5 (a) $1s^2 2s^2 2p^6 3s^2 3p^6$; (c) [Ar]$3d^5$. 7.7 (a) Sc_2O_3; (c) BaS. 7.9 The Cu^+ and Cd^{2+} ions do not have noble gas configurations, because they have filled 3d and 4d subshells, respectively. 7.11 The forces that bind Ar atoms together are very weak because there is no chemical bonding between the atoms. In RbCl, on the other hand, the Rb^+ ions interact strongly with Cl^- ions, forming a lattice with high stability due to the electrostatic attractive forces between ions of opposite sign. 7.13 The Y^{n-} anion is probably smaller than X^{n-}, allowing a closer approach of oppositely charged ions. 7.14 In BaCl and MgF_3 the formulas do not correspond to noble gas electron configurations about one or more ions; the formulations should be $BaCl_2$ and MgF_2. 7.16 Addition of an electron to the neutral atom results in larger total electron-electron repulsions. These are reduced to some extent by an increase in the volume of the ion, that allows the average distance between electrons to increase. 7.18 The $3p$ electron in K^+ experiences a larger effective nuclear charge, because the $4s$ electron removed in forming K^+ served to provide *a little* shielding of the nucleus from the $3p$ electrons (penetration effect). 7.20 (b) $Mg^{2+} < Na^+ < Br^-$.

7.23 (a) H—Si(—H)(—H)—H (e) $\ddot{\underset{..}{O}}$=$\ddot{N}$—$\ddot{\underset{..}{Cl}}$:

7.28 (b) BCl_3 contains 6 electrons in valence shell of B; (d) odd number of electrons in ClO_2, so one atom *must* not have an octet of electrons; (e) 10 electrons in valence shell of Te. 7.30 F is more electronegative than Cl and smaller; six Cl atoms cannot fit about a central sulfur atom. 7.33 (a) $\Delta H = 3D(H—H) + D(C—H) + D(C\equiv N) - 4D(C—H) - 3D(N—H) = -213$ kJ. 7.35 $NH(g) \longrightarrow N(g) + H(g)$, $\Delta H = 328$ kJ. 7.37 (a) $P < S < O$; (b) $Mg < Al < Si$; (c) $S < Br < Cl$; (d) $Si < C < N$. 7.39 (a) $C—S < N—O < B—F$. 7.42 Generally, electronegativity decreases with increasing size. Thus in nonmetals electronegativity decreases as one moves down a group. In metals the situation is more complex, because of the presence of the incompletely filled d orbitals and the f orbital effects. Among some metal families the electronegativity *increases* with increasing period, e.g., in the Group 2B metals. (Remember that electronegativities for these elements are the estimated values for the +2 ions.) 7.43 (a) N, +1; O, −2. (c) H, +1; P, +3; O, −2. (e) Cl, +3; F, −1. 7.45 (a) Sodium dihydrogen phosphate; (c) vanadium(V) oxide; (e) hypobromite ion. 7.47 (b) P_2O_5; (d) FeF_3. 7.49 (a) S is oxidized from 0 to +4; F is reduced from 0 to −1. 7.50 (a) zero; (c) +1. 7.52 (a) MgO; (b) N_2O, NO, N_2O_3, NO_2, N_2O_5. 7.55 (b) $Cl_2O_7(g) + H_2O(l) \longrightarrow 2HClO_4(aq)$; (d) $Cr_2O_3(s) + 6HCl(aq) \longrightarrow 2CrCl_3(aq) + 3H_2O(l)$; (e) $SrO(s) + H_2O(l) \longrightarrow Sr(OH)_2(aq)$. 7.57 (a) K^+, Ca^{2+}. 7.59 Mg^{2+} is smaller because of its larger charge, so there is a much larger attractive interaction between anions and cations. 7.62 (a) $E = -4.38 \times 10^{-18}$ J; on a molar basis, 264 kJ/mol. 7.64 The compound is most likely GeO_2 (m.p. GeO_2 = 1115°C). 7.67 [Kr] $4d^4$, MoO; [Kr] $4d^2$, MoO_2; [Kr], MoO_3.

7.69 (a)

$$\left[\begin{array}{c} :\ddot{O}: \\ | \\ :\ddot{S}-\ddot{S}-\ddot{O}: \\ | \\ :\ddot{O}: \end{array}\right]^{2-}$$

7.71 (a) $\Delta H = D(H{-}Br) + 2D(F{-}F) - 3D(Br{-}F) - D(H{-}F) = 366 + 2(155) - 3(237) - 567 = -602$ kJ. 7.73 (a) NaClO. 7.76 430 kJ/mol. 7.78 (b) $As_2O_5(s) + 3H_2O(l) \longrightarrow 2H_3AsO_4(aq)$. 7.80 Multiple oxidation states occur most often in elements that have incomplete subshells, as in the case of the transition elements, that possess from one to ten *d* electrons.

CHAPTER 8

8.1 (a) Trigonal planar; (b) tetrahedral; (c) trigonal bipyramidal; (d) octahedral. 8.3 (a) Trigonal bipyramidal. 8.5 (b) Tetrahedral; (d) bent. 8.8 Write the Lewis structures. Note that in NO_2^+ there are two multiple bonds (isoelectronic with CO_2), so we predict 180° bond angles. In NO_2, predict O—N—O angle larger than 120° because repulsive effect of one electron is less than that of a bonding pair. In NO_2^-, angle should be less than 120°. 8.9 (d) No, linear; (e) yes. 8.12 Q = 1.87×10^{-20} C; ionic character = 12%. 8.15 (a) sp^2; 120° bond angles in a plane; (c) d^2sp^3 or sp^3d^2; 90° bond angles. 8.17 (b) sp^3d; (d) sp^3; (g) sp^3. 8.19 (b) There are four π bonds altogether; two in the C≡C bond, one each in the C=O and C=C bonds. 8.21 (a) Bonding mo's are lower in energy than the starting atomic orbitals, the antibonding mo's are higher. (b) In the bonding mo's, electron density tends to concentrate in the space between the bonded atoms; in the antibonding mo's, it concentrates in the region in back of the atoms. 8.26 (a) NO (odd number of electrons); (b) O_2^+ would have one unpaired electron; (c) No; (d) No. 8.28 Addition of an electron to, or removal of an electron from, N_2 will reduce the bonding order from 3 to 2.5. Both N_2^- and N_2^+ should have longer N—N distances than N_2. 8.30 P_2 should have an energy level diagram similar to that for N_2, Table 8.7. However, the P_2 dissociation energy will be much lower, because P does not form strong π bonds. 8.32 (b) In bromine each Br is bonded to only one other atom. In C, the bonding forms a three-dimensional network. 8.35 The mo model predicts the presence of unpaired electrons, as observed. O—O bond dissociation energy and bond distance are also consistent with predicted O—O bond order. 8.36 In PH_3 the unshared electron pair exerts a greater repulsive effect on the other three electron pairs about P, forcing the H—P—H angle inward. In PH_4^+ all four electron pairs about P are equivalent. 8.39 (a) Yes; (b) yes; (c) no (planar trigonal); (d) yes; (e) no; (f) yes. 8.41 (c) sp^3d. 8.44 (a) $sp^3 \longrightarrow sp^2$; (b) $sp^3d \longrightarrow sp^3$; (c) $sp^3d^2 \longrightarrow sp^3d$. 8.46 (b) The bonding about each N should be trigonal planar. Overall, the molecule should be planar. (c) Nitrogen will employ $sp^2 + p$ hybrid orbitals. 8.49 (a) CH_5 is unstable because carbon has no vacant orbitals of accessible energy for adding the ninth electron. The octet of electrons is attained in CH_4. (c) OF_6 is unstable because there are no orbitals of accessible energy for accepting more than 8 electrons about O. In addition, there is not room about O for six F atoms at reasonable bonding distance.

8.51

Molecule or ion:	O_2^+	O_2	O_2^-	O_2^{2-}
Bond order:	2.5	2.0	1.5	1.0

$\longrightarrow$ Increasing bond length

8.54 NO should have the lowest ionization energy, N_2 the highest. 8.57 Strain is least in S_8. Bond angle strain is higher in the smaller rings, leading to higher chemical reactivity. 8.59 In AsF_2^+, six electrons in As valence shell; bent structure. In AsF_4^-, there are ten electrons in As valence orbitals. Predict seesaw structure.

CHAPTER 9

9.2 0.929 atm; (b) 3.62 atm; (c) 165 torr; (d) 709 mm Hg; (e) 90.6 kPa; (f) 79.5 kPa. 9.5 (a) 294 Pa
9.6 (a) Pressure of gas is 684 mm Hg. 9.8 Net pressure of O_2 is 708.0 mm Hg. 9.11 (a) False; pressure is directly proportional to number of moles of gas. (b) correct; (c) correct; (d) false; at constant volume, pressure is directly proportional to absolute temperature. 9.12 (a) P = 16.0 atm; (b) V = 0.431 L; (c) n = 0.0194 mol; (d) $T = 1.11 \times 10^3$ K. 9.14 978 g; V = 685 L. 9.17 (a) P_f = 1140 mm Hg. 9.19 (a) V = 1.50 L; (b) V = 0.458 L. 9.21 Would require 1.14 tanks. 9.24 (a) 6.8×10^{-3} atm; (b) 0.0355 atm. 9.27 Total pressure = 0.958 atm. 9.28 (a) d = 1.78 g/L. 9.29 (a) 40.5 g/mol. 9.30 (a) C_2N_2. 9.31 83.5 g/mol. 9.33 O_2 volume = 3.69 L. 9.34 Because water vapor pressure is 28 mm Hg, collect H_2 gas at 720 mm Hg. V = 1.34 L. 9.35 P = 0.215 atm. 9.41 (a) vessel A; (b) vessel B; (c) vessel B; (d) vessel A. 9.44 (a) u = 3.10×10^2 m/s. 9.46 (a) He. 9.48 (a) $r_{He}/r_{Ne} = (20.2/4.00)^{1/2} = 2.25$. 9.51 Deviations from ideal behavior are most evident at lowest temperatures and at high pressures. They result from finite sizes of molecules, and intermolecular attractive forces. 9.53 (a) UF_6; (b) $SiCl_4$; (c) SO_2. 9.55 (a) P = 0.966 atm; (b) P = 0.946 atm. 9.58 P = 38 kPa. 9.60 (a) 10.0 mm Hg; (b) Net height of water column is 9.67 m. 9.62 (a) $n = 8 \times 10^6$ mol H_2 = 2×10^7 g H_2; (b) 1.4×10^8 g H_2O. 9.64 1.5×10^{-4} mol/L; 8.8×10^{19} molecules/L. 9.68 n = 9.8×10^{-11} mol O_2; 4.8×10^{-9} g Mg. 9.70 n = 0.260 mol; 5.47 g CaH_2. 9.74 d = 0.596 g/L. 9.77 The balloon will contract because H_2 will diffuse out more rapidly than He will diffuse in.

CHAPTER 10

10.1 (a) The density of a liquid or solid is relatively insensitive to pressure, whereas the density of a gas is much lower and is proportional to pressure. (b) Diffusion is extremely slow or nonexistent in a solid, slow in a liquid, and rapid in a gas. (c) Liquids and solids are relatively incompressible, whereas gases are compressible. (d) Flow is nonexistent or slow in a solid, much more rapid in a liquid and even more rapid in a gas. 10.4 When liquid ethyl chloride boils, heat is required to overcome the attractive forces between molecules. This heat is extracted from the surface that contacts the liquid. 10.7 (a) Total heat required = 233 kJ. 10.9 1.8 g H_2O per min. 10.11 (a) No change; (b) increases with increasing temperature; (c) higher intermolecular forces mean lower vapor pressure; (d) no change. 10.13 Diethyl ether, about 23°C; ethyl alcohol, about 68°C; water, 90°C. 10.15 (a) Decreases; (b) increase; (c) increase; (d) increase; (e) increases; (f) increases; (g) increases; (h) unaffected.

10.17 $\Delta H_v = 60.3$ kJ/mol. 10.19 Intermolecular forces are much weaker; they are not directional; they do not have the same specific character as bonding forces, which reflect the valences of the atoms. 10.21 (a) Ion-ion; (b) dispersion force; (c) hydrogen bonding; (d) dipole-dipole. 10.23 $C_6H_5OH > CH_3NH_2 > H_2S$. 10.26 (a) The ionic substance, $FeCl_2$, is most likely to exist as a solid; (b) nonpolar, low molecular weight F_2 is most likely to exist as a gas. 10.28 Propane-1,3-diol > propanol > propane, based on hydrogen bonding interactions. 10.31 Critical temperature is abnormally high for HF because of hydrogen bonding forces. It increases in the series as $HCl < HBr < HI$ because of increasing dispersion forces. 10.32 (a) Molecular; (b) molecular; (c) metallic; (d) ionic; (e) covalent; (f) ionic; (g) atomic; (h) network polar covalent; (i) molecular; (j) ionic; (k) ionic; (l) molecular. 10.36 (a) One ($\frac{1}{8}$th of each of 8 corners); (b) two ($\frac{1}{8}$th of each of 8 corners, plus one in the center); (c) four ($\frac{1}{8}$th at each of 8 corners, plus $\frac{1}{2}$ of each of six faces); (d) one (an average of $\frac{1}{8}$ at each of 8 corners). 10.39 (a) $r = 1.28$ Å. 10.40 (a) $l = 5.387$ Å. 10.42 (a) Density = 1.45 g/cm^3. 10.43 (a) 108 amu; since this is four times the atomic mass of Al, the face-centered cubic cell contains four formula units (See Exercise 10.36). 10.45 (a) Coordination number is 12; (b) 6; (c) 8. 10.47 $d = 3.56$ Å. 10.52 (a) Water vapor would condense to form a solid at a pressure of around 4 mm Hg. At higher pressure, 5 atm or so, the solid would melt to form liquid water. This occurs because the melting point of ice, which occurs at 0°C at 1 atm, decreases with increasing pressure. 10.54 The melting point of ice decreases with increasing pressure because ice is less dense than water. As pressure is increased, the system responds by shifting in the direction of the denser phase. Solid CO_2 is more dense than $CO_2(l)$, so the opposite (and more common) behavior is seen. 10.56 $\Delta H = 14.7$ kJ. 10.58 The major source of intermolecular attractive energy in glycerol is hydrogen bonding. Increasing temperature diminishes the effects of hydrogen bonding and viscosity thus decreases. 10.61 $\Delta H = 24.3$ kJ/mol. Changing units of pressure would have no effect on the calculated ΔH_v. 10.63 All three molecules are planar-trigonal in shape and nonpolar. The increasing critical temperature reflects increasing London dispersion forces in the series. 10.65 H_2O_2 is not much more massive than H_2O. Thus, the melting and boiling points are abnormally high, just as they are for water, and for the same reason: hydrogen bonding. 10.68 In amorphous silica (SiO_2) the regular structure of quartz is disrupted. The loose, disordered structure that results should have a lower density than for quartz. 10.70 $r = 1.248$ Å. 10.73 Well-defined patterns should be seen for (a) sugar; (b) KBr; (d) pure iron; (e) ice.

CHAPTER 11

11.1 (a) Weight percentage = 1.96%. 11.2 (a) Mole fraction $CH_3OH = 3.94 \times 10^{-3}$. 11.3 (a) 8.40×10^{-3} *M*; (b) 0.222 *M*. 11.4 (a) 5.98 *m*. 11.5 (a) Weight percentage = 1.0%; mole fraction = 1.2×10^{-3}; molality = 0.065 *m*. 11.6 (a) Weight percentage H_2O = 41.1%; (b) mole fraction = 0.692; (c) 38.7 *m*; (d) 21.1 *M*. 11.9 (a) 0.0916 mol $Ca(NO_3)_2$; (c) 4.35×10^{-2} mol NaCl; (e) 0.025 mol HCl. 11.13 (a) 0.72 *N*; (b) 0.0350 *M*. 11.15 (a) Ion-dipole; (b) London dispersion forces; (c) F—H···O hydrogen bonding; (d) dipole-dipole (both molecules are quite polar). 11.17 $CH_4 < CH_3F < NaF < CH_3OH$. 11.21 (a) Na^+ (smaller); (b) Cl^- (smaller); (c) Fe^{3+} (smaller, higher charge; (d) Al^{3+} (smaller, higher charge). 11.23 $C_{He} = 1.8 \times 10^{-4}$ *M*; $C_{N_2} = 3.0 \times 10^{-4}$ *M*. 11.26 $CuCl_2$ should be more soluble at lower temperature. 11.28 (a) Strong electrolyte; (c) strong electrolyte; (e) nonelectrolyte. 11.31 (a) 0.10 *M* Na^+ + 0.10 *M* OH^-; (b) 0.35 *M* Ca^{2+} + 0.70 *M* Br^-; (c) 0.14 *M* C_2H_5OH; (d) 0.050 *M* K^+ + 0.20 *M* Na^+ + 0.050 *M* ClO_3^- + 0.10 *M* SO_4^{2-}. 11.33 (a) No; (b) yes; (c) no; (d) yes; (e) no; (f) no; (g) no (Remember that a colligative property is one that depends on the total concentration of solutes present. Thus, vapor pressure *lowering* is a colligative property, but vapor pressure itself is not.). 11.34 (a) Vapor pressure H_2O = 185.4 mm Hg. 11.35 (a) Boiling point = 79.9°C, freezing point = −117.0°C; (b) boiling point = 80.6°C, freezing point = 4.6°C. 11.37 (a) 23.8 atm. 11.38 Molecular weight = 267 g/mol. 11.42 (a) $CO_2(g) + OH^-(aq) \longrightarrow HCO_3^-(aq)$; (b) $NaH(s) + H_2O(l) \longrightarrow Na^+(aq) + H_2(g) + OH^-(aq)$; (c) $Zn(s) + Cu^{2+}(aq) \longrightarrow Zn^{2+}(aq) + Cu(s)$. 11.43 (a) Yes; (b) yes; (c) no; (d) no; (e) no; (f) no; (g) yes; (h) yes. 11.44 (a) $ZnS(s) + 2H^+(aq) \longrightarrow H_2S(g) + Zn^{2+}(aq)$; (b) $Ba^{2+}(aq) + CO_3^{2-}(aq) \longrightarrow BaCO_3(s)$; (c) $3H^+(aq) + PO_4^{3-}(aq) \longrightarrow H_3PO_4(aq)$ (H_3PO_4 is a relatively weak acid.); (d) $H^+(aq) + OH^-(aq)\ H_2O(l)$. 11.48 Look for a Tyndall effect by passing a beam of light through the liquid. 11.52 The added electrolyte changes the distribution of surface charges on the proteins suspended in milk. They thus coagulate and drop from colloidal suspension. 11.55 8.42 mL solution. 11.57 (a) The nitrogens of the caffeine molecule can engage in hydrogen bonding with H—O groups of the solvent, as can the carbonyl oxygen. In addition, the caffeine molecule as a whole is polar, so there is some attractive dipole-dipole interaction with water. (b) We can infer from the solubility dependence on temperature that the solution process is endothermic (ΔH positive). 11.59 Evaporation of I_2 and mixing of $Br_2(l)$ and $CCl_4(l)$ are either endothermic or have $\Delta H \sim 0$. They proceed because the system moves spontaneously toward states of larger disorder or randomness. Processes (c) and (d) both lead to more order; they nevertheless proceed because they result in a lower overall enthalpy for the system. 11.61 1.37×10^{-5} *M*. 11.63 2.0×10^{-6} atm. 11.66 420 g CH_3OH required. 11.69 Sample is about 80% lactose. 11.71 Molecular weight calculated from freezing point data = 1.8×10^2. Molecular formula is $C_9H_{13}O_3N$, for which formula weight is 183. 11.73 6.67×10^{-2} *M* Na^+; 0.133 *M* Cl^-; 3.33×10^{-2} *M* Ba^{2+}. 11.75 (a) $Al(OH)_3(s) + 3H^+(aq) \longrightarrow Al^{3+}(aq) + 3H_2O(l)$; (b) $Mg(OH)_2(s) + 2H^+(aq) \longrightarrow Mg^{2+}(aq) + 2H_2O(l)$; (c) $MgCO_3(s) + 2H^+(aq) \longrightarrow Mg^{2+}(aq) + H_2O(l) + CO_2(g)$.

CHAPTER 12

12.2 *Tropopause* (200 K); *stratopause* (270 K); *mesopause* (180 K). 12.4 The answer lies in Figure 12.2. Atmospheric pressure falls rapidly with increasing elevation. At 11 km the pressure is only about 250 mm Hg, and it declines to about 1 mm Hg at 50 km. Thus the molecules are much more densely packed at the lowest elevations. 12.6 3.49×10^{-19} J/molecule; $\lambda = 569$ nm. 12.9 Avg. molecular weight at surface = 28.0 g/mol. At 200 km, 19.3 g/mol. 12.11 O_3 in the stratosphere absorbs solar radiation from the 200 to 310 nm region. Such high energy radiation is injurious to health. 12.12 $NO(g) + O_3(g) \longrightarrow NO_2(g) + O_2(g)$; $\Delta H = 33.8 - 142.3 - 90.4 = -198.9$ kJ. $NO_2(g) + O(g) \longrightarrow O_2(g) + NO(g)$; $\Delta H = 90.4 - 247.5 -$

$33.8 = -190.9$ kJ. 12.16 6.7×10^{18} O_3 molecules/cm^3. 12.20 5.5×10^5 g sulfate/mi^2. 12.23 7.9% is in the form of carboxyhemoglobin. 12.25 $2.3 \times 10^{-6} M$ PO_4^{3-}. 12.30 2.5×10^9 g oil. 12.32 (a) Water hardness is due to Ca^{2+}, Mg^{2+} and Fe^{2+}. (b) These ions react with soaps to form insoluble materials. Also precipitation of carbonates can occur when "hard" water is heated (Equation 12.19). 12.34 26 ppm O_2. 12.39 Temperature maximum arises from the "ozone" machine: photodissociation of O_2, formation of O_3, then decomposition of O_3 to reform O_2. In effect, radiant energy is converted into thermal energy. 12.41 Photoionization requires much higher energy than photodissociation. Thus, in the lower atmosphere, photodissociation predominates over photoionization, because all the shorter wavelength radiation has been absorbed. At higher elevations, photoionization is extensive because the intensity of short wavelength radiation is high. 12.44 1×10^7 g phytoplankton per 0.1 km^3 of water (that is, in a 1 km^2 area 0.1 km in depth). 12.46 21 ppm O_2. 12.47 Add 0.35 mol $Ca(OH)_2$ per 10^3 L to remove bicarbonate. Add 0.15 mol Na_2CO_3 per 10^3 L to remove Ca^{2+} as $CaCO_3$.

CHAPTER 13

13.1

Time (min)	Time interval (min)	Concentration (M)	Concentration change	Rate (M/min)
0		1.85		
	79		0.18	2.3×10^{-3}
79		1.67		
	79		0.15	1.9×10^{-3}
158		1.52		
	158		0.22	1.4×10^{-3}
316		1.30		
	316		0.30	0.95×10^{-3}
632		1.00		

13.3 From the slopes of the graphs, the rates are: $t = 0$ s, 0.101 mm/s; $t = 5000$ s, 0.0348 mm/s; $t = 10{,}000$ s, 0.0138 mm/s. The rate constants are derived from these rates by dividing by the pressure of CH_3NC at each time. 13.4 (a) $-\Delta[NOCl]/\Delta t = +\Delta[NO]/\Delta t = +2\,\Delta[Cl_2]/\Delta t$. (b) $-\Delta[HI]/\Delta t = -\Delta[CH_3I]/\Delta t = +\Delta[CH_4]/\Delta t = +\Delta[I_2]/\Delta t$. 13.5 (a) $\Delta[CO_2]/\Delta t = +0.20\ M/s$; $\Delta[H_2O]/\Delta t = 0.40\ M/s$. 13.6 (a) rate $= k[N_2O_5] = (6.08 \times 10^{-4}/s)(0.100M) = 6.08 \times 10^{-5}\ M/s$. 13.8 (a) First order in A, first order in B, second order overall; (b) second order in A, zero order in B, second order overall; (c) first order in A, second order in B, third order overall.

13.10

Rate law	Effect of doubling [A]	Effect of doubling [B]
$k[A][B]$	Double rate	Double rate
$k[A]^2$	Fourfold increase	No effect
$k[A][B]^2$	Double rate	Fourfold increase

13.12 (a) Rate $= k[B]^2$; (b) zero order with respect to A, second order with respect to B, second order overall; (c) use any of the data sets in the table: $k = 7.0 \times 10^{-3}\ M/s$; (d) rate $= 1.8 \times 10^{-5}\ M/s$. 13.15 $k = 2.8 \times 10^{-4}/s$. 13.17 From slope of graph, $k = -2.30(-9.53 \times 10^{-6}/s) = 2.19 \times 10^{-5}/s$. 13.19 Graph of $\log [C_{12}H_{22}O_{11}]$ vs. time, $k = -2.30$ (slope) $= -2.30(-1.60 \times 10^{-3}/min) = 6.13 \times 10^{-3}/s$. 13.21 (a) $[N_2O_5] = 0.115$ mol N_2O_5; (b) $t = 330$ s. 13.25 (a) Reaction with (a), with smallest E_a, will be fastest. The slowest is (c), with largest E_a. (b) The reverse reaction has the smallest E_a in (b). It has the largest E_a in (a). Reverse reaction (b) is thus fastest, whereas reverse reaction (a) should be slowest. 13.27 $E_a = 47$ kJ/mol. 13.30 (a) $k_{873} = 1.19 \times 10^{-4}\ M/s$. (b) $k_{1073} = 1.28 \times 10^{-2}\ M/s$. 13.32 $E_a = 138$ kJ/mol. 13.33 (a) The first step is unimolecular, the second is bimolecular. (b) $2NO_2Cl(g) \longrightarrow 2NO_2(g) + Cl_2(g)$ (obtained by adding the two elementary processes). (c) Cl is an intermediate; it is formed and consumed in the overall reaction. 13.35 (a) The first step is bimolecular in the forward direction, unimolecular in the reverse direction. The second step is bimolecular. (b) First step: $-\Delta[NO]/\Delta t = k[NO][Br_2]$; reverse reaction: $-\Delta[NOBr_2]/\Delta t = k[NOBr_2]$; Second step: $-\Delta[NOBr_2]/\Delta t = k[NOBr_2][NO]$. (c) The chemical reaction is just the sum of the two steps: $2NO(g) + Br_2(g) \longrightarrow 2NOBr(g)$. (d) $NOBr_2$ is the intermediate. 13.37 (a) rate $= k[H_2O_2][I^-]$; (b) $2H_2O_2(aq) \longrightarrow 2H_2O(l) + O_2(g)$; (c) $IO^-(aq)$ is the intermediate. (d) A catalyst is present at the beginning of the reaction, whereas an intermediate is formed during reaction. They *are* alike in that neither appears in the balanced equation for the process. 13.41 The method of preparation and treatment of a catalyst will affect the surface area and the condition of the surface. That is, it will affect the numbers and kinds of active site at which reaction occurs. 13.43 (a) Concentrations of reactants; temperature; addition of a catalyst. (b) Increases in concentrations of reactants result in more frequent collisions per unit time. An increase in temperature results in an increase in average kinetic energy of the reactants, so a larger fraction of collisions produces reaction. A catalyst lowers the activation energy for reaction. 13.47 (a) Doubling [A] changes [A] and rate, but not k or [B]. (b) Addition of a catalyst changes rate and k. (c) $[A]^2$ increases by fourfold, [B] decreases by fourfold, rate and k are unchanged. (d) Increase in temperature increases k and rate. 13.49 rate $= 7 \times 10^{-8}\ M/s$; $\Delta[Cl^-]/\Delta t = 1.4 \times 10^{-7}\ M/s$. 13.52 (a) rate $= k[HgCl_2][C_2O_4^{2-}]^2$; (b) $k = 7.5 \times 10^{-3}\ M^2$-s; (c) rate $= 6.0 \times 10^{-6}$ M/s. 13.53 (a) $k = 4.28 \times 10^{-4}/s$; (b) [Urea] $= 0.059\ M$; (c) $t_{1/2} = 1.62 \times 10^3$ s. 13.55 A graph of log (% oxidation) vs. time is linear. The first order rate constant $k = -2.30$ (slope) $=$ 0.23/day. 13.56 The fact that the rate is no more than first order in HBr rules out the second or third steps as rate-determining. The first step must be the slow one. (b) We can't be sure that there is not some other set of equations leading to the same product that would be consistent with the observations. 13.62 From the graph, obtain an activation energy of 65 kJ/mol.

CHAPTER 14

14.1 (a) $K = \dfrac{[SO_2]^2[O_2]}{[SO_3]}$; (b) $K = [N_2][H_2O]^2$; (c) $K = \dfrac{[NO_2]^2}{[N_2O_4]}$; (d) $K = \dfrac{[H_2O]}{[H_2]}$; (e) $K = \dfrac{[N_2]^2[H_2O]^6}{[O_3]^3[NH_3]^4}$

14.3 $K_p = \dfrac{P^2_{Cl}}{P_{Cl2}} = \dfrac{(2.97 \times 10^{-2})^2}{1.00} = 8.82 \times 10^{-4}$

14.6 $K_c = \dfrac{[SO_3]}{[SO_2][O_2]^{1/2}} = 20.4$ (a) K_c for the reaction written is just the reciprocal of K_c written above: $K_c' = 1/K_c = 1/20.4 = 4.90 \times 10^{-2}$. (b) K_c'' is just the square of $K_c = (20.4)^2 = 416$. (c) $K_p = K_c''(RT)^{\Delta n}$, where Δn is the number of moles of gas product minus the number of moles of gas reactants. In this case $\Delta n = -1$. Thus, $K_p = 416(0.08201 \times 973)^{-1} = 5.21$. 14.8 (a) $K_p = K_c(RT)^{\Delta n}$; $\Delta n = 0$, so $K_p = K_c = 0.11$. 14.10 Moles NO reacting $= 0.030$; number moles H_2 remaining is 0.020; form 0.030 moles H_2O, 0.015 moles N_2. (b) $K_c = (0.015)(0.130)^2/(0.070)^2(0.020)^2 = 129$. 14.12 First calculate the number of moles of NOBr, and NO and Br_2 present: get

0.0293 mol NOBr, 0.103 mol NO and 0.0262 mol Br_2 (a) $K_c = 6.42 \times 10^{-2}$; (b) $K_p = 1.98$; (c) Total moles gas present is 0.158; $P = 0.968$ atm. 14.14 In each case compare the reaction quotient with $K_p = P_{NH_3}^2/P_{N_2} \times P_{H_2}^3 = 4.51 \times 10^{-5}$: (a) $Q = 2.7 \times 10^{-6}$; since $Q < K_p$, reaction shifts to right to attain equilibrium. 14.16 [HBr] = 26 M. 14.18 3.07×10^{-3} mol Br. 14.20 $P_{I_2} = P_{Br_2} = 1.2 \times 10^{-2}$ atm. 14.22 $[H_2S] = [NH_3] = 1.1 \times 10^{-2}$ *M*. 14.24 (a) Shift equilibrium to right; more CO will be formed. (b) No effect on equilibrium; so long as there is *any* C(s) present, the amount is not involved in the equilibrium. (c) Equilibrium is shifted to the right, direction in which reaction is endothermic. (d) An increase in pressure causes shift in equilibrium to the left. (e) No effect. (f) Removal of CO causes a shift to the right; more CO_2 reacts with C to form CO. 14.26 (a) No effect; (b) no effect; (c) increase equilibrium constant; (d) no effect. 14.29 $K = k_f/k_r = 1.48 \times 10^{-39}$. 14.34 (a) $K = \frac{[Ag(NH_3)_2^+]}{[Ag^+][NH_3]^2}$; (b) $K = [Ag^+][CrO_4^{2-}]$; (c) $K = \frac{[H^+][NO_2^-]}{[HNO_2]}$; (d) $K = \frac{[Zn^{2+}]}{[Cu^{2+}]}$; (e) $K = \frac{[NH_4^+][OH^-]}{[NH_3]}$. 14.36 (a) At equilibrium, $[SO_2] = 0.50\,M$; $[NO_2] = 1.0\,M$; $[SO_3] = 1.5\,M$; $[NO] = 3.0\,M$. (b) $K_c = 3.0$. 14.38 $K_c = 4.4 \times 10^{-2}$; $K_p = 4.0$. 14.41 (a) At equilibrium, $[NO] = 1.78 \times 10^{-2}\,M$, $[O_2] = 2.89 \times 10^{-2}\,M$; (b) $K_c = 0.53$. 14.45 (a) To the left; (b) to the right; (c) to the left; a volume increase would produce a lower pressure, and the system will respond by a shift in equilibrium toward the side that produces more moles of gas; (d) shift to the right; (e) shift to the left. 14.46 Increasing temperature causes a decrease in the value for *K* when the reaction is exothermic in the forward direction. 14.49 $CO(g) + 2H_2(g) \longrightarrow CH_3OH(l)$; $\Delta H = -238.6 - (-110.5) = -128.1$ kJ. The process is exothermic, therefore the equilibrium constant decreases with increasing temperature. Extent of reaction will increase with increase in pressure. 14.52 $[Br] = 3.2 \times 10^{-2}\,M$; (b) fraction of original Br_2 that dissociates is 0.016. 14.56 $Q = 8.3 \times 10^{-6}$. Since $Q > K_p$, system will shift to the left to attain equilibrium. Thus a catalyst that promoted attainment of equilibrium would result in a lower CO content in the exhaust.

CHAPTER 15

15.2 The hydrate consists of $[H_3O^+]$ and $[NO_3^-]$ ions. 15.3 6.55×10^3 mol HI/cm³ soln, 0.042 mol H_2O/cm³ soln. Thus, there are 6.4 water molecules for each pair of H^+ and I^- ions. HI is highly soluble because it undergoes *reaction* with water, to form $H^+(aq)$ and $I^-(aq)$ ions. 15.5 (b) HBr; (d) H_3PO_4. 15.6 (b) Br^-; (d) S^{2-}. 15.8 (a) Base: NH_2^-; conjugate acid: NH_3; acid: H_2O; conjugate base, OH^-; (d) base: CO_3^{2-}; conjugate acid: HCO_3^-; acid: $HC_2O_4^-$; conjugate base: $C_2O_4^{2-}$. 15.10 (a) Yes; (b) no; (c) yes; (d) no. 15.12 (a) pH = 4.52, acidic; (c) pH = 2.80, acidic. 15.14 (b) $2.2 \times 10^{-6}\,M$, acidic; (d) $4.68 \times 10^{-12}\,M$, basic. 15.17 (a) $[H_I^+]/[H_2^+] = 10$; (b) 2.0; (c) 100. 15.19 1.12×10^{-3} g KOH. 15.20 (b) pH = 10.78; (d) in this case $[H^+] = 0.10 + 10^{-3}\,M$; the second source is negligible in comparison with the first. pH = 1.00. 15.23 Acidic solutions: HF, $HClO_4$, $H_2C_2O_4$, H_3PO_3. Basic solutions: KCN, CaO, NaH, KOH. 15.25 (a) $HClO_3(aq) \rightleftharpoons H^+(aq) + ClO_3^-(aq)$; (b) $HSO_4^-(aq) \rightleftharpoons H^+(aq) + SO_4^{2-}(aq)$.

$$K_a = \frac{[H^+][ClO_3^-]}{[HClO_3]}; \quad K_a = \frac{[H^+][SO_4^{2-}]}{[HSO_4^-]}$$

15.26 $K_a = 1.4 \times 10^{-4}$. 15.28 (a) $[H^+] = 5.5 \times 10^{-5}\,M$. 15.29 0.13 g benzoic acid. 15.31 (a) 0.62% ionization. 15.34 Can ignore second ionization in comparison with the first. Use successive approximations or full quadratic form. pH = 2.24. 15.36 (b) $N_3^-(aq) + H_2O(l) \rightleftharpoons HN_3(aq) + OH^-(aq)$; $Kb = [HN_3][OH^-]/[N_3^-]$. 15.38 (a) pH = 8.96; (b) pH = 9.00; (c) pH = 10.36. 15.40 11% reaction of H_2NCH_3 with water. 15.41 (a) $K_b = 2.2\ 10^{-11}$. 15.44 pH = 10.15. 15.47 (a) Basic; (d) neutral. 15.49 (a) pH = 13.30; (b) pH = 10.00. 15.51 In general, acidity increases as electronegativity increases, because the X—H bond becomes more polar. 15.52 (a) NH_4^+; (d) H_2SO_3. 15.57 (b) Lewis acid: $FeBr_3$; Lewis base: Br^-. 15.59 (a) ZnI_2 (higher charge of Zn^{2+} as compared with Li^+). (d) $Fe(NO_3)_3$ (acetate ion is a base, NO_3^- is not basic.) 15.61 KCN > KNO_2 > $KClO_4$ > NH_4Br > $HClO_4$. KCN provides the most basic anion, $HClO_4$ is the strongest acid. 15.63 (b) 1.55; (d) K_b for CNO^- is $1 \times 10^{-14}/3.5 \times 10^{-4} = 2.9 \times 10^{-11}$. Solve for $[OH^-]$ in the usual way; pOH = 5.68, pH = 8.32. 15.65 (a) Using the full quadratic form, $[H^+] = 4.7 \times 10^{-3}\,M$; 9.0% ionization. 15.67 Solution will be blue, corresponding to an excess of the conjugate base form. 15.70 (a) NH_3 can't be a proton donor in water. If there were a base strong enough to remove a proton from NH_3 that base would preferentially react with the solvent, H_2O, because H_2O is a stronger proton donor than NH_3. The rule is that one cannot have a stronger base present (in this case, NH_2^-) than the conjugate base characteristic of the solvent, OH^-. (b) pH = 12.18. 15.72 pH is 6.34. 15.73 $[HSO_4^-] = 1.8 \times 10^{-2}\,M$; $[SO_4^{2-}] = 6.8 \times 10^{-3}\,M$; $[H^+] = 3.2 \times 10^{-2}\,M$. 15.75 pH = 11.34; $pK_a = 10.15$. 15.79 (a) pH = 5.39; (b) In strongly acidic solution glycine exists as $H_3NCH_2COOH^+$; in strongly basic solution it exists as $N_2NCH_2CO_2^-$. (c) At pH 10, there is somewhat more $H_2NCH_2CO_2^-$ present than the zwitterion, $^+H_3NCH_2COO^-$; at pH 2, about two-thirds of the molecules are in the form $^+H_3NCH_2COOH$, about one third in the zwitterionic form.

CHAPTER 16

16.1 (a) Increase; formate is a moderately strong base; (c) no change. 16.2 (a) 3.44; (b) 4.01. 16.5 $pK_a = 7.80$. 16.7 (a) $H^+(aq) + C_7H_5O_2^-(aq) \rightleftharpoons HC_7H_5O_2(aq)$; (b) *K* for reaction (a) is 1.5×10^4; (c) $[HC_7H_5O_2] = 0.10\,M$; $[H^+] = [C_7H_5O_2^-] = 2.5 \times 10^{-3}\,M$. 16.11 (a) True. 16.12 (a) $[HCO_3^-]/[H_2CO_3] = 11$; (b) $[HCO_3^-]/[H_2CO_3] = 5.5$. 16.15 (a) 7.22; (c) 9.13. 16.17 (a) pH = 4.56; (b) pH = 4.49; (c) pH = 4.64. 16.19 (b) 21.0 mL NaOH soln. 16.20 (a) 0.137 *M*; (c) 0.244 *M*. 16.21 (a) pH = 0.92; (c) pH = 3.30. 16.25 (a) $K_{sp} = [Cd^{2+}][S^{2-}]$; (c) $K_{sp} = [Ce^{3+}][F^-]^3$. 16.27 (a) 1.0×10^{-8}; (c) 8.8×10^{-12}. 16.29 (b) 9.8×10^{-3} g/L. 16.31 (a) $[Ag^+] = [I^-] = 9.1 \times 10^{-9}\,M$. 16.34 pH = 8.64. 16.36 $[Ca^{2+}]/[Fe^{2+}] = 88$. 16.38 (a) $[Cd^{2+}] = 2.5 \times 10^{-2}\,M$; (b) $[Cd^{2+}] = 2.5 \times 10^{-6}\,M$. 16.40 $Q = [Mn^{2+}][S^{2-}] = 3.0 \times 10^{-20}$. This ion product is smaller than K_{sp}, so no MnS(*s*) forms. 16.42 Molar solubility of $Ce(IO_3)_3$ is $1.8 \times 10^{-3}\,M$; (b) $NaIO_3$ concentration needs to be $1.2 \times 10^{-2}\,M$. 16.45 (a) $Al(OH)_3(s) + OH^-(aq) \rightleftharpoons$

$Al(OH)_4^-(aq)$; (c) $Cr(OH)_4^-(aq) + H^+(aq) \rightleftharpoons Cr(OH)_3(s) + H_2O(l)$. 16.47 $[Cu^{2+}] = 2 \times 10^{-12}\,M$. 16.50 $K = K_{sp} \times K_f = 2 \times 10^{11}$. 16.52 $[Cd^{2+}]/[Cd(SCN)_4^{2-}] = 2.5 \times 10^{-4}$. 16.55 PbS could precipitate from the pH 1 solution, because the ion product greatly exceeds K_{sp}. However, the quantity formed is very slight. 16.58 The greatest buffer capacity is possessed by solution (a); solution (b) has the least buffer capacity. 16.63 (b) pH = 1.30; (c) pH = 12.70; (d) pH = 4.44. 16.65 $pK_a = 4.68$. 16.68 Determine pH of a solution that gives equimolar concentrations of acid and conjugate base. (a) 12.4; (b) 3.5; (c) 9.3; (d) 4.2. The $HCNO/CNO^-$ comes closest to the desired pH. 16.70 pH = 7.24. 16.72 (a) pH = 5.97 at equivalence point. 16.75 If $[Mg^{2+}]$ is to be $3.0 \times 10^{-2}\,M$, then $[C_2O_4^{2-}] = 2.9 \times 10^{-3}\,M$. Thus, $[HC_2O_4^-] = 2.7 \times 10^{-2}\,M$. From K_b for $C_2O_4^{2-}$, 1.6×10^{-10}, determine that pH = 3.24. 16.77 (b) $MgSO_3(s) + 2H^+(aq) \rightleftharpoons SO_2(g) + Mg^{2+}(aq) + H_2O(l)$. 16.79 (a) $[HS^-] = 6 \times 10^{-6}\,M$; (b) $[HS^-] = 2 \times 10^{-3}\,M$. 16.81 (a) maximum allowable pH that will keep FeS from forming is about 3.9 to 4.0; (b) $[HC_2H_3O_2]/[C_2H_3O_2^-] = 5.6$. 16.84 The added contribution to $[OH^-]$ of the water due to dissolving of $Fe(OH)_3$ is small in comparison with equilibrium value of $1 \times 10^{-7}\,M$. Thus, $[Fe^{3+}][OH^-]^3 = [Fe^{3+}][1 \times 10^{-7}]^3 = 4 \times 10^{-38}$; $[Fe^{3+}] = 4 \times 10^{-17}\,M$.

CHAPTER 17

17.2 (a) 1 mol H_2, 1 atm pressure (larger volume). 17.3 (a) False. The essential question is whether the reaction proceeds far to the right before arriving at equilibrium. The position of equilibrium depends not only on the enthalpy change, but also on the entropy change in the process. (b) True. (c) False. Spontaneity relates to the *position* of equilibrium in a process, not the rate at which that equilibrium is attained. (d) True. (e) False. Such a process *might* be spontaneous, but not necessarily so. 17.6 (a) S° is higher for Hg(g) because it occupies a much larger volume. 17.8 (a) ΔS positive; (c) ΔS positive. 17.10 (b) $\Delta S^\circ = 2(240.4) - [2(210.6) + 205.0] = -145.3$ J/mol-K; (d) $\Delta S^\circ = 90.0 - [2(60.8) + 2(205.0)] = -441.5$ J/mol-K. ΔS° is negative in (b) because the number of moles of gas is lower in the products. It is negative in (d) for the same reason; a gas is converted to a solid. 17.12 There is no inconsistency. The second law states that in a spontaneous process there is an increase in entropy in the *universe*. While there may be a decrease in entropy of the system, the decrease is more than offset by an increase in entropy in the surroundings. 17.14 The absolute entropy at 0 K would be greater than zero, because each component would be distributed throughout a larger volume than in the pure substances, which serve as the reference point for zero entropy. 17.17 ΔG° applies only when all substances are in their standard states. When this is not the case, the free energy change is represented as ΔG. 17.18 (a) $\Delta G^\circ = 2(-394.4) + 0 - [(-464.8) + 2(-137.3)] = -49.4$ kJ; (c) $\Delta G^\circ = +146$ kJ. 17.20 (a) $\Delta H^\circ = -269.3$ kJ, $\Delta G^\circ = -218.1$ kJ, $\Delta S^\circ = -171.9$ J/K. Calculate ΔG° value from the ΔH° and ΔS° values, obtain -218.1 kJ at 298 K. 17.22 (a) $\Delta H^\circ = 229.7$ kJ; $\Delta S^\circ = 156$ J/K; (b) $\Delta G^\circ(900\text{ K}) = 229.7 \times 10^3 - 900(156) = 89$ kJ. 17.25 (b) ΔG becomes more negative. 17.26 (a) $\Delta G = \Delta G^\circ + 2.30RT \log Q = -33.4 + 2.30(8.314)(298) \log [4^2/(12)2^3] = -37.8$ kJ. 17.28 (a) $K = 1.6 \times 10^{14}$. 17.31 (a) The ethanol could be used as fuel in an internal combustion engine that could be used to do various kinds of work. 17.33 (a) ΔH for the process is zero if N_2 behaves as an ideal gas. 17.36 (a) When ice melts the individual molecules that were fixed in place in the solid are free to move throughout the volume of the liquid, and to take up various positions. 17.38 (b) ΔH is positive for this solution process; ΔS is positive, ΔG is negative. 17.40 Under these conditions, $\Delta S = \Delta H/T$. Both the gas and liquid phases are in their standard states; thus the entropy changes are ΔS° values. Values of ΔS° calculated are as follows: acetone, 92.1 J/mol-K; benzene, 87.0 J/mol-K; ammonia, 97.5 J/mol-K; water, 109 J/mol-K. The ΔS° values are significantly higher for the hydrogen-bonded substances, ammonia and especially water. Rupture of hydrogen bonding on vaporization makes an extra positive contribution to ΔS°. 17.42 (c) $\Delta G = -11.6$ kJ. 17.44 $T = 646$ K. 17.47 $\Delta G = -543$ kJ. 17.49 ΔG° values for reactions $4FeO + O_2 \longrightarrow 2Fe_2O_3$, $2Fe_3O_4 + \frac{1}{2}O_2 \longrightarrow 3Fe_2O_3$ and $3FeO + \frac{1}{2}O_2 \longrightarrow Fe_3O_4$ are -461, -194 and -249 kJ, respectively. From these data it is clear that the most stable oxide is Fe_2O_3. 17.51 (a) $\Delta G = +8.75$ kJ; (b) Because ΔG is positive, work must be done on the system (blood plasma plus muscle cells) to move the K^+ ions "uphill" in terms of the concentration gradient. ΔG represents the minimum amount of work needed to transfer a mole of K^+ ions from blood plasma at $5 \times 10^{-3}\,M$ to muscle cells fluids at 0.15 M, assuming constancy of concentrations.

CHAPTER 18

18.1 (a) $Al(OH)_3(s) + OH^-(aq) \longrightarrow Al(OH)_4^-(aq)$, no oxidation or reduction. (c) $ClO^-(aq) + 4OH^-(aq) \longrightarrow ClO_3^-(aq) + 2H_2O(l) + 4e^-$; oxidation. 18.3 (a) $3F_2(g) + Br_2(g) \longrightarrow 2BrF_3(g)$; bromine is oxidized, fluorine is reduced. (c) $2MnO_4^-(aq) + 6I^-(aq) + 4H_2O(l) \longrightarrow 2MnO_2(s) + 3I_2(s) + 8OH^-(aq)$. 18.6 (a) $5H_2C_2O_4(aq) + 2MnO_4^-(aq) + 6H^+(aq) \longrightarrow 2Mn^{2+}(aq) + 10CO_2(g) + 8H_2O(l)$; (b) 0.495% by weight. 18.8 (a) $2Au(s) + 8Br^-(aq) + 3IO_3^-(aq) + 3H_2O(l) \longrightarrow 2AuBr_4^-(aq) + 3I^-(aq) + 6OH^-(aq)$; $E^\circ = 0.858 + 0.49 = 1.35$ V. (b) $2Eu^{2+}(aq) + Sn^{2+}(aq) \longrightarrow 2Eu^{3+}(aq) + Sn(s)$; $E^\circ = 0.43 - 0.14 = 0.29$ V. 18.10 (a) The two half-reactions are: $Tl^{3+}(aq) + 2e^- \longrightarrow Tl^+(aq)$; $2(Cr^{2+}(aq) \longrightarrow Cr^{3+}(aq) + e^-)$. $E^\circ = E^\circ_r + 0.41 = 1.19$ V. $E^\circ_r = 0.78$ V. This is the standard potential for reduction of Tl^{3+} to Tl^+. 18.12 (a) Reaction at anode: $5Fe^{2+}(aq) \longrightarrow 5Fe^{3+}(aq) + 5e^-$. Reaction at cathode: $8H^+(aq) + MnO_4^-(aq) + 5e^- \longrightarrow Mn^{2+}(aq) + 4H_2O(l)$. Electrons move from the Pt electrode in the iron solution to the Pt electrode in the MnO_4^- solution. Anions migrate through the salt bridge from the cathode beaker to the anode beaker. The electrode in the iron-containing beaker has a negative sign, that in the MnO_4^- beaker has a positive sign. (b) Using Table 18.1, $E^\circ = (-0.77\text{ V}) + 1.51\text{ V} = 0.74$ V. 18.14 (a) In order of increasing reduction potential: $Cu^{2+}(aq) < O_2(g) < Cr_2O_7^{2-}(aq) < Cl_2(aq) < H_2O_2(aq)$. 18.17 The criterion for spontaneity is the value of E°. (a) Positive, spontaneous; (b) positive, spontaneous; (c) negative, nonspontaneous; (d) negative, nonspontaneous. 18.18 (a) True; (b) false. The free energy change is an *extensive* property of the system; it depends on the number of moles of substance reacting. (c) False; $\Delta G^\circ = -n\mathfrak{F}E^\circ$. The quantity $n\mathfrak{F}$ has units of coulombs. (d) True. 18.21 (a) $2Fe^{2+}(aq) + S_2O_6^{2-}(aq) + 4H^+(aq) \longrightarrow 2Fe^{3+}(aq) + 2H_2SO_3(aq)$ and $2Fe^{2+}(aq) + N_2O(aq) + 2H^+(aq) \longrightarrow 2Fe^{3+}(aq) + N_2(g) + H_2O(l)$. (b) For the first reaction, $E^\circ = -0.17$ V, $\Delta G^\circ = +33$ kJ, $K = 2 \times 10^{-6}$. For the second reaction, $E° = -2.54$ V, $\Delta G° = +490$ kJ, $K = 1 \times 10^{-86}$ 18.23 (a) $E° = +0.627$ V, $K = 2 \times 10^{21}$. 18.25 Any change that shifts the equilibrium to the left will make the reaction less spontaneous, thus will lower E. (a) Lowers E; (b) lowers E; (c) no effect on E; (d) no effect on E. 18.27 $E° = -0.010$ V; $[Pb^{2+}] = 1.6 \times 10^{-8}$ M. K_{sp} for $PbSO_4 = 1.6 \times 10^{-8}$. 18.29 pH = 1.23. 18.33 (a) In discharge: $Cd(s) + NiO_2(s) + 2H_2O(l) \longrightarrow Cd(OH)_2(s) + Ni(OH)_2(s)$. In charging, the reverse reaction occurs. (b) $E° = 1.25$ V. 18.35 Overall cell emf would be reduced, because the potential for oxidation of Cd is less positive than that for oxidation of Zn. 18.36 In aqueous solution electrolysis, water is reduced in preference to Mg^{2+}. The overall reactions in the two cases are $MgCl_2(l) \longrightarrow Mg(l) + Cl_2(g)$ and $2Cl^-(aq) + 2H_2O(l) \longrightarrow Cl_2(g) + H_2(g) + 2OH^-(aq)$. 18.39 (a) $E_{min} = 1.34$ V; (b) A larger voltage than E_{min} is required to overcome cell resistance, and to cause reaction to proceed at a reasonable rate. 18.40 96,487 C/mol e^-. 18.42 3.66 g Zn. 18.44 (a) 3.83 L Cl_2. 18.49 $E° = 1.201$ V; $w_{max} = 232$ kJ per mol Sn reacting. If 0.200 mol Sn is consumed, $w_{max} = 46.4$ kJ. 18.51 No. The potential for oxidation of Co to Co^{2+}, +0.28 V, is lower than that for oxidation of iron to Fe^{2+}, 0.44 V. To afford cathodic protection, a metal must be more readily oxidized than iron. 18.53 The pH is an important factor in corrosion. Note from Equation 18.50 that increased $[H^+]$ shifts the equilibrium to the right. When $[H^+]$ is depressed, reduction of O_2 is less spontaneous, and corrosion slows down. Added amines serve the function of keeping $[H^+]$ low. 18.55 (b) $3UO^{2+}(aq) + ClO_3^-(aq) \longrightarrow 3UO_2^{2+}(aq) + Cl^-(aq)$; (g) $2H_2SeO_3(aq) + H_2S(aq) \longrightarrow 2Se(s) + 2H_2O(l) + HSO_4^-(aq) + H^+(aq)$. 18.56 (a) $ZnS(s) + 4H_2O_2(aq) \longrightarrow Zn^{2+}(aq) + SO_4^{2-}(aq) + 4H_2O(l)$. 18.58 (a) $XeO_3(s) + 9I^-(aq) + 6H^+(aq) \longrightarrow Xe(g) + 3I_3^-(aq) + 3H_2O(l)$; (b) 0.723 L Xe gas; (c) 5.70×10^{-2} mol I_3^-. 18.60 (a) Zn is oxidized to Zn^{2+}, Ag_2O is reduced to $Ag(s)$; (b) anode is negatively charged, cathode is positive; (c) $Zn(s) + Ag_2O(s) + H_2O(l) \longrightarrow Zn^{2+}(aq) + 2Ag(s) + 2OH^-(aq)$ $E° = +1.10$ V. 18.62 Determine in each case whether $E°$ is positive (spontaneous) or negative (nonspontaneous): (a) spontaneous. 18.64 $E = +0.028$ V for cell reaction $Cr^{3+}(aq)\ (1.0\,M) \longrightarrow Cr^{3+}(aq)\ (0.040\,M)$. The cathode is in the compartment containing 1.0 M Cr^{3+}. 18.67 Add to $AgSCN(s) + e^- \longrightarrow Ag(s) + SCN^-(aq)$, $E° = +0.0895$ V, the reaction $Ag(s) \longrightarrow Ag^+(aq) + e^-$, $E° = -0.799$ V, get $AgSCN(s) \longrightarrow Ag^+(aq) + SCN^-(aq)$, $E° = -0.710$ V. $K = 1.0 \times 10^{-12}$. 18.69 (a) 5.9×10^6 C required for plating; (b) 9.8×10^5 amp. 18.71 Cathode reaction: $2H_2O(l) + 2e^- \longrightarrow H_2(g) + 2OH^-(aq)$ (note that OH^- is formed); anode reaction: $2H_2O(l) \longrightarrow 4H^+(aq) + O_2(g) + 4e^-$ (note that H^+ is formed). 18.73 460 h. 18.77 Mg acts as a sacrificial anode: $Mg(s) \longrightarrow Mg^{2+}(aq) + 2e^-$. Corresponding cathode reaction is reduction of water: $2H_2O(l) + 2e^- \longrightarrow H_2(g) + 2OH^-(aq)$. Mg replaces Fe as the anode material. 18.79 Oxygen plays role of reductant at the cathode in the electrochemical corrosion of iron. Secondly, it reacts with Fe^{2+} to form "rust," the hydrated oxide of iron (III). 18.82 1.37×10^6 C.

CHAPTER 19

19.2 (a) 8 protons, 9 neutrons; (b) 42 protons, 57 neutrons; (c) 55 protons, 81 neutrons; (d) 47 protons, 68 neutrons.
19.4 (a) ${}^{181}_{72}Hf \longrightarrow {}^{0}_{-1}e + {}^{181}_{73}Ta$;
(b) ${}^{226}_{88}Ra \longrightarrow {}^{222}_{86}Rn + {}^{4}_{2}He$;
(c) ${}^{205}_{82}Pb \longrightarrow {}^{0}_{1}e + {}^{205}_{81}Tl$; (d) ${}^{179}_{74}W + {}^{0}_{-1}e \longrightarrow {}^{179}_{73}Ta$
19.6 (a) ${}^{32}_{16}S + {}^{1}_{0}n \longrightarrow {}^{1}_{1}H + {}^{32}_{15}P$; (b) ${}^{7}_{4}Be + {}^{0}_{-1}e$ (orbital electron) $\longrightarrow {}^{7}_{3}Li$;
19.8 (i) ${}^{250}_{100}Fm \longrightarrow {}^{246}_{98}Cf + {}^{4}_{2}He$;
(ii) ${}^{246}_{98}Cf \longrightarrow {}^{242}_{96}Cm + {}^{4}_{2}He$;
(iii) ${}^{242}_{96}Cm \longrightarrow {}^{238}_{94}Pu + {}^{4}_{2}He$; (iv) ${}^{238}_{94}Pu \longrightarrow {}^{234}_{92}U + {}^{4}_{2}He$
19.11 If neutron/proton ratio is too high or low as compared with belt of stability in Figure 19.1, or if atomic number exceeds 83, nucleus will be radioactive. (b) Low neutron/proton ratio. (c) Low neutron/proton ratio. (e) High atomic number. 19.14 (a) No—low neutron/proton ratio, should be a positron emitter. (b) No—low neutron/proton ratio. (c) No—high neutron/proton ratio; should be a beta emitter. (d) No—high atomic number; should be an alpha emitter. 19.17 (a) High neutron/proton ratio; beta emission lowers this ratio by, in effect, converting a neutron into a proton. 19.19 1.56 mg ${}^{66}Ga$ remaining. 19.21 $t_{1/2} = 22.6$ yr. 19.23 $N_t/N_0 = 0.971$. 19.26 $k = 1.5 \times 10^{-10}$ yr^{-1}, $t = 1.9 \times 10^9$ yr. 19.30 $k = 1.79 \times 10^{-9}\ s^{-1}$; mole fraction of ${}^{3}H$ in the sample is 4.8×10^{-13}. 19.31 (a) Binding energy per nucleon is 8.25×10^{-3} amu; $\Delta E = 1.23 \times 10^{-12}$ J; (b) $\Delta E = 1.41 \times 10^{-12}$ J; (c) $\Delta E = 1.26 \times 10^{-12}$ J. 19.33 $\Delta E = 3.09 \times 10^{12}$ J per mol ${}^{6}Li$. 19.36 (a) $\Delta E = 1.70 \times 10^{12}$ J/mol; (b) $\Delta E = 3.15 \times 10^{11}$ J/mol; (c) $\Delta E = 1.77 \times 10^{12}$ J/mol. 19.38 From Figure 19.9, note that binding energy per nucleon is greatest for nuclei of mass numbers around 50. Thus ${}^{59}_{27}Co$ should possess the greatest mass defect per nucleon. 19.39 (a) ${}^{235}_{92}U + {}^{1}_{0}n \longrightarrow {}^{160}_{62}Sm + {}^{72}_{30}Zn + 4{}^{1}_{0}n$. 19.42 In fusion the idea is to jam two nuclei together. Because nuclei have positive charges, the very short distances needed to produce nuclear reaction require an input of enormous energy to overcome the repulsive potentials of the nuclei for one another. This same requirement is not present in nuclear fission. 19.45 Alpha particles have only a very short range. Thus, they damage biological tissue only when the alpha emitter has been ingested or inhaled. Plutonium is very poisonous under these conditions. 19.48 (a) Add ${}^{36}Cl$ to water as the chloride salt. Then dissolve ordinary CCl_3COOH. After a time, distill the volatile materials away from the salt; CCl_3COOH is volatile, and will distill with water. Count radioactivity in the volatile material. If chlorine exchange has occurred, there will be radioactivity. 19.53 $N_t/N_0 = 0.11$. 19.55 (b) ${}^{48}_{20}Ca + {}^{248}_{96}Cm \longrightarrow {}^{296}_{116}X$. 19.58 The most massive radionuclides will have the highest neutron/proton ratios. Thus they are most likely to decay by beta emission, which lowers this ratio. Least massive nuclei are most likely to decay by a process that increases the neutron/proton ratio, i.e., positron emission or electron capture. 19.60 5.4×10^3 yr. 19.62 This decay represents a change of 28 mass units. There must have been a total of seven alpha emissions. These by themselves would have caused a reduction of 14 in the atomic number. Since the series as a whole represents a decrease of 10 in atomic number, there must have been four beta emissions, each of which increases the atomic number by one. 19.65 Product nuclei are (a) ${}^{78}_{35}Br$; (b) ${}^{7}_{4}Be$; (c) ${}^{32}_{15}P$. 19.68 $\Delta E = 8.2 \times 10^{-14}$ J; using $\Delta E = hc/\lambda$, calculate $\lambda = 2.4 \times 10^{-3}$ nm.

CHAPTER 20

20.2 The first member in each family generally has higher electronegativity than the other members of the family. The atoms of the first element are smaller; they do not possess vacant d orbitals of the same major quantum number as the valence shell; they have generally higher ionization energies; they form stable π bonds. 20.5 (a) Nitrogen is most electronegative; (b) potassium is most metallic; (c) potassium also forms a positive ion most readily in the gas phase. However, $E°$ is most positive for Li. (d) Nitrogen has the smallest covalent atomic radius; (e) nitrogen also most readily forms π bonds. 20.11 (a) $NiO(s) + C(s) \longrightarrow CO(g) + Ni(s)$; (b) $2C_3H_7OH(l) + 9O_2(g) \longrightarrow 6CO_2(g) + 8H_2O(g)$; (c) $NaOCH_3(s) + H_2O(l) \longrightarrow NaOH(aq) + CH_3OH(aq)$. 20.14 (a) $Mg(s) + 2H^+(aq) \longrightarrow Mg^{2+}(aq) + H_2(g)$;

(b) $2H_2O(l) \xrightarrow{\text{electr.}} 2H_2(g) + O_2(g)$;

(c) $CH_4(g) + H_2O(g) \xrightarrow{1100°C} CO(g) + 3H_2(g)$;

(d) $C(s) + H_2O(g) \xrightarrow{1000°C} CO(g) + H_2(g)$.

20.17 (a) $2Al(s) + 6H^+(aq) \longrightarrow 2Al^{3+}(aq) + 3H_2(g)$;
(b) $Mg(s) + H_2O(g) \longrightarrow MgO(s) + H_2(g)$;
(c) $MnO_2(s) + H_2(g) \longrightarrow MnO(s) + H_2O(g)$;
(d) $CaH_2(s) + H_2O(l) \longrightarrow Ca(OH)_2(aq) + 2H_2(g)$.
20.21 (a) -2; (b) -1; (c) $+2$; (d) -2; (e) -1.

20.24 (a) $2HgO(s) \xrightarrow{\Delta} 2Hg(l) + O_2(g)$;
(b) $2Cu(NO_3)_2(s) \xrightarrow{\Delta} 2CuO(s) + 4NO_2(g) + O_2(g)$;
(c) $PbS(s) + 4O_3(g) \longrightarrow PbSO_4(s) + 4O_2(g)$;
(d) $2ZnS(s) + 3O_2(g) \longrightarrow 2ZnO(s) + 2SO_2(g)$;
(e) $2K_2O_2(s) + 2CO_2(g) \longrightarrow 2K_2CO_3(s) + O_2(g)$;
(f) $2Ag(s) + O_3(g) \longrightarrow Ag_2O(s) + O_2(g)$. 20.27 (a) HNO_2, $+3$; (b) N_2H_4, -2; (c) KN_3, $-\frac{1}{3}$; (d) NaCN, -3; (e) NH_4Cl, -3; (f) N_2O, $+1$; (g) NH_3, -3; (h) H_2NOH, -1; (i) NO, $+2$; (j) HNO_3, $+5$; (k) N_2O_5, $+5$; (l) Li_3N, -3. 20.30 (a) $Mg_3N_2(s) + 6H_2O(l) \longrightarrow 3Mg(OH)_2(s) + 2NH_3(aq)$; (b) $2NO(g) + O_2(g) \longrightarrow 2NO_2(g)$; (c) $4NH_3(g) + 3O_2(g) \longrightarrow 2N_2(g) + 6H_2O(g)$. 20.34 $N_2H_5^+(aq) \longrightarrow N_2(g) + 5H^+(aq) + 4e^-$, $E° = +0.23$ V. Reduction should occur for Sn^{2+} (marginal), Cu^{2+} and Ag^+. 20.36 (a) $E° = +1.06$ V; (b) $E° = -0.88$ V. 20.41 (a) CaC_2; (b) SiC; (c) NaCN; (d) $CaCO_3$; (e) $KHCO_3$. 20.44 $\Delta H° = -393.5$ kJ (graphite), -395.4 kJ (diamond).

20.46 (a) $ZnCO_3(s) \xrightarrow{\Delta} ZnO(s) + CO_2(g)$; (c) $C_2H_4(g) + 3O_2(g) \longrightarrow 2CO_2(g) + 2H_2O(g)$; (e) $NaCN(s) + H^+(aq) \longrightarrow Na^+(aq) + HCN(g)$; (g) $NaHCO_3(s) + H^+(aq) \longrightarrow Na^+(aq) + H_2O(l) + CO_2(g)$; (i) $C(s) + H_2O(g) \longrightarrow H_2(g) + CO(g)$. 20.48 (a) The CH_3 carbon is tetrahedrally surrounded; it employs an sp^3 hybrid orbital set. The other two carbons have linear geometry about them; they employ an sp hybrid orbital set. (b) The carbon in CN^- uses an sp hybrid orbital set. (c) The carbon in CS_2 uses an sp hybrid orbital set, consistent with the linear geometry of the molecule. (d) In C_2H_6, each carbon uses an sp^3 hybrid orbital set, and has an approximately tetrahedral environment of one C—C and three C—H bonds. 20.51 (a) $Mg_3N_2(s) + 6D_2O(l) \longrightarrow 2ND_3(aq) + 3Mg(OD)_2(s)$; (b) $SO_3(g) + D_2O(l) \longrightarrow 2D^+(aq) + SO_4^{2-}(aq)$; (c) $Na_2O(s) + D_2O(l) \longrightarrow 2Na^+(aq) + 2OD^-(aq)$. 20.54 (a) Mn_2O_7 (higher oxidation state of metal); (b) SnO_2; (c) SO_3; (d) SO_2; (e) Ga_2O_3; (f) SO_2. 20.57 (a) $H_2O_2(aq) + S^{2-}(aq) \longrightarrow 2OH^-(aq) + S(s)$; (b) $SO_2(g) + 2OH^-(aq) + H_2O_2(aq) \longrightarrow SO_4^{2-}(aq) + 2H_2O(l)$. 20.61 (A) $(CH_3)_2NNH_2 + 2N_2O_4 \longrightarrow 3N_2(g) + 4H_2O(g) + 2CO_2(g)$; (B) $(CH_3)HNNH_2 + (\frac{5}{4})N_2O_4 \longrightarrow (\frac{9}{4})N_2(g) + 3H_2O(g) + CO_2(g)$. In case (A) there are 9 mols of gas per mol $(CH_3)_2NNH_2$ plus two mols N_2O_4. Total mass of reactant is 244 g. Thus, there are 0.037 mol gas/1 g reactants. In case (B) there are 0.0393 mol gas/1 g reactant. Case (B) thus provides marginally better thrust. 20.65 You need to refer to Equations 20.57, 20.56 and 20.54. Obtain 1.9×10^{11} mol NH_3/yr if ignore recirculation of NO. Counting that, about 1.3×10^{11} mol NH_3/yr, corresponding to 7×10^{10} L at 45 atm. 20.69 The empirical formula of the hydrocarbon is C_3H_4. The formula of the carbide is Mg_2C_3. 20.72 (a) $Li_3N(s) + H_2O(l) \longrightarrow 3Li^+(aq) + 3OH^-(aq) + NH_3(aq)$; (c) $3NO_2(g) + H_2O(l) \longrightarrow NO(g) + 2H^+(aq) + 2NO_3^-(aq)$; (e) $4NH_3(g) + 5O_2(g) \longrightarrow 4NO(g) + 6H_2O(g)$; (g) $H_2CO_3(aq) \longrightarrow H_2O(l) + CO_2(g)$; (i) $CS_2(g) + 3O_2(g) \longrightarrow CO_2(g) + 2SO_2(g)$; (k) $2Na(s) + 2H_2O(l) \longrightarrow 2Na^+(aq) + 2OH^-(aq) + H_2(g)$; (m) $LiH(s) + H_2O(l) \longrightarrow Li^+(aq) + OH^-(aq) + H_2(g)$.

CHAPTER 21

21.1 (a) IO_3^- $(+5)$; (b) $HBrO_3$ $(+5)$; (c) BrF_3 $(+3)$; (d) NaClO $(+1)$; (e) HIO_2 $(+3)$; (f) KI_3 $(-\frac{1}{3})$; (g) $CaBr_2$ (-1); (h) XeO_3 $(+6)$; (i) $XeOF_4$ $(+6)$. 21.5 $I_2 < F_2 < Br_2 < Cl_2$. The lower than expected value for F_2 can be related to nonbonding repulsions between lone pairs on the two fluorines. 21.7 (a) Linear; (b) square-planar. 21.10 (a) $PBr_5(l) + 4H_2O(l) \longrightarrow H_3PO_4(aq) + 5H^+(aq) + 5Br^-(aq)$; (b) $IF_5(l) + 3H_2O(l) \longrightarrow H^+(aq) + IO_3^-(aq) + 5HF(aq)$; (c) $SiBr_4(l) + 4H_2O(l) \longrightarrow Si(OH)_4(s) + 4H^+(aq) + 4Br^-(aq)$. 21.12 (a) $Br_2(l) + 2OH^-(aq) \longrightarrow BrO^-(aq) + Br^-(aq) + H_2O(l)$; (c) $Br_2(l) + H_2O_2(aq) \longrightarrow 2Br^-(aq) + O_2(g) + 2H^+(aq)$; (e) $AlBr_3(s) + 3H_2O(l) \longrightarrow Al(OH)_3(s) + 3HBr(g)$. 21.15 Use the procedure described in Section 20.5: (a) $E° = +1.64$ V; (b) use the result from part (a): $E° = +1.57$ V. 21.20 (a) potassium thiosulfate; (b) aluminum sulfide; (c) sodium hydrogen selenite; (d) selenium hexafluoride (or selenium (VI) fluoride); (e) hydrogen selenide; (f) iron pyrite (iron persulfide); (g) sodium hydrogen sulfate (or sodium bisulfate); (h) sodium selenate. 21.25 (a) $SeO_2(s) + H_2O(l) \longrightarrow H_2SeO_3(aq) \rightleftharpoons H^+(aq) + HSeO_3^-(aq)$; (b) $ZnS(s) + 2H^+(aq) \longrightarrow Zn^{2+}(aq) + H_2S(g)$; (c) $8SO_3^{2-}(aq) + S_8(s) \longrightarrow 8S_2O_3^{2-}(aq)$;
(d) $Se(s) + 2H_2SO_4(l) \xrightarrow{\Delta} SeO_2(g) + 2SO_2(g) + 2H_2O(g)$;
(e) $SO_3(g) + H_2SO_4(l) \longrightarrow H_2S_2O_7(l)$;
(f) $H_2SeO_3(aq) + H_2O_2(aq) \longrightarrow 2H^+(aq) + SeO_4^{2-}(aq) + H_2O(l)$.
21.27 (a) $2MnO_4^-(aq) + 5H_2SO_3(aq) \longrightarrow 2MnSO_4(s) + 5SO_4^{2-}(aq) + 3H_2O(l) + 4H^+(aq)$; (b) $Cr_2O_7^{2-}(aq) + 3H_2SO_3(aq) + 2H^+(aq) \longrightarrow 2Cr^{3+}(aq) + 3SO_4^{2-}(aq) + 4H_2O(l)$; (c) $Hg_2^{2+}(aq) + H_2SO_3(aq) + H_2O(l) \longrightarrow 2Hg(l) + SO_4^{2-}(aq) + 4H^+(aq)$. 21.29 3.5 g FeS. 21.33 (a) Sodium phosphide; (b) phosphorous acid; (c) arsenic trifluoride; (d) phosphorus (V) oxide (sometimes also called diphosphorus pentoxide, because the empirical formula is P_2O_5); (e) arsenic acid.

21.36 (a)

(b) In PCl_4^+ the phosphorus employs an sp^3 hybrid orbital set; in PCl_6^-, an sp^3d^2 set. (c) The ionic form is stabilized in the solid state by the lattice energy. 21.40 Use the plane triangle formula $a^2 = b^2 + c^2 - 2bc\text{Cos}A$. Calculate P—P distances of 0.295 nm for P_4O_6, 0.283 nm for P_4O_{10}. The shorter distance in the latter case is due to both a sharper P—O—P angle and a shorter P—O distance, which is related to the higher effective nuclear charge on the phosphorus of higher oxidation state. 21.41 (a) SiO_2 (+4); (b) $GeCl_4$ (+4); (c) $NaBH_4$ (+3); (d) H_3BO_3 (+3); (e) $SnCl_2$ (+2). 21.43 In C_2H_4 and C_2H_2 the carbons are joined by π bonds as well as the C—C σ bond. Silicon does not form Si—Si π bonds of comparable stability, so compounds involving such bonds are unstable relative to other bonding arrangements. 21.45 (a) Carbon; (b) lead; (c) tin; (d) carbon; (e) carbon; (f) silicon. 21.47 (a) $GeCl_4(l) + Ge(s) \longrightarrow 2GeCl_2(l)$. (b) In $GeCl_2$ there are six electrons about Ge. According to the VSEPR model the three pairs are in a trigonal plane. The molecule should have a bent shape. 21.52 KSi_3AlO_8. 21.55 (a) SiO_4^{4-}; (b) SiO_3^{2-}; (c) SiO_3^{2-}. 21.58 (a) Tetraboric acid; (b) carborundum, or silicon carbide; (c) metaphosphoric acid; (d) xenon difluoride (or xenon (II) fluoride); (e) sodium sulfide; (f) potassium chlorate. 21.60 (b) $H_2SO_3(aq) + Br_2(l) + H_2O(l) \longrightarrow HSO_4^-(aq) + 2Br^-(aq) + 3H^+(aq)$; (c) $2K^+(aq) + 2OH^-(aq) + Br_2(l) + 5ClO^-(aq) \longrightarrow 5Cl^-(aq) + 2KBrO_3(s) + H_2O(l)$. 21.66 $\Delta H^\circ = 2\,\Delta H_f^\circ(P_2) - \Delta H_f^\circ(P_4) = 2(144.3\text{ kJ}) - 58.9\text{ kJ} = +229.7\text{ kJ}$. 21.70 (a) Assuming S_8 as the product: $8SO_2(g) + 16H_2S(g) \longrightarrow 3S_8(s) + 16H_2O(g)$; (b) 5.1×10^4 L H_2S; (c) 9.6×10^4 g S. (About 200 pounds S per ton of coal combusted; however, keep in mind that two thirds of this is presumably obtained also from coal.) 21.72 Add: $H_2(g)$ + X(std state) $\longrightarrow H_2X(g), \Delta H_f^\circ$; $2H(g) \longrightarrow H_2(g), \Delta H_f^\circ(H—H)$; $X(g) \longrightarrow$ X(std state), ΔH_3. Add to give $2H(g) + X(g) \longrightarrow H_2X(g)$, $\Delta H = \Delta H_f^\circ + \Delta H_f^\circ(H—H) + \Delta H_3$. Get −926, −733, −633 and −533 kJ for H_2X = H_2O, H_2S, H_2Se and H_2Te, respectively. The average bond dissociation energies in each case are just half these values, with positive signs.

CHAPTER 22

22.2 The two ions must have the same charge if it is to be a simple substitution, with no change in overall formula type. Secondly, the ionic radii must be closely similar, within about 15%. 22.4 The potassium ethyl xanthate molecule has a polar, ionic end and a relatively nonpolar organic end. The polar end interacts with the surface of the mineral through ion-ion electrostatic interactions. The nonpolar end protrudes into the solvent. These nonpolar ends tend to form a layer on bubble surfaces. 22.5 (a) $ZnCO_3(s) \longrightarrow ZnO(s) + CO_2(g)$; (c) $ZnS(s) + 2O_2(g) \longrightarrow Zn^{2+}(aq)^+ + SO_4^{2+}(aq)$; (e) $2UO_2(s) + 6CO_3^{2-}(aq) + O_2(g) + 2H_2O(l) \longrightarrow 2[UO_2(CO_3)_3]^{4-}(aq) + 4OH^-(aq)$. 22.6 580 g $Ni^{2+}(aq)$. 22.8 (a) $2HSeO_3(aq) + 2SO_2(aq) + H_2O(l) \longrightarrow Se(s) + 2SO_4^{2-}(aq) + 4H^+(aq)$; (c) $K(l) + TiCl_4(g) \longrightarrow Ti(s) + 4KCl(s)$; (e) assuming product gas is CO: $2Cu_2O(l) + CH_4(g) \longrightarrow 2H_2O(g) + 4Cu(l)$. $2H_2O(g) + 4Cu(l)$. 22.9 (a) $CaO(l) + CaAl_2O_3(l) \longrightarrow (AlO_2)_2(l)$. 22.10 Li_2O, MgO, Fe_2O_3, TiO_2, Re_2O_7. The oxides grow increasingly acidic with increase in oxidation number of the metal. 22.12 Purify electrochemically, using a soluble cobalt salt such as $CoSO_4 \cdot 7H_2O$ as electrolyte. Reduction of H_2O does not occur because of overvoltage. 22.14 The major reducing agent is CO, formed by partial oxidation of the coke (C) with which the furnace is charged: $Fe_2O_3(s) + 3CO(g) \longrightarrow 2Fe(l) + 3CO_2(g)$, $Fe_3O_4(s) + 4CO(g) \longrightarrow 3Fe(l) + 4CO_2(g)$. 22.15 Total SO_2 is about 4×10^3 kg. 22.17 In molten metals the metal atoms continue to be bonded to one another by metallic type bonds. In a slag, on the other hand, the bonding is that of a molten ionic material. The two types of materials are so different that they form immiscible layers. 22.19 (a) $\Delta G^\circ = -245$ kJ. 22.21 $SnO_2(s) + C(s) \longrightarrow Sn(l) + CO_2(g)$. Electrochemically, electrode reactions are $Sn(s) \longrightarrow Sn^{2+}(aq) + 2e^-$ at the anode, and the reverse of this reaction at the cathode. 22.24 A most important consideration is that molten iron itself not be oxidized too readily. On the other hand, the free energies of formation of the oxides we want to get rid of must be large and negative to permit nearly quantitative removal. 22.26 Sodium is metallic; each atom is bonded to many others. When the metal lattice is distorted, many bonds remain intact. In NaCl, the ionic forces are strong and the ions are arrayed in very regular arrays. The ionic forces tend to be broken along certain cleavage planes in the solid, and the substance does not tolerate much distortion before cleaving. 22.29 In the electron sea model the electrons move about in the metallic lattice while the atoms remain more or less fixed in position. Under the influence of an applied potential the electrons are free to move throughout the structure, giving rise to conductivity. 22.31 The variation in densities reflects shorter metal-metal bond distances. These shorter distances suggest in turn that the metal-metal bonding is increasing in the series. It appears that all the valence level electrons in these elements are involved in metallic bonding. 22.34 (a) Most likely is a simple acid-base reaction: $MoO_3(g) + MnO(l) \longrightarrow MnMoO_4(l)$; (c) $3Mn_3O_4(s) + 8Al(l) \longrightarrow 4Al_2O_3(s) + 9Mn(l)$. 22.35 E° values for oxidation of either Se or Te are negative. Thus, both elements accumulate as the free elements in the so-called anode slime, along with the noble metals. 22.38 (a) $NiO(s) + 2H^+(aq) \longrightarrow Ni^{2+}(aq) + H_2O(l)$; (c) $TiO_2(s) + C(s) + 2Cl_2(g) \xrightarrow{\Delta} TiCl_4(g) + CO_2(g)$. 22.39 3.1×10^5 kwh. 22.41 The carbonates of Ca^{2+}, Fe^{2+} and Mn^{2+} are insoluble in water. On contact with CO_2-saturated water the following reaction occurs: $MCO_3(s) + H_2O(l) + CO_2(aq) \longrightarrow M^{2+}(aq) + 2HCO_3^-(aq)$. 22.44 White tin is more metallic in character; higher conductivity, larger Sn—Sn distance (3.02 Å) than in grey tin (2.81 Å). 22.47 To have a ratio of valence electrons to atoms of 3 to 2 in the beta phases, Cu must be considered to provide one electron, Zn to provide two, Al three and Sn four. Similar results are found for the other phases.

CHAPTER 23

23.2 We can deduce that the 5*s* and 4*d* orbitals must be close in energy, because the 5*s* orbital occupancy varies between $5s^1$ and $5s^2$ in the series. 23.4 Only ionization energy is characteristic solely of isolated atoms. The electrode potential can be broken into a series of steps: vaporization of the metal, ionization to the +2 state, then solvation of the metal. Thus, in part it does depend on characteristics of the isolated atom. 23.7 (a) $NbCl_5$; (b) WCl_6; (c) $CoCl_3$; (d) $CdCl_2$; (e) Re_2Cl_7; (f) $HfCl_4$. 23.9 (a) $[Kr]4d^4$; (c) $[Kr]4d^8$. 23.10 (a) $2NH_4VO_3(s) \longrightarrow V_2O_5(s) + 2NH_3(g) + H_2O(g)$; (c) $V_2O_5(s) + 2Mg(s) + 10H^+(aq) \longrightarrow 2V^{3+}(aq) + 2Mg^{2+}(aq) + 5H_2O(l)$. 23.12 (a) Hydration causes E° for reduction to be more negative; (b) formation of a stable complex would cause E° for reduction to be more negative. 23.14 Copper(I) iodide is insoluble. When $Cu^{2+}(aq)$ is reduced in the presence of I^-, CuI(*s*) is formed: $Cu^{2+}(aq) + I^-(aq) + e^- \longrightarrow CuI(s)$. The reducing agent is iodide ion, forming I_2.

The reaction proceeds because the $E°$ values lead to a large equilibrium constant for the overall reaction, and because of the insolubility of CuI. 23.16 Hydrolysis of the $Fe(H_2O)_6^{3+}$ leads to a pH of 2.59. 23.18 Condensation reactions similar to those described for phosphates (page 705): $VO_4^{3-}(aq) + H^+(aq) \rightleftharpoons HVO_4^{2-}(aq)$; $2HVO_4^{2-}(aq) \rightleftharpoons V_2O_7^{4-}(aq) + H_2O(l)$; $V_2O_7^{4-}(aq) + H^+(aq) \rightleftharpoons HV_2O_7^{3-}(aq)$; $HV_2O_7^{3-}(aq) + HVO_4^{2-}(aq) \rightleftharpoons V_3O_{10}^{5-}(aq) + H_2O(l)$. 23.21 Oxides of metals in low oxidation states behave as bases; those of metals in high oxidation states behave as acids. (a) Basic; (b) acidic; (c) basic; (d) acidic. 23.23 (a) Octahedral; (b) tetrahedral; (c) octahedral; (d) octahedral. In a ferromagnet, the magnetic moments on the various sites interact with one another to form a much larger magnetic moment throughout the solid. Because of the interactions, the magnetic moment within the solid remains when the field is turned off. 23.31 (a) Mn_2O_7 (higher oxidation state of metal); (c) $FeCl_3$ (higher oxidation state of metal).

CHAPTER 24

24.1 (a) Four; (c) six. 24.2 (a) +4; (c) +2. 24.6 (a) Triamminetrinitrocobalt(III); (c) potassium tetrachloronickelate(II); (e) potassium pentachlorothiosulfatoiridate(IV). 24.8 $[Pt(NH_3)_4Cl_2]Cl_2 - Sr(NO_3)_2$; $K_3[Cr(C_2O_4)_3] - ScBr_3$; $[Co(NH_3)_4Cl_2]Cl - CsCl$. 24.9 (b) $[Co(NH_3)_4(N_3)_2]F$; (d) $[Co(en)_2Cl_2]Cl$.

24.11 (a)

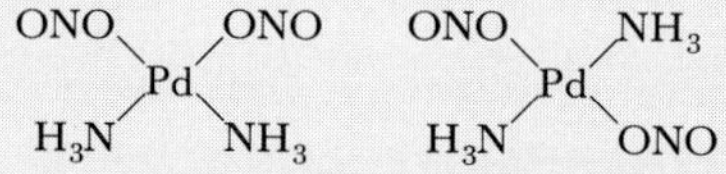

ONO, ONO / Pd / H_3N, NH_3 — *cis*

ONO, NH_3 / Pd / H_3N, ONO — *trans*

(b) $[Pd(NH_3)_2(ONO)_2]$, $[Pd(NH_3)_2(NO_2)_2]$

(c) $[V(N\frown N)_2Cl_2]^+$ mirror-image pair (N, N, N, N, Cl, Cl around V) — mirror plane

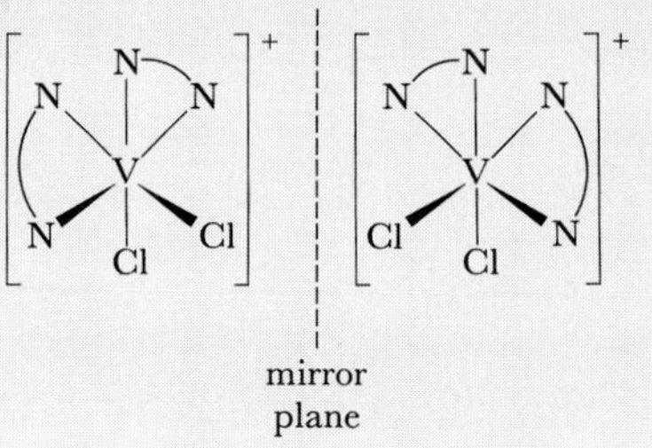

(d) $[Co(NH_3)_4(C_2O_4)]Cl_2$, $[Co(NH_3)_4Cl_2]C_2O_4$. 24.13 (a) Coordination sphere isomerism. 24.15 The observed isomerism could be due to linkage isomers, or to geometrical isomerism. It could not be due to coordination sphere isomerism, nor to optical isomerism. 24.18 Cobalt(III) complexes are generally inert. Thus the ions that form precipitates are outside the coordination sphere. Red complex is $[Co(NH_3)_5SO_4]Br$, pentaamminesulfatocobalt(III) bromide; violet complex is $[Co(NH_3)_5Br]SO_4$, pentaamminebromocobalt(III) sulfate. 24.20 $[Pt(NH_3)_4][CuCl_4]$; tetraammineplatinum(II) tetrachlorocuprate(II). $[Cu(NH_3)_4][PtCl_4]$; tetraamminecopper(II) tetrachloroplatinate(II). $[PtCl(NH_3)_3][CuCl_3(NH_3)]$ and $[CuCl(NH_3)_3][PtCl_3(NH_3)]$ are also possible, but it is doubtful whether such compounds could be prepared. 24.21 Blue to blue-violet. 24.22 (a) Seven; (c) three. 24.24 For emerald, λ_{max} = 650 nm or thereabouts; for ruby, about 530 nm; for amethyst, which is violet-colored, about 570 nm. 24.27 (a) Low-spin; (b) low-spin; (c) high-spin. 24.29 Use experimental procedure of Figure 23.8, using a weighed sample of known magnetism for calibration purposes. Complex is tetrahedral, d^7 (see Figure 24.32). 24.31 (a) No 4*d* electrons; (b) octahedral d^4: [1 |] [1 | 1 | 1] (c) octahedral d^6: [|] [↑↓ | ↑↓ | ↑↓]

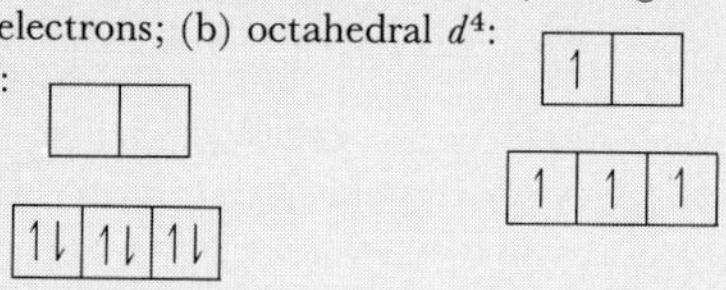

24.32 (a) One 5*d* electron; 0.4Δ; (c) Ir^{3+} has a $5d^6$ configuration. CFSE = 6 × 0.4Δ = 2.4Δ. 24.33 Iodide is a sufficiently good reducing agent to reduce Fe^{3+} to Fe^{2+}, with formation of I_2. Doesn't occur with Br^- or Cl^- because of their greater electronegativity, i.e., more negative potentials for oxidation of X^-. 24.37 The reaction that occurs produces an ion of higher charge: $[Co(NH_3)_4Br_2]^+(aq) + H_2O(l) \longrightarrow [Co(NH_3)_4(H_2O)Br]^{2+}(aq) + Br^-(aq)$. 24.40 Cobalt is present as Co^{3+}, thus the crystal field splitting is greater than for Mn^{2+}. Cobalt is d^6 and low-spin, manganese is d^5 and high-spin. 24.42 The absorption spectra of $[VF_6]^{3-}$, $[V(NH_3)_6]^{3+}$ and $[V(NO_2)_6]^{3-}$ should exhibit a trend toward absorption at shorter wavelength (higher energy). 24.44 (a) orange-yellow; (c) orange-red. 24.46 (a) $AgCl(s) + 2NH_3(aq) \longrightarrow [Ag(NH_3)_2]^+(aq) + Cl^-(aq)$; (b) $[Cr(en)_2Cl_2]^+(aq) + H_2O(l) \longrightarrow [Cr(en)_2(H_2O)Cl]^{2+}(aq) + Cl^-(aq)$ (The remaining Cl may be similarly replaced); (c) $Zn(OH)_2(s) + 4NH_3(aq) \longrightarrow [Zn(NH_3)_4]^{2+}(aq) + 2OH^-(aq)$; (d) $Co^{2+}(aq) + 4Cl^-(aq) \longrightarrow [CoCl_4]^{2-}(aq)$. 24.48 Longer wavelength absorption corresponds to a smaller splitting of the *d* orbital energies. (a) $[CoF_6]^{4-}$, because F^- is a weaker field ligand than CN^-. 24.50 See Figure 24.28. Application of pressure would result in shorter metal ion-oxide distances, thus would increase the ligand field splitting. Absorption would shift to higher energy.

CHAPTER 25

25.1 Alkane, C_6H_{14}; alkene, C_6H_{12}; alkyne, C_6H_{10}; aromatic, C_6H_6.

25.4 (b) $CH_3CH_2CH{=}CHCH_3$, $CH_3CH_2CH_2CH{=}CH_2$, $CH_3CH_2C(CH_3){=}CH_2$, $CH_3CH(CH_3)CH{=}CH_2$, $CH_3C(CH_3){=}CHCH_3$, cyclopentane with CH_3 (H_2C, CH_2, CH_2, H_2C, CH_2 ring), methylcyclobutane ($H_2C{-}C{-}CH_3$ / $H_2C{-}CH_2$), ethylcyclopropane (H, CH_2CH_3 on C; $CH_2{-}CH_2$)

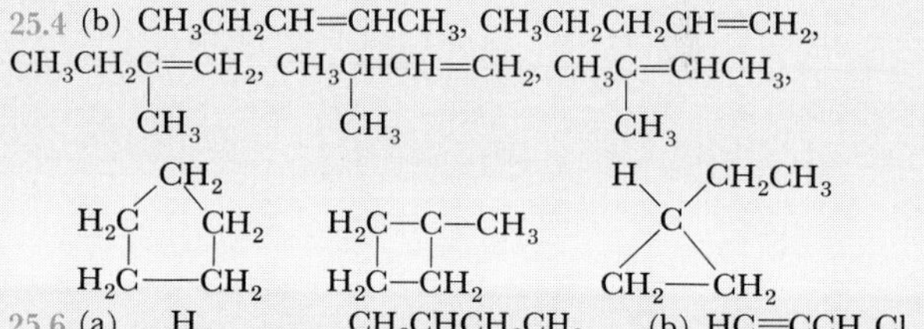

25.6 (a) H and H_3C on C=CH–$CH_2CHCH_2CH_3$ with CH_3 branch; (b) $HC{\equiv}CCH_2Cl$

(c) 1,2-dichlorobenzene (benzene ring with Cl, Cl)

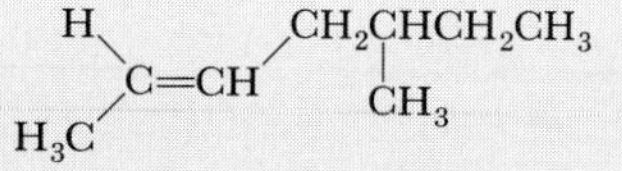

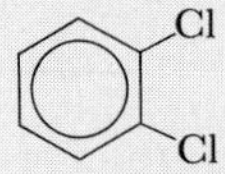

25.8 (a) 3,4-dimethylheptane; (b) *cis*-6-methyl-3-octene; (c) *para*-dibromobenzene; (d) 4,4-dimethyl-1-hexyne; (e) 2,4-dibromopentane; (f) 2,5-octadiene; (g) 2-phenyl-3,5-dichlorohexane. 25.9 (a) No;

(b) CH_3, CH_3 / C=C / Cl, Cl — *cis*; Cl, CH_3 / C=C / H_3C, Cl — *trans*

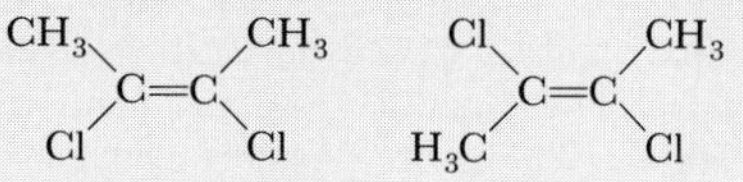

(c) no; (d) no. 25.13 Assuming that each component has the same effective octane number in the mixture, octane number = 72.5. 25.16 (a) $CH_3CH_2CH_2CH_2CH_3(g) + Cl_2(g) \longrightarrow CH_3CH_2CH_2CHClCH_3(g) + HCl(g)$; (b) $2CH_3CH{=}CH_2(g) + 9O_2(g) \longrightarrow 6CO_2(g) + 6H_2O(g)$.
25.18 (a) $CH_2{=}CHCH_2CH_3 + H_2 \longrightarrow CH_3CH_2CH_2CH_3$; (b) $CH_3CH{=}CHCH_3 + H_2O \xrightarrow{H_2SO_4} CH_3CH(OH)CH_2CH_3$.
25.19 (a) $CH_3\underset{\underset{\displaystyle Cl}{|}}{C}HCH_3$ (b) $(CH_3)_2\underset{\underset{\displaystyle Cl}{|}}{C}CH_2CH_3$

(c) $CH_3\overset{\overset{\displaystyle Cl}{|}}{\underset{\underset{\displaystyle Cl}{|}}{C}}CH_3$ or $CH_3\underset{\underset{\displaystyle Cl}{|}}{C}{=}CH_2$

25.21 (b) $CH_3CH_2CI{=}CH_2$ or $CH_3CH_2CI_2CH_3$. 25.23 (a) $\Delta H° = -1427.7$ kJ. 25.26 (a) Ketone; (b) carboxylic acid; (c) alcohol; (d) ester; (e) aldehyde; (f) ketone; (g) amide.

25.29 (a) $CH_3\underset{\underset{\displaystyle CH_3}{|}}{C}H\overset{\overset{\displaystyle O}{\|}}{C}{-}H$ (b) $CH_3CH_2\overset{\overset{\displaystyle O}{\|}}{C}{-}H$ or

$CH_3CH_2\overset{\overset{\displaystyle O}{\|}}{C}{-}OH$ (c) No possibility for mild oxidation here. (d) Oxidation here would probably go all the way to CO_2 and H_2O, although possibly CH_3COOH could be formed.

(e) $CH_3\overset{\overset{\displaystyle O}{\|}}{C}{-}OH$

25.30 (a) $CH_3\overset{\overset{\displaystyle O}{\|}}{C}{-}O\overset{\overset{\displaystyle CH_3}{|}}{\underset{\underset{\displaystyle CH_3}{|}}{C}}H$ (b) $CH_3\overset{\overset{\displaystyle O}{\|}}{C}{-}OCH_2CH_2CH_3$

(c) $H\overset{\overset{\displaystyle O}{\|}}{C}{-}OCH_2CH_3$

25.34 (a) $CH_3CH_2\overset{\overset{\displaystyle O}{\|}}{C}{-}H$ (b) $CH_3CH_2CH_2\overset{\overset{\displaystyle O}{\|}}{C}CH_3$

(c) $CH_3\underset{\underset{\displaystyle CH_3}{|}}{C}H\overset{\overset{\displaystyle O}{\|}}{C}CH_3$ (d) $CH_3CH_2\underset{\underset{\displaystyle CH_3}{|}}{C}H\overset{\overset{\displaystyle O}{\|}}{C}H$

25.38 (a) $nCH_2CHCl \longrightarrow {-}\!(CH_2CHCl){-}_n$
25.41 $(n+1)H_2NCH_2COOH \longrightarrow$

$H_2N{-}(CH_2\overset{\overset{\displaystyle O}{/\!/}}{C}{-}NH){-}_nCH_2COOH + nH_2O$
25.44 (a)$C_{20}H_{42}$;(b)$C_{18}H_{36}$;(c)$C_{12}H_{22}$. 25.48 (a)$CH_3C{\equiv}CH + HCl \longrightarrow CH_3CCl{=}CH_2$; $CH_3CCl{=}CH_2 + HBr \longrightarrow CH_3CClBrCH_3$; (b) $CH_2{=}CH_2 + Cl_2 \longrightarrow CH_2Cl{-}CH_2Cl$; (c) water is eliminated between the alcohol and acid functions to form ester linkages along a chain. 25.51 Cyclohexyne would not be stable because of the strain in bond angle about the alkynyl carbons, which would preferentially have 180° bond angles. 25.53 (a) $CH_3OCH_2CH_3$; (c) $CH_3C(O)CH_3$; (e) $CH_3CH_2CH_2CH_2CH_2OH$. 25.55 Empirical formula is C_2H_4O, with formula mass 44. From the gas density data, calculate molecular weight of 43.8 g/mol. Possibly acetaldehyde, CH_3CHO, which could be oxidized to acetic acid, CH_3COOH.

CHAPTER 26

26.2 All growth processes, which require synthesis of new cell materials, require energy. For example, transpiration of moisture in plants; in animals, action of involuntary muscles such as the heart, maintenance of uniform body temperature, synthesis of new cells, etc. 26.5 36 g O_2, corresponding to 25.4 L. 26.8 (a) An alpha amino acid contains an NH_2 function on the carbon adjacent to the carboxylic acid function. (b) In forming a protein, amino acids undergo a condensation reaction between the amino group of one molecule and the carboxylic acid group of another, to form the so-called amide link. 26.10 Two dipeptides are possible, $H_2NCH_2CONHCH(CH(CH_3)_2)COOH$ (glycylvaline), and $H_2NCH(CH(CH_3)_2)CONHCH_2COOH$ (valinylglycine). 26.13 Alanine, serine. 26.14 In the basic solution as the anion, $H_2NCH_2CO_2^-$; in acidic solution as the cation, $H_3NCH_2COOH^+$. 26.17 In glycine the α carbon atom has two hydrogens. It is therefore not chiral in the sense illustrated in Figure 26.4. 26.20 The biochemical properties are determined by all the aspects of protein configuration. The term *denaturation* refers to the loss of biochemical activity as a result of change in protein structure, by heating, change in solvent, addition of electrolytes or by other means. 26.23 1 – (d); 2 – (a); 3 – (e); 4 – (c); 5 – (b). 26.25 These enzymes are peptidases; their function is to hydrolyze the amide functions along the peptide chain. 26.27 Glucose exists in solution as a cyclic structure in which the aldehyde function on carbon 1 reacts with the OH group of carbon 5 to form what is called a hemiacetal. In the hemiacetal form carbon 1 carries an OH group. In α-glucose this OH group is on the same side of the ring as an OH group on the adjacent carbon 2 (illustrated in Equation 26.7). In the β form the OH groups are on opposite sides.

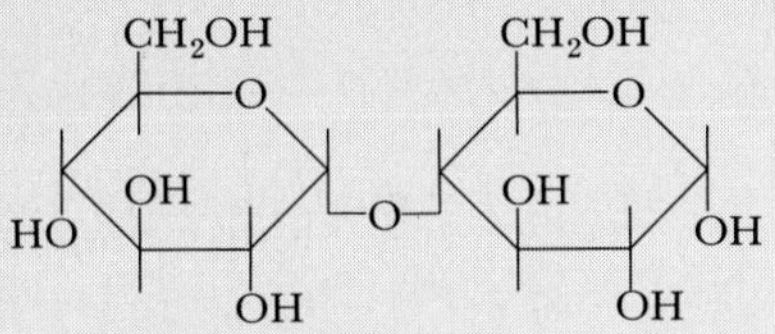

α-linkage

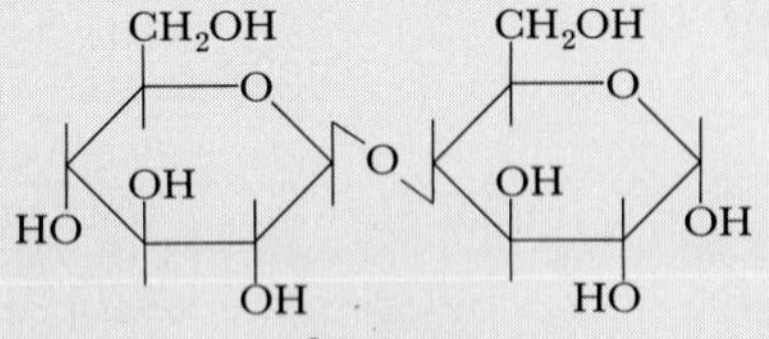

β-linkage

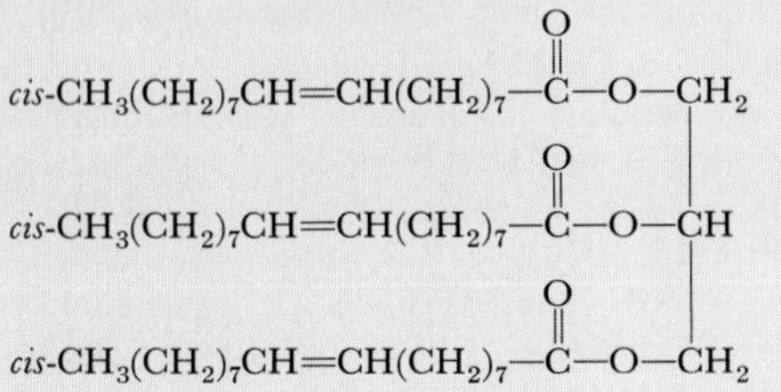

$$cis\text{-}CH_3(CH_2)_7CH{=}CH(CH_2)_7{-}\overset{\overset{\displaystyle O}{\|}}{C}{-}O{-}CH_2$$
$$cis\text{-}CH_3(CH_2)_7CH{=}CH(CH_2)_7{-}\overset{\overset{\displaystyle O}{\|}}{C}{-}O{-}CH$$
$$cis\text{-}CH_3(CH_2)_7CH{=}CH(CH_2)_7{-}\overset{\overset{\displaystyle O}{\|}}{C}{-}O{-}CH_2$$

26.34 $C_3H_5(O\overset{\overset{\displaystyle O}{\|}}{C}C_{13}H_{27})_3 + 3Na^+(aq) + 3OH^-(aq) \longrightarrow$

$$3C_{13}H_{27}\overset{\overset{\displaystyle O}{\|}}{C}{-}O^-(aq) + 3Na^+(aq) + C_3H_5(OH)_3$$

26.37 A nucleotide consists of nitrogen-containing aromatic compound, a sugar in the furanose (5-membered) ring form, and a phosphoric acid molecule. The structure of deoxycytidine monophosphate is:

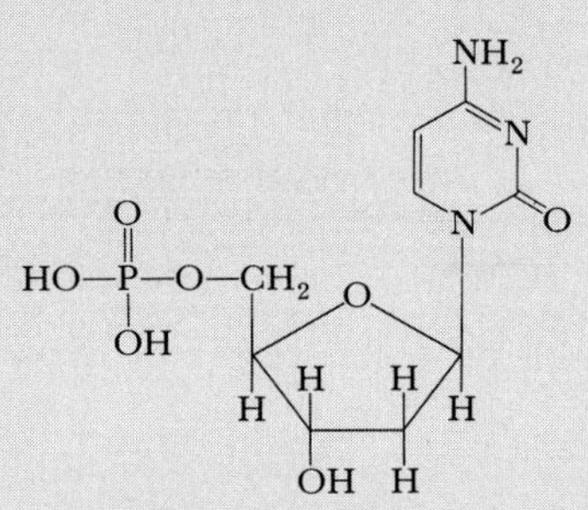

NH_2
N
O
N O
HO—P—O—CH_2 O
OH
H H
H H
OH H

26.39 —A—C—T—C—G—A—
—T—G—A—G—C—T— ⟵ Complementary strand

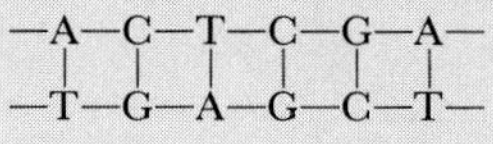

26.42 (a) None; (b) the carbon bearing the secondary OH has four different groups attached, and is thus chiral. (c) The carbon bearing the $-NH_2$ group and the carbon bearing the CH_3 group are both chiral. 26.46 glu-cys-gly is the only possible structure. 26.48 (a) An enzyme is thought to sometimes work by providing a kind of template on which the reactants can come together in a certain way to facilitate reaction. (b) A *substrate* is any substance that undergoes the reaction catalyzed by a particular enzyme. Enzyme *inhibition* occurs when the active site is blocked by some reagent which, while not itself a substrate, can effectively inactivate the active site. 26.51 The reaction is $2NH_2CH_2COOH(aq) \longrightarrow NH_2CH_2CONHCH_2COOH(aq) + H_2O(l)$; $\Delta G = (-488) + (-285.5) - 2(369) = -35.8$ kJ. 26.53 The fact that the rate doubles with a doubling of concentration of the sugar tells us that the fraction of enzyme tied up in an enzyme-substrate complex is small. Innositol is behaving like a competitor with sucrose for binding in the active site. 26.58 $[AMPOH^-]/[AMPO^{2-}] = 0.65$ when pH = 7.40.

Index

PHYSICAL AND CHEMICAL CONSTANTS

Atomic mass unit	1 amu = 1.66057×10^{-27} kg 6.022045×10^{23} amu = 1 g
Avogadro's number	$N = 6.022045 \times 10^{23}$/mol
Boltzmann's constant	$k = 1.38066 \times 10^{-23}$ J/K
Electron rest mass	$m_e = 5.48580 \times 10^{-4}$ amu $= 9.10953 \times 10^{-28}$ g
Electronic charge	$e = 1.6022 \times 10^{-19}$ coul
Faraday's constant	$\mathfrak{F} = Ne = 9.6485 \times 10^{4}$ coul/mol
Gas constant	$R = Nk = 8.3144$ J/K-mol $= 0.082057$ L-atm/K-mol
Neutron rest mass	$m_n = 1.00866$ amu $= 1.67495 \times 10^{-24}$ g
Pi	$\pi = 3.1415926536$
Planck's constant	$h = 6.6262 \times 10^{-34}$ J-sec
Proton rest mass	$m_p = 1.00728$ amu $= 1.67265 \times 10^{-24}$ g
Speed of light (in vacuum)	$c = 2.997925 \times 10^{8}$ m/sec